Algebra 2

Explorations and Applications

$r = 6.37 \times 10^6$

Authors

Senior Authors

Miriam A. Leiva   Richard G. Brown

Loring Coes III

Shirley Frazier Cooper

Joan Ferrini-Mundy

Amy T. Herman

Patrick W. Hopfensperger

Celia Lazarski

Stuart J. Murphy

Anthony Scott

Marvin S. Weingarden

McDougal Littell
A HOUGHTON MIFFLIN COMPANY
Evanston, Illinois ◆ **Boston** ◆ **Dallas**

Authors

Senior Authors

Richard G. Brown
Mathematics Teacher, Phillips Exeter Academy, Exeter, New Hampshire

Miriam A. Leiva
Cone Distinguished Professor for Teaching and Professor of Mathematics, University of North Carolina at Charlotte

Loring Coes III
Chair of the Mathematics Department, Rocky Hill School, E. Greenwich, Rhode Island

Shirley Frazier Cooper
Curriculum Specialist, Secondary Mathematics, Dayton Public Schools, Dayton, Ohio

Joan Ferrini-Mundy
Professor of Mathematics and Mathematics Education, University of New Hampshire, Durham, New Hampshire

Amy T. Herman
Mathematics Teacher, Atherton High School, Louisville, Kentucky

Patrick W. Hopfensperger
Mathematics Teacher, Homestead High School, Mequon, Wisconsin

Celia Lazarski
Mathematics Teacher, Glenbard North High School, Carol Stream, Illinois

Stuart J. Murphy
Visual Learning Specialist, Evanston, Illinois

Anthony Scott
Assistant Principal, Orr Community Academy, Chicago, Illinois

Marvin S. Weingarden
Supervisor of Secondary Mathematics, Detroit Public Schools, Detroit, Michigan

Editorial Advisors

Martha A. Brown
Mathematics Supervisor, Prince George's County Public Schools, Capitol Heights, Maryland

Diana Garcia
Mathematics Supervisor, Laredo Independent School District, Laredo, Texas

Sue Ann McGraw
Mathematics Teacher, Lake Oswego High School, Lake Oswego, Oregon

Editorial Advisors helped plan the concept, teaching approach, and format of the book. They also reviewed draft manuscripts.

ISBN: 0-395-86298-1   123456789—VH—01 00 99 98 97

Review Panel

Judy B. Basara — Curriculum Chair, St. Hubert's High School, Philadelphia, Pennsylviania

Dane Camp — Mathematics Teacher, New Trier High School, Winnetka, Illinois

Pamela W. Coffield — Mathematics Teacher, Brookstone School, Columbus, Georgia

Kathleen Curran — Mathematics Teacher, Ball High School, Galveston, Texas

Randy Harter — Mathematics Specialist, Buncombe County Schools, Asheville, North Carolina

William Leonard — Assistant Director of Mathematics, Fresno Unified School District, Fresno, California

Betty McDaniel — Coordinator of Mathematics, Florence School District, Florence, South Carolina

Roger O'Brien — Mathematics Supervisor, Polk County Schools, Bartow, Florida

Leo Ramirez — Mathematics Teacher, McAllen High School, McAllen, Texas

Michelle Rohr — Director of Mathematics, Houston Independent School District, Houston, Texas

May Samuels — Mathematics Chairperson, Weequahic High School, Newark, New Jersey

Betty Takesuye — Mathematics Teacher, Chaparral High School, Scottsdale, Arizona

Members of the review panel read and commented upon outlines, sample lessons, and research versions of the chapters.

Manuscript Reviewers

Dane Camp — Mathematics Teacher, New Trier High School, Winnetka, Illinois

Pamela W. Coffield — Mathematics Teacher, Brookstone School, Columbus, Georgia

Ravi Kamat — Mathematics Teacher, Roosevelt High School, Dallas, Texas

William Leonard — Assistant Director of Mathematics, Fresno Unified School District, Fresno California

Joseph E. Montaño — Mathematics Teacher, Martin High School, Laredo, Texas

Betty Takesuye — Mathematics Teacher, Chaparral High School, Scottsdale, Arizona

Straight Line Editorial Development, Inc. — Editorial Consultants, San Francisco, California

Manuscript Reviewers read and reacted to draft manuscript, focusing on its effectiveness from a teaching/learning viewpoint.

Student Advisors

Iruma Bello, Henry H. Filer Middle School, Hialeah, FL; Leslie Blaha, Walnut Springs Middle School, Westerville, OH; Tai-Ling Bloomfield, Alyeska Central School, Anchorage, AK; Gabriel Bonilla, G.W. Carver Middle School, Miami, FL; Anne Burke, Nichols School, Buffalo, NY; Eric de Armas, G.W. Carver Middle School, Miami, FL; Beth Donaldson, Bettendorf Middle School, Bettendorf, IA; Megan Foreman, Allen High School, Allen, TX; Clifton Gray, Riverside University High School, Milwaukee, WI; Colleen Kelly, Chaparral High School, Scottsdale, AZ; Tony Liberati, Hampton Middle School, Gibsonia, PA; Christian Maiden, Hathaway Brown School, Shaker Heights, OH; J.P. Marshall, Lincoln High School, Tallahassee, FL; Jeff Phi, Irving Junior High School, Kansas City, MO; Jaret Radford, Fondren Middle School, Houston, TX; Marisa A. Sharer, Chaparral High School, Scottsdale, AZ; Scott Terrill, Swartz Creek School, Swartz Creek, MI; Gabriela Zúñiga, San Marcos High School, San Marcos, TX.

Acknowledgment

The authors wish to thank Pamela W. Coffield and her students at Brookstone School in Columbus, Georgia, for using and providing comments on the Portfolio Projects in this book. Ms. Coffield's students are Kate Baker, Bo Bickerstaff, Lizzie Bowles, Bradford Carmack, Lucy Cartledge, Henry Dunn, Patrick Graffagnino, Charles Haines, Daniel McFall, Sravanthi Meka, Blake Melton, Jason Pease, Ann Phillips, Dorsey Staples, and Jeffrey Usman.

Contents

Modeling Using Algebra

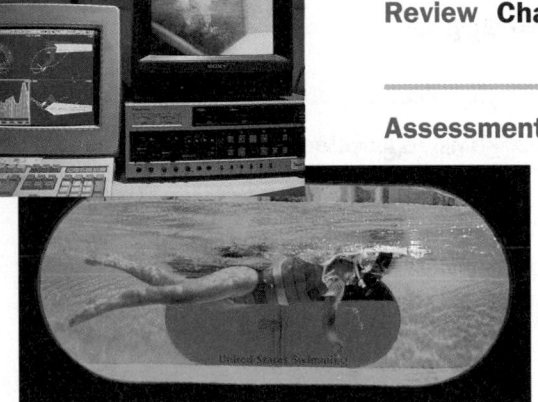

1.2 *Computer analysis of swimming technique*   13

Applications

Interview:
Twyla Lang
Business Owner
6, 26

─────── Connection ───────

Education	7	**Agriculture**	19
Languages	8	**Geometry**	20
Olympic Events	13	**Baseball**	27
History	14	**Games**	33

Additional applications include: business, manufacturing, industry, sports, entertainment, personal finance

Linear Functions

Portfolio Project
Gathering data   84

Applications

Interview:
Norbert Wu
Underwater Photographer
49, 54

Additional applications include: precision skating, geometry, sports, manufacturing, personal finance, temperature scales, oceanography, economics, health, business, physics

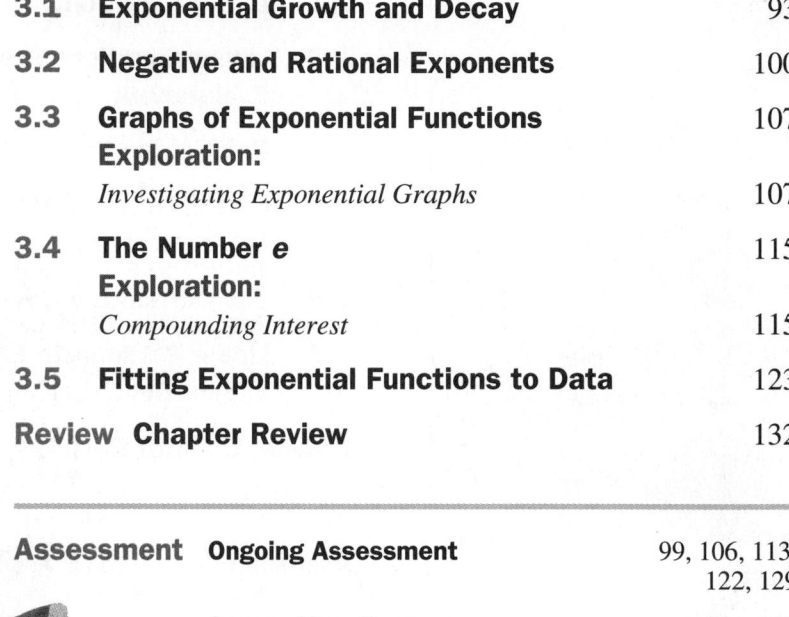

3.3 *Carbon dating* 113

Applications

Logarithmic Functions

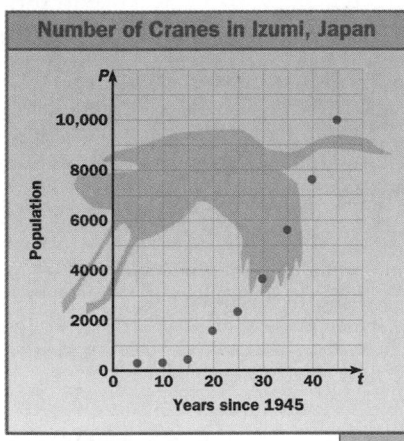

Number of Cranes in Izumi, Japan

4.5   *Using logarithms to model data*   168

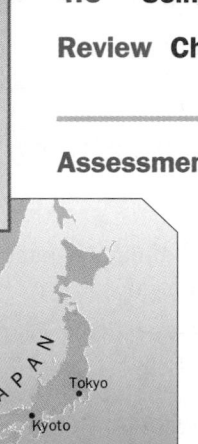

Applications

Interview:
Ednaly Ortiz
Archaeologist
146, 159

———— Connection ————

Transportation	147	**Radiology**	166
Astronomy	153	**Social Studies**	172
Acoustics	160	**Biology**	174

Additional applications include: business, forestry, recreation, mountain climbing, population, earth science, cooking, personal finance, ecology

Quadratic Functions

5.3 *Parabolic athletic fields*   204

Applications

Interview: *Mark Thomas*
               Automotive Engineer
               190, 220

━━━━━━━ Connection ━━━━━━━

Additional applications include: geometry, business, physics, sports science, government, driving, electricity

Investigating Data

6.3 *Archaeological sampling* 254

Systems

PACIFIC
OCEAN

7.5 *Comparing desert climates*   315

Applications

Interview:
Gina Oliva
Fitness Director
299, 316, 319, 326

——————— Connection ———————

Economics	295	**Physics**	313
Physics	301	**Meteorology**	320
Agriculture	307		

Additional applications include: personal finance,
advertising, social science, recreation, consumer
economics, finance, cooking, art, cartography,
manufacturing, medicine, geology, fitness

Radical Functions and Number Systems

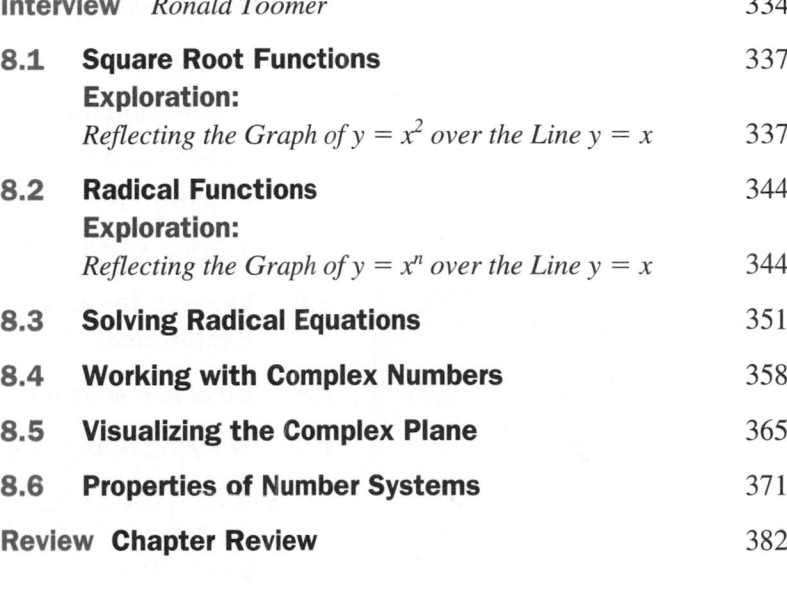

                                                        364, 370, 379

              Assess Your Progress   357, 379

              Journal writing   357, 379

              Portfolio Project:
              Walk This Way   380

              Chapter Assessment   384

8.3   *America's Cup boats*   356

Applications

Interview:
Ronald Toomer
Roller Coaster
Designer
342, 348

— **Connection** —

Oceanography	341	**Electronics**	363
Sports	349	**Mandelbrot Set**	369
Meteorology	354	**Dance**	377
Sports	356	**Games**	378

Additional applications include: biology, geometry,
astronomy, wind chill

Polynomial and Rational Functions

Planet	Mean distance from sun (millions of miles)	Apparent size of sun
Mercury	36.0	2.58
Venus	67.2	1.38
Earth	93.0	1.00
Mars	142.0	0.655
Jupiter	484.0	0.192
Saturn	885.0	0.105
Uranus	1780.0	0.0522
Neptune	2790.0	0.0333
Pluto	3660.0	0.0254

9.6 *Apparent size of the sun*   430

Applications

Interview:

Vera Rubin

Astronomer

430, 446

Additional applications include: biology, psychology, physiology, space travel, education, metallurgy

Sequences and Series

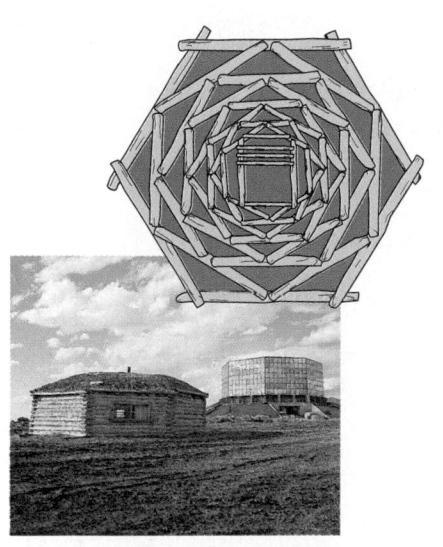

10.2 *Building hogans*   478

Applications

Interview:
Jhane Barnes
Textile Designer
471, 479, 502

Additional applications include: chemistry, biology, architecture, medicine, language arts, physics, economics, genealogy

Analytic Geometry

11.2 *Parabolic radio telescope* 525

Applications

Discrete Mathematics

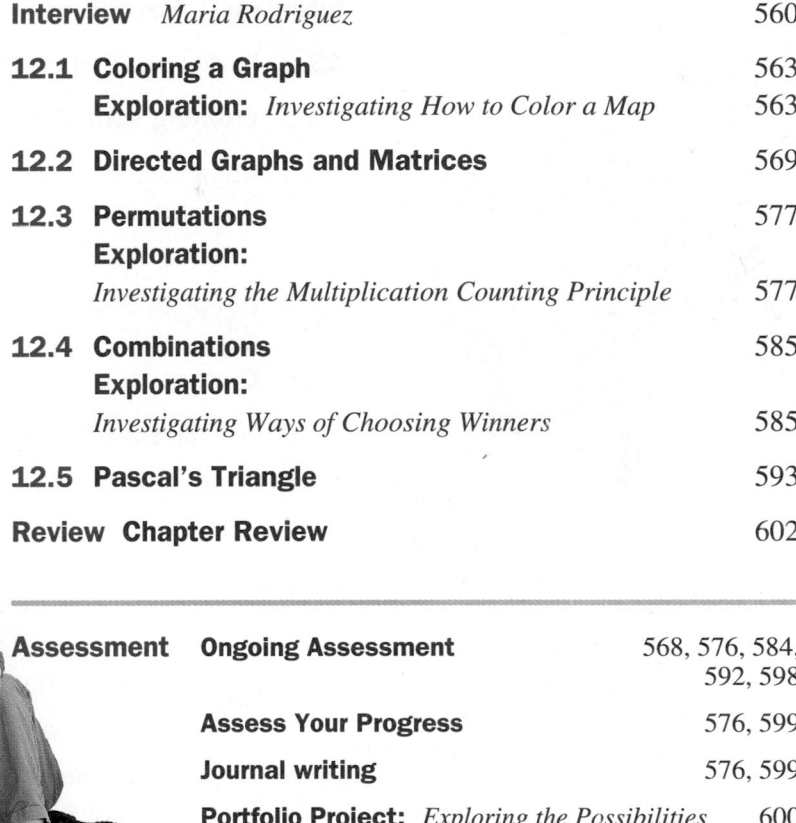

12.5 *Investigating Pascal's triangle* 593

Applications

Additional applications include: cartography, scheduling, chemistry, ecology, medicine, business, fashion, forensics, music, literature, hobbies, manufacturing, computers, advertising, fundraising, industry, recreation, government

Probability

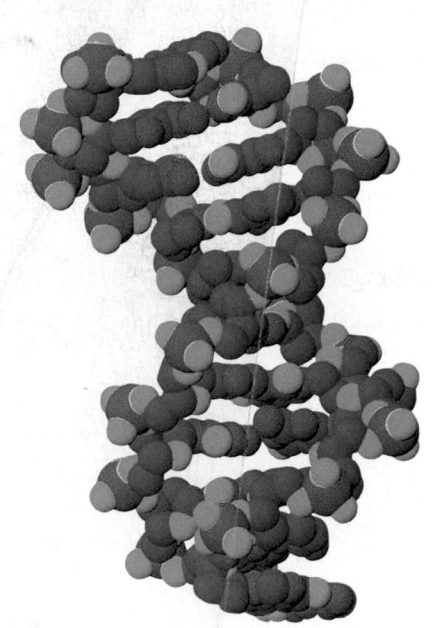

Applications

——————— Connection ———————

Additional applications include: sports, astronomy, manufacturing, physics, meteorology, movies, electronics, health, market research, architecture, driver's education, biology, agriculture

Triangle Trigonometry

14.6   *Pysanki of triangles*   696

Applications

Interview:
Johnpaul Jones
Zoo Designer
661, 668

———— Connection ————

Additional applications include: archaeology, exercising, parasailing, aviation, meteorology, sports, navigation, dancing, architecture, cooking

Trigonometric Functions

15.2 *Testing stereo speakers* 715

Applications

Additional applications include: amusement park rides, geometry, automotive mechanics, music, hot air ballooning, astronomy, engineering

Student Resources

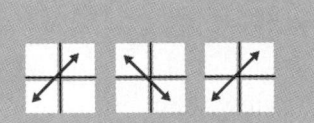

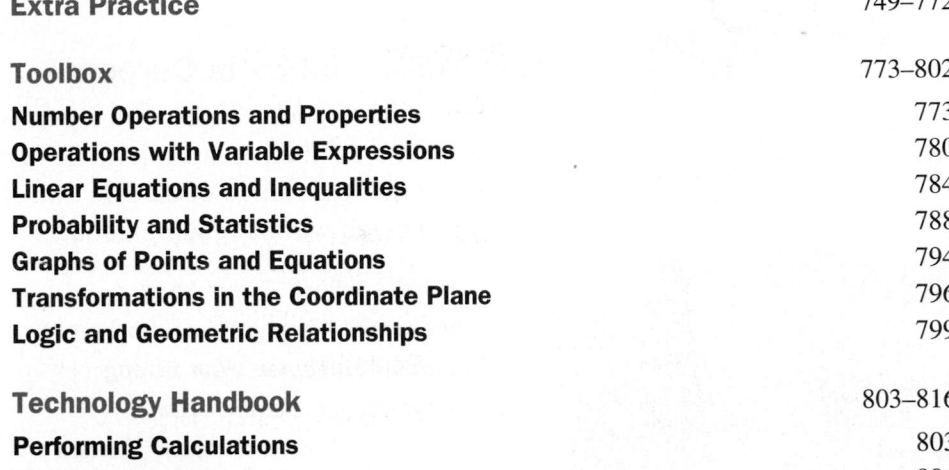

About *the* Interviews

Using Mathematics in Careers

Each chapter of this book starts with a personal interview with someone who uses mathematics in his or her career. You may be surprised by the wide range of careers that are included. These are the people you will be reading about:

- **Business Owner** *Twlya Lang*
- **Underwater Photographer** *Norbert Wu*
- **Aquarium Designer** *Finn Strong*
- **Archaeologist** *Ednaly Ortiz*
- **Automotive Engineer** *Mark Thomas*
- **Computer Visualization Specialist** *Donna Cox*
- **Fitness Director** *Gina Oliva*
- **Roller Coaster Designer** *Ronald Toomer*
- **Astronomer** *Vera Rubin*
- **Textile Designer** *Jhane Barnes*
- **Cartographer** *Kija Kim*
- **Electrical Engineer** *Maria Rodriguez*
- **Lawyer** *Robert Ward*
- **Zoo Designer** *Johnpaul Jones*
- **Record Producer** *Joe Lopez*

Ednaly Ortiz

Mark Thomas

Jhane Barnes

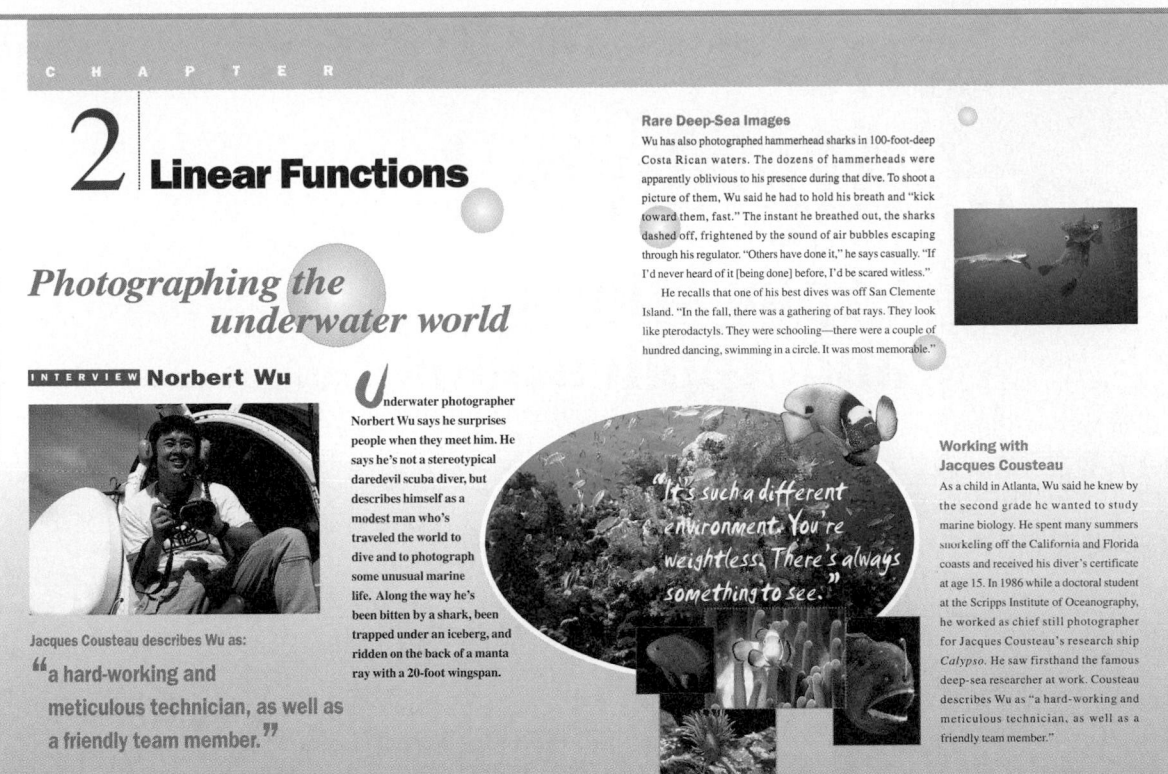

Applying What You've Learned

At the end of each interview, there are *Explore and Connect* questions that guide you in learning more about the career being discussed. Some of these questions involve research that is done outside of class. In addition, in each chapter there are *Related Examples and Exercises* that show how the mathematics you are learning is directly related to the career highlighted in the interview.

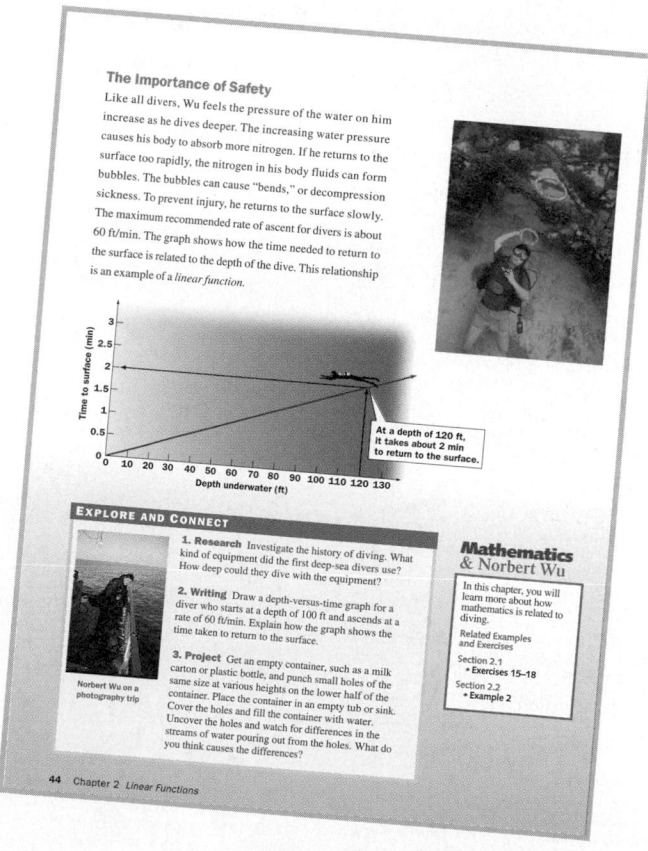

Welcome *to* Algebra 2

Explorations and Applications

GOALS OF THE COURSE

This book will help you use mathematics in your daily life and prepare you for success in future courses and careers.

In this course you will:
- Study the algebra concepts that are most important for today's students
- Apply these concepts to solve many different types of problems
- Learn how calculators and computers can help you find solutions

You will have a chance to develop your skills in:
- Reasoning and problem solving
- Communicating orally and in writing
- Studying and learning independently and as a team member

MATHEMATICAL CONTENT

This contemporary algebra course gives you a strong background in the types of mathematical reasoning and problem solving that will be important in your future.

The book emphasizes:
- Using functions, equations, and graphs to model problem situations
- Investigating connections of algebra to geometry, statistics, probability, and discrete mathematics

ACTIVE LEARNING

To learn algebra successfully, you need to get involved!

There will be many opportunities in this course for you to participate in:
- Explorations of mathematical concepts
- Cooperative learning activities
- Small-group and whole-class discussions

So don't sit back and be a spectator. If you join in and share your ideas, everyone will learn more.

Course Overview

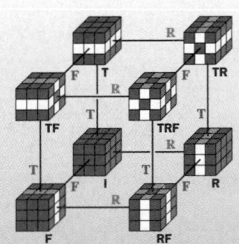

To get an overview of your course, turn to pages xxiv–xxxiii to see some of the types of problems you will solve and topics you will explore.

"What does algebra have to do with me?"

Applications and Connections

Algebra is about you and the world around you.

In this course you'll learn how algebra can help answer many different types of questions in daily life and in careers.

Section 5.1 Working with Simple Quadratic Functions

Learn how to...
- recognize and draw graphs of quadratic functions
- solve simple quadratic equations

So you can...
- examine the relationship between air resistance and speed when cycling, for example

If you have ever ridden a bicycle, you know that the air pushes against you and can slow you down even when there is no wind. The force that the air exerts on you, called *air resistance*, is given by the formula on the preceding page. (The formula applies to bicycles as well as cars.)

EXAMPLE 1 Application: Bicycling

Shirley Scott is cycling on a day with no wind. She is in a casual, upright position on her bicycle. In this situation, the drag coefficient is 1.1 and the frontal area is 5.5 ft². Write an equation that gives the air resistance R (in pounds) as a function of the bicycle's speed s (in miles per hour). Graph the function.

SOLUTION

Use the formula shown on the preceding page.

$$R = 0.00256(1.1)(5.5)s^2$$
$$= 0.0155s^2$$

Make a table of values, plot the points, and connect them with a smooth curve.

Substitute 1.1 for C_D and 5.5 for frontal area.

s	R
0	0
4	0.248
8	0.992
12	2.232
16	3.968
20	6.200

THINK AND COMMUNICATE

1. What happens to the air resistance when Shirley Scott doubles her speed?
2. What would you say is a reasonable domain for the air resistance function? What is the corresponding range?

5.1 Working with Simple Quadratic Functions   **185**

Recreation

How does increasing your speed on a bicycle affect air resistance?

(Chapter 5, page 185)

Business Management

How can past sales growth be used to estimate future sales of Kente cloth scarves?

(Chapter 1, page 6)

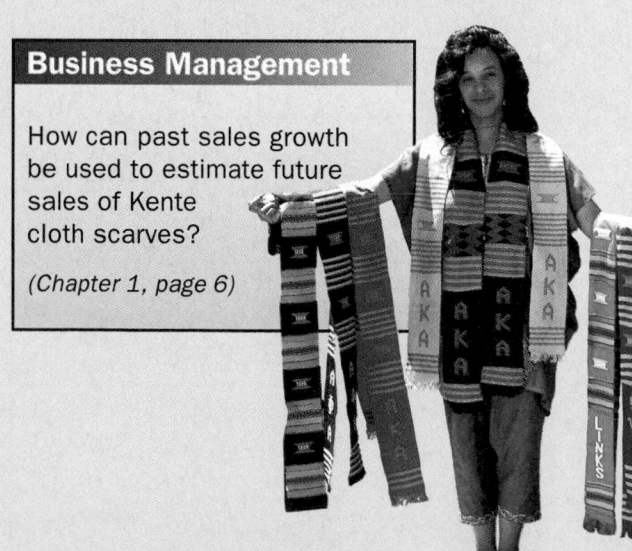

Year	Scarf sales
1992	100
1993	350
1994	650

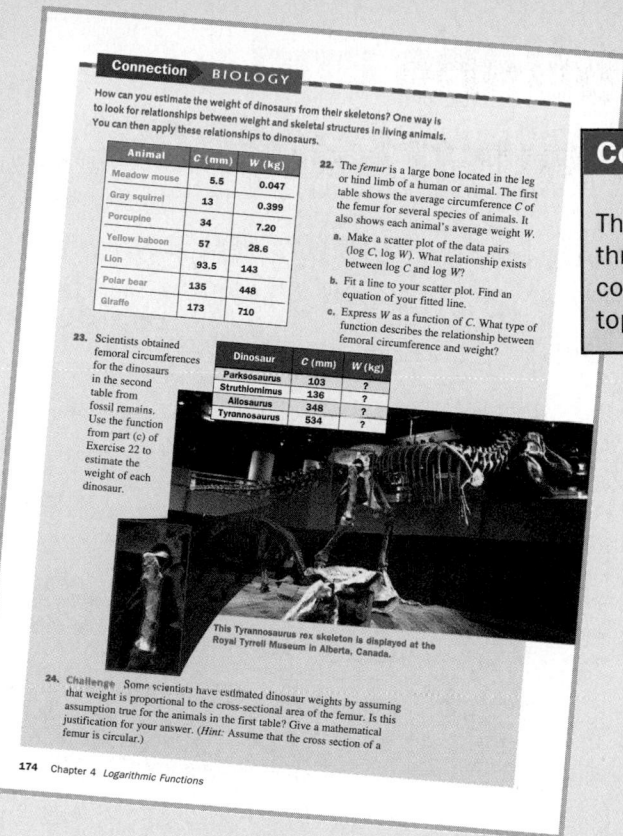

Connection BIOLOGY

How can you estimate the weight of dinosaurs from their skeletons? One way is to look for relationships between weight and skeletal structures in living animals. You can then apply these relationships to dinosaurs.

Animal	C (mm)	W (kg)
Meadow mouse	5.5	0.047
Gray squirrel	13	0.399
Porcupine	34	7.20
Yellow baboon	57	28.6
Lion	93.5	143
Polar bear	135	448
Giraffe	173	710

22. The *femur* is a large bone located in the leg or hind limb of a human or animal. The first table shows the average circumference C of the femur for several species of animals. It also shows each animal's average weight W.

a. Make a scatter plot of the data pairs (log C, log W). What relationship exists between log C and log W?

b. Fit a line to your scatter plot. Find an equation of your fitted line.

c. Express W as a function of C. What type of function describes the relationship between femoral circumference and weight?

23. Scientists obtained femoral circumferences for the dinosaurs in the second table from fossil remains. Use the function from part (c) of Exercise 22 to estimate the weight of each dinosaur.

Dinosaur	C (mm)	W (kg)
Parksosaurus	103	?
Struthiomimus	136	?
Allosaurus	348	?
Tyrannosaurus	534	?

This Tyrannosaurus rex skeleton is displayed at the Royal Tyrrell Museum in Alberta, Canada.

24. **Challenge** Some scientists have estimated dinosaur weights by assuming that weight is proportional to the cross-sectional area of the femur. Is this assumption true for the animals in the first table? Give a mathematical justification for your answer. (*Hint:* Assume that the cross section of a femur is circular.)

174 Chapter 4 *Logarithmic Functions*

Connections Exercises

These clusters of exercises, which appear throughout each chapter, focus on the connections of algebra to a particular topic, career, or branch of mathematics.

Oceanography

What is the water pressure 35,800 ft below the surface of the ocean?

(Chapter 2, page 49)

0 —
5,000 —
10,000 —
15,000 —
20,000 —
25,000 —
30,000 —
35,000 —
40,000 —

The Marianas Trench

Earth Science

How did the height of the Mississippi River change during a 1993 flood?

(Chapter 2, page 69)

Height above banks (in.)
10
9
8
7
6
5
4
3
2
1
0
0   5   10   15   20   25   30
Date in April, 1993

"Do we just sit back and listen?"

Explorations and Cooperative Learning

In this course you'll be an active learner.

Working individually and in groups, you'll investigate questions and then present and discuss your results. Here are some of the topics you'll explore.

Drawing a Parabola

How is a point on a parabola related to the focus and the directrix?

(Chapter 11, page 520)

Section

4.2 Using Inverses of Exponential Functions

Learn how to...
- find inverses of exponential functions
- evaluate logarithms

So you can...
- understand how an altimeter works, for example

You know that the inverse of a nonconstant linear function is also a linear function. Does this mean that the inverse of an exponential function is an exponential function? The answer is no. In fact, the inverse of an exponential function is a type of function you may not have seen before.

EXPLORATION
COOPERATIVE LEARNING

Investigating the Inverse of $f(x) = 2^x$

Work with a partner.
You will need: • a MIRA® transparent mirror
• graph paper

x	f(x)
-2	?
-1	?
0	?
1	?
2	?
3	?

1 Let $f(x) = 2^x$. Copy and complete the table.

2 Graph $y = f(x)$. Label the points on the graph corresponding to the ordered pairs in the table.

3 Graph the line $y = x$ in the same coordinate plane you used for graphing f.

4 Graph $y = f^{-1}(x)$ by using the MIRA® to draw the reflection of the graph of f over the line $y = x$. Label the points on the graph of f^{-1} that are reflections of labeled points on the graph of f.

Questions

1. Replace each ? with the number that makes the equation true.

a. $f^{-1}(8) = $?   b. $f^{-1}\left(\frac{1}{4}\right) = $?   c. $f^{-1}(1) = $?

2. Replace each ? with exponents or powers of 2.

a. For the function f, the input values are ? and the output values are ?.

b. For the function f^{-1}, the input values are ? and the output values are ?.

Investigating Inverse Functions

What type of function is the inverse of an exponential function?
(Chapter 4, page 148)

Portfolio Projects

These open-ended projects give you a chance to explore applications of the topics you have studied.

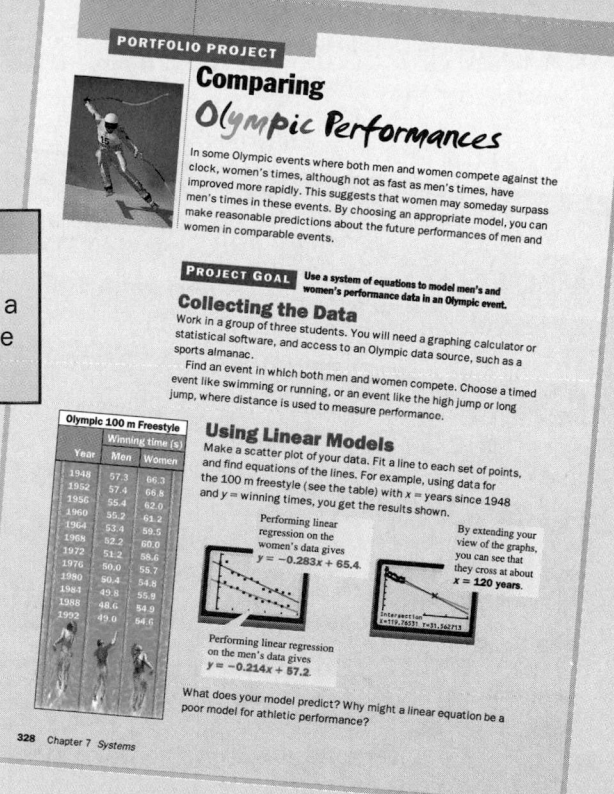

PORTFOLIO PROJECT

Comparing
Olympic Performances

In some Olympic events where both men and women compete against the clock, women's times, although not as fast as men's times, have improved more rapidly. This suggests that women may someday surpass men's times in these events. By choosing an appropriate model, you can make reasonable predictions about the future performances of men and women in comparable events.

PROJECT GOAL Use a system of equations to model men's and women's performance data in an Olympic event.

Collecting the Data

Work in a group of three students. You will need a graphing calculator or statistical software, and access to an Olympic data source, such as a sports almanac.

Find an event in which both men and women compete. Choose a timed event like swimming or running, or an event like the high jump or long jump, where distance is used to measure performance.

Olympic 100 m Freestyle

Year	Winning time (s)	
	Men	Women
1948	57.3	66.3
1952	57.4	66.8
1956	55.4	62.0
1960	55.2	61.2
1964	53.4	59.5
1968	52.2	60.0
1972	51.2	58.6
1976	50.0	55.7
1980	50.4	54.8
1984	49.8	55.9
1988	48.6	54.9
1992	49.0	54.6

Using Linear Models

Make a scatter plot of your data. Fit a line to each set of points, and find equations of the lines. For example, using data for the 100 m freestyle (see the table) with x = years since 1948 and y = winning times, you get the results shown.

Performing linear regression on the women's data gives $y = -0.283x + 65.4$.

By extending your view of the graphs, you can see that they cross at about $x = 120$ years.

Performing linear regression on the men's data gives $y = -0.214x + 57.2$.

What does your model predict? Why might a linear equation be a poor model for athletic performance?

328 Chapter 7 Systems

Finding an Exponential Model

Can you use an equation to model the temperature of cooling water?

(Chapter 4, page 167)

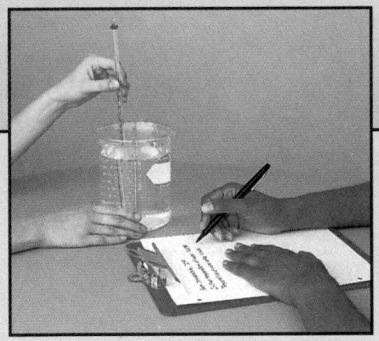

Modeling an Infinite Geometric Series

Is it possible for an infinite geometric series to have a sum?

(Chapter 10, page 497)

"*How can I visualize the problem?*"

Using Technology

Calculators and computers can help you see mathematical relationships.

In this course there are many opportunities to use technology to model problem situations, identify patterns, and find solutions.

Graphing Calculator

Can you find the *x*-intercepts of the graph of a quadratic equation just by looking at the equation?

(Chapter 5, page 199)

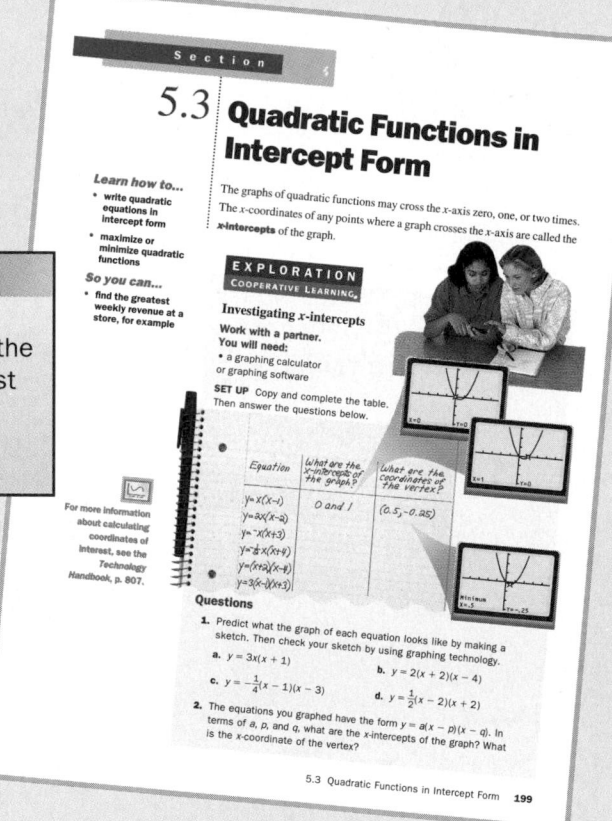

Section

5.3 Quadratic Functions in Intercept Form

Learn how to...
• write quadratic equations in intercept form
• maximize or minimize quadratic functions

So you can...
• find the greatest weekly revenue at a store, for example

The graphs of quadratic functions may cross the *x*-axis zero, one, or two times. The *x*-coordinates of any points where a graph crosses the *x*-axis are called the **x-intercepts** of the graph.

EXPLORATION
COOPERATIVE LEARNING

Investigating *x*-intercepts
Work with a partner.
You will need:
• a graphing calculator or graphing software

SET UP Copy and complete the table. Then answer the questions below.

Equation	What are the x-intercepts of the graph?	What are the coordinates of the vertex?
$y = x(x-1)$	0 and 1	$(0.5, -0.25)$
$y = 2x(x-2)$		
$y = -x(x+3)$		
$y = \frac{1}{2}x(x+4)$		
$y = (x+2)(x+4)$		
$y = 3(x-1)(x+3)$		

For more information about calculating coordinates of interest, see the *Technology Handbook*, p. 807.

Questions

1. Predict what the graph of each equation looks like by making a sketch. Then check your sketch by using graphing technology.
   a. $y = 3x(x + 1)$
   b. $y = 2(x + 2)(x - 4)$
   c. $y = -\frac{1}{4}(x - 1)(x - 3)$
   d. $y = \frac{1}{2}(x - 2)(x + 2)$

2. The equations you graphed have the form $y = a(x - p)(x - q)$. In terms of *a*, *p*, and *q*, what are the *x*-intercepts of the graph? What is the *x*-coordinate of the vertex?

5.3 Quadratic Functions in Intercept Form **199**

Matrix Operations

How much will each family pay for a catalog order of camping equipment?

(Chapter 1, pages 24-25)

```
[A][B]
  [[34 2000]
   [24 1230]
   [63 3160]]
```

	Tents	Sleeping bags	Back-packs	Cook sets
Graham	1	4	2	0
Piscitelli	0	2	3	1
Brewer	2	5	4	2

Connection · WIND ENERGY

In the book *Wind Energy Comes of Age*, Paul Gipe discusses the amount of energy generated from wind turbines in North America and Europe between 1980 and 1995. His data can be modeled using the equations

Europe: $y = 6.489(1.580)^x$

North America: $y = \dfrac{3500}{1 + 874e^{-0.882x}}$

where x = years since 1980 and y = gigawatt-hours of wind energy.
(*Note:* One gigawatt-hour equals 10^6 kilowatt-hours.)

Located in Scotland, this 135 ft tall wind turbine is one of the world's largest.

Technology For Exercises 22–25, use a graphing calculator or graphing software.

22. Graph the wind energy functions.

23. According to the models, what was the wind energy for each region in 1985? in 1990? in 1995?

24. Which region produced more wind energy in the 1980s?

25. When did Europe's wind energy production surpass North America's?

26. **Writing** Do you think an exponential function is a realistic model for European wind energy in the future? Why?

Wind turbines line California's Altamont Pass, where wind speeds average 13 mi/h to 18 mi/h.

CHEMISTRY Show that the two equations given for each radioactive element are equivalent by rewriting both in the form $A(t) = ab^t$.

27. **Polonium-210**
$A(t) = 100e^{-0.005t}$
$A(t) = 100\left(\dfrac{1}{2}\right)^{t/138}$

28. **Radium-226**
$A(t) = e^{-0.000825t}$
$A(t) = \left(\dfrac{1}{2}\right)^{t/1320}$

29. **Challenge** The radioactive element carbon-14 has a half-life of about 5730 years, so an equation describing its radioactive decay is:

$$A(t) = A_0\left(\dfrac{1}{2}\right)^{t/5730}$$

Write an equivalent equation that uses e as a base. (*Hint:* Look at the numbers in the exponents in Exercises 27 and 28.)

3.4 The Number e **121**

Technology Exercises

In these exercises you will be using graphing technology or spreadsheets to practice, apply, and extend what you have learned.

Spreadsheet

How much should you save each year for college, assuming that the cost of college continues to grow?

(Chapter 10, page 505)

College Planning

	A	B	C
1	Year	College cost	Savings
2	1	24876	1200
3	2	B2*1.059	C2*1.1+120

Graphing Technology

Why does wearing snowshoes keep you from sinking into deep snow?

(Chapter 9, page 426)

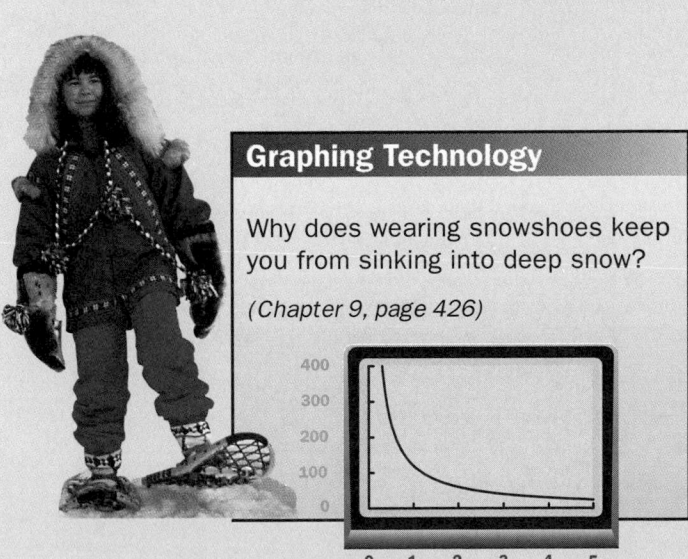

"Can I solve this problem with algebra?"

Integrating Math Topics

Sometimes you need to combine algebra with other math topics in order to find a solution.

In this course you'll see how you can solve problems by integrating algebra with geometry, statistics, probability, and discrete mathematics.

Discrete Mathematics

How is food energy transferred through an ecosystem?

(Chapter 12, page 569)

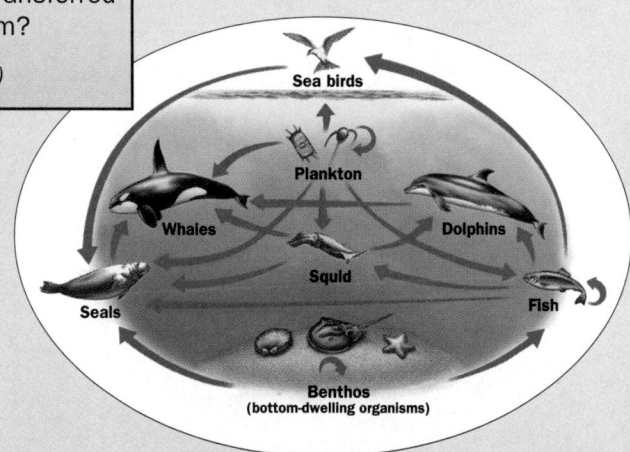

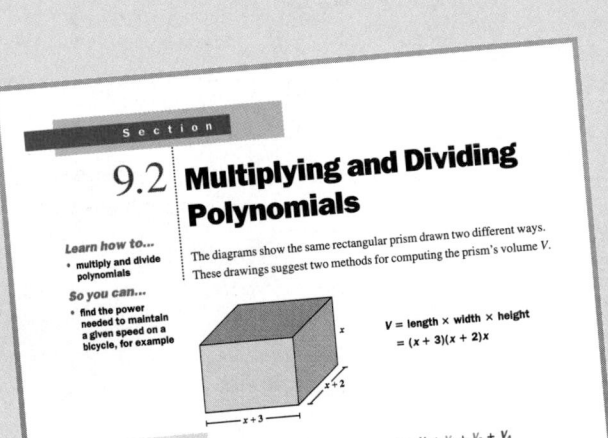

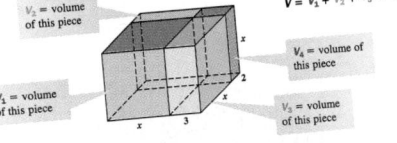

Geometry

How can thinking about the volume of a rectangular prism help you understand multiplying polynomials?

(Chapter 9, page 397)

Probability

What is the probability of a complete match between fingerprints from two randomly chosen people?

(Chapter 13, page 630)

Corresponding squares

Section

6.5 Describing the Variation of Data

Learn how to...
- find and interpret the range, interquartile range, and standard deviation of a data set

So you can...
- determine the variability of data, such as ozone readings

Life on Earth would not be possible without the ozone layer in the upper atmosphere filtering the ultraviolet rays of the sun. Recent studies have shown that the thickness of the ozone layer is decreasing. In the tropics, unlike other areas, the change of seasons has little effect on the thickness of the ozone layer. At non-tropical latitudes, the ozone layer is thickest in the spring and thinnest in the fall.

THINK AND COMMUNICATE

The readings below, which give the thickness of the ozone layer in Dobson units (DU), were taken in Fresno, California, for two 20-day periods.

March readings: 427, 466, 372, 299, 293, 284, 298, 284, 314, 302, 286, 296, 306, 308, 320, 318, 344, 345, 354, 381

November readings: 317, 330, 316, 296, 270, 271, 295, 277, 275, 275, 269, 275, 272, 267, 270, 291, 275, 268, 261, 296

The comparative box plot displays these readings.

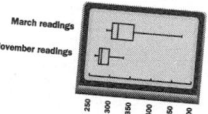

March readings

November readings

1. How does the graph show that the ozone over Fresno is generally thicker in the spring than in the fall?

2. In which month, *March* or *November*, would it be more likely to obtain a reading that is much different from the reading on the previous day? Why?

To determine the variability of a set of data, it is often helpful to find the *range*, the *interquartile range*, and any *outliers* of the data set. The **range** of a data set is the difference between the maximum and minimum data values. The **interquartile range (IQR)** of a data set is the difference between the upper quartile and the lower quartile. A data value can be considered an **outlier** if its distance from the nearer quartile is more than 1.5 times the interquartile range.

BY THE WAY

The mean ozone reading over the Antarctic dropped from 321 DU in 1956 to 117 DU in 1993. The ozone layer in this area is so thin that it is now referred to as "the Antarctic ozone hole."

266   Chapter 6 *Investigating Data*

Statistics

How do the changing seasons affect the thickness of the ozone layer?

(Chapter 6, page 266)

Analytic Geometry

What curves are produced by slicing through a double cone?

(Chapter 11, page 550)

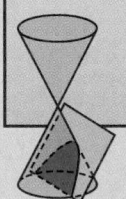 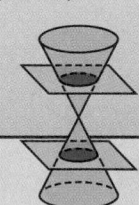 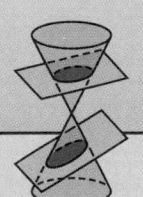 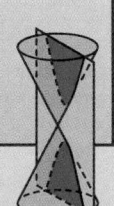

"When will I ever use this?"

Building for the Future

The skills you'll learn in this course will form a strong foundation for the future.

They'll prepare you for more advanced courses and increase your career opportunities.

3.5 | Exercises and Applications

Extra Practice
exercises on
page 753

Write an exponential function whose graph passes through each pair of points.

1. (0, 5), (1, 8) **2.** (1, 8), (2, 10) **3.** (0, 10), (1, 8) **4.** (5, 7), (6, 6)

5. (3, 3), (8, 7) **6.** (0, 6), (7, 9) **7.** (20, 4), (25, 3) **8.** (0, 17), (7, 15)

9. ASTRONOMY The *apparent magnitude* of a star is a measure of its apparent brightness. A star of magnitude 1 is 100 times brighter than a star of magnitude 6.

  a. Find an equation of the exponential decay function whose graph passes through (1, 100) and (6, 1).

  b. Graph the function from part (a).

  c. Copy and complete the table showing comparative brightness values for various sky objects.

Sky object	Apparent magnitude	Brightness value
Uranus	6	1
Aldebaran	1	100
Vega	0	?
Sirius	−1.5	?
Full moon	−12.5	?
Sun	−26.7	?

  d. **Research** Use a book about astronomy to find the names and magnitudes of some stars. Plot them on your graph from part (b). Calculate their brightness values and add them to your table from part (c).

10. SAT/ACT Preview Which of the equations represents an exponential decay function whose graph passes through (2, 1)?

  A. $y = (0.5)^x$ **B.** $y = (0.5x)^2$ **C.** $y = \frac{1}{4}(0.5)^{-x}$ **D.** $y = 4(0.5)^x$ **E.** $y = \frac{1}{4} \cdot 2^x$

11. Open-ended Problem The table shows the increase in the number of host sites on the Internet computer network.

Year	1988	1989	1990	1991	1992	1993	1994
Years since 1988	0	1	2	3	4	5	6
Internet hosts (1000s)	56	159	313	617	1136	2056	3864

  a. Choose any two years and write an exponential function to represent the growth.

  b. Check your function by seeing how its predictions compare to the other data values in the table.

  c. **Technology** Enter all the data into a graphing calculator or statistical software and find an exponential equation that gives a good fit. Compare the results to your results from part (a).

126 Chapter 3 *Exponential Functions*

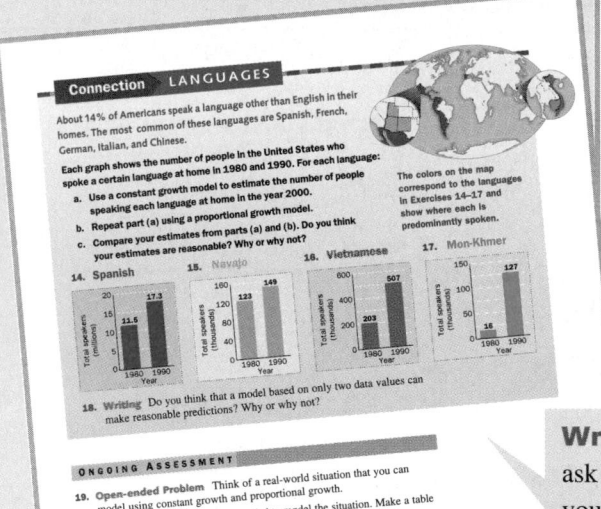

Connection LANGUAGES

About 14% of Americans speak a language other than English in their homes. The most common of these languages are Spanish, French, German, Italian, and Chinese.

Each graph shows the number of people in the United States who spoke a certain language at home in 1980 and 1990. For each language:

  a. Use a constant growth model to estimate the number of people speaking each language at home in the year 2000.

  b. Repeat part (a) using a proportional growth model.

  c. Compare your estimates from parts (a) and (b). Do you think your estimates are reasonable? Why or why not?

The colors on the map correspond to the languages in Exercises 14–17 and show where each is predominantly spoken.

14. Spanish **15.** Navajo **16.** Vietnamese **17.** Mon-Khmer

18. Writing Do you think that a model based on only two data values can make reasonable predictions? Why or why not?

ONGOING ASSESSMENT

19. Open-ended Problem Think of a real-world situation that you can model using constant growth and proportional growth.

  a. **Research** Gather the data needed to model the situation. Make a table or bar graph to display the data you find.

  b. Model the situation using constant growth and proportional growth. Make a prediction about the future using each model. Which model do you think gives the better estimate? Why?

SPIRAL REVIEW

20. Make a scatter plot using the precipitation data. *(Toolbox, page 795)*

For each equation, make a table of values for *x* and *y* when *x* = 0, 1, 2, and 3. *(Toolbox, page 795)*

21. $y = -3x + 6$ **22.** $y = 4.5(2^x)$

Simplify. Give answers with the appropriate number of significant digits. *(Toolbox, page 789)*

23. 3.2 + 0.15 **24.** 42.1 × 3.6

Average Precipitation	Average number of rainy days per year	Average yearly precipitation (inches)
Olympia, WA	164	51.0
Honolulu, HI	100	23.5
Charleston, WV	151	42.4
Madison, WI	118	30.8

8 Chapter 1 *Modeling Using Algebra*

Acoustical Engineering, page 160

Cartography, page 512

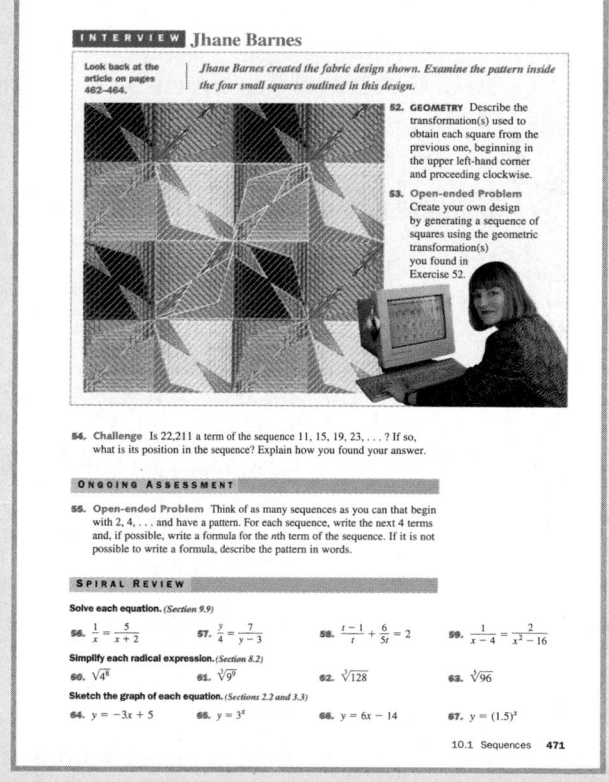

Textile Design, page 471

1 Modeling Using Algebra

Starting a *new tradition*

Twyla Lang

"**T**ravel is so broadening," author Sinclair Lewis wrote. Twyla Lang agrees wholeheartedly. "My parents always told me that travel was educational," Lang says. "Since I was a little girl, they encouraged me to visit countries all over the world so that I could learn about other cultures and broaden my horizons. Well, travel used to be part of my education, but now it's part of my business."

" No matter where I travel, everybody understands when it comes to numbers. Math truly is the universal language."

Showing Cultural Pride

In 1989, while still a college student, Lang founded a company that imports and sells jewelry, purses, and clothing from Africa. Her main business revolves around the sale of Kente cloth scarves made in Ghana and Togo. Kente cloth is a handwoven fabric made from cotton, rayon, or silk. "In Africa, it's a symbol of royalty, unity, and achievement," Lang says.

Some Ghanaians and Togolese wear wraps of Kente cloth on special occasions.

Lang sells the scarves to students graduating from high school or college. "Students wear Kente cloth as a way of showing pride in their achievements," she explains. "The idea is to add a touch of cultural diversity to the graduation ceremony."

Getting Started

Lang's company sprung from her habit of collecting gifts and souvenirs from her far-ranging trips. "People always seemed interested in what I brought back, so I thought maybe I could make a living selling these things," she says. She traveled throughout Africa, gathering samples and testing out their appeal, before settling on the Kente products from Ghana and Togo.

The business started small, initially with just a suitcase full of merchandise. But now Lang receives two shipments a month and sells hundreds of items each year.

1

Planning for the Future

Lang's company has experienced dramatic growth in recent years. Continued success will depend on careful planning, among other things. She uses information about her sales to create a *model* that can help her predict what the future may hold for her business.

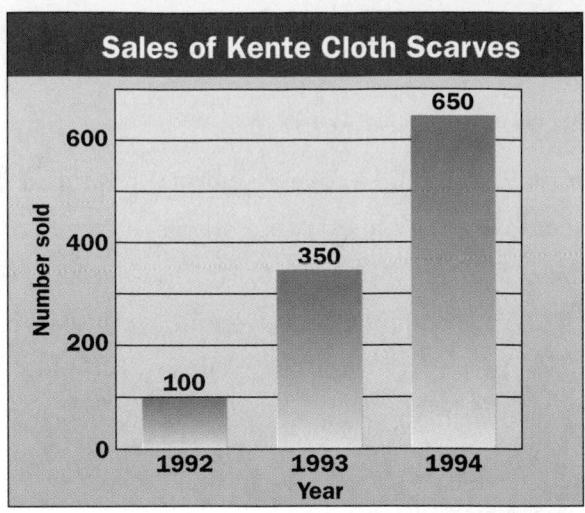

Sales of Kente Cloth Scarves

Note: Lang's actual sales figures are confidential.

"When I draw up a business plan, I chart past sales in order to see what I might expect in the future," she says. "An easy way to do that is with a graph. I can look at what I sold in the past and try to make realistic goals for the future."

Twyla Lang displays some of her Kente cloth products.

1. Research Investigate how a company in your community started. How long has it been in business? Did the owner start the company? Did the owner hold other positions in the company? Write a profile of the company based on your research.

2. Writing Based on the graph shown above, what prediction would you make for 1995 sales? Explain.

3. Project Lang was able to adapt the traditional use of Kente cloth in a way that made it meaningful for American students. Learn more about a tradition from another culture. Think of a way that this tradition can be adapted to be meaningful in American life.

Mathematics & Twyla Lang

In this chapter, you will learn more about how mathematics is used to make decisions in business.

Related Exercises

Section 1.1
• Exercises 2 and 3

Section 1.4
• Exercises 14 and 15

1.1 Modeling Growth with Graphs and Tables

Learn how to...

- **model real-world situations with tables and graphs**
- **compare different models describing the same situation**

So you can...

- **make predictions about future trends, such as growth in sales**

Three years ago the Carter High School Volunteer Club began selling T-shirts. The graph shows the amount of money the club raised each year. The club members want to know what amount they might expect to raise in the school year ahead. How much do you think the club will raise?

EXAMPLE 1   Application: Business

Toolbox p. 790
Mean, Median, and Mode

a. Find the average (mean) yearly increase in the amount raised.

b. Use your answer to part (a) and the initial amount raised to estimate the amount the club raises each year after the initial year. Then estimate the amount the club will raise in the school year ahead.

SOLUTION

a. Find the increase for each year. $5000 - 4000 = \mathbf{1000}$

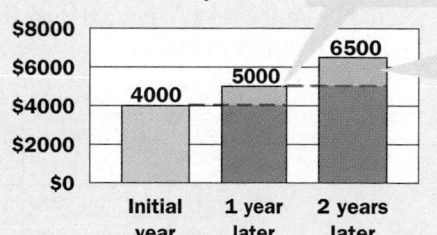

$6500 - 5000 = \mathbf{1500}$

The average yearly increase is $\dfrac{1000 + 1500}{2}$, or \$1250.

Solution continued on next page.

For more information about using a spreadsheet, see the *Technology Handbook,* **p. 816.**

S O L U T I O N *continued*

b. Make a table. You may want to use a spreadsheet.

	Volunteer Club	
	A	**B**
1	Year	Amount raised ($)
2	Initial year	4000
3	1 year later	5250
4	2 years later	6500
5	3 years later	7750

+1250
+1250
+1250

The club can expect to raise about $7750 in the school year ahead.

When you describe a situation using graphs, tables, or equations, you are making a **mathematical model**. Different assumptions can lead to different models.

EXAMPLE 2   **Application: Business**

Toolbox p. 775
Percent Change

Suppose that the amount of money raised from T-shirt sales grows by about the same *percent* from year to year. Create a model for this situation. Then use the model to estimate the amount the club will raise in the year ahead.

S O L U T I O N

Step 1 Find the average yearly percent change.

$$\frac{5000 - 4000}{4000} = 25\%$$

$$\frac{6500 - 5000}{5000} = 30\%$$

The average yearly percent change is $\frac{25 + 30}{2}$, or 27.5%.

Step 2 Find the *growth factor*. The growth factor is 100% of the previous year plus the average yearly percent change:

$$100\% + 27.5\% = 127.5\%, \text{ or } 1.275.$$

Step 3 Use the initial amount and the growth factor to estimate the amount the club will raise in the school year ahead.

	Volunteer Club	
	A	**B**
1	Year	Amount raised ($)
2	Initial year	4000
3	1 year later	5100
4	2 years later	6502.5
5	3 years later	8290.6875

×1.275
×1.275
×1.275

The club will raise about $8300 in the school year ahead.

The model in Example 1 is based on an assumption that sales will increase by the same *amount* each year. This is an example of a *constant growth* model. The model in Example 2 is based on an assumption that sales will increase by the same *percent* each year. This is an example of a *proportional growth* model.

One way to see which model is more accurate is to compare the values from each model to the actual data.

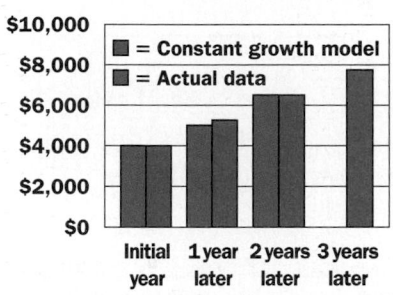
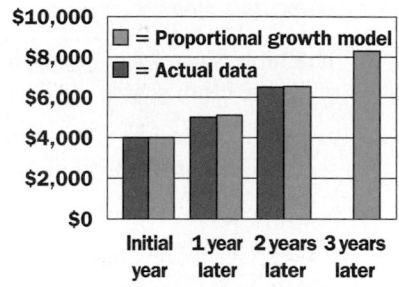

THINK AND COMMUNICATE

1. Does each model above accurately estimate the actual data? Explain.

2. Compare the predictions for the year ahead. What do you think is a reasonable estimate for the amount the club might raise in the school year ahead? Why?

☑ CHECKING KEY CONCEPTS

1. Use a constant growth model to estimate the number of people earning a bachelor's degree in 2000.

2. Repeat Question 1 using a proportional growth model.

3. Are your estimates in Questions 1 and 2 reasonable? Why?

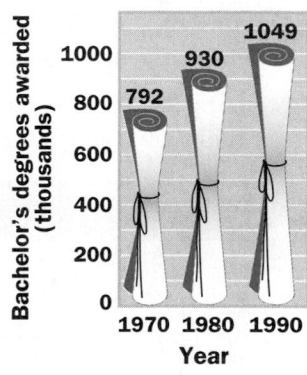

1.1 Exercises and Applications

Extra Practice exercises on page 749

1. The table shows the average daily hospital cost in the United States.

Average Daily Hospital Cost			
Year	Daily charge ($)	$ Increase	% Increase
1975	134	—	—
1980	245	245 − 134 = 111	111/134 = 83%
1985	460	?	?
1990	687	?	?

a. Copy and complete the table.

b. Use a constant growth model to estimate the daily cost in 1995.

c. Use a proportional growth model to estimate the daily cost in 1995. How does this estimate compare to the one you found in part (b)?

Twyla Lang

Look back at the article on pages xxxiv–2.

Twyla Lang estimates her company's future sales so that she can stock her inventory and decide how much she can afford to spend on advertising. Her estimates depend on the assumptions she makes about the company's growth.

2. The table shows the approximate sales of Kente cloth scarves from 1992 to 1994.

Year	Scarf sales
1992	100
1993	350
1994	650

   a. Use a constant growth model to estimate the sales of Kente cloth scarves in 1995.

   b. Repeat part (a) using a proportional growth model.

   c. What do you think is a reasonable estimate of the sales of Kente cloth scarves for 1995? Why?

3. Estimate the sales of Kente cloth scarves for the year 2000. Do you think your estimate is accurate? Why or why not?

4. **Writing** What assumptions do you make about a situation when you use a constant growth model? a proportional growth model? What other factors may influence how well a model predicts future trends?

5. **Cooperative Learning** Work with a partner. Use the graph showing the minimum wage in the United States since 1974.

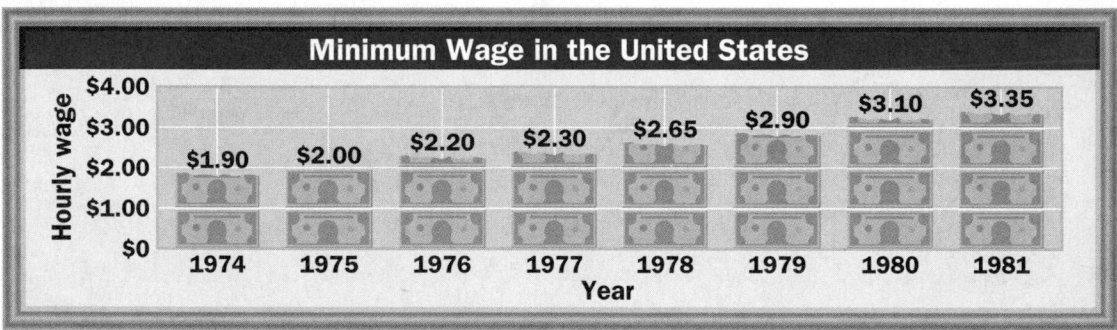

Minimum Wage in the United States

a. Each of you should estimate the minimum wage in 1994. One person should use a constant growth model and the other should use a proportional growth model.

b. In 1994 the minimum wage was $4.25. Which estimate from part (a) is closest to the actual data? Why might your estimate be off?

c. How would your estimate for 1994 change if you looked at data only for 1979–1981? only for 1974–1976?

6. **Challenge** Suppose you have data that *decrease* from year to year.

   a. How would this change be reflected in a constant growth model?

   b. How would this change be reflected in a proportional growth model?

BY THE WAY

There was no change in the minimum wage between January of 1981 and April of 1990. The minimum wage was raised to $3.80 in 1990.

A college education may be costly, but it may be more expensive *not* to go to college. In the early 1990s the average salary for a graduate of a four-year college was about 85% more than the average salary for someone without a college degree.

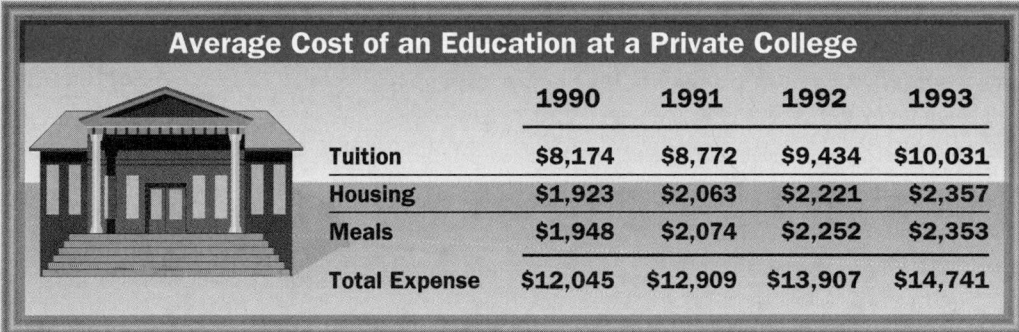

Average Cost of an Education at a Private College				
	1990	**1991**	**1992**	**1993**
Tuition	$8,174	$8,772	$9,434	$10,031
Housing	$1,923	$2,063	$2,221	$2,357
Meals	$1,948	$2,074	$2,252	$2,353
Total Expense	$12,045	$12,909	$13,907	$14,741

For each expense, estimate the cost in 1998. Describe the model you use.

7. tuition **8.** housing **9.** meals **10.** total expenses

11. Which 1990–1993 expense increased by the most dollars? by the largest percentage?

12. a. Suppose you begin college right after you graduate from high school. Estimate the total cost for your first year at a private college.

b. What do you think an education at a private college would cost you, assuming you complete college in four years? Do you think your predictions are reasonable? Why or why not?

c. What are some reasons you should use caution when you use a model to predict far into the future? Explain your reasoning.

The graph shows the average annual income for people in the United States in 1990 based on the highest degree earned.

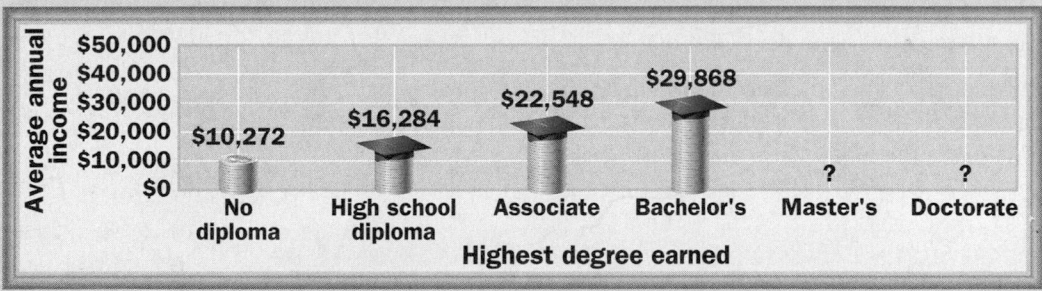

13. a. What is the average percent increase in annual income for each increase in the level of degree earned?

b. Estimate the annual incomes for a person with a master's degree and a person with a doctorate.

c. The average annual salary is about $38,532 for a person with a master's degree and about $54,540 for a person with a doctorate. Are these figures close to your estimates from part (b)? Why might the actual figures and your estimates be different?

About 14% of Americans speak a language other than English in their homes. The most common of these languages are Spanish, French, German, Italian, and Chinese.

Each graph shows the number of people in the United States who spoke a certain language at home in 1980 and 1990. For each language:

a. Use a constant growth model to estimate the number of people speaking each language at home in the year 2000.

b. Repeat part (a) using a proportional growth model.

c. Compare your estimates from parts (a) and (b). Do you think your estimates are reasonable? Why or why not?

The colors on the map correspond to the languages in Exercises 14–17 and show where each is predominantly spoken.

14. Spanish **15.** Navajo **16.** Vietnamese **17.** Mon-Khmer

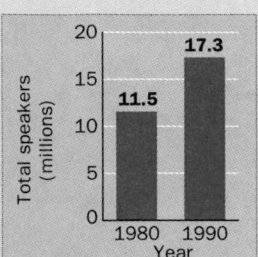 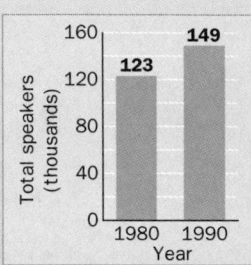 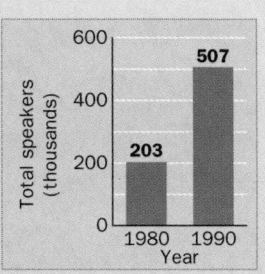 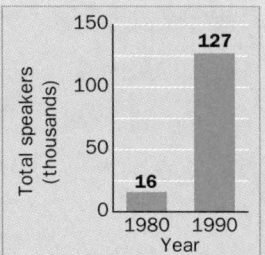

18. Writing Do you think that a model based on only two data values can make reasonable predictions? Why or why not?

ONGOING ASSESSMENT

19. Open-ended Problem Think of a real-world situation that you can model using constant growth and proportional growth.

a. **Research** Gather the data needed to model the situation. Make a table or bar graph to display the data you find.

b. Model the situation using constant growth and proportional growth. Make a prediction about the future using each model. Which model do you think gives the better estimate? Why?

SPIRAL REVIEW

20. Make a scatter plot using the precipitation data. *(Toolbox, page 795)*

For each equation, make a table of values for *x* and *y* when *x* = 0, 1, 2, and 3. *(Toolbox, page 795)*

21. $y = -3x + 6$ **22.** $y = 4.5(2^x)$

Simplify. Give answers with the appropriate number of significant digits. *(Toolbox, page 789)*

23. $3.2 + 0.15$ **24.** 42.1×3.6

Average Precipitation		
	Average number of rainy days per year	Average yearly precipitation (inches)
Olympia, WA	164	51.0
Honolulu, HI	100	23.5
Charleston, WV	151	42.4
Madison, WI	118	30.8

1.2 Using Functions to Model Growth

Learn how to...

- **write a function to model a set of data**

So you can...

- **make predictions about a situation, such as plastics production in the United States**

What do nylon stockings, polyester pants, and vinyl siding have in common? Each item contains plastic in both the product and its name! Nylon, polyester, and vinyl are just some of the types of plastics that have been developed since the 1930s. Look at the graph of plastics production in the United States from 1961 to 1991.

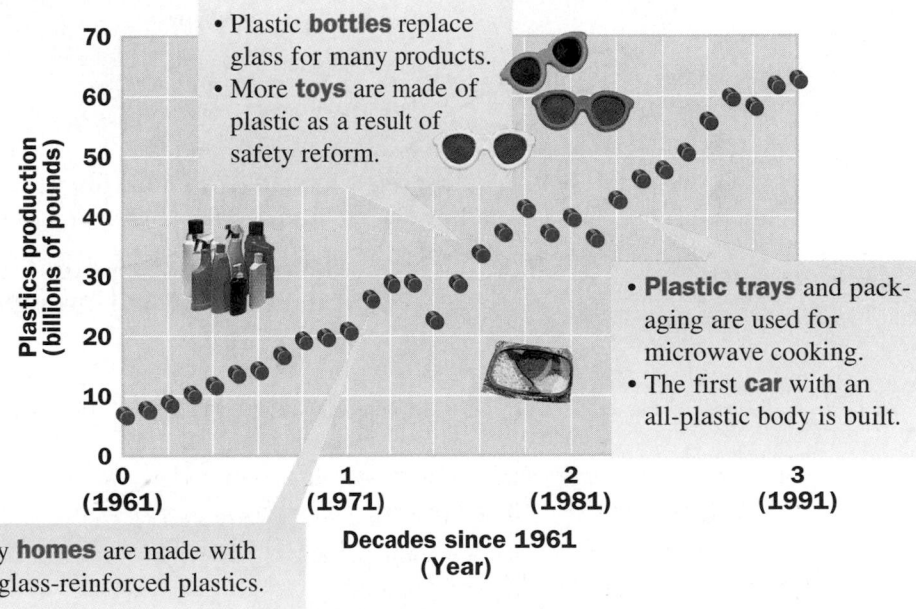

- Plastic **bottles** replace glass for many products.
- More **toys** are made of plastic as a result of safety reform.

- **Plastic trays** and packaging are used for microwave cooking.
- The first **car** with an all-plastic body is built.

- Many **homes** are made with fiberglass-reinforced plastics.

Decades since 1961 (Year)

THINK AND COMMUNICATE

1. Copy and complete the table.

Decades since 1961	Plastics production (billions of pounds)	Change per decade	Percent change per decade
0 (1961)	7	—	—
1 (1971)	21	?	?
2 (1981)	40	?	?
3 (1991)	63	?	?

2. What was the average change?

3. What was the average percent change?

4. How many decades since 1961 is 2001? Estimate the plastics production in 2001.

EXAMPLE 1   Application: Manufacturing

Write an equation to model the plastics production since 1961 using:

a. the average change from *Think and Communicate* Question 2.

b. the average percent change from *Think and Communicate* Question 3.

SOLUTION

Let x = the number of decades since 1961.
Let y = the plastics production (in billions of pounds).

a. You can find y by starting with the plastics production in 1961 and adding the constant growth rate x times.

$$y = \begin{pmatrix} \text{amount of plastic} \\ \text{produced in 1961} \end{pmatrix} + \begin{pmatrix} \text{constant} \\ \text{growth rate} \end{pmatrix} \cdot x$$

In 1961 there were 7 billion pounds of plastics produced.

The average constant growth rate is 18.7.

$$y = 7 + 18.7x$$

b. You can find y by starting with the plastics production in 1961 and multiplying by the proportional growth factor x times.

$$y = \begin{pmatrix} \text{amount of plastic} \\ \text{produced in 1961} \end{pmatrix} \begin{pmatrix} \text{proportional} \\ \text{growth factor} \end{pmatrix}^x$$

In 1961 there were 7 billion pounds of plastics produced.

Each decade about 116% more plastics are produced. The proportional growth factor is 216%.

$$y = 7(2.16)^x$$

A **function** pairs each input value, x, with exactly one output value, y. The two equations from Example 1 are functions.

A model based on constant growth is called a **linear** function.

A model based on proportional growth is called an **exponential** function.

$y = 7 + 18.7x$	
x (input)	y (output)
0	7
1	25.7
2	44.4
3	63.1

$y = 7(2.16)^x$	
x (input)	y (output)
0	7
1	15.1
2	32.6
3	70.5

Since you input an x-value to represent a decade of your choice, x is called the **independent variable**. Since the plastics production *depends* on the decade you choose, y is called the **dependent variable**.

EXAMPLE 2   Application: Manufacturing

a. Use a graphing calculator or graphing software to graph the equations from Example 1 and the plastics production data for 1961, 1971, 1981, and 1991 in the same coordinate plane.

b. Does each equation reasonably model the data? Why or why not?

c. Use each model to estimate the plastics production in 2001.

For more information about graphing equations and graphing data points, see the *Technology Handbook,* **p. 804 and p. 814.**

SOLUTION

a. Graph $y = 7 + 18.7x$, $y = 7(2.16)^x$, and the data points for 1961, 1971, 1981, and 1991.

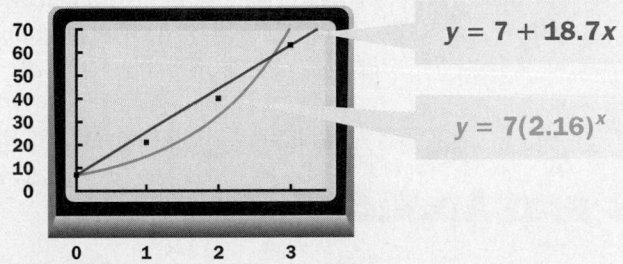

b. Both equations reasonably model the data, although the linear equation appears to lie a little closer to the data points.

c. The year 2001 is four decades after 1961. Trace along each graph until $x = 4$.

Linear Model

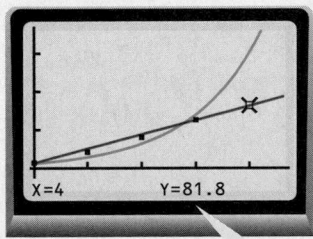

X=4   Y=81.8

Exponential Model

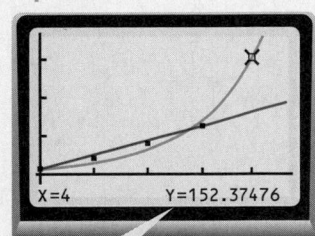

X=4   Y=152.37476

Read the value of
y when $x = 4$.

A low estimate of the plastics production in 2001, based on the linear model, is about 81 billion pounds. A high estimate, based on the exponential model, is about 152 billion pounds.

THINK AND COMMUNICATE

5. What would be a mid-range estimate for part (c) of Example 2?

6. Which function grows faster, the *exponential function* or the *linear function*? What does this mean when you use these functions to make long-term predictions?

7. Which function do you think more accurately predicts the growth of plastics production in the future? Why?

INDUSTRY The table shows the number of people working in the cellular telephone industry from 1987 to 1992. The average change was 5440 workers per year and the average percent change was 37.4% per year.

Year	Number of workers
1987	7,147
1988	11,400
1989	15,927
1990	21,382
1991	26,327
1992	34,348

1. Define the dependent and independent variables in this situation.

2. Write a linear function to model the number of people working in the cellular telephone industry since 1987.

3. Write an exponential function to model the situation.

4. Estimate the number of people working in the cellular telephone industry in 2000. Explain how you arrived at your estimate.

1.2 **Exercises and Applications**

Extra Practice exercises on page 749

SPORTS Jon-Paul and Marya each used an equation to model the number of youth softball teams, in thousands, in the United States. Refer to each model to answer Exercises 1–8.

Marya
Suppose x is number of years since 1980 and y is the number of youth softball teams in the United States, in thousands. An equation that models the situation is $y = 18 (1.10)^x$.

Jon-Paul
Let x represent the number of years since 1985 and y represent the number of youth softball teams in the United States, in thousands. An equation that models the situation is $y = 3x + 31$.

1. Which student created an exponential model? a linear model?

2. Which variable is the independent variable? the dependent variable?

3. What year is represented by an x-value of 0 in Jon-Paul's model? in Marya's model? Why are these values different?

4. How many youth softball teams were there in 1980? in 1985? How did you determine each value?

5. What is the average change in the number of youth softball teams? Which model did you use to find this information?

6. What is the average percent change in the number of youth softball teams each year? Which model did you use to find this information?

7. Use Marya's model to find the number of youth softball teams in 1988.

8. Use Jon-Paul's model to find the number of youth softball teams in 1995.

The table shows the winning times, in seconds, in the 400 m Olympic freestyle swimming event from 1960 to 1992. Modern training methods are one reason why the winning times have improved so greatly.

Year	Men	Women
1960	258.3	290.6
1964	252.2	283.3
1968	249.0	271.8
1972	240.27	259.44
1976	231.93	249.89
1980	231.31	248.76
1984	231.23	247.10
1988	226.95	243.85
1992	225.00	247.18

(Top) Janet Evans of the United States swims in the women's 400 m freestyle at the 1988 Olympics in Seoul. (Above) An Olympic swimmer trains for a competition by swimming against a pump-generated current. While she swims, a computer analyzes her stroke and oxygen intake.

9. **Writing** To analyze the swimming data, Fredric wants to let x be the number of years since 1960. Do you think this is a good definition for the independent variable? Why or why not?

10. a. Use a linear function to model the women's winning swim times. Why is the growth rate negative?

   b. Use an exponential function to model the women's times. (*Hint:* If the average percent change is -5%, the growth factor is $1.00 + (-0.05)$, or 0.95.)

   c. **Technology** Use a graphing calculator or graphing software to graph the data points and your equations from parts (a) and (b). Does each equation reasonably model the data? Why or why not?

   d. Estimate the women's winning swim times in 1996 and 2000.

11. a. Repeat Exercise 10 using the data for the men's winning times in the 400 m Olympic freestyle swimming event.

   b. **Writing** Use your models from Exercise 10 and part (a) of Exercise 11 to estimate the year in which the women's winning time will be better than the men's winning time. Do you think your estimate is a reasonable one? Explain.

   c. **Challenge** What x-value would you use in your models from part (a) to estimate the men's winning time in the 1956 Olympics? Why? Use this value to estimate the men's winning time in 1956.

12. **Writing** Do you think that the winning swim times will continue to decrease? Why or why not?

13. Cooperative Learning Work with a partner. You will need graphing calculators or graphing software.

a. Create a linear model and an exponential model of the growth of the United States national debt for your data set. One of you should use the 1950–1990 data and the other should use the 1990–1994 data.

b. 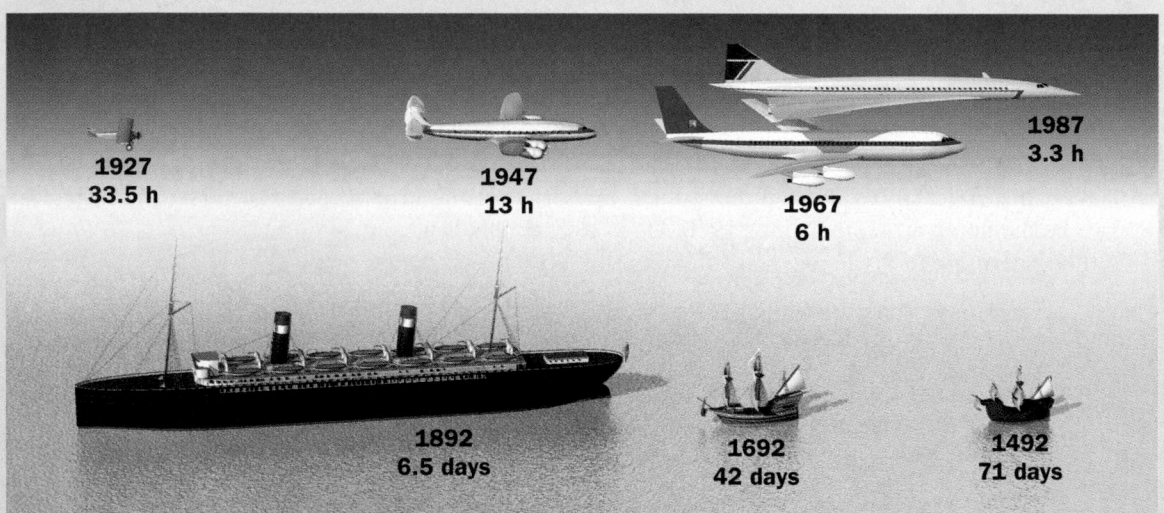 **Technology** Each of you should graph your equations and the data points in the same coordinate plane. With your partner, decide which equation is a better model for each data set.

c. Estimate the national debt in 2000. Compare your estimates.

d. **Writing** Which model do you think gives a better estimate of the national debt in 1996? in 2010? Explain your reasoning.

National Debt	
Year	Trillions of $
1950	0.26
1960	0.29
1970	0.38
1980	0.91
1990	3.27
1991	3.68
1992	4.08
1993	4.44
1994	4.70

Connection HISTORY

Christopher Columbus sailed across the Atlantic Ocean in 71 days. In 1927 it took Charles Lindbergh 33.5 hours to fly across the Atlantic Ocean. As transportation has improved, the amount of time needed to cross the Atlantic Ocean has changed.

1927
33.5 h

1947
13 h

1967
6 h

1987
3.3 h

1892
6.5 days

1692
42 days

1492
71 days

Use the information in the diagram above for Exercises 14 and 15.

14. a. Write a linear function and an exponential function to model the amount of time it takes to cross the Atlantic Ocean by sea. What interval of time does each x-value represent?

b. **Technology** Use a graphing calculator or graphing software to graph the data points and your equations from part (a) in the same coordinate plane. Does each equation reasonably model the data?

c. Use your model to estimate the amount of time it would take to cross the Atlantic Ocean by sea in 1992. A powerboat crossed the Atlantic Ocean in about 2.5 days in 1992. How close is your estimate?

15. Repeat Exercise 14 using the data for the amount of time it takes to cross the Atlantic Ocean by air. Use your model to estimate the amount of time it will take to cross the Atlantic Ocean by air in 2007.

16. Writing When do you think it is best to model a situation using a linear model? an exponential model?

17. The table shows the revenue from book sales in the United States. *(Section 1.1)*

a. Use a constant growth model to estimate the revenue in 1993.

b. Repeat part (a) using a proportional growth model.

Year	Revenue (billions)
1989	$18.0
1990	$19.0
1991	$20.1

Evaluate each expression when $a = 5$ and $b = -4$. *(Toolbox, page 780)*

18. $-(a + b)^2$ **19.** $-a - b$ **20.** $a - 2b + 14$

21. The table shows the number of major hurricanes from 1900 to 1980. Find the mean, median, and mode of the data. *(Toolbox, page 790)*

Period	1901–1920	1921–1940	1941–1960	1961–1980
Major hurricanes	13	13	18	10

ASSESS YOUR PROGRESS

VOCABULARY

mathematical model (p. 4)   **independent variable** (p. 10)
function (p. 10)   **dependent variable** (p. 10)

1. Writing What calculation do you perform when you model a situation using constant growth? proportional growth? *(Section 1.1)*

2. ENTERTAINMENT The table shows the average cost of a movie ticket from 1951 to 1991. *(Section 1.2)*

a. Write a linear function to model the situation.

b. Write an exponential function to model the situation.

c. **Technology** Use a graphing calculator or graphing software to graph the data points and your equations from parts (a) and (b) in the same coordinate plane. Does one model fit the data better? If so, which one? Explain your choice.

d. Estimate the price of a movie ticket in 2001.

3. Journal When do you think it is best to model a situation using a table? an equation? a graph? Which method do you prefer? Why?

1.3 Modeling with Matrices

Learn how to...

- **organize information in a matrix**
- **add matrices**
- **multiply a matrix by a scalar**

So you can...

- **use matrices to model data, such as characteristics of people**

Did you know that about ten percent of the population is left-handed? Scientists who studied subjects in works of art dating from more than 5000 years ago found that the percentage of left-handed people in the world has stayed fairly constant over time.

EXPLORATION

COOPERATIVE LEARNING

Organizing Data in Matrices

Work with half of your class.

1 Find out which hand each person in your group uses to write. Record the results in a *matrix* like the one shown.

$$\begin{array}{c} \\ \text{Males} \\ \text{Females} \end{array} \overset{\text{Left} \quad \text{Right} \quad \text{Both}}{\begin{bmatrix} ? & ? & ? \\ ? & ? & ? \end{bmatrix}}$$

Use the "Both" column if a person writes well with both hands.

2 Give your matrix to the other group. Create a new matrix by adding the number in each position of your matrix to the number in the same position in the other group's matrix. What does the new matrix represent?

3 How many females in your class are right-handed? How many males in your class are left-handed?

4 How do you think you could use your data to predict what a similar matrix for your whole school might be?

A **matrix** is a group of numbers arranged in rows and columns. Each number in a matrix is called an **element**. The **dimensions of a matrix** have the form $r \times c$ where r is the number of rows and c is the number of columns.

Matrix A has 2 rows and 3 columns. Its dimensions are **2 × 3** (read "two by three").

$$A = \begin{bmatrix} 3 & 1 & -4 \\ 0 & 7 & 15 \end{bmatrix}$$

You write $a_{2,1}$ to represent an element in the second row and first column of matrix A.

EXAMPLE 1

The groups in one of Kim Timpf's algebra classes got the matrices shown when they did the Exploration. Find the class total.

Group A:

$$\begin{matrix} & \text{Left} & \text{Right} & \text{Both} \end{matrix}$$

$$\begin{matrix} \text{Males} \\ \text{Females} \end{matrix} \begin{bmatrix} 2 & 8 & 0 \\ 1 & 5 & 0 \end{bmatrix} = A$$

Group B:

$$\begin{matrix} & \text{Left} & \text{Right} & \text{Both} \end{matrix}$$

$$\begin{matrix} \text{Males} \\ \text{Females} \end{matrix} \begin{bmatrix} 0 & 5 & 0 \\ 2 & 9 & 0 \end{bmatrix} = B$$

SOLUTION

The class total is the sum $A + B$. You can add the two matrices because the corresponding entries in each matrix represent the same characteristic. For example, $a_{1,1}$ and $b_{1,1}$ both represent left-handed males.

$$A + B = \begin{bmatrix} 2 & 8 & 0 \\ 1 & 5 & 0 \end{bmatrix} + \begin{bmatrix} 0 & 5 & 0 \\ 2 & 9 & 0 \end{bmatrix}$$

$$= \begin{bmatrix} 2+0 & 8+5 & 0+0 \\ 1+2 & 5+9 & 0+0 \end{bmatrix}$$

$$= \begin{bmatrix} 2 & 13 & 0 \\ 3 & 14 & 0 \end{bmatrix}$$

Add each element in matrix A to the element in the same position in matrix B.

THINK AND COMMUNICATE

1. What is the position of the 5 in matrix B of Example 1? What is the value of $a_{1,3}$ in matrix A of Example 1?

2. Based on the solution to Example 1, state how many students in Kim Timpf's class are:
   a. right-handed females   **b.** left-handed males

3. Can you add matrices G and H? Explain why or why not.

$$G = \begin{bmatrix} 3 & 9 & -8 \\ -4 & 1 & 1 \end{bmatrix} \qquad H = \begin{bmatrix} 6 & 4 \\ -2 & 0 \\ 3 & 5 \end{bmatrix}$$

EXAMPLE 2

Find $C - D$ when $C = \begin{bmatrix} 4 & -2 \\ 7 & 0 \end{bmatrix}$ and $D = \begin{bmatrix} 1 & 1 \\ 0 & 2 \end{bmatrix}$.

SOLUTION

$$C - D = \begin{bmatrix} 4 & -2 \\ 7 & 0 \end{bmatrix} - \begin{bmatrix} 1 & 1 \\ 0 & 2 \end{bmatrix}$$

$$= \begin{bmatrix} 4-1 & -2-1 \\ 7-0 & 0-2 \end{bmatrix}$$

$$= \begin{bmatrix} 3 & -3 \\ 7 & -2 \end{bmatrix}$$

Subtract corresponding elements.

Scalar Multiplication

You can multiply a matrix by a number, or **scalar**. This process, known as **scalar multiplication**, is shown in the following example.

EXAMPLE 3

Use the solution from Example 1 to estimate the totals for all 160 of Kim Timpf's algebra students.

SOLUTION

The elements of $A + B$ are *counts* of students. There are 32 students in all.

Toolbox p. 785
Ratios and Proportions

Step 1 Convert the counts to *ratios* by multiplying each element of $A + B$ by the scalar $\frac{1}{32}$.

$$\frac{1}{32}\begin{bmatrix} 2 & 13 & 0 \\ 3 & 14 & 0 \end{bmatrix} = \begin{bmatrix} \frac{2}{32} & \frac{13}{32} & \frac{0}{32} \\ \frac{3}{32} & \frac{14}{32} & \frac{0}{32} \end{bmatrix}$$

Step 2 Estimate the counts for all 160 of Kim Timpf's students by multiplying each element of the matrix from Step 1 by the scalar 160.

$$160\begin{bmatrix} \frac{2}{32} & \frac{13}{32} & \frac{0}{32} \\ \frac{3}{32} & \frac{14}{32} & \frac{0}{32} \end{bmatrix} = \begin{bmatrix} 160\left(\frac{2}{32}\right) & 160\left(\frac{13}{32}\right) & 160\left(\frac{0}{32}\right) \\ 160\left(\frac{3}{32}\right) & 160\left(\frac{14}{32}\right) & 160\left(\frac{0}{32}\right) \end{bmatrix}$$

$$= \begin{bmatrix} 10 & 65 & 0 \\ 15 & 70 & 0 \end{bmatrix}$$

THINK AND COMMUNICATE

BY THE WAY

People have a preferred foot, as well as a preferred hand. Most people use a particular foot to kick a ball or to pick up an object with their toes.

4. What is an estimate for the number of right-handed males in Kim Timpf's classes? for the number of left-handed females?

5. What assumption do you make when "scaling up" from the 32 students in Example 1 to the 160 students in Example 3?

☑ CHECKING KEY CONCEPTS

For Questions 1–6, use the matrices below to evaluate each matrix expression.

$$A = \begin{bmatrix} 1 & 3 \\ 6 & 2 \end{bmatrix} \quad B = \begin{bmatrix} 4 & 1 \\ 0 & 3 \end{bmatrix} \quad C = \begin{bmatrix} 5 & -2 & 1 \\ 0 & -8 & 6 \\ -3 & 4 & -7 \end{bmatrix} \quad D = \begin{bmatrix} -1 & 2 & 1 \\ 6 & -6 & 4 \\ 2 & -8 & 3 \end{bmatrix}$$

1. $A + B$

2. $C + D$

3. $A - B$

4. $2B$

5. $C + 3D$

6. $3C - 2D$

1.3 Exercises and Applications

Extra Practice
exercises on
page 749

For Exercises 1–3, use the matrix $A = \begin{bmatrix} -1 & 2 & 0 & 11 \\ 3 & 6 & -9 & 4 \end{bmatrix}$.

1. What are the dimensions of A?

2. What is the value of $a_{1,2}$? the value of $a_{2,1}$?

3. **Writing** Let $B = \begin{bmatrix} 1 & 8 & -3 \\ 4 & 2 & 0 \end{bmatrix}$. Can you add matrix A to matrix B? If so, describe the method you use. If not, explain why not.

For Exercises 4–11, use matrices *P, Q, R,* and *S* to evaluate each matrix expression. If an operation cannot be performed, write *undefined*.

$$P = \begin{bmatrix} 7 & 11 \\ -15 & 3 \end{bmatrix} \qquad Q = \begin{bmatrix} 1 & 0 \\ 4 & -\frac{11}{2} \end{bmatrix} \qquad R = \begin{bmatrix} 8 & -2 \\ 19 & 6 \\ 1 & -13 \end{bmatrix} \qquad S = \begin{bmatrix} 6 & 8 \\ 0 & 12 \\ 10 & -4 \end{bmatrix}$$

4. $P + Q$

5. $R + S$

6. $Q + R$

7. $S - R$

8. $2P$

9. $-4P + 2Q$

10. $3R - \frac{1}{2}S$

11. $-Q + 5S$

Connection ▸ AGRICULTURE

Farmers in the United States harvest crops from more than 300 million acres of land. Almost 25% of the crops are grown in Illinois, Iowa, and Kansas. The table shows the number of bushels, in thousands, harvested for four leading crops.

	Corn		Wheat		Soybeans		Oats	
	1990	1991	1990	1991	1990	1991	1990	1991
Illinois	1,320,800	1,177,000	91,200	44,800	354,900	341,250	11,560	6,600
Iowa	1,562,400	1,427,400	3,375	1,700	323,900	350,325	40,800	21,250
Kansas	188,000	206,250	472,000	363,000	46,800	43,700	6,600	5,830

12. **a.** Create a 3 × 4 matrix to represent the data for 1990. Label this matrix A.

   b. Create a 3 × 4 matrix using the data for 1991. Label this matrix B.

   c. For matrices A and B, what does each row represent? What does each column represent?

13. Write an expression in terms of A and B to represent the total amount of each crop produced in each state during 1990 and 1991. Then evaluate your expression.

14. Calculate the matrix $B - A$. What does this matrix represent? What does a negative number in matrix $B - A$ represent?

15. How would the dimensions of each matrix change if Nebraska and Missouri were added to the list of states? if barley were added to the list of crops?

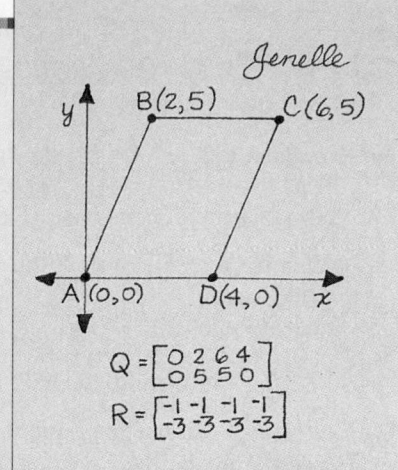

Have you ever used a computer drawing program to enlarge a figure or change its position? The computer can use scalar multiplication and matrix addition to find the coordinates of the changed figure.

Jenelle wrote the coordinates of quadrilateral *ABCD* in matrix *Q*.

16. What does each column of matrix *Q* represent?

17. a. Find $Q + R$.

b. Matrix $Q + R$ defines a new quadrilateral *EFGH*. Sketch *EFGH*. How is *EFGH* related to *ABCD*?

18. a. Find $2Q$.

b. Matrix $2Q$ defines a new quadrilateral *IJKL*. Sketch *IJKL*. How is *IJKL* related to *ABCD*?

For Exercises 19–21, the coordinates of a geometric figure are given in the black matrix, and a transformation is indicated in red. Describe each transformation. Then sketch the original figure and its image to check your answer.

19. $3\begin{bmatrix} 0 & 1 & 4 \\ 0 & 2 & 0 \end{bmatrix}$
    20. $\dfrac{1}{2}\begin{bmatrix} 0 & 8 & 2 & -2 \\ 2 & 4 & -8 & -2 \end{bmatrix}$
    21. $\begin{bmatrix} 1 & 3 & 5 & 3 \\ 3 & 6 & 3 & -3 \end{bmatrix} + \begin{bmatrix} -1 & -1 & -1 & -1 \\ 3 & 3 & 3 & 3 \end{bmatrix}$

22. Suppose *A* is any matrix.

a. Describe the matrix $A - A$.

b. How are the matrices $A + A$ and $2A$ related?

Use matrices *A*, *B*, and *C* for Exercises 23–25.

$$A = \begin{bmatrix} 3 & 1 & -2 \\ -1 & 5 & 6 \\ 4 & 13 & 0 \end{bmatrix} \quad B = \begin{bmatrix} -9 & 10 & -3 \\ 0 & 6 & 1 \\ 14 & 7 & -8 \end{bmatrix} \quad C = \begin{bmatrix} 7 & -2 & 0 \\ 15 & 1 & 19 \\ -4 & 12 & 2 \end{bmatrix}$$

23. Show that $A + B = B + A$.

24. Show that $A + (B + C) = (A + B) + C$.

25. Show that $r(A + B) = rA + rB$ where r is any scalar.

26. PERSONAL FINANCE Roger and Meredith Stone each have a savings account and a checking account. The money they had in each account at the end of May is represented by matrix *A*. Matrix *D* represents the total deposits Roger and Meredith made in June, and matrix *W* represents the total withdrawals each made in June.

	Savings	Checking			Savings	Checking			Savings	Checking
Roger	3000	540	$= A$		300	100	$= D$		0	-120	$= W$
Meredith	9500	1200			0	250			-1500	-1000	

a. Write an expression to show how to calculate the amount of money in each account at the end of June.

b. Simplify the expression you wrote in part (a) to write a matrix showing the amount of money in each account at the end of June.

27. Challenge Find X if X is a matrix and $2X + \begin{bmatrix} 9 & -2 \\ 4 & 7 \end{bmatrix} = \begin{bmatrix} 3 & 10 \\ -6 & 12 \end{bmatrix}$.

28. SAT/ACT Preview What value of a makes the equation true?

$$\begin{bmatrix} 2 & -5 \\ -1 & 4 \end{bmatrix} + \begin{bmatrix} 1 & 0 \\ 2a & -5 \end{bmatrix} = \begin{bmatrix} 3 & -5 \\ -9 & -1 \end{bmatrix}$$

A. 5   **B.** -5 **C.** -10 **D.** -4 **E.** 2

ONGOING ASSESSMENT

29. Cooperative Learning Work in a group of four. Each of you should choose a class (freshman, sophomore, junior, or senior) and ask ten to twenty people from the class how they most often get to school. Complete a matrix like the one below for each class.

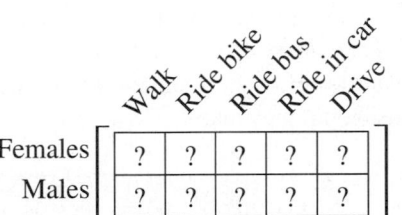

a. Find out the total number of students in each class in your school. Scale each matrix up to estimate how many males and females in each class use each form of transportation.

b. Add the four matrices you found in part (a) to create a matrix estimating how many students in your school use each form of transportation.

c. Why wouldn't you add the matrices first and then scale up?

SPIRAL REVIEW

ENTERTAINMENT The table shows the number of people, in millions, subscribing to cable television systems from 1960 to 1990. *(Section 1.2)*

30. Write a linear function to model the situation.

31. Write an exponential function to model the situation.

32. Estimate the number of cable television subscribers in 2000. Explain how you arrived at your estimate.

Decades since 1960	Cable users (millions)
0	0.65
1	4.5
2	16
3	50

Simplify. *(Toolbox, page 778)*

33. $4(-5) + 7(3)$ **34.** $16 \div 8 + 12 \div 3$ **35.** $15 \cdot 2 + 3 \cdot 6$

Tell whether each property is true for real numbers a, b, and c. Give an example to support your answer. *(Toolbox, page 774)*

36. $a + b = b + a$ **37.** $a - b = b - a$ **38.** $ab = ba$

39. $a \cdot \dfrac{1}{a} = 1$ **40.** $b \cdot 1 = b$ **41.** $|a| + |b| = |a + b|$

1.4 Multiplying Matrices

Have you ever gone canoeing? Depending on the route you choose, you may need to carry the canoes and supplies over land to get to another waterway or to avoid an obstacle. This process is called *portage*.

Learn how to...

- multiply matrices

So you can...

- solve problems involving matrix multiplication, such as planning a canoe trip

---

EXAMPLE 1 Application: Canoeing

A group of canoers arrives at the Canoe Centre at Opeongo Campgrounds in Algonquin Provincial Park. There are about seven hours of daylight left, six of which they can use to travel to a campsite. They can canoe 1 km in 10 min, but they need 20 min to portage 1 km. They would like to camp at one of the four sites shown on the map. Which route should they take?

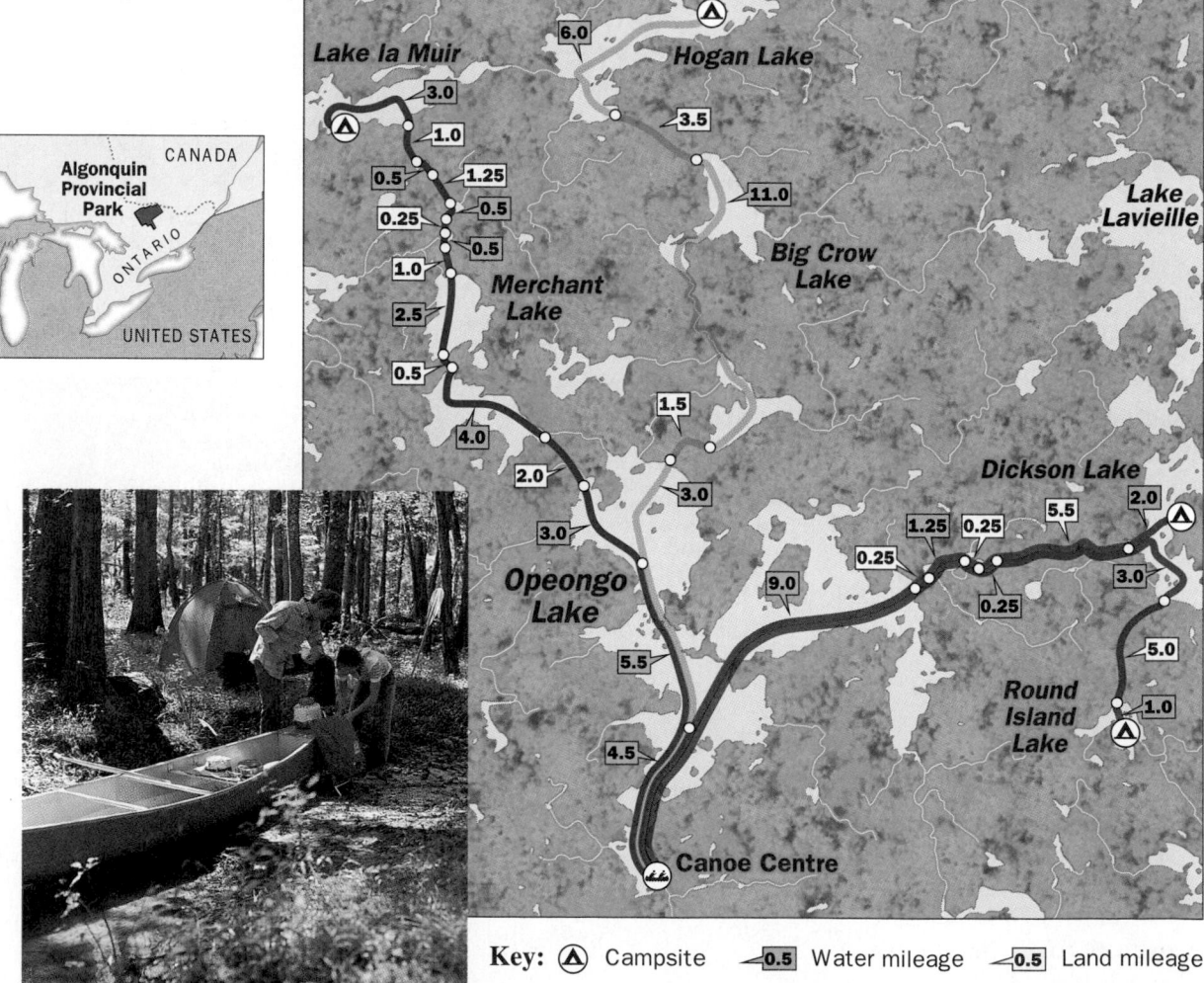

Key: ▲ Campsite   ◄0.5 Water mileage   ◄0.5 Land mileage

Let matrix A show the canoeing and portaging distances, in kilometers, along each route. Let matrix B show the number of minutes per kilometer that the group can travel using each method of travel.

$$\begin{array}{c} \phantom{\text{Route to Lake la Muir}} \begin{array}{cc} \text{Canoe} & \text{Portage} \\ \text{(km)} & \text{(km)} \end{array} \\ \begin{array}{r} \text{Route to Lake la Muir} \\ \text{Route to Hogan Lake} \\ \text{Route to Dickson Lake} \\ \text{Route to Round Island Lake} \end{array} \begin{bmatrix} 24 & 6 \\ 30 & 5 \\ 17 & 6 \\ 19 & 11 \end{bmatrix} = A \end{array}$$

$$\begin{array}{c} \begin{array}{c} \text{Time} \\ \text{(min/km)} \end{array} \\ \begin{array}{r} \text{Canoe} \\ \text{Portage} \end{array} \begin{bmatrix} 10 \\ 20 \end{bmatrix} = B \end{array}$$

Use matrices A and B to create a matrix of total travel times to each lake.

$$\begin{array}{c} \text{Time (min)} \\ \begin{array}{r} \text{Route to Lake la Muir} \\ \text{Route to Hogan Lake} \\ \text{Route to Dickson Lake} \\ \text{Route to Round Island Lake} \end{array} \begin{bmatrix} 24(10) + 6(20) \\ 30(10) + 5(20) \\ 17(10) + 6(20) \\ 19(10) + 11(20) \end{bmatrix} = \begin{bmatrix} 360 \\ 400 \\ 290 \\ 410 \end{bmatrix} \end{array}$$

Since the group has 6 h, or 360 min, they have time to get to either Lake la Muir or Dickson Lake.

Matrix Multiplication

Example 1 illustrates a process called **matrix multiplication**.

$$\begin{bmatrix} 24 & 6 \\ 30 & 5 \\ 17 & 6 \\ 19 & 11 \end{bmatrix} \begin{bmatrix} 10 \\ 20 \end{bmatrix} = \begin{bmatrix} 24(10) + 6(20) \\ 30(10) + 5(20) \\ 17(10) + 6(20) \\ 19(10) + 11(20) \end{bmatrix} = \begin{bmatrix} 360 \\ 400 \\ 290 \\ 410 \end{bmatrix}$$

Dimensions: $4 \times \underline{2}$ $\underline{2} \times 1$ 4×1

This element, which is in the second row and first column of the product matrix, is the sum of the products of the corresponding elements in the second row of A and the first column of B: $400 = 30(10) + 5(20)$.

For the product to be defined, the number of elements in a **row of A** must equal the number of elements in a **column of B**.

The product matrix has the same number of rows as A and the same number of columns as B.

When you multiply matrices with labels, the column labels of the first matrix must have the same attributes as the row labels of the second matrix.

These labels are the same.

$$\begin{array}{r} \text{Route to Lake la Muir} \\ \text{Route to Hogan Lake} \\ \text{Route to Dickson Lake} \\ \text{Route to Round Island Lake} \end{array} \begin{array}{c} \text{Canoe Portage} \\ \begin{bmatrix} 24 & 6 \\ 30 & 5 \\ 17 & 6 \\ 19 & 11 \end{bmatrix} \end{array} \times \begin{array}{c} \\ \text{Canoe} \\ \text{Portage} \end{array} \begin{bmatrix} \text{Time} \\ 10 \\ 20 \end{bmatrix} = \begin{array}{r} \text{Route to Lake la Muir} \\ \text{Route to Hogan Lake} \\ \text{Route to Dickson Lake} \\ \text{Route to Round Island Lake} \end{array} \begin{bmatrix} \text{Time} \\ 360 \\ 400 \\ 290 \\ 410 \end{bmatrix}$$

EXAMPLE 2 Application: Camping

The Graham, Piscitelli, and Brewer families order supplies for an upcoming camping trip from the catalog shown. The number of items needed by each family is given in the table. Find the shipping weight and total cost for each family's order.

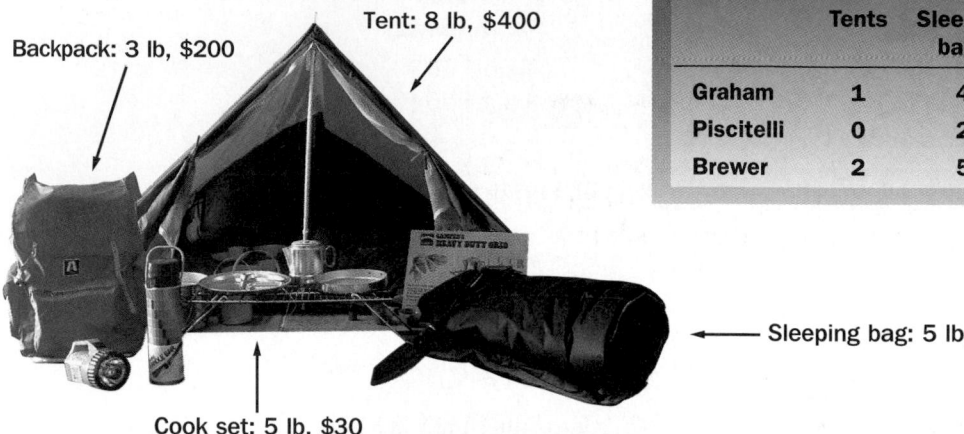

Backpack: 3 lb, $200

Tent: 8 lb, $400

Sleeping bag: 5 lb, $300

Cook set: 5 lb, $30

	Tents	Sleeping bags	Back-packs	Cook sets
Graham	1	4	2	0
Piscitelli	0	2	3	1
Brewer	2	5	4	2

SOLUTION

Let matrix A show each family's camping supplies order, and let matrix B show the shipping weight and cost of each item.

$$\begin{array}{c} \text{Graham} \\ \text{Piscitelli} \\ \text{Brewer} \end{array} \begin{bmatrix} 1 & 4 & 2 & 0 \\ 0 & 2 & 3 & 1 \\ 2 & 5 & 4 & 2 \end{bmatrix} = A \qquad \begin{array}{c} \text{Tents} \\ \text{Sleeping bags} \\ \text{Backpacks} \\ \text{Cook sets} \end{array} \begin{bmatrix} 8 & 400 \\ 5 & 300 \\ 3 & 200 \\ 5 & 30 \end{bmatrix} = B$$

The product AB will be a 3×2 matrix showing the shipping weight and total cost for each family. Find AB.

Method 1 Use paper and pencil.

$$AB = \begin{bmatrix} 1 & 4 & 2 & 0 \\ 0 & 2 & 3 & 1 \\ 2 & 5 & 4 & 2 \end{bmatrix} \begin{bmatrix} 8 & 400 \\ 5 & 300 \\ 3 & 200 \\ 5 & 30 \end{bmatrix}$$

$$= \begin{bmatrix} 1(8) + 4(5) + 2(3) + 0(5) & 1(400) + 4(300) + 2(200) + 0(30) \\ 0(8) + 2(5) + 3(3) + 1(5) & 0(400) + 2(300) + 3(200) + 1(30) \\ 2(8) + 5(5) + 4(3) + 2(5) & 2(400) + 5(300) + 4(200) + 2(30) \end{bmatrix}$$

$$= \begin{bmatrix} 34 & 2000 \\ 24 & 1230 \\ 63 & 3160 \end{bmatrix}$$

The order placed by the Graham family weighs 34 lb and costs $2000. The order placed by the Piscitelli family weighs 24 lb and costs $1230. The order placed by the Brewer family weighs 63 lb and costs $3160.

Method 2 Use a graphing calculator or software with matrix calculation capabilities.

Use the matrix feature to enter matrices *A* and *B*. Then find *AB*.

```
[A][B]
  [[34  2000]
   [24  1230]
   [63  3160]]
```

For more information about matrix calculations, see the *Technology Handbook*, pp. 811–812.

The order placed by the Graham family weighs 34 lb and costs $2000. The order placed by the Piscitelli family weighs 24 lb and costs $1230. The order placed by the Brewer family weighs 63 lb and costs $3160.

☑ CHECKING KEY CONCEPTS

Find each product. If the matrices cannot be multiplied, state that the product is *undefined*.

$$P = \begin{bmatrix} 2 & 0 \\ 5 & 1 \end{bmatrix} \quad Q = \begin{bmatrix} 4 & 3 & -1 \\ -2 & 5 & 1 \end{bmatrix} \quad R = \begin{bmatrix} 1 & 2 & -1 \\ 4 & 3 & -3 \\ 5 & 1 & -1 \end{bmatrix} \quad S = \begin{bmatrix} 8 & -10 & 0 \\ 3 & -6 & 4 \\ -2 & 1 & 5 \end{bmatrix}$$

1. a. *PQ* **2. a.** *QR* **3. a.** *RS* **4. a.** *QS*

   b. *QP*   **b.** *RQ*   **b.** *SR*   **b.** *SQ*

5. Which pair of products in Questions 1–4 were both defined? Were any of the products commutative?

1.4 Exercises and Applications

Extra Practice exercises on page 750

1. In Example 2, suppose each family also orders some compasses. How would the dimensions of matrices *A*, *B*, and *AB* change?

Tell whether *LM* and *ML* are defined. If so, give the dimensions of the product matrices. If not, explain why not.

2. *L* is any 4 × 2 matrix.
M is any 1 × 4 matrix.

3. *L* is any 8 × 6 matrix.
M is any 6 × 6 matrix.

4. *L* is any 3 × 3 matrix.
M is any 3 × 1 matrix.

5. *L* is any 4 × 5 matrix.
M is any 3 × 4 matrix.

For Exercises 6–13, use matrices *A*, *B*, *C*, *D*, and *E* to find each product. If the matrices cannot be multiplied, state that the product is *undefined*.

$$A = \begin{bmatrix} 2 \\ -4 \\ 6 \end{bmatrix} \quad B = \begin{bmatrix} 3 & 1 \\ 5 & 0 \end{bmatrix} \quad C = \begin{bmatrix} 2 & 0 & 7 \\ -1 & -2 & 0 \\ 3 & 7 & 4 \end{bmatrix} \quad D = \begin{bmatrix} 0 & 3 & -2 \\ 2 & 5 & 1 \end{bmatrix} \quad E = \begin{bmatrix} 1 & -2 \\ 5 & 3 \end{bmatrix}$$

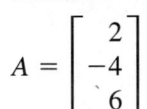 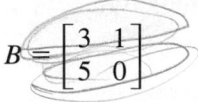 

6. *CA*   **7.** *BE*   **8.** *CD*   **9.** *DA*

10. *AB*   **11.** *ED*   **12.** *DC*   **13.** *BA*

Look back at the
article on pages
xxxiv–2.

In addition to receiving orders through the mail, Twyla Lang sells her products at conventions and trade shows. Along with the Kente scarves, Twyla's company offers other items, such as earrings and shirts.

14. The Kente scarves that Lang sells come in four different styles. Suppose matrix A shows the number of each style of scarf sold for 1992–1994.

$$\begin{array}{c} \\ 1992 \\ 1993 \\ 1994 \end{array} \begin{array}{cccc} \text{Class} & \text{Plain} & \text{Sorority} & \text{2-color sorority} \\ \left[\begin{array}{cccc} 0 & 0 & 100 & 0 \\ 250 & 100 & 15 & 20 \\ 450 & 85 & 65 & 75 \end{array}\right] \end{array} = A$$

a. The costs of the different scarves are shown in the photo below. Create a matrix C to represent the cost of each style of Kente scarf.

b. Find AC. What does this matrix represent?

Class year scarf: **$25**
Day 1 sales: **40**
Day 2 sales: **32**

Shirts: **$12**
Day 1 sales: **65**
Day 2 sales: **45**

Plain Kente scarf: **$20**
Day 1 sales: **45**
Day 2 sales: **52**

Earrings: **$15**
Day 1 sales: **25**
Day 2 sales: **18**

2-color sorority scarf: **$25**
Day 1 sales: **12**
Day 2 sales: **24**

Sorority Kente scarf: **$30**
Day 1 sales: **28**
Day 2 sales: **16**

15. Suppose the sales for each item from one show are those shown above.

   a. Create matrix S to show the sales for each item and matrix T to show the cost of each item. What are the dimensions of each matrix?

   b. Find ST. What does this matrix represent?

16. Cooperative Learning Work with a partner. Use matrices A, B, C, and D.

$$A = \begin{bmatrix} 4 & 0 \\ -2 & 1 \end{bmatrix} \qquad B = \begin{bmatrix} 1 & 0 \\ 0 & 1 \end{bmatrix} \qquad C = \begin{bmatrix} -2 & -3 \\ 3 & 5 \end{bmatrix} \qquad D = \begin{bmatrix} -5 & -3 \\ 3 & 2 \end{bmatrix}$$

   a. Multiply all possible pairs of matrices. (You do not need to multiply a matrix by itself.) Is the product always commutative?

   b. For real numbers, the number 1 is called the *multiplicative identity* because $1 \cdot n = n$ and $n \cdot 1 = n$. Which matrix acts as a multiplicative identity for matrices A, B, C, and D? Give an example.

   c. For real numbers not equal to zero, n and $\frac{1}{n}$ are *multiplicative inverses* because $n\left(\frac{1}{n}\right) = 1$ (the multiplicative identity). Which two matrices are multiplicative inverses of each other? Explain your reasoning.

On August 5, 6, and 7, 1994, the San Francisco Giants played the Houston Astros at the Houston Astrodome. The Astrodome has many different types of seating available, as shown in the diagram. The attendance for the general seating at each game is given in the table.

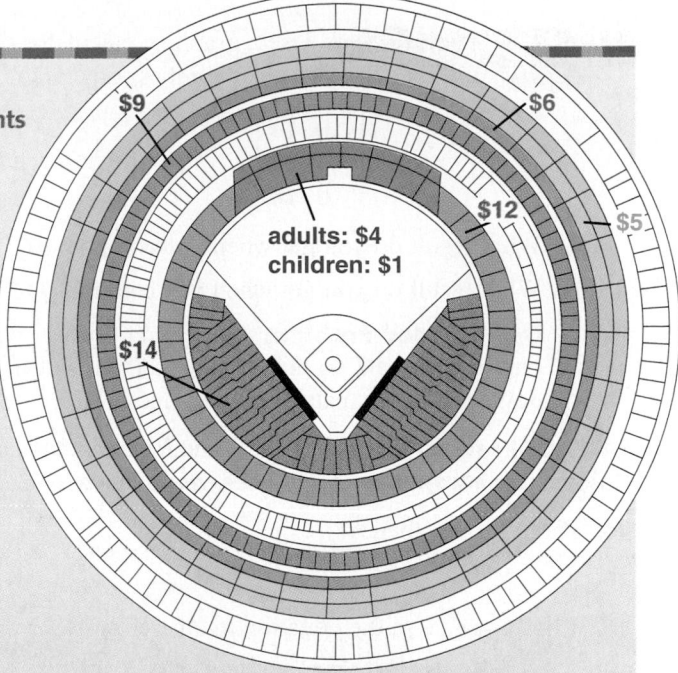

$9 $6 $12 $5 $14

adults: $4
children: $1

	Pavilion (adults)	Pavilion (children)	Upper Reserved	Upper Box	Loge	Mezzanine	Field Box
8/5/94	2,339	952	11,518	3,579	4,309	10,496	7,641
8/6/94	2,137	1,223	12,167	3,381	7,436	11,476	7,683
8/7/94	2,289	984	1,609	1,403	2,997	9,003	7,535

17. a. Create matrix A showing the attendance for each type of seating by date. Create matrix B showing the cost for each type of seating.

    b. Find AB. What does this matrix represent?

18. **Writing** Which type of seating has the most consistent revenue? Which has the most varied revenue? Explain your reasoning.

19. In addition to the general seating, the Astrodome has executive seats. The attendance for these seats is given in the table below. Create matrix C showing the attendance for each type of seating by date. Create matrix D showing the cost for each type of seating. Find CD.

	Skybox ($10)	Club Level ($15)	Bullpen Room ($15)	Star Deck ($17)	Star Suite ($30)	Owner's Club ($50)	Diamond Level ($100)
8/5/94	1325	47	65	2715	947	101	94
8/6/94	1768	52	50	2696	1140	80	94
8/7/94	1225	44	50	2718	721	51	93

20. Find the sum of AB and CD. What does this matrix represent?

21. Which game had the largest revenue?

22. **Writing** Which type of seat do you think is most popular? least popular? Explain your reasoning.

23.  **Technology** Use a graphing calculator or computer software with matrix calculation capabilities. Enter each matrix.

$$S = \begin{bmatrix} 10 & 14 \\ -3 & 23 \\ 0 & 18 \end{bmatrix} \qquad T = \begin{bmatrix} 12 & -26 & 6 & 36 \\ -45 & 25 & 0 & 24 \end{bmatrix}$$

a. What result do you get when you compute ST?

b. What result do you get when you compute TS? What does this result mean?

c. **Open-ended Problem** Create additional rows or columns to change matrices S and T so that you avoid the result you got in part (b). Write down your new matrices. Then find TS.

SAT/ACT Preview Suppose that *M* and *N* are matrices and that *M* ≠ *N*. Tell if each statement is *always true, sometimes true,* or *never true.*

24. If MN exists, then NM exists.

25. $MN = NM$

26. $M + N = N + M$

27. $3M + 3N = 3(M + N)$

28. Challenge Use the equation to answer parts (a) and (b).

$$\begin{bmatrix} 3 & -1 \\ 2 & 1 \end{bmatrix} \begin{bmatrix} x \\ y \end{bmatrix} = \begin{bmatrix} 1 \\ 9 \end{bmatrix}$$

a. Multiply the matrices on the left. Then equate corresponding elements to write two equations in terms of x and y.

b. Solve the equations from part (a) for x and y. Use matrix multiplication to check your work.

ONGOING ASSESSMENT

29. a. Writing Explain how you can use the dimensions of two matrices to determine if their product exists. If the product does exist, how are you able to determine the dimensions of the product matrix before you complete the multiplication?

b. **Open-ended Problem** Create a poster that demonstrates, step by step, the process for multiplying two 2 × 2 matrices. Use color-coding.

SPIRAL REVIEW

For Exercises 30–32, use matrices *A* and *B*. *(Section 1.3)*

$$A = \begin{bmatrix} 1 & 0 \\ 2 & 4 \end{bmatrix} \qquad B = \begin{bmatrix} 2 & 2 \\ 1 & 3 \end{bmatrix}$$

30. Find $A + B$.   **31.** Find $-\frac{3}{2}A$.   **32.** Find $4A + 2B$.   **33.** Find $B - A$.

Express each of the following as a decimal between 0 and 1. *(Toolbox, page 788)*

34. a 30% chance   **35.** a 1 in 4 chance   **36.** a 99% chance   **37.** a 1 in 100 chance

Solve each inequality. *(Toolbox, page 787)*

38. $2x + 1 < 5$ **39.** $-(x + 1) \geq 8$ **40.** $\frac{1}{2}x \geq 3$ **41.** $-5x + 1 \leq 4$

1.5 Using Simulations as Models

Have you ever played a game that imitated life? Some games allow players to experiment with topics such as real estate finance or life decisions. These games are examples of *simulations*. A **simulation** is an experiment that you use to model a situation and make predictions. You can use coins, dice, spinners, and calculators with a random number feature to simulate events.

EXPLORATION
COOPERATIVE LEARNING

Simulating a Coupon Giveaway

Work with a partner.
You will need:

- a die

SET UP Suppose a music store gives each customer a discount coupon. The amount of the discount is known when a sales clerk removes the seal on the coupon to reveal either a 10%, 15%, or 30% discount.

1 Suppose there is an equal chance of getting each discount. Let each face of a die represent the type of discount you receive.

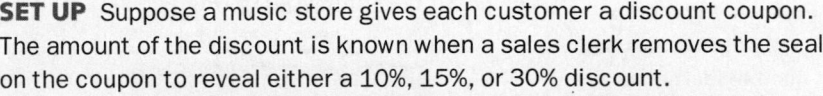

Roll of the die	⚀ or ⚁	⚂ or ⚃	⚄ or ⚅
Discount	10%	15%	30%

Roll the die until you get a 5 or a 6, and then record the number of rolls it took you to get a 5 or a 6. Repeat the simulation 10 times.

2 What was the greatest number of coupons you needed to have to get a 30% discount? What was the least number? the average number?

3 The store manager decides that in the next coupon order, she will request that half of the coupons give a 10% discount, a third give a 15% discount, and a sixth give a 30% discount. Which numbers on the die will you use to represent a discount of 10%? 15%? 30%? Why?

4 Run a simulation for the new coupon values. About how many coupons do you need to collect to get a 30% discount?

EXAMPLE 1

Mr. and Mrs. Skotzke are planning to have children. Create a simulation to estimate the number of children they need to have before their family includes at least one girl and one boy. Repeat the simulation 5 times. About how many children do you think will be in the Skotzke family?

SOLUTION

Assume that the chances of having a girl are the same as the chances of having a boy.

Method 1 Use a coin.

Let "heads up" represent the birth of a girl and "tails up" represent the birth of a boy. Flip the coin until at least 1 head (H) and 1 tail (T) land face up.

Simulation	1	2	3	4	5
Toss results	HHHT (GGGB)	HHT (GGB)	HHT (GGB)	TH (BG)	TTTTH (BBBBG)
Number of children	4	3	3	2	5

The average number of children is $\dfrac{4+3+3+2+5}{5} = 3.4$. Based on these results, the Skotzke family will have either 3 or 4 children.

Method 2 Use a calculator.

For more information about the random number feature on a calculator, see the *Technology Handbook*, p. 814.

The random number feature on a calculator gives a random decimal between 0 and 1. Let the birth be of a girl if the number is less than 0.5. Let the birth be of a boy if the number is 0.5 or greater.

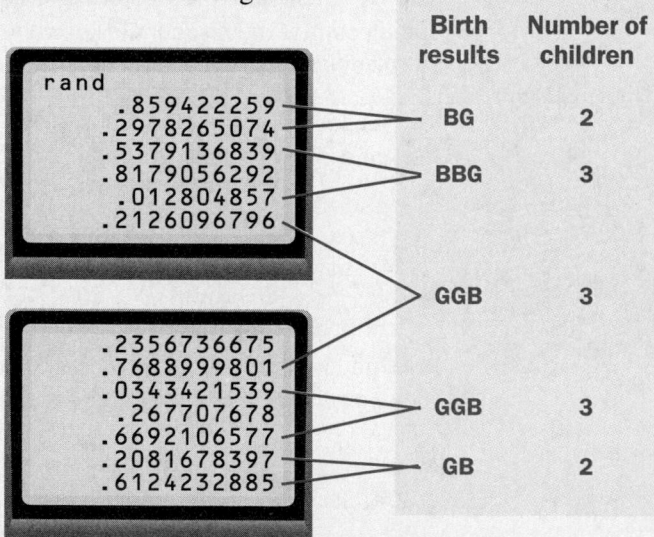

The average number of children is $\dfrac{2+3+3+3+2}{5} = 2.6$. Based on these results, the Skotzke family will have either 2 or 3 children.

EXAMPLE 2

Emily used a program to repeat the simulation from Example 1.

Get N, the number of repetitions. →	`:Disp "NO. OF SIMULATIONS":Input N`
T totals children for all simulations. →	`:0→T`
S, the simulation counter, runs from 1 to N. →	`:For (S,1,N)`
B and G are the boy and girl counters. →	`:0→B:0→G`
	`:Lbl 1`
A child is born; update either B or G. →	`:If rand>0.5:Then:B+1→B`
	`:Else:G+1→G:End`
Repeat until at least 1 boy and 1 girl. →	`:If B=0 or G=0:Goto 1`
Add the boy and girl counts to the total. →	`:T+B+G→T:End`
Report results. →	`:Disp "AVERAGE NUMBER OF CHILDREN IS", T/N`

a. Find the average number of children the Skotzke's will have if you use Emily's program to run the simulation 250 times.

b. Writing Which method do you think is most accurate: Method 1 from Example 1, Method 2 from Example 1, or Emily's method? Explain.

For more information about programming, see the *Technology Handbook*, p. 815.

SOLUTION

a.
```
prgmCHILDREN
HOW MANY TIMES?
?250
AVERAGE NUMBER
OF CHILDREN IS
          2.98
```

According to these results, the Skotzke family will have **3 children**.

b. Emily's method seems most reliable. Emily got her results after repeating the simulation 250 times. In Example 1 the simulations were repeated only 5 times for each method. Because there were so few results for the methods of Example 1, any unusual results had more of an effect on the average.

THINK AND COMMUNICATE

1. Tell what each line does in the program from Example 2.

a. `If rand>0.5:Then:B+1→B:Else:G+1→G`

b. `If B=0 or G=0:Goto 1`

2. Do you think your results would be much different if you ran Emily's program 500 times? 1000 times? Explain your reasoning.

3. Research has shown that the chance of a baby being a girl is 48.8% and the chance of it being a boy is 51.2%. How would you simulate this situation? Do you think you would get different results? Explain.

☑ CHECKING KEY CONCEPTS

Suppose there is one of six bonus stickers in each pack of baseball cards and there is an equal chance of getting each sticker.

1. Describe a simulation to find the number of packs of baseball cards you need to buy in order to get a particular sticker. In your simulation, how do you decide if a pack contains the desired sticker?

2. Describe a simulation to find the number of packs of baseball cards you need to buy in order to collect all six stickers.

3. Suppose that there is an 80% chance of getting a sticker of an individual player and a 20% chance of getting a sticker of a team. Describe a simulation to find the number of packs of baseball cards you need to buy in order to get a sticker of a team.

1.5 | Exercises and Applications

Extra Practice exercises on page 750

For Exercises 1–3, describe how you would simulate each situation.

1. guessing the correct answer on at least 7 of 10 true/false questions

2. choosing a bulb for a red tulip from a bin if one in six of the bulbs is for a red tulip

3. getting a hit in a baseball game if the batter averages one base hit every three times at bat

4. a. Simulate the situation from Exercise 3 to find the number of hits the player gets in a game if he bats 4 times. Repeat the simulation 30 times. Record the number of hits the player gets in each game.

   b. In how many games did the player get no hits? 1 hit? 2 hits? 3 hits? 4 hits? Do you think your results are reasonable? Why or why not?

5. SAT/ACT Preview Which simulation techniques are appropriate for randomly choosing between two equally likely events, A and B?

   I. Toss a coin. Choose A if you get heads, B if you get tails.
   II. Use the spinner. Choose A if you get red, B if you get blue.
   III. Roll a die. Choose A if you get an even number, B if you get an odd number.

   A. I only   **B.** I and III only   **C.** I, II, and III   **D.** none of these

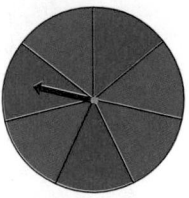

6. a. Do the simulation you described in *Checking Key Concepts* Question 1. Repeat your simulation 10 times.

   b. What is the greatest number of packs of baseball cards you need to buy to get the desired sticker? What is the least number? the average number?

7. Repeat Exercise 6 for the simulation you described in *Checking Key Concepts* Question 2.

8. Repeat Exercise 6 for the simulation you described in *Checking Key Concepts* Question 3.

In many societies people play games that involve tossing some sort of die or dice. One such game, played by the Paiute people of what is now Nevada, involves tossing twelve sticks painted on one side. A tossed stick will land with the painted side up about half the time.

9. To play, one player tosses all twelve sticks. If *exactly* five sticks land with the painted side up, the player gets a point. The players take turns tossing the sticks until one player gets fifty points.

   a. Simulate the Paiute game, but only go to five points. Describe how you decide whether a stick lands with the painted side up or down. Record the results of each player's toss and the number of points each player has.

   b. **Cooperative Learning** How many turns did your game take? Compare your results to those found by other students. What do you think is the average number of turns it takes to play the Paiute game to fifty points?

   c. Do you think each player has an equal chance of winning? Explain.

10. **Writing** Explain why a player who has exactly seven sticks land with the painted side *down* will get a point.

Sarah Winnemucca was an educator, interpreter, and spokeswoman for Native American Rights. In 1883, she wrote *Life Among the Paiutes.*

---

11. **Cooperative Learning** Work in a group of three students. The table shows the chance of 0, 1, 2, or 3 people arriving at a bus stop in a given minute. Suppose that a bus arrives about once every 6 min.

    a. Describe a way to simulate the number of people who arrive at the bus stop. Describe a way to simulate whether a bus arrives.

    b. Simulate the activity at the bus stop for a 60 min period. Use a table like the one below to record your results.

Number of people arriving in a given minute	Chance of the given number of people arriving
0	20%
1	40%
2	30%
3	10%

Minute of simulation	1	2	3	4	5	...	60
Number of people who arrive	1	3	0	2	0	...	3
Bus arrives	0	0	0	1	0	...	0

    c. How many people arrived at the bus stop in 60 min? How many buses arrived? What was the average number of people who got on each bus? (Assume that any people who arrive in the same minute as a bus get on that bus.)

    d. About how many minutes did a person have to wait before a bus arrived? What was the longest waiting time? the shortest?

12. 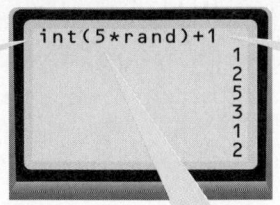 **Technology** Genna used a formula on her calculator to generate the numbers 1, 2, 3, 4, and 5 randomly.

This function cuts off the decimal part of a positive number.

```
int(5*rand)+1
              1
              2
              5
              3
              1
              2
```

Put the number you want to start with here.

a. What formula would you use to generate the numbers 4, 5, 6, 7, and 8 randomly? the numbers 0–9 randomly?

Put the number of numbers you want here.

b. Use Genna's method to repeat Exercise 6 if there are 8 stickers available.

 Technology For Exercises 13 and 14, use the given program or adapt it for use on your graphing calculator or computer.

a. Enter the program and run it at least 10 times. Enter a different number of simulations each time. Record each set of results.

b. What results, if any, are unusual? Explain.

c. What do you think is a reasonable result? Why?

13. This program simulates a best-of-seven championship series between two evenly matched teams. It tells how many times each team won the series and the average number of games the teams played in each series.

14. This program simulates the average number of times a basketball player makes both baskets when the player is given two free throw shots. Enter a player's free throw percentage as a number between 0 and 100.

```
:Disp "HOW MANY TIMES?"
:Input N
:0→T:0→C:0→D
:For (S,1,N)
:0→A:0→B:0→G
:Lbl 1
:G+1→G
:If rand>0.5:Then:A+1→A
:Else:B+1→B:End
:If A≠4 and B≠4:Goto 1
:If A=4:C+1→C:If B=4:D+1→D
:T+G→T:End
:Disp "TIMES A WON",C
:Disp "TIMES B WON",D
:Disp "AVG. NO. OF GAMES",T/N
```

```
:Disp "WHAT IS PLAYER'S
FREE-THROW PERCENTAGE?"
:Input P
:Disp "HOW MANY TIMES?":Input N
:0→C
:For (S,1,N)
:0→B
:For (X,1,2)
:If int (100*rand)+1≤P
:Then:B+1→B:End
:If B=2:Then:C+1→C:End
:End:End
:Disp "NO. OF 2 PT SHOTS",C
:Disp "PERCENT 2 PT SHOTS"
:Disp int(100*(C/N))
```

15. Challenge Change the program in Exercise 13 to simulate the winner of a best-of-seven championship series in which one team is twice as likely to win each game as the other team. Then repeat Exercise 13.

16. Open-ended Problem Simulate a situation of your choice. Describe the situation and the simulation. Record your results. Repeat your simulation enough times so that you have confidence in your results.

SPIRAL REVIEW

Multiply. *(Section 1.4)*

17. $\begin{bmatrix} 4 & 2 \\ 1 & 0 \end{bmatrix}\begin{bmatrix} 3 \\ 1 \end{bmatrix}$
  18. $\begin{bmatrix} 2 & 1 & 7 \\ 0 & 8 & 4 \\ 3 & 6 & 2 \end{bmatrix}\begin{bmatrix} 4 & 10 \\ 2 & 8 \\ 3 & 1 \end{bmatrix}$
  19. $\begin{bmatrix} 2 & -3 \\ -1 & 2 \\ 0 & -4 \end{bmatrix}\begin{bmatrix} 5 & 2 & -1 \\ 1 & 3 & -4 \end{bmatrix}$

For each equation, make a table of values for *x* and *y*. Use −3, −2, −1, 0, 1, 2, and 3 as values for *x*. Then use your points to graph each equation. *(Toolbox, page 795)*

20. $y = 4x$
   21. $y = -x$
   22. $y = 3x - 1$
   23. $y = -2x + 5$

ASSESS YOUR PROGRESS

VOCABULARY

matrix, element (p. 16)
dimensions of a matrix (p. 16)
scalar (p. 18)

scalar multiplication (p. 18)
matrix multiplication (p. 23)
simulation (p. 29)

For Exercises 1–10 use the matrices below. If the operation cannot be performed, write "undefined." *(Sections 1.3 and 1.4)*

$$A = \begin{bmatrix} 2 & 0 & -1 \\ -3 & 1 & 4 \end{bmatrix} \quad B = \begin{bmatrix} -1 & 6 & 2 \\ 0 & 3 & -3 \end{bmatrix} \quad C = \begin{bmatrix} 3 & -2 & 2 \\ 5 & -1 & -3 \\ 0 & 4 & 1 \end{bmatrix} \quad D = \begin{bmatrix} 1 & 0 \\ 4 & -2 \\ 3 & -1 \end{bmatrix}$$

1. $A + B$ **2.** $3C$ **3.** $C + 2D$ **4.** $C - 4D$ **5.** $-2A - B$

6. AC **7.** BD **8.** BA **9.** CD **10.** CB

11. Suppose there is a one in six chance that any employee of a 20-person company will leave the company in a given year. *(Section 1.5)*

  a. Simulate how long each employee will work for the company. Describe your simulation. Record your results in a table.

Employee	1	2	3	4	...	20
Simulation results	?	?	?	?	...	?
Number of years an employee stays	?	?	?	?	...	?

  b. What is the average number of years an employee stays with the company? How many of the current 20 employees are still with the company after 5 years? after 10 years?

12. Journal What do you think are the most difficult steps of a simulation? Why? What methods of simulation do you like best? Why?

Predicting Basketball Accuracy

In 1979 the National Basketball Association instituted a *three-point rule*. According to this rule, a basket shot from a point outside the three-point "line" is worth three points, while a basket shot from a closer distance is worth two points. Why do you think the rules were changed?

The **three-point line** is two inches wide. A shot counts for only two points if a player's foot touches the three-point line.

PROJECT GOAL Do an experiment to find how your ability to make a basket changes with distance.

Investigating Your Basketball Ability

For this investigation, you will need a piece of wadded-up paper to be your ball, a wastebasket, a yardstick, and graph paper. You may find it helpful to have a graphing calculator or graphing software.

1. **STAND** 0 yards from the wastebasket. Shoot your paper ball into the basket 10 times and record the number of shots that go into the basket.

2. **STEP** back one yard and record your results for 10 shots from this distance. Continue stepping back one yard until you reach a distance where none of your 10 shots go into the basket.

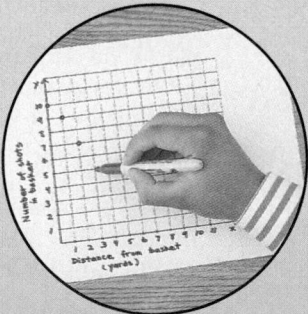

3. **MAKE** a scatter plot of your data. Let the horizontal axis show the distance from the basket, and let the vertical axis show the number of shots that went into the basket.

Making a Model

• **DESCRIBE** your graph. Do you think the data decreases *linearly*? *exponentially*? At what distance did the number of shots that went into the basket decrease the most?

• **ORGANIZE** your data in a table like the one shown. Find the average change and the average percent change.

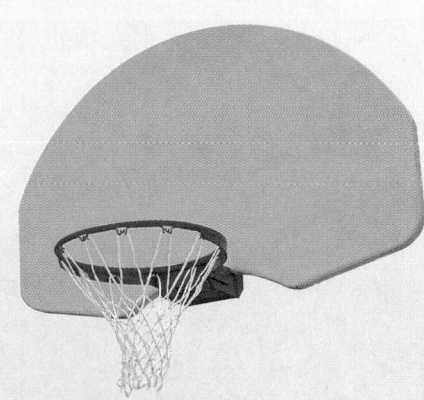

Distance from basket (yd)	Number of shots that went into basket	Change	Percent change
0	?	—	—
1	?	?	?
2	?	?	?
⋮	⋮	⋮	⋮

• **MODEL** your data with a linear function and with an exponential function.

• **GRAPH** your linear function, exponential function, and data points in the same coordinate plane. Which model do you think most accurately represents your data? Why?

Writing a Report

Write a report about your experiment that includes your data, graphs, models, and explanations. You may also want to investigate and report on these things:

• How do you think your results would change if you repeated the experiment 5 times? 10 times? 50 times? Explain your reasoning. Then repeat the experiment at least 5 times. How do your results compare to your prediction?

• Repeat the experiment using a real basketball and hoop. How could a basketball player use this experiment to decide whether to try for a two-point shot or a three-point shot?

• Why do you think the three-point rule was created? What other sports have rules that take the *likelihood* of scoring points into account? Describe the method of scoring for each sport.

Self-Assessment
Describe any difficulties you had in finding a linear and an exponential model to fit your data. How did you resolve these difficulties?

1 Review

What study techniques have you tried before? Write two brief paragraphs starting with these phrases:

- **To study for a mathematics test I usually...**
- **A study technique that has not helped me in the past is...**

VOCABULARY

mathematical model (p. 4)
function (p. 10)
independent variable (p. 10)
dependent variable (p. 10)
matrix (p. 16)
element (p. 16)

dimensions of a matrix (p. 16)
scalar (p. 18)
scalar multiplication (p. 18)
matrix multiplication (p. 23)
simulation (p. 29)

SECTIONS 1.1 *and* 1.2

Mathematical models help you understand and make predictions about real-world situations, such as the federal funding for research and development in transportation.

Years since 1990	Funding (billions of $)	Change	Percent change
0	1.05	—	—
1	1.23	0.18	17%
2	1.52	0.29	24%
3	1.78	0.26	17%
		0.24	19%

averages

Linear Model
$y = 1.05 + 0.24x$

x (input)	y (output)
0	1.05
1	1.29
2	1.53
3	1.77

Exponential Model
$y = 1.05(1.19)^x$

x (input)	y (output)
0	1.05
1	1.25
2	1.49
3	1.77

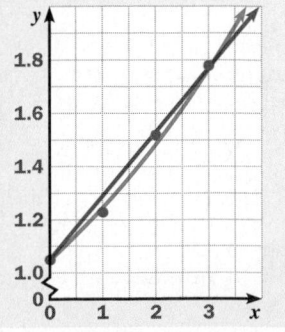

A **matrix** is a group of numbers arranged in rows and columns. For example, matrix I represents the current inventory of wooden toys and art in a small craft and souvenir store. Matrix S represents sales for one week, and matrix A represents additional inventory that has come in during the week.

$$\begin{array}{c} \text{Toys} \quad \text{Art} \\ \begin{array}{c} \text{New} \\ \text{Antique} \end{array} \begin{bmatrix} 15 & 9 \\ 10 & 21 \end{bmatrix} = I \end{array} \qquad \begin{array}{c} \text{Toys} \quad \text{Art} \\ \begin{array}{c} \text{New} \\ \text{Antique} \end{array} \begin{bmatrix} 4 & 1 \\ 2 & 7 \end{bmatrix} = S \end{array} \qquad \begin{array}{c} \text{Toys} \quad \text{Art} \\ \begin{array}{c} \text{New} \\ \text{Antique} \end{array} \begin{bmatrix} 5 & 2 \\ 3 & 0 \end{bmatrix} = A \end{array}$$

At the end of the week, the stock is $I - S + A$.

$$I - S + A = \begin{bmatrix} 15 - 4 + 5 & 9 - 1 + 2 \\ 10 - 2 + 3 & 21 - 7 + 0 \end{bmatrix} = \begin{array}{c} \text{Toys} \quad \text{Art} \\ \begin{bmatrix} 16 & 10 \\ 11 & 14 \end{bmatrix} \begin{array}{l} \text{New} \\ \text{Antique} \end{array} \end{array}$$

You can sometimes multiply two matrices to find information. For example, matrix A represents the amount of labor and paint required for a body shop to repair damage to different parts of a car. Matrix B represents the amount the body shop charges for each hour of labor and each pint of paint.

$$\begin{array}{c} \text{Labor} \quad \text{Paint} \\ \text{(hours)} \ \text{(pints)} \\ \begin{array}{r} \text{Door} \\ \text{Hood} \\ \text{Rear bumper} \end{array} \begin{bmatrix} 4.0 & 1.2 \\ 3.0 & 1.0 \\ 2.5 & 1.0 \end{bmatrix} = A \end{array} \qquad \begin{array}{c} \text{Charge (\$)} \\ \begin{array}{c} \text{Labor} \\ \text{Paint} \end{array} \begin{bmatrix} 32 \\ 18 \end{bmatrix} = B \end{array}$$

To determine the total cost to repair each car part, multiply A by B.

$$AB = \begin{bmatrix} 4.0 & 1.2 \\ 3.0 & 1.0 \\ 2.5 & 1.0 \end{bmatrix} \begin{bmatrix} 32 \\ 18 \end{bmatrix} = \begin{bmatrix} (4.0)32 + (1.2)18 \\ (3.0)32 + (1.0)18 \\ (2.5)32 + (1.0)18 \end{bmatrix} = \begin{array}{c} \text{Charge (\$)} \\ \begin{bmatrix} 149.6 \\ 114 \\ 98 \end{bmatrix} \begin{array}{l} \text{Door} \\ \text{Hood} \\ \text{Rear bumper} \end{array} \end{array}$$

A **simulation** is an experiment that models a situation and can be used to make predictions. For example, suppose Zack is selecting from a tray of spring rolls. One half of the rolls contain only vegetables and one half contain meat. He wants to get at least one vegetable roll. How many rolls should he take? To simulate this situation, toss a coin repeatedly, letting heads (H) represent meat rolls and tails (T) represent vegetable rolls. Stop when you get a T.

Simulation	1	2	3	4
Toss results	HT	T	HHHHT	T
Number of rolls	2	1	5	1

The average is
$$\frac{2 + 1 + 5 + 1}{4} = 2.25,$$
so he should take two or three rolls.

Assessment

VOCABULARY QUESTIONS

For Questions 1 and 2, complete each paragraph.

1. A(n) _?_ pairs every input value with exactly one output value. The input variable, x, is called the _?_. The output variable, y, is called the _?_ because its value depends upon what value you choose for x.

2. A(n) _?_ is a group of numbers arranged in rows and columns. Each number is called a(n) _?_.

SECTIONS 1.1 *and* 1.2

3. The table shows United States Census figures for the population of California.

Year	Population (millions)
1970	**20.0**
1980	**23.7**
1990	**29.8**

  a. What was the average change in population per decade?

  b. What was the average percent change in population per decade?

  c. Use a constant growth (linear) model to estimate California's population in the year 2000.

  d. Repeat part (c) using a proportional growth (exponential) model.

  e. **Writing** Compare your answers to parts (c) and (d). Which do you think is a more reasonable estimate for the population of California in the year 2000? Why?

4. The graph shows the amount Americans spent on dental care for 1986–1989.

  a. Write a linear function to model the amount spent on dental care. Let x = the number of years since 1986.

  b. Repeat part (a) using an exponential function to model the data.

Dental Care in U.S.

Amount spent (billions of dollars)

40, 30, 24.7, 27.1, 29.4, 31.6, 20, 10, 0

1986 1987 1988 1989

  c. Make a table showing the actual amount spent, the amount predicted by your linear function, and the amount predicted by your exponential function for each of the four years shown in the graph.

  d. Based on your answer to part (c), which model do you think is more reasonable—the *linear model* or the *exponential model*? Explain.

SECTIONS 1.3 *and* 1.4

For Exercises 5–10, use matrices *S, T, U,* and *V* to evaluate each matrix expression. If an operation cannot be performed, write *undefined*.

$$S = \begin{bmatrix} 1 & 0 \\ -2 & 1 \end{bmatrix} \quad T = \begin{bmatrix} 5 & 2 \\ 0 & -1 \\ 7 & 11 \end{bmatrix} \quad U = \begin{bmatrix} 13 \\ -6 \end{bmatrix} \quad V = \begin{bmatrix} 3 & -4 \\ -2 & 1 \\ 0 & 20 \end{bmatrix}$$

5. $T + V$ **6.** $U + S$ **7.** $4V - T$

8. TU **9.** $3S$ **10.** ST

For Exercises 11–13, use the matrix $P = \begin{bmatrix} 8 & 6 & 5 & 1 \\ -2 & 17 & 0 & -4 \end{bmatrix}.$

11. What are the dimensions of P?

12. What is the value of the element $p_{2,1}$?

13. If Q is a 3×2 matrix, what are the dimensions of the product QP?

SECTION 1.5

14. The makers of ABC, a new juice drink, have devised a contest for consumers to win a free bottle of ABC. One of the letters A, B, or C is printed on the liner of each bottle cap. One half of the caps have an A, one third of the caps have a B, and one sixth of the caps have a C. If you collect all 3 letters, you win a free bottle of ABC.

a. Describe how you would simulate this giveaway using a die.

b. The table shows some results from a simulation to determine how many bottle caps a person needs to collect to win a free bottle. Based on these results, how many bottle caps (on average) would a person need in order to win a free bottle?

Simulation	1	2	3	4	5
Results	B, A, A, A, A, A, C	A, A, A, A, A, A, C, C, B	A, A, B, C	B, A, C	A, B, B, A, B, A, A, B, A, A, B, A, C

PERFORMANCE TASK

15. Interview a local businessperson. See if the person is willing to share information about sales, inventory, pricing, special promotions, and so on. Use the information you collect to develop mathematical models involving functions, matrices, or simulations. Prepare a report to present to the class.

2 | Linear Functions

Photographing the underwater world

INTERVIEW Norbert Wu

Jacques Cousteau describes Wu as:

" a hard-working and meticulous technician, as well as a friendly team member. "

Underwater photographer Norbert Wu says he surprises people when they meet him. He says he's not a stereotypical daredevil scuba diver, but describes himself as a modest man who's traveled the world to dive and to photograph some unusual marine life. Along the way he's been bitten by a shark, been trapped under an iceberg, and ridden on the back of a manta ray with a 20-foot wingspan.

Rare Deep-Sea Images

Wu has also photographed hammerhead sharks in 100-foot-deep Costa Rican waters. The dozens of hammerheads were apparently oblivious to his presence during that dive. To shoot a picture of them, Wu said he had to hold his breath and "kick toward them, fast." The instant he breathed out, the sharks dashed off, frightened by the sound of air bubbles escaping through his regulator. "Others have done it," he says casually. "If I'd never heard of it [being done] before, I'd be scared witless."

He recalls that one of his best dives was off San Clemente Island. "In the fall, there was a gathering of bat rays. They look like pterodactyls. They were schooling—there were a couple of hundred dancing, swimming in a circle. It was most memorable."

"It's such a different environment. You're weightless. There's always something to see."

Working with Jacques Cousteau

As a child in Atlanta, Wu said he knew by the second grade he wanted to study marine biology. He spent many summers snorkeling off the California and Florida coasts and received his diver's certificate at age 15. In 1986 while a doctoral student at the Scripps Institute of Oceanography, he worked as chief still photographer for Jacques Cousteau's research ship *Calypso*. He saw firsthand the famous deep-sea researcher at work. Cousteau describes Wu as "a hard-working and meticulous technician, as well as a friendly team member."

The Importance of Safety

Like all divers, Wu feels the pressure of the water on him increase as he dives deeper. The increasing water pressure causes his body to absorb more nitrogen. If he returns to the surface too rapidly, the nitrogen in his body fluids can form bubbles. The bubbles can cause "bends," or decompression sickness. To prevent injury, he returns to the surface slowly. The maximum recommended rate of ascent for divers is about 60 ft/min. The graph shows how the time needed to return to the surface is related to the depth of the dive. This relationship is an example of a *linear function*.

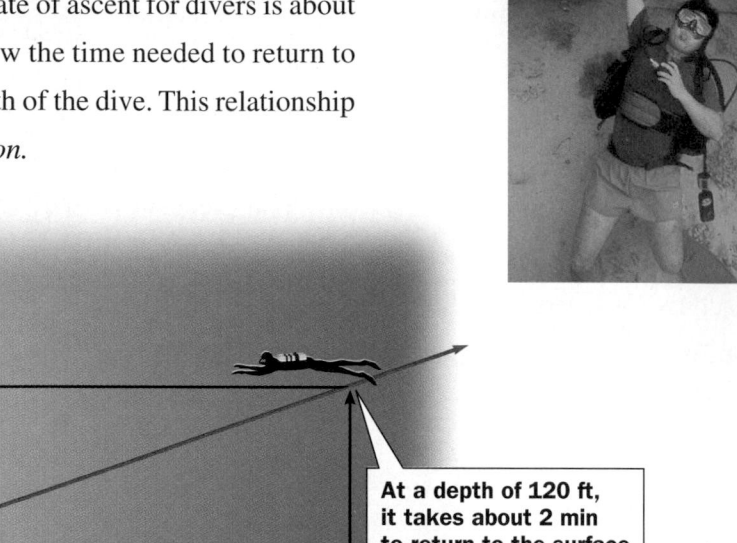

At a depth of 120 ft, it takes about 2 min to return to the surface.

EXPLORE AND CONNECT

1. Research Investigate the history of diving. What kind of equipment did the first deep-sea divers use? How deep could they dive with the equipment?

2. Writing Draw a depth-versus-time graph for a diver who starts at a depth of 100 ft and ascends at a rate of 60 ft/min. Explain how the graph shows the time taken to return to the surface.

3. Project Get an empty container, such as a milk carton or plastic bottle, and punch small holes of the same size at various heights on the lower half of the container. Place the container in an empty tub or sink. Cover the holes and fill the container with water. Uncover the holes and watch for differences in the streams of water pouring out from the holes. What do you think causes the differences?

Norbert Wu on a photography trip

Mathematics
& Norbert Wu

In this chapter, you will learn more about how mathematics is related to diving.

Related Examples and Exercises

Section 2.1
• Exercises 15–18

Section 2.2
• Example 2

2.1 Direct Variation

When precision skaters perform a maneuver called *the wheel,* they form a revolving line. A skater's speed depends upon the skater's distance from the center of rotation. How are speed and distance mathematically related in this situation? Consider the following table of data.

Learn how to...

- recognize direct variation

- write and use direct variation equations

So you can...

- analyze data and make predictions about skating speeds, for example

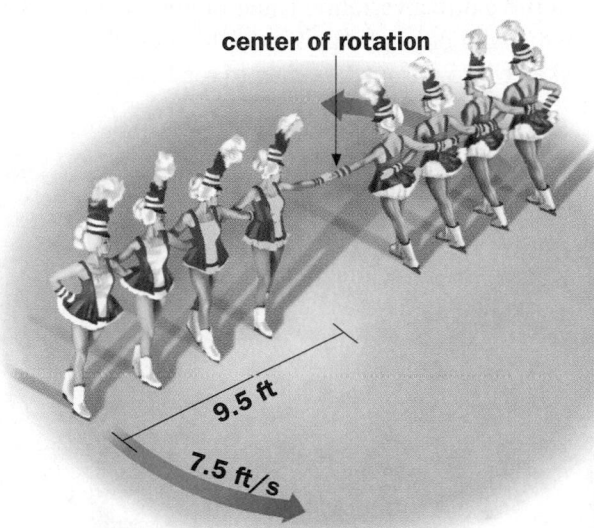

center of rotation

9.5 ft

7.5 ft/s

Distance from center of rotation (ft)	Speed (ft/s)
2	1.6
4.5	3.5
7	5.5
9.5	7.5
12	9.4

Toolbox p. 785
Ratios and Proportions

EXAMPLE 1   Application: Precision Skating

Use a spreadsheet to see if there is a common speed-to-distance ratio for the data given in the table.

SOLUTION

Calculate the ratios $\dfrac{\text{speed}}{\text{distance}}$ using a spreadsheet.

For more information about using a spreadsheet, see the *Technology Handbook*, p. 816.

	Precision Skating Data		
	A	B	C
1	Distance *d*	Speed *s*	*s/d* ratio
2	2	1.6	0.8
3	4.5	3.5	0.77777778
4	7	5.5	0.78571429
5	9.5	7.5	0.78947368
6	12	9.4	0.78333333

The ratios are approximately equal.

There is a common ratio of about 0.79 ft/s of speed per 1 ft of distance from the center of rotation.

When the ratio of two variables is constant, the variables are in **direct variation**. This means that if $\frac{y}{x} = a$ for some nonzero constant a, then:

$$y = ax$$

You can say that **y** *varies directly* with **x**, or **y** is *proportional* to **x**.

The number **a** is the **constant of variation**.

EXAMPLE 2

Write a direct variation equation for the data in Example 1, and use a graphing calculator or graphing software to graph the equation. Then predict the speed of a skater 19.5 ft from the center of rotation.

SOLUTION

In Example 1, you found that $\frac{s}{d} = 0.79$, so $s = 0.79d$.

A graph of this equation is a line. To check the reasonableness of the graph, compare it to a scatter plot of the data.

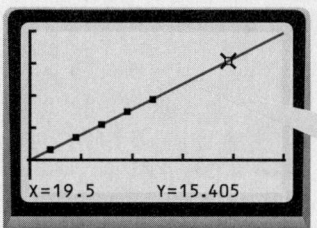

X=19.5   Y=15.405

On a graphing calculator or graphing software, graph **y = 0.79x** instead of $s = 0.79d$.

Notice how well the graph fits the plotted data. For a skater 19.5 ft from the center of rotation, $d = \mathbf{19.5}$ and $s = 0.79(\mathbf{19.5}) \approx 15.41$. The skater's speed is about 15.4 ft/s.

THINK AND COMMUNICATE

Moira

$\dfrac{Y}{19.5} = \dfrac{1.6}{2}$

$Y = \dfrac{1.6}{2}(19.5) \approx 15.6$

The skater's speed is about 15.6 ft/s.

1. If (x_1, y_1) and (x_2, y_2) are two data pairs from a direct variation, what can you say about $\frac{y_1}{x_1}$ and $\frac{y_2}{x_2}$? Explain.

2. Moira found the speed of the skater in Example 2 a different way.

    a. Describe the method Moira used.

    b. What explains the difference between Moira's answer and the answer given in Example 2?

3. Notice that the graph in Example 2 includes the origin. What does this mean in terms of the rotating line of skaters? Why does this make sense?

The graph of $y = ax$ is a line that passes through the origin. In a previous course you learned that the ratio $\frac{\text{vertical change}}{\text{horizontal change}}$ gives the **slope** of a line. For a direct variation graph, the slope is just the constant of variation, a.

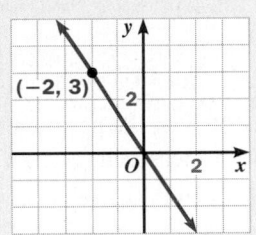

The graph of $y = 2x$ is a line that passes through the origin.

When x increases by **1** unit, y increases by **2** units. The slope of the line is $\frac{2}{1}$, or 2.

EXAMPLE 3

Write an equation for the direct variation graph shown below.

SOLUTION

The graph passes through the point $(-2, 3)$, so the constant of variation is:

$$a = \frac{3}{-2} = -\frac{3}{2} = -1.5$$

An equation of the graph is $y = -1.5x$.

$(-2, 3)$

☑ CHECKING KEY CONCEPTS

1. Tell whether y varies directly with x. Explain your reasoning.

   a. $y = 3.2x$ **b.** $y = -3.2x$ **c.** $y = 3.2x + 1$

2. Tell whether r and s are proportional. Explain your reasoning.

   a. $\dfrac{r}{s} = \dfrac{2}{5}$ **b.** $r = 8s$ **c.** $r + s = 2$

3. For each graph, tell whether y varies directly with x. If so, give an equation for the graph.

   a.
   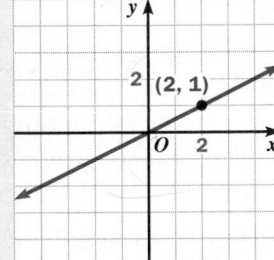

   b.
   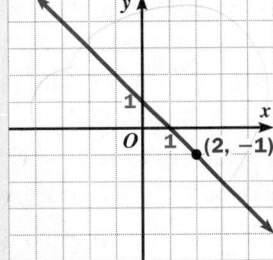

4. Graph $y = 3x$. Describe the change in y when:

   a. x increases by 1 unit   **b.** x doubles   **c.** x is halved

2.1 Exercises and Applications

Extra Practice exercises on page 750

For each table, tell whether there is a common ratio. If there is, write an equation relating the two variables.

1.

Calories in Lean Hamburger	
Serving size (oz)	Calories
2	125
2.5	156
3	188
3.5	219
4	250

2.

Income for a Software Company	
Sales (billion $)	Profit (billion $)
1.18	0.28
1.84	0.46
2.76	0.71
3.75	0.95
4.65	1.15

3.

Importer's Prices of Diamonds	
Size (carats)	Price (dollars)
$\frac{1}{3}$	199
$\frac{1}{2}$	399
$\frac{3}{4}$	499
1	599
2	1499

4. **GEOMETRY** The length of a side of a square is s meters.

   a. Find formulas in terms of s for the perimeter P and area A of the square. In what units are P and A measured?

   b. Does P vary directly with s? Does A vary directly with s? Explain your answers.

For each equation, tell whether y varies directly with x. If so, graph the equation.

5. $y = \dfrac{5}{2x}$

6. $y = \dfrac{5x}{2}$

7. $y = 5x + 2$

8. $y = \dfrac{2}{5}x$

9. $y = x + 4$

10. $y = \dfrac{4}{x}$

11. $y = 4$

12. $y = 4x$

13. **SPORTS** At the 1994 Winter Olympics in Hamar, Norway, speed skaters were required to complete 1.25 laps around the track for the 500 m race. Find a direct variation equation that relates the number of laps to the length of a race in meters. Then find the number of laps required to complete the 3000 m race.

14. **MANUFACTURING** Machinists often cut hexagonal shapes from round stock. Many times the machinist knows the distance f between the "flats" or sides of the hexagon. The direct variation equation $d = 1.1547f$ is then used to determine the distance d between opposite corners of the hexagon.

   a. Find the diameter of the round stock needed to cut a hex nut measuring 1.3750 in. between the flats.

   b. Use what you know about the geometry of a regular hexagon to derive the direct variation equation relating d and f.

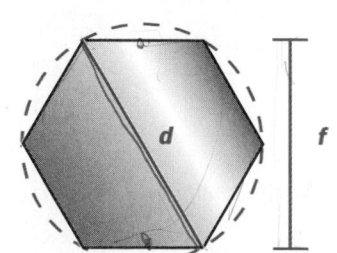

Norbert Wu

Look back at the
article on pages
42–44.

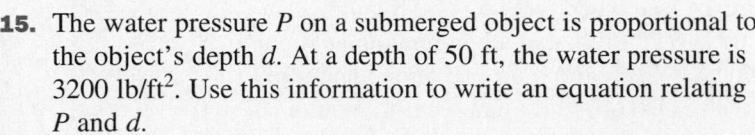

The deeper Norbert Wu dives, the more pressure the water exerts on him. This is true of anything else in the water, including animals and submarines.

15. The water pressure P on a submerged object is proportional to the object's depth d. At a depth of 50 ft, the water pressure is 3200 lb/ft^2. Use this information to write an equation relating P and d.

16. Using standard gear, a scuba diver can go to a maximum depth of about 130 ft. What is the water pressure on a scuba diver at that depth?

17. Some of the fish that Norbert Wu has photographed live at a depth of 3000 ft. (These fish have to be brought to the surface with a net.) What is the water pressure on the fish living at that depth?

18. The deepest ocean descent on record was achieved in 1960 by Dr. Jacques Piccard of Switzerland and Lt. Donald Walsh of the United States Navy. They used a small research submarine called a bathyscaph to descend 35,800 ft into the Pacific Ocean's Marianas Trench. At that depth, what was the water pressure on the hull of the bathyscaph?

0
5,000
10,000
15,000
20,000
25,000
30,000
35,000
40,000

The Marianas Trench

For each graph, tell whether _y_ varies directly with _x_. If so, give an equation for the graph.

19.

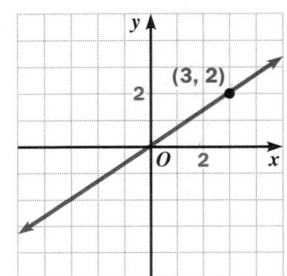

(3, 2)

2

O 2 *x*

20.

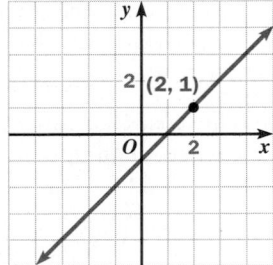

2 (2, 1)

O 2 *x*

21.

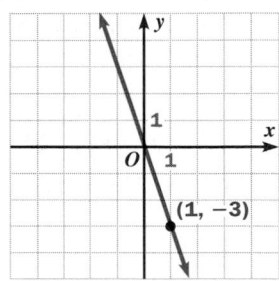

1

O 1 *x*

(1, −3)

The number of calories (Cal) you use when exercising is proportional to the amount of time you exercise.

22. **Visual Thinking** For a 150 lb person, the graphs of calories used when walking at 2.5 mi/h, when walking at 3.25 mi/h, and when running at 10 mi/h are shown. Which graph do you think applies to each type of exercise? Explain your reasoning.

23. Alex and Robert both weigh 150 lb. For 30 min, Alex walks at 3.25 mi/h while Robert walks at 2.5 mi/h. About how many more calories does Alex use than Robert?

24. Suppose Joe, who weighs 150 lb, has just eaten a snack containing 300 calories. He wants to "walk it off." About how long will it take if he walks at 2.5 mi/h? at 3.25 mi/h?

25. **a.** Find an equation for the running graph. What is the constant of variation? (Include its units.)

    b. Do you think the constant would be different for a person who weighs less than 150 lb? If so, do you expect it to be more or less than the constant in part (a)? Why?

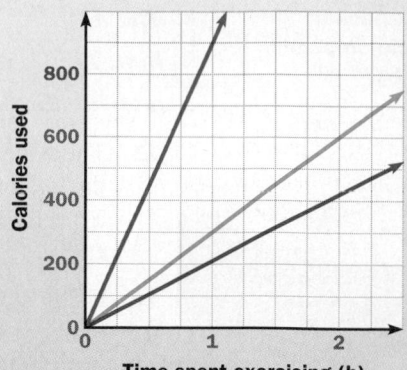

26. **Cooperative Learning** Work in a group of six or more students. Your group will need a stopwatch or a watch with a second hand.

    a. Starting with one person and adding another at each stage of the experiment, time how long it takes to pass a "wave" (an up-and-down motion of the arms) along a row of people.

    b. Do your data suggest that "wave" time varies directly with the number of people? If so, write an equation and compare its graph to a scatter plot of your data.

27. **Challenge** The equation $u = 0.33t$ gives the electrical usage u, measured in kilowatt-hours (kW·h), as a function of the time t, in hours, that Betty's television set is on. Also, the equation $C = 0.0918u$ gives the cost of electricity C, in dollars, as a function of Betty's electrical usage u.

    a. For Betty's television, can you say that the cost C of using it varies directly with the time t it is used? If so, what equation relates these two variables?

    b. Explain how you obtained the constant of variation for your equation in part (a).

28. **PERSONAL FINANCE** Sonya keeps a log of her car's fuel economy. Each time she stops for gasoline, she records the mileage from the car's odometer and the number of gallons of gas needed to fill the fuel tank. The beginning of Sonya's log is shown.

a. **Spreadsheets** Enter Sonya's data in a spreadsheet. Include two other columns, one labeled "Distance driven since last fuel stop (mi)" and one labeled "Fuel economy (mi/gal)." Have the spreadsheet perform the appropriate calculations for these two columns.

b. Does part (a) suggest a direct variation? If so, write an equation relating d, the distance driven, and g, the gas used.

c. The fuel tank of Sonya's car holds about 13.5 gal. How far can Sonya travel on a full tank of gas?

Odometer reading (mi)	Gas used since last fuel stop (gal)
18	Car is new with a full tank of gas.
232	8.2
534	11.6
621	3.4
871	9.5
1033	6.2
1308	10.7

ONGOING ASSESSMENT

29. **Writing** Find an example of direct variation in everyday life. Create a table of data for the two variables and use your data to find the constant of variation. Write a brief explanation of why one variable varies directly with the other.

SPIRAL REVIEW

30. When Leah took her little brother Johnny shopping, Johnny noticed a machine that dispenses small toys in plastic capsules. The machine holds five types of toys, including a high-bounce ball that Johnny would like to have. It costs $.25 to get a toy from the machine. If the five types of toys are equally likely to drop from the machine, how much would it cost, on average, to get the toy that Johnny wants? Use a simulation to answer this question. *(Section 1.5)*

31. The linear function $C = 3 + 0.25n$ gives a checking account's monthly cost C, in dollars, in terms of the number of checks written n. Tell what the numbers 3 and 0.25 mean in this situation. *(Section 1.2)*

32. Which pair(s) of matrices can be added together? Which pair(s) can be multiplied? *(Sections 1.3, 1.4)*

$$A = \begin{bmatrix} 1 & 2 \\ 3 & 4 \end{bmatrix} \qquad B = \begin{bmatrix} 2 \\ 3 \end{bmatrix} \qquad C = \begin{bmatrix} 4 & 2 \\ 1 & 0 \end{bmatrix}$$

2.2 | Linear Equations and Slope-Intercept Form

Learn how to...

- write an equation of a line in slope-intercept form
- graph a linear equation in slope-intercept form

So you can...

- model events, such as rising water levels

You've probably noticed that when you get into a bathtub, the water level rises. In the Exploration you will investigate a similar situation and develop a linear model.

EXPLORATION
COOPERATIVE LEARNING

Measuring Water Height

Work with a group of three or four students.
You will need:

- a glass container about half filled with water
- twenty marbles of the same size
- a centimeter ruler

A narrow container with straight sides works best.

Measure the water height to the nearest **0.1 cm**. Be sure to put your eyes at the level of the water before measuring.

Number of marbles	Water height (cm)
0	?
5	?
10	?
15	?
20	?

1 Copy the table. Measure and record the water height before you add any marbles to the container.

2 Add five marbles at a time to the container. Measure and record the water height each time.

3 Make a scatter plot of your data. Put the number of marbles on the horizontal axis and the water height on the vertical axis. What do you notice about your data points?

4 Calculate the change in water height per marble.

5 Let m = the number of marbles and h = the water height. Use your data to write an equation of the form:

water height	=	starting height	+	change in water height per marble	×	number of marbles

In doing the Exploration, Maria and Anthony obtained the graph shown.

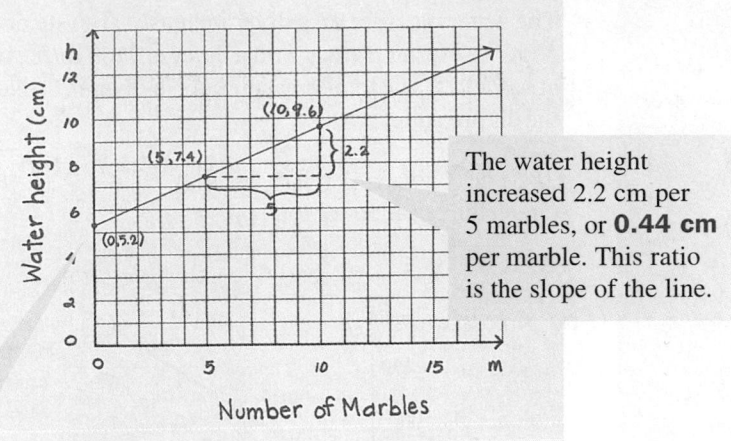

The water height started at **5.2 cm**. This number is called the **vertical intercept** because it indicates where the graph crosses the vertical axis.

The water height increased 2.2 cm per 5 marbles, or **0.44 cm** per marble. This ratio is the slope of the line.

An equation for Maria and Anthony's graph is $h = 5.2 + 0.44m$, or $h = 0.44m + 5.2$. This equation is in **slope-intercept form** because it fits this pattern:

$$y = ax + b$$

The number **a** tells you the slope.

The number **b** tells you the vertical intercept.

WATCH OUT!

You may remember the slope-intercept form as $y = mx + b$ from a previous course. The letter a is used in place of m here because that is what you will see on some common graphing calculators.

Notice how the slope of Maria and Anthony's graph is calculated:

$$\text{slope} = \frac{\text{vertical change}}{\text{horizontal change}} = \frac{9.6 - 7.4}{10 - 5} = \frac{2.2}{5} = 0.44$$

In general, if (x_1, y_1) and (x_2, y_2) are two points on a line, then the slope a is:

$$a = \frac{\text{vertical change}}{\text{horizontal change}} = \frac{y_2 - y_1}{x_2 - x_1}$$

EXAMPLE 1

Find the slope-intercept equation of the line through the points $(-4, 1)$ and $(0, 2)$.

SOLUTION

Step 1 The line crosses the y-axis at $(0, 2)$, so $b = 2$ is the vertical intercept, or y-intercept.

Step 2 To find the slope a, use the points $(-4, -1)$ and $(0, 2)$ on the line:

$$a = \frac{2 - (-1)}{0 - (-4)} = \frac{3}{4}$$

Substituting $a = \frac{3}{4}$ and $b = 2$ in $y = ax + b$ gives the slope-intercept equation:

$$y = \frac{3}{4}x + 2$$

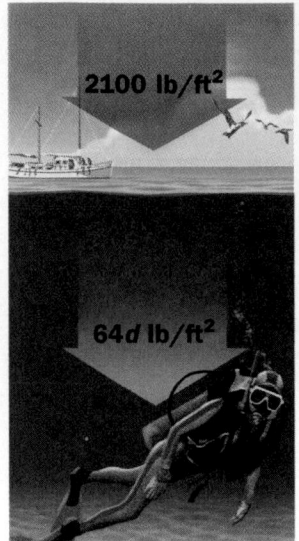

2100 lb/ft²

64*d* lb/ft²

EXAMPLE 2 Interview: Norbert Wu

The water pressure on a diver increases at a rate of 64 lb/ft² for every 1 ft of depth *d*. The air pressure on a diver is 2100 lb/ft², which is constant for anyone at sea level. The total pressure *P* on a diver is the sum of the water pressure and the air pressure:

$$P = 64d + 2100$$

Graph the equation for total pressure.

SOLUTION

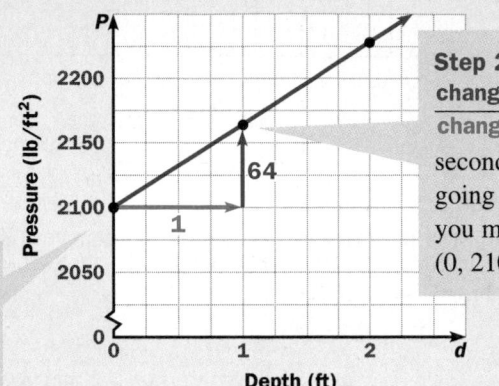

Step 2 Use the slope, $\dfrac{\text{change in } P}{\text{change in } d} = \dfrac{64}{1}$, to plot a second point on the line by going up **64** units when you move right **1** unit from (0, 2100).

Step 1 Plot the point **(0, 2100)** where the line crosses the *P*-axis.

THINK AND COMMUNICATE

1. Does the total pressure on a diver vary directly with depth? Explain.

2. How is the graph in Example 2 geometrically related to the graph of the equation for water pressure only (see Exercise 15 on page 49)?

☑ CHECKING KEY CONCEPTS

Find the slope of the line passing through each pair of points.

1. (1, 3) and (3, 7)   2. (−4, 2) and (5, −3)   3. (−1, 6) and (1, 6)

State the slope and *y*-intercept of each line with the given equation.

4. $y = 8x − 5$ 5. $y = \dfrac{1}{4} − \dfrac{3}{7}x$ 6. $y = 1$

Find the slope-intercept equation of each line.

7.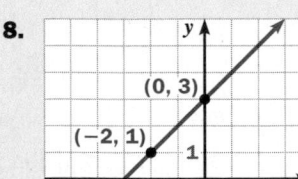

8.

2.2 Exercises and Applications

Extra Practice
exercises on
pages 750–751

State the slope and *y*-intercept of each line with the given equation.

1. $y = 3x + 7$ **2.** $y = -\frac{2}{3}x + \frac{1}{3}$ **3.** $y = x$ **4.** $y = 2 - 5x$

Find the slope-intercept equation of each line.

5.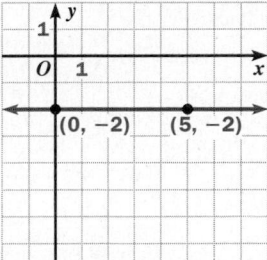

6.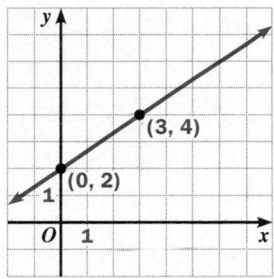

7.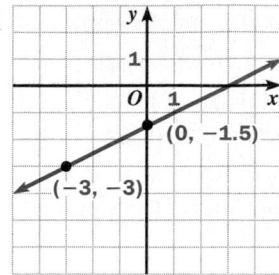

For Exercises 8–13, model each situation with an equation and a graph. Be sure to identify the independent and dependent variables.

8. An amusement park charges $4.00 admission and $1.00 per ride.

9. An empty 0.75 gal pitcher weighs 1.2 lb. Water weighs 8.3 lb/gal.

10. A film club membership costs $10.00 and $2.00 per movie.

11. A test is worth 100 points with 5 points deducted for each wrong answer.

12. A candle is 8 in. tall and burns at a rate of 1.5 in./h.

13. The gas tank of a van holds 16 gal. The van uses $\frac{1}{15}$ gal/mi.

14. Open-ended Problem Describe a situation that could be modeled by the equation $y = 0.75x + 4.25$.

Graph each equation.

15. $y = 1 - 2x$ **16.** $y = -4 + \frac{2}{3}x$ **17.** $y = 4$

18. $y = -3 - 3x$ **19.** $y = 7x - 12$ **20.** $y = x$

21. Writing Describe how the sign of the slope affects the graph in Exercises 15–20.

22. TEMPERATURE SCALES The drawing shows the temperatures at which water freezes and boils in both degrees Celsius and Fahrenheit. The Fahrenheit temperature *F* is a linear function of the Celsius temperature *C*.

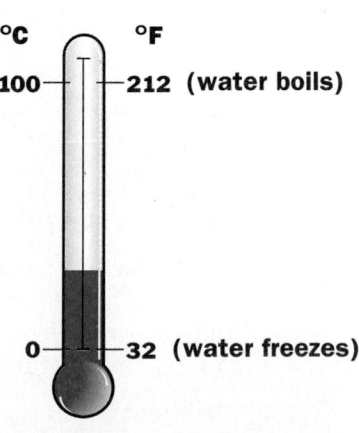

a. Plot the two data pairs (C, F) in a coordinate plane, and draw a line through the plotted points.

b. Find the slope-intercept equation of the line.

c. Find the Fahrenheit temperature equivalent to 56°C.

d. The equation from part (b) gives *F* as a function of *C*. Rewrite this equation so that it expresses *C* as a function of *F*.

e. Use your equation from part (d) to find the Celsius temperature equivalent to 77°F.

When European explorers first saw Mount Kenya and Mount Kilimanjaro in the 1840s, they sent back reports of snow-covered mountains in central Africa. European geographers of the time thought that the explorers had really seen white quartz, because they thought that snow could not exist near the equator.

6 km — ?
5 km — ?
4 km — ?
3 km — −4.5°C
2 km — 2.0°C
1 km — 8.5°C
0 km — 15.0°C

Mount Kilimanjaro in Tanzania is the highest mountain in Africa.

23. The scale above gives the standard air temperature at different altitudes.

   a. Plot the data in a coordinate plane. Draw a line through the points.

   b. Calculate the change in temperature per 1 km increase in altitude.

   c. Write an equation to model the temperature at different altitudes. What is the slope? What is the vertical intercept?

   d. What is the predicted temperature on top of Mount Kilimanjaro, which is 5.9 km high? Is snow possible at this temperature?

24. The boiling temperature of water is 100°C at sea level. For every 1 km increase in altitude, the boiling temperature decreases by about 3.5°C. (The boiling temperature of water decreases with altitude because the air pressure decreases.)

   a. Write the slope-intercept equation giving boiling temperature as a function of altitude.

   b. What is the boiling temperature of water in Mexico City, which is at altitude 2.2 km?

   c. What is the boiling temperature of water on top of Mount Kenya, which is 5.2 km high?

   d. **Research** At the top of Mount Everest (8.85 km high), water boils at about 69°C. This makes cooking food difficult. Use a cookbook to learn about the problems of cooking at high altitude, or use a chemistry book to find out what factors affect the boiling point of water. Write a brief report.

BY THE WAY

Mount Kilimanjaro has several distinct vegetation zones, including semiarid scrub at the base, a dense cloud forest, an alpine desert, and a moss and lichen community at the summit.

25. **Challenge** In this exercise, you will use algebra to show that the slope of the line $y = ax + b$ is a.

    a. Let (x_1, y_1) and (x_2, y_2) be any two distinct points on the line $y = ax + b$. Write an expression for the slope of the line in terms of x_1, y_1, x_2, and y_2.

    b. Use the fact that $y_1 = ax_1 + b$ and $y_2 = ax_2 + b$ to write your expression from part (a) in terms of only x_1 and x_2.

    c. Show that your expression from part (b) equals a.

26. **Technology** Miguel Santos wants to rent a car for a day while on vacation. Carla's Cars charges $22 for an economy car plus $.08 per mile driven. Friendly Rentals charges $38 for the same model of car with no additional charge per mile.

    a. For each company, find the slope-intercept equation giving the cost C to rent a car as a function of the distance d driven.

    b. Use a graphing calculator or graphing software to graph the two equations from part (a) in the same coordinate plane.

    c. **Writing** Discuss the circumstances for which each company offers Miguel the better deal.

27. **SAT/ACT Preview** If $A = -4(x + 1)$ and $B = -4x - 3$, choose the statement that is true for all values of x.

    A. $A > B$ **B.** $A < B$ **C.** $A = B$

    D. relationship cannot be determined

28. **GEOMETRY** *Without* plotting the points $(-2, -1)$, $(1, 5)$, and $(3, 10)$, determine whether the points lie on the same line. Explain your reasoning.

ONGOING ASSESSMENT

29. **Cooperative Learning** Work with a partner. You will need a set of identical paper cups. Measure the height of a single cup. Then add cups to form a stack, and measure the height of the stack each time a cup is added. Develop a linear model relating the height of the stack to the number of cups added to the first cup in the stack.

SPIRAL REVIEW

30. A typist types at a rate of 50 words per minute. Write a direct variation equation that relates the number of words typed to the amount of time spent typing. *(Section 2.1)*

31. Which of the following expressions is equivalent to $-2x + 6$? *(Toolbox, page 781)*

    A. $8 - 2(x + 1)$ **B.** $2 - 2(x - 3)$ **C.** $-2(x + 3)$

Solve each proportion. *(Toolbox, page 785)*

32. $\dfrac{x - 3}{3 + x} = 2$ 33. $\dfrac{4}{5 - 2x} = \dfrac{1}{2}$ 34. $12 = \dfrac{x + 2}{4x - 2}$

2.3 Point-Slope Form and Function Notation

Learn how to...

- **write the point-slope equation of a line**
- **use function notation**
- **find the domain and range of a linear function**

So you can...

- **make predictions about a runner's finishing time, for example**

You probably find it harder to run in a headwind (a wind blowing against your direction of motion) than to run in still air or with the wind at your back. In fact, the time required for a runner to finish a 1 mi race increases an average of 1.5 s for every 1 mi/h of headwind speed.

EXAMPLE 1

Lisa runs a 1 mi race in a 12 mi/h headwind. She finishes the race in 5:20 (5 min 20 s). Find an equation giving Lisa's finishing time as a function of headwind speed.

SOLUTION

Since Lisa's finishing time t increases 1.5 s for each 1 mi/h increase in headwind speed h, the graph giving t as a function of h is a line with slope 1.5. Also, $t = 5(60) + 20 = 320$ when $h = 12$, so the line passes through the point (12, 320).

To find an equation of the line, let (h, t) be any point on the line other than (12, 320).

The slope between these points should equal 1.5.

$$\frac{t - 320}{h - 12} = 1.5$$

$$t - 320 = 1.5(h - 12)$$

$$t = 320 + 1.5(h - 12)$$

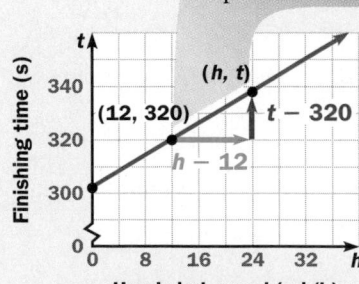

The equation $t = 320 + 1.5(h - 12)$ gives Lisa's finishing time t as a function of headwind speed h.

THINK AND COMMUNICATE

1. In Example 1, predict Lisa's finishing time in still air.

2. Write the equation $t = 320 + 1.5(h - 12)$ in the slope-intercept form $t = ah + b$. What does the vertical intercept tell you?

The Point-Slope Equation of a Line

In Example 1, the equation

$$t = 320 + 1.5(h - 12)$$

is based on knowing the point (12, 320) and the slope 1.5. In general, if you know that a line passes through a point (x_1, y_1) and has slope a, you can write the following **point-slope equation** of the line:

$$y = y_1 + a(x - x_1)$$

You can also use a point-slope equation when you know two points.

EXAMPLE 2

Write a point-slope equation of the line through the points $(-4, 7)$ and $(3, 1)$.

SOLUTION

Step 1 Find the slope a:

$$a = \frac{1 - 7}{3 - (-4)} = \frac{-6}{7} = -\frac{6}{7}$$

Step 2 Write a point-slope equation using either $(-4, 7)$ or $(3, 1)$ as the point (x_1, y_1). If you choose $(-4, 7)$, your equation is:

$$y = 7 + \left(-\frac{6}{7}\right)(x - (-4))$$

$$y = 7 - \frac{6}{7}(x + 4)$$

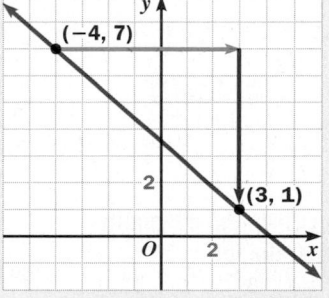

A point-slope equation of the line is $y = 7 - \frac{6}{7}(x + 4)$.

THINK AND COMMUNICATE

3. **a.** Write a point-slope equation of the line in Example 2 using (3, 1) for (x_1, y_1). Compare your answer with the equation obtained using $(-4, 7)$. Do the two equations look the same?

   b. Write the two equations in slope-intercept form. What do you notice?

4. A line has equation $y = -2 + \frac{1}{3}(x + 1)$.

   a. What is the line's slope?

   b. Name a point through which the line passes.

Function Notation

The **function notation** $y = f(x)$ tells you that y is a function of x. If there is a rule relating y to x, such as $y = 2x + 3$, then you can also write:

$$f(x) = 2x + 3$$

The name of the function is **f**. Other letters are also used to name functions, especially **g** and **h**.

$f(x)$, read "**f of x**," represents the value of the function at x.

The **domain** of a function f is the set of values of x for which f is defined. The **range** of a function f is the set of all values of $f(x)$, where x is in the domain of f.

For instance, if the function in Example 1 applies to headwind speeds up to 20 mi/h, then the domain of the function is $0 \le h \le 20$. The corresponding range of the function is $302 \le t \le 332$.

EXAMPLE 3

Suppose the function $f(x) = \frac{3}{2}x - 2$ is defined for $x \ge 2$.

a. Find $f(6)$.　　　　　　　　　　**b.** Find the domain and range of f.

SOLUTION

a. $f(6) = \frac{3}{2}(6) - 2 = 7$

b. **Method 1**
Use algebra.
The domain is $x \ge 2$. You can obtain the range from the domain.

Transform the inequality so that the left side becomes $\frac{3}{2}x - 2$.

$$x \ge 2$$
$$\frac{3}{2}x \ge \frac{3}{2}(2)$$
$$\frac{3}{2}x - 2 \ge 3 - 2$$
$$f(x) \ge 1 \quad \longleftarrow \text{ This is the range of } f.$$

Method 2
Look at a graph.

The range consists of the y-coordinates of the points on the graph. The **range** of f is $y \ge 1$.

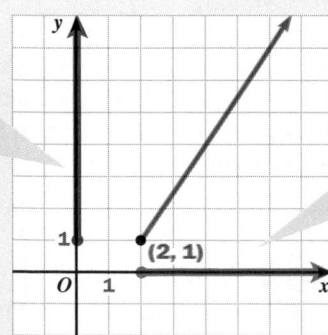

The domain consists of the x-coordinates of the points on the graph. The **domain** of f is $x \ge 2$.

(2, 1)

For Questions 1–3, write a point-slope equation of the line passing through the given point and having the given slope.

1. point: (2, 3)

slope = 1

2. point: (−4, 6)

slope = $-\dfrac{1}{4}$

3. point: (0, 3)

slope = $-\dfrac{3}{2}$

For Questions 4–6, write a point-slope equation of the line passing through the given points.

4. (1, 3) and (3, −5)

5. (−2, 3) and (−3, 2)

6. (0, 8) and (4, 1)

7. Suppose the function $f(x) = 4x + 3$ is defined for $x \geq -2$.

  a. Find $f(-1)$.

  b. Find the domain and range of f.

2.3 Exercises and Applications

Extra Practice exercises on page 751

For Exercises 1–4:

a. Write a point-slope equation of the line passing through the given point and having the given slope.

b. Graph the equation.

c. Write the equation in slope-intercept form.

1. point: (3, 2)

slope = 1

2. point: (−1, 5)

slope = $-\dfrac{1}{4}$

3. point: (3, 1)

slope = $-\dfrac{2}{3}$

4. point: (−5, −2)

slope = $\dfrac{4}{5}$

5. OCEANOGRAPHY The speed of sound in the ocean depends on the temperature, salinity, and depth of the water. However, in the open ocean below approximately 1000 m, temperature and salinity are fairly constant and the speed of sound increases linearly at a rate of 17 m/s for every 1000 m of depth.

  a. The speed of sound at a depth of 1000 m is about 1484 m/s. Write a point-slope equation giving the speed of sound s in meters per second as a function of the depth d in meters.

  b. Use the equation you found in part (a) to predict the speed of sound at a depth of 2000 m. Then use this value to determine the distance between the red and black submarines if it takes 3.2 s for sound to travel from one to the other.

  c. The Marianas Trench, at a depth of 10,924 m, is the deepest point in Earth's oceans. What is the domain and range of the function you found in part (a) in Earth's oceans? Explain.

BY THE WAY

Understanding how the speed of sound varies in water is essential for mapping the sea floor with sonar, and for communicating underwater.

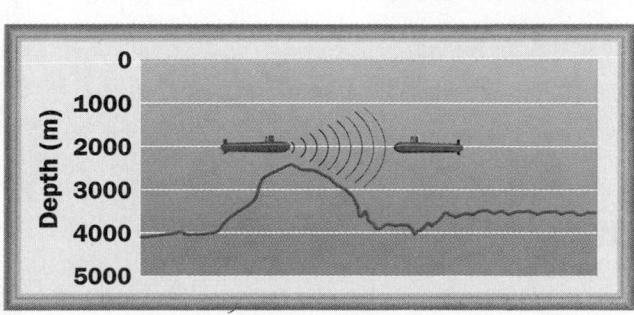

Sunrise and sunset occur earliest at the eastern edge of a time zone. They occur four minutes later for each degree west you travel within that time zone, if you remain at roughly the same latitude.

6. If the sun rises at 6:05 A.M. in Philadelphia, then for any location at about the same latitude as Philadelphia and at longitude *l* within the Eastern time zone, sunrise occurs at the time *t* given by:

$$t = 6:05 + (:04)(l - 75)$$

Cities near 40°N latitude	Longitude (°W)
Philadelphia, PA	75
Indianapolis, IN	86

   a. Use the equation to predict the time the sun will rise in Indianapolis.

   b. The Eastern time zone includes longitudes from about 67°W to 90°W. State the domain and range of the given function.

7. a. Suppose the sun sets at 6:17 P.M. in Sweetwater, Texas. Write a point-slope equation giving sunset time as a function of longitude for locations at about 32°N within the Central time zone.

Cities near 32°N latitude	Longitude (°W)
Sweetwater, TX	100
Marshall, TX	94

   b. Use the equation to predict the time the sun will set in Marshall, Texas.

8. **Open-ended Problem** Use an almanac to find the longitude for your town. Check a newspaper to find today's sunset time for your town. Write an equation giving sunset times for locations along the same latitude as your town and in your time zone. Find another town at about the same latitude and predict the sunset time there today.

9. Anton receives a New York newspaper in Akron, Ohio, located at 82°W longitude. He notices that the sunrise time given for New York City, located at the same latitude as Akron, is 32 min earlier than the sunrise time printed in his Akron newspaper. Determine the longitude of New York City.

10. **Challenge** Draw a graph of sunrise times across the United States at 40°N latitude if you know that sunrise occurs in Philadelphia at 6:15 A.M. on a given day. (*Note:* Your graph should include the loss of 1 h each time you pass into a new time zone going west. At 40°N latitude, the United States extends from about 74°W to 124°W longitude.)

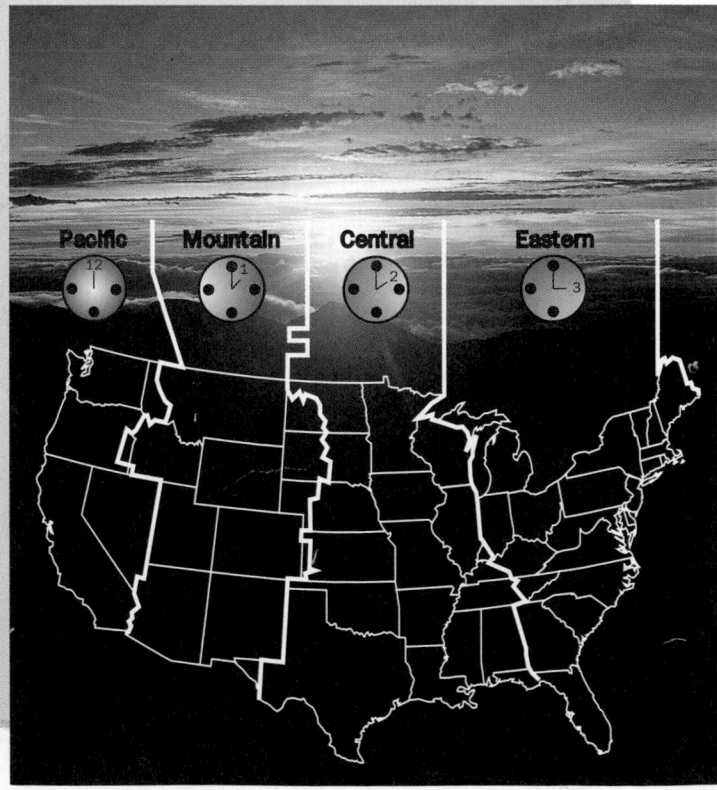

For each function f whose graph is shown, find the domain and range.

11.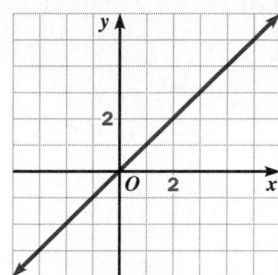

12.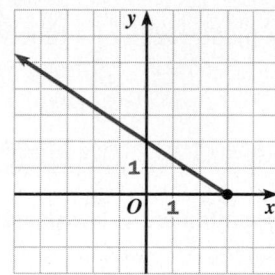

13.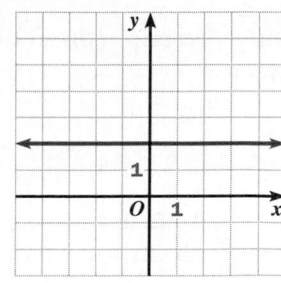

14. **BIOLOGY** The weight of Jersey cattle increases roughly linearly with age. A typical Jersey cow weighs 150 lb at 3 months and 540 lb at 15 months. Write a point-slope equation giving the weight of a Jersey cow as a function of age. Predict the weight of a Jersey cow at 20 months.

15. **SPORTS** Although a headwind hurts a runner's performance, a tailwind helps a runner move more quickly. Studies have shown that the time required for a runner to finish a 1 mi race decreases an average of 0.5 s for every 1 mi/h of tailwind speed.

    a. Running with an 8 mi/h tailwind, Alicia finishes a 1 mi race in 6:30 (6 min 30 s). Find a point-slope equation giving Alicia's finishing time t in seconds as a function of tailwind speed w in miles per hour.

    b. Predict Alicia's finishing time in still air.

16. **SAT/ACT Preview** Which ordered pair is a solution to the equation $y = 15(x - 3) + 7$?

    A. $(-3, 7)$ **B.** $(0, 0)$ **C.** $(3, 15)$ **D.** $(7, 3)$ **E.** $(3, 7)$

Find a point-slope equation of each line.

17.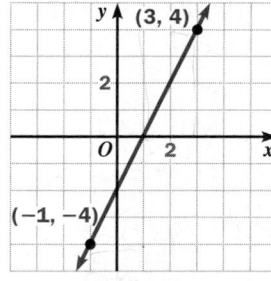

18.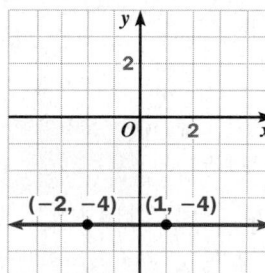

19.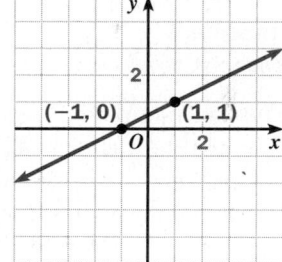

20. **SPORTS** A 120 lb in-line skater burns 7.4 Cal/min skating at 10 mi/h. For every extra 20 lb of weight, a skater skating at 10 mi/h burns an extra 0.9 Cal/min.

    a. Write a point-slope equation giving c, the number of calories burned per minute, as a function of w, the weight of the skater.

    b. **Open-ended Problem** What is a reasonable domain of the function you found in part (a)? What is the range?

For Exercises 21–24, find the domain and range of each function.

21. $f(x) = 5x - 2$ for $x > 0$

22. $f(x) = 2x - 1$ for $x \le 0.5$

23. $f(x) = 7.2x + 320$ for $x \ge 14$

24. $f(x) = 18$ for $0 \le x \le 5$

25. Writing Describe how to find the domain and range of a function from its graph.

For Exercises 26–28, find the slope-intercept equation of the line passing through each pair of points. *(Section 2.2)*

26. $(0, 2)$ and $(4, 5)$ **27.** $(0, -3)$ and $(3, -5)$ **28.** $(-1, -2)$ and $(0, -1)$

29. Michael is studying the number of families with preschoolers in his neighborhood for a social studies project. His data appear in the table. Write a linear equation to model the number of families with preschoolers each year. *(Section 1.2)*

Year	Families
1993	11
1994	10
1995	8
1996	8

Solve each equation. *(Toolbox, page 784)*

30. $x + \dfrac{3}{2} = \dfrac{7}{8}$

31. $20 - x = 3x + 4$

32. $5n = 125$

33. $\dfrac{s}{8} = 7s + 11$

ASSESS YOUR PROGRESS

VOCABULARY

direct variation (p. 46)   **point-slope equation** (p. 59)
constant of variation (p. 46)   **function notation** (p. 60)
slope (p. 47)   **domain** (p. 60)
vertical intercept (p. 53)   **range** (p. 60)
slope-intercept form (p. 53)

1. a. The table shows the Loquasto family's driving time and distance traveled for each day of their car trip from Boston to Denver. Show that the distance traveled each day varies directly with driving time.

Time (h)	Distance (mi)
8.5	468
6	328
7	385
6.5	360
8	438

   b. From Denver, the Loquastos plan to drive to Salt Lake City, a distance of 493 mi. Estimate how long it will take them. *(Section 2.1)*

2. Graph the line $y = 3 - \dfrac{2}{3}x$. *(Section 2.2)*

3. Suppose the domain of the function $f(x) = 5 - 2x$ is $x \le 3$. What is the range? Graph the function. *(Section 2.3)*

4. Journal How are the point-slope and slope-intercept forms of a linear equation related? How are they different?

2.4 Fitting Lines to Data

You know that the more pages a book has, the thicker it is. The Exploration will help you decide if the relationship between page count and thickness is linear.

Learn how to...

- draw a line of fit and find its equation

So you can...

- make predictions about bicycle production, for example

EXPLORATION
COOPERATIVE LEARNING

Measuring the Thickness of Books

Work with a group of three or four students.
You will need:
- a set of encyclopedias
- a metric ruler
- graph paper and a pencil

1 Choose ten of the volumes from a set of encyclopedias. For each, find the number of pages *P* and measure the volume's thickness *T* to the nearest 0.1 cm. Record your data in a table.

2 Plot the data pairs (*P*, *T*) in a coordinate plane.

Toolbox p. 795
Making a Scatter Plot

3 Use the ruler to draw a *line of fit*.

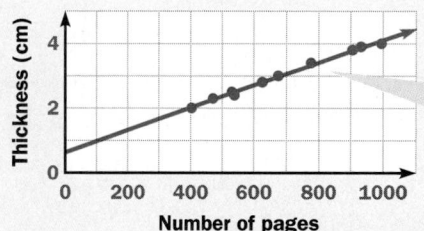

A **line of fit** is a line that lies as close as possible to all the points in a scatter plot. It does not have to pass through any of them.

Use your graph to answer the following questions.

1. Find the slope of your line. What does the slope represent in this situation? (*Hint:* Think about the relationship between the number of pages and the number of pieces of paper.)

2. a. Find an equation of your line of fit. What is the significance of the *T*-intercept?

b. Measure the thickness of a volume not included in your data set. Use the equation to predict how many pages it has. Then use the book to check your prediction.

Lines of fit are often used to model data, such as data obtained in scientific studies, opinion polls, and sporting events, and to predict what happens between and beyond plotted data points. The "eyeball method" used in the Exploration is the simplest way to determine a line of fit.

EXAMPLE 1   Application: Manufacturing

The table gives the number of bicycles produced worldwide for various years between 1965 and 1990.

Year	1965	1970	1975	1980	1985	1990
Number of bicycles (in millions)	21	36	43	62	79	95

a. Show that the data have a linear relationship by making a scatter plot and drawing a line of fit.

b. Find an equation of the fitted line.

c. Use the equation to predict the number of bicycles produced in the year 2000.

SOLUTION

a. Let t be the time (in years) since 1965, and let b be the number of bicycles (in millions) produced.

t	b
0	21
5	36
10	43
15	62
20	79
25	95

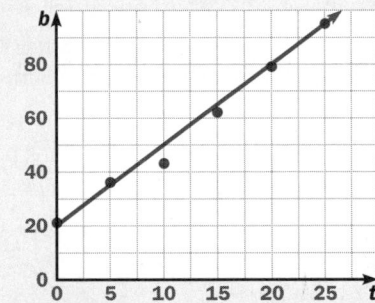

b. Choose any two points that *appear* to lie on the fitted line, for example (5, 35) and (20, 80). The slope of the line is:

$$\frac{80 - 35}{20 - 5} = \frac{45}{15} = 3$$

Therefore a point-slope equation of the line is:

$$y = 35 + 3(x - 5)$$

c. The year 2000 is 35 years after 1965, so substitute $x = 35$ into the equation from part (b):

$$y = 35 + 3(35 - 5)$$
$$= 125$$

You can predict that about 125 million bicycles will be produced in the year 2000.

1. Does a line of fit have to pass through at least one of the data points?

2. **a.** If two different students eyeballed a line of fit for the same data, would you expect the lines of fit to be the same? Explain.

   b. How would you know which student's line was a better fit for the data?

The Least-Squares Line

Different people tend to "eyeball" different lines of fit for a set of data points. To avoid this problem, mathematicians have developed a standard line of fit called the *least-squares regression line*, or just the **least-squares line.**

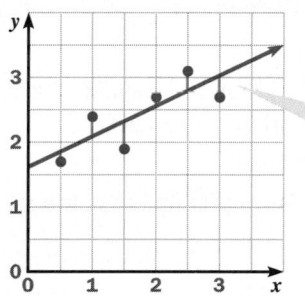

The sum of the squares of the vertical distances shown in green is smallest for the least-squares line.

Most graphing calculators and statistical software will find an equation of the least-squares line for you.

EXAMPLE 2   Application: Manufacturing

Refer to the bicycle data in Example 1.

a. Use a graphing calculator or statistical software to find an equation of the least-squares line for the data.

b. Use the equation from part (a) to predict the number of bicycles produced in the year 2000.

For more information about linear regression, see the *Technology Handbook*, p. 814.

SOLUTION

a. Enter the data and have the calculator or software perform a linear regression. You should get results like those shown. An equation of the least-squares line is $y = 2.96x + 19$.

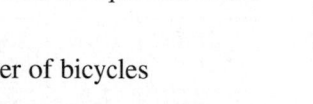
```
LinReg
 y=ax+b
 a=2.96
 b=19
 r=.9939507756
```

b. Substitute $x = 2000 - 1965 = 35$ into the equation from part (a):

$$y = 2.96(35) + 19$$
$$= 122.6$$

You can predict that about 123 million bicycles will be produced in the year 2000.

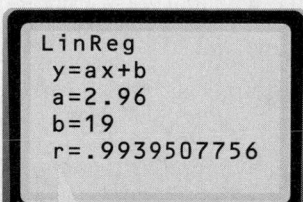

For now, ignore the variable r in the display.

> **WATCH OUT!**
>
> On some calculators the roles of a and b are reversed: $y = a + bx$. Be sure to check your calculator's instruction manual.

☑ CHECKING KEY CONCEPTS

1. The table gives the grams of fat and the number of calories in 100 g of various salad dressings. The lowest in fat and calories is low-fat French, and the highest in fat and calories is mayonnaise.

Fat (g)	4.3	42.3	50.2	52.3	60.0	79.9
Calories	96	435	502	504	552	718

  a. Show that the data have a linear relationship by making a scatter plot and drawing a line of fit.

  b. Find an equation of the line.

  c. Use the equation to predict the number of calories in a salad dressing that has 30 g of fat.

2. a. **Technology** Use a graphing calculator or statistical software to find an equation of the least-squares line for the salad dressing data in Question 1.

  b. Use the equation from part (a) to predict the number of calories in a salad dressing that has 30 g of fat.

2.4 | **Exercises and Applications**

Extra Practice exercises on page 751

1. ECONOMICS If a person works an eight-hour day, how much of that time goes just toward paying taxes? The table shows how this figure has changed from 1958 to 1994.

Year	1958	1964	1970	1976	1982	1988	1994
Hours per day for taxes	2.2	2.3	2.5	2.6	2.7	2.7	2.8

  a. Make a scatter plot of the data and draw a line of fit.

  b. Find an equation of the line. What does the line's slope represent?

  c. Predict the number of hours per day that a person living in 2010 will have to work just to pay taxes.

  d. **Writing** Does the model ever predict that a person's entire workday will go to pay taxes? Discuss the limitations of the model.

2. **Technology** The table shows the winning times for men in the Boston Marathon for various years between 1955 and 1995. Use a graphing calculator or statistical software to find an equation of the least-squares line for winning time as a function of year.

Year	1955	1960	1965	1970	1975	1980	1985	1990	1995
Time (min)	138.37	140.90	136.55	130.50	129.92	132.18	134.08	128.32	129.37

In 1993 some of the worst flooding in U.S. history occurred when the Mississippi and Missouri rivers overflowed their banks after reaching record crests. The flooding was caused by downpours in the Midwest.

3. The table gives the height of the Mississippi River above its banks in Hannibal, Missouri, for one week in April, 1993.

   a. Show that the data have a linear relationship by making a scatter plot and drawing a line of fit.

   b. Find an equation of the line of fit. Predict the height of the river on April 1st and on April 30th. Which prediction do you think is more accurate?

Date in April	Height above banks (in.)
4	2.1
5	2.4
6	2.8
7	3.2
8	3.8
9	4.4
10	4.6

4. The scatter plot shows the height of the Mississippi River above its banks in Hannibal, Missouri, for the entire month of April, 1993.

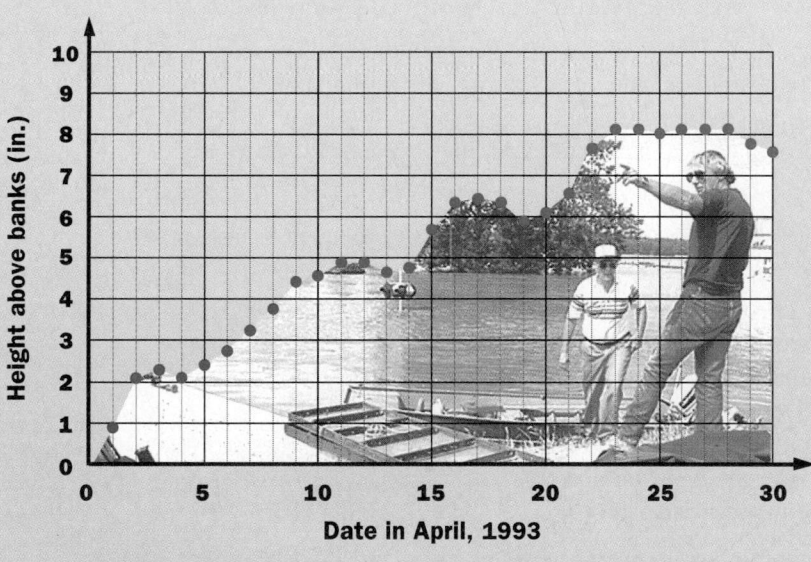

Date in April, 1993

   a. **Visual Thinking** Does the height of the water still appear to increase roughly linearly over time? Use a pencil or ruler to eyeball a line of fit for the entire month of April, and for April 4–10. What do you notice?

   b. Using the graph, find an equation for a new line of fit for the height of the water as a function of time for the month of April.

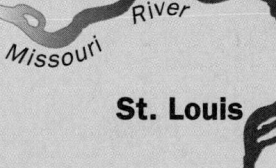

Lines of fit have been used in court to interpret data about election results.

Year	District	Democratic machine vote	Republican machine vote	Democratic absentee vote	Republican absentee vote
1982	2	47,767	21,340	551	205
1982	4	44,437	28,533	594	312
1982	8	55,662	13,214	338	115
1986	2	39,034	23,363	609	316
1986	4	52,817	16,541	666	306
1986	8	48,315	11,605	477	171
1990	2	27,543	26,843	660	509
1990	4	39,193	27,664	482	831
1990	8	34,598	8,551	308	148

5. The table shows the results in three Philadelphia voting districts for state senate elections. "Machine vote" means votes cast at official polling places.

   a. **Technology** Let m = the difference between the Democratic and the Republican machine vote. Let a = the difference between the Democratic and the Republican absentee vote. Use a graphing calculator or statistical software to add two columns to the table for m and a.

   b. Make a scatter plot of the data, with m on the horizontal axis and a on the vertical axis. Find a line of fit.

6. **Visual Thinking** In a special runoff election in District 2 in 1993, the Democratic machine vote was 19,127, and the Republican machine vote was 19,691. The Democratic absentee vote was 1396, and the Republican absentee vote was 371. Find m and a for this election and include the new point on your scatter plot. What do you notice?

7. **Writing** The special runoff election was challenged in court. The Republicans charged that many of the absentee ballots were fraudulent. The Democrats argued that they had done a good job turning out the absentee vote. The judge awarded the seat to the Republican. Why do you think the judge made this decision?

8. **Writing** The data point for the 1990 election in District 4 is also far off the line of fit. Why do you think this election was not challenged? Should it have been?

BALLOTS

9. **BIOLOGY** The table gives the weight and oxygen consumption of four harbor seals.

Weight (kg)	26.8	31.3	35.6	41.0
Oxygen consumption (mL/min)	230	266	287	332

   a. Show that the data have a linear relationship by drawing a scatter plot and a line of fit.

   b. Find an equation of the line of fit.

   c. Predict the oxygen consumption of a 25 kg harbor seal.

10. **HEALTH** The table gives average weights for men and women of different heights. The data are for people of medium frame, aged 30–39 years.

   a. Make two scatter plots, one for men and one for women, in the same coordinate plane. Put height on the horizontal axis and weight on the vertical axis, and use different symbols or colors to distinguish the two sets of data.

   b. For each scatter plot, draw a line of fit and find its equation.

   c. Predict the average weights of men 80 in. tall and women 76 in. tall.

   d. **Writing** Compare the slopes of the two lines of fit. Did you expect them to be the same? Explain.

Women		Men	
Height (in.)	Weight (lb)	Height (in.)	Weight (lb)
60	120	64	145
62	126	66	153
64	132	68	161
66	139	70	170
68	146	72	179
70	154	74	188
72	164	76	199

ONGOING ASSESSMENT

11. **Open-ended Problem** Look through a newspaper or almanac and find a set of real-world data that appears to be linear. Make a scatter plot of the data, draw a line of fit, and find an equation of the line. Use your equation to make predictions.

SPIRAL REVIEW

12. a. The table gives the number of eighth graders, in millions, enrolled in public schools in the United States. Make a scatter plot of the data. *(Toolbox, page 795)*

   b. Does it make sense to fit a line to the data?

Year	Eighth graders (millions)
1960	2.70
1970	3.60
1980	3.09
1990	2.98

Write a point-slope equation of the line passing through the given point and having the given slope. *(Section 2.3)*

13. point: (7, 3)
    slope = −4.2

14. point: (12, 3)
    slope = 2

15. point: (−2, −3)
    slope = 0.5

For each equation, state whether y varies directly with x. *(Section 2.1)*

16. $y = \frac{1}{2}x$

17. $y = \frac{1}{3}x$

18. $y = 2(x + 1)$

19. $y = -3x$

2.5 The Correlation Coefficient

Learn how to...

- **interpret a correlation coefficient**
- **recognize positive and negative correlation**

So you can...

- **determine the strength of a linear relationship, such as between longitude and the highest January temperature**

Is there a strong linear relationship between a city's location and the highest January temperature in that city? Examine the two scatter plots.

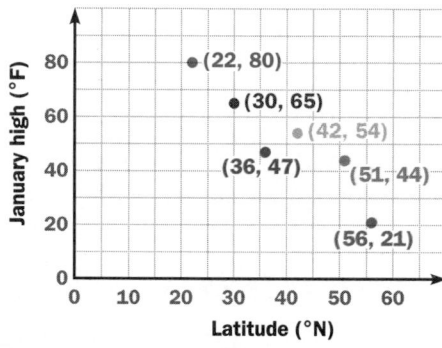

January High versus Longitude

(31, 65) (88, 80)
(12, 54)
(0, 44) (140, 47)
(38, 21)

January high (°F) vs Longitude (°E)

January High versus Latitude

(22, 80)
(30, 65)
(42, 54)
(36, 47) (51, 44)
(56, 21)

January high (°F) vs Latitude (°N)

THINK AND COMMUNICATE

1. For which scatter plot do the data points "line up" better?

2. Do you think the longitude of a city or the latitude of a city more accurately predicts the highest January temperature in that city? Explain why your answer makes sense.

A **correlation coefficient**, denoted r, is a number between -1 and $+1$ that measures how well data points line up. There is a **positive correlation** $(0 < r \leq 1)$ between the variables x and y when y tends to increase as x increases. There is a **negative correlation** $(-1 \leq r < 0)$ between the variables when y tends to decrease as x increases.

Perfect Positive Correlation	**Perfect Negative Correlation**
$r = +1$	$r = -1$

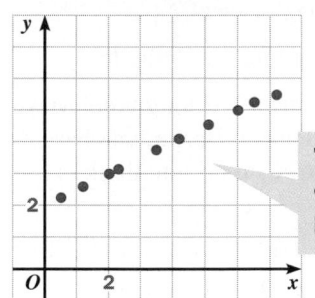

The points lie on a line with **positive slope**.

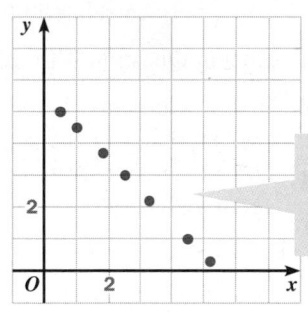

The points lie on a line with **negative slope**.

Below are examples of correlations between -1 and $+1$. If $|r|$ is near 1, the data points almost lie on a line. If $|r|$ is near 0, the data points tend not to lie on any line.

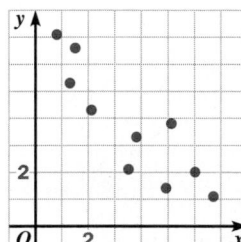

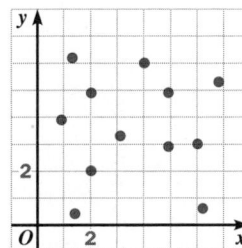

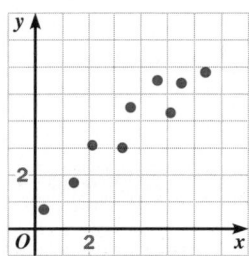

−1

$r \approx -0.88$
strong negative correlation

$r \approx 0$
weak correlation

$r \approx 0.93$
strong positive correlation

+1

The formula for the correlation coefficient r is complicated, but graphing calculators and statistical software often find r when they compute the least-squares line.

EXAMPLE 1 **Application: Meteorology**

Refer to the scatter plots on the previous page.

a. Use a graphing calculator or statistical software to find the correlation coefficient for the scatter plot of highest January temperature versus longitude.

b. Repeat part (a) for the scatter plot of highest January temperature versus latitude.

SOLUTION

a. Enter the data pairs (longitude, January high) into a graphing calculator or statistical software. The top calculator screen shows that the correlation coefficient r is about 0.20.

b. For the data pairs (latitude, January high), the bottom calculator screen shows that the correlation coefficient r is about -0.93, which is close to -1.

```
LinReg
 y=ax+b
 a=.0761472156
 b=47.91175173
 r=.2009759978
```

```
LinReg
 y=ax+b
 a=-1.457596095
 b=109.4083791
 r=-.93144244
```

3. In Example 1, are highest January temperature and longitude positively or negatively correlated? Is this correlation strong or weak?

4. Repeat Question 3 for highest January temperature and latitude.

5. Berlin, Germany, is located at 53°N latitude and 13°E longitude. Make two estimates of the highest January temperature in Berlin, one based on latitude and the other on longitude. Which estimate do you think is more accurate? Why?

Correlation and Causation

You've probably noticed that there is a strong positive correlation between the amount of time you study for a test and the grade you receive. Plenty of study time is one cause of high test scores. Correlation does not always imply causation, however.

The Secret of Long Life? Be Dour and Dependable

Score one for those pious voices of prudence: being cautious and somewhat dour is a key to longevity, according to a 60-year study of more than 1000 men and women.

Those who were conscientious as children were 30 percent less likely to die in any given year of adulthood than their most free-wheeling peers.

"We don't really know why conscientious people live longer—it's not as simple as wearing your sweater when it's cold outside," said Dr. Howard S. Friedman, who did the research.

EXAMPLE 2

The article suggests that being conscientious causes people to live longer. Explain what might account for this correlation.

SOLUTION

There are probably many factors, such as leading a healthy lifestyle, setting long-term life goals, and avoiding risks, that are highly correlated with being conscientious. These are the factors that actually cause people to live longer.

☑ CHECKING KEY CONCEPTS

For Questions 1–3, match each scatter plot to one of the following values of r: 0.6, −0.3, −0.9.

1. 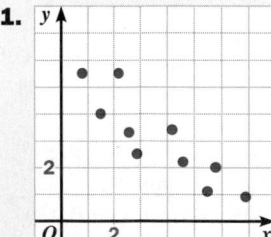   **2.** 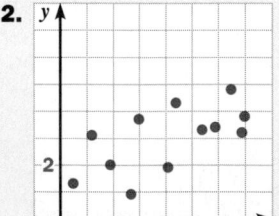   **3.**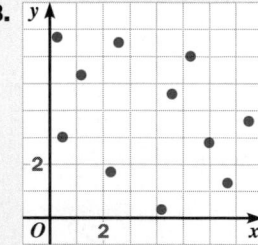

4. Brahim found a strong positive correlation between a baseball team's season record and home attendance. Can you say that a high season record causes a high home attendance? Explain.

2.5 | Exercises and Applications

Extra Practice
exercises on
pages 751–752

For Exercises 1–3, estimate the correlation coefficient for the given scatter plots.

1.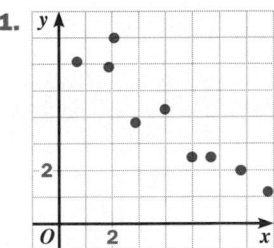

2.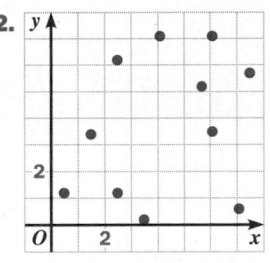

3.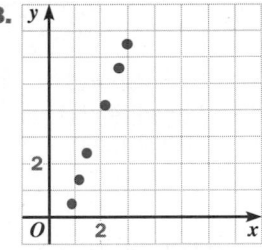

4. Writing Can the points (1, 2) and (4, 5) both lie on a scatter plot with correlation coefficient $r = -1$? Explain your answer.

For Exercises 5–8, tell whether you would expect the correlation between the two quantities to be positive, negative, or about zero.

5. the height and weight of a person

6. the age of a car and its value

7. the shoe size and salary of an adult

8. the outside temperature and parka sales

9. Writing The correlation coefficient for variables u and v is 0.43, while the correlation coefficient for variables x and y is -0.92. Elvia says that u and v are more strongly correlated than x and y since $0.43 > -0.92$. Write a sentence or two explaining Elvia's mistake.

10. Writing There is a strong positive correlation between the number of mailboxes in a city and the amount of light pollution at night in that city. This certainly does not mean that mailboxes cause light pollution. Explain what might account for the correlation.

11. Challenge If variables x and y are positively correlated and variables y and z are negatively correlated, what can be said of the correlation between x and z? Explain your reasoning.

12. BUSINESS The bar graphs give information about sales figures and profits for a major United States company.

a. 📈 **Technology** Use a graphing calculator or statistical software to find the equation of the least-squares line and the correlation coefficient for $x =$ total sales and $y =$ total profits.

b. Use the equation for the least-squares line to predict the total profits when the total sales for the company are 6.0 billion dollars.

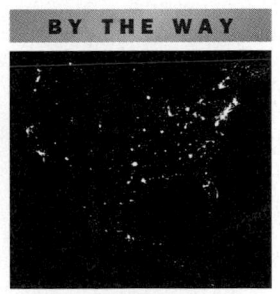

BY THE WAY

Many people consider artificial light a form of pollution because it makes the environment unfit for certain activities, such as astronomy.

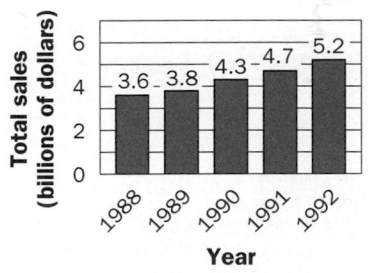

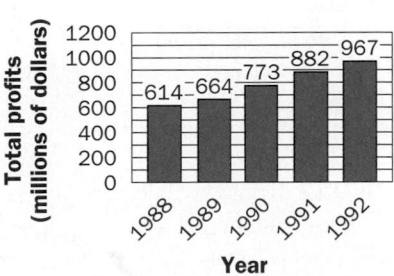

Large animals have many advantages over small animals, including being able to move more rapidly.

13. **Technology** The table gives the average body lengths and the highest observed flying speeds of various animals.

   a. Use a graphing calculator or statistical software to make a scatter plot of the data. Estimate the correlation coefficient from your scatter plot.

   b. Find the correlation coefficient of flying speed versus body length.

   c. Find the correlation coefficient of body length versus flying speed. Compare this coefficient with the answer to part (b). Does the value of the correlation coefficient depend on whether body length is the independent or dependent variable?

Species	Length (cm)	Flying speed (m/s)
Horsefly	1.3	6.6
Ruby-throated hummingbird	8.1	11.2
Dragonfly	8.5	10.0
Willow warbler	11	12.0
Flying fish	34	15.6
Whimbrel	41	23.2
Common pintail	56	22.8

14. The two scatter plots show the highest observed running speeds and swimming speeds versus body lengths of various animals.

   a. Which scatter plot shows the stronger correlation? Explain what this tells you.

   b. What variables besides body length might be correlated with speed?

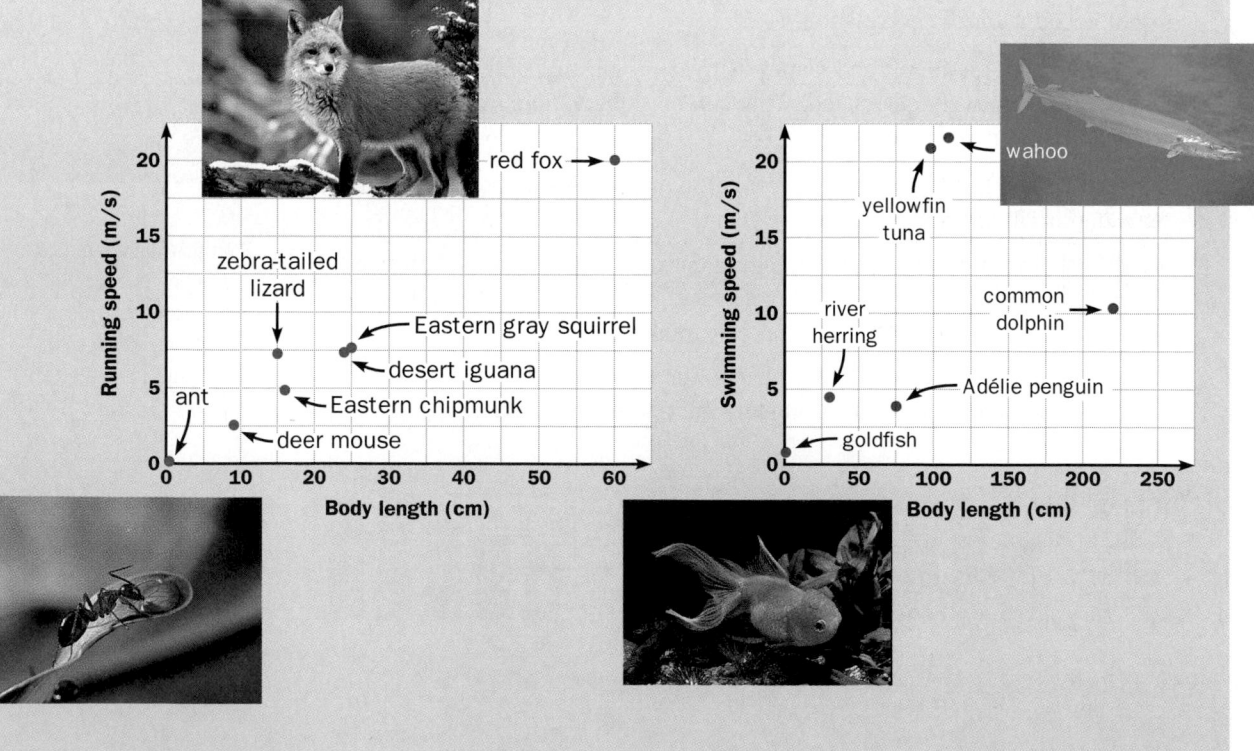

15. GOVERNMENT The table lists six political parties in Mexico, the percent of votes cast for each party in the 1991 election, and the size of their delegations to the governing body called the *Cámara Federal de Diputados*.

a. **Technology** Use a graphing calculator or statistical software to find the correlation coefficient for the data.

b. Is the correlation between percent of votes and delegation size positive or negative? Is it strong or weak?

c. **Research** Read about the United States Senate and House of Representatives in an encyclopedia or some other source. Is there a strong correlation between a state's population and its number of senators? between a state's population and its number of representatives? Explain.

Political party	Percent of votes cast	Total seats
PRI	61.46	320
PAN	17.72	89
PRD	8.26	41
PFCRN	4.36	23
PARM	2.15	15
PPS	1.80	12

ONGOING ASSESSMENT

16. Cooperative Learning Work in a group to collect data on the ages and odometer readings of cars. Make a scatter plot of the data and estimate the correlation coefficient. Write a brief explanation of why the variables are correlated.

SPIRAL REVIEW

17. The table shows the percent of women marrying who are within various age groups. For example, 37.1% of the women marrying in 1980 were between 20 and 24 years old. *(Section 2.4)*

Year	20–24 years old	25–29 years old	65 years and older
1980	37.1	18.7	1.0
1985	34.4	22.1	1.0
1988	31.5	24.1	1.0

a. For each age group make a scatter plot of percent of women marrying versus year. Then draw a line of fit.

b. Find an equation for each line.

c. Predict the percent of women marrying who will be in each age group in 1990 and 2000.

Write an equation of the line with the given slope and y-intercept. *(Section 2.2)*

18. slope $= 4$
y-intercept $= 7$

19. slope $= 0$
y-intercept $= -2$

20. slope $= -3$
y-intercept $= 1$

Solve each linear equation. *(Toolbox, page 784)*

21. $3 + 2t = 5$

22. $\dfrac{t - 7}{3} = 2$

23. $3(t - 2) = 6$

2.6 Linear Parametric Equations

Learn how to...

- **write and graph a pair of linear parametric equations**

- **rewrite a pair of parametric equations as a single equation**

So you can...

- **model situations, such as the motion of a plane**

When an airplane takes off or lands, it moves both horizontally (forward) and vertically (up or down). To analyze motion in two dimensions, you can use *parametric equations*.

EXPLORATION
COOPERATIVE LEARNING

Landing a Plane

Work with a partner. You will need:
- a graphing calculator with parametric mode

SET UP Imagine you are piloting a small airplane at **12,000 ft**, preparing to land. Once you begin your descent to the runway, your altitude changes at a rate of **–15 ft/s**. Your horizontal speed is **200 ft/s**.

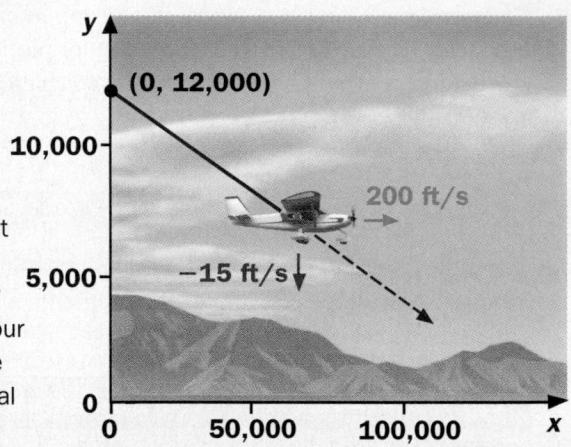

1 The plane is initially at the point (0, 12,000). Complete the equations for *x* and *y*.

The plane's horizontal position at time *t* (in seconds) is given by:

$$x = 0 + \underline{?}\, t$$

The plane's vertical position at time *t* (in seconds) is given by:

$$y = \underline{?} + \underline{?}\, t$$

2 Enter your equations from Step 1 into a graphing calculator set in parametric mode. Then graph the equations, adjusting the viewing window and *t*-values so you can see where the graph crosses the *x*-axis.

3 Use your calculator's trace feature to determine how long it takes the plane to reach ground level. What is the value of *x* at this time? At what horizontal distance from the airport should the airplane begin its descent? (*Note:* 1 mi = 5280 ft)

For more information about parametric mode, see the *Technology Handbook*, p. 810.

In the Exploration, the airplane's x- and y-coordinates were not related through an equation of the form $y = f(x)$. Instead, you expressed x and y as separate functions of t:

$$x = g(t) \qquad\qquad y = h(t)$$

The variable t is called a **parameter**.

The equations $x = g(t)$ and $y = h(t)$ are called **parametric equations**.

EXAMPLE 1 Application: Physics

Tom lives directly west across a river from his grandfather's house. The river flows south at 3 mi/h and is 0.5 mi wide. If Tom tries to row east across the river at 2 mi/h, how far from his grandfather's house will he land?

SOLUTION

Let x = Tom's east-west position and y = Tom's north-south position, in miles.

Step 1 Draw a sketch and show a coordinate system with the origin at Tom's house.

Step 2 Write parametric equations for x and y.

$$x = 2t \qquad y = -3t$$

Step 3 Find t when $x = 0.5$.

$0.5 = 2t$, so $t = 0.25$

Step 4 Find y when $t = 0.25$.

$y = -3(0.25)$, so $y = -0.75$

Tom will land 0.75 mi downstream from his grandfather's house.

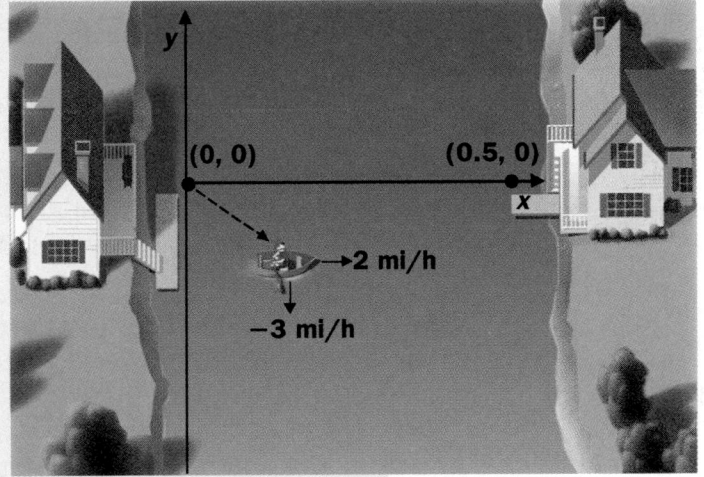

EXAMPLE 2

Graph the parametric equations from Example 1 for $0 \le t \le 0.25$.

SOLUTION

Use a table to plot a few points.

t	$x = 2t$	$y = -3t$
0	0	0
0.1	0.2	−0.3
0.2	0.4	−0.6
0.25	0.5	−0.75

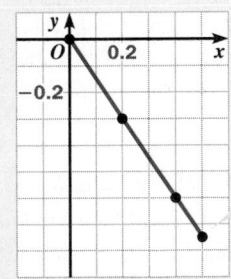

For $0 \le t \le 0.25$, $0 \le x \le 0.5$ and $-0.75 \le y \le 0$.

THINK AND COMMUNICATE

1. In Example 1, what would the parametric equations be if the origin were located at Tom's grandfather's house?

2. In Example 2, what are the domain and range of the function $x = 2t$? of the function $y = -3t$?

3. What does the graph in Example 2 represent?

EXAMPLE 3

Using the parametric equations from Example 1, express *y* as a function of *x*.

SOLUTION

Your goal is to eliminate the parameter t. To do this, first solve $x = 2t$ for t:

$$2t = x$$
$$t = 0.5x$$

Then substitute $0.5x$ for t in $y = -3t$:

$$y = -3t$$
$$= -3(0.5x)$$
$$= -1.5x$$

Bear in mind that a restriction on t creates a restriction on x. For the equation $y = -1.5x$, the restriction on x is $0 \leq x \leq 0.5$ because $0 \leq t \leq 0.25$.

THINK AND COMMUNICATE

4. What information do the parametric equations give you that is missing from the equation you found in Example 3?

5. Consider the graph of the function $y = 4 + x$ for $x \geq -1$. Find a pair of parametric equations having the same graph. Is more than one answer possible?

☑ CHECKING KEY CONCEPTS

For each pair of parametric equations, find *x* and *y* when *t* = 2.

1. $x = 2t - 1$
   $y = -5t - 2$
   $t \geq 0$

2. $x = 0.5t$
   $y = t + 1$
   $0 \leq t \leq 12$

3. $x = 12 - t$
   $y = 14$
   $t \geq 5$

4–6. Express y as a function of x using the equations in Questions 1–3. State any restriction on x.

Exercises and Applications

*Extra Practice
exercises on
page 752*

Graph each pair of parametric equations for the given restriction on *t*.

1. $x = 3t$
$y = t$
$t \geq 0$

2. $x = -1 - t$
$y = -3 + 2t$
$t \leq 0$

3. $x = -5 + 5t$
$y = 4t$
$0 \leq t \leq 1$

4. $x = 1 + t$
$y = -2 + t$
no restriction on t

5–8. Express y as a function of x using the equations in Exercises 1–4. State any restriction on x.

9. Challenge Find a pair of parametric equations to describe the graph of $y = 2x - 3$ for $x \geq 2$.

Connection ▶ RECREATION

Immediately after jumping from a plane, a parachutist falls with increasing speed. After about 9 s, however, the speed of a parachutist in a flat stable position levels off at about 110 mi/h, or 160 ft/s. This speed is called *terminal velocity*. The parachutist is in *free fall* until the parachute is opened.

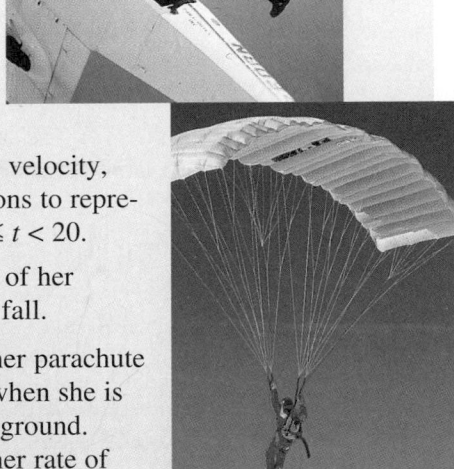

10. Johanna has just reached terminal velocity at 6000 ft above the ground. She descends at 160 ft/s and the wind blows her horizontally at 12 ft/s.

a. Let $t = 0$ correspond to the time when she reaches terminal velocity, 6000 ft above the ground. Write a pair of parametric equations to represent her motion for $0 \leq t < 20$.

b. Sketch a graph of her descent in free fall.

11. Johanna deploys her parachute at time $t = 20$ s, when she is 2800 ft above the ground. Almost instantly her rate of descent changes to 10 ft/s.

a. When does she land?

b. She is still traveling horizontally at 12 ft/s. Write a pair of parametric equations to model her descent with the open parachute. What are the restrictions on t?

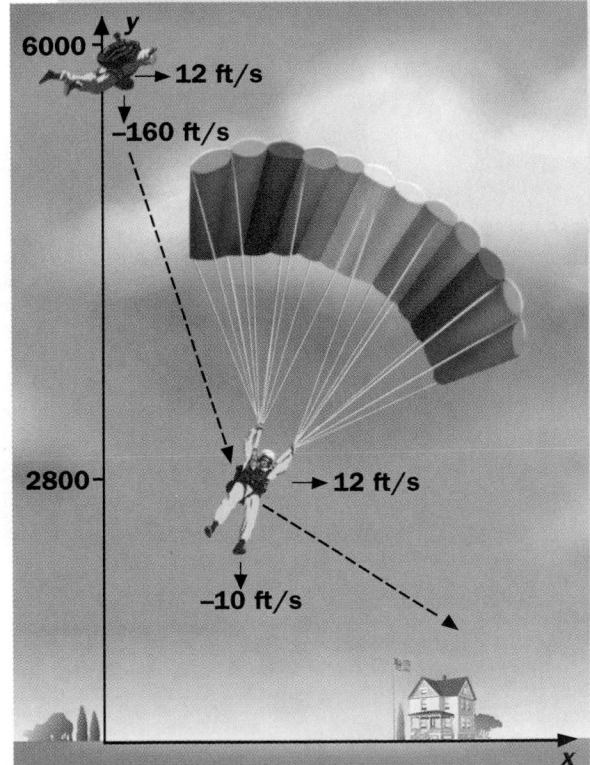

12. An object moves according to the parametric equations $x = 3t$ and $y = 4t$ where t is in seconds and x and y are in meters.

a. Draw a graph of the object's path for $0 \leq t \leq 1$.

b. What is the object's speed along its path in meters per second? (*Hint:* Find the distance traveled and divide it by the travel time.)

c. Another object moves according to the parametric equations $x = 6t$ and $y = 8t$ for $0 \leq t \leq 0.5$. What is the relationship between the path of this object and the path of the first object? What is the relationship between their speeds?

d. **Writing** Write a paragraph about a situation that could be modeled by these parametric equations.

Connection ▶ PERFORMING ARTS

When you direct a performance, it is important to know where the performers will be at different times.

13. The drama club is putting on a play set in a haunted house. The script calls for two characters to back into each other. Glen starts at $(0, 15)$, and his path is described by the equations $x = 2.5t$ and $y = 15 - t$, with t in seconds and x and y in feet. Neepa starts at $(40, 10)$, and her path is described by the equations $x = 40 - 3t$ and $y = 10$.

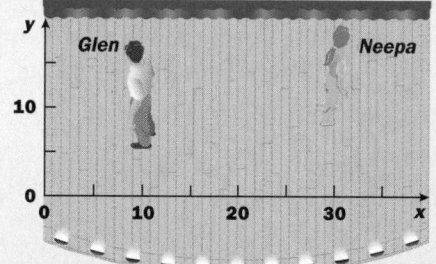

a. 📈 **Technology** Use a graphing calculator or a table of values to plot Glen and Neepa's paths.

b. Will they bump into each other? How do you know?

c. Could you have answered part (b) if you were given Glen and Neepa's paths with y expressed as a function of x? Explain.

14. Yuki is choreographing a modern dance performance on the same stage as the play in Exercise 13. She wants three dancers to run past each other at the same instant. The dancers' movements for $0 \leq t \leq 3$ are described by the equations in the table below, with x and y in feet and t in seconds.

Anita	$x = 5t$	$y = 10 - \frac{5}{3}t$
Bao	$x = 27 - 4t$	$y = 8$
Claire	$x = 5t$	$y = 16 - \frac{5}{3}t$

a. What is the position of each dancer at $t = 0$?

b. 📈 **Technology** Enter the equations for the three dancers into a graphing calculator set in parametric mode. Do the dancers pass each other at the same time? (*Hint:* To pass, they should be at different points on the same line.)

15. **Open-ended Problem** Think of a situation you can model with two or three objects following paths defined by parametric equations.

   a. Write up your problem. Some questions to consider are: What are the restrictions on each object's movement? Where are the objects initially? Do you want the objects to meet or to miss each other?

   b. **Writing** Write a paragraph or short story about what happens in the problem you wrote.

16. Explain the difference between correlation and causation. *(Section 2.5)*

17. Sales of computer books at The Science Bookshop are growing at an average of 15% a year. In 1990 the bookshop sold 560 computer books. *(Section 1.2)*

   a. Write an equation to model the growth in computer book sales.

   b. How many computer books did the bookshop sell in 1994?

Write a point-slope equation of the line through each pair of points. *(Section 2.3)*

18. $(13, 8)$ and $(-1, 1)$ 19. $(-3, 7)$ and $(2, -5)$ 20. $(-3, 2)$ and $(1, 4)$

ASSESS YOUR PROGRESS

VOCABULARY

least-squares line (p. 67)   **negative correlation** (p. 72)
correlation coefficient (p. 72)   **parameter** (p. 79)
positive correlation (p. 72)   **parametric equations** (p. 79)

Exercises 1 and 2 refer to the table, which shows the number of English-language newspapers published in the United States.

Year	Newspapers
1980	1745
1982	1711
1984	1688
1986	1657
1988	1642
1990	1611

1. Show that the data have a linear relationship by making a scatter plot and drawing a line of fit. *(Section 2.4)*

2. **Technology** Use a graphing calculator or statistical software to find the correlation coefficient for the data. Is the correlation strong or weak? positive or negative? *(Section 2.5)*

3. a. Graph the parametric equations $x = 2 + t$ and $y = 1 - t$ for $t \geq 0$.

   b. Express y as a function of x and give the function's domain. *(Section 2.6)*

4. **Journal** Find a real-life example of correlation and explain whether it involves causation.

Stretching a Rubber Band

When you put fruit on a grocery store scale, the scale's dial tells you the weight. If you could look inside the scale, you would see a spring that controls the movement of the dial. This spring stretches as you add more fruit.

A rubber band behaves in a similar way: the harder you pull on it, the longer it stretches. These situations both involve a pulling force that increases length. Have you ever wondered how the force used to stretch a spring or a rubber band is related to its length?

PROJECT GOAL Perform an experiment to study the relationship between pulling force and rubber band length.

Conducting an Experiment

Work with a partner. Here are a few suggestions for carrying out the experiment.

1. GATHER a rubber band, 2 large paper clips, a paper cup, 20 marbles, a ruler (preferably transparent), graph paper, and a pencil.

2. SET UP the equipment. Bend the paper clips to make S-shaped hooks. Using the hooks and rubber band, suspend the paper cup from the edge of a desk or table as shown.

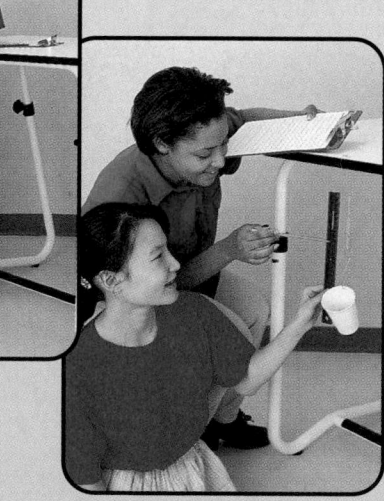

3. MEASURE the length of the rubber band to the nearest 0.1 cm. After one partner adds marbles to the cup four at a time and measures the rubber band's length, the other partner records the data. Copy and complete the table.

m = number of marbles in cup	0	4	8	12	16
l = length of rubber band (cm)	?	?	?	?	?

Analyzing the Data

- **DRAW** a scatter plot of the data pairs (m, l) in a coordinate plane. The plotted points should almost line up. Use the ruler or a graphing calculator to obtain a line of fit.

- **WRITE** an equation for your line. Use the equation to find l when $m = 20$.

- **CHECK** the prediction by putting 20 marbles in the cup and measuring the rubber band length. How did your predicted value for $m = 20$ compare to your experimental value?

Writing a Report

Write a report about your experiment. Include a paragraph on each of these points:

- goals of the experiment
- descriptions of your procedure
- tables and graphs of your data
- conclusions based on your results

You may want to extend your report by investigating and reporting on some other ideas.

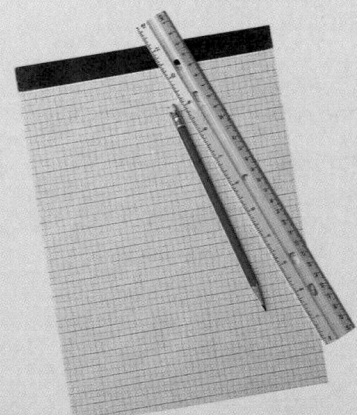

- Use your graph from the project to sketch a predicted graph for this experiment: You start with 20 marbles in the cup and remove four at a time. Then conduct the experiment to check your graph.

- Repeat the project using various types of rubber bands. Do some rubber bands stretch more than others? What factors do you think contribute to the stretchability of the rubber bands?

- Research Hooke's law from a physics book and explain how it applies to this project.

Self-Assessment

Were you satisfied with your prediction of rubber band length based on your line of fit? If so, what factors influenced your success? If not, how can you adjust your methods to improve your results?

2 | Review

STUDY TECHNIQUE

Go back through the sections in this chapter. For each section, write two questions that could appear on the chapter test. One should be a short-answer question. One should be a more involved question about general ideas. Exchange your questions with another student and answer the set you receive.

VOCABULARY

direct variation (p. 46)
constant of variation (p. 46)
slope (p. 47)
vertical intercept (p. 53)
slope-intercept form (p. 53)
point-slope equation (p. 59)
function notation (p. 60)
domain (p. 60)

range (p. 60)
least-squares line (p. 67)
correlation coefficient (p. 72)
positive correlation (p. 72)
negative correlation (p. 72)
parameter (p. 79)
parametric equations (p. 79)

SECTIONS | 2.1, 2.2, *and* 2.3

The variables x and y are in **direct variation** if $y = ax$ for some nonzero constant a. A direct variation graph is a straight line passing through the origin.

For example, a 39 oz box of laundry detergent can clean 13 loads of clothing. The number of loads l that you can clean varies directly with d, the amount of detergent in ounces, so $l = ad$.

To find a, the **constant of variation**, rewrite $l = ad$ as $a = \dfrac{l}{d}$.

$$a = \frac{13}{39} = \frac{1}{3}$$

The direct variation equation is $l = \dfrac{1}{3}d$.

The slope of the graph is the constant of variation, $\dfrac{1}{3}$.

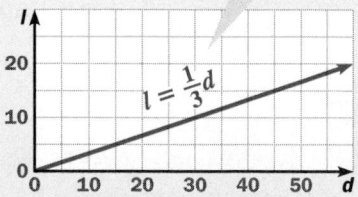

To find the **slope** of the line that passes through two points (x_1, y_1) and (x_2, y_2), use this formula:

$$\text{slope} = \frac{\text{vertical change}}{\text{horizontal change}} = \frac{y_2 - y_1}{x_2 - x_1}$$

For example, the line shown passes through $(-2, -1)$ and $(0, 3)$. The slope a is:

$$a = \frac{-1 - 3}{-2 - 0} = 2$$

The **slope-intercept equation** of a line is $y = ax + b$, and a **point-slope equation** is $y = y_1 + a(x - x_1)$. For the line shown, the equations are:

Slope-intercept $\quad y = 2x + 3$

Point-slope $\quad y = -1 + 2(x + 2)$

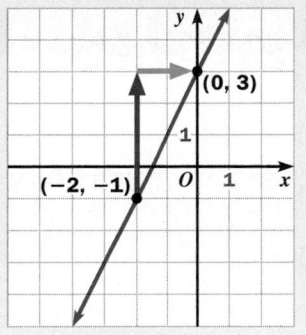

SECTIONS $\ 2.4$ *and* 2.5

For a set of paired data, you can draw a line of fit by hand, or use a graphing calculator or statistical software to find the **least-squares line**.

For example, the table gives six data pairs, which are plotted below.

x	y
0	2.5
2	2.6
4	3.6
6	5.4
8	5.5
10	7.4

```
LinReg
 y=ax+b
 a=.5
 b=2
 r=.9689381826
```

The scatter plot shows that the plotted points "line up," which implies a **strong correlation**.

The least-squares line has slope 0.5 and vertical intercept 2, so an equation is **$y = 0.5x + 2$**.

SECTION $\ 2.6$

Use **parametric equations** to solve problems about two-dimensional motion.

For example, the graph of the parametric equations $x = t + 1$ and $y = 3t - 2$ for $1 \leq t \leq 3$ is shown.

You can express y as a function of x by eliminating the **parameter** t. For the graph shown, $y = 3x - 5$ for $2 \leq x \leq 4$.

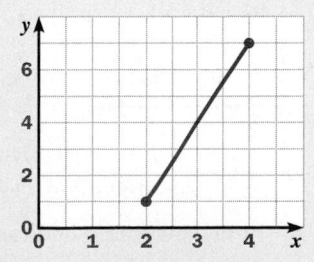

2 Assessment

VOCABULARY QUESTIONS

For Questions 1–3, complete each paragraph.

1. The equation $y = ax$ tells you that x and y are in __?__ .

2. The __?__ form for the equation of a line is $y = ax + b$. It has this name because a is the __?__ of the line, and b is the line's __?__ .

3. When you enter data pairs into a graphing calculator or statistical software, you can obtain an equation for a line of fit called the __?__ . You can also obtain the value of r, called the __?__ , which measures how well this line fits the data points.

SECTIONS 2.1, 2.2, *and* 2.3

For each equation, tell whether *y* varies directly with *x*. If so, graph the equation.

4. $y = \dfrac{4}{3x}$　　　　5. $y = \dfrac{4x}{3}$　　　　6. $y = 4x + 3$

7. Suppose p varies directly with s. When $s = 12$, $p = 5$.

   a. Find an equation relating p and s.　　**b.** Find p when $s = 27$.

Give the slope and *y*-intercept of the line described by each equation.

8. $y = 6x$　　　　9. $y = 2 - 5x$　　　　10. $y = -3$

Graph each equation.

11. $y = \dfrac{5}{3}x - 4$　　　　12. $y = -x$　　　　13. $y = 2.4$

Find a point-slope equation of each line.

14. 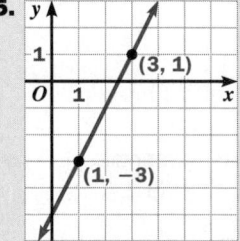　　15.　　16.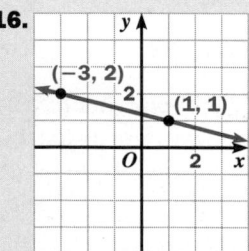

17. Suppose the function $f(x) = -2x + 5$ is defined for $x \le 3$.

   a. Find the domain and range of f.　　**b.** Graph the function.

18. SPORTS The table gives the 1993 populations of six states and the total number of professional baseball, basketball, football, and hockey teams in each state at that time.

State	Population (millions)	Number of sports teams
CA	30	16
GA	6	3
IL	11	5
MA	6	4
NY	18	9
PA	12	7

  a. Show that the data have a linear relationship by making a scatter plot.

  b. Is the correlation between the population and the number of teams *positive* or *negative*? Is it *strong* or *weak*?

  c. **Technology** Use a graphing calculator or statistical software to find the correlation coefficient for the data.

  d. Fit a line to the data. Find an equation for your line of fit.

  e. Use the equation to predict the number of professional sports teams in New Jersey, with population 8 million. How does your prediction compare with New Jersey's actual figure of two professional teams?

19. Open-ended Problem Do you think there is a correlation between the amount of time a person spends reading each week and the amount of time the person spends watching television each week? Explain your reasoning. If there is a correlation, is it *positive* or *negative*? Why? Is there causation? Explain.

SECTION 2.6

For each of Exercises 20–22:

a. Graph the pair of parametric equations for the given restriction on *t*.

b. Express *y* as a function of *x*, and state any restriction on *x*.

20. $x = -t$
$y = 5t - 3$
$t \le 0$

21. $x = 2 + t$
$y = 3t + 4$
$0 \le t \le 3$

22. $x = \frac{1}{2}t + 1$
$y = -3t + 6$
$2 \le t \le 6$

PERFORMANCE TASK

23. Design and carry out an experiment that you think will result in linear data. Make a scatter plot of the data, draw a line of fit, and find an equation of the line. Use your equation to make predictions. In addition to the graph, equation, and predictions, your report should include a description of the experiment and an assessment of how well a linear model fits your data.

3 Exponential Functions

A river running through it

Finn Strong

When Finn Strong was two years old, he started pouring water over rock piles. At the age of 12, he built a rock garden with plants and fish in the living room of his family's home. The garden included a waterfall that tumbled into a fish tank. At the age of 24, Strong started a company to sell a new kind of fish tank. He invented the "river tank" to show what a river might look like if you put a window along one side. Inside the fish swim through rapids, waterfalls, and eddies. Strong says, "They're much more active than fish in normal tanks."

> **"If you have a crazy idea that everyone says can't be done, you still ought to try."**

The Rainmaker

The water in the tank stays clear because it is constantly flowing, cleansing itself like an actual river. The continual churning of the water helps provide more oxygen for the fish as well. "People put chemicals in our tanks when they're not supposed to," Strong says. "The key is to set it up and leave it alone." Evaporation posed a problem for Strong's river tank because water had to be added frequently. Strong solved this problem by designing a rainmaking device. With this device, water has to be added only once every one to two weeks. Beads of water condense on the surfaces of the rainmaker until they periodically fall as rain.

"Everything runs according to nature."

A Natural Interdependency

Frogs, toads, lizards, salamanders, snails, and plants live in the water and on the riverbanks of Strong's river tank. "The tank shows the interdependence of plants and animals," Strong says. The fish can't live without the plants, because the plants keep the water clean. The plants, in turn, live off the fish wastes. Even the bacteria have a role, eating nutrients that would otherwise cloud the water and make it toxic. The idea, Strong adds, is "to provide the right ingredients so that everything runs according to nature."

Raindrops Keep Falling

How does the rainmaker work? The amount of water vapor that air can hold depends on the temperature. At higher temperatures, air can hold much more water. The graph shows how the maximum amount of water vapor that air can hold depends on the temperature. In Strong's river tank, the warm moist air near the surface of the water rises to the top. Raindrops form when the air touches the cool surfaces of the rainmaker.

The relationship between temperature and the amount of water vapor the air can hold can be modeled by an *exponential function*. The graph of an exponential function rises slowly at first and then takes off dramatically. For example, the graph shows that when the air is cold, a small change in temperature has a small effect on the amount of water vapor the air can hold. At higher temperatures, a small increase in temperature produces a large increase in the amount of water vapor the air can hold.

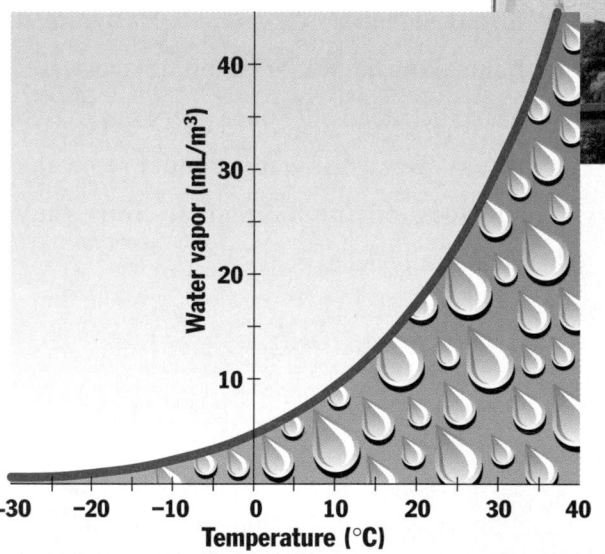

EXPLORE AND CONNECT

Strong points out a shelf that he designed for lizards. They can keep dry by staying in the upper areas of the tank.

1. Writing In many parts of the world, hot weather is often accompanied by high humidity. Cold weather is often accompanied by dry air. How does the graph help explain these phenomena?

2. Project In some ways, Finn Strong's river tank and rainmaking device are a model of the *hydrologic cycle*. This phrase is used to describe the constant movement of water between the atmosphere and the surface of Earth. Find out more about the hydrologic cycle. Draw a diagram to show the cycle.

3. Research When you listen to a weather report, you may hear the forecaster talk about *relative humidity*. Find out what is meant by relative humidity.

Mathematics & Finn Strong

In this chapter, you will learn more about how mathematics is related to the water content of air and the oxygen content of water.

Related Exercises

Section 3.3
• Exercises 23 and 24

Section 3.5
• Exercises 17–19

3.1 Exponential Growth and Decay

Learn how to...

- **describe growth and decay using tables of data and equations**

So you can...

- **use exponential equations to model real-life situations, such as making books**

Did you know that this book was made by printing on both sides of large sheets of paper and folding them many times? Each large sheet forms a group of pages called a *signature,* which is trimmed and then bound with other signatures. An exponential function relates the number of times a sheet is folded to the number of pages in the signature.

EXAMPLE 1 **Application: Bookmaking**

Each time you fold a sheet of paper used in bookmaking, you double the number of sheets of paper in the signature. Each sheet of paper, called a *leaf,* has a front and a back, so there are two pages for every leaf.

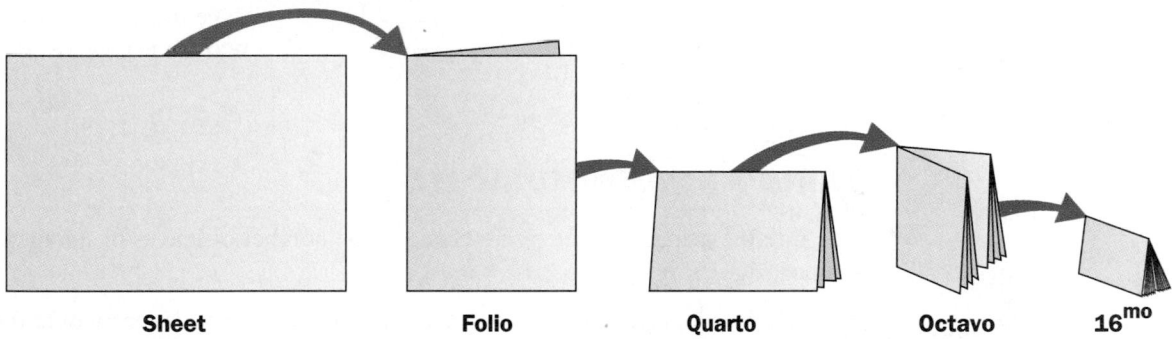

Sheet **Folio** **Quarto** **Octavo** **16**^{mo}

a. Make a table showing the number of leaves and pages when a sheet of paper is folded many times. The names of page sizes are given in the table.

b. Write equations that show how the number of leaves and the number of pages depend on the number of folds.

Name of page size	Number of folds	Number of leaves	Number of pages
—	0		
folio	1		
quarto	2		
octavo	3		
16mo	4		
32mo	5		
64mo	6		

SOLUTION

a.

Name of page size	Number of folds	Number of leaves	Number of pages
—	0	1	2
folio	1	2	4
quarto	2	4	8
octavo	3	8	16
16mo	4	16	32
32mo	5	32	64
64mo	6	64	128

b. Let n = the number of folds.

Each fold doubles the number of leaves, so the number of leaves L is given by this equation:

$$L = 2^n$$

Each leaf has two pages, so the number of pages P is given by this equation:

$$P = 2 \cdot L$$
$$= 2 \cdot 2^n$$
$$= 2^{n+1}$$

> Since $2 = 2^1$ and since 2^1 and 2^n have the same base, you can **add** the exponents.

THINK AND COMMUNICATE

1. Are the names of page sizes related to the number of leaves or the number of pages after folding?

2. Suppose a piece of paper is folded 7 times. How many leaves would there be? What might the page size be named?

Toolbox p. 776
Exponents and Powers

To rewrite $2 \cdot 2^n$ as 2^{n+1} in part (b) of Example 1, you use the product property of exponents. That property, as well as two others you learned in a previous course, are stated below.

Properties of Exponents

For any $b > 0$ and positive integers m and n:

Examples:

Product Property: $b^m \cdot b^n = b^{m+n}$ $2^3 \cdot 2^4 = 2^7$

Quotient Property: $\dfrac{b^m}{b^n} = b^{m-n}$ $\dfrac{5^6}{5^4} = 5^2$

Power Property: $\left(b^m\right)^n = b^{mn}$ $\left(3^6\right)^2 = 3^{12}$

When paper is folded to make a signature for a book, each fold is perpendicular to the previous fold. You can use this information to see how the page width, height, and area are affected by folding.

EXAMPLE 2 Application: Bookmaking

a. The "Royal" paper size is 46 cm wide and 60 cm high. Make a table of the width, height, and area of the page sizes made by folding a sheet of Royal-sized paper in half many times.

b. Write an equation for the page area as a function of the number of folds.

SOLUTION

a.

Name of page size	Number of folds	Page width (cm)	Page height (cm)	Page area (cm²)
ROYAL	0	46	60	2760
folio	1	30	46	1380
quarto	2	23	30	690
octavo	3	15	23	345
16 mo	4	11.5	15	172.5
32 mo	5	7.5	11.5	86.25
64 mo	6	5.75	7.5	43.125

b. The original page area is 2760 cm². Each fold halves the page size, so the page area A is given by:

$$A = 2760\left(\frac{1}{2}\right)^n$$

EXAMPLE 3

Suppose a stack of 400 leaves of a mathematics textbook measures 1 in.

a. What is the thickness of one leaf?

b. Write an equation for the thickness T of the stack of leaves when a sheet is folded n times.

SOLUTION

a. $\dfrac{1 \text{ in.}}{400 \text{ leaves}} = 0.0025$ in./leaf

b. $T = 0.0025\left(2^n\right)$

To find T, multiply the total number of leaves by the thickness per leaf.

THINK AND COMMUNICATE

3. Suppose a stack of 900 leaves of a history textbook measures 2 in. How does the thickness equation given in part (b) of Example 3 change?

4. Compare the equations from Examples 2 and 3. How are they similar?

BY THE WAY

The word *paper* comes from *papyrus*, a writing material made by ancient Egyptians from the stem of the papyrus plant. Paper, however, consists of dissolved fibers of wood, cotton, or linen and is made by a process invented in China about the second century B.C.

The equations from Examples 2 and 3, $A = 2760\left(\frac{1}{2}\right)^n$ and $T = 0.0025\left(2^n\right)$, both have this general form:

amount after x events $\qquad$ $y = ab^x$ $\qquad$ number of events

original amount $\qquad$ proportional growth or decay factor

There is an important difference in their graphs, however.

WATCH OUT!

In both graphs, curves are *not* drawn through the points, because it does not make sense to consider non-integral values for the number of times paper is folded.

Exponential Growth

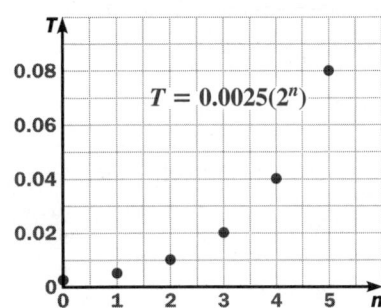

When $a > 0$ and $b > 1$, the graph of $y = ab^x$ *rises* from left to right.

Exponential Decay

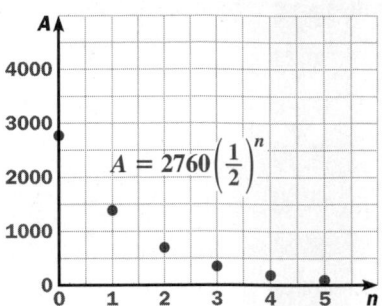

When $a > 0$ and $0 < b < 1$, the graph of $y = ab^x$ *falls* from left to right.

THINK AND COMMUNICATE

Use the equations for stack thickness and page area given with the graphs above.

5. What is the original amount for the stack thickness function? for the page area function?

6. What is the proportional growth factor for the stack thickness function?

7. What is the proportional decay factor for the page area function?

8. What "event" is represented by the variable n?

☑ CHECKING KEY CONCEPTS

Tell how many leaves would be formed if a sheet of paper could be folded the given number of times.

1. 7 $\qquad$ **2.** 8 $\qquad$ **3.** 9 $\qquad$ **4.** 10

5. The "Demy" paper size is 38 cm wide and 51 cm high.

    a. Make a table of the width, height, and area of the page sizes made by folding a sheet of Demy-sized paper.

    b. Write an exponential decay equation for the page area as a function of the number of folds.

3.1 Exercises and Applications

Extra Practice exercises on page 752

Tell how many folds are needed in a sheet of paper to create the given number of leaves.

1. 16   **2.** 64   **3.** 8   **4.** 128

Write each expression as a power of 2.

5. $2 \cdot 2 \cdot 2 \cdot 2$ **6.** $4 \cdot 32$ **7.** $2 \cdot 2^5$ **8.** $2^5 \cdot 2^7$

Evaluate each expression when $x = 5$.

9. $5(2^x)$ **10.** $200(2^x)$ **11.** $6400\left(\frac{1}{2}\right)^x$ **12.** $128\left(\frac{1}{2}\right)^x$

Tell whether each equation represents growth that is *linear*, *exponential*, or *neither*.

13. $y = (3^2)x$ **14.** $y = 3(2^x)$ **15.** $y = 3x^2$ **16.** $y = 2(3^x)$

17. Technology Use a graphing calculator or graphing software to graph the equation from part (b) of Example 3. Use the trace feature to find how many folds are needed to make a stack of pages taller than you are.

18. Visual Thinking Fold a piece of paper in half four times and unfold it. How can you number the small rectangles on both sides so that when you fold the paper again and trim it along three edges, you get a small book with 32 correctly numbered pages? (*Hint:* One side of one possible answer is shown. How should the other side be numbered?)

24	9	16	17
25	8	1	32
28	5	4	29
21	12	13	20

19. Investigation Try folding a sheet of paper in half six times. What do you notice? Why do you think signatures of 64 leaves or more are not commonly used in bookmaking? Look at some hardcover books to see how many folds were used to create the signatures.

20. Spreadsheets You can use a spreadsheet to make a table of the width, height, and area of the page sizes made by folding a sheet of paper. An example, using the "Demy" paper size, is shown.

a. The "Pott" paper size is 31 cm wide and 39 cm high. Use a spreadsheet to make a table of the width, height, and area of the page sizes made by folding a sheet of Pott-sized paper.

b. Write an exponential decay equation for the page area as a function of the number of folds.

c. Graph the equation from part (b).

Bookmaking Page Sizes

	C9		=0.5*D8	

	A	B	C	D	E
1	Name of page size	Number of folds	Page width (cm)	Page height (cm)	Page area (cm^2)
2	DEMY	0	38	51	1938
3	folio	1	25.5	38	969
4	quarto	2	19	25.5	484.5
5	octavo	3	12.75	19	242.25
6	16mo	4	9.5	12.75	121.125
7	32mo	5	6.375	9.5	60.5625
8	64mo	6	4.75	6.375	30.28125
9	128mo	7	3.1875	4.75	15.140625

This is half of the amount in cell D8.

This is the same as the amount in cell C8.

21. Cooperative Learning Here is a way to discover a word someone is thinking of just by asking 20 questions:

Ask your friend to think of a word. Open a dictionary at the halfway point and ask a question such as, "Is your word before or after *lifesaver* in a dictionary?"

Suppose your friend is thinking of the word *electricity*. Your friend says, "Before."

Look in the middle of the first half and ask another question: "Is your word before or after *diesel*?" Your friend says, "After."

Divide the indicated section in half and continue asking questions this way until you either find the word or use up 20 questions.

a. Try the dictionary game with a friend or family member. Are you able to discover the word in fewer than 20 questions?

b. About how many words are in the dictionary you used in part (a)?

c. How many words would a dictionary have to include to require you to use more than 20 questions?

d. Writing Explain how the dictionary game works. Use powers of $\frac{1}{2}$ or powers of 2 in your explanation.

22. MYTHOLOGY The Mongol people known as Kalmyks in Central Asia believed the proportions of the universe were fixed by mathematical rules. In their tales, a central mountain called Sumeru rises 80,000 leagues above the surface of the world-ocean. Around it are seven circular mountain chains, the first and innermost being 40,000 leagues high, the next 20,000 leagues high, and so on.

a. Find the height of the seventh mountain chain.

b. Suppose the mountain chains continue beyond the seventh. Write an equation for the height of the *n*th mountain chain.

Connection ▸ GRAPHIC DESIGN

In many countries, the International Standards Organization "A series" of paper sizes is used, with metric measurements as shown.

AO paper size

	A	B	C	D
ISO A Series Paper Sizes				
1	Name of paper size	Paper width (mm)	Paper height (mm)	Paper area (m^2)
2	A0	841	1189	1.000
3	A1	594	841	0.500
4	A2	420	594	0.249
5	A3	297	420	0.125
6	A4	210	297	0.062
7	A5	148	210	0.031
8	A6	105	148	0.016

Write an exponential decay equation for each variable. Use the numbers in the paper size names as the independent variable.

23. paper width   **24.** paper height   **25.** paper area

26. Challenge What is the ratio of paper height to paper width for any sheet in the "A series"? Explain why this ratio allows you to write exponential decay functions for paper width and paper height. Is there any other ratio that would allow you to write such functions?

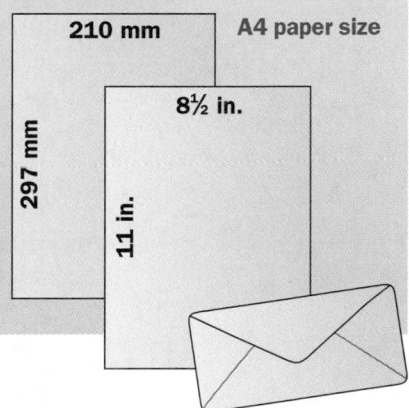

210 mm   A4 paper size

297 mm

8½ in.

11 in.

The A0 paper size is 16 times larger than the A4 paper size.

ONGOING ASSESSMENT

27. GEOMETRY The edge length of each red triangle in the pattern is twice the edge length of the triangle above it.

a. Make a table showing edge length, perimeter, and area.

b. Write exponential growth equations for each column of the table.

c. Graph the equations from part (b).

SPIRAL REVIEW

For each pair of parametric equations, express *y* as a function of *x*. *(Section 2.6)*

28. $x = 0.5t$
$y = 3(t - 1)$

29. $x = t + 1$
$y = t + 2$

30. $x = 4t + 1$
$y = t$

Tell whether each equation shows direct variation. *(Section 2.1)*

31. $y = 2.5x$ **32.** $y + 2 = x + 2$ **33.** $y = 2x - 1$

34. Write a linear equation that gives the number of decades, *d*, in *y* years. Is this an example of direct variation? *(Section 2.1)*

3.2 Negative and Rational Exponents

Learn how to...

- **evaluate expressions that use negative and rational exponents**

So you can...

- **describe situations involving continuous exponential growth or decay, such as population increases**

Every ten years the federal government does a census to estimate the number of U.S. residents. It is not practical to do a census every year, but by using a model to describe the growth, you can estimate the population between census years.

EXAMPLE 1   Application: Census

Find an exponential model for the U.S. census data given in the spreadsheet.

U.S. Population Data			
D3		=C3/C2	
	A	**B**	**C**	**D**
1	Census year	Decades since 1800	Population (millions)	Proportional growth rate
2	1800	0	5.31	
3	1810	1	7.24	1.363
4	1820	2	9.64	1.331
5	1830	3	12.87	1.335
6	1840	4	17.07	1.326
7	1850	5	23.19	1.359

SOLUTION

Column D in the spreadsheet calculates the proportional growth factors between consecutive census years. The average of these proportional growth factors is about 1.34.

An exponential model for the data is

$$P = 5.31(1.34)^d$$

where P is the population in millions and d is the number of decades since 1800.

Negative Exponents

You can use the model from Example 1 to predict the population of the United States a decade before the 1800 census. Consider a pictograph of the first few population values predicted by the model.

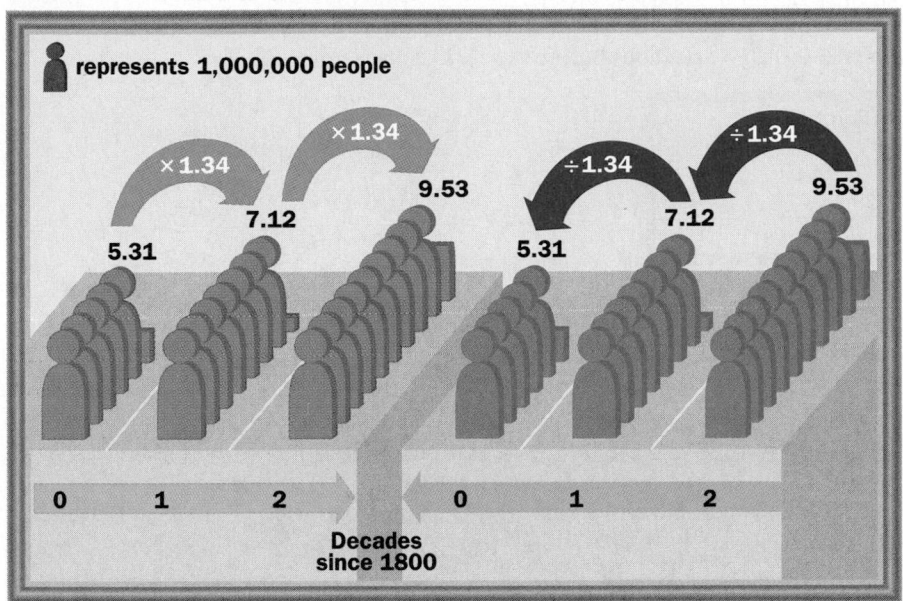

Moving forward in time means **multiplying** the population values by the proportional growth factor.

Moving backward in time means **dividing** the population values by the proportional growth factor.

THINK AND COMMUNICATE

1. Divide 5.31 by 1.34 to predict the U.S. population in 1790, one decade *before* the 1800 census.

2. **a.** What value of d would you use in the model $P = 5.31(1.34)^d$ to get the population in 1790?

   b. Use a calculator to evaluate $5.31(1.34)^d$ using the value of d from part (a). Do you get the same result as in Question 1?

You have seen that $5.31(1.34)^{-1} = \dfrac{5.31}{1.34}$. This implies that $1.34^{-1} = \dfrac{1}{1.34}$, which is an example of the following property of exponents.

The Meaning of b^{-n}

For any base $b > 0$ and any positive integer n:

$$b^{-n} = \frac{1}{b^n}$$

Example:

$$2^{-3} = \frac{1}{2^3} = \frac{1}{8}$$

EXAMPLE 2

Predict the U.S. population in 1780.

SOLUTION

Use the model $P = 5.31(1.34)^d$ from Example 1.

Since 1780 is two decades before the 1800 census, $d = -2$.

You can also get this result by evaluating $\dfrac{5.31}{1.34^2}$.

The predicted population in 1780 is about 2.96 million.

Rational Exponents

You can use the model from Example 1 to predict the U.S. population *between* census years, too.

Just as **1.34** is the population's 10-year proportional growth factor, **(1.34)$^{1/2}$** is the population's 5-year growth factor.

Year	1800	1805	1810
Decades since 1800	0	$\frac{1}{2}$	1
Population (millions)	$5.31(1.34)^0$	$5.31(1.34)^{1/2}$	$5.31(1.34)^1$

THINK AND COMMUNICATE

3. a. You can use a calculator to find $(1.34)^{1/2}$. What do you get? Store this result in memory.

   b. Multiply 5.31 by the number stored in memory. What does the result represent?

   c. Multiply the result from part (b) by the number stored in memory. What does the result represent?

4. a. Assume that the power property, $\left(b^m\right)^n = b^{mn}$, applies to rational as well as integral values of m and n. Apply the property to $\left[(1.34)^{1/2}\right]^2$. What do you get?

   b. Describe the mathematical relationship between $(1.34)^{1/2}$ and 1.34.

Toolbox p. 799
Logical Statements

The Meaning of $b^{1/n}$

For any base $b > 0$ and any positive integer n, $b^{1/n}$ is the number c whose nth power is b.

$b^{1/n} = c$ **if and only if** $b = c^n$

Example:

$8^{1/3} = 2$ because $8 = 2^3$

By thinking of $b^{m/n}$ as $\left(b^{1/n}\right)^m$, you can see that it is possible to raise a base to any rational power, such as $\frac{2}{3}$ or -2.4. The properties of exponents listed in Section 3.1 also apply to all rational exponents.

Toolbox p. 779
Rational Numbers and Irrational Numbers

EXAMPLE 3

Use the properties of exponents to simplify each expression.

a. $5^{2/3} \cdot 5^{5/3}$ **b.** $\dfrac{2^{7/3}}{2^{2/3}}$ **c.** $\left(6^{1/3}\right)^{3/5}$

SOLUTION

a. $5^{2/3} \cdot 5^{5/3} = 5^{(2/3\,+\,5/3)}$ **b.** $\dfrac{2^{7/3}}{2^{2/3}} = 2^{(7/3\,-\,2/3)}$ **c.** $\left(6^{1/3}\right)^{3/5} = 6^{(1/3\,\cdot\,3/5)}$

$\qquad\qquad = 5^{7/3}$ $= 2^{5/3}$ $= 6^{1/5}$

EXAMPLE 4

Predict the U.S. population in 1776.

SOLUTION

Use the model $P = 5.31(1.34)^d$.

Since 1776 is 24 years, or 2.4 decades, before 1800, $d = -2.4$.

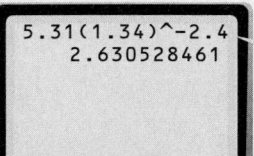
```
5.31(1.34)^-2.4
    2.630528461
```

A calculator will accept negative rational exponents as well as positive ones.

The predicted population in 1776 is about 2.63 million.

BY THE WAY

In 1776, Philadelphia had a population of 40,000, which was more than the populations of Boston and New York combined.

EXAMPLE 5

Use the equation $P = 5.31(1.34)^d$ to write an equation that gives the U.S. population in terms of y years after 1800 instead of d decades.

SOLUTION

The number of decades is one tenth the number of years: $d = \frac{1}{10}y$.

$P = 5.31(1.34)^d$

$\quad = 5.31(1.34)^{(1/10)y}$ ⟵ Substitute $\frac{1}{10}y$ for d.

$\quad = 5.31\left[(1.34)^{1/10}\right]^y$ ⟵ $b^{mn} = \left(b^m\right)^n$

$\quad = 5.31(1.03)^y$ ⟵ $(1.34)^{1/10} \approx 1.03$

THINK AND COMMUNICATE

5. What is the significance of the number 1.03 in Example 5?

6. Evaluate the function $P = 5.31(1.03)^y$ when $y = 10$. How does the result help you check the solution to Example 5?

☑ CHECKING KEY CONCEPTS

Evaluate each expression. Explain the significance of the result in terms of the population model in this section.

1. $5.31(1.34)^0$

2. $5.31(1.34)^{-2}$

3. $5.31(1.34)^2$

4. $5.31(1.34)^{-1.5}$

5. Explain why the values predicted by the model $P = 5.31(1.34)^d$ do not exactly match the population data in the spreadsheet for Example 1.

Evaluate each expression.

6. $64^{1/2}$

7. $64^{1/3}$

8. $64^{3/2}$

9. $64^{2/3}$

10. $100^{1/2}$

11. $100^{1/3}$

3.2 Exercises and Applications

Extra Practice exercises on page 752

1. Use the model from Example 1. Explain why the population in 1805 is not halfway between the population in 1800 and the population in 1810.

Use the model $P = 5.31(1.34)^d$ to estimate the population for each year.

2. 1830

3. 1850

4. 1870

Use the model $P = 5.31(1.03)^y$ to estimate the population for each year.

5. 1812

6. 1799

7. 1950

8. The actual U.S. population in 1950 was 151.3 million. Does this agree with the prediction in Exercise 7? What may account for any lack of agreement?

9. Writing Explain why it does not make sense to use negative or non-integral exponents in the examples in Section 3.1. Explain why it *does* make sense to use such exponents in the examples in this section.

Simplify.

10. 3^{-2}

11. 2^{-4}

12. 5^{-3}

13. 7^{-1}

Simplify using the properties of exponents.

14. $5^{1/2} \cdot 5^{1/2}$

15. $6^{1/2} \cdot 6^{3/2}$

16. $3^{3/2} \cdot 3^{-7/2}$

17. $6^{2/3} \cdot 6^{-2/3}$

18. $\dfrac{2^{5/2}}{2^{1/2}}$

19. $\dfrac{4^{13/5}}{4^{3/5}}$

20. $\dfrac{25^{3/4}}{25^{1/4}}$

21. $\dfrac{4^{7/3}}{2^{11/3}}$

22. $\left(7^4\right)^{1/2}$

23. $\left(25^3\right)^{1/6}$

24. $\left(8^{2/3}\right)^{3/2}$

25. $\left(16^0\right)^{3/7}$

The Central Artery, a highway through Boston, was opened in the late 1950s. It was designed to accommodate 75,000 vehicles per day. The table shows how the traffic has increased over time. By 1989, this stretch of highway was more crowded than any interstate in America.

Year	Vehicles per day (thousands)
1959	75
1969	100
1979	140
1989	190

26. Find the average growth factor per decade.

27. Write an equation modeling the Central Artery traffic as a function of the number of decades after 1959.

In the 1990s Boston planners began building a larger highway underground to replace the Central Artery. The project was called "The Big Dig."

Use the equation from Exercise 27 to estimate the traffic for each year.

28. 1964   **29.** 1974   **30.** 1984

31. Use the method of Example 5 to rewrite the equation from Exercise 27 in terms of years rather than decades after 1959.

32. Use the new equation from Exercise 31 to predict the traffic in 2001. Why do you think Boston planners are replacing the Central Artery?

33. Open-ended Problem Write several numerical examples that confirm that the properties of exponents in Section 3.1 can be extended to include:

a. negative integer exponents   **b.** fractional exponents

34. SAT/ACT Preview Which of the following is *true*?

A. $2^{1/3} \cdot 2^3 = 2^1$ **B.** $5^{1/2} + 5^{1/2} = 5^1$

C. $4^{-2} \cdot 4^2 = 8^0$ **D.** $6^{3/2} - 6^{1/2} = 6^1$

E. none of these

35. SNOWMAKING Artificial snow is made by combining air and water in a ratio that depends on the outdoor temperature. The table shows the rate of air flow needed per gallon of water as a function of temperature.

a. Write an exponential equation for the rate of air flow needed per gallon of water as a function of temperature.

b. Use the equation to predict the rate of air flow needed at 5°F, at 15°F, and at 25°F.

Temperature (°F)	Air flow per gallon of water (ft³/min)
0	3.0
10	4.7
20	9.8

The musical note "concert A" has a frequency of 440 vibrations per second. The A note that is one octave higher has a frequency of 880 vibrations per second. The frequencies of the notes between these two A's follow an exponential pattern:

$$F = 440 \cdot 2^{x/12}$$

36. Explain why the growth factor between two adjacent notes is $2^{1/12}$.

37. Copy and complete the table of frequencies shown below. Give answers to the nearest tenth.

Note	A	A#	B	C	C#	D	D#	E	F	F#	G	G#	A
x	0	1	2	3	4	5	6	7	8	9	10	11	12
F	440	?	?	?	?	?	?	?	?	?	?	?	880

The bassoon can play an A with a frequency of 110 vibrations per second, which is two octaves lower than concert A.

38. Open-ended Problem The interval between two notes whose frequencies have a ratio of 2:3 is called a "perfect fifth." Find two notes whose interval is close to a perfect fifth.

The 12-note scale in Exercises 36–38 is called a half-tone scale. Some composers have used scales that include more than 12 notes.

39. The quarter-tone scale has 24 notes. Its frequency equation is $F = 440 \cdot 2^{x/24}$. Complete the table of frequency equations for sixth-, eighth-, and sixteenth-tone scales.

Julián Carrillo

Name of scale	Notes in scale	Frequency equation
half-tone	12	$F = 440 \cdot 2^{x/12}$
quarter-tone	24	$F = 440 \cdot 2^{x/24}$
sixth-tone	36	$F = 440 \cdot 2^{x/?}$
eighth-tone	48	$F = 440 \cdot 2^{x/?}$
sixteenth-tone	96	$F = 440 \cdot 2^{x/?}$

40. The Mexican composer Julián Carillo used the sixteenth-tone scale. Find the frequencies of the first 10 notes in his 96-note scale. Start with the frequency 440.

ONGOING ASSESSMENT

41. Cooperative Learning Work with another student. One of you should graph $y = 8^{x/3}$ for $x = 0, 1, 2,$ and 3. The other should graph $y = 2^x$ for $x = 0, 1, 2,$ and 3. Compare your graphs. What do you notice?

SPIRAL REVIEW

Evaluate each expression. *(Section 3.1)*

42. $7\left(2^3\right)$ **43.** $3.2(0.5)^4$ **44.** $4.8\left(2^1\right)$

Compare each graph to the graph of $y = 3 + 2x$. *(Section 2.2)*

45. $y = 6 + 2x$ **46.** $y = 3 + 6x$ **47.** $y = 1 + 2(x + 1)$

48. Carry out a simulation to estimate how many people are needed on average before two have the same birth month. *(Section 1.5)*

3.3 Graphs of Exponential Functions

Learn how to...

- **draw graphs of exponential functions**

- **interpret how different values of *a* and *b* affect the graph of *y* = *ab^x***

So you can...

- **determine the doubling time for an investment or the half-life of caffeine, for example**

In this section you will see how different values of a and b affect the graphs of exponential functions in the form $y = ab^x$.

EXPLORATION
COOPERATIVE LEARNING

Investigating Exponential Graphs

Work in a group of four students.
You will need:
- a graphing calculator

1 In the same coordinate plane, graph $y = 8^x$, $y = 4^x$, $y = 2^x$, and $y = 1^x$. Sketch the results. What do you notice about the *y*-intercepts?

2 In the same coordinate plane, graph $y = \left(\frac{1}{8}\right)^x$, $y = \left(\frac{1}{4}\right)^x$, and $y = \left(\frac{1}{2}\right)^x$. Sketch the results. How are they related to the graphs from Step 1?

3 Choose a positive value for *b* and graph $y = b^x$ and $y = \left(\frac{1}{b}\right)^x$. What do you notice about the graphs? Draw the graphs from your group on one piece of paper. Which graphs represent exponential growth? Which represent exponential decay?

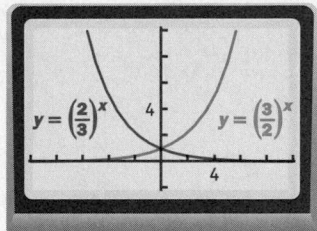

4 In the same coordinate plane, graph $y = 8 \cdot 2^x$, $y = 4 \cdot 2^x$, $y = 2 \cdot 2^x$, and $y = 1 \cdot 2^x$. Sketch the results. What do you notice?

5 Choose a positive value for *a* and graph $y = a \cdot 3^x$ and $y = -a \cdot 3^x$. What do you notice about the graphs? Draw the graphs from your group on one piece of paper.

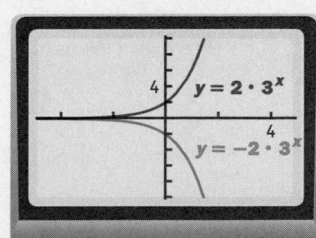

6 Discuss how the values of *a* and *b* affect the graph of $y = ab^x$.

Doubling Time

When something is growing exponentially, the time it takes for an initial amount to double, called the **doubling time**, is one way of describing its growth.

EXAMPLE 1 Application: Banking

Suppose you have $300 invested in a bank that offers 5% interest compounded annually.

a. Write an equation in the form $y = ab^x$ for the amount in your account after x years.

b. How many years will it take to double your money?

For more information about finding intersections, see the *Technology Handbook*, p. 808.

SOLUTION

a. $y = 300(1.05)^x$

The interest rate is 5%, so the **growth factor** is 1.05.

b. You want to find the value of x for which $300(1.05)^x = 2 \cdot 300$.

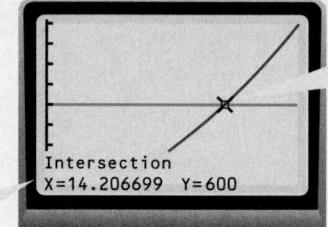

Intersection
X=14.206699   Y=600

Find the intersection of the graphs of $y = 600$ and $y = 300(1.05)^x$.

It will take about **14.2 years** to double your money if the interest rate is 5%.

You saw in the Exploration that an exponential function in the form $y = ab^x$ represents *growth* when a is positive and b is greater than 1. Such a growth function can always be rewritten in this form:

$$y = a \cdot 2^{x/d}$$

d is the doubling time.

For instance, the equation in Example 1 could be rewritten this way:

$$y = 300 \cdot 2^{x/14.2}$$

THINK AND COMMUNICATE

Toolbox p. 781
Simplifying Variable Expressions

1. Show that the equations $y = 300(1.05)^x$ and $y = 300 \cdot 2^{x/14.2}$ are equivalent equations by:

   a. comparing their graphs   **b.** evaluating $2^{1/14.2}$

2. Joey says, "If each year the bank increases the amount by 5%, then in 14.2 years the total increase is only 71%. That's not double!" Explain what is wrong with his reasoning.

Half-Life

A function in the form $y = ab^x$ represents *decay* when a is positive and b is between 0 and 1. A decay function can always be rewritten in this form:

$$y = a \cdot \left(\frac{1}{2}\right)^{x/h} \qquad \text{h is the \textit{half-life}.}$$

The **half-life** is the amount of time needed for the function to be half its initial value.

EXAMPLE 2   **Application: Physiology**

A 5 oz cup of coffee has about 120 mg of caffeine. Caffeine is eliminated from the bloodstream at a rate of about 12% per hour in adults.

a. Write an equation of the form $y = ab^x$ for the amount of caffeine in the bloodstream after x hours.

b. How many hours will it take for half the caffeine to be eliminated?

c. Write an equation of the form $y = a \cdot \left(\frac{1}{2}\right)^{x/h}$ for the amount after x hours.

SOLUTION

a. $y = 120(0.88)^x$

> The **decay factor** is $1 - 0.12$, or 0.88.

b. You want to find the value of x for which
$$120(0.88)^x = \frac{120}{2}.$$

> Find the intersection of the graphs of $y = 60$ and $y = 120(0.88)^x$.

It will take about **5.4 h** for half the caffeine to be eliminated.

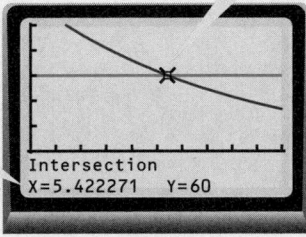

Intersection
X=5.422271   Y=60

c. $y = 120 \cdot \left(\frac{1}{2}\right)^{x/5.4}$

> Use **5.4** as the **half-life**.

THINK AND COMMUNICATE

3. Check part (c) of Example 2 by showing that $0.88 \approx 0.5^{1/5.4}$.

4. After how many hours will the amount of caffeine be 30 mg? 15 mg?

5. Suppose the initial amount of money in Example 1 was $500 instead of $300. How long would $500 take to double?

6. Suppose the initial amount of caffeine in Example 2 was 80 mg instead of 120 mg. What would the half-life be?

Exponential Growth and Decay

Exponential Growth

The function $y = ab^x$ represents exponential growth when $a > 0$ and $b > 1$.

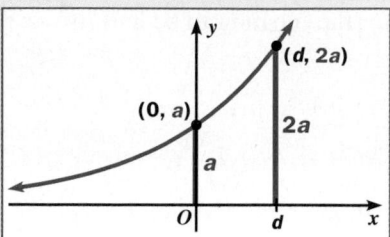

The graph shows that the initial amount a is doubled when $x = d$, the doubling time. Replacing b with $2^{1/d}$ gives:

$$y = a \cdot 2^{x/d}$$

Exponential Decay

The function $y = ab^x$ represents exponential decay when $a > 0$ and $0 < b < 1$.

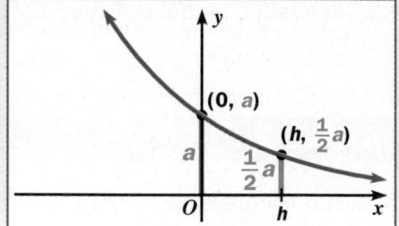

The graph shows that the initial amount a is halved when $x = h$, the half-life. Replacing b with $\left(\dfrac{1}{2}\right)^{1/h}$ gives:

$$y = a \cdot \left(\frac{1}{2}\right)^{x/h}$$

☑ CHECKING KEY CONCEPTS

Tell whether each equation is an example of *exponential growth* or *exponential decay*.

1. $y = (3.2)^x$ **2.** $y = (0.6)^x$ **3.** $y = \left(\dfrac{4}{3}\right)^x$

Find the *y*-intercept of the graph of each equation.

4. $y = 3^x$ **5.** $y = 3 \cdot 2^x$ **6.** $y = 2 \cdot 3^x$

7. $y = -4^x$ **8.** $y = -7 \cdot 6^x$ **9.** $y = -2.5(0.2)^x$

Find the doubling time or half-life of each function.

10. $y = 3.4 \cdot 2^{x/7}$ **11.** $y = 7 \cdot \left(\dfrac{1}{2}\right)^{x/4}$ **12.** $y = 4(2.7)^x$

3.3 Exercises and Applications

Extra Practice exercises on page 752

1. a. In the same coordinate plane, graph $y = 10^x$, $y = 5^x$, $y = 4^x$, and $y = (1.25)^x$.

b. Which of the graphs represents the fastest growth? the slowest growth?

c. What are the *y*-intercepts of the graphs?

2. a. Sketch the reflections of the graphs in Exercise 1 over the y-axis.

   b. Write equations for the graphs you sketched in part (a).

   c. 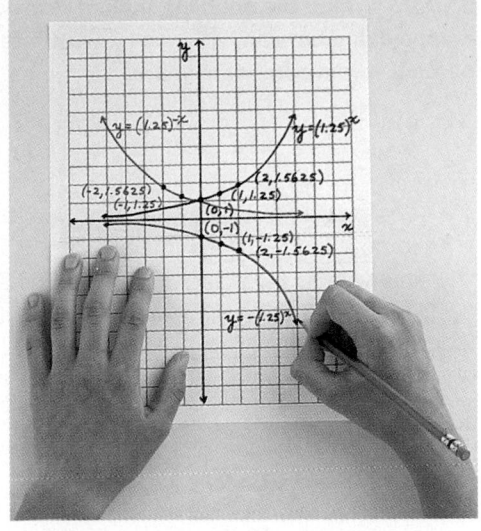 **Technology** Check your equations in part (b) by graphing them using a graphing calculator or graphing software.

3. a. Sketch the reflections of the graphs in Exercise 1 over the x-axis.

   b. Write equations for the graphs you sketched in part (a).

   c. **Technology** Check your equations in part (b) by graphing them using a graphing calculator or graphing software.

For each function in Exercises 4–7, do the following:

a. Find the y-intercept of the graph.

b. Tell whether the graph represents *exponential growth* or *exponential decay*.

c. Sketch the graph.

> **Toolbox p. 797**
> *Reflecting a Figure; Line Symmetry*

4. $y = 4(1.05)^x$ **5.** $y = (4.5)^x$ **6.** $y = 6^x$ **7.** $y = 2.5(0.8)^x$

8. a. Explain why the y-intercept of $y = ab^x$ is a when $b > 0$.

   b. **Investigation** What does the graph of $y = ab^x$ look like when $b < 0$ and the domain of the function is all integers? What happens if the domain is real numbers? $\left(\textit{Hint: } \text{Try evaluating the function when } x = \dfrac{1}{2}.\right)$

9. a. For what value of b is the graph of $y = ab^x$ a horizontal line? Describe the line.

   b. For what value of a is the graph of $y = ab^x$ a horizontal line? Describe the line.

10. Writing What does the graph of $y = ab^{-x}$ look like for different values of a and b? How does it compare to the graph of $y = ab^x$? How does it compare to the graph of $y = a\left(\dfrac{1}{b}\right)^x$?

Match the graphs and the equations.

11. $y = -3 \cdot 2^x$

12. $y = 3 \cdot 2^x$

13. $y = 3(0.5)^x$

14. $y = -3 \cdot 2^{-x}$

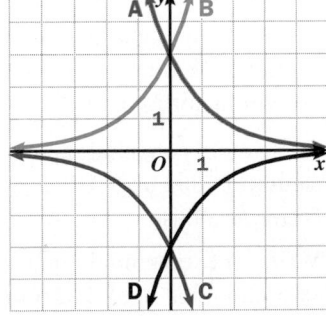

15. Challenge You may have noticed in the Exploration that the graph of $y = 8 \cdot 2^x$ looks like the graph of $y = 2^x$ translated to the left by 3 units. Explain why this is so.

BANKING Find the doubling time of money in a bank account earning interest compounded annually at each interest rate. Round each doubling time to the nearest whole number of years.

16. 8%

17. 6%

18. 4%

19. 9%

20. 12%

21. 18%

22. **Cooperative Learning** One way to estimate how long it will take money to double at various interest rates is called the "Rule of 72." Discuss your answers to Exercises 16–21 and what you think the "Rule of 72" is. Write the rule in a way that you can remember.

INTERVIEW **Finn Strong**

Look back at the article on pages 90–92.

The "rainmaker" on top of Finn Strong's fish tank is cooler than the air near the water's surface. Water vapor in the air that touches the rainmaker condenses into rain droplets, because air that is cool can hold less water vapor than air that is warm. The table shows the amount of water vapor that can exist in a cubic meter of air at various temperatures.

Temperature (°C)	Saturation (mL of water per m³ of air)
0	4.847
5	6.797
10	9.399
15	12.83
20	17.30
25	23.05
30	30.38
35	39.63

23. **a.** Find the growth factors for the saturation column of the table. What is the average growth factor associated with a 5-degree difference in temperature?

   b. Write an exponential equation for the saturation that uses the average 5-degree growth factor as a base.

24. **a.** Write an exponential equation for the saturation that uses a 1-degree growth factor as a base.

   b. Write an exponential equation for the saturation that uses 2 as a base.

   c. What change of temperature allows you to double the amount of water in a cubic meter of air?

When a plant or animal dies, it stops acquiring carbon-14 and carbon-12 from the atmosphere. Because carbon-14 is an unstable radioactive isotope, it decays over time, with a half-life of about 5730 years. The fraction of the original amount A of carbon-14 that remains in a sample after t years is given by $A = \left(\dfrac{1}{2}\right)^{t/5730}$.

Tell what fraction of the original carbon-14 remains in a sample after each number of years.

25. 2500   **26.** 5000

27. 10,000   **28.** 20,000

29. As a graduate student, Kerry Sieh discovered a way to document earthquakes that happened in the past. Using carbon dating on the peat found in streambeds that had been offset by earthquakes, he found that about 64% of the original carbon-14 remained. When was the earthquake that disturbed the stream?

Kerry Sieh walked a 250 mile stretch of the San Andreas fault to search for streambeds.

30. Chess historians were puzzled when archaeologists originally dated these animal-bone chess pieces to the first century A.D., 500 years before chess was invented in India. More recent carbon dating revealed that about 87.6% of the original carbon-14 remains in the pieces. What is a better estimate of their age?

For Exercises 31 and 32, assume exponential growth or decay.

31. A company wants to double its sales in 7 years. By what percent must sales increase each year to achieve this goal?

32. A neighborhood committee wants to cut the amount of littering in half in 5 years. By what percent must littering be reduced each year to achieve this goal?

33. Open-ended Problem   Write an equation representing an exponential growth or decay situation with a factor other than 2 or $\dfrac{1}{2}$. Show how the initial amount and the growth factor appear in your equation. Graph the equation and determine the doubling time or half-life.

The equation $P = 11.7(1.02)^x$ models the population P of Chile in millions, with $x =$ years since 1985. Find P for each year. *(Section 3.2)*

34. 1990   **35.** 1980   **36.** 1984

37. What is the doubling time for the population of Chile? *(Section 3.3)*

Simplify. *(Section 3.2)*

38. $8^{1/3}$ **39.** 3^{-2} **40.** $16^{3/2}$

41. The table shows the water content and calories of several soups. *(Section 2.4)*

    a. Graph the data and draw a line of fit.

    b. Find an equation for the line of fit.

    c. Make a prediction about the number of calories in vegetable beef soup, which is 91.9% water.

Soup	Percent water	Calories
Bean with pork	84.4	67
Beef noodle	93.2	28
Minestrone	89.5	43
Split pea	85.4	59
Tomato	90.5	36

ASSESS YOUR PROGRESS

VOCABULARY

exponential growth (p. 96)   **doubling time** (p. 108)
exponential decay (p. 96)   **half-life** (p. 109)

Evaluate each expression when $n = 4$. *(Section 3.1)*

1. 2^n **2.** $\left(\dfrac{1}{2}\right)^n$ **3.** $3 \cdot 2^{n-1}$

4. SPORTS A National Collegiate Athletic Association Basketball Championship starts with 64 teams in the first round. Each subsequent round eliminates half the teams. Write an equation for the number of teams that remain after n rounds. *(Section 3.1)*

Simplify using the properties of exponents. *(Section 3.2)*

5. $\dfrac{2^{5/4}}{2^{1/2}}$ **6.** $7^{3/4} \cdot 7^{1/4}$ **7.** $\left(4^{1/3}\right)^{3/2}$

A city has been reducing its traffic accident rate by 2% a year. In 1993 there were 278 traffic accidents. Estimate the number of accidents in each year. *(Section 3.2)*

8. 1995   **9.** 2000   **10.** 1985

11. A mail order company wants to reduce complaints by 5% a year. If the company is successful, when will the company be receiving half as many complaints as the company is now? *(Section 3.3)*

12. Journal Explain how the values of a and b affect the doubling time or half-life of an exponential function $y = ab^x$.

3.4 The Number *e*

In Section 3.3 you used an exponential function to describe compounding interest annually in a bank account. Most banks offer interest that is compounded more than once a year. When interest is compounded *n* times per year for *t* years at an interest rate *r* (expressed as a decimal), a principal of *P* dollars grows to the amount *A* given by this formula:

$$A = P\left(1 + \frac{r}{n}\right)^{nt}$$

EXPLORATION
COOPERATIVE LEARNING

Compounding Interest

**Work in a group of four students.
You will need:**
- a scientific calculator

1 The table shows the value of one dollar after one year of compounding. Copy and complete the table. Each of you should complete one column for one of the interest rates.

Compounding	n	Formula	r=0.05	r=0.10	r=0.50	r=1.00
annually	1	$\left(1+\frac{r}{1}\right)^1$	1.05			
semiannually	2	$\left(1+\frac{r}{2}\right)^2$		1.1025		
quarterly	4	$\left(1+\frac{r}{4}\right)^4$			1.6018	
monthly	12	$\left(1+\frac{r}{12}\right)^{12}$				2.6130
daily	365	$\left(1+\frac{r}{365}\right)^{365}$				
hourly						
every minute						
every second						

2 Describe how increasing the frequency of compounding interest affects the value of a dollar that is invested for one year.

3 If a bank compounded interest more often than every second, would this make much difference? Explain.

The last column of the table in the Exploration shows what happens to a dollar when 100% interest is compounded more and more frequently. As the frequency of compounding increases, the amount in the account approaches $2.718. . . . Because the number 2.718. . . is special in mathematics, it is given a name: *e*.

> ### The Number *e*
>
> As *n* increases, $\left(1 + \dfrac{1}{n}\right)^n$ approaches *e*.
>
> $$e = 2.718281828. . .$$

THINK AND COMMUNICATE

Use a calculator to evaluate each power of *e*.

1. $e^{0.05}$ **2.** $e^{0.10}$ **3.** $e^{0.50}$ **4.** e^1

5. Compare your answers to Questions 1–4 with the last row of the table from the Exploration. What do you notice?

6. What does your answer to Question 5 suggest is the value of $\left(1 + \dfrac{r}{n}\right)^n$ as *n* gets very large?

7. When compounding takes place without interruption, it is called *continuous compounding*. Explain why it is reasonable that the formula for continuous compounding of interest is $A = Pe^{rt}$.

8. a. If $r = 0$, what is the value of e^r?

    b. If $r > 0$, what can you say about the value of e^r?

    c. If $r > 0$, does the graph of $A = Pe^{rt} = P\left(e^r\right)^t$ represent *exponential growth* or *exponential decay*? Why does this make sense?

Because $e > 1$, $y = e^x$ is an exponential growth function. Also, because $e^{-1} = \dfrac{1}{e} < 1$, $y = e^{-x}$ is an exponential decay function.

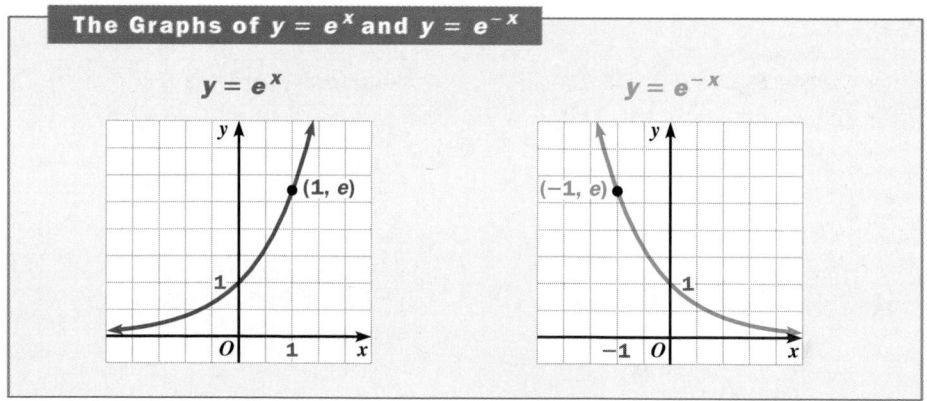

Exponential functions with the base *e* are often used to describe continuous growth or decay, such as continuously compounded interest or radioactive decay, as shown in Examples 1 and 2 on the next page.

EXAMPLE 1 Application: Banking

Suppose a bank offers 3.5% interest compounded continuously. If an account starts with $1000, what will its value be after 1 year, 5 years , and 10 years? Use the formula $A = Pe^{rt}$.

SOLUTION

Evaluate the formula $A = Pe^{rt}$ when $P = 1000$, $r = 0.035$, and $t = 1, 5$, and 10.

```
1000e^(.035*1)
        1035.619709
1000e^(.035*5)
        1191.246217
1000e^(.035*10)
        1419.067549
```

After 1 year, the account's value is about **$1035.62**.

After 5 years, the account's value is about **$1191.25**.

After 10 years, the account's value is about **$1419.07**.

EXAMPLE 2 Application: Chemistry

Polonium-210 is a radioactive element. An initial amount of 100 micrograms (μg) decays to an amount $A(t)$ in t days according to this formula:

$$A(t) = 100e^{-0.005t}$$

a. Use a graphing calculator or graphing software to graph the decay function.

b. What is the half-life of polonium-210? Write an equation for the amount of polonium-210 using $\frac{1}{2}$ as a base instead of e.

Polonium was discovered by Marie Curie in 1898. She named it after her native country, Poland.

SOLUTION

a.

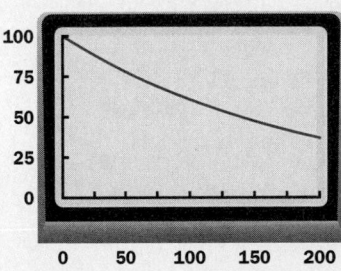

b. Find the value of t when $A(t) = 50$.

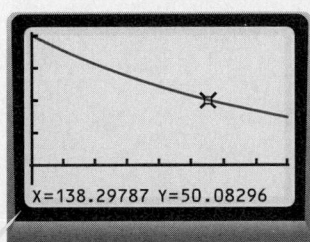

X=138.29787 Y=50.08296

The **half-life** of polonium-210 is about **138 days**.

An equation for the decay of 100 μg of polonium-210 is:

$$A(t) = 100\left(\frac{1}{2}\right)^{t/138}$$

Logistic Growth

In many situations, exponential growth does not continue forever, because environmental or other conditions limit the growth. One kind of limited growth, called **logistic growth**, begins with rapid growth but eventually levels off. Equations for logistic growth are often written using e.

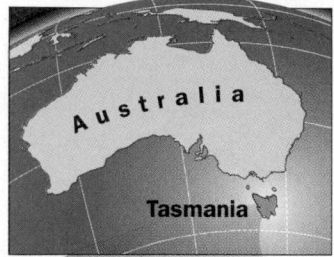

EXAMPLE 3 **Application: Biology**

About 187,000 sheep inhabited the island of Tasmania in 1819. As European settlers began to develop the land for sheep farming, this population increased until it stabilized about 70 years later.

The sheep population $P(t)$ can be modeled by a logistic function, where t is the number of years since 1819:

$$P(t) = \frac{1{,}670{,}000}{1 + 7.915e^{-0.131t}}$$

a. Graph the sheep population function using a graphing calculator or graphing software. Evaluate the function when $t = 0$. Does your answer make sense?

b. Estimate the date when the sheep population reached 1,500,000.

c. Investigate the value of $P(t)$ for large values of t. According to the model, at what value did the sheep population eventually stabilize?

SOLUTION

a. The graphing calculator screen shows the graph of

$$y = \frac{1{,}670{,}000}{1 + 7.915e^{-0.131x}}.$$

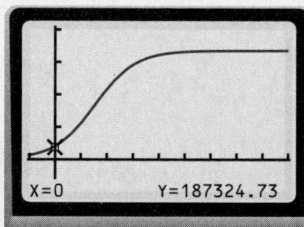

When $t = 0$, $e^{-0.131t} = e^0 = 1$. Therefore,

$$P(0) = \frac{1{,}670{,}000}{1 + 7.915(1)} \approx 187{,}000.$$

This makes sense, because the sheep population in 1819 was 187,000.

b. Find the intersection of the graph from part (a) and the line

$$y = 1{,}500{,}000.$$

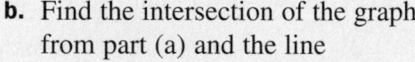

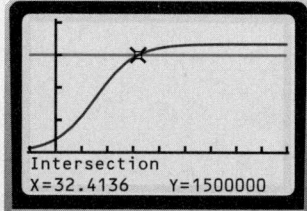

The sheep population reached 1,500,000 about 32 years after 1819, in the year 1851.

c. The expression $e^{-0.131t}$ is an example of exponential decay. When t is large, this expression approaches 0. Therefore, for large values of t:

$$P(t) \approx \frac{1{,}670{,}000}{1 + 7.915(0)} = 1{,}670{,}000.$$

According to the model, the sheep population on Tasmania stabilized at about 1,670,000 sheep.

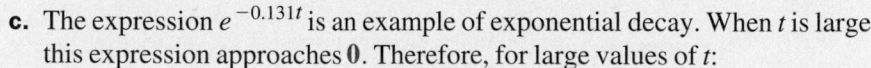

☑ CHECKING KEY CONCEPTS

1. Suppose you deposit $1000 in the account described in the advertisement. What amount would be in the account after 1 year? What do you think *effective annual yield* means?

2. Suppose Neighborhood Bank offered continuous compounding instead of daily compounding. Find the value of a $1000 deposit after 1 year. How does this result compare with the value from Question 1?

3. Radium-226 is a radioactive element that decays after t years according to this formula:

$$A(t) = A_0 e^{-0.000525t}$$

Suppose A_0 is 1 gram. Find the amount of radium-226 left after:

   a. 2640 years
   b. 3960 years
   c. 5280 years

4. Estimate the half-life of radium-226 and write a decay equation using $\frac{1}{2}$ as a base instead of e.

5. Use the logistic function $L(t) = \dfrac{2000}{1 + 19e^{-0.1t}}$.

   a. Find $L(0)$.
   b. Investigate the value of $L(t)$ for large values of t.
   c. For what value of t is $L(t) = 1000$?
   d. For what value of t is $L(t) = 4000$?

NEIGHBORHOOD BANK

INTRODUCING THE
— *PASSPORT SAVINGS* —
ACCOUNT

| *3.20%* | *3.25%* |
| Interest Compounded Daily* | Effective Annual Yield |

*Based on a Minimum Daily Balance of $500

Extra Practice exercises on page 753

3.4 Exercises and Applications

Suppose a bank compounds interest 360 times a year. For each interest rate, find the *effective annual yield*. (The effective annual yield is the annual interest rate that would yield the same amount after one year.)

1. 2.50%   2. 5.00%   3. 7.50%   4. 10.00%

Suppose a bank offers interest compounded continuously. Use the formula $A = Pe^{rt}$ to find the value of $1000 after 10 years at each interest rate.

5. 3.75%   6. 6.25%   7. 9.50%   8. 11.00%

Find the value of $\left(1 + \dfrac{1}{n}\right)^n$ for each value of n. Round each answer to six decimal places.

9. 1   10. 10   11. 10^2 12. 10^3
13. 10^4 14. 10^5 15. 10^6 16. 10^7

17. Cooperative Learning Work in a group of four students to investigate the interest rates that double your money in a year if interest is compounded n times a year. Each of you should choose a different value of n from the table shown below.

a. 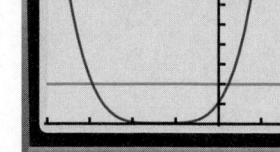 **Technology** Graph $y = 2$ and $y = \left(1 + \dfrac{x}{n}\right)^n$ for your value of n. Sketch your graphs.

b. For your value of n, what interest rate doubles your money in a year?

c. Combine your results to complete the table. Discuss the results.

d. **Writing** Describe how your group's graphs compare to the graph of $y = e^x$.

e. If interest is compounded continuously, what interest rate will double your money in one year?

Doubling Your Money in a Year		
Compounding	n	Interest Rate
annually	1	
semiannually	2	
quarterly	4	
monthly	12	
daily	365	

Connection FARMING

Data gathered in the 1920s suggested that the number of eggs E a Leghorn chicken produces per year declines exponentially with its age t in years, according to this function:

$$E(t) = 179.2e^{-0.12t}$$

18. Graph the egg production function.

19. Estimate how many years it takes for a chicken's egg production to be cut in half. Use your estimate to write an egg production function in the form $E(t) = a \cdot \left(\dfrac{1}{2}\right)^{t/h}$.

20. Write an egg production function in the form $E(t) = ab^t$. What is the yearly proportional decay factor?

21. By the 1990s certain breeds of 1-year-old chickens were laying about 250 eggs per year. Compare the egg production of one of these chickens to the egg production of a 1-year-old Leghorn from the 1920s. By what percent does a modern chicken out-produce the Leghorn?

In the book *Wind Energy Comes of Age*, Paul Gipe discusses the amount of energy generated from wind turbines in North America and Europe between 1980 and 1995. His data can be modeled using the equations

Europe: $y = 6.489(1.580)^x$

North America: $y = \dfrac{3500}{1 + 874e^{-0.852x}}$

where x = years since 1980 and y = gigawatt-hours of wind energy. (*Note:* One gigawatt-hour equals 10^6 kilowatt-hours.)

Located in Scotland, this 135 ft tall wind turbine is one of the world's largest.

 Technology **For Exercises 22–25, use a graphing calculator or graphing software.**

22. Graph the wind energy functions.

23. According to the models, what was the wind energy for each region in 1985? in 1990? in 1995?

24. Which region produced more wind energy in the 1980s?

25. When did Europe's wind energy production surpass North America's?

26. **Writing** Do you think an exponential function is a realistic model for European wind energy in the future? Why?

Wind turbines line California's Altamont Pass, where wind speeds average 13 mi/h to 18 mi/h.

CHEMISTRY Show that the two equations given for each radioactive element are equivalent by rewriting both in the form $A(t) = ab^t$.

27. **Polonium-210**

$$A(t) = 100e^{-0.005t}$$

$$A(t) = 100\left(\frac{1}{2}\right)^{t/138}$$

28. **Radium-226**

$$A(t) = e^{-0.000525t}$$

$$A(t) = \left(\frac{1}{2}\right)^{t/1320}$$

29. **Challenge** The radioactive element carbon-14 has a half-life of about 5730 years, so an equation describing its radioactive decay is:

$$A(t) = A_0\left(\frac{1}{2}\right)^{t/5730}$$

Write an equivalent equation that uses e as a base. (*Hint:* Look at the numbers in the exponents in Exercises 27 and 28.)

30. **BIOLOGY** The temperature at which red-eared slider turtle eggs are incubated affects the percent that are hatched as males. A logistic decay function that models this phenomenon is:

$$M(T) = \frac{100}{1 + e^{4.99T - 11.2}}$$

where $M(T) =$ the percent of males and $T =$ the incubation temperature measured in degrees Celsius above $27°$.

a.  **Technology** Use a graphing calculator or graphing software to graph the function.

b. At what temperature will about half the turtles be male?

c. At what temperature will about 10% of the turtles be female?

O N G O I N G   A S S E S S M E N T

31. **Cooperative Learning** Work with a partner.

Radon-222 gas decays according to the function $A(t) = A_0 e^{-0.18t}$, where t is measured in days.

a. Graph the decay function for radon-222.

b. Work with your partner to make a table showing the number of days it takes for $A(t)$ to reach 90%, 80%, 70%, . . . , 10% of its initial value.

S P I R A L   R E V I E W

Exercises 32 and 33 refer to the graph of $y = ab^x$. *(Section 3.3)*

32. For what values of a and b will the graph show exponential decay?

33. For what values of a and b will the graph show exponential growth?

Exercises 34–36 refer to the table that shows public spending, in billions of dollars, on maternal and child health care programs in the United States over 7 years. *(Section 2.4)*

Years since 1985	Public funds (billions)
0	1.3
1	1.4
2	1.6
3	1.7
4	1.8
5	1.9
6	2.0

34. Show that the data have a linear relationship by making a scatter plot and drawing a line of fit.

35. **Technology** Use a graphing calculator or statistical software to find an equation of the least-squares line that fits the data.

36. Predict public spending on maternal and child health care in the year 2012.

Multiply. *(Section 1.4)*

37. $\begin{bmatrix} 5 & 2 \\ 0 & 8 \end{bmatrix}\begin{bmatrix} 3 & -4 & 0 \\ 12 & 6 & -5 \end{bmatrix}$

38. $\begin{bmatrix} 14 & 2 & 1 \\ 0 & 2 & 3 \\ 11 & 7 & 6 \end{bmatrix}\begin{bmatrix} 1 \\ 3 \\ 1 \end{bmatrix}$

3.5 Fitting Exponential Functions to Data

Learn how to...

- write exponential functions that fit sets of data

So you can...

- make predictions about exponential growth and decay situations, such as learning curves and computer passwords

When data appear to grow or decay exponentially, you can try to find an exponential function that is a "good fit." You will need to find appropriate values of a and b for an equation of the form $y = ab^x$.

EXAMPLE 1   **Application: Psychology**

Psychologists studying how people learn timed a young piano student as she played Chopin's "Minute Waltz." On her second try, she took 2.367 min to perform the piece. On her third she took 2.067 min.

Suppose the student's learning curve is exponential. Write an equation that relates the duration of her performance to the number of times she has played the piece.

SOLUTION

You want to find an exponential function whose graph passes through the points $(2, 2.367)$ and $(3, 2.067)$.

Step 1 Substitute the coordinates of each point into the general equation $y = ab^x$.

$$2.067 = ab^3$$
$$2.367 = ab^2$$

Step 2 Divide the two equations to solve for b.

$$\frac{2.067}{2.367} = \frac{ab^3}{ab^2}$$
$$0.873 \approx b$$

Step 3 To solve for a, use the value of b from Step 2 in either equation.

$$2.367 = a(0.873)^2$$
$$\frac{2.367}{(0.873)^2} = a$$
$$3.104 \approx a$$

An equation that describes the piano student's performance duration y on the xth attempt is

$$y = 3.104(0.873)^x.$$

THINK AND COMMUNICATE

1. Predict how many times the student must play the "Minute Waltz" before she will be able to play it in about one minute.

2. Explain why an exponential decay model is not useful after the student has mastered playing the waltz in one minute.

3. How would you find an exponential equation that passes through two points if the x-coordinates of the points are *not* exactly 1 unit apart?

EXAMPLE 2   Application: Banking

To use an automated teller machine, you need to enter a password. Longer passwords are better because they are more difficult for other people to guess.

If a password uses the letters A–Z and is four letters long, there are 456,976 possible passwords. If the password is seven letters long, there are 8,031,810,176 possible passwords. Write an exponential equation that relates the number of passwords to the password length.

SOLUTION

You need to find an exponential function whose graph passes through the points (4, 456,976) and (7, 8,031,810,176).

Step 1 Write two equations. $8,031,810,176 = ab^7$ — Use substitution as in Example 1.

$$456,976 = ab^4$$

Step 2 Solve for b. $\dfrac{8,031,810,176}{456,976} = b^3$ — Dividing the equations eliminates a.

$$17,576 = b^3$$

If $b^n = c$, then $b = c^{1/n}$. — $(17,576)^{1/3} = b$

$$26 = b$$

Step 3 Solve for a. $456,976 = a \cdot 26^4$ — To solve for a, use **26** as the value of b in either equation.

$$\dfrac{456,976}{26^4} = a$$

$$1 = a$$

The number y of all possible passwords when x letters are used is $y = 1 \cdot 26^x$, or $y = 26^x$.

THINK AND COMMUNICATE

4. Explain why the base 26 makes sense if the password includes only the letters A–Z.

5. What would the exponential function be if a password includes a mix of the letters A–Z, the letters a–z, and the numbers 0–9?

You can also use a graphing calculator or statistical software to find an exponential equation that is a good fit to a set of data.

EXAMPLE 3   **Application: Biology**

The table shows that the mean weight of Atlantic cod from the Gulf of Maine increases with age. Find an exponential function that models the data.

Atlantic Cod from the Gulf of Maine	
Age (y)	Weight (kg)
1	0.751
2	1.079
3	1.702
4	2.198
5	3.438
6	4.347
7	7.071
8	11.518

SOLUTION

First enter the data values.

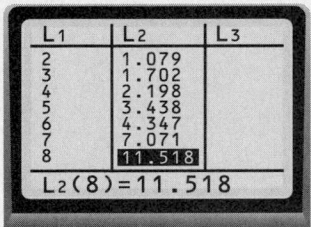

Then use the exponential regression feature of a graphing calculator.

```
ExpReg
 y=a*b^x
 a=.508948569
 b=1.459711295
 r=.99728485
```

If the correlation coefficient **r is close to 1** (or −1), then **exponential growth** (or decay) is a good model.

An exponential function that models the Atlantic cod weight data is $y = 0.509(1.460)^x$ where x is age in years and y is weight in kilograms.

☑ CHECKING KEY CONCEPTS

Write an exponential function whose graph passes through each pair of points.

1. (0, 5), (1, 3)   **2.** (0, 10), (2, 20)   **3.** (2, 48), (5, 3072)

4. HEALTH Tuberculosis cases in the United States fell exponentially from about 84,000 in 1953 to about 22,000 in 1985. Write an exponential function that models this decay.

3.5 Exercises and Applications

Extra Practice exercises on page 753

Write an exponential function whose graph passes through each pair of points.

1. (0, 5), (1, 8)

2. (1, 8), (2, 10)

3. (0, 10), (1, 8)

4. (5, 7), (6, 6)

5. (3, 3), (8, 7)

6. (0, 6), (7, 9)

7. (20, 4), (25, 3)

8. (0, 17), (7, 15)

9. ASTRONOMY The *apparent magnitude* of a star is a measure of its apparent brightness. A star of magnitude 1 is 100 times brighter than a star of magnitude 6.

a. Find an equation of the exponential decay function whose graph passes through (1, 100) and (6, 1).

b. Graph the function from part (a).

c. Copy and complete the table showing comparative brightness values for various sky objects.

Sky object	Apparent magnitude	Brightness value
Uranus	6	1
Aldebaran	1	100
Vega	0	?
Sirius	−1.5	?
Full moon	−12.5	?
Sun	−26.7	?

d. Research Use a book about astronomy to find the names and magnitudes of some stars. Plot them on your graph from part (b). Calculate their brightness values and add them to your table from part (c).

10. SAT/ACT Preview Which of the equations represents an exponential decay function whose graph passes through (2, 1)?

A. $y = (0.5)^x$ **B.** $y = (0.5x)^2$ **C.** $y = \frac{1}{4}(0.5)^{-x}$ **D.** $y = 4(0.5)^x$ **E.** $y = \frac{1}{4} \cdot 2^x$

11. Open-ended Problem The table shows the increase in the number of host sites on the Internet computer network.

Year	1988	1989	1990	1991	1992	1993	1994
Years since 1988	0	1	2	3	4	5	6
Internet hosts (1000s)	56	159	313	617	1136	2056	3864

a. Choose any two years and write an exponential function to represent the growth.

b. Check your function by seeing how its predictions compare to the other data values in the table.

c. ▨ **Technology** Enter all the data into a graphing calculator or statistical software and find an exponential equation that gives a good fit. Compare the results to your results from part (a).

12. **Cooperative Learning** Work with another student.

   a. Each of you should write an exponential function that your partner cannot see.

   b. Find two points that are on the graph of your exponential function. Give their coordinates to your partner.

   c. Find the exponential function that passes through the two points your partner gives you.

   d. Check your answers by revealing the original exponential functions.

13. **Investigation** Suppose a computer program tests passwords at a rate of 10,000 characters per second. Make a table showing how long it will take to go through all the passwords of various lengths if the password uses:

   a. only the digits 0–9

   b. only the letters A–Z

   c. the letters A–Z, a–z, and the digits 0–9

14. **SPORTS** The table shows the breathing rate (in liters of air per minute) of bicyclists traveling at various speeds on two types of bikes.

Touring bike speed (mi/h)	Breathing rate (L/min)	Racing bike speed (mi/h)
0	7	0
1.8	9	3.2
6	13	7.2
8.3	18	10.5
12	29	14.5
16	50	19
18.5	72	22
21	93	25
22.5	115	27

   a. Write an exponential function to show how the breathing rate of a bicyclist on a racing bike depends on speed.

   b. Write an exponential function to show how the breathing rate of a bicyclist on a touring bike depends on speed.

   c. **Technology** Use a graphing calculator or graphing software to graph the two functions from parts (a) and (b) in the same coordinate plane.

   d. For a given speed, who breathes at a faster rate, a bicyclist on a racing bike or a bicyclist on a touring bike? Explain why your answer makes sense.

15. **Challenge** Explain why the growth or decay factor for an exponential function whose graph passes through (p, q) and (r, s) is $\left(\dfrac{s}{q}\right)^{1/(r - p)}$.

16. GEOMETRY If you fold the corners of a square piece of paper to the center of the square, you get a smaller square. You can perform the folding process *x* times to get smaller and smaller squares.

a. Copy and complete the table.

b. Write an exponential function for each column of the table.

x	Area	Side length	Perimeter
0	1	1	4
1	0.5	?	?
2	?	?	?
3	?	?	?
4	?	?	?

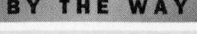
To make the candy dish, flip the square over before folding a second time.

  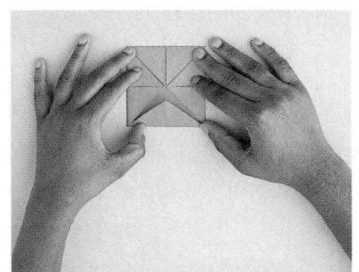

INTERVIEW Finn Strong

Look back at the article on pages 90–92.

In Finn Strong's fish tanks, the fish need oxygen to survive. The amount of oxygen that water can hold varies exponentially with the temperature and depends on the water's chlorinity (a measure related to its saltiness).

17. Use the table to write an exponential function that fits the data for fresh water (with 0% chlorinity), such as the water in Finn Strong's tanks.

18. Use the table to write an exponential function that fits the data for ocean water (with 2% chlorinity).

19. Open-ended Problem For any given temperature, describe how the chlorinity of the water affects the amount of oxygen the water can hold. Predict table values for water of 0.5% and 1.5% chlorinity. Justify your predictions.

Oxygen Absorbed by Water (mL/L)			
Temperature (°C)	0%	Chlorinity 1%	2%
0	10.29	9.13	7.97
10	8.02	7.19	6.35
15	7.22	6.50	5.79
20	6.57	5.95	5.31
30	5.57	5.01	4.46

20. Writing Suppose you have to explain to a friend who missed school what you have learned about exponential modeling. Write a page reminding your friend what modeling is, explaining how you determine if a relationship is exponential, and suggesting some ways to find an exponential function that describes a set of data.

SPIRAL REVIEW

21. BIOLOGY The growth of the bacteria *Mycobacterium tuberculosis* can be modeled by the equation $P(t) = P_0 e^{0.116t}$, where $P(t)$ is the population after t hours and P_0 is the population when $t = 0$. *(Section 3.4)*

   a. At 1:00 P.M. there are 30 *Mycobacterium tuberculosis* bacteria in a sample. Write a function for the number of bacteria after 1:00 P.M.

   b. What is the population at 5:00 P.M.?

   c. What was the population at noon?

   d. How would you find the population at 3:45 P.M.?

22. Use the equation $y = 8 - 2x$. *(Section 2.2)*

   a. Find y when $x = -3$.   **b.** Find x when $y = 5$.

23. a. Graph the parametric equations $x = 2 - t$ and $y = 3t + 1$ for $t \geq 0$. *(Section 2.6)*

   b. Express y as a function of x. State any restriction on x.

ASSESS YOUR PROGRESS

VOCABULARY

e (p. 116)   **logistic growth** (p. 118)

Find the value of an investment of $2000 after four years if the interest rate is 4.5% and interest is compounded as specified. *(Section 3.4)*

   1. annually   **2.** weekly   **3.** continuously

4. Graph $y = 2^x$, $y = e^x$, and $y = 3^x$ in the same coordinate plane. What do you notice?

5. PHYSIOLOGY Your visual *near point* is the closest point at which your eyes can see an object distinctly. Your near point moves farther away from you as you grow older, as shown by the data in the table. *(Section 3.5)*

Age (years)	Near point (cm)
10	10
20	12
30	15
40	25
50	40
60	100

   a. Choose two data pairs and write an exponential function to represent how near point distance varies with age.

   b. **Technology** Enter the data into a graphing calculator or statistical software and perform an exponential regression. What equation do you get?

   c. Predict the near point distance in centimeters at age 80.

6. Journal Compare exponential and logistic growth. Give some examples of situations in which one model is better than the other.

The Way a Ball Bounces

Have you ever watched a ball bounce repeatedly? Due to the loss of energy each time a bouncing ball hits the floor, the ball never rebounds to the same height from which it fell.

The bounciness, or "bounce factor," of a ball can be determined by comparing the ball's rebound height to the original height from which the ball was dropped. Then, using the bounce factor of the ball, you can predict the height of the ball after any number of bounces.

PROJECT GOAL **Create a model that relates the bounce height of a ball to the number of times the ball has bounced.**

Conducting an Experiment

Work with a partner to design and carry out an experiment in which you measure the bounce height of a ball after dropping the ball from varying heights. You will need a ball and two metersticks or a tape measure. Here are some guidelines for conducting your experiment.

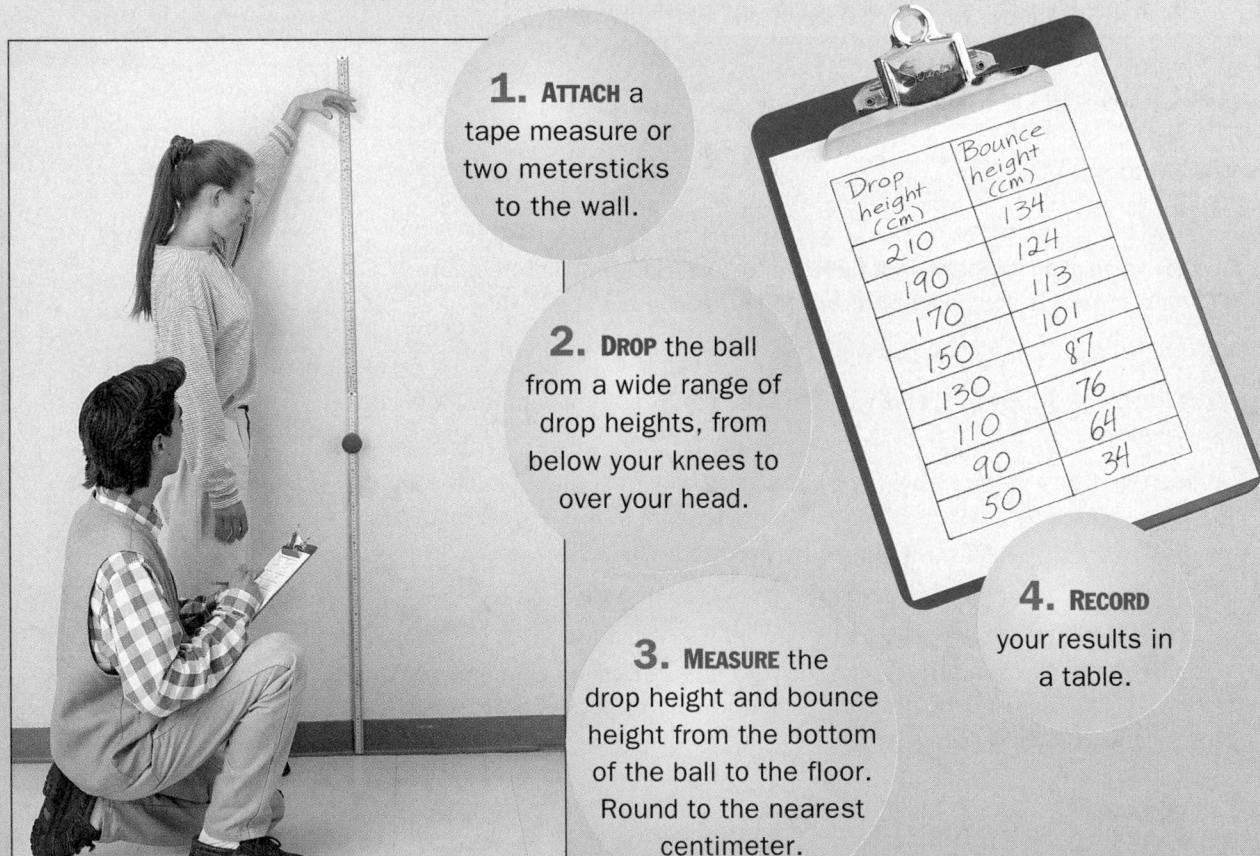

1. ATTACH a tape measure or two metersticks to the wall.

2. DROP the ball from a wide range of drop heights, from below your knees to over your head.

3. MEASURE the drop height and bounce height from the bottom of the ball to the floor. Round to the nearest centimeter.

4. RECORD your results in a table.

Drop height (cm)	Bounce height (cm)
210	134
190	124
170	113
150	101
130	87
110	76
90	64
50	34

Bounce Factor		
C10		= SUM (C2:C9)/8
	A	B	C
	Drop height (cm)	Bounce height (cm)	Bounce factor
1			
2	210	134	0.64
3	190	124	0.65
4	170	113	0.66
5	150	101	0.67
6	130	87	0.67
7	110	76	0.69
8	90	64	0.71
9	50	34	0.68
10	Mean bounce factor:		0.67

Analyzing the Data

• **CALCULATE** the bounce factor of your ball for each drop height. The bounce factor of your ball is equal to the ratio $\frac{\text{bounce height}}{\text{drop height}}$.

• **FIND** the mean of the bounce factors. Use this figure as the average bounce factor of your ball.

Predicting Exponential Decay

Suppose you dropped your ball from the height of the top of your head, and watched it fall and rebound, over and over.

• **MAKE** a table of predicted rebound heights. Start with the ball's initial height, and calculate the ball's height after one bounce, after two bounces, and so on. Continue for several bounces.

• **FIND** an equation that describes rebound height as a function of the number of bounces. Determine the proper values for a and b to substitute into the exponential decay model $y = ab^x$. What does a stand for in your model? What does b stand for?

• **PREDICT** how many bounces it will take until the ball's rebound height is less than the height of your knee. Test your prediction.

Writing a Report

Write a report about your experiment. State the project goal, describe your procedures, and present data that support the conclusions you make. Be sure to explain the role that drop height and bounce factor have in your model and evaluate how well your exponential decay model predicts bounce height.

   You may wish to extend your report to include data about other types of balls, or do some research about the official rules concerning the bounciness of a ball in a sport like tennis, basketball, or football.

Self-Assessment

Does the model that you developed for this project make sense to you? If so, in what ways? If not, where did you get confused? What would help make things clearer for you?

Review

STUDY TECHNIQUE

Review your homework from this chapter. Find one exercise from each section that gave you the most difficulty. Write and solve new exercises like the ones you chose or choose similar exercises from the text to solve. Exchange your solutions with another student and check each other's work.

VOCABULARY

exponential growth (p. 96)   **half-life** (p. 109)
exponential decay (p. 96)   **e** (p. 116)
doubling time (p. 108)   **logistic growth** (p. 118)

SECTIONS | 3.1 *and* 3.2

You can simplify expressions involving exponents by using the properties of exponents and the definitions of negative and rational exponents.

Properties of Exponents

For any $b > 0$:   **Examples:**

Product Property: $b^m \cdot b^n = b^{m+n}$ $4^3 \cdot 4^2 = 4^5$

Quotient Property: $\dfrac{b^m}{b^n} = b^{m-n}$ $\dfrac{7^{1.8}}{7^{0.8}} = 7^1 = 7$

Power Property: $\left(b^m\right)^n = b^{mn}$ $3^{2x} = \left(3^2\right)^x = 9^x$

Negative and Rational Exponents

For any base $b > 0$:

Negative Exponent   **Example:**

$b^{-n} = \dfrac{1}{b^n}$ $2^{-5} = \dfrac{1}{2^5} = \dfrac{1}{32}$

Rational Exponent   **Example:**

$b^{1/n} = c$ if and only if $b = c^n$ $81^{1/2} = 9$ because $81 = 9^2$

SECTION 3.3

Exponential growth and **exponential decay** can be modeled by the equation $y = ab^x$.

Suppose you invested $100 at 4% interest compounded yearly. You can model the growth of the investment using $y = 100(1.04)^x$.

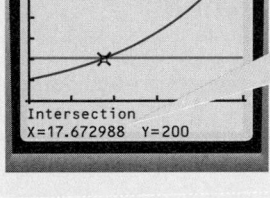

Intersection
X=17.672988   Y=200

Because the investment has a **doubling time** of about 17.7 years, you can rewrite $y = 100(1.04)^x$ as $y = 100 \cdot 2^{x/17.7}$.

The trade-in value of an $8000 used car decreases by 14% per year. You can model the decay in the car's value using $y = 8000(0.86)^x$.

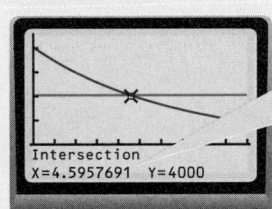

Intersection
X=4.5957691   Y=4000

Because the car's value has a **half-life** of about 4.6 years, you can rewrite $y = 8000(0.86)^x$ as $y = 8000\left(\dfrac{1}{2}\right)^{x/4.6}$.

SECTIONS 3.4 and 3.5

The irrational number e, approximately equal to 2.718, is often used as the base of exponential functions.

$A(t)$ gives the amount in t years when $100 is invested at a 4% annual rate compounded continuously.

$$A(t) = 100e^{0.04t}$$
$$A(2) = 100e^{0.04(2)}$$
$$\approx 108.3$$

Evaluating $A(2)$ gives the amount in 2 years.

The graph shows the **logistic growth** of a population modeled by:

$$P(t) = \frac{1000}{1 + 9e^{-0.05t}}$$

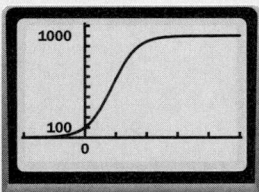

At time $t = 0$, the population is **100**. According to the model, the population stabilizes at **1000**.

You can fit an exponential function to paired data by finding an exponential curve that passes through two representative data points or by performing exponential regression on all the data points. For example, the table shows how the sales of home fax machines increased since 1989.

Using these two data points, you get $y = 306(1.14)^x$ for an exponential model.

x (years)	y (thousands)
1	350
2	400
3	520
4	800

Using all the data points, you can perform exponential regression on a graphing calculator or graphing software to get $y = 247(1.32)^x$.

Assessment

VOCABULARY QUESTIONS

For Questions 1 and 2, complete each paragraph.

1. The graph of the function $y = ab^x$ describes __?__ when $b > 1$ or __?__ when $0 < b < 1$.

2. As n gets larger and larger, the value of $\left(1 + \dfrac{1}{n}\right)^n$ approaches __?__ .

SECTIONS 3.1 *and* 3.2

Simplify using the properties of exponents.

3. $4^2 \cdot 4^{-3/2}$

4. $\dfrac{6^{3/4}}{6^{9/4}}$

5. $\left(7^{-1/2}\right)^4$

6. $\dfrac{9^x}{3^{2x}}$

7. **AGRICULTURE** The number of eggs E a Leghorn chicken produces per year declines exponentially with its age y in years. A function that models the data is:

$$E = 179.2(0.89)^y$$

   a. Rewrite the equation to give the number of eggs as a function of age w in weeks.

   b. Use your equation from part (a) to estimate egg production for a 104-week-old hen.

SECTION 3.3

8. A computer systems analyst earned a salary of \$30,000 in 1990 and averaged 5% more per year after that time.

   a. Write an equation for s, the annual salary, as a function of n, the number of years since 1990.

   b. Find the doubling time for the analyst's salary.

9. The radioactive isotope Germanium-71 decays at a rate of about 6% per day. A scientist has an initial amount of 10 g.

   a. Write an equation of the form $y = a \cdot \left(\dfrac{1}{2}\right)^{x/h}$ for the number of grams remaining after x days.

   b. How many grams are left after 2 days?

10. Suppose a bank offers interest compounded continuously. Use the formula $A = Pe^{rt}$ to find the value of \$2000 after 15 years at each rate.

a. 3.00%   **b.** 5.75%   **c.** 11.50%

11. The function $E(t) = 50e^{-0.03t}$ gives the average number of typing errors a student makes on a typing test after t days of typing instruction.

a. Rewrite the function in the form $E(t) = ab^t$.

b. Find the average number of errors after 20 days of typing instruction.

12. Writing Describe a situation for which a logistic function would provide a good model. Include a rough sketch in your answer.

Write an exponential function whose graph passes through each pair of points.

13. $(3, 7), (8, 15)$ **14.** $(2, 8), (5, 9)$ **15.** $(3, 9), (1, 2)$

16. **Technology** The United States Department of Health and Human Services estimates that national health care expenses will grow exponentially according to the data in the table.

a. Enter the data into a graphing calculator or statistical software and perform an exponential regression. What equation do you get?

b. Estimate the expenses in 2040 using the equation from part (a).

National Health Care Expenses	
Decades since 1990	Expenses (trillions of dollars)
0	0.7
1	1.8
2	3.8
3	7.8
4	16.0

17. ECONOMICS The Consumer Price Index (CPI) for consumers in the United States compares the cost of goods and services with their cost at another time. Assume that the value of the CPI increases exponentially over time. The CPI was 130.7 in 1990 and 136.2 in 1991.

a. Write an exponential function to model the value of the CPI.

b. Use your model to estimate the value of the CPI in 1988 and in 1992.

PERFORMANCE TASK

18. The table gives the percent of the world population in urban areas for decades since 1800. Use the data to write exponential models based on:

- the average proportional growth rate
- an estimate of the doubling time or half-life
- two representative data pairs
- exponential regression

Compare your models and use one that gives a good fit to the data to make predictions about the percent of the world population living in urban areas in 2030.

You can also use other data that appear to be exponential. A good source for data is an almanac.

World Population	
Decades since 1800	Percent in urban areas
0	2
5	4
10	9
15	21

Cumulative Assessment

C H A P T E R S $1-3$

CHAPTER 1

SPORTS The table shows the winning times for the men's 400 m dash in the Olympics from 1980 to 1992. Use the table in Questions 1–3.

Year	1980	1984	1988	1992
Time (s)	44.60	44.27	43.87	43.50

1. Let x be the number of 4-year intervals since 1980. Find a linear function and an exponential function to model the winning times.

2. Estimate the winning time in 1996.

3. **Writing** Which model do you think gives a better estimate of the winning time in 1996? in 2096? Explain your reasoning.

For Questions 4–7, use the matrices below to evaluate each matrix expression. If an expression cannot be evaluated, write *undefined*.

$$A = \begin{bmatrix} 2 & -7 & 4 \\ 3 & 0 & -1 \end{bmatrix} \quad B = \begin{bmatrix} -9 & 0 & 4 \\ 3 & 6 & 1 \end{bmatrix} \quad C = \begin{bmatrix} 8 & -4 \\ 4 & 5 \end{bmatrix} \quad D = \begin{bmatrix} 7 & 2 \\ -1 & -3 \end{bmatrix}$$

4. $3A + 2B$ 5. BD 6. CD 7. $A - B$

8. **Open-ended Problem** Suppose that in a family with three children, each child is as likely to be a girl as a boy. Simulate the number of girls and boys in 20 different families, each with three children. Describe the method you used for your simulation. How many of the families have 3 girls? 2 girls? 1 girl? no girls?

CHAPTER 2

For each equation, tell whether y varies directly with x. If so, graph the equation.

9. $y = 3x - 1$ 10. $y = \dfrac{4}{5x}$ 11. $y = \dfrac{4}{5}x$

12. Write an equation that relates the number of meters, m, to the number of centimeters, c. Tell whether m varies directly with c. Explain your answer.

Find the slope-intercept equation of the line passing through each pair of points.

13. $(1, 3)$ and $(4, 7)$ 14. $(2, 5)$ and $(6, 9)$ 15. $(3, 7)$ and $(5, 8)$

Graph each equation.

16. $y = 3x - 7$ 17. $y = -2x + 9$ 18. $y = \dfrac{1}{2}x + 8$

19. **MANUFACTURING** American women's shoe sizes 7 and 9 correspond to European women's shoe sizes 38 and 40.

   a. Write a point-slope equation that gives the European shoe size, E, as a linear function of the American shoe size, A.

   b. Find the European size that corresponds to an American size 10.

20. **Technology** Refer to the table used in Questions 1–3. Use a graphing calculator or statistical software to find the equation of the least-squares line and the correlation coefficient for the data. Is the correlation *positive* or *negative*? *strong* or *weak*?

21. **Open-ended Problem** Give an example of two real-world quantities for which the correlation would be positive and strong.

22. a. Graph the parametric equations $x = 10 - t$ and $y = 5 + t$ for $t \le 0$.

   b. Express y as a function of x and give the function's domain.

   c. **Writing** Explain why you could find a different pair of parametric equations for the graph in part (a).

CHAPTER 3

Simplify using the properties of exponents.

23. $\dfrac{9^{5/8}}{9^{1/8}}$

24. $\left(6^{-9}\right)^{1/3}$

25. $3^5 \cdot 3^{-5}$

26. $16^{1/8} \cdot 16^{5/8}$

27. a. Sketch the graphs of $y = 100(2^x)$ and $y = 100\left(\frac{1}{2}\right)^x$ in the same coordinate plane.

   b. **Writing** Compare the graphs in part (a).

   c. **Open-ended Problem** Describe a real-life situation that could be modeled by each equation in part (a).

28. **BANKING** Find the number of years needed to double the value of a $500 investment if the 7.2% annual interest is compounded:

   a. annually   b. monthly   c. continuously

29. **CHEMISTRY** The amount that remains of a 50 g sample of the radio-active element carbon-11 after t seconds is modeled by the formula $A(t) = 50e^{-0.0347t}$.

   a. Find the amount remaining after 30 s.

   b. Find the half-life h and use it to write a formula in the form
   $$A(t) = a \cdot \left(\frac{1}{2}\right)^{t/h}.$$

30. **Technology** Use the table from Questions 1–3. Let $x =$ the number of years since 1980. Enter the data into a graphing calculator or statistical software and find an exponential equation that gives a good fit. Predict the winning time in 1996 and compare it with the answer to Question 3.

4 Logarithmic Functions

Digging for new ideas

INTERVIEW **Ednaly Ortiz**

" I went to the symposium and everyone was surprised I was a high school student!"

It's not often that a young person impresses a society of professional scientists. By the time Ednaly Ortiz Camacho entered college, she had done that several times at presentations in Spain and throughout the United States. Ortiz earned international fame while still in high school for her soil studies on archaeological sites in her native Puerto Rico.

As a young girl, Ortiz often accompanied her father, a history teacher, on archaeological digs. Now her work provides archaeologists with new methods for detecting a historical human presence at excavations—even if no artifacts are found. "When people ask me what I do," she says, "I tell them it's not only an excavation for artifacts, it's also a study involving chemistry, geology, and topography."

Examining the Soil

Ortiz set out to determine the impact of these cultures on the soil by conducting a chemical analysis of soil samples. It was difficult to find a laboratory to work with her, she says, because she was only 14 years old. Ortiz concluded through chemical testing of the soil that erosion of the artifacts affected the soil's pH level. The high concentrations of calcium, due to the breakdown of the mollusk shells and animal bones, made the soil more alkaline. She compared the soil to other soil from pits where no cultural evidence was found. "In areas where the Elenoid or Chicoid people did not live," she says, "the components were normal."

New Evidence from Old Cultures

Her four-year, four-phase project began in 1989 at a site near the northeastern shore of Puerto Rico. Armed with trowels, sifters, brushes, and other tools, Ortiz and her fellow freshman classmates uncovered pieces of ceramic bowls, stone weights, pestles, shellfish remains, and the bones of an extinct rabbit-like animal called *jutía*. The artifacts belonged to two pre-Columbian cultures, the Elenoid and Chicoid peoples, who lived from around A.D. 900 to 1500. Both groups of people grew crops and fished for food, which explains the abundance of shell remains. It was the first time evidence of these native people had been discovered in this area of Puerto Rico.

The pH Scale

When Ortiz measured the pH level of soil, she really measured the concentration of hydrogen ions in the soil (pH stands for "potential of hydrogen").

The pH scale runs from 0 to 14 where 7 is neutral, alkalinity increases as the numbers increase, and acidity increases as the numbers decrease. Since each whole-number increase in pH represents a ten-fold decrease in the concentration of hydrogen ions, the pH scale is an example of a *logarithmic scale*.

> ❝ **What I do is not only an excavation for artifacts, it's also a study involving chemistry, geology, and topography.** ❞

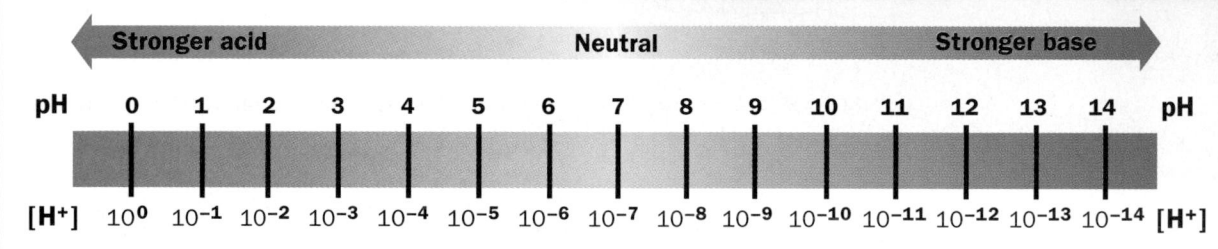

Stronger acid — Neutral — Stronger base

| pH | 0 | 1 | 2 | 3 | 4 | 5 | 6 | 7 | 8 | 9 | 10 | 11 | 12 | 13 | 14 | pH |

| $[H^+]$ | 10^0 | 10^{-1} | 10^{-2} | 10^{-3} | 10^{-4} | 10^{-5} | 10^{-6} | 10^{-7} | 10^{-8} | 10^{-9} | 10^{-10} | 10^{-11} | 10^{-12} | 10^{-13} | 10^{-14} | $[H^+]$ |

EXPLORE AND CONNECT

Ednaly Ortiz stands near the shore of Ojo del Buey, the site of her investigation into pre-Columbian cultures.

1. Project The acidity of many common foods and drinks can be measured using the pH scale. Ask a science teacher for litmus paper and use it to measure the pH of various foods and drinks in your house. Make a chart comparing the acidity of the items you test.

2. Writing A soft drink has a pH of 3.0 and black coffee has a pH of 5.0. Which is more acidic? How many times greater is the concentration of hydrogen ions in the soft drink? Explain your reasoning.

3. Research Find out what plants are best suited for the pH level of the soil in your area. What do gardeners or farmers do to change the pH level of their soil?

Mathematics
& Ednaly Ortiz

In this chapter, you will learn more about how mathematics relates to archaeology.

Related Exercises

Section 4.1
• **Exercises 13 and 14**

Section 4.3
• **Exercises 18 and 19**

4.1 Using Inverses of Linear Functions

Learn how to...

- **graph and find equations for inverses of linear functions**

So you can...

- **solve problems about bicycling and home construction, for example**

Most bicycles today have multiple gears to help cyclists ride comfortably over different terrain. By choosing the right gear, a cyclist can conserve energy and ride longer distances.

Lower gears are used for going up hills.

Higher gears are used on level ground.

THINK AND COMMUNICATE

The table shows the distance a bicycle travels for different numbers of pedal rotations when the bicycle is in first gear. Use the table and graph for Questions 1–4.

Number of pedal rotations	Distance traveled (meters)
0	0
1	3.1
2	6.2
3	9.3

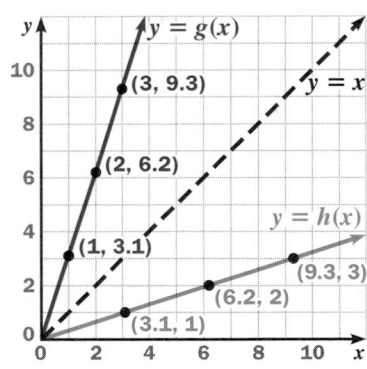

Graph showing $y = g(x)$, $y = x$, and $y = h(x)$, with points (1, 3.1), (2, 6.2), (3, 9.3) on g and (3.1, 1), (6.2, 2), (9.3, 3) on h.

1. Replace each $\underline{?}$ with *number of pedal rotations* or *distance traveled*.

   a. g gives $\underline{?}$ as a function of $\underline{?}$. **b.** h gives $\underline{?}$ as a function of $\underline{?}$.

2. **a.** Which function would you use to find the distance traveled in 2.5 pedal rotations? Why?

   b. Which function would you use to find the number of pedal rotations needed to travel 8 m? Why?

3. Complete these statements: If the point (a, b) is on the graph of g, then the point $\underline{?}$ is on the graph of h. So if $g(a) = b$, then $h(\underline{?}) = \underline{?}$.

4. How are the graphs of g and h geometrically related to the line $y = x$?

In the *Think and Communicate* questions, the function h is the *inverse* of the function g. In general, the **inverse** of a function f, denoted f^{-1}, is a function that satisfies this property:

$$f^{-1}(b) = a \text{ if and only if } f(a) = b.$$

You read f^{-1} as "f inverse."

WATCH OUT!

The symbol -1 in f^{-1} is *not* an exponent. In general, $f^{-1}(x) \neq \dfrac{1}{f(x)}$.

A point (b, a) is on the graph of f^{-1} if and only if (a, b) is on the graph of f.

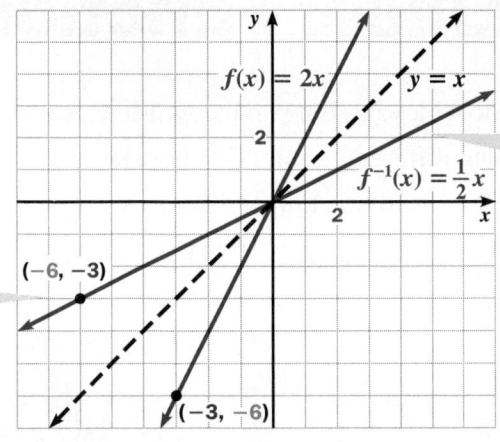

The graph of f^{-1} is the reflection of the graph of f over the line $y = x$.

Given the graph of a function f, you can find the graph of f^{-1}.

Toolbox p. 797
Reflecting a Figure; Line Symmetry

EXAMPLE 1

Let $f(x) = -3x + 6$. Graph $y = f(x)$ and $y = f^{-1}(x)$ in the same coordinate plane.

SOLUTION

Note that f is a linear function with $f(0) = 6$ and $f(1) = 3$.

Step 1 Graph $y = f(x)$. The graph of f is a line passing through the points $(0, 6)$ and $(1, 3)$.

Step 2 Draw the line $y = x$.

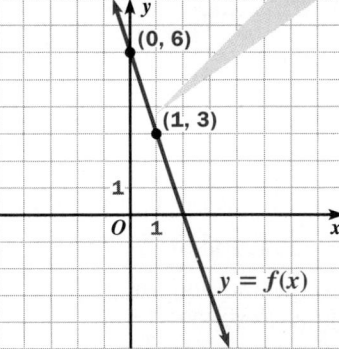

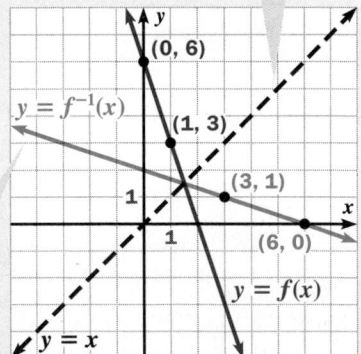

Step 3 Graph $y = f^{-1}(x)$ by reflecting the graph of f over the line $y = x$. Since the points $(0, 6)$ and $(1, 3)$ are on the graph of f, the points $(6, 0)$ and $(3, 1)$ are on the graph of f^{-1}.

THINK AND COMMUNICATE

5. Is the inverse of a nonconstant linear function always a nonconstant linear function? Explain.

6. Does a constant function have an inverse? Why or why not? (*Hint:* What is the reflection of the graph of a constant function over the line $y = x$?)

Given an equation for a function f, you can find an equation for f^{-1}.

EXAMPLE 2

Let $g(x) = 7x - 4$. Find an equation for g^{-1}.

SOLUTION

Step 1 Replace $g(x)$ with y in the equation for g.

$$y = 7x - 4$$

Step 2 Solve for x.

$$y + 4 = 7x$$

$$\frac{1}{7}(y + 4) = \frac{1}{7}(7x)$$

$$\frac{1}{7}y + \frac{4}{7} = x$$

> **Toolbox p. 788**
> *Rewriting Equations and Formulas*

Step 3 Switch x and y so that x is the independent variable of the inverse function.

$$\frac{1}{7}x + \frac{4}{7} = y$$

Step 4 Replace y with $g^{-1}(x)$.

$$\frac{1}{7}x + \frac{4}{7} = g^{-1}(x)$$

An equation for g^{-1} is $g^{-1}(x) = \frac{1}{7}x + \frac{4}{7}$.

THINK AND COMMUNICATE

7. How can you use graphs to verify that $y = \frac{1}{7}x + \frac{4}{7}$ is the inverse of $y = 7x - 4$?

8. Use the function f in Example 1.

   a. Find an equation for f^{-1} using the fact that its graph is a line passing through the points $(6, 0)$ and $(3, 1)$.

   b. Find an equation for f^{-1} using the method in Example 2. Compare this equation with the equation you found in part (a). Which method do you prefer? Why?

EXAMPLE 3 | Application: Business

Jaime Ramirez is a builder who is developing a subdivision of a small town. He sells housing lots in the subdivision for $84,000 each. He also builds houses on the lots for $60 per square foot of floor space.

The cost of a lot is $84,000.

The cost per square foot is $60.

a. Find an equation giving the cost C of building a house as a function of the house's size S in square feet.

b. Find the inverse of the function from part (a).

c. How large a house can be built for $225,000?

SOLUTION

a. Note that:

$$\begin{matrix} \text{Cost of} \\ \text{building a house} \end{matrix} = \begin{matrix} \text{Cost of} \\ \text{a lot} \end{matrix} + \left(\begin{matrix} \text{Cost per} \\ \text{square foot} \end{matrix} \times \begin{matrix} \text{Number of} \\ \text{square feet} \end{matrix} \right)$$

$$C = 84{,}000 + 60S$$

An equation is $C = 84{,}000 + 60S$.

b. Solve the equation in part (a) for S.

$$C - 84{,}000 = 60S$$

$$\frac{1}{60}(C - 84{,}000) = \frac{1}{60}(60S)$$

$$\frac{1}{60}C - 1400 = S$$

The inverse is $S = \frac{1}{60}C - 1400$.

c. Use the inverse to find S when $C = 225{,}000$.

$$S = \frac{1}{60}(225{,}000) - 1400$$

$$= 3750 - 1400$$

$$= 2350$$

A house with 2350 ft^2 of floor space can be built.

> **WATCH OUT!** ▶
>
> You could write the functions in Example 3 as:
> $f(x) = 84{,}000 + 60x$
> and
> $f^{-1}(x) = \frac{1}{60}x - 1400$.
> However, in real-life applications you should use variables like C and S that are easily associated with the quantities they represent.

THINK AND COMMUNICATE

9. Replace each $\underline{?}$ with *cost* or *square footage*.

  a. The equation $C = 84{,}000 + 60S$ gives $\underline{?}$ as a function of $\underline{?}$.

  b. The equation $S = \frac{1}{60}C - 1400$ gives $\underline{?}$ as a function of $\underline{?}$.

10. If you want to graph the two functions in Example 3 on a graphing calculator, you have to enter them as $y = 84{,}000 + 60x$ and $y = \frac{1}{60}x - 1400$. For each function, tell what x and y represent.

Complete each statement.

1. If $f(2) = 6$, then $f^{-1}(6) = \underline{\ ?\ }$. **2.** If $h^{-1}(c) = d$, then $h(d) = \underline{\ ?\ }$.

For each pair of functions whose graphs are shown, tell whether g is the inverse of f.

3.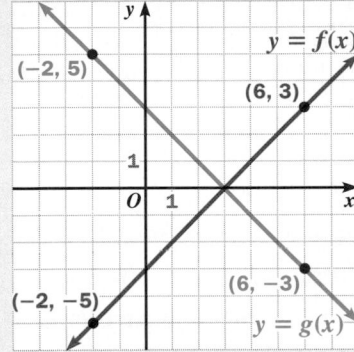

4.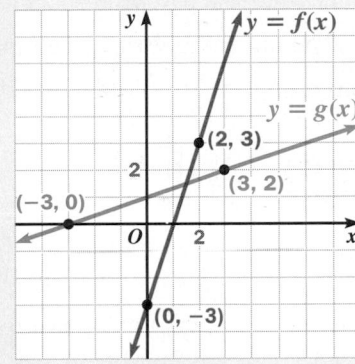

4.1 **Exercises and Applications**

Extra Practice
exercises on
page 753

For each function:

a. Graph the function and its inverse in the same coordinate plane.

b. Find an equation for the inverse.

1. $f(x) = 3x$

2. $f(x) = -2x$

3. $y = -\dfrac{1}{4}x$

4. $y = \dfrac{1}{5}x$

5. $g(x) = 4x + 7$

6. $h(x) = -5x + 1$

7. $y = \dfrac{1}{3}x - 2$

8. $y = -\dfrac{3}{2}x - \dfrac{1}{2}$

9. SAT/ACT Preview Which function is the inverse of $y = 6x - 11$?

A. $y = -6x + 11$ **B.** $y = \dfrac{1}{6x - 11}$ **C.** $y = \dfrac{1}{6}x - \dfrac{11}{6}$ **D.** $y = \dfrac{1}{11 - 6x}$ **E.** $y = \dfrac{1}{6}x + \dfrac{11}{6}$

10. FORESTRY A forester needs to know the diameter of a tree in order to determine the amount of wood it contains. Because a tree's diameter is difficult to measure directly, a forester will often calculate the diameter from the measured circumference.

   a. Open-ended Problem Explain how you could measure a tree's circumference and why measuring the circumference is easier than measuring the diameter.

   b. Write an equation giving a tree's circumference C as a function of its diameter d.

   c. Find the inverse of the function from part (b).

   d. A forester measures the circumference of a tree to be 75 in. Find the tree's diameter.

11. Look back at Example 3. Find the domains and ranges of the functions
$C = 84{,}000 + 60S$ and $S = \frac{1}{60}C - 1400$. What do you notice?

12. **RECREATION** One kind of bicycle travels 7.7 m per pedal rotation when in tenth gear.

   a. Find an equation giving the distance d that the bicycle travels in tenth gear as a function of the number n of pedal rotations.

   b. Find the inverse of the function from part (a).

   c. Complete this statement: The inverse gives _?_ as a function of _?_ .

   d. How many pedal rotations are needed to ride the bicycle 200 m?

INTERVIEW Ednaly Ortiz

Look back at the article on pages 138–140.

Archaeologists such as Ednaly Ortiz often make excavation grids like the one below on sites where they are searching for artifacts. An excavation grid consists of squares of earth separated by partitions called balks.

13. **Writing** What do you think excavation grids are used for?

14. Suppose a group of 10 people is excavating a section of a large field. The group creates an excavation grid such that the area of each grid square and its surrounding balks is 25 m². Each person can excavate a grid square in 20 days.

Letters are used to identify a square's **east-west** position.

Numbers are used to identify a square's **north-south** position.

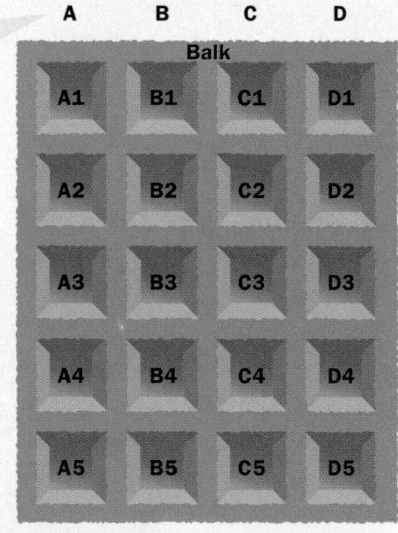

   a. Find an equation giving the area A that the group can excavate in n days.

   b. Find the inverse of the function from part (a).

   c. How many days will it take the group to excavate 125 m² of the field?

This excavation site, located in Maine, has been divided into grid squares.

15. **Challenge** Consider the general linear function $f(x) = ax + b$.

   a. Find an equation for f^{-1}. (Your equation will involve a and b.)

   b. Use the equation from part (a) to find the inverse of $f(x) = -8x + 3$.

   c. How does the equation from part (a) show that a constant function does *not* have an inverse?

   d. What must be true about a and b if $f(x) = f^{-1}(x)$ for all real numbers x?

The table shows the average fuel economy of cars in the United States for the years from 1980 to 1990.

16.  **Technology** Let x = the number of years since 1980. Let y = the average fuel economy.

   a. Enter the data pairs (x, y) into a graphing calculator or statistical software. Make a scatter plot of the data. What do you notice about the plotted points?

   b. Find an equation of the least-squares line for the scatter plot.

   c. Predict the average fuel economy in the year 2000.

17. a. Find the inverse of the function from part (b) of Exercise 16. What quantity does the independent variable of the inverse represent? What quantity does the dependent variable represent?

   b. Use your answer to part (a) to predict the year in which the average fuel economy will reach 30 mi/gal.

Year	Fuel economy (mi/gal)
1980	15.46
1981	15.94
1982	16.65
1983	17.14
1984	17.83
1985	18.20
1986	18.27
1987	19.20
1988	19.87
1989	20.31
1990	21.02

ONGOING ASSESSMENT

18. Writing Write a note to a friend explaining the relationship between the graph of a function f and the graph of f^{-1}. Also list the steps needed to find an equation for f^{-1} given an equation for f.

SPIRAL REVIEW

19. **Technology** The table shows the volume of a landfill for the years from 1990 to 1994. *(Section 3.5)*

   a. Enter the data pairs (x, y) into a graphing calculator or statistical software. Find an exponential function that models the data.

   b. Predict the volume of the landfill in 1998.

   c. Estimate the volume of the landfill in 1987.

x = years since 1990	y = volume (m³)
0	50,000
1	63,000
2	79,000
3	102,000
4	124,000

Find a point-slope equation of the line passing through the given points. *(Section 2.3)*

20. $(1, 2)$ and $(2, 5)$ **21.** $(-3, 4)$ and $(0, -2)$ **22.** $(-6, -3)$ and $(4, 9)$

23. Suppose Chen deposits $500 in a savings account earning 4% annual interest compounded continuously. Find the amount of money in Chen's account after 3 years, assuming he makes no more deposits or withdrawals. *(Section 3.4)*

4.2 Using Inverses of Exponential Functions

Learn how to...

- **find inverses of exponential functions**

- **evaluate logarithms**

So you can...

- **understand how an altimeter works, for example**

You know that the inverse of a nonconstant linear function is also a linear function. Does this mean that the inverse of an exponential function is an exponential function? The answer is no. In fact, the inverse of an exponential function is a type of function you may not have seen before.

EXPLORATION
COOPERATIVE LEARNING

Investigating the Inverse of $f(x) = 2^x$

Work with a partner.
You will need: • a MIRA® transparent mirror
• graph paper

x	f(x)
−2	?
−1	?
0	?
1	?
2	?
3	?

1 Let $f(x) = 2^x$. Copy and complete the table.

2 Graph $y = f(x)$. Label the points on the graph corresponding to the ordered pairs in the table.

3 Graph the line $y = x$ in the same coordinate plane you used for graphing f.

4 Graph $y = f^{-1}(x)$ by using the MIRA® to draw the reflection of the graph of f over the line $y = x$. Label the points on the graph of f^{-1} that are reflections of labeled points on the graph of f.

Questions

1. Replace each _?_ with the number that makes the equation true.

    a. $f^{-1}(8) = \underline{?}$ **b.** $f^{-1}\left(\frac{1}{4}\right) = \underline{?}$ **c.** $f^{-1}(1) = \underline{?}$

2. Replace each _?_ with *exponents* or *powers of 2*.

    a. For the function f, the input values are _?_ and the output values are _?_.

    b. For the function f^{-1}, the input values are _?_ and the output values are _?_.

Logarithmic Functions

The inverse of an exponential function is called a **logarithmic function**. You write the inverse of the exponential function $f(x) = b^x$ as $f^{-1}(x) = \log_b x$. The number b, which must be positive and not equal to 1, is called the *base* of the logarithmic function. The expression $\log_b x$ is called the **base-b logarithm** of x.

For example, the inverse of $y = 3^x$ is $y = \log_3 x$. The graphs of these functions are shown.

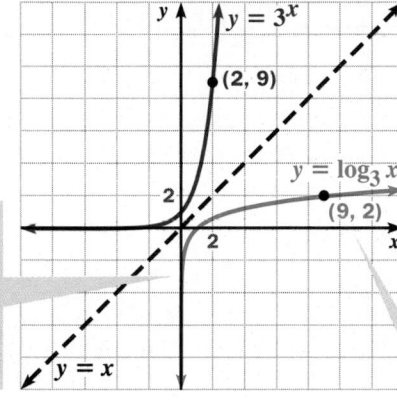

The graph of $y = \log_b x$ is the reflection of the graph of $y = b^x$ over the line $y = x$.

The domain of $y = \log_b x$ is $x > 0$. The range is all real numbers.

Since $3^2 = 9$, it follows that $\log_3 9 = 2$.

Since $f^{-1}(x) = \log_b x$ is the inverse of $f(x) = b^x$, you know that:

$$f^{-1}(N) = k \text{ if and only if } f(k) = N$$

Therefore:

power of b

exponent

$$\log_b N = k \text{ if and only if } b^k = N$$

base

This equation is in **logarithmic form**.

This equation is in **exponential form**.

The exponential and logarithmic equations are equivalent. This means that an equation given in one form can be rewritten in the other form.

EXAMPLE 1

a. Write $4^3 = 64$ in logarithmic form.

b. Write $\log_5 \frac{1}{25} = -2$ in exponential form.

SOLUTION

a. $\log_4 64 = 3$

b. $5^{-2} = \frac{1}{25}$

EXAMPLE 2

Evaluate each logarithm.

a. $\log_6 36$

b. $\log_{16} 8$

SOLUTION

a. Let $\log_6 36 = k$.

Then: $6^k = 36$ ←—— Write the equation in exponential form. ——→ **b.** Let $\log_{16} 8 = k$.

Then: $16^k = 8$

$6^k = 6^2$ ←—— Make the bases the same. ——→ $(2^4)^k = 2^3$

$2^{4k} = 2^3$

$k = 2$ ←—— Equate exponents. ——→ $4k = 3$

$k = \dfrac{3}{4}$

$\log_6 36 = 2$

$\log_{16} 8 = \dfrac{3}{4}$

THINK AND COMMUNICATE

1. To evaluate $\log_6 36$ mentally, you should ask yourself the question, "What power of 6 equals 36?" What question should you ask to evaluate each of these logarithms mentally?

   a. $\log_2 8$
   b. $\log_7 \dfrac{1}{7}$
   c. $\log_9 3$

2. Mentally evaluate each logarithm in Question 1.

3. Explain why $\log_6 (-36)$ is not defined.

4. **a.** Evaluate these logarithms: $\log_2 1$, $\log_3 1$, $\log_4 1$.

   b. In general, what is the value of $\log_b 1$?

5. **a.** Evaluate these logarithms: $\log_2 2$, $\log_3 3$, $\log_4 4$.

   b. In general, what is the value of $\log_b b$?

6. **a.** Complete this equation: $\log_b b^k = \underline{\ ?\ }$.

   b. Explain why your answer to part (a) is consistent with your results from Questions 4 and 5.

Most calculators have LOG and LN keys for evaluating common and natural logarithms.

Common and Natural Logarithms

Two kinds of logarithms are used so often that they are given special names and symbols. The **common logarithm** of a positive number N is the base-10 logarithm of N and is denoted **log N** (rather than $\log_{10} N$). The **natural logarithm** of N is the base-e logarithm of N and is denoted **ln N** (rather than $\log_e N$).

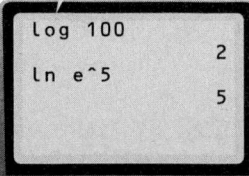

Log 100

2

Ln e^5

5

EXAMPLE 3 | **Application: Mountain Climbing**

Mountain climbers use an instrument called an *altimeter* to help them navigate. An altimeter finds a climber's height above sea level by measuring the air pressure. The height h and air pressure P are related by the function

$$P = 101{,}300e^{-h/8005}$$

where h is in meters and P is in pascals.

a. Find the inverse of the given function.

b. In 1975, Junko Tabei of Japan became the first woman to reach the summit of Mount Everest, the highest mountain in the world. What was the reading on Tabei's altimeter when the air pressure was 60,000 pascals?

Junko Tabei stands on "the top of the world." She had to wear an oxygen mask due to the low level of oxygen at that high altitude.

SOLUTION

a. Solve the equation $P = 101{,}300e^{-h/8005}$ for h.

$$\frac{P}{101{,}300} = e^{-h/8005}$$

$$\ln \frac{P}{101{,}300} = -\frac{h}{8005}$$ Write the equation in logarithmic form.

$$-8005 \ln \frac{P}{101{,}300} = h$$

The inverse is $h = -8005 \ln \dfrac{P}{101{,}300}$.

b. Use the inverse to find h when $P = 60{,}000$.

$$h = -8005 \ln \frac{60{,}000}{101{,}300}$$ Substitute **60,000** for *P*.

```
ln(60000/101300)
      -.523741849
```

$$\approx (-8005)(-0.5237)$$

$$\approx 4190$$ Use a calculator to evaluate the logarithm.

The reading on Junko Tabei's altimeter was about 4190 m above sea level.

7. In Example 3, does a decrease in air pressure indicate an *increase* or *decrease* in height above sea level? How do you know?

8. In the solution to part (a) of Example 3, why would it be incorrect to write $\log \dfrac{P}{101{,}300} = -\dfrac{h}{8005}$?

☑ CHECKING KEY CONCEPTS

Find the inverse of each function.

1. $f(x) = 2^x$

2. $g(x) = \left(\dfrac{1}{3}\right)^x$

3. $y = e^x$

Write each equation in logarithmic form.

4. $3^4 = 81$

5. $8^0 = 1$

6. $10^{-2} = \dfrac{1}{100}$

Write each equation in exponential form.

7. $\log_6 216 = 3$

8. $\log \dfrac{1}{10} = -1$

9. $\log_{27} 3 = \dfrac{1}{3}$

Evaluate each logarithm.

10. $\log_2 32$

11. $\log_{125} 25$

12. $\ln e^3$

Use a calculator to find the value of each logarithm to the nearest hundredth.

13. $\log 146$

14. $\ln 15$

15. $\log 0.32$

4.2 Exercises and Applications

Extra Practice exercises on page 753

1. Look back at the graph of $y = \log_3 x$ on page 149. For what value(s) of x is $\log_3 x$:

a. positive?

b. negative?

c. zero?

d. undefined?

Write each equation in logarithmic form.

2. $3^2 = 9$

3. $2^3 = 8$

4. $10^4 = 10{,}000$

5. $e^0 = 1$

6. $5^{-1} = \dfrac{1}{5}$

7. $64^{-1/3} = \dfrac{1}{4}$

8. $(0.3)^2 = 0.09$

9. $\left(\dfrac{4}{5}\right)^3 = \dfrac{64}{125}$

Write each equation in exponential form.

10. $\log_4 16 = 2$

11. $\log_2 16 = 4$

12. $\log_3 \dfrac{1}{9} = -2$

13. $\log_2 \dfrac{1}{32} = -5$

14. $\log 1 = 0$

15. $\log_{1/6} \dfrac{1}{36} = 2$

16. $\log_{0.2} 0.008 = 3$

17. $\log_{64} 16 = \dfrac{2}{3}$

Evaluate each logarithm.

18. $\log_3 27$

19. $\log_7 49$

20. $\ln e^{-6}$

21. $\log \dfrac{1}{1000}$

22. $\log_{1/2} \dfrac{1}{8}$

23. $\log_9 1$

24. $\log_4 8$

25. $\log_{81} 27$

26. Writing Explain why the base b of $y = \log_b x$ cannot be 1.

Star	Apparent magnitude
Sirius	−1.5
Tau Ceti	3.5
Barnard's Star	9.5
Luyten 726-8	12.5
Wolf 359	13.5

This time-lapse photograph shows the motion of Barnard's Star over a four-year period.

Recall from Section 3.5 that the *apparent magnitude* of a star is a measure of how bright the star looks to a person on Earth. The greater a star's apparent magnitude, the *less* bright the star appears. The table shows the apparent magnitudes of several stars.

You can see stars with apparent magnitudes of up to 6.5 without a telescope. To see a star with magnitude $M > 6.5$, you need a telescope having an objective lens or mirror whose diameter is at least D mm, where:

$$D = 10^{(M - 2)/5}$$

27. What minimum lens or mirror diameter must a telescope have for you to be able to see Barnard's Star?

28. Find the inverse of $D = 10^{(M - 2)/5}$.

29. A telescope's *limiting magnitude* is the apparent magnitude of the dimmest star that can be seen with the telescope.

a. A typical amateur telescope might have a lens or mirror diameter of about 75 mm. Use your answer to Exercise 28 to find the limiting magnitude of such a telescope.

BY THE WAY

Barnard's Star, located in the constellation of Ophiuchus, moves across the sky faster than any other star.

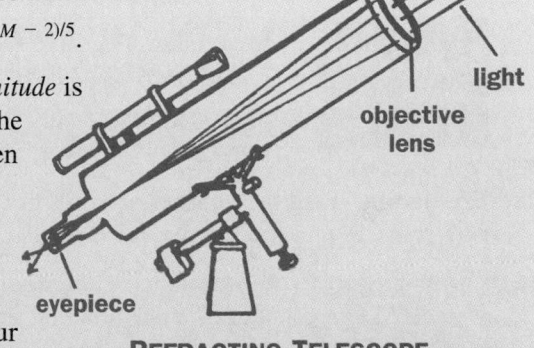

light

objective lens

eyepiece

REFRACTING TELESCOPE

b. One of the largest optical telescopes in the world is located on Mount Semirodriki in Russia. This telescope has a mirror whose diameter is 236 in. Find the telescope's limiting magnitude. (*Hint:* 1 in. = 25.4 mm)

c. Which of the stars in the table can be seen with the telescope in part (a)? Which can be seen with the telescope in part (b)?

30. Writing Suppose the apparent magnitude of star A is less than the apparent magnitude of star B. Does star A necessarily give off more light than star B? Explain.

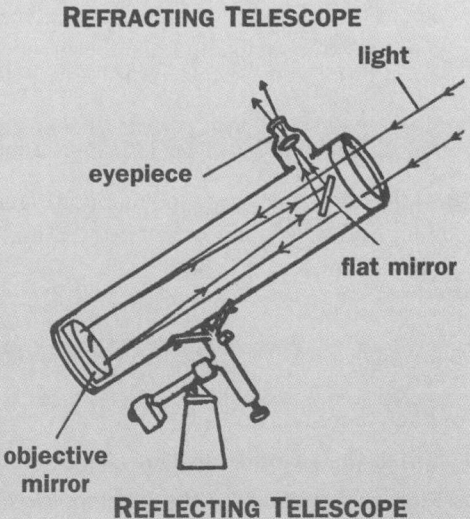

light

eyepiece

flat mirror

objective mirror

REFLECTING TELESCOPE

Use a calculator to find the value of each logarithm to the nearest hundredth.

31. $\log 12$

32. $\log 257$

33. $\ln 6.5$

34. $\ln \dfrac{3}{8}$

For each function:

a. Graph the function and its inverse in the same coordinate plane.

b. Find an equation for the inverse.

35. $f(x) = 4^x$

36. $h(x) = 10^x$

37. $y = \left(\dfrac{1}{2}\right)^x$

38. $y = 2 \cdot 3^x$

39. POPULATION The population P of Nepal (in millions) can be modeled by the function

$$P = 19.1e^{0.025t}$$

where t is the number of years since 1990.

a. Find the inverse of the given function. What information can you obtain with the inverse?

b. Predict the year in which the population of Nepal will reach 30 million.

c. Estimate the year in which the population of Nepal was 15 million.

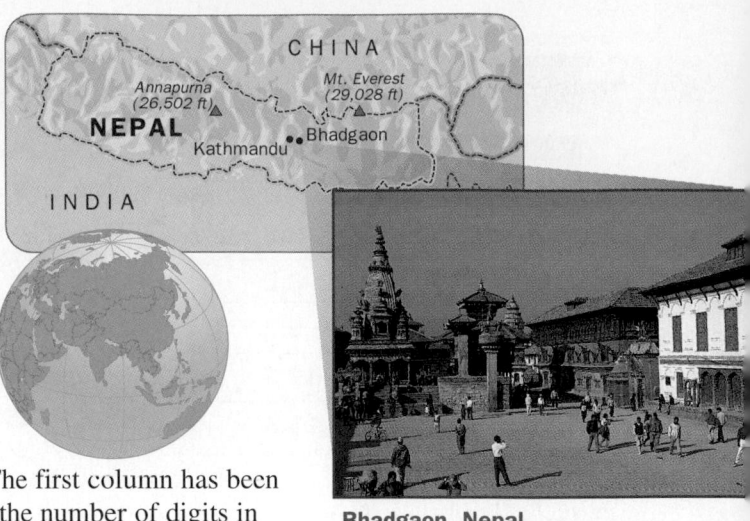

Bhadgaon, Nepal

40. Investigation Copy and complete the table. (The first column has been done for you.) What is the relationship between the number of digits in a positive integer N and the common logarithm of N?

N	7	10	52	100	613	1,000	4,849	10,000	91,770
Number of digits in N	1	?	?	?	?	?	?	?	?
$\log N$	0.85	?	?	?	?	?	?	?	?

41. Challenge Find the number of digits in 3^{1000}. (*Hint:* Express 3 as a power of 10, and use the result from Exercise 40.)

42. Open-ended Problem Choose an exponential function of the form $y = ae^{kx}$ from Chapter 3. Your function should model something from real life. Find the inverse of your function. Describe what information the inverse gives, and use the inverse to solve a problem.

ONGOING ASSESSMENT

43. Writing List three facts about logarithmic functions. Write down a logarithmic function $y = \log_b x$, and evaluate your function for at least two values of x.

SPIRAL REVIEW

For each function:

a. Graph the function and its inverse in the same coordinate plane.

b. Find an equation for the inverse. *(Section 4.1)*

44. $f(x) = 5x$ **45.** $g(x) = -\dfrac{1}{3}x$ **46.** $h(x) = 2x - 10$ **47.** $y = \dfrac{3}{8}x + \dfrac{7}{8}$

48. Let $A = \begin{bmatrix} 2 & 5 \\ -8 & 0 \end{bmatrix}$ and $B = \begin{bmatrix} -1 & 3 \\ -5 & 6 \end{bmatrix}$. Find each matrix. *(Sections 1.3 and 1.4)*

     a. $A + B$ **b.** $B - A$ **c.** $2A + 3B$ **d.** AB

Tell whether each equation represents growth that is *linear*, *exponential*, or *neither*. *(Section 3.1)*

49. $y = 7x^2$ **50.** $y = 7(2^x)$ **51.** $y = 2(7^x)$ **52.** $y = (7^2)x$

4.3 Working with Logarithms

In Chapter 3, you learned rules for simplifying products, quotients, and powers of powers. As you'll see in the Exploration, there are corresponding rules for logarithms.

Learn how to...

- **use the properties of logarithms**

So you can...

- **understand logarithmic scales, such as the Richter scale for earthquakes**

EXPLORATION
COOPERATIVE LEARNING

Investigating Properties of Logarithms

Work with a partner.
You will need:
● a scientific calculator

SET UP Copy the table. Use a calculator to complete the table. Round each logarithm to four decimal places.

N	1	2	3	4	5	6	7	8	9	10	20	30	40	50	60	70	80	90	100	1000
log N	?	?	?	?	?	?	?	?	?	?	?	?	?	?	?	?	?	?	?	?

Questions

1. a. Use your table to find log 2 + log 4.

   b. Look for a value of N in your table for which log N equals your answer to part (a). Complete this equation: log 2 + log 4 = log _?_ .

2. Use your table to complete these equations.

   a. log 8 + log 5 = log _?_   **b.** log 4 + log 20 = log _?_

   c. log 7 + log 10 = log _?_   **d.** log 9 + log 10 = log _?_

3. Generalize your results from Questions 1 and 2:
   log M + log N = log _?_ .

4. Use your table to complete these equations.

   a. log 8 − log 2 = log _?_   **b.** log 50 − log 5 = log _?_

   c. log 60 − log 10 = log _?_   **d.** log 90 − log 30 = log _?_

5. Generalize your results from Question 4: log M − log N = log _?_ .

6. Use your table to complete these equations.

   a. 2 log 3 = log _?_   **b.** 2 log 10 = log _?_   **c.** 3 log 10 = log _?_

7. Generalize your results from Question 6: k log M = log _?_ .

You can use the following properties to rewrite logarithms of products, quotients, and powers.

Properties of Logarithms

Let M, N, and b be positive numbers with $b \neq 1$. Then:

Product Property: $\qquad \log_b MN = \log_b M + \log_b N$

Quotient Property: $\qquad \log_b \dfrac{M}{N} = \log_b M - \log_b N$

Power Property: $\qquad \log_b M^k = k \log_b M$

EXAMPLE 1

Write $\log_3 \dfrac{p^4 q^5}{r^{1/2}}$ in terms of $\log_3 p$, $\log_3 q$, and $\log_3 r$.

SOLUTION

$$\log_3 \frac{p^4 q^5}{r^{1/2}} = \log_3 p^4 q^5 - \log_3 r^{1/2}$$

Use the **quotient property**.

$$= \log_3 p^4 + \log_3 q^5 - \log_3 r^{1/2}$$

Use the **product property**.

$$= 4 \log_3 p + 5 \log_3 q - \frac{1}{2} \log_3 r$$

Use the **power property**.

EXAMPLE 2

Write $\log_a 48 - 3 \log_a 2 + \log_a 5$ as a logarithm of a single number.

SOLUTION

$$\log_a 48 - 3 \log_a 2 + \log_a 5 = \log_a 48 - \log_a 2^3 + \log_a 5$$

Use the **power property**.

$$= \log_a 48 - \log_a 8 + \log_a 5$$

$$= \log_a \frac{48}{8} + \log_a 5$$

Use the **quotient property**.

$$= \log_a 6 + \log_a 5$$

$$= \log_a (6 \cdot 5)$$

Use the **product property**.

$$= \log_a 30$$

THINK AND COMMUNICATE

1. Use a calculator to evaluate $\log_a 48 - 3 \log_a 2 + \log_a 5$ and $\log_a 30$ when $a = 10$. Are your results consistent with Example 2?

EXAMPLE 3   Application: Earth Science

The *Richter scale* is used to rate the severity of earthquakes. An earthquake is assigned a *Richter magnitude* based on the amount of energy it releases. The Richter magnitude M and energy E are related by the equation

$$M = \frac{2}{3} \log \frac{E}{10^{11.8}}$$

where E is measured in units called *ergs*. Suppose the difference in the Richter magnitudes of two earthquakes, A and B, is 1. How many times as much energy does earthquake A release as earthquake B?

BY THE WAY

The smallest earthquakes that can be felt have Richter magnitudes of about 2.0. The most powerful earthquake ever recorded had a Richter magnitude of 8.9. It occurred in Japan on March 2, 1933.

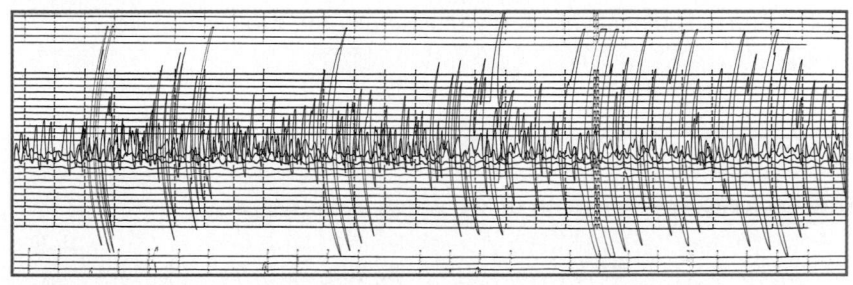

A seismograph draws a graph like this to show the movement of the earth during an earthquake. Scientists use these graphs to determine the amount of energy released.

SOLUTION

Let E_A and M_A be the energy and Richter magnitude for earthquake A. Let E_B and M_B be the energy and Richter magnitude for earthquake B. Then:

$$M_A - M_B = 1$$

The difference in magnitudes is 1.

Use the given equation.

$$\frac{2}{3} \log \frac{E_A}{10^{11.8}} - \frac{2}{3} \log \frac{E_B}{10^{11.8}} = 1$$

$$\frac{2}{3}\left(\log \frac{E_A}{10^{11.8}} - \log \frac{E_B}{10^{11.8}} \right) = 1$$

Toolbox p. 782
Factoring Variable Expressions

$$\log \frac{E_A}{10^{11.8}} - \log \frac{E_B}{10^{11.8}} = \frac{3}{2}$$

Use the quotient property.

$$\log \left(\frac{E_A/10^{11.8}}{E_B/10^{11.8}} \right) = \frac{3}{2}$$

$$\log \left(\frac{E_A}{10^{11.8}} \cdot \frac{10^{11.8}}{E_B} \right) = \frac{3}{2}$$

$$\log \frac{E_A}{E_B} = \frac{3}{2}$$

Write the equation in exponential form.

$$\frac{E_A}{E_B} = 10^{3/2}$$

$$E_A = 10^{3/2} \cdot E_B \approx 32E_B$$

Earthquake A releases about 32 times as much energy as earthquake B.

The Richter scale is an example of a *logarithmic scale*. In general, *adding* 1 on a logarithmic scale corresponds to *multiplying* a related quantity (such as energy) by some number. Logarithmic scales are also used to measure acidity and loudness. (See Exercises 18, 19, and 26–28.)

THINK AND COMMUNICATE

2. Summarize the result in Example 3 by completing this statement: *Adding* 1 on the Richter scale corresponds to *multiplying* the energy by __?__ .

3. In 1982, an earthquake measuring 6.0 on the Richter scale struck Yemen. In 1993, an earthquake measuring 8.0 on the Richter scale struck Guam. Compare the amounts of energy released by the two earthquakes.

☑ CHECKING KEY CONCEPTS

Write each expression in terms of $\log_2 p$, $\log_2 q$, and $\log_2 r$.

1. $\log_2 p^3$ **2.** $\log_2 pq^5$ **3.** $\log_2 \dfrac{p^2}{r^4}$ **4.** $\log_2 \dfrac{p^7 q^{10}}{r^{3/8}}$

Write as a logarithm of a single number or expression.

5. $5 \log_a 2$ **6.** $\log_b \dfrac{1}{8} + 3 \log_b 4$

7. $9 \log_a x - 2 \log_a y$ **8.** $2 \ln 3 + \ln 6 - \dfrac{3}{4} \ln 81$

4.3 | **Exercises and Applications**

Extra Practice exercises on page 754

Write each expression in terms of $\log_7 p$, $\log_7 q$, and $\log_7 r$.

1. $\log_7 p^4$ **2.** $\log_7 q^{1/7}$ **3.** $\log_7 q^2 r^3$ **4.** $\log_7 \dfrac{p^{2/5}}{q^8}$

5. $\log_7 \dfrac{1}{q}$ **6.** $\log_7 p^6 q^3 r^{1/4}$ **7.** $\log_7 \dfrac{p^9 q^{11}}{r^7}$ **8.** $\log_7 \dfrac{r^2}{pq^{1/3}}$

9. Writing Tamara expressed the product property of logarithms in words to help her understand what it means. Express the quotient and power properties in words.

Tamara
Product Property:
$\log_b MN = \log_b M + \log_b N$

The logarithm of a product of two numbers equals the sum of the logarithms of the numbers.

Write as a logarithm of a single number or expression.

10. $4 \log_a 3$

11. $-\dfrac{1}{2} \log_a 49$

12. $\log_3 p + 2 \log_3 q$

13. $6 \log_5 u - 10 \log_5 v$

14. $2 \log_b 10 - \dfrac{2}{3} \log_b 125$

15. $12 \log x^2 + \dfrac{4}{5} \log x^{10}$

16. $3 \ln p - \ln q - 7 \ln r$

17. $\dfrac{1}{3} \log_b 27 + 2 \log_b 6 - \dfrac{1}{2} \log_b 144$

Ednaly Ortiz

Look back at the article on pages 138–140.

Archaeologists like Ednaly Ortiz use the pH scale to rate the acidity of soils. The pH of a soil is given by the equation

$$pH = -\log [H^+]$$

where $[H^+]$ is the soil's hydrogen ion concentration in moles per liter. The greater a soil's pH, the more alkaline (or less acidic) the soil is.

An *ion* is an electrically charged atom or group of atoms. A *mole* of ions contains about 6.02×10^{23} ions.

18. The diagram shows the different soil layers found in one of Ortiz's excavation pits, as well as the hydrogen ion concentration in each layer.

  a. Find the pH of each soil layer.

  b. Ortiz found that the presence of human artifacts increases a soil's alkalinity. Which of the soil layers in the diagram do you think contained the most artifacts? Explain.

19. a. Suppose the difference in pH of two soils, A and B, is 1. Compare the hydrogen ion concentrations in the soils.

  b. **Writing** Explain why the pH scale is a logarithmic scale.

$[H^+] = 4.0 \times 10^{-7}$ moles/L

$[H^+] = 1.6 \times 10^{-7}$ moles/L

$[H^+] = 2.0 \times 10^{-6}$ moles/L

20. EARTH SCIENCE Look back at Example 3.

  a. On June 9, 1994, a powerful earthquake struck La Paz, Bolivia. The earthquake released about $10^{24.1}$ ergs of energy. Find the Richter magnitude of the earthquake to the nearest tenth.

  b. On March 27, 1964, an earthquake in Alaska released about twice as much energy as the La Paz earthquake. Find the Richter magnitude of the Alaska earthquake to the nearest tenth. Compare this magnitude with the magnitude of the La Paz earthquake.

21. Open-ended Problem For each statement, find positive numbers M, N, and b (with $b \neq 1$) that show that the statement is false in general.

  a. $\log_b (M + N) = \log_b M + \log_b N$

  b. $\log_b (M - N) = \log_b M - \log_b N$

Let $x = \log_a 5$ and $y = \log_a 10$. Write each expression in terms of x and y.

22. $\log_a 50$ **23.** $\log_a 2$ **24.** $\log_a 100$ **25.** $\log_a 250$

The loudness of a sound is measured on the *decibel scale* and depends on the sound's *intensity*, or power per unit area. The loudness L and intensity I are related by the equation $L = 10 \log \dfrac{I}{I_0}$ where L is measured in decibels (dB), I is measured in watts per square meter (W/m^2), and $I_0 = 10^{-12}$ W/m^2 is the intensity of a barely audible sound.

Sound	Intensity (W/m^2)
Falling pin	10^{-11}
Quiet conversation	10^{-6}
Subway train	10^{-3}
Jet at takeoff	1

An acoustical engineer studies the noise created by an industrial machine.

26. The table shows the intensities of some common sounds. Find the loudness of each sound.

27. a. By what percent must the intensity of a sound increase in order for the loudness to be raised by 1 dB?

   b. **Writing** Explain why the decibel scale is a logarithmic scale.

28. a. Suppose the intensity of sound from a stereo speaker is doubled. By how many decibels does the loudness increase?

   b. **Challenge** Suppose the intensity of sound from a speaker is increased by a factor of n. By how many decibels does the loudness increase? (Your answer will involve n.)

29. **SAT/ACT Preview** Which of the following does $2 \log_3 6 - \log_3 4$ equal?

   A. 2   **B.** 3   **C.** 9   **D.** 48   **E.** none of these

30. **Technology** Graph $y = \log x^2$ and $y = 2 \log x$ using a graphing calculator or graphing software. Are the graphs of the two functions the same? If not, why not? Do the graphs contradict the power property of logarithms? Explain.

31. Antonio proved the product property of logarithms as shown. Justify each step of Antonio's proof.

32. Prove the power property: $\log_b M^k = k \log_b M$. (*Hint:* Let $x = \log_b M$. Then $M = b^x$, so that $\log_b M^k = \log_b (b^x)^k$.)

33. Use the product and power properties to prove the quotient property:
$$\log_b \frac{M}{N} = \log_b M - \log_b N.$$
$\left(\textit{Hint:} \text{ Write } \dfrac{M}{N} \text{ as } MN^{-1}.\right)$

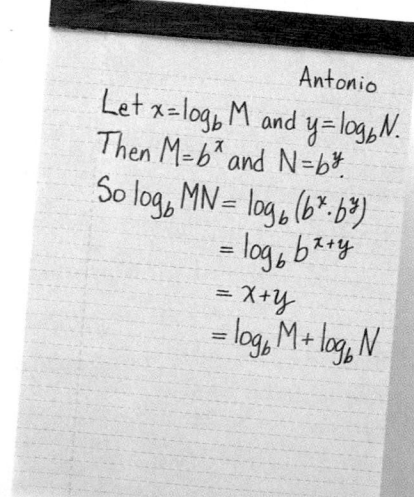

Antonio

Let $x = \log_b M$ and $y = \log_b N$.
Then $M = b^x$ and $N = b^y$.
So $\log_b MN = \log_b (b^x \cdot b^y)$
$= \log_b b^{x+y}$
$= x + y$
$= \log_b M + \log_b N$

34. Writing An encyclopedia gives the equation relating the energy and Richter magnitude of an earthquake in this form:

$$\log E = 11.8 + 1.5M$$

Show that this equation is equivalent to the equation in Example 3. (*Hint:* Note that $11.8 = \log 10^{11.8}$.)

35. Write $4^2 = 16$ in logarithmic form. *(Section 4.2)*

36. Write $\log 1000 = 3$ in exponential form. *(Section 4.2)*

37. **Technology** Use a graphing calculator or graphing software to find the solution of $30e^{0.1x} = 75$ to the nearest hundredth. *(Section 3.4)*

ASSESS YOUR PROGRESS

VOCABULARY

inverse of a function (p. 142)   **common logarithm** (p. 150)
logarithmic function (p. 149)   **natural logarithm** (p. 150)
base-b logarithm (p. 149)

For each function:

a. Graph the function and its inverse in the same coordinate plane.

b. Find an equation for the inverse. *(Section 4.1)*

  1. $f(x) = 2x$ **2.** $g(x) = 3x + 2$ **3.** $h(x) = \dfrac{1}{4}x - 1$ **4.** $y = -6x - 5$

Evaluate each logarithm. *(Section 4.2)*

  5. $\log_2 4$ **6.** $\log_4 2$ **7.** $\log_3 \dfrac{1}{81}$ **8.** $\ln e^{5/11}$

9. POPULATION The population P of Argentina (in millions) can be modeled by the function $P = 33.9e^{0.0109t}$ where t is the number of years since 1994. *(Section 4.2)*

  a. Find the inverse of the given function.

  b. Predict the year in which Argentina's population will reach 35 million.

10. Write $\log \dfrac{p^3 q^4}{r^5}$ in terms of $\log p$, $\log q$, and $\log r$. *(Section 4.3)*

11. Write $\log_b 18 - 2 \log_b 3 + \log_b 4$ as a logarithm of a single number. *(Section 4.3)*

12. Journal Explain what a logarithmic scale is. Give two examples of a logarithmic scale used in real life.

4.4 Exponential and Logarithmic Equations

Learn how to...

- **solve exponential and logarithmic equations**

So you can...

- **find how long it takes fudge to cool, for example**

The notecard below shows a recipe for fudge. Notice that before adding powdered milk and vanilla, you have to let the fudge mixture cool.

> FUDGE
> Melt ¼ cup butter in large saucepan over medium heat. Add 1½ cups boiling water. Mix in 3 cups sugar, ⅔ cup cocoa, and ⅛ teaspoon cream of tartar. Stir until mixture boils. Cover 3 min. Uncover, lower heat, and cook until mixture reaches 236°F. Remove from heat. After mixture has cooled to 110°F, add 6 tablespoons powdered milk and 1 teaspoon vanilla. Beat until creamy. Pour into 8 in. X 8 in. buttered pan.

EXAMPLE 1   Application: Cooking

The equation $T = 164e^{-0.041t} + 72$ gives the temperature T of a fudge mixture t minutes after it begins cooling from 236°F. How long does it take for the mixture to cool to 110°F?

SOLUTION

Find t when $T = 110$.

$$164e^{-0.041t} + 72 = 110$$

> Substitute **110** for T in the given equation.

$$164e^{-0.041t} = 38$$

$$e^{-0.041t} \approx 0.232$$

$$-0.041t \approx \ln 0.232$$

> Write the equation in logarithmic form.

$$t \approx \frac{\ln 0.232}{-0.041}$$

$$t \approx 35.6$$

It takes about 36 min for the mixture to cool from 236°F to 110°F.

THINK AND COMMUNICATE

1. **Technology** Use a graphing calculator or graphing software to graph $y = 164e^{-0.041x} + 72$.

   a. Check the solution of Example 1 by using the trace or intersection feature to find x when $y = 110$.

   b. To what temperature will the fudge mixture eventually cool? How do you know?

EXAMPLE 2

Solve $3^x = 20$.

SOLUTION

$$3^x = 20$$
$$\log 3^x = \log 20$$
$$x \log 3 = \log 20$$

Take the common logarithm of both sides.

Use the power property.

$$x = \frac{\log 20}{\log 3}$$
$$x \approx 2.727$$

To the nearest hundredth, the solution is 2.73.

THINK AND COMMUNICATE

2. Use natural logarithms to solve the equation in Example 2.

3. Compare the steps you would use to find decimal solutions of $e^x = 5$ and $2^x = 5$. Which equation is easier to solve? Why?

In Example 2, the solution of $3^x = 20$ was expressed as a quotient of common logarithms:

$$x = \frac{\log 20}{\log 3}$$

You can also express the solution as a single base-3 logarithm by writing $3^x = 20$ in logarithmic form:

$$x = \log_3 20$$

So $\log_3 20 = \dfrac{\log 20}{\log 3}$. This illustrates the *change-of-base formula*.

Change-of-Base Formula

Let M, b, and c be positive numbers with $b \neq 1$ and $c \neq 1$. Then:

$$\log_b M = \frac{\log_c M}{\log_c b}$$

You can use the change-of-base formula and a graphing calculator to evaluate any logarithm or graph any logarithmic function.

To evaluate **log₂ 6**, calculate the quotient
$$\frac{\log 6}{\log 2} \text{ or } \frac{\ln 6}{\ln 2}.$$

To graph **y = log₂ x**, enter the equation
$$y = \frac{\log x}{\log 2} \text{ or } y = \frac{\ln x}{\ln 2}.$$

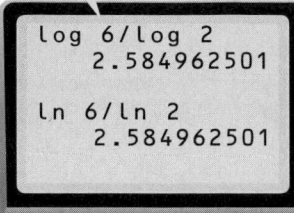

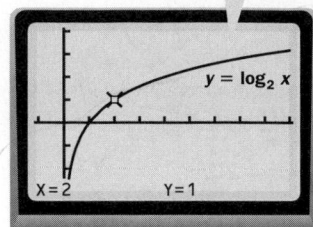

EXAMPLE 3

Solve $\log_2 (5x + 4) - \log_2 (x - 1) = 3$.

SOLUTION

Method 1

Use a graphing calculator or graphing software.

Graph $y = \log_2 (5x + 4) - \log_2 (x - 1)$ and $y = 3$ in the same coordinate plane. Use the change-of-base formula to express the first equation in terms of common logarithms when entering it into the calculator or software:

$$y = \frac{\log (5x + 4)}{\log 2} - \frac{\log (x - 1)}{\log 2}$$

The solution is 4.

Find the *x*-coordinate of the point where the graphs intersect.

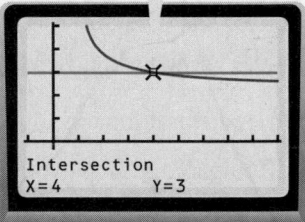

Method 2

Use algebra.

$$\log_2 (5x + 4) - \log_2 (x - 1) = 3$$

Use the quotient property.

$$\log_2 \frac{5x + 4}{x - 1} = 3$$

Write the equation in exponential form.

$$\frac{5x + 4}{x - 1} = 2^3$$

Multiply both sides by $x - 1$.

$$5x + 4 = 8(x - 1)$$
$$5x + 4 = 8x - 8$$
$$-3x + 4 = -8$$
$$-3x = -12$$
$$x = 4$$

The solution is 4.

Check
$$\log_2 (5 \cdot 4 + 4) - \log_2 (4 - 1) \stackrel{?}{=} 3$$
$$\log_2 24 - \log_2 3 \stackrel{?}{=} 3$$
$$\log_2 \frac{24}{3} \stackrel{?}{=} 3$$
$$\log_2 8 \stackrel{?}{=} 3$$
$$3 = 3 ✔$$

When solving a logarithmic equation algebraically, it is possible to obtain a solution that does *not* satisfy the original equation. Such a solution is called an **extraneous solution**. Be sure to check your solutions to eliminate any that are extraneous.

THINK AND COMMUNICATE

4. Use Kim's work.

a. Substitute Kim's solution into the original equation. Explain why the solution is extraneous.

b. 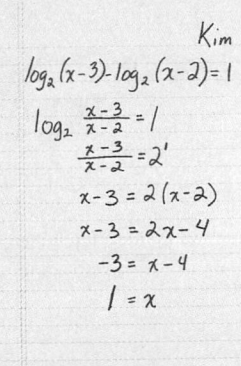 **Technology** Use a graphing calculator or graphing software to graph $y = \log_2 (x - 3) - \log_2 (x - 2)$ and $y = 1$. How do the graphs show that Kim's equation has no solution?

Kim

$\log_2 (x-3) - \log_2 (x-2) = 1$

$\log_2 \frac{x-3}{x-2} = 1$

$\frac{x-3}{x-2} = 2^1$

$x - 3 = 2(x-2)$

$x - 3 = 2x - 4$

$-3 = x - 4$

$1 = x$

☑ CHECKING KEY CONCEPTS

Solve each equation. Round your answers to the nearest hundredth.

1. $e^x = 3$ **2.** $2^y + 7 = 60$ **3.** $5^t = 4$ **4.** $9 \cdot 10^{-0.2x} = 2$

Evaluate each logarithm. Round your answers to the nearest hundredth.

5. $\log_2 5$ **6.** $\log_6 4$ **7.** $\log_8 14.2$ **8.** $\log_3 \frac{7}{5}$

Solve each equation. Be sure to check your solutions.

9. $\log_2 x = 5$ **10.** $\log_4 (11x + 20) = 3$

11. $\log_2 u - \log_2 (u - 6) = 2$ **12.** $\log_3 (5y + 2) - \log_3 (y - 4) = 1$

4.4 Exercises and Applications

Extra Practice exercises on page 754

Solve each equation. Round your answers to the nearest hundredth.

1. $e^x = 7$ **2.** $2e^{-0.3t} = 1.6$ **3.** $10^{4y} = 5$ **4.** $3^x = 8$

5. $\left(\frac{1}{2}\right)^x = \frac{3}{4}$ **6.** $6e^k - 13 = 29$ **7.** $1 + 8(0.43)^{2u} = 3$ **8.** $4 \cdot 7^w = 3 \cdot 2^w$

9. Open-ended Problem Write an exponential equation. Exchange equations with a classmate, and solve your classmate's equation.

10. Look back at Example 1. How long does it take for the fudge mixture to cool to 150°F?

11. Challenge Find the solution of $ab^x = cd^x$ in terms of a, b, c, and d.

Evaluate each logarithm. Round your answers to the nearest hundredth.

12. $\log_2 9$ **13.** $\log_9 2$ **14.** $\log_5 7$ **15.** $\log_{16} 75$

16. $\log_7 \frac{1}{8}$ **17.** $\log_{1/3} 6$ **18.** $\log_4 0.93$ **19.** $\log_{0.1} 0.4$

If X-rays of a fixed wavelength strike a material x cm thick, then the intensity $I(x)$ of the X-rays transmitted through the material is given by $I(x) = I_0 e^{-\mu x}$ where I_0 is the initial intensity and μ is a number that depends on the type of material and the wavelength of the X-rays.

Material	Value of μ
Aluminum	0.43
Copper	3.2
Lead	43

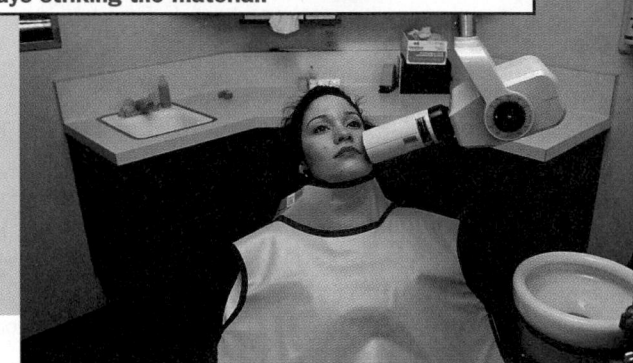

x cm

$I(x) = I_0 e^{-\mu x}$

I_0 is the initial intensity of the X-rays striking the material.

20. The table shows values of μ for various materials. These μ-values apply to X-rays of medium wavelength.

   a. Find the thickness of aluminum shielding that reduces the intensity of X-rays to 30% of their initial intensity. (*Hint:* Find the value of x for which $I(x) = 0.3I_0$.)

   b. Repeat part (a) for copper shielding.

   c. Repeat part (a) for lead shielding.

21. Writing Your dentist puts a lead apron on you before taking X-rays of your teeth to protect you from harmful radiation. Based on your results from Exercise 20, explain why lead is a better material to use than aluminum or copper.

Solve each equation. Be sure to check your solutions.

22. $\log_3 x = -2$

23. $\ln \dfrac{x}{5} = 4$

24. $\log_6 (7t + 13) = 1$

25. $3 \log_8 (u - 1) = 2$

26. $\log_2 (x - 1) - \log_2 (x - 8) = 3$

27. $\log_4 (3y - 4) = -1 + \log_4 (y + 6)$

28. $\log_5 (9x + 2) - \log_5 (3x + 8) = 2$

29. $\log (\log w) = 0$

30. PERSONAL FINANCE When you take out a loan to buy a house, you must repay a portion of the loan with interest each month. The amount $A(n)$ you still owe after making n monthly payments is given by

$$A(n) = \left(A_0 - \frac{P}{r}\right)(1 + r)^n + \frac{P}{r}$$

where A_0 is the original amount of the loan, P is the monthly payment, and r is the monthly interest rate expressed as a decimal. (The monthly rate is the annual rate divided by 12.)

   a. Kachina Tome borrows $140,000 at 12% annual interest to buy a house. She plans to pay back the loan over 30 years, or 360 months. Calculate Kachina's monthly payment. (*Hint:* Use the fact that $A(360) = 0$.)

   b. Write an equation giving the amount Kachina still owes after making n monthly payments.

   c. After how many months will Kachina have paid back half the money she borrowed? Does your answer surprise you?

31. **Visual Thinking** For parts (a)–(c), use a graphing calculator or graphing software to graph the two equations in the same coordinate plane.

   a. $y = \log_2 x$
   $y = \log_{1/2} x$

   b. $y = \log_3 x$
   $y = \log_{1/3} x$

   c. $y = \log_4 x$
   $y = \log_{1/4} x$

   d. How are the graphs of $y = \log_b x$ and $y = \log_{1/b} x$ related?

   e. Write an equation expressing $\log_{1/b} x$ in terms of $\log_b x$.

ONGOING ASSESSMENT

32. **Cooperative Learning** The equation $T = 164e^{-0.041t} + 72$ in Example 1 is a special instance of *Newton's law of cooling*. This law states that the temperature T of a heated substance t minutes after it begins cooling is given by this equation:

$$T = (T_0 - T_R)e^{-kt} + T_R$$

T_0 is the substance's initial temperature.

k is a constant.

T_R is room temperature.

In this activity, you will find an equation that models the temperature of cooling water. Work in a group of four students. You need a transparent glass, an outdoor thermometer, and a watch. Follow these steps:

Step 1 Record the room temperature T_R from the thermometer.

Step 2 Fill the glass with hot tap water, and measure the water's initial temperature T_0.

Step 3 Wait 10 min and measure the water's new temperature T_{10}.

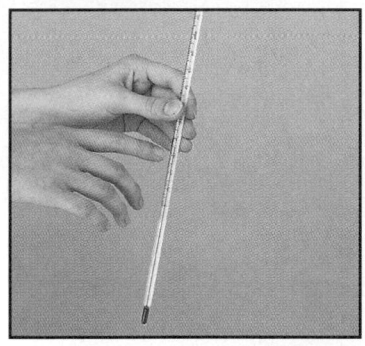

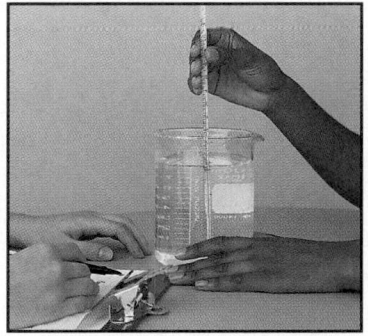

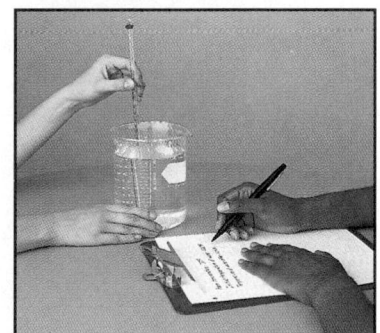

   a. Substitute your values of T_0 and T_R into the general cooling equation above. Then use the fact that $T = T_{10}$ when $t = 10$ to find k.

   b. Use your values of T_0, T_R, and k to write an equation giving the water's temperature T after t minutes.

   c. Use your equation from part (b) to predict the temperature of the water 20 min after you measured the initial temperature T_0. Check your prediction by actually measuring this temperature with the thermometer. Compare the predicted and actual temperatures.

SPIRAL REVIEW

Write as a logarithm of a single number or expression. *(Section 4.3)*

33. $2 \log_a 3 + \dfrac{1}{2} \log_a 49$

34. $6 \ln u + \ln v - 8 \ln w$

Graph each function. *(Section 3.3)*

35. $y = 3^x$

36. $y = 3^{-x}$

37. $y = -3^x$

4.5 Using Logarithms to Model Data

Learn how to...

• use logarithms to find exponential and power functions that model data

So you can...

• make predictions about real-life quantities, such as animal populations and weights

If you travel to Izumi, Japan, during the winter, you'll see thousands of cranes in the wetlands near the ocean. This wasn't always the case, however. There were fewer than 300 cranes in Izumi at the end of World War II. The table and scatter plot show how the number of cranes increased from 1945 to 1990.

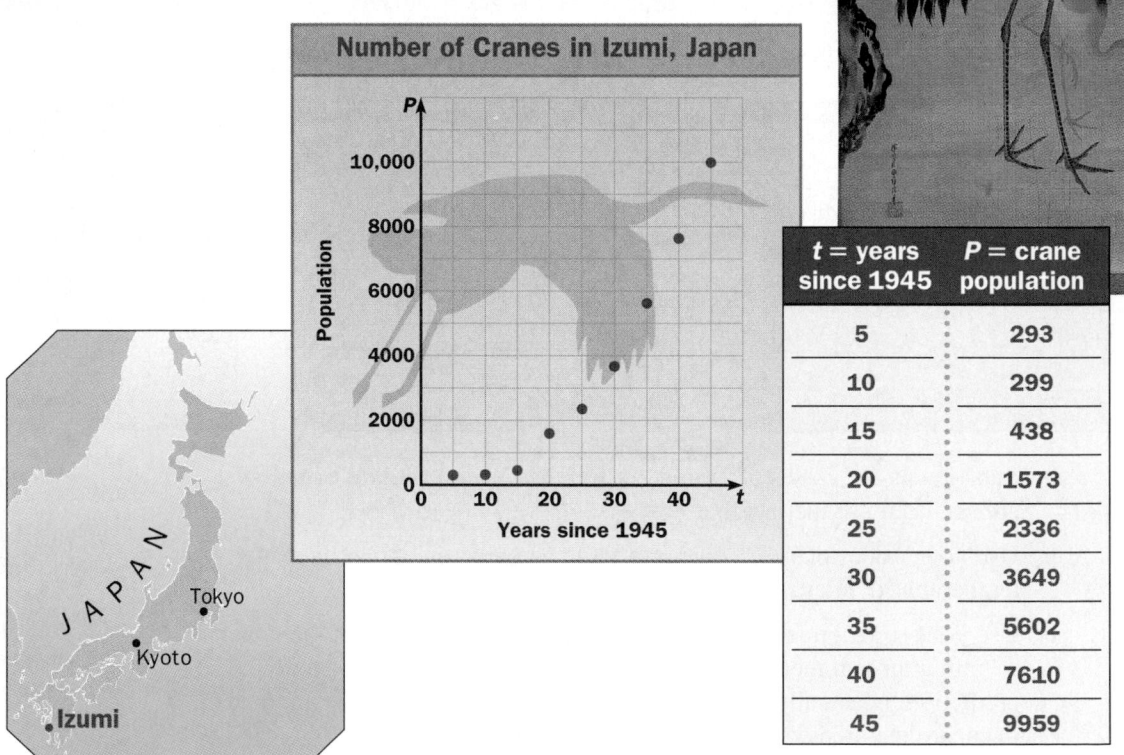

Number of Cranes in Izumi, Japan

Population vs. Years since 1945

t = years since 1945	P = crane population
5	293
10	299
15	438
20	1573
25	2336
30	3649
35	5602
40	7610
45	9959

THINK AND COMMUNICATE

1. Look at the scatter plot. Do you think a linear function is a good model for the crane data? Why or why not?

2. What type of function do you think best models the crane data? Explain your reasoning.

You can use logarithms to find a function that models the crane data.

Use the table on page 168.

a. Make a scatter plot of the data pairs $(t, \log P)$. What relationship exists between t and $\log P$?

b. Find an equation giving P as a function of t. What type of function is this equation?

SOLUTION

a. Make a table of data pairs $(t, \log P)$. Plot the pairs in a coordinate plane. The scatter plot suggests that there is a linear relationship between t and $\log P$.

	Crane Data		
	A	**B**	**C**
1	t	P	$\log P$
2	5	293	2.47
3	10	299	2.48
4	15	438	2.64
5	20	1573	3.20
6	25	2336	3.37
7	30	3649	3.56
8	35	5602	3.75
9	40	7610	3.88
10	45	9959	4.00

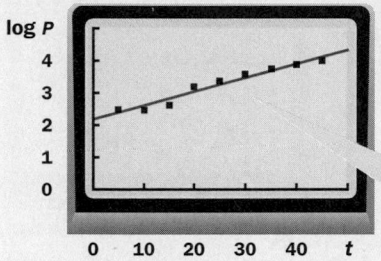

An equation of this fitted line is
$\log P = 0.0431t + 2.18$.

b. Use the equation of the fitted line.

$$\log P = 0.0431t + 2.18$$

$$P = 10^{0.0431t + 2.18}$$

Write the equation in exponential form.

$$P = (10^{2.18})(10^{0.0431t})$$

$$P = (10^{2.18})(10^{0.0431})^t$$

$$P = 151(1.10)^t$$

An equation is $P = 151(1.10)^t$. This equation is an exponential function.

THINK AND COMMUNICATE

3.   **Technology**   Use a graphing calculator or graphing software to graph the data pairs (t, P) on page 168 and the function $P = 151(1.10)^t$ in the same coordinate plane. Do you think the given function is a good model for the crane data? Explain.

4. Predict the number of cranes in Izumi in 1998.

5. Based on the exponential model in Example 1, by about what percent did the Izumi crane population increase each year from 1945 to 1990?

EXAMPLE 2 Application: Biology

The table shows how the weight of a chicken embryo inside an egg changes over time.

a. Make a scatter plot of the data pairs $(\log d, \log W)$. What relationship exists between $\log d$ and $\log W$?

b. Find an equation giving W as a function of d.

d = days after egg is laid	W = weight of embryo (grams)
1	0.0002
4	0.05
8	1.15
12	5.07
16	15.98
20	30.21

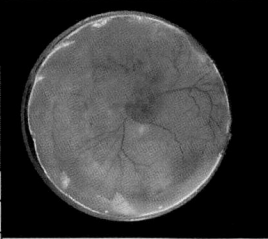

Chicken embryo after 4 days of incubation

SOLUTION

a. Make a table of data pairs $(\log d, \log W)$. Plot the pairs in a coordinate plane. The scatter plot suggests that there is a linear relationship between $\log d$ and $\log W$.

Embryo Data

	A	B	C	D
	d	W	$\log d$	$\log W$
1				
2	1	0.0002	0.00	−3.70
3	4	0.05	0.60	−1.30
4	8	1.15	0.90	0.06
5	12	5.07	1.08	0.71
6	16	15.98	1.20	1.20
7	20	30.21	1.30	1.48

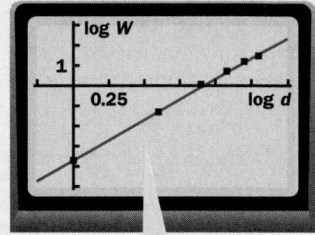

An equation of this fitted line is
$\log W = 4.04 \log d - 3.69$.

b. Use the equation of the fitted line.

$$\log W = 4.04 \log d - 3.69$$

$$\log W = \log d^{4.04} - 3.69 \qquad \text{Use the power property.}$$

$$\log W - \log d^{4.04} = -3.69$$

$$\log \frac{W}{d^{4.04}} = -3.69 \qquad \text{Use the quotient property.}$$

$$\frac{W}{d^{4.04}} = 10^{-3.69} \qquad \text{Write the equation in exponential form.}$$

$$W = 0.000204 d^{4.04}$$

An equation is $W = 0.000204 d^{4.04}$.

The function $W = 0.000204 d^{4.04}$ in Example 2 is called a *power function*. A **power function** has the form $y = ax^b$ where a and b are constants.

THINK AND COMMUNICATE

6. A chick hatches 21 days after the egg is laid. Estimate the weight of a chick when it hatches.

7. Tell whether each function is a power function.

   a. $y = 0.3(1.6)^x$ **b.** $y = 0.3x^{1.6}$ **c.** $y = \dfrac{0.3}{x + 1.6}$ **d.** $y = 5x^{-2}$

Using Exponential and Power Functions to Model Data

1. You can use an **exponential function** to model a set of data pairs (x, y) if there is a **linear relationship** between x and $\log y$.

2. You can use a power function to model a set of data pairs (x, y) if there is a **linear relationship** between $\log x$ and $\log y$.

☑ CHECKING KEY CONCEPTS

For each table:

a. Make a scatter plot of the data pairs (x, $\log y$).

b. Make a scatter plot of the data pairs ($\log x$, $\log y$).

c. Tell whether an *exponential function* or a *power function* is a better model for the data. Find an equation giving y as a function of x.

1.

x	y
1	2.4
2	2.8
3	3.3
4	3.9
5	4.7
6	5.5

2.

x	y
5	6.3
10	15.7
15	26.7
20	39.0
25	52.2
30	66.3

3.

x	y
2	1.25
4	1.05
6	0.94
8	0.87
10	0.82
12	0.79

4.5 Exercises and Applications

Extra Practice exercises on page 754

1. Writing Describe how you can use logarithms to determine if an exponential function is a good model for a set of data pairs (x, y). Describe how you can determine if a power function is a good model.

Write y as a function of x.

2. $\log y = 0.06x + 1.3$ **3.** $\log y = 0.2x + 3.8$ **4.** $\log y = -0.35 + 0.45x$

5. $\log y = -0.04x - 0.091$ **6.** $\ln y = x + 2$ **7.** $\ln y = 0.83 - 3.27x$

4.5 Using Logarithms to Model Data **171**

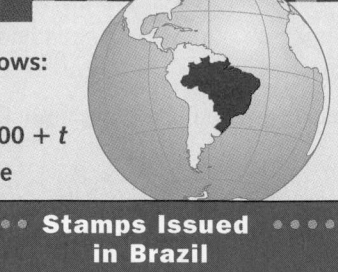

Use the table. The variables in the table are defined as follows:

t = number of years since 1900

N = total varieties of stamps issued through year 1900 + t

D = varieties of stamps issued in the previous decade

8. a. Make a scatter plot of the data pairs $(t, \log N)$. What relationship exists between t and $\log N$?

b. Fit a line to your scatter plot. Find an equation of your fitted line.

c. Find an equation giving N as an exponential function of t.

d. Predict the number of different stamps Brazil will have issued by the year 2010.

9. a. Copy and complete the third column of the table. The first entry has been done for you.

b. **Challenge** Show that D is an exponential function of t *without* using logarithms. (*Hint:* Use the equation from part (c) of Exercise 8 and the fact that $D(t) = N(t) - N(t - 10)$.)

c. Make a scatter plot of the data pairs $(t, \log D)$. Use your scatter plot to find an equation giving D as an exponential function of t. Compare this function with the one you found in part (b).

Stamps Issued in Brazil		
t	N	D
0	145	—
10	173	28
20	216	?
30	299	?
40	477	?
50	658	?
60	868	?
70	1125	?
80	1668	?
90	2186	?

Use the numbers in the second column to find this value:

$173 - 145 = 28$

Write y as a function of x.

10. $\log y = 0.4 \log x + 0.8$ **11.** $\log y = 0.72 \log x + 2.34$ **12.** $\log y = -0.3 + 4.85 \log x$

13. $\log y = -0.61 \log x - 0.039$ **14.** $\ln y = 3.7 \ln x + 3.7$ **15.** $\ln y = 1.48 - \ln x$

16. **Technology** Some graphing calculators and statistical software will find an exponential or power function for a set of data pairs (x, y). They may also give a correlation coefficient r that tells you how well a line fits the data points $(x, \log y)$ for exponential functions or $(\log x, \log y)$ for power functions. The closer $|r|$ is to 1, the better the fit is. The screen shows the exponential function and correlation coefficient given by one calculator for the crane data on page 168.

a. Use a graphing calculator or statistical software to find a power function that models the crane data. What is the correlation coefficient?

b. Compare the correlation coefficients for the exponential and power functions. Based on these coefficients, which type of function is the better model?

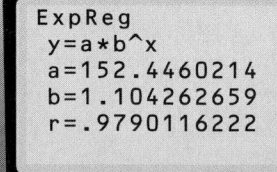

```
ExpReg
y=a*b^x
a=152.4460214
b=1.104262659
r=.9790116222
```

17. Use the table of chicken embryo data in Example 2.

a. Make a scatter plot of the data pairs (d, W). Is it clear from the scatter plot that a power function is a better model for the data than an exponential function? Explain.

b. Make a scatter plot of the data pairs $(d, \log W)$. Fit a line to your scatter plot.

c. **Writing** Compare the scatter plot from part (b) with the scatter plot of the data pairs $(\log d, \log W)$ in Example 2. Explain why these scatter plots show that a power function is a better model for the chicken embryo data than an exponential function.

18. **Spreadsheets** You can use a spreadsheet to compare the rates at which different types of functions increase.

a. Consider the linear function $y = 2x$, the exponential function $y = 2^x$, the logarithmic function $y = \log_2 x$, and the power function $y = x^2$. Use a spreadsheet to find the values of these functions for $x = 1, 2, 3, \ldots, 12$. Rank the functions in order from fastest to slowest rate of increase.

	A	B	C	D	E
	x	y = 2*x	y = 2^x	y = log(x)/log(2)	y = x^2
1					
2	1	?	?	?	?
3	2	?	?	?	?
4	3	?	?	?	?
5	4	?	?	?	?
6	5	?	?	?	?
7	6	?	?	?	?
8	7	?	?	?	?
9	8	?	?	?	?
10	9	?	?	?	?
11	10	?	?	?	?
12	11	?	?	?	?
13	12	?	?	?	?

Comparing Functions

b. Repeat part (a) several times using functions of the form $y = bx$, $y = b^x$, $y = \log_b x$, and $y = x^b$ where b is a number greater than 2.

c. **Writing** Write a paragraph summarizing your results from parts (a) and (b). Your paragraph should include a general comparison of the rates of increase of linear, exponential, logarithmic, and power functions.

19. **Cooperative Learning** Work with a partner. One person should choose an exponential function f to generate a table of 10 ordered pairs $(x, f(x))$. The other person should use logarithms to "discover" an equation for f. Reverse roles with your partner, and repeat the activity using a power function.

20. **Research** Choose a two-column table of data from a magazine, almanac, or other source. Find an exponential function and a power function that model the data. Decide which function is the better model.

21. **SAT/ACT Preview** Suppose $\log y = 2 \log x + 1$. Which equation gives y as a function of x?

A. $y = 2x + 1$ **B.** $y = x^2 + 1$ **C.** $y = 10^{2x+1}$ **D.** $y = 10x^2$ **E.** none of these

How can you estimate the weight of dinosaurs from their skeletons? One way is to look for relationships between weight and skeletal structures in living animals. You can then apply these relationships to dinosaurs.

Animal	C (mm)	W (kg)
Meadow mouse	5.5	0.047
Gray squirrel	13	0.399
Porcupine	34	7.20
Yellow baboon	57	28.6
Lion	93.5	143
Polar bear	135	448
Giraffe	173	710

22. The *femur* is a large bone located in the leg or hind limb of a human or animal. The first table shows the average circumference C of the femur for several species of animals. It also shows each animal's average weight W.

   a. Make a scatter plot of the data pairs ($\log C$, $\log W$). What relationship exists between $\log C$ and $\log W$?

   b. Fit a line to your scatter plot. Find an equation of your fitted line.

   c. Express W as a function of C. What type of function describes the relationship between femoral circumference and weight?

23. Scientists obtained femoral circumferences for the dinosaurs in the second table from fossil remains. Use the function from part (c) of Exercise 22 to estimate the weight of each dinosaur.

Dinosaur	C (mm)	W (kg)
Parksosaurus	103	?
Struthiomimus	136	?
Allosaurus	348	?
Tyrannosaurus	534	?

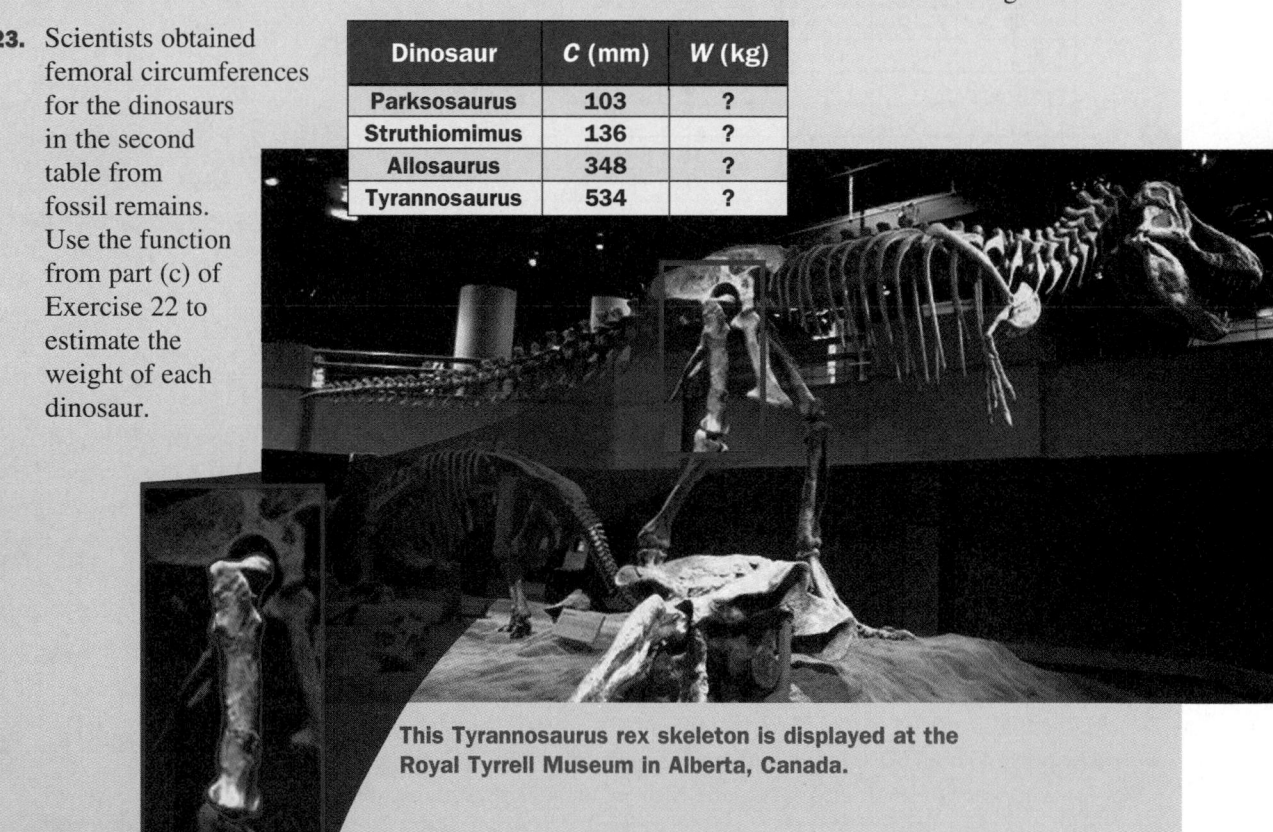

This Tyrannosaurus rex skeleton is displayed at the Royal Tyrrell Museum in Alberta, Canada.

24. Challenge Some scientists have estimated dinosaur weights by assuming that weight is proportional to the cross-sectional area of the femur. Is this assumption true for the animals in the first table? Give a mathematical justification for your answer. (*Hint:* Assume that the cross section of a femur is circular.)

25. **Open-ended Problem** The table shows the number of African-American elected officials in the United States for various years. Find at least two models for the data. Discuss how well each of your models fits the data. Use your models to predict the number of African-American elected officials for future years.

Year	1982	1984	1986	1988	1990	1992
Number of officials	5115	5654	6384	6793	7335	7517

SPIRAL REVIEW

Evaluate each logarithm. Round your answers to the nearest hundredth.
(Section 4.4)

26. $\log_3 12$ 27. $\log_8 7$ 28. $\log_5 0.36$ 29. $\log_2 \dfrac{5}{9}$

30. **Writing** Explain the relationship between the slope of the least-squares line and the sign of the correlation coefficient for a set of data points (x, y).
(Section 2.5)

Evaluate each expression when $x = 3$ and $x = -2$. *(Toolbox, page 780)*

31. x^2 32. $5x^2$ 33. $-4x^2$ 34. $\dfrac{1}{2}x^2$

ASSESS YOUR PROGRESS

VOCABULARY

extraneous solution (p. 165)   **power function** (p. 170)

Solve each equation. Round your answers to the nearest hundredth. *(Section 4.4)*

1. $e^x = 6$ 2. $4^t = 23$ 3. $10(1.09)^{2k} + 7 = 51$

Solve each equation. Be sure to check your solutions. *(Section 4.4)*

4. $\log_4 x = 3$ 5. $\log_2 (w - 9) - \log_2 (w - 2) = 3$

6. **BIOLOGY** The table shows the average weight W (in grams) of several rats t days after birth for $t = 1, 2, \ldots, 6$. *(Section 4.5)*

   a. Find an exponential function that models the data.

   b. Find a power function that models the data.

   c. Which of the functions from parts (a) and (b) is the better model? Explain your reasoning.

t	W
1	5.64
2	6.21
3	7.07
4	8.37
5	9.74
6	10.7

7. **Journal** Discuss how logarithms can be used to model data.

PORTFOLIO PROJECT

Investigating Zipf's Law

Bumba
population: 75,156
rank: 19
CONGO
Zaire
BASIN
Bukavu
population: 158,920
rank: 11
Kinshasa
population: 2,222,981
rank: 1
Kananga
population: 460,091
rank: 3
Lubumbashi
population: 596,297
rank: 2
MITUMBA MTNS
Lubumbashi
Kinshasa
log R
1.6
1.2
0.8
0.4
O
4.8
5.0
5.2
5.4
5.6
5.8
6.0
6.2
6.4
log P

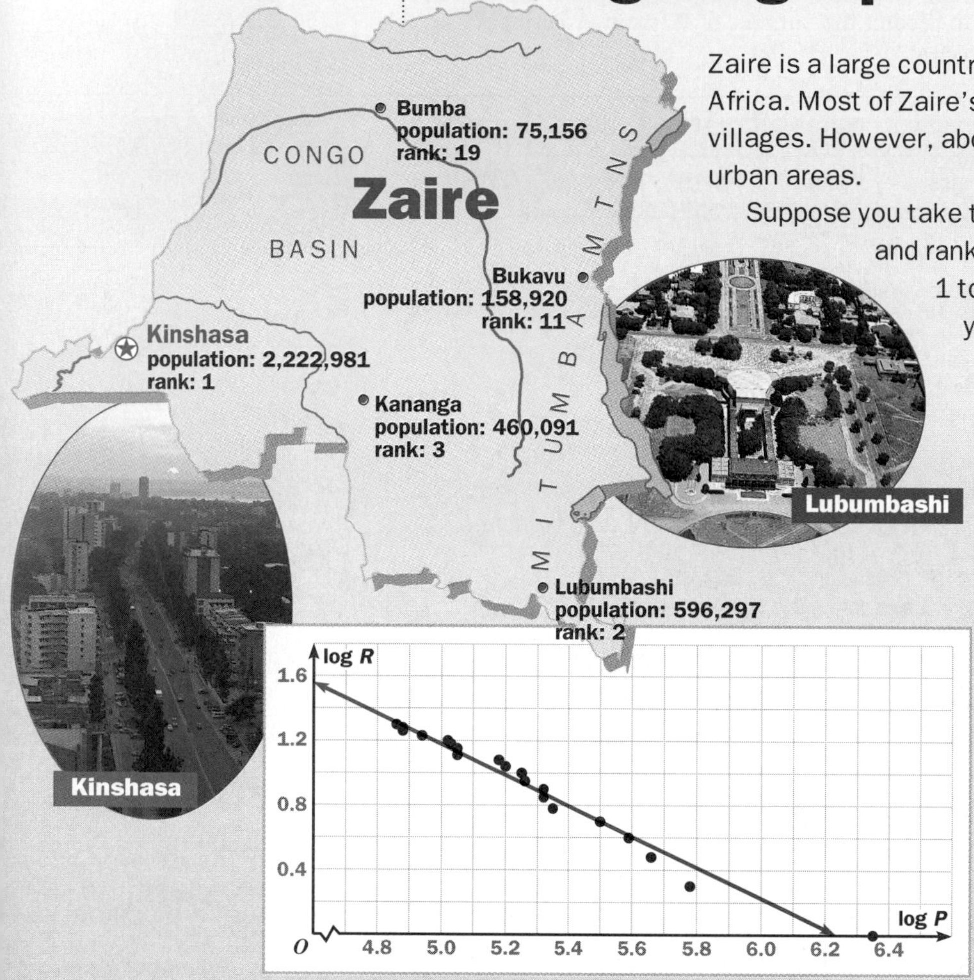

Zaire is a large country located in the middle of Africa. Most of Zaire's people live in small rural villages. However, about 40% of the people live in urban areas.

Suppose you take the 20 largest cities in Zaire and rank them by population from 1 to 20 using data from 1992. If you plot the logarithm of each city's rank R against the logarithm of its population P, you obtain the graph shown.

Since the data points ($\log P$, $\log R$) lie approximately on a line, you can use a power function to model the relationship between P and R. One good model is given by:

$$R = 871{,}000P^{-0.952}$$

This function illustrates *Zipf's law*, which says that rank is a power function of population for the cities of any country.

PROJECT GOAL **Your goal is to use Zipf's law to model the relationship between population and rank for cities in a country.**

Collecting the Data

The Twenty Largest Cities in Zaire
City | Population | Rank
Kinshasa 2,222,981 1
Lubumbashi 596,297 2
Kananga 460,091 3
Mbuji-Mayi 391,845 4
Kisangani

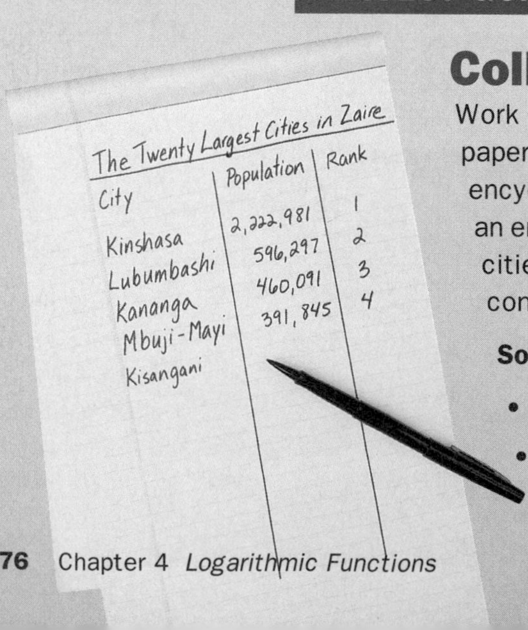

Work in a group of three students. Your group needs graph paper or a graphing calculator and access to a set of encyclopedias. Choose a country that interests you. Use an encyclopedia to make a list of the country's largest cities, their populations, and their ranks. Your list should contain at least 10 cities.

Some possibilities:

- "The Twenty Largest Cities in India"
- "Chinese Cities Having at Least One Million People"

Analyzing the Data

1. MAKE a scatter plot of the data pairs (log *P*, log *R*) for your cities. Fit a line to the plotted points, and find an equation of the line.

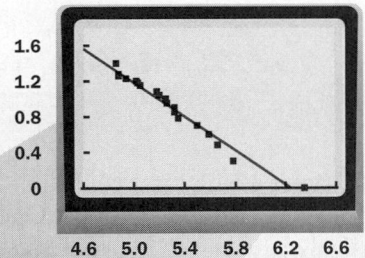

```
LinReg
 y=ax+b
 a=-.9521288837
 b=5.940205715
 r=-.9876633883
```

2. EXPRESS rank as a power function of population, using the equation of your fitted line.

3. PREDICT the rank of some city not used in your scatter plot. How well does the predicted rank match the actual rank?

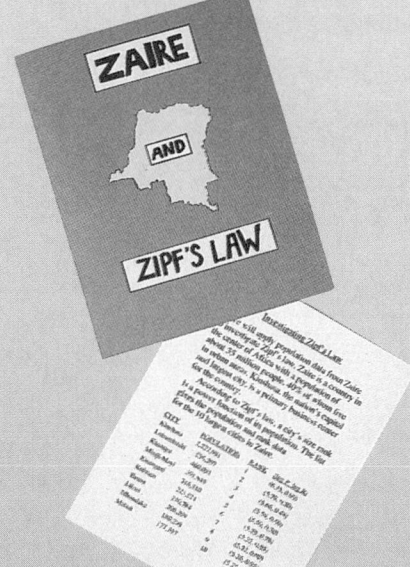

Writing a Report

Write a report summarizing your results. Your report should include:

- a statement of the goal of your project

- a brief description of your chosen country and its people

- a map of your country

- the population and rank data you collected

- a scatter plot and fitted line for the data pairs (log *P*, log *R*)

- an equation giving rank as a power function of population

- an assessment of how well Zipf's law describes the relationship between population and rank for cities in your country

Self-Assessment

In your report, explain how this project increased your understanding of logarithms and their role in mathematical modeling. How well did you work with your partners as a group? What could you do to improve your project?

4 | Review

Compare what you learned in Chapter 4 with what you learned in Chapter 3. Make a list of the important ideas you learned in Chapter 3 and then revisited and extended in Chapter 4. Also list any new ideas discussed in Chapter 4.

VOCABULARY

inverse (p. 142)
logarithmic function (p. 149)
base-*b* logarithm (p. 149)
common logarithm (p. 150)

natural logarithm (p. 150)
extraneous solution (p. 165)
power function (p. 170)

SECTIONS | 4.1 *and* 4.2

Given the graph of a function f, you can find the graph of f^{-1}.

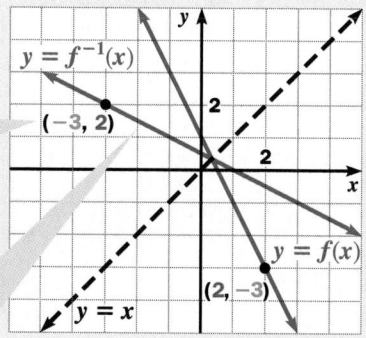

$y = f^{-1}(x)$

$(-3, 2)$

$y = f(x)$

$(2, -3)$

$y = x$

A point (b, a) is on the graph of f^{-1} if and only if (a, b) is on the graph of f.

The graph of f^{-1} is the reflection of the graph of f over the line $y = x$.

You can find the **inverse** of a function $y = f(x)$ by solving for x and switching x and y.

$$f(x) = -2x + 1$$
$$y = -2x + 1$$
$$-\frac{1}{2}y + \frac{1}{2} = x \qquad \text{Solve for } x.$$
$$y = -\frac{1}{2}x + \frac{1}{2} \qquad \text{Switch } x \text{ and } y.$$
$$f^{-1}(x) = -\frac{1}{2}x + \frac{1}{2}$$

The inverse of an exponential function is a **logarithmic function**.

	Exponential form	Logarithmic form	
base 10	$y = 10^x$ $100 = 10^2$	$\log y = x$ $\log 100 = 2$	common log
base e	$y = e^x$ $20 \approx e^3$	$\ln y = x$ $\ln 20 \approx 3$	natural log
base b $(b > 0, b \neq 1)$	$y = b^x$ $81 = 3^4$	$\log_b y = x$ $\log_3 81 = 4$	base-b log

SECTIONS 4.3 *and* 4.4

You can use properties of logarithms to simplify logarithmic expressions and to solve exponential and logarithmic equations.

Let M, N, b, and c be positive numbers with $b \neq 1$ and $c \neq 1$. Then:

Product Property
$\log_b MN = \log_b M + \log_b N$

$$\log 4 + \log 25 = \log 100$$
$$= 2$$

Quotient Property
$\log_b \dfrac{M}{N} = \log_b M - \log_b N$

$$\log_2 x - \log_2 8 = 1$$
$$\log_2 \frac{x}{8} = 1$$
$$\frac{x}{8} = 2^1$$
$$x = 16$$

Power Property
$\log_b M^k = k \log_b M$

$$5^x = 8$$
$$\log 5^x = \log 8$$
$$x \log 5 = \log 8$$
$$x = \frac{\log 8}{\log 5} \approx 1.29$$

Change-of-Base Property
$\log_b M = \dfrac{\log_c M}{\log_c b}$

This property is often used to change base-b logarithms to common or natural logarithms.

$$\log_3 5 = \frac{\log 5}{\log 3} \approx 1.46$$
$$\log_4 7 = \frac{\ln 7}{\ln 4} \approx 1.40$$

SECTION 4.5

You can use exponential functions or **power functions** to make predictions. For example, you can predict the United States per capita personal income I for some number of years t in the future. The table at the right shows the data for six years, and the table below gives an exponential and a power model for the data.

	Exponential Model	Power Model
Equation of the fitted line	$\log I = 0.026t + 1.01$	$\log I = 0.431 \log t + 0.834$
Function	$I = 10.2(1.06)^t$	$I = 6.82t^{0.431}$

t (years since 1980)	I (thousands of dollars)
5	13.9
6	14.6
7	15.6
8	16.6
9	17.7
10	18.6

In general, a set of data pairs (x, y) can be modeled by:

- an exponential function $y = ab^x$ if the points $(x, \log y)$ lie on or close to a line.
- a power function $y = ax^b$ if the points $(\log x, \log y)$ lie on or close to a line.

4 Assessment

VOCABULARY QUESTIONS

For Questions 1 and 2, complete each paragraph.

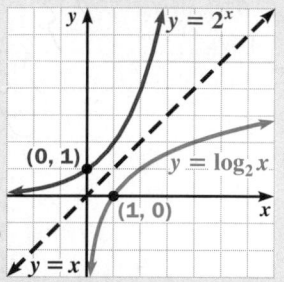

1. The function $y = \log_2 x$ is the __?__ of the function $y = 2^x$. The graph of $y = \log_2 x$ is the __?__ of the graph of $y = 2^x$ over the line $y = x$.

2. The expressions $\log 12$, $\ln 12$, and $\log_5 12$ represent the __?__ logarithm, the __?__ logarithm, and the __?__ logarithm of 12. The __?__ formula can be used to evaluate $\log_5 12$.

SECTIONS 4.1 *and* 4.2

For each function:

a. Graph the function and its inverse in the same coordinate plane.

b. Find an equation for the inverse.

3. $f(x) = -\dfrac{1}{2}x$ 4. $y = -x + 3$ 5. $g(x) = 10^{x/2}$ 6. $y = e^{-x}$

7. **a.** Look at your answers to Questions 3–6. What is interesting about the equation in Question 4?

   b. Writing Explain why the equation in Question 4 has the characteristic you noted in part (a).

8. A 10 in. tall cylindrical candle burns at a rate of 0.25 in./h.

   a. Write an equation giving height as a function of time.

   b. Find the inverse of the function in part (a).

   c. When is the height of the candle 6 in.?

Write each equation in logarithmic form.

9. $2^6 = 64$ 10. $e^{-1} = \dfrac{1}{e}$ 11. $(0.5)^{-3} = 8$

Write each equation in exponential form.

12. $\log_4 2 = 0.5$ 13. $\ln 1 = 0$ 14. $\log_{1/9} 27 = -\dfrac{3}{2}$

Evaluate each logarithm.

15. $\log_{27} 3$ 16. $\ln e^{-2}$ 17. $\ln e^4$ 18. $\log 0.01$

19. If $0 < x < 1$, what can you say about $\log x$? Explain your answer.

20. Without using a calculator, explain why $2 < \log_3 15 < 3$.

Write each expression in terms of $\log_5 p$, $\log_5 q$, and $\log_5 r$.

21. $\log_5 pq^2$ **22.** $\log_5 \dfrac{5}{r}$ **23.** $\log_5 \dfrac{r^{1/4}}{p^3 q}$

Write as a logarithm of a single number or expression.

24. $\dfrac{1}{4} \ln 16$ **25.** $3 \log_a u + 2 \log_a v$ **26.** $3 \log_b 5 - \dfrac{1}{2} \log_b 25$

27. Open-ended Problem Find positive numbers M, N, and b (with $b \neq 1$) to show that, in general, $\log_b MN \neq \log_b M \cdot \log_b N$.

28. CHEMISTRY The formula used to calculate pH is $\text{pH} = -\log [\text{H}^+]$ where $[\text{H}^+]$ is the concentration of hydrogen ions in moles per liter. Find the pH of acid rain, for which $[\text{H}^+] = 3 \times 10^{-5}$ moles/L. Round your answer to the nearest hundredth.

Solve each equation. Round each answer to the nearest hundredth.

29. $4e^x = 7.2$ **30.** $12(0.6)^{2x} = 15$ **31.** $2 \log_4 (t + 3) = 6$

Evaluate each logarithm. Round each answer to the nearest hundredth.

32. $\log_9 42$ **33.** $\log_{1/6} 3$ **34.** $\log_3 0.7$

SECTION 4.5

35. The table shows the median price for single-family homes in the United States.

Years since 1980	Price (thousands)
5	75.5
9	93.1
10	95.5
11	100.3

  a. **Technology** Find an exponential function and a power function that model the data.

  b. Which of the functions from part (a) is the better model? Explain your choice.

  c. Use the model you named in part (b) to predict the median price for single-family homes in the United States in 1992. Compare this with the actual median price of $103,700.

PERFORMANCE TASK

36. Cooperative Learning Work with a partner. One of you should write several functions for exponential growth. The other should write several functions for exponential decay. (See Chapter 3 for ideas.)

    Exchange your functions and use logarithms to find the doubling times for the growth functions or the half-lives for the decay functions. Generalize your results using $y = ab^x$ with $b > 1$ (for growth) and with $0 < b < 1$ (for decay). Compare your generalizations and describe any similarities.

Soon Yi

$y = 50(1.02)^x$

$100 = 50(1.02)^x$

$2 = (1.02)^x$

$\ln 2 = x \ln (1.02)$

$\dfrac{\ln 2}{\ln (1.02)} = x$

$35.0 \approx x$

5 | Quadratic Functions

Engineering a paper car

INTERVIEW **Mark Thomas**

Mark Thomas had not thought about engineering until a high school guidance counselor suggested it would be a good career for someone like him who did well in mathematics and science. "At the time, I didn't have the faintest idea as to what engineering was," Thomas admits. "But I decided to check it out anyway." Now, as a senior project engineer for a large auto manufacturer, Thomas is giving students an early introduction to the field he has grown to love. With the help of his coworkers Vincent Lyons and Elliot Lyons, Thomas has started the Paper Vehicle Project, a program that is offering Detroit-area high school students hands-on experience in engineering. The challenge for the students is to design and build a steerable, movable car made entirely out of paper and capable of supporting at least 300 lb.

> "Never say never, no matter how big the challenge. If you work hard, you can do whatever it takes to make it happen."

Mark Thomas stands with *Predator,* **a paper vehicle constructed by students at Western High School in Detroit, Michigan.**

Teamwork and Communication

"At first, they don't have any idea how to do it," Thomas explains. "But eventually they realize that a big, complex task can be accomplished through teamwork and by breaking it down into smaller, more manageable chunks."

In addition to gaining technical skills, the students learn how to communicate their ideas and collaborate successfully. "That's essential in any field," Thomas says, "whether they go on to become lawyers or janitors. The popularity of the program is obvious," he adds, "given that 90% of our kids come to school for three hours on Saturday mornings to build their cars. That's pretty amazing!"

Students must spend many hours preparing to build a car before they can actually begin. They use CAD (Computer-Aided Design) software to develop their car design.

Basic Engineering

Still, the cars won't run on enthusiasm alone. Students learn enough mathematics, physics, and basic engineering to make their vehicles worth more than the paper they're constructed from. Students compute velocity by measuring how far a car travels in a given time. They also calculate *drag coefficients* for vehicles of different sizes and shapes to see how a car's design affects its performance. A drag coefficient, C_D, is a number that describes the aerodynamics of a car—the lower the number, the more aerodynamic the car.

Air Resistance

Aerodynamics is concerned with a force called *air resistance* that pushes against a car when it moves. This force depends on several variables, all of which are constant for a given car in a given place, except for the speed of the air relative to the car. The formula shows how air resistance is calculated. Air resistance is measured in pounds, frontal area in square feet, and relative speed in miles per hour.

The **frontal area** is determined by the shape of the car.

Relative speed is a combination of the speed of the car and the speed of the air. In still air, the relative speed is just the car's speed.

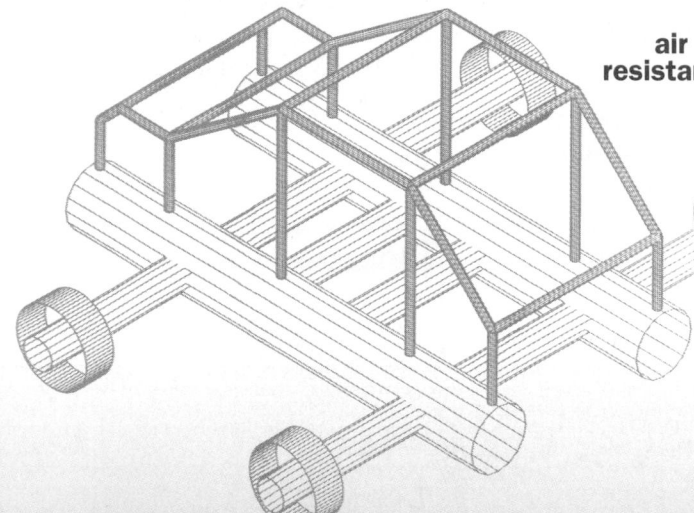

$$\text{air resistance} = 0.00256 \times C_D \times \text{frontal area} \times \left(\text{relative speed} \right)^2$$

Because air resistance depends on the square of the relative speed, the functional relationship between these two variables is *quadratic*. With quadratic functions, small changes in the independent variable can cause large changes in the dependent variable.

EXPLORE AND CONNECT

The Lyons brothers lend a helping hand to Mark Thomas as he completes an inspection of another paper vehicle.

1. Research Engineers work in many different fields and have different specialties. Find out more about one of these specialties (some examples are civil, environmental, electrical, mechanical, and nuclear engineering) and how mathematics is used in the work.

2. Project Gather pictures of old and new cars and trucks. Rank them according to their aerodynamics. Explain your ranking.

3. Writing What do you think car designers do to minimize air resistance? Which parts of the formula for air resistance can they control?

Mathematics & Mark Thomas

In this chapter, you will learn more about air resistance and how the speed of a car affects fuel economy.

Related Exercises

Section 5.1
• **Exercises 22 and 23**

Section 5.5
• **Exercises 32–34**

5.1 Working with Simple Quadratic Functions

Learn how to...

- recognize and draw graphs of quadratic functions

- solve simple quadratic equations

So you can...

- examine the relationship between air resistance and speed when cycling, for example

If you have ever ridden a bicycle, you know that the air pushes against you and can slow you down even when there is no wind. The force that the air exerts on you, called *air resistance*, is given by the formula on the preceding page. (The formula applies to bicycles as well as cars.)

EXAMPLE 1   Application: Bicycling

Shirley Scott is cycling on a day with no wind. She is in a casual, upright position on her bicycle. In this situation, the drag coefficient is 1.1 and the frontal area is 5.5 ft². Write an equation that gives the air resistance R (in pounds) as a function of the bicycle's speed s (in miles per hour). Graph the function.

SOLUTION

Use the formula shown on the preceding page.

$$R = 0.00256(\mathbf{1.1})(\mathbf{5.5})s^2$$
$$= 0.0155s^2$$

Substitute **1.1** for C_D and **5.5** for **frontal area**.

Make a table of values, plot the points, and connect them with a smooth curve.

$R = 0.0155s^2$	
s	R
0	0
4	0.248
8	0.992
12	2.232
16	3.968
20	6.200

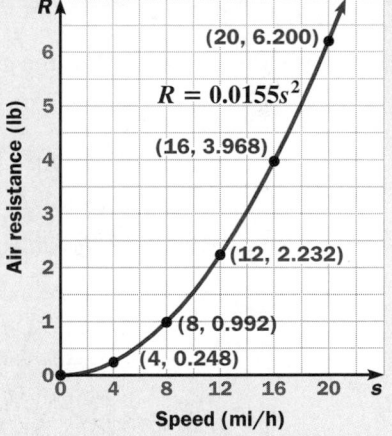

THINK AND COMMUNICATE

1. What happens to the air resistance when Shirley Scott doubles her speed?

2. What would you say is a reasonable domain for the air resistance function? What is the corresponding range?

The air resistance function in Example 1 is an example of a **quadratic function** having the general form $y = ax^2$. The graph of the quadratic function $y = x^2$ with a domain of all real numbers is shown below.

Toolbox p. 797
Reflecting a Figure; Line Symmetry

The graph of a quadratic function is called a **parabola**.

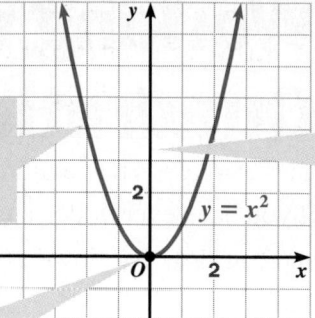

A parabola has a **line of symmetry**. This means that the part of the parabola on one side of the line is a reflection of the part on the other side.

$y = x^2$

The point where the line of symmetry crosses the parabola is called the **vertex**.

THINK AND COMMUNICATE

3. For the graph of the function $y = x^2$, what is the vertex? What is the line of symmetry?

4. What is the range of the function $y = x^2$?

5. The vertex of a parabola is sometimes called the *turning point*. Explain the significance of this name.

6. Why did the graph in Example 1 lack symmetry?

EXAMPLE 2

Use the graph of $y = x^2$ to sketch the graphs of the related functions $y = 2x^2$, $y = \frac{1}{2}x^2$, and $y = -x^2$. (In this case, $y = x^2$ is called a *parent* function.)

SOLUTION

Choose a point such as $(1, 1)$ on the graph of $y = x^2$. Then plot the points with the same x-coordinate on the graphs of the other functions.

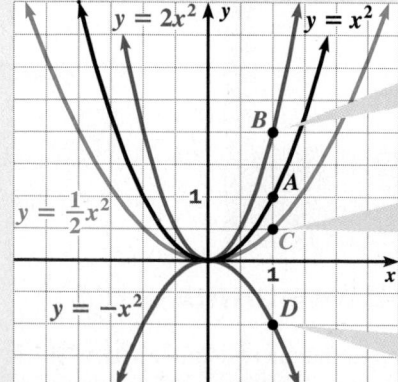

The y-coordinate of $B(1, 2)$ is twice the y-coordinate of $A(1, 1)$.

The y-coordinate of $C\left(1, \frac{1}{2}\right)$ is one half of the y-coordinate of $A(1, 1)$.

The y-coordinate of $D(1, -1)$ is the opposite of the y-coordinate of $A(1, 1)$.

THINK AND COMMUNICATE

7. You can think of the graphs of $y = 2x^2$ and $y = \frac{1}{2}x^2$ as being the result of vertical stretches and vertical shrinks of the graph of $y = x^2$. Explain.

> **Toolbox p. 798**
> *Dilating a Figure*

8. Complete each statement using *vertical stretch* or *vertical shrink*.

   a. When $0 < a < 1$, the graph of $y = ax^2$ is the result of a $\underline{?}$ of the graph of $y = x^2$.

   b. When $a > 1$, the graph of $y = ax^2$ is the result of a $\underline{?}$ of the graph of $y = x^2$.

9. Complete this statement: The graph of $y = -3x^2$ is the *reflection over the x-axis* of the graph of $y = \underline{?}$.

> If $y = 16$, then $x = 4$ or $x = -4$.

You know that for every input x the function $y = x^2$ gives a unique output y. But what happens when you know y and want to find x?

The graph shows that two x-values are paired with each positive y-value. The x-values are called the *square roots* of y.

A number x is a **square root** of a number y if it satisfies the equation $x^2 = y$.

$$\text{If } x^2 = y, \text{ then } x = \begin{cases} 0, \text{ if } y = 0. \\ \pm\sqrt{y}, \text{ if } y > 0. \end{cases}$$

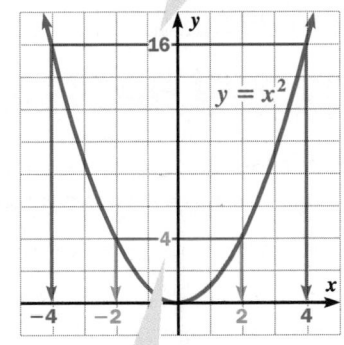

> Read "plus-or-minus square root of *y*."

> If $y = 4$, then $x = 2$ or $x = -2$.

EXAMPLE 3

According to legend, the Italian scientist Galileo Galilei dropped objects of different weights from the top of the Leaning Tower of Pisa, about 180 ft high. The equation $d = 16t^2$ gives the distance d, in feet, an object falls as a function of time t, in seconds. How many seconds would it take an object to fall 180 ft?

SOLUTION

Use $d = 16t^2$ to find t when $d = 180$.

180 ft

$$d = 16t^2$$

Substitute **180** for *d*.

$$180 = 16t^2$$

$$\frac{180}{16} = t^2$$

Divide by 16 to get t^2 alone on one side of the equation.

$$11.25 = t^2$$

$$\pm\sqrt{11.25} = t$$

A calculator gives this decimal approximation for $\sqrt{11.25}$.

$$\pm 3.35 \approx t$$

Since t represents time, use the positive solution. An object would fall 180 ft in about 3.4 s.

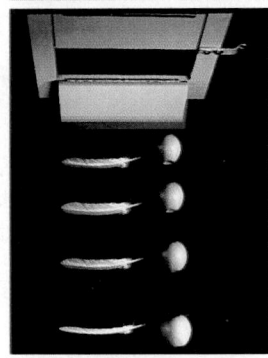

BY THE WAY

The experiment shows that objects of different weights fall the same distance in the same amount of time.

☑ **CHECKING KEY CONCEPTS**

1. Sketch the graph of $y = 3x^2$. Label the parabola, the line of symmetry, and the vertex.

2. Graph $y = -5x^2$ and $y = \frac{1}{5}x^2$ in the same coordinate plane. Compare these graphs with the graph of $y = x^2$.

Solve each equation. Give solutions to the nearest tenth when necessary.

3. $x^2 = 25$

4. $x^2 = 8$

5. $x^2 + 3 = 15$

6. $55 = 1.1x^2$

7. $200 = 0.8x^2$

8. $-0.15 = -0.05x^2$

5.1 | **Exercises and Applications**

Extra Practice exercises on page 754

Match each graph with its equation.

1.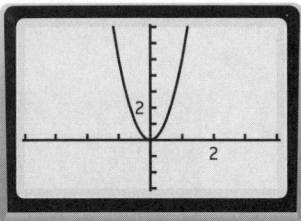

2.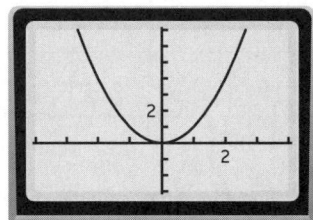

3.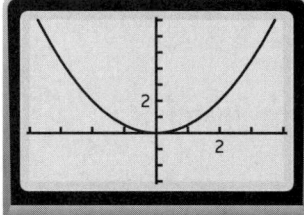

A. $y = x^2$

B. $y = 5x^2$

C. $y = 0.5x^2$

4. **Open-ended Problem** Explain why the graph shown at the right is not symmetric about the y-axis. Then write an equation whose graph *is* symmetric about the y-axis. Tell why your equation is symmetric about the y-axis.

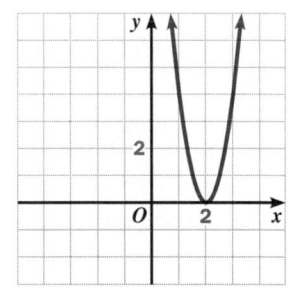

5. **Challenge** Prove that the graph of $y = ax^2$ is symmetric about the line $x = 0$. (*Hint:* If (x, y) is any point on the graph, what are the coordinates of the point's reflection over the y-axis? What must be true about this image point?)

Solve each equation. Give solutions to the nearest tenth when necessary.

6. $x^2 = 16$

7. $x^2 = 12$

8. $8x^2 = 24$

9. $36x^2 = 9$

10. $3x^2 = 27$

11. $\frac{1}{3}x^2 - 27 = 0$

12. $-4x^2 = -64$

13. $-5x^2 = -35$

Each graph has an equation of the form $y = ax^2$. Find a.

14.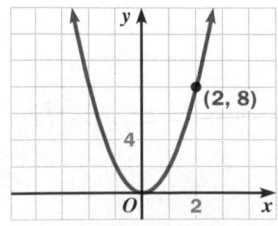

(2, 8)

15.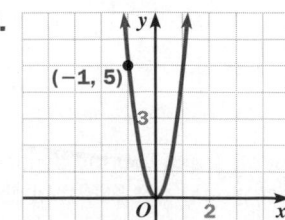

(−1, 5)

16.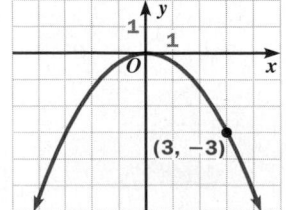

(3, −3)

188 Chapter 5 *Quadratic Functions*

17. **GEOMETRY** An equation for the volume of a cylinder is $V = \pi r^2 h$.

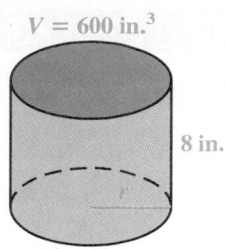

$V = 600 \text{ in.}^3$

8 in.

   a. Suppose $h = 8$ in. and $V = 600$ in.^{3} Write an equation you could use to find r.

   b. Find r. Use $\pi \approx 3.14$.

18. **Challenge** In Section 3.2, you learned about the meaning of rational powers. In this section, you learned about the meaning of square roots.

   a. Use a calculator to compare the decimal values of $16^{1/2}$ and $\sqrt{16}$, $5^{1/2}$ and $\sqrt{5}$, and $(0.1)^{1/2}$ and $\sqrt{0.1}$. What do you notice?

   b. **Writing** State a general rule about $\frac{1}{2}$ powers and square roots. Then give a convincing argument in support of your rule.

Connection ▶ **BICYCLING**

In Example 1 on page 185, the values of the drag coefficient and the frontal area determined the air resistance on a bicyclist in a particular riding position. These values change for different riding positions and bicycle designs.

For Exercises 19–21, use the information about air resistance shown with the photographs below for various bicycling positions.

19. Write an equation that gives the air resistance R on a bicyclist traveling at speed s in each situation shown (assuming still air).

20.  **Technology** Use a graphing calculator or graphing software to graph each equation from Exercise 19 in the same coordinate plane. How do the graphs show which rider experiences the least air resistance at a given speed?

With complete fairing: $C_D = 0.11$, frontal area = 4.56 ft^2

A *fairing* is an external surface that reduces drag.

21. What is the speed of a crouched, racing bicyclist who experiences 4 lb of air resistance?

Upright position: $C_D = 1.1$, frontal area = 5.5 ft^2

Racing crouch position: $C_D = 0.83$, frontal area = 3.9 ft^2

Partial fairing and racing crouch: $C_D = 0.7$, frontal area = 4.1 ft^2

Closely following another bicycle: $C_D = 0.5$, frontal area = 3.9 ft^2

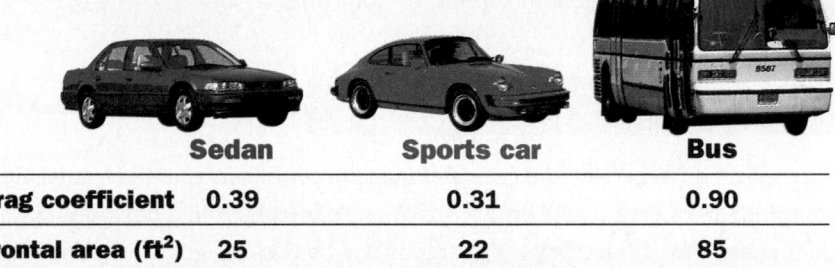

INTERVIEW Mark Thomas

Look back at the article on pages 182–184.

Automotive engineers like Mark Thomas know that the more aerodynamic a vehicle is, the less power the vehicle's engine must provide to overcome air resistance. As the formula on page 184 shows, the air resistance that a vehicle encounters depends on the vehicle's drag coefficient, its frontal area, and the square of the relative speed. The table below gives typical values of the drag coefficient and frontal area for several types of vehicles.

	Sedan	Sports car	Bus
Drag coefficient	0.39	0.31	0.90
Frontal area (ft²)	25	22	85

Use the table above and the formula on page 184 for Exercises 22 and 23.

22. a. Write an equation giving the air resistance R on a sedan as a function of the sedan's speed s.

 b. Write an equation giving the air resistance on a sports car as a function of its speed.

 c. Find the air resistance on both a sedan and a sports car at 30 mi/h and at 60 mi/h. What do you notice?

23. a. Suppose the air resistance on a bus is 200 lb. About how fast is the bus traveling?

 b. By what factor should the bus reduce its speed in order to reduce the air resistance by 50%?

ASTRONOMY The surface area of a sphere with radius r units is given by the formula S.A. $= 4\pi r^2$. Use this formula in Exercises 24 and 25.

24. a. Earth's radius is about 4000 mi. What is its surface area?

 b. What fraction of Earth's surface is in daylight at any given moment?

 c. What is the surface area of the part of Earth that is in daylight at any given moment?

25. Sunlight travels about 93,000,000 mi to reach Earth's surface. Because light from the sun radiates out in all directions, you can think of the available sunlight at 93,000,000 mi as the surface of a sphere with a radius equal to that distance.

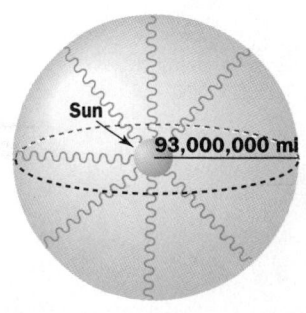

 a. Find the fraction of available sunlight that hits Earth's surface at any given moment.

 b. How large would Earth's radius have to be for Earth to receive just 1% of the available sunlight?

26. SAT/ACT Preview If $A = \sqrt{x}$ and $B = \sqrt{y}$ for positive integers x and y, and $x > y$, then:

A. $A > B$ **B.** $B > A$

C. $A = B$ **D.** relationship cannot be determined

27. The spreadsheet shows the breaking weights w, in pounds, for polyester ropes of various maximum diameters d, in inches. In theory, the breaking weight for a rope is proportional to the rope's cross-sectional area A: $w = kA$.

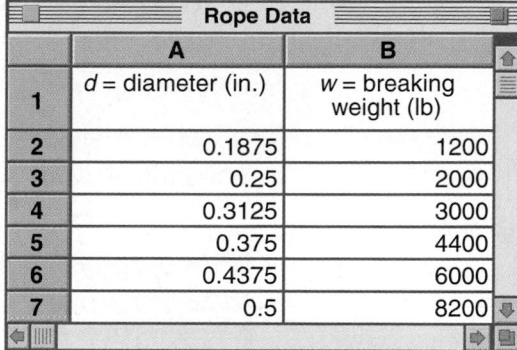

	A	B
	Rope Data	
	d = diameter (in.)	w = breaking weight (lb)
1		
2	0.1875	1200
3	0.25	2000
4	0.3125	3000
5	0.375	4400
6	0.4375	6000
7	0.5	8200

a. Explain why w should be proportional to d^2 if w is proportional to A.

b. 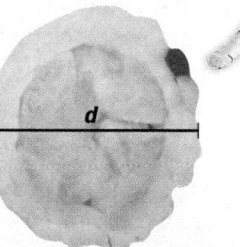 **Spreadsheets** Copy the spreadsheet shown and create a third column that gives the ratio $\dfrac{w}{d^2}$ for each size of rope. Are the ratios about equal? What is the average ratio?

c. Open-ended Problem Use the average ratio to write an equation relating w and d. If someone wants to use your equation to find the breaking weight of a rope, would you still use the average ratio? Explain.

Cross-section of rope

28. Open-ended Problem The *period* of a pendulum is the time the pendulum takes to swing back and forth. The period t (in seconds) of a pendulum of length l (in meters) is given by $l = 0.25t^2$. Create a word problem that can be solved by using this quadratic function. Solve the problem.

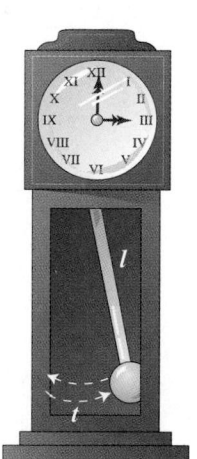

29. Suppose $\log y = 2 \log x + 1.3$. Write y as a function of x. *(Section 4.5)*

30. Graph each equation. *(Section 2.2)*

a. $y = 3x - 1$ **b.** $y = 3x$ **c.** $y = 3x + 2$

31. How are the graphs in Exercise 30 alike? How are they different? *(Section 2.2)*

32. Suppose Maia invests $1200 in a bank account that offers 7% interest compounded continuously. *(Section 3.4)*

a. How much money does Maia have in the account after five years? after ten years?

b. Find the doubling time for the investment.

5.2 Translating Parabolas

You know that the value of a in the equation $y = ax^2$ affects the width of the graph and tells you whether the parabola opens up or down. In the Exploration, you will see how introducing other constants affects the graph.

Learn how to...

- graph equations in the form $y = a(x - h)^2 + k$

- solve quadratic equations

So you can...

- solve problems, such as finding how long it takes a coconut to fall to the ground

EXPLORATION
COOPERATIVE LEARNING

Moving Parabolas Around

Work with a partner.
You will need:
- a graphing calculator or graphing software

SET UP Copy and complete the table. Then answer the questions below.

> **Toolbox p. 796**
> *Translating a Figure*

Equation 1	Equation 2	How is the graph of equation 2 geometrically related to the graph of equation 1?
$y = x^2$	$y = (x - 1)^2$	translated 1 unit to the right
$y = x^2$	$y = x^2 + 1$	?
$y = 2x^2$	$y = 2(x + 3)^2$	?
$y = 2x^2$	$y = 2x^2 - 4$	?
$y = -\frac{1}{2}x^2$	$y = -\frac{1}{2}(x + 1)^2 + 3$	?
$y = -\frac{1}{2}x^2$	$y = -\frac{1}{2}(x - 2)^2 + 1$	?

Questions

1. Predict what the graph of each equation below looks like by making a sketch. Then check your sketch by using a graphing calculator or graphing software.

   a. $y = \frac{1}{3}x^2 + 5$ **b.** $y = -4(x - 1)^2$ **c.** $y = (x + 4)^2 - 3$

2. **a.** How is the graph of $y = ax^2 + k$ geometrically related to the graph of $y = ax^2$ when k is positive? when k is negative?

   b. How is the graph of $y = a(x - h)^2$ geometrically related to the graph of $y = ax^2$ when h is positive? when h is negative?

The equations that you graphed in the Exploration have the general form $y = a(x - h)^2 + k$. To graph an equation in this form, you can start with the graph of the simpler equation $y = ax^2$ and translate it *h* **units horizontally** and *k* **units vertically**.

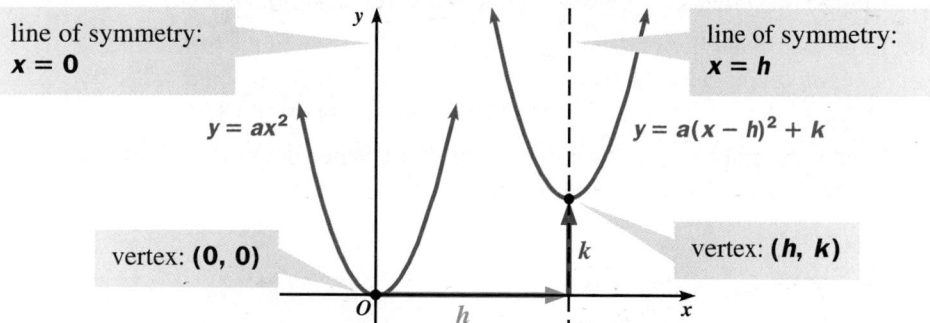

line of symmetry:
x = 0

line of symmetry:
x = h

$y = ax^2$

$y = a(x - h)^2 + k$

vertex: **(0, 0)**

vertex: **(h, k)**

Because you can read the vertex (as well as other information about the graph) directly from the equation $y = a(x - h)^2 + k$, the equation is called the **vertex form** of a quadratic equation.

Vertex Form of a Quadratic Equation

The graph of $y = a(x - h)^2 + k$ is a parabola that:

- has its vertex at (*h*, *k*),
- has the line *x = h* as its line of symmetry,
- opens up if *a* > 0 and opens down if *a* < 0.

EXAMPLE 1

Describe the graph of $y = -2(x + 3)^2 + 4$ and then sketch it.

SOLUTION

Rewrite the equation as $y = -2(x - (-3))^2 + 4$. The vertex is $(-3, 4)$. The line of symmetry is $x = -3$. The parabola opens down.

For *x*-values **1** unit to the **left** or **right** of the line of symmetry, the corresponding *y*-value is $y = -2(\pm 1)^2 + 4 = 2$.

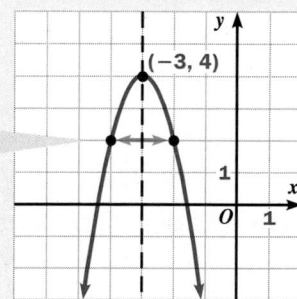

EXAMPLE 2 Application: Physics

The people living on many islands in the South Pacific harvest coconuts by climbing coconut palm trees and cutting the stems of the fruit. The coconuts then fall to the ground. Suppose a coconut falls from a height of 80 ft.

a. Express the height of the falling coconut as a function of the time t since it began falling. Use the fact that the distance d, in feet, an object falls in t seconds is given by $d = 16t^2$. (See Example 3 on page 187.)

b. Sketch the graph of the equation from part (a). When does the coconut hit the ground?

SOLUTION

a. The diagram shows that the relationship between the distance fallen, d, and the height above the ground, h, is given by:

$$d + h = 80$$

Since $d = 16t^2$, substitute and solve for h:

$$16t^2 + h = 80$$
$$h = 80 - 16t^2$$
$$= -16t^2 + 80$$

80 ft

b. The graph of the height function has its vertex at (0, 80) and opens down.

Height of a Falling Coconut

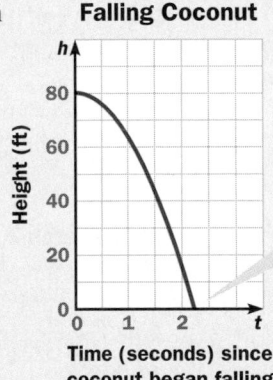

Time (seconds) since coconut began falling

The coconut hits the ground when $h = 0$. The graph suggests that this happens at time $t \approx 2.2$ s.

THINK AND COMMUNICATE

1. Substitute 0 for h in the equation $h = -16t^2 + 80$. Solve for t. Does your answer agree with the solution of part (b) of Example 2?

2. Does the graph in Example 2 represent the *path* of the falling coconut? Explain.

3. The formula $d = 16t^2$ gives the distance an object falls in feet. Similarly, the formula $d = 4.9t^2$ gives the distance an object falls in meters. If a coconut falls from a height of 25 m, express the height of the falling coconut as a function of the time t since it began falling.

Minimum and Maximum Values

Because the vertex (h, k) represents the *lowest* point on the graph of $y = a(x - h)^2 + k$ when $a > 0$, k is called the **minimum value** of the quadratic function.

Likewise, because the vertex (h, k) represents the *highest* point on the graph of $y = a(x - h)^2 + k$ when $a < 0$, k is called the **maximum value** of the quadratic function.

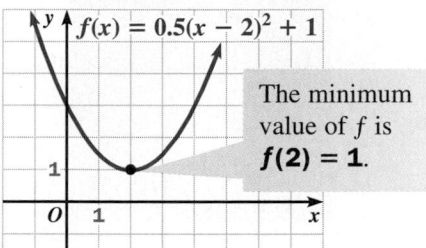

The minimum value of f is $f(2) = 1$.

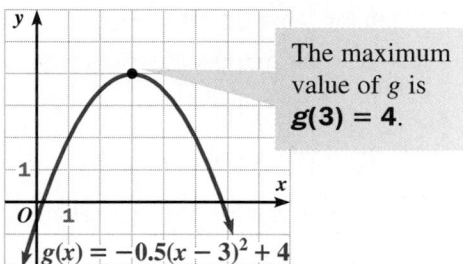

The maximum value of g is $g(3) = 4$.

EXAMPLE 3

Nyasha throws a ball straight up into the air. The ball reaches a maximum height of 8 ft above the ground in 0.5 s.

a. Find an equation giving the ball's height h as a function of time t.

b. When does the ball hit the ground?

SOLUTION

a. If you measure time from the moment the ball reaches its maximum height, a graph of the height function looks like this:

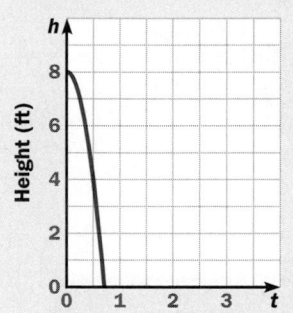

Time (seconds) since maximum height was reached

But if you measure time from the moment the ball leaves Nyasha's hands, a graph of the height function looks like this:

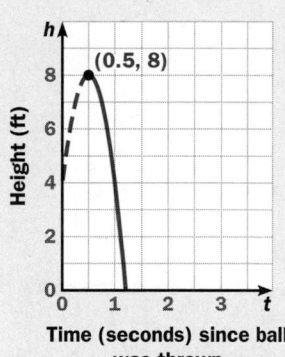

Time (seconds) since ball was thrown

8 ft

An equation for this graph is $h = -16t^2 + 8$. This is just like the falling coconut's equation in Example 2.

The previous graph has been translated 0.5 units to the right. The equation is now $h = -16(t - 0.5)^2 + 8$.

Solution continued on next page.

SOLUTION *continued*

b. Let $h = 0$ and solve for t:

$$h = -16(t - 0.5)^2 + 8$$
$$0 = -16(t - 0.5)^2 + 8$$
$$-8 = -16(t - 0.5)^2$$
$$0.5 = (t - 0.5)^2$$
$$\pm 0.707 \approx t - 0.5$$
$$0.5 \pm 0.707 \approx t$$
$$-0.207, 1.207 \approx t$$

The ball hits the ground in about 1.2 s.

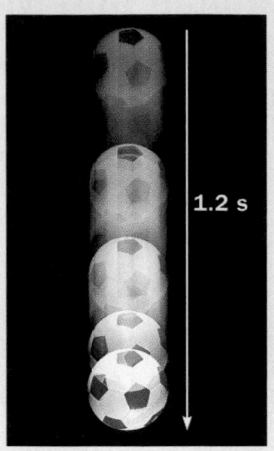

1.2 s

☑ CHECKING KEY CONCEPTS

1. Suppose the parabola with equation $y = -2x^2$ is translated 1 unit horizontally and -2 units vertically. Write the new equation.

Tell whether the graph of each function *opens up* or *opens down*. What is the maximum value or minimum value of the function?

2. $f(x) = 0.12(x - 5)^2 + 3$ **3.** $f(x) = -0.12(x + 5)^2 - 2$

Solve each equation. Give solutions to the nearest tenth.

4. $4(x - 5)^2 + 20 = 100$ **5.** $-3(x + 2)^2 + 34 = 25$

5.2 | **Exercises and Applications**

Extra Practice exercises on page 755

Match each quadratic equation with one of the graphs below.

1. $y = 2x^2 + 3$ **2.** $y = 2(x + 1)^2$ **3.** $y = -2(x - 1)^2 + 3$

A. 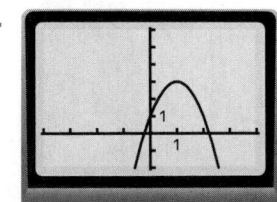   **B.** 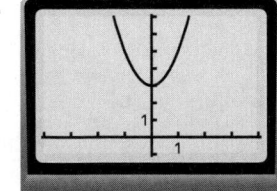   **C.**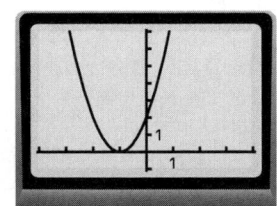

4. Writing Describe the effect that each change has on the graph of each original equation.

  a. changing $y = 5(x - 4)^2 + 1$ to $y = 5(x - 4)^2 - 2$

  b. changing $y = 5(x - 4)^2 + 1$ to $y = 5(x + 4)^2 + 1$

  c. changing $y = 5(x - 4)^2 + 1$ to $y = -5(x - 4)^2 + 1$

The amount a doctor earns each year may depend on the doctor's age. The table gives doctors' average gross incomes and net incomes in 1991 for various age ranges. Use the table for Exercises 5–8.

5. a. Make a scatter plot of the data. Use one color or symbol for the gross incomes and a different color or symbol for the net incomes.

   b. Writing Examine your scatter plot and describe any patterns you see. How do the gross and net incomes compare overall?

6. The data points appear to lie on curves rather than lines. Suppose the curves are parabolas.

   a. What would you say is the approximate location of the vertex of each parabola?

   b. Make a sketch of each parabola on your scatter plot.

7. **Spreadsheets** Alisa modeled the gross income data by using the spreadsheet shown. She let x = age and y = income. She then assumed that the vertex of the parabola for gross income was at (50, 300,000), and so she created two new variables: $u = x - 50$ and $v = y - 300,000$.

   a. Look at the ratios $\dfrac{v}{u^2}$ in the last column of Alisa's spreadsheet. What do you notice?

Robin A. Winthrob, M.D.

Incomes (dollars)

Age (years)	Gross	Net
30–34	175,370	111,200
35–39	227,040	137,610
40–44	270,050	155,680
45–49	301,470	164,410
50–54	295,930	167,720
55–59	281,580	156,100
60–64	229,090	127,130
65–69	167,710	92,280

For modeling purposes, use the middle age in each range: 32, 37, 42, 47, and so on.

b. If you ignore the two middle ratios (which are more variable than the rest), what average ratio do you get?

c. If a is your average ratio from part (b), then you can write $\dfrac{v}{u^2} = a$, or $v = au^2$.
   What equation do you get when you substitute $x - 50$ for u and $y - 300,000$ for v?

d. Writing Graph your equation from part (c) and comment on the reasonableness of the model for the gross income data.

Doctors' Gross Incomes

	C	D	E
1	$u = x - 50$	$v = y - 300000$	v/u^2
2	−18	−124630	−384.66049
3	−13	−72960	−431.71598
4	−8	−29950	−467.96875
5	−3	1470	163.333333
6	2	−4070	−1017.5
7	7	−18420	−375.91837
8	12	−70910	−492.43056
9	17	−132290	−457.75087

Alisa copied the gross income data from the table, putting the ages (the x-values) in Column A and the gross incomes (the y-values) in Column B, not shown here.

8. Open-ended Problem Use a method like the one described in Exercise 7 to develop a quadratic model for the net income data.

What is the maximum value or minimum value of each function?

9. $f(x) = -4(x + 4)^2 + 1$ **10.** $f(x) = 0.3(x + 1)^2 + 4$ **11.** $f(x) = 3(x - 1)^2 - 4$

12. $f(x) = -0.4(x + 4)^2 - 1$ **13.** $f(x) = 5(x - 3)^2 - 2$ **14.** $f(x) = -2(x + 2)^2 + 6$

Describe the graph of each function. Make a sketch of each graph.

15. $y = -3(x + 3)^2 - 3$ **16.** $y = 6(x - 1)^2$ **17.** $y = -(x + 2)^2$

18. $y = \frac{1}{2}(x - 2)^2 - 5$ **19.** $y = -2(x + 4)^2 + \frac{1}{2}$ **20.** $y = -\frac{1}{3}(x - 3)^2 - 4$

Write an equation in the form $y = a(x - h)^2 + k$ for each parabola shown.

21. 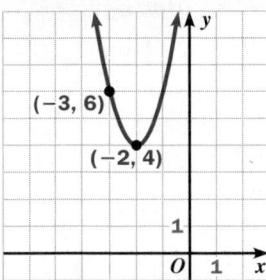   **22.** 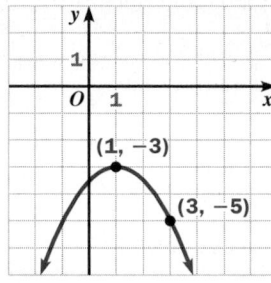   **23.**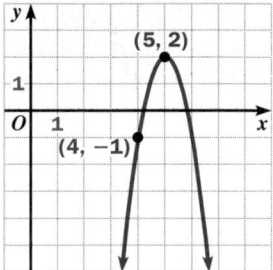

24. Use the equation $h = -16(t - 0.5)^2 + 8$ from part (b) of Example 3 on page 196. Find the possible times the ball is at each height.

   a. 5 ft   **b.** 3 ft   **c.** 1 ft

25. BIOLOGY Many birds drop clams or other shellfish in order to break the shell and get to the food inside. Crows along the west coast of Canada use this method to break the shells of whelks. Suppose a crow drops a whelk from a height of 17 ft.

   a. Write a quadratic equation for this situation.

   b. How long will it take the whelk to reach the ground?

26. Challenge Show that the line $x = h$ is the line of symmetry for the graph of $y = a(x - h)^2 + k$. (*Hint:* Consider the x-values $x = h \pm b$ where b is an arbitrary number.)

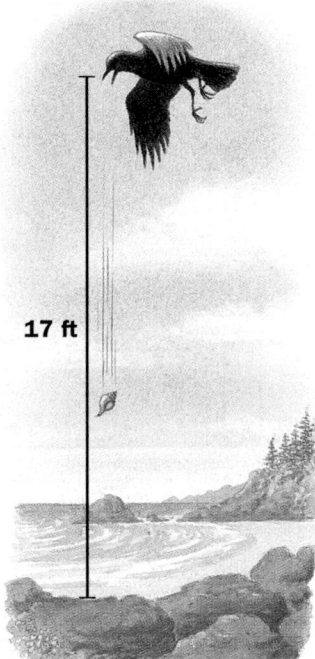

17 ft

ONGOING ASSESSMENT

27. Writing Describe how to graph an equation in vertex form. Include sketches illustrating your description.

SPIRAL REVIEW

Solve each equation. (*Section 5.1*)

28. $x^2 = 0.25$ **29.** $5x^2 = 50$ **30.** $2x^2 - 7 = 11$

Rewrite each expression in factored form. (*Toolbox, page 782*)

31. $3x + 6$ **32.** $35x - 21y$ **33.** $5x^2 + 2x$

Find the inverse of each function. (*Sections 4.1 and 4.2*)

34. $f(x) = \ln x$ **35.** $g(x) = 3x + 15$ **36.** $h(x) = 2^x$

5.3 Quadratic Functions in Intercept Form

Learn how to...

- **write quadratic equations in intercept form**

- **maximize or minimize quadratic functions**

So you can...

- **find the greatest weekly revenue at a store, for example**

The graphs of quadratic functions may cross the *x*-axis zero, one, or two times. The *x*-coordinates of any points where a graph crosses the *x*-axis are called the **x-intercepts** of the graph.

EXPLORATION
COOPERATIVE LEARNING

Investigating *x*-intercepts

Work with a partner.
You will need:
- a graphing calculator or graphing software

SET UP Copy and complete the table. Then answer the questions below.

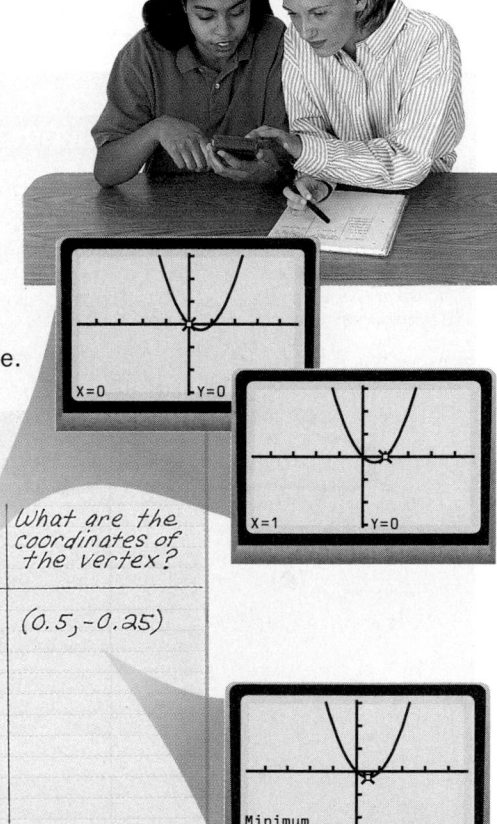

For more information about calculating coordinates of interest, see the *Technology Handbook*, p. 807.

Equation	What are the x-intercepts of the graph?	What are the coordinates of the vertex?
$y = x(x-1)$	0 and 1	$(0.5, -0.25)$
$y = 2x(x-2)$		
$y = -x(x+3)$		
$y = -\frac{1}{2}x(x+4)$		
$y = (x+2)(x-4)$		
$y = 3(x-1)(x+3)$		

Questions

1. Predict what the graph of each equation looks like by making a sketch. Then check your sketch by using graphing technology.

   a. $y = 3x(x + 1)$ **b.** $y = 2(x + 2)(x - 4)$

   c. $y = -\frac{1}{4}(x - 1)(x - 3)$ **d.** $y = \frac{1}{2}(x - 2)(x + 2)$

2. The equations you graphed have the form $y = a(x - p)(x - q)$. In terms of *a*, *p*, and *q*, what are the *x*-intercepts of the graph? What is the *x*-coordinate of the vertex?

Because you can read x-intercepts directly from an equation in the form $y = a(x - p)(x - q)$, it is called the **intercept form** of a quadratic function. As you saw in the Exploration, there is a relationship between the graph of the function and the three factors a, $x - p$, and $x - q$.

One x-intercept is **p**. This is where the factor $x - p$ equals 0.

The other x-intercept is **q**. This is where the factor $x - q$ equals 0.

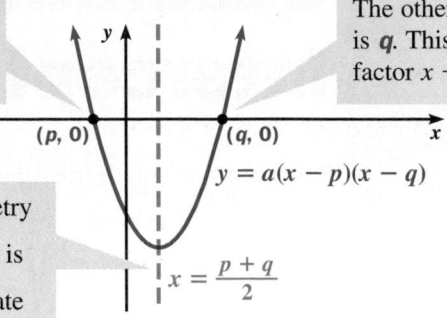

The line of symmetry is $x = \dfrac{p + q}{2}$. This is also the x-coordinate of the vertex.

$(p, 0)$ $(q, 0)$ x

$y = a(x - p)(x - q)$

$x = \dfrac{p + q}{2}$

EXAMPLE 1

Graph $y = (3x + 6)(1 - x)$.

Toolbox p. 782
Factoring Variable Expressions

SOLUTION

Step 1 Rewrite $y = (3x + 6)(1 - x)$ in the form $y = a(x - p)(x - q)$.

$$y = (3x + 6)(1 - x)$$
$$= [3(x + 2)][-1(x - 1)]$$
$$= -3(x + 2)(x - 1)$$
$$= -3(x - (-2))(x - 1)$$

Step 2 Identify the x-intercepts and the coordinates of the vertex.

The x-intercepts are **−2** and **1**.

The x-coordinate of the vertex is:

$$x = \frac{-2 + 1}{2} = -0.5$$

The y-coordinate of the vertex is:

$$y = [3(-0.5) + 6][1 - (-0.5)]$$
$$= 4.5 \cdot 1.5$$
$$= 6.75$$

Step 3 Graph the function.

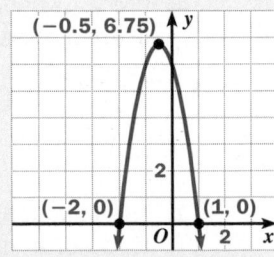

$(-0.5, 6.75)$

$(-2, 0)$ $(1, 0)$

THINK AND COMMUNICATE

1. In Step 1 of Example 1, Joanna found the x-intercepts by setting each factor, $3x + 6$ and $1 - x$, equal to 0. Does this method work? Explain.

2. What is the maximum value of the function in Example 1?

EXAMPLE 2

Write an equation for the parabola shown below.

SOLUTION

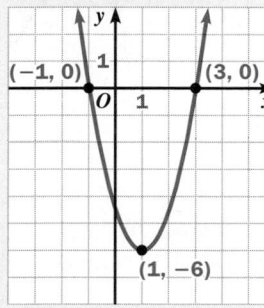

From the graph, you see that the x-intercepts are -1 and 3, and the vertex is $(1, -6)$. Use the intercept form $y = a(x - p)(x - q)$.

$$y = a(x - p)(x - q)$$

Substitute -1 and 3 for p and q.

$$y = a(x - (-1))(x - 3)$$

$$y = a(x + 1)(x - 3)$$

Substitute 1 for x and -6 for y to find a.

$$-6 = a(1 + 1)(1 - 3)$$

$$-6 = -4a$$

$$\frac{3}{2} = a$$

Note that **a is positive.** This agrees with the fact that the graph opens up.

An equation in intercept form is $y = \frac{3}{2}(x + 1)(x - 3)$.

EXAMPLE 3   Application: Business

A store manager receives 100 dresses with a suggested retail price of $70.00. The projected sales are 20 dresses per week. Based on past experience, the store manager knows that sales will increase by 2 dresses per week for every $5.00 decrease in price.

a. Find a function that gives the store's weekly revenue as a function of the number of $5.00 price decreases.

b. What price maximizes weekly revenue?

SOLUTION

a. Let $x = $ the number of $5 decreases in price.

Then $70 - 5x = $ the cost of each dress after x decreases in price,

and $20 + 2x = $ the number of dresses sold after x decreases in price.

The weekly revenue R is the product of the cost per dress, $70 - 5x$, and the number sold, $20 + 2x$. So $R = (70 - 5x)(20 + 2x)$.

Without any price decreases, the weekly revenue is **($70)(20) = $1400**.

Solution continued on next page.

b. The number of price decreases that will maximize weekly revenue is the *x*-coordinate of the vertex of the graph of $R = (70 - 5x)(20 + 2x)$.

Step 1 Find the *x*-intercepts.

$$R = (70 - 5x)(20 + 2x)$$

Substitute **0** for **R**. ➤ $0 = (70 - 5x)(20 + 2x)$

The product of the factors equals 0 if and only if at least one of the factors equals 0.

$$0 = 70 - 5x \quad \text{or} \quad 0 = 20 + 2x$$

$$5x = 70 \qquad\qquad -2x = 20$$

$$x = 14 \qquad\qquad x = -10$$

The *x*-intercepts are 14 and -10.

Step 2 Find the *x*-coordinate of the vertex.

The *x*-coordinate of the vertex is $x = \dfrac{14 + (-10)}{2} = 2$.

The store manager will maximize weekly revenue by decreasing the price of a dress by 2($5), or $10, making the price $70 − $10, or $60.

THINK AND COMMUNICATE

3. In Example 3, if the manager sells the dresses at $60, what will the store's weekly revenue be?

4. Does maximizing weekly revenue also maximize *total* revenue from selling the dresses? Why would a store manager discount the price?

☑ CHECKING KEY CONCEPTS

1. Use the function $y = 6(x + 7)(x - 6)$.

   a. What are the *x*-intercepts of the function's graph?

   b. What are the coordinates of the graph's vertex?

   c. Sketch the graph.

2. Write an equation for the parabola shown.

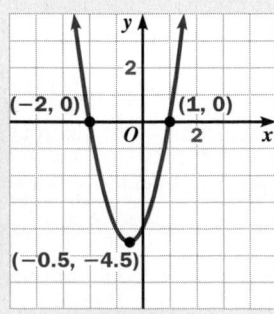

3. What is the maximum value of the function $f(x) = (4x + 12)(6 - 2x)$?

5.3 Exercises and Applications

Extra Practice
exercises on
page 755

For each equation in Exercises 1–6:

  a. Find the *x*-intercept(s).   **b. Find the vertex.**   **c. Sketch the graph.**

1. $y = 4(x - 1)(x + 3)$ **2.** $y = -3(x + 1)(x - 5)$ **3.** $y = -(x - 4)(x + 4)$

4. $y = (2x + 4)(3x + 3)$ **5.** $y = (6 - x)(5x + 20)$ **6.** $y = 0.2(x - 3)(x - 3)$

7. Writing Cheryl and Juan each wrote different equations for the graph
shown. Are the students' answers equivalent? Explain why or why not.

Write an equation for each graph.

8.

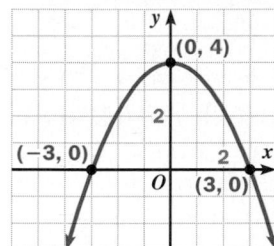

9.

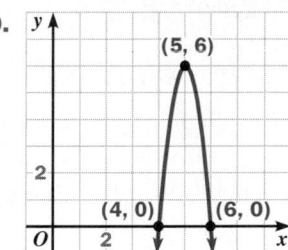

10.

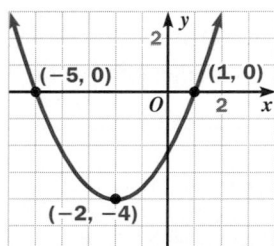

11. Cooperative Learning Work with two other students. You may want to
use a graphing calculator or graphing software. Each student should choose
one of these three equations:

    $y = (x + 2)(3x - 12)$ $y = (0.5x + 1)(x - 4)$ $y = (2x + 4)(4 - x)$

  a. Graph your equations in the same coordinate plane. How are the three
graphs alike? How are they different?

  b. Write your equations in intercept form. How are the three equations alike?
How are they different?

12. BUSINESS A magazine publisher polled its subscribers to see how much
more they would be willing to pay for a year's subscription to the magazine.
The magazine currently has 20,000 subscribers, and the subscription price is
$16. Based on the results of the poll, the publisher knows that, on average,
for every $4 increase in the price of a magazine subscription, 1000 people
will decide not to renew their subscriptions.

  a. Write a quadratic function that gives the revenue R as a function of x, the
number of subscription price increases.

  b. What subscription price will give the maximum revenue?

  c. What is the maximum revenue for this situation?

Engineers and designers often need to consider the effects of weather on objects and surfaces to be constructed. For example, the surfaces of some roads and athletic fields are shaped like parabolas to allow rain to run off to either side.

13. The diagram shows the parabolically shaped surface of a football field with synthetic turf.

18 in.

160 ft

   a. Use the fact that football fields are 160 ft wide. Find an equation that gives the height h of a field at a distance d from a sideline.

   b. How much higher is the field surface 15 ft from a sideline than at the sideline?

14. The cross-sectional sketch of a two-lane road below shows that each lane is 12 ft wide and the surface at the center line of the road is 3 in. higher than at the edges of the road.

   a. Suppose you set up a coordinate plane with the origin at one side of the road shown. Then one of the x-intercepts of the road surface is 0. What is the other x-intercept? What are the coordinates of the vertex? Write an equation for the cross section of the road surface.

   b. Suppose you use the same x-axis as in part (a), but you put the y-axis on the line of symmetry for the road surface. What are the coordinates of the vertex now? What are the x-intercepts? Write an equation for the road surface.

   c. How is the graph of the equation in part (a) geometrically related to the graph of the equation in part (b)?

   d. Use either the equation from part (a) or the equation from part (b) to find how much higher the road surface is at 4 ft from an edge of the road than at the edge of the road.

3 in.

24 ft

15. Open-ended Problem Make a sketch of a parabola that has each given number of x-intercepts. Write an equation for each graph.

   a. two   **b.** one   **c.** none

16. Challenge A gardener has 50 ft of fencing and plans to create a rectangular garden using a side of a house as one edge of the garden. What dimensions maximize the area of the garden?

17. SAT/ACT Preview If $y = (3x + 2)^2$, then $y = \underline{\ ?\ }$.

   A. $9x + 4$ **B.** $9x^2 + 4$ **C.** $9x^2 + 6x + 4$

   D. $9x^2 + 12x + 4$ **E.** none of these

18. Writing Given a quadratic function in the form $y = a(x - p)(x - q)$, describe all you know about its graph from the values of a, p, and q. Give an example.

SPIRAL REVIEW

Graph each function in the same coordinate plane. Describe the relationships among the four graphs. *(Section 5.2)*

19. $y = x^2$

20. $y = (x - 1)^2$

21. $y = (x - 1)^2 - 1$

22. $y = -(x - 1)^2 - 1$

Solve each equation. Give solutions to the nearest tenth when necessary. *(Sections 5.1 and 5.2)*

23. $16t^2 = 400$

24. $3(x - 1)^2 = 12$

25. $5\left(x + \dfrac{1}{2}\right)^2 - 10 = 0$

Write as a logarithm of a single number or expression. *(Section 4.3)*

26. $-\dfrac{1}{3}\log_b 8$

27. $4 \log_2 x + \dfrac{1}{2}\log_2 x^4$

28. $6 \log_a p - 3 \log_a q$

ASSESS YOUR PROGRESS

VOCABULARY

quadratic function (p. 186) **vertex form** (p. 193)
parabola (p. 186) **minimum value** (p. 195)
line of symmetry (p. 186) **maximum value** (p. 195)
vertex (p. 186) **x-intercept** (p. 199)
square root (p. 187) **intercept form** (p. 200)

For Exercises 1–3:

a. Graph each function in the same coordinate plane.

b. Solve for *x* when *y* = 12. Give solutions to the nearest tenth when necessary. *(Section 5.1)*

1. $y = x^2$

2. $y = 4x^2$

3. $y = \dfrac{1}{4}x^2$

Write an equation in the form $y = a(x - h)^2 + k$ using the given coordinates of the vertex and of another point on the graph. *(Section 5.2)*

4. vertex: $(3, 7)$; $(0, -20)$

5. vertex: $(-1, 4)$; $(-5, 12)$

6. vertex: $(5, -2)$; $(7, 6)$

7. vertex: $(-6, -1)$; $(2, 7)$

Rewrite each function in intercept form. Then graph each function. *(Section 5.3)*

8. $y = (3x + 27)(x + 4)$

9. $y = (6x - 18)(4 - x)$

10. $y = (4x - 12)(x + 3)$

11. $y = (10 - 5x)(2x + 4)$

12. Journal Describe how you can find the coordinates of the vertex and the *x*-intercepts of the graph of a quadratic function given in intercept form. How can you do the same for a function given in vertex form?

5.4 Completing the Square

You can write a quadratic function in many different forms. You have already used the vertex form and the intercept form. The **standard form** of a quadratic function is $y = ax^2 + bx + c$. A process called *completing the square* is sometimes used to change quadratic functions in standard form to vertex form. In the Exploration, you will model this process with algebra tiles.

Learn how to...

- complete the square to write quadratic functions in vertex form

So you can...

- find maximums or minimums, such as the maximum height of an object thrown into the air

EXPLORATION
COOPERATIVE LEARNING

Using Algebra Tiles to Complete the Square

Work with a partner.
You will need:
- algebra tiles

1 Use tiles to model the expression $x^2 + 6x$.

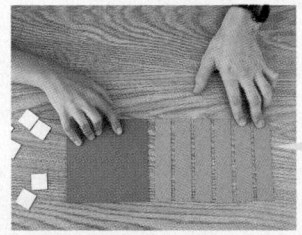

You will need one x^2-tile and six x-tiles.

2 If possible, arrange the tiles in a square. Your arrangement may have a "hole."

You want the length and width of your "square" to be equal.

3 Determine the number of 1-tiles needed to fill the hole.

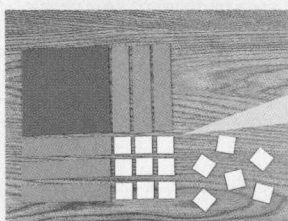

By adding nine 1-tiles, you see that $x^2 + 6x + 9 = (x + 3)^2$.

Questions

1. Copy and complete the table by following the procedure described on the previous page.

Expression	Number of 1-tiles needed to complete the square	Expression written as a square
$x^2 + 6x +$?	9	$x^2 + 6x + 9 = (x + 3)^2$
$x^2 + 4x +$?	?	?
$x^2 + 8x +$?	?	?
$x^2 + 12x +$?	?	?

2. Look for patterns in the last column of your table. Consider the general statement $x^2 + bx + c = (x + d)^2$.

   a. How is d related to b in each case?

   b. How is c related to d in each case?

   c. How could you obtain the numbers in the second column of your table directly from the coefficients of x in the expressions from the first column?

EXAMPLE 1

Write $y = x^2 + 3x + 2$ in vertex form.

SOLUTION

Use the method of completing the square.

$$y = x^2 + 3x + 2$$
$$= (x^2 + 3x +\ \underline{?}\ ) + 2 -\ \underline{?}$$
$$= \left(x^2 + 3x + \frac{9}{4}\right) + 2 - \frac{9}{4}$$
$$= \left(x + \frac{3}{2}\right)^2 - \frac{1}{4}$$

Whatever number you **add** to complete the square, you must also **subtract** to avoid changing the equation.

The number you want is the square of half of the coefficient of x: $\left(\frac{3}{2}\right)^2 = \frac{9}{4}$.

In vertex form, the equation is $y = \left(x + \frac{3}{2}\right)^2 - \frac{1}{4}$.

THINK AND COMMUNICATE

1. **Technology** Use a graphing calculator or graphing software to graph $y = x^2 + 3x + 2$ and $y = \left(x + \frac{3}{2}\right)^2 - \frac{1}{4}$ in the same coordinate plane. What do you notice?

2. The graph of the equation in Example 1 has what point as its vertex?

Projectiles

You know that a dropped object falls a distance of $16t^2$ feet in t seconds. (See Example 3 on page 187.) When an object is not simply released but is thrown or launched, it is called a *projectile*. What happens to a projectile that is launched with some initial vertical velocity v_0 (measured in feet per second) at some initial height h_0 (measured in feet)?

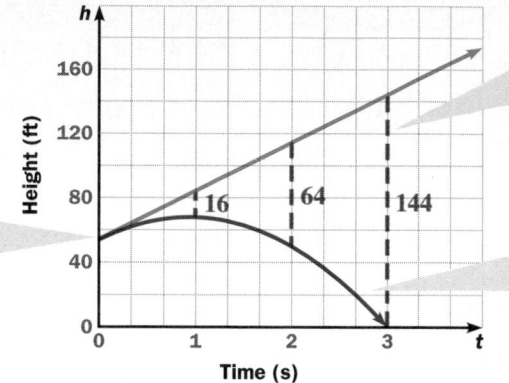

Without gravity to pull the projectile down, its height h would increase according to the equation $h = v_0 t + h_0$.

With gravity, the projectile falls **$16t^2$** feet in t seconds.

So the projectile's height at any time t is given by $h = -16t^2 + v_0 t + h_0$.

The function $h = -16t^2 + v_0 t + h_0$ is a model for the height of a projectile, in feet, as a function of time, in seconds. The model has its limitations, since it does not take into account factors such as air resistance, which can affect the projectile's vertical movement.

THINK AND COMMUNICATE

3. Why is the coefficient of t^2 negative in $h = -16t^2 + v_0 t + h_0$ but positive in $d = 16t^2$ (from Example 3 on page 187)?

4. Consider a baseball thrown upward and a leaf thrown upward. Which situation is more accurately modeled by $h = -16t^2 + v_0 t + h_0$? Why?

5. Use the projectile graph shown in red above.

    a. Estimate the object's initial height and maximum height.

    b. What is the domain of the object's height function? What is the range?

    c. Does the graph tell you anything about the object's *horizontal* position at time t? Explain.

    d. How would the graph of the object's height function change if the object's initial vertical velocity were increased?

16 ft

4 ft

EXAMPLE 2   **Application: Physics**

Shelly throws her keys up in the air, releasing them from a height of 4 ft above the ground, with an initial vertical velocity of 32 ft/s.

a. What maximum height do the keys reach?

b. Her brother Mark is standing on a balcony above her. If his outstretched arms are 16 ft above the ground, at what time(s) can he catch the keys?

SOLUTION

a. The maximum height of the keys is at the vertex of the parabola described by the equation $h = -16t^2 + 32t + 4$. Use the method of completing the square to rewrite the equation in vertex form.

$$h = -16t^2 + 32t + 4$$
$$= -16(t^2 - 2t) + 4$$
$$= -16(t^2 - 2t + 1) + 4 - (-16)$$
$$= -16(t - 1)^2 + 20$$

By putting a **1** inside the parentheses to complete the square, you have added **−16 · 1**, or **−16**, to the equation. So you must also subtract **−16**.

The vertex is at $(1, 20)$. So the keys are at a maximum height of 20 ft after 1 s.

b. Use the equation $h = -16(t - 1)^2 + 20$ from part (a). Find t when $h = 16$.

$$-16(t - 1)^2 + 20 = 16 \quad\quad \text{Substitute \textbf{16} for } \textbf{\textit{h}}.$$
$$-16(t - 1)^2 = -4$$
$$(t - 1)^2 = \frac{1}{4}$$
$$t - 1 = \pm\frac{1}{2}$$
$$t = 1 \pm \frac{1}{2}$$
$$t = \frac{1}{2}, \frac{3}{2}$$

The keys will be at a height of 16 ft after 0.5 s and again after 1.5 s.

☑ CHECKING KEY CONCEPTS

1. Find the number that completes the square for each expression.

   a. $x^2 + 14x + \underline{?}$ **b.** $x^2 + 5x + \underline{?}$ **c.** $x^2 + 0.20x + \underline{?}$

2. Write each equation in vertex form.

   a. $y = 3x^2 + 6x + 1$ **b.** $y = x^2 + \frac{3}{4}x + 1$ **c.** $y = -5x^2 + 25x + 2$

3. What is the maximum value of the function $f(x) = -2x^2 + 8x + 3$?

4. Hamish throws a baseball straight up with a velocity of 24 ft/s from an initial height of 6 ft.

   a. What equation describes the height of the ball as a function of time?

   b. What is the ball's maximum height? When does the ball reach that height?

   c. If Hamish doesn't catch the ball, when does it hit the ground?

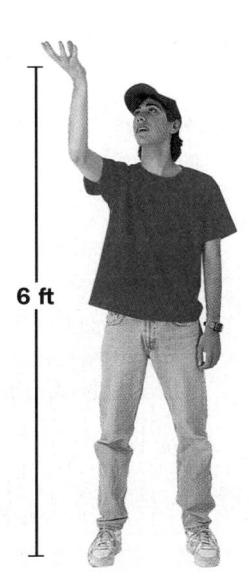

6 ft

5.4 Exercises and Applications

Extra Practice
exercises on
page 756

Write each function in vertex form.

1. $y = x^2 + 4x + 3$ 2. $y = -x^2 + 8x - 1$ 3. $y = x^2 - 7x - 12$ 4. $y = 2x^2 + 7x + 3$

5. $y = x^2 + 6x$ 6. $y = 8x^2 + 2x + 9$ 7. $y = 2x^2 - 5x - 3$ 8. $y = -x^2 + 4x - 7$

9. Carol throws a softball upward from a height of 3 ft above the ground and with an initial velocity of 24 ft/s.

   a. Write an equation to model this situation.

   b. What is the maximum height the ball reaches?

   c. When does the ball reach its maximum height?

   d. When does the ball hit the ground?

10. Andrew tosses a rock into an empty well from a height of 54 ft above the bottom of the well. Write an equation giving the rock's height above the bottom of the well as a function of time in the following situations.

    a. He throws the rock upward with an initial velocity of 32 ft/s.

    b. He drops the rock, with no initial velocity.

    c. He throws the rock downward, with an initial velocity of −32 ft/s.

    d. Find the coordinates of the vertex of each of the parabolas you found in parts (a)–(c). What do you notice?

Match each equation with its graph. Explain your choices.

11. $y = -x^2 - 3x + 5$ 12. $y = x^2 - 3x - 2$ 13. $y = 2x^2 + 4x + 2$

A.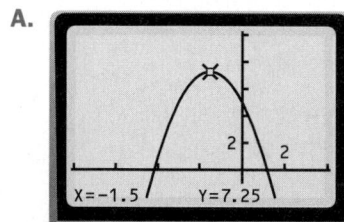

B.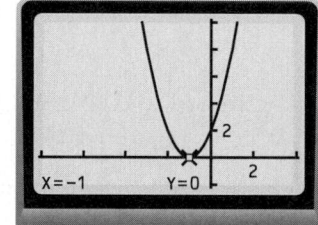

C.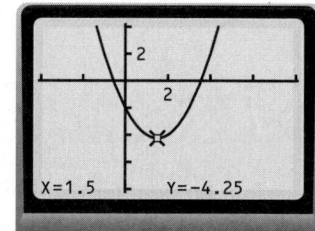

14. **PHYSICS** To jump, a frog pushes itself into the air using its long hind legs. Suppose this jumping frog's body is $\frac{1}{3}$ ft above the ground when the frog's outstretched hind legs leave the ground. At this point, you can treat the frog as a projectile.

    a. Suppose the frog jumps with a vertical velocity of 4 ft/s. Write an equation to model the height of the frog's body above the ground.

    b. What maximum height does the frog's body reach?

$\frac{1}{3}$ ft

To determine the energy cost of motion (walking, running, and so on) for a human or an animal, biologists often measure the rate at which oxygen is consumed. As you might expect, oxygen consumption depends on the level and type of exertion.

15. The table shows the rate of oxygen consumed by a test subject running at a speed of 16 km/h using various stride lengths. The graph shows the data from the table and a parabola that fits the data. The parabola has the equation $y = 0.001292x^2 - 0.3784x + 31.56$.

Stride length (cm)	Oxygen consumption (L/min)
135	4.03
149	3.87
169	4.52

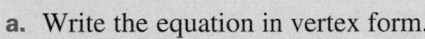
stride length

a. Write the equation in vertex form. What are the coordinates of the vertex of the parabola? What does this vertex mean to the runner?

b. **Writing** At 16 km/h, the test subject freely chose the stride length of 149 cm when allowed to run in a way that felt natural. What do you notice about this stride length and the x-coordinate of the vertex from part (a)? What conclusion might you draw from this observation?

16.  **Technology** Use a graphing calculator or statistical software that can perform quadratic least-squares regression.

a. Find an equation of a parabola that is a good fit for the data below. Here, the test subject described in Exercise 15 ran at 14 km/h.

Stride length (cm)	119	135	153
Oxygen consumption (L/min)	3.55	3.35	3.75

b. The subject's "natural" stride length in this case is 135 cm. Does this confirm your conclusion from part (b) of Exercise 15? Explain.

17. The graph shows the oxygen consumption of a parakeet flying level at various speeds. The data can be modeled by the function $y = 0.06x^2 - 4.0x + 87$, where x is the speed in kilometers per hour and y is the oxygen consumption in milliliters of oxygen per gram of body mass per hour.

a. Write the equation in vertex form.

b. What is the minimum value of the function $y = 0.06x^2 - 4.0x + 87$? What does it mean for the parakeet?

18. **Challenge** Estimate the oxygen consumption of a 35 g parakeet flying level at 20 km/h for 5 min.

BY THE WAY

Olympic long-distance runner Paavo Nurmi of Finland won six gold medals in the 1920s. Nurmi's unusually long stride length was the most economical for *him*, but not for the other runners who tried to copy it.

20 km/h

When distances are measured in feet and time in seconds, you can determine the position (x, y) of a projectile at any time t using this pair of parametric equations:

$$x = v_x t + x_0$$

$$y = -16t^2 + v_y t + y_0$$

In these equations, v_x and v_y are the initial velocities in the horizontal and vertical directions, and x_0 and y_0 are the initial horizontal and vertical positions of the projectile. (*Note:* For help in using parametric equations, look back at Section 2.6.)

15 ft

19. Suppose a basketball player throws the ball as shown above. The initial velocity in the horizontal direction is 20 ft/s and the initial velocity in the vertical direction is 16 ft/s.

   a. **Open-ended Problem** Where would you place the coordinate axes in this situation? Draw a sketch.

   b. Write a pair of parametric equations to model the *trajectory*, or path, of the basketball.

   c. Find the y-coordinate of the ball when the x-coordinate of the ball equals the x-coordinate of the hoop. Does the player make the shot?

20. Suppose the basketball player in Exercise 19 jumps up 1 ft before throwing the ball.

   a. How would your parametric equations change?

   b. Would the player make the basket?

1 ft

21. Complete the square on the parametric equation for the height of the basketball in part (b) of Exercise 19. What maximum height does the ball reach? When does it reach that height?

22. **Cooperative Learning** Work with a partner to find other initial horizontal and vertical velocities that are likely to result in a basket for the player in Exercise 19. You may wish to use a graphing calculator or graphing software to help you examine the possibilities graphically.

23. **Challenge** Using the parametric equations from part (b) of Exercise 19, express y as a function of x. Graph your function in an xy-plane to confirm that the graph is the trajectory of the basketball.

24. Challenge The Buckingham Fountain in Chicago shoots water to a maximum height of 125 ft above the fountain jet. Find the initial vertical velocity of the water. (*Hint:* Remember that the water falling from the maximum height drops $16t^2$ feet in t seconds.)

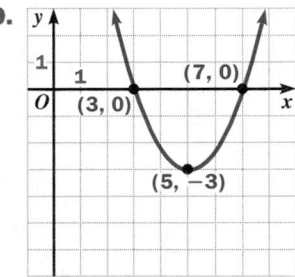

125 ft

State whether each function has a *maximum value* or a *minimum value*. Then find that value.

25. $y = -16x^2 + 32x + 4$

26. $y = 3x^2 - 6x + 5$

27. $y = -x^2 - 6x + 3$

28. $y = 3x^2 - 12x + 2$

29. $y = -x^2 - 6x + 2$

30. $y = x^2 - x - 1$

31. $y = 2x^2 - x + \dfrac{1}{3}$

32. $y = \dfrac{2}{3}x^2 - x$

33. $y = -x^2 + 8x + 5$

34. $y = -3x^2 + 9x - 4$

35. a. Write the functions $y = x^2 + x$, $y = x^2 + 2x$, and $y = x^2 + 3x$ in vertex form. Examine the graph of each function and note the following characteristics:

  • the x-intercepts   • the coordinates of the vertex

b. Investigation Using integer values of b, investigate the graphs of functions of the form $y = x^2 + bx$ and describe the effect of b on the characteristics noted in part (a).

c. Challenge The vertices of the graphs of $y = x^2 + bx$ all lie on a single parabola. Find an equation for that parabola.

ONGOING ASSESSMENT

36. Open-ended Problem Write any quadratic function in standard form. Put the function in vertex form.

SPIRAL REVIEW

Write an equation in intercept form for each parabola shown. (*Section 5.3*)

37.

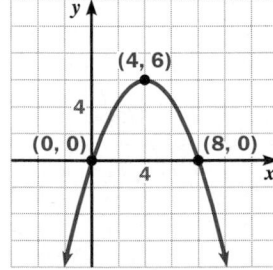

(4, 6), (0, 0), (8, 0)

38.

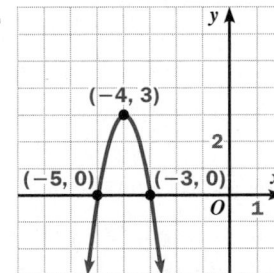

(−4, 3), (−5, 0), (−3, 0)

39.

(7, 0), (3, 0), (5, −3)

Solve each equation. Give solutions to the nearest tenth when necessary. (*Section 5.1*)

40. $200 = 4x^2$

41. $12x^2 = 48$

42. $32x^2 = 50$

43. $\dfrac{2}{9}x^2 = \dfrac{1}{2}$

Find the inverse of each function. (*Section 4.1*)

44. $f(x) = 2x - 5$

45. $f(x) = 2.5x - 7.8$

46. $g(x) = x + 7$

47. $h(x) = 4 - 2x$

5.5 Using the Quadratic Formula

Learn how to...

- **solve equations using the quadratic formula**

- **use the discriminant to determine how many solutions an equation has**

So you can...

- **find how long it takes a projectile to reach the ground, for example**

Michael and his algebra class want to use what they know about quadratic functions while playing baseball. Suppose Michael hits a baseball with a vertical velocity of 60 ft/s from a height of 4 ft. He wants to know how high the ball will travel and how long it will take the ball to hit the ground in the outfield.

Michael can use the height model for a projectile, $h = -16t^2 + v_0t + h_0$. For the baseball he hit, the equation is:

$$h = -16t^2 + 60t + 4$$

The questions below are answered using both the equation for Michael's baseball and a general equation, $y = ax^2 + bx + c$.

Question 1: *How high will the baseball go?*

Use the method of completing the square to find the vertex of the graph of each equation.

MICHAEL'S BASEBALL	**GENERAL EQUATION**
$h = -16t^2 + 60t + 4$	$y = ax^2 + bx + c$
$= -16\left(t^2 + \left(-\dfrac{15}{4}\right)t\right) + 4$	$= a\left(x^2 + \dfrac{b}{a}x\right) + c$
$= -16\left(t^2 - \dfrac{15}{4}t + \dfrac{225}{64}\right) + 4 - \left(-16 \cdot \dfrac{225}{64}\right)$	$= a\left(x^2 + \dfrac{b}{a}x + \dfrac{b^2}{4a^2}\right) + c - \left(a \cdot \dfrac{b^2}{4a^2}\right)$
$= -16\left(t - \dfrac{15}{8}\right)^2 + \dfrac{241}{4}$	$= a\left(x + \dfrac{b}{2a}\right)^2 + \dfrac{4ac - b^2}{4a}$

So the vertex of the height function's graph is at $\left(\dfrac{15}{8}, \dfrac{241}{4}\right)$. Michael's baseball will reach a maximum height of $\dfrac{241}{4}$ ft, or $60\dfrac{1}{4}$ ft.

THINK AND COMMUNICATE

1. The vertex of the graph of $y = ax^2 + bx + c$ has what x-coordinate?

2. The function $y = ax^2 + bx + c$ has what maximum or minimum value? How can you tell from the equation whether the value is a maximum or a minimum?

Question 2: *How long will it take the baseball to reach the ground?*

The ground is at height $h = 0$. Find t when $h = 0$. Use the completed-square form from Question 1.

MICHAEL'S BASEBALL

$$h = -16\left(t - \frac{15}{8}\right)^2 + \frac{241}{4}$$

$$0 = -16\left(t - \frac{15}{8}\right)^2 + \frac{241}{4}$$

$$-\frac{241}{4} = -16\left(t - \frac{15}{8}\right)^2$$

$$\frac{241}{64} = \left(t - \frac{15}{8}\right)^2$$

$$\pm\sqrt{\frac{241}{64}} = t - \frac{15}{8}$$

$$\frac{15}{8} \pm \sqrt{\frac{241}{64}} = t$$

$$\frac{15}{8} \pm \frac{\sqrt{241}}{8} = t$$

$$\frac{15 \pm \sqrt{241}}{8} = t$$

GENERAL EQUATION

$$y = a\left(x + \frac{b}{2a}\right)^2 + \frac{4ac - b^2}{4a}$$

$$0 = a\left(x + \frac{b}{2a}\right)^2 + \frac{4ac - b^2}{4a}$$

$$\frac{b^2 - 4ac}{4a} = a\left(x + \frac{b}{2a}\right)^2$$

$$\frac{b^2 - 4ac}{4a^2} = \left(x + \frac{b}{2a}\right)^2$$

$$\pm\sqrt{\frac{b^2 - 4ac}{4a^2}} = x + \frac{b}{2a}$$

$$-\frac{b}{2a} \pm \sqrt{\frac{b^2 - 4ac}{4a^2}} = x$$

$$-\frac{b}{2a} \pm \frac{\sqrt{b^2 - 4ac}}{2a} = x$$

$$\frac{-b \pm \sqrt{b^2 - 4ac}}{2a} = x$$

THINK AND COMMUNICATE

3. Find decimal values for $\dfrac{15 + \sqrt{241}}{8}$ and $\dfrac{15 - \sqrt{241}}{8}$. Which value is the answer to Question 2?

4. What are the two solutions for the general equation?

You can use the vertex that you found in Question 1 and the solutions when $h = 0$ or $y = 0$ from Question 2 to graph each equation.

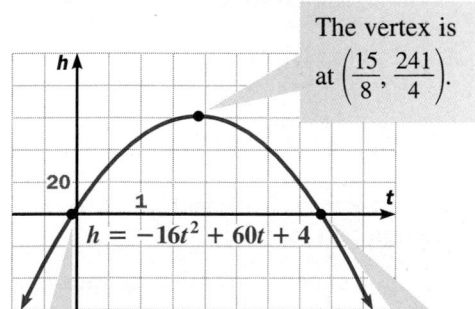

The vertex is at $\left(\dfrac{15}{8}, \dfrac{241}{4}\right)$.

$h = -16t^2 + 60t + 4$

One *t*-intercept is −0.066.

The other *t*-intercept is 3.82.

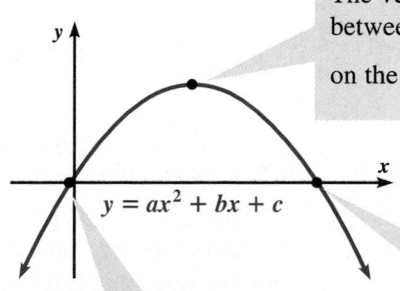

The vertex lies halfway between the *x*-intercepts, on the line $x = -\dfrac{b}{2a}$.

$y = ax^2 + bx + c$

The other *x*-intercept is $-\dfrac{b}{2a} + \dfrac{\sqrt{b^2 - 4ac}}{2a}$.

One *x*-intercept is $-\dfrac{b}{2a} - \dfrac{\sqrt{b^2 - 4ac}}{2a}$.

The solution for the general equation in Question 2 is called the *quadratic formula*. You can use the **quadratic formula** to find the solutions of any quadratic equation in the form $ax^2 + bx + c = 0$ directly, without having to complete the square each time.

The Quadratic Formula

The solutions of any quadratic equation in the form
$ax^2 + bx + c = 0$ are:

$$x = \frac{-b + \sqrt{b^2 - 4ac}}{2a} \quad \text{and} \quad x = \frac{-b - \sqrt{b^2 - 4ac}}{2a}$$

---

70 ft/s

EXAMPLE 1

Suppose Michael hits the baseball with a vertical velocity of 70 ft/s instead of 60 ft/s.

a. How long does it take the baseball to reach the ground?

b. When is the baseball at its highest point?

SOLUTION

The height model for Michael's baseball is $h = -16t^2 + 70t + 4$.

a. Solve $-16t^2 + 70t + 4 = 0$ using the quadratic formula.

$$t = \frac{-70 \pm \sqrt{70^2 - 4(-16)(4)}}{2(-16)}$$

$$= \frac{-70 \pm \sqrt{5156}}{-32}$$

Substitute **−16** for *a*, 70 for *b*, and **4** for *c* in the quadratic formula.

$$\approx 4.43, \ -0.06$$

The baseball will return to the ground in about 4.4 s.

b. The greatest height is at time $t = -\dfrac{b}{2a}$.

$$t = -\frac{70}{2(-16)} \approx 2.19$$

Substitute **−16** for *a* and **70** for *b*.

The baseball is at its highest point about 2.2 s after it is hit.

THINK AND COMMUNICATE

5. Why is the value $t \approx -0.06$ not a solution in part (a) of Example 1?

6. a. Is it possible for the value of $b^2 - 4ac$, the expression under the radical sign in the quadratic formula, to be negative? Explain why or why not.

b. Can you find the square root of a negative number?

c. Discuss whether you can use the quadratic formula to find the solutions of an equation when $b^2 - 4ac$ is negative.

Using the Discriminant

The value $b^2 - 4ac$ in the quadratic formula is called the **discriminant**. You can use the sign of the discriminant to determine the number of solutions that a quadratic equation has. You can also tell how many x-intercepts the graph of the related function $y = ax^2 + bx + c$ has.

Value of the discriminant	Quadratic equation $ax^2 + bx + c = 0$	Related function $y = ax^2 + bx + c$ (with examples)
$b^2 - 4ac > 0$	Two solutions: $$x = \frac{-b + \sqrt{b^2 - 4ac}}{2a}$$ $$x = \frac{-b - \sqrt{b^2 - 4ac}}{2a}$$	Two x-intercepts
$b^2 - 4ac = 0$	One solution: $$x = -\frac{b}{2a}$$	One x-intercept
$b^2 - 4ac < 0$	No solution	No x-intercept

EXAMPLE 2

Find the number of x-intercepts that the graph of each function has.

a. $y = x^2 - 3x - 28$ **b.** $y = x^2 - 6x + 9$ **c.** $y = 2x^2 + 3x + 6$

SOLUTION

	Equation when $y = 0$	Value of $b^2 - 4ac$	Number of x-intercepts
a.	$x^2 - 3x - 28 = 0$	121	2
b.	$x^2 - 6x + 9 = 0$	0	1
c.	$2x^2 + 3x + 6 = 0$	-39	none

1. Suppose a soccer player standing on a sideline throws the ball onto the field from an initial height of 5 ft and with an initial vertical velocity of 16 ft/s.

   a. How long will it take the ball to reach the ground?

   b. How long will it take the ball to reach its maximum height?

   c. What is the ball's maximum height?

2. Determine whether the graph of $y = 3x^2 + 6x + 6$ has *two x-intercepts*, *one x-intercept*, or *no x-intercepts*.

5.5 Exercises and Applications

Extra Practice exercises on page 756

For Exercises 1–9, find the solution(s) of each quadratic equation. Use the quadratic formula.

1. $2x^2 + 8x - 13 = 0$
2. $3x^2 + 18x + 5 = 0$
3. $5x^2 - 2x - 3 = 0$
4. $4x^2 - 9x + 1 = 0$
5. $x^2 + 2x - 15 = 0$
6. $x^2 + 6x + 9 = 0$
7. $2x^2 + 5x + 3 = 9$
8. $x^2 = 6x - 2$
9. $3x^2 - 8 = 4x$

BY THE WAY

In Russia, circus artists must complete years of rigorous training at a professional school in Moscow before they can join one of the many circuses that perform around the country.

10. A circus performer is launched up off a springboard. The height in feet that the performer reaches in t seconds is given by $h = -16t^2 + 18t$.

    a. After how many seconds will the performer land on the ground?

    b. After how many seconds will the performer reach maximum height?

    c. What maximum height does the performer reach?

11. **SPORTS SCIENCE** A researcher testing javelins found that the horizontal distance d (in feet) that a particular javelin traveled when launched at a speed of 100 ft/s and at an angle a (in degrees) could be modeled by $d = -0.251a^2 + 16.5a + 53.4$.

    a. Find the angle at which the javelin should be launched to maximize the horizontal distance it travels.

    b. What maximum horizontal distance does the javelin travel?

Tell whether each equation has *two solutions*, *one solution*, or *no solution*.

12. $x^2 + 12x + 3 = 0$
13. $4x^2 - 12x + 9 = 0$
14. $7x^2 - x + 2 = 0$
15. $24x^2 - 14x + 5 = 0$
16. $16x^2 + 40x + 25 = 0$
17. $3x^2 - 21 = 0$

Find the *x*-intercepts, if any, for the graph of each quadratic function.

18. $y = x^2 + 1$
19. $y = x^2 + 3x + 4$
20. $y = 3x^2 + 5x$
21. $y = -2x^2 + 4x - 2$
22. $y = 6x^2 - 7x + 2$
23. $y = -0.1x^2 + 0.2x + 0.1$

24. **Challenge** Suppose the *x*-intercepts of the graph of the quadratic function $y = x^2 + bx + c$ are $\dfrac{1 - \sqrt{3}}{2}$ and $\dfrac{1 + \sqrt{3}}{2}$. Find b and c.

For a science fair project, a student used audio and visual stimuli to test the reaction times of people of various ages. His data led to the following models where x is a person's age and y is the reaction time (measured in an arbitrary time unit):

Audio stimulus: $y = 15.0008 - 0.3185x + 0.0051x^2$

Visual stimulus: $y = 22.0036 - 0.2287x + 0.005x^2$

25. Visual Thinking Examine the graphs of the two models. What general conclusions can you draw about reaction times?

26. a. Find the age at which the minimum value of each model occurs.

    b. Writing Compare your answers to part (a) and comment on their significance.

27. Suppose a person's reaction time to the visual stimulus is 30 time units. About how old is the person?

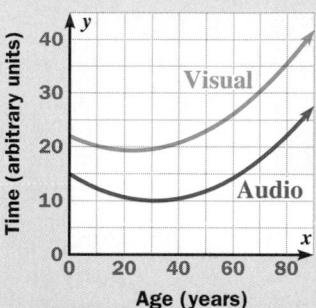

Reaction Times to Audio and Visual Stimuli

28. Suppose a person's reaction time to the audio stimulus is 12 time units. About how old is the person?

29. Open-ended Problem What do the models predict for *your* reaction times to audio and visual stimuli?

30. Suppose a parabola has vertex $(-1, 5)$ and crosses the x-axis at $(-6, 0)$ and $(4, 0)$.

    a. Make a sketch of the parabola.

    b. Write an equation for the parabola in intercept form, $y = a(x - p)(x - q)$.

    c. Write an equation for the parabola in vertex form, $y = a(x - h)^2 + k$.

    d. Compare your equations. Are the values for a the same?

    e. Put each equation in standard form. Are the equations the same?

    f. Let $y = 0$ in your equation from part (e). Solve the resulting equation for x using the quadratic formula. Do the solutions agree with the information you were given about the parabola?

31. GOVERNMENT Each year, money not spent by the federal hospital insurance program is put in a trust fund. Reports issued in 1995 indicated that this trust fund was in danger of going bankrupt. The function $y = -3.59x^2 + 5.09x + 135$ described the projected year-end balance y for each year x since 1994. In what year was the fund projected to go bankrupt?

BY THE WAY

The federal hospital insurance (HI) program pays for inpatient hospital care for those age 65 or over, and for the long-term disabled. The HI program is financed primarily by payroll taxes.

INTERVIEW Mark Thomas

Look back at the article on pages 182–184.

The cars built by students in Thomas's Paper Vehicle Project do not need fuel to run, but cars and trucks on the roads and highways do. Fuel consumption can be harmful to the environment and expensive as well, so it makes sense to drive at speeds that use fuel most efficiently. The fuel efficiency of an average car is given by

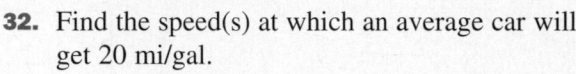
$$y = -0.0177x^2 + 1.48x + 3.39$$

where x is the speed in miles per hour and y is the fuel economy in miles per gallon.

32. Find the speed(s) at which an average car will get 20 mi/gal.

33. Challenge Determine the range of speeds that yield 32 mi/gal or better.

34. What speed yields the best fuel economy?

ONGOING ASSESSMENT

35. Open-ended Problem Experiment with different values of a, b, or c to answer parts (a)–(f). For some parts, there may be more than one possible answer.

a. Find b if the graph of $y = x^2 + bx - 2$ has an x-intercept of 1.

b. Find c if the graph of $y = 4x^2 + 4x + c$ is tangent to the x-axis.

c. Find a if the graph of $y = ax^2 + 8x + 2$ is tangent to the x-axis.

d. Find a if the graph of $y = ax^2 + x + 1$ has two x-intercepts.

e. Find b if the graph of $y = x^2 + bx + 3$ has two x-intercepts.

f. Find c if the graph of $y = 3x^2 - 6x + c$ has no x-intercept.

SPIRAL REVIEW

Use the method of completing the square to find the vertex of the graph of each equation. *(Section 5.4)*

36. $y = x^2 - 8x + 1$

37. $y = 3x^2 + 6x$

38. $y = x^2 + x - \dfrac{3}{4}$

Solve each equation. *(Section 5.1)*

39. $2x^2 - 50 = 0$

40. $\dfrac{1}{3}x^2 + 1 = 4$

41. $-\dfrac{1}{8}x^2 = -18$

Solve each equation. Round your answers to the nearest hundredth. *(Section 4.4)*

42. $5^{2x} = 12$

43. $e^{2x-1} = 37$

44. $\log_2(x + 1) - \log_2(x - 1) = 1$

5.6 Factoring Quadratics

Learn how to...

- **factor quadratic expressions**
- **solve quadratic equations using factoring**

So you can...

- **find the maximum speed a car can be traveling to be able to stop within a given distance, for example**

To find the x-intercepts of the graph of a quadratic function such as $y = x^2 + 7x + 12$, it would be helpful to rewrite the function as a product of factors. The Exploration will show you how this can be done.

EXPLORATION
COOPERATIVE LEARNING

Using Algebra Tiles to Factor

Work with a partner.
You will need:
- algebra tiles

1 Use tiles to model the expression $x^2 + 7x + 12$.

> You will need one x^2-tile, seven x-tiles, and twelve 1-tiles.

length: $x + 4$

2 Arrange the loose tiles to form a complete rectangle. Read the dimensions of the rectangle.

width: $x + 3$

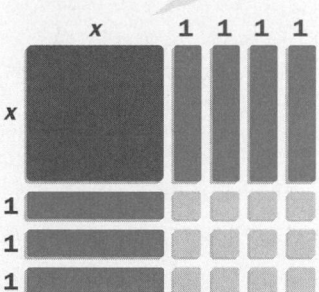

3 Write the original expression as a product of factors.

$$x^2 + 7x + 12 = (x + 3)(x + 4)$$

> The area of a rectangle ...

> ... equals the product of its width and length.

Rewrite each expression as a product of factors.

1. $x^2 + 5x + 4$ **2.** $x^2 + 6x + 8$ **3.** $x^2 + 4x + 4$

When **factoring** $x^2 + bx + c$, your job is to find (if possible) integers m and n such that $x^2 + bx + c = (x + m)(x + n)$. One way is to use algebra tiles as you did in the Exploration. Another, more general way is to consider relationships among b, c, m, and n:

$$x^2 + bx + c = (x + m)(x + n)$$

Compare
corresponding
coefficients.
$$= x^2 + mx + nx + mn$$
$$= x^2 + (m + n)x + mn$$

THINK AND COMMUNICATE

1. Complete each statement with either "b" or "c."

    a. Whatever m and n are, their product must equal _?_ .

    b. Whatever m and n are, their sum must equal _?_ .

2. In the Exploration you saw that $x^2 + 7x + 12 = (x + 3)(x + 4)$. Do the relationships in Question 1 hold in this case?

EXAMPLE 1

Write the function $y = x^2 + 2x - 24$ as a product of factors.

SOLUTION

To write $x^2 + 2x - 24$ as $(x + m)(x + n)$, you need to find integers m and n such that $mn = -24$ and $m + n = 2$. Make a table of possibilities as shown at the right.

Factors of −24	Sum of factors
−1, 24	23
1, −24	−23
−2, 12	10
2, −12	−10
−3, 8	5
3, −8	−5
−4, 6	2
4, −6	−2

These factors give you the sum you want.

A factorization of $y = x^2 + 2x - 24$ is $y = (x - 4)(x + 6)$.

THINK AND COMMUNICATE

3. In Example 1, why would it be difficult to use algebra tiles to factor $x^2 + 2x - 24$?

4. a. **Technology** Use a graphing calculator or graphing software to graph $y = x^2 + 2x - 24$ and $y = (x - 4)(x + 6)$ in the same coordinate plane. What do you notice?

    b. If you rewrite $y = (x - 4)(x + 6)$ as $y = (x - 4)(x - (-6))$, you know that the x-intercepts of the function's graph are _?_ and _?_ . Does your graph of $y = (x - 4)(x + 6)$ from part (a) confirm this?

To factor $ax^2 + bx + c$ when $a \neq 1$, you need to find (if possible) integers k, l, m, and n where:

$$ax^2 + bx + c = (kx + m)(lx + n)$$
$$= klx^2 + knx + lmx + mn$$
$$= klx^2 + (kn + lm)x + mn$$

Note that kn and lm are factors of $klmn$, which is the product of kl (the coefficient of x^2) and mn (the constant).

THINK AND COMMUNICATE

5. Use "a," "b," and "c" to complete this statement: If $ax^2 + bx + c$ can be written as the product of two factors, then __?__ must be the sum of factors of the product of __?__ and __?__.

6. Does the statement you completed in Question 5 apply to $ax^2 + bx + c$ when $a = 1$? Explain.

EXAMPLE 2

Write $y = 4x^2 - 20x + 21$ as a product of factors.

SOLUTION

Step 1 Find the product of the coefficient of x^2 and the constant:

$$4 \cdot 21 = 84$$

List the factors of the product and select the pair whose sum is the coefficient of x.

Factors of 84	Sum of factors
1, 84	85
2, 42	44
3, 28	31
4, 21	25
6, 14	20
7, 12	19

Factors of 84	Sum of factors
−1, −84	−85
−2, −42	−44
−3, −28	−31
−4, −21	−25
−6, −14	−20
−7, −12	−19

The pair of factors −6 and −14 gives the sum you want.

Step 2 Write the function as a product of factors.

$$y = 4x^2 - 20x + 21$$
$$= 4x^2 - 6x - 14x + 21$$
$$= 2x(2x - 3) - 7(2x - 3)$$
$$= (2x - 3)(2x - 7)$$

Separate into two groups of two terms and factor each group.

Pull out the common factor between the two groups.

Written as a product of factors, the function $y = 4x^2 - 20x + 21$ becomes $y = (2x - 3)(2x - 7)$.

THINK AND COMMUNICATE

7. Would it make any difference in Step 2 of Example 2 if you wrote $4x^2 - 20x + 21$ as $4x^2 - 14x - 6x + 21$? Show what happens if you do.

8. What are the x-intercepts of the graph of the function in Example 2?

EXAMPLE 3 **Application: Driving**

On dry asphalt, the distance d in feet needed to stop a car traveling at speed s in miles per hour is given by this quadratic function:

$$d = 0.05s^2 + 1.1s$$

The car travels **0.05s^2** feet after the brakes are applied.

The car travels **1.1s** feet during the time it takes the driver to react to an emergency and apply the brakes.

Suppose a driver has just rounded a curve and sees a stop sign 78 ft ahead. At what maximum speed should the car be traveling in order to come to a stop in time?

── 78 ft ───────────────────────────────▶ STOP

SOLUTION

Let $d = 78$ and solve for s.

$$78 = 0.05s^2 + 1.1s$$
$$0 = 0.05s^2 + 1.1s - 78$$
$$0 = 0.05(s^2 + 22s - 1560)$$
$$0 = 0.05(s + 52)[s + (-30)]$$
$$s + 52 = 0 \quad \text{or} \quad s - 30 = 0$$
$$s = -52 \text{ or} \qquad s = 30$$

> You need to find a factor pair whose product is -1560 and whose sum is 22. Use **52** and **−30**.

The car should be traveling at 30 mi/h or less to stop within 78 ft.

☑ CHECKING KEY CONCEPTS

Write each quadratic function as a product of factors.

1. $y = x^2 - 18x + 81$ **2.** $y = 2x^2 + 7x + 3$ **3.** $y = 6x^2 + 10x - 4$

4. What are the x-intercepts of the graph of the function $y = 5x^2 - 21x + 18$?

5. In Example 3, at what maximum speed should the car be traveling in order to come to a stop within 36 ft?

5.6 | **Exercises and Applications**

*Extra Practice
exercises on
page 756*

Write each quadratic function as a product of factors.

1. $y = x^2 - 5x + 4$ **2.** $y = x^2 + 5x - 24$ **3.** $y = x^2 - 3x - 4$

4. $y = 2x^2 - 7x - 15$ **5.** $y = 9x^2 + 12x + 4$ **6.** $y = 6x^2 + 33x - 63$

7. $y = 4x^2 + 3x - 10$ **8.** $y = 24x^2 + 14x - 3$ **9.** $y = 36x^2 - 100$

10. DRIVING On wet concrete, the distance d in feet needed to stop a car traveling
at speed s in miles per hour is given by the function $d = 0.055s^2 + 1.1s$. At
what maximum speed should a car be traveling on wet concrete to stop
at a traffic light 44 ft ahead?

11. GEOMETRY The photo shows videocassette tape wound on a spool.

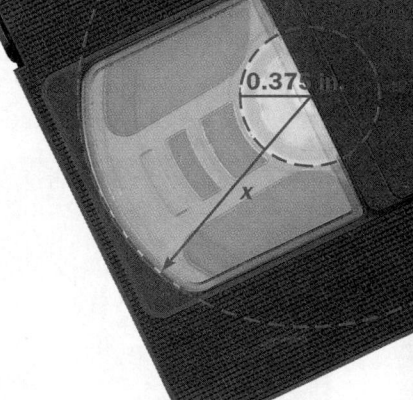

  a. Write a function giving the area of the top surface of the reel of tape
(without the spool). (*Hint:* Find the area between the two dashed circles
shown on the photo.) Express your function both in standard form and as
a product of factors.

  b. Give a geometric argument to explain why the area you found in part (a)
should equal the length of the tape times the thickness of the tape.

  c. The thickness of videocassette tape is about 0.0005 in. Use this fact and
your answers to parts (a) and (b) to express the length of tape wound on
the spool as a function of the distance from the center of the spool to the
outer edge of the tape.

Find the x-intercept(s) of the graph of each function.

12. $y = x^2 + 8x - 20$ **13.** $y = x^2 + 12x + 36$ **14.** $y = x^2 - 9x - 90$

15. $y = 3x^2 - 12x + 9$ **16.** $y = 4x^2 + 5x - 6$ **17.** $y = 6x^2 + 17x - 45$

18. $y = 0.2x^2 - 1.4x - 1.6$ **19.** $y = 25x^2 + 110x - 75$ **20.** $y = 15x - 15x^2$

21. Cooperative Learning Work with two other students. The sum of the first
x terms of the number pattern $10 + 8 + 6 + \cdots$ is given by the function
$S(x) = 11x - x^2$. How many terms does it take to get the sum of zero?

  a. One student should use a graph to answer the question. Another student
should use algebra. The third student should use arithmetic.

  b. Compare the three answers. Which method was easiest to use? Explain.

  c. Make a conjecture about a function that gives the sum of the first x terms
of the pattern $12 + 10 + 8 + 6 + \cdots$. Test your conjecture using one of
the methods in part (a).

22. Challenge Consider the following quadratic functions:

    $y = 2x^2 + 3x - 1$ $y = 2x^2 - x - 1$ $y = 8x^2 + 2x - 3$ $y = 3x^2 - 4x + 4$

  a. Write each function as a product of factors, if possible.

  b. Let $y = 0$ and find the discriminant of each equation. What kind of number
is each discriminant?

  c. Make a conjecture about the relationship between factorable quadratic
functions and discriminants. Test your conjecture using other quadratic functions.

The surface of a spinning liquid forms a *paraboloid*, a three-dimensional figure created by rotating a parabola about its line of symmetry. By spinning reflective liquids, astronomers have started to make telescopes having "liquid mirrors" with surfaces smoother than glass. A cross section of the surface of a spinning liquid is described by the equation

$$y = \frac{2\pi^2 f^2}{g} x^2 - \frac{\pi^2 f^2 R^2}{g}$$

where f is the spinning frequency in revolutions per second, R is the container's radius in méters, and $g = 9.8$ m/s^2 (the acceleration due to gravity).

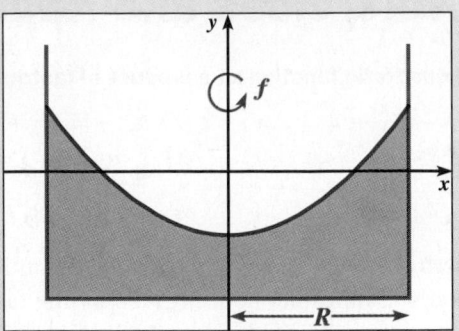

23. What is the equation for the cross-sectional surface of the liquid before it is spun? What does the equation tell you about the placement of the x-axis relative to the liquid?

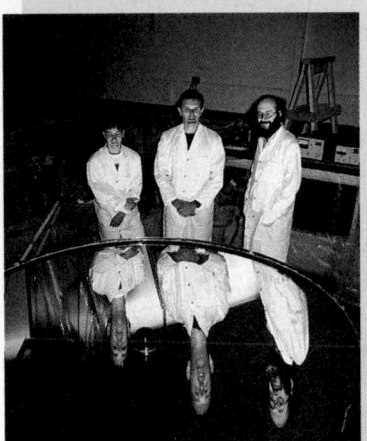

24. Suppose a reflective liquid is put in a container of radius $\sqrt{2}$ m and spun with a frequency of 0.5 revolutions per second.

   a. Write an equation describing a cross section of the surface.

   b. Rewrite your equation from part (a) as a product of factors.

   c. What are the x-intercepts of the graph of the function in part (b)? What are the coordinates of the vertex? Make a sketch of the graph.

   d. **Open-ended Problem** Using the same coordinate plane as in part (c), show the cross sections that result when the liquid is spun at other frequencies. What do you notice?

25. **Challenge** Show that no matter how fast the container is spun, the x-intercepts of the graph of $y = \dfrac{2\pi^2 f^2}{g} x^2 - \dfrac{\pi^2 f^2 R^2}{g}$ are $\pm \dfrac{R}{\sqrt{2}}$.

Ph.D. student Luc Girard, Prof. Ermanno F. Borra, and Dr. Robert Content are reflected in a liquid mirror at Université Laval in Quebec City, Canada.

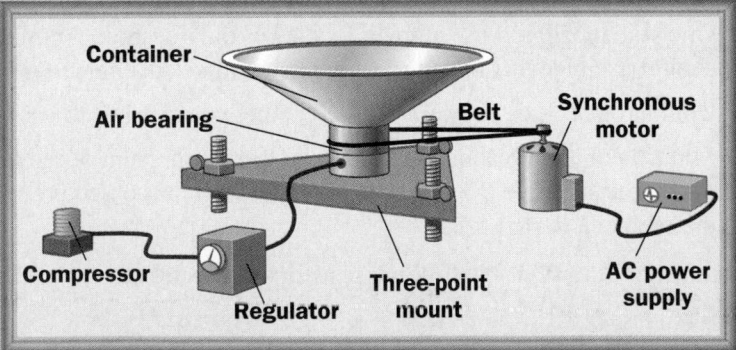

NASA built a liquid mercury mirror based on the design above. It is being used in a telescope for locating space debris that may damage spacecraft or space stations.

Use factoring and the properties of logarithms to solve for x. Be sure to check your answers.

26. $\log(x^2 + 15x) = 2$

27. $\log_3(5x^2 - 30x + 45) - \log_3 5 = 2$

28. $\log_2 x + \log_2(x - 2) = 3$

29. $\log_5(6x - 1) + \log_5(x + 3) = \log_5 25$

ONGOING ASSESSMENT

30. Cooperative Learning Work with three other students. Each of you should choose one of these four quadratic functions:

A. $y = x^2 + 5x + 6$

B. $y = 2x^2 + 5x + 6$

C. $y = 2x^2 + 5x + 1$

D. $y = x^2 + 0.8x + 0.3$

a. Writing Describe the methods you could use to find the x-intercepts of the graph of *any* quadratic function.

b. Which method would each of you use to find the x-intercepts of the graph of the function you chose? Explain your choice.

SPIRAL REVIEW

Solve each equation. *(Section 5.5)*

31. $x^2 - 8x + 3 = 0$

32. $2x^2 + 3x - 2 = 0$

33. $5x^2 - 2x - 4 = 0$

34. $1.2x^2 + 3.4x + 2.4 = 0$

35. Use a constant growth model to estimate the median weekly earnings for families in 1995. *(Section 1.1)*

Sketch the graph of each function. *(Section 5.2)*

36. $y = 4(x + 7)^2 - 3$

37. $y = -2(x - 1)^2 + 5$

38. $y = \frac{1}{3}(x - 3)^2 - 4$

39. $y = -\frac{1}{2}x^2 + 1$

Median Weekly Earnings for U.S. Families with Wage Earners

ASSESS YOUR PROGRESS

VOCABULARY

standard form (p. 206)   discriminant (p. 217)
quadratic formula (p. 216)   factoring (p. 222)

Write each equation in vertex form. *(Section 5.4)*

1. $y = 15x^2 + 9x$

2. $y = x^2 + 12x + 7$

3. $y = 4x^2 - 20x - 3$

Find the x-intercepts, if any, for the graph of each quadratic function. *(Section 5.5)*

4. $y = 4x^2 + 8x + 3$

5. $y = x^2 - 2x + 5$

6. $y = 2x^2 - 20$

Write each quadratic function as a product of factors. *(Section 5.6)*

7. $y = x^2 + 7x + 12$

8. $y = 3x^2 - 5x - 2$

9. $y = 2x^2 + 13x - 7$

10. Journal Explain the advantages and disadvantages of writing quadratic functions in standard, intercept, and vertex form.

Investigating the Flow of Water

Often you will see a town using a tall water tower as a means of creating enough pressure to deliver water throughout the town. The water tower functions on the simple principle that as the height of a column of water increases, the weight of the water in the column increases, which in turn causes the pressure at the base of the column to increase.

You can model this situation using a can filled with water. As you let water flow out of a hole in the bottom, the pressure at the hole decreases. What effect do you think decreasing pressure will have on the time it takes the water to flow out of the can?

PROJECT GOAL Conduct an experiment to find out how the water level in a can affects the time it takes the water to flow from a hole in the can's bottom.

Conducting an Experiment

1. PUNCH a hole in the bottom of the can.

Work with a partner to plan and perform an experiment in which you measure the height of water in a can as the water flows from a hole in the bottom. You will need a large can (like a coffee can), a nail, a hammer, a metric ruler, and a watch. Here are some guidelines for conducting your experiment.

3. RECORD the water's initial height, h_0, in millimeters.

2. COVER the hole in the can's bottom with your finger and fill the can with water.

Testing a Quadratic Model

A theoretical model based on physics applies to the situation you have investigated. The model involves the following constants (all measured in millimeters): h_0, the water's initial height; D, the diameter of the can; and d, the diameter of the hole. The model says that the water's height h, in millimeters, is a quadratic function of the time t, in seconds, after the water begins running out of the container:

$$h(t) = \left(\sqrt{h_0} - \frac{70d^2}{D^2}t\right)^2$$

- **MEASURE** the diameters of the can and the hole.

- **SUBSTITUTE** your values of D, d, and h_0 into the model.

- **CALCULATE** $h(t)$ for the t-values in your data table. Compare your calculated water-height values with the actual data from your experiment. Are there differences? What might explain them?

- **GRAPH** the function $h(t)$ along with your data using a graphing calculator or graphing software. Is the graph a good model for your data?

PROJECT GOAL

TIME (s)	WATER HEIGHT (mm)	PREDICTED HEIGHT (mm)
0		
10		
20		
30		
40		
50		
60		
70		

Writing a Report

Write a report about your experiment. Describe your procedures, and present a comparison of your data to the model's predictions. Include the following:

- a statement of the goal of your project

- a data table that compares your experimental water-height data to the values predicted by the quadratic model

- an evaluation of how well the quadratic model predicts water height

- any ideas you may have that could explain differences between the data and the model

To extend your project, you may wish to investigate other factors that could affect the flow of liquid from a container. For example, what if you used a wider can or made a smaller hole? What if you used a different liquid, like pancake syrup?

Self-Assessment

In your report, include a description of any difficulties you had. For example, was it difficult to get precise measurements? How might you alter the experiment to improve the results?

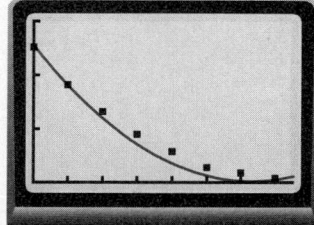

5. ORGANIZE your data in a table. Then make a scatter plot. Put time on the horizontal axis and water height on the vertical axis.

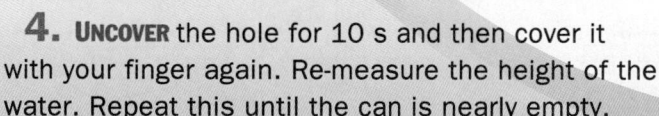

4. UNCOVER the hole for 10 s and then cover it with your finger again. Re-measure the height of the water. Repeat this until the can is nearly empty.

5 Review

STUDY TECHNIQUE

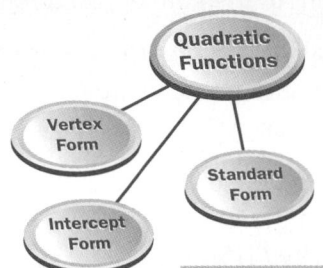

A *concept map* is a diagram that highlights the connections between ideas. Drawing a concept map for a chapter or a section can help you focus on the important ideas and on how they are related. Draw a concept map for this chapter, showing key concepts and vocabulary terms.

VOCABULARY

quadratic function (p. 186)
parabola (p. 186)
line of symmetry (p. 186)
vertex (p. 186)
square root (p. 187)
vertex form (p. 193)
minimum value (p. 195)

maximum value (p. 195)
x-intercept (p. 199)
intercept form (p. 200)
standard form (p. 206)
quadratic formula (p. 216)
discriminant (p. 217)
factoring (p. 222)

SECTIONS | 5.1, 5.2, *and* 5.3

The graph of a **quadratic function** is a **parabola**. A quadratic function can be expressed in several forms. In each of the following forms, the value of a affects the width of the parabola and the direction it opens. If $a > 0$, the function has a **minimum value**; if $a < 0$, the function has a **maximum value**.

Vertex form: $y = a(x - h)^2 + k$
vertex: (h, k)

line of symmetry: $x = h$

Intercept form: $y = a(x - p)(x - q)$
x-intercepts: p and q

line of symmetry: $x = \dfrac{p + q}{2}$

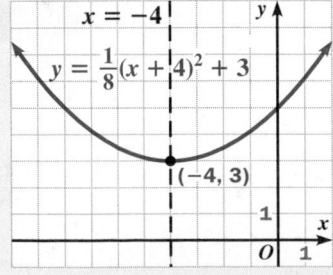

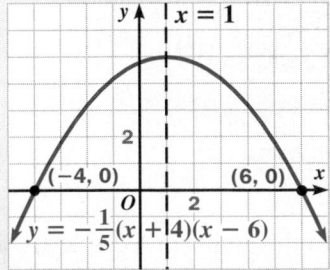

SECTIONS 5.4, 5.5, and 5.6

The equation $y = ax^2 + bx + c$ is the **standard form** of a quadratic function. To find the maximum value or minimum value of a quadratic function in standard form, or to find the coordinates of the **vertex** of the function's graph, rewrite the equation in vertex form by completing the square.

Add the square of half the coefficient of x.

$$\left(\frac{-6}{2}\right)^2 = (-3)^2 = 9$$

$$y = -2x^2 + 12x - 10$$
$$= -2(x^2 - 6x + \mathbf{9}) - 10 + \mathbf{2(9)}$$
$$= -2(x - 3)^2 + \mathbf{8}$$

The vertex is **(3, 8)**.

Add **2(9)** to balance **−2(9)**.

To find the **x-intercept(s)** of the graph of a quadratic function in standard form or the solutions of a quadratic equation in the form $ax^2 + bx + c = 0$, you can complete the square or use the **quadratic formula**. For example, consider $y = -2x^2 + 12x - 10$.

Method 1: Complete the square.
Rewrite $y = -2x^2 + 12x - 10$ as $y = -2(x - 3)^2 + 8$. Let $y = 0$.

$$-2(x - 3)^2 + 8 = 0$$
$$-2(x - 3)^2 = -8$$
$$(x - 3)^2 = 4$$
$$x - 3 = \pm 2$$
$$x = 3 \pm 2 = 1, 5$$

Method 2: Use the quadratic formula.
Let $y = 0$.

$$-2x^2 + 12x - 10 = 0$$

Substitute -2 for a, 12 for b, and -10 for c into

$$x = \frac{-b \pm \sqrt{b^2 - 4ac}}{2a}.$$

$$x = \frac{-12 \pm \sqrt{12^2 - 4(-2)(-10)}}{2(-2)}$$
$$= \frac{-12 \pm 8}{-4} = 1, 5$$

The **discriminant** of the quadratic formula, $b^2 - 4ac$, tells you how many solutions the equation $ax^2 + bx + c = 0$ has and how many x-intercepts the graph of $y = ax^2 + bx + c$ has, as shown in the table at the right.

Some quadratic functions of the form $y = ax^2 + bx + c$ can be written as a product of factors by finding two numbers whose product is ac and whose sum is b. For example, to write $y = 3x^2 + 7x - 6$ as a product of factors, find two numbers whose product is $3(-6) = -18$ and whose sum is 7.

$b^2 - 4ac$	Solutions/ x-intercepts
> 0	2
= 0	1
< 0	0

Factors of −18	Sum of factors
1, −18	−17
−1, 18	17
2, −9	−7
−2, 9	7
3, −6	−3
−3, 6	3

$$y = 3x^2 + 7x - 6$$
$$= 3x^2 - 2x + 9x - 6$$
$$= x(3x - 2) + 3(3x - 2)$$
$$= (x + 3)(3x - 2)$$

Write **7x** as the sum **−2x + 9x**.

Factor.

The factors **−2** and **9** have the needed sum, **7**.

5 | Assessment

VOCABULARY QUESTIONS

For Questions 1 and 2, complete each paragraph.

1. The equation $y = 2(x + 1)^2 - 3$ is a _?_ function in _?_ form. The function has a _?_ value of -3. The graph of the equation is a _?_ with $(-1, -3)$ as the _?_ and with $x = -1$ as the _?_ .

2. The equation $y = ax^2 + bx + c$ is the _?_ form of a quadratic function. The expression $b^2 - 4ac$ is called the _?_ , and its value can be used to find the number of _?_ that the graph of $y = ax^2 + bx + c$ has.

SECTIONS 5.1, 5.2, and 5.3

3. Graph $y = -0.5x^2$.

4. The graph of an equation of the form $y = ax^2$ passes through the point $(-2, 12)$. Find the value of a.

5. The area of a circular garden is 30 ft^2. How much fencing is needed to enclose the garden? (*Hint*: Use the formulas $A = \pi r^2$ and $C = 2\pi r$.)

6. **Writing** Explain how to obtain the graph of $y = -(x - 2)^2 + 1$ from the graph of $y = -x^2$. Then describe the graph of $y = -(x - 2)^2 + 1$.

7. **Open-ended Problem** Find a quadratic function that has 0 as its minimum value and $x = -4$ as the line of symmetry for its graph.

8. a. Write an equation in the form $y = a(x - h)^2 + k$ for a parabola that has vertex $(-1, -7)$ and passes through the point $(1, 1)$.

    b. Find the coordinates of the points, to the nearest hundredth, where the parabola from part (a) intersects the x-axis.

9. Rewrite $y = (2 - x)(2 + 2x)$ in intercept form. Then graph the equation and label the vertex.

10. Find an equation for each graph.

a.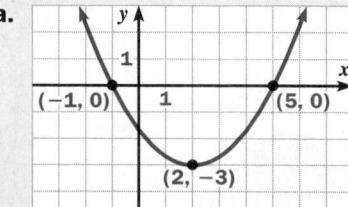

b.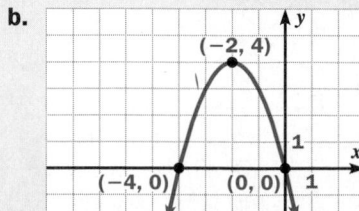

11. The equation $h = -16t^2 + 20t + 6$ gives the height h, in feet, of a basketball as a function of time t, in seconds.

    a. What is the maximum height the ball reaches?

    b. At what time does the ball hit the ground? What method did you use to find the answer?

SECTIONS 5.4, 5.5, *and* 5.6

Write each function in vertex form.

12. $y = 2x^2 + 9x + 3$ **13.** $y = -x^2 + 6x + 3$ **14.** $y = \frac{1}{4}x^2 - 2x + 11$

15. **ELECTRICITY** In a 120 volt electrical circuit that has a resistance of 16 ohms, the power P, in watts, is given by $P = 120I - 16I^2$, where I is the amount of current flowing, in amperes.

    a. Write the equation in vertex form.

    b. What are the coordinates of the vertex of the graph of the equation? What do they mean in terms of the situation?

Find the *x*-intercepts, if any, for the graph of each quadratic function.

16. $y = 4x^2 - 20x + 25$ **17.** $y = 2x^2 + 3x + 5$ **18.** $y = 3x^2 - 16x - 12$

19. **PHYSICS** The equation $y = 0.00139x^2 + 0.0304x + 1.25$ gives the time y, in seconds, that it takes a car to accelerate to a given speed x, in miles per hour. To the nearest integer, what speed can the car reach in 4 s?

Write each quadratic function as a product of factors.

20. $y = x^2 - 6x + 8$ **21.** $y = 6x^2 - 7x - 5$ **22.** $y = 4x^2 + 27x + 18$

PERFORMANCE TASK

23. The game of Leapfrog involves moving pegs along a line of holes. The game begins with a group of n pegs of the same color at one end of the line, a group of n pegs of another color at the other end, and one hole between the two groups. The goal of the game is to interchange the two groups of pegs by moving individual pegs in one of the following ways:

    (1) Move a peg into an adjacent hole.

    (2) Move a peg of one color over an adjacent peg of the other color to get the peg into a hole on the other side.

Play the game with various numbers of pegs. (You may wish to use coins instead of pegs, and a line of squares drawn on paper instead of a line of holes drilled in wood.) Make a table relating n, the number of pegs of each color, to m, the *minimum* number of moves needed to interchange the colors. Then find a functional relationship between m and n. (*Hint*: m is a quadratic function of n.)

C H A P T E R

6 Investigating Data

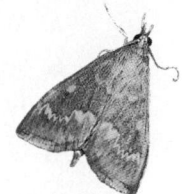

Visualizing the data

INTERVIEW **Donna Cox**

"There is natural beauty in mathematics and physics."

As a child growing up in Oklahoma, Donna Cox found herself drawn to the seemingly divergent worlds of art and science. By the time she started college, she still could not decide which subject to pursue. Instead, she "kept bouncing back and forth between the two." In graduate school, she finally discovered computer graphics—a field that draws almost equally on art, science, and mathematics.

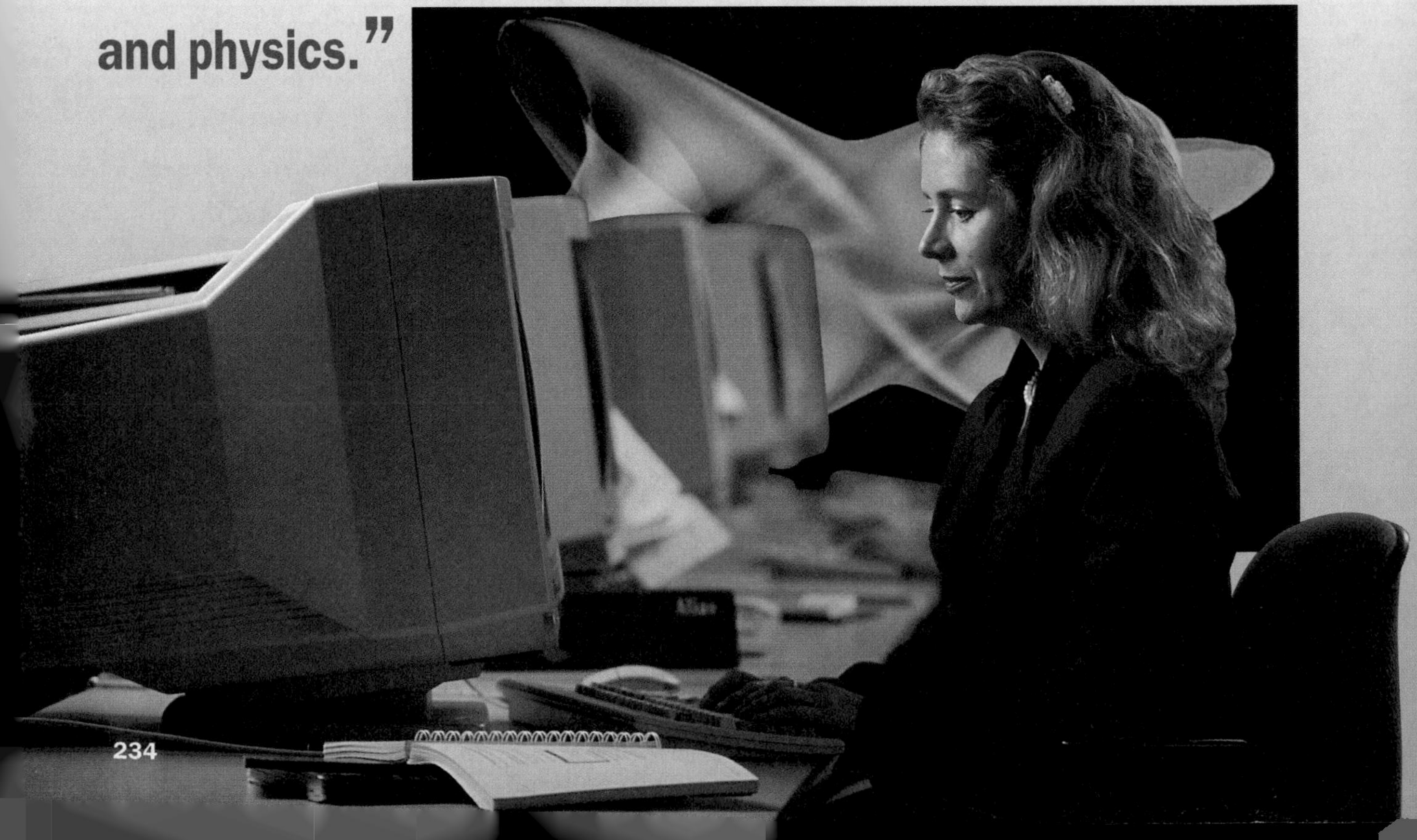

Bringing the Data to Life

Now, as an art professor at the University of Illinois and an associate director at the National Center for Supercomputing Applications, Cox specializes in *scientific visualizations*—using pictures or animated movies to display the information contained in raw, numerical data.

She joins forces with scientists to figure out ways of making data more readily accessible. Cox calls these cooperative ventures "Renaissance teams," a term she coined in 1986. The term refers to the Renaissance era, a period of notable intellectual and artistic achievement from the 14th through the 16th centuries.

European corn borer adult

A Renaissance Team

A good example is the work Cox did with David Onstad, an entomologist at the University of Illinois, and Edward Kornkven, a computer scientist at the South Dakota School of Mines and Technology. Onstad created a mathematical model of the larval and adult forms of the European corn borer, an insect that damages corn plants. He wanted to see how the larval and the adult populations were affected by an insect disease and how this, in turn, affected the corn plants.

Kornkven programmed these interactions on a supercomputer, which produced billions of numbers covering reams and reams of paper. Obviously, there was no shortage of printed output, but extracting useful information from it was a tedious process.

European corn borer larva

Donna Cox's team created the graphic at the left (*behind Cox*) using mathematics and an experimental computer program. The image that resulted, named *Etruscan Venus*, shows a mathematical figure known as a *Klein bottle* in three-dimensional space.

Art Is Communication

The image below is an example of what Cox produced using Onstad and Kornkven's data. The colors of the spikes refer to the health status of larval or adult corn borers, whereas the heights of the spikes refer to the relative numbers of these insects on each plant. In this image, color is used to present *categorical data* and height is used to present *numerical data*.

Cox says, "Artists can make a significant contribution to these technical projects, because art is, in essence, a form of communication. Taking an abstraction such as scientific data and putting it in a visual (or audio) form is something we can do quite well."

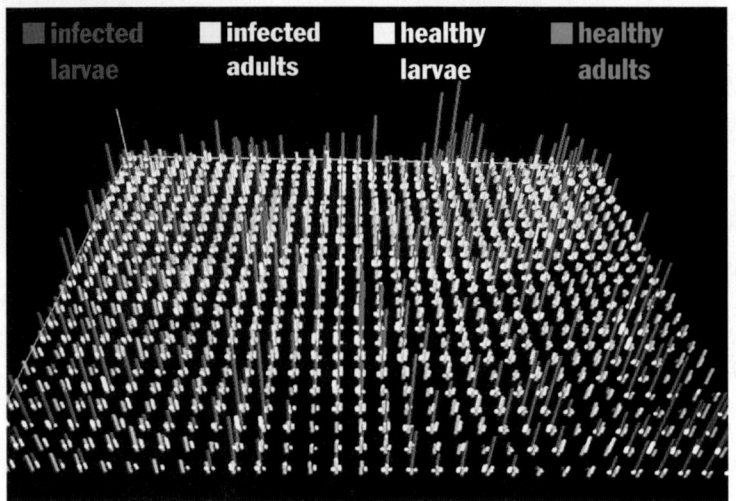

infected larvae   infected adults   healthy larvae   healthy adults

This image is of a cornfield where each group of spikes represents a single corn plant. Disease-infected corn borer larvae and adults do minimal damage to the plants, so only those plants with tall white and blue spikes are being significantly damaged.

"**Art is, in essence, a form of communication.**"

EXPLORE AND CONNECT

Donna Cox works on her computer.

1. Writing Examine the cornfield shown in the computer image above. What do you notice about the distributions of the various categories of corn borers?

2. Research Find out what insects are of concern to farmers near you. What kinds of data do the farmers and agricultural specialists use to make decisions about controlling the insects?

3. Project Use an almanac to find the average monthly temperature and the average monthly rainfall for five cities. Choose cities from different regions of the United States. Transform your numerical data into a visual display.

Mathematics & Donna Cox

In this chapter, you will learn more about ways to describe data used in agriculture.

Related Exercises

Section 6.3
• Exercises 13–19

Section 6.5
• Exercises 14–16

6.1 Types of Data

Learn how to...

- organize information and classify data

So you can...

- analyze data, such as pizza delivery times

Have you ever seen an ad like the one shown? Did you ever wonder how many pizzas are actually given away? The owner of the Hayden Pizza Shop is thinking of placing the ad at the right. The owner uses this method to make a decision:

- The staff keeps a journal like the one shown below.

- The owner makes a graph of all the data gathered by the staff.

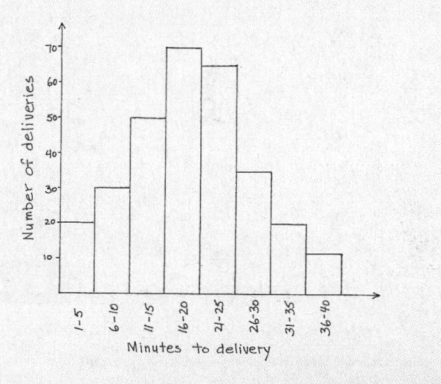

THINK AND COMMUNICATE

1. Look at the journal kept by the staff. Why do you think the owner wants to know the day of delivery? the pizza size?

2. Look at the graph made by the owner. About how long does it take to deliver most pizzas? How many deliveries take longer than 30 minutes?

3. If you were the owner of the Hayden Pizza Shop, would you place the ad shown above? Use the owner's graph to support your answer.

The method used by the owner of the Hayden Pizza Shop can be summarized into five steps. These steps will help you gather data and make conclusions from the data in later sections of this chapter.

> ### Statistical Investigations
>
> 1. Consider the problem or question.
> 2. Collect and record the data.
> 3. Describe and display the data.
> 4. Interpret the data.
> 5. Summarize the findings and make conclusions.

In the journal kept by the staff of the Hayden Pizza Shop, two types of data were gathered. The distances and delivery times are examples of *numerical data*. **Numerical data** are counts or measurements. The days of the week and pizza sizes are examples of *categorical data*. **Categorical data** are names or labels.

Toolbox p. 791
Data Displays

Numerical data are often displayed in a *histogram* or a *box plot*.

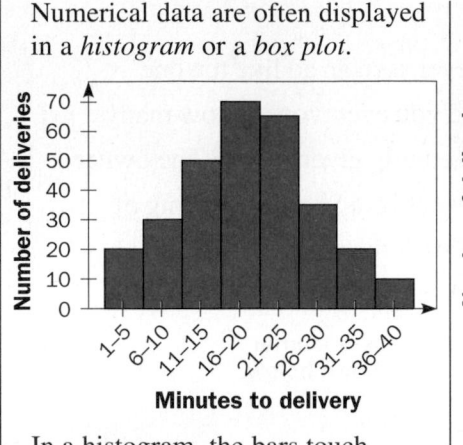

Minutes to delivery

In a histogram, the bars touch and the horizontal axis is a number line.

Categorical data are often displayed in a bar graph or a circle graph.

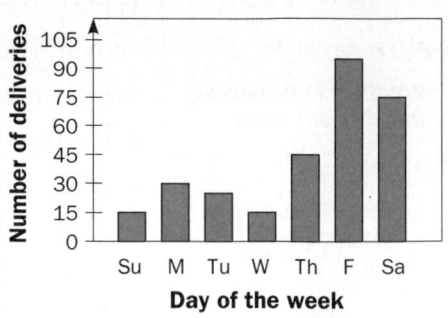

Day of the week

In a bar graph, the bars do not touch and the horizontal axis is not a number line.

The vertical axis for both a histogram and a bar graph is a number line from which you can read the height of each bar.

EXAMPLE 1

Tell whether the data that can be gathered about each variable are *categorical* or *numerical*. Then describe the categories or numbers.

a. blood type   **b.** height of an adult   **c.** grade point average

SOLUTION

a. The data are *categorical*. Blood types are classified as O+, O−, A+, A−, B+, B−, AB+, and AB−.

b. The data are *numerical*. Most adult heights range from 4 ft to 7 ft.

c. The data are *numerical*. A grade point average usually ranges between 0.0 and 4.0.

Some data may appear to be numerical when they are really categorical. For example, you can describe pizza sizes as 8 in., 12 in., and 16 in., but these numbers are not obtained by actually measuring pizzas. Instead, they represent categories, just as small, medium, and large do.

THINK AND COMMUNICATE

Tell whether the data that can be gathered about each variable are *categorical* or *numerical*. Explain your reasoning.

4. shoe size   **5.** words per minute   **6.** month of the year

EXAMPLE 2 | **Application: Medicine**

At the Foley High School blood drive, 200 people each donated a pint of blood. The number of pints of each blood type collected is shown in the table.

Blood type	O+	O−	A+	A−	B+	B−	AB+	AB−
Number of donors	77	6	74	14	11	12	6	0

a. Use the results above to estimate the percent of people in the United States population with each blood type.

b. Compare the estimates from part (a) with those made by the American Red Cross, shown at the right. Does the school collection reflect the percent of people in the general population with each blood type? Explain.

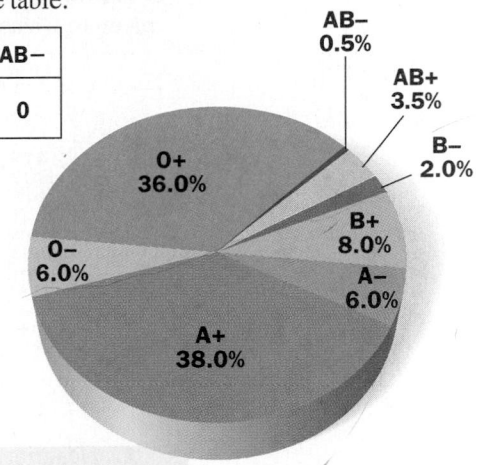

SOLUTION

a. Find the percent of the 200 donors with each blood type. A spreadsheet may be helpful.

	A	B	C	D	E	F	G	H	I
					Blood Donors				
1	Blood type	O+	O−	A+	A−	B+	B−	AB+	AB−
2	Number of donors	77	6	74	14	11	12	6	0
3	Percent of donors	38.5	3	37	7	5.5	6	3	0

b. Almost all of the estimates from the school collection are within 3 percentage points of the estimates for the general population. The school estimate for the number of people in the general population with B− blood was 4 percentage points higher than the estimate made by the American Red Cross.

   Since the American Red Cross estimates are based on donations from a much larger group of people, their estimates more closely reflect the figures for the general population. Therefore, the proportion of people with B− blood seems to be higher in the school than in the general population.

The people who donated at the Foley High School blood drive represent a *sample* of the entire *population*.

A complete group is a **population**. For the American Red Cross data, the population consists of the people who live in the United States.

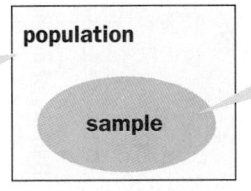

A part of a group is a **sample**. There were 200 people in the sample at the Foley High School blood drive.

The graphs show the results of a survey on pet ownership. The survey was given to a class of 32 students.

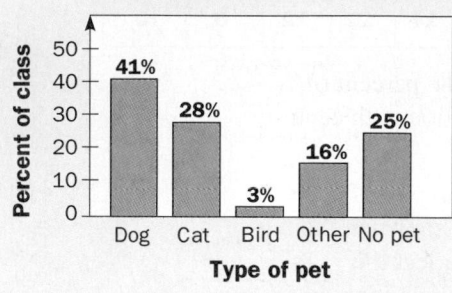

What type of pet do you own?

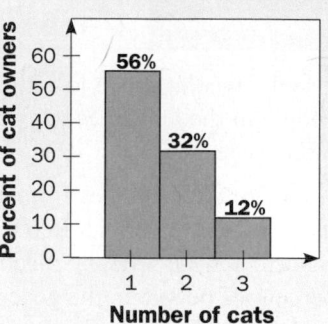

How many cats do you own?

1. Describe each graph and tell whether the data shown are *categorical* or *numerical*.

2. Use the graph of pet ownership to estimate the number of dog owners you might expect to find in a sample of 50,000 households.

3. Repeat Question 2 for the number of cat owners.

4. Why do the percents in the first graph total more than 100%?

5. Use your answer to Question 3 and the graph of cat ownership to estimate the number of cats you might expect to find in a sample of 50,000 households.

6.1 Exercises and Applications

Extra Practice exercises on page 757

Tell whether the data that can be gathered about each variable are *categorical* or *numerical*. Then describe the categories or numbers.

1. the calories in a flavor of yogurt

2. the ZIP code of a city

3. a person's favorite color

4. the scores at a golf tournament

For each graph, tell if it is an appropriate way to display the data. If not, tell what type of display would be better. Explain your reasoning.

5.

Dottie's Dress Shop Orders

6. School SAT Scores
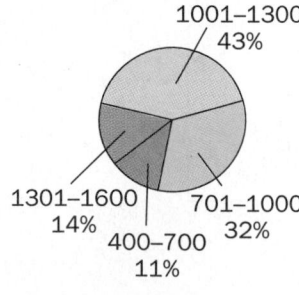

7. Top 1993 Buyers of U.S. Exports
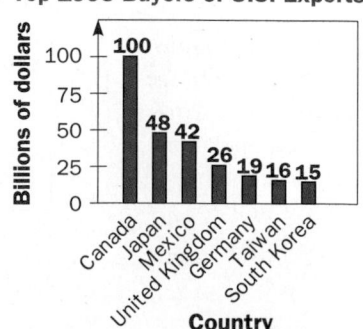

Connection ▸ SPORTS

Sports teams often provide data about their players. The data tell spectators about each athlete, such as a player's performance statistics, position, and experience. Data for the 1994 Stanford Women's NCAA Volleyball Championship team are given in the table.

Player	Year in school	Position	Height
Lisa Sharpley	1	S-OH	6 ft 0 in.
Cary Wendell	3	S-OH	6 ft 0 in.
Shelly Foster	2	DS	5 ft 5 in.
Marnie Triefenbach	3	OH	6 ft 0 in.
Anne Wicks	4	MB	6 ft 2 in.
Paula McNamee	2	MB-OH	6 ft 0 in.
Barbara Ifejika	1	MB	6 ft 2 in.
Nikki Otto	2	MB	6 ft 0 in.
Eileen Murfee	2	MB	6 ft 2 in.
Maureen McLaren	3	OH	6 ft 1 in.
Catherine Juillard	2	S-DS	5 ft 8 in.
Debbie Lambert	1	OH	6 ft 0 in.
Denise Rotert	4	OH-DS	5 ft 10 in.
Kristin Folkl	1	OH	6 ft 2 in.
Wendy Hromadka	3	OH	5 ft 11 in.

The table lists four positions: outside hitter (OH), middle blocker (MB), setter (S), and defensive specialist (DS).

Stanford's Lisa Sharpley (*left*) sets up a spike for teammate Barbara Ifejika at the NCAA Division 1 Women's Volleyball Championship on December 17, 1994, in Austin, Texas. Stanford defeated UCLA in the best-of-five series.

8. What type of data is represented by a player's year in school? by a player's position? by a player's height? Explain.

9. The graph displays the heights, in inches, of the women on the team.

   a. What type of graph is shown?

   b. Use the graph to estimate the mean height of the team in inches.

   c. Convert the heights in the table to inches. Then calculate the mean of the heights. How accurate was your estimate from part (b)?

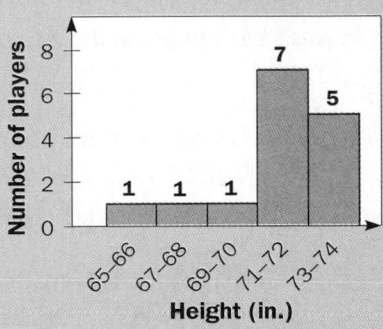

10. a. Make a bar graph to show the number of players at each position.

   b. Writing Does it make sense to find the mean of the number of players at each position? Why or why not?

11. Writing What type of graph would you use to display the data for each player's year in school? Why?

12. Open-ended Problem Suppose you were doing a survey about people's recycling habits. What numerical data could you collect about the subject? What categorical data could you collect? What type of graph would you use to display each type of data? Explain your choice.

For Exercises 13–17:

a. Describe the population and the sample.

b. Tell whether the data that can be gathered in each situation are *categorical* or *numerical*.

c. Tell what type of graph you would use to display the data.

13. A car manufacturer tests 100 cars of the same model to find the fuel efficiency (in miles per gallon).

14. The cast of the play wants to estimate how many programs to order. Cast members ask students from each grade level if they plan to attend the play.

15. A ratings company polls radio listeners between the ages of 12 and 25 to find the most popular radio station in a city.

16. After a severe storm, an insurance company selects 50 homes to assess the average amount of money each homeowner will need for repairs.

17. Supporters of a political proposal want to know where they need to increase their campaign efforts. On a survey of voters, they ask people to write down their voting district.

18. How many exercises are in the "population" of this exercise set? in the "sample" your teacher assigned?

19. Cooperative Learning Work in a group of four students.

a. Each of you should write one of the variables listed below on a piece of paper so that there is one piece of paper for each variable.

- your eye color
- whether you have a driver's license
- your age
- how many classes you are taking

As a group, discuss whether the data that can be gathered about each variable are *categorical* or *numerical*.

b. Pass each paper to the right until each of you has recorded your data on each piece of paper. When you get your original paper back, summarize the data. Use your summary to predict the number of students in your class who would respond in the same way. Explain how you made your prediction.

c. As a class, use a table like the one shown to record each group's results for each variable. Find the totals and compare them with the predictions you made in part (b). Were your predictions close? Why or why not?

d. Use your class results to make predictions about the entire school population. Do you think your predictions are reasonably accurate? Why or why not?

Class Results for Eye Color			
Blue eyes	**Brown eyes**	**Hazel eyes**	**Green eyes**
Group 1 — 1	3	0	0
Group 2 — 1	2	1	0
Group 3 — 0	3	0	1
Group 4 — 1	2	1	0
Group 5 — 2	2	0	0
Group 6 — 1	2	0	1
Total — 6	14	2	2

Aiyana, Kelly, and Marcus are college students completing a degree in music. Each person contacts some of the orchestras in his or her region of the country to find the average starting salary for musicians in an orchestra.

Aiyana's Results		Kelly's Results		Marcus's Results	
Orchestra	Salary	Orchestra	Salary	Orchestra	Salary
Dallas	$46,500	Boston	$59,500	Atlanta	$43,750
Houston	$42,000	Buffalo	$32,000	Florida	$22,750
Los Angeles	$58,250	Chicago	$59,500	New Orleans	$16,500
Oregon	$27,500	Cincinnati	$52,250	North Carolina	$29,500
Phoenix	$22,250	Cleveland	$56,500	Louisville	$21,250
Utah	$31,750	Detroit	$52,500	New Jersey	$19,500
San Diego	$25,000	Indianapolis	$39,750	New York Opera	$26,000
Average	$36,179	Average	$50,286	Average	$25,607

20. What type of data did each person collect? What type of graph do you think would be appropriate to display each person's data?

21. From what population are the samples taken? Which person's sample do you think best represents the population? Why?

22. a. Use each person's sample to estimate the percent of orchestras that pay more than $40,000 as a starting salary.

   b. Repeat part (a) using the salaries from all three samples. Which estimate do you think is most reliable? Why?

   c. There are about 42 major orchestras in the United States. Estimate how many orchestras pay more than $40,000 as a starting salary.

ONGOING ASSESSMENT

23. Cooperative Learning Work in a group of four students. Your group will be conducting a survey about a subject of your choice. In each section of this chapter, you will complete a part of the project.

   In this section you should decide on a subject for your survey. Keep in mind that you will need to gather both categorical and numerical data about your subject. Write down the subject and any questions that you would like to consider. Describe the population you want to study.

SPIRAL REVIEW

Find the x-intercept(s) of the graph of each function. (*Section 5.6*)

24. $y = 2x^2 + 2x - 12$ **25.** $y = x^2 + 2x - 3$ **26.** $y = 6x^2 + x - 1$ **27.** $y = 36 + 13x + x^2$

Graph each equation. (*Section 5.3*)

28. $y = 2(x + 1)(x + 2)$ **29.** $y = (x - 3)(x + 3)$ **30.** $y = 4(x - 7)(x + 2)$ **31.** $y = -(x + 2)(x - 5)$

6.2 Writing Survey Questions

What makes a person a good journalist, lawyer, or scientist? The answer may be the way in which people in these fields ask questions. Knowing which questions to ask may help a person get a lead on a story, prevent an innocent person from going to jail, or discover a cure for a life-threatening disease. In this section, you will learn how the way you ask a question can affect the response you receive.

Learn how to...

- **write good survey questions**

So you can...

- **get more accurate responses to your surveys**

- **judge the accuracy of responses to other people's surveys**

EXPLORATION
COOPERATIVE LEARNING

Comparing Ways of Asking a Question

Work with half the class.

1 One half of the class should answer the questions in Survey I, and the other half of the class should answer the questions in Survey II. The data you collect for this Exploration may be invalid if you read the questions on the other group's survey. To help you avoid this temptation, Survey II has been printed upside down.

Survey I

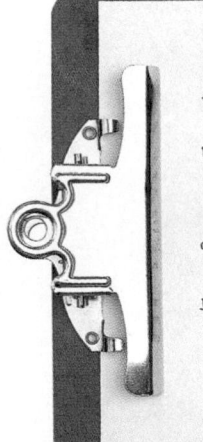

A. Do you think it is important to get good grades in school?

B. Do you think the long hours a student works at an after-school job make it more difficult for that student to get good grades?

C. Do you think it is a good idea for a student to have a job?

D. If a student does have a job, what do you think is the maximum number of hours he or she should work in a week?

Survey II

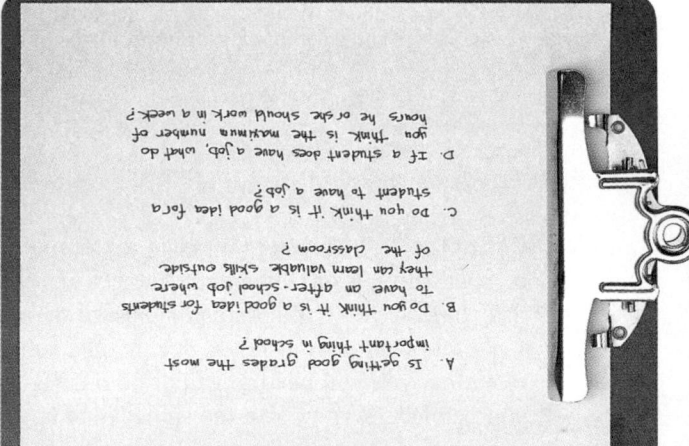

A. Is getting good grades the most important thing in school?

B. Do you think it is a good idea for students to have an after-school job where they can learn valuable skills outside of the classroom?

C. Do you think it is a good idea for a student to have a job?

D. If a student does have a job, what do you think is the maximum number of hours he or she should work in a week?

2 Record the percent of people in your group who responded "yes" to survey questions A, B, and C. Also record the percent of people in your group who responded "no" to questions A, B, and C. Finally, find the average of the numbers given in response to question D.

Questions

1. Compare the questions on your group's survey with the questions on the other group's survey. Do the questions ask about the same information? Do you think your group's results will be similar to those for the other group? Why or why not?

2. Share the results your group recorded in Step 2 with the other group. In both surveys, question A asked about the importance of good grades. According to the results of the two groups, does it appear that one group thinks good grades are more important than the other group does? Do you think this is really the case? Why or why not?

3. Compare the results for the other three questions. Describe any differences. Why do you think these differences occurred?

4. Suppose questions C and D on each survey in the Exploration were the first two questions on each survey. Do you think the results from these questions would be different? Why or why not?

When a question produces responses that do not accurately reflect the opinions or actions of the respondents, the question is said to be a **biased question**.

A biased question may encourage the respondent to answer in a particular way, may be perceived as too sensitive to respond to truthfully, or may not provide the respondent with enough information to give an accurate opinion. Bias may also be introduced through the order in which the questions are asked.

BY THE WAY

A fictitious law was included in a 1981 study. Thirty percent of the people involved in the study claimed to know of this made-up law.

EXAMPLE 1

Tell why each question may be biased.

a. "Many national parks are being heavily damaged by acid rain. Do you favor government funding to help prevent acid rain?"

b. "Do you agree with the amendments to the Clean Air Act?"

c. Police officers ask mall visitors, "Do you wear your seat belt regularly?"

SOLUTION

a. This is an example of a *leading question*. Respondents may think a "no" response means they are not in favor of supporting the national parks. In this way, the question encourages the respondent to answer "yes."

b. This question assumes that a respondent is familiar with the amendments to the Clean Air Act. Responses by people unfamiliar with the amendments could lead to misleading conclusions.

c. Many motorists may answer untruthfully because a police officer is asking the question, especially if the law requires seat belt use. The data collected might not accurately represent the percent of people who wear seat belts.

THINK AND COMMUNICATE

1. Question B from Survey I of the Exploration is an example of a leading question. What answer does the wording of each version of the question encourage? Why? How can you rewrite the question so that it is not biased?

2. Discuss how you would word an unbiased survey question on each of the following topics:

   a. the need to repave the school's parking lot

   b. the need for a new teen center in your hometown

   c. the need for your state to encourage businesses to hire teens

Some surveys have respondents choose their answer to a question from a list of options. People who conduct surveys, called *pollsters*, have found that results are more accurate when people are given a list of choices than when an open-ended question is asked. Also, because it is easier to place a check mark next to a response than it is to write out a response, more people are likely to complete the survey. A multiple-choice method also makes it easier for the pollster to analyze the results.

EXAMPLE 2   Application: Market Research

Due to rising printing costs, the staff of the school newspaper is considering charging $.50 for each issue of the weekly paper. The staff would like to know how this would affect the number of issues of the paper students read. Write a group of unbiased questions that will help the staff make a decision.

SOLUTION

NEWSPAPER SURVEY

1. How often do you read the weekly school newspaper?

☐ 4 times a month   ☐ 2 or 3 times a month
☐ once a month   ☐ occasionally
☐ never

2. Rising printing costs will require that we charge $.50 for each paper if we continue to publish it weekly. Would you pay $.50 for each weekly issue, or would you prefer a monthly paper with no charge?

☐ weekly issue for $.50
☐ monthly issue at no charge

3. How often would you buy the paper if it is published weekly and costs $.50 a copy?

☐ 4 times a month   ☐ 2 or 3 times a month
☐ once a month   ☐ occasionally
☐ never

Ask this question to determine if the respondent currently reads the paper.

Ask this question to see if students will pay $.50 for the paper.

Ask this question to see if the number of issues a student reads will change when a fee is charged.

☑ CHECKING KEY CONCEPTS

Tell why each question may be biased. Then describe what changes you would make to improve the question.

1. "Which city council candidate's platform do you support?"

2. The last question on a test asks students, "How long did you study for this test?"

3. "Aerosol products containing CFCs damage Earth's ozone layer. Do you think aerosol products containing CFCs should be banned?"

4. "How often do you exercise?"

6.2 Exercises and Applications

Extra Practice exercises on page 757

For Exercises 1–6, tell why each question may be biased. Then describe what changes you would make to improve the question.

1. "How much do you weigh?"

2. "A survey of the voters in the state shows that 85% would support a candidate who favors a tax decrease. Do you favor a tax decrease?"

3. "Which one of the following meals would you eliminate from the menu of our Japanese restaurant?

  ___ yakitori   ___ chasoba   ___ udon"

4. "Are you in favor of replacing the baseball stadium that has been used by hometown teams since 1905 with a large, new sports complex?"

5. "Do you think the defendant in the Carter case was given a fair trial?"

6. "What is your grade point average?"

7. Look back at the surveys in the Exploration on page 244. Rewrite the questions so that they are not biased.

Use the senior class survey at the right for Exercises 8–11.

8. What information is the survey trying to obtain?

9. **Writing** Which questions are worded well? Which are worded poorly? Are there any additional questions you would ask? Explain your reasoning.

10. Rewrite any questions that you think are poorly worded.

11. Tell what type of data each question will generate. Explain how the data will help the planning committee make decisions about the banquet.

Do you know which Japanese dish this is?

Please complete this questionnaire regarding the senior class banquet and return it to your homeroom teacher by Friday.

• Are you a male or a female?

  _____

• Do you think a senior banquet is a good idea?

  ___ yes ___ no
  ___ no opinion

• Do you think formal attire should be worn at the banquet?

  ___ yes ___ no
  ___ no opinion

• How much would you be willing to pay to attend the banquet?

  _____

Andrew Jackson

Public polling first became popular in the United States in 1824, when it was used to predict the outcome of the presidential election between Andrew Jackson and John Quincy Adams. Since then, politicians and journalists have questioned people about their political beliefs, candidate preferences, and voting habits.

During the 1992 presidential election, a candidate distributed a questionnaire asking voters about their beliefs. Some of the questions from that survey are shown below.

John Quincy Adams

Please darken box completely.

1. Should the President have the Line Item Veto to eliminate waste?
   YES☐   NO☐

2. Do you want a Constitutional Balanced Budget Amendment, with emergency funding limited exclusively to National Defense?
   YES☐   NO☐

3. Should laws be passed to eliminate all possibilities of special interests giving huge sums of money to candidates?
   YES☐   NO☐

12. a. Which questions on this survey do you think are biased? Why?

b. Rewrite the questions so they are not biased.

13. Writing In the original survey, 97% of the respondents answered "yes" to question 1. When the question was rewritten to ask "Should the President have the Line Item Veto, or not?", only 57% of the respondents answered "yes."

a. What do you think caused people to support question 1 so strongly in the original questionnaire?

b. What do the results indicate voters are really concerned about? Explain.

14. Question 3 was revised in a new survey that asked, "Should laws be passed to prohibit interest groups from contributing to campaigns, or do groups have a right to contribute to the candidate they support?"

a. Do you think the revised question is biased? Why or why not?

b. Match each set of results with the question you think produced those results. Explain your choice.

Question 3 responses
YES, pass a law 40%   NO 55%

Question 3 responses
YES, pass a law 80%   NO 17%

c. Why do you think the results do not add up to 100%?

15. **Research** Find out about the questioning techniques used by people in the fields of law, journalism, science, or medicine. Give examples of the types of questions used by people in at least one of these fields. Do people in these fields ever use biased questions? Do they always avoid asking open-ended questions? Explain.

Write a survey about each topic. Include at least three unbiased questions on each survey. Tell what population should receive each survey.

16. The manager of a restaurant wants to know whether to add a popular special to the regular menu.

17. The transportation department is trying to decide whether it should expand the number of hours train service is provided.

18. A crafts group is trying to decide what day of the week it should hold its regular meetings so that the most people can attend.

19. A local television station wants to know what percent of viewers watch its programs on Thursday nights.

ONGOING ASSESSMENT

20. **Open-ended Problem** Work with the members of your group from the *Ongoing Assessment* exercise from Section 6.1.

    a. Write a survey to gather information about the topic you chose in Section 6.1. Your survey should include these things:

       • an opening statement that tells the purpose of the survey

       • directions on how to complete the survey

       • at least five well-written questions about the topic, including at least two questions that ask for categorical data and at least two questions that ask for numerical data

    b. After you write your survey, explain what information you want to obtain from each question. Then give your survey to another group to review. Use the comments you receive to improve your survey.

SPIRAL REVIEW

Tell whether the data that can be gathered about each variable are *categorical* or *numerical*. Then describe the categories or numbers. *(Section 6.1)*

21. car mileage   22. gender   23. museum membership

Match each scatter plot with one of the following values of the correlation coefficient *r*: 0.64, −0.95, 0.25. *(Section 2.5)*

24.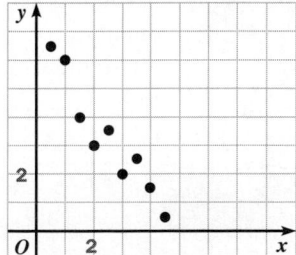

25.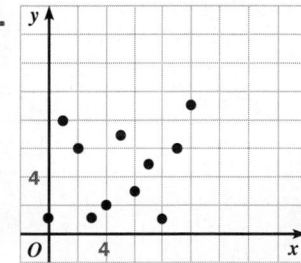

26.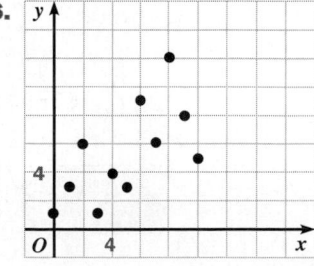

6.3 Collecting Data from Samples

It is often too difficult, time-consuming, and expensive to survey everyone in a population. As a result, pollsters often gather data from a sample of the population. Some ways of selecting a sample are shown below.

Self-selected sample

Let people volunteer.

Systematic sample

Use a pattern, such as selecting every other person.

Convenience sample

Choose people who are easy to reach, such as those in the front row.

Random sample

Use a method that gives everyone an equally likely chance of being selected, such as drawing names out of a hat.

Stratified random sample

Divide the population into groups and randomly select people from each group.

Cluster sample

Choose people as a group rather than as individuals.

Although there are many ways of sampling a population, a random sample is preferred since it is most likely to produce a representative sample of the population. A sample that overrepresents or underrepresents part of the population is a **biased sample**.

THINK AND COMMUNICATE

1. Which of the sampling methods described on the previous page is *least* likely to produce a representative sample of the population? Explain.

2. Think of some other ways to choose a sample and describe them.

EXAMPLE 1   Application: Market Research

The managers of a movie theater chain want to find out how many movies people in the community usually see in a theater each month. Each manager suggests a method for gathering the data. Identify the type of sample each method describes. Tell if the sample is biased. Explain your reasoning.

a. Have the ticket sellers at each theater survey customers when they purchase their tickets.

b. Place an ad in the local paper and ask people to mail in their responses.

c. Randomly select phone numbers from the phone book and call people to ask their opinions.

SOLUTION

a. This is a convenience sample. The sample is biased because it underrepresents people who seldom or never attend movies in a theater.

b. This is a self-selected sample. The sample is biased because it underrepresents people who do not read the paper. Bias is also introduced because the people who respond to the survey are more likely to enjoy movies than those who do not take the time to respond.

c. This is a random sample. The method of selecting people is not biased, but because people who do not have a phone or who have an unlisted number are not included, the sample is biased.

THINK AND COMMUNICATE

3. Which method in Example 1 do you think is best? Why?

For each sample in Example 1, tell how these factors might influence the survey results.

4. time of day the survey is taken

5. day of the week the survey is taken

6. location of the theater

7. time of year the survey is taken

EXAMPLE 2

The Science Club at Park High School wants to give a survey about recycling to 50 of the 900 students in the school. The club has a list of names of all the students. Describe how the club can select a random sample.

SOLUTION

There are many ways to select a random sample. Here are two methods.

Method 1 Use a physical model.

Put each name on a piece of paper and place it in a hat. Mix up the pieces of paper and pull out 50 names without looking at them.

Method 2 Use a calculator.

Number the students' names from 1 to 900. Then use a calculator to generate random numbers from 1 to 900. Most calculators will produce a random number between 0 and 1. To generate other random numbers, you can use a formula as shown on the calculator screen below.

This function generates a random number between 0 and 1.

Put the number of people you want to survey here.

Put the smallest number you want to generate here.

This function cuts off the decimal part of a positive number.

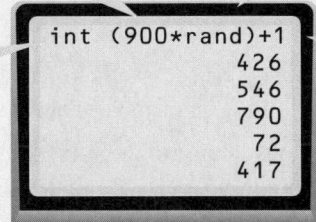

```
int (900*rand)+1
            426
            546
            790
             72
            417
```

Press the ENTER key until you generate 50 different numbers. Then give the survey to the students assigned to the numbers you generate.

In Example 2, the population of the school may be better represented by finding the percent of students in each grade and then choosing a random sample having the same percent of students from each grade.

THINK AND COMMUNICATE

8. Is it possible that the 50 students selected in Example 2 could all be seniors? Do you think that would be a good sample? Why or why not?

9. Suppose the school in Example 2 has 22% freshmen, 28% sophomores, 26% juniors, and 24% seniors. How many students from each grade should be in the sample so that it is a stratified random sample?

✓ CHECKING KEY CONCEPTS

Identify the type of sample and describe the population of the survey. Then tell if the sample is biased. Explain your reasoning.

1. A grocery store wants to survey customers about how the store can improve. A survey is given to every fifth person entering the store.

2. A news program asks viewers to phone in a vote for or against increasing funding for the development of electric cars.

3. The student council wants to survey students about school activities. It randomly selects 25 students from each grade to answer a survey.

4. A concert promoter wants to ask people at a jazz concert what other jazz musicians they would like to see perform. The promoter surveys the people sitting in Section D of the concert hall.

Call-in surveys became popular after the 1980 presidential debate when a television network asked viewers to decide which candidate had "won" the debate.

6.3 Exercises and Applications

Extra Practice exercises on page 757

1. Suppose the homeroom teachers in a school are asked to send four students to participate on a school festival committee. Here is how some teachers chose the four students.

Ms. Rose mixes the names of all the girls in a box and chooses two without looking. Then she does the same for the boys' names.

Mr. Champine's homeroom is working in groups of four. He sends students from one of the groups.

Mrs. Santanella puts the names of all the students in a box, mixes the names, and pulls out four names without looking.

Mrs. Kim chooses the four students in the first row, closest to where she is standing.

a. For each homeroom, tell if the teacher used a *random sample*, a *convenience sample*, a *cluster sample*, or a *stratified random sample*.

b. How would you choose the four students? Why?

Identify the type of sample and describe the population of the survey. Then tell if the sample is biased. Explain your reasoning.

2. The school board wants to find out how voters feel about a proposed addition to the high school. Each board member randomly selects 30 names from the phone book and calls each number during a weekday afternoon.

3. A local sports station wants to find out how many hours per week people in the viewing area watch sporting events on television. The station surveys people at the nearby sports stadium.

4. The alumni club at Pioneer High School wants to set up an employment referral network among alumni. The club mails a survey to all graduates from the past 20 years and asks them to complete and return it.

5. A taxicab company wants to know if its customers are satisfied with the service. Each driver surveys every tenth customer during the day.

6. **BIOLOGY** A biologist wants to estimate the number of deer in a state park. She gathers a random sample of 30 deer and tags each one. Then she releases the tagged deer back into the park. After the deer have mixed with the other deer, she gathers a second random sample of 30 deer and finds that there are 8 tagged deer in this sample.

   a. What percent of the deer in the second sample are tagged? Use this percent to estimate the number of deer in the entire park.

   b. **Writing** Do you think your estimate from part (a) is accurate? Why or why not? What could you do to get a better estimate?

Connection   ARCHAEOLOGY

When it becomes too expensive to excavate an entire site, archaeologists resort to sampling portions of the site. Artifacts on the surface of a site may influence where to excavate.

The images below are of an excavation site, where the 42 red squares on each image show the portions to be excavated. For each image, tell what sampling method is suggested.

7.
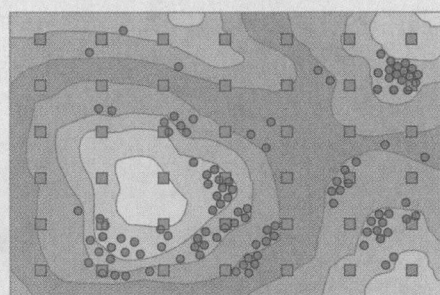

Archaeological dig at Port au Choix National Historic Park in Newfoundland, Canada

8.
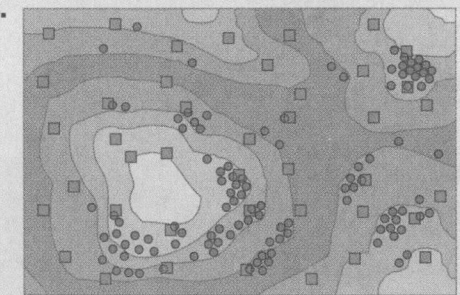

9.
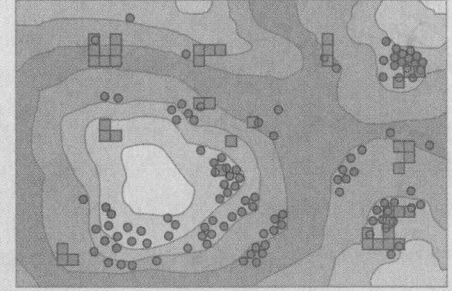

10. **Writing** If there are no artifacts on the surface of the site, which sampling method shown above seems the most reasonable? Why?

11. a. Each blue dot in the images above represents the location of a buried artifact. If you assume that all the artifacts that lie in or partially in the red squares will be found, which sampling method finds the most artifacts?

    b. Do you think this method will always be the best? Why or why not?

12. **Open-ended Problem** Suppose you are in charge of an archaeological dig and must use sampling to stay on budget. Describe the procedure that you would use in your attempt to find the most artifacts. Explain why you think that your procedure is the best.

Look back at the
article on pages
234–236.

*Computer images like the ones below, which were created by Donna Cox,
are used to study the ecology of crops and the insects that attack them.
Each image below shows a cornfield with 800 plants.*

The **black squares**
show plants with no
larvae.

The green squares
show plants with
healthy larvae. These
larvae damage crops.

The **red squares**
show plants with
infected larvae. These
larvae do minimal
damage to crops.

Other colors show
plants with a mixture
of healthy and
infected larvae.

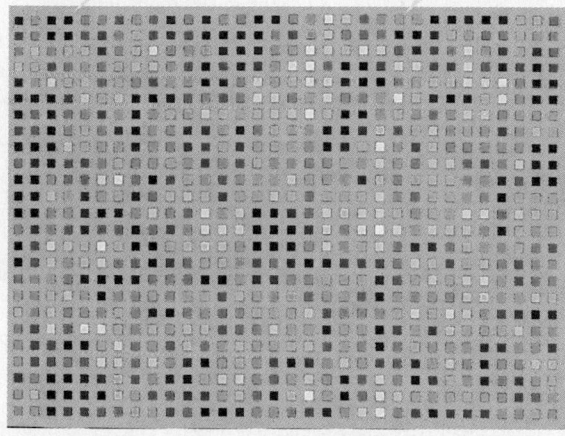

Field 1

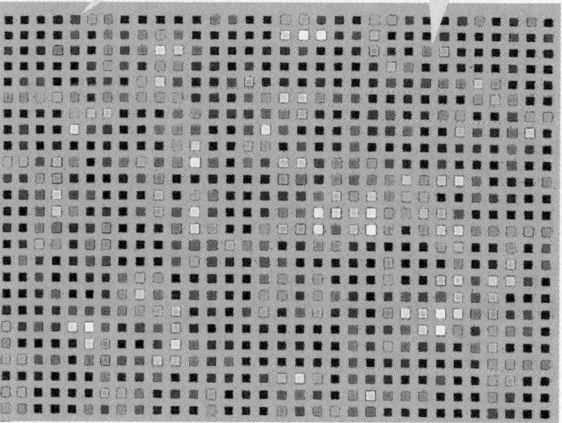

Field 2

13. **Visual Thinking** For each field, visually estimate the percent of the
    plants being damaged by healthy larvae. Which field will probably
    produce more corn? Why?

**Use the indicated method to select a sample of 10 plants from each field. Note
the color associated with each plant in your samples.**

14. cluster   15. systematic   16. random

17. a. For each of your samples in Exercises 14–16, count the number of
       plants being damaged by healthy larvae. Convert the counts to percents.

    b. For each field, do your sample percents from part (a) agree with each
       other? Do they agree with your visual estimate from Exercise 13? If
       not, explain any discrepancies.

18. **Open-ended Problem** Obtain another sample from each field. You
    may want to choose one of the sampling methods in Exercises 14–16 but
    this time use a sample with more than 10 plants, or you may want to use
    some other sampling method. Once again calculate the percent of plants
    being damaged by healthy larvae. Compare the results with those from
    Exercise 17.

19. **Writing** Suppose a farmer asks for your help in determining the damage
    being done to the farmer's corn crop by European corn borer larvae.
    Describe the steps you would take to judge the health of the crop without
    examining every plant.

20. **Cooperative Learning** Work with the members of your group from the *Ongoing Assessment* exercise from Section 6.1.

    a. Describe how you will choose a random sample of at least 30 people to take the survey you wrote in Section 6.2.

    b. **Writing** Describe how you will contact people to respond to your survey. You should consider these things:

    • Will you interview people in person? interview people over the phone? distribute a written questionnaire? use some other method?

    • When will you distribute your survey? Where?

    • Is there any information you should record about each respondent?

    • How will you know when you have responses from enough people? How will you ensure that someone does not take the survey more than once?

    c. Distribute your survey from Section 6.2. Keep the surveys or a record of people's responses for your work in later sections.

For Exercises 21–23:

a. Describe any bias in the question. *(Section 6.2)*

b. Tell whether the data that can be gathered from each question are *categorical* or *numerical*. Then tell what type of graph you would use to display the data. *(Section 6.1)*

21. "A recent study says that people who read books regularly have a better vocabulary than people who seldom read books. How many books do you read in a month?"

22. "Do you support the decisions made at last Monday's board meeting?"

23. "What is your favorite flavor of ice cream?"

24. Find the mean, the median, and the mode of the data in the table below. *(Toolbox, page 790)*

City	Bern	Brasília	Buenos Aires	London	Madrid	Mexico City	Paris	Seoul	Tokyo
Price of 1 lb of cheese ($)	6.15	1.28	0.45	2.32	4.23	2.76	2.85	5.40	6.32

25. **GOVERNMENT** Use the table at the right. *(Section 3.5)*

    a. Choose any two periods and write an exponential function to represent the growth. Check your function by comparing its predictions with the other data values in the table.

    b. **Technology** Use a graphing calculator or statistical software to perform exponential regression. Compare the equation with your results from part (a).

Period	Number of new stamps issued
1839–1868	88
1869–1898	205
1899–1928	354
1929–1958	476
1959–1988	1277

ASSESS YOUR PROGRESS

VOCABULARY

numerical data (p. 238)
categorical data (p. 238)
population (p. 239)
sample (p. 239)
biased question (p. 245)
self-selected sample (p. 250)

systematic sample (p. 250)
convenience sample (p. 250)
random sample (p. 250)
stratified random sample (p. 250)
cluster sample (p. 250)
biased sample (p. 251)

Tell whether the data that can be gathered about each variable are *categorical* or *numerical*. Then describe the categories or numbers. *(Section 6.1)*

1. savings account balance

2. the month of a person's birthday

3. a person's coat size

4. figure skating scores

For each graph, tell if it is an appropriate way to display the data. If not, tell what type of display would be better and why. *(Section 6.1)*

5. **Household Water Use**

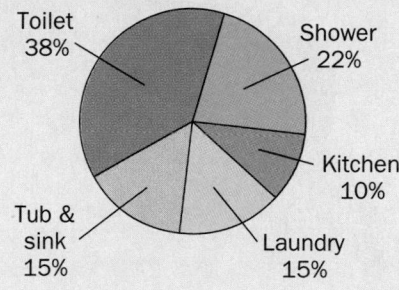

6. **Parker High School Enrollment**

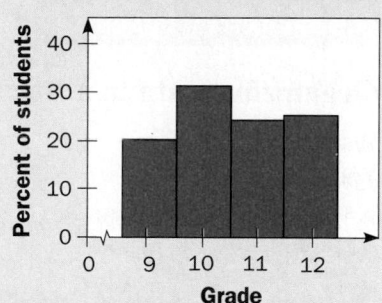

Tell why each question may be biased. Then describe what changes you would make to improve the question. *(Section 6.2)*

7. "College tuition costs are on the rise. Do you favor government spending on student financial aid?"

8. Your dentist asks, "Have you been flossing regularly?"

For Exercises 9 and 10, identify the type of sample and describe the population of the survey. Then tell if the sample is biased. Explain. *(Sections 6.1 and 6.3)*

9. A veterinarian wants to know where to set up a new practice within a county. She randomly telephones 25 people from four towns within the county to ask how many pets are in each household.

10. A department store wants to know customers' opinions of its merchandise and service. The store offers a chance of winning a gift certificate through a random drawing if the customer fills out a survey.

11. **Writing** Describe two ways to obtain a sample of 50 people from your community to ask about the need for more community parks. *(Section 6.3)*

12. **Journal** Is it ever appropriate to ask a biased question? Explain.

6.4 Displaying and Analyzing Data

Learn how to...

- **make histograms and box plots**

So you can...

- **create appropriate displays for numerical data**

Newspapers and magazines often display data in graphs so that readers can easily understand the information presented. Sometimes a poorly presented display can lead people to draw wrong conclusions about the data. It is important to know how to make good data displays so that you present information accurately.

EXPLORATION

COOPERATIVE LEARNING

Organizing Data in a Box Plot

Work with a partner.
You will need:
- scissors
- graph paper

1 Count the number of letters in your first name, in your last name, and in your first and last names combined. Share your results with your class. Write down each person's responses. Use the data for the total number of letters in a person's first and last names for this Exploration. Save the rest of the data to use in Exercises 4–6.

2 List the data in order from least to greatest on a strip of graph paper. Put one number in each square. Then cut the paper so that there are no extra squares to the left or right of the data.

| 7 | 7 | 8 | 9 | 9 | 9 | 9 | 9 | 10 | 10 | 11 | 11 | 11 | 12 | 12 | 12 | 12 | 12 | 13 | 13 | 13 | 13 | 13 | 14 | 14 | 15 | 15 | 16 | 17 |

3 Fold the strip of paper in half. Then fold the paper in half again. Open the paper up and draw a line on each fold.

| 7 | 7 | 8 | 9 | 9 | 9 | 9 | 9 | 10 | 10 | 11 | 11 | 11 | 12 | 12 | 12 | 12 | 12 | 13 | 13 | 13 | 13 | 13 | 14 | 14 | 15 | 15 | 16 | 17 |

Questions

1. What does the middle fold tell you about the data? What do the other folds represent?

2. Below what number of letters do 25% of the names fall? 75% of the names?

In the Exploration you completed the first steps of making a *box-and-whisker plot*, or simply a *box plot*. A **box plot** is a display that shows the median, the *quartiles*, and the *extremes* of a data set.

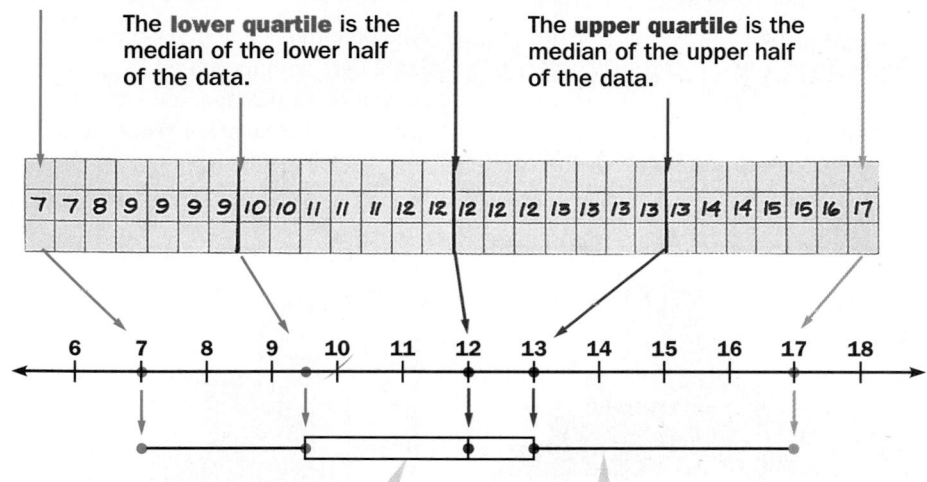

The **lower extreme** is the least data value.

The **lower quartile** is the median of the lower half of the data.

The **median** divides the data set into two halves.

The **upper quartile** is the median of the upper half of the data.

The **upper extreme** is the greatest data value.

The box in a box plot shows where the **middle 50%** of the data fall.

Each "whisker" in a box plot shows where the **extreme 25%** of the data fall.

EXAMPLE 1   Application: Economics

Make a box plot of the data in the table.

SOLUTION

Step 1 Find the median, the quartiles, and the extremes of the data and plot these points just below a number line.

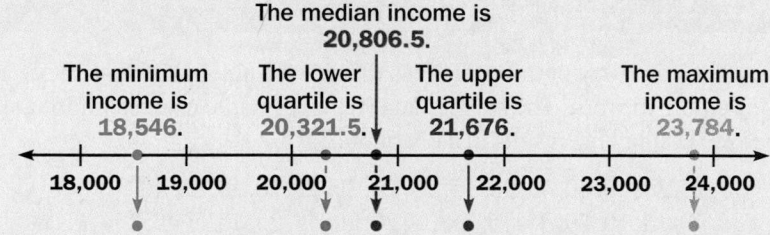

The median income is 20,806.5.

The minimum income is 18,546.

The lower quartile is 20,321.5.

The upper quartile is 21,676.

The maximum income is 23,784.

Step 2 Make the box plot.

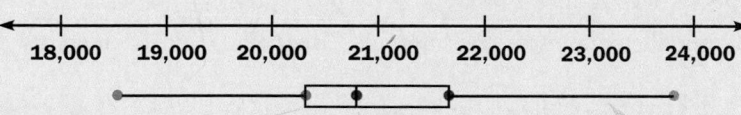

Draw a box from the lower quartile to the upper quartile.

Then draw a vertical line segment through the median.

Draw line segments from the box to the extremes.

Average 1994 Per Capita Personal Income for Midwestern States	
State	**Income ($)**
Illinois	23,784
Indiana........	20,378
Iowa..........	20,265
Kansas........	20,896
Michigan	22,333
Minnesota	22,453
Missouri.......	20,717
Nebraska	20,488
North Dakota...	18,546
Ohio	20,928
South Dakota...	19,577
Wisconsin	21,019

Histograms

A **histogram**, like a box plot, is a graph that displays numerical data. The horizontal axis is a number line divided into intervals of equal width. The vertical axis shows the *frequency* of the data items that fall within each interval.

When the frequency is a count of the data within each interval, the vertical axis shows the **absolute frequency**.

When the frequency gives the count of the data within each interval as a *percent* of all the data, the vertical axis shows the **relative frequency**.

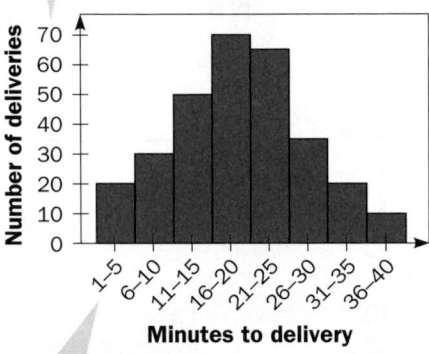

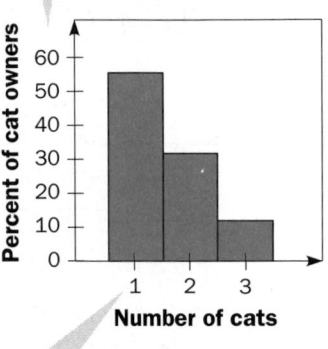

Here, each interval covers a period of 5 min.

Here, each interval includes a single integer.

A histogram displays the overall shape of the distribution of data values. The histogram on the left above shows a **symmetric distribution** because its shape is approximately symmetrical about a vertical line passing through the interval with the greatest frequency. The histogram on the right shows a **skewed distribution** because it is not symmetric.

EXAMPLE 2   Application: Transportation

A company took a survey of its employees to learn the amount of time they spend commuting to work. The survey data (reported to the nearest 5 min) are shown below. Display the results in a histogram.

15, 25, 30, 40, 60, 40, 25, 15, 15, 20, 30, 75, 20, 15, 65, 40,
30, 20, 15, 10, 20, 35, 45, 35, 30, 20, 10, 55, 25, 15, 5, 45, 20,
10, 5, 15, 30, 30, 50, 15, 5, 10, 45, 30, 20, 20, 10, 30, 20, 30

SOLUTION

Step 1 Group the data into equal intervals. (Between 5 and 10 intervals is best.)

The data values range from 5 to 75. Choose intervals of equal width that will allow you to cover the entire range of values, such as 1–10, 11–20, 21–30, 31–40, 41–50, 51–60, 61–70, and 71–80.

Step 2 Organize the data into a frequency table.

Interval	Tally	Frequency	Relative frequency
1-10	HHT III	8	16%
11-20	HHT HHT HHT II	17	34%
21-30	HHT HHT II	12	24%
31-40	HHT	5	10%
41-50	IIII	4	8%
51-60	II	2	4%
61-70	I	1	2%
71-80	I	1	2%
Total		50	100%

Find the relative frequency by dividing each absolute frequency by the total number of data values.

$$\frac{2}{50} = 0.04 = 4\%$$

Step 3 Draw the histogram.

For each interval, draw a rectangle with a width equal to the interval width and a height equal to the frequency of the data within the interval. You may draw either type of histogram shown below.

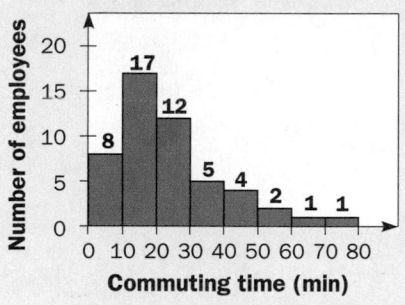

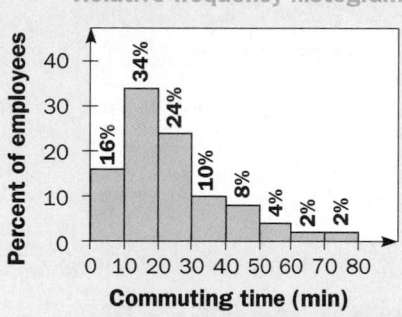

You can use technology to draw histograms and box plots. The displays below both show the same data taken from an auto magazine survey of new car prices.

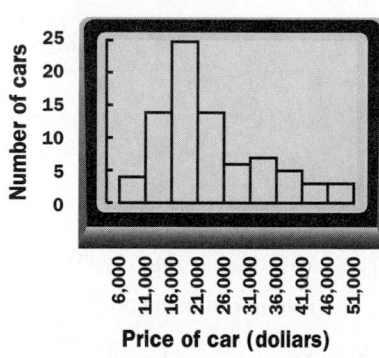

For more information about drawing histograms and box plots, see the *Technology Handbook*, p. 813.

THINK AND COMMUNICATE

1. What visual impressions do you get from each graph of car prices?

2. Estimate the median car price. How does each graph show that there are car prices much greater than the median?

3. Do you think the mean car price is *about the same*, *greater than*, or *less than* the median price? Explain.

The weights, to the nearest quarter pound, of fish gathered in one harvest at a fish farm are listed below. Use the data to answer Questions 1 and 2.

2.25, 1.0, 0.75, 1.5, 2.0, 2.75, 4.0, 1.0, 1.25, 1.75, 1.5, 2.5, 3.25,
5.0, 3.5, 1.0, 2.0, 1.75, 2.25, 1.0, 1.75, 1.25, 2.25, 2.5, 1.25, 3.0,
1.0, 1.5, 2.0, 1.25, 2.75, 1.25, 1.75, 2.25, 1.5, 1.25, 3.75, 1.0,
4.75, 1.5, 3.0, 1.75, 1.0, 3.25, 4.5, 1.25, 2.25, 3.75, 1.5

1. Make a histogram of the data. Does your histogram display *relative frequencies* or *absolute frequencies*? Is the distribution *symmetric* or *skewed*?

2. a. Make a box plot of the data.

   b. Above what weight are 50% of the harvest? 25% of the harvest?

The weights of the fish gathered in another harvest are shown below.

1.5, 0.75, 1.25, 1.75, 1.0, 2.25, 1.5, 1.25, 1.0, 1.75, 2.75, 1.25,
1.5, 1.75, 2.0, 1.75, 1.25, 1.5, 1.0, 3.0, 2.0, 2.25, 1.25, 1.5, 1.0,
1.75, 2.5, 1.5, 1.0, 1.25, 1.75, 1.5, 2.0, 1.5, 3.25, 1.75, 1.25, 1.5,
2.25, 1.75, 2.0, 1.25, 1.5, 1.75, 2.5, 2.0, 1.5, 2.0, 1.25, 1.5

3. Make a box plot of the data. Compare this box plot with the box plot you made in Question 2. Which harvest was better? Why?

6.4 | **Exercises and Applications**

Extra Practice exercises on page 758

1. Writing What information does a box plot tell you about a set of data? What information does a histogram tell you?

For each histogram, tell whether the distribution is *symmetric* or *skewed*.

2.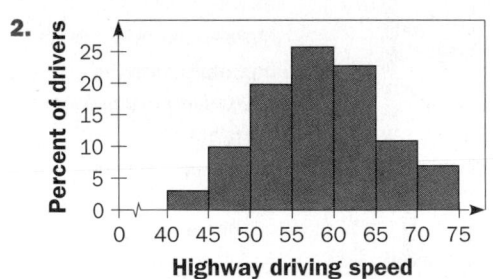
Highway driving speed

3.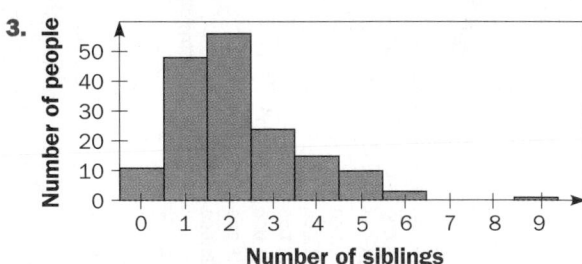
Number of siblings

Use the data you gathered in the Exploration on page 258 to answer Exercises 4–6.

4. Make a box plot of the number of letters in the first names of the students in your class. Identify the medians, quartiles, and extremes of the data.

5. Repeat Exercise 4 using the number of letters in the last names of the students in your class.

6. Compare the box plots you made in Exercises 4 and 5. Describe the relationship between the medians, quartiles, and extremes.

The highest elevation in the United States is Mt. McKinley, in Alaska. Native Americans living near the mountain call it *Denali*, which means "the great one." The table below gives the highest elevation in each of the 50 states.

Mt. McKinley

HIGHEST ELEVATIONS PER STATE (FEET)						
West of the Mississippi				**East of the Mississippi**		
AK 20,320	MN 2,301	TX 8,749		IN 1,257	NC 6,684	
AZ 12,633	MO 1,772	UT 13,528		KY 4,139	OH 1,549	
AR 2,753	MT 12,799	WA 14,410		ME 5,267	PA 3,213	
CA 14,494	NE 5,426	WY 13,804		MD 3,360	RI 812	
CO 14,433	NV 13,140		AL 2,405	MA 3,487	SC 3,560	
HI 13,796	NM 13,161		CT 2,380	MI 1,979	TN 6,643	
ID 12,662	ND 3,506		DE 442	MS 806	VT 4,393	
IA 1,670	OK 4,973		FL 345	NH 6,288	VA 5,729	
KS 4,039	OR 11,239		GA 4,784	NJ 1,803	WV 4,861	
LA 535	SD 7,242		IL 1,235	NY 5,344	WI 1,951	

7. a. Find the median, the upper and lower quartiles, and the extremes of the elevation data for all 50 states.

b. Make a box plot. Between what numbers do 50% of the data fall?

c. If you removed Alaska's highest elevation from the data set, how would the box plot change?

8. a. Make a comparative box plot for the highest elevations east and west of the Mississippi River.

b. The Appalachian Mountains are east of the Mississippi River, and the Rocky Mountains are west. Which mountain range has the higher mountains? How can you tell from the comparative box plot?

9. a. Make a histogram to display the elevation data for all 50 states. Is the distribution *symmetric* or *skewed*?

b. Open-ended Problem Estimate the "middle" of the distribution.

c. Find the mean and median of the elevation data. How do these values compare with your estimate from part (b)?

d. Do you think the *mean* or the *median* is a better measure of the "middle" of the distribution? Why?

10. Research Use an almanac or a set of encyclopedias to find the lowest elevation in each of the 50 states. Display the data in either a box plot or a histogram. Describe what the graph shows about the data. Compare the lowest elevation data with the highest elevation data given here.

BY THE WAY

Denali is a word in a Native American language that is part of a group of related languages called *Athabaskan*. Speakers of Athabaskan languages live in parts of Canada, the United States, and Mexico.

For each graph, tell why the graph may misrepresent the data. Then suggest a way to improve the graph.

11.

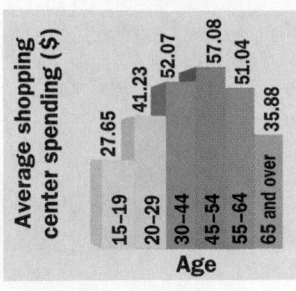

12.

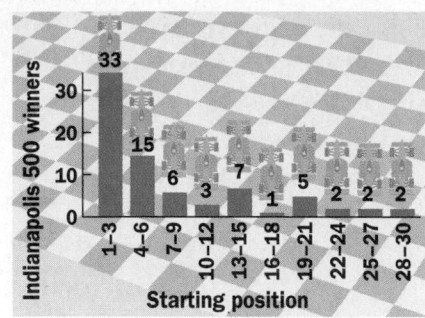

13. Writing If you are given only an absolute frequency histogram, can you create a relative frequency histogram? Can you create an absolute frequency histogram given only a relative frequency histogram? Explain.

Connection ▸ HISTORY

European settlers began arriving in what is now the United States in the sixteenth century. In 1787, Delaware, Pennsylvania, and New Jersey became the first former colonies to achieve statehood. The box plots compare the year in which a present-day state was settled by Europeans and the year in which it achieved statehood.

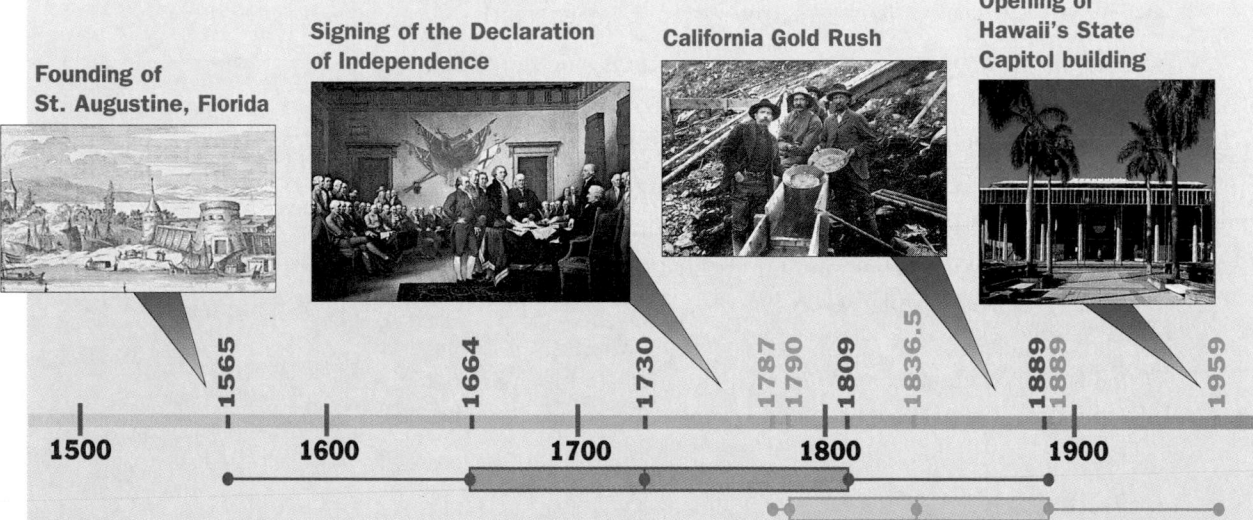

Founding of St. Augustine, Florida

Signing of the Declaration of Independence

California Gold Rush

Opening of Hawaii's State Capitol building

1565   1664   1730   1787 1790 1809   1836.5   1889 1889   1959

1500   1600   1700   1800   1900

14. By what year did half of the present-day states have European settlements? By what year did half of the present-day states achieve statehood?

15. In 1788, eight former colonies became states—the largest number achieving statehood in one year. This number is twice the next closest number to achieve statehood in one year. What effect does this have on the statehood box plot?

16. Research Find out the year your state was settled by Europeans and the year it achieved statehood. In which quarter of the data is the year your state was settled? In which quarter of the data is the year your state achieved statehood?

1994 Recording Industry Sales by Age						
Age group (years)	10–14	15–19	20–24	25–29	30–34	35–39
Percent of all sales	7.0	16.8	15.4	12.6	11.8	11.5

17. Make a histogram to display the data in the table from the Recording Industry Association of America. Does the histogram show a *symmetric distribution* or a *skewed distribution*? Explain.

18. Make another histogram of the data. Combine the data so that each interval covers a period of 10 years. How does the choice of intervals change the appearance of the distribution?

19. Writing People who are 40 years old or older accounted for 24% of the total recording industry sales in 1994.

    a. People who are 10 and older accounted for what percent of the recording industry sales in 1994? Why do you think this total is not equal to 100%?

    b. Carly wants to compare spending by people under age 20, people between 20 and 30, and people over 30. Can she use a histogram? Explain.

BY THE WAY

The soundtrack from Walt Disney's "The Lion King" and Ace of Base's "The Sign" tied for the best-selling album in 1994, at about 7 million copies each.

ONGOING ASSESSMENT

20. Cooperative Learning Work with the members of your group from the *Ongoing Assessment* exercise from Section 6.1.

    a. Organize the data you gathered in Section 6.3. Decide how you will display the data. You should use at least one histogram or box plot and at least one circle graph or bar graph.

    b. Make a frequency table for each set of numerical data. Find the mean and median of the data. If you make a histogram, decide if it will display absolute frequencies or relative frequencies.

    c. Draw the graphs to display your data. What conclusions can you make about your data?

SPIRAL REVIEW

Identify the type of sample and describe the population of the survey. Then tell if the sample is biased. Explain your reasoning. *(Section 6.3)*

21. The members of the drama club are trying to decide whether they should perform a fall play. The group surveys every fifth person entering the door at the spring play.

22. A sports club wants to find out which football team is the most popular. The manager asks people to call in a vote for their favorite team.

Simplify. *(Toolbox, page 777)*

23. $\sqrt{54}$ **24.** $\sqrt{200}$ **25.** $\sqrt{\dfrac{300}{3}}$ **26.** $\sqrt{\dfrac{880}{5}}$

6.5 Describing the Variation of Data

Learn how to...

- find and interpret the range, interquartile range, and standard deviation of a data set

So you can...

- determine the variability of data, such as ozone readings

Life on Earth would not be possible without the ozone layer in the upper atmosphere filtering the ultraviolet rays of the sun. Recent studies have shown that the thickness of the ozone layer is decreasing. In the tropics, unlike other areas, the change of seasons has little effect on the thickness of the ozone layer. At non-tropical latitudes, the ozone layer is thickest in the spring and thinnest in the fall.

THINK AND COMMUNICATE

The readings below, which give the thickness of the ozone layer in Dobson units (DU), were taken in Fresno, California, for two 20-day periods.

March readings: 427, 466, 372, 299, 293, 284, 298, 284, 314, 302, 286, 296, 306, 308, 320, 318, 344, 345, 354, 381

November readings: 317, 330, 316, 296, 270, 271, 295, 277, 275, 275, 269, 275, 272, 267, 270, 291, 275, 268, 261, 296

The comparative box plot displays these readings.

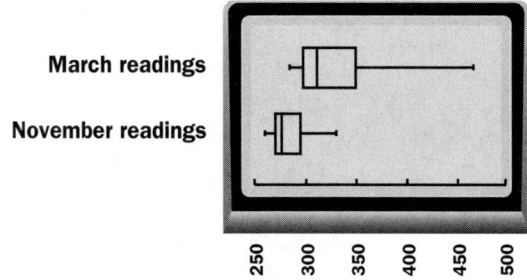

1. How does the graph show that the ozone over Fresno is generally thicker in the spring than in the fall?

2. In which month, *March* or *November*, would it be more likely to obtain a reading that is much different from the reading on the previous day? Why?

To determine the variability of a set of data, it is often helpful to find the *range*, the *interquartile range*, and any *outliers* of the data set. The **range** of a data set is the difference between the maximum and minimum data values. The **interquartile range (IQR)** of a data set is the difference between the upper quartile and the lower quartile. A data value can be considered an **outlier** if its distance from the nearer quartile is more than 1.5 times the interquartile range.

EXAMPLE 1 **Application: Earth Science**

Use the ozone readings on the previous page.

a. Find the range and the interquartile range for each set of readings.

b. Identify the outliers in each data set.

SOLUTION

a. First find the minimum and maximum values (shown in gold below) and the upper and lower quartiles (shown in green) for each data set.

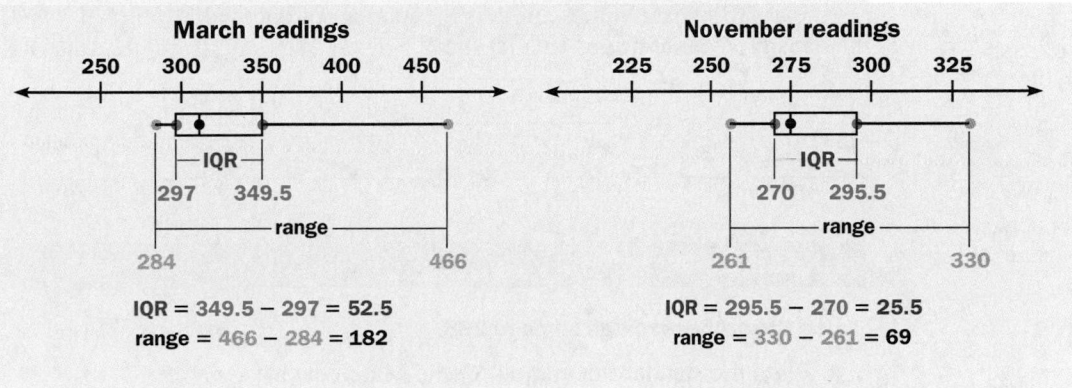

IQR = 349.5 − 297 = **52.5**   IQR = 295.5 − 270 = **25.5**
range = 466 − 284 = **182**   range = 330 − 261 = **69**

b. Use each IQR from part (a) to set bounds (shown in blue) for the outliers.

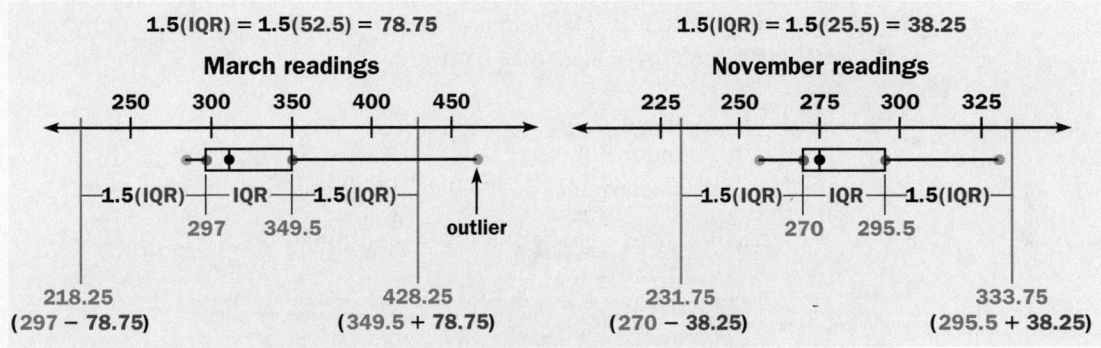

Any values in the March data set that are less than 218.25 or greater than 428.25 are outliers. The ozone reading of 466 is an outlier.

Any values in the November data set that are less than 231.75 or greater than 333.75 are outliers. Since all data values are between 231.75 and 333.75, there are no outliers.

THINK AND COMMUNICATE

3. For each data set in Example 1, compare the range and interquartile range. How do these numbers describe the variability of each data set?

4. A scientist analyzing the March ozone data decides to recalculate the mean, median, range, and interquartile range without the 466 reading. Why do you think the scientist would do this?

Standard Deviation

The range and the interquartile range measure the variability of a data set using only the quartiles and extremes of the data. Another measure of variability, called *standard deviation*, uses the square of the deviation of every data value from the mean.

Standard Deviation

For a data set $x_1, x_2, x_3, \ldots, x_n$, where $\bar{x}$ (read "x bar") is the mean of the data values and n is the number of data values, the **standard deviation** σ (read "sigma") is:

$$\sigma = \sqrt{\frac{(x_1 - \bar{x})^2 + (x_2 - \bar{x})^2 + \cdots + (x_n - \bar{x})^2}{n}}$$

EXAMPLE 2   Application: Earth Science

Use the ozone readings on page 266.

a. Find the standard deviation for the March data.

b. Find the standard deviation for the November data.

SOLUTION

You can find the standard deviation of a set of data by using a table or by using a graphing calculator or statistical software.

a. Use a table.

Step 2 Subtract the mean from each data value.
Example:
427 − 329.85 = 97.15

Step 3 Square each difference from Step 2.
Example:
(97.15)^2 = 9438.1225

Step 1 Find the mean of the data.
$$\bar{x} = \frac{6597}{20} = 329.85$$

Ozone reading	reading − $\bar{x}$	(reading − $\bar{x}$)^2
427	97.15	9,438.1225
466	136.15	18,536.8225
372	42.15	1,776.6225
299	−30.85	951.7225
⋮	⋮	⋮
345	15.15	229.5225
354	24.15	583.2225
381	51.15	2,616.3225
6597		46,288.55

Step 4 Sum the squares.

$$\sigma = \sqrt{\frac{46{,}288.55}{20}} \approx 48.1$$

Step 5 Divide the sum by the number of data values and take the square root.

The standard deviation is about 48.1.

SOLUTION

b. Use a graphing calculator or statistical software.

First enter the data.   Then have the calculator or computer find the standard deviation. It is usually shown as σ_x or σ_n.

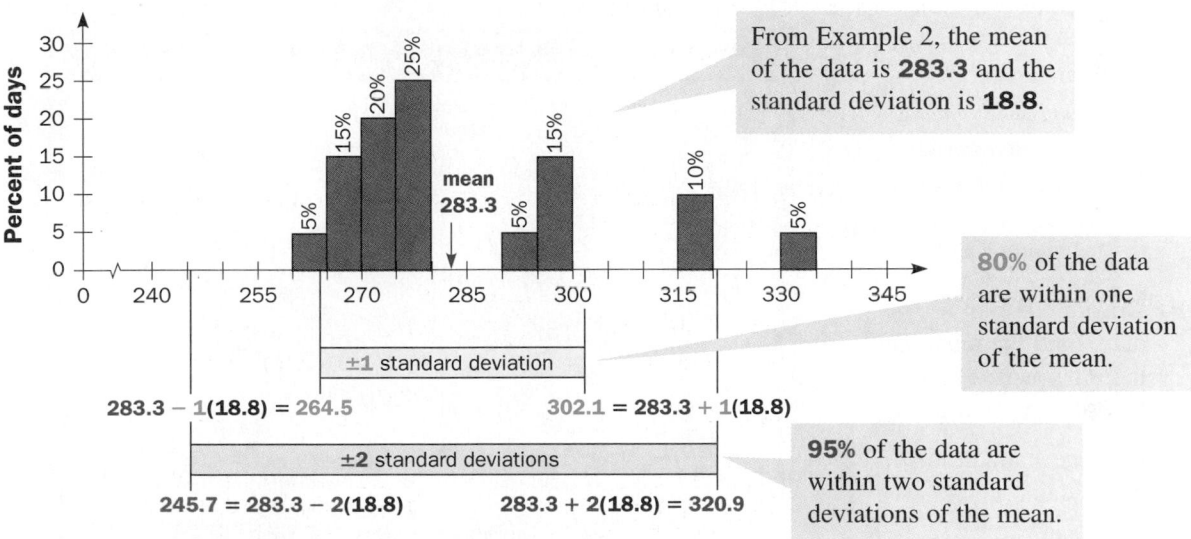

```
L1      L2      L3
317    -----   -----
330
316
296
270
271
295
L1=(317,330,316...
```

```
1-Var Stats
x̄=283.3
Σx=5666
Σx²=1612232
Sx=19.26846346
σx=18.78057507
↓n=20
```

The standard deviation is about **18.8**.

WATCH OUT!

In the display shown, S_x is the standard deviation for a sample. This book will use the standard deviation for a population, σ_x, regardless of whether the data come from a sample or an entire population.

Statisticians often describe a distribution of data by the percent of the data that fall within a certain number of standard deviations of the mean. The histogram below shows the percent of data from the November ozone readings within one and two standard deviations of the mean.

From Example 2, the mean of the data is **283.3** and the standard deviation is **18.8**.

$283.3 - 1(18.8) = 264.5$ $302.1 = 283.3 + 1(18.8)$

$245.7 = 283.3 - 2(18.8)$ $283.3 + 2(18.8) = 320.9$

80% of the data are within one standard deviation of the mean.

95% of the data are within two standard deviations of the mean.

☑ CHECKING KEY CONCEPTS

MEDICINE Doctors often take a sample of blood from a patient to help with a diagnosis. The white blood cell counts from 27 patients are shown at the right.

1. Make a box plot of the data. Are there any outliers?

2. Find the mean and the standard deviation of the data.

3. Identify the values that are within one standard deviation of the mean. What percent of the values are within one standard deviation?

4. A white blood cell count greater than 10,000 indicates the possibility of a bacterial infection. How many standard deviations from the mean is 10,000?

White blood cell counts
5620, 5730, 5750, 6210,
6390, 6750, 6900, 7030,
7230, 7450, 7600, 7710,
7730, 7850, 8090, 8370,
8630, 8880, 9060, 9240,
9380, 9440, 9700, 9890,
10,250, 10,900, 11,070

6.5 | **Exercises and Applications**

Extra Practice exercises on page 758

The box plot at the right shows the number of minutes a restaurant's customers sit at their table before they place their orders. Use the box plot to find each value asked for in Exercises 1–3.

1. the range   **2.** the interquartile range   **3.** an outlier

One way to estimate the standard deviation of a set of data is to divide the range by 4. Use this rule to estimate the standard deviation for each set of data. Then find the standard deviation and compare your results.

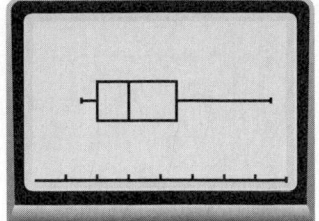

0 2 4 6 8 10 12 14 16

4. 10, 12, 13, 16, 18   **5.** 15, 22, 30, 40, 55, 75

6. 6, 9, 13, 18, 22, 28, 32   **7.** 1, 5, 15, 25

8. Writing Can a data set have a standard deviation of zero? Explain.

Connection ▶ **BIOLOGY**

In 1995 there were more than 900 plant and animal species protected under the United States Endangered Species Act. The map shows the number of endangered species in each state.

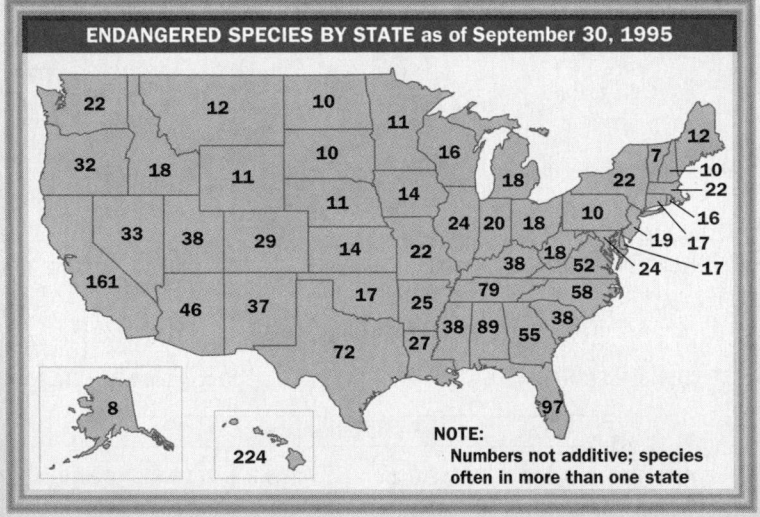

ENDANGERED SPECIES BY STATE as of September 30, 1995

NOTE:
Numbers not additive; species often in more than one state

9. Draw a box plot of the data. Find the range and the interquartile range. Identify any outliers.

10. Find the standard deviation of the data.

11. About what percent of the data are within one standard deviation of the mean? within two standard deviations of the mean?

12. *Chebyshev's rule* says that the proportion of data within k standard deviations of the mean, where $k > 1$, is at least $1 - \dfrac{1}{k^2}$.

    a. Use Chebyshev's rule to estimate the percent of data within two standard deviations of the mean. How does the estimate from Chebyshev's rule compare with your answer to the second question in Exercise 11?

    b. Use Chebyshev's rule to estimate the percent of data within three standard deviations of the mean.

13. About how many standard deviations from the mean is the data value for California?

BY THE WAY

The bluefin tuna, one of the largest fish in the Atlantic, was placed on the 1994 list of the ten most endangered species. The bluefin tuna population has decreased by 80% since 1974.

Donna Cox

Look back at the article on pages 234–236.

The matrix below gives typical cornfield data that Donna Cox might use to create a visual display. Each element of the matrix corresponds to a corn plant and has two values related to corn borer larvae, as described.

The numbers in the **shaded column** tell the total number of corn borer larvae on each plant.

	C1		C2		C3		C4		C5		C6		C7		C8		C9	
R1	3	1	7	4	0	0	4	1	8	3	5	3	9	6	8	7	11	5
R2	4	2	0	0	0	0	6	4	4	1	3	2	0	0	6	5	10	10
R3	0	0	0	0	0	0	3	2	3	0	3	3	5	2	1	0	9	7
R4	0	0	0	0	0	0	0	0	1	0	0	0	2	2	0	0	3	3
R5	0	0	0	0	3	0	0	0	0	0	0	0	7	3	9	5	5	1
R6	5	3	6	6	0	0	3	3	6	3	0	0	0	0	0	0	0	0
R7	0	0	0	0	0	0	0	0	0	0	2	0	0	0	0	0	1	0
R8	0	0	0	0	0	0	0	0	0	0	0	0	4	2	8	6	0	0
R9	0	0	0	0	0	0	0	0	0	0	0	0	0	0	3	0	0	0
R10	3	0	0	0	0	0	0	0	0	0	0	0	0	0	0	0	0	0

The numbers in the **unshaded column** tell how many of the larvae on each plant are infected.

14. Using only the plants that actually have larvae on them, make a comparative box plot of the data in the shaded and unshaded columns of the matrix.

15. Find the median, the range, and the interquartile range of each box plot you made in Exercise 14. How do corresponding values compare with one another?

16. Challenge How would your box plots and statistics change if you used all the data?

Use the graphs below to answer Exercises 17–19.

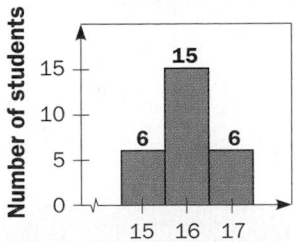

Ages of students in a class

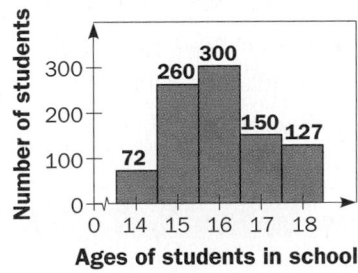

Ages of students in school

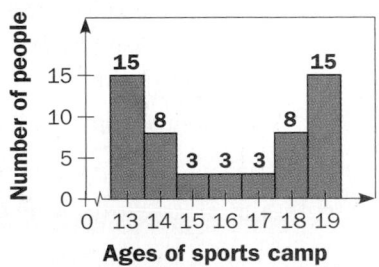

Ages of sports camp participants and staff

17. Find the mean of each data set. How do the means compare?

18. Visual Thinking Which set of data do you think has the largest standard deviation? Why?

19. Find the standard deviation of each data set. How do the standard deviations compare with one another?

The Pan-African Film and Television Festival was first held in the winter of 1969 to spotlight African-made films. The week-long festival is now held every other year in Burkina Faso, Africa. Countries outside of Africa also participate.

Year	1969	1970	1972	1973	1976	1979	1981	1983
Number of films	23	40	36	51	75	78	78	69
Number of African countries	5	9	18	23	17	16	16	25
Total number of countries	7	9	23	29	26	26	27	37

20. **a.** Make a box plot using the total number of countries participating in each of the first 8 festivals. Find the range and the interquartile range. Are there any outliers?

   b. Find the mean and standard deviation of the data. Are any values more than one standard deviation from the mean? more than two standard deviations from the mean?

21. Repeat Exercise 20 for the number of African countries participating in the festival each year.

22. Repeat Exercise 20 for the number of films in the festival.

In downtown Ouagadougou, Burkina Faso, a sculpture representing reels of movie film and camera lenses celebrates the Pan-African Film and Television Festival.

ONGOING ASSESSMENT

23. **Cooperative Learning** Work with the members of your group from the *Ongoing Assessment* exercise from Section 6.1. Complete parts (a)–(c) for each set of numerical data you gathered.

   a. Draw a box plot of the data. Find the range and the interquartile range. Identify any outliers.

   b. Calculate the standard deviation for the data. How many values are within one standard deviation of the mean? within two standard deviations of the mean? within three standard deviations of the mean?

   c. What conclusions can you make about your data?

SPIRAL REVIEW

24. A housing committee took a survey of the amount of money people pay to rent an apartment. Use the results of the survey shown in the table at the right to make a histogram of the data. Describe the distribution of the data. *(Section 6.4)*

Solve each inequality. *(Toolbox, p. 787)*

25. $x + 5 \leq 13$ 26. $-4x - 6 \geq 18$ 27. $3x + 2 < 8$

Apartment rent ($)
295, 600, 225, 280, 430, 290, 310, 200, 300, 725, 350, 575, 260, 375, 450, 400, 380, 250, 175, 325, 420, 425, 130, 475, 300, 220, 340, 550, 375, 625, 330, 525, 180, 325, 360, 350

6.6 Making Decisions from Samples

Learn how to...

- **find the margin of error for a sample proportion**

So you can...

- **make decisions about opinion poll results**

Although statistics such as mean and standard deviation do not apply to categorical data, you can find the *proportion* of data in a particular category. For example, if 7 out of 10 people respond "yes" to a survey question, then the **sample proportion** of "yes" responses is $\frac{7}{10}$, or 70%. In the Exploration, you will investigate what happens when you take many samples from the same population and look at a distribution of sample proportions.

EXPLORATION
COOPERATIVE LEARNING

Making a Sampling Distribution

Work with a partner.
You will need:
- a calculator with a random number generator

A person who drives 10,000 miles a year in a fuel-efficient car, rather than a gas-guzzler, will save how much in gasoline costs for one year?

A. Under $100
B. Several hundred dollars
C. About $1000
D. Several thousand dollars

SET UP In 1990, the Consumer Federation of America designed a test of consumer knowledge. The group gave the test to 1139 adults at shopping centers. At the left is one of the questions on the test.

1 Write the answer that you think is correct on a piece of paper. Collect all the papers and write the responses on the board, numbering them for easy reference. Your teacher will provide the correct answer.

2 With your partner, use the random number generator on your calculator to choose 6 different random samples of 5 responses from the board. For each sample, find the proportion of correct responses. Write each sample proportion as a percent.

3 In Step 2, did you get the same proportion from each sample? Based on your results from Step 2, would you say that the percent correct for the entire class is closest to *0%, 20%, 40%, 60%, 80%,* or *100%*? Explain.

4 As a class, make a histogram of all the sample proportions. Do they tend to cluster around some value? If so, what is that value?

5 Calculate the *population proportion* (that is, the percent of correct answers for the entire class). Compare this with what you found in Step 4.

In doing the Exploration, one class made the histogram below. It displays a **sampling distribution** for many samples taken from the same population.

Sample Size = 5

Number of samples

Sample proportion

Notice the variation in the sample proportions.

Notice that the results cluster around the class's population proportion of 56%.

Each sample in the histogram above consists of 5 responses, so the **sample size** in this case is 5. The sampling distributions below show what happens when the sample size is increased.

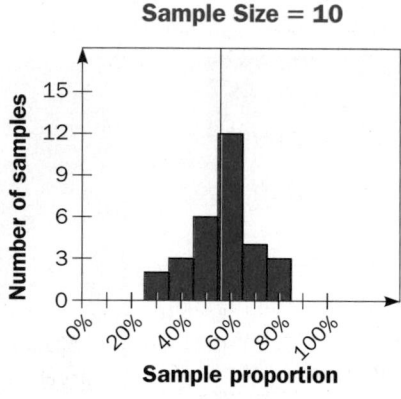

Sample Size = 10

Number of samples

Sample proportion

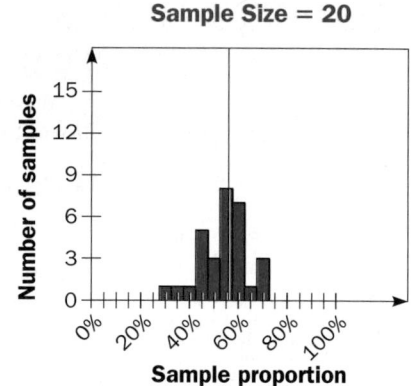

Sample Size = 20

Number of samples

Sample proportion

THINK AND COMMUNICATE

1. What do you observe about the clustering of the sample proportions in each histogram?

2. **a.** What would happen if you used a sample size greater than 20?

   b. What would happen if you made the sample size as large as the population size?

WATCH OUT!

The margin of error formula is most accurate for sample proportions between 40% and 60%. For other proportions, this approximation overestimates the true margin of error.

People often use **margin of error** to indicate the expected variability in sampling. For example, a sample proportion with a margin of error of ±3% is very likely, but not certain, to be within 3 percentage points of the population proportion.

> **Margin of Error Formula**
>
> When a random sample of size n is taken from a large population, the sample proportion has a margin of error approximated by this formula:
>
> $$\text{margin of error} \approx \pm \frac{1}{\sqrt{n}}$$

EXAMPLE 1   Application: Consumer Economics

In the test of consumer knowledge conducted by the Consumer Federation of America, 58% of the 1139 adults who participated answered the question stated in the Exploration on page 273 correctly.

a. Find the margin of error for the sample proportion.

b. Find an interval that is likely to contain the proportion of all adult Americans who would give the correct answer to the question.

SOLUTION

a. Find the margin of error using $n = 1139$.

$$\text{margin of error} \approx \pm \frac{1}{\sqrt{1139}} \approx \pm 0.0296 \approx \pm 3\%$$

The margin of error for the sample proportion is about $\pm 3\%$.

b.

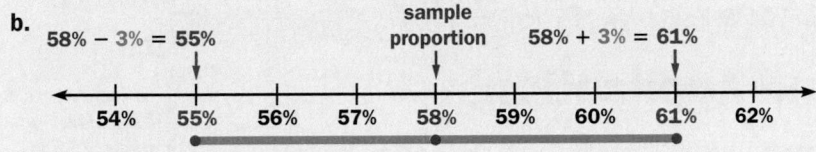

The proportion of people who would answer the question correctly is likely to be between 55% and 61%.

EXAMPLE 2   Application: Polling

Based on the newspaper report at the right, is it reasonable to assume that Costa will win the election?

> **In a telephone poll, we asked local voters which candidate for mayor they plan to vote for in the upcoming election.**
>
> | Candidate | Percent of votes |
> |-----------|------------------|
> | Costa | 54% |
> | Kwan | 46% |
>
> **Margin of error: $\pm 5\%$**

SOLUTION

The diagram shows the range of possible voting results for each candidate.

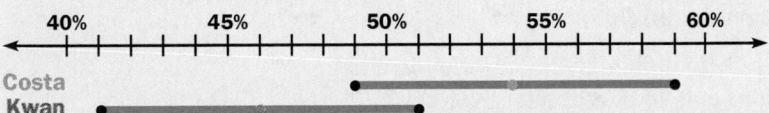

The margin of error for the poll makes it possible that Kwan might get as much as 51% of the vote, and Costa might get as little as 49% of the vote. If this were the case, Kwan would win. Based on this poll, it is not reasonable to assume that Costa will win the election.

We asked 500 people which candidate they planned to vote for in the upcoming election.

Candidate	Number of votes
Mirdik	280
Wong	220

POLLING For Questions 1–5, use the poll results at the left.

1. Find the sample proportion for each candidate.

2. Estimate the margin of error for this poll.

3. Find an interval that is likely to contain the percent of *all* voters who would vote for Mirdik.

4. Find an interval that is likely to contain the percent of *all* voters who would vote for Wong.

5. If 1000 people, instead of 500 people, had been included in the sample, what would happen to the margin of error?

6.6 Exercises and Applications

Extra Practice exercises on page 758

1. **EMPLOYMENT** In a survey of 2990 adult Americans, respondents were asked about their jobs. In response to one question, 53.3% said they have a full-time job. Find an interval that is likely to contain the proportion of all adult Americans who have a full-time job.

2. **CONSUMER ECONOMICS** A question from the Consumer Federation of America's test of consumer knowledge is shown at the right. When asked this question, 36% of 1139 adult Americans gave the correct answer. Find an interval that is likely to contain the proportion of all adult Americans who would give the correct answer to this question.

living well & eating right

The truth about what's in your food...

TAKE THE FOOD LABEL CHALLENGE

The ingredients on food labels are listed:
A. by nutritional importance, from most to least.
B. by weight, from most to least.
C. alphabetically.
D. in any order the manufacturer chooses.

3. **EDUCATION** A poll reported that 86% of Americans believe that financial need should be a consideration in distributing federal student aid. Is it possible to find the margin of error for this sample proportion? If so, find it. If not, explain why you cannot.

4. Does your class proportion from the Exploration on page 273 fall within the interval found in part (b) of Example 1? If not, what might account for this?

5. **Open-ended Problem** A few days before the mayoral election in a large city, a local newspaper took a random sample of 850 likely voters. Of those surveyed, 453 planned to vote for Timothy Marden and 397 planned to vote for Dipak Johari. If you were a reporter for the newspaper, how would you report the results of this survey?

6. **a.** If the margin of error for a poll is ±4%, find the size of the sample. (*Hint:* Solve for n in the formula for margin of error.)

   b. If the margin of error is ±2%, find the size of the sample. How does this sample size compare with the one you found in part (a)?

Boston Garden was the home sports arena for Boston's professional basketball and hockey teams. On February 26, 1993, the *Boston Globe* published the article below about a deal to build a new arena.

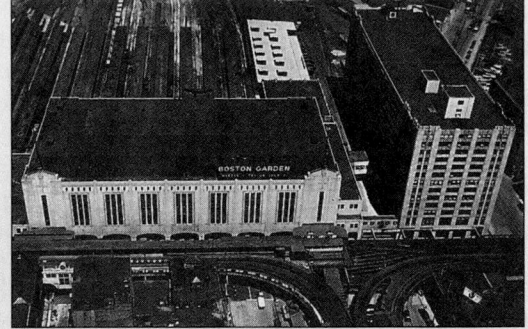

Last Friday and Sunday, Boston Globe readers were asked their opinion of the new Boston Garden deal.... As of noon yesterday the Globe had received 1119 responses from readers. Of those, 841 (approximately 75 percent) said they were in favor of the deal and 278 (25 percent) were against it, though the results are far from scientific.

BY THE WAY

Boston Garden was built in 1928. In 1995, it was replaced by the FleetCenter.

7. Why did the *Boston Globe* say the results of this mail-in survey were "far from scientific"?

8. a. The *Boston Globe* also published the results of a survey based on a random sample of 400 registered voters. In this survey, 71% of the voters favored the plans for the new sports arena. Find the margin of error for this sample proportion.

b. Writing How did the results of the random sample compare with the results of the voluntary sample? Would you expect the two surveys to have the same results? Explain why or why not.

9. The plans for the new sports arena were a popular topic on many talk radio shows in the Boston area. In the survey described above, respondents were also asked whether they listen to talk radio. Of those surveyed, 63% of the men and 54% of the women said "yes." Is it reasonable to conclude that more men than women listen to talk radio, given a ±5% margin of error? Explain.

10. In a poll of 12,300 high school students asked about life in the 22nd century, 80% predicted that Americans will be working in space in the year 2100. Also, 24% predicted that people will learn how to control the weather. Find intervals that are likely to contain the proportion of all high school students who would make these predictions.

11. a. **Technology** Graph the function $y = \dfrac{1}{\sqrt{x}}$ on a graphing calculator or graphing software. Use a viewing window with $0 \le x \le 2500$ and $0 \le y \le 0.25$. Use the graph to complete the table.

b. How are sample size and margin of error related?

c. What sample size is needed for the margin of error to be zero?

12. Writing One organization that regularly conducts public opinion polls uses a random sample of 1000 people for most of its polls. Given that the margin of error decreases as the sample size increases, why do you think the organization does not use a larger sample size?

Sample size	Margin of error
100	?
200	?
400	?
800	?
1600	?
2400	?

13. Sherille Rudbek asked 100 students, "How many hours did you work at a job last week?" The mean number of hours was 18, with a standard deviation of 5.5 hours. In order to find the margin of error for her numerical data, she needed to use a different formula than the one given on page 274. For numerical data, the margin of error for the population mean is given by

$$\text{margin of error} = \pm \frac{2\sigma}{\sqrt{n}}$$

where σ = standard deviation and n = sample size. Find the margin of error for Sherille Rudbek's survey.

14. The formula for the margin of error for a sample proportion given on page 274 can be stated precisely as follows:

$$\text{margin of error} = \pm 2 \sqrt{\frac{\hat{p}(1 - \hat{p})}{n}}$$

where $\hat{p}$ = sample proportion and n = sample size.

a. Find the margin of error if the sample proportion is 0.20 and the sample size is 500.

b. Show that $2 \sqrt{\dfrac{\hat{p}(1 - \hat{p})}{n}} = \dfrac{1}{\sqrt{n}}$ when $\hat{p} = 0.50$.

c. **Technology** Let $n = 1000$. Use a graphing calculator or graphing software to graph the two formulas for margin of error with $0 \le \hat{p} \le 1$. Describe what the graphs tell you about the relationship between the two formulas.

d. **Challenge** Give a mathematical argument that shows why $\dfrac{1}{\sqrt{n}} \ge 2 \sqrt{\dfrac{\hat{p}(1 - \hat{p})}{n}}$ for all n.

ONGOING ASSESSMENT

15. **Cooperative Learning** Work with the members of your group from the *Ongoing Assessment* exercise from Section 6.1. Use the data you have collected from your survey.

a. Draw conclusions about the categorical data you have collected. Give sample proportions and margins of error.

b. Summarize all of your work in a report. Your report should include the following items:

• a title page

• a copy of your questionnaire

• a description of how you obtained your sample

• the results of the survey, including any graphs that you used to organize and display your data

• a summary of any calculations that you performed, such as mean, median, standard deviation, and margin of error

• a written summary of your findings and conclusions, including comments on any problems that you encountered in doing the project

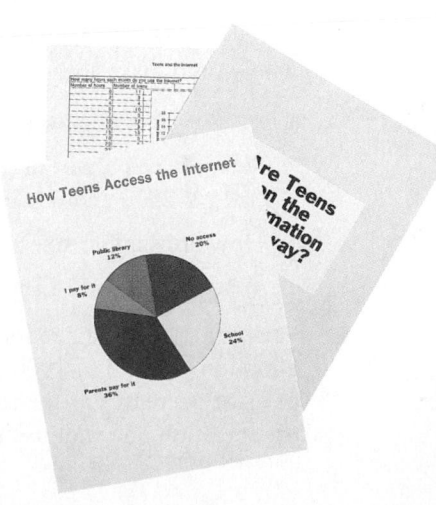

Find the mean and the standard deviation of each data set. *(Section 6.5)*

16. 1, 2, 2, 3, 5, 18, 20, 22, 23, 27

17. 4, 5, 5, 6, 8, 8, 8, 9, 9, 10, 12, 12, 14

Write a point-slope equation of the line passing through the given points. *(Section 2.3)*

18. $(6, 3)$ and $(-3, 3)$ **19.** $(-4, 1)$ and $(1, 4)$ **20.** $(-3, 5)$ and $(6, -1)$

ASSESS YOUR PROGRESS

VOCABULARY

box plot (p. 259)
lower, upper extremes (p. 259)
lower, upper quartiles (p. 259)
histogram (p. 260)
absolute frequency (p. 260)
relative frequency (p. 260)
symmetric distribution (p. 260)
skewed distribution (p. 260)

range (p. 266)
interquartile range (IQR) (p. 266)
outlier (p. 266)
standard deviation (p. 268)
sample proportion (p. 273)
sampling distribution (p. 274)
sample size (p. 274)
margin of error (p. 274)

The table at the right shows the ages of children attending a summer camp. Use the data for Exercises 1–3. *(Sections 6.4 and 6.5)*

Ages of camp participants			
8	4	10	4
7	8	5	5
6	8	7	9
6	6	10	6
7	5	7	6
5	5	8	6
10	8	8	6
6	9	4	6

1. a. Make a box plot of the data.

   b. Above what age are 50% of the children? 25% of the children?

2. a. *Writing* Suppose four more children register for the camp. Their ages are 11, 9, 8, and 9. Do you think the inclusion of these data will have a significant effect on the distribution of the ages? Why or why not?

   b. Make a box plot that includes the new data. Compare this box plot with the one you made in Exercise 1. How did the inclusion of these data affect the distribution of the ages?

3. a. Find the range and the interquartile range for the set of camp participants, including the four children from Exercise 2. Identify any outliers in the data set.

   b. Find the standard deviation for the ages of camp participants.

4. A superintendent of a city school system wanted to find out whether the majority of voters in the city favored a proposed building addition to the high school. The superintendent surveyed 500 voters and found that 260 favored the building addition. *(Section 6.6)*

   a. Find an interval that would most likely contain the percent of all voters who favor the building addition.

   b. Based on the results from part (a), can the superintendent conclude that the voters will pass the building proposal?

5. *Journal* Explain how you can use the topics you have studied in this chapter to interpret the results of a survey or study that you read about in a newspaper or magazine.

Survey Says...

Newspapers and magazines often report the results of public opinion surveys. Some surveys are sponsored by independent groups and reported by the press; others are sponsored by news organizations themselves to accompany the articles they publish.

It is important to analyze critically the results of a survey. Does the report fail to note the sampling method used or the sample size? Were the questions worded in ways that might bias the results? Have data been displayed in a misleading manner?

PROJECT GOAL Analyze the presentation and conclusions of a public opinion survey published in a newspaper or magazine.

Analyzing a Survey

Skim through some newspapers or magazines to find an opinion survey that interests you. Analyze the presentation and conclusions of the report.

Here are some questions to consider:

• Who conducted the survey? Does the person or group have a stake in its outcome?

Casa Suntuoso, your FAVORITE!!

Which pizza do you prefer?

CASA SUNTUOSO   70%
PIZZA PRIMO   30%

*Based on a survey of 1000 Casa Suntuoso customers

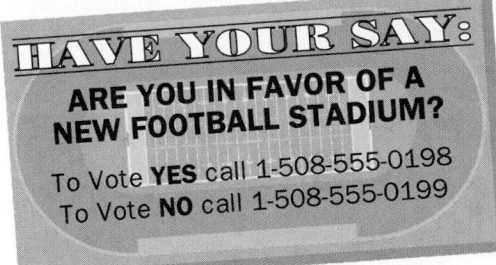

HAVE YOUR SAY:

ARE YOU IN FAVOR OF A NEW FOOTBALL STADIUM?

To Vote **YES** call 1-508-555-0198
To Vote **NO** call 1-508-555-0199

• How was the survey done? Does the sampling method seem reasonable? Is the sample representative of the population being studied?

• What is the sample size? Is it large enough to give meaningful results?

• How are the questions worded? Does the wording seem biased? If so, how would you improve the questions?

POLL SHOWS MARCIO IN FRONT IN MAYORAL RACE

Who will get your vote for mayor?

MARCIO	41%
DEHORS	39%
NOT SURE	20%

From a telephone poll of 1000 city residents. Margin of error is +/–3%.

FIZZEE   OTHER BRAND

MORE PEOPLE LIKE FIZZEE

• What is the survey's margin of error? When you take the margin of error into account, are the conclusions drawn from the survey accurate?

• Does the report have any graphs? Are they accurately drawn? Are they in any way misleading?

Writing a Report

Write a report summarizing your work. Include a copy of the survey and your analysis of it. List any information that you think is missing from the survey, and suggest ways the survey can be improved. You may wish to extend your report as follows:

• Write a letter to the sponsor of the survey to ask any questions you still have about it.

• Is the publisher simply reporting the results of a survey from another organization? If so, get the original report from that organization, and compare it with the version that you found.

• In your own community, conduct the same survey that you analyzed. Compare your results with the original findings.

Self-Assessment

Describe any difficulties you had in analyzing your survey. Do you feel that you understand the elements of a well-done public opinion survey? If not, what things are still unclear?

6 Review

The steps for performing statistical investigations are listed on page 237. Use what you learned in this chapter to discuss these steps in detail with a partner. If necessary, go back through the chapter to look up any details that are not clear to you.

VOCABULARY

numerical data (p. 238)
categorical data (p. 238)
population (p. 239)
sample (p. 239)
biased question (p. 245)
self-selected sample (p. 250)
systematic sample (p. 250)
convenience sample (p. 250)
random sample (p. 250)
stratified random sample (p. 250)
cluster sample (p. 250)
biased sample (p. 251)
box plot (p. 259)
lower, upper extremes (p. 259)

lower, upper quartiles (p. 259)
histogram (p. 260)
absolute frequency (p. 260)
relative frequency (p. 260)
symmetric distribution (p. 260)
skewed distribution (p. 260)
range (p. 266)
interquartile range (IQR) (p. 266)
outlier (p. 266)
standard deviation (p. 268)
sample proportion (p. 273)
sampling distribution (p. 274)
sample size (p. 274)
margin of error (p. 274)

SECTIONS | 6.1, 6.2, *and* 6.3

Suppose you want to gather information about the students in your school. The complete group of students is the **population**. The part of the population that you survey is the **sample**. There are many ways you can choose a sample of the students in your school. For example:

- **self-selected:** Students volunteer.

- **convenience:** Survey the students whose lockers are near yours.

- **stratified random:** Divide student names by grade and randomly choose names from each grade.

- **systematic:** Survey every fourth student in the lunch line.

- **cluster:** Survey the students in your homeroom.

- **random:** Draw names out of a hat and survey the selected students.

You should avoid asking **biased questions** in your survey so that you get responses that accurately reflect the opinions or actions of the respondents.

Data that are names or labels are **categorical data**. Categorical data are often displayed in a bar graph or a circle graph. Data that are counts or measurements are **numerical data**. Numerical data, such as the fuel efficiency of various models of cars (see table), are often displayed in a *box plot* or a *histogram*.

Fuel efficiency (mi/gal)
30, 25, 41, 29, 22, 29, 18, 23, 26, 23, 29, 24, 25, 28, 25, 20, 17, 19, 17, 17, 18, 20, 13, 23, 12, 17, 21, 17, 18, 26, 29, 29, 26, 22, 17, 22, 26

A **box plot** displays the median, the **quartiles**, and the **extremes** of a data set. The "box" of a box plot shows the middle half of the data.

A **histogram** displays the overall shape of a distribution of numerical data. The distribution may be **symmetric** or **skewed**.

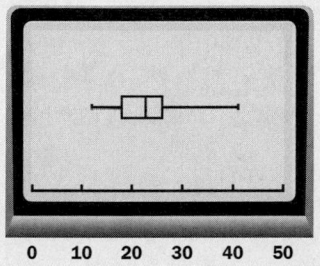

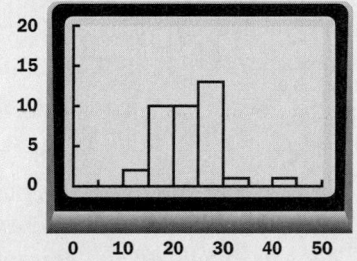

Some measures of variability within a data set are the **range** and **interquartile range**. Another measure of variability is the **standard deviation**. If the data values are $x_1, x_2, \ldots, x_n$, and the mean of the data is $\bar{x}$, then the standard deviation, σ, is given by this formula:

$$\sigma = \sqrt{\frac{(x_1 - \bar{x})^2 + (x_2 - \bar{x})^2 + \cdots + (x_n - \bar{x})^2}{n}}$$

Measures of variability for the fuel efficiency data are shown below.

IQR = 26 − 18 = 8

range = 41 − 12 = 29

```
1-Var Stats
x̄=22.78378378
Σx=843
Σx²=20385
Sx=5.720990275
σx=5.643150046
↓n=37
```

The standard deviation of the fuel efficiency data is about 5.6.

SECTION 6.6

When a random sample of size *n* is taken from a large population, the **sample proportion**, which is the fraction or percent of responses that fall into a particular category, has a **margin of error** approximated by the formula

$$\text{margin of error} \approx \pm \frac{1}{\sqrt{n}}$$

where *n* is the size of the sample used in the survey. For example, many surveys have a **sample size** of about 1000, which gives a margin of error of about

$\pm \dfrac{1}{\sqrt{1000}} \approx \pm 0.0316$, or $\pm 3\%$.

6 | **Assessment**

VOCABULARY QUESTIONS

For Questions 1–3, complete each paragraph.

1. Numerical data are often displayed in a(n) _?_ or a(n) _?_. _?_ data are often displayed in a bar graph or circle graph.

2. Selecting every tenth name from a list produces a(n) _?_ sample. Drawing names out of a hat produces a(n) _?_ sample.

3. When you can read the actual count of the data within each interval of a histogram, the vertical axis shows _?_. The vertical axis shows _?_ when the count is expressed as a percent of all the data.

SECTIONS 6.1, 6.2, *and* 6.3

Tell whether the data that can be gathered about each variable are *categorical* or *numerical*. Then describe the categories or numbers.

4. the species of animals at a zoo   5. semester grades

6. the amount of electricity a household uses each day

For each graph, tell if it is an appropriate way to display the data. If not, tell what type of display would be better. Explain your reasoning.

7. **Amount of Television Watched by Tenth Graders Each Day**

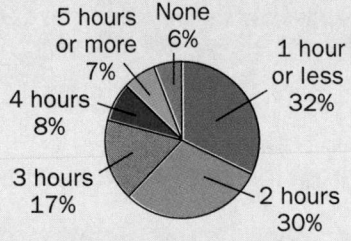

8. **Percent of U.S. Births, by Month, in 1991**

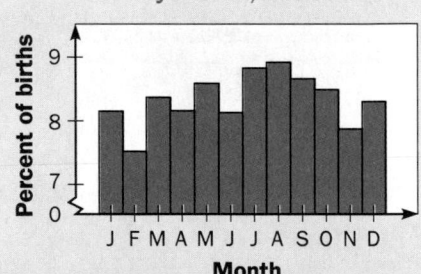

Tell why each question may be biased. Then describe what changes you would make to improve the question.

9. "Have you ever cheated on a test?"

10. "Some people want to censor the material that can be transmitted through the Internet, thus violating people's First Amendment right to freedom of speech. Do you think restrictions should be placed on the material that can be transmitted through the Internet?"

Identify the type of sample and describe the population of the survey. Then tell if the sample is biased. Explain your reasoning.

11. A marketing group uses a computer to generate phone numbers.

12. A teacher collects the homework of every third student.

13. The principal sends the student council officers to represent the school at a district fundraising event.

SECTIONS 6.4, 6.5, *and* 6.6

SPORTS In 1969, four new teams joined major league baseball. In addition, the height of the pitching mound was lowered, giving batters an advantage. Use the data in the table to answer Questions 14–17.

Average Number of Home Runs per Game	
1950–1968	**1969–1993**
1.23, 1.37,	1.15, 1.28,
1.42, 1.50,	1.36, 1.36,
1.57, 1.66,	1.39, 1.41,
1.67, 1.67,	1.44, 1.46,
1.67, 1.70,	1.47, 1.48,
1.70, 1.72,	1.51, 1.55,
1.78, 1.80,	1.56, 1.57,
1.81, 1.82,	1.60, 1.60,
1.85, 1.85,	1.60, 1.61,
1.91	1.63, 1.71,
	1.73, 1.76,
	1.78, 1.81,
	2.12

14. a. Make two box plots to compare the average number of home runs hit per game for the 1950–1968 seasons and the 1969–1993 seasons.

    b. Between what numbers do 50% of the data fall in the box plot of the 1950–1968 data?

    c. Repeat part (b) using the box plot of the 1969–1993 data.

15. Make a histogram of each data set. Describe the distribution of the data.

16. a. Find the range, interquartile range, and any outliers of each data set.

    b. Find the standard deviation of each data set.

17. **Writing** Do you think the changes made in 1969 had an effect on the average number of home runs hit per game? Explain your reasoning.

18. The cheerleaders at a high school took a random sample of 80 students and found that 60 students would buy a spirit T-shirt.

    a. Find the sample proportion of students who would buy a T-shirt.

    b. Find the margin of error. Use the margin of error to find an interval for the proportion of the entire student body that would buy a T-shirt.

19. **Writing** In what ways can errors in data collection, data displays, and data analysis affect the conclusions made from a set of data?

PERFORMANCE TASK

20. The manager of a restaurant wants to gather data about the restaurant's service, customers, and food. Write a report to the manager suggesting a way to gather, display, and analyze the data. Include these things:

• Give examples of categorical and numerical data that may be collected. Describe how you would display each type of data.

• Write a sample survey of unbiased questions to distribute to customers. Then describe how you would distribute the survey.

• Give examples of how the standard deviation, range, interquartile range, outliers, and margin of error can be used to analyze the data.

Cumulative Assessment

CHAPTERS $4-6$

CHAPTER 4

For each function:

a. Graph the function and its inverse in the same coordinate plane.

b. Find an equation for the inverse.

1. $f(x) = 6x$ **2.** $g(x) = 4^x$ **3.** $h(x) = -\dfrac{2}{5}x + \dfrac{1}{5}$

Write each equation in logarithmic form.

4. $5^0 = 1$ **5.** $81^{-1/2} = \dfrac{1}{9}$ **6.** $2^5 = 32$

Evaluate each logarithm. Round decimal answers to the nearest hundredth.

7. $\log 0.01$ **8.** $\ln e^{32}$ **9.** $\log_{1/2} 6$

10. Write $\log_4 \dfrac{p^6}{q^{2/3}}$ in terms of $\log_4 p$ and $\log_4 q$.

11. Write $-\dfrac{3}{4}\log_a 81 + \dfrac{1}{2}\log_a 9$ as a logarithm of a single number.

Solve each equation. Round your answers to the nearest hundredth.

12. $4^x = 15$ **13.** $\log_4 x = -3$ **14.** $2\ln(x+1) = 6$

Write y as a function of x.

15. $\log y = 0.8x + 1$ **16.** $\ln y = 0.6 - 0.4 \ln x$

17. The table shows the number of households, in millions, in the continental United States with television sets for various years from 1955 through 1985.

a. Find an exponential function that models the data.

b. Find a power function that models the data.

c. **Writing** Which of the functions from parts (a) and (b) is the better model? Explain your reasoning.

Years since 1950	Millions of households
5	32.0
10	45.2
15	53.8
20	60.1
25	69.6
30	76.3
35	84.9

CHAPTER 5

Describe the graph of each function. Make a sketch of each graph.

18. $y = -\dfrac{1}{2}x^2$ **19.** $y = -\dfrac{1}{2}(x-1)^2$ **20.** $y = -\dfrac{1}{2}(x+2)(x-3)$

Solve each equation. Give solutions to the nearest tenth.

21. $-3x^2 = -15$ **22.** $7(x+4)^2 - 28 = 0$ **23.** $2x^2 - 15x + 21 = 0$

24. **PHYSICS** The equation $h = -4.9(t-1.1)^2 + 7$ gives the height h of a ball in meters t seconds after it is thrown straight up. When does it hit the ground?

25. Write an equation in the form $y = a(x - h)^2 + k$ for a parabola that has its vertex at $(-4, 1)$ and passes through the point with coordinates $(-2, 9)$.

26. Write $y = -2x^2 - 4x + 6$ in vertex form.

27. Write an equation for the parabola shown at the right.

28. **Open-ended Problem** Describe two ways to show that the equation $5x^2 + x + 8 = 0$ has no solution.

29. Find the x-intercepts, if any, for the graph of $y = 0.1x^2 + 0.4x + 0.4$.

30. The equation $y = -0.05x^2 + 2x + 5$, where the distances x and y are measured in feet, describes the path of a stream of water from a fire hose. What is the maximum height of the stream of water?

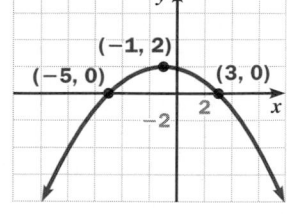

Write each quadratic function as a product of factors.

31. $y = x^2 - 64$ **32.** $y = 3x^2 + x - 10$ **33.** $y = 9x^2 - 24x + 16$

34. **Writing** What information can you get about the graphs of the functions in Questions 31–33 by writing each function as a product of factors?

CHAPTER 6

35. A public radio station sends surveys to its contributors to determine which programs are most popular.

   a. Describe the population of the survey.

   b. Identify the type of sample. Then tell if the sample is biased. Explain.

   c. Tell whether the data that will be gathered are *numerical* or *categorical*.

   d. Tell what type of graph you would use to display the data.

36. **Writing** Describe three ways in which a survey question may be biased. Give an example of each.

37. **Open-ended Problem** Margery Gupta is trying to decide whether to run for mayor. She plans to use an opinion poll to determine the level of support for her candidacy. Explain how she could choose a sample.

SPORTS For Questions 38 and 39, use the table showing the number of wins per team in the National Hockey League in the 1994–1995 season.

38. a. Make a histogram to display the data for all the teams.

   b. Is the distribution shown on your histogram *symmetric* or *skewed*?

39. a. Identify the median, the quartiles, and the extremes of each data set.

   b. Make a comparative box plot of the data for the two conferences.

   c. For each set of data, find the range and the interquartile range, and identify any outliers.

   d. Find the standard deviation for each set of data. How do the standard deviations compare?

Eastern Conference				
27	22	20	19	18
22	15	22	9	28
29	30	17	22	

Western Conference				
16	24	24	17	33
17	16	28	19	21
18	16			

7 Systems

Seeing how to exercise

(American Manual Alphabet for "exercise")

Gina Oliva

When Gina Oliva took her first aerobics class in 1981, she instantly loved it. "When I find something I love, I want to share it with other people," Oliva says. "I also saw that I could do it—that I had a feel for the rhythm." This is no small matter because Oliva is almost completely deaf; she can hear only low-frequency music with a very strong bass line. Yet she picked up the basic aerobics steps quickly and soon decided to "bring the joy of exercise to both deaf and hearing people."

"I've devoted my life to making exercise more accessible to deaf people."

Innovative Hand and Arm Gestures

Oliva was well-positioned to work with deaf people, given that she was already on the staff of Gallaudet University in Washington, D.C., the world's leading learning institution for the deaf. The challenge was coming up with a system to help students follow an aerobics class even when they cannot hear the instructor's commands.

Oliva developed a set of approximately one hundred visual cues—hand and arm gestures that enable teachers to communicate the entire sequence of steps without saying a single word. Like pantomime, most of the visual cues are designed to resemble the dance-exercise movements. She also uses American Sign Language to convey some of the cues not easily represented by gesture.

"Math can help you become more fit."

Visual Cues for Everyone

Equipped with this new "vocabulary," Oliva has taught classes and trained aerobics instructors all over the country. She stresses that the visual cues she invented are for everyone, both hearing and deaf alike.

"Among hearing people, one of the biggest problems is voice injuries to aerobics instructors trying to be heard over the music," she says. "Even when the teacher uses a microphone, many students can't hear the instructions. It helps if you can see a signal at the same time you hear something."

Getting a Good Workout

In addition to aerobics, Oliva teaches her students how to determine their target heart rate zone—the range of heart rates that indicates a good aerobic workout. Oliva explains, "In a way, math can help you become more fit. It can tell you the amount of exertion you need to get a good workout without overdoing it."

To figure out your target heart rate zone, subtract your age from 220. Multiply that number by 0.60 to get your minimum target heart rate and by 0.80 to get your maximum target heart rate. As a general rule during exercise, try to keep your heart rate between these two values. This requirement can be expressed by a *system of inequalities*, whose graph is shown.

Find your heart rate by counting your pulse for 6 s and multiplying that number by 10. To feel your pulse, place your index and middle fingers on your wrist. Begin counting the beats within 15 s of stopping an activity.

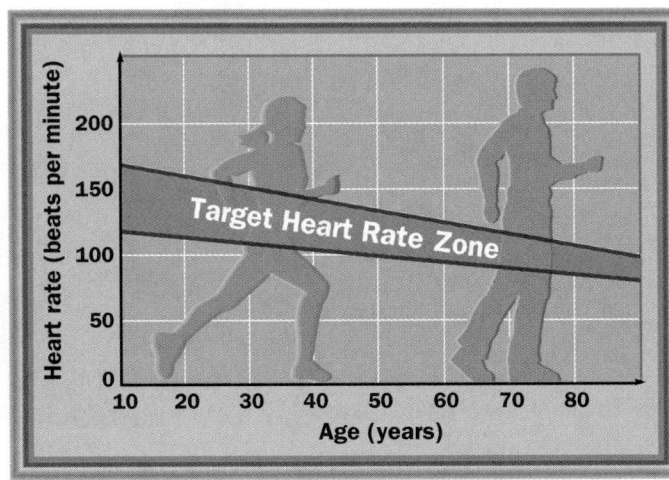

EXPLORE AND CONNECT

Gina Oliva prepares to start an aerobics session.

1. Writing Calculate your own minimum target heart rate and maximum target heart rate using the guidelines above. Then use the graph to explain how these values change as you get older.

2. Research What are some ways that other sports are adapted to make them more accessible to handicapped individuals? You may want to contact the Disabled Sports USA organization in Washington, D.C., for more information. Report your findings to your class.

3. Project Try engaging in different physical activities like walking, running, or bicycling. After 5 min of an activity, find your heart rate. Is this rate within your target heart rate zone? Compare the results for each activity.

Mathematics
& Gina Oliva

In this chapter, you will learn more about how mathematics is related to fitness.

Related Examples and Exercises

Section 7.2
• Example 2

Section 7.5
• Example 1
• Exercises 14–18

Section 7.6
• Exercises 15–17

7.1 Systems of Equations

Learn how to...
- use technology to find points of intersection of graphs
- use substitution to solve systems of equations

So you can...
- solve problems about personal finance and advertising, for example

The political party that controls, or has the most members in, the U.S. Senate can more easily accomplish its legislative goals. The graph shows the number of Democrats and Republicans in the U.S. Senate during fifteen Congresses.

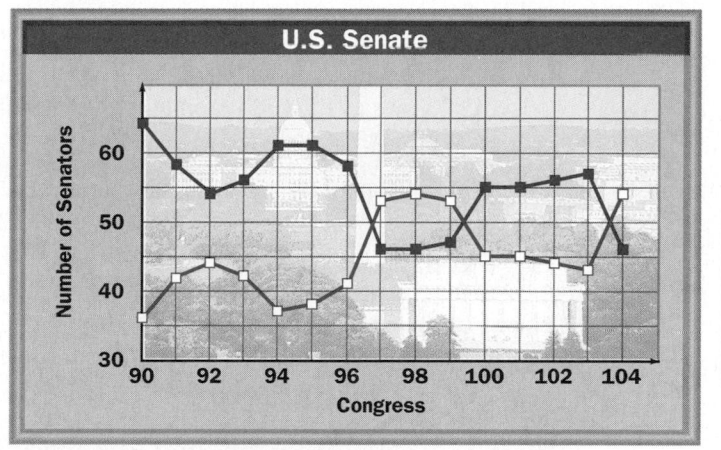

THINK AND COMMUNICATE

~~~~~~~~~~~~~~~~~~~~~~~~~~~~~ ⇒stions.

1. During which Congresses were the Democrats in control? the Republicans in control?

2. How does the graph show when changeovers in control took place? How many changeovers were there?

A point where two graphs intersect often has significance. Sometimes you may want to use technology to find an intersection point.

EXAMPLE 1   Application: Personal Finance

When buying a car, Tom Fiore needs to finance $12,000 over four years. The car dealer offers him a 2.9% annual interest rate if he finances the full amount *or* a 9.9% annual interest rate if he takes a $1500 up-front discount. The amount A owed on each loan after p monthly payments is given by:

$$A = 109{,}688 - 97{,}688(1.00242)^p \quad \longleftarrow \text{ first loan option}$$
$$A = 32{,}218 - 21{,}718(1.00825)^p \quad \longleftarrow \text{ second loan option}$$

In how many months will the amounts needed to pay off the two loans be equal?

SOLUTION

Use a graphing calculator or graphing
software to graph the equations and find
where they intersect.

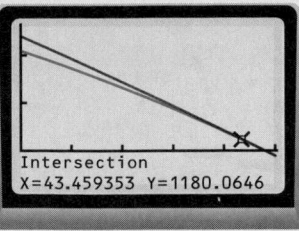

Intersection
X=43.459353 Y=1180.0646

> Tom Fiore will owe the
> same amount on both loans
> in about **43 months**.

THINK AND COMMUNICATE

3. Compare the two A-values when $p = 0$. What do these numbers represent?

4. If Tom decides to sell his car while he is still making payments on it, he'd
like to owe as little as possible. Under which loan plan would Tom owe the
least after 43 months?

Two or more equations involving the same variables, such as the equations
$A = 109{,}688 - 97{,}688(1.00242)^p$ and $A = 32{,}218 - 21{,}718(1.00825)^p$ from
Example 1, form a **system of equations**.

Solving a system means finding the coordinates of the point(s) where the
graphs of the equations intersect. When one equation is already solved for one
of the variables, you can use substitution to solve the system algebraically.

FOR
EVERY
SPORT

EXAMPLE 2   Application: Advertising

A footwear company is preparing to market a new kind of sports shoe. The
company plans to spend $28 million on advertising. Past experience shows
that the amount spent on television advertising should be 3.5 times as much
as the amount spent on magazine advertising. Determine how much money
should be spent on each type of advertising.

SOLUTION

Step 1 Write a system of equations. Let $t =$ the amount spent on television
advertising, in millions, and let $m =$ the amount spent on magazine
advertising, in millions.

$$t + m = 28 \quad \longleftarrow \text{ total money spent on advertising}$$

$$t = 3.5m \quad \longleftarrow \begin{array}{l}\text{how money is to be split between}\\ \text{television and magazine ads}\end{array}$$

Step 2 Solve the system using substitution.

> Substitute **3.5m** for
> **t** and solve for m.

$$3.5m + m = 28 \qquad\qquad t = 3.5m$$

$$4.5m = 28 \qquad\qquad \approx 3.5(6.22)$$

> Substitute
> **6.22** for m
> to find t.

$$m \approx 6.22 \qquad\qquad \approx 21.78$$

The company should spend about $21.78 million on television advertising and
about $6.22 million on magazine advertising.

EXAMPLE 3

Use substitution to solve this system of equations.

$$5x - 2y = 4$$
$$4x - y = 5$$

SOLUTION

Step 1 Solve one equation for one of the variables. In this case, it is convenient to solve the second equation for y.

$$4x - y = 5$$
$$y = 4x - 5$$

Step 2 Solve the system using substitution.

$$5x - 2(4x - 5) = 4$$
$$5x - 8x + 10 = 4$$
$$-3x = -6$$
$$x = 2$$

Substitute $4x - 5$ for y and solve for x.

$$y = 4x - 5$$
$$= 4(2) - 5$$
$$= 3$$

Substitute 2 for x to find y.

The solution of the system is $(x, y) = (2, 3)$.

Check
$$5(2) - 2(3) \stackrel{?}{=} 4 \qquad 4(2) - 3 \stackrel{?}{=} 5$$
$$10 - 6 \stackrel{?}{=} 4 \qquad 8 - 3 \stackrel{?}{=} 5$$
$$4 = 4 ✔ \qquad 5 = 5 ✔$$

☑ CHECKING KEY CONCEPTS

1. Explain what it means to solve a system of equations.

 **Technology** Use a graphing calculator or graphing software to graph each pair of equations and find the point(s) of intersection.

2. $y = 2x^2$
$y - x = 6$

3. $y = 3x + 10$
$2y = 12x + 11$

4. $y = 2^x$
$y = x + 5$

Use substitution to solve each system of equations.

5. $y = 13x$
$y + 12x = 75$

6. $2x + y = 14$
$x - 3y = 5$

7. $x + 2y = 11$
$4x + 2y = 17$

7.1 Exercises and Applications

Extra Practice exercises on page 758

Solve each system of equations.

1. $2x - y = 0$
$4x + y = 12$

2. $y = 1.2x$
$y + 2.8x = 5$

3. $x = y - 1$
$3x + y = 13$

4. $-2x + y = 2$
$4x - y = 9$

5. $y + 2x = 15$
$2y - 6x = 10$

6. $3y + x = 10$
$2x + 5y = 19$

7. $y = 2.6x$
$2y + 3x = 12$

8. $3.1x + y = 0$
$3y + 2.5x = 17$

9. **Technology** Johanna wanted to solve the system

$$y = 3x + 4$$
$$y = 4x - 1$$

by setting up tables of values. Her calculator displayed the y_1-values (for $y = 3x + 4$) and the y_2-values (for $y = 4x - 1$) for $x = 1, 2, \ldots, 7$.

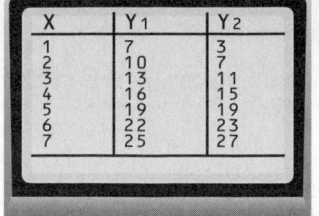

X	Y1	Y2
1	7	3
2	10	7
3	13	11
4	16	15
5	19	19
6	22	23
7	25	27

a. What is the solution of the system? Explain your reasoning.

b. Open-ended Problem Describe other ways that Johanna could solve the system.

10. SOCIAL SCIENCE People who study demographics have noted shifts in the distribution of the U.S. population over time. To see the changes more clearly, demographers look at the country in terms of four regions: the Northeast, the Midwest, the South, and the West.

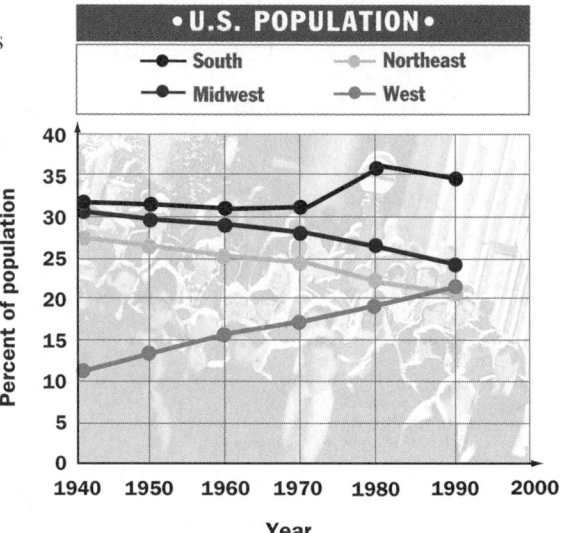

a. What region has had the largest percent of the U.S. population since 1940?

b. What region has been steadily growing since 1940?

c. What was significant about the decade 1980–1990?

d. What predictions might you make about the decade 1990–2000?

11. SPORTS A baseball player's batting average (the number of hits per at-bat) is .280 at the beginning of a game. The player gets 3 hits during 5 at-bats. He ends the game with a .300 batting average. How many times has the player batted this season?

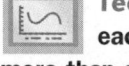 **Technology** Use a graphing calculator or graphing software to solve each system of equations. (The graphs of the equations may intersect more than once.)

12. $y = 4(2^x)$
$y = 34.6 + 17x$

13. $y = 103(1.08)^x$
$y = 90 + 15x$

14. $y = x^2$
$y = 7$

15. $y = x^2 - 1$
$y = 3x$

16. $y = 3x^2 - 4x + 5$
$y = -5x + 10$

17. $y = x^2 + 2x - 2$
$y = 7 + 10x$

18. $y = x^2 + 2x - 2$
$y = 4x + 4$

19. $y = x^2 - 16$
$y = -x^2 + 6$

20. $y = 2x + 5$
$y = 14x - 17$

21. $y = -2(3^x)$
$y = -53$

22. $y = x^2 - 3x$
$y = 5x + 8$

23. $y = x^2 + x - 1$
$y = 2x^2 - 6$

24. **Technology** Margot and Claudia are modeling the world's population P, in billions. The world's population in 1990 was 5.33 billion.

• Margot predicts a 1.7% annual increase after 1990, giving the exponential equation $P = 5.33(1.017)^x$, where $x =$ the number of years since 1990.

• Claudia predicts an annual increase of 130 million people, giving the linear equation $P = 5.33 + 0.13x$.

Use a graphing calculator or graphing software to determine the years for which the two models will give the same predicted populations.

25. Challenge Find the time *t* between successive alignments of the hands of a clock. (*Hint:* The hour hand rotates *d*° at a rate of 30°/h in time *t*, while the minute hand rotates (360 + *d*)° at a rate of 360°/h in the same amount of time.)

Connection ECONOMICS

Although appliances can be expensive, they often pay for themselves after a certain number of uses.

26. Writing What do you think the phrase "pay for itself" means for an appliance?

27. A breadmaker costs $199. The ingredients and electricity to make one loaf of bread with the machine cost $.79. A loaf of bread of similar quality costs $1.59 at a grocery store.

   a. Write two equations giving the total cost *C* in terms of the number of loaves *l* made or purchased for each situation.

   b. Solve the system of equations to find the number of loaves of bread you have to make before the breadmaker pays for itself.

28. An electric clothes dryer costs $300 and uses $.43 worth of electricity for a load that takes 1 h to dry. A laundromat charges $1.50 to use a dryer for an hour.

   a. Write equations representing the total cost *C* for drying *l* loads of laundry in each situation.

   b. How many loads of laundry do you have to dry before the dryer pays for itself?

29. A noodle ringer for rolling homemade noodles costs $40. The ingredients for each batch of homemade noodles cost $.25. A batch of store-bought noodles costs $.50.

   a. Write equations representing the total cost *C* for *b* batches of noodles in each situation.

   b. How many batches of noodles do you have to make before the noodle ringer pays for itself?

30. Research Find out how much a home water purifier costs and how much a bottle of purified water costs. Determine how many gallons of water you would have to purify for the purifier to pay for itself.

31. Open-ended Problem Choose an appliance and determine how many times you have to use it before it pays for itself.

32. a. Visual Thinking A brainteaser like the one shown below appeared in an ad from a company boasting of its problem-solving abilities. Write an equation for each of the first three pictures.

    b. Challenge Use the substitution method several times to determine how many oranges will balance the bunch of grapes in the final picture.

ONGOING ASSESSMENT

33. Cooperative Learning Work with two other students.

    a. Each of you should take a piece of paper and write a linear equation using the variables x and y. One of the variables should have the coefficient 1 or -1.

    b. Pass your paper to the person on your right. Write a second linear equation on the paper you receive.

    c. Pass the paper you have to the person on your right once again. Solve the system of equations on the paper you receive.

    d. Pass papers to the right one last time; you should receive your original paper. Check the solution shown by substituting it into the two equations. If it doesn't check, help the person who solved the system find where errors were made.

SPIRAL REVIEW

Solve each equation. If all numbers are solutions, write *all numbers*. If there is no solution, write *no solution*. *(Toolbox, page 786)*

34. $2x + 1 = -2 - 4x$ **35.** $4x - 1 = 4x + 3$ **36.** $2x + 1 = x - 3$ **37.** $x - 1 = x + 1$

38. The table shows the distance from the sun, d, and the period of revolution, p, of five planets. (Distances are given in *astronomical units* (AU), where 1 AU = the distance between Earth and the sun. Periods are given in Earth years.) *(Section 4.5)*

    a. Make a scatter plot of the data pairs $(\log d, \log p)$. What relationship exists between $\log d$ and $\log p$?

    b. Find an equation giving p as a function of d.

    c. Make a prediction about the period of Saturn if you know that its distance from the sun is 9.541 AU.

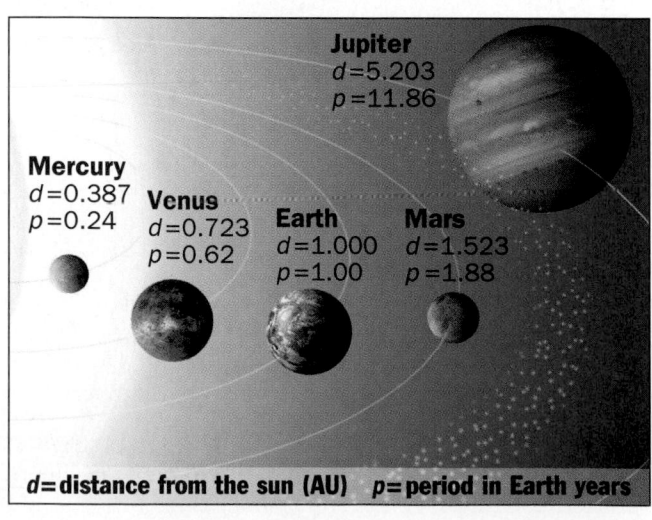

Jupiter
$d=5.203$
$p=11.86$

Mercury
$d=0.387$
$p=0.24$

Venus
$d=0.723$
$p=0.62$

Earth
$d=1.000$
$p=1.00$

Mars
$d=1.523$
$p=1.88$

$d=$ distance from the sun (AU)   **$p=$ period in Earth years**

7.2 Linear Systems

In Chapter 2, you worked with linear equations in "$y =$" form, such as $y = 2x + 3$. When x and y appear together on the same side of the equation, as in $-2x + y = 3$, the equation is said to be in *standard form*. In the Exploration, you'll investigate systems of linear equations in standard form.

Learn how to...
- solve linear systems by adding equations

So you can...
- solve problems involving two unknown values, such as boat speed and current speed

EXPLORATION
COOPERATIVE LEARNING

Investigating Ways that Lines Intersect

Work with a partner.
You will need:
- a graphing calculator or graphing software

Be sure to use parentheses when the coefficient of x is a fraction.

1 Graph this system:

$$2x - 3y = 5$$
$$3x + 4y = -12$$

Your calculator or software may require that you solve each equation for y first.

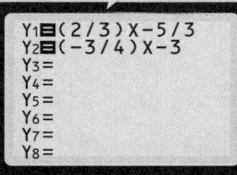

```
Y₁◼(2/3)X-5/3
Y₂◼(-3/4)X-3
Y₃=
Y₄=
Y₅=
Y₆=
Y₇=
Y₈=
```

2 Do the lines intersect in a single point? If so, find the approximate coordinates of the point. If not, describe the geometric relationship between the lines.

3 Repeat Steps 1 and 2 for each of the following systems of equations.

$$2x - 3y = 5 \qquad\qquad 2x - 3y = 5$$
$$4x - 6y = 1 \qquad\qquad -6x + 9y = -15$$

Questions

1. In what ways can two lines in a plane intersect? What does this mean in terms of the number of solutions that a system of linear equations can have?

2. If a system of linear equations is to have a *unique* solution, what must be true about the slopes of the lines?

When the coefficients of one of the variables in a system of linear equations are the *same*, you can use *subtraction* to solve the system algebraically. Likewise, you can use *addition* when the coefficients of one of the variables are *opposites* of each other.

EXAMPLE 1   Application: Recreation

TACHOMETER READING: 3000 rev/min		
	Downstream	**Upstream**
TIME (min:sec)	3:12	3:22
SPEED (knots)	18.8	17.8

Jared's boat does not have a speedometer, so he timed how long it took to travel 1 nautical mile (nm) with a river's current (downstream) and 1 nm against it (upstream) while keeping his boat's engine running at a constant 3000 rev/min (as measured on the boat's tachometer).

Use Jared's log to determine the speed of his boat in still water when the engine runs at 3000 rev/min, and find the speed of the current.

The speed of a boat is often measured in nautical miles per hour, or knots: 1 knot ≈ 1.15 mi/h.

SOLUTION

Step 1 Write a system of equations. Use the fact that the boat's speed s is increased by the water's speed w when traveling with the current and decreased by the water's speed when traveling against the current.

$s + w = 18.8$ ⟵ combined downstream speed

$s - w = 17.8$ ⟵ combined upstream speed

Step 2 Solve the system using addition. Add corresponding sides of each equation to eliminate the variable w.

$$s + w = 18.8$$
$$\underline{s - w = 17.8}$$
$$2s = 36.6$$
$$s = 18.3$$

Substitute **18.3** for s.

$$18.3 + w = 18.8$$
$$w = 0.5$$

The speed of the boat is 18.3 knots. The speed of the current is 0.5 knots.

THINK AND COMMUNICATE

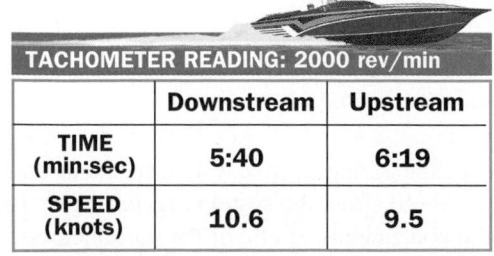

TACHOMETER READING: 2000 rev/min		
	Downstream	**Upstream**
TIME (min:sec)	5:40	6:19
SPEED (knots)	10.6	9.5

1. In Example 1, explain how traveling 1 nm in 3:22 translates into a speed of 17.8 knots.

2. Jared repeated his speed experiment, this time keeping his boat's engine running at a constant 2000 rev/min. Use the information shown in Jared's log to determine the speed of his boat in still water when the engine runs at 2000 rev/min.

Sometimes you can eliminate a variable by multiplying an equation by a constant before adding or subtracting.

EXAMPLE 2 **Interview: Gina Oliva**

Gina Oliva likes to combine jogging and walking to exercise. One day she jogged and walked for 1 h and covered 4.2 mi. Her jogging speed was 5 mi/h, and her walking speed was 3 mi/h. Find her time spent jogging and her time spent walking.

SOLUTION

Step 1 Write a system of equations involving jogging time j and walking time w, both in hours.

Step 2 Multiply one of the equations by a constant. In this case, it is convenient to multiply the first equation by 5.

total time ⟶ $j + w = 1$ — × 5→ $5j + 5w = 5$

total distance ⟶ $5j + 3w = 4.2$ ⟶ $5j + 3w = 4.2$

Step 3 Solve the system using subtraction. Subtract corresponding sides of each equation to eliminate j.

$$5j + 5w = 5$$
$$\underline{5j + 3w = 4.2}$$
$$2w = 0.8$$
$$w = 0.4$$

Substitute **0.4** for **w**.

$$j + 0.4 = 1$$
$$j = 0.6$$

Gina Oliva spent 0.4 h (24 min) walking and 0.6 h (36 min) jogging.

THINK AND COMMUNICATE

3. Describe Sheila's method for solving the system in Example 2.

4. Can the system in Example 2 be solved by the method of substitution described in Section 7.1? Explain.

5. How would the system of equations in Example 2 change if j and w were measured in minutes?

As you saw in the Exploration on page 297, a system of linear equations can have no solution or infinitely many solutions. When a system has *no solution*, it is called an **inconsistent system**. When it has *infinitely many solutions*, it is called a **dependent system**.

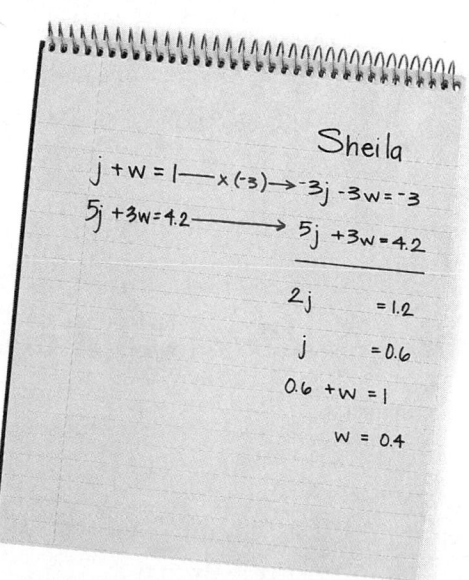

Sheila

$j + w = 1$ — × (-3)→ $-3j - 3w = -3$

$5j + 3w = 4.2$ ⟶ $5j + 3w = 4.2$

$2j = 1.2$

$j = 0.6$

$0.6 + w = 1$

$w = 0.4$

EXAMPLE 3

Solve each system of equations.

a. $x - y = 5$
$x - y = -3$

b. $3x + 6y = 9$
$2x + 4y = 6$

SOLUTION

a. Use subtraction.

$$x - y = 5$$
$$\underline{x - y = -3}$$
$$0 \neq 8$$

> The equation $0 = 8$ is never true.

Values that satisfy the equation $x - y = 5$ never satisfy the equation $x - y = -3$. This is an *inconsistent system*. It has no solution.

b. Multiply both sides of the first equation by 2, and multiply both sides of the second equation by 3. Then subtract.

$$3x + 6y = 9 \longrightarrow \times 2 \longrightarrow 6x + 12y = 18$$
$$2x + 4y = 6 \longrightarrow \times 3 \longrightarrow \underline{6x + 12y = 18}$$
$$0 = 0$$

> The equation $0 = 0$ is always true.

Values that satisfy the equation $3x + 6y = 9$ always satisfy the equation $2x + 4y = 6$. This is a *dependent system*. It has infinitely many solutions.

☑ CHECKING KEY CONCEPTS

Describe at least one way to solve each system of equations.

1. $3x + 5y = 19$
$5x - 5y = -3$

2. $2x + 4y = 6$
$5x + 4y = 15$

3. $y = 7x + 2$
$y = 2x - 4$

4. $x + 2y = 6$
$3x - 4y = 2$

5. $2x + 4y = 28$
$3x + 6y = 42$

6. $3x - 4y = 55$
$y = 2x$

Write a system of equations that would best be solved by each method.

7. addition

8. subtraction

9. multiplication and addition

7.2 Exercises and Applications

Extra Practice exercises on page 759

Solve each system of equations.

1. $5x - 3y = 13$
$7x + 3y = 11$

2. $12y - 2x = 21$
$3x + 12y = -4$

3. $6x + 2y = 5$
$8x + 2y = 3$

4. $4x + 5y = 12$
$4x - 5y = 6$

5. $y - x = 13$
$2x - 3y = 1$

6. $x + 6y = -1$
$3x - 3y = 6$

7. $3x + 7y = 6$
$2x + 9y = 4$

8. $8x + 9y = 15$
$5x - 2y = 17$

Solve each system of equations. State whether the system has *one solution*, *infinitely many solutions*, or *no solution*.

9. $3x + 7y = 2$
   $6x + 14y = 4$

10. $4x + 5y = 3$
    $6x + 9y = 9$

11. $6x + 11y = 12$
    $12x + 22y = 10$

12. $0.3x + 0.7y = 1.2$
    $6x + 14y = 24$

13. $0.4x + 2.1y = 40$
    $3.6x + 6.3y = 22$

14. $27x - 42y = 102$
    $45x - 70y = 170$

15. $27x - 42y = 102$
    $45x - 70y = 150$

16. $9x + 10y = 11$
    $10x + 9y = 11$

Connection ▸ PHYSICS

A *lever* is a rod that tilts on a *fulcrum*. Any weight applied to a lever creates *torque*, defined as the product of the weight and the distance between the fulcrum and the point where the weight is applied. For a lever *not* to move when force is applied, any torque in a clockwise direction must equal the torque in a counterclockwise direction.

This man creates a counterclockwise torque equal to 60 lb · 8 ft, or 480 ft · lb (read "foot-pounds").

60 lb

This large stone creates a clockwise torque equal to 120 lb · 4 ft, or 480 ft · lb.

120 lb

⊢————— 8 ft —————⊣ ⊢——— 4 ft ———⊣

Because the counterclockwise and clockwise torques are equal, the lever is in balance.

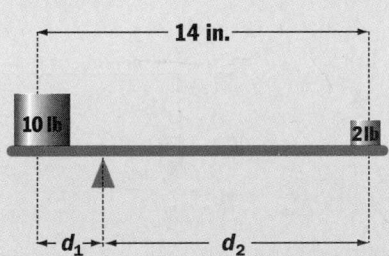

⊢——————— 14 in.———————⊣

10 lb   2 lb

←d₁→⊢————— d₂ —————⊣

17. In the situation shown at the left, the lever is in balance. Complete parts (a)–(c) to determine the distances d_1 and d_2 between the weights and the fulcrum.

   a. Write an equation representing the equal torques.

   b. Write an equation for the combined distances.

   c. Find d_1 and d_2.

18. Suppose a 100 ft bridge weighs 100 tons and has a support at each end. A 3 ton truck is 10 ft from the right end of the bridge.

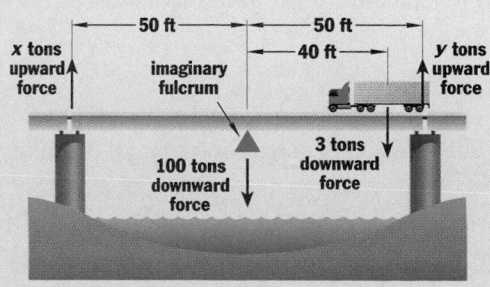

⊢———— 50 ft ————⊣⊢———— 50 ft ————⊣
                  ⊢——— 40 ft ———⊣

x tons upward force

imaginary fulcrum

y tons upward force

100 tons downward force

3 tons downward force

   a. You can think of the bridge as a type of lever. Imagine a fulcrum at the midpoint of the bridge, and think of the bridge's weight as being concentrated at the fulcrum. Write an equation representing the balanced torques that occur at each support and at the truck.

   b. For an object at rest, the downward forces on the object equal the upward forces. Write an equation representing the balanced vertical forces acting on the bridge. (*Note:* There are no distances involved.)

   c. Solve your system of equations to find the force, in tons, that each support must exert to keep the bridge from moving.

19. a. **Technology** Graph the equations $-x - y = 3$ and $3x - y = 1$ using a graphing calculator or graphing software. Where do the lines intersect?

b. **Writing** Add the two equations from part (a), solve for y, and graph the resulting equation on the same screen. Describe what you see.

c. **Challenge** Use the general equations $Ax + By = C$ and $Dx + Ey = F$ to give an algebraic argument for why the graph of the sum of the two equations passes through the same intersection point as the given equations. (*Hint:* If the coordinates (p, q) satisfy both $Ax + By = C$ and $Dx + Ey = F$, show that they must satisfy the sum of the two equations.)

20. CONSUMER ECONOMICS The octane number of gasoline indicates how well it fights *knock*, the noise an engine makes when fuel is not being burned properly. Octane numbers are based on a standard isooctane-and-heptane reference fuel. For example, 87-octane gasoline fights knock as well as a reference fuel that is 87% isooctane. Alexis decides to mix 87-octane and 93-octane gasolines to get 15 gal of 89-octane gasoline.

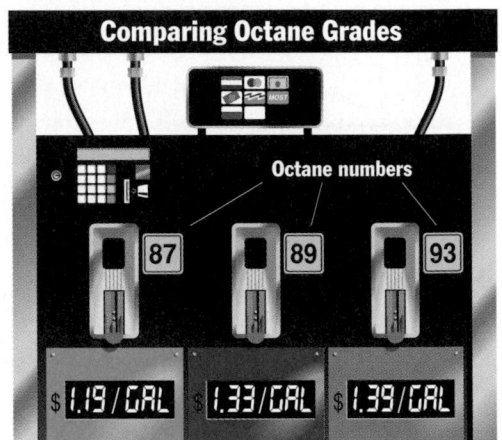

a. Write equations describing the total volume of gasoline and the volume of isooctane in the reference fuel.

b. Solve the system to determine how many gallons of each grade of gasoline are needed.

c. Use the diagram. Is it cheaper for Alexis to buy the 89-octane gasoline directly or to mix the other two grades herself? Explain.

21. FINANCE Suppose a university allots $335,000 to pay the salaries of a new research team. The team of 10 members will consist of researchers who will be paid $44,000 and technicians who will be paid $29,000. Determine how many researchers the university can afford to hire for the team.

ONGOING ASSESSMENT

22. Writing When is the addition method better than the substitution method? When is the substitution method better? Give an example of each.

SPIRAL REVIEW

Use substitution to solve each system of equations. (*Section 7.1*)

23. $y = 7x$
$y + 6x = 39$

24. $x + y = 8$
$2x - 3y = 6$

25. $x + 2y = 11$
$3x + 2y = 17$

26. $x + y = 5.8$
$2x + 3y = 14.4$

For Exercises 27–30, use matrices *A, B, C*, and *D* to evaluate each matrix expression. If an expression is not defined, write *undefined*. (*Section 1.3*)

$$A = \begin{bmatrix} 7 & 11 \\ -15 & 3 \end{bmatrix} \qquad B = \begin{bmatrix} 1 & 0 \\ 4 & -\frac{11}{2} \end{bmatrix} \qquad C = \begin{bmatrix} 8 & -2 \\ 19 & 6 \\ 1 & -13 \end{bmatrix} \qquad D = \begin{bmatrix} 6 & 8 \\ 0 & 12 \\ 10 & -4 \end{bmatrix}$$

27. $A + B$

28. $2A$

29. $D - C$

30. $-4A + 2B$

7.3 Solving Linear Systems with Matrices

Learn how to...

- solve systems of linear equations using matrices

So you can...

- solve problems, such as determining how many batches to prepare of two recipes so that all supplies are used

At the end of the gardening season, the last of the fruits and vegetables are often canned for use during the winter. A system of equations can help you divide your supplies among canning recipes so that you don't waste anything.

EXAMPLE 1   Application: Cooking

Betsy has harvested her garden before leaving on vacation. She has 42 qt of tomatoes and 6 qt of green peppers. She wants to use up all her tomatoes and peppers by canning tomato sauce and tomato juice.

   Use matrices to determine how many batches of sauce and juice she should make so that no tomatoes or peppers are wasted.

Tomato Juice
8 Qt tomatoes
0.5 Qt green peppers
2 onions
1 beet
1 hot pepper
1 garlic clove
celery leaves

Base for Tomato Sauce
4 Qt tomatoes
1 Qt green peppers
1 Qt celery
1 Qt onions

SOLUTION

Step 1 Write a system of equations. Let x = the number of batches of sauce, and let y = the number of batches of juice.

	Number of quarts needed for:		
	x batches of sauce	y batches of juice	
Tomatoes	$4x$	$8y$	$4x + 8y = 42$
Peppers	$1x$	$0.5y$	$1x + 0.5y = 6$

There are **42 qt** of tomatoes available.

There are **6 qt** of peppers available.

Solution continued on next page.

SOLUTION *continued*

Step 2 Rewrite the system of equations as a matrix equation.

$$\begin{bmatrix} 4 & 8 \\ 1 & 0.5 \end{bmatrix} \begin{bmatrix} x \\ y \end{bmatrix} = \begin{bmatrix} 42 \\ 6 \end{bmatrix}$$

This equation has the form $A \begin{bmatrix} x \\ y \end{bmatrix} = B.$

To solve this equation for x and y, you multiply both sides of the equation by the *inverse* of the coefficient matrix.

$$\begin{bmatrix} x \\ y \end{bmatrix} = \begin{bmatrix} 4 & 8 \\ 1 & 0.5 \end{bmatrix}^{-1} \begin{bmatrix} 42 \\ 6 \end{bmatrix}$$

This equation has the form $\begin{bmatrix} x \\ y \end{bmatrix} = A^{-1}B.$

For more information about inverse matrices, see the *Technology Handbook*, **p. 812.**

Use a graphing calculator or software with matrix calculation capabilities to find $A^{-1}B$.

```
[A]-1[B]
        [[4.5]
         [3  ]]
```

Betsy needs to make **4.5 batches of sauce** and **3 batches of juice**.

THINK AND COMMUNICATE

1. Confirm that $(x, y) = (4.5, 3)$ is the solution of the system of equations in Example 1.

2. In the solution of Example 1, what happens if you try to compute $B(A^{-1})$ instead of $A^{-1}B$? Why doesn't this work?

3. How would you write a matrix equation to represent the system below?

$$15x + 8y = 9$$
$$-3y = 8$$

Square matrices such as $\begin{bmatrix} 1 & 0 \\ 0 & 1 \end{bmatrix}$ and $\begin{bmatrix} 1 & 0 & 0 \\ 0 & 1 & 0 \\ 0 & 0 & 1 \end{bmatrix}$, which have 1's on the

diagonal running from upper left to lower right and 0's elsewhere, are called **identity matrices**. An identity matrix I has the special property that

$$AI = IA = A$$

for any matrix A having the same dimensions as I.

The **inverse matrix** of a square matrix A, written A^{-1}, is the matrix that, when multiplied by A, gives an identity matrix:

$$A^{-1}A = A(A^{-1}) = I$$

Not all square matrices have inverses. If a matrix A does not have an inverse, you will get an error message when you try to find A^{-1} on a graphing calculator or computer.

WATCH OUT! ▶

Just as $a \cdot a^{-1} = 1$ for any nonzero number a, $A \cdot A^{-1} = I$ for any invertible matrix A. However, the notation A^{-1} does *not* mean $\frac{1}{A}$.

THINK AND COMMUNICATE

4. **Technology** Use a graphing calculator or software with matrix calculation capabilities to show that the product of the coefficient matrix and its inverse in Example 1 is an identity matrix.

EXAMPLE 2

Find an equation of the parabola passing through the points $(-3, 1), (4, 4)$, and $(7, -1)$.

SOLUTION

You need to find values of a, b, and c such that the coordinates of each point satisfy $y = ax^2 + bx + c$.

Step 1 Substitute the coordinates of each point into $ax^2 + bx + c = y$.

Using $(-3, 1)$: $a(-3)^2 + b(-3) + c = 1 \longrightarrow 9a - 3b + c = 1$

Using $(4, 4)$: $a(4)^2 + b(4) + c = 4 \longrightarrow 16a + 4b + c = 4$

Using $(7, -1)$: $a(7)^2 + b(7) + c = -1 \longrightarrow 49a + 7b + c = -1$

Step 2 Write the system of equations in matrix form, and solve the system using a graphing calculator or software with matrix calculation capabilities.

$$\begin{bmatrix} 9 & -3 & 1 \\ 16 & 4 & 1 \\ 49 & 7 & 1 \end{bmatrix} \begin{bmatrix} a \\ b \\ c \end{bmatrix} = \begin{bmatrix} 1 \\ 4 \\ -1 \end{bmatrix}$$

$$\begin{bmatrix} a \\ b \\ c \end{bmatrix} = \begin{bmatrix} 9 & -3 & 1 \\ 16 & 4 & 1 \\ 49 & 7 & 1 \end{bmatrix}^{-1} \begin{bmatrix} 1 \\ 4 \\ -1 \end{bmatrix}$$

```
[A]-1[B]
[[-.2095238095]
 [.6380952381 ]
 [4.8         ]]
```

Rounding the coefficients to two significant digits, you get
$$y = -0.21x^2 + 0.64x + 4.8$$
as an equation of the parabola.

✓ CHECKING KEY CONCEPTS

Write each system of equations as a matrix equation, and solve using a graphing calculator or software with matrix calculation capabilities.

1. $14x - 13y = -6$
$-5x - y = -2$

2. $512x + 94y = 318$
$137x = -89$

3. $81.2x + 17.4y = 0.8$
$4.6x - 12.8y = 12.8$

4. Find an equation of the parabola passing through the points $(2, 5)$, $(3, -7)$, and $(5, 5)$.

5. Use matrices to solve this system for x, y, and z:

$$3x + 2y + 5z = 9$$
$$2x - 4y + 6z = 10$$
$$5x + 9y - 2z = 15$$

7.3 Exercises and Applications

*Extra Practice
exercises on
page 759*

 Technology Throughout the exercises, use a graphing calculator or
software with matrix calculation capabilities to solve the matrix equations.

Rewrite each of the systems in matrix form, and solve for *x* and *y*.

1. $8x - 3y = 32$
$7x - 9y = 24$

2. $24x - 24y = 89$
$67x - 19y = 22$

3. $5x - 12y = 13$
$14x - 7y = 34$

4. $51x - 25y = 98$
$13x - 19y = 34$

5. COOKING Charlie wants to use up 8 cups of buttermilk and 11 eggs by
baking rolls and muffins to freeze. A batch of rolls uses 2 cups of buttermilk and 3 eggs. A batch of muffins uses 1 cup of buttermilk and 1 egg.

   a. How many batches should Charlie make of each recipe?

   b. How would your answer to part (a) change if Charlie wanted to use
up 8 cups of buttermilk and 6 eggs?

6. The graph shows two parallel lines, $y = -2x + 1$ and $y = -2x - 2$.

   a. Can you solve this system of equations? Explain.

   b. Rewrite the equations as a matrix equation of the form $A \begin{bmatrix} x \\ y \end{bmatrix} = B$.

   c. Enter the matrices A and B into a graphing calculator or software
with matrix calculation capabilities. Try to find $A^{-1}B$. What happens?

   d. Open-ended Problem Write two equations that describe the same
line. Repeat parts (b) and (c) for this set of equations.

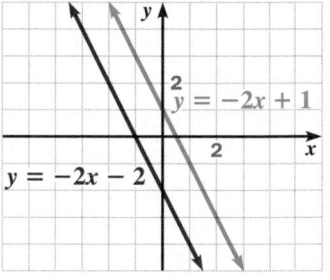

Use matrices to solve each system for *x* and *y*.

7. $y = 14x + 2$
$y = 31x - 3$

8. $y = 4.1x - 8.6$
$y = -6.2$

9. $y = 57x + 48$
$y = -12x$

10. $y = -4x + 8.3$
$y = -26x - 2.5$

11. $x = -y + 2$
$y = 2x - 7$

12. $2x - 14y = 57$
$15x + 8y = 18$

13. $31x + 8y = -7$
$2 - 17y = 0$

14. $7.2x - 8.8y = -31.2$
$0.1y - 2.4x = 8.1$

15. Challenge Consider this matrix equation:

$$\begin{bmatrix} a & b \\ c & d \end{bmatrix} \begin{bmatrix} 1 & 2 \\ 3 & 4 \end{bmatrix} = \begin{bmatrix} 1 & 0 \\ 0 & 1 \end{bmatrix}$$

   a. Use matrix multiplication to rewrite the left-hand side of the equation as
a single matrix.

   b. Use your matrices to write four equations involving *a*, *b*, *c*, and *d* by setting
corresponding entries equal. (*Hint:* The first equation is $a + 3b = 1$.)

   c. Without using technology, find *a*, *b*, *c*, and *d*.

   d. What is the matrix $\begin{bmatrix} a & b \\ c & d \end{bmatrix}$? Check your answer using technology.

Use matrices to solve each system for *x*, *y*, and *z*.

16. $5x - 4y - 8z = 10$
$7x - 11y - 2z = 13$
$-3x - y - 3z = 17$

17. $6x - z = 10$
$x - 2y - 11z = -3$
$7x - 9z = -1$

18. $3.1x + 0.9y - 8.3z = 1.8$
$0.1x - 0.2y - 1.0z = 4.0$
$8.5x + 2.6y + 5.0z = 5.0$

Find an equation of the parabola passing through each set of points.

19. $(-3, 4.5)$, $(-1.2, -2.0)$, and $(0.8, 1.6)$

20. $(-4, 17)$, $(6, 2)$, and $(13, 15)$

21. $(-1, 5)$, $(7, 2)$, and $(9, 5)$

22. $(-2, 3.5)$, $(3, 8)$, and $(6, 11.5)$

Connection ▶ AGRICULTURE

Soil scientists have conducted many experiments to determine how to use various fertilizers. The table shows the yield of hard red spring wheat (in kilograms per hectare) when different amounts of nitrogen fertilizer were applied to a field in the western United States.

Fertilizer (kg/ha)	Wheat yield (kg/ha)
99	1814
155	3562
211	3965
267	4368
323	4032

23. Use the last three entries from the table to find an equation for a parabola passing through those three points. What maximum yield does the model predict for the field?

24. Open-ended Problem Choose another set of three points from the table and find an equation of the parabola passing through them. How well do the two models agree? Is this a good way to fit a quadratic model to data? Why or why not?

25.  **Technology** If you have a calculator or software that will perform quadratic regression, find a quadratic model using all the data. Compare this model with the model you found in Exercise 23. Which model is better? Why?

26. ART Monica is making a mobile. She wants to suspend three objects from a lightweight rod, as shown. Suppose the weights of the objects are $w_1 = 2.5$ oz, $w_2 = 1.2$ oz, and $w_3 = 2.0$ oz.

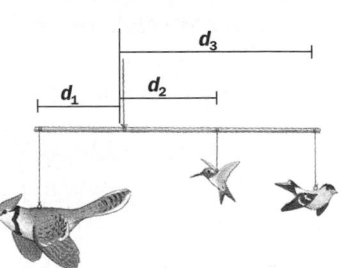

a. Write three equations involving the unknown distances d_1, d_2, and d_3 based on the following facts:

• The length of the rod is 14.5 in.

• To balance the mobile, she must position the objects so that $w_1d_1 = w_2d_2 + w_3d_3$.

• She wants the third object to be 1.5 times as far from the support as the second object is.

b. Write the system of equations as a matrix equation.

c. Find d_1, d_2, and d_3.

27. **Cooperative Learning** Work with a partner. Each of you should choose three points in the coordinate plane.

    a. Each of you should find an equation of the parabola passing through your three points.

    b. For each set of three points, no two points can be on the same vertical line. Why?

    c. Compare your answers to part (a). Is there always a parabola? What are the values of a, b, and c when the three points lie on a line?

SPIRAL REVIEW

Solve each system of equations. State whether the system has *one solution*, *infinitely many solutions*, or *no solution*. *(Sections 7.1 and 7.2)*

28. $2x - 2y = 6$
    $x - y = 1$

29. $2x + 2y = 5$
    $x + 3y = 0$

30. $-3x + 4y = 1$
    $6x - 8y = -2$

Solve each inequality for **x**. *(Toolbox, page 787)*

31. $2x + 4 \geq 1$

32. $5x - 3 < 7$

33. $12x > 6$

Write each quadratic function as a product of factors. *(Section 5.6)*

34. $y = x^2 - 2x - 8$

35. $y = 2x^2 - 9x - 5$

36. $y = x^2 - 4x + 3$

ASSESS YOUR PROGRESS

VOCABULARY

system of equations (p. 292)
inconsistent system (p. 299)
dependent system (p. 299)

identity matrices (p. 304)
inverse matrix (p. 304)

Solve each system of equations. *(Section 7.1)*

1. $y = 5x + 2$
    $y = 2x^2$

2. $3y = 9$
    $-5x - y = 6$

3. $3x + 5y = 3$
    $14x + 10y = 6$

Solve each system of equations. State whether the system has *one solution*, *infinitely many solutions*, or *no solution*. *(Section 7.2)*

4. $2x + 4y = 7$
    $5x - 3y = 1$

5. $8x + 2y = 4$
    $-4x - y = 2$

6. $-3x - y = -5$
    $x + y = -1$

7. **Technology** Use a graphing calculator or software with matrix calculation capabilities to find an equation of the parabola passing through the points $(-11, -7)$, $(-8, 2)$, and $(-3, 3)$. *(Section 7.3)*

8. **Journal** Explain the difference between an inconsistent system and a dependent system. Can nonlinear systems be inconsistent or dependent? Explain.

7.4 Inequalities in the Plane

For a point to be on the graph of $y = x + 3$, its y-coordinate must be 3 more than its x-coordinate. What do you think must be true about points on the graph of $y > x + 3$ or $y < x + 3$? In the Exploration, you will investigate such inequalities.

EXPLORATION
COOPERATIVE LEARNING

Guessing People's Ages

Work with a group of four or five students.
You will need:
- paper and pencil
- graph paper

1 Each person in your group should estimate the ages of the people shown. (Keep your estimates a secret; it will be more fun.)

2 Get the list of the actual ages from your teacher.

3 Make a scatter plot of your group's data. Use the horizontal axis for a = actual age and the vertical axis for e = estimated age.

Use your scatter plot to answer the following questions.

1. What does each vertical column of dots represent?

2. Do any plotted points fall on the line $e = a$? If so, what does this mean?

3. **a.** What does it mean if a plotted point falls below the line $e = a$? How would you represent this mathematically?

   b. What does it mean if a plotted point falls above the line $e = a$? How would you represent this mathematically?

4. Overall, what pattern do you see in your group's scatter plot? Compare your scatter plot with those from other groups.

5. If people were estimating your age, would you prefer that most points fell above the line $e = a$, below the line, or on the line? Do you think everyone prefers that region?

EXAMPLE 1 **Application: Consumer Economics**

The label on a package of ground beef (hamburger) tells you that the beef is "90% lean," which means that it contains no more than 10% fat (by weight). Write and graph an inequality that relates the acceptable amount of fat to the weight of the ground beef being sold.

SOLUTION

Let b = the weight of the ground beef, and let f = the weight of the fat. Then the inequality is $f \leq 0.1b$.

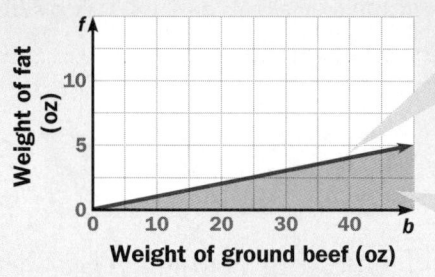

Step 1 Draw the line $f = 0.1b$.

Step 2 Shade below the line.

THINK AND COMMUNICATE

1. Use the graph from Example 1.

    a. For the grocer preparing ground beef, what does the shaded region represent?

    b. What does the region above the line represent? What inequality defines this region?

The graph of $y = ax + b$ divides the coordinate plane into two regions called **half-planes**. Each region is defined by an inequality.

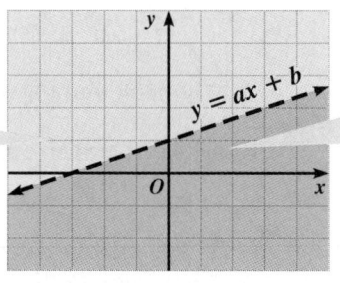

The inequality $y > ax + b$ defines the half-plane **above** the line.

The inequality $y < ax + b$ defines the half-plane **below** the line.

When a line is vertical, it has an equation of the form $x = k$. Vertical lines also define half-planes.

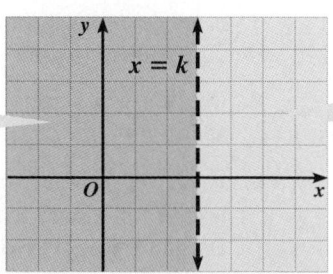

The graph of $x < k$ is the half-plane to the **left** of the line $x = k$.

The graph of $x > k$ is the half-plane to the **right** of the line $x = k$.

EXAMPLE 2

Graph each inequality.

a. $y \geq 2x - 1$

b. $y < 2x - 1$

SOLUTION

a.

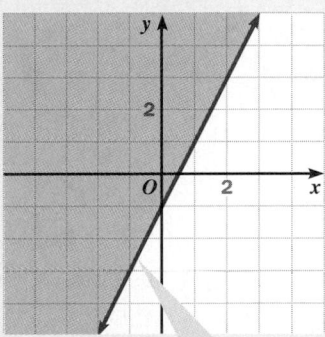

Graph $y = 2x - 1$ as a **solid** line since it is included in the graph of the inequality.

b.

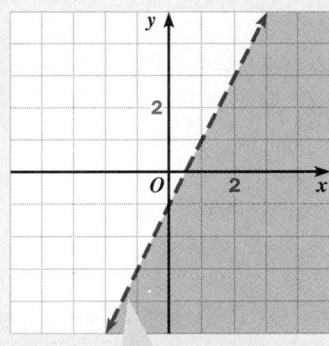

Graph $y = 2x - 1$ as a **dashed** line since it is not included in the graph of the inequality.

EXAMPLE 3

Graph the inequality $2x - 3y > 4$.

SOLUTION

Solve the inequality for y.

$$2x - 3y > 4$$
$$-3y > -2x + 4$$
$$y < \frac{2}{3}x - \frac{4}{3}$$

Remember that an inequality sign **changes direction** when each side is multiplied or divided by a negative number.

Graph $y = \frac{2}{3}x - \frac{4}{3}$ as a dashed line and shade below it.

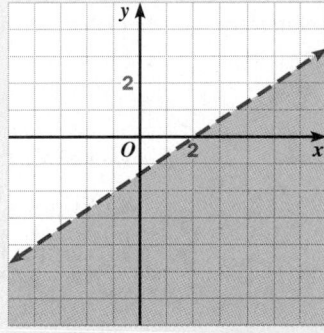

Toolbox p. 787
Solving Linear Inequalities

THINK AND COMMUNICATE

2. a. Do the coordinates $(3, -3)$ satisfy the inequality in Example 3? Do the coordinates $(0, 0)$ satisfy the inequality?

b. How does checking the coordinates of a point help you graph an inequality?

EXAMPLE 4

Find an inequality that defines the shaded region shown below.

SOLUTION

The line through the points $(0, 3)$ and $(4, 1)$ has the equation $y = \left(\dfrac{1-3}{4-0}\right)x + 3$, or $y = -\dfrac{1}{2}x + 3$. The region is below and does not include the line, so the inequality is $y < -\dfrac{1}{2}x + 3$.

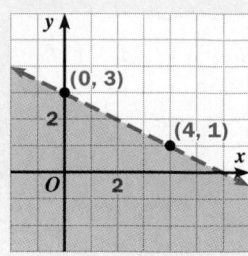

☑ CHECKING KEY CONCEPTS

Graph each inequality.

1. $y < 3x$ **2.** $y > -2x + 4$ **3.** $y \le 5x + 5$

4. $x < 7$ **5.** $3x + 2y > 1$ **6.** $2y \le x - 4$

7. During a "red tag" sale, a department store offers discounts of up to 25% on all tagged merchandise. Write and graph an inequality that relates the discounted price to the normal price of tagged merchandise.

7.4 **Exercises and Applications**

Extra Practice exercises on page 759

1. A day-care center provides at least one supervisor for every three infants. Write and graph an inequality that relates the number of supervisors on duty to the number of infants at the center.

Graph each inequality.

2. $y > 5x$ **3.** $y > 3x - 2$ **4.** $y < -4x + 5$ **5.** $y < -1.2x$

6. $y > 4.5x + 3$ **7.** $x \ge 2.5$ **8.** $4x + 6y > 24$ **9.** $3.1x - 4.6y < 33$

10. $5x + 2y \ge 0$ **11.** $y > 8.5$ **12.** $x + y \ge 6$ **13.** $-5x + 10y < -15$

14. Challenge Graph the quadratic inequality $y > x^2 + 3$.

Find an inequality that defines each shaded region.

15.
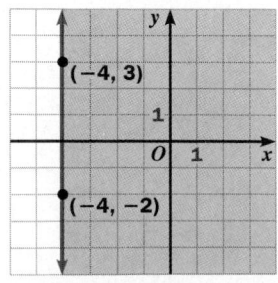

16.
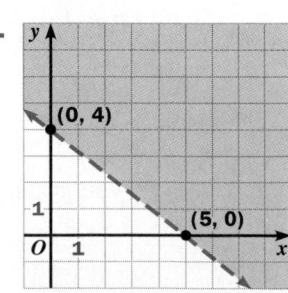

17.
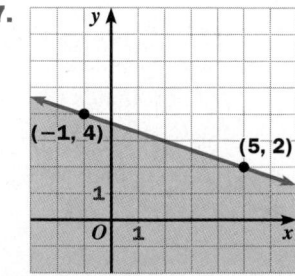

An object immersed in a liquid will be buoyed by a force equal to the weight of the liquid that the object displaces. Since 1 ft^3 of salt water weighs 64 lb, a submerged object having a volume of 1 ft^3 and weighing more than 64 lb will sink, while one having the same volume and weighing less than 64 lb will float.

BY THE WAY

In 1971, a 2400-ton barge was raised in the Gulf of Mexico using the "sphere injection" method.

18. Divers use weight belts to control their buoyancy. In order to float on the surface of the ocean, divers must weigh less than the amount of water they displace. Write and graph an inequality that relates the volume of salt water a floating diver displaces to the weight of the diver.

19. An interesting form of ocean salvage involves pumping hollow spheres into sunken vessels in order to supply buoyancy. To become buoyant, each ton to be raised requires at least 80 specially designed spheres of diameter 11 in. Write and graph an inequality that expresses the relationship between the number of spheres needed to raise a vessel and the weight of the vessel.

20. Lifting bags are sometimes attached to a sunken object and inflated in order to increase the object's buoyancy. Shellie is using a lifting bag to raise an object from the bottom of a freshwater lake. She knows that 1 ft^3 of fresh water weighs 62.4 lb. Write and graph an inequality that relates the volume of water the lifting bag displaces to the weight of the submerged object.

21. Open-ended Problem Why do you think that you float more easily in salt water than in fresh water?

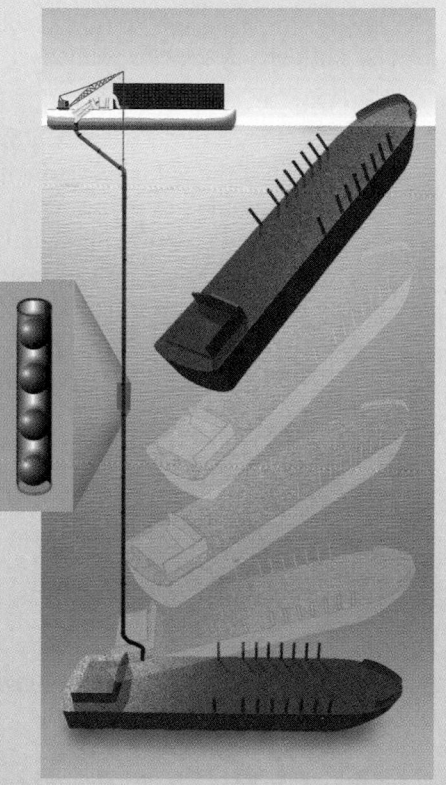

22. CARTOGRAPHY In order to make the symbols on maps legible to the average person, cartographers set lower limits for the width and height of the symbols they use. The minimum size of a symbol depends linearly on the distance at which it will be viewed.

a. The minimum width (in inches) of a legible symbol is 0.007 times the viewing distance (in feet). Write and graph an inequality that relates these two variables.

b. The minimum height (in inches) of a legible symbol is 0.03 times the viewing distance (in feet). Write and graph an inequality that relates these two variables.

23. AGRICULTURE On an ostrich farm, female ostriches lay up to 70 eggs per year. Write and graph an inequality that relates the number of eggs produced per year to the number of female ostriches on the farm.

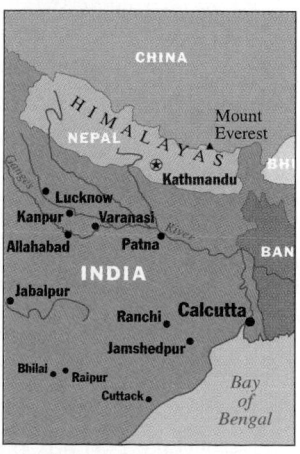

24. **MANUFACTURING** For many farmers in the Tai Lake Valley region of China, silkworm cocoons provide the main source of income. Unfortunately, the industry is being threatened by fluoride pollution from nearby brick factories.

a. To form its cocoon, each silkworm produces a single filament that stretches up to 1.5 km long. Write an inequality that relates the total possible length of silk filament produced to the number of silkworms.

b. Mature silkworms, on average, can tolerate 40 micrograms of fluoride in each gram of mulberry leaves they consume. Write an inequality that relates the mass of fluoride a silkworm can safely consume to the mass of mulberry leaves consumed.

25. **MEDICINE** A comparison of blood pressures in a patient's arm and ankle can help a doctor find hidden health risks. The doctor may consider it a warning sign if the ankle's systolic blood pressure is less than 90% of the arm's systolic blood pressure. Write and graph an inequality that relates the acceptable blood pressure in the ankle to the blood pressure in the arm. Systolic arm blood pressures, measured in millimeters of mercury (mm Hg), typically range from 100 mm Hg to 130 mm Hg.

26. **SAT/ACT Preview** Which inequality is equivalent to $4x - 3y > 6$?

A. $y < -\frac{4}{3}x - 2$ B. $y > -\frac{4}{3}x + 2$ C. $y < \frac{4}{3}x - 2$

D. $y > \frac{4}{3}x + 2$ E. none of these

ONGOING ASSESSMENT

27. **Open-ended Problem** Find an example of an inequality involving two variables in everyday life. Write and graph your inequality.

SPIRAL REVIEW

Solve each system of equations. *(Section 7.2)*

28. $3x + 4y = 17$
    $x + 7y = 17$

29. $3.1x - 5.8y = 19$
    $2.6x + 1.8y = 51$

30. $23x + 45y = 91$
    $26x - 28y = 24$

Use the following data on the number of passengers who entered various stations on two subway lines during one day in Boston. *(Section 6.4)*

Orange Line: 6043; 8368; 5714; 9512; 4743; 2029; 9703; 19,480; 3393; 4886; 15,171; 2769; 5552; 2798; 4167; 2441; 2962; 10,936

Red Line: 9990; 5865; 8744; 14,724; 7877; 8163; 7208; 4636; 16,278; 2960; 3087; 7576; 723; 3777; 1157; 9161; 5114; 4122; 7133; 5920; 5024

31. a. Make two box plots to compare the data for the Orange Line and the Red Line.

    b. **Writing** How do the two plots compare?

7.5 Systems of Inequalities

Learn how to...

- **graph a system of linear inequalities**

- **find the system of linear inequalities that describes a given graph**

So you can...

- **represent and interpret situations involving inequalities, such as target heart rate**

Hot deserts in North America include three general types: the arid Mojave desert, the grassier Chihuahuan Desert, and the diverse Sonoran Desert. What distinguishes the three deserts climatically is the amount of precipitation that falls in each season. The graph below shows the mean annual precipitation and the percent of precipitation that falls in winter for various sites in the three North American hot deserts.

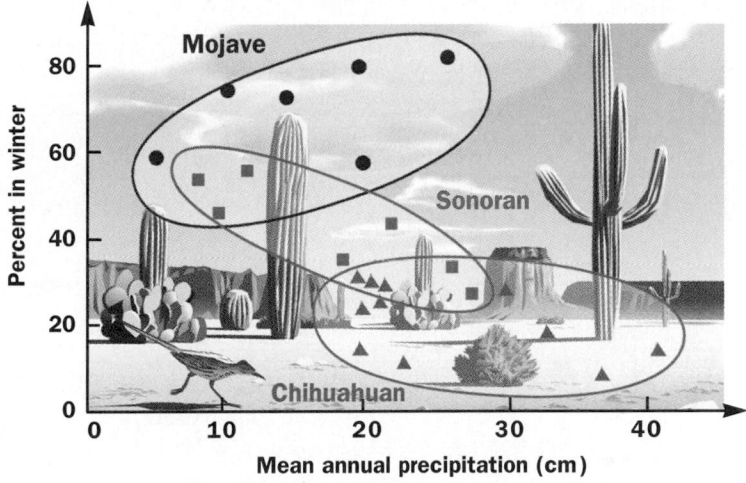

THINK AND COMMUNICATE

Use the graph above to answer the following questions.

1. If a North American hot desert site gets 30 cm of rain, and 20% of that in the winter, which desert do you think it is in? What if the site gets 25 cm of rain, and 30% of that in the winter?

2. How does the graph show that the climate in the Sonoran Desert is sometimes like the climate in the Mojave Desert?

3. Which two deserts are the most different climatically? How does the graph show this?

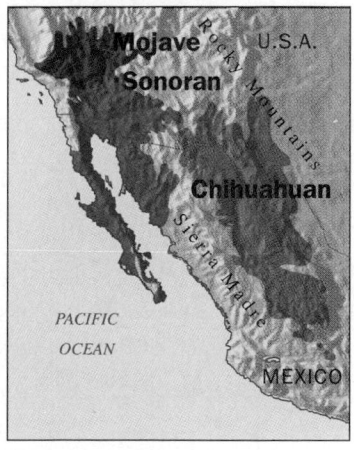

A region of a coordinate plane can be described using a **system of inequalities**. For example, the first quadrant of the *xy*-plane is described by this system of inequalities: $x > 0$ and $y > 0$.

EXAMPLE 1   Interview: Gina Oliva

Look back at the opening interview on pages 288–290. To find a person's target heart rate, subtract the person's age from 220. The target heart rate, in heartbeats per minute, is between 60% and 80% of this number.

a. Write a system of inequalities using a for age and h for target heart rate.

b. Graph the system of inequalities.

SOLUTION

a.

Inequality 1: h should be at least 60% of $(220 - a)$.	**Inequality 2:** h should be no more than 80% of $(220 - a)$.
$h \geq 0.6(220 - a)$, or $h \geq -0.6a + 132$	$h \leq 0.8(220 - a)$, or $h \leq -0.8a + 176$

b.

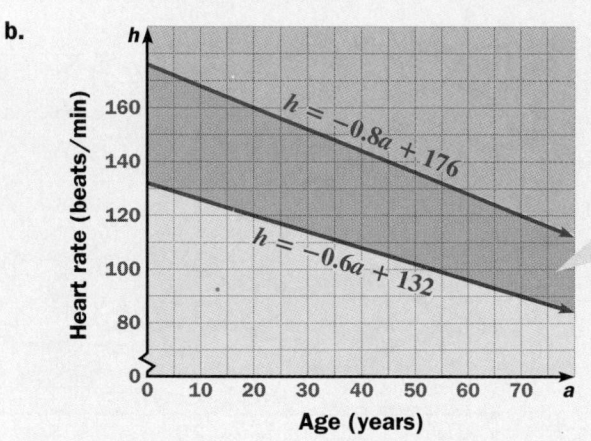

The possible values for the target heart rate are contained in the **overlap** of the red region and the blue region.

EXAMPLE 2

The Junior-Senior Prom Committee must have five to eight representatives drawn from the junior and senior classes. The committee must include at least two members of the junior class and at least two members of the senior class. Write and graph a system of inequalities that represents the possible committees.

SOLUTION

Step 1 Write inequalities to represent the restrictions in the problem. Let $x =$ the number of juniors, and let $y =$ the number of seniors.

The committee has at least five members.	The committee has at most eight members.	There are at least two juniors.	There are at least two seniors.
$x + y \geq 5$, or $y \geq -x + 5$	$x + y \leq 8$, or $y \leq -x + 8$	$x \geq 2$	$y \geq 2$

Step 2 Graph the region described by the system of inequalities.

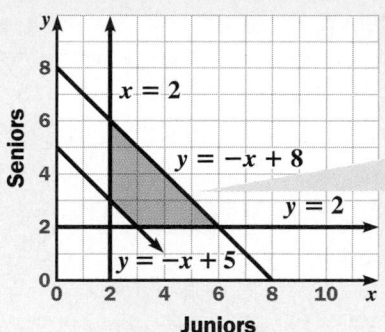

All possible combinations of juniors and seniors on the committee are shown in the shaded region.

THINK AND COMMUNICATE

4. Does *every* point (x, y) in the shaded region in Example 2 represent a committee consisting of x juniors and y seniors? Explain.

5. Redraw the solution of Example 2 using discrete points rather than a shaded region.

EXAMPLE 3

Find a system of inequalities defining the shaded region shown below.

SOLUTION

Inequality 1: An equation of the line through the points $(0, 5)$ and $(6, 3)$ is

$$y = \left(\frac{3 - 5}{6 - 0}\right)x + 5, \text{ or } y = -\frac{1}{3}x + 5.$$

The region is below and includes the

line, so the inequality is $y \le -\frac{1}{3}x + 5.$

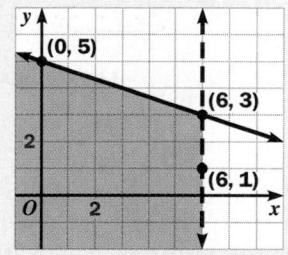

Inequality 2: An equation of the line through the points $(6, 3)$ and $(6, 1)$ is $x = 6$. The region is to the left of and does not include the line, so an inequality is $x < 6$.

☑ CHECKING KEY CONCEPTS

Graph each system of inequalities.

1. $y \ge x + 3$
$x \le 5$

2. $y > 0$
$y < 10$

3. $y \ge -3x - 3$
$y < x + 1$
$x \le 15$

4. Writing What must be true about the coordinates of any point in the graph of a system of inequalities?

7.5 Exercises and Applications

Extra Practice exercises on page 759

Graph each system of inequalities.

1. $y \geq -x - 1$
$y \geq 4x - 12$

2. $x \leq 5$
$x \geq -2$

3. $x \leq -1$
$3y \leq -x - 2$

4. $y \geq 1$
$y \leq -3x + 8$

5. $y \geq 5x$
$y \leq 2x - 4$

6. $y \leq 5$
$y \geq \frac{1}{2}x + 1$

7. $y < x + 3$
$y < 4$

8. $y \geq -3x + 3$
$y \leq \frac{1}{4}x + 4$

9. GEOLOGY Although "Old Faithful," a geyser in Yellowstone National Park, is known for its regular eruptions, the lengths of eruptions and the times between eruptions do vary. The scatter plot shows the length of an eruption, l, and the time until the next eruption, t, over a four-day period.

a. Open-ended Problem Imagine a parallelogram that encloses the plotted points. Find the inequalities that describe the interior of the parallelogram.

b. Writing If an eruption lasts 2 min, what range of time would you expect to wait until the next eruption? Explain how you can use your graph from part (a) to answer this question.

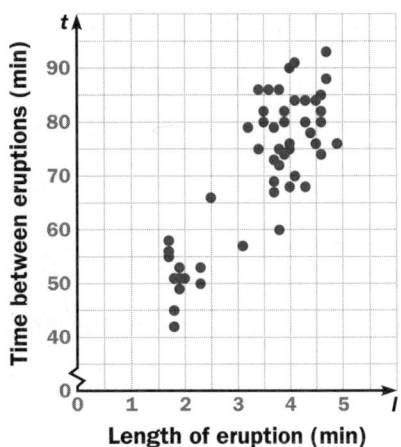

Length of eruption (min)

BY THE WAY

A geyser is a spring that intermittently throws hot water up with explosive force. The only geyser group in the United States is in Yellowstone National Park.

10. George has to write a research paper on the work of a well-known author. His bibliography must include at least 12 books or articles, and at least 4 of the references must be original works by the author (primary sources). The other references can be discussions of the author's work or background material (secondary sources).

a. Write inequalities to represent the number of primary sources, p, and the number of secondary sources, s, that George can list in his bibliography.

b. Graph the system of inequalities you found in part (a).

Write a system of inequalities defining each shaded region.

11.

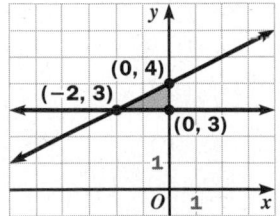

12.

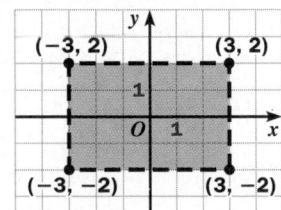

13.

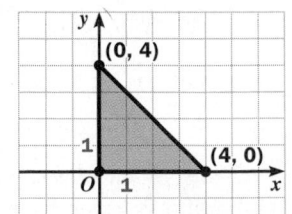

Look back at the article on pages 288–290.

Many people go to aerobics classes like Gina Oliva's to maintain a recommended weight. The table shows the recommended weights for people of various heights and ages.

14. Use the data for people aged 19–34. Make a scatter plot with height on the horizontal axis and weight on the vertical axis. For each height, use one color or symbol to plot the lowest recommended weight, and a second color or symbol to plot the highest recommended weight.

15. Using your scatter plot from Exercise 14, draw a line of fit for the lowest recommended weight. Draw a line of fit for the highest recommended weight. Shade the recommended weight region.

16. Write inequalities to describe your graph from Exercise 15.

17. a. Repeat Exercises 14–16 using the data for people aged 35 and over.

    b. **Writing** Compare the region for people aged 19–34 with the region for people aged 35 and over.

18. **Research** Recommended weight also depends on gender, build, and medical history. Use an almanac or other source to find the recommended weights for a group of people, and repeat Exercises 14–16 for the new data.

Healthy Weight Ranges for Men and Women

Height (in.)	Age 19–34 years Weight (lb)	Age 35 years and over Weight (lb)
60	97–128	108–138
61	101–132	111–143
62	104–137	115–148
63	107–141	119–152
64	111–146	122–157
65	114–150	126–162
66	118–155	130–167
67	121–160	134–172
68	125–164	138–178
69	129–169	142–183
70	132–174	146–188
71	136–179	151–194
72	140–184	155–199

Weather is defined as the condition of the atmosphere during a brief period. *Climate* is the characteristic weather in an area over a long period of time. Climate variables include the variation in temperature from month to month. The table gives the average monthly high and low temperatures in Tokyo, Japan.

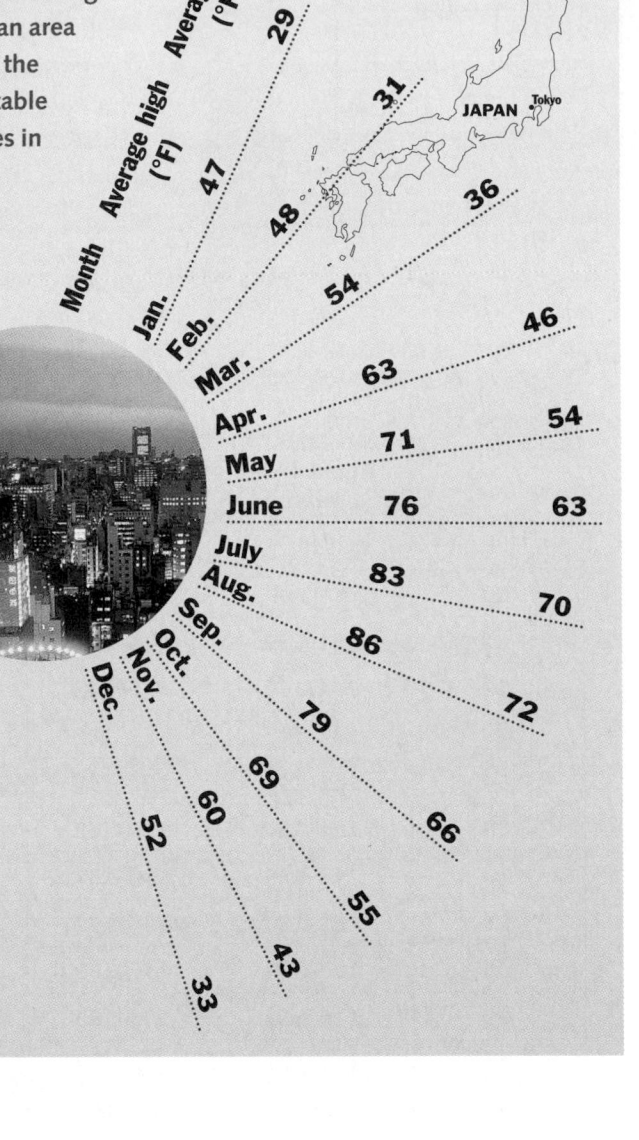

Month	Average high (°F)	Average low (°F)
Jan.	47	29
Feb.	48	31
Mar.	54	36
Apr.	63	46
May	71	54
June	76	63
July	83	70
Aug.	86	72
Sep.	79	66
Oct.	69	55
Nov.	60	43
Dec.	52	33

19. Use the data in the table to create a graph showing the range of average temperatures in Tokyo. Put time, measured in months since January, on the horizonal axis, and temperature, measured in degrees Fahrenheit, on the vertical axis.

20. In what months would you expect the temperature to be above 50°? below 60°?

21. Open-ended Problem Use an almanac, atlas, or other reference book to find average monthly high and low temperatures for another location. Make a graph, and compare it with your graph from Exercise 19.

BY THE WAY

The climate of the Japanese islands is moderated by the sea. Winters are milder and precipitation is heavier than for locations on the Asian mainland at the same latitude.

Graph each system of inequalities.

22. $y < 7$
$x + y \geq 2$
$y \leq x$

23. $y \leq -2x$
$y \leq \frac{2}{3}x + 5$

24. $y < \frac{3}{2}x + 6$
$y \leq \frac{1}{2}x + 6$
$y \leq x + 8$

25. $x - 2y \geq -2$
$2x + y \geq 0$
$2x - y \leq 8$

26. $y \geq \frac{1}{2}x + 1$
$y \leq -1$

27. $y \leq \frac{3}{2}x + 4$
$y \geq -x - 6$
$x \leq 6$

28. $y \geq x$
$y \leq 2x$
$y \geq -3x - 1$

29. $y \geq \frac{1}{4}x + 3$
$y \leq x + 8$
$y \leq 6$

30. SAT/ACT Preview Which one of the following points is included in the region defined by $-2 \leq x < 8$ and $4 < y \leq 12$?

A. $(-2, 7)$ **B.** $(-2, 4)$ **C.** $(-2, -2)$ **D.** $(8, 7)$ **E.** $(8, 12)$

Write a system of inequalities defining each shaded region.

31.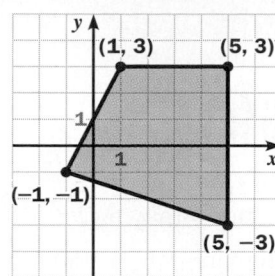

32.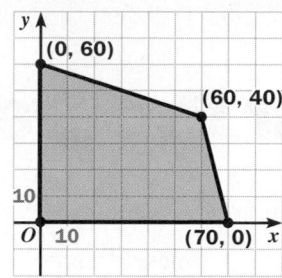

33.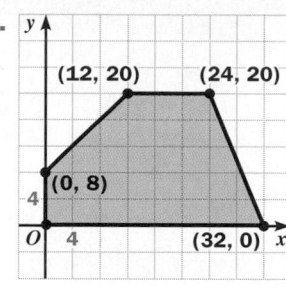

34. Challenge Graph the region defined by the inequalities $y \leq \frac{1}{2}x^2 + 6$ and $y \geq 3x^2 - 4$.

35. CONSUMER ECONOMICS An international telephone call from the United States to most countries in Asia costs between \$1.56 and \$5.58 for the first minute. Each additional minute costs between 89% and 95% as much as the first minute.

a. Write inequalities to represent the relationship between the cost of the first minute, f, and the cost of each additional minute, a.

b. Graph the system of inequalities you found in part (a). Put the first-minute cost on the horizontal axis and the additional-minute cost on the vertical axis. Label the lines on your graph.

c. The first minute of a call to Singapore costs \$1.73. What might you expect to pay for each additional minute?

ONGOING ASSESSMENT

36. Cooperative Learning Work with a partner. One of you should use an almanac to find the normal monthly temperature and precipitation for the area where you live. The other should find the normal monthly temperature and precipitation for another area. Each of you should then draw a scatter plot of temperature and precipitation for your chosen area using the same scale. Draw a polygon enclosing the plotted points, and find the inequalities defining the interior of the polygon. Compare the two graphs.

SPIRAL REVIEW

Graph each inequality. *(Section 7.4)*

37. $5x + 3y < 18$ **38.** $x - 4y > 8$ **39.** $-2y \geq 5$ **40.** $-7x - 2y \leq 12$

Solve each system of equations. *(Section 7.1)*

41. $2x + y = 15$
$x + y = 10$

42. $y = 3.2x$
$1.5x - 0.5y = 0.8$

43. $-x - y = 11$
$x + 3y = -3$

44. $5x + 4y = 7$
$3x = 9$

Find the x-intercepts of the graph of each equation. *(Section 5.3)*

45. $y = 3(x - 1)(x + 4)$ **46.** $y = (6x + 3)(x - 2)$ **47.** $y = (x + 1)(x - 1)$

7.6 Linear Programming

Quilting—the art of stitching layers of fabric together—may have originated with the Chinese. Patchwork quilting began in the 1700s when colonial women pieced together scraps of material to make blankets. There are now about 15.5 million quilters and 3000 quilting stores in the United States.

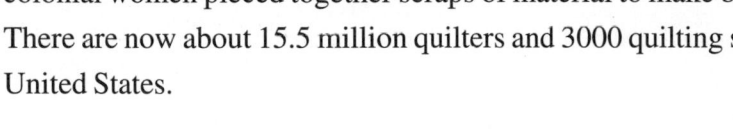

EXAMPLE 1   Application: Business

Elizabeth Ferris owns a quilting store. In addition to quilting supplies, she stocks queen-size and baby quilts. She wants to order up to 50 quilts to display and sell this fall. She knows she should have at least as many queen-size quilts as baby quilts, and she wants to have at least 10 baby quilts. Make a graph to show the possible combinations of quilts she can order.

SOLUTION

Step 1 Express each *constraint* as an inequality.

Let x = the number of queen-size quilts, and let y = the number of baby quilts.

Constraint 1: She will order up to 50 quilts.	**Constraint 2:** She should have at least as many queen-size quilts as baby quilts.	**Constraint 3:** She wants to have at least 10 baby quilts.
$x + y \leq 50$, or $y \leq -x + 50$	$x \geq y$, or $y \leq x$	$y \geq 10$

Step 2 Graph the constraints.

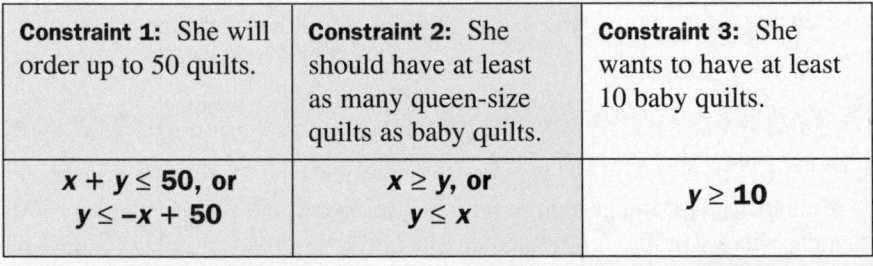

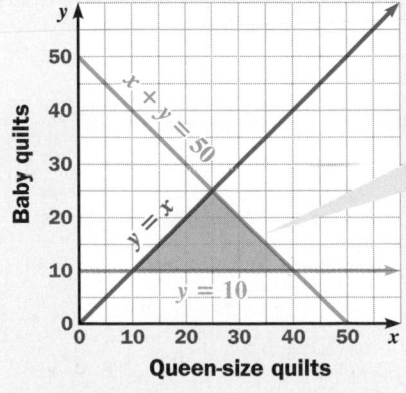

The shaded region contains all possible combinations of quilts that Elizabeth Ferris can order.

In Example 1, suppose Elizabeth Ferris prices her quilts so as to make a profit of $125 on every queen-size quilt and $25 on every baby quilt. Then an equation giving her total profit P on the quilts is:

$$P = 125x + 25y$$

The graph from Example 1, which is called a **feasible region**, is shown at the right along with several *profit lines* drawn in red. Each line represents a particular value of P.

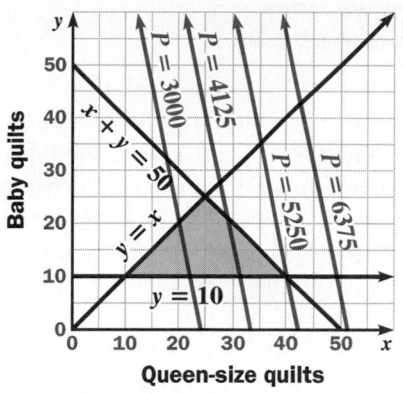

THINK AND COMMUNICATE

1. Confirm that the point (30, 15) is in the feasible region shown above. Which profit line is it on? Name another point in the feasible region that is on the same profit line.

2. For each profit line, name a point with integer coordinates that is on the line and in the feasible region, if possible.

 a. $P = 3000$ **b.** $P = 5250$ **c.** $P = 6375$

3. At what point in the feasible region is the profit greatest? Explain.

You often want to find the best solution to a problem, such as the number of quilts that will maximize profit. The **corner-point principle** states that for a polygonal feasible region, the best solution will be at a corner point of the feasible region. The method of finding the best solution is called **linear programming**.

<table>
<tr><td>**EXAMPLE 2**</td><td>**Application: Business**</td></tr>
</table>

Suppose Elizabeth Ferris makes a $100 profit on a queen-size quilt and a $35 profit on a baby quilt. Given the constraints in Example 1, how many quilts of each type should she order to maximize her profit?

SOLUTION

Step 1 Write an equation describing the profit she earns.

$$P = 100x + 35y$$

Step 2 Check the corner points of the feasible region in the profit equation.

Corner point	Value of P
(10, 10)	$P = 100(10) + 35(10) = 1350$
(25, 25)	$P = 100(25) + 35(25) = 3375$
(40, 10)	$P = 100(40) + 35(10) = 4350$

For the maximum profit, $4350, she should order **40 queen-size quilts** and **10 baby quilts**.

EXAMPLE 3   Application: Consumer Economics

Julian and Yoshi are making punch for a party using papaya juice and apple juice. They want to make between 12 qt and 15 qt of the punch. They know from experience that there should be at least twice as much, but not more than three times as much, apple juice as papaya juice.

Apple juice costs $2.49 per quart. Papaya juice costs $1.53 per quart. How much apple juice and papaya juice should Julian and Yoshi buy to minimize the cost of the punch?

SOLUTION

Step 1 Write inequalities to represent the constraints. Let $x =$ the number of quarts of apple juice, and let $y =$ the number of quarts of papaya juice.

Contraint 1: Make 12 qt to 15 qt of punch.	**Contraint 2:** Use at least twice as much apple juice as papaya juice.	**Contraint 3:** Use no more than three times as much apple juice as papaya juice.
$12 \leq x + y \leq 15$, or $y \geq 12 - x$ and $y \leq 15 - x$	$x \geq 2y$, or $y \leq \frac{1}{2}x$	$x \leq 3y$, or $y \geq \frac{1}{3}x$

Step 2 Graph the feasible region and find the corner points. You can solve a system of two equations to find each corner point. For example, the graphs of $y = 12 - x$ and $y = \frac{1}{3}x$ intersect at (9, 3).

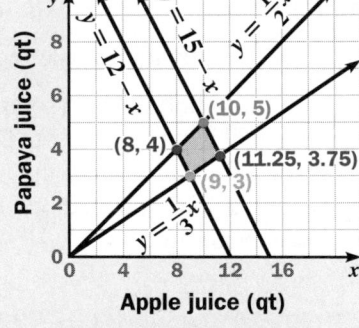

Step 3 Write an expression for the cost function and check the corner points.

$$C = 2.49x + 1.53y$$

Corner point	Value of C
(8, 4)	$C = 2.49(8) + 1.53(4) = 26.04$
(10, 5)	$C = 2.49(10) + 1.53(5) = 32.55$
(11.25, 3.75)	$C = 2.49(11.25) + 1.53(3.75) = 33.75$
(9, 3)	$C = 2.49(9) + 1.53(3) = 27.00$

Julian and Yoshi should purchase 8 qt of apple juice and 4 qt of papaya juice, for a minimum cost of $26.04.

THINK AND COMMUNICATE

4. In Example 3, suppose the cost lines were parallel to an edge of the feasible region. Does the corner-point principle still work? Explain.

Jenn and Kate are making oatmeal and chocolate chip cookies for a bake sale. They want to make 120 to 144 cookies. They want at least 24 of each kind.

1. Write constraints for this situation, and graph the feasible region. Label the edges and corner points.

2. Suppose they make a profit of $.20 per oatmeal cookie and $.15 per chocolate chip cookie. How many of each type of cookie should they make to maximize their profit?

3. Suppose their cost is $.20 per oatmeal cookie and $.25 per chocolate chip cookie. How many of each type of cookie should they make to minimize their cost?

7.6 | **Exercises and Applications**

Extra Practice exercises on page 760

Graph each feasible region.

1. $y \le 14$
$y \ge \frac{1}{2}x + 2$
$x \ge 1$

2. $y \ge 2$
$y \le x$
$y \le 22 - 2x$

3. $y \ge 5 - x$
$y \ge \frac{1}{3}x + 1$
$y \le 3 + x$
$y \le 15 - 2x$

4. $y \ge 0$
$y \le 4x$
$y \ge 2x - 14$
$y \le 8$

5–8. For each of the feasible regions you graphed in Exercises 1–4, find the maximum profit for the profit function $P = 5x + 4y$.

9–12. For each of the feasible regions you graphed in Exercises 1–4, find the minimum cost for the cost function $C = 3x + 3y$.

13. BUSINESS A potter is preparing to make serving bowls and plates. A serving bowl uses 5 lb of clay. A plate uses 4 lb of clay. She has 40 lb of clay and wants to make at least 4 serving bowls.

a. Let x = the number of serving bowls, and let y = the number of plates. Write the constraints for this situation in terms of x and y.

b. Graph the feasible region. Label the corner points.

c. If the profit on a serving bowl is $35 and the profit on a plate is $30, how many bowls and plates should she make to maximize her profit?

14. ART Laura is making a necklace 15 in. to 18 in. long. She will use two types of beads: brightly painted wooden ovals, 0.75 in. long and $.65 apiece, and small animal shapes, 0.50 in. long and $1.00 apiece. She wants to have at least three times as many ovals as animals, but not more than six times as many ovals as animals.

a. Write inequalities to represent the constraints for this situation.

b. Graph the feasible region. Label the corner points.

c. If the cord to string the beads costs $.75, how many of each type of bead should she buy to minimize her cost?

Look back at the article on pages 288–290.

Fitness instructors like Gina Oliva suggest that you vary your workout with different exercises, to work different muscle groups and to avoid boredom with one routine. For Exercises 15–17, use the table below.

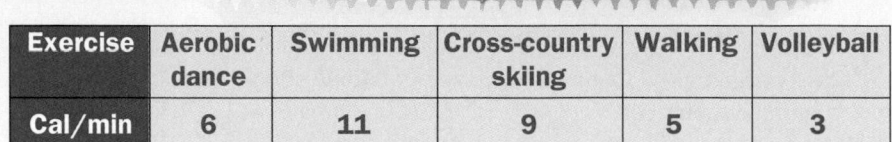

Exercise	Aerobic dance	Swimming	Cross-country skiing	Walking	Volleyball
Cal/min	6	11	9	5	3

15. Aaron wants to combine aerobics and swimming. He wants to work out for 3 h to 5 h (180 min to 300 min) a week. He wants to spend at least 1 h on each activity. He plans to burn at least 1200 Cal a week with this exercise program.

    a. Write a system of inequalities to express Aaron's constraints. Let s = the number of minutes spent swimming and a = the number of minutes spent doing aerobics.

    b. Graph the feasible region. Label the corner points.

    c. How much time should Aaron spend on swimming and on aerobics if he wants to minimize the time spent exercising, but burn as many calories as possible in that time?

    d. How much time should he spend on each exercise if he wants to maximize the number of calories burned?

16. Len wants to combine aerobics and cross-country skiing. He plans to exercise 6 h to 8 h a week this winter. He will spend at least 1 h but no more than 3 h each week doing aerobics and at least 4 h skiing.

    a. Write a system of inequalities to express Len's constraints.

    b. Graph the feasible region. Label the corner points.

    c. How much time should Len spend on each activity if he wants to maximize the number of calories he burns?

17. **Open-ended Problem** Create a linear programming problem about combining two types of exercise. Then exchange your problem with someone in your class and solve each other's problems.

For Exercises 18–21, find the minimum cost or maximum profit for the feasible region shown.

18. $P = 3x + 5y$ **19.** $P = 5x + 3y$

20. $C = 3x + 5y$ **21.** $C = 5x + 3y$

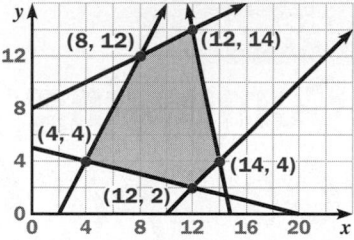

ONGOING ASSESSMENT

22. Writing Explain what a linear programming problem is and how to solve it.

SPIRAL REVIEW

For each function, find the inverse. *(Section 4.1)*

23. $f(x) = 2x + 7$ **24.** $g(x) = x$ **25.** $h(x) = 15x - 3$

Tell why each question might be biased. Then describe what changes you would make to improve the question. *(Section 6.2)*

26. "How much money do you donate to charity each year?"

27. "Given that state taxes were increased twice in the past decade, would you favor another state tax increase?"

ASSESS YOUR PROGRESS

VOCABULARY

half-planes (p. 310)   corner-point principle (p. 323)
system of inequalities (p. 315)   linear programming (p. 323)
feasible region (p. 323)

Graph each inequality. *(Section 7.4)*

1. $2x + 4y \leq -16$ **2.** $3x - y > 5$ **3.** $-15x + 6y > 12$

4. PERSONAL FINANCE As a general rule, banks allow people to spend up to 28% of their gross incomes on a home mortgage. Write and graph an inequality for this situation. *(Section 7.4)*

Graph each system of inequalities. *(Section 7.5)*

5. $y \geq 2x + 1$ **6.** $y \geq -2x + 1$ **7.** $y \leq 2x + 1$
    $y \leq 4x$ $y < 3x - 6$ $y < 4x$

Graph the feasible region and find the maximim profit if $P = 12x + 15y$. *(Section 7.6)*

8. $x \geq 10$ **9.** $3x + y \leq 13$ **10.** $0 \leq y \leq 12$
    $y \leq 18$ $x + 4y \geq 8$ $x - y \leq 8$
    $2x - y \leq 8$ $x \geq 0$ $x \geq 2$

11. Journal How are a system of linear equations and a system of linear inequalities alike? How are they different?

Comparing
Olympic Performances

In some Olympic events where both men and women compete against the clock, women's times, although not as fast as men's times, have improved more rapidly. This suggests that women may someday surpass men's times in these events. By choosing an appropriate model, you can make reasonable predictions about the future performances of men and women in comparable events.

PROJECT GOAL   Use a system of equations to model men's and women's performance data in an Olympic event.

Collecting the Data

Work in a group of three students. You will need a graphing calculator or statistical software, and access to an Olympic data source, such as a sports almanac.

Find an event in which both men and women compete. Choose a timed event like swimming or running, or an event like the high jump or long jump, where distance is used to measure performance.

Olympic 100 m Freestyle		
	Winning time (s)	
Year	Men	Women
1948	57.3	66.3
1952	57.4	66.8
1956	55.4	62.0
1960	55.2	61.2
1964	53.4	59.5
1968	52.2	60.0
1972	51.2	58.6
1976	50.0	55.7
1980	50.4	54.8
1984	49.8	55.9
1988	48.6	54.9
1992	49.0	54.6

Using Linear Models

Make a scatter plot of your data. Fit a line to each set of points, and find equations of the lines. For example, using data for the 100 m freestyle (see the table) with x = years since 1948 and y = winning times, you get the results shown.

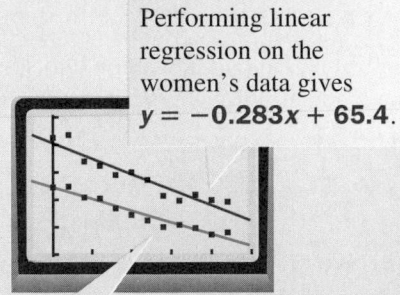

Performing linear regression on the women's data gives $y = -0.283x + 65.4$.

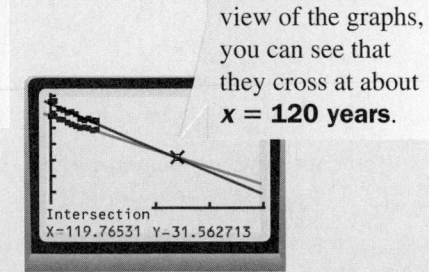

By extending your view of the graphs, you can see that they cross at about $x = 120$ years.

Performing linear regression on the men's data gives $y = -0.214x + 57.2$.

What does your model predict? Why might a linear equation be a poor model for athletic performance?

Using Exponential Models

Suppose you want to model winning times using an exponential decay equation of the form $y = ab^x$ (where $0 < b < 1$). You know y approaches 0 as x increases. Since you wouldn't expect performance times to decrease to 0, it makes sense to choose a reasonable lower bound for the data.

Try using a decay equation of the form $y = ab^x + c$, where the number c is the lower bound. (If you think your data have an upper bound, use a model of the form $y = c - ab^x$.) Suppose, for instance, that you don't expect men's or women's times for the 100 m freestyle to go below 40 s.

STEP 1 Find an exponential model that fits the data points (x, y') where $y' = y - 40$.

STEP 2 In the equation from Step 1, substitute $y - 40$ for y' and solve for y to obtain a model that fits the data points (x, y).

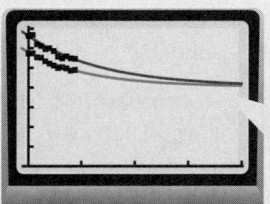

These are the data points (x, y) representing the men's times.

The function $y = 17.7(0.983)^x + 40$ models the original data points. The graph does not fall below the line $y = 40$.

Performing exponential regression on the data points (x, y') gives $y' = 17.7(0.983)^x$.

STEP 3 Repeat Steps 1 and 2 for the women's times, and graph the models for both data sets to see if they intersect.

Over a span of 200 years, the graphs don't intersect but do get closer and closer.

Would you get different results if you changed the lower bound? What if you used different lower bounds for men and women?

Writing a Report

Write a report summarizing your results. Include these items: a table and a scatter plot of your data sets, graphs of your linear and exponential models, a comparison of future perform-ances based on your models, and an evaluation of the limitations of each model. To extend your project, you may wish to compare results with other groups, or investigate the topic of human limits in sports.

Self-Assessment

Describe how comfortable you are modeling data and making predictions. What aspects of modeling are you unclear or unsure about?

7 | Review

Form a study group with six members. Have each group member take one section of the chapter and write a "how to" synopsis for the major concepts in that section. Have each member present his or her synopsis to the group, and discuss.

VOCABULARY

system of equations (p. 292)
inconsistent system (p. 299)
dependent system (p. 299)
identity matrices (p. 304)
inverse matrix (p. 304)

half-planes (p. 310)
system of inequalities (p. 315)
feasible region (p. 323)
corner-point principle (p. 323)
linear programming (p. 323)

SECTIONS | 7.1, 7.2, *and* 7.3

To solve a **system of equations**, you need to find the values of the variables that make both of the equations true. For example, you can use any of the four methods shown below to solve the system $3x - y = 7$ and $5x + 2y = -3$.

Use technology. Solve each equation for y and graph.

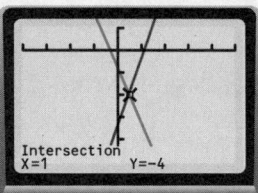

Intersection
X=1   Y=-4

Use substitution. Solve the first equation for y and substitute into the second equation.

$$5x + 2(3x - 7) = -3 \qquad 3(1) - y = 7$$
$$11x = 11 \qquad\qquad -y = 4$$
$$x = 1 \qquad\qquad y = -4$$

Use addition. Multiply the first equation by 2 and add the two equations.

$$6x - 2y = 14$$
$$5x + 2y = -3$$
$$11x = 11 \rightarrow x = 1$$
$$5(1) + 2y = -3 \rightarrow y = -4$$

Use matrices. Write the system as a matrix equation and solve using an **inverse matrix**.

$$\begin{bmatrix} 3 & -1 \\ 5 & 2 \end{bmatrix}\begin{bmatrix} x \\ y \end{bmatrix} = \begin{bmatrix} 7 \\ -3 \end{bmatrix}$$

$$\begin{bmatrix} x \\ y \end{bmatrix} = \begin{bmatrix} 3 & -1 \\ 5 & 2 \end{bmatrix}^{-1}\begin{bmatrix} 7 \\ -3 \end{bmatrix} = \begin{bmatrix} 1 \\ -4 \end{bmatrix}$$

The graphs of inequalities and **systems of inequalities** are regions of the coordinate plane.

Example: A town's citizens want to plant maple trees and spruce trees in the town square. They want at least 60 trees, and they have enough space for up to 100 trees. To minimize leaf pickup, they want at least 1.5 times as many spruce trees as maples. To provide fall color, they want at least 15 maples. You can follow the steps below to represent this situation graphically.

Step 1 Write a system of inequalities using s for the number of spruce trees and m for the number of maple trees.

$$m + s \leq 100 \qquad s \geq 1.5m$$
$$m + s \geq 60 \qquad m \geq 15$$

Step 2 Graph one of the inequalities.

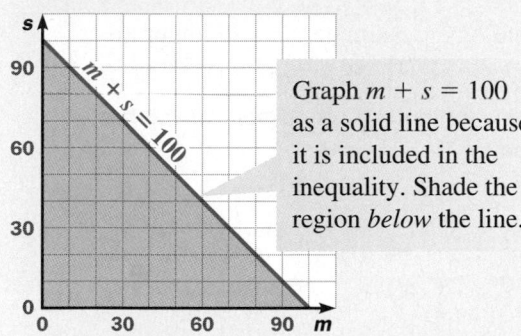

Graph $m + s = 100$ as a solid line because it is included in the inequality. Shade the region *below* the line.

Step 3 Graph the other inequalities in the system.

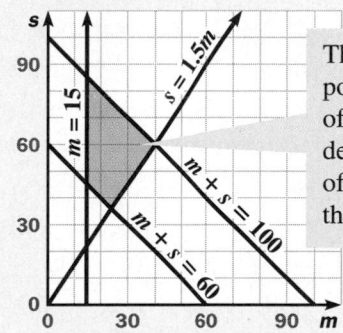

The coordinates of the points in the *overlap* of all the **half-planes** defined by a system of inequalities satisfy the system.

Linear programming involves finding the best solution to a problem subject to various constraints. For example, suppose that in the example above, each maple tree costs $40 and each spruce tree costs $60. The town's citizens hope to minimize the cost of the planting.

Step 4 Write a cost function for the situation: $C = 40m + 60s$.

Step 5 Find the corner points of the **feasible region** from Step 3.

Step 6 Check the coordinates of the corner points of the feasible region in the cost equation.

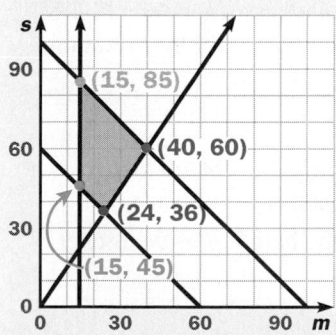

Corner point	Value of C
(15, 45)	$C = 40(15) + 60(45) = 3300$
(15, 85)	$C = 40(15) + 60(85) = 5700$
(24, 36)	$C = 40(24) + 60(36) = 3120$
(40, 60)	$C = 40(40) + 60(60) = 5200$

For a minimum cost of $3120, the town's citizens should plant 24 maple trees and 36 spruce trees.

7 Assessment

VOCABULARY QUESTIONS

For Questions 1 and 2, complete each sentence.

1. A system of linear equations is called a(n) _?_ system if it has no solutions and a(n) _?_ system if it has infinitely many solutions.

2. Multiplying a square matrix by its _?_ , if it has one, gives a(n) _?_ .

SECTIONS 7.1, 7.2, *and* 7.3

3. **Technology** Use technology to estimate the solution of the system of equations $y = 26(1.07)^x$ and $y = 34(1.035)^x$.

Solve each system of equations. State whether the system has *one solution*, *infinitely many solutions*, or *no solution*.

4. $7x - 3y = -5$
$x + 2y = 4$

5. $4x - 2y = 12$
$2x - y = 3$

6. $0.5x + 3.7y = 2.02$
$0.4x - y = 0$

7. $25x + 10y = 20$
$10x + 4y = 20$

8. $3x + 5y = 2$
$6x - 4y = 11$

9. $2x - 9y = 2$
$7x - 12y = -6$

10. **Writing** Describe what a solution of a system of equations means both in terms of a graph and in terms of algebra.

11. **CONSUMER ECONOMICS** Suppose a 15-watt fluorescent bulb costs $12 to buy and $.0018 per hour to run. A standard bulb with the same light output costs $.50 to buy and $.0072 per hour to run.

   a. Write a system of equations representing the cost of each bulb as a function of hours of use.

   b. Solve the system. What does the solution mean?

   c. **Writing** If a fluorescent bulb lasts 10,000 h and an incandescent bulb lasts 1000 h, will it be cost-effective to use fluorescent bulbs? Explain.

12. **PHYSICS** Estimates of the height of a model rocket for various times after the end of the burn phase are given in the table.

   a. **Technology** Using matrices and a graphing calculator or software with matrix calculation capabilities, find an equation of the form $h = at^2 + bt + c$ that describes the height h of the rocket t seconds after the burn ends.

   b. What is the rocket's maximum height?

Time after burn(s)	Height (ft)
1	250
3	235
5	100

SECTIONS 7.4, 7.5, *and* 7.6

Graph each inequality.

13. $y \leq -4x$　　　　**14.** $4x + 3y < 10$　　　　**15.** $3.5x - 7y \geq 1.75$

Write a system of inequalities defining each shaded region.

16. 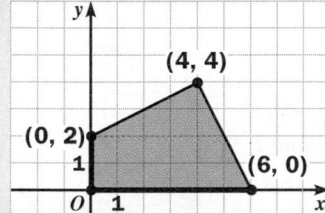　　　　**17.**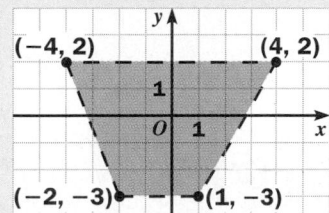

18. FITNESS The President's Council on Physical Fitness program gives awards to children who meet certain standards for a group of exercises. For example, the number of "curl-ups" a boy between the ages of 6 and 14 must complete in a minute to qualify towards an award is modeled by the equations below, where x is age and y is the number of curl-ups.

Presidential Award	National Award
$y = 2.82x + 16.39$	$y = 2.62x + 8.5$

a. Graph the system $y \geq 2.62x + 8.5$ and $y \leq 2.82x + 16.39$ for $6 \leq x \leq 14$.

b. What does the system in part (a) represent for this situation?

19. Mai Wong wants to invest a total of $12,000 to $20,000 in two mutual funds: a conservative balanced fund and an aggressive growth fund. She wants at least two-thirds as much money in the balanced fund as in the growth fund, but she wants no more than $10,000 in the balanced fund.

a. Write inequalities to represent the constraints for this situation.

b. Graph the feasible region. Label each corner point.

c. The balanced fund is expected to grow by 6% and the growth fund by 18% over a year. What investments should Mai Wong make to maximize her gain?

PERFORMANCE TASK

20. From your own experience, describe a situation involving two variables. For example, perhaps you have had to decide how to split your time between two activities, or you belong to a club that has tried to raise funds by making and selling two items.

a. Write a system of equations or a system of inequalities to represent your situation.

b. Show how you can use the system in part (a) to resolve the situation.

8 Radical Functions, Number Systems

Designs for radical rides

Ronald Toomer

In the fast-moving world of amusement park rides, there's a legend and a classic joke. The legend is Ronald Toomer, the man who has designed and built ninety of the world's biggest and scariest roller coasters. The joke is that Toomer personally never rides these things. He says, "They just don't appeal to me. I have a terrible time with motion sickness and would much rather design them."

"It's just a matter of keeping things moving."

Keeping Things Moving

In the 1960s Toomer was working as a mechanical engineer on missile and rocket systems when he heard about an engineering company that needed help constructing its first roller coaster. He signed up for temporary work on that project. "It's been a long temporary job, lasting about 30 years," says Toomer, who is now president of the company.

The company has been so successful that each year about 200 million people ride its coasters. The key to a popular ride is pretty simple, Toomer says. "We try to keep it going as fast as we can, all the time, while throwing in a few changes of pace. Basically, it's just a matter of keeping things moving."

"For some crazy reason, people like to be scared."

Figuring Out the Fun

His company built *Desperado,* the world's largest roller coaster, located near Las Vegas, Nevada. The ride can theoretically reach a velocity of about 82 mi/h. However, this velocity is reduced both by friction between the wheels and the track and by air resistance against the passenger cars.

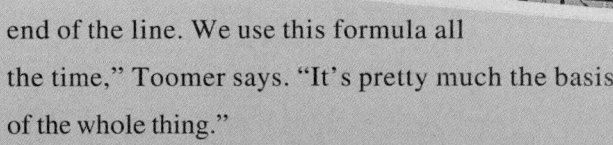

Toomer calculates the velocity of a passenger car at the bottom of a hill by using a formula that includes values for the height of the drop and the acceleration due to gravity. "We keep doing the same calculation for the hills until the coaster finally reaches the end of the line. We use this formula all the time," Toomer says. "It's pretty much the basis of the whole thing."

At the top of
a hill, $v = 0$.

A Radical Relationship

The formula Toomer uses, shown below, gives the relationship between a roller coaster's velocity and the height of its drop. This formula is an example of a *radical function*. In the hands of a master engineer like Toomer, this function can lead to a hair-raising good time.

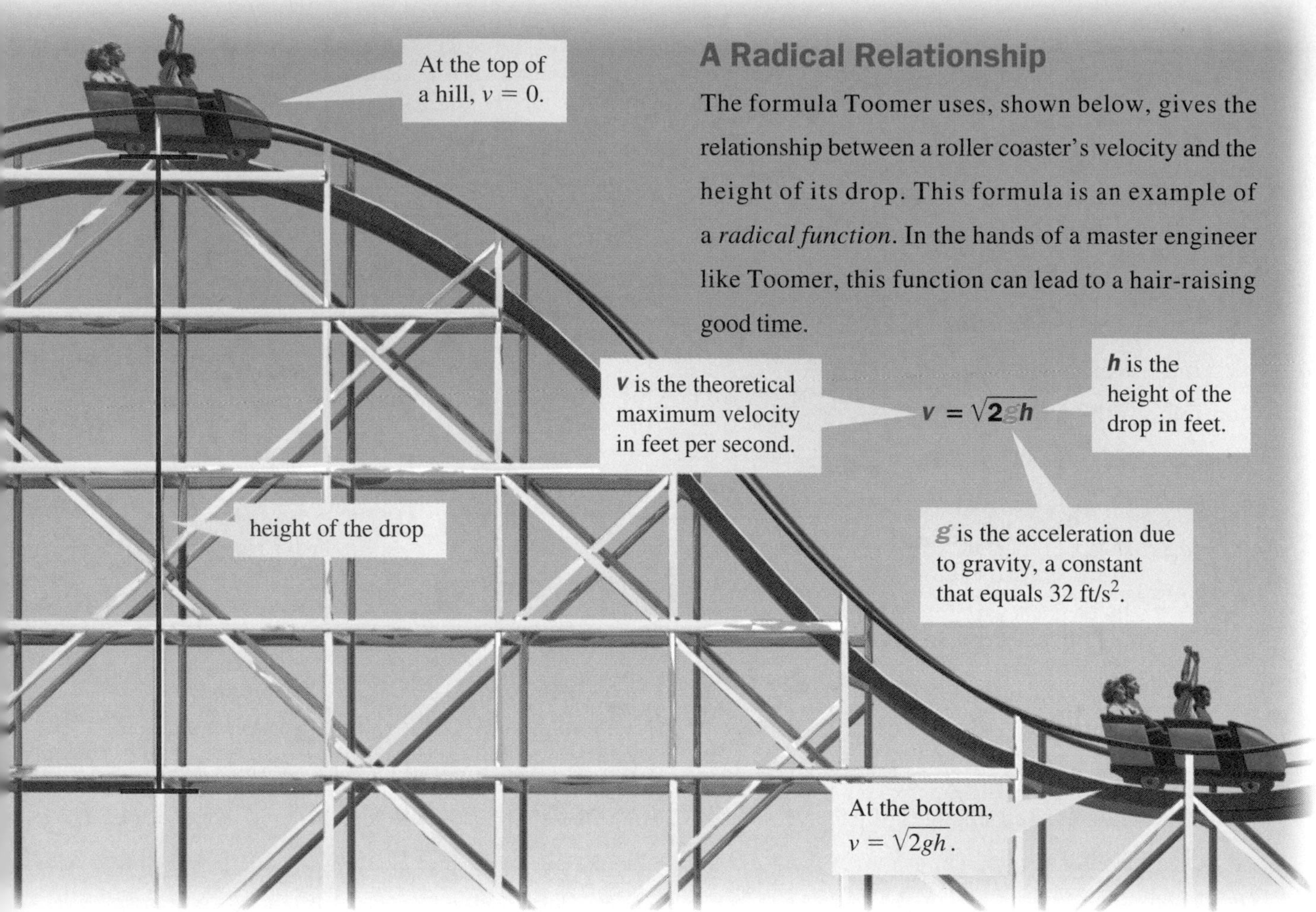

v is the theoretical maximum velocity in feet per second.

$$v = \sqrt{2gh}$$

h is the height of the drop in feet.

g is the acceleration due to gravity, a constant that equals 32 ft/s².

height of the drop

At the bottom, $v = \sqrt{2gh}$.

EXPLORE AND CONNECT

Ronald Toomer designs a roller coaster.

1. Writing Roller coasters have motor-driven chains that pull a car up the first and highest hill. The energy a car gains from moving up the first hill keeps it moving for the rest of the ride. Why do you think successive hills must decrease in height?

2. Project Draw a sketch of a roller coaster with three hills. For each hill in your model, measure the height of the drop to the nearest inch. Convert the model heights to life-size heights using 1 in. = 25 ft. Calculate the velocity of a passenger car at the bottom of each drop, assuming that $v = 0$ at the top of each drop.

3. Research Contact a nearby amusement park. Ask for an employee who can tell you the height of the first drop for the park's largest roller coaster. Use the given height to calculate the velocity of the roller coaster at the bottom of the drop.

Mathematics & Ronald Toomer

In this chapter, you will learn more about how mathematics is related to amusement park rides.

Related Exercises

Section 8.1
• Exercises 17–19

Section 8.2
• Exercises 29–32

8.1 Square Root Functions

In Chapter 4 you saw that reflecting the graph of a function over the line $y = x$ can produce another function, called an *inverse*. For example, the inverse of an exponential function is a logarithmic function.

What happens when the graph of a quadratic function is reflected over the line $y = x$?

Learn how to...

- **restrict the domain of a function to obtain an inverse function**
- **graph and evaluate square root functions**

So you can...

- **understand the dynamics of pole vaulting, for example**

EXPLORATION
COOPERATIVE LEARNING

Reflecting the Graph of $y = x^2$ over the Line $y = x$

Work with a partner. You will need:
- graph paper
- a MIRA® transparent mirror, tracing paper, or a ruler

---

1 Graph $y = x^2$ and $y = x$ in the same coordinate plane.

---

2 Sketch the reflection of the graph of $y = x^2$ over the line $y = x$. Remember to reflect not only the points above the line but also the points below the line. Here are three ways of reflecting the graph:

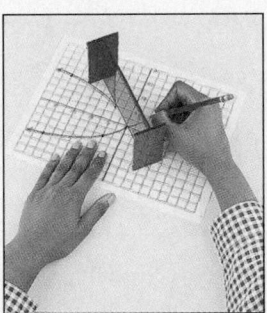  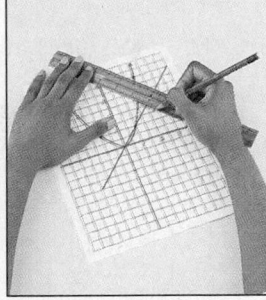

You can place a MIRA® along the line $y = x$ and sketch the reflection you see.

You can fold tracing paper along the line $y = x$ and trace on the other side of the paper.

You can hold a ruler perpendicular to the line $y = x$ and measure equal distances on each side.

Questions

1. Does the graph you get represent a function? Explain why or why not.

2. Consider the function $y = x^2$ with a restricted domain of $x \geq 0$. What does the graph look like? Is the reflection of the graph over the line $y = x$ the graph of a function?

To find the inverse of $f(x) = x^2$, you might be tempted to do the following:

$$y = x^2 \longleftarrow \text{ Substitute } y \text{ for } f(x).$$

$$\pm\sqrt{y} = x \longleftarrow \text{ Solve for } x \text{ (see page 187).}$$

$$\pm\sqrt{x} = y \longleftarrow \text{ Switch } x \text{ and } y.$$

The equation $y = \pm\sqrt{x}$ does *not* define a function, however, because a positive input results in *two* outputs. For example, if $x = 9$, then $y = \pm\sqrt{9} = \pm 3$.

To get around this problem, you can restrict the domain of $f(x) = x^2$. Two ways of doing this are shown below.

For more information about graphing inverses, see the *Technology Handbook*, p. 809.

1. Restrict the domain of $f(x) = x^2$ to $x \geq 0$.

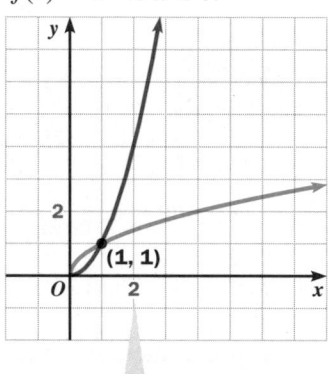

The inverse of $f(x) = x^2$ for $x \geq 0$ is $f^{-1}(x) = \sqrt{x}$.

2. Restrict the domain of $f(x) = x^2$ to $x \leq 0$.

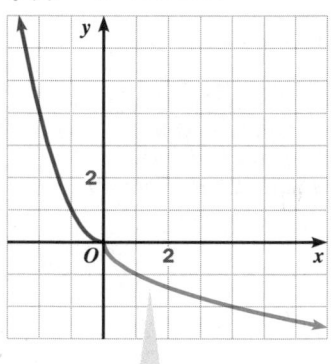

The inverse of $f(x) = x^2$ for $x \leq 0$ is $f^{-1}(x) = -\sqrt{x}$.

THINK AND COMMUNICATE

1. **Technology** Use a graphing calculator or graphing software to graph the equations shown.

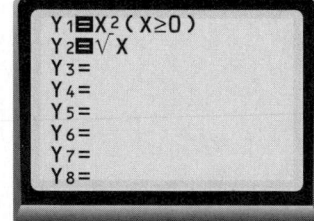

 a. Use trace to convince yourself that a point (b, a) is on the graph of $y = \sqrt{x}$ whenever the point (a, b) is on the graph of $y = x^2$ for $x \geq 0$.

 b. Repeat part (a) using $y = x^2$ for $x \leq 0$ and $y = -\sqrt{x}$.

2. Would you be able to find an inverse of $f(x) = x^2$ if you restricted the domain of f to $-1 \leq x \leq 1$? Explain.

3. Why is it *not* necessary to restrict the domain of a nonconstant linear function or an exponential function when finding its inverse?

EXAMPLE 1   Application: Sports

In 1993, Ukrainian athlete Sergei Bubka set a world record in pole vaulting, clearing a bar that was 20 ft 2 in. above the ground. The equation

$$h = \frac{v^2}{64}$$

gives the height h (in feet) that a pole vaulter's center of gravity is raised as a function of running velocity v (in feet per second) at launch.

a. Write an equation for the vaulter's velocity as a function of the height the vaulter's center of gravity is raised.

b. Graph the equation from part (a). State the domain and range.

c. Suppose Bubka's center of gravity while running was 3 ft 2 in. above the ground. Find the running velocity he needed to clear the bar placed 20 ft 2 in. above the ground.

BY THE WAY

There are no length limits on poles for vaulting, but poles longer than 16 ft are considered too heavy to use because they slow down runners.

SOLUTION

a. Solve the equation for v.

$$h = \frac{v^2}{64}$$

$$64h = v^2$$

$$\sqrt{64h} = v$$

$$8\sqrt{h} = v$$

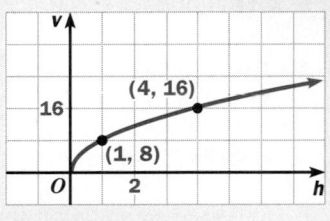

Use the **positive square root**, because velocity here must be nonnegative.

An equation is $v = 8\sqrt{h}$.

b.

The domain is $h \geq 0$.

The range is $v \geq 0$.

c. Bubka needed to raise his center of gravity from 3 ft 2 in. to 20 ft 2 in. above the ground. The difference in these heights is 17 ft.

$$v = 8\sqrt{17} \quad \longleftarrow \quad \text{Substitute 17 for } h.$$

$$\approx 32.985$$

Bubka needed to run about 33 ft/s for his world record vault.

THINK AND COMMUNICATE

4. **Technology** Verify that $\sqrt{64h} = 8\sqrt{h}$ by using a graphing calculator or graphing software to compare the graphs of $y = \sqrt{64x}$ and $y = 8\sqrt{x}$.

Toolbox p. 777
Square Roots

5. Compare the graphs of $y = 8\sqrt{x}$ and $y = \sqrt{x}$. How are they geometrically related?

6. Compare the graphs of $y = 8 + \sqrt{x}$ and $y = \sqrt{x}$. How are they geometrically related?

EXAMPLE 2

Compare the graphs of $y = \sqrt{x}$ and $y = \sqrt{x-2} + 1$. How are they geometrically related? State the domain and range of the second function.

SOLUTION

WATCH OUT!

Use parentheses on a graphing calculator to indicate what is under the square root sign.

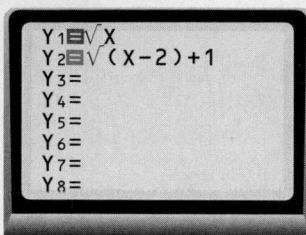

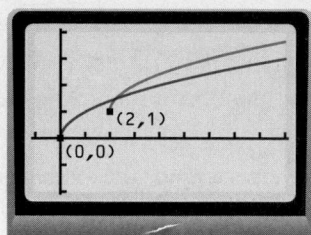

The second graph is the first graph translated 2 units to the right and 1 unit up. The domain of the second function is $x \geq 2$. The range is $y \geq 1$.

☑ CHECKING KEY CONCEPTS

1. If you measure the height that a pole vaulter's center of gravity is raised in meters and the pole vaulter's running velocity in meters per second, an equation relating height and velocity is $h = \dfrac{v^2}{19.6}$.

   a. Write an equation for the velocity as a function of height.

   b. Graph the equation from part (a). State the domain and range.

   c. Suppose a pole vaulter's center of gravity is 1 m above the ground when running. Find the running velocity needed to clear a bar 5 m above the ground.

Graph each function and compare it with the graph of $f(x) = \sqrt{x}$.

2. $f(x) = 3\sqrt{x}$ 3. $f(x) = \sqrt{x} - 3$ 4. $f(x) = \sqrt{x-3}$

8.1 Exercises and Applications

Extra Practice exercises on page 760

1. Use the tools you used in the Exploration on page 337.

   a. Graph $y = x^2 + 1$ and $y = x$ in the same coordinate plane.

   b. Reflect the graph of $y = x^2 + 1$ over the line $y = x$.

   c. Does the graph in part (b) represent a function? Explain.

   d. Write equations for two functions that together create the graph in part (b). What domain restrictions must you apply to the functions?

In Example 1, you saw that you can rewrite $\sqrt{64h}$ as $8\sqrt{h}$ because 64 is a perfect square. Find an equivalent expression for each of the following. Assume $x \geq 0$.

2. $\sqrt{4x}$ **3.** $\sqrt{8x}$ **4.** $\sqrt{16x}$

5. $\sqrt{25x}$ **6.** $\sqrt{50x}$ **7.** $\sqrt{100x}$

Early photographers captured the Krakatau eruption, one week after the eruptions started.

Connection ▸ OCEANOGRAPHY

When the volcanic island of Krakatau exploded on August 27, 1883, ocean waves caused by the disturbance traveled great distances. The map at the bottom of the page shows how many minutes it took the first wave to reach various places in the Sunda Strait, between Sumatra and Java.

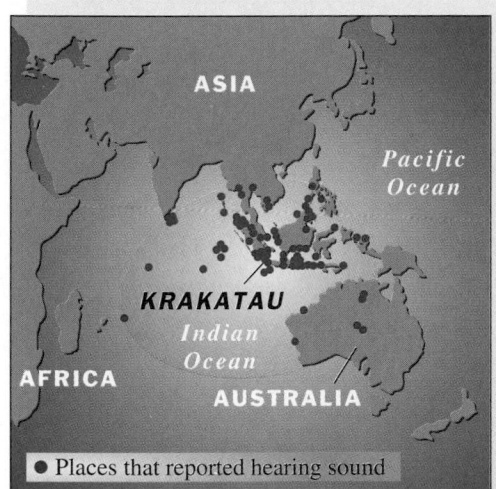

● Places that reported hearing sound

Sounds of the explosions from Krakatau were heard over a region covering 1/14th of the globe.

8. Estimate the speeds of the waves from Krakatau to Princes Island, Telok Betong, and Jakarta. Give your answers in kilometers per minute.

9. The speed of a wave depends on the depth of the water it travels through, according to the equation

$$s = \sqrt{35.28d}$$

where s = the speed of the wave in kilometers per minute and d = the ocean depth in kilometers. Use this equation to estimate the average ocean depths from Krakatau to Princes Island, Telok Betong, and Jakarta. Convert your answers to meters.

10. Waves from Krakatau's eruption reached Port Elizabeth, South Africa, in 15 h 12 min, after traveling 7546 km.

a. Find the average wave speed.

b. Find the average depth of the Indian Ocean between Krakatau and Port Elizabeth.

This 1979 explosion occurred on the island Anak Krakatau ("son of" Krakatau), which was formed after Krakatau exploded.

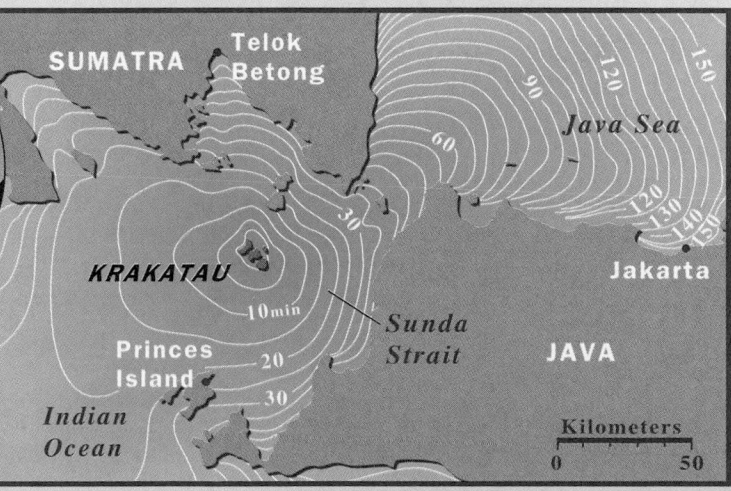

Match each function with its graph.

11. $f(x) = \sqrt{x - 1} + 2$

12. $f(x) = \sqrt{x + 1} - 2$

13. $f(x) = -\sqrt{3 - x}$

14. $f(x) = -\sqrt{x - 3}$

15. $f(x) = \sqrt{x} + 2$

16. $f(x) = \sqrt{x + 2}$

A.
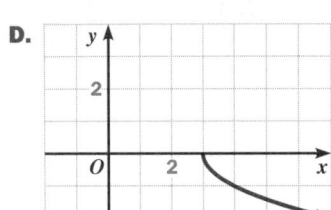

B.
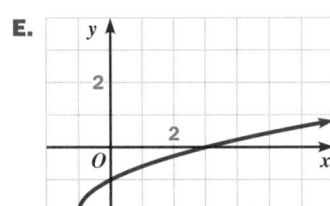

C.
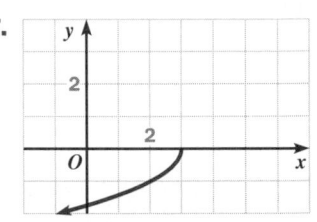

D.

E.

F.

INTERVIEW # Ronald Toomer

Look back at the article on pages 334–336.

Amusement park ride designers such as Ronald Toomer use physics to predict the velocity of a ride. In the Demon Drop ride, you plunge dozens of feet before your car follows a curve to slow down. If you ignore friction and air resistance, potential energy at the top is transformed into kinetic energy at the bottom, according to the equation

kinetic energy ▷ $\frac{1}{2}mv^2 = mgh$ ◁ potential energy

where m is mass, v is velocity, g is the acceleration due to gravity, and h is the height of the drop.

Over 10 million people have ridden the Demon Drop at Cedar Point in Sandusky, Ohio.

17. Solve the equation for *v*. Does the Demon Drop car drop faster when people are in it or when it is empty? Explain.

18. The Demon Drop ride involves a drop of 60 ft. Using $g = 32$ ft/s², find the velocity of the car after it falls 60 ft.

19. By what factor would you have to change the height of the Demon Drop ride in order to double the car's velocity at the bottom?

Graph each function. State the domain and range.

20. $y = \sqrt{x} + 3$ **21.** $y = -\sqrt{x} - 3$ **22.** $y = \sqrt{x+3}$ **23.** $y = -\sqrt{x+3}$

24. $y = -\sqrt{x}$ **25.** $y = 3 - \sqrt{x}$ **26.** $y = \sqrt{3x}$ **27.** $y = \sqrt{3x-3}$

28. $y = \sqrt{x-1}$ **29.** $y = \sqrt{x-1} + 3$ **30.** $y = \sqrt{x+1} - 3$ **31.** $y = \sqrt{x+3} - 1$

Technology Use a graphing calculator or graphing software to graph the left and right sides of each equation as separate functions. Decide whether each statement is *always true, sometimes true,* or *never true.*

32. $\sqrt{x^2 + 9} = x + 3$ **33.** $\sqrt{x-2} = \sqrt{x} - 2$ **34.** $\sqrt{x^2} = x$ **35.** $\left(\sqrt{x}\right)^2 = x$

36. $\sqrt{16x} = 4\sqrt{x}$ **37.** $\sqrt{-x} = -\sqrt{x}$ **38.** $\sqrt{x^4} = x^2$ **39.** $\sqrt{\dfrac{x}{9}} = \dfrac{\sqrt{x}}{3}$

40. Writing Is the statement $\sqrt{a+b} = \sqrt{a} + \sqrt{b}$ true for all nonnegative values of a and b? Explain your answer.

ONGOING ASSESSMENT

41. Cooperative Learning Work in a group of four students.

a. Discuss how the values of a, h, and k affect the graph of $y = a(x - h)^2 + k$.

b. Make conjectures about how the values of a, h, and k affect the graph of $y = a\sqrt{x - h} + k$. Then test your conjectures by graphing the equation for several values of a, h, and k, some positive and some negative.

c. Discuss how you can put equations like $y = 3\sqrt{2x - 4} + 1$ and $y = 3 - 2\sqrt{4x + 1}$ in the form of $y = a\sqrt{x - h} + k$ for graphing. Then work out several examples.

SPIRAL REVIEW

42. Chu is ordering T-shirts and sweatshirts for a fundraiser. He wants to order between 200 and 300 shirts altogether, and he wants at least 100 T-shirts and no more than 150 sweatshirts. *(Section 7.6)*

a. Write a system of inequalities and graph the feasible region. Label the coordinates of the vertices of the feasible region.

b. If the T-shirts cost $5 each and the sweatshirts cost $9 each, find the number of each that Chu should order to minimize his cost.

c. If the profit on a T-shirt is $3 and the profit on a sweatshirt is $6, find the number of each that Chu should order to maximize his profit.

Simplify. *(Section 3.2)*

43. $100^{1/2}$ **44.** $16^{1/4}$ **45.** $125^{1/3}$ **46.** $49^{1/2}$

Graph each inequality. *(Section 7.4)*

47. $y > x$ **48.** $2x - y \geq 1$ **49.** $150x + 40y \leq 200$ **50.** $-5x + 2y < 25$

8.2 Radical Functions

Learn how to...

- **evaluate radical expressions**

- **graph radical functions**

So you can...

- **solve problems involving radical functions, such as finding the length of a fish based on its weight**

If you restrict the domain of $y = x^2$ to values of $x \geq 0$, then the reflection of the graph of the restricted function over the line $y = x$ represents another function, the inverse $y = \sqrt{x}$. In the Exploration, you will see when you need to restrict the domains of other power functions of the form $y = x^n$ to create **radical functions** of the form $y = \sqrt[n]{x}$.

EXPLORATION
COOPERATIVE LEARNING

Reflecting the Graph of $y = x^n$ over the Line $y = x$

Work with a partner.
You will need:
• a graphing calculator or graphing software

SET UP Adjust the viewing window so that it shows $-3 \leq x \leq 3$ and $-2 \leq y \leq 2$.

1 Graph the line $y = x$ and the function $y = x^3$. Then draw the reflection of the function's graph over the line. Does the reflection represent a function?

2 Repeat Step 1 using the functions $y = x^4$, $y = x^5$, and $y = x^6$. (Clear the screen each time.)

Use your calculator or software's draw-inverse feature.

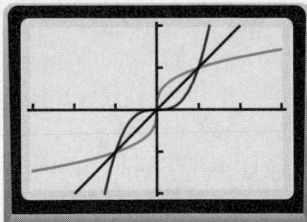

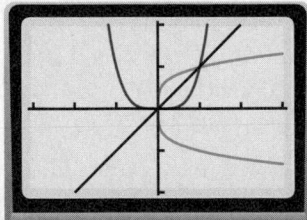

Questions

1. Describe how the graphs of $y = x^n$ when n is odd are different from the graphs when n is even. How are their reflections different?

2. Which functions of the form $y = x^n$ must have their domains restricted in order for the reflections of their graphs over the line $y = x$ to represent functions? How would you restrict the domains of the original functions?

Radical functions of the form

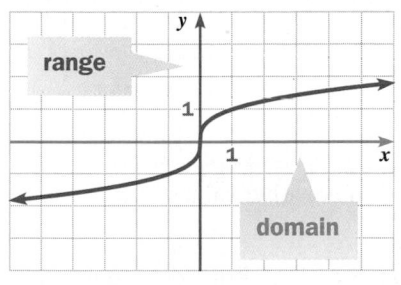

The positive integer n is called the *index* of the radical.

$$y = \sqrt[n]{x}$$

Read this as "the nth root of x."

have a domain and range of *all* numbers when n is *odd* and a domain and range of *nonnegative* numbers when n is *even*.

Example: $y = \sqrt[3]{x}$ This is a cube root function.

Example: $y = \sqrt[4]{x}$ This is a fourth root function.

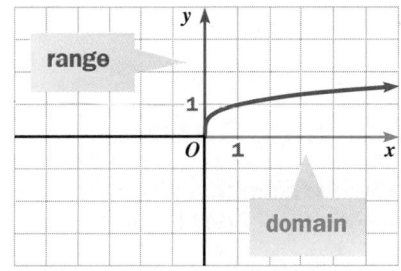

As you saw in the Exploration, a radical function is an inverse of a power function with a suitably restricted domain.

EXAMPLE 1   Application: Biology

A conservationist estimates the weight of a northern pike (in pounds) by cubing its length (in inches) and dividing by 3500.

a. Write an equation for the pike's weight w as a function of its length l.

b. Write an equation for the inverse of the function in part (a).

c. Graph the inverse function. Estimate the length of a 3 lb pike.

SOLUTION

a. $w = \dfrac{1}{3500}l^3$

b. $\dfrac{1}{3500}l^3 = w$

   $l^3 = 3500w$

   $l = \sqrt[3]{3500w}$

Because weight is a cubic function of length, length is a cube root function of weight.

c.

Use the trace feature of a graphing calculator.

X=3   Y=21.897596

A 3 lb northern pike is about 22 in. long.

Because radical functions are inverses of power functions, you can use the
following definition to help you evaluate $y = \sqrt[n]{x}$ for given values of x.

The Meaning of $\sqrt[n]{b}$

The nth root of a positive
number b is the positive number c
whose nth power is b.

$$\sqrt[n]{b} = c \text{ if and only if } b = c^n$$

Examples:

$\sqrt{25} = 5$ because $25 = 5^2$

$\sqrt[3]{343} = 7$ because $343 = 7^3$

$\sqrt[4]{81} = 3$ because $81 = 3^4$

THINK AND COMMUNICATE

1. Can the definition given above be extended to negative values of b if n is
   even? if n is odd? Explain.

2. Look back at the definition of $b^{1/n}$ given on page 102. What do you notice?
   What can you conclude about $\sqrt[n]{b}$ and $b^{1/n}$?

You can extend the properties of exponents to radical expressions as long as the
values of a and b below are not both negative.

Properties of Exponents	Examples with Fractional Exponents	Examples with Radical Expressions
$(ab)^m = a^m b^m$	$(ab)^{1/2} = a^{1/2} \cdot b^{1/2}$	$\sqrt{ab} = \sqrt{a} \cdot \sqrt{b}$
$\left(\dfrac{a}{b}\right)^m = \dfrac{a^m}{b^m}$	$\left(\dfrac{a}{b}\right)^{1/5} = \dfrac{a^{1/5}}{b^{1/5}}$	$\sqrt[5]{\dfrac{a}{b}} = \dfrac{\sqrt[5]{a}}{\sqrt[5]{b}}$
$\left(b^m\right)^n = b^{mn} = b^{nm} = \left(b^n\right)^m$	$\left(b^2\right)^{1/3} = b^{2/3} = \left(b^{1/3}\right)^2$	$\sqrt[3]{b^2} = \left(\sqrt[3]{b}\right)^2$

EXAMPLE 2

Simplify each radical expression. Assume $b \ge 0$ in part (c).

a. $\sqrt[3]{-24}$ b. $\sqrt[4]{810}$ c. $\sqrt[3]{b^6}$

SOLUTION

a. $\sqrt[3]{-24} = \sqrt[3]{-8 \cdot 3}$

$\quad = \sqrt[3]{-8} \cdot \sqrt[3]{3}$

$\quad = -2\sqrt[3]{3}$

b. $\sqrt[4]{810} = \sqrt[4]{81 \cdot 10}$

$\quad = \sqrt[4]{81} \cdot \sqrt[4]{10}$

$\quad = 3\sqrt[4]{10}$

c. $\sqrt[3]{b^6} = \left(b^6\right)^{1/3}$

$\quad = b^{6 \cdot 1/3}$

$\quad = b^2$

$\sqrt[3]{-8} = -2$ because
$(-2)^3 = -8$.

☑ CHECKING KEY CONCEPTS

State the domain and range of each function.

1. $y = \sqrt[3]{x + 2}$

2. $y = \sqrt[4]{x - 1}$

3. $y = \sqrt[5]{x} + 7$

Evaluate each radical expression, or state that it is *undefined*.

4. $\sqrt{4}$

5. $\sqrt{-4}$

6. $\sqrt[3]{27}$

7. $\sqrt[3]{-27}$

8. $\sqrt[6]{64}$

9. $\sqrt[3]{-64}$

Simplify each radical expression.

10. $\sqrt{45}$

11. $\sqrt[3]{270}$

12. $\sqrt[4]{2500}$

13. GEOMETRY The equation $V = \frac{1}{3}\pi r^3$ gives the volume V of a cone whose height and base radius are both r.

    a. Write an equation for the radius as a function of the volume.

    b. Find the radius of such a cone with a volume of 1000 cm³.

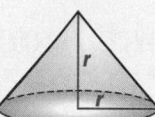

8.2 Exercises and Applications

Extra Practice exercises on page 760

Use the equation from Example 1 to find the length, to the nearest half inch, of a northern pike of each weight.

1. 1 lb

2. 2 lb

3. 4 lb

4. 5 lb

Evaluate each radical expression, or state that it is *undefined*.

5. $\sqrt{-25}$

6. $\sqrt{0}$

7. $\sqrt[3]{1000}$

8. $\sqrt[3]{-27}$

9. $\sqrt[3]{\dfrac{1}{8}}$

10. $\sqrt[3]{-64}$

11. $\sqrt[4]{625}$

12. $\sqrt[4]{-10{,}000}$

13. $\sqrt[4]{10{,}000}$

14. $\sqrt[5]{-1}$

15. $\sqrt[5]{32}$

16. $\sqrt[5]{-\dfrac{1}{32}}$

17. GEOMETRY The equation $V = \frac{1}{6}\pi d^3$ gives the volume V of a sphere as a function of its diameter d.

    a. Write an equation for the diameter as a function of the volume.

    b. Find the diameter of a sphere with a volume of 1000 cm³.

    c. Challenge Use part (a) to show that an equation for the radius r of a sphere as a function of its volume is $r = \sqrt[3]{\dfrac{3V}{4\pi}}$.

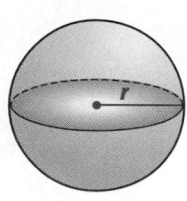

State the domain and range of each function.

18. $y = \sqrt{x + 1}$

19. $y = \sqrt[3]{x - 2}$

20. $y = \sqrt[4]{x} + 5$

21. $y = 3\sqrt[6]{x}$

In Exercises 22–24, assume all variables are restricted to nonnegative values.

22. Express using fractional exponents.

   a. $\sqrt[3]{b}$ **b.** $\sqrt[4]{d^3}$ **c.** $\sqrt[5]{s^2}$ **d.** $\sqrt[6]{t^5}$

23. Express using radical notation.

   a. $z^{1/2}$ **b.** $p^{2/3}$ **c.** $q^{3/2}$ **d.** $y^{7/10}$

24. Simplify each expression.

   a. $\sqrt{75}$ **b.** $\sqrt[3]{-432}$ **c.** $\sqrt[4]{3m^{12}}$ **d.** $\sqrt[5]{64g^{10}}$

Use fractional exponents and properties of exponents to prove each statement. Assume $x \geq 0$.

25. $\sqrt[6]{x} \cdot \sqrt[6]{x} = \sqrt[3]{x}$ 26. $\dfrac{\sqrt[3]{x^2}}{\sqrt{x}} = \sqrt[6]{x}$ 27. $x^{m/n} = \sqrt[n]{x^m}$ 28. $\left(\sqrt[n]{x}\right)^n = x$

INTERVIEW Ronald Toomer

Look back at the article on pages 334–336.

When designing a roller coaster, Ronald Toomer can use the formula

$$v = \sqrt{2gh}$$

to find the maximum velocity of a roller coaster car, which occurs at the bottom of the first hill. For the exercises, ignore energy losses due to friction and air resistance.

Ronald Toomer designed this roller coaster, the Magnum XL-200 on Lake Erie in Sandusky, Ohio.

29. In customary units, $g = 32$ ft/s². Use this fact to show that $v = \sqrt{2gh}$ can be simplified to $v = 8\sqrt{h}$.

30. Find the maximum velocity of each of the roller coaster cars in the table, using the data for the greatest vertical drop distances.

Roller coaster	Drop (ft)
American Eagle (Gurnee, IL)	147
Hercules (Allentown, PA)	157
Magnum XL-200 (Sandusky, OH)	205
Desperado (Jean, NV)	225

31. Use the results of Exercise 30 to graph $v = 8\sqrt{h}$. Identify the points on the graph that represent the roller coasters shown in the table.

32. Suppose you want to design a roller coaster whose maximum velocity is 100 mi/h. Convert this velocity from miles per hour to feet per second. Use your graph or the equation to find the drop for the first hill.

A scientist studying rowing found that a crew's speed s (in meters per second) depends on the number of rowers n as follows:

$$s = 4.62\sqrt[9]{n}$$

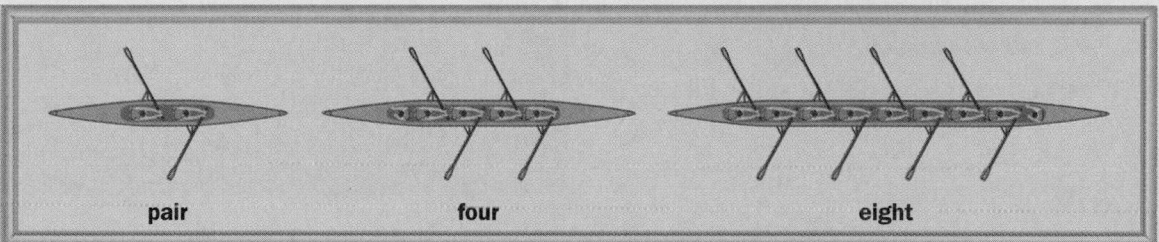

pair   four   eight

33. For $n = 1, 2, 4$, and 8 rowers, find s to the nearest hundredth of a meter per second. Does doubling the number of rowers double the speed?

34. Crew races are typically 2000 m long. Find the time in minutes each boat shown above takes to complete a race.

35. Suppose a new kind of boat is designed to accommodate 6 rowers. Predict its speed and race time.

36. Open-ended Problem When is the square root of a number greater than the number itself? Use graphs to show your reasoning. How is the answer different for cube roots? Make a generalization about other even roots and odd roots.

37. SAT/ACT Preview Suppose $A = \sqrt[3]{\sqrt{x}}$ and $B = \sqrt[6]{x}$ for $x \geq 0$. Which of the following is true?

  A. $A > B$ **B.** $B > A$ **C.** $A = B$ **D.** relationship cannot be determined

38. BIOLOGY Scientists can compare an animal's metabolic rate M (its daily heat production in Calories) to its weight W (in kilograms). For many species, the results can be modeled by this equation:

$$M = 70\sqrt[4]{W^3}$$

a. Rewrite the equation using a fractional exponent.

b. Calculate the missing metabolic rates in the table.

c. Solve the equation for W.

d. Calculate the missing weights in the table.

Animal	Weight (kg)	Metabolic rate (Cal)
Mouse	?	3.6
Cat	3	?
Rabbit	?	191
Dog	14	?
Goat	?	800
Chimpanzee	38	?
Sheep	?	1160
Pony	253	?
Bull	?	12,100
Elephant	3672	?

39. ASTRONOMY Saturn has many satellites. The table shows the distances of some of these from Saturn, and the number of days each satellite takes to orbit around Saturn.

Satellite	Distance (100,000 km)	Orbit time (days)
Mimas	1.86	0.942
Enceladus	2.38	1.370
Tethys	2.95	1.888
Dione	3.77	2.737
Rhea	5.27	4.518
Titan	12.22	15.94
Hyperion	14.84	21.28
Iapetus	35.62	79.33
Phoebe	129.30	?

a. Use the methods of Section 4.5 to find a power function that shows how orbit time t is a function of the distance d. Express the power function using a fractional exponent.

b. Rewrite the power function from part (a) using radical notation.

c. Predict the orbit time of the moon Phoebe.

O N G O I N G   A S S E S S M E N T

40. Cooperative Learning Work with another student. You will need a round balloon and a tape measure.

a. Blow up your balloon using equal shallow breaths. Keep track of the number of breaths you use.

b. When the balloon looks like a sphere, record its volume V (in number of breaths) and have your partner measure its circumference C. Make a table showing V, C, and $\sqrt[3]{V}$ for your balloon.

c. Add another breath and measure the balloon's new circumference. Repeat until you have at least four data values in your table.

d. Find an equation of the form $C = a\sqrt[3]{V}$ that models your data.

S P I R A L   R E V I E W

Graph each function. State the domain and range. *(Section 8.1)*

41. $y = \sqrt{x} + 1$ **42.** $y = -\sqrt{2x}$ **43.** $y = \sqrt{x + 4} - 2$

44. An object falls according to the equation $d = 16t^2$, where d is the distance in feet, and t is the time in seconds. *(Section 5.1)*

a. How far does an object fall in 3 s?

b. How long does it take an object to fall 210 ft?

Solve. Check to eliminate any extraneous solutions. *(Sections 4.4 and 5.6)*

45. $\log_2 (x + 6) + \log_2 (x - 6) = 6$ **46.** $\log_6 (x + 1) + \log_6 x = 1$

8.3 Solving Radical Equations

Learn how to...

- **solve equations with radical expressions**

So you can...

- **solve problems involving free fall, wind chill, and hurricanes, for example**

In *Danny Dunn and the Fossil Cave*, by J. Williams and R. Abrashkin, the young hero Danny Dunn gets separated from his friends while exploring a cave. He finds himself in a narrow passageway near a crevice.

All at once, there was nothing under his right foot.

Nothing! Even as the thought flashed through his mind, his foot went down into empty air and he threw out his right hand to try to find some support. His flashlight went whirling away, and vanished into a huge black space, and in the same instant he realized that he was not going to fall. He was wedged tightly between the rocks by the folds of his jacket, and this held him securely in place.

He was too frightened even to yell. . . . At the same time, his mind was busy counting: *one, two, three, four, five* . . . and then he heard the faint clatter as his flashlight hit bottom somewhere far below.

EXAMPLE 1

It takes time for Danny Dunn's flashlight to fall down the crevice, and it takes time for the sound to travel up to his ears. The combined time is 5 s.

a. Recall from Example 3 on page 187 that the equation $d = 16t^2$ gives the distance d (in feet) that an object drops in t seconds. Use the equation to find the time it takes the flashlight to drop d feet.

b. Write an expression for the time it takes the sound to travel up d feet. Assume the speed of sound in the cave is 1116 ft/s.

c. Combine the results in parts (a) and (b) to write an equation that states the total time is 5 s. Solve the equation to estimate the crevice depth.

SOLUTION

a. Solve $d = 16t^2$ for t.

$$16t^2 = d$$

$$t^2 = \frac{1}{16}d$$

$$t = \sqrt{\frac{1}{16}d}$$

$$t = \frac{1}{4}\sqrt{d}$$

Since time is positive, use the positive square root.

b. Use the formula
distance = rate × time:

$$d = 1116t$$

$$\frac{1}{1116}d = t$$

Solution continued on next page.

SOLUTION *continued*

c. Since the sum of flashlight travel time and sound travel time is 5 s, an equation using the expressions from parts (a) and (b) is

$$\frac{1}{4}\sqrt{d} + \frac{1}{1116}d = 5.$$

Graph $y = \frac{1}{4}\sqrt{x} + \frac{1}{1116}x$ and $y = 5$ on a graphing calculator or graphing software, and find the intersection.

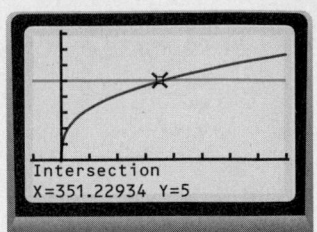

Intersection
X=351.22934 Y=5

The crevice is about 350 ft deep.

THINK AND COMMUNICATE

1. Explain why the expression $\sqrt{\frac{1}{16}d}$ can be rewritten as $\frac{1}{4}\sqrt{d}$.

2. If the crevice were 700 ft deep instead of 350 ft deep, would Danny Dunn have had to wait 10 s instead of 5 s to hear his flashlight?

Example 1 showed a graphical method for solving radical equations. The following examples show an algebraic method. This method sometimes produces extraneous solutions, however.

EXAMPLE 2

Solve $\sqrt{2x - 3} = x - 3$.

SOLUTION

$$\sqrt{2x - 3} = x - 3$$

$$\left(\sqrt{2x - 3}\right)^2 = (x - 3)^2$$

Square both sides to eliminate the radical expression.

$$2x - 3 = x^2 - 6x + 9$$

$$0 = x^2 - 8x + 12$$

Rearrange terms and solve using factoring.

$$0 = (x - 6)(x - 2)$$

$$x - 6 = 0 \ \text{ or } \ x - 2 = 0$$

These are the possible solutions.

$$x = 6 \ \text{ or } \ \ \ \ x = 2$$

Check

$$\sqrt{2 \cdot 6 - 3} \overset{?}{=} 6 - 3$$

$$\sqrt{9} \overset{?}{=} 3$$

$$3 = 3 ✔$$

The solution x = 6 does check.

Check

$$\sqrt{2 \cdot 2 - 3} \overset{?}{=} 2 - 3$$

$$\sqrt{1} \overset{?}{=} -1$$

$$1 \neq -1$$

The solution x = 2 does not check.

The solution of the equation $\sqrt{2x - 3} = x - 3$ is 6.

3. Which solution in Example 2 is the extraneous solution?

4. [📈] **Technology** Solve $\sqrt{2x - 3} = x - 3$ using technology. Do you get extraneous solutions when you use technology?

5. Suppose you are solving $\sqrt{2x + 3} = x - 3$ instead of $\sqrt{2x - 3} = x - 3$. You get the equation $0 = x^2 - 8x + 6$. What would you do next?

EXAMPLE 3

Solve $\sqrt[3]{x - 5} + 4 = -3$.

SOLUTION

$$\sqrt[3]{x - 5} + 4 = -3$$

$$\sqrt[3]{x - 5} = -7$$

$$\left(\sqrt[3]{x - 5}\right)^3 = (-7)^3$$

$$x - 5 = -343$$

$$x = -338$$

Isolate the radical on one side.

Cube both sides of the equation.

Check

$$\sqrt[3]{-338 - 5} + 4 \overset{?}{=} -3$$

$$\sqrt[3]{-343} + 4 \overset{?}{=} -3$$

$$-7 + 4 \overset{?}{=} -3$$

$$-3 = -3 ✔$$

☑ CHECKING KEY CONCEPTS

Solve by raising both sides of each equation to the same power.

1. $\sqrt{x - 3} = 3$ **2.** $\sqrt{x + 5} = 4$ **3.** $\sqrt{2x + 7} = 1$

4. $\sqrt[3]{5x + 4} = 4$ **5.** $\sqrt[4]{4x + 1} = 3$ **6.** $\sqrt[5]{3x - 7} = 2$

Solve. Check to eliminate extraneous solutions.

7. $\sqrt{x + 13} = x + 7$ **8.** $\sqrt{3(x - 7)} = x - 7$ **9.** $\sqrt{3(x + 9)} = x + 3$

8.3 **Exercises and Applications**

Extra Practice exercises on page 761

Solve.

1. $\sqrt{x} = 7$ **2.** $\sqrt{x - 1} = 7$ **3.** $\sqrt{x + 3} = 8$ **4.** $\sqrt{x + 4} = 0$

5. $\sqrt{2x} = 6$ **6.** $\sqrt{3x - 2} = 5$ **7.** $\sqrt{4x + 4} = 8$ **8.** $\sqrt{5(x + 3)} = 10$

9. $5\sqrt{x + 3} = 10$ **10.** $5 + \sqrt{x + 3} = 10$ **11.** $10 + \sqrt{x + 3} = 5$ **12.** $5\sqrt{\dfrac{x - 9}{4}} = 15$

Solve by raising both sides of each equation to the same power.

13. $\sqrt[3]{2x - 3} = 7$ **14.** $\sqrt[4]{\frac{1}{2}x + 7} = 4$ **15.** $\sqrt[5]{7x + 2} = -3$ **16.** $\sqrt[6]{\frac{2}{3}x} - 4 = 0$

17. Visual Thinking The graphs of $y = x - 1$, $y = x$, and $y = x + 1$ are shown with the graph of $y = \sqrt{x}$. Use the graphs to predict the number of solutions of each equation. Then find the solutions.

a. $\sqrt{x} = x - 1$ **b.** $\sqrt{x} = x$ **c.** $\sqrt{x} = x + 1$

18. Writing Explain why the equation $\sqrt{3x + 4} = -5$ has no solution, but the equation $\sqrt[3]{3x + 4} = -5$ *does* have a solution.

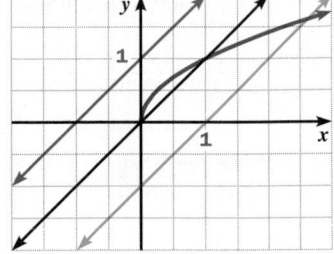

Connection ▶ **METEOROLOGY**

In a hurricane, the mean sustained wind velocity v, measured in meters per second, is given by

$$v = 6.3\sqrt{1013 - p}$$

where p is the air pressure, measured in millibars (mb), at the center of the hurricane.

The map shows the air pressure and wind velocity at various points along the path of Hurricane Hugo, which passed through the Caribbean in 1989.

19. Estimate the mean sustained wind velocities at 0 h (midnight) on September 13th, 14th, 15th, and 16th.

20. Estimate the air pressures at 12 h (noon) on September 13th, 14th, 15th, and 16th.

21. What happens to wind velocity in a hurricane when air pressure decreases?

Path of Hurricane Hugo, September 1989

Darker color indicates greater storm intensity.

0 h Sept. 16
923 mb

0 h Sept. 14
984 mb

0 h Sept. 15
962 mb

0 h Sept. 13
994 mb

12 h Sept. 16
61.7 m/s

12 h Sept. 15
64.3 m/s

12 h Sept. 14
43.7 m/s

12 h Sept. 13
30.9 m/s

22. **WIND CHILL** The *wind chill index* gives you an idea of how cold it feels outside when wind increases the rate of heat loss from your skin. The index is the *equivalent temperature* that would have the same cooling effect if the wind velocity were only 6 km/h.

When the wind velocity is less than 100 km/h, an equation for wind chill W (in degrees Celsius), based on temperature T (in degrees Celsius) and wind velocity v (in kilometers per hour), is:

$$W = 33 - (0.0393)(33 - T)(12.36 + 6.13\sqrt{v} - 0.32v)$$

a. Suppose $T = 4°C$ and $v = 40$ km/h. Find W.

b. Suppose $W = -15°C$ and $v = 20$ km/h. Find T.

c. **Technology** Suppose $W = -25°C$ and $T = -8°C$. Find v using a graphing calculator or graphing software.

d. **Cooperative Learning** Conditions are extremely dangerous for humans when the wind chill index is less than $-30°C$, because exposed flesh can freeze in less than a minute. Make a wind chill index table that shows several combinations of temperature and wind velocity that produce such dangerous conditions.

Solve by using the method of Example 2. Check to eliminate extraneous solutions.

23. $\sqrt{x - 3} = x - 5$

24. $\sqrt{5x + 39} = x + 3$

25. $\sqrt{13x + 90} = x + 10$

26. $\sqrt{14x - 5} = x + 2$

27. $\sqrt{3x + 7} = x + 1$

28. $\sqrt{x + 5} = x + 5$

29. $\sqrt{2x + 7} = x - 4$

30. $\sqrt{5x + 6} = x + 2$

31. $\sqrt{-(3x + 2)} = x + 4$

32. $\sqrt{9x - 2} = x + 2$

33. $2 - x = \sqrt{44 - x}$

34. $\sqrt{2x + 15} = x + 6$

35. **Open-ended Problem** Choose one problem from Exercises 23–34 that has two solutions. Use a graphing calculator or draw sketches to explain why there are two solutions.

36. a. Solve some examples of $\sqrt{x - k} = x - k$ using different values of k.

b. **Challenge** Express the two solutions of $\sqrt{x - k} = x - k$ in terms of k.

 Technology **Solve using a graphing calculator or graphing software. Check your answers.**

37. $\sqrt{x} + 2 = x$

38. $\sqrt[3]{x - 1} = \frac{1}{4}(x - 1)$

39. $\sqrt[4]{x} - 2 = x - 3$

40. a. Lisa tried solving the equation $\sqrt{x - 5} = \sqrt{x} - 1$ algebraically. Part of her work is shown. Complete the solution.

b. Try solving $\sqrt{2x + 1} = \sqrt{x} + 1$ in a similar way.

c. What makes solving equations like those in parts (a) and (b) different from solving the other radical equations in this section?

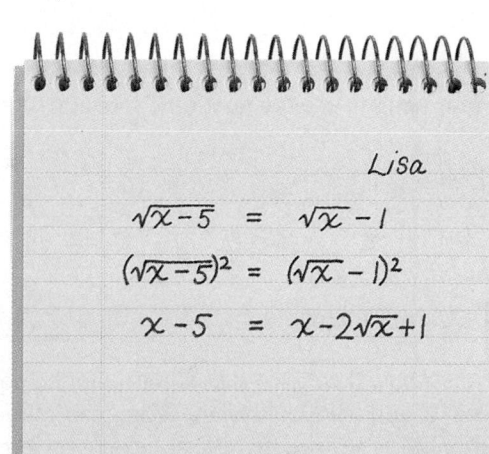

Lisa

$\sqrt{x - 5} = \sqrt{x} - 1$

$(\sqrt{x - 5})^2 = (\sqrt{x} - 1)^2$

$x - 5 = x - 2\sqrt{x} + 1$

In the America's Cup competition, boats must meet this rule to qualify:

$$\frac{L + 1.25\sqrt{S} - 9.8\sqrt[3]{D}}{0.679} \leq 24.000$$

where L is boat length (in meters), S is the sail area (in square meters), and D is the displacement (in cubic meters). The expression on the left side of the inequality is the rating of the boat. Boat designers try to make this value as close to 24 as possible.

In 1995, a women's team participated in the America's Cup races. Nearly 700 women asked for applications to try out for the 23 places on the team.

For each boat, find the value of the unknown variable that yields a rating of 24, qualifying the boat for the America's Cup race. Give answers to three decimal places.

	Boat	L (m)	S (m^2)	D (m^3)
41.	Argoknot	?	327.334	24.244
42.	Barnicle	21.870	?	22.440
43.	Coralgrazer	20.950	277.300	?

44. **Open-ended Problem** Suppose you are designing a boat for the America's Cup.

   a. Using the table below, find values for L, S, and D that meet the qualifying rule and fall within the ranges given. Give answers to three decimal places.

   b. How close to 24 is the rating of the boat you have designed?

Variable	Minimum	Your boat	Maximum
L (m)	20.600	?	22.000
S (m^2)	250.000	?	330.000
D (m^3)	15.610	?	24.390

Solve. You may need to use the quadratic formula. Check to eliminate extraneous solutions.

45. $\sqrt{6x + 4} = x + 2$ 46. $\sqrt{5x + 16} = x + 4$ 47. $\sqrt{x} = x$ 48. $\sqrt{x} = -x$

49. $\sqrt{3 - 3x} = x$ 50. $x - 3 = \sqrt{8 - 2x}$ 51. $\sqrt{3x + 5} = 2x$ 52. $x = \sqrt{x + 1}$

ONGOING ASSESSMENT

53. **Writing** Show a step-by-step solution to the equation $\sqrt{ax + b} = c$, where a, b, and c are constants. What restrictions must you place on a, b, or c in order to be sure that the solution you get is not extraneous?

Simplify each radical expression. *(Section 8.2)*

54. $\sqrt[3]{324}$ **55.** $\sqrt[2]{50}$ **56.** $\sqrt[3]{72}$ **57.** $\sqrt[4]{400}$

Tell whether each equation has *two solutions*, *one solution*, or *no solution*.
(Section 5.5)

58. $5x^2 + 3x + 2 = 0$ **59.** $-x^2 + 3x - 2 = 0$ **60.** $2x^2 + 6x + \dfrac{9}{2} = 0$ **61.** $-4x^2 - 3x - 5 = 0$

Solve each system of equations. *(Section 7.1)*

62. $y = x^2 + 2x + 1$
     $y = x + 1$

63. $y = 4x + 2$
     $y = \dfrac{1}{2}x$

64. $y = 5x + 2$
     $y = -3x$

65. $y = x^2 - x + 3$
     $y = 4 - x$

ASSESS YOUR PROGRESS

VOCABULARY

radical functions (p. 344)

1. **a.** Graph the function $y = (x - 1)^2$ and its reflection over the line $y = x$.
   (Section 8.1)

   b. Write equations for two functions that together create the reflected graph you found in part (a). Include the domain restrictions.

   c. Why must you use two functions to describe the graph?

SPORTS In the sport of powerlifting, the weights lifted in the squat, bench press, and deadlift are combined. Senior men's world record data can be modeled by the equation

$$T = 47.3\sqrt[3]{w^2}$$

where *T* is the total weight lifted and *w* is the athlete's weight class, both measured in kilograms. *(Section 8.2)*

2. Rewrite the equation using a fractional exponent.

3. Complete the table of the total weights predicted by the equation. Give answers to the nearest half kilogram.

Weight class *w* (kg)	52	56	60	67.5	75	82.5	90	100	110	125
Predicted total *T* (kg)	?	?	?	?	?	?	?	?	?	?

Solve. *(Section 8.3)*

4. $\sqrt{4(-x + 2)} = x - 3$ 5. $\sqrt{3x + 1} = x$ 6. $\sqrt{x^2 - 8} = 4$

7. **Journal** Explain how the relationship between power functions and radical functions is similar to and different from the relationship between exponential functions and logarithmic functions.

8.4 Working with Complex Numbers

Learn how to...

- add, subtract, multiply, and divide complex numbers

- find complex solutions to equations that have no real solutions

So you can...

- solve problems involving complex numbers, such as finding the impedance of an electrical circuit

Calvin and Hobbes

by Bill Watterson

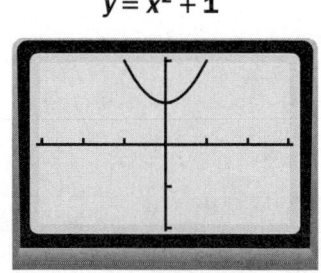

So far in this book the numbers you have used for the domain and range of functions have been **real numbers**, which include both rational numbers, such as 3 and $-\frac{1}{2}$, and irrational numbers, such as $\sqrt{2}$ and π. In this section you will see how to extend this set of numbers to include other numbers, called *imaginary numbers*, which together with real numbers form the system of *complex numbers*.

THINK AND COMMUNICATE

Use the graphs to answer Questions 1–3.

$$y = x^2 - 1$$

$$y = x^2 + 1$$

1. What does the graph on the left tell you about solutions of $x^2 - 1 = 0$?

2. What does the graph on the right tell you about solutions of $x^2 + 1 = 0$?

3. Apply the quadratic formula to the equations in Questions 1 and 2. What do you notice?

The equation $x^2 + 1 = 0$ has no *real* solutions, but it does have solutions if you define the square roots of negative numbers.

Imaginary Numbers

The letter i is used for the square root of -1, the fundamental unit in the system of *imaginary numbers*.

$$i = \sqrt{-1}$$
$$i^2 = -1$$

The square root of any negative number is called a **pure imaginary number**. Every pure imaginary number can be written in the form bi, where $b \neq 0$.

Examples:

$$\sqrt{-2} = \sqrt{(-1)(2)} = \sqrt{-1}\,\sqrt{2} = i\sqrt{2}$$
$$\sqrt{-3} = \sqrt{(-1)(3)} = \sqrt{-1}\,\sqrt{3} = i\sqrt{3}$$
$$\sqrt{-4} = \sqrt{(-1)(4)} = \sqrt{-1}\,\sqrt{4} = 2i$$

WATCH OUT!

The rule $\sqrt{ab} = \sqrt{a}\,\sqrt{b}$ applies only when a and b are both nonnegative, or when one is nonnegative and the other is negative. The rule does *not* apply if a and b are *both* negative.

EXAMPLE 1

Show that i is a solution of $x^2 + 1 = 0$.

SOLUTION

Substitute i for x in $x^2 + 1 = 0$.

$$i^2 + 1 \overset{?}{=} 0$$
$$\left(\sqrt{-1}\right)^2 + 1 \overset{?}{=} 0$$
$$-1 + 1 = 0 \quad ✔$$

THINK AND COMMUNICATE

4. Show that $-i$ is also a solution of $x^2 + 1 = 0$.

5. Find the pure imaginary solutions of $x^2 + 4 = 0$.

6. Express each radical in the form bi.

   a. $\sqrt{-9}$ **b.** $\sqrt{-25}$ **c.** $\sqrt{-200}$

Complex Numbers

Any number of the form $a + bi$ (where a and b are real numbers and $b \neq 0$) is an **imaginary number**. The imaginary numbers together with the real numbers form the set of **complex numbers**.

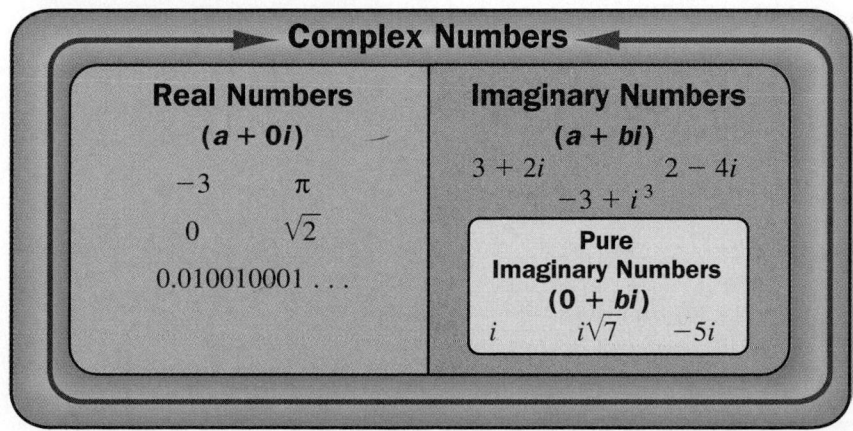

Adding and Subtracting Complex Numbers

You can add or subtract complex numbers by adding or subtracting their real parts and their pure imaginary parts.

$$(a + bi) + (c + di) = (a + c) + (b + d)i$$

$$(a + bi) - (c + di) = (a - c) + (b - d)i$$

EXAMPLE 2

Perform the indicated operation.

a. $(3 - 4i) + (5 + 9i)$ **b.** $(7 + 2i) - (8 - 3i)$

SOLUTION

a. $(3 - 4i) + (5 + 9i) = (3 + 5) + (-4 + 9)i = 8 + 5i$

b. $(7 + 2i) - (8 - 3i) = (7 - 8) + (2 - (-3))i = -1 + 5i$

Multiplying Complex Numbers

You can multiply complex numbers as you would multiply binomials.

$$\begin{aligned}(a + bi)(c + di) &= ac + bci + adi + bdi^2 \\ &= ac + (bc + ad)i - bd \\ &= (ac - bd) + (bc + ad)i\end{aligned}$$

THINK AND COMMUNICATE

7. Justify each step in the multiplication of $a + bi$ and $c + di$ shown above.

8. Write a rule for multiplying $(a - bi)(c - di)$.

EXAMPLE 3

Perform the indicated operation.

a. $(3 + 2i)(4 + 7i)$ **b.** $(6 + 5i)^2$ **c.** $(2 + 9i)(2 - 9i)$

SOLUTION

a.
$$\begin{aligned}(3 + 2i)(4 + 7i) &= 12 + 8i + 21i + 14i^2 \\ &= 12 + 29i - 14 \\ &= -2 + 29i\end{aligned}$$

b.
$$\begin{aligned}(6 + 5i)^2 &= (6 + 5i)(6 + 5i) \\ &= 36 + 30i + 30i + 25i^2 \\ &= 36 + 60i - 25 \\ &= 11 + 60i\end{aligned}$$

Use the definition of squaring.

c. $(2 + 9i)(2 - 9i) = 4 + 18i - 18i - 81i^2$

$$= 4 + 81$$

$$= 85$$

In this case, multiplying two complex numbers results in a real number.

Whenever you multiply numbers of the form $a + bi$ and $a - bi$, called **complex conjugates**, the result is a real number. You can use complex conjugates to divide complex numbers.

EXAMPLE 4

Use complex conjugates to write $\dfrac{3 + i}{2 - 3i}$ in $a + bi$ form.

SOLUTION

$$\frac{3 + i}{2 - 3i} = \frac{3 + i}{2 - 3i} \cdot \frac{2 + 3i}{2 + 3i}$$

Multiply the numerator and denominator by the complex conjugate of the denominator.

$$= \frac{(3 + i)(2 + 3i)}{(2 - 3i)(2 + 3i)}$$

$$= \frac{6 + 2i + 9i + 3i^2}{4 - 6i + 6i - 9i^2}$$

$$= \frac{6 + 11i - 3}{4 + 9}$$

The denominator is a real number.

$$= \frac{3 + 11i}{13}$$

$$= \frac{3}{13} + \frac{11}{13}i$$

Rewrite the fraction in $a + bi$ form.

EXAMPLE 5

Solve the equation $x^2 - 10x + 41 = 0$. Check the solutions.

SOLUTION

$$x = \frac{-b \pm \sqrt{b^2 - 4ac}}{2a}$$

Substitute **1** for a, -10 for b, and **41** for c.

$$= \frac{-(-10) \pm \sqrt{(-10)^2 - 4(1)(41)}}{2(1)}$$

$$= \frac{10 \pm \sqrt{100 - 164}}{2}$$

$$= \frac{10 \pm \sqrt{-64}}{2}$$

$$= \frac{10 \pm 8i}{2}$$

$$= 5 \pm 4i$$

Check

Substitute $5 + 4i$ for x in $x^2 - 10x + 41 = 0$.

$$(5 + 4i)^2 - 10(5 + 4i) + 41 \stackrel{?}{=} 0$$

$$(25 + 40i + 16i^2) - 50 - 40i + 41 \stackrel{?}{=} 0$$

$$25 + 40i - 16 - 50 - 40i + 41 \stackrel{?}{=} 0$$

$$(25 - 16 - 50 + 41) + (40i - 40i) \stackrel{?}{=} 0$$

$$0 + 0 = 0 \ \checkmark$$

The check of the other solution is left to you.

9. Show that the solution $5 - 4i$ in Example 5 also checks.

10. Without graphing, tell whether the graph of $y = x^2 - 10x + 41$ intersects the x-axis. How do you know?

☑ CHECKING KEY CONCEPTS

Express in *bi* form.

1. $\sqrt{-7}$ 2. $\sqrt{-32}$ 3. $\sqrt{-300}$ 4. $\sqrt{-16}$

Perform the indicated operation.

5. $(5 - 3i) + (9 + 8i)$ 6. $(1 + 6i) + (7 - 12i)$

7. $(10 + 9i) - (5 + 3i)$ 8. $(9 + 6i) - (10 + 7i)$

9. $4(3 + 2i)$ 10. $6i(5 + 3i)$

11. $(7 + i)(2 + 3i)$ 12. $(4 + 7i)^2$

13. $\dfrac{1 + 2i}{3 + 4i}$ 14. $\dfrac{1 + 3i}{2 + 4i}$

Solve and check the solutions.

15. $x^2 - 14x + 53 = 0$ 16. $x^2 - 2x + 26 = 0$

8.4 **Exercises and Applications**

Extra Practice exercises on page 761

Express in *bi* form.

1. $\sqrt{-36}$ 2. $\sqrt{-40}$ 3. $-\sqrt{-28}$ 4. $\sqrt{-64} + \sqrt{-121}$

Write a quadratic equation with the given solutions.

5. $4i, -4i$ 6. $10i, -10i$ 7. $i\sqrt{2}, -i\sqrt{2}$ 8. $3i\sqrt{3}, -3i\sqrt{3}$

Add or subtract.

9. $(5 + 3i) + (7 + 2i)$ 10. $(4 + 3i) + (2 + i)$ 11. $(12 + 5i) + (6 - 2i)$ 12. $(3 + 7i) + (-2 - i)$

13. $(2 + i) + (-4 + 3i)$ 14. $(4 - 2i) + (3 - 7i)$ 15. $(12 + 7i) - (2 + 3i)$ 16. $(1 + 3i) - (3 + i)$

17. $(-7 + 5i) - (4 + i)$ 18. $(5 + 11i) - (3 - 2i)$ 19. $(-8 - i) - (-8 - i)$ 20. $(2 - 5i) - (-3 + i)$

Multiply.

21. $4(5i)$ 22. $-2i(6i)$ 23. $5(2 + 3i)$ 24. $-8(12 - 3i)$

25. $-9(5 - 2i)$ 26. $6i(-2 + 4i)$ 27. $(3 + i)(7 + 2i)$ 28. $(5 + 2i)(7 - 3i)$

29. $(2 - i)(3 + 4i)$ 30. $(10 - 5i)(5 - 2i)$ 31. $(3 + 4i)(3 - 4i)$ 32. $(6 - 3i)^2$

Use complex conjugates to write each quotient in $a + bi$ form.

33. $\dfrac{2 + 5i}{8 + i}$ 34. $\dfrac{3 - 2i}{2 - i}$ 35. $\dfrac{-3 + i}{4 + 3i}$ 36. $\dfrac{2 - 5i}{i}$

Electrical circuit components such as resistors, inductors, and capacitors oppose the flow of current in different ways. The opposition is called *impedance*, denoted by *Z*. The value of *Z* is a real number *R* for a resistor of *R* ohms (Ω), a pure imaginary number *Li* for an inductor of *L* ohms, and a pure imaginary number *–Ci* for a capacitor of *C* ohms. (See the table at the right for examples.)

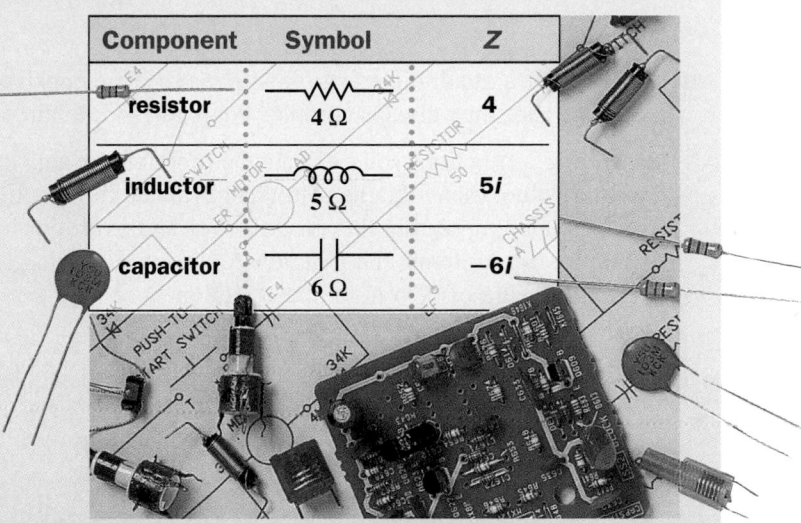

Component	Symbol	Z
resistor	4 Ω	4
inductor	5 Ω	5*i*
capacitor	6 Ω	–6*i*

Impedance in a Series Circuit	**Impedance in a Parallel Circuit**
A series circuit has only one pathway. To find the impedance *Z* in a series circuit, add the impedance for each component in the circuit.	A parallel circuit has more than one pathway that includes the alternating current source. To find the impedance *Z* in a parallel circuit with two pathways, find the impedance in each pathway, Z_1 and Z_2, and then apply this formula: $$Z = \frac{Z_1 Z_2}{Z_1 + Z_2}$$

Example:

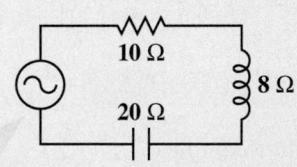

This is the symbol for an alternating current source.

In the series circuit shown, the impedance is:

$$Z = 10 + 8i - 20i$$
$$= 10 - 12i$$

Example:

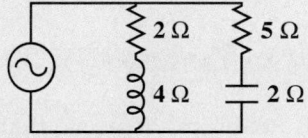

In the parallel circuit shown, the impedance is:

$$Z = \frac{(2 + 4i)(5 - 2i)}{(2 + 4i) + (5 - 2i)}$$
$$= \frac{158}{53} + \frac{76}{53}i$$

37. Justify that $\frac{(2 + 4i)(5 - 2i)}{(2 + 4i) + (5 - 2i)} = \frac{158}{53} + \frac{76}{53}i$. Show each step.

Find the impedance *Z* for each circuit.

38. series circuit

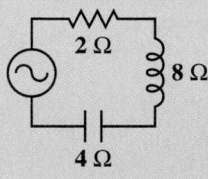

39. parallel circuit

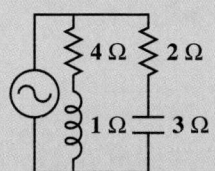

40. series circuit

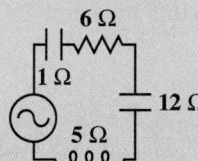

41. parallel circuit

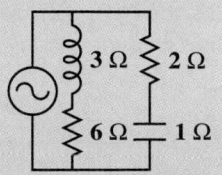

Solve using the quadratic formula. Check your solutions.

42. $x^2 - 4x + 13 = 0$ **43.** $x^2 - 14x + 50 = 0$ **44.** $x^2 + 6x + 11 = 0$

45. $x^2 + 12 = -6x$ **46.** $2x^2 + 20x = -82$ **47.** $x(x + 1) = -1$

48. Writing If a quadratic equation with real coefficients has two imaginary solutions, are they always complex conjugates? Explain your answer.

49. To *iterate* a function, you evaluate the function for a given starting value, then take the result and evaluate the function for that value, repeating this process over and over. For example, if you iterate the function $f(z) = z^2 + 1$ using a starting z-value of 0, you get these results:

$$0$$
$$0^2 + 1 = 1$$
$$1^2 + 1 = 2$$
$$2^2 + 1 = 5$$
$$5^2 + 1 = 26$$
$$\text{and so on.}$$

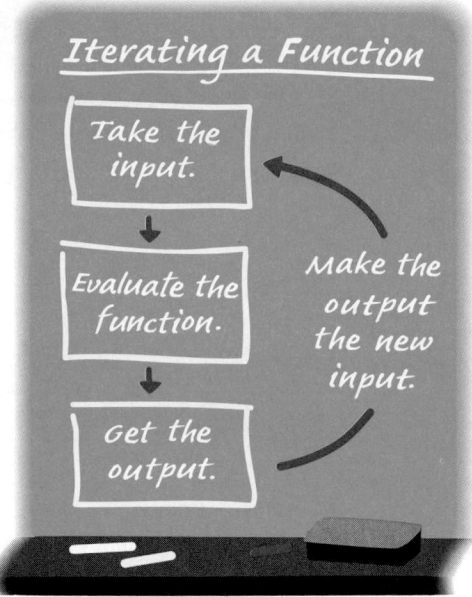

Iterating a Function

Take the input.

↓

Evaluate the function.

Make the output the new input.

↓

Get the output.

a. Iterate the function $f(z) = z^2 + 2$ starting with $z = 0$.

b. Iterate the function $f(z) = z^2 - 1$ starting with $z = 0$.

c. Iterate the function $f(z) = z^2 + i$ starting with $z = 0$.

d. Iterate the function $f(z) = z^2 + 1 + i$ starting with $z = 0$.

e. Open-ended Problem Choose any complex number c and iterate the function $f(z) = z^2 + c$ three times starting with $z = 0$. Write the results in $a + bi$ form.

ONGOING ASSESSMENT

50. Open-ended Problem Write a quadratic equation that has complex solutions of the form $a + bi$, where a and b are nonzero. Solve the equation. Show by substitution that the solutions check.

SPIRAL REVIEW

Solve. *(Section 8.3)*

51. $\sqrt{x - 2} = 8$ **52.** $\sqrt[3]{x + 2} = 3$ **53.** $\sqrt[4]{2x - 5} = 2$

Write an exponential function whose graph passes through each pair of points.
(Section 3.5)

54. $(2, 5), (3, 10)$ **55.** $(4, 8), (5, 10)$ **56.** $(2, 2), (5, 6)$

Use the Pythagorean theorem to find the length of the hypotenuse of each right triangle. *(Toolbox, page 801)*

57.

58.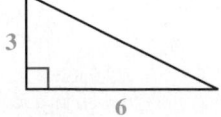

59.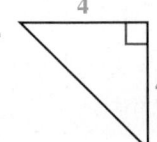

8.5 Visualizing the Complex Plane

Learn how to...

- **plot complex numbers in the complex plane**

- **calculate the magnitude of a complex number**

So you can...

- **understand the mathematics used to generate images of the Mandelbrot set, for example**

Some of the most astonishing images that computers have created are representations of the *Mandelbrot set*, a closed region of the plane with an extremely complicated border. In this section you will learn how repeatedly adding and multiplying complex numbers creates these images.

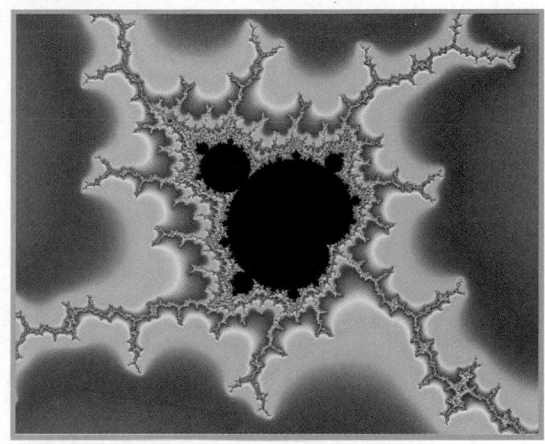

Images of the Mandelbrot set are plotted in the **complex plane**, where each point (a, b) represents the complex number $a + bi$.

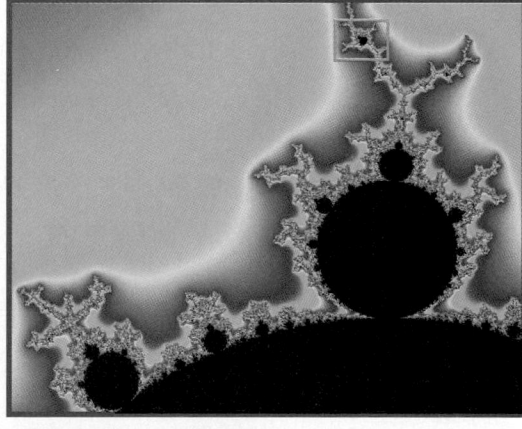

This point is the graph of $-1 + i$.

The complex plane has a horizontal **real axis**.

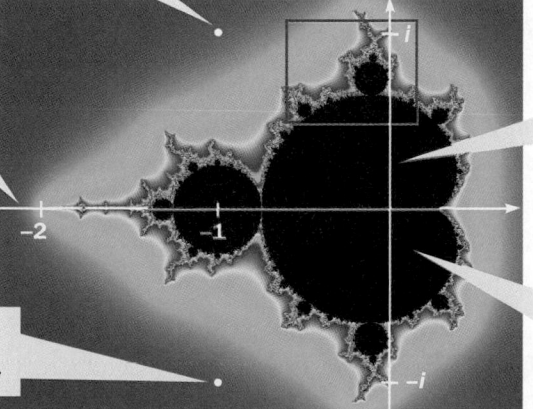

The complex plane has a vertical **imaginary axis**.

This point is the graph of $-1 - i$.

The **Mandelbrot set** is shown in black. It lies inside a circle of radius 2 centered at the origin of the complex plane.

THINK AND COMMUNICATE

1. In what quadrant would you plot the complex number $3 - 4i$?

2. In what quadrant would you plot the complex number $-4 + 3i$?

3. Which is farther from the origin, the graph of $3 - 4i$ or $-4 + 3i$? Explain.

Magnitude of a Complex Number

The **magnitude** of a complex number $a + bi$, denoted by $|a + bi|$, is the distance from $(0, 0)$ to (a, b).

You can use the Pythagorean theorem to find this distance:

$$|a + bi| = \sqrt{a^2 + b^2}$$

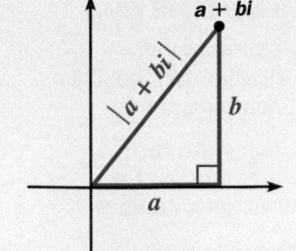

EXAMPLE 1

Plot $3 + i$, $2 + 3i$, and their sum in the complex plane. Find their magnitudes.

SOLUTION

First find the sum algebraically:

$$(3 + i) + (2 + 3i) = 5 + 4i$$

The numbers are plotted in the complex plane at the right.

Then calculate the magnitudes:

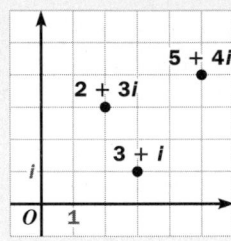

$$|3 + i| = \sqrt{3^2 + 1^2} = \sqrt{9 + 1} = \sqrt{10}$$

$$|2 + 3i| = \sqrt{2^2 + 3^2} = \sqrt{4 + 9} = \sqrt{13}$$

$$|5 + 4i| = \sqrt{5^2 + 4^2} = \sqrt{25 + 16} = \sqrt{41}$$

THINK AND COMMUNICATE

4. **a.** Draw a quadrilateral whose vertices are $(0, 0)$ and the three points in the diagram from Example 1. What is special about the quadrilateral?

   b. Repeat the construction with the sum of two different complex numbers of your own choice. Is the same kind of geometric figure formed?

5. **a.** In Example 1, is $|(3 + i) + (2 + 3i)| = |3 + i| + |2 + 3i|$ true?

   b. In general, is the *magnitude of the sum* of two complex numbers equal to the *sum of the magnitudes* of the complex numbers?

6. Under what circumstances is the magnitude of the sum of two complex numbers equal to the sum of the magnitudes of the complex numbers?

EXAMPLE 2

Plot $1 + i$, $1 + 3i$, and their product in the complex plane. Find their magnitudes.

SOLUTION

First find the product algebraically:

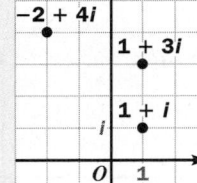

$$(1 + i)(1 + 3i) = 1 + i + 3i + 3i^2$$
$$= 1 + 4i - 3$$
$$= -2 + 4i$$

The numbers are plotted in the complex plane at the right.

Then calculate the magnitudes:

$$|1 + i| = \sqrt{1^2 + 1^2} = \sqrt{2}$$

$$|1 + 3i| = \sqrt{1^2 + 3^2} = \sqrt{10}$$

$$|-2 + 4i| = \sqrt{(-2)^2 + 4^2} = \sqrt{4 + 16} = \sqrt{20}$$

THINK AND COMMUNICATE

7. Draw a quadrilateral whose vertices are $(0, 0)$ and the three points in the diagram from Example 2. Is there anything special about it? Is this the same as the type of figure formed in *Think and Communicate* Question 4?

8. a. How are the magnitudes of the complex numbers in Example 2 related?

b. In general, do you think the *magnitude of the product* of two complex numbers is equal to the *product of the magnitudes* of the complex numbers? Test your conjecture with another pair of complex numbers.

☑ CHECKING KEY CONCEPTS

Plot each pair of complex numbers and their sum in the complex plane. Find their magnitudes.

1. $1, i$ **2.** $1 + i, -1 + i$ **3.** $2 + 3i, 3 + 2i$

4. $-4 - 3i, 2 - i$ **5.** $3 + 5i, 3 - 5i$ **6.** $-4 + 7i, 4 - 7i$

Plot each pair of complex numbers and their product in the complex plane. Find their magnitudes.

7. $1 + i, 2i$ **8.** $3i, -2 + 2i$ **9.** $-4, 2 + i$

10. $3i, -2i$ **11.** $3 + 4i, 3 - 4i$ **12.** $-1 + 3i, 4 - i$

13. Plot the numbers $i, i^2, i^3, i^4, i^5, \ldots$ in the complex plane. What do you notice?

8.5 Exercises and Applications

Extra Practice exercises on page 761

Name the two complex numbers plotted in each complex plane. Each pair has the same magnitude. Find that magnitude and name two other numbers that have the same magnitude.

1.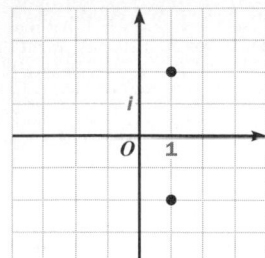

2.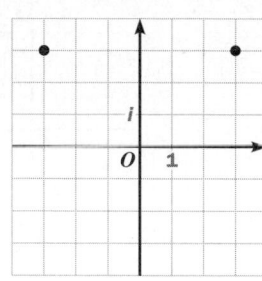

3.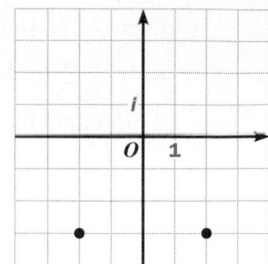

Plot each pair of complex numbers and their sum in the complex plane. Find their magnitudes.

4. $3, -2i$

5. $-4 + 3i, -3 + 4i$

6. $7i, -3i$

7. $2 + i, -2 + i$

8. $-3 + i, 2 - 4i$

9. $2 + i, -5 + 3i$

10. $3 + 2i, -5 - 3i$

11. $8 - 4i, -8 + 4i$

Plot each pair of complex numbers and their product in the complex plane. Find their magnitudes.

12. $i, 4 + 2i$

13. $-2i, 3 + 5i$

14. $-4i, -2i$

15. $5i, -i$

16. $-2, 1 + i$

17. $3 + i, 3 + 2i$

18. $5 - 6i, 5 + 6i$

19. $i, -1$

A complex number and its *opposite* have a sum of 0. Find the opposite of each complex number.

20. $4 + i$

21. $-5 + 2i$

22. $3 - 6i$

23. $5 + 6i$

24. $-1 - i$

25. $-(4 - 7i)$

26. Plot the complex numbers and their opposites from Exercises 20−25. How are opposites geometrically related in the complex plane?

27. What is the opposite of the complex number $a + bi$?

28. a. Open-ended Problem Choose a complex number and its opposite. Square both numbers and plot the results in the complex plane. What do you notice?

   b. How are the magnitudes of a complex number and its square related?

   c. What do you think is true of the *square roots* of a complex number?

Recall that the *complex conjugate* of *a + bi* is *a − bi*. Plot each number and its complex conjugate in the complex plane. Then plot their sum.

29. $5 + 3i$

30. $6 - 2i$

31. $5i$

32. $-3 - 4i$

33. $-1 + 7i$

34. -8

35. How are complex conjugates geometrically related in the complex plane?

36. Visual Thinking Show that the square of any complex number and the square of its conjugate are also conjugates. Use a complex plane diagram to illustrate.

In Exercise 49 of Section 8.4, you iterated the function $f(z) = z^2 + c$ for a complex value of c, starting with $z = 0$. If you plot the results in the complex plane, they form the *critical orbit* for that number c.

The complex numbers whose critical orbits settle into predictable cycles (visiting the same point or points in the complex plane repeatedly) form the *Mandelbrot set*. Complex numbers whose critical orbits escape to infinity (eventually moving far away from the origin) are *not* in the Mandelbrot set.

37. **Spreadsheets** You can use a spreadsheet to calculate the critical orbit of a complex number c for the function $f(z) = z^2 + c$.

  a. Explain why the cell formulas shown are correct.

  b. Create a spreadsheet like the one shown, and use the fill-down feature to iterate the function 1500 times. Test your spreadsheet by entering the c-values from Exercise 49 of Section 8.4. Do the results agree with your earlier work?

  c. What happens in your spreadsheet when the magnitudes of numbers in a critical orbit become very large? That is, how does the spreadsheet tell you a value of c is *not* in the Mandelbrot set?

Enter these cell formulas, then fill down:
B8 = B7^2 − D7^2 + B4
D8 = 2*B7*D7 + D4

Mandelbrot Set Calculator				
B18		=B17^2−D17^2+B4		

	A	B	C	D	E
1	Mandelbrot Set Calculator: $f(z)=z^2 + c$				
2					
3			c		
4		1	+	1	i
5					
6	Iteration		z		
7	0	0	+	0	i
8	1	1	+	1	i
9	2	1	+	3	i
10	3	−7	+	7	i
11	4	1	+	−97	i
12	5	−9407	+	−193	i
13	6	88454401	+	3631103	i
14	7	7.811E+15	+	6.42E+14	i
15	8	6.0599E+31	+	1E+31	i
16	9	3.5715E+63	+	1.22E+63	i
17	10	1.128E+127	+	8.7E+126	i
18	11	5.169E+253	+	2E+254	i
19	12	#NUM!	+	#NUM!	i

The spreadsheet shows that $1 + i$ is *not* in the Mandelbrot set.

Use your spreadsheet to tell whether each complex number *is* or *is not* in the Mandelbrot set.

38. 0

39. i

40. $-1 + 0.5i$

41. $-1.2 + 0.1i$

42. $-0.1 + 0.8i$

43. $-0.3 + 0.7i$

44. $-0.6 - 0.5i$

45. $-0.5 - 0.6i$

46. **Investigation** You may have noticed that some critical orbits eventually settle into a cycle that repeatedly visits one, two, three, or more points in the complex plane. Examine these cycles.

  a. Look at the orbits of these c-values, which lie in region A: $-0.5 + 0i$, $0.1 + 0.4i$, $-0.6 + 0.2i$. What do you notice?

  b. Look at the orbits of these c-values, which lie in region B: $-1 + 0i$, $-0.9 + 0.1i$, $-1.1 + 0.2i$. What do you notice?

  c. Look at the orbits of these c-values, which lie in region C: $-0.2 + 0.7i$, $-0.05 + 0.75i$. What do you notice?

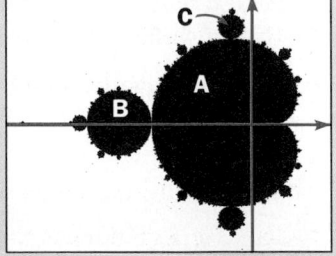

Look back at the graph of the Mandelbrot set on page 365.

  d. **Open-ended Problem** Find a point in the Mandelbrot set that has a critical orbit with a cycle that visits 4 or more points.

47. **Visual Thinking** Examine the critical orbit of any complex number and the critical orbit of its complex conjugate. How do the results help explain the symmetry of the Mandelbrot set?

48. Writing Show that multiplying a complex number $a + bi$ by the complex number i is equivalent to rotating the point that represents $a + bi$ in the complex plane by 90° counterclockwise around the origin. Use an algebraic argument to explain why this is so.

49. Open-ended Problem Plot two complex numbers and their product in the complex plane. Draw line segments connecting the points to the origin. Measure the angles these segments make with the positive real axis. What do you notice? Make a conjecture and test it.

50. SAT/ACT Preview What is the square of $3 + 2i$?

A. $9 + 4i$ **B.** $25i^2$ **C.** 5 **D.** $5 + 12i$ **E.** $13 + 12i$

51. Challenge Prove that the magnitude of the product of two complex numbers is equal to the product of the magnitudes of the two numbers. That is, show that this equation is always true:

$$|(a + bi)(c + di)| = |a + bi| \cdot |c + di|$$

ONGOING ASSESSMENT

52. a. Open-ended Problem Find two complex numbers whose sum is a pure imaginary number.

  b. Open-ended Problem Find two complex numbers whose product is a pure imaginary number.

SPIRAL REVIEW

Use the information in the table to find the mean and standard deviation of each data set. *(Section 6.5)*

53. recycled waste

54. incinerated waste

55. landfilled waste

Write a quadratic equation with the given solutions. *(Section 8.4)*

56. $i, -i$

57. $3i, -3i$

58. $i\sqrt{13}, -i\sqrt{13}$

Solid Waste Management in New England (1991)

State	Recycled (%)	Incinerated (%)	Landfilled (%)
Connecticut	15	65	20
Maine	17	45	38
Massachusetts	29	47	24
New Hampshire	5	23	72
Rhode Island	15	0	85
Vermont	20	8	72

Evaluate each matrix expression. *(Sections 1.3 and 1.4)*

59. $\begin{bmatrix} 1 & 4 \\ 10 & 7 \end{bmatrix} + \begin{bmatrix} 6 & -1 \\ 3 & -2 \end{bmatrix}$

60. $\begin{bmatrix} 6 & -1 \\ 3 & -2 \end{bmatrix} + \begin{bmatrix} 1 & 4 \\ 10 & 7 \end{bmatrix}$

61. $\begin{bmatrix} 1 & 5 \\ 9 & 2 \end{bmatrix}\begin{bmatrix} 6 & 3 \\ 1 & 0 \end{bmatrix}$

62. $\begin{bmatrix} 6 & 3 \\ 1 & 0 \end{bmatrix}\begin{bmatrix} 1 & 5 \\ 9 & 2 \end{bmatrix}$

8.6 | Properties of Number Systems

Learn how to...

- **identify the number systems to which a number belongs**

- **evaluate whether group properties hold for a set and an operation**

So you can...

- **see structural similarities in groups, such as country dancing, puzzle cube rotations, and the multiplication of complex roots of 1**

In Peter Høeg's book *Smilla's Sense of Snow,* Smilla Qaavigaaq Jaspersen compares the development of number systems to human development while her neighbor, a mechanic, prepares dinner:

> ". . . the number system is like human life. First you have the natural numbers. The ones that are whole and positive. The numbers of a small child. But human consciousness expands. The child discovers a sense of longing, and do you know what the mathematical expression is for longing?"
>
> He adds cream and several drops of orange juice to the soup.
>
> "The negative numbers. The formalization of the feeling that you are missing something. And human consciousness expands and grows even more, and the child discovers the in between spaces. Between stones, between pieces of moss on the stones, between people. And between numbers. And do you know what that leads to? It leads to fractions. Whole numbers plus fractions produce rational numbers. And human consciousness doesn't stop there. It wants to go beyond reason. It adds an operation as absurd as the extraction of roots. And produces irrational numbers."
>
> He warms French bread in the oven and fills the pepper mill.
>
> "It's a form of madness. Because the irrational numbers are infinite. They can't be written down. They force human consciousness out beyond the limits. And by adding irrational numbers to rational numbers, you get real numbers."
>
> I've stepped into the middle of the room to have more space. It's rare that you have a chance to explain yourself to a fellow human being. Usually you have to fight for the floor. And this is important to me.
>
> "It doesn't stop. It never stops. Because now, on the spot, we expand the real numbers with imaginary square roots of negative numbers. These are numbers we can't picture, numbers that normal human consciousness cannot comprehend. And when we add the imaginary numbers to the real numbers, we have the complex number system. The first number system in which it's possible to explain satisfactorily the crystal formation of ice. It's like a vast, open landscape. The horizons. You head toward them and they keep receding."

THINK AND COMMUNICATE

Toolbox p. 779
Rational Numbers and Irrational Numbers

1. Why does Smilla call the natural numbers "the numbers of a small child"?

2. How does a mathematician use the word *irrational*? What other meaning of *irrational* does Smilla use?

3. Mathematicians sometimes develop number systems in order to solve equations. For each equation, tell what type of number solves it.

   a. $x + 6 = 2$ **b.** $2x = 3$ **c.** $x^2 = 2$ **d.** $x^2 + x + 1 = 0$

4. Diagram the relationships among the sets of numbers that Smilla mentions.

Smilla Jaspersen's description of the complex number system shows that it includes real and imaginary numbers. The real number system, in turn, includes rational and irrational numbers.

The diagram below is one way to show how various important number systems are "nested." A number in any box belongs to the number systems of all the boxes that surround it.

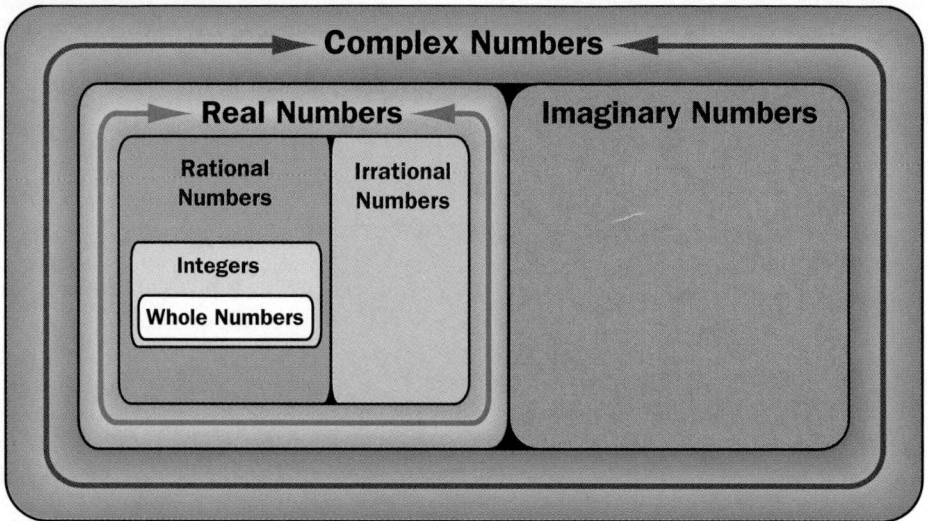

EXAMPLE 1

Identify the number systems to which each number belongs.

a. $3 + 4i$ **b.** $\sqrt{3}$ **c.** -56

SOLUTION

a. imaginary numbers
complex numbers

b. irrational numbers
real numbers
complex numbers

c. integers
rational numbers
real numbers
complex numbers

EXAMPLE 2

If possible, give an example of a number that satisfies each description.

a. integral and real   **b.** integral but not whole   **c.** rational and imaginary

SOLUTION

a. 3   **b.** -5 **c.** not possible

Mathematicians sometimes study number systems by investigating whether certain properties hold when numbers are combined using operations. For example, if an operation that combines two numbers from a number set yields a number that is also in the set, the number set is *closed* under that operation.

A number set and operation form a **group** if the set is closed under the operation and the *identity, inverse,* and *associative properties* hold. If the *commutative property* also holds for a group, it is called a **commutative group**. The group properties are summarized in the tables on this page and the next.

◦◦◦◦ Group Properties of Real Numbers under Addition ◦◦◦◦			
Property	**Definition**	**Example**	**Summary**
Closure	If *a* and *b* are real numbers, then *a* + *b* is a real number.	4 and -12 are real; $4 + (-12) = -8$, which is also real	If you add any two real numbers, the sum is also a real number.
Identity	If *a* is a real number, then *a* + 0 = 0 + *a* = *a*. 0 is the identity for addition.	$5 + 0 = 5$ $0 + \pi = \pi$	If you add 0 to any real number, the result is the number you started with.
Inverse	If *a* is a real number, then there is a real number $-a$ such that *a* + (−*a*) = 0. Note: −*a* is called the *opposite*, or *additive inverse*, of *a*.	$7 + (-7) = 0$ $-\sqrt{2} + \sqrt{2} = 0$	Every real number has an opposite that is also real. When you add a number and its opposite, the sum is 0, the identity for addition.
Associative	If *a*, *b*, and *c* are real numbers, then (*a* + *b*) + *c* = *a* + (*b* + *c*).	$(3 + 7) + 4 = 3 + (7 + 4)$ $\left(1 + \frac{1}{3}\right) + \frac{2}{3} = 1 + \left(\frac{1}{3} + \frac{2}{3}\right)$	When you add three real numbers, you get the same answer whether you add the sum of the first and second to the third or you add the first to the sum of the second and third.
Commutative	If *a* and *b* are real numbers, then *a* + *b* = *b* + *a*.	$3 + 4 = 4 + 3$ $e + 1 = 1 + e$	The order in which you add two real numbers does not change the sum.

EXAMPLE 3

Tell whether each number set is closed under the given operation. If so, tell whether the number set and operation *form a group, form a commutative group,* or *do not form a group.* Explain.

a. odd integers; +   **b.** even integers; +

SOLUTION

a. The set of odd integers is *not closed* under addition because $1 + 1 = 2$, which is not an odd integer. Because the closure property does not hold, the set of odd integers *does not form a group* under the operation of addition.

b. The set of even integers is *closed* under addition, because when you add two even numbers, the result is always even.

   Because the *identity, inverse, associative,* and *commutative properties* all hold as well, the set of even integers *forms a commutative group* under addition.

Group Properties of Real Numbers under Multiplication

Property	Definition	Example	Summary
Closure	If *a* and *b* are real numbers, then *ab* is a real number.	4 and -12 are real; $4(-12) = -48$, which is also real	If you multiply any two real numbers, the product is also a real number.
Identity	If *a* is a real number, then $a \cdot 1 = 1 \cdot a = a$. **1 is the identity for multiplication.**	$5 \cdot 1 = 5$ $1 \cdot \pi = \pi$	If you multiply any real number by 1, the result is the number you started with.
Inverse	If *a* is a nonzero real number, then there is a real number $\frac{1}{a}$ such that $a\left(\frac{1}{a}\right) = 1$. Note: $\frac{1}{a}$ is called the *reciprocal*, or *multiplicative inverse*, of *a*.	$7\left(\frac{1}{7}\right) = 1$ $-\sqrt{2}\left(-\frac{1}{\sqrt{2}}\right) = 1$	Every nonzero real number has a reciprocal. When you multiply a number and its reciprocal, the product is 1, the identity for multiplication.
Associative	If *a*, *b*, and *c* are real numbers, then $(ab)c = a(bc)$.	$(3 \cdot 7) \cdot 4 = 3 \cdot (7 \cdot 4)$ $\left(\frac{1}{2} \cdot \frac{3}{4}\right) \cdot \frac{4}{3} = \frac{1}{2} \cdot \left(\frac{3}{4} \cdot \frac{4}{3}\right)$	When you multiply three real numbers, you get the same answer whether you multiply the product of the first and second by the third or you multiply the first by the product of the second and third.
Commutative	If *a* and *b* are real numbers, then $ab = ba$.	$3 \cdot 4 = 4 \cdot 3$ $2 \cdot e = e \cdot 2$	The order in which you multiply two real numbers does not change the product.

EXAMPLE 4

Tell whether each number set is closed under the given operation. If so, tell whether the number set and operation *form a group, form a commutative group,* or *do not form a group.* Explain.

a. positive integers; $\times$ **b.** positive rational numbers; $\times$

SOLUTION

a. The set of positive integers is *closed* under multiplication, because the product of any two positive integers is a positive integer.

   However, the *inverse property does not hold* for the set of positive integers under multiplication, because the multiplicative inverse of 3, for example, is $\frac{1}{3}$, which is not an integer. Because the inverse property does not hold, the set of positive integers *does not form a group* under multiplication.

b. The set of positive rational numbers is *closed* under multiplication, because the product of any two positive rational numbers is a positive rational number.

   All the other group properties also hold, so the set of positive rational numbers *forms a commutative group* under multiplication.

5. Does the set of numbers -1, 0, and 1 form a commutative group under addition? under multiplication? Explain.

6. Does the set of integers form a commutative group under addition? under multiplication? Explain.

☑ CHECKING KEY CONCEPTS

Identify the number systems to which each number belongs.

1. 0

2. $-\sqrt{16}$

3. $5i$

If possible, give an example of a number that satisfies each description.

4. rational and even

5. negative and imaginary

6. irrational and odd

7. real and complex

Tell whether each number set is closed under the given operation. If so, tell whether the number set and operation *form a group, form a commutative group*, or *do not form a group*. Explain.

8. multiples of 3; $+$

9. multiples of 3; $\times$

10. -1, 1; $\times$

11. -2, 0, 2; $+$

8.6 Exercises and Applications

Extra Practice exercises on page 762

Identify the number systems to which each number belongs.

1. $5 - i\sqrt{2}$

2. $\dfrac{4}{7}$

3. $-\pi$

If possible, give an example of a number that satisfies each description.

4. rational and odd

5. real and imaginary

6. irrational and even

7. whole and complex

Tell whether each number set is closed under the given operation. If so, tell whether the number set and operation *form a group, form a commutative group*, or *do not form a group*. Explain.

8. negative integers; $+$

9. negative integers; $\times$

10. rational numbers; $+$

11. rational numbers; $\times$

12. irrational numbers; $+$

13. irrational numbers; $\times$

14. complex numbers; $+$

15. complex numbers; $\times$

16. multiples of 7; $+$

17. multiples of 7; $\times$

18. -10, -5, 0, 5, 10; $+$

19. -10, -5, 0, 5, 10; $\times$

20. Writing A *subset* is a set whose members are in another set. If a group property holds for a set of numbers, does it hold for any subset of those numbers? Use the results of Exercises 8–19 to explain your answer.

21. Consider the set of 2 × 2 matrices whose elements are real numbers.

    a. What group properties hold for this set under matrix addition?

    b. What group properties hold for this set under matrix multiplication?

22. Consider the operation of exponentiation using the set of real numbers.

    a. Is exponentiation commutative? That is, for all real values of a and b (not both zero), is $a^b = b^a$? Explain.

    b. Is exponentiation associative? That is, for all real values of a, b, and c (not all zero), is $\left(a^b\right)^c = a^{(b^c)}$? Explain.

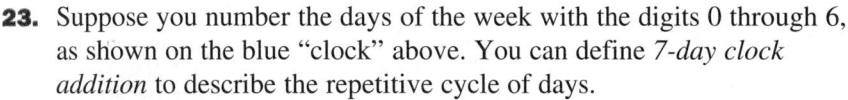

23. Suppose you number the days of the week with the digits 0 through 6, as shown on the blue "clock" above. You can define *7-day clock addition* to describe the repetitive cycle of days.

    a. Suppose today is Friday (day 5). What day will it be six days from now? In 7-day clock addition, $5 + 6 = \underline{\ ?\ }$.

    b. Copy and complete the addition table for the 7-day clock. What group properties hold for this table?

+	0	1	2	3	4	5	6
0	0	1	2	3	4	5	6
1	1	2	3	4	5	6	0
2	2	3	4	5	6	0	1
3	?	?	?	?	?	?	?
4	?	?	?	?	?	?	?
5	?	?	?	?	?	?	?
6	?	?	?	?	?	?	?

For each table, tell whether the given group property holds. If the group property does not hold, give a counterexample.

24. closure

+	0	2	4	6
0	0	2	4	6
2	2	4	6	8
4	4	6	8	0
6	6	8	0	2

25. identity

†	α	β	γ	δ
α	β	γ	δ	α
β	δ	α	β	γ
γ	γ	δ	α	β
δ	α	β	γ	δ

26. commutativity

*	W	X	Y	Z
W	W	X	Z	Y
X	X	Z	Y	W
Y	Z	Y	W	X
Z	Y	W	X	Z

Copy and complete each table so that it shows the group properties of closure, identity, inverse, and commutativity.

27.

×	A	B	C	D
A	D	C	B	A
B	C	?	A	?
C	B	A	?	?
D	A	?	?	?

28.

+	0	1	2	3
0	0	1	2	3
1	?	2	?	0
2	?	3	0	1
3	?	?	?	2

29.

@	□	●	▽	▲
□	?	?	●	?
●	□	●	▽	▲
▽	●	?	?	?
▲	?	?	?	●

30. Name the identity for each table in Exercises 27–29.

31. Use your answers to Exercises 27–29. How is symmetry across the diagonal of a table related to whether the operation is commutative?

32. Open-ended Problem Do you think the associative property holds for the tables you completed in Exercises 27–29? Give examples or counterexamples to support your conjecture.

In Western square dancing, the pattern known as "right and left grand" is performed by four couples as shown below.

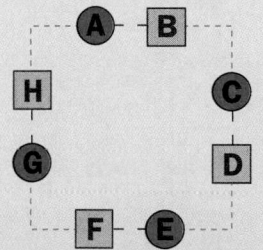

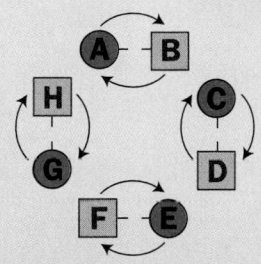

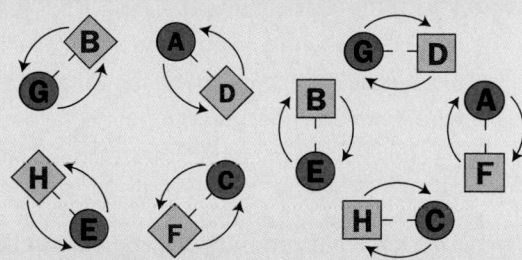

Female dancers A, C, E, and G face male dancers B, D, F, and H.

Partners join right hands and pull by each other to switch positions.

Dancers continue along the circle, giving a left hand to the next person, a right hand to the next, a left hand to the next, and so on.

33. a. Have a group of eight students demonstrate the "right and left grand" pattern twice. Imagine viewing the dancers from above. Which dancers move clockwise? counterclockwise?

   b. How many dancers does dancer A pass before she meets her partner again? before she returns to her original place?

The original order of dancers is ABCDEFGH.
After 1 move the order is BADCFEHG.
After 2 moves the order is GDAFCHEB.

⊕	0	1	2	3	4	5	6	7
ABCDEFGH 0	0	1	2	3	4	5	6	7
BADCFEHG 1	1	2	3	4	5	6	7	0
GDAFCHEB 2	2	3	4	5	6	7	0	1
3								
4								
5								
6								
7								

34. Copy and complete the left column of the table, showing how the order changes as dancers move around the circle.

35. Let 0–7 correspond to the 8 orders of dancers you listed in Exercise 34. Using the digits 0–7 for the orders, complete the table.

36. What group properties does this addition table have?

A *field* consists of a set of elements and two operations (such as × and +) that satisfy all the commutative group properties as well as the distributive property:

$$a \times (b + c) = (a \times b) + (a \times c)$$

37. Tell whether each set of numbers forms a field under the operations of standard multiplication and addition.

   a. integers   **b.** rational numbers   **c.** real numbers

38. Do the numbers 0 and 1, and the operations of multiplication and addition, as defined in the tables at the right, form a field? Explain.

×	0	1
0	0	0
1	0	1

+	0	1
0	0	1
1	1	0

If you start with an unmixed puzzle cube and restrict your moves to turning the middle layers 180°, you will always get one of the 8 cube patterns shown.

If you denote 180° turns of the middle layers parallel to the top, right, and front faces of the cube by T, R, and F, then the 8 cube patterns can be named I, T, R, F, TR, TF, RF, and TRF. (The letter I, for identity, names the unmixed cube.)

Use a puzzle cube to complete the following exercises. If you don't have a cube, you can follow the paths that connect cubes in the diagram to see the result of any combination of moves.

Name the cube pattern you get if you start at I and follow each sequence of moves.

39. F, R

40. R, F

41. T, R, F

42. T, R, F, T, F

43. T, R, F, T, R, F

44. T, T

45. T, R, T, F

46. T, T, R, T, F, F, R, R, R, R, T, T, F

47. Writing Based on your answers to Exercises 39–46, describe how you can name the cube pattern you get after *any* sequence of T's, R's, and F's.

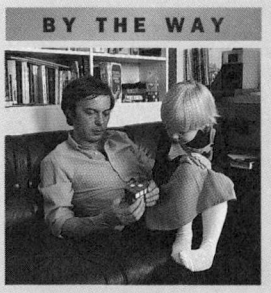
You can think of I, T, R, F, TR, TF, RF, and TRF as naming the 8 cube patterns or as naming sequences of moves. *Composition* (∘) is the operation that combines one sequence of moves with another sequence of moves. For example, T ∘ TR = R because if you start at I and move T followed by T and R, you end up at the cube pattern R.

48. Copy and complete the table.

49. Is the set of 8 elements closed under the operation of composition? Explain.

50. Is there an identity element? If so, what is it?

51. Which elements in the set have an inverse element? What do you notice about their inverses?

∘	I	T	R	F	TR	TF	RF	TRF
I	I	T	R	F	TR	TF	RF	TRF
T	T	I	TR	TF	R	F	TRF	RF
R	R	TR	I	RF	?	?	?	?
F	F	TF	RF	I	?	?	?	?
TR	?	?	?	?	?	?	?	?
TF	?	?	?	?	?	?	?	?
RF	?	?	?	?	?	?	?	?
TRF	?	?	?	?	?	?	?	?

52. Does the associative property hold? Give examples or counterexamples to support your conjecture.

53. Does the commutative property hold? Give examples or counterexamples to support your answer.

54. Based on your answers to Exercises 49–53, do the 8 cube patterns form a group under the operation of composition?

55. There are four solutions to the equation $x^4 = 1$ in the complex number system. The solutions are 1, i, -1, and $-i$.

×	1	i	−1	−i
1	1	?	?	?
i	?	−1	?	?
−1	?	?	?	?
−i	?	?	?	?

a. Copy and complete the table showing the multiplication of these four complex numbers.

b. Do the numbers 1, i, -1, and $-i$ form a commutative group under multiplication? Explain.

Plot each pair of complex numbers and their product in the complex plane. *(Section 8.5)*

56. $2 + i$, $2 + 2i$ **57.** $-3 + i$, $-2 - i$ **58.** $5i$, $-6i$

Write each quadratic function as a product of factors. *(Section 5.6)*

59. $y = 2x^2 + x - 15$ **60.** $y = 16x^2 - 4x - 12$ **61.** $y = 4x^2 - 49$

Simplify. *(Toolbox, page 781)*

62. $(4x + 5) + (6x - 9)$ **63.** $(9x - 7) - (4x + 8)$ **64.** $(x + 2) - (7 - x)$

ASSESS YOUR PROGRESS

VOCABULARY

real number (p. 358)	**magnitude** (p. 366)
imaginary number (p. 359)	**group** (p. 373)
pure imaginary number (p. 359)	**commutative group** (p. 373)
complex number (p. 359)	**closure property** (pp. 373, 374)
complex conjugate (p. 361)	**identity property** (pp. 373, 374)
complex plane (p. 365)	**inverse property** (pp. 373, 374)
imaginary axis (p. 365)	**associative property** (pp. 373, 374)
real axis (p. 365)	**commutative property** (pp. 373, 374)

Perform the indicated operation. *(Section 8.4)*

1. $(4 + 2i) - (5 - 3i)$ **2.** $(7 + 3i)(7 - 3i)$ **3.** $(8 - 5i)^2$

4. Plot two numbers that are complex conjugates in the complex plane. Find their sum and their product and plot them as well. *(Section 8.5)*

5. Find the magnitude of $-12 + 3i$.

Tell whether each number set is closed under the given operation. If so, tell whether the number set and operation *form a group, form a commutative group*, or *do not form a group*. Explain. *(Section 8.6)*

6. negative real numbers; $+$ **7.** real numbers a where $0 \le a \le 1$; $\times$

8. Journal Compare the operations of addition, subtraction, multiplication, and division on real numbers and complex numbers.

Walk This Way

Try walking across a flat, open space at an increasing speed. You'll notice that at some point you'll feel the urge to switch to a run. Your body recognizes when it's more efficient to run than to walk.

Now think about a small child trying to keep up with an adult who's walking briskly. With shorter legs than the adult, the child often has to run. Obviously, how fast someone can walk (without breaking into a run) has something to do with leg length.

PROJECT GOAL Examine the relationship between leg length and walking speed.

Doing an Experiment

Work with a partner to design and carry out an experiment in which you measure the leg length, in inches, of ten subjects, and then time the subjects as they walk. Be sure to choose subjects with a variety of leg lengths. You will need a tape measure and a stopwatch. Here are some guidelines for doing your experiment.

1. MEASURE each subject's leg length from hip to heel.

2. CHOOSE a straight, flat, 60 ft course that is free of obstacles.

3. INSTRUCT each subject to walk the 60 ft course as fast as he or she can. With each step, the walker must be sure that the rear foot does not leave the ground before the forward foot lands.

4. TIME each subject as he or she walks the course. Divide the distance traveled (60 ft) by each subject's time to calculate his or her speed in feet per second.

Using a Model

A person's maximum walking speed s, in feet per second, is given by the following model:

$$s = \frac{\sqrt{gl}}{12}$$

where g, the acceleration due to gravity, is 384 in./s^2, and l is leg length in inches.

• **CALCULATE** the predicted maximum speed for each of your subjects according to the model. You may find a spreadsheet helpful.

	A	B	C	D	E
	Name	Leg length (in.)	Time (s)	Actual speed (ft/s)	Predicted speed (ft/s)
2	Lester	40	5.9	10.2	10.33
3	Cindy	34	6.4	9.38	9.52
4	Gena	31	6.7	8.96	9.09

Walking Speed Data

• **PLOT** the predicted speeds on the same graph as the actual values. Use different colors or symbols for each data set.

• **COMPARE** the actual values with the predicted values. Are there differences? What might explain them?

5. **ORGANIZE** your data in a table, and then make a scatter plot. Put leg length on the horizontal axis and walking speed on the vertical axis.

Writing a Report

Write a report summarizing your results. It should include all your data as well as a comparison of the data with the predictions of the model. You can also extend your project to explore one of these ideas:

• The acceleration due to gravity varies from planet to planet. Suppose you were walking on Mars or on the moon. How fast could you walk? How does gravity affect walking speed?

• The model above is based on the assumption that the *centripetal force* acting on a walker's hips cannot exceed the walker's weight.

Centripetal force is given by $\frac{ms^2}{l}$, and weight is given by mg, where m is mass. Derive the model from this information. (You'll need to take units of measurement into account.)

Self-Assessment

In your report, describe any difficulties you had gathering data for this project. Did you feel that your data were in agreement with the model? If not, what might have gone wrong?

Review

Look through the chapter and identify your trouble spots: questions that you had difficulty answering or solved incorrectly on homework assignments. Review the material covering these topics, and then try to answer the questions again. If you're still having difficulty, seek the help of another student or your teacher.

VOCABULARY

radical functions (p. 344)
real numbers (p. 358)
imaginary number (p. 359)
pure imaginary number (p. 359)
complex numbers (p. 359)
complex conjugates (p. 361)
complex plane (p. 365)
imaginary axis (p. 365)
real axis (p. 365)

magnitude (p. 366)
group (p. 373)
commutative group (p. 373)
closure property (pp. 373, 374)
identity property (pp. 373, 374)
inverse property (pp. 373, 374)
associative property (pp. 373, 374)
commutative property (pp. 373, 374)

SECTIONS 8.1, 8.2, *and* 8.3

Radical functions of the form $y = \sqrt[n]{x}$ have a domain and range of all real numbers when n is odd, and a domain and range of nonnegative numbers when n is even.

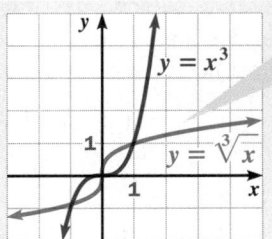

$y = \sqrt[3]{x}$ is the inverse of $y = x^3$.

You can solve radical equations by graphing or by using algebra. For example, you can use the equation $t = \sqrt{\dfrac{h}{4}}$, which gives the time (in seconds) of a jump that reaches a height of h ft, to find the height of a jump that lasts 0.95 s.

Graph $y = 0.95$ and $y = \sqrt{\dfrac{x}{4}}$ in the same coordinate plane.

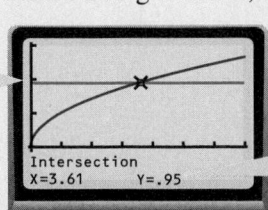

Intersection
X=3.61   Y=.95

A jump that lasts 0.95 s reaches a height of about 3.6 ft.

$$0.95 = \sqrt{\dfrac{h}{4}}$$

$$(0.95)^2 = \dfrac{h}{4}$$

$$4(0.95)^2 = h$$

$$3.6 \approx h$$

SECTIONS | 8.4 *and* 8.5

The letter i is used for the square root of -1, the fundamental unit in the system of **imaginary numbers**: $i = \sqrt{-1}$.

A **pure imaginary number** has the form bi where $b \neq 0$.

Examples:

$$\sqrt{-4} = 2i$$
$$5 + \sqrt{-9} = 5 + 3i$$

Any number of the form $a + bi$ where a and b are real numbers and $b \neq 0$ is an **imaginary number**.

The real numbers and imaginary numbers together form the set of **complex numbers**, which can be graphed in a **complex plane**.

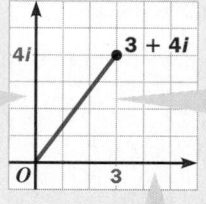

imaginary axis

real axis

The **magnitude** of a complex number is its distance from $(0, 0)$. The magnitude of $3 + 4i$ is 5.

Add or subtract complex numbers by adding or subtracting their real parts and their imaginary parts.

Multiply complex numbers as you would multiply binomials.

Divide complex numbers using a **complex conjugate** to get an answer in $a + bi$ form.

Operation	Example
Addition or subtraction	$(7 + 5i) - (4 + 2i) = (7 - 4) + (5i - 2i)$ $= 3 + 3i$
Multiplication	$(2 + 3i)(5 + 2i) = (2 \cdot 5) + (2 \cdot 2i) + (3i \cdot 5) + (3i \cdot 2i)$ $= 10 + 4i + 15i + (-6) = 4 + 19i$
Division	$\dfrac{2 - i}{1 + 3i} = \dfrac{2 - i}{1 + 3i} \cdot \dfrac{1 - 3i}{1 - 3i}$ $= \dfrac{2 - 6i - i + 3i^2}{10} = \dfrac{2 - 7i - 3}{10} = -\dfrac{1}{10} - \dfrac{7}{10}i$

• • • • **Examples of Operations with Complex Numbers** • • • •

SECTION | 8.6

The diagram shows the relationships among the various number systems. A number belongs to all the number systems linked to its left.

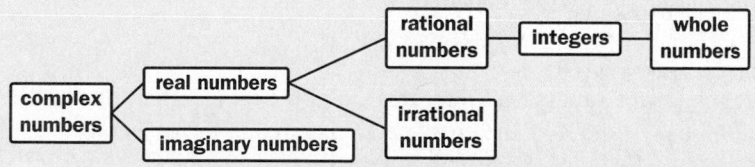

The properties of **closure**, **identity**, **inverse**, **associativity**, and **commutativity** are **group properties**. These properties apply to addition and multiplication on the set of real numbers (see pages 373 and 374).

Assessment

VOCABULARY QUESTIONS

For Questions 1–3, complete each statement.

1. The inverse of a power function is a(n) __?__.

2. A complex number can be graphed in the __?__. The distance of a complex number from the origin is called its __?__.

3. The __?__ property of addition says that the sum of any two real numbers is another real number.

SECTIONS 8.1, 8.2, and 8.3

State the domain and range of each function.

4. $y = \sqrt{x} - 4$

5. $y = 1 + \sqrt{x}$

6. $y = \sqrt{x - 2} + 3$

7. $y = \sqrt[3]{x - 3}$

8. $y = 1 + \sqrt[4]{x + 5}$

9. $y = \sqrt[6]{x} + 2$

Express using fractional exponents.

10. $\sqrt[9]{z}$

11. $\sqrt[5]{y^3}$

12. $\sqrt[4]{y^7}$

Express using radical notation.

13. $t^{4/3}$

14. $b^{3/8}$

15. $p^{5/7}$

Solve.

16. $\sqrt{4x - 7} = 3$

17. $\sqrt[3]{2x + 1} = 3$

18. $\sqrt[4]{x - 1} = 1$

19. **Writing** Show the steps you would use to solve $\sqrt[n]{ax + b} = c$, where a, b, and c are nonnegative real numbers. What is an expression for x in terms of a, b, and c?

A falling object reaches **terminal velocity** when the force of gravity is balanced by the drag force of the medium through which it falls.

20. **PHYSICS** The *terminal velocity* v (in meters per second) of a spherical object falling through the air can be estimated by using the equation

$$v = \sqrt{\frac{2mg}{0.6A}}$$

where m = the object's mass (in kilograms), g = the acceleration due to gravity (9.8 m/s^2), and A = the cross-sectional area of the object (in square meters). Estimate the terminal velocity of each object.

a. baseball: $m = 0.145$ kg, $A = 0.0042$ m^2

b. golf ball: $m = 0.046$ kg, $A = 0.0014$ m^2

c. hailstone: $m = 0.00048$ kg, $A = 0.000079$ m^2

SECTIONS 8.4 *and* 8.5

Perform the indicated operation.

21. $(6 + 2i) + (-3 - 4i)$ **22.** $(7 - 6i) + (2 + 3i)$ **23.** $-3i(5i)$

24. $(4 - 5i) - (-8 + 4i)$ **25.** $(3 - i)(-2 + 4i)$ **26.** $\dfrac{8 - i}{5 + 2i}$

Plot each number in the complex plane and find its magnitude. Give two other numbers that have the same magnitude.

27. $1 + 4i$ **28.** $7 + 3i$ **29.** $4 + 4i$

30. $2 - i$ **31.** $5 + 2i$ **32.** $-3 + i$

Plot each pair of numbers and their product in the complex plane.

33. $1 + 4i, 2 - i$ **34.** $7 + 3i, 5 + 2i$ **35.** $-3 + i, 4 + 4i$

SECTION 8.6

Tell whether each number set is closed under the given operation. If so, tell whether the number set and operation *form a group*, *form a commutative group*, or *do not form a group*. Explain.

36. multiples of 5; + **37.** $-1, 0, 1$; + **38.** $2, 1, \dfrac{1}{2}$; $\times$

For each table, tell whether the given group property holds. If the group property does not hold, give a counterexample.

39. commutativity

✝	#	@	*	%
#	#	@	%	*
@	@	%	*	#
*****	%	*	#	&
%	*	#	&	%

40. identity

Δ	A	B	C	D
A	A	D	C	B
B	C	B	A	D
C	D	C	B	A
D	B	A	D	C

41. closure

×	3	5	7	9
3	9	5	1	7
5	5	5	5	5
7	1	5	9	3
9	7	5	3	1

PERFORMANCE TASK

42. a. Draw a three-dimensional coordinate system with two horizontal x-axes, one real and the other imaginary, and one vertical y-axis.

b. Graph $y = x^2 + 1$ using real values of x as the domain. Then graph $y = x^2 + 1$ using only pure imaginary values of x as the domain.

c. Explain how you can use your graph to solve equations like $x^2 + 1 = 0$, $x^2 + 1 = 5$, and $x^2 + 1 = -8$.

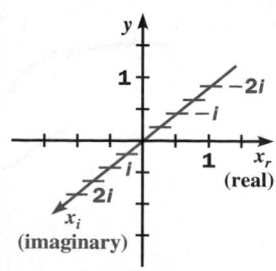

9 Polynomial and Rational Functions

Stellar mysteries in the data

INTERVIEW **Vera Rubin**

As a child, Vera Rubin used to lie awake studying the stars outside her bedroom window. She picked out the constellations and the planets, and saw the night sky as a "great mystery waiting to be solved." She decided to become an astronomer, even though, she says, "teachers in those days just didn't know what to do with girls who were interested in science." Rubin was not discouraged, however, and went on to make an important discovery about the universe.

> "Sometimes all it takes to solve a puzzle is to look at it through new eyes."

Vera Rubin holds a picture of herself as a young woman using her college telescope.

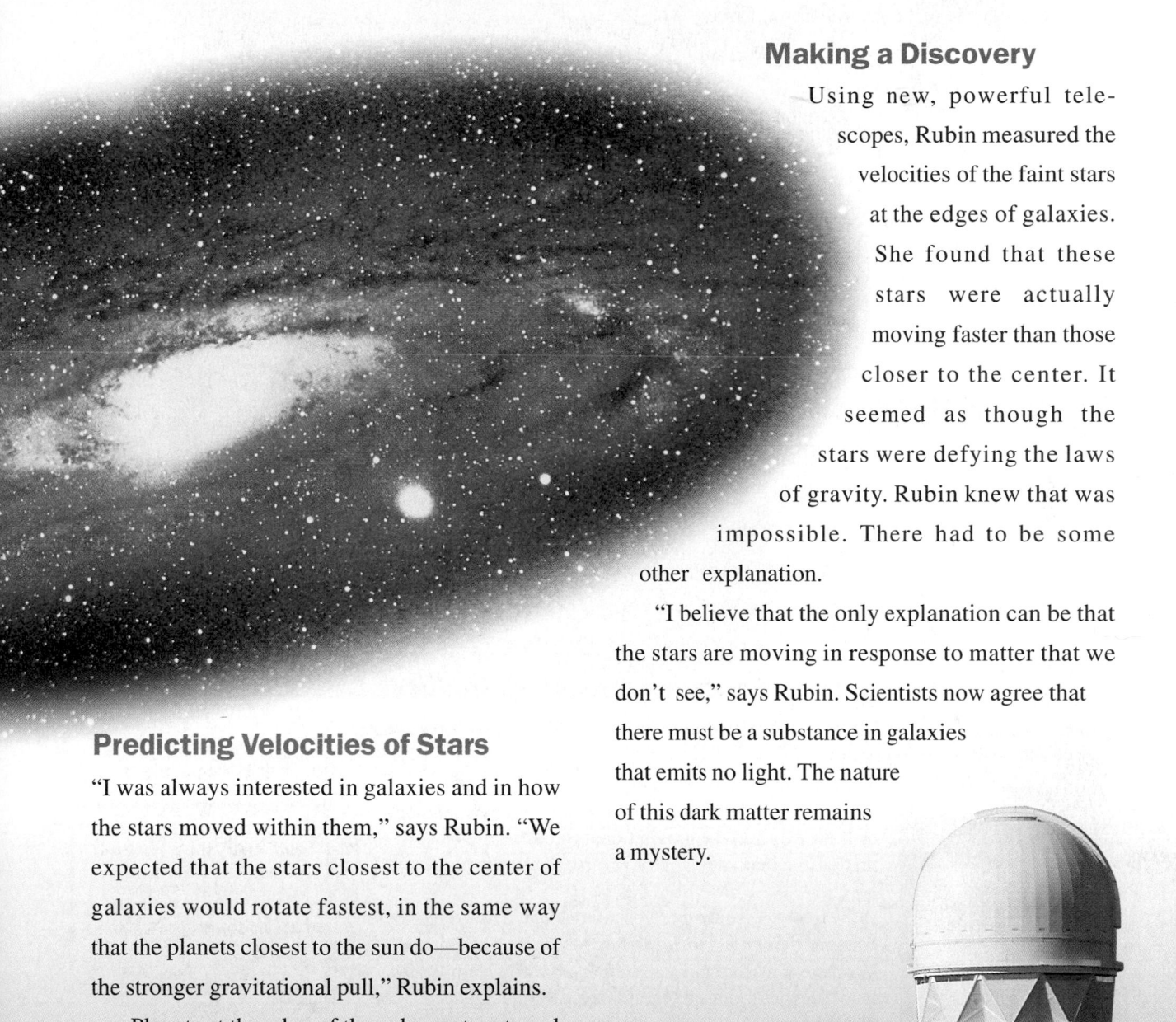

Making a Discovery

Using new, powerful telescopes, Rubin measured the velocities of the faint stars at the edges of galaxies. She found that these stars were actually moving faster than those closer to the center. It seemed as though the stars were defying the laws of gravity. Rubin knew that was impossible. There had to be some other explanation.

"I believe that the only explanation can be that the stars are moving in response to matter that we don't see," says Rubin. Scientists now agree that there must be a substance in galaxies that emits no light. The nature of this dark matter remains a mystery.

Predicting Velocities of Stars

"I was always interested in galaxies and in how the stars moved within them," says Rubin. "We expected that the stars closest to the center of galaxies would rotate fastest, in the same way that the planets closest to the sun do—because of the stronger gravitational pull," Rubin explains.

Planets at the edge of the solar system travel much more slowly, because the gravitational pull of the sun is weaker. Astronomers expected the stars far from the center of a galaxy to travel slower because the gravitational pull, which comes from matter farther away, is weaker.

Kitt Peak National Observatory

Gravitational Force

Central to Rubin's work is the idea that any two objects exert a gravitational force on each other. This force, typically measured in newtons (N), changes with the distance between the objects.

Consider, for example, the probe *Voyager 2*, launched by NASA in the late 1970s to explore the outer planets of our solar system. Although the gravitational force between the probe and Earth was 9143 N at Earth's surface, the force decreased as the probe moved away from Earth. As the graph shows, gravitational force is a function of distance. This relationship is an example of a *rational function*.

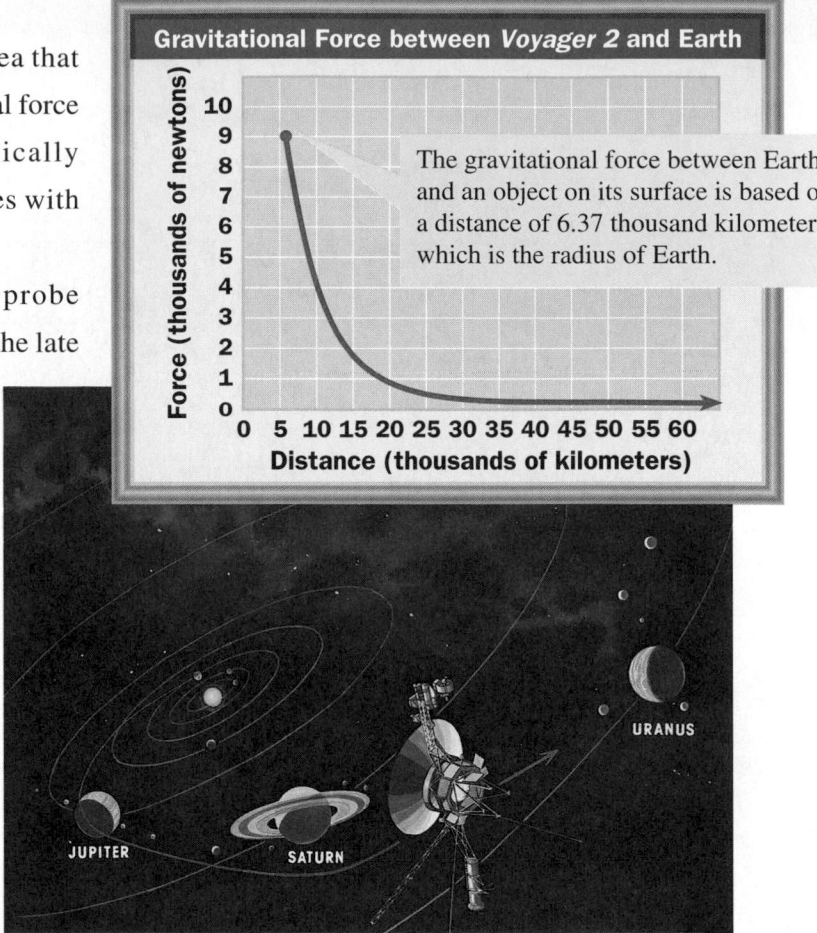

Gravitational Force between *Voyager 2* and Earth

The gravitational force between Earth and an object on its surface is based on a distance of 6.37 thousand kilometers, which is the radius of Earth.

JUPITER   SATURN   URANUS

EXPLORE AND CONNECT

Vera Rubin accepts the National Medal of Science from President Clinton.

1. Research Use an encyclopedia to find out if astronomers expect the universe to continue expanding, or if they expect the gravitational forces acting on matter to eventually stop or reverse the expansion.

2. Writing Use the graph shown above to describe the change in force as *Voyager 2* moves from 10 thousand kilometers to 20 thousand kilometers away from Earth. By what factor does the gravitational force between them decrease?

3. Project According to Einstein's theory of general relativity, gravity exists because objects "curve" the space around them. Do this activity to model the curvature of space by an object:

Hold a flat bed sheet above the ground and have a person pull tightly at each corner of the sheet. Roll a tennis ball across the sheet and observe the ball's movement. Next, place a basketball on the sheet's center and roll the tennis ball across one more time. What happens to the movement of the tennis ball?

Mathematics
& Vera Rubin

In this chapter, you will learn more about how mathematics is related to astronomy.

Related Exercises

Section 9.6
• **Exercises 13 and 14**

Section 9.8
• **Exercises 5 and 6**

9.1 | **Polynomials**

Many early civilizations used drawings of everyday objects to represent numbers. Some of the symbols used by the Egyptian and Aztec peoples are shown below. Notice that the Egyptian numeration system was based on powers of 10, while the Aztec system was based on powers of 20.

Learn how to...

- **recognize, evaluate, add, and subtract polynomials**

So you can...

- **understand numeration systems of ancient cultures, for example**

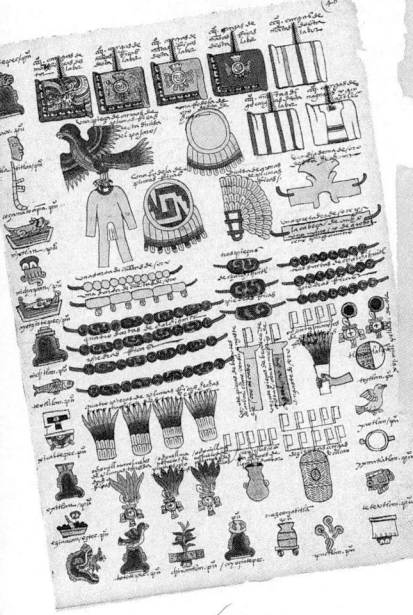

This page is from an Aztec *codex*, an illustrated history of their culture.

EGYPTIAN SYMBOLS

Lotus flower	Coiled rope	Arch	Stroke
$= 10^3$	$= 10^2$	$= 10$	$= 1$

This Egyptian painting is from the tomb of Tashep-Khonsu.

AZTEC SYMBOLS

Maize doll	Maize plant	Flag	Maize seed
$= 20^3$	$= 20^2$	$= 20$	$= 1$

You can use the information above to find the decimal form of the numbers represented by these groups of symbols:

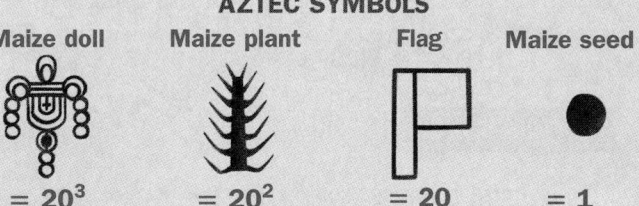

$$2(10^3) \quad + \quad 5(10^2) \quad + \quad 4(10) \quad + \quad 6 \quad = \quad 2{,}546$$

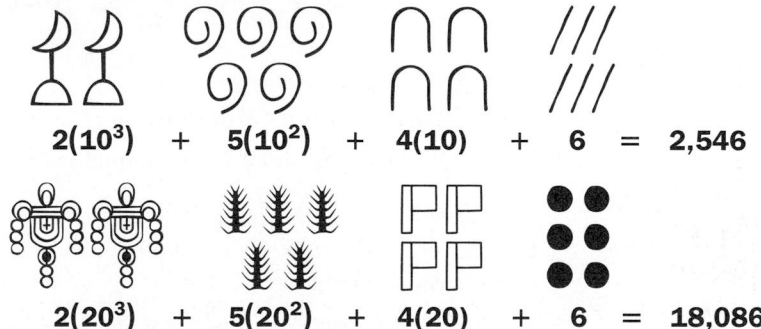

$$2(20^3) \quad + \quad 5(20^2) \quad + \quad 4(20) \quad + \quad 6 \quad = \quad 18{,}086$$

THINK AND COMMUNICATE

1. Compare the Egyptian numeration system with the decimal system used in the United States today. How are the systems alike? How are they different?

2. What is the maximum number of maize seeds that can be used in the Aztec representation of a number? Explain.

The expressions

$$2(10^3) + 5(10^2) + 4(10) + 6$$

and

$$2(20^3) + 5(20^2) + 4(20) + 6$$

on the previous page are similar in form to the *polynomial* below:

A **polynomial** is a term or a sum of terms like $2x^3$. Each term is the product of a real-number coefficient and a variable with a whole-number exponent.

The exponent of this variable is 1, since $x^1 = x$.

$$2x^3 + 5x^2 + 4x + 6$$

The **degree** of a polynomial is the greatest exponent. This polynomial has degree 3.

This term is called the **constant term** of the polynomial. Note that 6 is the same as $6x^0$, since $x^0 = 1$.

A polynomial whose exponents decrease from left to right is said to be in **standard form**. For example, the polynomial shown above is in standard form.

EXAMPLE 1

Tell whether each expression is a polynomial. If so, write the polynomial in standard form and state its degree. If not, explain why not.

a. $x^2 - 8x^3 - 2 + 3x^4 + 7x$ **b.** $4x^{-3} + 9x + x^{1/2} + 10$

c. $7r^4 - 2r^3 + 5r^2 + \dfrac{1}{r+1}$ **d.** $-6.4m + \dfrac{2}{7}m^3 - \pi m^5 + \sqrt{3}\,m^2$

SOLUTION

a. The expression *is* a polynomial.

The standard form of the polynomial is $3x^4 - 8x^3 + x^2 + 7x - 2$.

The degree of the polynomial is 4.

b. The expression *is not* a polynomial, because the exponents -3 and $\dfrac{1}{2}$ are not whole numbers.

c. The expression *is not* a polynomial, because $\dfrac{1}{r+1}$ is not the product of a real number and a whole-number power of r.

d. The expression *is* a polynomial.

The standard form of the polynomial is $-\pi m^5 + \dfrac{2}{7}m^3 + \sqrt{3}\,m^2 - 6.4m$.

The degree of the polynomial is 5.

Synthetic Substitution

Look at the group of Aztec symbols shown.

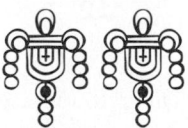  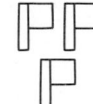 

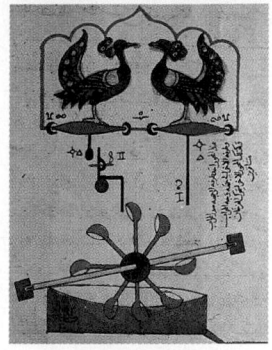
You can find the decimal form of the number these symbols represent by evaluating the polynomial

$$2x^3 + x^2 + 3x + 7$$

when $x = 20$. One way to do this is simply to substitute 20 for x in the polynomial. However, it is more efficient to first rewrite the polynomial this way:

$$2x^3 + x^2 + 3x + 7 = (2x^2 + x + 3)x + 7$$
$$= ((2x + 1)x + 3)x + 7$$

To evaluate the polynomial for any value of x, follow these steps:

$$((2x + 1)x + 3)x + 7$$

Step 1 Multiply 2 by x.
Step 2 Add 1.
Step 3 Multiply by x.
Step 4 Add 3.
Step 5 Multiply by x.
Step 6 Add 7.

Steps 1–6 illustrate a procedure called **synthetic substitution**. The next example shows an easy way to perform this procedure.

EXAMPLE 2

Evaluate $2x^3 + x^2 + 3x + 7$ when $x = 20$.

SOLUTION

Write the polynomial's coefficients in a row. Perform Steps 1–6 above.

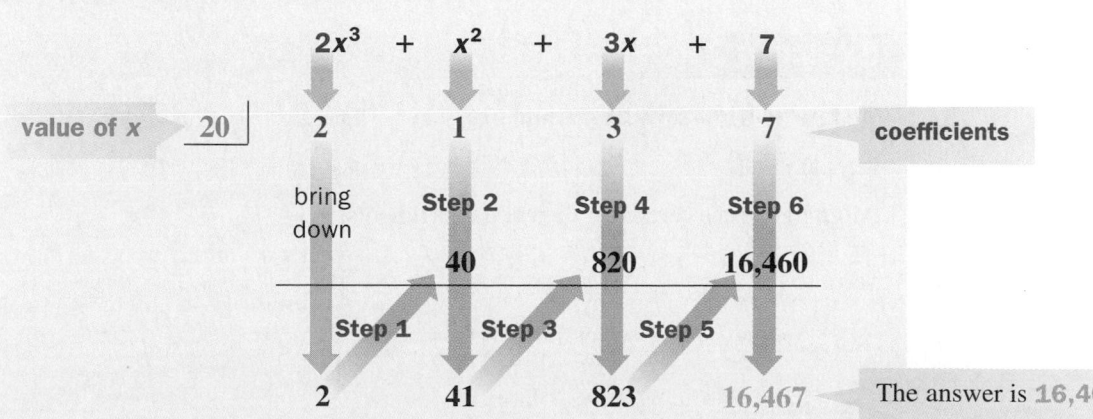

The answer is 16,467.

When writing the row of coefficients in synthetic substitution, you must include a zero coefficient for each "missing" term of the polynomial. For example, you would write these coefficients for the polynomial $4x^5 - 7x^3 - x^2 + 12$:

$$4 \qquad 0 \qquad -7 \qquad -1 \qquad 0 \qquad 12$$

This zero represents the missing x^4-term.

This zero represents the missing x-term.

THINK AND COMMUNICATE

3. Use synthetic substitution to evaluate $4x^5 - 7x^3 - x^2 + 12$ when $x = -2$. Would you get the same answer if you left out the zeros?

Adding and Subtracting Polynomials

Look at the two Egyptian numbers shown. You can find the decimal form of their sum by adding coefficients of terms containing the same power of 10.

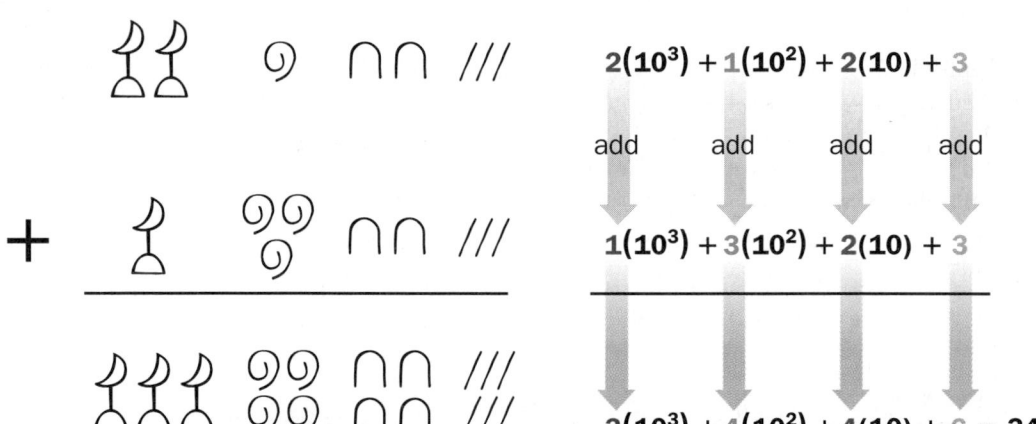

$$2(10^3) + 1(10^2) + 2(10) + 3$$

add   add   add   add

$$1(10^3) + 3(10^2) + 2(10) + 3$$

$$3(10^3) + 4(10^2) + 4(10) + 6 = 3446$$

To add two polynomials, you add coefficients of terms containing the same power of the variable, such as $2x^3$ and $5x^3$. Such terms are called **like terms**.

EXAMPLE 3

Add $3x^4 + 8x^3 - x^2 - 2x + 6$ and $9x^4 - 4x^2 + 2x - 1$.

SOLUTION

Align like terms vertically, then add coefficients.

$$
\begin{array}{r}
3x^4 + 8x^3 - x^2 - 2x + 6 \\
9x^4 \qquad\;\; - 4x^2 + 2x - 1 \\
\hline
12x^4 + 8x^3 - 5x^2 + 0x + 5
\end{array}
$$

The sum is $12x^4 + 8x^3 - 5x^2 + 5$.

Subtracting polynomials is similar to subtracting real numbers. To subtract a real number, you add the opposite of the number. To subtract a polynomial, you add the opposite of each term of the polynomial.

EXAMPLE 4

Subtract $-3x^2 - 5x - 7 + 6x^3$ from $7x^3 - 5x^2 + x - 4$.

SOLUTION

Align like terms vertically.

$$7x^3 - 5x^2 + \ x - 4$$
$$-(6x^3 - 3x^2 - 5x - 7)$$

Write the **opposite** of each term, then add.

$$7x^3 - 5x^2 + \ x - 4$$
$$-6x^3 + 3x^2 + 5x + 7$$
$$\overline{x^3 - 2x^2 + 6x + 3}$$

The difference is $x^3 - 2x^2 + 6x + 3$.

☑ CHECKING KEY CONCEPTS

Tell whether each expression is a polynomial. If so, write the polynomial in standard form and state its degree. If not, explain why not.

1. $1 + x^{-1} + x^{-2} + x^{-3}$ **2.** $5 + 2x^4 - 4x^3 + x - 9x^2$

3. $-1.6y^2 - 3.8y^4 + 0.22y - 0.94y^5$ **4.** $4t^{4/5} + 3t^3 + 2t^2 + t$

Use synthetic substitution to evaluate each polynomial for the given value of the variable.

5. $x^3 - 4x^2 + 6x - 2;\ x = 3$ **6.** $-2t^5 + 3t^3 + 5t^2 - 8;\ t = -1$

Find each polynomial sum or difference.

7. $(6x^3 + x^2 + 3x + 4) + (2x^3 + 7x^2 + x + 5)$

8. $(-9n^5 + 3n^4 - 2n^2 - 2n - 8) - (-n^5 + 5n^4 + 10n^3 - 2n^2 + 4)$

9.1 | Exercises and Applications

Extra Practice exercises on page 762

Tell whether each expression is a polynomial. If so, write the polynomial in standard form and state its degree. If not, explain why not.

1. $5x^{2.9}$ **2.** $5x^3$

3. $6t + t^2 - 1$ **4.** $6t + \dfrac{1}{t^2} - 1$

5. $-\dfrac{3}{2}u^4 + 5u^7 + \pi - 0.6u^{10} - \sqrt{2}\,u^8$ **6.** $-9u^2 + 7u + 2\sqrt[3]{u} + 4$

7. $1 + 2^r + 3^r + 4^r$ **8.** $1 + r^2 + r^3 + r^4$

Use synthetic substitution to evaluate each polynomial for the given value of the variable.

9. $x^2 + 4x + 3$; $x = 10$

10. $2x^3 - 3x^2 - 7x + 5$; $x = 2$

11. $t^4 + t^3 - 4t^2 - 8t - 10$; $t = -2$

12. $-3x^4 + 6x^3 + x^2 + 5x - 2$; $x = -3$

13. $2n^3 - 6n + 1$; $n = 5$

14. $-w^5 - 7w^4 - 2w^2 - 9$; $w = -4$

15. $6x^4 + 7x^3 + 5x - 3$; $x = \dfrac{1}{3}$

16. $-5u^7 + 9u^5 - 8u^2 + 4u + 11$; $u = \sqrt{2}$

17. BIOLOGY The average weight w of a rainbow trout l in. long can be modeled by the equation

$$w = 0.0005l^3$$

where w is measured in pounds.

a. Is $0.0005l^3$ a polynomial? If so, what is the polynomial's degree?

b. Find the average weight of a rainbow trout 20 in. long.

c. By what factor does average weight for a rainbow trout increase when length doubles?

Use the information about Egyptian and Aztec numeration systems on page 389 to find the decimal form of each number.

18.

19.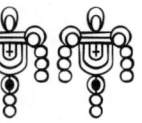

Add.

20. $(4x^2 + x + 3) + (3x^2 + 5x + 1)$

21. $(6x^3 + 8x^2 - 2x + 4) + (10x^3 + x^2 + 11x + 9)$

22. $(-t^4 + 7t^3 + 18t - 13) + (4t^4 - 2t^3 - 6t^2 + 5t - 20)$

23. $(3.9y^5 + 0.5y^4 - 4.8y^3 - 2y^2 + 7.6y) + (5.2y^5 - 2.7y^4 - 8.8y^2 + 0.9y + 4)$

24. $\left(\dfrac{3}{8}x^2 - \dfrac{5}{3}x^3 + \dfrac{7}{12} + \dfrac{1}{2}x^4 - \dfrac{2}{5}x\right) + \left(-\dfrac{1}{6}x^3 + \dfrac{2}{9} - \dfrac{5}{2}x^4 + \dfrac{11}{4}x^2\right)$

25. $\left(3\sqrt{2}\,m^3 - 6\sqrt{5}\,m^2 + 9m + \sqrt{3}\right) + \left(-\sqrt{3} + 14m - \sqrt{5}\,m^2 - 7\sqrt{2}\,m^3\right)$

Subtract.

26. $(2x^2 + 5x + 3) - (x^2 + 2x + 1)$

27. $(8x^3 - 4x^2 - 7x + 2) - (9x^3 + 6x^2 - 7x - 2)$

28. $(-3u^4 - 10u^3 + 19u^2 - 12) - (-16u^3 + 14u^2 + 5u + 4)$

29. $\left(\dfrac{3}{4}v - \dfrac{2}{7}v^2 + \dfrac{1}{3}v^3 - \dfrac{11}{12}\right) - \left(\dfrac{4}{3}v^3 - \dfrac{3}{5}v + \dfrac{13}{18} - \dfrac{5}{14}v^2\right)$

30. $\left(4\sqrt{3} + 5y - 10\sqrt{7}\,y^2 + 9y^4 - 3\sqrt{3}\,y^6\right) - \left(4\sqrt{7}\,y^2 - \sqrt{3}\,y^6 + 5\sqrt{3} - y\right)$

31. $(0.65t^5 - 0.7t^4 + 1.09t^2 - 0.3) - (-3.82 - 1.5t^2 + 2.1t + 0.36t^5 + 0.08t^4)$

For each positive integer n, you can make a type of pyramid called a *tetrahedron* with n layers of dots. The number of dots $T(n)$ in the nth tetrahedron is called the *nth tetrahedral number.*

 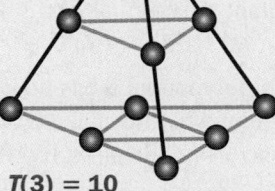 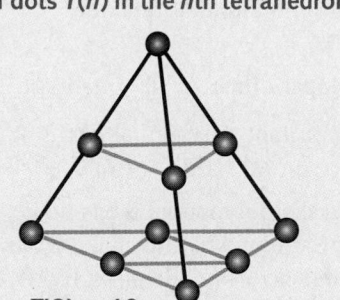 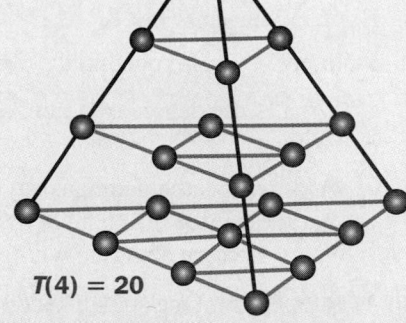

$T(1) = 1$ $T(2) = 4$ $T(3) = 10$ $T(4) = 20$

In Exercises 32–34, you will develop an equation for $T(n)$.

32. Consider the polynomial function $f(x) = x^2 + 3x + 1$. The diagram below shows how to compute *1st differences* and *2nd differences* for $f(x)$.

x	1	2	3	4	5	6
$f(x)$	5	11	19	?	?	?

$11 - 5$ $19 - 11$

1st differences   6   8   ?   ?   ?

$8 - 6$

2nd differences   2   ?   ?   ?

     a. Copy and complete the diagram.

     b. What do you notice about the 2nd differences?

33. Now consider these polynomial functions:

$$g(x) = 3x^3 - 5x^2 + x + 10$$
$$h(x) = 2x^4 - 7x^3 - 4x^2 - 6x + 1$$

     a. Find 1st, 2nd, 3rd, and 4th differences for $g(x)$ and $h(x)$ using the x-values 1, 2, 3, . . . , 7.

     b. **Writing** Based on your results from part (a) and Exercise 32, what can you say about the kth differences for a polynomial function of degree k?

34. a. **Visual Thinking** Show that the 5th and 6th tetrahedral numbers, $T(5)$ and $T(6)$, are 35 and 56, respectively.

     b. It can be shown that $T(n)$ is a polynomial function of n. Use the values of $T(1)$, $T(2)$, . . . , $T(6)$ to compute various differences for $T(n)$. Based on your results, what is the degree of $T(n)$?

     c. **Technology** Look back at Section 7.3, where you used matrices to fit quadratic functions to data. Use an extension of this matrix method to find an equation for $T(n)$. Verify that the first six tetrahedral numbers given by your equation are correct.

35. SAT/ACT Preview If $A = x^4 + x^2 + 1$, $B = x^5 + x^3 + x$, and $0 < x < 1$, then:

A. $A > B$ **B.** $B > A$ **C.** $A = B$ **D.** relationship cannot be determined

36. Tell whether each type of function is a polynomial function. Explain your answers.

  a. linear   **b.** exponential   **c.** logarithmic   **d.** quadratic   **e.** radical

37. a. What is the degree of a nonzero constant polynomial like 2 or -3? Explain.

  b. Writing Mathematicians say that the polynomial 0 has no degree, or that the degree of 0 is undefined. Explain why this makes sense. (*Hint:* Is there a unique whole number n for which $0 = 0x^n$?)

38. PSYCHOLOGY One social scientist found that the percent P of people who describe themselves as "highly annoyed" by noise from traffic, airports, and other sources can be modeled by the equation

$$P = 0.8553L - 0.0401L^2 + 0.00047L^3$$

where L is the average loudness of the noise in decibels (dB).

  a. Is $0.8553L - 0.0401L^2 + 0.00047L^3$ a polynomial? If so, write the polynomial in standard form and state its degree.

  b. The noise level of city traffic is typically about 70 dB. At this level, what percent of people are highly annoyed by the noise?

ONGOING ASSESSMENT

39. Open-ended Problem Recall from geometry that the formula $V = \frac{4}{3}\pi r^3$ gives the volume V of a sphere as a polynomial function of its radius r. List three other polynomial formulas from geometry, and tell what information each formula gives.

SPIRAL REVIEW

Tell whether each set of numbers is closed under the given operation. If so, tell whether the number set and operation *form a group*, *form a commutative group*, or *do not form a group*. Explain. *(Section 8.6)*

40. whole numbers; $+$

41. whole numbers; $\times$

42. real numbers; $+$

43. odd numbers; $\times$

Evaluate each logarithm. *(Section 4.2)*

44. $\log_2 8$ **45.** $\log 10{,}000$ **46.** $\log_5 \dfrac{1}{25}$ **47.** $\log_9 27$

Multiply. *(Toolbox, page 782)*

48. $(x + 1)(x + 2)$ **49.** $(x - 3)(x + 4)$ **50.** $(2y + 5)(2y - 5)$ **51.** $(6n + 1)^2$

9.2 Multiplying and Dividing Polynomials

Learn how to...
- multiply and divide polynomials

So you can...
- find the power needed to maintain a given speed on a bicycle, for example

The diagrams show the same rectangular prism drawn two different ways. These drawings suggest two methods for computing the prism's volume V.

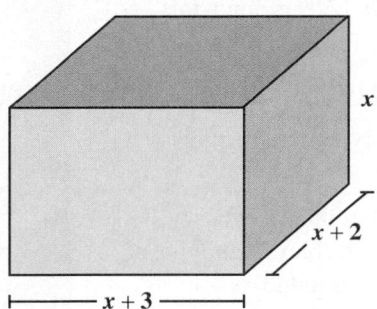

x

$x + 2$

$x + 3$

V = length × width × height

$= (x + 3)(x + 2)x$

V_2 = volume of this piece

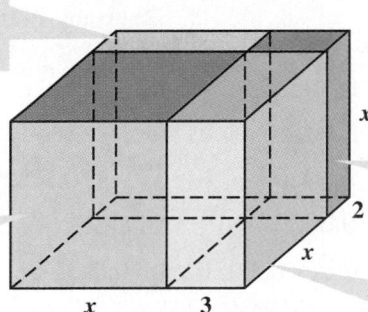

x

2

x

x 3

V_1 = volume of this piece

V_4 = volume of this piece

V_3 = volume of this piece

$V = V_1 + V_2 + V_3 + V_4$

THINK AND COMMUNICATE

1. Express the volume of each piece of the prism shown above in terms of x.

   a. V_1 **b.** V_2 **c.** V_3 **d.** V_4

2. Use your results from Question 1 to complete this equation with a polynomial in standard form:

   $$(x + 3)(x + 2)x = \underline{\ ?\ }$$

3. **a.** Use the distributive property to find the product $(x + 3)(x + 2)$.

   b. Use the distributive property again to find the product of x and your answer to part (a).

   c. How does your answer to part (b) compare with your answer to Question 2?

Toolbox p. 774
Properties of Real Numbers

Polynomial Multiplication

You can use the distributive property to write a product of polynomials as a single polynomial.

EXAMPLE 1 Application: Bicycling

The power P, measured in horsepower (hp), required to keep a certain kind of bicycle moving at s mi/h is given by

$$P = 0.00267sF$$

where F is the force of road and air resistance in pounds. On level ground, this force is a quadratic function of speed:

$$F = 0.0116s^2 + 0.789$$

a. Write P as a polynomial function of s alone.

b. How much power must a cyclist provide to keep the bicycle moving at 15 mi/h on level ground?

SOLUTION

a. Use the equation for P.

> Substitute **$0.0116s^2 + 0.789$** for **F**.

$$P = 0.00267sF$$
$$= 0.00267s(\mathbf{0.0116s^2 + 0.789})$$

> Use the distributive property.

$$= 0.00267s(0.0116s^2) + 0.00267s(0.789)$$
$$\approx 0.0000310s^3 + 0.00211s$$

> Simplify.

The function is $P = 0.0000310s^3 + 0.00211s$.

b. Substitute 15 for s in the equation from part (a).

$$P = 0.0000310(\mathbf{15}^3) + 0.00211(\mathbf{15})$$
$$\approx 0.136$$

A cyclist must provide about 0.136 hp.

THINK AND COMMUNICATE

4. a. How much power must a cyclist provide to keep the bicycle in Example 1 moving at 30 mi/h on level ground?

b. Compare your answer to part (a) with the answer to part (b) of Example 1. Does it take twice as much power to double speed? Explain.

c. For a certain moped, the power from the engine must be 1.35 hp in order to travel 30 mi/h on level ground. Which vehicle—the moped or the bicycle in Example 1—uses energy more efficiently? Explain.

When you use the distributive property to multiply two polynomials, you multiply each term of one polynomial by each term of the other polynomial. You can do this easily using the vertical format shown in Example 2.

EXAMPLE 2

Multiply $2x - 3$ and $4x^2 - 5x + 8$.

SOLUTION

Align like terms vertically.

$$4x^2 - 5x + 8 \quad\longleftarrow\quad \text{Write the polynomial with more terms first.}$$

$$\underline{2x - 3}$$

$$8x^3 - 10x^2 + 16x \qquad\longleftarrow\quad \text{This is } 2x(4x^2 - 5x + 8).$$

$$\underline{\quad - 12x^2 + 15x - 24} \qquad\longleftarrow\quad \text{This is } -3(4x^2 - 5x + 8).$$

$$8x^3 - 22x^2 + 31x - 24 \qquad\longleftarrow\quad \text{Add like terms.}$$

The product is $8x^3 - 22x^2 + 31x - 24$.

THINK AND COMMUNICATE

5. a. Multiply $2x - 3$ and $4x^2 - 5x + 8$ using a horizontal format:

$$(2x - 3)(4x^2 - 5x + 8) = 2x(4x^2 - 5x + 8) - 3(4x^2 - 5x + 8) = \underline{\ ?\ }$$

b. Compare the horizontal format in part (a) with the vertical format in Example 2. Which format do you prefer? Why?

Polynomial Division

In elementary school, you learned how to use long division to write a fraction like $\frac{173}{3}$ as a mixed number:

$$
\begin{array}{r}
57 \\
3\overline{)\,173} \\
\end{array}
$$

Subtract 5×3. $\quad 15$

$\quad\overline{\quad 23}$

Subtract 7×3. $\quad 21$

$\quad\overline{\quad\quad 2}$

$$\frac{\text{dividend}}{\text{divisor}} \qquad \text{quotient} \qquad \frac{\text{remainder}}{\text{divisor}}$$

$$\frac{173}{3} = 57 + \frac{2}{3} = 57\frac{2}{3}$$

You can use a similar procedure to divide one polynomial by another as shown in Example 3 on the next page.

EXAMPLE 3

Divide $6x^3 - 5x^2 + 9$ by $2x + 1$.

SOLUTION

$$\frac{6x^3}{2x} = 3x^2 \qquad \frac{-8x^2}{2x} = -4x \qquad \frac{4x}{2x} = 2$$

$$
\begin{array}{r}
3x^2 - 4x + 2 \\
2x + 1 \overline{\smash{\big)}\, 6x^3 - 5x^2 + 0x + 9}
\end{array}
$$

Subtract $3x^2(2x + 1)$. $\longrightarrow$ $\underline{6x^3 + 3x^2}$

$-8x^2 + 0x$

Insert the "missing" x-term by using 0 as the coefficient.

Subtract $-4x(2x + 1)$. $\longrightarrow$ $\underline{-8x^2 - 4x}$

$4x + 9$

Subtract $2(2x + 1)$. $\longrightarrow$ $\underline{4x + 2}$

7

Stop when either the degree of the remainder is less than the degree of the divisor, or the remainder is 0.

Therefore, $\dfrac{6x^3 - 5x^2 + 9}{2x + 1} = 3x^2 - 4x + 2 + \dfrac{7}{2x + 1}$.

$$\frac{\textbf{dividend}}{\textbf{divisor}} = \textbf{quotient} + \frac{\textbf{remainder}}{\textbf{divisor}}$$

THINK AND COMMUNICATE

6. a. Explain why the word equation at the end of Example 3 is equivalent to this equation:

dividend = divisor × quotient + remainder

 b. Use the word equation in part (a) to check the answer to Example 3.

☑ CHECKING KEY CONCEPTS

Multiply.

 1. $(3x - 5)(x + 2)$ **2.** $(x^3 + 2x)(4x^4 - 7x^2)$

 3. $(u + 1)(u^2 - 5u - 1)$ **4.** $(2t - 9)(-3t^2 + 8t + 2)$

Divide.

 5. $\dfrac{x^2 + 6x + 10}{x + 2}$ **6.** $\dfrac{6w^2 + 7w + 5}{3w - 1}$

 7. $\dfrac{y^3 + 4y^2 - 20y + 1}{y - 3}$ **8.** $\dfrac{8x^3 - 12x - 7}{2x + 3}$

9.2 Exercises and Applications

Extra Practice exercises on page 762

Multiply.

1. $(4x - 1)(2x + 7)$

2. $(5x + 3)^2$

3. $-2t^3(3t^2 - 6t - 1)$

4. $y(y - 1)(y - 2)$

5. $(u + 1)(u^2 + u + 1)$

6. $(v - 2)(-v^2 + 8v - 9)$

7. $(3m + 4)(2m^2 - 13m + 4)$

8. $(x + 2)^3$

9. $(-10n^2 + n + 6)(-4n - 7)$

10. $(2x - 1)(3x + 5)(4x + 3)$

11. $(w - 3)^4$

12. $(r^2 - 5r + 2)(2r^2 + 8r - 3)$.

Connection ▶ HISTORY

According to Chinese legend, an emperor named Yu found a mystical turtle while walking along the Lo River. A drawing of the turtle is shown. The square array gives the number of dots on each plate of the turtle's shell. The array is called a *magic square* because the entries in each row, column, and diagonal add up to the same number, called a *magic constant*.

The entries in this 3 × 3 magic square are the integers from 1 to 9. In general, the entries in an $n \times n$ magic square are the integers from 1 to n^2.

Portrait of Emperor Yu

13. **a.** Verify that the square array shown is a magic square. What is the magic constant?

   b. In general, the magic constant M for an $n \times n$ magic square is given by this equation:

   $$M = \frac{1}{2}n^3 + \frac{1}{2}n$$

   Find M when $n = 3$, and compare this value with the magic constant you found in part (a).

14. **a.** Use the equation for M in Exercise 13 to find an equation giving the sum S of *all* the entries in an $n \times n$ magic square.

   $\left(\textit{Hint: Note that } S = \dfrac{\text{sum of the numbers}}{\text{in each row}} \times \dfrac{\text{number}}{\text{of rows}}\right)$

   b. Use your equation from part (a) to find the sum of all the entries in the given 3 × 3 magic square. Check your answer by computing the sum directly.

15. **Cooperative Learning** Work with a partner to create a 4 × 4 magic square. Verify that the equation for M in Exercise 13 applies to your square.

BY THE WAY

The 3 × 3 magic square on the shell of Emperor Yu's turtle is called the *Lo shu*, which means "Lo River writing."

The diagram shows one version of a dessert called Baked Alaska. You make this dessert by covering a hemisphere-shaped mound of ice cream with a 1 in. layer of meringue and then browning the meringue in an oven. The amount of meringue needed depends on the radius of the ice cream mound.

16. Use the diagram.

    a. Express V_I as a polynomial function of r. (*Hint*: Recall from geometry that the volume of a sphere is given by $V = \frac{4}{3}\pi r^3$. Use 3.14 for π.)

    b. Express V_{I+M} as a polynomial function of r.

    c. Use your results from parts (a) and (b) to express V_M as a polynomial function of r. Write the polynomial in standard form.

    d. How much meringue is needed to make Baked Alaska if the radius of the ice cream mound is 4 in.?

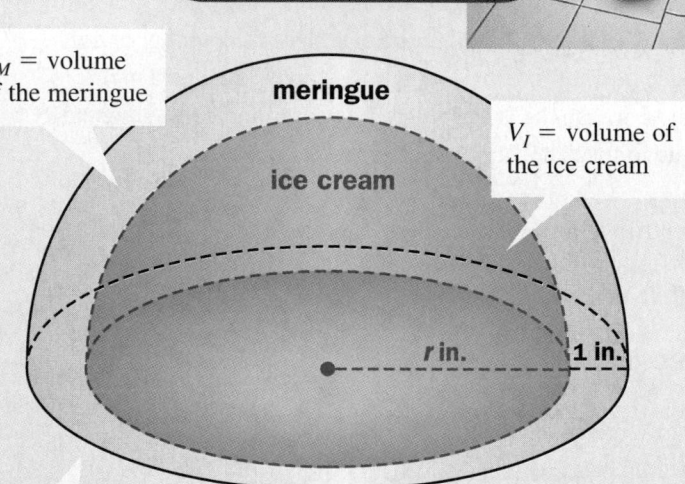

V_M = volume of the meringue

meringue

V_I = volume of the ice cream

ice cream

r in. 1 in.

V_{I+M} = combined volume of the ice cream and meringue

17. The function you found in part (c) of Exercise 16 gives the volume of meringue needed in cubic inches. It is more convenient to measure meringue in cups.

    a. Use the fact that 1 in.$^3 \approx 0.0693$ cups to express V_M in cups as a function of r in inches.

    b. How many cups of meringue are needed if the radius of the ice cream mound is 5 in.?

    c. **Challenge** Suppose you want the thickness of the meringue to be something other than 1 in. Express the number of cups of meringue needed in terms of r and the desired meringue thickness h (where r and h are in inches).

18. SAT/ACT Preview Suppose $(2x + a)(bx^2 + cx + d) = 2x^3 + 7x^2 - 19x + 6$. Which of the following equals $ac + 2d - 3b$?

A. 13   **B.** 14   **C.** 16   **D.** -19 **E.** -22

Divide.

19. $\dfrac{x^2 + 3x + 6}{x + 1}$

20. $\dfrac{2x^2 - 2x + 5}{x - 3}$

21. $\dfrac{3t^2 + 7t}{t - 2}$

22. $\dfrac{u^2 + 4}{u + 2}$

23. $\dfrac{y^3 + 8y^2 + 14y - 3}{y + 4}$

24. $\dfrac{4v^3 + 16v^2 - v - 43}{2v + 5}$

25. $\dfrac{12x^3 - 19x - 12}{2x - 3}$

26. $\dfrac{w^3 - 6w^2 + 1}{w - 1}$

27. $\dfrac{2x^4 - x^3 + x^2 + 10x - 6}{x + 2}$

28. $\dfrac{-6x^4 + 7x^3 + 11x + 4}{3x + 1}$

29. $\dfrac{8n^3 + 26n^2 - 15n + 9}{2n^2 + 7n - 5}$

30. $\dfrac{r^4 - 4r^3 - 2r^2 + 7r + 13}{r^2 - r + 3}$

31. Writing Let $p(x)$ and $d(x)$ be polynomials with degrees m and n, respectively, where $m > n$.

    a. What is the degree of $p(x) \cdot d(x)$? How do you know?

    b. Suppose that when $p(x)$ is divided by $d(x)$, the quotient is $q(x)$ and the remainder is $r(x)$. What is the degree of $q(x)$? What can you say about the degree of $r(x)$? Explain.

ONGOING ASSESSMENT

32. Open-ended Problem Write two polynomials $p(x)$ and $d(x)$ such that $\dfrac{p(x)}{d(x)} = 4x^2 - x + 6$. Explain how you found your polynomials, and verify that they satisfy the given equation.

SPIRAL REVIEW

Use synthetic substitution to evaluate each polynomial for the given value of the variable. *(Section 9.1)*

33. $2x^2 + 6x + 11; x = 3$

34. $x^3 - 9x^2 - 5x + 5; x = 2$

35. $3y^4 + 2y^3 - 7y + 1; y = -2$

36. $-4m^3 + 13m^2 + 5m - 10; m = \dfrac{1}{4}$

 Technology Use a graphing calculator or software with matrix calculation capabilities to solve each system of equations. *(Section 7.3)*

37. $2x - y = 5$
$\ x + 3y = 6$

38. $-70x + 10y = 19$
$\ -10x + 2y = 3$

39. $3x + 5y - 9z = 26$
$\ 4x + 7y + 2z = -7$
$\ 6x - 9y - 8z = 3$

Graph each equation. *(Sections 2.2 and 5.4)*

40. $y = 3x - 2$

41. $y = -2x + 6$

42. $y = \dfrac{1}{3}x + 4$

43. $y = x^2 - 6x + 4$

44. $y = -3x^2 - 12x - 8$

45. $y = 2x^2 - 5x - 7$

9.3 Exploring Graphs of Polynomial Functions

You already know what the graphs of some polynomial functions look like. For example, the graph of $y = 2x + 5$ is a line, and the graph of $y = 4x^2 - 8x - 3$ is a parabola. You'll look at graphs of higher-degree polynomial functions in the Exploration.

EXPLORATION
COOPERATIVE LEARNING

Looking for Patterns in Graphs

Work with a partner.
You will need: • a graphing calculator or graphing software

SET UP Adjust the viewing window for your calculator or software so that the intervals $-5 \leq x \leq 5$ and $-20 \leq y \leq 20$ are shown on the axes. Complete Steps 1 and 2 for each of these functions:

- $y = 4x^2 - 8x - 3$
- $y = x^3 - x^2 - 3x + 1$
- $y = x^4 + 2x^3 - 5x^2 - 7x + 3$
- $y = 2x^5 + 6x^4 - 2x^3 - 14x^2 + 5$
- $y = 3x^6 - 13x^4 + 15x^2 + x - 17$
- $y = x^7 - 8x^5 + 18x^3 - 6x$

1 Describe what happens to the graph as x takes on large positive and large negative values.

Example: $y = 2x^2 - 2x - 11$

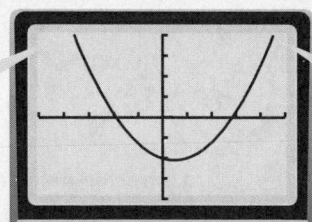

As x takes on large negative values, the graph rises.

As x takes on large positive values, the graph rises.

2 Find the number of *turning points*. Give the approximate coordinates of each turning point.

Example: $y = 2x^2 - 2x - 11$

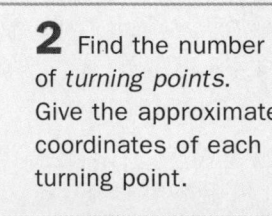

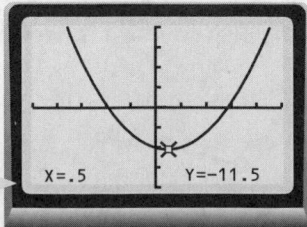

X=.5   Y=-11.5

The graph has one turning point. Its coordinates are $(0.5, -11.5)$.

Questions

1. How do the graphs of the even-degree polynomial functions behave as *x* takes on large positive values? as *x* takes on large negative values?

2. How do the graphs of the odd-degree polynomial functions behave as *x* takes on large positive values? as *x* takes on large negative values?

3. a. How is the number of turning points on the graph of each function in the Exploration related to the function's degree?

b. Decide if the relationship from part (a) is always true by graphing other polynomial functions. If necessary, modify your statement of the relationship so that it applies to *all* polynomial functions.

An important characteristic of a polynomial function *f* is its *end behavior*. The phrase **end behavior** refers to what happens to $f(x)$ as *x* takes on large positive and large negative values.

You can describe end behavior with the symbols $+\infty$ and $-\infty$, which stand for "positive infinity" and "negative infinity," respectively. The diagram explains how these symbols are used.

To indicate that *x* takes on larger and larger negative values, you write "$x \to -\infty$."

To indicate that $f(x)$ takes on larger and larger positive values, you write "$f(x) \to +\infty$."

To indicate that $f(x)$ takes on larger and larger negative values, you write "$f(x) \to -\infty$."

To indicate that *x* takes on larger and larger positive values, you write "$x \to +\infty$."

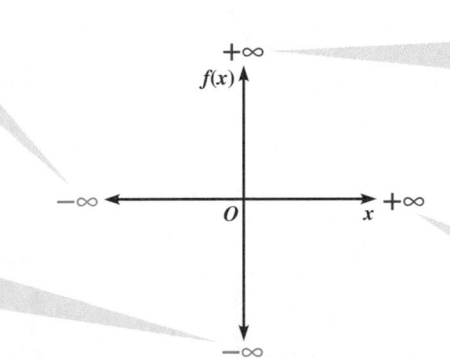

EXAMPLE 1

Graph $f(x) = 2x^3 + x^2 - 3x + 4$ using a graphing calculator or graphing software. Describe the end behavior of *f*.

SOLUTION

The graph of *f* is shown. Use infinity notation to describe the end behavior.

As $x \to -\infty$, $f(x) \to -\infty$. This sentence is read, "As *x* approaches negative infinity, $f(x)$ approaches negative infinity."

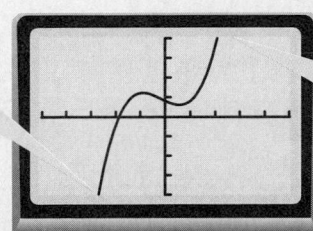

As $x \to +\infty$, $f(x) \to +\infty$.

The end behavior of a polynomial function depends on the function's degree and the sign of its *leading coefficient*. The **leading coefficient** is the coefficient of the highest power of the variable. For example, the leading coefficient of $f(x) = 2x^3 + x^2 - 3x + 4$ is **2**. The graphs below illustrate the four types of end behavior for nonconstant polynomial functions.

Type 1: As $x \rightarrow +\infty$, $f(x) \rightarrow +\infty$.
As $x \rightarrow -\infty$, $f(x) \rightarrow -\infty$.

Example:

$$f(x) = 4x^3 - 2x^2 - 5x + 3$$

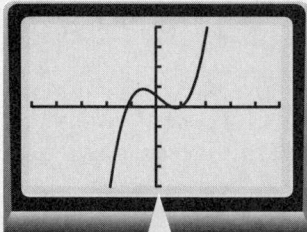

This type applies to functions whose **degree** is **odd** and whose **leading coefficient** is **positive**.

Type 2: As $x \rightarrow +\infty$, $f(x) \rightarrow -\infty$.
As $x \rightarrow -\infty$, $f(x) \rightarrow +\infty$.

Example:

$$f(x) = -1x^5 + 8x^3 - 10x$$

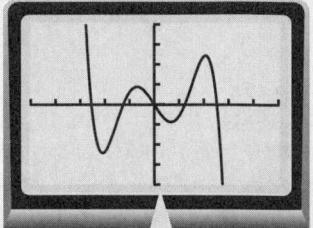

This type applies to functions whose **degree** is **odd** and whose **leading coefficient** is **negative**.

Type 3: As $x \rightarrow +\infty$, $f(x) \rightarrow +\infty$.
As $x \rightarrow -\infty$, $f(x) \rightarrow +\infty$.

Example:

$$f(x) = 1x^4 + 3x^3 + x^2 - 7x - 10$$

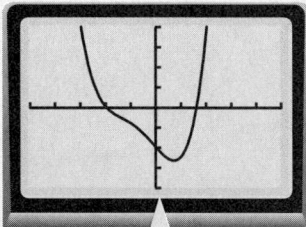

This type applies to functions whose **degree** is **even** and whose **leading coefficient** is **positive**.

Type 4: As $x \rightarrow +\infty$, $f(x) \rightarrow -\infty$.
As $x \rightarrow -\infty$, $f(x) \rightarrow -\infty$.

Example:

$$f(x) = -2x^6 + 9x^4 - 9x^2 + x + 7$$

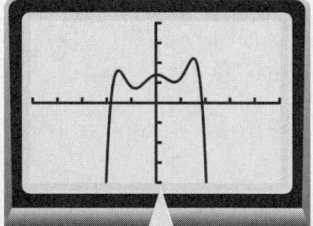

This type applies to functions whose **degree** is **even** and whose **leading coefficient** is **negative**.

THINK AND COMMUNICATE

1. What is the leading coefficient of $f(x) = 7x^3 - x^2 + 2x - 5x^4 + 4$?

2. a. Use infinity notation to describe the end behavior of the function $h(x) = 3x^6 - 6x^4 + 5x - 7$ *without* actually graphing it.

b. **Technology** Graph $y = h(x)$ using a graphing calculator or graphing software. Does the graph support your answer to part (a)?

A **turning point** of the graph of a function is a point that is higher or lower than all nearby points. The graph of $g(x) = -2x^3 - 3x^2 + 12x + 5$, shown below, has two turning points. In general, the graph of a polynomial function of degree n has at most $n - 1$ turning points.

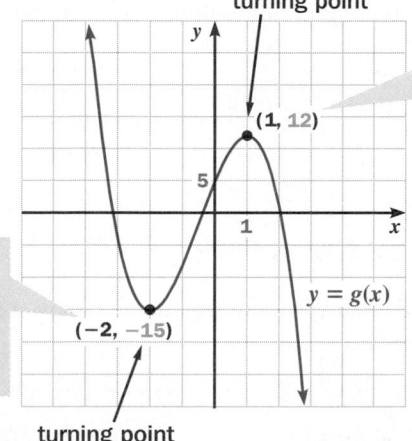

turning point

The *y-coordinate* of a turning point *higher* than all nearby points is a **local maximum**.

(1, 12)

5

1

$y = g(x)$

The *y-coordinate* of a turning point *lower* than all nearby points is a **local minimum**.

(−2, −15)

turning point

EXAMPLE 2 Application: Physiology

During a normal five-second respiratory cycle in which a person inhales and then exhales, the volume V of air brought into the person's lungs can be modeled by the function

$$V = 0.027t^3 - 0.27t^2 + 0.675t$$

where volume is measured in liters and t is the number of seconds after the person starts inhaling. When is this volume at a maximum? What is the maximum volume?

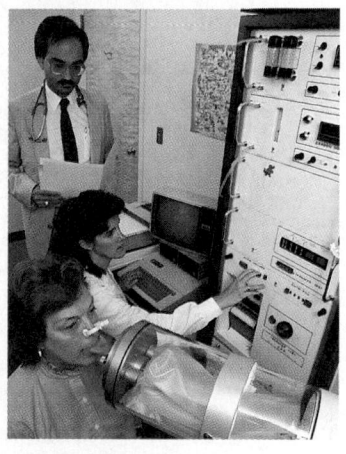

A doctor observes a patient taking a pulmonary function test.

SOLUTION

Use a graphing calculator or graphing software to graph the function $y = 0.027x^3 - 0.27x^2 + 0.675x$. Adjust the viewing window so that the interval $0 \le x \le 5$ is shown on the x-axis.

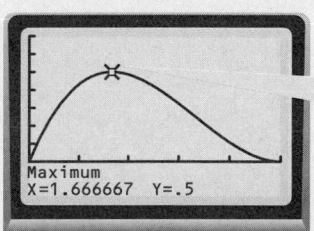

The volume reaches a **maximum** at this turning point. The coordinates of the point are about (1.7, 0.5).

The volume of air brought into the lungs is at a maximum after about 1.7 s. The maximum volume is about 0.5 L.

THINK AND COMMUNICATE

3. In Example 2, the volume given by $V = 0.027t^3 - 0.27t^2 + 0.675t$ does not include the "residual volume" of air present in the lungs even after a person exhales. This volume is about 2.5 L.

    a. Write an equation for the total volume of air in the lungs during a five-second respiratory cycle.

    b. Find the maximum total volume of air in the lungs *without* graphing the function from part (a). Explain how you obtained your answer.

☑ CHECKING KEY CONCEPTS

For Questions 1–4, use what you know about end behavior to match each polynomial function with its graph.

    1. $f(x) = 3x^4 - 10x^2 - x + 3$ **2.** $f(x) = -2x^3 + 3x^2 + 2x + 3$

    3. $f(x) = x^5 - 6x^3 + 5x + 3$ **4.** $f(x) = -4x^6 + 7x^5 + 8x^2 - 8x + 3$

A.
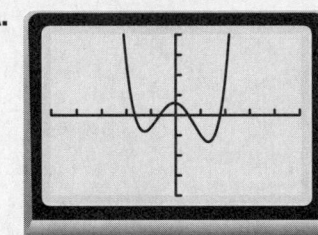

B.
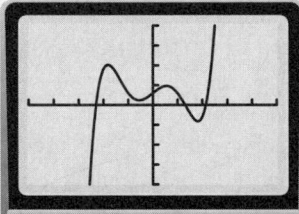

C.
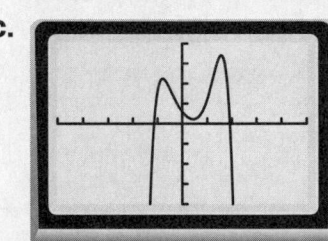

D.
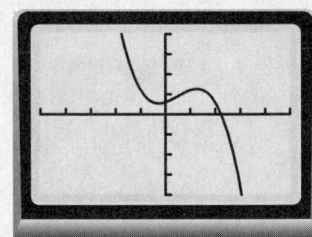

5–8.   **Technology** Use a graphing calculator or graphing software to graph each function in Questions 1–4. Find all local maximums and local minimums.

9.3 Exercises and Applications

*Extra Practice
exercises on
page 762*

1. Writing In your own words, explain what is meant by the "end behavior" of a polynomial function. How is the end behavior related to the function's degree and leading coefficient?

2. Look back at the graph of $g(x) = -2x^3 - 3x^2 + 12x + 5$ near the top of the previous page. Why does it make sense to call 12 a "local maximum," but not a "maximum," of the function g?

---

408   Chapter 9 *Polynomial and Rational Functions*

Technology Use a graphing calculator or graphing software to graph each polynomial function. For each function:

a. Describe the end behavior using infinity notation.

b. Find all local maximums and local minimums.

3. $f(x) = 3x^2 + 6x - 4$

4. $f(x) = -5x^2 + 19x - 8$

5. $f(x) = -x^3 + x^2 - 5x + 2$

6. $f(x) = 2x^3 - 7x$

7. $g(x) = 4x^5 + 7x^4 - 9x^3 - 15x^2$

8. $g(x) = x^4 + x^3 - 6x^2 - x + 4$

9. $h(x) = -2x^4 + 5$

10. $h(x) = -2x^5 + 3x^4 + x^2 + 6$

11. $f(x) = x^6 - 7x^4 + 8x^2 + 7$

12. $f(x) = -5x^6 + x^5 + 21x^4 - 3x^3 - 22x^2 + 9$

Connection   SWIMMING

The graph shows how the speed of a swimmer doing the breaststroke changes over the course of one complete stroke. An equation for the graph is:

$$s = -241.0t^7 + 1062t^6 - 1871t^5 + 1647t^4 - 737.4t^3 + 143.9t^2 - 2.432t$$

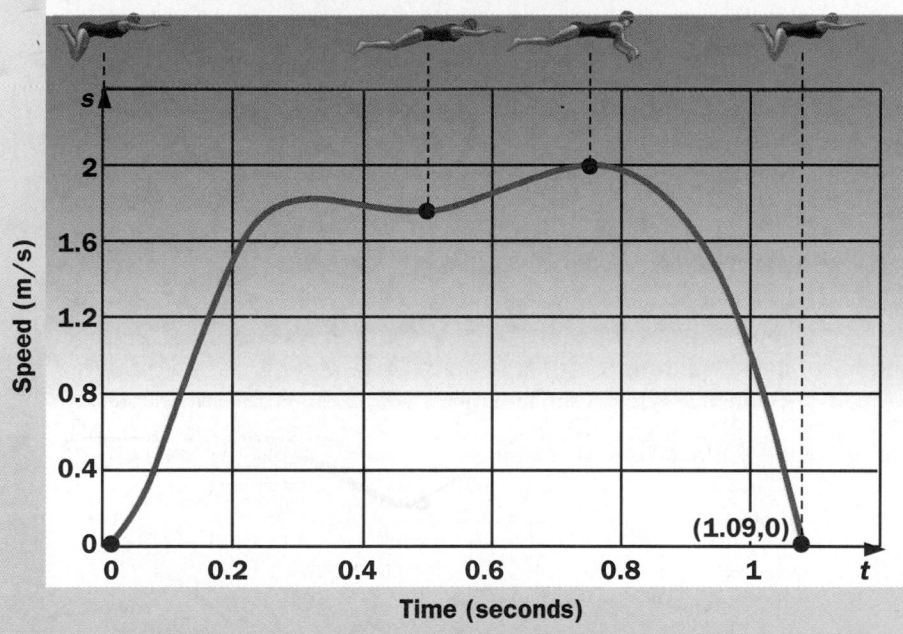

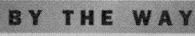

BY THE WAY

Swim meets are held in both long-course pools, which are 50 m in length, and short-course pools, which are 25 m in length. Dai Guohong of China holds the women's short-course world records in the 100 m and 200 m breaststroke.

13. a. Use the graph. Estimate the coordinates of all turning points. For each turning point, tell whether a *local maximum* or *local minimum* occurs at the point.

   b. **Technology** Use a graphing calculator or graphing software to graph the given equation and check your answers to part (a). At what time is the swimmer's speed greatest? What is the greatest speed?

14. Visual Thinking Extend the graph shown so that it gives the speed of the swimmer over the course of three complete strokes.

15. Open-ended Problem What do you think is the average speed of the swimmer during one complete stroke? Explain your reasoning.

The photograph shows a reconstruction of a dwelling called a *neesquttow* once built by the Wampanoag, a Native American people. A *neesquttow* was constructed with poles covered by strips of bark.

16. **Challenge** Suppose 600 ft^2 of bark is to be used to make a *neesquttow* in the shape of a half-cylinder. You can find the dimensions of the roomiest *neesquttow* that can be built with this amount of bark.

   a. Use the diagram. Write an equation giving the surface area S of the *neesquttow* in terms of its radius r and length l. Assume the *neesquttow* has a rectangular opening at one end as shown. (*Hint:* The surface area of a cylinder is given by the formula $S = 2\pi r^2 + 2\pi rl$.)

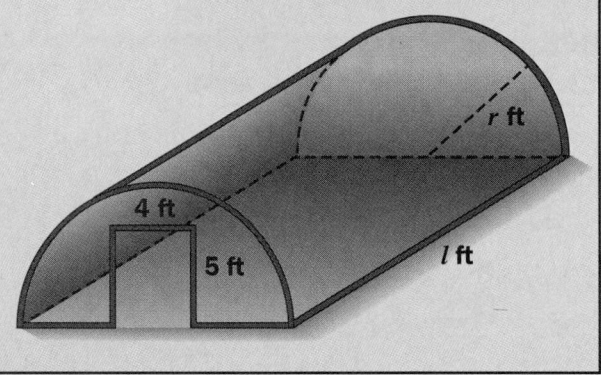

   b. Knowing that $S = 600$, use your equation from part (a) to solve for l in terms of r.

   c. Use your expression for l from part (b) to write an equation giving the volume V of the *neesquttow* as a polynomial function of r. (*Hint:* The volume of a cylinder is given by the formula $V = \pi r^2 l$.)

   d. **Technology** Graph your equation for V using a graphing calculator or graphing software. Find the radius r that maximizes the volume of the *neesquttow*. Use this r-value to find the length l that maximizes the volume. What is the maximum volume?

17. **Writing** Is the "best" *neesquttow* that can be built with a given amount of bark necessarily the one having maximum volume? What factors besides volume do you think the Wampanoag considered when building a *neesquttow*?

BY THE WAY

The Wampanoag live in what is now Rhode Island and southern Massachusetts. They shared the first Thanksgiving feast with the Pilgrims in 1621.

18. a. Find the domain and range of each function whose graph is shown on page 406.

   b. In general, what can you say about the domain and range of an odd-degree polynomial function? of an even-degree polynomial function?

ONGOING ASSESSMENT

19. **Open-ended Problem** Write a polynomial function f of degree 5 such that $f(x) \to -\infty$ as $x \to +\infty$. Use a graphing calculator or graphing software to find the coordinates of each turning point on the function's graph.

Write a point-slope equation of the line passing through the given points.
(Section 2.3)

20. $(2, 1)$ and $(3, 3)$ **21.** $(-5, -3)$ and $(4, 0)$ **22.** $(6, 6)$ and $(8, -2)$

Solve each equation. *(Sections 5.5 and 5.6)*

23. $x^2 - 7x + 12 = 0$ **24.** $6x^2 + x - 15 = 0$ **25.** $2x^2 + 3x - 1 = 0$

ASSESS YOUR PROGRESS

VOCABULARY

polynomial (p. 390)	**like terms** (p. 392)
degree (p. 390)	**end behavior** (p. 405)
constant term (p. 390)	**leading coefficient** (p. 406)
standard form (p. 390)	**turning point** (p. 407)
synthetic substitution (p. 391)	**local maximum/minimum** (p. 407)

1. Write the polynomial $-13x + 8x^4 + 9 + x^3 - 2x^5$ in standard form and state its degree. *(Section 9.1)*

2. Use synthetic substitution to evaluate $3t^4 - 4t^2 + t + 7$ when $t = 2$. *(Section 9.1)*

3. Add $2x^3 - 5x^2 - 6x + 3$ and $x^3 + 2x^2 + 6x - 10$. *(Section 9.1)*

4. Subtract $3y^4 - 5y^3 + y + 8$ from $7y^4 + 4y^3 - 8y^2 - 1$. *(Section 9.1)*

5. Multiply $4u + 7$ and $2u^2 - u + 5$. *(Section 9.2)*

6. Divide $4x^3 - x^2 + 8$ by $x + 2$. *(Section 9.2)*

7. Describe the end behavior of $f(x) = -7x^4 + 3x^3 - 5x + 1$ using infinity notation. *(Section 9.3)*

8. **PACKAGING** A manufacturer of packaging materials makes open-top boxes by cutting an x in. by x in. square from each corner of an 18 in. by 24 in. piece of cardboard as shown. *(Section 9.3)*

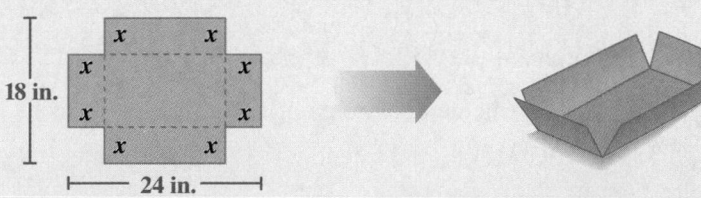

    a. Write an equation giving the volume V of the box as a function of x.

    b. **Technology** Use a graphing calculator or graphing software to graph the equation from part (a). Find the value of x that maximizes the volume of the box. What is the maximum volume?

9. **Journal** Choose two of the four basic operations—addition, subtraction, multiplication, and division—and explain how to perform the operations on polynomials.

9.4 Solving Cubic Equations

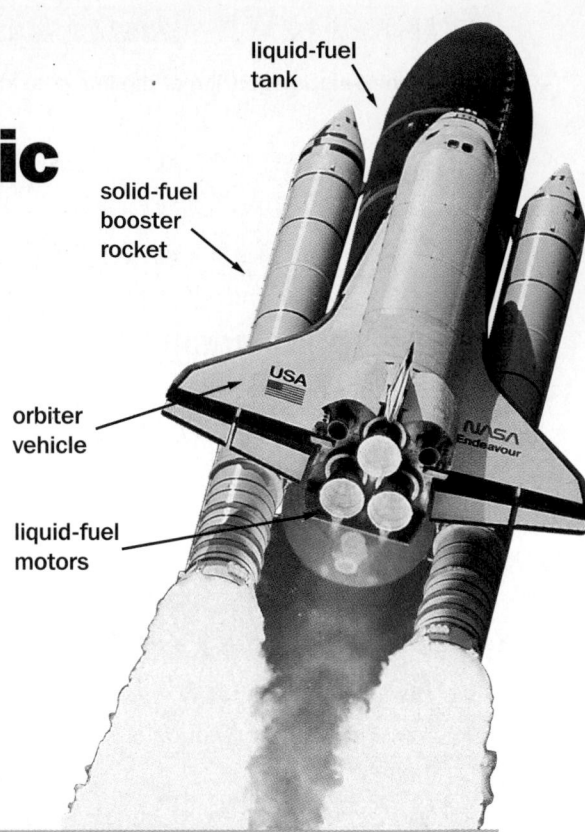

liquid-fuel tank

solid-fuel booster rocket

orbiter vehicle

liquid-fuel motors

Learn how to...

- **solve cubic equations**

- **find equations for graphs of cubic functions**

- **find zeros of cubic functions**

So you can...

- **analyze the flight of the space shuttle after launch, for example**

The space shuttle is powered by three liquid-fuel motors in its tail and two external solid-fuel booster rockets. When the shuttle reaches a speed of about 3000 mi/h, the booster rockets fall off and return to Earth. The liquid-fuel motors continue firing and propel the shuttle into orbit.

EXAMPLE 1 Application: Space Travel

The speed of the space shuttle t seconds after launch can be approximated using the function

$$s(t) = 0.000559t^3 + 0.0313t^2 + 13.6t$$

where $s(t)$ is measured in miles per hour. How long after launch do the booster rockets fall off?

SOLUTION

The booster rockets fall off when $s(t) = 3000$, so you need to solve this equation:

$$0.000559t^3 + 0.0313t^2 + 13.6t = 3000$$

Method 1 Use a graphing calculator or graphing software.

Graph $y = 0.000559x^3 + 0.0313x^2 + 13.6x$ and $y = 3000$ in the same coordinate plane.

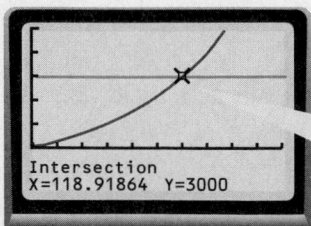

Intersection
X=118.91864 Y=3000

Find the x-coordinate of the point where the graphs intersect. The x-coordinate is about **119**.

The booster rockets fall off about 119 s (or about 2 min) after launch.

Method 2 Use a spreadsheet.

Step 1 Have the spreadsheet calculate the shuttle's speed at ten-second intervals from $t = 50$ to $t = 150$.

	A	B	C	D	E	F	G	H	I	J	K	L
					Shuttle Speeds 1							
1	t	50	60	70	80	90	100	110	120	130	140	150
2	$s(t)$	828	1049	1297	1575	1885	2232	2619	3049	3525	4051	4631

Since 3000 is between these two numbers, the desired t-value is between **110** and **120**.

Step 2 Have the spreadsheet calculate the shuttle's speed at one-second intervals from $t = 110$ to $t = 120$.

	A	B	C	D	E	F	G	H	I	J	K	L
					Shuttle Speeds 2							
1	t	110	111	112	113	114	115	116	117	118	**119**	120
2	$s(t)$	2619	2660	2701	2743	2785	2828	2871	2915	2959	3004	3049

Notice that $s(\mathbf{119}) \approx 3000$.

The booster rockets fall off about 119 s (or about 2 min) after launch.

The function $s(t) = 0.000559t^3 + 0.0313t^2 + 13.6t$ in Example 1 is called a *cubic function*. A **cubic function** is a polynomial function of degree 3.

Cubic functions may be written in standard form, $f(x) = ax^3 + bx^2 + cx + d$, or in *intercept form*, $f(x) = a(x - p)(x - q)(x - r)$. The next example shows why the phrase "intercept form" is used.

EXAMPLE 2

Let $f(x) = 2(x - 1)(x - 3)(x - 4)$. Find the x-intercepts of the graph of f.

SOLUTION

To find the x-intercepts, solve the equation $f(x) = 0$.

$$2(x - 1)(x - 3)(x - 4) = 0$$

Divide both sides by 2.

$$(x - 1)(x - 3)(x - 4) = 0$$

$$x - 1 = 0 \quad \text{or} \quad x - 3 = 0 \quad \text{or} \quad x - 4 = 0$$

$$x = 1 \quad \text{or} \quad x = 3 \quad \text{or} \quad x = 4$$

The product of the factors equals 0 if and only if at least one of the factors equals 0.

The x-intercepts are 1, 3, and 4.

If f is a polynomial function, then each solution of the equation $f(x) = 0$ is called a **zero** of f. For cubic functions in the form $f(x) = a(x - p)(x - q)(x - r)$, the zeros are p, q, and r. Each zero is an x-intercept of the graph of f provided the zero is a real number.

THINK AND COMMUNICATE

1. What are the zeros of $f(x) = (x - 1)(x^2 + 1)$? What are the x-intercepts of the graph of f? Is every zero an x-intercept? Explain.

2. In general, if $x - k$ is a factor of a polynomial $f(x)$, what can you say about $f(k)$?

EXAMPLE 3

Find an equation for the cubic function g whose graph is shown.

SOLUTION

The x-intercepts of the graph of g are $-1, 2$, and 4. Therefore, an equation for g has the form

$$g(x) = a(x - (-1))(x - 2)(x - 4),$$

or

$$g(x) = a(x + 1)(x - 2)(x - 4).$$

To find the value of a, use the fact that $g(0) = 24$:

$$a(0 + 1)(0 - 2)(0 - 4) = 24$$

$$8a = 24$$

$$a = 3$$

An equation for g is $g(x) = 3(x + 1)(x - 2)(x - 4)$.

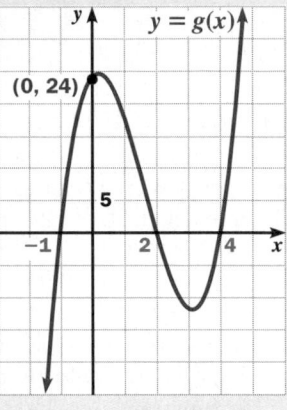

THINK AND COMMUNICATE

3. In general, if $f(x)$ is a polynomial and $f(k) = 0$, what can you say about $x - k$?

Examples 2 and 3 illustrate the following result, known as the *factor theorem*.

> ### Factor Theorem
>
> Let $f(x)$ be a polynomial.
> Then $x - k$ is a factor of $f(x)$ if and only if $f(k) = 0$.

Suppose you want to find the zeros of $f(x) = ax^3 + bx^2 + cx + d$, a cubic function given in standard form. The factor theorem says that if you can find one zero k, then you can write $f(x)$ in the form

$$f(x) = (x - k) \cdot q(x)$$

where $q(x) = \dfrac{f(x)}{x - k}$ is a polynomial of degree 2. You can then find the remaining zeros by solving the quadratic equation $q(x) = 0$, since any zero of q is also a zero of f.

There is a simple test you can use to determine whether a given polynomial function has any integral zeros.

Finding Integral Zeros of Polynomial Functions

Let f be a polynomial function with integral coefficients. Then the only possible integral zeros of f are the divisors of the constant term.

Example: Let $f(x) = 3x^3 + x^2 - 19x + 10$. The possible integral zeros of f are the divisors of **10**: ± 1, ± 2, ± 5, and ± 10.

THINK AND COMMUNICATE

4. **a.** Use the function $f(x) = 3x^3 + x^2 - 19x + 10$. Show that f has only one integral zero by using synthetic substitution to evaluate $f(x)$ at each divisor of the constant term.

   b. Let k = the integral zero you found in part (a). Divide $f(x)$ by $x - k$. Compare the coefficients of the quotient with the numbers in the last row of the synthetic substitution array for k. What do you notice?

EXAMPLE 4

Find the zeros of $f(x) = x^3 + 4x^2 - 5x - 8$.

SOLUTION

Step 1 List the possible integral zeros of f.
The possible integral zeros are the divisors of -8: ± 1, ± 2, ± 4, and ± 8.

Step 2 Use synthetic substitution to find one zero from among the divisors.
Try $x = 1$:   Try $x = -1$:

$$\begin{array}{r|rrrr} 1 & 1 & 4 & -5 & -8 \\ & & 1 & 5 & 0 \\ \hline & 1 & 5 & 0 & -8 \end{array} \qquad \begin{array}{r|rrrr} -1 & 1 & 4 & -5 & -8 \\ & & -1 & -3 & 8 \\ \hline & 1 & 3 & -8 & 0 \end{array}$$

Since $f(1) = -8$, 1 *is not* a zero.   Since $f(-1) = 0$, -1 *is* a zero.

Step 3 Find the quotient polynomial $q(x) = \dfrac{f(x)}{x - (-1)} = \dfrac{f(x)}{x + 1}$.

As you saw in *Think and Communicate* Question 4, the coefficients of $q(x)$ are the red numbers in the synthetic substitution array for $x = -1$:

$$q(x) = 1x^2 + 3x - 8$$

Step 4 Find the remaining zeros of f by solving the equation $q(x) = 0$.

$$x^2 + 3x - 8 = 0$$

$$x = \frac{-3 \pm \sqrt{3^2 - 4(1)(-8)}}{2(1)} = \frac{-3 \pm \sqrt{41}}{2} \qquad \text{Use the quadratic formula.}$$

The zeros of f are -1, $\dfrac{-3 + \sqrt{41}}{2}$, and $\dfrac{-3 - \sqrt{41}}{2}$.

THINK AND COMMUNICATE

5. Use a calculator to verify that $\dfrac{-3 + \sqrt{41}}{2}$ and $\dfrac{-3 - \sqrt{41}}{2}$ are zeros of the function $f(x) = x^3 + 4x^2 - 5x - 8$ in Example 4.

6. Write the function f in Example 4 in intercept form.

7. Can you use the procedure described in Example 4 to find the zeros of $f(x) = x^3 + x^2 - 9x + 2$? Why or why not? How can you approximate the zeros using a graphing calculator or graphing software? using a spreadsheet?

☑ CHECKING KEY CONCEPTS

 Technology Use technology to approximate all real solutions of each equation to the nearest hundredth.

1. $x^3 + x^2 + 5x = 4$ **2.** $-2.28x^3 + 9.14x^2 - 5.07 = 0$

Find an equation for the cubic function whose graph has the given *x*-intercepts and passes through the given point.

3. *x*-intercepts: 1, 2, 3; $(0, -12)$ **4.** *x*-intercepts: $-2, -1, 2$; $(1, 18)$

Find the zeros of each function.

5. $f(x) = (x - 2)(x - 4)(x - 5)$ **6.** $f(x) = 4(x + 9)(x + 3)(x - 3)$

7. $g(x) = x^3 - 7x^2 + 4x + 12$ **8.** $h(x) = 3x^3 + 6x^2 - 22x + 8$

9.4 | Exercises and Applications

Extra Practice exercises on page 762

 Technology Use technology to approximate all real solutions of each equation to the nearest hundredth.

1. $x^3 - 2x^2 + x = 12$ **2.** $-x^3 + x^2 + x = 10$

3. $2x^3 + 3x^2 - 7x - 3 = 0$ **4.** $3x^3 - 5x^2 - 66x + 120 = 0$

5. $-0.5x^3 + 2.1x^2 + 1.3x = 4$ **6.** $0.45x^3 - 2.07x^2 + 0.12x + 3.84 = 0$

Find an equation for each cubic function whose graph is shown.

7.
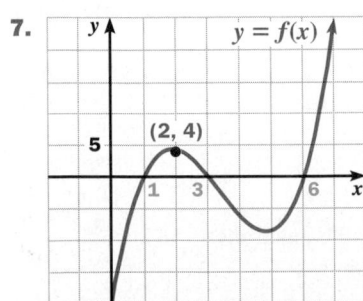

8.
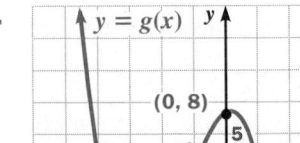
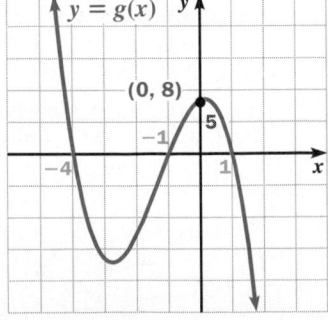

9.
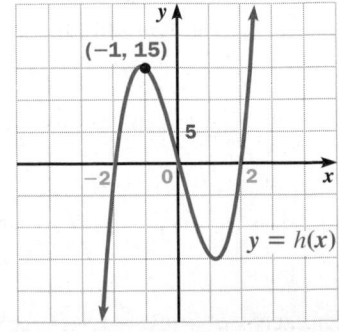

Unlike cars, most motorboats do not have speedometers that tell how fast they are moving. Instead, the speed of a motorboat can be estimated using the boat's *tachometer* and a graph called a *speed curve*. A tachometer shows the propeller speed in revolutions per minute (rev/min). A speed curve gives the boat's speed as a function of the propeller speed.

tachometers

10. **Technology** A speed curve for a certain motorboat traveling in still water is shown. An equation of the curve is:

$$b = 0.00547p^3 - 0.224p^2 + 3.60p - 11.0$$

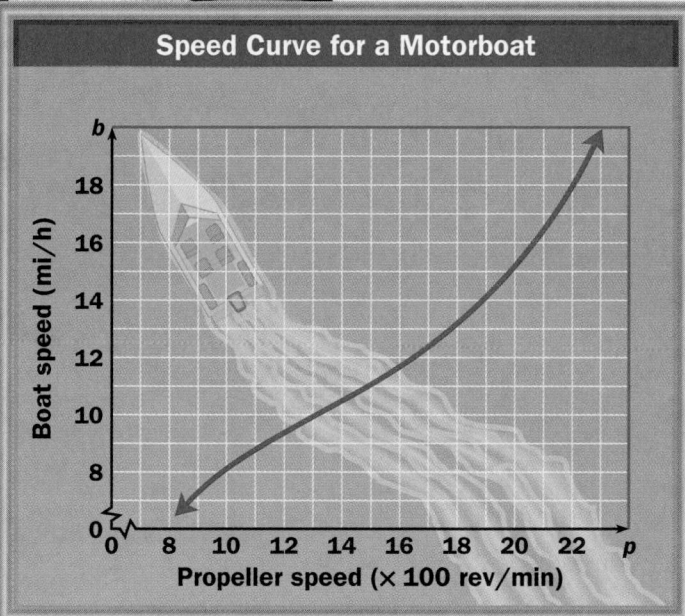

Speed Curve for a Motorboat

Boat speed (mi/h)

Propeller speed (× 100 rev/min)

a. Suppose the boat's tachometer reads 1400 rev/min. How fast is the boat traveling?

b. At what tachometer reading does the boat's speed reach 15 mi/h?

c. What minimum tachometer reading should be maintained in order to cross a lake 2 mi wide in at most 10 min?

11. Suppose the motorboat in Exercise 10 is traveling against a current of 3 mi/h. Sketch a new speed curve for the boat, and find an equation of this curve. How is the new speed curve geometrically related to the speed curve shown?

12. Open-ended Problem Describe how you would go about creating a speed curve for a motorboat.

Find the zeros of each function.

13. $f(x) = 2(x - 3)(x + 10)(x - 7)$

14. $f(x) = (3x + 1)(2x - 5)(x + 8)$

15. $g(x) = x^3 - 4x^2 - x + 4$

16. $g(x) = x^3 + 8x^2 + 17x + 10$

17. $h(x) = x^3 + 8x^2 - 10x - 9$

18. $h(x) = -x^3 + 6x^2 - 2x$

19. $f(x) = 2x^3 - 16x^2 + 19x + 30$

20. $f(x) = 3x^3 + 17x^2 + 16x - 24$

21. EDUCATION For the years from 1985 to 1991, the number of women (in thousands) enrolled in graduate school in the United States can be modeled by the function

$$W = 1.97t^3 - 14.1t^2 + 50.4t + 701$$

where t is the number of years since 1985. Use technology to estimate the year when the number of female graduate students reached 803 thousand.

Write each function in intercept form.

22. $f(x) = (2x + 4)(x - 5)(x - 6)$

23. $g(x) = (2x + 3)(3x + 4)(4x + 5)$

24. $h(x) = x^3 - 3x^2 - 4x + 12$

25. $f(x) = x^3 + 5x^2 - 18x - 28$

BY THE WAY

The first college in the United States to admit women was Oberlin Collegiate Institute in Oberlin, Ohio. The college opened in 1833 with 44 students, 15 of whom were women.

26. SAT/ACT Preview Suppose $x - 2$ is a factor of $f(x) = 3x^3 - x^2 + cx - 8$. What is the value of c?

A. -18 **B.** -6 **C.** 0 **D.** 3 **E.** 4

27. HISTORY Bhaskara was an Indian mathematician who lived during the twelfth century A.D. In his book *Siddhanta Siromani* (*The Gem of Mathematics*), he solved the cubic equation $x^3 - 6x^2 = -12x + 35$ using these steps:

$$x^3 - 6x^2 + 12x - 8 = 27$$

Add $12x - 8$ to both sides so that each side becomes a perfect cube.

Write each side as the cube of a number or an expression.

$$(x - 2)^3 = 3^3$$

$$x - 2 = 3$$

Take the cube root of both sides.

$$x = 5$$

a. Use Bhaskara's method to solve the equation $x^3 + 3x^2 = -3x + 7$.

b. Writing Can Bhaskara's method be used to solve *any* cubic equation? If not, what types of equations are solvable with his method?

c. Research Find a book on the history of mathematics. Discuss some of the contributions made by mathematicians of various cultures to solving polynomial equations.

28. Let $f(x) = ax^3 + bx^2 + cx + d$ where a, b, c, and d are integers.

a. Challenge Prove that if k is an integral zero of f, then k is a divisor of the constant term, d. (*Hint:* Start with the equation $f(k) = 0$. Solve this equation for d, and show that d can be written as the product of k and another integer.)

b. Prove or disprove the converse of the statement in part (a): If k is a divisor of d, then k is a zero of f.

ONGOING ASSESSMENT

29. Cooperative Learning Work with a partner. One person should choose three integers and write a cubic function f in standard form whose zeros are the chosen integers. The other person should use the methods from this section to "discover" the zeros of f. Reverse roles with your partner and repeat the activity.

SPIRAL REVIEW

Describe the end behavior of each function using infinity notation. (*Section 9.3*)

30. $f(x) = 2x^3$ **31.** $g(x) = -x^4 + 2x^3$ **32.** $h(x) = -8x^7 - x^6 + 2x^5 - 14x^2$

Simplify. (*Section 3.2*)

33. 5^{-2} **34.** 2^{-5} **35.** $9^{1/2}$ **36.** $32^{4/5}$

Solve using the quadratic formula. Check your solutions. (*Section 8.4*)

37. $x^2 + 16 = 0$ **38.** $x^2 - 2x + 2 = 0$ **39.** $x^2 = 4x - 13$ **40.** $3x^2 + 6x = -7$

9.5 Finding Zeros of Polynomial Functions

Learn how to...

- **find zeros of higher-degree polynomial functions**

So you can...

- **solve problems about beam deflection, for example**

You can find the number of real zeros of a polynomial function by counting the x-intercepts of the function's graph. As you'll see in the Exploration, the number of real zeros is related to the function's degree.

EXPLORATION
COOPERATIVE LEARNING

Counting Real Zeros

Work with a partner.
You will need:
- a graphing calculator or graphing software

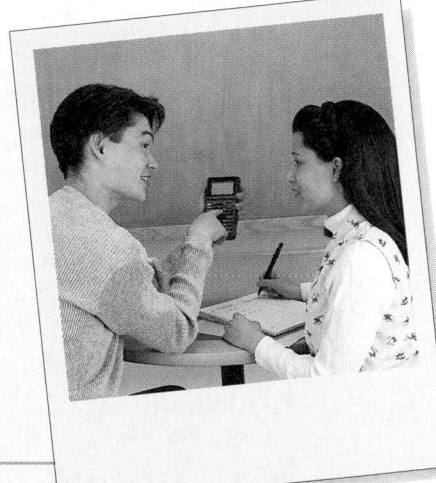

SET UP Adjust the viewing window for your calculator or software so that the intervals $-5 \le x \le 5$ and $-20 \le y \le 20$ are shown on the axes.

This function has **three** real zeros.

1 Graph the following functions. Tell how many real zeros each function has.

- $y = x^3 + x^2 - 7x - 5$
- $y = x^4 - 10x^2 + 9$
- $y = x^5 - 7x^3 - x^2 + 8x$
- $y = x^6 - 11x^4 + 29x^2 - x - 8$

◀ **Example:** $y = x^3 - 2x^2 - 5x + 6$

2 For each function in Step 1, how is the number of real zeros related to the function's degree?

3 Repeat Step 1 using these functions:

- $y - x^3 + x^2 + 3x + 5$
- $y = x^4 + 2x^3 - x^2 + 6x + 7$
- $y = x^5 - x^4 - 6x^3 + 4x^2 + 8x$
- $y = x^6 - 3x^4 - 7x^2 + 10$

Is the relationship you found in Step 2 true for these functions? Explain.

4 Graph several other polynomial functions having different degrees. If necessary, modify your answer for Step 2 so that the stated relationship between a polynomial function's degree and its number of real zeros is *always* true.

The following result gives a maximum for the number of real zeros of a polynomial function.

Real Zeros of Polynomial Functions

A polynomial function of degree n has at most n real zeros.

For example, a cubic function may have 1, 2, or 3 real zeros, as shown below.

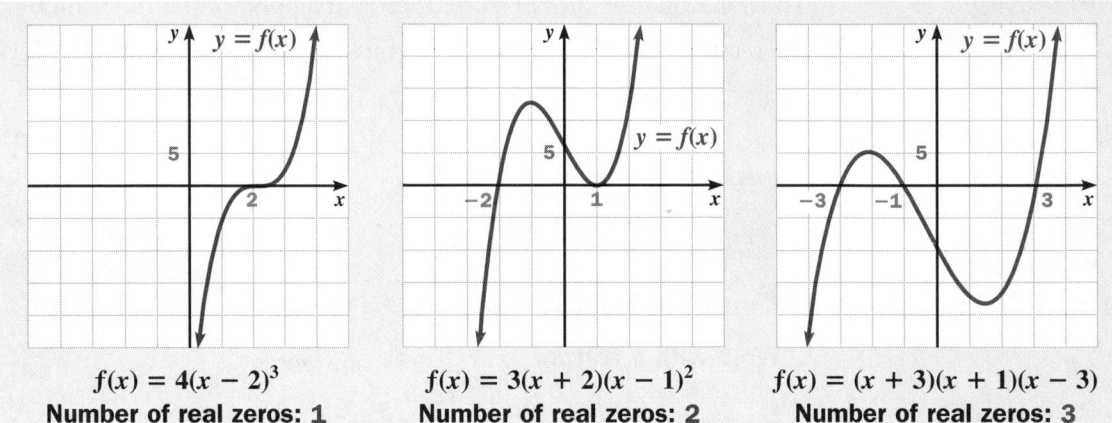

$f(x) = 4(x - 2)^3$
Number of real zeros: 1

$f(x) = 3(x + 2)(x - 1)^2$
Number of real zeros: 2

$f(x) = (x + 3)(x + 1)(x - 3)$
Number of real zeros: 3

The second of the three functions, $f(x) = 3(x + 2)(x - 1)^2$, has a *double zero* at $x = 1$. A number k is a **double zero** of a polynomial function f if $(x - k)^2$ is a factor of $f(x)$. Similarly, k is a **triple zero** of f if $(x - k)^3$ is a factor of $f(x)$. The first function, $f(x) = 4(x - 2)^3$, has a triple zero at $x = 2$.

The next example shows that a polynomial function may have imaginary zeros as well as real zeros.

EXAMPLE 1

Find the zeros of $f(x) = x^3 - x^2 - 7x + 15.$

SOLUTION

Find one zero of f by checking the divisors of the constant term, 15. You will find that -3 is a zero:

$$
\begin{array}{r|rrrr}
-3 & 1 & -1 & -7 & 15 \\
   &   & -3 & 12 & -15 \\
\hline
   & 1 & -4 & 5 & 0
\end{array}
$$

The quotient polynomial is $q(x) = \dfrac{f(x)}{x + 3} = 1x^2 - 4x + 5$. Find the remaining zeros of f by solving the equation $q(x) = 0$:

$$x^2 - 4x + 5 = 0$$

Use the quadratic formula. $\longrightarrow$ $x = \dfrac{-(-4) \pm \sqrt{(-4)^2 - 4(1)(5)}}{2(1)} = \dfrac{4 \pm \sqrt{-4}}{2} = \dfrac{4 \pm 2i}{2} = 2 \pm i$

The zeros of f are -3, $2 + i$, and $2 - i$.

In Example 1, you saw that the cubic function $f(x) = x^3 - x^2 - 7x + 15$ has three zeros in the set of complex numbers. The next theorem, known as the *fundamental theorem of algebra*, extends this result.

Fundamental Theorem of Algebra

A polynomial function of degree n has exactly n complex zeros, provided each double zero is counted as 2 zeros, each triple zero is counted as 3 zeros, and so on.

BY THE WAY

The fundamental theorem of algebra was first proved in 1799 by Carl Friedrich Gauss, one of the greatest mathematicians of all time. Gauss, a child prodigy, is said to have detected an error in his father's bookkeeping when he was only three years old.

THINK AND COMMUNICATE

1. Explain why the fundamental theorem of algebra implies the property stated on the previous page: A polynomial function of degree n has at most n real zeros.

Rational Zeros of Polynomial Functions

In Section 9.4, you learned a procedure for finding the possible integral zeros of a polynomial function. The following more general result lets you find all possible rational zeros (that is, zeros of the form $\frac{p}{q}$ where p and q are integers and $q \neq 0$).

Rational Zeros Theorem

Let f be a polynomial function with integral coefficients. Then the only possible rational zeros of f are $\frac{p}{q}$ where p is a divisor of the constant term of $f(x)$ and q is a divisor of the leading coefficient.

For example, you can use the theorem above to find the possible rational zeros of $f(x) = 4x^3 - 7x^2 + 2x - 3$:

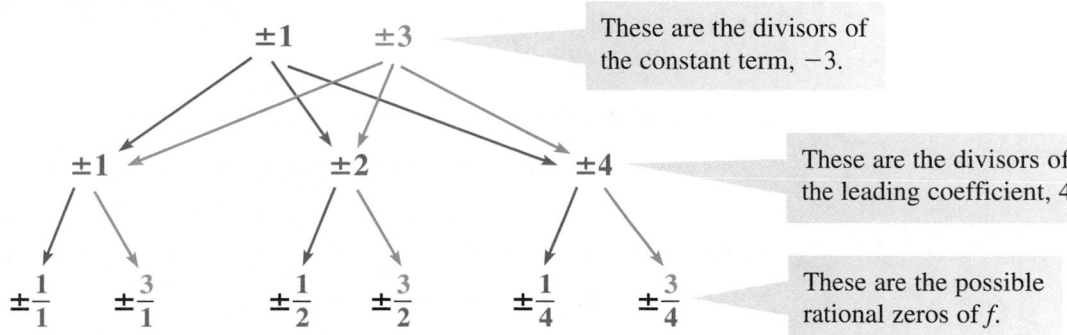

These are the divisors of the constant term, -3.

These are the divisors of the leading coefficient, 4.

These are the possible rational zeros of f.

Note that the possible rational zeros include the possible integral zeros: ± 1 and ± 3.

EXAMPLE 2

Find the zeros of $f(x) = 9x^4 - 6x^3 + 19x^2 - 12x + 2$.

SOLUTION

Step 1 Find the possible rational zeros of f.
The divisors of the constant term, 2, are ± 1 and ± 2.
The divisors of the leading coefficient, 9, are ± 1, ± 3, and ± 9.

So the possible rational zeros of f are ± 1, ± 2, $\pm\frac{1}{3}$, $\pm\frac{2}{3}$, $\pm\frac{1}{9}$, and $\pm\frac{2}{9}$.

Step 2 Use synthetic substitution to find one zero of f from the list in Step 1.
To save space, you can use a "short form" of synthetic substitution in which
the coefficients of $f(x)$ are listed only once and all additions are done mentally.

	9	−6	19	−12	2	← coefficients of $f(x)$
1	9	3	22	10	12	← $f(1) = 12$
−1	9	−15	34	−46	48	← $f(-1) = 48$
2	9	12	43	74	150	← $f(2) = 150$
−2	9	−24	67	−146	294	← $f(-2) = 294$
$\frac{1}{3}$	9	−3	18	−6	0	← $f\left(\frac{1}{3}\right) = 0$

possible rational zeros

One zero of f is $\frac{1}{3}$.

Step 3 Find a second zero of f by finding a zero of the quotient polynomial
$q_1(x) = \dfrac{f(x)}{x - \dfrac{1}{3}} = 9x^3 - 3x^2 + 18x - 6$.

Continue checking the possible rational zeros of f listed in Step 1, starting
with $\frac{1}{3}$ $\left(\text{since } \frac{1}{3} \text{ may be a double zero}\right)$.

	9	−3	18	−6	← Use the coefficients of $q_1(x)$.
$\frac{1}{3}$	9	0	18	0	← $q_1\left(\frac{1}{3}\right) = 0$

So $\frac{1}{3}$ is a double zero of f.

Step 4 Find the remaining zeros of f by solving the equation $q_2(x) = 0$
where $q_2(x)$ is the new quotient polynomial: $q_2(x) = \dfrac{q_1(x)}{x - \dfrac{1}{3}} = 9x^2 + 0x + 18$.

$$9x^2 + 18 = 0$$
$$9x^2 = -18$$
$$x^2 = -2$$
$$x = \pm\sqrt{-2} = \pm i\sqrt{2}$$

The zeros of f are $\frac{1}{3}$ (a double zero), $i\sqrt{2}$, and $-i\sqrt{2}$.

THINK AND COMMUNICATE

2. In Example 2, how do you know that the quotient polynomial $q_1(x)$ has degree 3?

3. In Step 3 of Example 2, it was assumed that any rational zero of q_1 must be in the list of possible rational zeros of f. Explain why this is true.

4. Refer to the functions f and q_2 in Example 2.

    a. Complete this equation: $f(x) = \underline{\ ?\ } \cdot q_2(x)$. How does the equation show that each zero of q_2 is also a zero of f?

    b. Verify directly that the zeros of q_2 are zeros of f by evaluating $f(x)$ when $x = i\sqrt{2}$ and when $x = -i\sqrt{2}$.

5. Does the degree of the function f in Example 2 equal the number of *distinct* complex zeros of f? Does this contradict the fundamental theorem of algebra? Explain.

6. Use the rational zeros theorem to show that $h(x) = 3x^3 + 4x^2 + x - 5$ has no rational zeros. Does this mean that h has no *real* zeros? If not, how can you approximate any real zeros of h?

✓ CHECKING KEY CONCEPTS

 Technology Use a graphing calculator or graphing software to approximate all real zeros of each function to the nearest hundredth.

1. $y = x^3 - 5x^2 + 11x - 10$ **2.** $y = x^4 + 4x^3 - 2x^2 - 11x + 1$

3. $y = 2x^5 + 7x^4 - 3x^3 - 19x^2 + 7$ **4.** $y = -3x^6 + 12x^4 - 4x^2 - 8$

For each function in Questions 5–8:

a. List the possible rational zeros of the function.

b. Find all real and imaginary zeros of the function. Identify any double or triple zeros.

    5. $f(x) = 2x^3 - 5x^2 - 4x + 3$ **6.** $f(x) = 4x^3 + 3x^2 + 12x + 9$

    7. $g(x) = x^4 - 6x^3 + 14x^2 - 16x + 8$ **8.** $h(x) = 3x^4 + 8x^3 + 6x^2 - 1$

9.5 Exercises and Applications

Extra Practice exercises on page 763

 Technology Use a graphing calculator or graphing software to approximate all real zeros of each function to the nearest hundredth.

1. $y = x^3 + x^2 - 9x + 3$

2. $y = 2x^3 + 3x^2 - 3x + 5$

3. $y = x^4 - x^3 - 4x^2 + 2x - 12$

4. $y = 3x^4 - 14x^2 - 2x + 7$

5. $y = -6x^5 - 18x^4 + x^3 + 27x^2 - 4$

6. $y = x^5 - 6x^4 + 5x^3 + 15x^2 - 9x - 8$

7. $y = 2x^6 - 16x^4 + 32x^2 - x - 10$

8. $y = -4x^6 + 13x^4 + x^3 - 11x^2 - 2x + 10$

For each function in Exercises 9–18:

a. List the possible rational zeros of the function.

b. Find all real and imaginary zeros of the function. Identify any double or triple zeros.

9. $f(x) = x^3 - 5x^2 + 17x - 13$

10. $f(x) = 4x^3 + 12x^2 + x + 3$

11. $f(x) = 3x^3 + 10x^2 + 4x - 8$

12. $f(x) = 30x^3 - x^2 - 6x + 1$

13. $g(x) = x^4 - 2x^3 - 21x^2 + 22x + 40$

14. $g(x) = 4x^4 - 4x^3 - 11x^2 + 6x + 9$

15. $g(x) = 16x^4 + 8x^3 + 17x^2 + 8x + 1$

16. $g(x) = -6x^4 + 19x^3 - 6x^2 - 41x + 20$

17. $h(x) = -x^5 + 2x^4 + 10x^3 + 8x^2 - x - 2$

18. $h(x) = 2x^5 - 5x^4 - 40x^3 - x^2 - 52x - 84$

Connection ▶ PHYSICS

A *beam* is an oblong piece of metal, wood, or other material used as a horizontal support. When a force is applied to a beam, the beam is *deflected*, or bent, from its original position. The deflection can often be modeled by a polynomial function.

19. **Technology** One example of a beam is a diving board. The first diagram shows a 150 lb person standing at the end of an aluminum diving board. The deflection of the board, in inches, at a point x in. from its support can be modeled by this function:

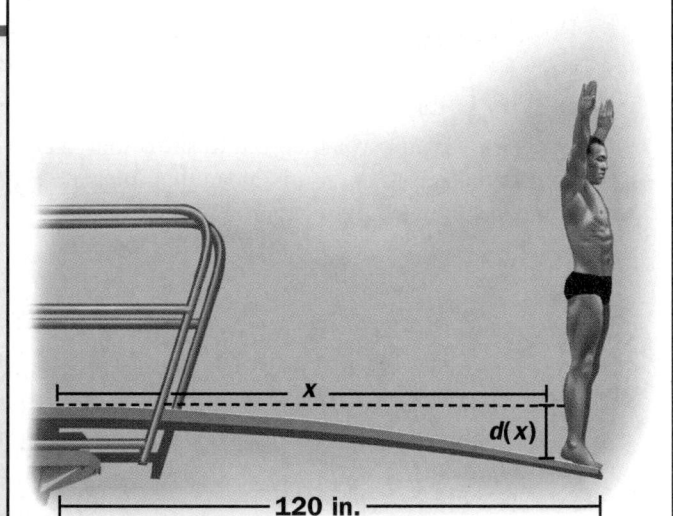

$$d(x) = (-4.752 \times 10^{-7})x^3 + (1.711 \times 10^{-4})x^2$$

Use a graphing calculator or graphing software to find the portion of the diving board where the deflection exceeds 1 in.

20. **Technology** Another example of a beam is a bookshelf. The second diagram shows a wooden bookshelf loaded with 180 lb of books. The deflection of the bookshelf, in inches, at a point x in. from its left end can be modeled by this function:

$$d(x) = (2.724 \times 10^{-7})x^4 - (3.269 \times 10^{-5})x^3 + (9.806 \times 10^{-4})x^2$$

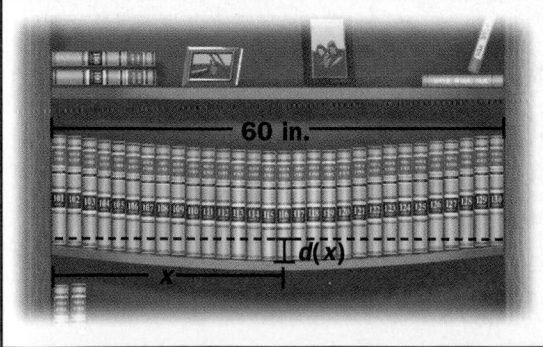

a. Use a graphing calculator or graphing software to graph $y = d(x)$. Approximate all real zeros of $y = d(x)$ in the interval $0 \le x \le 60$. Explain why your answers make sense in the given situation.

b. At what point on the bookshelf is the deflection at a maximum? What is the maximum deflection?

c. Find the portion of the bookshelf where the deflection exceeds 0.1 in.

21. SAT/ACT Preview Let $f(x) = 15x^4 + lx^3 + mx^2 + nx + 12$ where l, m, and n are integers. Which of the following numbers *cannot* be a zero of f?

A. $\dfrac{12}{5}$ **B.** 6 **C.** $-\dfrac{2}{3}$ **D.** $\dfrac{5}{4}$ **E.** $-\dfrac{1}{15}$

22. Investigation For parts (a)–(d), identify the double zeros of each function.

a. $f(x) = (x + 2)(x - 2)^2$ **b.** $g(x) = -3(x + 2)^2(x - 1)$

c. $h(x) = x^2(x - 3)^2$ **d.** $f(x) = -(x - 1)^2(x - 4)^2$

e. [graph icon] **Technology** Use a graphing calculator or graphing software to graph the functions in parts (a)–(d). Describe the behavior of each graph when the value of x is close to a double zero.

ONGOING ASSESSMENT

23. a. Writing Explain why an odd-degree polynomial function must have at least one real zero.

b. Open-ended Problem Give an example of an even-degree polynomial function that has no real zeros.

SPIRAL REVIEW

Find an equation for the cubic function whose graph has the given x-intercepts and passes through the given point. *(Section 9.4)*

24. x-intercepts: 1, 2, 4; $(0, -8)$ **25.** x-intercepts: $-7, -5, 0$; $(-1, 48)$

For each equation, tell whether y varies directly with x. *(Section 2.1)*

26. $y = 3x$ **27.** $y = 3x + 3$ **28.** $y = \dfrac{x}{3}$ **29.** $y = \dfrac{3}{x}$

ASSESS YOUR PROGRESS

VOCABULARY

cubic function (p. 413)   **double zero** (p. 420)
zero (p. 413)   **triple zero** (p. 420)

1. ENERGY For a certain windmill, the power produced when the wind speed is s m/s can be modeled by the function

$$P(s) = -0.699s^3 + 19.8s^2 - 120s + 220$$

where $P(s)$ is in kilowatts (kW) and $4 \le s \le 16$. At what wind speed does the power produced by the windmill reach 300 kW? *(Section 9.4)*

Find the zeros of each function. *(Sections 9.4 and 9.5)*

2. $f(x) = 5(x - 1)(x + 2)(x - 3)$ **3.** $f(x) = x^3 - 14x + 8$

4. $g(x) = 6x^3 - x^2 - 9x - 10$ **5.** $h(x) = 2x^4 - 5x^3 + 5x^2 - 20x - 12$

6. Journal Discuss the relationship between the factors and zeros of a polynomial function f. Explain how the rational zeros theorem can help you find the zeros of f.

9.6 Inverse Variation

Learn how to...

- recognize inverse variation

- write and use inverse variation equations

So you can...

- find the pressure that a person exerts when walking on snow, for example

If you've ever walked through deep snow, you can appreciate the value of snowshoes. Snowshoes prevent you from sinking by distributing your weight over a large area. This reduces the pressure you exert on the snow.

EXAMPLE 1   Application: Physics

For a 120 lb person, the average pressure P exerted on the snow beneath the person's footwear is given by

$$P = \frac{120}{A}$$

where A is the area of the bottom of the footwear in square feet and P is measured in pounds per square foot.

a. Use the pictures below. Compare the pressure that a 120 lb person exerts when wearing the snowshoes with the pressure the person exerts when wearing the boots.

b. Graph $P = \dfrac{120}{A}$ using a graphing calculator or graphing software. Describe what happens to pressure as area gets very small and very large.

The area **A** of the soles of a pair of boots is **A = 0.4 ft²**.

The area **A** of a pair of snowshoes is **A = 4 ft²**.

SOLUTION

a. For the boots, $P = \dfrac{120}{0.4} = 300$ lb/ft².

For the snowshoes, $P = \dfrac{120}{4} = 30$ lb/ft².

The pressure exerted when the snowshoes are worn is only one tenth of the pressure exerted when the boots are worn.

b.

As **area** gets very small, pressure gets very large.

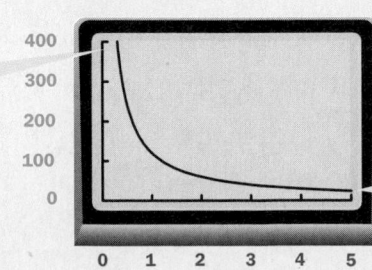

As **area** gets very large, pressure gets very small.

THINK AND COMMUNICATE

1. **a.** In Example 1, what happens to the pressure exerted on the snow when the area of the footwear is doubled? tripled? increased by a factor of *n*?

   b. What can you say about the product of pressure and area?

2. 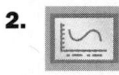 **Technology** Use a graphing calculator or graphing software to graph $y = \dfrac{120}{x}$. Adjust the viewing window so that the intervals $-5 \le x \le 5$ and $-400 \le y \le 400$ are shown on the axes. Compare this graph with the graph in Example 1. What do you notice?

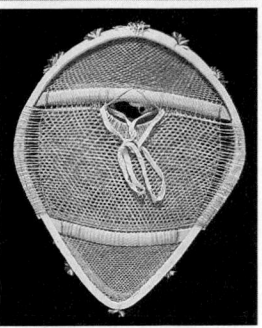
Two variables *x* and *y* are said to show **inverse variation** if they are related by an equation of this form:

You can say that *y* **varies inversely** with *x*, or that *y* is **inversely proportional** to *x*.

$$y = \frac{a}{x}$$

The number *a*, which must be nonzero, is called the **constant of variation**.

In Example 1, for instance, the equation $P = \dfrac{120}{A}$ indicates that *P* varies inversely with *A*. The constant of variation is 120.

The graph of an inverse variation equation is called a **hyperbola**. A hyperbola consists of two pieces, called *branches*. For $a > 0$, the graph of $y = \dfrac{a}{x}$ looks like this:

The **y-axis** is a vertical *asymptote* of the graph. An **asymptote** is a line that a graph approaches more and more closely.

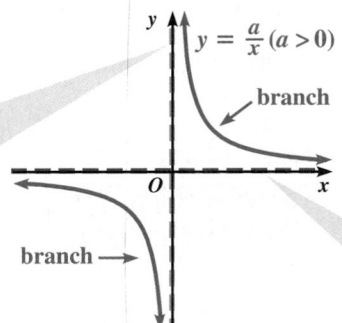

branch

branch

The **x-axis** is a horizontal asymptote of the graph.

Note that if *x* and *y* represent positive quantities (such as area and pressure), the graph of $y = \dfrac{a}{x}$ is only the first-quadrant branch of a hyperbola.

THINK AND COMMUNICATE

3. 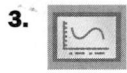 **Technology** Use a graphing calculator or graphing software to graph $y = \dfrac{a}{x}$ for several negative values of *a*. Compare the graph of $y = \dfrac{a}{x}$ where $a < 0$ with the graph of $y = \dfrac{a}{x}$ where $a > 0$.

4. Explain why the graph of $y = \dfrac{a}{x}$ (where $a \ne 0$) never crosses either the *x*- or *y*-axis.

An inverse variation equation $y = \dfrac{a}{x}$ can also be written this way:

$$xy = a$$

Therefore, y varies inversely with x if and only if the product of x and y is constant.

EXAMPLE 2   Application: Biology

The rate r at which a bird flaps its wings depends on the bird's wing length l. The pictures show values of l and r for different species of birds.

  a. Show that r varies approximately inversely with l.

  b. Write an equation giving r as a function of l.

Merganser

$l = 28.9$ cm
$r = 4.6$ flaps
    per second

Widgeon

$l = 26.2$ cm
$r = 5.1$ flaps per second

Peregrine falcon

$l = 30.9$ cm
$r = 4.3$ flaps per second

Egyptian vulture

$l = 49.5$ cm
$r = 2.7$ flaps
    per second

SOLUTION

  a. Find the product lr for each species of bird. A spreadsheet may be helpful.

	A	B	C
	l	r	lr
2	26.2	5.1	133.6
3	28.9	4.6	132.9
4	30.9	4.3	132.9
5	49.5	2.7	133.7

Bird Data

Each product is about 133.

Since the products lr are nearly constant, r varies approximately inversely with l.

  b. From part (a), the relationship between l and r can be modeled by the equation $lr = 133$, or $r = \dfrac{133}{l}$.

THINK AND COMMUNICATE

  5. The wing length of a cormorant is about 35.0 cm. About how many times per second does a cormorant flap its wings?

EXAMPLE 3   **Application: Home Repair**

The force F needed to loosen a bolt with a wrench is inversely proportional to the turning radius r.

The wrench shown can loosen the bolt with 250 lb of force. How much force is needed when a wrench with a 10 in. turning radius is used?

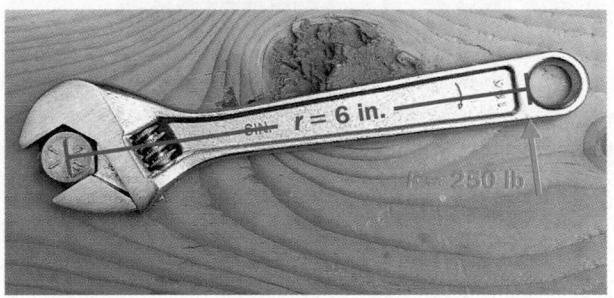

$r = 6$ in.

250 lb

SOLUTION

The product of **turning radius** and **force** is constant. Therefore:

$$\textbf{10 in.} \times \begin{array}{c}\text{force needed with}\\\text{a 10 in. radius}\end{array} = \textbf{6 in.} \times \begin{array}{c}\text{force needed with}\\\text{a 6 in. radius}\end{array}$$

F is the unknown force.

$$10F = 6 \cdot 250$$
$$10F = 1500$$
$$F = 150$$

The force needed is 150 lb.

THINK AND COMMUNICATE

6. Write an equation giving the force F needed to loosen the bolt in Example 3 as a function of the wrench's turning radius r.

☑ CHECKING KEY CONCEPTS

Tell whether y varies inversely with x. If so, state the constant of variation.

1. $y = \dfrac{5}{x}$　　　**2.** $y = \dfrac{x}{5}$　　　**3.** $\dfrac{y}{x} = -0.3$　　　**4.** $xy = -0.3$

Tell whether y varies inversely with x. If so, write an equation giving y as a function of x.

5.

x	y
1	3
2	6
3	9
4	12

6.

x	y
1	24
2	12
3	8
4	6

7.

x	y
1	4
2	3
3	2
4	1

8. The time it takes Teri Petersen to drive home from college is inversely proportional to her average speed. When Teri drives at an average speed of 45 mi/h, it takes her 6 h to get home. How fast must Teri drive if she wants to get home in 5 h?

9.6 Exercises and Applications

Extra Practice exercises on page 763

Technology For each equation, tell whether *y* varies inversely with *x*. If so, state the constant of variation and graph the equation using a graphing calculator or graphing software.

1. $y = \dfrac{2}{x}$ **2.** $xy = 10$ **3.** $y = -\dfrac{7}{2}x$ **4.** $y = -\dfrac{7}{2x}$

5. $y = \dfrac{2}{x + 1}$ **6.** $x = -\dfrac{1.6}{y}$ **7.** $3xy = 15$ **8.** $xy^3 = 15$

Tell whether you think the two quantities show inverse variation. Explain your reasoning.

9. the possible length and width of a rectangular garden having an area of 72 ft²

10. the number of college students sharing an apartment, and each student's rent

11. the distance driven in a car and the amount of gasoline used

12. the possible radius and height of a can holding 1 L of juice

---

INTERVIEW **Vera Rubin**

Look back at the article on pages 386–388.

Astronomers like Vera Rubin use a measurement called **apparent size** *to describe how big an object in space looks from different locations. For example, the apparent size of the sun from the planet Mercury is about 2.58. This means that the sun appears about 2.58 times as wide on Mercury as it does on Earth.*

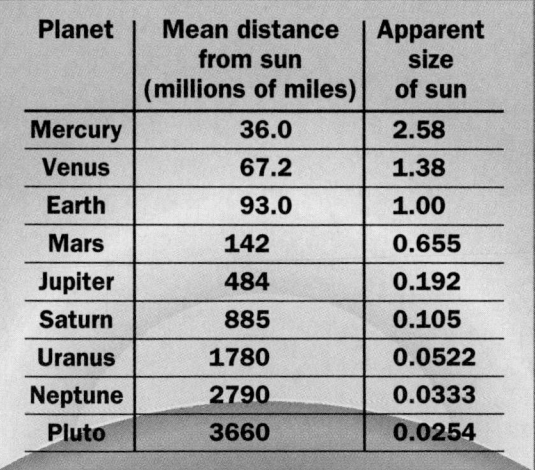

Planet	Mean distance from sun (millions of miles)	Apparent size of sun
Mercury	36.0	2.58
Venus	67.2	1.38
Earth	93.0	1.00
Mars	142	0.655
Jupiter	484	0.192
Saturn	885	0.105
Uranus	1780	0.0522
Neptune	2790	0.0333
Pluto	3660	0.0254

13. Use the table.

  a. Show that the apparent size of the sun varies inversely with the distance from where it is observed.

  b. Write an equation giving the sun's apparent size *s* from a point *d* million miles away.

  c. Ceres is an asteroid located between the orbits of Mars and Jupiter. The mean distance of Ceres from the sun is about 257 million miles. What is the apparent size of the sun from Ceres?

14. Visual Thinking Draw a circle representing the sun as seen from Earth. Use the diameter of this circle and the apparent sizes in the table to draw eight more circles representing the sun as seen from each of the other eight planets.

On May 11, 1970, a powerful tornado struck Lubbock, Texas. The diagram below shows the horizontal cross section of the tornado at ground level. Notice that a tornado consists of a central **core region** and a surrounding free vortex region.

The May 11, 1970, tornado caused major damage to Lubbock, Texas.

15. In the core region, the rotational wind speed s varies *directly* with the distance d from the tornado's center. Use the information in the diagram to estimate the rotational wind speed 700 ft from the center of the Lubbock tornado.

16. In the free vortex region, s varies *inversely* with d. Use the information in the diagram to estimate the rotational wind speed 2000 ft from the center of the Lubbock tornado.

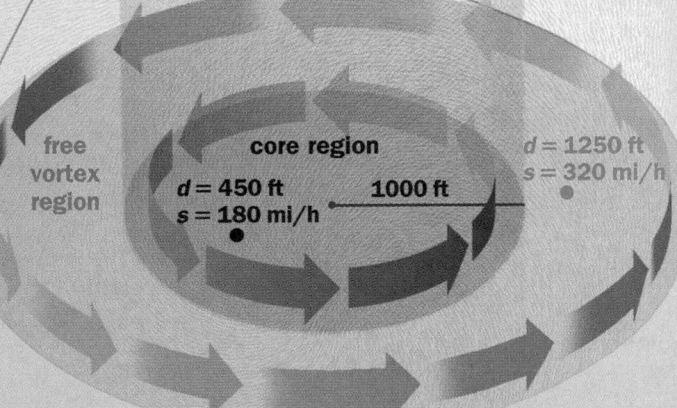

free vortex region

core region

$d = 450$ ft
$s = 180$ mi/h

1000 ft

$d = 1250$ ft
$s = 320$ mi/h

17. a. Write an equation giving s as a function of d in the core region of the Lubbock tornado. What is the function's domain?

b. Write an equation giving s as a function of d in the free vortex region of the Lubbock tornado. What is this function's domain?

c. Challenge Draw a graph showing s as a function of d for $0 \le d \le 4000$.

18. Use the equations and graph from Exercise 17.

a. How far from the center of the Lubbock tornado did the maximum rotational wind speed occur? What is the significance of this distance? What was the maximum speed?

b. How far from the center of the tornado was the rotational wind speed 300 mi/h?

19. Open-ended Problem Draw a graph of a function having $x = 3$ as a vertical asymptote and $y = -2$ as a horizontal asymptote.

20. SAT/ACT Preview Suppose A $= \dfrac{a}{x}$ and B $= ax$ where a is a positive constant. Which of the following is true for all positive values of x?

A. A > B   **B.** B > A   **C.** A = B   **D.** relationship cannot be determined

21. HYDRAULICS The time t required to fill a container with water from a hose is given by the equation

$$t = \frac{V}{As}$$

where V is the volume of the container, A is the cross-sectional area of the hose, and s is the speed of the water from the hose.

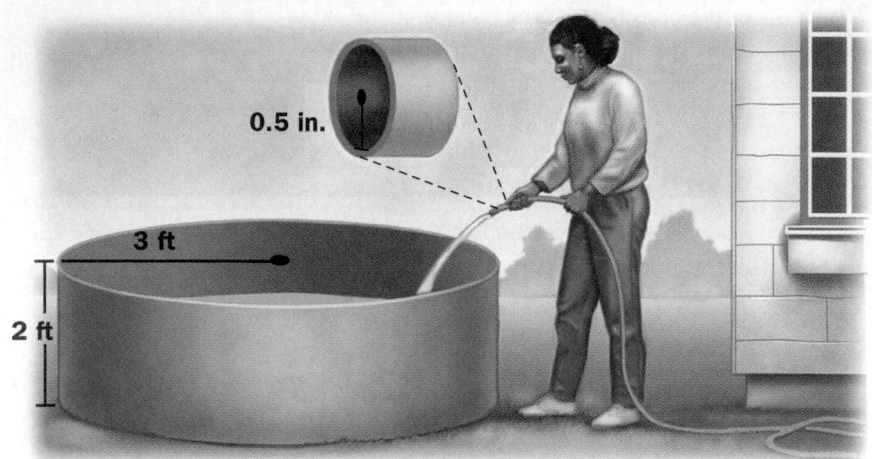

0.5 in.

3 ft

2 ft

a. Use the diagram. Find an equation giving the time required to fill the swimming pool as a function of the water speed. How long will it take to fill the pool if the water speed is 10 ft/s? (*Hint:* First convert the units of the hose radius to feet.)

b. In part (a), does time vary inversely with water speed? If so, what is the constant of variation?

c. **Cooperative Learning** Working with a partner, devise an experiment in which you use the equation $t = \dfrac{V}{As}$ to estimate the speed of water from a garden hose. If you have a garden hose available, perform the experiment and write a report summarizing your results.

ONGOING ASSESSMENT

22. Writing Explain how inverse variation differs from direct variation. Give real-life examples illustrating each type of variation.

SPIRAL REVIEW

Find all real and imaginary zeros of each function. Identify any double or triple zeros. *(Section 9.5)*

23. $f(x) = 6x^3 - 23x^2 - 5x + 4$ **24.** $g(x) = 15x^3 - x^2 + 3x - 2$

25. $h(x) = -4x^4 - 12x^3 - 5x^2 + 16x - 5$ **26.** $f(x) = 18x^5 + 63x^4 + 82x^3 + 48x^2 + 12x + 1$

Describe how the graph of each equation is related to the graph of $y = 5x^2$. *(Section 5.2)*

27. $y = 5(x - 1)^2$ **28.** $y = 5x^2 + 3$ **29.** $y = 5(x - 1)^2 + 3$ **30.** $y = 5(x + 1)^2 - 3$

9.7 Working with Simple Rational Functions

Learn how to...

- **identify important features of translated hyperbolas**

- **find equations of translated hyperbolas**

So you can...

- **understand population models, for example**

In Chapter 5, you learned how to translate parabolas by introducing new constants in their equations. As you'll see in the Exploration, you can do the same thing with hyperbolas.

EXPLORATION
COOPERATIVE LEARNING

Investigating Translated Hyperbolas

Work with a partner.
You will need: ● a graphing calculator or graphing software

1 Graph the following equations. For each graph, identify the asymptotes and tell how the graph is related to the graph of $y = \frac{1}{x}$.

- $y = \frac{1}{x - 1}$
- $y = \frac{1}{x - 3}$
- $y = \frac{1}{x + 2}$
- $y = \frac{1}{x + 4}$

Example: $y = \frac{1}{x - 2}$

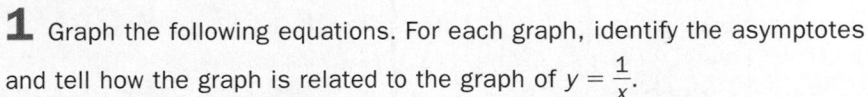

The line $x = 2$ is a vertical asymptote.

The line $y = 0$ is a horizontal asymptote.

The graph is the hyperbola $y = \frac{1}{x}$ translated **2 units right**.

2 Repeat Step 1 using these equations.

- $y = \frac{1}{x} + 1$
- $y = \frac{1}{x} + 3$
- $y = \frac{1}{x} - 2$
- $y = \frac{1}{x} - 4$

3 Predict how the graph of each equation below is related to the graph of $y = \frac{2}{x}$. Use your calculator or software to check your predictions.

- $y = \frac{2}{x - 4}$
- $y = \frac{2}{x + 1}$
- $y = \frac{2}{x} + 3$
- $y = \frac{2}{x} - 5$

THINK AND COMMUNICATE

1. How is the graph of $y = \dfrac{a}{x - h}$ geometrically related to the graph of

$y = \dfrac{a}{x}$ when h is positive? when h is negative?

2. How is the graph of $y = \dfrac{a}{x} + k$ geometrically related to the graph of

$y = \dfrac{a}{x}$ when k is positive? when k is negative?

You can obtain the graph of $y = \dfrac{a}{x - h} + k$ by translating the graph of $y = \dfrac{a}{x}$
horizontally h units and **vertically k units**.

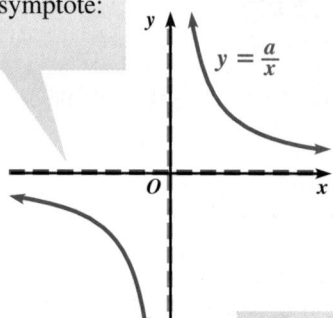

horizontal asymptote:
y = 0

vertical asymptote:
x = 0

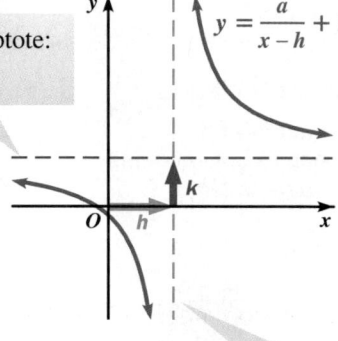

horizontal asymptote:
y = k

vertical asymptote:
x = h

EXAMPLE 1

Find an equation for the function f whose graph is the hyperbola shown.

SOLUTION

A vertical asymptote of the graph is $x = 3$,
and a horizontal asymptote is $y = 1$.
Therefore, an equation for f has this form:

$$f(x) = \frac{a}{x - 3} + 1$$

To find the value of a, use the fact
that $f(-1) = 0$:

$$\frac{a}{-1 - 3} + 1 = 0$$

$$\frac{a}{-4} = -1$$

$$a = 4$$

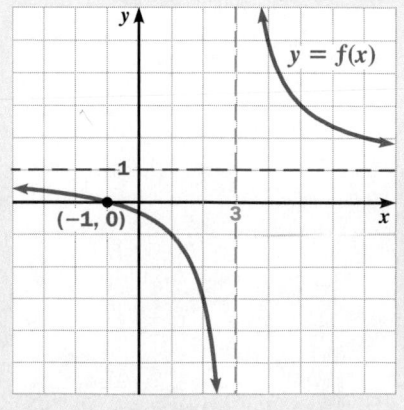

An equation for f is $f(x) = \dfrac{4}{x - 3} + 1$.

Rational Functions

Notice that the function $f(x) = \dfrac{4}{x - 3} + 1$ in Example 1 can be written this way:

Toolbox p. 783
Adding and Subtracting Rational Expressions

$$f(x) = \frac{4}{x - 3} + 1$$

$$= \frac{4}{x - 3} + \frac{x - 3}{x - 3}$$

$$= \frac{4 + x - 3}{x - 3}$$

$$= \frac{x + 1}{x - 3} \quad \longleftarrow \quad \text{polynomial}$$
$$\phantom{= \frac{x + 1}{x - 3}} \quad \longleftarrow \quad \text{polynomial}$$

This is an example of a *rational function*. A **rational function** has the form $f(x) = \dfrac{p(x)}{q(x)}$ where $p(x)$ and $q(x)$ are polynomials. In this section, $p(x)$ and $q(x)$ will always be constant or first-degree polynomials.

EXAMPLE 2

Let $g(x) = \dfrac{-6x + 7}{2x + 1}$. Find the asymptotes of the graph of g, and tell how the graph is related to a hyperbola with equation of the form $y = \dfrac{a}{x}$.

SOLUTION

Step 1 Divide the denominator of $g(x)$ into the numerator.

$$\begin{array}{r} -3 \\ 2x + 1 \overline{\smash{\big)} -6x + 7} \\ \underline{-6x - 3} \\ 10 \end{array}$$

So $\dfrac{-6x + 7}{2x + 1} = -3 + \dfrac{10}{2x + 1}$.

Step 2 Write $g(x)$ in the form $g(x) = \dfrac{a}{x - h} + k$.

$$g(x) = \frac{-6x + 7}{2x + 1}$$

$$= \frac{10}{2x + 1} - 3$$

$$= \frac{\dfrac{1}{2}}{\dfrac{1}{2}} \cdot \frac{10}{2x + 1} - 3$$

$$= \frac{5}{x + \dfrac{1}{2}} - 3$$

$$= \frac{5}{x - \left(-\dfrac{1}{2}\right)} + (-3)$$

A vertical asymptote is $x = -\dfrac{1}{2}$. A horizontal asymptote is $y = -3$. The graph of g is the graph of $y = \dfrac{5}{x}$ translated $\dfrac{1}{2}$ **unit left** and **3 units down**.

THINK AND COMMUNICATE

3. In Example 2, how do you know that the direction of the translation is left and down?

4. a. What are the domain and range of the function *g* in Example 2?

   b. Generalize your answer to part (a) by identifying the domain and range of any function that can be written in the form $f(x) = \dfrac{a}{x - h} + k$.

EXAMPLE 3   Application: Population Growth

In 1960, three scientists from the University of Illinois published a paper claiming that the world population *P* in some future year *t* can be predicted with the equation

$$P = \frac{179}{2026 - t}$$

where *P* is in billions and $t < 2026$. This population model is often called the "doomsday" model.

a. Graph the scientists' equation using a graphing calculator or graphing software. What is the predicted world population in the year 2000?

b. Writing According to the doomsday model, what general trend in world population will occur in future years? Why is the name "doomsday" appropriate?

SOLUTION

a.

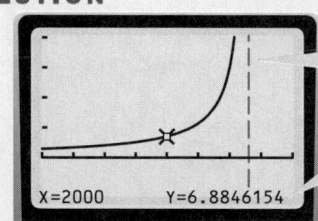

A vertical asymptote is **t = 2026**.

The predicted population in the year 2000 is about **6.88 billion**.

X=2000   Y=6.8846154

b.

The doomsday model predicts that world population will increase more and more rapidly as time passes. As the year 2026 approaches, the population will become infinitely large. Because Earth's resources will not be able to sustain such a population, the name "doomsday" is appropriate.

5. Do you think the doomsday model is realistic? Why or why not?

6. The world population in 1990 was about 5.29 billion. How does this figure compare with the prediction from the doomsday model?

☑ **CHECKING KEY CONCEPTS**

Match each equation with its graph.

1. $y = \dfrac{3}{x + 4} + 2$

2. $y = \dfrac{3}{x + 4} - 2$

3. $y = \dfrac{3}{x - 4} + 2$

4. $y = \dfrac{3}{x - 4} - 2$

A.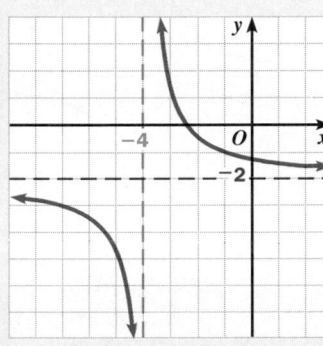

B.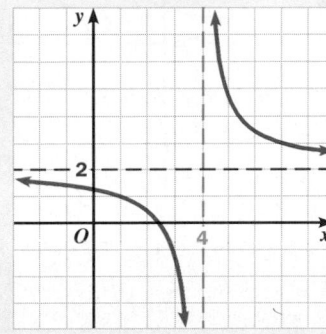

C.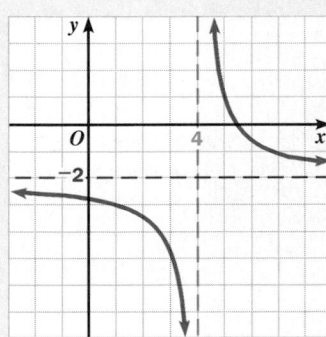

D.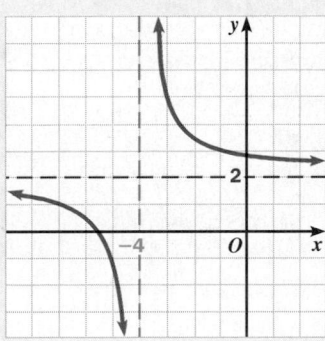

9.7 | **Exercises and Applications**

Extra Practice exercises on page 763

Tell whether each function is a rational function. Explain your answers.

1. $f(x) = \dfrac{2x + 3}{5x - 1}$

2. $f(x) = \dfrac{\sqrt{x - 4}}{x}$

3. $g(t) = \dfrac{t^2 + 1}{2^t + 1}$

4. $h(t) = \dfrac{t^2 + 1}{6t^3 - t^2 - 12}$

5. Challenge Show that the equation $xy - x + 2y = 5$ defines a rational function.

Find an equation of each hyperbola.

6.
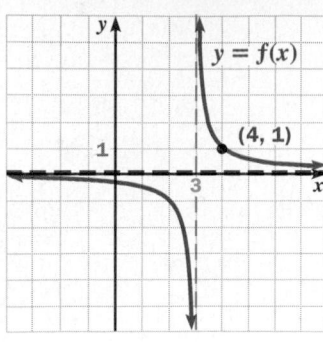

$y = f(x)$

(4, 1)

1

3

7.
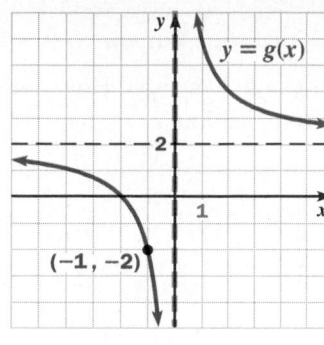

$y = g(x)$

2

1

(−1, −2)

8.
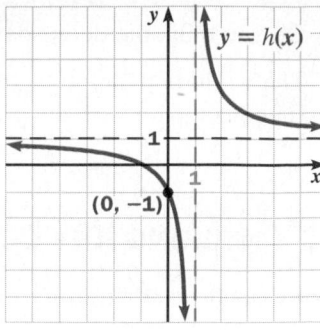

$y = h(x)$

1

1

(0, −1)

9.
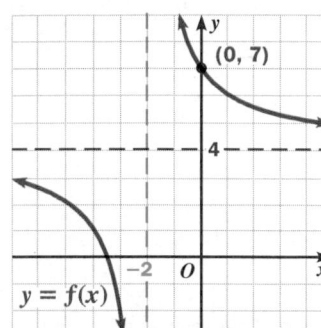

(0, 7)

4

−2   O

$y = f(x)$

10.
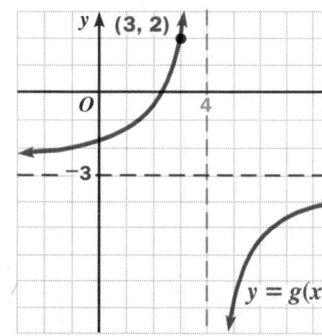

(3, 2)

O   4

−3

$y = g(x)$

11.
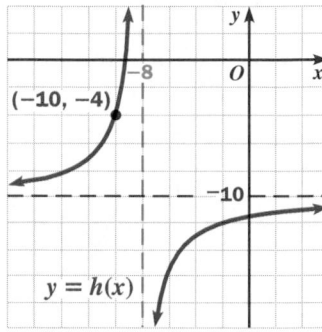

−8   O

(−10, −4)

−10

$y = h(x)$

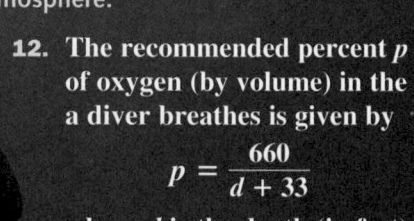

Connection   SCUBA DIVING

You may be surprised to learn that oxygen under high pressure can be toxic to the human body. This is a concern for scuba divers, who often experience high pressure caused by the weight of the water above them. In order to breathe safely, divers working in deep water must fill their scuba tanks with air containing less oxygen than Earth's atmosphere.

12. The recommended percent p of oxygen (by volume) in the air a diver breathes is given by

$$p = \frac{660}{d + 33}$$

where d is the depth (in feet) at which the diver is working.

a. [icon] **Technology** Graph the given equation using a graphing calculator or graphing software. What is the recommended percent of oxygen for a diver working at a depth of 100 ft? 200 ft? 300 ft?

b. What value does the recommended percent of oxygen approach as a diver's depth increases?

13. **Research** Find out the percent of oxygen (by volume) in Earth's atmosphere. Compare this figure with the one given by the equation in Exercise 12 for a diver at the surface.

When a bolt of lightning strikes, you don't hear thunder right away. The time t (in seconds) it takes for the thunder to reach your ears is given by

$$t = \frac{d}{1.09T + 1050}$$

where d is your distance from the lightning (in feet) and T is the air temperature (in degrees Fahrenheit).

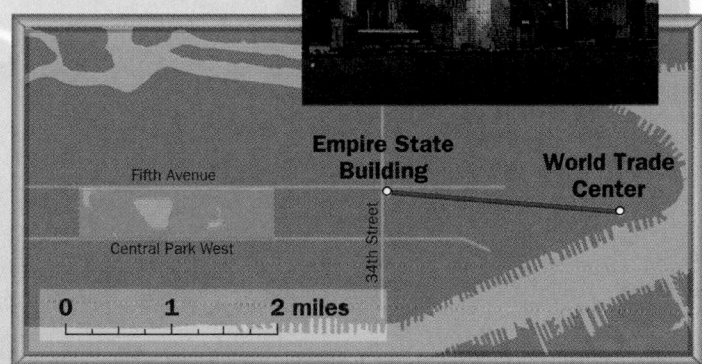

Fifth Avenue

Central Park West

34th Street

Empire State Building

World Trade Center

0   1   2 miles

14. The World Trade Center in New York City is often struck by lightning during storms. Suppose you are at the Empire State Building when lightning strikes the World Trade Center. (See the map.)

a. Write an equation giving the time it takes to hear the thunder as a function of air temperature. (*Note:* 1 mi = 5280 ft)

b. **Technology** Graph the equation from part (a) using a graphing calculator or graphing software. How does the time it takes to hear the thunder change as air temperature increases? What is this time when the temperature is 30°F? 60°F? 90°F?

15. **Writing** Explain how you can use the equation

$$t = \frac{d}{1.09T + 1050}$$

to estimate how far away lightning is.

For each function:

a. Find the asymptotes of the function's graph.

b. Tell how the function's graph is related to a hyperbola with equation of the form $y = \dfrac{a}{x}$.

16. $f(x) = \dfrac{2x + 5}{x - 1}$

17. $f(x) = \dfrac{8x + 14}{2x - 3}$

18. $f(x) = \dfrac{21x + 16}{3x + 7}$

19. $f(x) = \dfrac{3x + 11}{x + 2}$

20. $g(x) = \dfrac{-x + 5}{x + 4}$

21. $g(x) = \dfrac{-20x + 3}{4x + 1}$

22. $h(x) = \dfrac{-4x + 19}{7x - 21}$

23. $h(x) = \dfrac{-30x - 1}{5x - 4}$

24. **CONSUMER ECONOMICS** Tanya wants to order several pizzas from Marco's Pizza Palace and have them delivered to her graduation party. The price of each pizza is $8, and there is a $3 delivery charge regardless of the number of pizzas ordered.

a. Write an equation giving Tanya's total cost C if she orders p pizzas.

b. Use your answer from part (a) to write an equation giving the *average* cost A of a pizza if p pizzas are ordered.

c. Copy and complete the table. What value does the average cost of a pizza approach as the number of pizzas ordered increases? What does this value tell you about the effect of the delivery charge on large orders?

d. Write your equation from part (b) in the form $A = \dfrac{a}{p - h} + k$. How does this form show you what value the average cost approaches as the number of pizzas ordered increases?

p	A
2	?
5	?
10	?
20	?
50	?
100	?

25. Open-ended Problem Write a rational function of the form $f(x) = \dfrac{ax + b}{cx + d}$.
Find the asymptotes of the graph of your function, and make a sketch of the graph.

Tell whether y varies inversely with x. If so, state the constant of variation.
(Section 9.6)

26. $\dfrac{y}{x} = 3$ **27.** $y = \dfrac{1.5}{x}$ **28.** $4xy = -28$ **29.** $x^2y = 2$

30. Ben took 11 tests in his algebra class. The scores he received were 75, 83, 84, 66, 75, 70, 84, 91, 80, 84, and 77. Find each statistic for Ben's test data. *(Sections 6.4 and 6.5)*

  a. mean   **b.** median   **c.** range   **d.** standard deviation

Divide. *(Section 9.2)*

31. $\dfrac{x^2 + 8x + 10}{x + 1}$ **32.** $\dfrac{3x^2 + x + 7}{x^2 - 3x + 2}$ **33.** $\dfrac{8x^3 - 6x^2 - x - 4}{2x - 1}$

ASSESS YOUR PROGRESS

VOCABULARY

inverse variation (p. 427)   **asymptote** (p. 427)
constant of variation (p. 427)   **rational function** (p. 435)
hyperbola (p. 427)

1. Use the table. Tell whether y varies inversely with x. If so, write an equation giving y as a function of x. *(Section 9.6)*

x	−2	−0.5	1.5	3
y	6	24	−8	−4

2. The time it takes a person to type a handwritten essay varies inversely with the person's typing rate. Matt, who types at a rate of 40 words per minute, can type a five-page essay in 30 min. If Matt takes a typing class that increases his rate to 60 words per minute, how long will it take him to type a five-page essay? *(Section 9.6)*

3. Find an equation of the hyperbola that has asymptotes $x = 1$ and $y = -3$ and that passes through the point with coordinates $(0, -5)$. *(Section 9.7)*

4. Let $g(x) = \dfrac{6x + 13}{3x + 2}$. Find the asymptotes of the graph of g, and tell how the graph is related to a hyperbola with equation of the form $y = \dfrac{a}{x}$. *(Section 9.7)*

5. **Journal** If $y = \dfrac{a}{x - h} + k$, can you say that y varies inversely with x? Explain.

9.8 | Working with General Rational Functions

So far, the graph of each rational function you have seen has been a hyperbola with a single vertical asymptote. In the Exploration, you'll work with rational functions having more complicated graphs.

Learn how to...

- **identify important features of graphs of rational functions**

- **describe the end behavior of rational functions**

So you can...

- **analyze how the acidity level in your mouth changes when you eat sugary foods, for example**

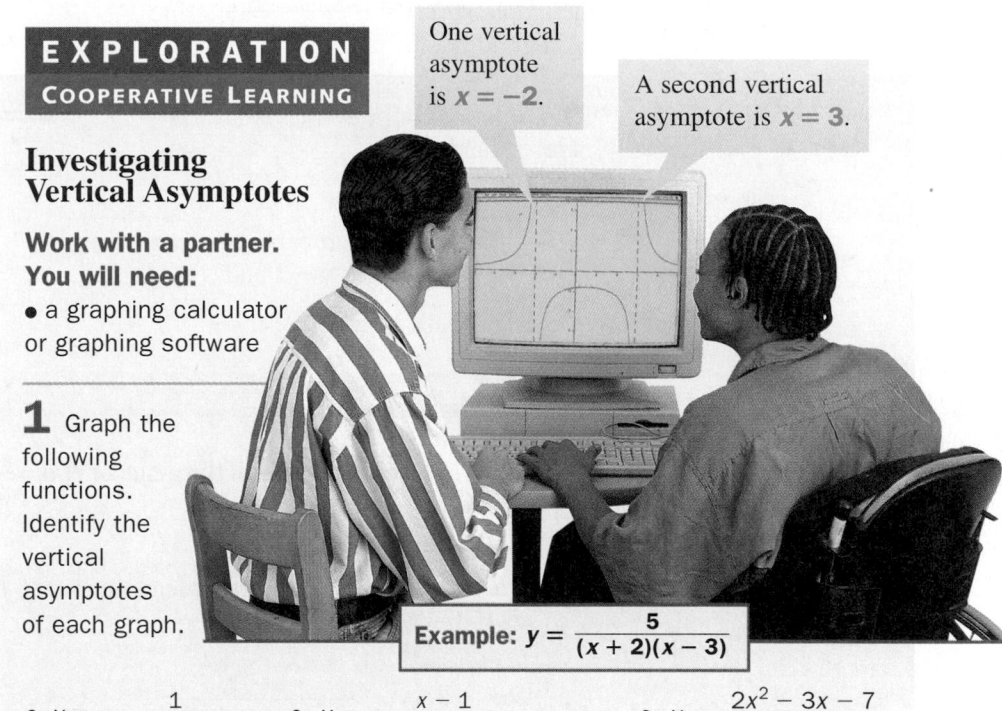

One vertical asymptote is $x = -2$.

A second vertical asymptote is $x = 3$.

EXPLORATION

COOPERATIVE LEARNING

Investigating Vertical Asymptotes

Work with a partner. You will need:

- a graphing calculator or graphing software

1 Graph the following functions. Identify the vertical asymptotes of each graph.

Example: $y = \dfrac{5}{(x + 2)(x - 3)}$

- $y = \dfrac{1}{(x - 1)(x - 3)}$ • $y = \dfrac{x - 1}{(x - 2)(x - 5)}$ • $y = \dfrac{2x^2 - 3x - 7}{(x + 1)(x - 4)}$

- $y = \dfrac{4}{x(x + 3)(x - 3)}$ • $y = \dfrac{-x^2 - 6x + 2}{(x + 3)(x + 1)(x - 4)}$ • $y = \dfrac{x^4 + x^3 - 5x^2}{(x + 1)^2(x - 2)}$

2 Predict the vertical asymptotes of the graph of each function below. Use your calculator or software to check your predictions.

- $y = \dfrac{-3x^2}{(x + 3)(x - 2)}$ • $y = \dfrac{7x + 2}{(x + 5)(x + 1)(x - 3)}$ • $y = \dfrac{x^3}{x^2 - 16}$

3 Based on your results from Steps 1 and 2, how are the vertical asymptotes of a rational function's graph related to the function's equation?

The following result makes it easy to identify the vertical asymptotes of a rational function's graph.

Finding Vertical Asymptotes for Rational Functions

Let $p(x)$ and $q(x)$ be polynomials having no common factors.

Then the graph of $f(x) = \dfrac{p(x)}{q(x)}$ has a vertical asymptote at each

real solution of $q(x) = 0$.

For example, you can use the summary above to find the vertical asymptotes for

the graph of $f(x) = \dfrac{x + 2}{(x + 3)(x - 1)}$.

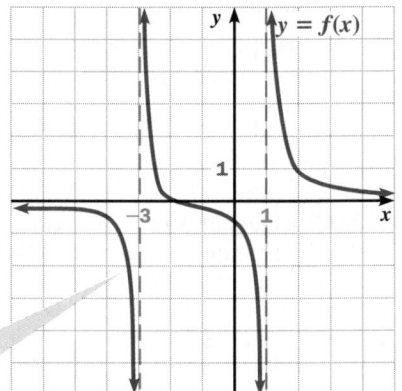

The solutions of $(x + 3)(x - 1) = 0$ are −**3** and **1**.

The vertical asymptotes are $x = -3$ and $x = 1$.

EXAMPLE 1

Find the vertical asymptotes of the graph of $f(x) = \dfrac{x^2 - 1}{2x^2 + 5x - 12}$.

SOLUTION

Step 1 Write the numerator and denominator of $f(x)$ as a product of factors to see if they have common factors.

$$f(x) = \frac{x^2 - 1}{2x^2 + 5x - 12}$$

$$= \frac{(x - 1)(x + 1)}{(2x - 3)(x + 4)}$$

The numerator and denominator have **no common factors**. Therefore, the vertical asymptotes occur where the denominator is 0.

Step 2 Solve the equation $(2x - 3)(x + 4) = 0$.

$$(2x - 3)(x + 4) = 0$$

$$2x - 3 = 0 \quad \text{or} \quad x + 4 = 0$$

$$x = \frac{3}{2} \quad \text{or} \quad x = -4$$

The vertical asymptotes are $x = \dfrac{3}{2}$ and $x = -4$.

THINK AND COMMUNICATE

1. a. **Technology** Graph $g(x) = \dfrac{x + 3}{x^2 + x - 6}$ using a graphing calculator or graphing software. Identify the vertical asymptotes.

b. Solve the equation $x^2 + x - 6 = 0$. Does the graph of g have a vertical asymptote at each solution? If not, does this contradict the boxed statement at the top of the previous page? Explain.

2. a. **Technology** Graph $h(x) = \dfrac{5}{x^2 + 1}$ using a graphing calculator or graphing software. Identify any vertical asymptotes.

b. Solve the equation $x^2 + 1 = 0$. How do your solutions support your answer to part (a)?

End Behavior of Rational Functions

As x takes on large positive and negative values, a rational function $y = f(x)$ can behave in one of several different ways. You can describe this end behavior using the infinity notation introduced in Section 9.3.

EXAMPLE 2

Use infinity notation to describe the end behavior of each function.

a. $f(x) = \dfrac{2x - 1}{x^2 - 4}$ **b.** $g(x) = \dfrac{6x^2 + 2x - 13}{3x^2 + x - 10}$ **c.** $h(x) = \dfrac{-x^3 + 5}{x^2}$

SOLUTION

Graph each function using a graphing calculator or graphing software.

a.

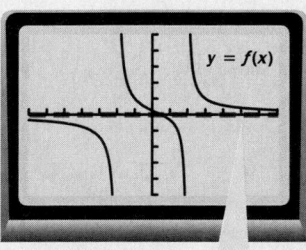

As $x \to \pm\infty$, $f(x) \to 0$.

b.

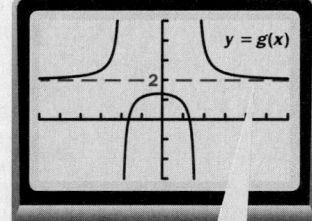

As $x \to \pm\infty$, $g(x) \to 2$.

c.

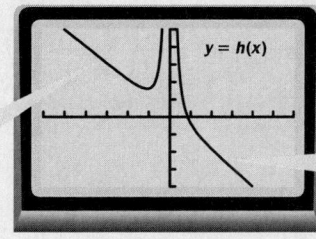

As $x \to -\infty$, $h(x) \to +\infty$.

As $x \to +\infty$, $h(x) \to -\infty$.

You can also determine a rational function's end behavior by comparing the degrees of the numerator and denominator.

Determining the End Behavior of Rational Functions

Let $f(x) = \dfrac{p(x)}{q(x)}$ where $p(x)$ is a polynomial of degree m and $q(x)$ is a polynomial of degree n. Let a be the leading coefficient of $p(x)$, and let b be the leading coefficient of $q(x)$.

- If **$m < n$**, then $f(x) \to 0$ as $x \to \pm\infty$. The graph of f has a horizontal asymptote at $y = 0$.

- If **$m = n$**, then $f(x) \to \dfrac{a}{b}$ as $x \to \pm\infty$. The graph of f has a horizontal asymptote at $y = \dfrac{a}{b}$.

- If **$m > n$**, then f has the same end behavior as the polynomial function $y = \dfrac{a}{b}x^{m-n}$. The graph of f has no horizontal asymptote.

EXAMPLE 3   Application: Dentistry

When you eat sugary foods, the amount of acid in your mouth temporarily increases. This causes your mouth's pH to decrease. (See page 140 for a description of the pH scale.) The pH level t minutes after eating can be approximated with this equation:

$$\text{pH} = \frac{65t^2 - 204t + 2340}{10t^2 + 360}$$

a. Tooth decay can occur if the pH level in your mouth falls too low (and remains low for a period of time). How long after sugary foods are eaten is the lowest pH level reached? What is the lowest pH?

b. As time passes (that is, as $t \to +\infty$), what value does the pH level approach?

SOLUTION

a. Graph the equation using a graphing calculator or graphing software.

Find the coordinates of the lowest point on the graph. These coordinates are **(6, 4.8)**.

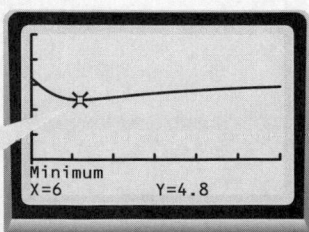

Minimum
X=6   Y=4.8

The lowest pH level is reached after 6 min. The lowest pH is 4.8.

b. In the equation for pH, the numerator **$65t^2 - 204t + 2340$** has the same degree as the denominator **$10t^2 + 360$**. So the pH level approaches $\dfrac{65}{10}$, or 6.5, as time passes.

3. Use the equation $pH = \dfrac{65t^2 - 204t + 2340}{10t^2 + 360}$ to find the pH level when $t = 0$. What do you notice? Explain why your answer makes sense.

4. Show that the rules for end behavior at the top of the previous page apply to the rational functions in Example 2.

5. Find a polynomial function that has the same end behavior as the function $f(x) = \dfrac{-15x^4 + x^3 + 4x^2 + 4}{3x^2 + 11x + 1}$.

☑ CHECKING KEY CONCEPTS

For each function:

a. Find the vertical asymptotes of the function's graph.

b. Describe the function's end behavior using infinity notation.

1. $f(x) = \dfrac{2}{(x + 7)(x - 4)}$ **2.** $f(x) = \dfrac{8x^2 + 2x - 15}{2x^2 + 13x + 11}$

3. $g(x) = \dfrac{3x^2}{4x - 17}$ **4.** $h(x) = \dfrac{-x^5 + x + 1}{x^3 - 2x}$

9.8 Exercises and Applications

Extra Practice exercises on page 764

Use what you know about vertical asymptotes to match each function with its graph.

1. $y = \dfrac{3}{x^2 - 9}$ **2.** $y = \dfrac{x - 8}{x^2 - 2x - 3}$ **3.** $y = \dfrac{-2x}{x^2 + 2x - 3}$

A. 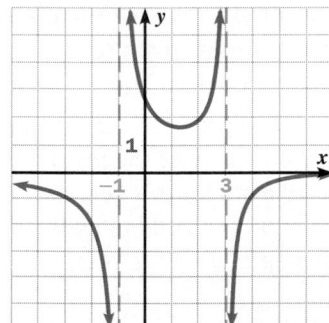   **B.** 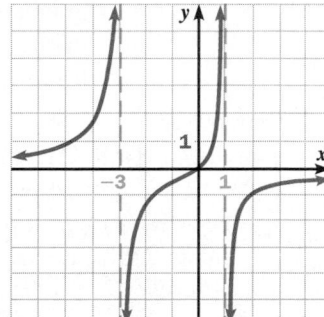   **C.**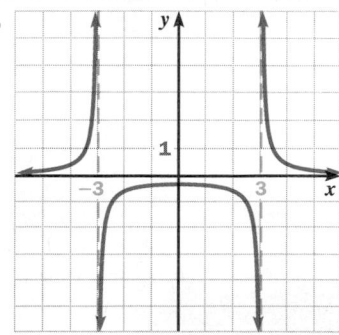

4. SAT/ACT Preview Let $f(x) = \dfrac{p(x)}{q(x)}$ be a rational function such that $f(x) \to 3$ as $x \to \pm\infty$. If A = the degree of $p(x)$ and B = the degree of $q(x)$, then:

A. $A > B$ **B.** $B > A$ **C.** $A = B$ **D.** relationship cannot be determined

Look back at the article on pages 386–388.

As an astronomer, Vera Rubin uses gravity to help explain the motion of planets, stars, and other objects in space. Of all these objects, Earth is the source of the gravity that affects you most directly. The gravitational force F, measured in newtons (N), that Earth exerts on a relatively small object is given by

$$F = \frac{(4.0 \times 10^{14})m}{d^2}$$

where m is the object's mass (in kilograms) and d is the distance (in meters) of the object from the center of Earth.

5. The space probe *Voyager 2* was launched in 1977 to explore the outer planets of the solar system. In order to leave Earth, *Voyager 2* had to overcome Earth's gravity. The mass of *Voyager 2* is 930 kg.

   a. Write an equation giving Earth's gravitational force on *Voyager 2* as a function of the distance *d* of *Voyager 2* from Earth's center.

   b. 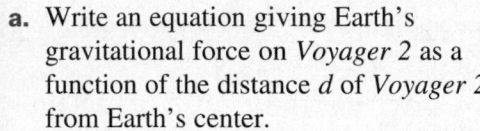 **Technology** Graph the function from part (a) using a graphing calculator or graphing software. What are the function's domain and range? (*Hint:* The radius of Earth is about 6.37×10^6 m.) What value did Earth's gravitational force on *Voyager 2* approach as the probe's distance from Earth became greater and greater?

   c. The average distance between the center of Earth and the center of the moon is about 3.84×10^8 m. When *Voyager 2* was this far from Earth's center, what was Earth's gravitational force on the probe? How does this force compare with the force required to lift an algebra book (about 20 N)?

6. Suppose two space probes, probe A and probe B, are traveling away from Earth. The mass of probe B is twice the mass of probe A, and probe B is twice as far from Earth's center as probe A is. On which probe is the pull of Earth's gravity greater? Give a mathematical justification for your answer.

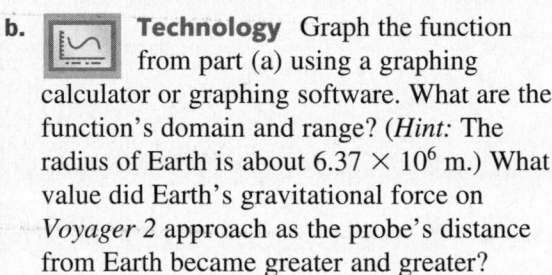

Earth

d

2d

probe A
(mass = *m*)

probe B
(mass = 2*m*)

For each function:

a. Find the vertical asymptotes of the function's graph.

b. Describe the function's end behavior using infinity notation.

7. $f(x) = \dfrac{1}{(x + 5)(x - 2)}$

8. $g(x) = \dfrac{4x^2}{12x^2 + x - 1}$

9. $h(x) = \dfrac{-10x^2 - 13x + 3}{5x - 9}$

10. $f(x) = \dfrac{x - 5}{x^3 - 5x^2 + 4x}$

11. $g(x) = \dfrac{-x^3 + 1}{x^3 - x^2 - 14x + 24}$

12. $h(x) = \dfrac{6x^4 + 11}{x^2 + 1}$

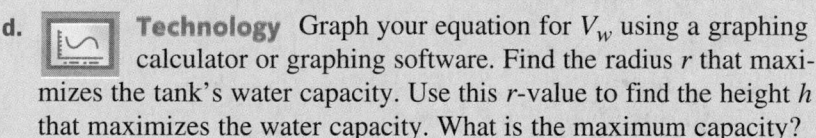

Connection   ENGINEERING

An *impluvium* is a type of dwelling found in western Africa. One version of an impluvium consists of several houses that surround a courtyard and share a single funnel-shaped roof. The roof is used to direct rainwater into a concrete tank or into clay jars.

13. Challenge Suppose you are making a cylindrical tank for the courtyard of an impluvium. You have 100 ft³ of concrete. The sides and base of the tank should be 1 ft thick. The tank's inner radius r and inner height h should be chosen so that the tank holds as much water as possible. You can determine the values of r and h as follows.

a. Write an equation giving the volume V_c of concrete needed to make the tank in terms of r and h. (*Hint:* The volume of concrete is the difference of the volumes of two cylinders.)

b. Knowing that $V_c = 100$, use your equation from part (a) to solve for h in terms of r.

c. Use your expression for h from part (b) to write an equation giving the tank's water capacity V_w as a rational function of r.

d. 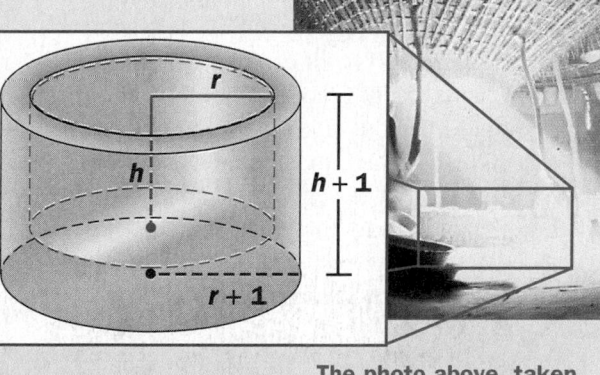 **Technology** Graph your equation for V_w using a graphing calculator or graphing software. Find the radius r that maximizes the tank's water capacity. Use this r-value to find the height h that maximizes the water capacity. What is the maximum capacity?

The photo above, taken by Jean-Paul Bourdier, shows an impluvium courtyard.

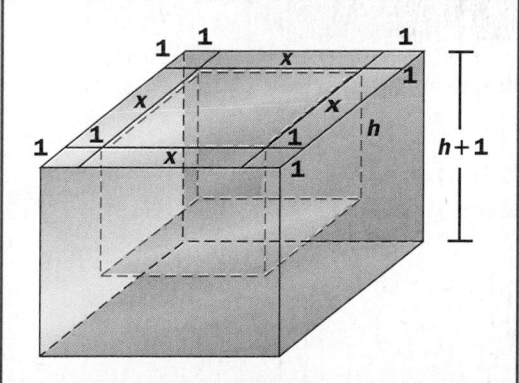

14. Cooperative Learning Suppose you want to make a tank like the one at the left for an impluvium. Work with a partner to find the dimensions x and h that maximize the tank's water capacity, assuming you can use 100 ft³ of concrete. Compare the maximum capacity for this type of tank with the maximum capacity for a cylindrical tank. Which type of tank is better for storing water? Explain.

15. CHEMISTRY The density of water depends on the temperature of the water. The density D and temperature T are related by the equation

$$D = \frac{k_0 + k_1T + k_2T^2 + k_3T^3 + k_4T^4 + k_5T^5}{1 + k_6T}$$

where D is measured in kilograms per cubic meter, T is measured in degrees Celsius, and the constants $k_0, k_1, \ldots, k_6$ are given in the table.

k_0	9.9983952×10^2
k_1	1.6945176×10^1
k_2	$-7.9870401 \times 10^{-3}$
k_3	$-4.6170461 \times 10^{-5}$
k_4	1.0556302×10^{-7}
k_5	$-2.8054253 \times 10^{-10}$
k_6	1.6879850×10^{-2}

a. **Technology** Graph the equation for D using a graphing calculator or graphing software. Adjust the viewing window so that the intervals $0 \le T \le 10$ and $999.5 \le D \le 1000.5$ are shown on the axes. At what temperature is water most dense? What is the maximum density?

b. **Writing** For a typical liquid, density increases steadily as the temperature decreases to the liquid's freezing point. Is this true for water? Explain. (*Note:* The freezing point of water is 0°C.)

c. **Writing** How does your answer to part (b) explain why lakes freeze at the top, rather than at the bottom?

16. GARDENING Brian Keegan is making a rectangular vegetable garden adjacent to his house as shown. The vegetables he wants to plant require 200 ft² of earth in which to grow. Brian plans to put fencing on three sides of the garden to keep out animals that might eat the vegetables. To save money, he wants to use the least amount of fencing possible. Find the dimensions x and y that minimize the length of fencing needed. (*Hint:* Use a procedure similar to the one given in Exercise 13.)

ONGOING ASSESSMENT

17. Open-ended Problem Write a rational function $f(x)$ so that the graph of f has vertical asymptotes at $x = -1$ and $x = 2$, and so that $f(x) \to -\infty$ as $x \to \pm\infty$.

SPIRAL REVIEW

Tell how the graph of each function is related to a hyperbola with equation of the form $y = \dfrac{a}{x}$. (*Section 9.7*)

18. $f(x) = \dfrac{1}{x - 2} + 4$ **19.** $f(x) = \dfrac{5}{x + 6} - 3$ **20.** $g(x) = \dfrac{2x + 9}{x + 1}$ **21.** $h(x) = \dfrac{-14x + 3}{2x - 1}$

Solve each proportion. (*Toolbox, page 785*)

22. $\dfrac{x}{10} = \dfrac{3}{2}$ **23.** $\dfrac{3}{8} = \dfrac{9}{4x}$ **24.** $\dfrac{4}{3k - 4} = \dfrac{7}{2}$ **25.** $\dfrac{2n - 3}{5} = \dfrac{n + 2}{6}$

9.9 Solving Rational Equations

Learn how to...

- solve rational equations

So you can...

- solve problems that involve mixing metals, for example

When two or more metals are melted, mixed together, and allowed to harden, the material they form is called an *alloy*. Some alloys, such as steel (a mixture of iron and carbon), are used mainly for industrial purposes. Other alloys, including those made from gold and silver, are highly valued for their beauty.

| EXAMPLE 1 | Application: Metallurgy |

"Green gold," a gold-and-silver alloy named for its greenish color, is used to make jewelry. Green gold is 75% gold and 25% silver by weight. A naturally occurring alloy called *electrum* is 80% gold and 20% silver by weight. How much pure silver should be mixed with 12 oz of electrum to make green gold?

SOLUTION

Let x = the weight in ounces of pure silver to be mixed with the electrum. The value of x should be chosen so that this equation is satisfied:

This is the fraction of the electrum-and-silver mixture that is gold.

$$\frac{\text{weight of gold in mixture}}{\text{total weight of mixture}} = 0.75$$

Green gold is 75% gold.

The weight of the gold is 80% of the electrum's weight.

$$\frac{(0.80)(12)}{12 + x} = 0.75$$

The total weight is the weight of the electrum, 12, plus the weight x of the silver added.

$$\frac{9.6}{12 + x} = 0.75$$

Multiply both sides by $12 + x$.

$$(12 + x)\left(\frac{9.6}{12 + x}\right) = (12 + x)(0.75)$$

$$9.6 = 9 + 0.75x$$

$$0.6 = 0.75x$$

$$0.8 = x$$

The amount of silver that should be mixed with the electrum is 0.8 oz.

THINK AND COMMUNICATE

1. a. Solve the problem in Example 1 by starting with this equation:

$$\frac{\text{weight of silver in mixture}}{\text{total weight of mixture}} = 0.25$$

b. How does this method compare with the method shown in Example 1?

The equation $\frac{9.6}{12 + x} = 0.75$ in Example 1 is called a *rational equation*. A **rational equation** contains only polynomials or quotients of polynomials.

EXAMPLE 2

Solve the rational equation $\frac{x}{x + 1} + \frac{2}{x - 3} = \frac{4}{(x + 1)(x - 3)}.$

SOLUTION

Step 1 Multiply each side of the equation by the least common denominator (LCD) of the fractions. Solve the resulting equation for x.

The LCD is
$(x + 1)(x - 3)$.

$$(x + 1)(x - 3)\left(\frac{x}{x + 1} + \frac{2}{x - 3}\right) = (x + 1)(x - 3)\left(\frac{4}{(x + 1)(x - 3)}\right)$$

$$x(x - 3) + 2(x + 1) = 4$$

$(x + 1)(x - 3)\left(\dfrac{x}{x + 1}\right) = x(x - 3)$

$(x + 1)(x - 3)\left(\dfrac{2}{x - 3}\right) = 2(x + 1)$

$$x^2 - 3x + 2x + 2 = 4$$

$$x^2 - x - 2 = 0$$

$$(x + 1)(x - 2) = 0$$

$$x + 1 = 0 \quad \text{or} \quad x - 2 = 0$$

$$x = -1 \quad \text{or} \quad x = 2$$

Step 2 Check the solutions. Use the *original* equation.

Check

For x = −1:

$$\frac{-1}{-1 + 1} + \frac{2}{-1 - 3} \stackrel{?}{=} \frac{4}{(-1 + 1)(-1 - 3)}$$

$$\frac{-1}{0} + \frac{2}{-4} \stackrel{?}{=} \frac{4}{0}$$

You can't divide by **0**, so −1 *is not* a solution.

Check

For x = 2:

$$\frac{2}{2 + 1} + \frac{2}{2 - 3} \stackrel{?}{=} \frac{4}{(2 + 1)(2 - 3)}$$

$$\frac{2}{3} + \frac{2}{-1} \stackrel{?}{=} \frac{4}{(3)(-1)}$$

$$-\frac{4}{3} = -\frac{4}{3} \checkmark$$

The only solution is 2.

Example 2 shows that rational equations can have extraneous solutions. You should always check your solutions to eliminate any that are extraneous.

THINK AND COMMUNICATE

2. Solve the equation $\dfrac{6x - 1}{2x + 5} = 3$. What happens? What does this tell you about the equation's solutions?

3. What happens when you try to solve $\dfrac{3x - 6}{x - 2} = 3$? What are the solutions of this equation? (*Hint:* Be careful. Remember that a solution must satisfy the *original* equation.)

☑ CHECKING KEY CONCEPTS

Solve each equation.

1. $\dfrac{20 + x}{4 + x} = 3$

2. $\dfrac{2}{x - 1} = \dfrac{x}{6}$

3. $\dfrac{3}{t} + \dfrac{8}{t - 5} = 1$

4. $\dfrac{u}{u - 3} + \dfrac{1}{u - 4} = \dfrac{1}{(u - 3)(u - 4)}$

5. **METALLURGY** *Sterling silver* and *jewelry silver* are both silver-and-copper alloys. Sterling silver is 92.5% silver and 7.5% copper by weight. Jewelry silver is 80% silver and 20% copper by weight. How much pure silver must be mixed with 10 oz of jewelry silver to make sterling silver?

9.9 Exercises and Applications

Extra Practice exercises on page 764

Solve each equation.

1. $\dfrac{16 - x}{5 + x} = 2$

2. $\dfrac{1}{x} + \dfrac{8}{5x} = 1$

3. $\dfrac{x}{x - 3} = 4 + \dfrac{3}{x - 3}$

4. $\dfrac{6}{x(x - 6)} - \dfrac{1}{x - 6} = 0$

5. $\dfrac{1}{x - 2} - \dfrac{8}{x - 1} = -3$

6. $\dfrac{2}{x - 10} + \dfrac{3x}{x - 4} = 5$

7. $\dfrac{t^2 - 7t - 8}{(3t - 4)(t + 2)} + \dfrac{3t + 7}{t + 2} = 3$

8. $\dfrac{4}{y + 1} - \dfrac{1}{y} = 1$

9. $\dfrac{u - 2}{(u + 3)(u + 2)} = \dfrac{3}{u + 3} - \dfrac{2}{u + 2}$

10. $\dfrac{v - 3}{v - 5} + \dfrac{4}{10 - 2v} = 1$

11. $\dfrac{5}{r - 1} - \dfrac{3}{r + 1} = \dfrac{2}{r}$

12. $\dfrac{s + 1}{s} + \dfrac{14}{s - 7} = \dfrac{3s - 7}{s^2 - 7s}$

13. **Challenge** The ancient Greeks thought that the rectangles with the most pleasing appearance were those for which the ratio of the width w to the length l equals the ratio of l to $l + w$. The ratio of width to length for these "ideal" rectangles is called the *golden ratio*. Find the value of the golden ratio. (*Hint:* Write a proportion involving w and l. Express your proportion in terms of $x = \dfrac{w}{l}$, and solve the proportion for x.)

The double arrow means that the reaction occurs in both directions.

When you mix hydrogen and iodine gases together, a third gas, called *hydrogen iodide*, is formed. Each molecule of hydrogen (H_2) reacts with one molecule of iodine (I_2) to produce two molecules of hydrogen iodide (HI). Simultaneously, a "reverse reaction" occurs, in which the newly-formed hydrogen iodide breaks down into hydrogen and iodine. This process is illustrated above.

H_2 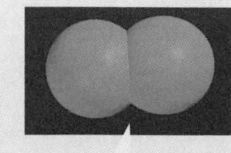 I_2 $\rightleftharpoons$ 2HI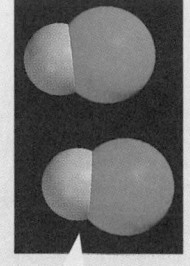

An H_2 molecule consists of two hydrogen atoms.

An I_2 molecule consists of two iodine atoms.

An HI molecule consists of one hydrogen atom and one iodine atom.

Eventually, the amounts of the three gases stabilize. When this happens, the gas mixture is said to be in *equilibrium*. At equilibrium, the amounts of the gases are related by the *law of mass action*,

$$\frac{(\text{amount of HI})^2}{(\text{amount of } H_2)(\text{amount of } I_2)} = K$$

where K is a constant that depends on the temperature of the gases.

14. Suppose 1 mole of hydrogen gas reacts with 2 moles of iodine gas to produce hydrogen iodide. (*Note:* 1 mole $\approx 6.023 \times 10^{23}$ molecules) The temperature of the gas mixture is 430°C. Assume the mixture has reached equilibrium.

   a. **Writing** Let x = the amount of hydrogen (in moles) that is converted into hydrogen iodide. Use the illustration above to explain why x moles of iodine are also converted.

   b. Copy the table and replace each "?" with an expression involving x. (All amounts are in moles.)

Molecule	H_2	I_2	HI
Initial amount	1	2	0
Net change	$-x$	$-x$	?
Equilibrium amount	?	?	?

   c. At a temperature of 430°C, the value of K is 54.3 in the law of mass action. Use this K-value and the equilibrium amounts in the table to find the value of x. (*Hint:* You will obtain two x-values, but only one of the values makes sense in this situation.) At equilibrium, how much of the gas mixture is hydrogen? iodine? hydrogen iodide?

15. Suppose 5 moles of hydrogen react with 3 moles of iodine at 430°C. Find the equilibrium amounts of hydrogen, iodine, and hydrogen iodide.

The bottle shown contains iodine in its gaseous and solid forms.

16. **Open-ended Problem** Choose a real-world application of a rational function from Section 9.7 or 9.8. Write a question based on this application that can be answered by solving a rational equation. Exchange questions with a classmate, and answer your classmate's question.

SPIRAL REVIEW

Perform the indicated operation. *(Section 8.4)*

17. $(5 - 7i) + (8 + 3i)$ 18. $(2 + i) - (6 - 11i)$ 19. $(3 + i)^2$

 Technology Use a graphing calculator or software with matrix calculation capabilities to find an equation of the parabola passing through each set of points. *(Section 7.3)*

20. $(1, 2)$, $(2, 8)$, and $(3, 18)$ 21. $(1, 0)$, $(2, 2)$, and $(3, 6)$

ASSESS YOUR PROGRESS

VOCABULARY

rational equation (p. 450)

For each function:

a. Find the vertical asymptotes of the function's graph.

b. Describe the function's end behavior using infinity notation. *(Section 9.8)*

1. $f(x) = \dfrac{2x - 7}{(x - 1)(x - 5)}$

2. $g(x) = \dfrac{-x^2 - 2x + 8}{x + 3}$

3. $h(x) = \dfrac{x^4}{x^2 + 3x - 10}$

4. $f(x) = \dfrac{3x^3 + 1}{x^3 - 9x}$

Solve each equation. *(Section 9.9)*

5. $\dfrac{5}{3} + \dfrac{2x + 9}{x + 4} = 4$

6. $\dfrac{w}{w - 5} = \dfrac{2}{w - 3} + \dfrac{4}{(w - 3)(w - 5)}$

7. **SPORTS** The 1994 major league baseball season ended prematurely because of a players' strike. When the season ended, Tony Gwynn of the San Diego Padres had the league's highest batting average (the ratio of hits to at-bats). Gwynn had 165 hits in 419 at-bats, for an average of .394. If there had been no strike and Gwynn had continued playing, how many consecutive hits would he have needed to raise his average to .400? *(Section 9.9)*

Tony Gwynn

> **BY THE WAY**
>
> If Tony Gwynn had batted at least .400 in 1994, he would have been the first major league player to do so since 1941, when Ted Williams batted .406.

8. **Journal** Explain how the end behavior of a rational function $f(x) = \dfrac{p(x)}{q(x)}$ is related to the degrees of the polynomials $p(x)$ and $q(x)$.

The Shape of Things

Grocery store shelves are filled with containers having many different sizes and shapes, and made of many different materials. When designing containers, food packagers must consider serving size, the cost of materials, and other factors. Packages must be durable, attractive, and easy to handle.

For liquid products that can assume the shape of any container, selecting the shape with the least surface area for a specified volume may reduce the cost of the material used to make the container.

PROJECT GOAL Determine the best shape and dimensions of a container that will hold a specified volume of liquid.

Designing an Efficient Container

Work in groups of three. Suppose you are part of a package design team assigned to design a container that will hold 500 cm³ of tomato sauce. The most desirable package design will minimize material costs, yielding the most efficient container. Here's a possible plan of action for your team:

1. CONSIDER a cylindrical can. Since many food containers are cylindrical, this is a logical shape to consider first. Your team must determine the radius and height of a 500 cm³ cylindrical can with the least surface area.

STEP 1 Express surface area as a function of base radius only. (*Hint:* The fact that the volume must be 500 cm³ will allow you to replace *h* with an expression involving *r*.)

STEP 2 Find the dimensions of the cylinder that minimizes surface area. One method is to examine a graph using a graphing calculator or graphing software. Another method is to examine a table of values using a spreadsheet like the one shown.

	A	B	C
	Cylinder Surface Areas		
1	Radius	Height	Surface area
2	3.00	17.684	389.684
3	3.10	16.570	382.931
4	3.20	15.542	376.649
5	3.30	14.615	371.266

2. INVESTIGATE other shapes for your tomato sauce container. Are there other geometric shapes that can hold 500 cm^3 using less surface area than the cylinder you identified in Part 1?

With other team members, discuss the advantages and disadvantages of using unusual container shapes, such as a cone, pyramid, prism, or sphere. Analyze their surface areas as you did in Part 1.

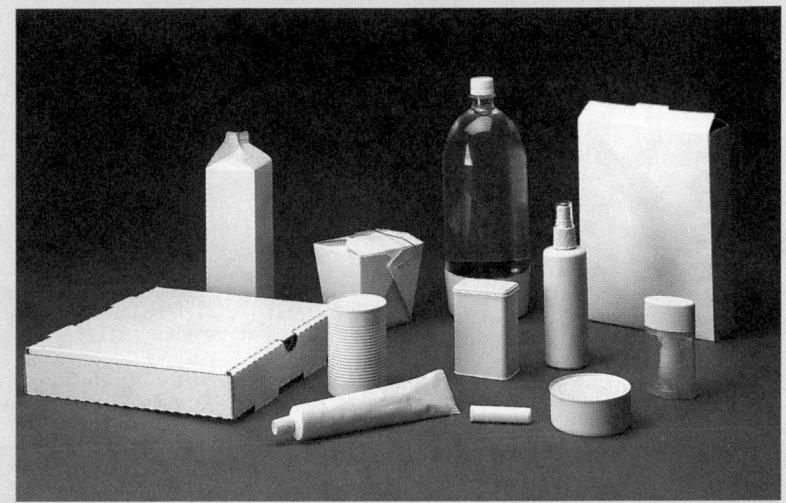

Presenting a Report

Present your report as a recommendation to the management of your company. Describe your procedures, and provide data that support the conclusions you make. You may also wish to extend your project. Here are a few ideas:

• Based on your findings for a cylindrical can with a volume of 500 cm^3, make a generalization about the ratio of radius to height for a cylindrical can of any specified volume if the surface area is to be minimized.

• Visit a grocery store. Can you find cylindrical containers that have the radius-to-height ratio described above? Based on the examples you see in the store, what are some other factors that should be considered in packaging a product?

• Consider the problems of packaging objects with unusual shapes. How would you package a football, a boomerang, or a dozen coat hangers?

• Most products are shipped in large quantities after packaging. What issues need to be considered when making larger packages from smaller ones? For example, how would you package 24 cans of tomato sauce so that they can be shipped?

Self-Assessment

How did the team divide up the tasks? What problems, if any, did you have setting up and analyzing the equations you used for the project? In what ways, if any, did the results of the project surprise you?

9 Review

STUDY TECHNIQUE

Reread the chapter, writing down questions about the ideas you don't understand. If any questions are still unanswered at the end of the chapter, ask a parent, friend, or teacher to help you with them.

VOCABULARY

polynomial (p. 390)	**cubic function** (p. 413)
degree (p. 390)	**zero** (p. 413)
constant term (p. 390)	**double zero** (p. 420)
standard form (p. 390)	**triple zero** (p. 420)
synthetic substitution (p. 391)	**inverse variation** (p. 427)
like terms (p. 392)	**constant of variation** (p. 427)
end behavior (p. 405)	**hyperbola** (p. 427)
leading coefficient (p. 406)	**asymptote** (p. 427)
turning point (p. 407)	**rational function** (p. 435)
local maximum/minimum (p. 407)	**rational equation** (p. 450)

SECTIONS 9.1 and 9.2

A **polynomial** is a term or a sum of terms such that each term is the product of a real-number coefficient and a variable with a whole-number exponent. An example of a polynomial is $3x^4 - 5x^2 + 6x - 1$. The **degree**, or greatest exponent, of this polynomial is 4. The polynomial is in **standard form** because its exponents decrease from left to right.

You can use **synthetic substitution** to evaluate a polynomial for a given value of the variable. An example is shown at the right.

Evaluate $3x^4 - 5x^2 + 6x - 1$ when $x = 2$.

```
3    0   −5    6   −1      Repeatedly
     6   12   14   40      multiply by 2,
                          and then add.
─────────────────────
3    6    7   20   39  ◄─── answer
```

You can also add, subtract, multiply, and divide polynomials. An example is shown at the right.

Add $x^4 + 2x^3 + 4x^2 - 7$ and $3x^4 - 5x^2 + 6x - 1$.

$$
\begin{array}{r}
x^4 + 2x^3 + 4x^2 \qquad\quad - 7 \\
3x^4 \qquad\quad - 5x^2 + 6x - 1 \\
\hline
4x^4 + 2x^3 - \ x^2 + 6x - 8
\end{array}
$$

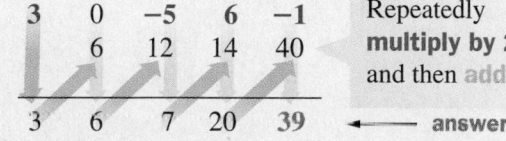 Add **like terms**.

◄─── answer

Four important features of a polynomial function and its graph are **end behavior, turning points, local maximums and minimums,** and **zeros**.

$f(x) = 3(x + 1)(x - 2)^2$

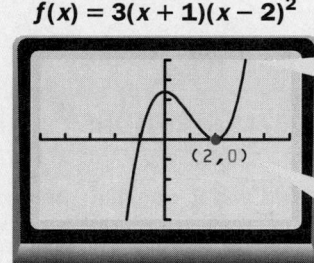

As $x \to +\infty$, $f(x) \to +\infty$.
As $x \to -\infty$, $f(x) \to -\infty$.

You can obtain information about a polynomial function's zeros using the *factor theorem*, the *fundamental theorem of algebra*, and the *rational zeros theorem*.

This is a **turning point** of the graph. Its *x*-coordinate is a **zero** of the function. Its *y*-coordinate is a **local minimum**.

Two variables x and y show **inverse variation** if $y = \dfrac{a}{x}$, or $xy = a$, for some nonzero number a. For example, the time it takes to travel a fixed distance varies inversely with speed.

The graph of $y = \dfrac{a}{x}$ is a **hyperbola**. The graph of $y = \dfrac{a}{x - h} + k$ is the graph of $y = \dfrac{a}{x}$ translated h units horizontally and k units vertically.

$y = \dfrac{1}{x - 2} + 3$

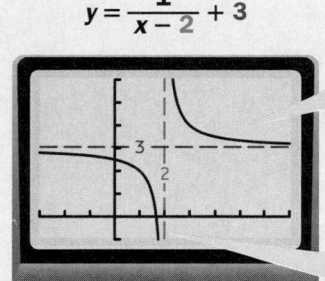

This graph is the graph of $y = \dfrac{1}{x}$ shifted **2 units right** and **3 units up**.

A **rational function** has the form $f(x) = \dfrac{p(x)}{q(x)}$ where $p(x)$ and $q(x)$ are polynomials.

The **asymptotes** of the graph are $x = 2$ and $y = 3$.

A rational function $f(x) = \dfrac{p(x)}{q(x)}$ has a vertical asymptote at each real solution of $q(x) = 0$, provided $p(x)$ and $q(x)$ have no common factors.

$f(x) = \dfrac{2x^2}{x^2 - 1}$

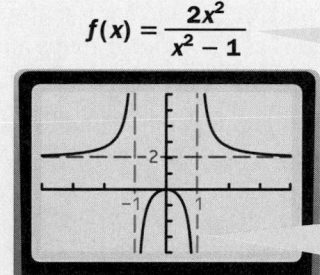

The numerator $2x^2$ has the same **degree** as the denominator $1x^2 - 1$. So $f(x) \to \dfrac{2}{1} = 2$ as $x \to \pm\infty$.

The end behavior of $f(x) = \dfrac{p(x)}{q(x)}$ is determined by the degrees and leading coefficients of $p(x)$ and $q(x)$.

The graph has vertical asymptotes where $x^2 - 1 = 0$: at $x = 1$ and $x = -1$.

A **rational equation**, such as $\dfrac{3x}{x - 4} + \dfrac{2}{x - 10} = 5$, contains only polynomials or quotients of polynomials. When solving a rational equation, be sure to check your solutions to eliminate any that are extraneous.

9 Assessment

VOCABULARY QUESTIONS

For Questions 1 and 2, complete each paragraph.

1. Let $f(x) = 5x^3 - x^2 + 9x - 7$. For this function, the _?_ is 3, the _?_ is 5, and the _?_ is -7.

2. If $y = \dfrac{16}{x}$, then x and y show _?_ . The number 16 is called the _?_ .

SECTIONS 9.1 *and* 9.2

3. Use synthetic substitution to evaluate $-2t^4 + 5t^3 + t - 8$ when $t = 3$.

4. Add $x^3 + 6x^2 - 10x + 3$ and $-3x^3 + 7x^2 - x - 2$.

5. Subtract $9u^2 - 8u^3 + 20 + 4u^4$ from $-1 + 11u + 2u^2 - 5u^3 + 8u^4$.

6. Multiply $3y - 1$ and $4y^2 - 2y + 7$.

7. Divide $10x^3 + 3x^2 - 5$ by $2x + 3$.

SECTIONS 9.3, 9.4, *and* 9.5

8. Describe the end behavior of $f(x) = -4x^5 + 2x^4 + 6x^2 + 12$ using infinity notation.

9. Find all local maximums and minimums of $g(x) = x^4 - 7x^2 - 4x + 9$.

10. **Writing** Explain why odd-degree polynomial functions can have only *local* maximums and *local* minimums, but even-degree polynomial functions can have maximums and minimums.

11. **POPULATION** For the period 1890–1990, the American Indian, Eskimo, and Aleut population P (in thousands) can be modeled by

$$P = 0.00496t^3 - 0.432t^2 + 11.3t + 212$$

where t is the number of years since 1890. In what year did the population reach 502 thousand?

12. Find an equation for the cubic function whose graph is shown at the left.

Find all real and imaginary zeros of each function.

13. $f(x) = x^3 - 10x^2 + 25x - 4$ **14.** $g(x) = 4x^4 + 4x^3 + x^2 + 30x + 36$

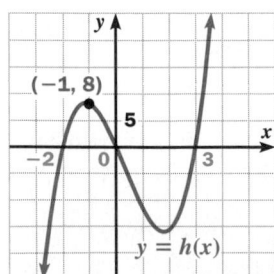

$(-1, 8)$

$y = h(x)$

15. HOME REPAIR On some tubes of caulking, the diameter of the round nozzle opening from which the caulking flows can be adjusted by the user. As the diameter d increases, the length l of caulking obtained from the tube decreases, as shown in the table.

  a. Does l vary inversely with d? Explain.

  b. Find the area A of each nozzle opening whose diameter is given in the table. Does l vary inversely with A? Explain.

  c. Write an equation giving l as a function of A.

d (in.)	l (in.)
$\frac{1}{8}$	1440
$\frac{1}{4}$	360
$\frac{3}{8}$	160
$\frac{1}{2}$	90

16. Let $f(x) = \dfrac{30x + 17}{6x + 1}$. Find the asymptotes of the graph of f, and tell how the graph is related to a hyperbola with equation of the form $y = \dfrac{a}{x}$.

17. Let $h(x) = \dfrac{x^4}{4x^2 + 11x - 3}$. Find the vertical asymptotes of the graph of h, and describe the end behavior of h using infinity notation.

18. SPORTS Janice is a member of her high school's varsity basketball team. So far this season, she has made 26 of the 80 three-point shots she has attempted, for a shooting percentage of 32.5%. How many consecutive three-point shots must she make to raise her shooting percentage to 40%?

19. Solve the equation $\dfrac{2}{x(x - 2)} - \dfrac{1}{x - 2} = 1$.

PERFORMANCE TASK

20. This problem was posed 700 years ago by the Chinese mathematician Ch'in Chiu-shao:

    There is a circular town of unknown diameter having four gates. A tall tree stands 3 li from the north gate as shown. When you exit the south gate and turn east, you must walk 9 li before you can see the tree. Find the diameter of the town. (*Note:* 1 li ≈ 0.33 mi)

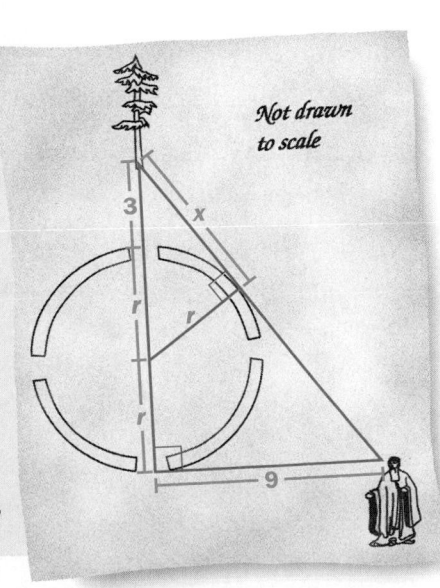

Not drawn to scale

  a. Show that the radius r of the town must satisfy the polynomial equation $4r^4 + 12r^3 + 9r^2 - 486r - 729 = 0$.

  b. Find all real and imaginary solutions of the equation in part (a). Explain why only one of the solutions makes sense in the given situation. What is this solution? What is the diameter of the town?

Cumulative Assessment
CHAPTERS 7–9

CHAPTER 7

Solve each system of equations. Tell whether the system has *one solution*, *infinitely many solutions*, or *no solution*.

1. $3x - 2y = 12$
$5x + 2y = 4$

2. $7x + y = 7$
$5x + 3y = 15$

3. $x - y = 6$
$2x - 3y = 8$

4. **Open-ended Problem** Write a system of equations that has infinitely many solutions and another system that has no solution. Use graphs and algebraic methods to show that your choices of systems are correct.

5. Rewrite the system at the right in matrix form and solve for x, y, and z.

6. **Writing** Explain how to decide which half-plane to shade when you graph an inequality such as $7x + 5y > 35$.

$$\begin{aligned} 0.3x - 2.1y + 4.5z &= 12.4 \\ 5.2x + 0.2y - 3.7z &= 4.0 \\ -1.8x - 3.0y - 1.6z &= 9.2 \end{aligned}$$

Graph each inequality or system of inequalities.

7. $y \leq -0.4x + 0.6$

8. $x > -4$
$y \leq 5$

9. $2x - 3y \geq -12$
$y < -\frac{2}{3}x - 2$

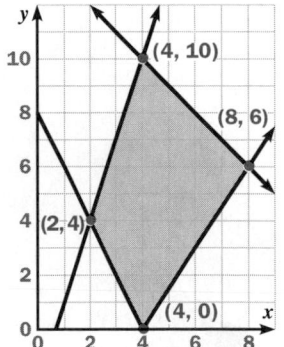

10. a. Write a system of inequalities defining the feasible region shown.

b. For the feasible region in part (a), find the minimum cost for the cost function $C = 5x + 8y$.

11. Daniel can spend up to $100 on tapes and CDs. He wants to buy at least as many CDs as tapes, and at least 1 tape. A tape costs $10 and a CD costs $15. How many CDs and how many tapes should Daniel buy to maximize the number of items bought?

CHAPTER 8

12. OCEANOGRAPHY The equation $l = \dfrac{\pi v^2}{4.9}$ shows how the velocity v of a deep water wave (in meters per hour) is related to the wavelength l (in meters).

a. Write an equation for the velocity as a function of the wavelength.

b. Find the domain and range of the function you wrote in part (a).

c. Find the velocity of a deep water wave with wavelength 10 m.

13. Graph the function $y = -\sqrt{x - 2} + 2$. State the domain and range.

14. Writing Explain why $\sqrt[3]{-729}$ is a real number, but $\sqrt{-729}$ is not.

Solve. Check to eliminate extraneous solutions.

15. $2\sqrt{2x - 1} = 6$ **16.** $\sqrt[3]{2x - 1} = 6$ **17.** $x^2 + 32 = 0$

Perform the indicated operation. Plot the result in the complex plane and find its magnitude.

18. $(7 + i) + (1 - i)$ **19.** $(7 + i) - (1 - i)$ **20.** $(7 + i)(1 - i)$

21. Use complex conjugates to write the quotient $\dfrac{7 + i}{7 - i}$ in $a + bi$ form.

22. Tell which group properties hold for the whole numbers under multiplication. If a group property does not hold, give a counterexample.

23. Open-ended Problem Give an example of a geometry problem whose solution is an irrational number.

CHAPTER 9

Perform the indicated operation.

24. $\dfrac{2v^2 + 6v - 5}{v - 2}$ **25.** $(1.2x^3 + 3x - 5) + (-x^2 + 7x - 8)$

26. Use synthetic substitution to evaluate $8x^3 - 7x^2 + 3x - 5$ for $x = \sqrt{3}$.

27. Writing Explain why $2\sqrt{x}$ is not a polynomial, but $x\sqrt{2}$ is a polynomial.

28. **Technology** Use a graphing calculator or graphing software to graph $f(x) = x^4 - 5x^3 + 5x^2$.

   a. Describe its end behavior.

   b. Find all local maximums and local minimums of the function.

29. Find an equation for the cubic function whose graph has x-intercepts -2, 1, and 4 and passes through the point with coordinates $(2, 4)$.

For each function, list the possible rational zeros, and find all real and imaginary zeros. Identify any double or triple zeros.

30. $f(x) = 6x^3 - 23x^2 + 9x + 5$ **31.** $g(x) = x^4 - 2x^3 - 3x^2 - 50x - 50$

32. Open-ended Problem Find a polynomial function that has degree 4 and a triple zero at $x = 1$.

33. GEOMETRY Does the base length of a triangle with area 20 cm^2 vary inversely with the triangle's height? If so, state the constant of variation.

For each function, find the vertical asymptotes of the function's graph, and describe the function's end behavior using infinity notation.

34. $f(x) = \dfrac{2x - 1}{x + 5}$ **35.** $g(x) = \dfrac{x}{x^2 + x - 6}$

36. Find an equation of the hyperbola that has asymptotes $x = 2$ and $y = -3$ and that passes through the point with coordinates $(1, 5)$.

37. Solve the rational equation $\dfrac{1}{x - 1} + \dfrac{2}{x + 1} = 1$.

C H A P T E R

10 Sequences and Series

Fractals for fashions

INTERVIEW **Jhane Barnes**

"Everything I create is ultimately visual. And fractals are very visual."

Imagine starting your own business while still a college student. That's what Jhane Barnes did while she was enrolled in New York's Fashion Institute of Technology. Starting with a loan from a professor, she built a business that has grown to become both a commercial and a critical success. In 1980, at the age of 25, Barnes was the youngest person, and the first female, to win a Coty American Fashion Critics' Award for menswear.

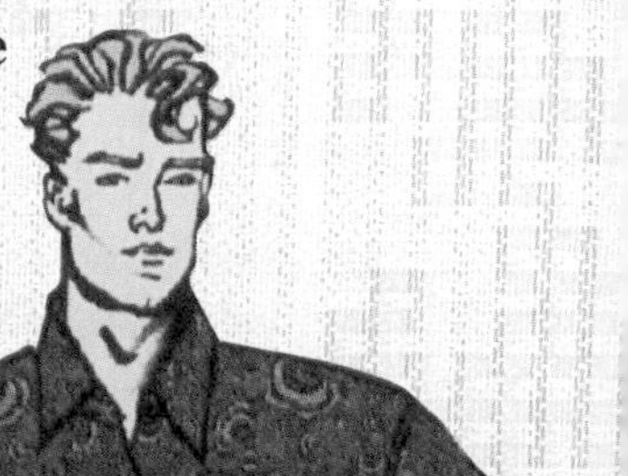

The design in the background is the pattern Jhane Barnes calls "Dragon."

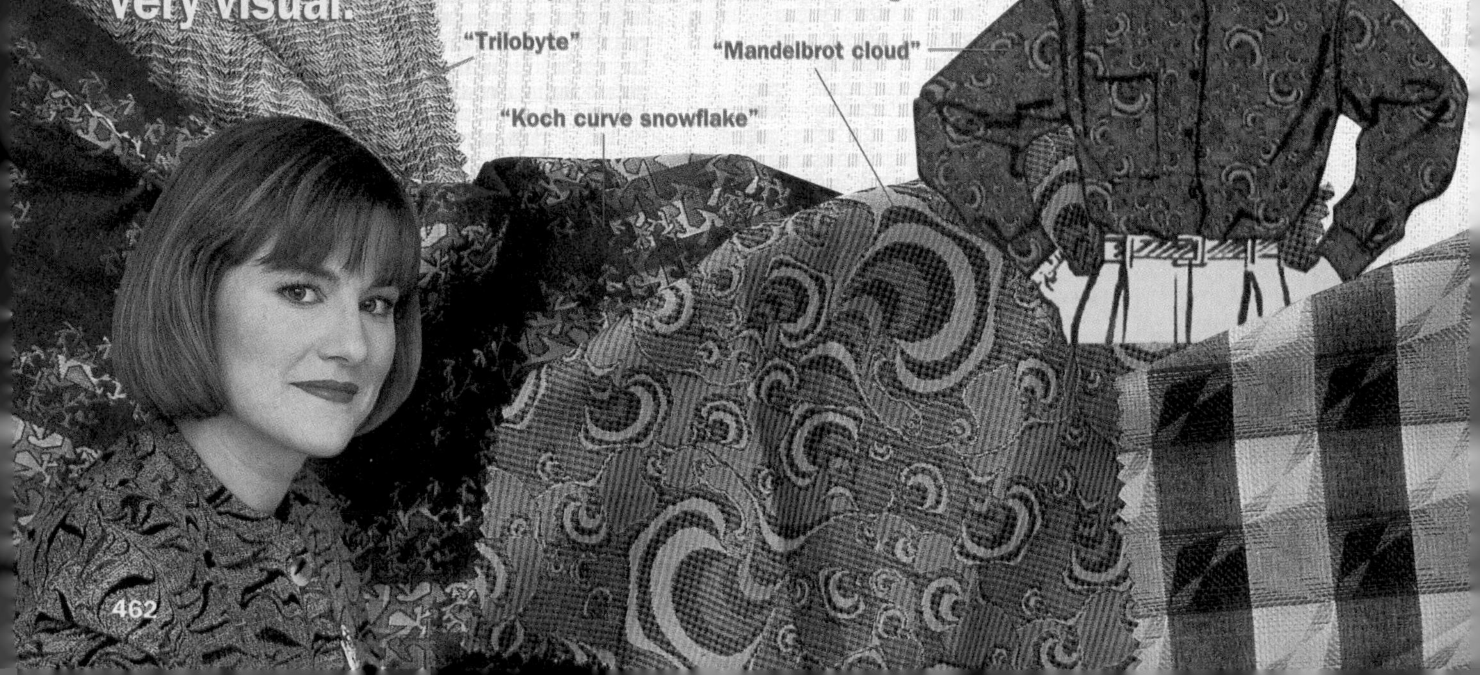

"Trilobyte"
"Mandelbrot cloud"
"Koch curve snowflake"

Creating Magical Patterns

Barnes designs textiles that are immediately recognizable for their unique use of color and texture. Many of her designs start with a basic element, to which geometric transformations such as rotations and translations are repeatedly applied. In some cases the basic element is as simple as a swirl; in other cases it is a complex figure known as a *fractal*. A close look at a fractal reveals patterns that repeat on an ever-diminishing scale.

Programs with Fashion Finesse

Barnes uses computer programs to transfer her designs to a weaving loom. Only certain designs are transferable, because the loom's frame must hold vertical and horizontal threads known as *warp* and *weft*. "I'm constrained by fabrics," Barnes says. "I can only do things that can be made with warp and weft thread."

Using a computer to control the loom is what led Barnes to discover that a computer could also help her with fabric design.

"Symmetry star"

Working with several programmers, Barnes develops the computer code that creates her most intricate designs. "Instead of thinking of patterns only in a visual sense, I now break them down to their simplest elements and think of them mathematically."

Designing with Mathematics

Barnes can go through hundreds of pattern ideas a day for a tie, a jacket, or a pair of socks. Her creative genius shines through in her color choices, and in knowing which patterns will adapt to which fabrics. The outcome, she says, is like magic.

"I realized that if patterns can be generated mathematically, so can the arrangement of color. Many of my designs demonstrate how the sequence of color enhances the movement and flow of the pattern," says Barnes.

A Sierpinski Pattern

Barnes designed one textile using a fractal known as a *Sierpinski triangle*, named after Waclaw Sierpinski, the Polish mathematician who introduced it in 1916. This fractal is developed in stages, beginning with a solid triangle. At the next stage, the original triangle is divided into four congruent, smaller triangles and the center one is removed. This process is repeated to create successive stages involving smaller and smaller triangles. As the illustration shows, these stages form a *sequence* of triangles, each more elaborate than the one before.

Stage 1

Stage 2

Stage 3

Fabrics created by Jhane Barnes are also used in interior design.

EXPLORE AND CONNECT

Jhane Barnes works on one of her fabric designs using a computer.

1. Project A Sierpinski carpet is like a Sierpinski triangle except that it starts with a shaded square. At each stage of creating the carpet, you must divide all shaded squares from the previous stage into nine congruent, smaller squares and shade all but the middle square. Show three stages of this sequence.

2. Writing Give a written description of the stages in this sequence:

Ask someone who has not seen this sequence to try drawing it from your description of it.

3. Research Use an encyclopedia to find out about weaving. Describe the use of warp and weft in plain and satin weaves. How do the patterns of warp and weft differ in each of the weaves?

Mathematics
& Jhane Barnes

In this chapter, you will learn more about how mathematics is related to fashion.

Related Exercises

Section 10.1
• **Exercises 52 and 53**

Section 10.2
• **Exercises 35–37**

Section 10.5
• **Exercises 38 and 39**

10.1 Sequences

The sights and sounds of everyday life are full of interesting patterns. Some are easy to detect, while others demand careful study. Here are just a few examples. Think about what they have in common.

A. Part of a Philippine folk dance

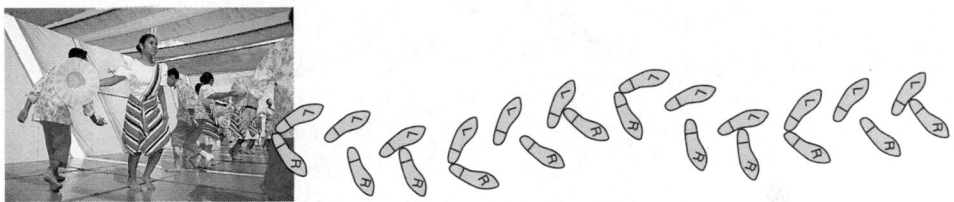

B. A pattern completion exercise from a standardized test

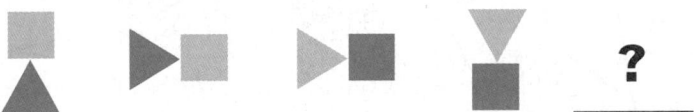

C. Part of the sheet music for the national anthem of Mexico, words by Francisco González Bocanegra and music by Jaime Nunó

ce - ro a - pre-stad y el bri - dón.

THINK AND COMMUNICATE

1. Explain why order is important in each situation shown.

2. Describe any patterns you see. Use them to predict what comes next when the pattern is continued.

In each situation shown above, the elements are ordered in time or space. Each set of ordered elements forms a **sequence**, or ordered list. The elements of a sequence are called its **terms**. In mathematics, the terms of a sequence are numbers or geometric figures.

EXAMPLE 1

Find the next 4 terms of each sequence, if possible. If not, explain why.

a. 13, 10, 7, 4, . . .

b. daily closing prices of a stock:

$$18\tfrac{3}{8},\ 18\tfrac{1}{4},\ 17,\ 17\tfrac{5}{8},\ \ldots$$

SOLUTION

a.

$$13,\quad 10,\quad 7,\quad 4, \ldots$$
$$-3\quad -3\quad -3$$

Notice that each term after the first is 3 less than the term before it. The next 4 terms of the sequence are $1, -2, -5, -8$.

b. The price of a stock at the close of the trading day depends on many factors, most of which cannot be predicted. Therefore, it is not possible to find the next 4 terms of the sequence.

When a number sequence has a pattern, you may be able to write a rule for the relationship between the terms of a sequence and their position in the sequence. For example, consider the sequence of positive even numbers:

terms of the sequence: 2, 4, 6, 8, . . .

position in the sequence: 1 2 3 4 . . .

Each term t is simply 2 times its position number n. This means that t is a function of n, and you can write:

For any sequence, the **domain** consists of the **counting numbers** 1, 2, 3, ...

$$t(n) = 2n$$

For any sequence, the **range** consists of the **terms** of the sequence.

To distinguish sequences from other functions, formulas for sequences are written in *subscript notation*, where t_n represents the nth term of the sequence:

Read this as "t sub n."

$$t_n = 2n$$

EXAMPLE 2

Find the first 4 terms of the sequence $t_n = n^2 - 3$.

SOLUTION

Substitute $1, 2, 3,$ and 4 for n in the formula.

$$t_1 = 1^2 - 3 \qquad t_2 = 2^2 - 3 \qquad t_3 = 3^2 - 3 \qquad t_4 = 4^2 - 3$$
$$= -2 \qquad\qquad = 1 \qquad\qquad = 6 \qquad\qquad = 13$$

The first 4 terms of the sequence are $-2, 1, 6, 13$.

EXAMPLE 3 **Application: Chemistry**

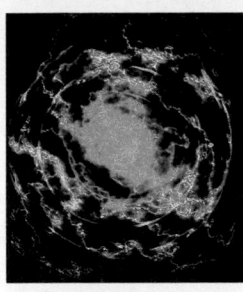

In an atom, the electrons can be found in energy levels outside the nucleus. An atom can have from 1 to 7 energy levels, and there is a maximum number of electrons that each level can hold. For the first 4 levels, these numbers are 2, 8, 18, 32.

a. Write a formula for the *n*th term of the sequence.

b. Determine the maximum number of electrons in the outermost energy level.

Representation of the energy levels of a helium atom

SOLUTION

a. Look for a relationship between each term and its position number.

Energy level	1	2	3	4
Maximum number of electrons	2	8	18	32

$$1 \cdot 2 \qquad 2 \cdot 4 \qquad 3 \cdot 6 \qquad 4 \cdot 8$$

Each term t_n is the product of *n* and a positive even number, so:

$$t_n = n \cdot 2n$$
$$= 2n^2$$

b. Level 7 is the outermost level. Find t_7.

$$t_7 = 2 \cdot 7^2$$

Substitute 7 for n.

$$= 98$$

The maximum number of electrons in the outermost energy level is 98.

A sequence like 2, 8, 18, 32, . . . , 98 that has a last term is a **finite sequence**. A sequence like 2, 4, 6, 8, . . . that continues without end is an **infinite sequence**.

EXAMPLE 4

Tell whether each sequence is *finite* or *infinite*. Explain.

a. the nonnegative multiples of 3   **b.** the dates of all Sundays this month

SOLUTION

a. The nonnegative multiples of 3 are produced by multiplying 3 by the whole numbers:

$$3 \cdot 0 = 0, \quad 3 \cdot 1 = 3, \quad 3 \cdot 2 = 6, \quad \ldots, \quad 3n, \quad \ldots$$

This sequence continues without end, so it is infinite.

b. No month has more than 5 Sundays, so the sequence is always finite.

For each sequence:

a. **Find the next 4 terms.**   b. **Write a formula for t_n.**

1. $1, \dfrac{1}{4}, \dfrac{1}{9}, \dfrac{1}{16}, \ldots$ **2.** $8, 14, 20, 26, \ldots$ **3.** $1, \sqrt{2}, \sqrt{3}, 2, \ldots$

Find the 7th term of each sequence.

4. $t_n = 2n + 3$ **5.** $t_n = 4^{n+1}$ **6.** $t_n = 5n^3 - 6$

10.1 Exercises and Applications

**Extra Practice
exercises on
page 764**

Find the next 4 terms of each sequence.

1. $4, 16, 64, 256, \ldots$

2. $800, 400, 200, 100, \ldots$

3. $1, 9, 3, 18, 5, 36, \ldots$

4. $1, 0, -1, 5, 1, 10, \ldots$

5. $5, 55, 555, 5555, \ldots$

6. $121, 11{,}311, 1{,}114{,}111, \ldots$

7. $43, 63, 93, 133, \ldots$

8. $7, -7, 7, -7, \ldots$

9. $\sqrt{3}, 3, \sqrt{27}, 9, \ldots$

Tell whether it is possible to find the next 4 terms of each sequence. Explain.

10. balances in a checkbook: $500.00, $467.32, $421.84, $471.84, \ldots

11. Presidential election years in the U.S. since 1790: 1792, 1796, 1800, 1804, \ldots

12. days on which it rains in June: 3, 7, 8, 15, \ldots

13. house numbers along one side of a street: 901, 903, 905, 907, \ldots

14. **BIOLOGY** Frogs belong to a class of animals called *amphibians*. After tadpoles hatch from frogs' eggs, they undergo a sequence of changes in body structure, called *metamorphosis,* before becoming adult frogs. The major stages in the metamorphosis of a frog are shown out of order.

A.

B.

C.
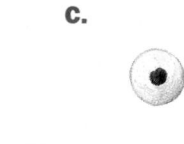

D.
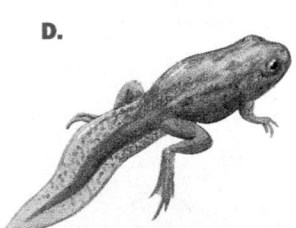

E.
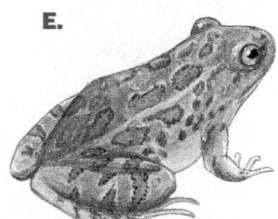

F.
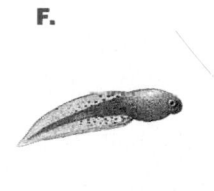

a. Put the pictures in the correct order to show the sequence of events that occurs during the metamorphosis of a frog.

b. **Writing** Describe the sequence of events shown.

15. **Cooperative Learning** Work with another student.

   a. Each of you should cut out a comic strip from a newspaper. (Do not show your comic strip to your partner.)

   b. Cut out the individual frames of the strip, mix them up, and pass them to your partner.

   c. Each of you should try to put the frames in the correct sequence.

Find the 2nd, 5th, and 12th terms of each sequence.

16. $t_n = 2n + 1$

17. $t_n = n^3$

18. $t_n = \dfrac{3}{n}$

19. $t_n = \dfrac{10n}{3n + 2}$

20. $t_n = 3n - 2^n$

21. $t_n = (3n - 2)^2$

22. $t_n = 2n(5 - n)$

23. $t_n = (2n)^2$

24. $t_n = (4n)^{1/n}$

25. $t_n = (-1)^{n + 3}$

26. $t_n = 6^{n - 2}$

27. $t_n = \log 8n$

Write a formula for t_n.

28. $1, 8, 15, 22, \ldots$

29. $43, 49, 55, 61, \ldots$

30. $7, 49, 343, 2401, \ldots$

31. $\dfrac{3}{5}, \dfrac{3}{25}, \dfrac{3}{125}, \dfrac{3}{625}, \ldots$

32. $-3, 3, -3, 3, \ldots$

33. $103, 97, 91, 85, \ldots$

34. $1, 2\sqrt{2}, 3\sqrt{3}, 8, \ldots$

35. $1, -\dfrac{1}{3}, \dfrac{1}{5}, -\dfrac{1}{7}, \dfrac{1}{9}, \ldots$

36. $12, 42, 92, 162, \ldots$

37. $\dfrac{\sqrt{2}}{2}, 1, \sqrt{2}, 2, \ldots$

38. $6, 30, 150, 750, \ldots$

39. $84, 42, 21, 10.5, \ldots$

Connection ▶ MUSIC

The *acoustics,* or sound quality, in a concert hall depends in part on the stereo effect that occurs when sound reaches listeners' ears from the sides of the hall. To improve acoustics, architects often install *diffusers* with wells of varying depths that scatter sound in all directions.

   In one type of diffuser, the depths of the wells are the terms of this sequence: the remainders when n^2 (for $n = 0, 1, 2, \ldots$) is divided by a prime number p. A shorthand way to write this is $t_n = (n^2)_{\mathrm{mod}\ p}$.

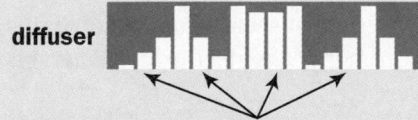

diffuser

wells

40. Find the first $2p$ terms of the sequence for $p = 5$, $p = 11$, and $p = 13$.

41. Use your results in Exercise 40 to make a conjecture about the terms of $t_n = (n^2)_{\mathrm{mod}\ p}$.

42. The Michael Fowler Centre opened in 1983 in Wellington, New Zealand. It has diffusers with well depths $t_n = (n^2)_{\mathrm{mod}\ 7}$. Find the terms of the sequence and sketch the pattern of the wells in a diffuser.

The Michael Fowler Centre, Wellington, New Zealand

Tell whether each sequence is *finite* or *infinite*. Explain.

43. a pattern around the rim of a pottery vase

44. the digits in the decimal form of π

45. the positive odd numbers

46. the squares of the counting numbers

47. the digits in the decimal form of $\frac{1}{128}$

48. the positive integer powers of 10 less than one billion

Connection · ASTRONOMY

Johann Elert Bode

In 1766 the German astronomer Johann Daniel Titius announced a rule for computing the approximate distances of the planets from the sun in astronomical units (AU). One astronomical unit is the mean distance between Earth and the sun. Today, Titius's rule is known as *Bode's law*, for Johann Elert Bode, who made it famous.

According to Bode's law, the distances of the first four planets from the sun are given by the sequence in the table.

Planet	Distance (AU)
Mercury	0.4 = 4/10
Venus	0.7 = (4 + 3)/10
Earth	1.0 = (4 + 3 · 2)/10
Mars	1.6 = (4 + 3 · 2 · 2)/10

49. Write a formula for Bode's law. (*Hint:* Begin with Venus.)

50. a. When Bode's law was developed, only six of the nine planets had been discovered. The other two known planets were Jupiter, about 5 AU from the sun, and Saturn, about 10 AU from the sun. Use these distances to find each planet's position in the sequence for Bode's law.

b. In 1781, the planet Uranus was discovered. Uranus fits into Bode's sequence immediately after Saturn. Use Bode's law to find the approximate distance of Uranus from the sun in astronomical units.

c. Research Find the 5th term in Bode's sequence. Use an encyclopedia or astronomy book to confirm that there is no planet at this distance from the sun. What astronomical object *is* located at this distance? When was it discovered?

1 AU

51. The last two planets, Neptune and Pluto, were not discovered until 1846 and 1930, respectively. Neptune is located about 30 AU from the sun and Pluto is about 39 AU from the sun. Do Neptune and Pluto fit into Bode's sequence? Explain.

Look back at the
article on pages
462–464.

*Jhane Barnes created the fabric design shown. Examine the pattern inside
the four small squares outlined in this design.*

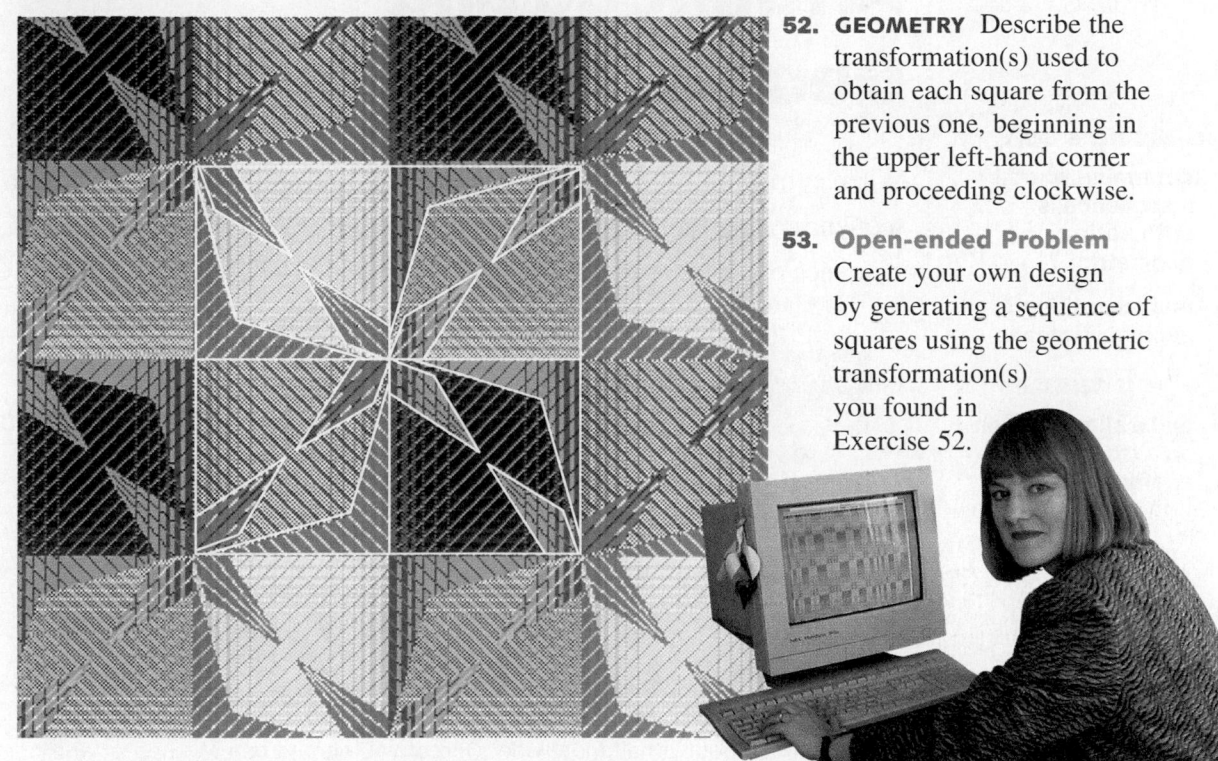

52. GEOMETRY Describe the
transformation(s) used to
obtain each square from the
previous one, beginning in
the upper left-hand corner
and proceeding clockwise.

53. Open-ended Problem
Create your own design
by generating a sequence of
squares using the geometric
transformation(s)
you found in
Exercise 52.

54. Challenge Is 22,211 a term of the sequence 11, 15, 19, 23, . . . ? If so,
what is its position in the sequence? Explain how you found your answer.

ONGOING ASSESSMENT

55. Open-ended Problem Think of as many sequences as you can that begin
with 2, 4, . . . and have a pattern. For each sequence, write the next 4 terms
and, if possible, write a formula for the nth term of the sequence. If it is not
possible to write a formula, describe the pattern in words.

SPIRAL REVIEW

Solve each equation. *(Section 9.9)*

56. $\dfrac{1}{x} = \dfrac{5}{x + 2}$

57. $\dfrac{y}{4} = \dfrac{7}{y - 3}$

58. $\dfrac{t - 1}{t} + \dfrac{6}{5t} = 2$

59. $\dfrac{1}{x - 4} = \dfrac{2}{x^2 - 16}$

Simplify each radical expression. *(Section 8.2)*

60. $\sqrt{4^8}$

61. $\sqrt[3]{9^9}$

62. $\sqrt[3]{128}$

63. $\sqrt[5]{96}$

Sketch the graph of each equation. *(Sections 2.2 and 3.3)*

64. $y = -3x + 5$

65. $y = 3^x$

66. $y = 6x - 14$

67. $y = (1.5)^x$

10.2 Arithmetic and Geometric Sequences

Learn how to...

- determine whether a sequence is arithmetic or geometric

- find an arithmetic or geometric mean

So you can...

- find a slide position on a trombone, for example

When cutting up vegetables, experienced cooks usually begin by making a sequence of parallel cuts. Then they gather all the pieces and make another sequence of cuts through all the pieces at once. Do you know why?

EXPLORATION
COOPERATIVE LEARNING

Investigating Methods for Cutting Paper

Work with a partner.
You will need:
- scissors

SET UP Turn an $8\frac{1}{2}$ in. by 11 in. sheet of paper sideways. Cut the paper into strips about a half inch wide. One of you should use Method A, and the other should use Method B. Each time you make a cut, record the total number of cuts and the total number of pieces of paper in a table.

| No. of | No. of pieces | |
cuts	A	B
1	2	2
2	4	3
⋮	⋮	⋮

Method A

1 Cut the paper in half.

2 Stack the halves. Cut the stack in half.

3 Continue stacking and cutting.

Method B

1 Cut a thin strip from one end.

2 Continue cutting off strips.

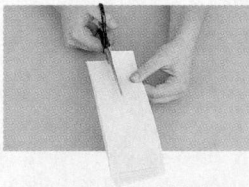

Answer Questions 1–3 for each method.

1. How does the number of pieces change with each cut after the first?

2. Find the next 3 terms of the sequence of the number of pieces.

3. Write a formula for the nth term of the sequence in terms of n and the first term of the sequence.

The sequence formed by Method A in the Exploration is a *geometric sequence*. In a **geometric sequence**, the ratio of any term to the term before it is constant. The constant ratio is called the **common ratio**. For example, consider this sequence:

The common ratio is 5.

$$2, \quad 10, \quad 50, \quad 250, \quad \ldots, \quad 2 \cdot 5^{n-1}$$

$\times 5 \quad \times 5 \quad \times 5$

To find the *n*th term, **multiply** the first term by the common ratio $(n - 1)$ times.

The sequence formed by Method B in the Exploration is an *arithmetic sequence*. In an **arithmetic sequence**, the difference between any term and the term before it is constant. The constant difference is called the **common difference**. For example, consider this sequence:

The common difference is 5.

$$2, \quad 7, \quad 12, \quad 17, \quad \ldots, \quad 2 + 5(n - 1)$$

$+ 5 \quad + 5 \quad + 5$

To find the *n*th term, **add** the common difference to the first term $(n - 1)$ times.

Arithmetic and Geometric Sequences

Arithmetic Sequence

General formula:

$$t_n = t_1 + d(n - 1)$$

common difference

Example: $t_n = 2 + 5(n - 1)$

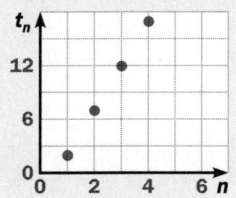

Geometric Sequence

General formula:

$$t_n = t_1 \cdot r^{n-1}$$

common ratio

Example: $t_n = 2 \cdot 5^{n-1}$

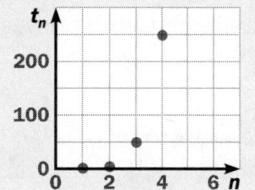

THINK AND COMMUNICATE

1. How are arithmetic sequences and linear functions related? Which elements of the general formula for an arithmetic sequence correspond to the slope, the dependent variable, and the independent variable of a linear function? Explain.

2. How are geometric sequences and exponential functions related? Which elements of the general formula for a geometric sequence correspond to the proportional growth factor, the dependent variable, and the independent variable of an exponential function? Explain.

3. What effect does a common ratio *between 0 and 1* have on a geometric sequence? on its graph?

4. What effect does a *negative* common difference have on an arithmetic sequence? on its graph?

EXAMPLE 1

Tell whether each sequence is *arithmetic, geometric,* or *neither*.

a. $-14, -42, -126, -378, \ldots$ **b.** $1, 19, 49, 91, \ldots$ **c.** $31, 27, 23, 19, \ldots$

SOLUTION

Look for a common difference or a common ratio.

a. $-14, \quad -42, \quad -126, \quad -378, \ldots$

differences: $-28 \quad -84 \quad -252$

ratios: $3 \quad\quad 3 \quad\quad 3$

The common ratio is 3. The sequence is geometric.

b. $1, \quad 19, \quad 49, \quad 91, \ldots$

differences: $18 \quad\quad 30 \quad\quad 42$

ratios: $19 \quad 2.579 \quad 1.857$

There is neither a common difference nor a common ratio.
The sequence is neither geometric nor arithmetic.

c. $31, \quad 27, \quad 23, \quad 19, \ldots$

differences: $-4 \quad\quad -4 \quad\quad -4$

ratios: $0.871 \quad 0.852 \quad 0.826$

The common difference is -4. The sequence is arithmetic.

EXAMPLE 2

Find the 11th term of each sequence.

a. the sequence in part (a) of Example 1

b. the sequence in part (c) of Example 1

SOLUTION

a. The sequence $-14, -42, -126, -378, \ldots$ is geometric with a common
ratio of 3.

Step 1 Write a formula for t_n.

$$t_n = -14\left(3^{n-1}\right)$$

> Substitute -14 for t_1 and 3 for r in the general formula.

Step 2 Find t_{11}.

$$t_{11} = -14\left(3^{11-1}\right)$$
$$= -14(59{,}049)$$
$$= -826{,}686$$

> Substitute 11 for n.

The 11th term of the sequence is $-826{,}686$.

b. The sequence 31, 27, 23, 19, . . . is arithmetic with a common difference of -4.

> **Step 1** Write a formula for t_n.
>
> $$t_n = 31 - 4(n - 1)$$

Substitute **31** for t_1 and -4 for d in the general formula.

> **Step 2** Find t_{11}.
>
> $$t_{11} = 31 - 4(11 - 1)$$
> $$= 31 - 40 = -9$$

Substitute **11** for n.

The 11th term of the sequence is -9.

EXAMPLE 3 Application: Music

To play different notes on a trombone, you change the *effective length* of the tube by moving the slide in and out. The sequence of lengths at different slide positions is geometric. Find the effective length at 4th position on this trombone.

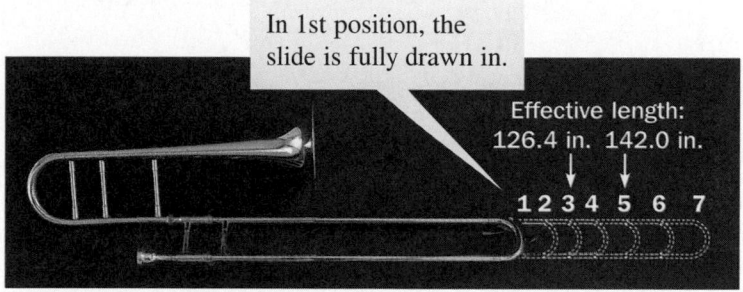

In 1st position, the slide is fully drawn in.

Effective length: 126.4 in. 142.0 in.

1 2 3 4 5 6 7

SOLUTION

Use the fact that a geometric sequence has a common ratio.

The ratios of successive terms are equal.

$$\frac{t_4}{t_3} = \frac{t_5}{t_4}$$

Substitute **126.4** for t_3 and **142.0** for t_5.

$$\frac{t_4}{126.4} = \frac{142.0}{t_4}$$

Use the means-extremes property.

$$(t_4)^2 = 126.4 \cdot 142.0 = 17{,}948.8$$

$$t_4 = \sqrt{17{,}948.8} \approx 134.0$$

Since length is always positive, find the **positive square root**.

Toolbox p. 785
Ratios and Proportions

The effective length at 4th position is about 134.0 in.

The term between any two terms of a geometric sequence is called the **geometric mean** of the given terms. In Example 3, you found the geometric mean of two positive terms. In general, for any two positive terms a and b:

$$\textbf{geometric mean of } \textit{a} \textbf{ and } \textit{b} = \sqrt{ab}$$

The term between any two terms of an arithmetic sequence is called the **arithmetic mean** of the given terms. The arithmetic mean is simply the *average* or *mean* of the given terms. For any two terms a and b:

$$\textbf{arithmetic mean of } \textit{a} \textbf{ and } \textit{b} = \frac{a + b}{2}$$

For each sequence:

a. Tell whether it is *arithmetic, geometric,* or *neither.*

b. Find the 11th term.

   1. 11, 15, 19, 23, . . .   **2.** 3, 15, 75, 375, . . .   **3.** 53, 47, 41, 35, . . .

   4. $63, 21, 7, 2\frac{1}{3}, \ldots$ **5.** 1, 4, 9, 16, . . .   **6.** $-3, -1, 1, 3, \ldots$

For each pair of numbers:

a. Find the arithmetic mean.   b. Find the geometric mean.

   7. 4, 25   **8.** 5, 45   **9.** $\sqrt{2}, 3\sqrt{2}$

10.2   **Exercises and Applications**

Extra Practice exercises on page 764

Tell whether each sequence is *arithmetic, geometric,* or *neither.*

   1. 8, 16, 24, 32, . . .   **2.** $-4, -16, -64, -256, \ldots$ **3.** 36, 49, 64, 81, . . .

   4. 32, 33, 34, 35, . . .   **5.** 243, 81, 27, 9, . . .   **6.** $1, -1, 1, -1, \ldots$

   7. 0, 14, 52, 254, . . .   **8.** 129, 88, 47, 6, . . .   **9.** $\frac{1}{3}, \frac{1}{9}, \frac{1}{27}, \frac{1}{81}, \ldots$

  10. $\sqrt{3}, 2\sqrt{3}\ 4\sqrt{3}, 7\sqrt{3}, \ldots$ **11.** $-0.61, -0.64, -0.67, -0.70, \ldots$ **12.** $i, -2, -4i, 8, \ldots$

13. For each arithmetic sequence in Exercises 1–12, find:

   a. the common difference   b. the 14th term

14. For each geometric sequence in Exercises 1–12, find:

   a. the common ratio   b. the 14th term

15. BIOCHEMISTRY When scientists analyze DNA for medical diagnosis, criminal law, evolutionary biology, and gene research, they often need larger samples than they are given. The *polymerase chain reaction (PCR)* process amplifies a tiny sample of DNA by making multiple copies of the DNA molecules. In each cycle of the process, the number of DNA molecules doubles.

   a. Is the sequence of the number of DNA molecules after each cycle *arithmetic, geometric,* or *neither*? Explain.

   b. Write a formula for the *n*th term of the sequence, beginning with one molecule of DNA.

   c. How many cycles are needed to produce a billion copies?

   d. One cycle takes about 6.5 min. About how long does it take to produce a billion copies?

BY THE WAY

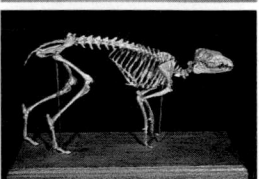

The PCR process was discovered in 1983 by biochemist Kary Mullis. One of its uses has been to study the genes of dinosaurs and other animals that lived millions of years ago.

16. Use the information about trombone slide positions in Example 3 on page 475.

    a. Find the common ratio of the sequence of effective lengths.

    b. **Writing** Explain what the common ratio means in this situation.

For each pair of numbers:

a. Find the arithmetic mean.   **b. Find the geometric mean.**

17. 9, 64   **18.** 5, 19   **19.** 16, 5   **20.** $\dfrac{7}{8}, \dfrac{1}{7}$

21. 3.7, 2.3   **22.** 7, 33   **23.** 4, 0.04   **24.** $8i, 32i$

25. **BIOLOGY** Domestic honeybees construct a honeycomb by adding ring after ring of hexagonal cells around a single hexagonal cell.

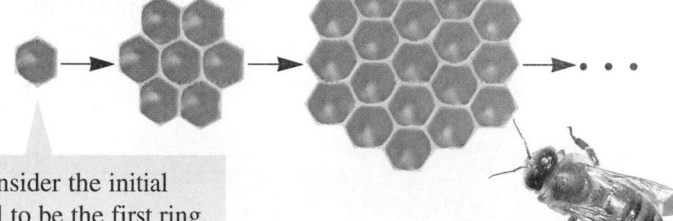

    a. **Visual Thinking** Is the sequence of the number of cells in each ring *arithmetic, geometric,* or *neither*? Explain.

    b. Write a formula for the *n*th term of the sequence.

Consider the initial cell to be the first ring.

Connection ▶ GRAPHIC DESIGN

The *weight* of a typeface refers to the thickness of the strokes used in the letters. Sometimes type designers use numbers to represent typeface weights. The weights of the typeface in the table form a geometric sequence.

Light 215	The quick brown fox jumps over the lazy red dog.
Regular _?_	The quick brown fox jumps over the lazy red dog.
Semibold 422	The quick brown fox jumps over the lazy red dog.
Bold _?_	**The quick brown fox jumps over the lazy red dog.**
Black 830	**The quick brown fox jumps over the lazy red dog.**

26. Complete the table of typeface weights.

27. What is the common ratio for this geometric sequence?

28. For this typeface, suppose you want to create an Ultra Light weight one step lighter in weight than Light. What number should you use to describe it?

29. For this typeface, suppose you want to create an Extra Black weight one step heavier than Black. What number should you use to describe it?

30. arithmetic sequence: $t_3 = 5$ and $t_9 = -7$ **31.** geometric sequence: $t_4 = 15$ and $t_7 = 120$

32. ARCHITECTURE The Navajo people of the western United States used a technique called *corbeling* to build dome-shaped log houses. In a corbeled house, or *hogan*, the logs are stacked in layers. In each layer, the ends of the logs meet at the midpoints of the logs below.

This is the view looking down onto a corbeled roof.

a. Suppose the logs in each layer of a corbeled roof form a regular hexagon. Write a formula for the *n*th term of the sequence of log lengths, starting with *x*.

b. Is the sequence of log lengths *arithmetic, geometric,* or *neither*? Explain.

Connection ▶ SPORTS

The track shown is shaped like a rectangle with semicircles on either end. The track has 8 lanes that are each 1.22 m wide. The lanes are numbered in ascending order from the inside. In a 400 m race, the runners must stay in their lanes. (Assume that the distance around each lane is measured along the center of the lane.)

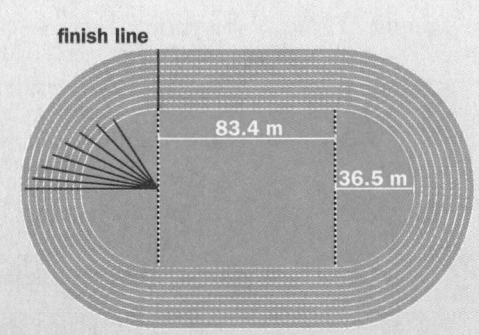

finish line

83.4 m

36.5 m

33. a. Each red radius on the diagram is the radius of the curve for one of the lanes. Tell whether the sequence formed by these radii is *arithmetic, geometric,* or *neither.*

b. Write a formula for the sequence in part (a).

c. According to the rules of the International Amateur Athletic Federation (IAAF), world records can be set only on tracks that have a curve radius of at most 50 m in the outside lane. Does the track shown meet the IAAF requirement?

34. a. On the track shown, each lane except the first is longer than 400 m. For each lane, find how much longer.

b. Tell whether the sequence in part (a) is *arithmetic, geometric,* or *neither.*

c. **Writing** Explain why the IAAF requires that the starting line be staggered when 400 m races are run on a track like the one shown.

INTERVIEW Jhane Barnes

Look back at the article on pages 462–464.

Jhane Barnes named this fabric pattern "Sierpinski triangle" for the fractal shape it contains. A fractal *is an irregular shape that has the property of* self-similarity, *so that any part looks like the whole fractal. The first four stages in the construction of a Sierpinski triangle are shown. The initial triangle is equilateral with sides 1 unit long. The cut-out triangles are formed by connecting the midpoints of the sides of the shaded triangles.*

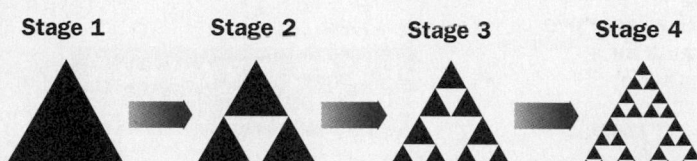

Stage 1   Stage 2   Stage 3   Stage 4

For each sequence described in Exercises 35–37:

a. **Find the first 4 terms.**

b. **Tell whether the sequence is *arithmetic, geometric,* or *neither.* Explain.**

c. **Write a formula for the *n*th term.**

35. the number of shaded triangles at each stage

36. the perimeter of the figure that remains at each stage (*Hint:* Add the perimeters of all the shaded triangles at each stage.)

37. the area of the figure that remains at each stage (*Hint:* Add the areas of all the shaded triangles at each stage.)

ONGOING ASSESSMENT

38. a. Graph the sequences $t_n = 3(2)^{n-1}$ and $u_n = 3(-2)^{n-1}$ in the same coordinate plane using two different symbols, such as "X" and "O," for your plotted points.

    b. Writing How are the graphs alike? How are they different?

    c. Writing Explain how the two sequences suggest the need to restrict the base *b* to positive numbers (other than 1) for an exponential function of the form $y = ab^x$.

SPIRAL REVIEW

Find the 15th term of each sequence. (*Section 10.1*)

39. $\sqrt{3}, 2\sqrt{3}, 3\sqrt{3}, 4\sqrt{3}, \ldots$ **40.** $0.9, 0.09, 0.009, 0.0009, \ldots$ **41.** $1, 6, 15, 28, \ldots$

Find the zeros of each function. (*Section 9.4*)

42. $f(x) = 6(x + 2)(x + 1)(x - 1)$ **43.** $f(x) = x^3 - 5x^2 - 2x + 10$

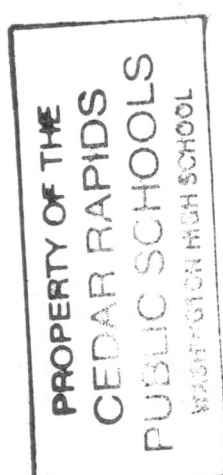

10.3 | Exploring Recursion

In Sections 10.1 and 10.2, you wrote formulas for the nth term, t_n, of a sequence. These formulas, which express t_n as a function of its position number n, are called **explicit formulas**. In this section you will explore another way to write a formula for the nth term of a sequence.

Learn how to...

- **find the terms of a sequence defined recursively**

- **write a recursive formula for a sequence**

So you can...

- **find out what happens to the level of medication in the bloodstream over time, for example**

EXPLORATION
COOPERATIVE LEARNING

Investigating Repeated Addition and Multiplication

Work with a partner.
You will need:
- graphing calculators

1 The calculator screen shows the procedure you should follow to generate a sequence. Choose various starting values and various constants (both positive and negative), and write down the first 6 terms of the sequences you generate.

Choose a starting value.

Add a constant to the "last answer."

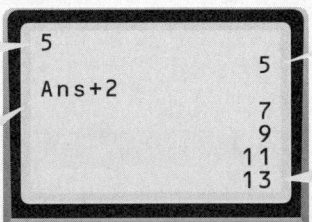

```
5
            5
Ans+2
            7
            9
           11
           13
```

Press ENTER.

By repeatedly pressing ENTER, you will continue to add the constant to the "last answer."

2 Repeat Step 1, but this time *multiply* by a constant instead of adding a constant. For example, use Ans * 2 instead of Ans + 2.

Questions

1. In each sequence, what is the relationship between any term (except the first) and the term before it?

2. Tell whether each sequence formed by repeatedly adding a constant is *arithmetic, geometric,* or *neither*.

3. Tell whether each sequence formed by repeatedly multiplying by a constant is *arithmetic, geometric,* or *neither*.

In the Exploration you generated the terms of a sequence by entering a starting value and repeatedly applying the same operation to your last answer. This process is called **recursion**.

Recursion gives you another way to write a formula for the *n*th term of a sequence. A **recursive formula** for a sequence has two parts. The first part assigns a starting value. The second part is a *recursion equation* for t_n as a function of t_{n-1}, the term before it.

Recursive Formulas

Arithmetic Sequence

General formula:

$$t_1 = \text{starting value}$$
$$t_n = t_{n-1} + d$$

Example: $t_1 = 5$
$$t_n = t_{n-1} + 2$$

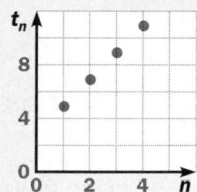

Geometric Sequence

General formula:

$$t_1 = \text{starting value}$$
$$t_n = r(t_{n-1})$$

Example: $t_1 = 5$
$$t_n = 2(t_{n-1})$$

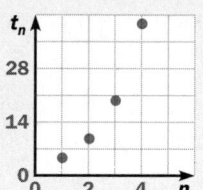

THINK AND COMMUNICATE

1. How are the general explicit formula (see page 473) and the general recursive formula for arithmetic sequences alike? different?

2. Answer Question 1 for geometric sequences.

EXAMPLE 1

Find the first 4 terms of each sequence.

a. $t_1 = -3;\ t_n = t_{n-1} + 8$

b. $t_1 = 1875;\ t_n = \frac{1}{5}(t_{n-1})$

SOLUTION

Each time you find a term of the sequence, use it to find the next term.

a. $t_1 = -3$

$t_2 = t_1 + 8 = -3 + 8 = 5$

$t_3 = t_2 + 8 = 5 + 8 = 13$

$t_4 = t_3 + 8 = 13 + 8 = 21$

The sequence is $-3, 5, 13, 21, \ldots$.

b. $t_1 = 1875$

$t_2 = \frac{1}{5}(t_1) = \frac{1}{5}(1875) = 375$

$t_3 = \frac{1}{5}(t_2) = \frac{1}{5}(375) = 75$

$t_4 = \frac{1}{5}(t_3) = \frac{1}{5}(75) = 15$

The sequence is $1875, 375, 75, 15, \ldots$.

EXAMPLE 2

Write a recursive formula for each sequence.

a. $1, 2.5, 6.25, 15.625, \ldots$ **b.** $2, 5, 26, 677, \ldots$ **c.** $4, 2.8, 1.6, 0.4, \ldots$

SOLUTION

a. The sequence is geometric with a common ratio of **2.5**. Use the general recursive formula for a geometric sequence.

Substitute **1** for the starting value. $t_1 = 1$ $t_n = 2.5(t_{n-1})$ Substitute **2.5** for *r*.

b. The sequence is neither arithmetic nor geometric. Look for a relationship between each term after the first and the term before it:

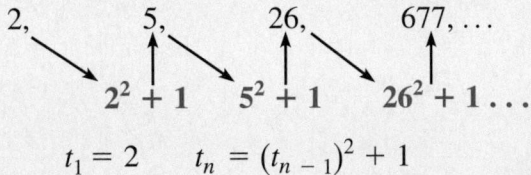

$$2, \quad 5, \quad 26, \quad 677, \ldots$$

$$2^2 + 1 \quad 5^2 + 1 \quad 26^2 + 1 \ldots$$

$$t_1 = 2 \qquad t_n = (t_{n-1})^2 + 1$$

c. The sequence is arithmetic with a common difference of **−1.2**. Use the general recursive formula for an arithmetic sequence.

Substitute **4** for the starting value. $t_1 = 4$ $t_n = t_{n-1} + (-1.2)$ Substitute **−1.2** for *d*.

EXAMPLE 3   Application: Medicine

One medication used to treat high blood pressure is supplied in 1.25 mg tablets. Patients take one tablet at the same time every day. By the time a patient takes the next dose, only about 32% of the medication is left in the bloodstream. What happens to the level of medication in a patient's bloodstream over time?

SOLUTION

Write a recursive formula to model the level of medication in the bloodstream after each dose.

$$t_1 = 1.25$$

$$t_n = 0.32(t_{n-1}) + 1.25$$

Method 1   Use the last-answer key on a calculator.

```
1.25
              1.25
.32*Ans+1.25
              1.65
             1.778
           1.81896
         1.8320672
```

```
       1.836261504
       1.837603681
       1.838033178
       1.838170617
       1.838214597
       1.838228671
       1.838233175
```

After several days, the amount of medication in the bloodstream levels off at about 1.84 mg.

Method 2 Use a graph.

Be sure your calculator is set for sequence graphing in dot mode.

Step 1 Enter the recursion equation.   **Step 2** Set the viewing window.

For more information about graphing sequences, see the *Technology Handbook*, p. 810.

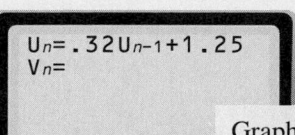

Un=.32Un-1+1.25
Vn=

Graph the sequence for values of *n* from 1 to 10.

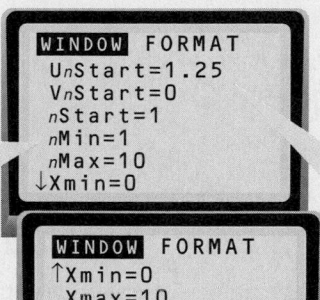

WINDOW FORMAT
UnStart=1.25
VnStart=0
nStart=1
nMin=1
nMax=10
↓Xmin=0

WINDOW FORMAT
↑Xmin=0
Xmax=10
Xscl=1
Ymin=-.2
Ymax=2
Yscl=.2

Enter the starting value of the sequence.

Step 3 Graph the recursive formula.

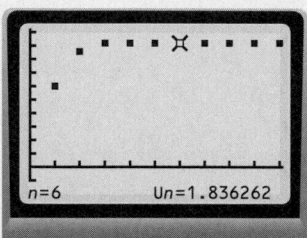

n=6   Un=1.836262

After several days, the amount of medication in the bloodstream levels off at about 1.84 mg.

THINK AND COMMUNICATE

3. Doctors sometimes begin a course of medication with a "booster" dose of 1.6 times the regular dose. What is the booster dose for the medication in Example 3? How does it compare with the solution to Example 3?

☑ CHECKING KEY CONCEPTS

Find the first 4 terms of each sequence.

1. $t_1 = 0$
$t_n = t_{n-1} + 6$

2. $t_1 = 8$
$t_n = -2(t_{n-1})$

3. $t_1 = -2$
$t_n = t_{n-1} - 5$

4. $t_1 = 3$
$t_n = 4(t_{n-1}) + 0.3$

5. $t_1 = 6.2$
$t_n = \frac{1}{2}(t_{n-1})$

6. $t_1 = 5\frac{1}{4}$
$t_n = 2(t_{n-1})^2 + 1$

Write a recursive formula for each sequence.

7. 4, 28, 196, 1372, …

8. 7, −14, 28, −56, …

9. 19, 13, 7, 1, …

10. 384, 96, 24, 6, …

11. $\frac{1}{3}, \frac{1}{9}, \frac{1}{81}, \frac{1}{6561}, \ldots$

12. 0.66, 0.69, 0.72, 0.75, …

10.3 Exercises and Applications

Extra Practice exercises on page 765

For each sequence:

a. Find the first 4 terms.

b. Write an explicit formula.

1. $t_1 = 4$
   $t_n = -\frac{1}{2}(t_{n-1})$

2. $t_1 = 7$
   $t_n = t_{n-1} + 3$

3. $t_1 = 4.5$
   $t_n = 6(t_{n-1})$

4. $t_1 = -1$
   $t_n = t_{n-1} - 6$

5. $t_1 = 9$
   $t_n = t_{n-1}$

6. $t_1 = -3$
   $t_n = t_{n-1} + 6$

7. $t_1 = 0$
   $t_n = t_{n-1} + 6$

8. $t_1 = 2i$
   $t_n = i(t_{n-1})$

9. **LANGUAGE ARTS** Linguists use the name *recursion* for the property of mental grammar that permits the creation of infinitely many new sentences by nesting grammatical patterns inside one another. Verna Aardema uses linguistic recursion in her book *Bringing the Rain to Kapiti Plain,* a retelling of a folktale of the Nandi culture of Kenya, in Africa. How is the linguistic recursion Verna Aardema used like mathematical recursion?

Here are the first four verses of *Bringing the Rain to Kapiti Plain*:

This is the cloud, all heavy with rain, that shadowed the ground on Kapiti Plain.

This is the grass, all brown and dead, that needed the rain from the cloud overhead— the big, black cloud, all heavy with rain, that shadowed the ground on Kapiti Plain.

These are the cows, all hungry and dry, who mooed for the rain to fall from the sky; to green-up the grass, all brown and dead, that needed the rain from the cloud overhead— the big, black cloud, all heavy with rain, that shadowed the ground on Kapiti Plain.

This is Ki-pat, who watched his herd as he stood on one leg, like the big stork bird; Ki-pat, whose cows were so hungry and dry, they mooed for the rain to fall from the sky; to green-up the grass, all brown and dead, that needed the rain from the cloud overhead—the big, black cloud, all heavy with rain, that shadowed the ground on Kapiti Plain.

Write a recursive formula for each sequence.

10. $-8, -10, -12, -14, \ldots$

11. $1.1, 2.2, 3.3, 4.4, \ldots$

12. $71, 65, 59, 53, \ldots$

13. $50, 25, 12.5, 6.25, \ldots$

14. $7, \frac{14}{3}, \frac{28}{9}, \frac{56}{27}, \ldots$

15. $3, 9, 81, 6561, \ldots$

16. $1.3, -2.7, -6.7, -10.7, \ldots$

17. $3, 3\sqrt{5}, 15, 15\sqrt{5}, \ldots$

18. $572, 57.2, 5.72, 0.572, \ldots$

19. $t_n = 3n - 1$

20. $t_n = 2^n$

21. $t_n = n$

22. Suppose a tree farm has 3800 trees. Each year the farmer harvests 20% of the trees and plants 900 new trees.

    a. Write a recursive formula for the total number of trees each year.

    b. What happens to the total number of trees over time?

    c. **Writing** What other factors may affect the number of trees on the farm? How do these factors affect the formula you wrote in part (a)?

Connection ▸ BIOLOGY

The chambered nautilus is a cone-shaped shellfish. To protect its soft body, it makes a shell that contains many chambers, or compartments. Each successive chamber is larger than the previous one but has the same shape as the animal's body. As the nautilus grows, it moves into the next larger chamber. The cross section of a nautilus shell is an *equiangular spiral* in which all the walls between the chambers intersect the outer boundary at congruent angles.

23. a. GEOMETRY Follow these steps to construct a figure that approximates an equiangular spiral.

    Step 1 Begin with a small unit square.
    Step 2 Form a rectangle by adding another unit square above it.
    Step 3 Form a larger rectangle by adding a square on the left side of the previous rectangle.
    Step 4 Form a still larger rectangle by adding a square below the previous rectangle.
    Step 5 Continue adding squares in this way, moving in a counterclockwise direction, until there are nine squares in all.
    Step 6 Starting at the lower right-hand vertex of the *second* square, draw a spiral that passes through one pair of opposite vertices of each square after the first.

    b. In terms of the length of a side of the unit square, find the lengths of the sides of the nine squares you drew in part (a).

24. The numbers you wrote in part (b) of Exercise 23 are the first 9 terms of an infinite sequence called the *Fibonacci sequence.* Write a recursive formula for this sequence. (*Hint:* Each term after the second is a function of the *two* terms immediately before it.)

(above left) A chambered nautilus *(above)* Cross section of a chambered nautilus shell, showing the chamber walls intersecting the outside of the shell at congruent angles

BY THE WAY

"Fibonacci" is the pen name of mathematician Leonardo of Pisa. Almost 800 years ago, he introduced the Western world to both Hindu-Arabic numerals and the mathematics developed in China, Arabia, and India.

Photographers who use manual 35mm cameras must adjust various camera settings to account for different lighting conditions.

25. To allow more (or less) light to enter the camera, the photographer changes the camera setting that increases (or decreases) the area of a circular opening. The camera settings, called *f-stops,* are inversely proportional to the diameter of the opening. Each f-stop lets in half as much light as the previous one. The amount of light is proportional to the area of the opening.

The photograph above was taken with sufficient light entering the camera. With too much light, the picture is too bright. With too little light, the picture is too dark.

    a. Tell whether the sequence of f-stops is *geometric, arithmetic,* or *neither.* (*Hint:* By what factor does the diameter change when the area is halved?)

    b. Use 2 as the first f-stop. Find the next 7 f-stops.

    c. Write a recursive formula for the sequence of f-stops.

26. The table gives factors for modifying the flash guide number (G) and the film speed number (ASA) in order to set the correct f-stop and exposure time for flash photography. Use the table to write a recursive formula for each sequence of factors:

    a. G—available light is principal source   **b.** G—flash is principal source

    c. ASA—available light is principal source   **d.** ASA—flash is principal source

Principal Light Source

Lighting ratio	Available light		Flash	
	Factor for G	Factor for ASA	Factor for G	Factor for ASA
2 : 1	$\sqrt{2}$	2	$\sqrt{2}$	2
3 : 1	$\sqrt{3}$	$\frac{3}{2}$	$\sqrt{\frac{3}{2}}$	3
4 : 1	$\sqrt{4}$	$\frac{4}{3}$	$\sqrt{\frac{4}{3}}$	4
5 : 1	$\sqrt{5}$	$\frac{5}{4}$	$\sqrt{\frac{5}{4}}$	5

For each sequence in Exercises 27 and 28, any term t_n after the second term is a function of the two terms, t_{n-1} and t_{n-2}, immediately before it in the sequence. Write a recursive formula for each sequence.

27. 6, 7, 26, 66, 184, 500, . . .   **28.** 2, 3, 2, 2, $\frac{4}{3}, \frac{8}{9}, \ldots$

Challenge Write a recursive formula that defines each operation.

29. a number raised to a positive integral power   **30.** a factorial of a positive integer

31. SAT/ACT Preview The third term of a sequence is 143. The recursion equation for the sequence is $t_n = 2(t_{n-1}) - 3$ for $n = 2, 3, \ldots$. What is the starting value for the sequence?

A. 140   **B.** 283   **C.** 38   **D.** 73   **E.** none of these

ONGOING ASSESSMENT

32. Open-ended Problem Create a sequence in which each term after the third term is a function of the three terms immediately before it. Write a recursive formula for your sequence and use it to find the first 12 terms.

SPIRAL REVIEW

Find the geometric mean of each pair of numbers. *(Section 10.2)*

33. 72, 6   **34.** 12, 26   **35.** $8\sqrt{6}, \sqrt{15}$

Tell whether the data that can be gathered about each variable are *categorical* **or** *numerical*. **Then describe the categories or numbers.** *(Section 6.1)*

36. class size   **37.** clothing size   **38.** pets

Evaluate each expression when $a = 7$ **and** $b = 2$. *(Toolbox, page 780)*

39. $5(a + 4b)$ **40.** $2a + 3(a - b)$ **41.** $\dfrac{-5(a + 2b)}{ab}$

ASSESS YOUR PROGRESS

VOCABULARY

sequence (p. 465)   **common difference** (p. 473)
term (p. 465)   **geometric mean** (p. 475)
finite sequence (p. 467)   **arithmetic mean** (p. 475)
infinite sequence (p. 467)   **explicit formula** (p. 480)
geometric sequence (p.473)   **recursion** (p. 481)
common ratio (p. 473)   **recursive formula** (p. 481)
arithmetic sequence (p. 473)

For each sequence, find the next three terms and tell whether the sequence is *arithmetic, geometric,* **or** *neither.* *(Sections 10.1 and 10.2)*

1. $-5, -7, -10, -14, \ldots$ **2.** $8, -16, 32, -64, \ldots$ **3.** $4, 13, 22, 31, \ldots$

Find the arithmetic mean and the geometric mean for each pair of numbers.
(Section 10.2)

4. 8, 22   **5.** 0.9, 62.5   **6.** $\sqrt{3}, 5\sqrt{3}$

Write an explicit formula and a recursive formula for each sequence.
(Sections 10.2 and 10.3)

7. $-7, -13, -19, -25, \ldots$ **8.** $3, 2, \dfrac{4}{3}, \dfrac{8}{9}, \ldots$ **9.** $19, 22, 25, 28, \ldots$

10. Journal What are some of the advantages and disadvantages of explicit formulas for sequences? of recursive formulas for sequences?

10.4 Sums of Arithmetic Series

Learn how to...

- use a formula to find the sum of an arithmetic series

- use sigma notation for series

So you can...

- find the number of seats in a theater, for example

Imagine the amazement of his elementary school teacher when young Carl Friedrich Gauss announced that he had added the integers from 1 to 100 in his head! This happened over 200 years ago in Germany. Gauss was a child prodigy who grew up to be one of the greatest mathematicians of all time. To see how Gauss did it, look at this:

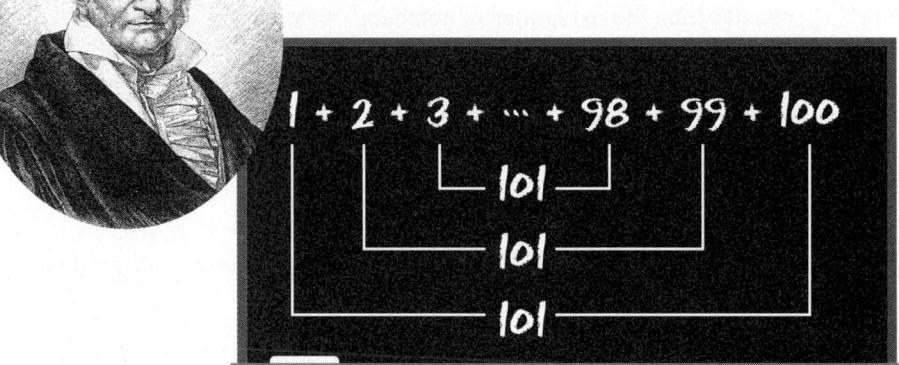

THINK AND COMMUNICATE

1. a. How many sums of 101 are there in the sum of the integers from 1 to 100? How do you know?

  b. Find the sum of the integers from 1 to 100. Explain how you found your answer.

2. Is the sequence 1, 2, 3, . . . , 100 *arithmetic, geometric,* or *neither*? Is it *finite* or *infinite*?

3. Use Gauss's method to find the sum of the terms of each sequence. Check your answers with a calculator.

  a. 1, 2, 3, . . . , 200   **b.** 5, 10, 15, . . . , 100

  c. 60, 58, 56, . . . , 0   **d.** −12, −11, −10, . . . , 12

When the terms of a sequence are added, the indicated sum is called a **series**:

sequence: 1, 2, 3, . . . , 100

series: 1 + 2 + 3 + · · · + 100

The indicated sum of an arithmetic sequence is called an **arithmetic series**.

The method Gauss used to find the sum of the integers from 1 to 100 leads to a general formula for the sum S_n of any finite arithmetic series with n terms.

Sum of a Finite Arithmetic Series

The general formula for the sum S_n of a finite arithmetic series $t_1 + t_2 + t_3 + t_4 + \cdots + t_n$ is:

sum of n terms

$$S_n = \frac{n}{2}(t_1 + t_n)$$

half the number of terms   first term   last term

THINK AND COMMUNICATE

4. Describe Gauss's formula in words.

5. Derive an alternative formula for the sum of a finite arithmetic series by substituting the explicit formula $t_1 + d(n - 1)$ for t_n.

EXAMPLE 1 **Application: Architecture**

The last row has 45 seats.

The seats in the theater shown are staggered to improve visibility. How many seats are there in the section shown? Use the information that accompanies the photo.

The first row has 22 seats.

SOLUTION

To find the total number of seats, find the sum of the series whose terms are the number of seats in each row.

Each row after the first has one more seat than the row before it.

Step 1 Find the related sequence.
The first four terms of the sequence are: 22, 23, 24, 25.
The sequence is arithmetic with a common difference of 1.

Step 2 Find the number of terms in the related sequence.
Use the formula for the nth term of an arithmetic sequence.

$$45 = 22 + 1(n - 1)$$

Solve for n. $45 = 22 + n - 1$ Substitute 45 for t_n, 22 for t_1, and 1 for d.

$$24 = n$$

Step 3 Find the sum of the series.
Use the formula for the sum of a finite arithmetic series.

$$S_{24} = \frac{24}{2}(22 + 45)$$

Substitute 24 for n, 22 for t_1, and 45 for t_n.

$$= 12 \cdot 67 = 804$$

There are 804 seats in the section shown.

Sigma Notation

You can write a series in compact form by using the summation symbol Σ, which is the capital Greek letter *sigma*. For example, the expression below represents the sum of the values of $3n + 10$ for integer values of n from 1 to 9.

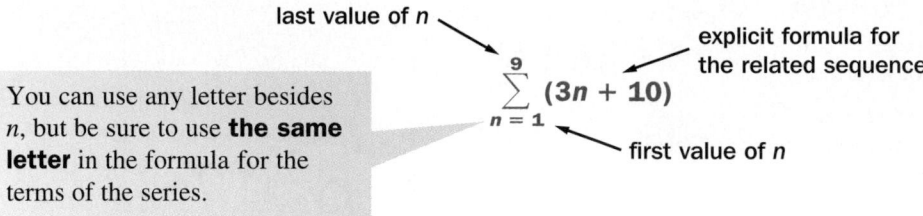

last value of n

explicit formula for the related sequence

$$\sum_{n=1}^{9} (3n + 10)$$

first value of n

You can use any letter besides n, but be sure to use **the same letter** in the formula for the terms of the series.

To write a series in *expanded form,* substitute each value of n into the formula. For the expression above, you have:

$$\sum_{n=1}^{9} (3n + 10) = (3 \cdot 1 + 10) + (3 \cdot 2 + 10) + \cdots + (3 \cdot 9 + 10)$$

$$= 13 + 16 + \cdots + 37$$

EXAMPLE 2

In Korea it is a tradition to celebrate three special birthdays over a person's lifetime: the *paegil* or hundredth day after birth, the *tol* or first birthday, and the *hwan'gap* or 60th birthday. For these celebrations, tables are set with stacks of fruit, rice cakes, sweetened bean cakes, and other special foods.

Jae Joon Hyun's mother stacks oranges for his *tol.* Use sigma notation to write the series that represents the number of oranges in the stack.

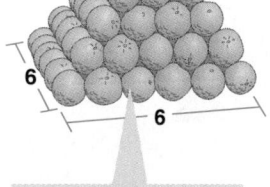

There are 6 layers.

SOLUTION

Step 1 Write an explicit formula for the sequence.
From the top, the layers contain these numbers of oranges:

$$1, \quad 4, \quad 9, \ldots, 36$$
$$1^2, \quad 2^2, \quad 3^2, \ldots, 6^2$$

Each term is the square of its position number, so the explicit formula is:

$$t_n = n^2$$

Step 2 Write the series in sigma notation.

$$\sum_{n=1}^{6} n^2$$

For each series:

a. Find the number of terms.

b. Find the sum.

1. $0 + 6 + 12 + 18 + \cdots + 120$ **2.** $1 + 9 + 17 + 25 + \cdots + 97$

3. $15 + 12 + 9 + 6 + \cdots + (-21)$ **4.** $100 + 99 + 98 + 97 + \cdots + 5$

Expand each series.

5. $\displaystyle\sum_{n=1}^{6} 4n^3$ **6.** $\displaystyle\sum_{k=1}^{7} (k + 5)$ **7.** $\displaystyle\sum_{n=1}^{5} (3n - 2)$ **8.** $\displaystyle\sum_{s=1}^{8} (2s^2 + 7)$

Write each series in sigma notation.

9. $3 + 12 + 48 + 192$ **10.** $(-14) + (-9) + (-4) + 1 + 6$

11. $26 + 24 + 22 + 20 + \cdots + 10$ **12.** $4 + 9 + 16 + 25 + \cdots + 81$

10.4 **Exercises and Applications**

Extra Practice exercises on page 765

Find the sum of each series.

1. $1 + 2 + 3 + 4 + \cdots + 80$ **2.** $0 + 4 + 8 + 12 + \cdots + 52$

3. $200 + 199 + 198 + 197 + \cdots + 100$ **4.** $20 + 14 + 8 + 2 + \cdots + (-76)$

5. $58 + 61 + 64 + 67 + \cdots + 349$ **6.** $11 + 4 + (-3) + (-10) + \cdots + (-73)$

7. Visual Thinking Cut out a block of unit squares from a piece of graph paper to represent the terms of a finite arithmetic series. Then cut out an identical block and put it together with the first block to form a rectangle.

Example: $4 + 7 + 10 + 13$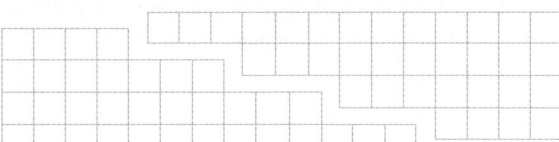

Repeat this procedure for three other arithmetic series. Then answer these questions:

a. What feature of the series does the length of each rectangle represent?

b. What feature of the series does the width of each rectangle represent?

c. What is the relationship between the area of each rectangle and the sum of the series?

The first four terms and the sum of a finite arithmetic series are given. Find the last term of each series.

8. $1 + 1\frac{1}{4} + 1\frac{1}{2} + 1\frac{3}{4} + \cdots + \underline{\ ?\ }$; $S_n = 51$ **9.** $1 + 2 + 3 + 4 + \cdots + \underline{\ ?\ }$; $S_n = 1275$

10. $0 + 11 + 22 + 33 + \cdots + \underline{\ ?\ }$; $S_n = 396$ **11.** $90 + 88 + 86 + 84 + \cdots + \underline{\ ?\ }$; $S_n = 948$

Expand each series.

12. $\displaystyle\sum_{m=1}^{4} 9m^2$ **13.** $\displaystyle\sum_{n=1}^{6} (n-12)$ **14.** $\displaystyle\sum_{n=1}^{5} (6n+1)$ **15.** $\displaystyle\sum_{k=1}^{7} (-2k^3+5)$

Write each series in sigma notation and find the sum.

16. $13 + 25 + 37 + 49 + \cdots + 217$

17. $0 + 2.3 + 4.6 + 6.9 + \cdots + 92$

18. $164 + 113 + 62 + 11 + \cdots + (-499)$

19. $8 + 11\frac{1}{2} + 15 + 18\frac{1}{2} + \cdots + 67\frac{1}{2}$

20. $(-184) + (-170) + (-156) + (-142) + \cdots + (-72)$

21. $7.1 + 6.9 + 6.7 + 6.5 + \cdots + 0.7$

Connection ▸ QUILTING

Quilters may create original patterns or choose from a variety of traditional patterns, including those used in the two quilts shown. A portion of each quilt has been enlarged to make the pattern easier to see.

22. The pattern shown at the right is called *around the world,* because the quilter sews groups of squares around a central square.

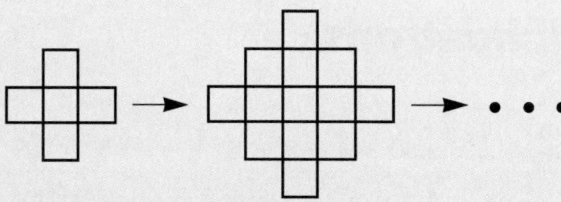
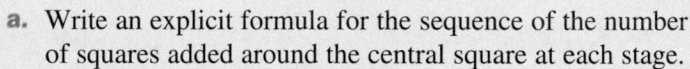

   a. Write an explicit formula for the sequence of the number of squares added around the central square at each stage.

   b. Tell whether the sequence is *arithmetic, geometric,* or *neither*. Explain.

   c. For the portion of the quilt that has been enlarged, the sequence has 7 terms. Write a formula for the sum of the sequence. Express your answer in sigma notation.

   d. Suppose the pattern were continued. How many stages would be needed to bring the total number of squares added around the central square to 480?

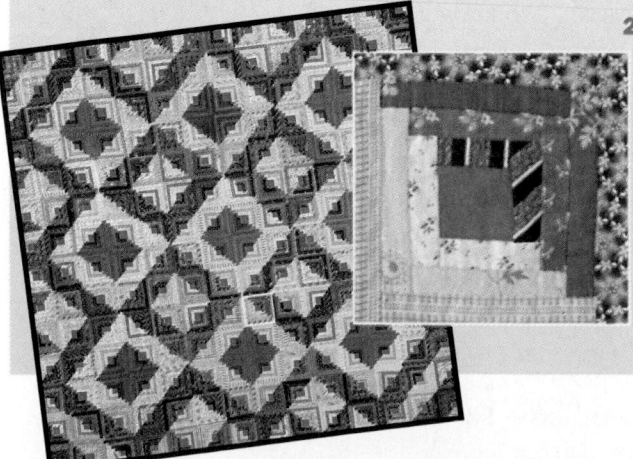

23. The pattern shown at the left is called a *log cabin.* It is created by cutting pieces from a long strip of cloth. All the pieces are 1 unit wide. The central piece is a unit square. Write a formula for the total length of cloth used as a function of the number of pieces cut from the cloth strip. Express your answer in sigma notation.

24. The *Fibonacci sequence* appears in many natural patterns (see Exercises 23 and 24 on page 485). The first and second terms of the sequence are both 1. Each term after the second term is the sum of the two terms immediately before it.

   a. Write the first 15 terms of the Fibonacci sequence.

   b. Find the sum of n terms of the Fibonacci sequence when $n = 2, 3, 4, \ldots, 13$.

   c. Add 1 to each sum you found in part (b) and compare the results with the terms of the Fibonacci sequence. Make a conjecture about the sum of the terms of the Fibonnaci sequence.

   d. **Challenge** Make a conjecture about the sum of any 10 consecutive terms of the Fibonacci sequence.

For Exercises 25 and 26, refer to Example 2 on page 490.

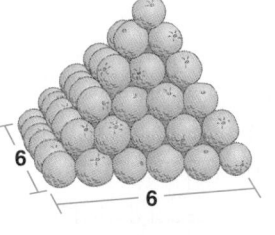

25. a. How many oranges are in the stack?

   b. Suppose the stack of oranges had 12 layers. Use sigma notation to write the series that represents the number of oranges in the stack.

26.  **Technology** Use a spreadsheet in part (a) and a graphing calculator or software with matrix calculation capabilities in part (b).

   a. Evaluate $\sum_{k=1}^{n} k^2$ for $n = 1, 2, 3, \ldots, 20$. These values form a *sequence of partial sums.*

Sums of Squares		
C8	= C7 + B8	
A	**B**	**C**
Position number	Term	Partial sum
1	1	1
2	4	5
3	9	14
4	16	30
5	25	55
6	36	91
7	49	140

   b. You know that $\sum_{k=1}^{n} k = \dfrac{n(n+1)}{2}$, which means that the sum of the first n positive integers is a *quadratic* function of n. Using the steps below, test the conjecture that the sum of the squares of the first n positive integers is a *cubic* function of n.

   Step 1 Write a system of four equations of this form using $n = 1, 2, 3, 4$:

   $$\sum_{k=1}^{n} k^2 = an^3 + bn^2 + cn + d, \text{ where } a, b, c, \text{ and } d \text{ are constants}$$

   Step 2 Solve the system using matrices (see Section 7.3).
   Step 3 Check to see whether the values you found for a, b, c, and d produce the correct sums when $n = 5, 6, 7, \ldots, 20$. (See part (a).)

27. SAT/ACT Preview What is the relationship between A and B for all values of n?

$$A = \sum_{k=1}^{n} (k + 4) \qquad B = \sum_{k=1}^{n} (k + 6)$$

A. $A = B$ **B.** $A < B$ **C.** $A > B$ **D.** relationship cannot be determined

28. PHYSICS When an object is in free fall, it experiences a constant acceleration of 32 ft/s² due to gravity. This means that the object's speed increases by 32 ft/s each second.

a. Starting with an initial speed of 0 ft/s, write the sequence of speeds for an object in free fall after $0, 1, 2, 3, \ldots, t$ seconds.

b. You can find the distance that the object falls during any second by multiplying the *average* speed by the time. For example, during the first second of fall, the object's average speed is $\dfrac{0 + 32}{2} = 16$ ft/s, so the object falls $(16 \text{ ft/s})(1 \text{ s}) = 16$ ft. Write the sequence of distances the object falls during each of the t seconds.

c. The total distance that an object falls is the sum of the terms of the sequence you wrote in part (b). Find the total distance. How does this compare with the function given in Example 3 on page 187?

ONGOING ASSESSMENT

29. Cooperative Learning Work with four other students.

For each step described below, each of you should write your answer on the paper you have and then pass it to the student on your right. When you have completed all the steps, pass the papers to the right one last time. Check the work on the paper you receive and correct any errors.

a. **Open-ended Problem** Make up a finite arithmetic sequence and write the terms of the sequence at the top of a blank sheet of paper.

b. Write an explicit formula for the nth term of the sequence on the paper you receive.

c. Using sigma notation, write an expression for the series related to the sequence on the paper you next receive.

d. Find the sum of the series on the paper you next receive.

SPIRAL REVIEW

Write a recursive formula for each sequence. *(Section 10.3)*

30. $12, 12i, -12, -12i, \ldots$ **31.** $14, 6, -2, -10, \ldots$

Multiply. *(Section 1.4)*

32. $\begin{bmatrix} 5 & -1 & 6 \\ 0 & 0 & 2 \\ -1 & 1 & 0 \end{bmatrix} \begin{bmatrix} 3 & 8 \\ 4 & 7 \\ -2 & 0 \end{bmatrix}$ **33.** $\begin{bmatrix} 8 & -3 \\ 7 & -2 \\ 1 & -3 \end{bmatrix} \begin{bmatrix} -3 & 0 & -1 \\ 6 & 9 & -9 \end{bmatrix}$ **34.** $\begin{bmatrix} 0 & 1 \\ 1 & 1 \end{bmatrix} \begin{bmatrix} 0 \\ 1 \end{bmatrix}$

10.5 Sums of Geometric Series

Learn how to...

- use a formula to find the sum of a finite geometric series

- use a formula to find the sum of an infinite geometric series, if the sum exists

So you can...

- find the number of games in an elimination tournament, for example

Many companies set up a telephone tree to notify employees when the business is shut down due to a hurricane, snow storm, or other emergency. Suppose a company sets up a telephone tree so that each employee (except those in the last level of the tree) calls three other employees. The first three levels of the tree look like this:

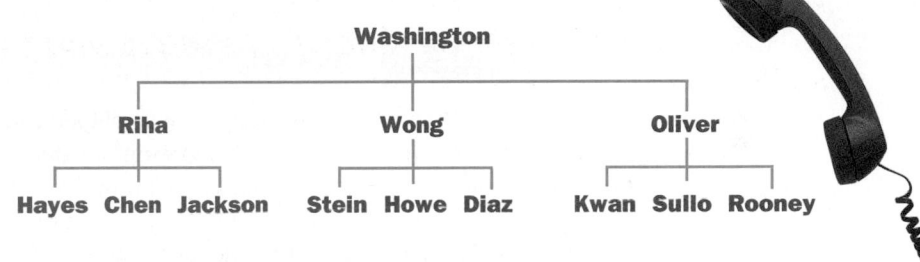

Washington

Riha   Wong   Oliver

Hayes Chen Jackson   Stein Howe Diaz   Kwan Sullo Rooney

THINK AND COMMUNICATE

1. **a.** Write the first 4 terms of the sequence for the number of employees at each level of the tree.

   b. Tell whether the sequence is *arithmetic, geometric,* or *neither.* Explain.

   c. Write an explicit formula for the sequence.

2. **a.** What does the related series represent in this situation?

   b. Write the related series in sigma notation, using k as the variable and n as the number of terms.

The indicated sum of a geometric sequence is called a **geometric series**. To find a general formula for the sum of a finite geometric series, first multiply the expanded series by the common ratio and then subtract the product from the original series:

$$S_n = t_1 + t_1 \cdot r + t_1 \cdot r^2 + t_1 \cdot r^3 + \cdots + t_1 \cdot r^{n-1}$$
$$r \cdot S_n = \qquad t_1 \cdot r + t_1 \cdot r^2 + t_1 \cdot r^3 + \cdots + t_1 \cdot r^{n-1} + t_1 \cdot r^n$$
$$\overline{S_n - r \cdot S_n = t_1 \qquad\qquad\qquad\qquad\qquad\qquad\qquad -t_1 \cdot r^n}$$

Subtract $r \cdot S_n$ from S_n.

THINK AND COMMUNICATE

3. Solve the equation above for S_n. Then check the formula using $n = 3$, $t_1 = 1$, and $r = 3$ to see whether it gives the number of employees shown in the telephone tree at the top of the page.

The general formula for the sum S_n of the finite geometric series
$t_1 + t_1 \cdot r + t_1 \cdot r^2 + \cdots + t_1 \cdot r^{n-1}$ where $r \neq 1$ is:

$$S_n = \frac{t_1(1 - r^n)}{1 - r}$$

THINK AND COMMUNICATE

4. Describe the general formula for the sum of a finite geometric series in your own words.

5. Why does the general formula for the sum of a finite geometric series apply only to series that have a common ratio that is not equal to 1? What sum would you get if $r = 1$?

EXAMPLE 1   Application: Sports

The four grand-slam tennis tournaments played each year are *elimination tournaments*. In both the men's and women's divisions, 128 players enter the first round of each tournament. In each round, pairs of players compete in matches. Players who win their matches move on to the next round. Players who lose are eliminated from the tournament.

a. How many rounds are there in a grand-slam tournament?

b. How many matches are played in a grand-slam tournament?

SOLUTION

Sketch a tree diagram of a simpler tournament, such as one with only four players.

In each round, the number of matches is **half** the number of players.

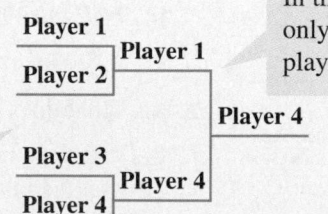

In the final round, only **1** match is played.

The number of matches in each round forms a geometric sequence with common ratio $\frac{1}{2}$. The number of matches in the first round is $128\left(\frac{1}{2}\right) = 64$.

a. Use the formula for the nth term of a geometric sequence: $t_n = t_1 \cdot r^{n-1}$.

$$1 = 64\left(\frac{1}{2}\right)^{n-1}$$

Substitute **1** for t_n, **64** for t_1, and $\frac{1}{2}$ for r.

$$\frac{1}{64} = \left(\frac{1}{2}\right)^{n-1}$$

$$\left(\frac{1}{2}\right)^6 = \left(\frac{1}{2}\right)^{n-1}$$

There are **7 rounds** in a grand-slam tournament.

$$6 = n - 1$$

$$7 = n$$

The exponents are equal.

BY THE WAY

The four grand-slam tournaments are the Australian Open, the French Open, Wimbledon, and the U.S. Open. Steffi Graf of Germany won all four in 1988.

b. The total number of matches is the sum of the number of matches in each round. Use the formula for the sum of a finite geometric series with $r \neq 1$.

$$S_7 = \frac{64\left(1 - \left(\frac{1}{2}\right)^7\right)}{1 - \frac{1}{2}} = 127$$

Substitute **64** for t_1, $\frac{1}{2}$ for r, and 7 for n.

In a grand-slam tournament, 127 matches are played.

Infinite Geometric Series

An infinite geometric series has the general form:

$$t_1 + t_1 \cdot r + t_1 \cdot r^2 + t_1 \cdot r^3 + t_1 \cdot r^4 + \cdots$$

Do you think it is possible for a series like this to have a sum?

EXPLORATION
COOPERATIVE LEARNING

Investigating an Infinite Geometric Series

Work with a partner.
You will need:

• scissors

SET UP Begin with a large square piece of paper. Let the length of a side of the square be 1 unit.

1 Cut the square in half to form two rectangles. Put one half aside and record its area as shown in the table below.

2 Cut the leftover piece in half to form two rectangles. Again put one half aside, placing it next to the piece set aside in the previous step. Express the combined area of the pieces put aside as an indicated sum.

3 Repeat Step 2 over and over until the leftover piece is too small to cut in half. Each time be sure to add the area of the piece put aside to the indicated sum that represents the combined area of the pieces put aside.

Questions

1. Suppose you can halve the leftover piece forever.

    a. Write the series that represents the combined area of the pieces put aside. Express your answer in expanded form.

    b. Is the series in part (a) *finite* or *infinite*? Is it *arithmetic*, *geometric*, or *neither*? Explain.

2. What would you say is the sum of the series in Question 1? Explain your reasoning.

Pieces put aside	Combined area
1	$\frac{1}{2}$
2	$\frac{1}{2} + ?$
3	$\frac{1}{2} + ? + ?$
$\vdots$	$\frac{1}{2} + ? + ? + \cdots$

In the Exploration you saw that the infinite geometric series

$$\frac{1}{2} + \frac{1}{4} + \frac{1}{8} + \frac{1}{16} + \cdots$$

appears to have a sum of 1. To see why this is so, consider the sum of the first n terms of this series:

$$S_n = \frac{\frac{1}{2}\left(1 - \left(\frac{1}{2}\right)^n\right)}{1 - \frac{1}{2}}$$

Substitute $\frac{1}{2}$ for t_1 and $\frac{1}{2}$ for r in the general formula for the sum of a finite geometric series.

$$= 1 - \left(\frac{1}{2}\right)^n$$

THINK AND COMMUNICATE

6. a. What happens to $\left(\frac{1}{2}\right)^n$ as $n \to \infty$? **b.** What happens to S_n as $n \to \infty$?

7. a. Describe what happens to $S_n = \dfrac{t_1(1 - r^n)}{1 - r}$ as $n \to \infty$ for *any* geometric series with $|r| < 1$.

b. Why is your conclusion in part (a) *not* true when $|r| > 1$?

Sum of an Infinite Geometric Series

The general formula for the sum S of the infinite geometric series
$t_1 + t_1 \cdot r + t_1 \cdot r^2 + t_1 \cdot r^3 + t_1 \cdot r^4 + \cdots$ where $|r| < 1$ is:

$$S = \frac{t_1}{1 - r}$$

You can use sigma notation to express the sum of an infinite geometric series. For example, you can write:

$$\frac{1}{2} + \frac{1}{4} + \frac{1}{8} + \frac{1}{16} + \cdots = \sum_{n=1}^{\infty} \frac{1}{2^n}$$

Use the symbol for **infinity** in place of the last value of n.

EXAMPLE 2

Find the sum of each infinite geometric series, if the sum exists.

a. $\displaystyle\sum_{n=1}^{\infty} -21(2)^{n-1}$

b. $5 + \dfrac{5}{3} + \dfrac{5}{9} + \dfrac{5}{27} + \cdots$

SOLUTION

For each series, first find the common ratio r. If $|r| < 1$, use the general formula to find the sum of the series.

a. The common ratio r is 2. Since $|r| > 1$, the series does not have a sum.

b. The common ratio r is $\frac{1}{3}$. Since $|r| < 1$, the series has a sum:

$$S = \frac{5}{1 - \frac{1}{3}}$$

$$= \frac{5}{\frac{2}{3}}$$

$$= 5\left(\frac{3}{2}\right)$$

$$= 7.5$$

> Substitute **5** for t_1 and $\frac{1}{3}$ for r in the general formula for the sum of an infinite geometric series.

EXAMPLE 3

Express 0.4545. . . as a fraction.

SOLUTION

Express the repeating decimal as an infinite series:

$$0.4545\ldots = 0.45 + 0.0045 + 0.000045 + 0.00000045 + \cdots$$

This series is geometric with $r = 0.01$. Since $|r| < 1$, the series has a sum:

$$S = \frac{0.45}{1 - 0.01}$$

$$= \frac{0.45}{0.99}$$

$$= \frac{45}{99} = \frac{5}{11}$$

> Substitute **0.45** for t_1 and **0.01** for r in the general formula for the sum of an infinite geometric series.

☑ CHECKING KEY CONCEPTS

Find the sum of each geometric series, if the sum exists. If a series does not have a sum, explain why not.

1. $2 + 6 + 18 + 54 + \cdots + 486$

2. $1 + 5 + 25 + 125 + \cdots$

3. $3 + (-6) + 12 + (-24) + \cdots + (-384)$

4. $200 + 50 + 12.5 + 3.125 + \cdots$

5. $\displaystyle\sum_{n=1}^{\infty} 4^{n-1}$

6. $\displaystyle\sum_{n=1}^{8} 7\left(\frac{1}{2}\right)^{n-1}$

7. $\displaystyle\sum_{n=1}^{\infty} \left(\frac{1}{3}\right)^{n-1}$

8. $\displaystyle\sum_{n=1}^{5} 3(-2)^{n-1}$

9–12. Write each series in Questions 1–4 in sigma notation.

10.5 Exercises and Applications

Extra Practice exercises on page 765

Find the sum of each series.

1. $4 + 8 + 16 + 32 + \cdots + 256$

2. $125 + 75 + 45 + 27 + \cdots + 5.832$

3. $2 + 2 \times 10 + 2 \times 10^2 + \cdots + 2 \times 10^{12}$

4. $\pi + 3\pi + 9\pi + 27\pi + \cdots + 6561\pi$

5. $12 + (-6) + 3 + \left(-\dfrac{3}{2}\right) + \cdots + \left(-\dfrac{3}{128}\right)$

6. $0.31 + 0.93 + 2.79 + 8.37 + \cdots + 225.99$

7. $\displaystyle\sum_{n=1}^{9} 43(0.3)^{n-1}$

8. $\displaystyle\sum_{n=1}^{5} -1\left(\dfrac{1}{3}\right)^{n-1}$

9. $\displaystyle\sum_{n=1}^{30} 5^{n-1}$

10. $\displaystyle\sum_{n=1}^{22} -2(-3)^{n-1}$

11. Suppose the company telephone tree on page 495 has 12 levels when it is completed. How many employees does the company have?

Find the sum of each series, if the sum exists. If a series does not have a sum, explain why not.

12. $300 + 150 + 75 + 37.5 + \cdots$

13. $6 + 4 + \dfrac{8}{3} + \dfrac{16}{9} + \cdots$

14. $-0.18 + 0.36 + (-0.72) + 1.44 + \cdots$

15. $88 + 22 + 5.5 + 1.375 + \cdots$

16. $1 + \left(-\dfrac{1}{3}\right) + \dfrac{1}{9} + \left(-\dfrac{1}{27}\right) + \cdots$

17. $12 + 16 + 21\dfrac{1}{3} + 28\dfrac{4}{9} + \cdots$

18. $\displaystyle\sum_{n=1}^{\infty} \dfrac{1}{2}(3)^{n-1}$

19. $\displaystyle\sum_{n=1}^{\infty} 5(0.2)^{n-1}$

20. $\displaystyle\sum_{n=1}^{\infty} -14\left(-\dfrac{1}{5}\right)^{n-1}$

21. $\displaystyle\sum_{n=1}^{\infty} \dfrac{5}{4}(-1)^{n-1}$

22. **Spreadsheets** Use a spreadsheet to explore the *sequence of partial sums*

$$S_1 = t_1, \quad S_2 = t_1 + t_2, \quad S_3 = t_1 + t_2 + t_3, \quad S_4 = t_1 + t_2 + t_3 + t_4, \quad \ldots$$

for an infinite geometric series $t_1 + t_2 + t_3 + t_4 + \cdots$.

Infinite Geometric Series		
C7	= C6 + B7	
	A	B	C
	A	**B**	**C**
1	First term =	0.5	
2	Common ratio =	0.5	
3			
4	Position number	Sequence	Partial sum
5	1	0.5	0.5
6	2	0.25	0.75
7	3	0.125	0.875

Set up your spreadsheet so that you can easily change the first term and the common ratio.

a. Do the partial sums of the series you investigated in the Exploration on page 497 appear to get closer and closer to any single number? If so, what is the number and how does it compare with the sum of the series?

b. Investigate the sequence of partial sums for three other infinite geometric series where $|r| < 1$. What do you notice?

c. Investigate the sequence of partial sums for three infinite geometric series where $|r| > 1$. What do you notice?

23. ECONOMICS In 1974 the Malaysian Tourist Development Corporation (MTDC) estimated the economic benefit of tourist brochures. The MTDC based its analysis on these assumptions:

- Each person or organization will spend 80.5% of each *ringgit,* or Malaysian dollar (M$), received in payment for goods or services.

- Malaysian dollars remain in circulation in Malaysia.

  - Each tourist receiving a brochure about the capital city Kuala Lumpur will spend an additional M$4.72 per person while in Malaysia.

  a. Write the first 4 terms of the series that represents the total spending generated from each additional ringgit spent by a tourist. (*Hint:* The person who receives M$1 spends 80.5% of M$1, the person who receives 80.5% of M$1 spends 80.5% of 80.5% of M$1, and so on.)

  b. Find the sum of the series you wrote in part (a).

c. What is the total additional spending that the MTDC predicted from each additional tourist brochure?

24. Investigation You can use a procedure like the one described in the Exploration on page 497 to find the sum of any series of the form

$$\sum_{n=1}^{\infty} \frac{1}{c^n} \text{ where } c = 2, 3, 4, \ldots.$$

a. Writing Describe how to modify the procedure in the Exploration for finding the sum of this series:

$$\frac{1}{3} + \frac{1}{9} + \frac{1}{27} + \frac{1}{81} + \cdots$$

b. Carry out the procedure you described in part (a) and explain why it shows that the sum of the series is $\frac{1}{2}$.

c. Modify the procedure once again to find $\sum_{n=1}^{\infty} \frac{1}{4^n}$.

d. Challenge Make a conjecture about an explicit formula for this

sequence: $\sum_{n=1}^{\infty} \frac{1}{2^n}, \sum_{n=1}^{\infty} \frac{1}{3^n}, \sum_{n=1}^{\infty} \frac{1}{4^n}, \ldots$. Explain your reasoning.

Express each repeating decimal as a fraction.

25. 0.888...   **26.** 0.181818...   **27.** 0.040404...

28. 0.3823823823...   **29.** 3.717171...   **30.** 0.13121312...

31. 1.059059...   **32.** 0.005005005...   **33.** −0.0453453...

34. 4.002626...   **35.** −10.8787...   **36.** 3.0005555...

37. a. Express the repeating decimal 0.9999. . . as a fraction. Are you surprised by the result?

   b. Writing Use your result in part (a) to explain why any integer can be expressed as a repeating decimal.

INTERVIEW Jhane Barnes

Look back at the article on pages 462–464.

Jhane Barnes likes to use fractals in her fabric designs. The fractal shown below is one of a class of Koch constructions, named for the Swedish mathematician Helge von Koch. It is formed by beginning with a unit square and repeatedly replacing the middle third of each boundary segment with a small square that is missing one side:

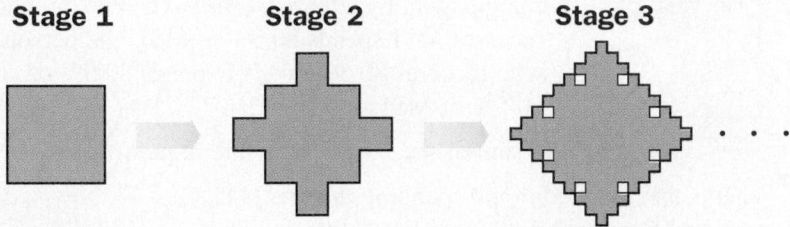

Stage 1   Stage 2   Stage 3

38. a. Consider the *number* of new squares added at each stage of this Koch construction. Write an explicit formula for the *n*th term of this sequence. (*Hint:* The sequence is geometric. Express t_n in the form $t_1 \cdot r^{n-1}$.)

   b. Consider the *area* of *each* of the new squares added at each stage of this Koch construction. Write an explicit formula for the *n*th term of this sequence. See the hint for part (a).

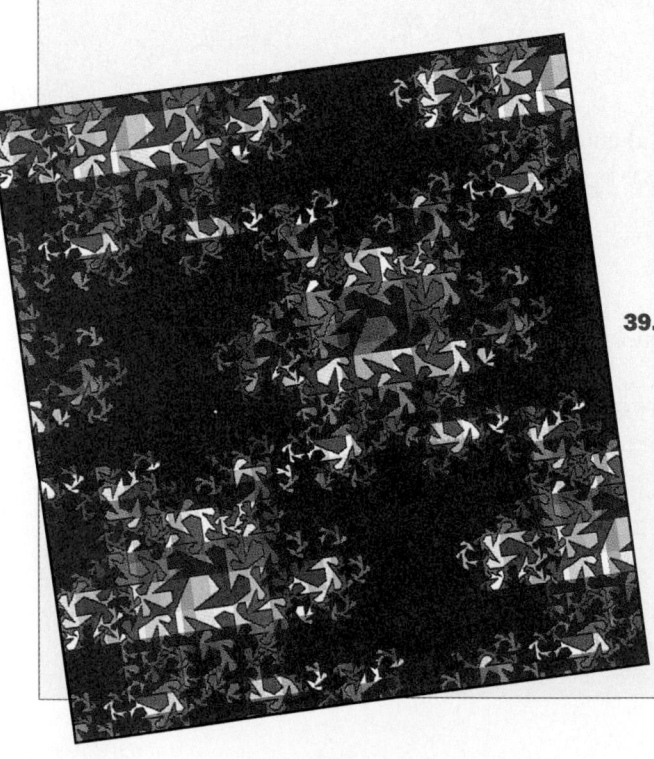

   c. Consider the *combined area* of *all* the new squares added at each stage of this Koch construction. Write an explicit formula for the *n*th term of this sequence. See the hint for part (a).

   d. Writing Find the total area enclosed by this Koch construction after infinitely many stages. Explain how you found your answer.

39. a. Consider the length added to the perimeter of this Koch construction at each stage. Write an explicit formula for the *n*th term of this sequence. See the hint for part (a) of Exercise 38.

   b. Writing Explain why the perimeter of this Koch construction after infinitely many stages is infinitely long.

"Koch curve snowflake"

Open-ended Problem Make up a geometric series of each type. Write each series in sigma notation. If the series has a sum, tell what the sum is.

40. finite series that has a sum

41. finite series that does not have a sum

42. infinite series that has a sum

43. infinite series that does not have a sum

Write each series in sigma notation and find the sum. *(Section 10.4)*

44. $0.6 + 0.67 + 0.74 + 0.81 + \cdots + 1.58$

45. $183 + 169 + 155 + 141 + \cdots + (-391)$

Solve each system of equations. *(Section 7.2)*

46. $2x - 3y = 9$
$x + y = 2$

47. $4y + 2x = -1$
$3x + 8y = 12$

48. $3x + 6y = -9$
$4x + 8y = -12$

49. $6x - y = 2$
$x + 5y = 16$

Find the slope of the line passing through each pair of points. *(Section 2.2)*

50. $(0, 0)$ and $(7, 9)$

51. $(0, 2)$ and $(5, 1)$

52. $(3, -2)$ and $(-1, 4)$

53. $(-1, -5)$ and $(1, 5)$

ASSESS YOUR PROGRESS

VOCABULARY

series (p. 488)   **geometric series** (p. 495)
arithmetic series (p. 488)

For each series:

a. Tell whether it is *arithmetic* or *geometric*.

b. Find the number of terms.

c. Find the sum. *(Sections 10.4 and 10.5)*

1. $640 + 320 + 160 + 80 + \cdots + 5$

2. $\left(-\dfrac{3}{8}\right) + \left(-\dfrac{1}{4}\right) + \left(-\dfrac{1}{8}\right) + 0 + \cdots + \dfrac{1}{2}$

3. $6 + 30 + 150 + 750 + \cdots + 93{,}750$

Find the sum of each geometric series, if the sum exists. *(Section 10.5)*

4. $\dfrac{1}{64} + \dfrac{1}{16} + \dfrac{1}{4} + 1 + \cdots$

5. $100 + 20 + 4 + 0.8 + \cdots$

Write each series in sigma notation and find the sum, if the sum exists.
(Sections 10.4 and 10.5)

6. $114 + 97 + 80 + 63 + \cdots + (-5)$ **7.** $3 + 2 + \dfrac{4}{3} + \dfrac{8}{9} + \cdots$

Expand each series and find the sum. *(Sections 10.4 and 10.5)*

8. $\displaystyle\sum_{n=1}^{9} (5n + 1)$

9. $\displaystyle\sum_{n=1}^{7} 0.3(7)^{n-1}$

10. $\displaystyle\sum_{n=1}^{\infty} (0.8)^n$

11. Journal Explain why infinite arithmetic series never have a sum but infinite geometric series sometimes have a sum.

The Future Is Now

A college education can be expensive, and the cost continues to rise as time passes. Planning ahead makes the task of providing the money needed to pay for college much easier. Paying for a college education can be difficult if you wait too long before starting a savings program.

Based on today's costs, you can estimate the cost of a college education at some point in the future. You can then explore various savings and investment plans to see how they can help pay for college.

PROJECT GOAL Estimate the future cost of a college education, and test the effectiveness of several plans for saving this amount of money.

Collecting the Data

Suppose you had 20 years to save enough money to pay for a college education. How could you determine the best method for achieving this goal?

Work in groups of three to collect the data necessary to develop good models. Each group member should do research in one of these areas:

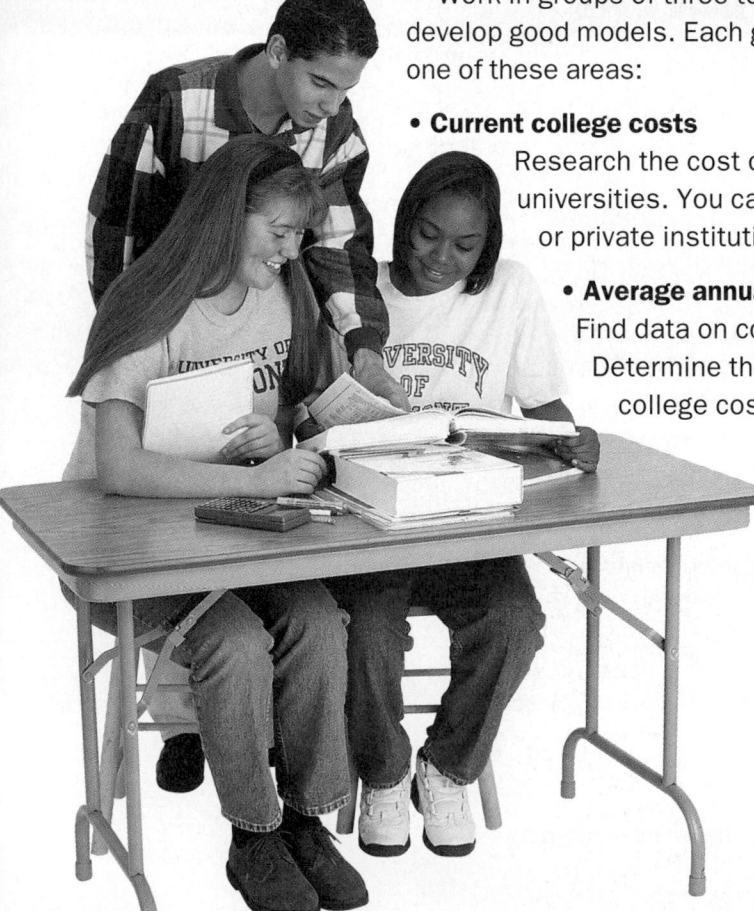

- **Current college costs**

  Research the cost of going to various 4-year colleges and universities. You can use either an average cost for all public or private institutions or the cost for a college of your choice.

- **Average annual percent increase in college costs**

  Find data on college costs over the past 10–15 years. Determine the average annual percent increase in total college costs.

- **Interest rates for various investments**

  Find out the average annual rate of return for various investments, such as savings accounts, certificates of deposit, and mutual funds.

  Two good sources for college data are the *Digest of Educational Statistics* and the *Statistical Abstract of the United States*. For investment data, try calling a bank or brokerage firm.

Projecting Costs and Savings

Use a spreadsheet to project future college costs and to explore different methods of saving for college. For example, the spreadsheet below shows college costs growing at 5.9% per year. It also shows a savings plan in which annual deposits of $1200 earn 10% annual interest. Will this plan yield enough savings to pay for college in 20 years?

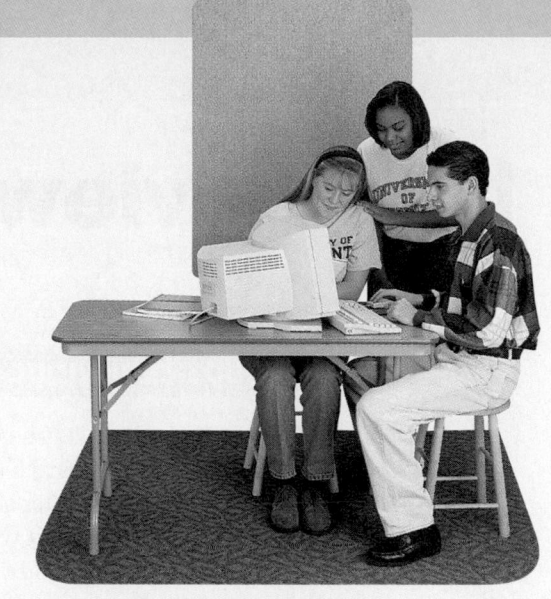

College Planning

	A	B	C
	Year	College cost	Savings
1	1	24876	1200
2	2	B2*1.059	C2*1.1+120
3			

Enter the amount of **money saved annually** in the first cell of this column. Use a recursive formula to calculate the effect of annual savings and deposits.

Enter the **current college cost** in the first cell of this column. Use a recursive formula to calculate costs for future years.

Use a spreadsheet to investigate scenarios like the one above, as well as other savings and investment programs, including those that involve:

• making only a large initial investment

• increasing your annual investment each year

• waiting until the final few years before starting to save

You can extend your project by investigating the cost differences between public and private institutions.

Writing a Report

Write a report summarizing your results. Include the following:

• all your data, estimates, and sources

• a description of each scenario you investigated and a printout of each spreadsheet you created

• a discussion of the feasibility of each scenario and which one makes the most sense for most people

Self-Assessment

What do you think are the limitations of the models you used in this project? Did you have any trouble setting up your spreadsheet models? What insights did you gain from this project?

Review

Taking a practice test is a good way to determine how ready you are for an actual test. One way to prepare for a test is to write one yourself. Writing a test requires you to review the chapter and become familiar with it. Create a test for this chapter, then exchange it with a classmate. After taking the tests, correct and discuss them. You should study any topics that gave you difficulty.

VOCABULARY

sequence (p. 465)
term (p. 465)
finite sequence (p. 467)
infinite sequence (p. 467)
geometric sequence (p. 473)
common ratio (p. 473)
arithmetic sequence (p. 473)
common difference (p. 473)

geometric mean (p. 475)
arithmetic mean (p. 475)
explicit formula (p. 480)
recursion (p. 481)
recursive formula (p. 481)
series (p. 488)
arithmetic series (p. 488)
geometric series (p. 495)

SECTIONS 10.1 *and* 10.2

A **sequence** is an ordered list of numbers or figures, called **terms**.

Arithmetic sequence

$$t_n = t_1 + d(n - 1)$$

An **explicit formula** gives t_n in terms of n.

Geometric sequence

$$t_n = t_1 \cdot r^{n-1}$$

nth term   first term   **common difference**

nth term   first term   **common ratio**

Examples of arithmetic and geometric sequences are given below.

A sequence is **finite** if it has a last term, and it is **infinite** if it continues without end.

Sequence	50, 47, 44, 41, . . .	1000, 100, 10, 1, . . . , 0.00000001
Type	infinite arithmetic	finite geometric
Explicit formula	$t_n = 50 + (-3)(n - 1)$	$t_n = 1000(0.1)^{n-1}$
10th term	$t_{10} = 50 + (-3)(9) = 23$	$t_{10} = 1000(0.1)^9 = 0.000001$

Suppose a, x, and b are consecutive terms of a sequence.

- If the sequence is arithmetic, then $x = \dfrac{a+b}{2}$.

  x is the **arithmetic mean** of a and b.

  Example: 50, 47, **44**, **41**, **38**, 35, . . . ← $41 = \dfrac{44+38}{2}$

- If the sequence is geometric, then $x = \sqrt{ab}$.

  x is the **geometric mean** of a and b.

  Example: 1000, 100, **10**, **1**, **0.1**, 0.01, . . . ← $1 = \sqrt{10 \cdot 0.1}$

SECTION | 10.3

Recursion is the process of obtaining each term of a sequence from the previous one by repeatedly performing the same operation(s). You can find **recursive formulas** for arithmetic and geometric sequences.

Arithmetic sequence	**Geometric sequence**
$t_1 =$ **starting value**	$t_1 =$ **starting value**
$t_n = t_{n-1} + d$	$t_n = r(t_{n-1})$
Example: The sequence 50, 47, 44, 41, . . . is defined recursively as $t_1 = 50$ and $t_n = t_{n-1} - 3$.	Example: The sequence 1000, 100, 10, 1, . . . is defined recursively as $t_1 = 1000$ and $t_n = 0.1(t_{n-1})$.

SECTIONS | 10.4 *and* 10.5

When the terms of a sequence are added, as in $t_1 + t_2 + \cdots + t_n$, the indicated sum is called a **series**. A series in *expanded form* can be expressed more compactly by using *sigma notation*. You can use formulas to find the sum S_n of a finite arithmetic series and a finite geometric series, each with n terms.

Arithmetic series	**Geometric series ($r \neq 1$)**
$S_n = \dfrac{n}{2}(t_1 + t_n)$	$S_n = \dfrac{t_1(1 - r^n)}{1 - r}$
Example: $9 + 22 + 35 + 48 + \cdots + 152$	Example: $1 + 2 + 4 + 8 + \cdots + 256$
$= \displaystyle\sum_{n=1}^{12} [9 + 13(n-1)]$	$= \displaystyle\sum_{n=1}^{9} 1(2^{n-1})$
$= \dfrac{12}{2}(9 + 152) = 966$	$= \dfrac{1(1 - 2^9)}{1 - 2} = 511$

If a geometric series is infinite and $|r| < 1$, then the sum is $S = \dfrac{t_1}{1-r}$.

Example: $32 + 16 + 8 + 4 + \cdots = \displaystyle\sum_{n=1}^{\infty} 32(0.5)^{n-1} = \dfrac{32}{1 - 0.5} = 64$

VOCABULARY QUESTIONS

For Questions 1 and 2, complete each paragraph.

1. When you divide any term by its preceding term in a(n) _?_ sequence, the quotient is a constant called the _?_ . The term between any two terms in a(n) _?_ sequence is the average, or _?_ , of the given terms.

2. If a sequence has a last term, it is a(n) _?_ sequence. When you add the terms of a sequence, the indicated sum is called a(n) _?_ .

SECTIONS 10.1 *and* 10.2

For each sequence, find the next 4 terms and tell whether the sequence is *arithmetic*, *geometric*, or *neither*.

3. $1, 8, 27, 64, \ldots$

4. $1, -1, 1, -1, \ldots$

5. $7, 2, -3, -8, \ldots$

6–8. Write a formula for t_n for each sequence in Questions 3–5.

Find the 12th term of each sequence.

9. $t_n = \dfrac{n}{n + 1}$

10. $t_n = \left(\dfrac{1}{3}\right)^n$

11. $t_n = 5n - 2$

For each pair of numbers:

a. Find the arithmetic mean.   **b. Find the geometric mean.**

12. $16, 25$

13. $7, 11$

14. $0.3, 1.2$

SECTION 10.3

For each sequence:

a. Find the first 4 terms.   **b. Write an explicit formula.**

15. $t_1 = 100$
    $t_n = -0.1(t_{n-1})$

16. $t_1 = 100$
    $t_n = -0.1 + t_{n-1}$

17. $t_1 = 3$
    $t_n = \sqrt{2}(t_{n-1})$

Write a recursive formula for the sequence given in each question.

18. Question 5

19. Question 10

20. Question 11

21. **GENEALOGY** Assume that a person has two parents, four grandparents, and so on. Write an explicit formula and a recursive formula for the number of ancestors the person has when you go back n generations. Is the sequence *arithmetic*, *geometric*, or *neither*? Explain.

22. Writing Explain how to find the 20th term when given a recursive formula for an arithmetic sequence and for a geometric sequence.

SECTIONS 10.4 *and* 10.5

For each series:

a. Tell whether it is *arithmetic* or *geometric*.

b. Find the number of terms.

c. Write the series in sigma notation.

d. Find the sum.

23. $20 + 12 + 4 + \cdots + (-100)$

24. $2 + 10 + 50 + \cdots + 156{,}250$

25. $8 + (-4) + 2 + \cdots + \left(-\dfrac{1}{4}\right)$

26. $7 + 7.5 + 8 + \cdots + 15$

27. ARCHITECTURE A concert hall has seating for 800 people. There are 21 seats in the first row. Each of the other rows has two more seats than the row in front of it. How many rows are in the concert hall? How many seats are in the last row?

28. Expand each series and find the sum.

a. $\displaystyle\sum_{n=1}^{5} \frac{1}{2}n^3$

b. $\displaystyle\sum_{n=1}^{\infty} 1000\left(-\frac{1}{4}\right)^{n-1}$

29. Open-ended Problem Write a geometric series that does not have a sum. Explain why the series has no sum.

30. Express $0.0175175175\ldots$ as a fraction.

PERFORMANCE TASK

31. The Tower of Hanoi puzzle involves moving disks from one post to another. The puzzle begins with a group of n disks of increasing diameter (from top to bottom) on one of three wooden posts. The goal of the puzzle is to transfer all the disks from this starting post to one of the other two posts following these rules:

(1) Only one disk can be moved at a time.

(2) No disk can be placed on top of a smaller disk.

Solve the puzzle with various numbers of disks. (You may wish to use coins of different sizes instead of disks, and three locations A, B, and C on a sheet of paper to represent the three posts.) Make a table relating n, the number of disks, to M_n, the *minimum* number of moves needed to transfer all the disks to another post. Then find an explicit formula for M_n. (*Hint:* Try applying a recursive procedure to moving the disks.)

11 Analytic Geometry

Taking maps into the 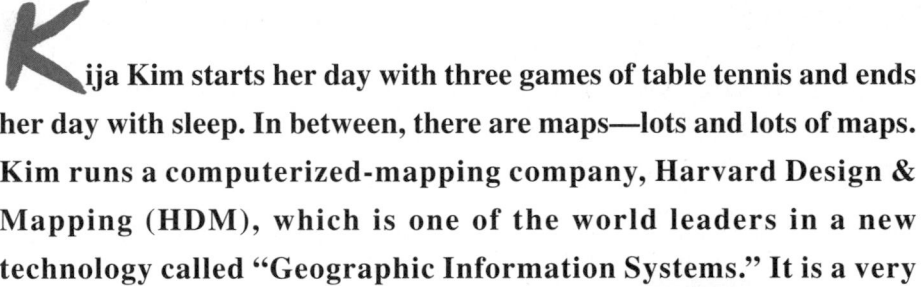 FUTURE

INTERVIEW **Kija Kim**

Kija Kim starts her day with three games of table tennis and ends her day with sleep. In between, there are maps—lots and lots of maps. Kim runs a computerized-mapping company, Harvard Design & Mapping (HDM), which is one of the world leaders in a new technology called "Geographic Information Systems." It is a very competitive arena, which explains her morning pastime. "Table tennis builds up my competitiveness and gets me mentally ready," she says.

"Ten years ago, we relied on rulers to measure distances. Now computers do the job instantly."

> ## "Geography combines the other subjects I like—physical science, social science, and mathematics."

Leading the Field

Kim began considering a profession in geography early in life, while she was still a high school student in South Korea. She did not work with computers until many years later, after she had moved to the United States. "During my first course in computer programming, I started thinking about how to automate mapmaking processes," she recalls. "That's when it all began."

Now her company designs maps for a variety of government and business applications. "When I first began applying computers to *cartography*, or mapmaking, people didn't think you could build a business around that idea. Now it's one of the fastest-growing areas in the whole computer field."

Creating Thinking Maps

HDM's computerized maps can display a vast amount of information compared with normal maps. For example, one of HDM's "intelligent maps" provides a detailed picture of all the water flowing into a Massachusetts reservoir, shows the industrial activity in the region, locates the toxic waste sites, and determines how both affect the water quality.

Another computerized map, shown on the next page, predicts what would happen if a break occurred in the Sudbury Dam, which is located on the Sudbury River near Framingham, Massachusetts. "Our software can tell you in 30 seconds whether you're in a potential flood zone," Kim notes. "Your insurance company could take two weeks getting that information."

Automating Calculations

Computers can provide information more quickly and accurately than old-fashioned cartographic techniques. "Just ten years ago, we relied on rulers and other tools to determine the position of objects on maps and to measure the distance between them. Now computers do that job instantly," Kim says.

A useful technique for identifying points and finding distances is to superimpose a coordinate grid onto a map. For example, the coordinate grid shown below allows you to locate a point on Sudbury Dam and another point on a nearby office building using coordinates. To calculate the distance between these two points, you can use the *distance formula*, which you will learn about in this chapter.

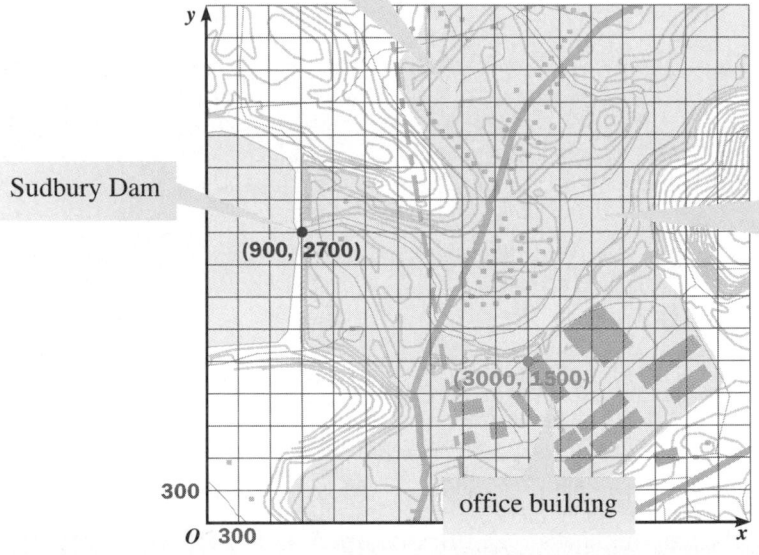

A potential flood zone is shown in gray.

Sudbury Dam

(900, 2700)

(3000, 1500)

The Sudbury River is shown in blue.

office building

300

O 300

Kija Kim and an assistant review the details of a map.

1. Writing Using the coordinates, given in feet, on the map above, find the east-west distance and the north-south distance between Sudbury Dam and the office building. How can you use these distances to calculate the straight-line distance between the dam and the building? Explain.

2. Research Find two maps of your state and compare them. How does each map represent the cities and landmarks? What scale does each map use for distances? Why is a scale important for reading a map?

3. Project Work in a group of four students to make a map of a public area, such as a local park or the school grounds. Identify at least 10 landmarks in the area. Choose a scale and make a map of the area, identifying the landmarks using coordinates.

Mathematics
& Kija Kim

In this chapter, you will learn more about how mathematics is related to making maps.

Related Exercises

Section 11.1
• **Exercises 22 and 23**

Section 11.3
• **Exercises 10–14**

11.1 Distances, Midpoints, and Lines

Learn how to...

- **find the distance between two points in a coordinate plane**
- **find the coordinates of the midpoint of a line segment in a coordinate plane**

So you can...

- **find the distance between two points on a map, for example**

Given any right triangle with sides of length a, b, and c (with c being the length of the hypotenuse), the Pythagorean theorem states that $c^2 = a^2 + b^2$. Ancient writings from Egypt, China, Babylonia, India, and Greece show that this relationship was known and used by many civilizations thousands of years ago.

EXPLORATION
COOPERATIVE LEARNING

Finding Distances and Midpoints

Work with a partner.
You will need:
- centimeter graph paper
- a metric ruler

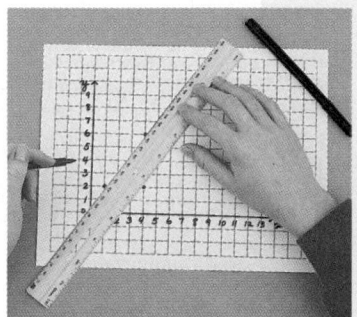

1 Draw coordinate axes on a piece of graph paper. Plot the points (1, 2), (4, 2), and (4, 6). Find the distance in centimeters between (1, 2) and (4, 6). Verify your measurement using the Pythagorean theorem applied to the right triangle formed by the three plotted points.

2 Explain how you can use the Pythagorean theorem to find the distance between any two points in a coordinate plane. Then choose two points and find the distance between them.

3 On the same graph, use your ruler to estimate the coordinates of the point halfway between (1, 2) and (4, 6).

4 Find the mean of the x-coordinates of the points (1, 2) and (4, 6). Also find the mean of the y-coordinates of these points. How are these means related to the coordinates of the point you found in Step 3?

5 Explain how you can find the midpoint of the line segment connecting any two points in a coordinate plane. Then choose two points and find the coordinates of the midpoint of the line segment connecting them.

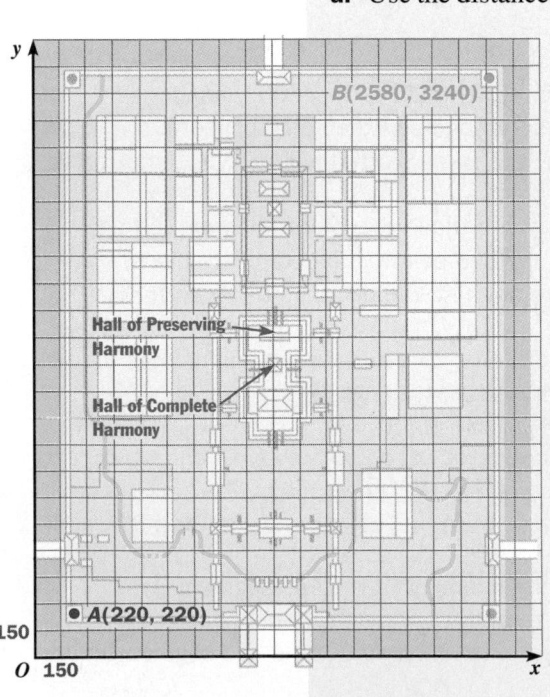
You can find the distance between any two points in a coordinate plane
by finding the length of the hypotenuse of a right triangle. You can find
the coordinates of the midpoint of a segment by finding the mean of the
x-coordinates and the mean of the *y*-coordinates of the segment's endpoints.

Distance and Midpoint Formulas

The distance d between two points (x_1, y_1) and (x_2, y_2) is:

$$d = \sqrt{(x_2 - x_1)^2 + (y_2 - y_1)^2}$$

The midpoint of the line segment joining the points (x_1, y_1) and (x_2, y_2)
has coordinates:

$$\left(\frac{x_1 + x_2}{2}, \frac{y_1 + y_2}{2} \right)$$

EXAMPLE 1   Application: Architecture

The Palace Museum is located in the city of Beijing in China. A moat and a
35 ft high wall completely surround nearly 1000 buildings with 9000 rooms.

a. Find the distance between the watchtowers at point $A(220, 220)$ and
point $B(2580, 3240)$, where the coordinates are given in feet, on the map
of the Palace Museum shown below.

b. Find the coordinates of the midpoint of the line segment joining A and B.
What is located on the map at the midpoint of this line segment?

SOLUTION

a. Use the distance formula.

$$\begin{aligned}
d &= \sqrt{(x_2 - x_1)^2 + (y_2 - y_1)^2} \\
&= \sqrt{(2580 - 220)^2 + (3240 - 220)^2} \\
&= \sqrt{2360^2 + 3020^2} \\
&= \sqrt{14{,}690{,}000} \\
&\approx 3833
\end{aligned}$$

The distance between A and B is about 3830 ft.

b. Use the midpoint formula.

$$\left(\frac{x_1 + x_2}{2}, \frac{y_1 + y_2}{2} \right) = \left(\frac{220 + 2580}{2}, \frac{220 + 3240}{2} \right)$$

$$= (1400, 1730)$$

The midpoint of the segment connecting A and B has
coordinates (1400, 1730). This point is the center of
the terrace between the Hall of Preserving Harmony
and the Hall of Complete Harmony.

Parallel and Perpendicular Lines

When you solved systems of linear equations in Chapter 7, you found that some systems of equations are inconsistent because the graphs of the equations are parallel lines. Parallel lines never intersect because they have the same slope.

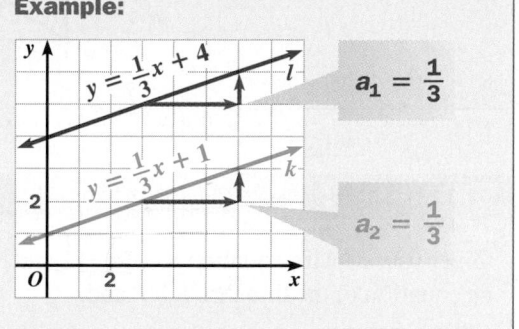

Parallel Lines

If line l has slope a_1 and line k has slope a_2, lines l and k are parallel if and only if $a_1 = a_2$.

Example:

$y = \frac{1}{3}x + 4$

$y = \frac{1}{3}x + 1$

$a_1 = \frac{1}{3}$

$a_2 = \frac{1}{3}$

E X P L O R A T I O N
COOPERATIVE LEARNING

Exploring Slopes of Perpendicular Lines

Work with a partner.
You will need: • graph paper   • a straightedge
• a protractor

1 Draw a line with a slope of 1. Choose a point on the line and label the point.

2 Use a protractor to draw another line perpendicular to the first line at the labeled point.

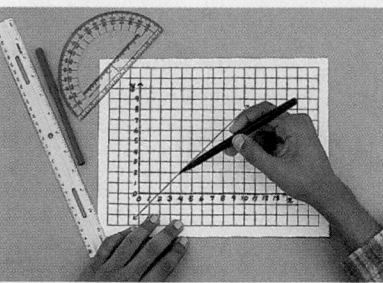

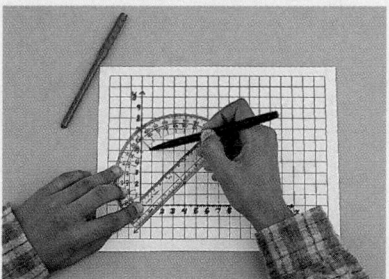

3 Find the slope of the perpendicular line and record it in a table like the one shown.

4 Repeat Steps 1–3 at least three more times, each time starting with a new line having a different slope, such as 2.

5 Make a conjecture about the slopes of perpendicular lines. Compare your conjecture with those of other groups.

Slope of line	Slope of perpendicular line
1	?
2	?
$-\frac{1}{3}$	?

If line l has slope a_1 and line k has slope a_2, lines l and k are perpendicular if and only if $a_1 = -\dfrac{1}{a_2}$ (that is, the slopes are negative reciprocals).

Example:

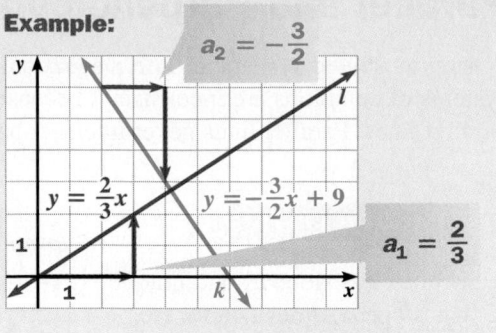

$a_2 = -\dfrac{3}{2}$

$y = \dfrac{2}{3}x$ $y = -\dfrac{3}{2}x + 9$

$a_1 = \dfrac{2}{3}$

EXAMPLE 2

Given a line l with equation $y = 3x + 2$ and point P with coordinates $(4, 2)$, find an equation of the line through P and:

a. parallel to l **b.** perpendicular to l

SOLUTION

a. Line l has a slope of 3. A line parallel to l will also have a slope of 3. An equation of the line through P and parallel to l is $y = ax + b$ where:

> Substitute **4** for x, **2** for y, and **3** for a.

$$2 = 3 \cdot 4 + b$$
$$2 = 12 + b$$
$$-10 = b$$

An equation of the line through P and parallel to l is $y = 3x - 10$.

b. A line perpendicular to l will have a slope of $-\dfrac{1}{3}$. An equation of the line through P and perpendicular to l is $y = ax + b$ where:

$$2 = -\frac{1}{3} \cdot 4 + b$$
$$2 = -\frac{4}{3} + b$$
$$\frac{10}{3} = b$$

> Substitute **4** for x, **2** for y, and $-\dfrac{1}{3}$ for a.

An equation of the line through P and perpendicular to l is $y = -\dfrac{1}{3}x + \dfrac{10}{3}$.

☑ CHECKING KEY CONCEPTS

For each pair of points, find:

a. the distance between the points

b. the coordinates of the midpoint of the line segment connecting the points

1. $(1, 5), (4, 8)$ **2.** $(-2, 6), (1, 4)$ **3.** $(-3, -2), (-6, 0)$

4. Given a line l with equation $y = 4x - 3$ and point P with coordinates $(-3, 5)$, find an equation of the line through P and:

    a. parallel to l **b.** perpendicular to l

11.1 **Exercises and Applications**

Extra Practice exercises on page 766

1. **a.** Find the midpoint M of segment $\overline{GH}$ for $G(4, -2)$ and $H(-1, -6)$.

   b. What is the slope of $\overline{GH}$?

   c. What is the slope of a line perpendicular to $\overline{GH}$ at M?

   d. Find an equation of the perpendicular bisector of $\overline{GH}$.

For each pair of points, find:

a. the distance between the points

b. an equation of the perpendicular bisector of the line segment connecting the points

2. $(1, 8), (6, 21)$ **3.** $(5, 4), (6.5, 2)$ **4.** $(-8, 0), (-5, 4)$ **5.** $(-11, -5), (3, -4)$

6. $(-4, 2), (3, -1)$ **7.** $(-3, -4), (2, 5)$ **8.** $(3.4, 4.6), (-2.5, 8.2)$ **9.** $(4.3, 0.4), (-6.4, -7.9)$

10. $(-12, 15), (4, -16)$ **11.** $\left(-4, \frac{1}{2}\right), \left(\frac{3}{4}, \frac{5}{8}\right)$ **12.** $\left(\frac{1}{3}, \frac{5}{6}\right), \left(-\frac{2}{5}, \frac{8}{3}\right)$ **13.** $(-3.5, 12.4), (-1.8, -9)$

14. GEOMETRY Verify that $\triangle ABC$ with vertices $A(1, 7)$, $B(6, 2)$, and $C(8, 6)$ is isosceles.

Find the value(s) of k so that the given points are n units apart.

15. $(6, k), (9, 11), n = 5$ **16.** $(-3, 2), (5, k), n = 17$ **17.** $(3, k), (-7, -10), n = 26$

18. $(3.5, 1.4), (k, 2.8), n = 5$ **19.** $(-3, 6), (k, 4), n = 6$ **20.** $(-4, 1), (k, 6), n = 12$

21. GEOMETRY Show that $Q(-5, 7)$, $R(0, 12)$, $S(5, 7)$, and $T(0, 2)$ are the vertices of a square. Explain your reasoning.

INTERVIEW **Kija Kim**

Look back at the article on pages 510–512.

Storm drains in the city shown carry rainwater from the streets to a local river. The water may pick up contaminants before reaching the river. To determine the impact that this water has on local recreation areas, city engineers start by finding the distance between a discharge pipe and a nearby recreation area.

22. Find the distance between the discharge pipe and the closest point of the recreation area shown on the map, where the coordinates are given in feet.

23. Estimate the distance between the discharge pipe and the farther end of the recreation area. Explain how you found your estimate.

24. FLAG MAKING According to the rules for creating California's state flag, the length of the bear is measured from the nose tip *N* to the rear of the right-hand paw *P*.

a. Find the length of the bear using the flag shown, where the coordinates are given in inches. Confirm that this length is about two thirds the height of the flag.

b. The midpoint *M* of $\overline{NP}$ should also be the midpoint of a segment drawn horizontally across the flag. Find the coordinates of *M*. Confirm that the bear is correctly positioned horizontally.

25. a. GEOMETRY Find the perimeter of quadrilateral *ABCD* with vertices $A(-2, 5)$, $B(3, 8)$, $C(3, 3)$, and $D(-2, 0)$.

b. Prove that *ABCD* is a parallelogram.

26. GEOMETRY Show that $\triangle PQR$ with vertices $P(3, 12)$, $Q(6, 5)$, and $R(-1, 2)$ is a right triangle. Which side is the hypotenuse? Which sides are perpendicular?

Connection ▶ LITERATURE

Harper Lee's novel *To Kill a Mockingbird* is set in the fictional town of Maycomb, Alabama, the county seat of Maycomb County. In the book, the exact location of the county seat is credited to Mr. Sinkfield, who ran an inn near the geographic center of the county.

Business was excellent when Governor William Wyatt Bibb, with a view to promoting the newly created county's domestic tranquillity, dispatched a team of surveyors to locate its exact center and there establish its seat of government. The surveyors, Sinkfield's guests, . . . showed him the probable spot where the county seat would be built. Had not Sinkfield made a bold stroke to preserve his holdings, Maycomb would have sat in the middle of Winston Swamp. . . . Sinkfield . . . induced them to bring forward their maps and charts, lop off a little here, add a bit there, and adjust the center of the county to meet his requirements.

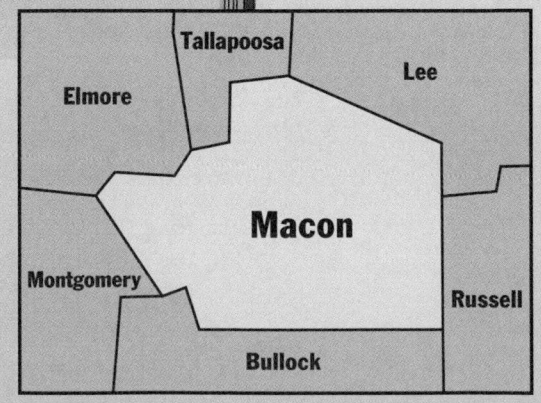

27. Writing Explain how you can approximate the center of a county using a coordinate map. Why does it make sense to locate the county seat at the center of the county?

28. Open-ended Problem Maycomb County is fictional. Use the map of Macon County in Alabama. Sketch a copy of the map. Locate the spot that you think is the center of the county. Explain your reasoning.

29. The pattern at the right was created by connecting the midpoints of the sides of one square to form a new square.

   a. Find the coordinates of the vertices of the red square.

   b. Find the coordinates of the vertices of the green square.

   c. Find the perimeters of the blue square, the red square, and the green square. Predict the perimeter of the next square in the pattern. Check your prediction.

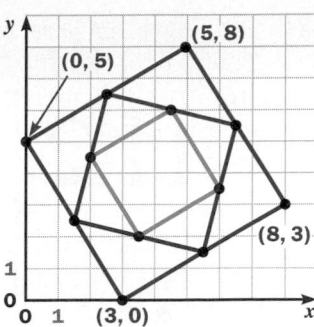

30. The second pattern at the right was created by connecting the points one quarter of the way along the sides of one square to form a new square.

   a. Explain how you can use the midpoint formula to find the coordinates of the vertices for each new square in this pattern.

   b. Find the coordinates of the vertices of the red square and the green square.

   c. Find the areas of the blue square, the red square, and the green square. Predict the area of the next square in the pattern. Check your prediction.

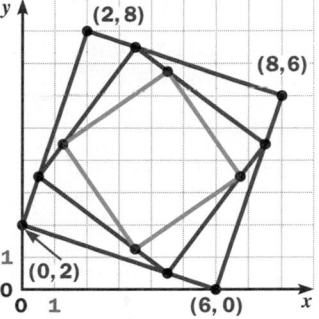

31. GEOMETRY Show that the midpoint of the hypotenuse of any right triangle is equidistant from each vertex. (*Hint:* Any right triangle can be placed on the coordinate plane so that its vertices are $(0, 0)$, $(2x, 0)$, and $(0, 2y)$.)

32. Challenge Use the distance formula to show that $M\left(\dfrac{x_1 + x_2}{2}, \dfrac{y_1 + y_2}{2}\right)$ is the same distance from point $P(x_1, y_1)$ as it is from point $Q(x_2, y_2)$.

ONGOING ASSESSMENT

33. Open-ended Problem Work in a group of 2 or 3 students. Create a game or puzzle that involves finding the lengths and midpoints of line segments and finding parallel and perpendicular lines. Write the rules for your game or puzzle. Suggest strategies for winning your game or provide a solution for your puzzle.

SPIRAL REVIEW

Write each geometric series in sigma notation. Find the sum of each series, if the sum exists. If a series does not have a sum, explain why not. (*Section 10.5*)

34. $512 + (-128) + 32 + (-8) + \cdots$ **35.** $6 + 18 + 54 + 162 + \cdots + 4374$

For Exercises 36 and 37:

a. Graph the feasible region.

b. Find the maximum profit for the profit function $P = 3x + 5y$.

c. Find the minimum cost for the cost function $C = 0.5x + 3y$. (*Section 7.6*)

36. $y \le 5$ **37.** $y \ge 3x + 1$

    $y \ge \dfrac{4}{5}x - \dfrac{3}{5}$ $y \le \dfrac{1}{2}x + 8$

    $y \le 3x - 5$ $y \ge -4x + 9$

Describe the graph of each function. Make a sketch of each graph. (*Section 5.2*)

38. $y = 2(x - 5)^2$ **39.** $y = 0.2(x - 3)^2 - 4$ **40.** $y = -3(x + 3)^2 - 3$

11.2 Parabolas

Learn how to...

- **find the focus and directrix of a parabola**

- **write an equation of a parabola**

So you can...

- **locate the antenna of a radio telescope dish, for example**

Many antennas are shaped so that a cross section of the antenna is a parabola. This shape improves the ability of the antenna to focus the signal that it is sending or receiving. The sender or receiver for the antenna is located at a point called the *focus* of the parabolic cross section. You can define a parabola using this point and a line called the *directrix*.

EXPLORATION
COOPERATIVE LEARNING

Using Focus-Directrix Paper to Draw a Parabola

Work with a partner.
You will need:

● focus-directrix paper

SET UP The circles on focus-directrix paper share a common center and the parallel lines are tangent to the circles.

1 Label as *d* the line that is tangent to the circle with radius 4 units and is *below* the center of the circle. Mark the single point where the line that is 2 units above *d* intersects the circle with radius 2 units.

2 Mark the two points where the line 3 units above *d* intersects the circle with radius 3 units.

3 Continue marking points as in Step 2, each time increasing by 1 unit the distance above *d* and the radius of the circle.

4 All the points you marked in Steps 2 and 3 should lie on a parabola. Sketch the parabola.

Questions

1. What is special about the point you marked in Step 1?

2. Line *d* is the *directrix* of the parabola, and the common center of the circles is the *focus* of the parabola. Based on Steps 2 and 3, how is a point on a parabola related to its focus and directrix?

3. What happens if you start with a directrix that lies *above* the focus?

You can define a parabola geometrically in terms of a focus and directrix.

A **parabola** is the set of all points in a plane that are the same distance from a point F, called the **focus**, and a line d, called the **directrix**.

The distance between a point on the parabola and the directrix is measured along a line perpendicular to the directrix.

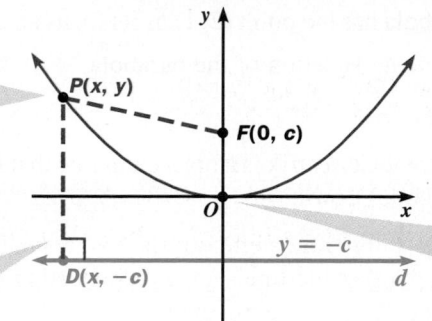

The **vertex** of the parabola lies halfway between the focus and the directrix.

In the graph shown above, the focus is the point $F(0, c)$ and the directrix is the horizontal line $y = -c$. If $P(x, y)$ is any point on the parabola, then you can derive an equation of the parabola as follows:

$$PF = PD$$

$D(x, -c)$ is the point of intersection of d and the line through P that is perpendicular to d.

$$\sqrt{(x - 0)^2 + (y - c)^2} = y - (-c)$$

$$x^2 + (y - c)^2 = (y + c)^2$$

Square both sides.

$$x^2 + y^2 - 2cy + c^2 = y^2 + 2cy + c^2$$

$$x^2 = 4cy$$

$$\frac{1}{4c}x^2 = y$$

THINK AND COMMUNICATE

1. What type of function does the equation $y = \frac{1}{4c}x^2$ represent? Does this surprise you? Explain.

2. What would happen to the equation of the parabola shown above if the parabola were translated h units horizontally and k units vertically?

Equation of a Parabola that Opens Up or Down

The graph of $y - k = \frac{1}{4c}(x - h)^2$ is a parabola that:

- has vertex at (h, k) and focus at $(h, k + c)$

- has directrix with equation $y = k - c$

- has a line of symmetry with equation $x = h$

- opens up if $c > 0$ and opens down if $c < 0$

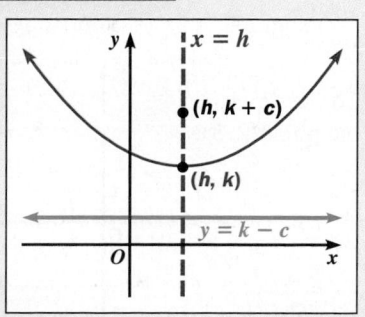

EXAMPLE 1

A parabola has the point (2, 1) as its focus and the line $y = 5$ as its directrix.

a. Write an equation of the parabola.   **b.** Graph the equation.

SOLUTION

a. Since the directrix is a horizontal line that lies above the focus, the parabola will open down.

Step 1 Find the vertex of the parabola. The vertex (h, k) is at the midpoint of the line segment connecting the focus (2, 1) and the point (2, 5) on the directrix.

$$(h, k) = \left(\frac{2 + 2}{2}, \frac{1 + 5}{2} \right)$$

Use the midpoint formula.

$$= (2, 3)$$

Step 2 Find the value of c. Use the fact that the directrix is the line $y = k - c$.

Substitute **5** for y and **3** for k.

$$5 = 3 - c$$

$$-2 = c$$

Note that a negative c-value agrees with the fact that the parabola opens down.

> **Check:**
>
> The focus is located at $(h, k + c)$.
>
> $(2, 1) \overset{?}{=} (2, 3 + (-2))$
>
> $1 \overset{?}{=} 3 + (-2)$
>
> $1 = 1 ✔$

Step 3 Find an equation of the parabola using the general form of an equation of a parabola that opens up or down.

$$y - k = \frac{1}{4c}(x - h)^2$$

Substitute **2** for h, **3** for k, and **−2** for c.

$$y - 3 = \frac{1}{4(-2)}(x - 2)^2$$

$$y - 3 = -\frac{1}{8}(x - 2)^2$$

An equation of the parabola is $y - 3 = -\frac{1}{8}(x - 2)^2$.

b. Plot the vertex and at least one other point. Use symmetry to sketch the graph.

The vertex is the point (2, 3).

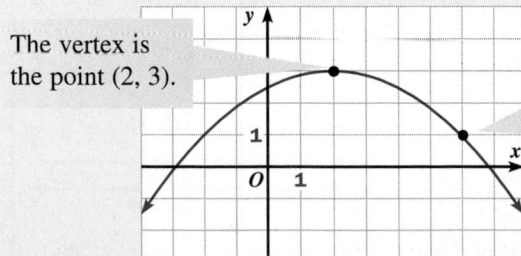

When $x = 6$,

$$y = -\frac{1}{8}(6 - 2)^2 + 3 = 1.$$

Plot the point (6, 1).

EXAMPLE 2 Application: Astronomy

Astronomers use radio telescopes to collect data from space. A main component of a radio telescope is a large dish with a parabolic cross section. The dish captures radio waves from space and reflects them to a receiver at the focus of the dish. The dish for the Fortaleza radio telescope in Brazil has a width of 14.2 m and a depth of 2.16 m. How far from the vertex is the receiver located?

SOLUTION

Step 1 Sketch a parabolic cross section of the radio telescope in a coordinate plane so that the vertex of the parabola is at (0, 0). Then an equation of the parabola is $y = \frac{1}{4c}x^2$, and the focus is c meters above the vertex.

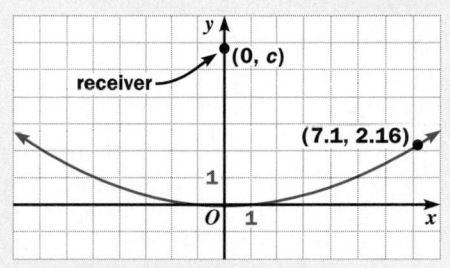

Step 2 Choose a point on the rim of the dish, as shown. The x-coordinate of this point is equal to half the diameter of the dish, 7.1 m. The y-coordinate of this point is equal to the depth of the dish, 2.16 m.

$2.16 = \frac{1}{4c}(7.1)^2$ Substitute **7.1** for x and **2.16** for y into the equation of the parabola.

$2.16c = \frac{1}{4}(50.41)$

$2.16c = 12.6025$ Multiply both sides by c and solve for c.

$c \approx 5.83$

The receiver is about 5.8 m from the vertex of the dish.

Parabolas that Open Right or Left

So far, you have seen parabolas that open either up or down. If a parabola opens to the left or right, its equation changes, as stated below.

Equation of a Parabola that Opens Right or Left

The graph of $x - h = \frac{1}{4c}(y - k)^2$ is a parabola that:

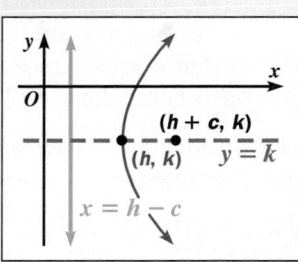

- has vertex at (h, k) and focus at $(h + c, k)$
- has directrix with equation $x = h - c$
- has a line of symmetry with equation $y = k$
- opens right if $c > 0$ and opens left if $c < 0$

EXAMPLE 3

An equation for a parabola is $x = -\frac{1}{3}(y + 1)^2 + 2$. Find the vertex, focus, and directrix of the parabola. Sketch the parabola with its focus and directrix.

SOLUTION

Step 1 Recognize the general form of the equation. The given equation involves squaring y, so its graph is a parabola that opens left or right.

Step 2 Find the values of h and k. You can rewrite the equation as

$$x - 2 = -\frac{1}{3}(y - (-1))^2,$$

so $h = 2$ and $k = -1$.

Multiply both sides by c and solve for c.

Step 3 Find the value of c.

$$\frac{1}{4c} = -\frac{1}{3}$$

$$\frac{1}{4} = -\frac{1}{3}c$$

$$-\frac{3}{4} = c$$

Because $c < 0$, the parabola opens to the left.

Step 4 Find the vertex, focus, and directrix.

Vertex: $(h, k) = (2, -1)$

Focus: $(h + c, k) = \left(2 + \left(-\frac{3}{4}\right), -1\right)$

$$= \left(1\frac{1}{4}, -1\right)$$

Directrix: $x = h - c$

$$= 2 - \left(-\frac{3}{4}\right)$$

$$= 2\frac{3}{4}$$

Step 5 Sketch the parabola with its focus and directrix.

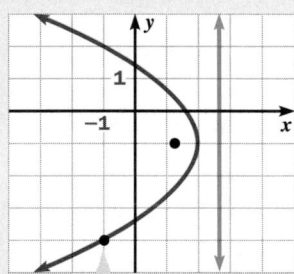

When $y = -4$, $x = -1$, so the parabola passes through the point $(-1, -4)$.

☑ CHECKING KEY CONCEPTS

1. The parabolas at the right all have the same vertex. Each parabola is the same color as its directrix and focus. Explain how the distance between the focus and the directrix affects the shape of the parabola.

Find an equation of the parabola with focus F and directrix d.

2. $F(0, 4)$; $d: y = -2$

3. $F(1, 3)$; $d: y = 0$

4. $F(4, -3)$; $d: y = 5$

5. $F(-2, -1)$; $d: x = 4$

11.2 Exercises and Applications

Extra Practice exercises on page 766

1. **Writing** Describe how you can find the vertex of a parabola and tell whether the parabola opens up, down, left, or right if you know its focus and directrix.

Match each equation with its graph.

2. $y - 4 = 2(x - 1)^2$

3. $x - 1 = \frac{1}{4}(y - 2)^2$

4. $x - 1 = 2(y - 2)^2$

A.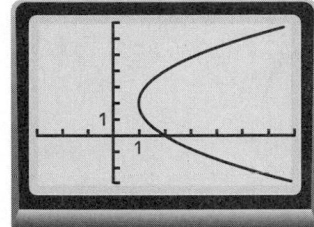

B.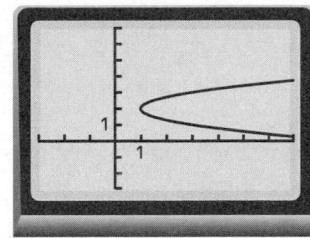

C.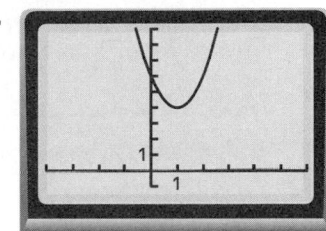

For Exercises 5–10:

a. Find an equation of the parabola with the given characteristics.

b. Graph your equation from part (a).

5. vertex $(0, 0)$; directrix $y = 3$

6. vertex $(-3, 0)$; directrix $y = -2$

7. vertex $(0, 0)$; directrix $y = -1$

8. vertex $(0, 0)$; focus $(0, -4)$

9. focus $(-4, 6)$; vertex $(2, 6)$

10. vertex $(-3, 3)$; directrix $x = 5$

ASTRONOMY Draw a sketch of a cross section of each radio telescope dish in a coordinate plane. Find the distance of the receiver from the vertex of each dish.

11. The submillimeter telescope at Steward Observatory in Tucson, Arizona, has a width of 10 m and a depth of 3.5 m.

12. The Hobart telescope in Tasmania has a width of 85 ft and a depth of 12.96 ft.

13. The Parkes telescope at the Australia Telescope National Facility has a width of 64 m and a depth of 9.2 m.

14. **a.** Solve the equation $x = 2(y + 1)^2 - 3$ for y.

   b. 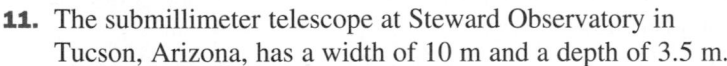 **Technology** Graph the equation from part (a) using a graphing calculator or graphing software. Explain what you must do to obtain a graph of the entire parabola.

The Parkes telescope is located in the state of New South Wales in southeastern Australia.

Name the vertex, focus, and directrix of a parabola with the given equation. Sketch the parabola with its focus and directrix. You may want to use a graphing calculator or graphing software to check your work.

15. $y = 3x^2$

16. $x = -\frac{1}{2}y^2$

17. $y = \frac{1}{8}(x + 1)^2 + 4$

18. $y = \frac{1}{16}(x - 3)^2 - 2$

19. $x = 4(y - 1)^2 + 2$

20. $x = \frac{1}{12}(y - 4)^2$

21. $x^2 = y - 2$

22. $x = 2y^2 + 5$

23. $y^2 = 4x + 3$

24. a. Investigation Complete the following steps and describe the results.

Step 1 Use a sheet of lined paper. Turn the paper so that the lines are vertical. Mark a point near the center of the paper.

Step 2 Fold the paper up from the bottom so that the bottom edge passes through the point you marked in Step 1.

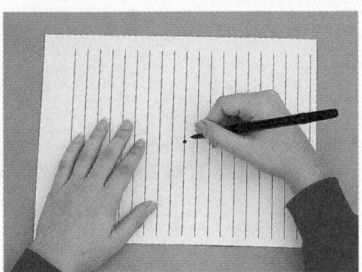
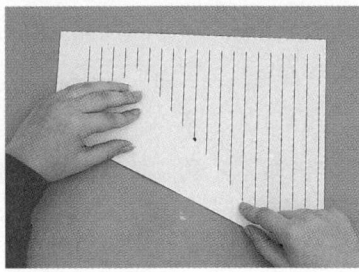

Step 3 Repeat Step 2 about 20 times, each time matching a different point along the bottom edge with the marked point.

b. Explain how you know that the figure traced out by the folds is a parabola. What are the focus and directrix of the parabola?

25. a. Challenge Use your parabola from Exercise 24. Trace along one of the tangent lines you folded. Mark the point where this line is tangent to the parabola. On both sides of the paper, draw a vertical line through the marked point. Fold the paper on the tangent. What happens to the part of the vertical line below the parabola?

b. The vertical line in part (a) is reflected in the same way that a light ray would be reflected by the parabola. How does this help to explain how the focus got its name?

26. Writing Explain why an equation whose graph is a parabola that opens up or down is a function. Explain why an equation whose graph is a parabola that opens left or right is not a function.

Connection ▶ **ACOUSTICS**

Many science museums have an exhibit that allows two people to whisper to each other over long distances. One person speaks into a ring attached to a parabolic dish while another person listens at the ring attached to a second dish across the room from the first.

27. At the Ann Arbor Hands-On Museum in Ann Arbor, Michigan, the "whisper dishes" are 49 in. wide and 8 in. deep. On graph paper, sketch a cross section of one of the parabolic dishes. Put the vertex of the dish at the origin. Show the location of the ring on your sketch, and explain why it is in the correct position.

28. When the dishes directly face each other, they can be used to transmit messages. Use your results from Exercise 25 to draw a sketch that shows how messages are transmitted. Explain your reasoning.

29. a. Open-ended Problem Choose the coordinates of a point and an equation of a horizontal or vertical line. Give all of the information that you can about the parabola with the chosen point as its focus and the chosen line as its directrix.

  b. Write an equation of the parabola from part (a). Explain the steps you need to take in order to sketch a graph of your equation. Include a sketch in your explanation.

SPIRAL REVIEW

For each pair of points, find:

a. **the distance between the points**

b. **the coordinates of the midpoint of the line segment connecting the points**
  (Section 11.1)

30. $(9, -1), (14, -5)$ **31.** $(-8, 5), (16, -11)$ **32.** $(7, 2), (-3, 8)$

Solve each system of equations. *(Section 7.1)*

33. $x - 3y = 0$ **34.** $y = 4.5x$ **35.** $2x = y - 4$
    $x + y = 5$ $y + 4.5x = 18$ $x + y = 15$

ASSESS YOUR PROGRESS

VOCABULARY

parabola (p. 521)   **directrix** (p. 521)
focus (p. 521)   **vertex** (p. 521)

For each pair of points, find:

a. **the distance between the points**

b. **the coordinates of the midpoint of the line segment connecting the points**
  (Section 11.1)

1. $(-3, 8), (5, -1)$ **2.** $(-6, -3), (7, 1)$

3. Given a line l with equation $y = -\dfrac{4}{5}x - 6$ and point P with coordinates $(5, -3)$, find an equation of the line through P and:

  a. parallel to l **b.** perpendicular to l *(Section 11.1)*

Name the vertex, focus, and directrix of each parabola with the given equation.
(Section 11.2)

4. $x + 3 = (y - 1)^2$ **5.** $y + 6 = \dfrac{1}{20}(x + 5)^2$

Find an equation of each parabola with the given characteristics. *(Section 11.2)*

6. vertex $(3, 1)$; directrix $y = -2$ **7.** vertex $(-2.5, 0)$; directrix $x = 1$

8. Journal How are the new ideas about lines and parabolas in Sections 11.1 and 11.2 related to what you already knew about lines and parabolas when you started studying this chapter?

11.3 Circles

Learn how to...

- **write an equation of a circle**

- **graph an equation of a circle**

So you can...

- **locate points on a circle, such as the location of a ship just within range of a lighthouse beam**

The Pharos of Alexandria, Egypt, is thought to be the first lighthouse ever constructed. Built in about 280 B.C., this lighthouse is considered one of the Seven Wonders of the Ancient World. A fire burning at the top of the lighthouse provided light. The use of reflectors increased the visibility of the light so that it could be seen from a distance of 35 miles. Modern lighthouses, like Boston Light shown above, are powered by electricity.

THINK AND COMMUNICATE

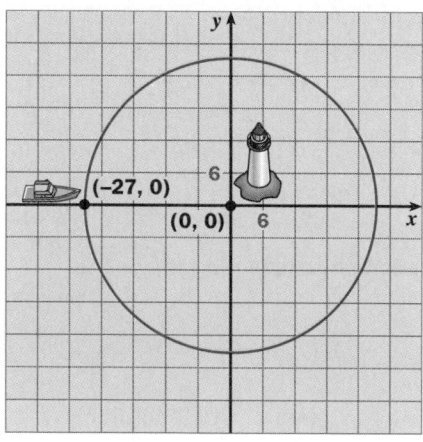

1. The ship shown on the map, where the coordinates are given in miles, has just come within range of the beam from Boston Light. Find the range of the beam.

2. Use the distance formula to find an equation for the distance between the lighthouse and any ship with coordinates (x, y) just within range of the beam.

3. Can a ship with coordinates (24, 12) see the beam? Explain.

The set of points that lie at the outer range of the lighthouse beam form a *circle*. A **circle** is the set of all points in a plane that are the same distance from the **center** of the circle. The **radius** of a circle is the distance between the center of the circle and any point on the circle. Any line containing the center of a circle is a line of symmetry for the circle.

The distance between the center $C(h, k)$ and any point $P(x, y)$ on the circle shown is r. You can use the distance formula to express this relationship as an equation:

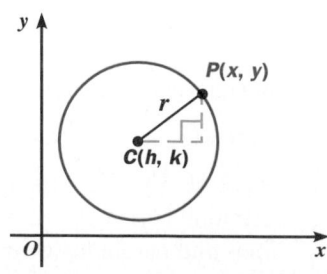

$$r = \sqrt{(x - h)^2 + (y - k)^2}$$
$$r^2 = (x - h)^2 + (y - k)^2$$

An equation of a circle with center $C(h, k)$ and radius r is:

$$(x - h)^2 + (y - k)^2 = r^2$$

EXAMPLE 1

Write an equation of the circle with center (1, −4) and radius 5.

SOLUTION

$$(x - h)^2 + (y - k)^2 = r^2$$ ← Use the standard form of an equation of a circle.

$$(x - 1)^2 + (y - (-4))^2 = 5^2$$

$$(x - 1)^2 + (y + 4)^2 = 25$$ ← Substitute **1** for **h**, **−4** for **k**, and **5** for **r**.

EXAMPLE 2

Graph the circle with equation $(x + 3)^2 + (y - 2)^2 = 36$.

SOLUTION

Method 1 Use graph paper.

The center is $(-3, 2)$. ← **In the given equation, $h = -3$ and $k = 2$.**

The radius is 6. ← **In the equation, $r^2 = 36$, so $r = 6$.**

Step 1 Plot the center point $(-3, 2)$.

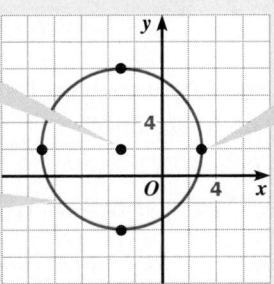

Step 2 The radius is 6. Plot the points that are 6 units above, below, to the left of, and to the right of the center.

Step 3 Sketch the circle through the four points.

Method 2 Use a graphing calculator or graphing software.

Solve the equation for y.

$$(x + 3)^2 + (y - 2)^2 = 36$$

$$(y - 2)^2 = 36 - (x + 3)^2$$

$$y - 2 = \pm\sqrt{36 - (x + 3)^2}$$

$$y = 2 \pm \sqrt{36 - (x + 3)^2}$$

There are two equations to graph: $y = 2 + \sqrt{36 - (x + 3)^2}$ represents the top half of the circle, and $y = 2 - \sqrt{36 - (x + 3)^2}$ represents the bottom half.

Solution continued on next page.

For more information
about squaring the
viewing window, see
the *Technology
Handbook*, p. 805.

SOLUTION *continued*

Enter the equations into a graphing
calculator or graphing software.

```
Y₁🔳2+√(36−(X+3)²
)
Y₂🔳2−√(36−(X+3)²
)
Y₃=
Y₄=
Y₅=
Y₆=
```

Set an appropriate square
window and graph the circle.

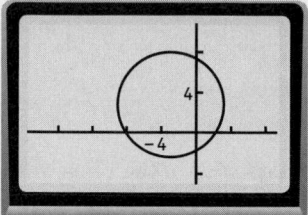

EXAMPLE 3

Find the points of intersection of the graphs of $(x + 4)^2 + (y - 1)^2 = 25$ and
$(x - 2)^2 + (y - 3)^2 = 9$.

SOLUTION

Method 1 Solve the system of equations.

Step 1 Expand both equations and subtract like terms.

$$
\begin{aligned}
x^2 + 8x + 16 + y^2 - 2y + 1 &= 25 \\
\underline{x^2 - 4x + 4 + y^2 - 6y + 9} &= \underline{9} \\
12x + 12 + 4y - 8 &= 16
\end{aligned}
$$

$$12x - 12 = -4y \qquad \text{Solve for } y.$$

$$-3x + 3 = y$$

Step 2 Substitute the equation from Step 1 into
one of the original equations and solve.

Substitute
−3x + 3 for **y**.

$$(x - 2)^2 + (-3x + 3 - 3)^2 = 9$$

$$x^2 - 4x + 4 + 9x^2 = 9 \qquad \text{Write the equation in standard form.}$$

$$10x^2 - 4x - 5 = 0$$

Use the quadratic
formula.

$$x = \frac{-(-4) \pm \sqrt{(-4)^2 - 4(10)(-5)}}{2(10)}$$

$$x = \frac{2 \pm 3\sqrt{6}}{10}$$

$$x \approx 0.93, \; -0.53$$

Step 3 Find the coordinates of the points of intersection.

When $x = \dfrac{2 + 3\sqrt{6}}{10}, y = -3\left(\dfrac{2 + 3\sqrt{6}}{10}\right) + 3 \approx 0.20.$

When $x = \dfrac{2 - 3\sqrt{6}}{10}, y = -3\left(\dfrac{2 - 3\sqrt{6}}{10}\right) + 3 \approx 4.60.$

The circles intersect at about $(0.93, 0.20)$ and at about $(-0.53, 4.60)$.

Method 2 Use a graphing calculator or graphing software.

Step 1 Solve the equations for y.
Enter the equations into the
calculator or software.

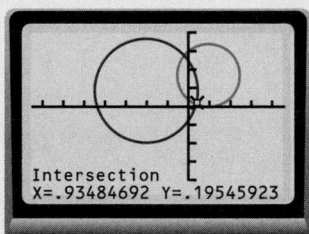

Y1☰1+√(25−(X+4)²
)
Y2☰1−√(25−(X+4)²
)
Y3☰3+√(9−(X−2)²)

Y4☰3−√(9−(X−2)²)

Step 2 Find the point(s) of intersection.

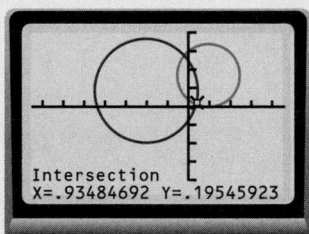

Intersection
X=.93484692 Y=.19545923

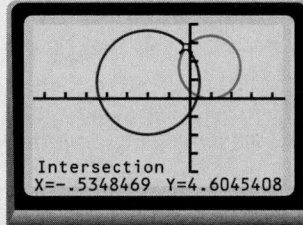

Intersection
X=−.5348469 Y=4.6045408

The graphs intersect at about $(0.93, 0.20)$ and at about $(-0.53, 4.60)$.

THINK AND COMMUNICATE

4. Why do you have to enter two equations to graph a circle on a graphing calculator?

5. The circles in Example 3 intersect at two points. Can two circles intersect at only one point? at more than two points? Explain.

☑ CHECKING KEY CONCEPTS

For each equation of a circle, identify the center and the radius.

1. $(x - 6)^2 + (y - 4)^2 = 144$
2. $(x + 1)^2 + (y - 3)^2 = 10$
3. $(x + 5)^2 + y^2 = 36$
4. $(x + 3)^2 + (y + 2)^2 = 18$

For each graph, write an equation of the circle.

5.

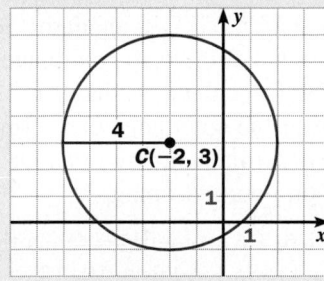

6.

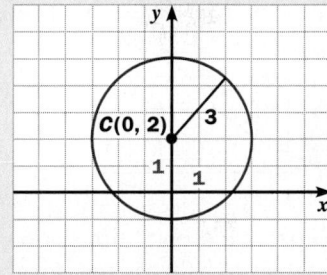

Graph each equation.

7. $(x - 5)^2 + (y + 1)^2 = 4$
8. $x^2 + (y - 3)^2 = 25$

9. Find the point(s) of intersection of the graphs of $x^2 + (y - 1)^2 = 9$ and $x^2 + y^2 = 4$.

11.3 Exercises and Applications

Extra Practice exercises on page 766

For each equation of a circle, identify the center and the radius. Then graph the equation.

1. $(x - 4)^2 + (y - 2)^2 = 25$

2. $x^2 + y^2 = 16$

3. $x^2 + (y - 1)^2 = 4$

4. $(x - 3)^2 + (y + 5)^2 = 6^2$

5. $(x + 1)^2 + (y + 1)^2 = 10$

6. $(x + 6)^2 + y^2 = 1$

7. $x^2 + (y + 3)^2 = 2$

8. $x^2 - 10x + 25 + y^2 = 4$

9. $x^2 + y^2 - 2y + 1 = 8$

INTERVIEW Kija Kim

Look back at the article on pages 510–512.

The maps that Kija Kim's company makes are used by a variety of other companies and organizations. Kim's company created the computerized map below for a Massachusetts fire department. The map illustrates the estimated fire engine response time from the fire station, which is located at the origin.

A fire engine leaving the station can reach a fire that is within 0.6 mi from the station within 3 min.

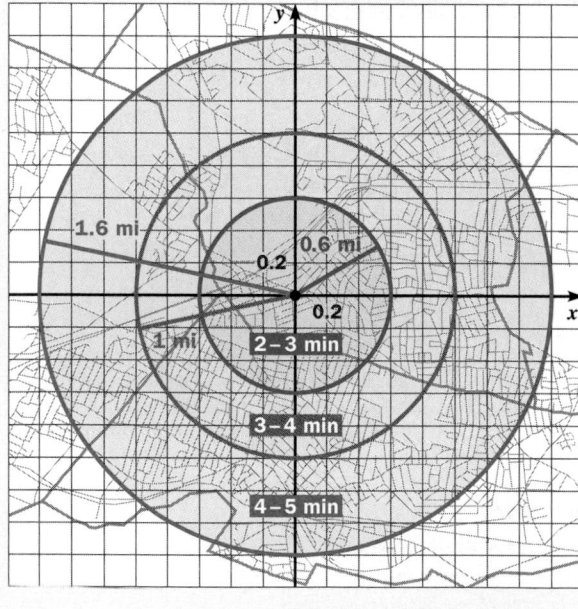

10. Write an equation of the circle defining the region that a fire engine leaving the station can reach within 3 min. What is the radius of this circle?

11. Repeat Exercise 10 for the circle defining the region that a fire engine can reach within 4 min.

12. Repeat Exercise 10 for the circle defining the region that a fire engine can reach within 5 min.

13. Suppose the station receives an alarm from a building located at the map coordinates (0.9, 1.2). The fire chief says that the building is near the outer edge of the ring defining the 4–5 min response time. Do you agree? Explain your reasoning.

14. Suppose a second fire station located at (1.6, 1.6) sometimes sends fire engines to help the fire department referred to in Exercises 10–13.

    a. Suppose the second station assists with fires within a 1 mi radius. Write an equation of the circle defining the region that the second station covers.

    b. Would the second station assist with a fire at the building in Exercise 13?

Write an equation of the circle with the given center C and radius r.

15. $C(0, 0)$; $r = 7$

16. $C(2, 5)$; $r = 3$

17. $C(-1, 6)$; $r = 3$

18. $C(-3, 2)$; $r = \sqrt{6}$

19. $C(0, \sqrt{3})$; $r = 11$

20. $C(-4, -2\sqrt{2})$; $r = 3$

21. $C(a, b)$; $r = 11$

22. $C(a, -b)$; $r = 2k$

23. $C(-a, 2b)$; $r = c^2$

24. A diameter of a circle has endpoints $(-2, -3)$ and $(4, 1)$. Find an equation of the circle.

Connection COMMUNICATIONS

A cellular phone network uses towers like the one shown to transmit calls. When a person makes a call, a tower forwards the signal to a switching station. The switching station transfers the call to the regional telephone office, which then processes the call. The range of a tower can be defined by a circle.

25. Suppose a cellular phone service company is installing towers in a town. The current tower locations are shown on the map, where the coordinates are given in miles.

  a. Tower A is located at the origin and has a range of 2 mi. Find an equation of the circle that defines the tower's range.

  b. Write an equation of the circle that defines the range of tower B.

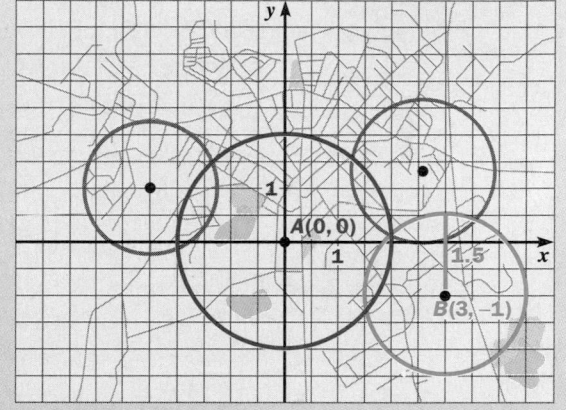

26. The company determines that additional towers should be installed to provide more complete service in the town.

  a. Open-ended Problem Select a location for a new tower in a region that does not have coverage. Determine a reasonable range for the tower.

  b. Write an equation of the circle that defines the tower's range.

Use substitution to determine if each given point is on the circle with equation $(x + 1)^2 + (y - 5)^2 = 16$.

27. $(-1, 1)$ **28.** $(3, 4)$ **29.** $(-5, 5)$ **30.** $(-3.4, 1.8)$

31. Writing Is the center of a circle a point on the circle? Why or why not?

32. Challenge The point $(-4, -1)$ is on the circle with equation $(x - 1)^2 + (y + 3)^2 = 29$. Write an equation of the line that is tangent to the circle and passes through the point $(-4, -1)$. (*Hint:* The tangent is perpendicular to the radius with an endpoint at $(-4, -1)$.)

33. Challenge Find an equation of the circle that passes through the points $(3, 6)$, $(-1, -2)$, and $(6, 5)$.

Find the point(s) of intersection of each pair of circles with the given equations.

34. $(x - 3)^2 + y^2 = 24$
$x^2 + y^2 = 9$

35. $x^2 + (y - 1)^2 = 16$
$x^2 + (y + 1)^2 = 28$

36. $(x + 3)^2 + (y + 2)^2 = 18$
$(x + 2)^2 + (y + 1)^2 = 11$

The NAVSTAR Global Positioning System (GPS) enables people, such as hikers, to locate their position on Earth using a handheld device that communicates with a group of satellites. Suppose three satellites are transmitting signals to a receiver held by a hiker. Each satellite determines a circle on which the hiker is positioned. The intersection of the three circles gives the coordinates of the hiker's location.

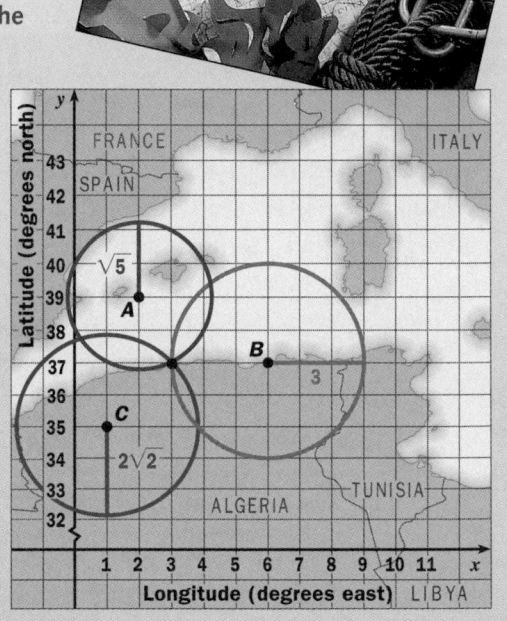

37. Write an equation of each circle shown on the map.

38.   **Technology** Use a graphing calculator or graphing software.

   a. Find the points of intersection of circles A and B.

   b. Find the points of intersection of circles A and C.

   c. Find the points of intersection of circles B and C.

39. a. Use your answers to Exercise 38 to find the coordinates for the position of the hiker.

   b. **Challenge** Suppose the hiker lives near Barcelona, Spain, at coordinates (3, 42). Use the fact that the hiker's current location and the hiker's home are at the same longitude to find how far from home, in miles, the hiker is. (*Hint:* The radius of Earth is about 3963 mi.)

ONGOING ASSESSMENT

40. Cooperative Learning Work with a partner. You will need graph paper.

   a. Each of you should draw axes on a piece of graph paper and sketch a circle so that its center is not at the origin. Label the coordinates of the center and a point on your circle. Give your paper to your partner.

   b. Write an equation of the circle you receive. Then graph your equation. Does your graph match the one your partner sketched? If not, check your work and correct any errors.

SPIRAL REVIEW

Find an equation of each parabola with the given focus and directrix. Sketch the parabola with its focus and directrix. *(Section 11.2)*

41. focus (3, 1)
   directrix $y = -1$

42. focus (−2, −1)
   directrix $x = 2$

43. focus (4, −3)
   directrix $y = 1$

Graph each function. State the domain and range. *(Section 8.1)*

44. $y = \sqrt{x + 2}$

45. $y = \sqrt{x - 1} + 4$

46. $y = \sqrt{x} + 3$

11.4 Ellipses

At one time, astronomers thought that the orbit of a planet around the sun was a circle or a combination of circles. In 1609, the German astronomer Johannes Kepler realized that the orbit of a planet follows an elliptical path, with the sun located at one focus of the elliptical orbit. In the Exploration, you will see how to draw an *ellipse*.

Learn how to...
- write an equation of an ellipse
- graph an equation of an ellipse

So you can...
- describe an elliptical object, such as the Oval Office in the White House

EXPLORATION
COOPERATIVE LEARNING

Drawing an Ellipse

Work with a partner.
You will need:
- graph paper
- a 10-inch piece of string
- two pushpins
- a piece of cardboard

1 Draw axes on a piece of graph paper. Label the points (–4, 0) and (4, 0). Place the graph paper on the cardboard. Use the pins to hold one end of the string at (–4, 0) and the other end at (4, 0).

2 Use a pencil to pull the string until it is taut. Move the pencil above and below the *x*-axis, keeping the string taut, until you have sketched a closed geometric figure called an *ellipse*.

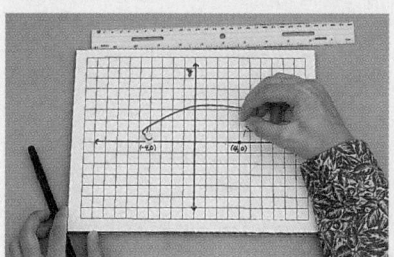

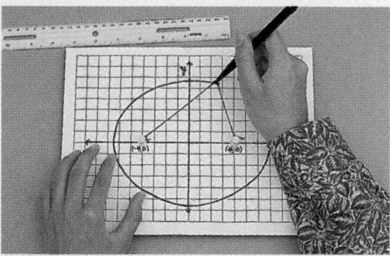

Questions

1. If *P* and *Q* are any two points on the figure you drew, how do the lengths F_1P and F_2P compare with the lengths F_1Q and F_2Q?

2. Give a definition for the figure you drew using the relationship between F_1P, F_2P, and the length of the string.

3. Move the pins closer together and repeat Steps 1 and 2. How did the ellipse change? What do you think happens to the shape of the ellipse when the pins are moved even closer together?

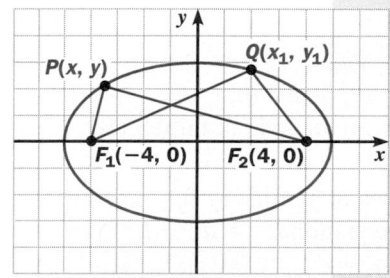

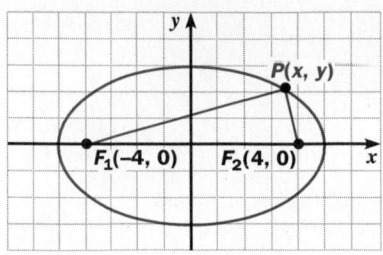

An **ellipse** is the set of all points in a plane such that the sum of the distances between any point and two fixed points F_1 and F_2, called the **foci**, is constant. (*Foci* is the plural of *focus*.) The **center** of the ellipse is the midpoint of $\overline{F_1F_2}$.

You can use the definition of an ellipse to find an equation of an ellipse. The ellipse in the Exploration has its center at the origin and has foci $F_1(-4, 0)$ and $F_2(4, 0)$. The string is 10 inches long. Let $P(x, y)$ be any point on the ellipse.

$$PF_1 + PF_2 = 10$$

Use the distance formula.

$$\sqrt{(x + 4)^2 + y^2} + \sqrt{(x - 4)^2 + y^2} = 10$$

$$\sqrt{(x - 4)^2 + y^2} = 10 - \sqrt{(x + 4)^2 + y^2}$$

Square both sides of the equation.

$$(x - 4)^2 + y^2 = 100 - 20\sqrt{(x + 4)^2 + y^2} + (x + 4)^2 + y^2$$

$$(x^2 - 8x + 16) + y^2 = 100 - 20\sqrt{(x + 4)^2 + y^2} + (x^2 + 8x + 16) + y^2$$

$$20\sqrt{(x + 4)^2 + y^2} = 100 + 16x$$

$$5\sqrt{(x + 4)^2 + y^2} = 25 + 4x$$

Square both sides of the equation.

$$25\left[(x + 4)^2 + y^2\right] = (25 + 4x)^2$$

$$25\left(x^2 + 8x + 16 + y^2\right) = 625 + 200x + 16x^2$$

Combine like terms and simplify.

$$9x^2 + 25y^2 = 225$$

Divide each term by 225.

$$\frac{x^2}{25} + \frac{y^2}{9} = 1$$

So an equation of the ellipse is $\frac{x^2}{25} + \frac{y^2}{9} = 1$. This equation can be generalized as:

$$\frac{x^2}{a^2} + \frac{y^2}{b^2} = 1$$

The foci are at $(c, 0)$ and $(-c, 0)$.

The **major axis** of an ellipse is the line segment that contains the foci and has the two vertices as its endpoints.

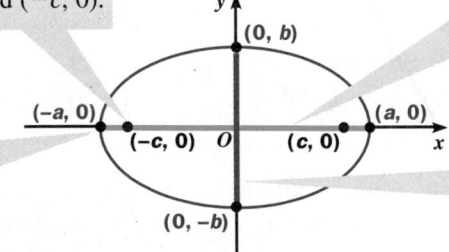

The points $(a, 0)$ and $(-a, 0)$ are the *vertices* of the ellipse. The **vertices** are the points of intersection of the ellipse and the line containing its foci.

The segment with endpoints $(0, b)$ and $(0, -b)$ is the **minor axis** of this ellipse.

The lines containing the axes of an ellipse are lines of symmetry for the ellipse. Also, the values of a, b, and c are related by the equation $c^2 = a^2 - b^2$. You will investigate this relationship in the exercises.

The summary below describes ellipses whose vertices are aligned either vertically or horizontally and whose centers are at $C(h, k)$.

Standard Forms of an Equation of an Ellipse

An equation of a *horizontal ellipse* (an ellipse with a horizontal major axis) with center $C(h, k)$ is:

$$\frac{(x - h)^2}{a^2} + \frac{(y - k)^2}{b^2} = 1 \quad (a > b > 0)$$

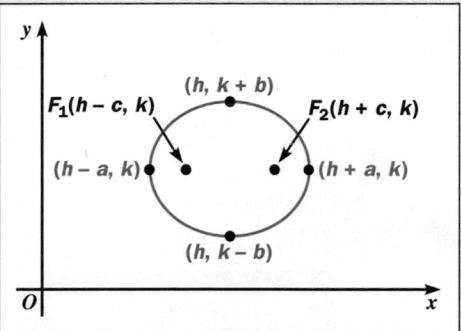

Foci: $F_1(h - c, k)$ and $F_2(h + c, k)$ where $c > 0$ and $c^2 = a^2 - b^2$

Length of major axis: $2a$

Length of minor axis: $2b$

An equation of a *vertical ellipse* (an ellipse with a vertical major axis) with center $C(h, k)$ is:

$$\frac{(x - h)^2}{b^2} + \frac{(y - k)^2}{a^2} = 1 \quad (a > b > 0)$$

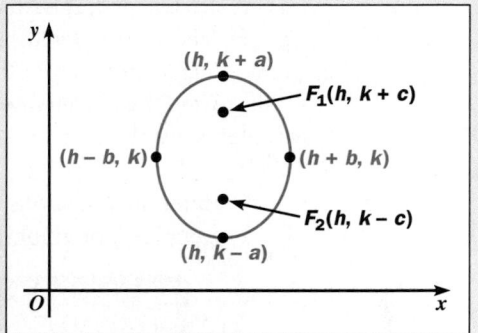

Foci: $F_1(h, k + c)$ and $F_2(h, k - c)$ where $c > 0$ and $c^2 = a^2 - b^2$

Length of major axis: $2a$

Length of minor axis: $2b$

EXAMPLE 1

Graph the equation $\frac{(x + 2)^2}{16} + \frac{(y - 3)^2}{25} = 1.$

SOLUTION

Method 1 Use graph paper.

In the given equation, $h = -2$ and $k = 3$. The center is $C(-2, 3)$.

Since the denominator of the term involving y is greater than the denominator of the term involving x, the ellipse is vertical. The major axis is parallel to the y-axis.

Since $a^2 = 25$, $a = \sqrt{25} = 5$. The major axis has length $2a = 10$.

Since $b^2 = 16$, $b = \sqrt{16} = 4$. The minor axis has length $2b = 8$.

Step 1 Plot the center point $C(-2, 3)$.

Step 2 Plot the endpoints of the major axis. They are **5 units** above and below the center.

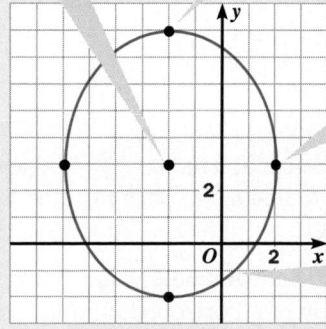

Step 3 Plot the endpoints of the minor axis. They are **4 units** to the left and right of the center.

Step 4 Sketch the ellipse through the four points plotted in Steps 2 and 3.

Solution continued on next page.

Method 2 Use a graphing calculator or graphing software.

Solve the equation for *y*.

$$\frac{(x+2)^2}{16} + \frac{(y-3)^2}{25} = 1$$

$$\frac{(y-3)^2}{25} = 1 - \frac{(x+2)^2}{16}$$

$$(y-3)^2 = 25 - \frac{25(x+2)^2}{16}$$

There are two equations to graph. The equation with + represents the top half of the ellipse, and the equation with − represents the bottom half.

$$y - 3 = \pm\sqrt{25 - \frac{25(x+2)^2}{16}}$$

$$y = 3 \pm\sqrt{25 - \frac{25(x+2)^2}{16}}$$

Enter the equations into a graphing calculator or graphing software.

Set an appropriate square window and graph the ellipse.

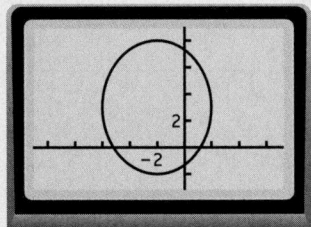

EXAMPLE 2

Write an equation of the ellipse with center (6, 8), one vertex at (1, 8), and one focus at (2, 8).

SOLUTION

Step 1 Determine if the ellipse is *horizontal* or *vertical*. Since the *y*-coordinates of the center, vertex, and focus are the same, the ellipse is horizontal.

Step 2 Find *a*.

$a = 6 - 1 = 5$

Subtract the *x*-coordinates of the center and the vertex to find *a*.

Step 3 Find *c*.

$c = 6 - 2 = 4$

The distance between the center and the focus is *c*.

Step 4 Find *b*.

$b^2 = 5^2 - 4^2 = 9$

$b = 3$

Substitute the values of *a* and *c* into the equation $b^2 = a^2 - c^2$.

Step 5 Substitute the values for *h*, *k*, *a*, and *b* into the standard form for an equation of a horizontal ellipse.

$$\frac{(x-6)^2}{25} + \frac{(y-8)^2}{9} = 1$$

Substitute 6 for *h*, 8 for *k*, 5 for *a*, and 3 for *b*.

EXAMPLE 3   Application: Architecture

The Oval Office in the White House is in the shape of an ellipse. Write an equation of the ellipse.

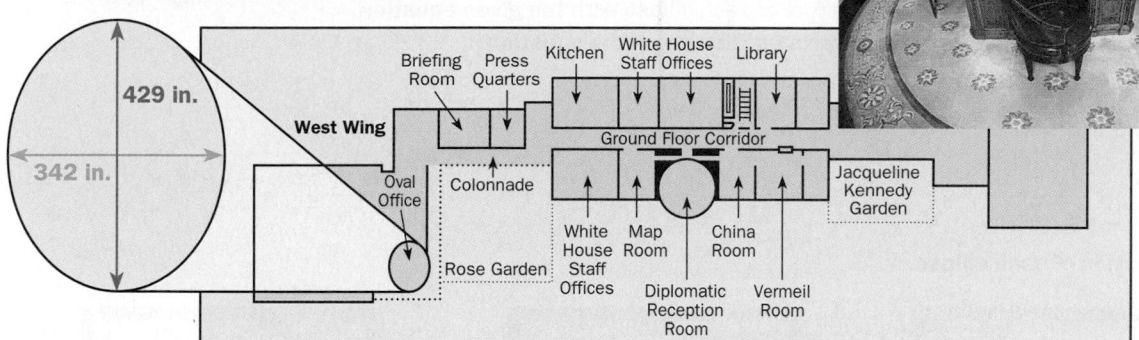

SOLUTION

Step 1 Find a.

$2a = 429$

$a = 214.5$

The major axis is 429 in. long.

Step 2 Find b.

$2b = 342$

$b = 171$

The minor axis is 342 in. long.

Step 3 Write an equation of the ellipse. Suppose the center of the Oval Office is positioned at the origin and the major axis lies on the y-axis.

$$\frac{x^2}{b^2} + \frac{y^2}{a^2} = 1 \quad \Longrightarrow \quad \frac{x^2}{29{,}241} + \frac{y^2}{46{,}010.25} = 1$$

Use the equation of a vertical ellipse.

The coordinates of points on this ellipse are measured in inches.

☑ CHECKING KEY CONCEPTS

Use the ellipse shown to answer Questions 1–3.

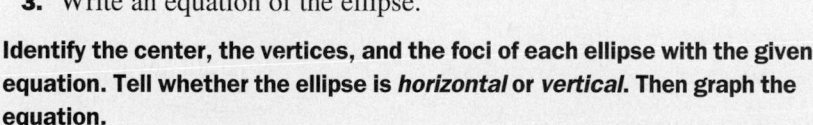

1. What are the coordinates of the center? the foci?

2. What is the length of the major axis? the minor axis?

3. Write an equation of the ellipse.

Identify the center, the vertices, and the foci of each ellipse with the given equation. Tell whether the ellipse is *horizontal* or *vertical*. Then graph the equation.

4. $\dfrac{(x - 6)^2}{16} + \dfrac{(y - 4)^2}{49} = 1$

5. $\dfrac{(x + 1)^2}{64} + \dfrac{(y - 3)^2}{4} = 1$

6. Discuss the graph of $\dfrac{x^2}{a^2} + \dfrac{y^2}{b^2} = 1$ when $a = b$. What can you say about the foci of this ellipse?

Extra Practice
exercises on
page 766

Identify the center, the vertices, and the foci of each ellipse with the given equation. Tell whether the ellipse is *horizontal* or *vertical*. Then graph the equation.

1. $\dfrac{x^2}{64} + \dfrac{y^2}{25} = 1$

2. $\dfrac{x^2}{4} + \dfrac{y^2}{49} = 1$

3. $\dfrac{x^2}{8} + \dfrac{(y-1)^2}{2} = 1$

4. $\dfrac{(x-3)^2}{16} + \dfrac{(y+3)^2}{9} = 1$

5. $\dfrac{(x+8)^2}{25} + \dfrac{(y+1)^2}{100} = 1$

6. $\dfrac{(x-5)^2}{81} + y^2 = 1$

Write an equation of each ellipse.

7.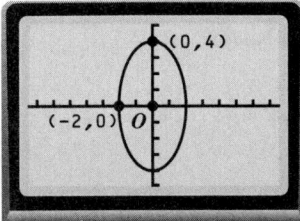

8.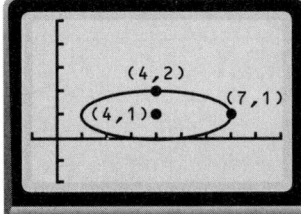

9.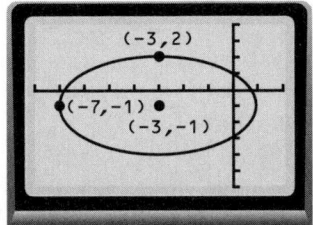

Connection   ART

Artist William Wainwright used mathematics to design a miniature golf hole in the shape of an ellipse for a museum exhibit entitled "Strokes of Genius: Mini Golf by Artists." The starting position of the ball is at one focus and the cup is at the other focus. Wainwright positioned a barrier at the center of the ellipse.

10. The length of the major axis of the elliptical frame of the hole is 12 ft. The length of the minor axis is 8 ft.

   a. Find an equation of the elliptical frame. Assume the ellipse is centered at the origin and the major axis is horizontal.

   b. Estimate the location of the foci.

11. Suppose you position the ellipse in a coordinate plane so that the origin is located at the starting position of the ball.

   a. What are the coordinates of the center of the barrier?

   b. What are the coordinates of the cup?

   c. Write an equation for the elliptical frame in this position.

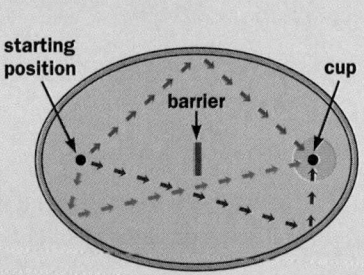

12. Suppose the barrier is removed. What is the shortest path from the starting position of the ball to the cup? What is the length of the path?

13. With the barrier in place, what is the shortest path with a single putt from the starting position of the ball to the cup? What is the length of the path?

A *lithotripter* is an elliptical medical instrument used to disintegrate a kidney stone within a patient's body. Shock waves, initiated by a generator located at one focus of the instrument, reflect off the sides of a three-dimensional elliptical reflecting dish and converge at the other focus. The patient is positioned so that the kidney stone is at the other focus.

The shape of a cross section of the reflecting dish on a certain lithotripter can be represented by the equation $\dfrac{x^2}{196} + \dfrac{y^2}{63.6804} = 1$, **where x and y are measured in centimeters.**

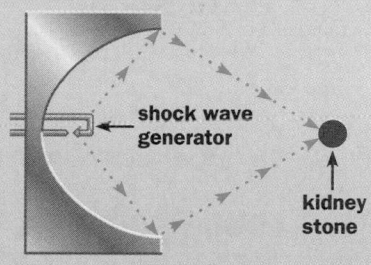

14. What are the coordinates of the shock wave generator? of the kidney stone?

15. Graph the equation.

Scientists have determined that a deeper reflecting dish operates more effectively. A different model lithotripter is 1.5 times as wide (along the x-axis) as the model discussed in Exercises 14 and 15.

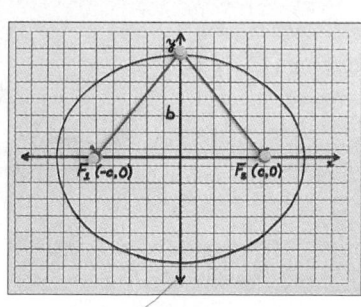

shock wave generator

kidney stone

16. Write an equation of a cross section of this reflecting dish. How is this equation different from the equation you used in Exercises 14 and 15?

17. Graph the equation you wrote in Exercise 16.

18. For this lithotripter, what are the coordinates of the shock wave generator? of the kidney stone?

19. Writing The ends of the string in the figure at the right are at $(-c, 0)$ and $(c, 0)$. Suppose that the length of the string is $2a$.

a. Explain why the vertices of the ellipse must be $(-a, 0)$ and $(a, 0)$.

b. Explain why the endpoints of the minor axis must be $(0, -b)$ and $(0, b)$, where $b^2 = a^2 - c^2$.

Write an equation of the ellipse with the given characteristics.

20. center $(0, 0)$; vertex $(-4, 0)$; y-intercept -3

21. center $(5, -3)$; $a = 3$; $b = 2$; major axis is vertical

22. center $(6, 2)$; vertex $(6, 7)$; focus $(6, -1)$

23. center $(-2, 4)$; vertex $(-9, 4)$; minor axis is 6 units long

24. CARPENTRY A standard vent pipe with a diameter of 3.75 in. needs to be installed on a roof with a slope of $\frac{4}{3}$. The hole that needs to be cut into the roof to fit the pipe is in the shape of an ellipse.

a. Use the slope of the roof to find k.

b. What is the length of the major axis? the minor axis?

c. Write an equation of the ellipse that should be cut from the roof to fit the pipe.

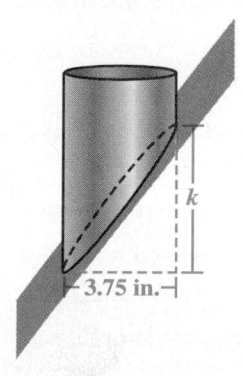

11.4 Ellipses **541**

25. Open-ended Problem Choose the length of the major and minor axes of an ellipse. Write an equation of your ellipse. Graph the equation. Find the vertices and foci.

Write an equation of the circle with the given center C and radius r. Then sketch its graph. *(Section 11.3)*

26. $C(3, 0); r = 5$ **27.** $C(-2, -1); r = 6$ **28.** $C(4, -3); r = \sqrt{8}$

Find the slope-intercept equation of each line. *(Section 2.2)*

29. 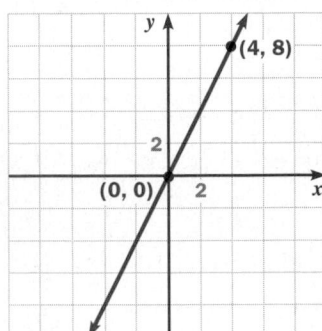   **30.** 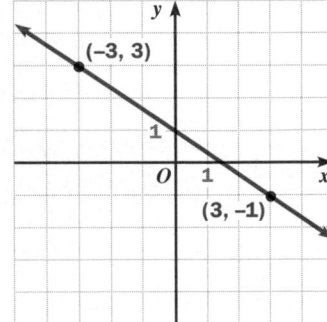   **31.**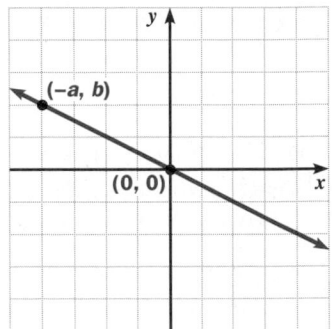

ASSESS YOUR PROGRESS

VOCABULARY

circle (p. 528)   **foci** (p. 536)
center (pp. 528, 536)   **vertices** (p. 536)
radius (p. 528)   **major axis** (p. 536)
ellipse (p. 536)   **minor axis** (p. 536)

For each equation of a circle, identify the center and the radius. *(Section 11.3)*

1. $(x + 1)^2 + (y - 5)^2 = 16$ **2.** $(x - 3)^2 + y^2 = 25$

3. Find the point(s) of intersection of the equations in Exercises 1 and 2. *(Section 11.3)*

Identify the center, the vertices, and the foci of each ellipse with the given equation. Tell whether the ellipse is *horizontal* or *vertical*. *(Section 11.4)*

4. $\dfrac{x^2}{64} + \dfrac{(y - 5)^2}{36} = 1$ **5.** $\dfrac{(x + 3)^2}{100} + \dfrac{(y - 5)^2}{196} = 1$

Write an equation of each figure described. Then graph the equation. *(Sections 11.3 and 11.4)*

6. circle with center $(0, -5)$ and radius 4

7. ellipse with center $(-1, 2)$, a vertex at $(-1, -3)$, and a focus at $(-1, 6)$

8. Journal Discuss the geometric and algebraic relationships between circles and ellipses.

11.5 Hyperbolas

The Fermi National Accelerator Laboratory in Batavia, Illinois, is the site of several large geometric sculptures. One of these, named *Acqua Alle Funi*, is a three-sided sculpture in the shape of a *hyperbola*.

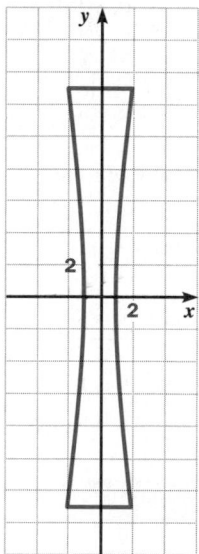

THINK AND COMMUNICATE

By superimposing a coordinate system on one side of the sculpture, as shown at the left, you can describe the vertical edges of the sculpture by the equation

$$x^2 - \frac{3y^2}{169} = 1, \text{ where } x \text{ and } y \text{ are measured in feet.}$$

1. How is the equation like the equation of an ellipse? How is it different?

2. **a.** The narrowest part of the sculpture corresponds to the x-intercepts of the graph of the given equation. What are the x-intercepts?

   b. How wide is the sculpture at its narrowest width?

3. The height of the sculpture is 26 ft. How wide is the sculpture at the top?

A **hyperbola** is the set of all points in a plane such that the difference of the distances between a point and each of two **foci**, F_1 and F_2, is constant. The **center** of the hyperbola is the midpoint of $\overline{F_1F_2}$.

The sculpture's equation is of this form:

$$\frac{x^2}{a^2} - \frac{y^2}{b^2} = 1$$

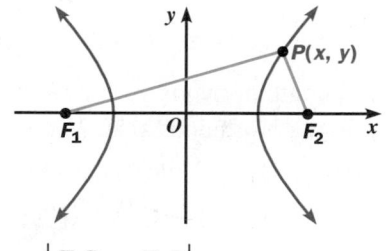

$$|F_1P - F_2P| = \text{a constant}$$

The hyperbola has the lines $y = \frac{b}{a}x$ and $y = -\frac{b}{a}x$ as asymptotes. The asymptotes contain the diagonals of a rectangle with dimensions $2a$ and $2b$.

The **major axis** has endpoints at the vertices, so its length is $2a$.

The **minor axis** is perpendicular to the major axis and has length $2b$.

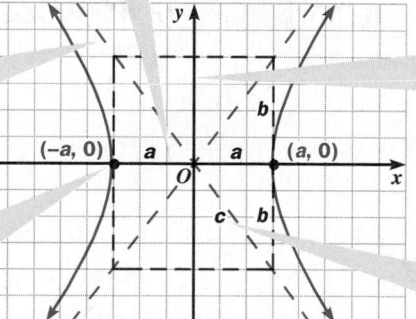

The **vertices** of the hyperbola are $(a, 0)$ and $(-a, 0)$.

By the Pythagorean theorem, $c^2 = a^2 + b^2$.

The lines containing the axes of a hyperbola are lines of symmetry.

The summary below describes hyperbolas whose vertices are aligned either vertically or horizontally and whose centers are at $C(h, k)$.

Standard Forms of an Equation of a Hyperbola

An equation of a *horizontal hyperbola* (a hyperbola with a horizontal major axis) with center $C(h, k)$ is:

$$\frac{(x - h)^2}{a^2} - \frac{(y - k)^2}{b^2} = 1 \quad (a > 0, b > 0)$$

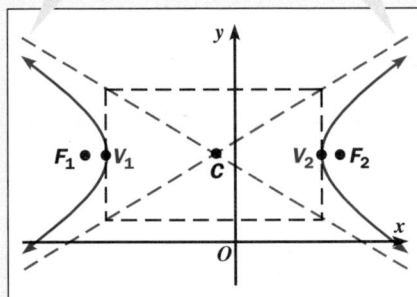

$$y = -\frac{b}{a}(x - h) + k \qquad y = \frac{b}{a}(x - h) + k$$

Foci: $F_1(h - c, k)$ and $F_2(h + c, k)$ where $c > 0$ and $c^2 = a^2 + b^2$

Vertices: $V_1(h - a, k)$ and $V_2(h + a, k)$

Asymptotes: $y = \pm\frac{b}{a}(x - h) + k$

Length of major axis: $2a$
Length of minor axis: $2b$

An equation of a *vertical hyperbola* (a hyperbola with a vertical major axis) with center $C(h, k)$ is:

$$\frac{(y - k)^2}{a^2} - \frac{(x - h)^2}{b^2} = 1 \quad (a > 0, b > 0)$$

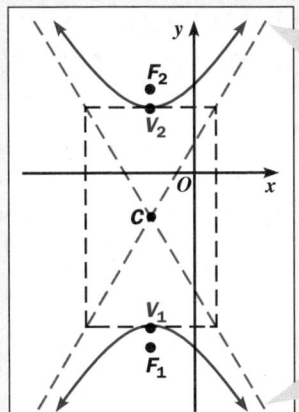

$$y = \frac{a}{b}(x - h) + k$$

$$y = -\frac{a}{b}(x - h) + k$$

Foci: $F_1(h, k - c)$ and $F_2(h, k + c)$ where $c > 0$ and $c^2 = a^2 + b^2$

Vertices: $V_1(h, k - a)$ and $V_2(h, k + a)$

Asymptotes: $y = \pm\frac{a}{b}(x - h) + k$

Length of major axis: $2a$
Length of minor axis: $2b$

EXAMPLE 1

Graph the hyperbola with equation $\dfrac{(x + 1)^2}{25} - \dfrac{(y - 3)^2}{9} = 1$.

SOLUTION

Method 1 Use graph paper. In the given equation, $h = -1$ and $k = 3$.

The center is $C(-1, 3)$.

WATCH OUT!

The major axis of a hyperbola is $\overline{V_1 V_2}$. It is not necessarily the longer axis.

Since the term involving y is subtracted from the term involving x, the hyperbola is horizontal. The major axis is parallel to the x-axis.

Since $a^2 = 25$, $a = \sqrt{25} = 5$. The major axis has length $2a = 10$.

Since $b^2 = 9$, $b = \sqrt{9} = 3$. The minor axis has length $2b = 6$.

The vertices are $V_1(-6, 3)$ and $V_2(4, 3)$.

The hyperbola has asymptotes $y = -\frac{3}{5}(x + 1) + 3$ and $y = \frac{3}{5}(x + 1) + 3$.

Step 1 Plot the center $C(-1, 3)$ and the vertices $V_1(-6, 3)$ and $V_2(4, 3)$.

Step 2 Draw the rectangle centered at $(-1, 3)$ and having a horizontal dimension of 10 and a vertical dimension of 6.

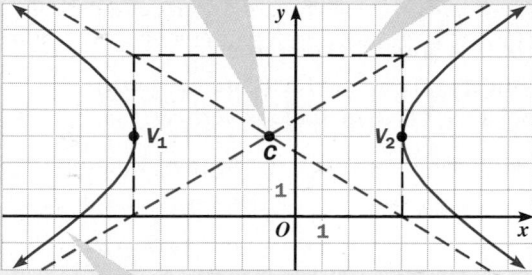

Step 3 Sketch the asymptotes

$$y = -\frac{3}{5}(x + 1) + 3$$

and $y = \dfrac{3}{5}(x + 1) + 3$,

which are the diagonals of the rectangle.

Step 4 Sketch the hyperbola through the vertices, extending toward the asymptotes.

Method 2 Use a graphing calculator or graphing software.

Solve the equation for y.

$$\frac{(x + 1)^2}{25} - \frac{(y - 3)^2}{9} = 1$$

$$\frac{(y - 3)^2}{9} = \frac{(x + 1)^2}{25} - 1$$

$$(y - 3)^2 = \frac{9(x + 1)^2}{25} - 9$$

$$y - 3 = \pm\sqrt{\frac{9(x + 1)^2}{25} - 9}$$

$$y = 3 \pm \sqrt{\frac{9(x + 1)^2}{25} - 9}$$

There are two equations to graph: $y = 3 + \sqrt{\dfrac{9(x + 1)^2}{25} - 9}$ represents the top

half of the hyperbola, and $y = 3 - \sqrt{\dfrac{9(x + 1)^2}{25} - 9}$ represents the bottom half of the hyperbola.

Enter the equations into a graphing calculator or graphing software.

Set an appropriate square window and graph the hyperbola.

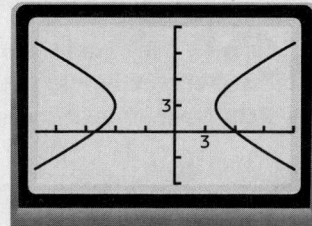

EXAMPLE 2

Write an equation of the hyperbola with center $(-2, 4)$, one vertex at $(-2, 1)$, and one focus at $(-2, 9)$.

SOLUTION

Step 1 Determine if the hyperbola is horizontal or vertical.
Since the x-coordinates of the center, vertex, and focus are the same, the major axis is parallel to the y-axis. The hyperbola is vertical.

Step 2 Find a.	**Step 3** Find c.	**Step 4** Find b.
$a = 4 - 1 = 3$	$c = 9 - 4 = 5$	$5^2 = 3^2 + b^2$
		$b^2 = 25 - 9 = 16$
		$b = 4$

The distance between the center and a vertex is a.

The distance between the center and a focus is c.

Use the fact that $c^2 = a^2 + b^2$.

Step 5 Substitute the values for h, k, a, and b into the standard form of an equation of a vertical hyperbola.

$$\frac{(y - k)^2}{a^2} - \frac{(x - h)^2}{b^2} = 1$$

$$\frac{(y - 4)^2}{9} - \frac{(x + 2)^2}{16} = 1$$

Substitute -2 for h, 4 for k, 3 for a, and 4 for b.

✓ CHECKING KEY CONCEPTS

Use the hyperbola shown to answer Questions 1–3.

1. What are the coordinates of the center? the foci?

2. Find equations of the asymptotes.

3. Write an equation of the hyperbola.

4. Graph $\dfrac{(x - 2)^2}{4} - \dfrac{(y + 1)^2}{25} = 1$ and $\dfrac{(x - 2)^2}{25} - \dfrac{(y + 1)^2}{4} = 1$. How are the graphs alike? How are they different?

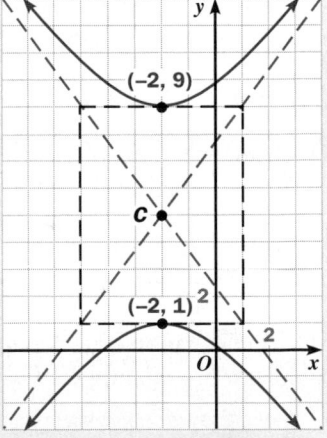

Identify the center, the vertices, and the foci of each hyperbola with the given equation. Tell whether the hyperbola is *horizontal* or *vertical*. Find equations of the asymptotes. Then graph the equation.

5. $\dfrac{(x - 6)^2}{16} - \dfrac{(y - 4)^2}{49} = 1$

6. $\dfrac{(x + 1)^2}{64} - \dfrac{(y - 3)^2}{4} = 1$

11.5 | **Exercises and Applications**

Extra Practice exercises on page 767

Identify the center, the vertices, and the foci of each hyperbola with the given equation. Tell whether the hyperbola is *horizontal* or *vertical*. Find equations of the asymptotes. Then graph the equation.

1. $\dfrac{x^2}{36} - \dfrac{y^2}{25} = 1$

2. $\dfrac{y^2}{4} - \dfrac{x^2}{49} = 1$

3. $\dfrac{x^2}{81} - \dfrac{(y-3)^2}{9} = 1$

4. $\dfrac{(x-1)^2}{16} - \dfrac{(y+3)^2}{64} = 1$

5. $\dfrac{(y+7)^2}{25} - \dfrac{(x-5)^2}{100} = 1$

6. $\dfrac{(x-2)^2}{81} - y^2 = 1$

7. $\dfrac{y^2}{3^2} - \dfrac{(x+4)^2}{7^2} = 1$

8. $4x^2 - 25y^2 = 100$

9. $\dfrac{y^2}{36} - \dfrac{x^2}{25} = 3$

10. Investigation Use a piece of bifocal paper. Let the centers of each set of concentric circles represent the foci, F_1 and F_2, of a hyperbola.

    a. Plot points on the paper that satisfy the equation $|F_1P - F_2P| = 2$. Then connect the points to sketch the curve.

    b. Use a different color pen to plot points on the same paper that satisfy the equation $|F_1P - F_2P| = 4$. Then connect the points to sketch the curve.

    c. How does the shape of the hyperbola change as you increase the value of $|F_1P - F_2P|$?

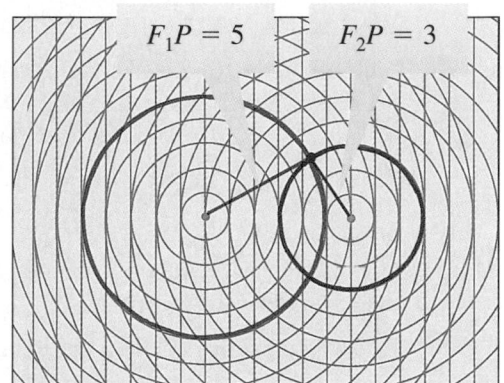

Match each graph with its equation.

11. $\dfrac{(x+3)^2}{9} - \dfrac{(y-1)^2}{25} = 1$

12. $\dfrac{(y+3)^2}{9} - \dfrac{(x-1)^2}{25} = 1$

13. $\dfrac{(x+3)^2}{25} - \dfrac{(y-1)^2}{9} = 1$

14. $\dfrac{(y+3)^2}{25} - \dfrac{(x-1)^2}{9} = 1$

A.

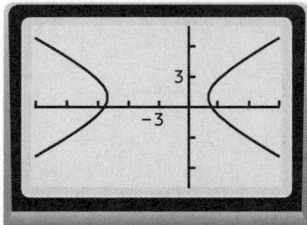

B.

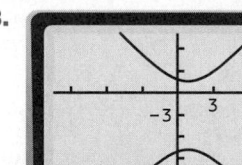

C.

D.

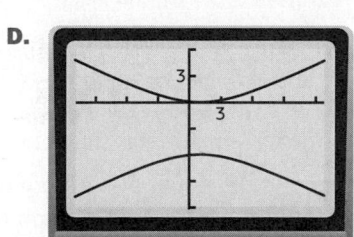

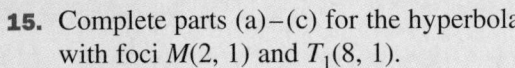

Connection ▸ NAVIGATION

Hyperbolic navigation allows ships to locate their position using radio transmitters. A master transmitter, M, and two secondary transmitters, T_1 and T_2, send signals to a receiver on the ship. The time it takes for the ship to receive each signal determines the ship's position along two hyperbolas. The intersection of the two hyperbolas gives the coordinates of the ship's location. All three transmitters are located at foci of the hyperbolas.

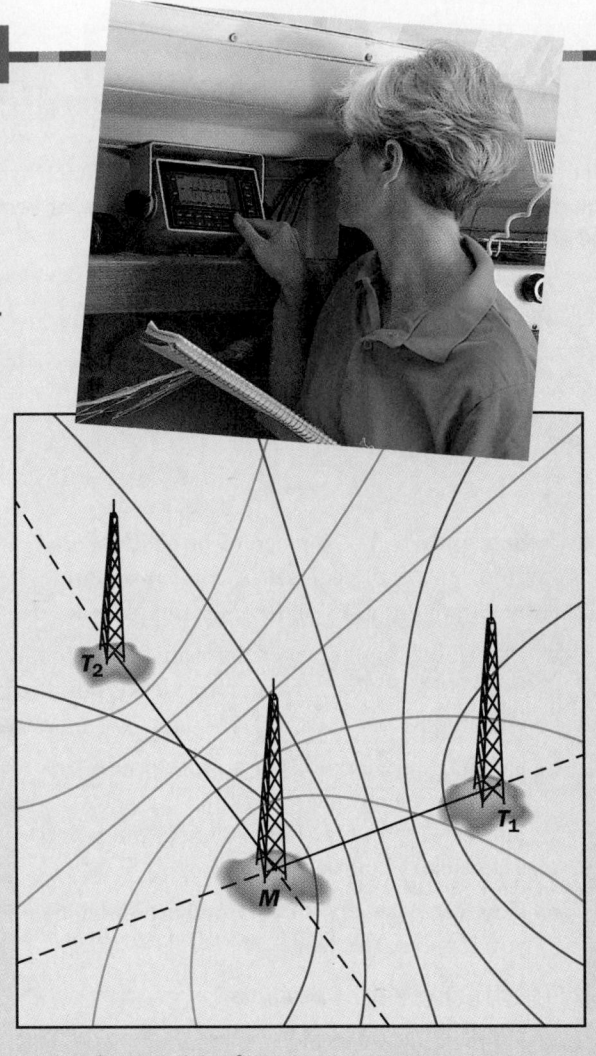

15. Complete parts (a)–(c) for the hyperbola with foci $M(2, 1)$ and $T_1(8, 1)$.

    a. What is the center of the hyperbola?

    b. Suppose the difference in the distances between a ship and the transmitters at M and T_1 is 2. (This means that the ship received a signal from one of the transmitters 2 units of time sooner than it received a signal from the other transmitter.) Find the coordinates of the vertices of the hyperbola.

    c. Write an equation of the hyperbola.

16. Repeat Exercise 15 for the hyperbola with foci M and $T_2(2, 9)$. In part (b), suppose that the difference in the distances between the ship and the transmitters at M and T_2 is 6.

17.   **Technology** Graph the equations you wrote in part (c) of Exercises 15 and 16. The ship is located at the point where the branch of the hyperbola in Exercise 15 that is closer to T_1 intersects the branch of the hyperbola in Exercise 16 that is closer to T_2. Find the coordinates of the ship.

18. Writing How is the hyperbola with equation $\dfrac{x^2}{a^2} - \dfrac{y^2}{b^2} = 1$ like the hyperbola with equation $\dfrac{y^2}{a^2} - \dfrac{x^2}{b^2} = 1$? How are they different? How do equations of the asymptotes of each graph compare?

19. Cooperative Learning Work in a group of three students.

    a. Graph each equation.

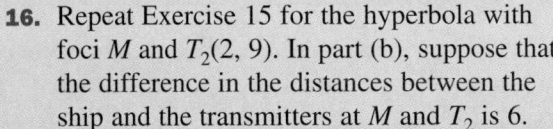

$$\frac{x^2}{4} - \frac{y^2}{4} = 1 \qquad\qquad \frac{x^2}{9} - \frac{y^2}{4} = 1 \qquad\qquad \frac{x^2}{16} - \frac{y^2}{4} = 1$$

    b. What happens to the graph of each equation as the value of a increases?

    c. What do you think will happen to the graph of each equation as the value of a stays the same and the value of b increases? Graph three equations with the same value for a and different values for b to test your prediction.

20. a. Graph the equation $xy = 4$.

    b. What is the center of the graph? What are equations of the asymptotes?

    c. **Writing** How is the hyperbola with equation $xy = 4$ like the other hyperbolas in this section? How is it different?

Write an equation for each hyperbola with the given characteristics. Then sketch its graph.

21. asymptotes: $y = \pm \dfrac{4}{3}x$; focus $(5, 0)$

22. asymptotes: $y = \pm \dfrac{5}{12}x$; focus $(0, 13)$

23. center $(0, 0)$; vertex $(6, 0)$; focus $(10, 0)$

24. center $(-5, 2)$; vertex $(-5, -1)$; focus $(-5, 7)$

25. center $(1, -4)$; vertex $(7, -4)$; focus $(-9, -4)$

26. a. Show that $\dfrac{x^2}{a^2} - \dfrac{y^2}{b^2} = 1$ can be rewritten as $y = \pm \dfrac{b}{a}\sqrt{x^2 - a^2}$.

    b. Explain why $\pm \dfrac{b}{a}\sqrt{x^2 - a^2} \approx \pm \dfrac{b}{a}x$ for large positive values of x.

    c. How does your answer to part (b) explain what happens to the graph of $\dfrac{x^2}{a^2} - \dfrac{y^2}{b^2} = 1$ as $x \to \infty$?

27. Challenge Suppose that the foci of a hyperbola are $F_1(-5, 0)$ and $F_2(5, 0)$ and the difference in the distances between any point $P(x, y)$ and these foci is 8. Derive an equation of the form $\dfrac{x^2}{a^2} - \dfrac{y^2}{b^2} = 1$ to describe this hyperbola.

ONGOING ASSESSMENT

28. Writing Graph the equations $\dfrac{x^2}{16} - \dfrac{y^2}{9} = 1$ and $\dfrac{x^2}{16} + \dfrac{y^2}{9} = 1$ in the same coordinate plane. How are the graphs alike? How are they different?

SPIRAL REVIEW

Write an equation of the ellipse with the given characteristics. Then sketch its graph. *(Section 11.4)*

29. center $(-6, -1)$; $a = 4$; $b = 1$; major axis is horizontal

30. center $(1, 2)$; vertex $(1, -4)$; focus $(1, 7)$

31. center $(2, -3)$; vertex $(15, -3)$; focus $(-10, -3)$

Write the equation of each function in vertex form and in intercept form.
(Sections 5.4 and 5.5)

32. $y = x^2 - 8x - 20$ **33.** $y = x^2 - 12x + 32$ **34.** $y = 2x^2 - 5x - 3$

11.6 Identifying Conics

Consider two cones placed together so that they have a common vertex and axis of symmetry. If you cut across the cones with a plane at different places and different angles, you can produce all of the curves that you have studied in this chapter. The name **conic section**, or simply *conic*, applies to these curves because they are all slices of a double cone.

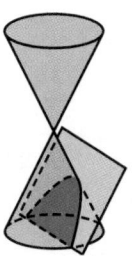 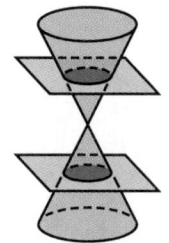 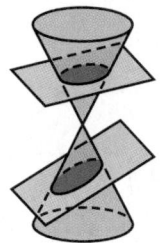 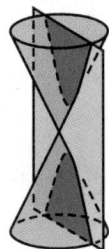

THINK AND COMMUNICATE

1. For each plane, use the diagrams above to tell what conic you get when the plane intersects the double cone and does *not* pass through the vertex. Also tell what conic you get when the plane *does* pass through the vertex.

   a. a horizontal plane   **b.** a vertical plane

The general equation for all conics is $Ax^2 + Bxy + Cy^2 + Dx + Ey + F = 0$. For any conic that has an axis of symmetry parallel to the *x*-axis or the *y*-axis, $B = 0$. In this section, you will work only with cases where $B = 0$.

EXAMPLE 1

For each equation of a conic, rewrite the equation in standard form, identify the conic, and state its important characteristics.

a. $x^2 + y^2 + 6x - 4y - 12 = 0$ **b.** $x^2 - y^2 + 6x + 8y - 32 = 0$

SOLUTION

a. Complete the square in x and y.

$$x^2 + y^2 + 6x - 4y - 12 = 0$$
$$(x^2 + 6x + \underline{\ ?\ }) + (y^2 - 4y + \underline{\ ?\ }) = 12 + \underline{\ ?\ } + \underline{\ ?\ }$$
$$(x^2 + 6x + 9) + (y^2 - 4y + 4) = 12 + 9 + 4$$
$$(x + 3)^2 + (y - 2)^2 = 25$$

This is an equation of the circle with center $C(-3, 2)$ and radius 5.

b. Complete the square in x and y.

$$x^2 - y^2 + 6x + 8y - 32 = 0$$
$$(x^2 + 6x + \underline{\ ?\ }) - (y^2 - 8y + \underline{\ ?\ }) = 32 + \underline{\ ?\ } - \underline{\ ?\ }$$
$$(x^2 + 6x + 9) - (y^2 - 8y + 16) = 32 + 9 - 16$$
$$(x + 3)^2 - (y - 4)^2 = 25$$
$$\frac{(x + 3)^2}{25} - \frac{(y - 4)^2}{25} = 1$$

This is an equation of the horizontal hyperbola with center $C(-3, 4)$, vertices $V_1(-8, 4)$ and $V_2(2, 4)$, and asymptotes $y = \pm 1(x + 3) + 4$.

You can use the coefficients of the x^2- and y^2-terms in the general equation of a conic to identify the conic without completing the square.

Identifying Conics

Provided $B = 0$, the graph of $Ax^2 + Bxy + Cy^2 + Dx + Ey + F = 0$ is:
- a parabola if $A = 0$ or $C = 0$, but not both
- a circle if $A = C$ and neither A nor C is 0
- an ellipse if A and C have the same sign ($AC > 0$), and $A \neq C$
- a hyperbola if A and C have opposite signs ($AC < 0$)

Degenerate Conics

The graph of the equation

$$Ax^2 + Bxy + Cy^2 + Dx + Ey + F = 0$$

can be the graph of either a conic section or a *degenerate conic*. A **degenerate conic** occurs when a plane intersects a double cone at its vertex.

Conic	Degenerate case
Ellipse or circle	Point
Parabola	Line
Hyperbola	Pair of intersecting lines

EXAMPLE 2

Identify and graph the degenerate conic $4x^2 - y^2 = 0$.

SOLUTION

$$4x^2 - y^2 = 0$$
$$(2x - y)(2x + y) = 0$$
$$2x - y = 0 \quad \text{or} \quad 2x + y = 0$$
$$y = 2x \quad \text{or} \quad y = -2x$$

The graphs of these equations are intersecting lines. They represent a degenerate hyperbola.

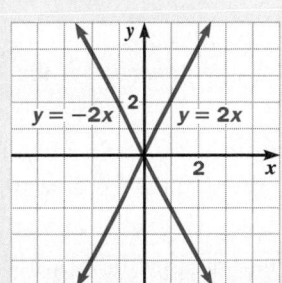

For each equation of a conic, rewrite the equation in standard form, identify the conic, and state its important characteristics.

1. $x^2 + y^2 - 6x + y + 3 = 0$ **2.** $12x^2 + 6y^2 - 36 = 0$

3. Identify and graph the degenerate conic $4x^2 + y^2 = 0$.

11.6   Exercises and Applications

Extra Practice exercises on page 767

For each equation of a conic, rewrite the equation in standard form, identify the conic, and state its important characteristics.

1. $9x^2 - 4y^2 + 90x - 16y + 173 = 0$ **2.** $4x^2 + 8y^2 - 8x - 32 = 0$

3. $-6x^2 - 12x + y - 6 = 0$ **4.** $9x^2 + y^2 - 18x + 6y - 12 = 0$

5. $4y^2 + 4x^2 + 12y - 16 = 0$ **6.** $y^2 - 4x^2 - 8x - 18y + 17 = 0$

7. **LANGUAGE ARTS** The slant angle s of a cone and the tilt angle t of a plane that cuts the cone determine which conic is created.

  a. If t (where $0° < t < 90°$) "falls short" of s (that is, $t < s$), an ellipse is created. What is created when t equals s? What is created when t exceeds s?

  b. **Research** Look up the definitions of "hyperbole," "parable," and "ellipsis." Describe how these words relate to exceeding, equaling, or falling short of reality.

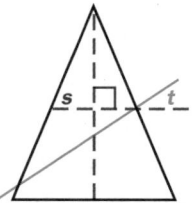

Connection ▶ AVIATION

Major Charles E. Yeager, USAF pilot, was the first person to fly faster than the speed of sound.

When an aircraft flies at the speed of sound or faster, it leaves behind a conical pressure wave, or shock wave. This wave causes a loud "sonic boom" that is heard simultaneously at all points of intersection of the cone with the ground.

8. Consider an aircraft flying parallel to the ground and faster than the speed of sound. What conic section is generated when the shock wave intersects the ground?

9. What conic section(s) can be generated if the plane is ascending?

10. Suppose the sonic boom is heard along a circle on the ground. In what direction is the aircraft traveling?

11. Is it possible for the sonic boom to be heard along a straight line? Explain.

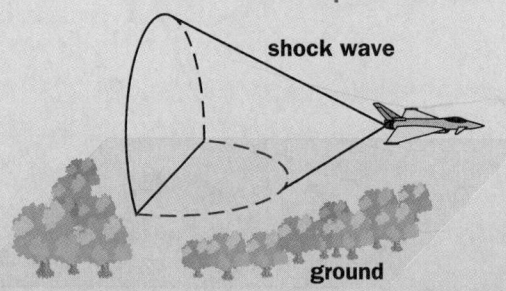

shock wave

ground

Identify and graph each equation of a degenerate conic.

12. $x^2 + y^2 = 0$ **13.** $(x + y)^2 = 0$ **14.** $-3x^2 + y^2 - 30x - 6y - 66 = 0$

15. $x^2 + 3y^2 - 4x + 4 = 0$ **16.** $x^2 - 9y^2 = 0$ **17.** $-x^2 + 4y^2 - 2x - 8y + 3 = 0$

ONGOING ASSESSMENT

18. Cooperative Learning Work with a partner. Hold a flashlight perpendicular to a wall and observe the shape of the light. Change the angle of the flashlight and describe the new shape of the light. What other conics can you create with a flashlight?

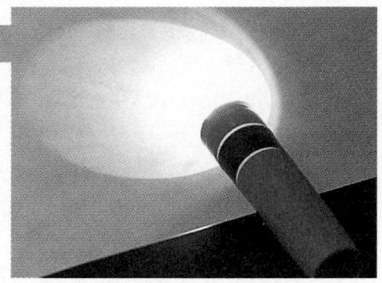

SPIRAL REVIEW

Identify the center, the vertices, and the foci of each hyperbola with the given equation. Tell whether the hyperbola is _horizontal_ or _vertical_. Find equations of the asymptotes. Then graph the equation. *(Section 11.5)*

19. $\dfrac{(x + 1)^2}{49} - \dfrac{(y - 3)^2}{25} = 1$ **20.** $\dfrac{y^2}{16} - \dfrac{(x + 2)^2}{36} = 1$

Graph each system of inequalities. *(Section 7.5)*

21. $y \ge 2x$
$y \le -3x - 1$

22. $y \le 3x + 2$
$y \ge -2x + 7$

23. $y < -0.5x + 1$
$2y \le -x + 6$

ASSESS YOUR PROGRESS

VOCABULARY

hyperbola (p. 543) **major axis** (p. 543)
foci (p. 543) **minor axis** (p. 543)
center (p. 543) **conic section** (p. 550)
vertices (p. 543) **degenerate conic** (p. 551)

Identify the center, the vertices, and the foci of each hyperbola with the given equation. Tell whether the hyperbola is _horizontal_ or _vertical_. Find equations of the asymptotes. Then graph the equation. *(Section 11.5)*

1. $\dfrac{x^2}{81} - \dfrac{(y - 5)^2}{36} = 1$ **2.** $\dfrac{(y - 2)^2}{25} - \dfrac{(x + 2)^2}{4} = 1$

For each equation of a conic, rewrite the equation in standard form, identify the conic, and state its important characteristics. *(Section 11.6)*

3. $x^2 + y^2 + 14x - 8y + 1 = 0$ **4.** $x^2 - y^2 + 4x + 10y - 30 = 0$

Identify and graph each equation of a degenerate conic. *(Section 11.6)*

5. $x^2 + 36y^2 = 0$ **6.** $x^2 - 36y^2 = 0$

7. Journal Compare the conic sections from this chapter. Describe any similarities you notice in the graphs or equations. Which conics do you think are most closely related? Why?

Designing a Logo

Do you recognize the logo shown? It's the symbol for the Olympic Games, which include thousands of participants from around the world.

The five rings symbolize Africa, Asia, Australia, Europe, and the Americas.

The interlocking of the rings symbolizes the meeting of athletes from all over the world.

At least one of the colors of the Olympic logo appears in the flag of every nation.

A *logo* is a symbol that visually represents an organization, such as a club, a charity, or a company. A good logo will often suggest or explain something about the organization it represents. In this project, you will design a logo for an organization that interests you.

PROJECT GOAL   **Use a graphing calculator or graphing software to design a logo composed of conic sections.**

Drawing with a Graphing Calculator

Paloma is designing a logo for her mother's company, *Orbital Transmissions*, which sells satellite dishes. For her design, Paloma decides to feature a planet with an orbiting satellite. These are the steps she takes to make her logo.

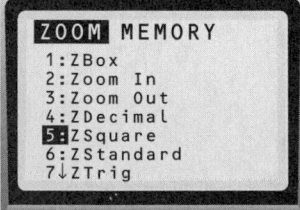

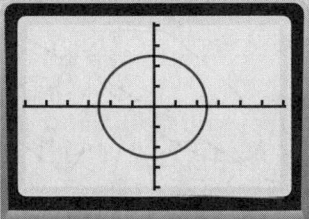

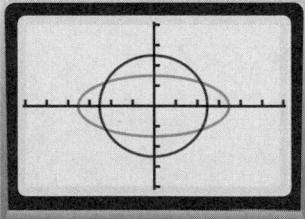

1. **USE** a square window so that circles are not distorted.

2. **ENTER** equations to draw a circle, leaving enough room for the rest of the drawing.

3. **ENTER** equations to draw an ellipse with a minor axis that is shorter than the radius of the circle and a major axis that is longer.

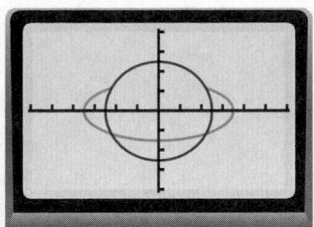

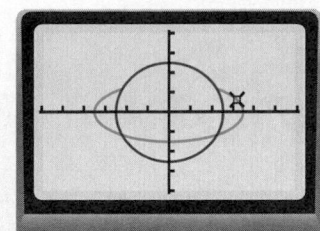

4. RESTRICT the domain of the equation for the upper half of the ellipse so that the ellipse appears to go behind the circle.

5. PLOT a point on the ellipse to give the appearance of an orbiting satellite.

Making Your Own Logo

Describe the organization for which you will design a logo. Then use a graphing calculator or graphing software to create the logo using conic sections. Include points and lines if you wish. Your logo should meet these criteria:

• The logo should be simple and easy to read.

• It should be distinctive and easily understood.

• It should feature information about the organization's functions, services, or products.

Transfer your logo to a poster. Use color to highlight parts of your logo.

Presenting Your Design

Present your logo to the class. Explain what the parts of your logo represent and how you created your design. Show all the equations you used as well as any restrictions on the variables. Explain how your logo meets the criteria given above.

You can extend your project by exploring these ideas:

• Try using a graphing calculator or graphing software to duplicate an actual logo for which conic sections are part of the design.

• Make a list of places where simple visual symbols like the ones shown are used as a substitute for words.

• Try using a graphing calculator or graphing software to draw simple pictures (not necessarily logos) that include conic sections.

Self-Assessment
Describe any problems that you had while creating your logo. How has your understanding of conic sections improved as a result of doing this project?

11 Review

Without looking at the book or your notes, write a list of important concepts from this chapter. Then look through the chapter and your notes and compare them with your list. Did you miss anything?

VOCABULARY

parabola (p. 521)
focus, foci (pp. 521, 536, 543)
directrix (p. 521)
vertex, vertices (pp. 521, 536, 543)
circle (p. 528)
center (pp. 528, 536, 543)
radius (p. 528)

ellipse (p. 536)
major axis (pp. 536, 543)
minor axis (pp. 536, 543)
hyperbola (p. 543)
conic section (p. 550)
degenerate conic (p. 551)

SECTION 11.1

For any two points $A(x_1, y_1)$ and $B(x_2, y_2)$:
- the distance AB between the two points is $AB = \sqrt{(x_2 - x_1)^2 + (y_2 - y_1)^2}$
- the midpoint M of $\overline{AB}$ has coordinates $\left(\dfrac{x_1 + x_2}{2}, \dfrac{y_1 + y_2}{2} \right)$

For example, if A has coordinates $(3, 1)$ and B has coordinates $(-2, 5)$, then:

$$AB = \sqrt{(-2 - 3)^2 + (5 - 1)^2} \qquad M = \left(\frac{3 + (-2)}{2}, \frac{1 + 5}{2} \right)$$

$$= \sqrt{41}$$

$$\approx 6.4 \qquad\qquad\qquad = \left(\frac{1}{2}, 3 \right)$$

Two lines are parallel if and only if they have the same slope and different y-intercepts. Two lines are perpendicular if and only if the slope of one line is the negative reciprocal of the slope of the other line.

SECTIONS 11.2, 11.3, 11.4, *and* 11.5

A **parabola** with **vertex** $V(h, k)$ and a distance of $2c$ between its **focus** F and **directrix** d has equation $y - k = \dfrac{1}{4c}(x - h)^2$ if it opens up or down, and has equation $x - h = \dfrac{1}{4c}(y - k)^2$ if it opens left or right.

A **circle** with **center** (h, k) and **radius** r has equation $(x - h)^2 + (y - k)^2 = r^2$.

An equation of an **ellipse** with **major axis** of length $2a$, **minor axis** of length $2b$, and center $C(h, k)$ is written in one of two ways:

- A *horizontal* ellipse has equation $\dfrac{(x - h)^2}{a^2} + \dfrac{(y - k)^2}{b^2} = 1$.

- A *vertical* ellipse has equation $\dfrac{(x - h)^2}{b^2} + \dfrac{(y - k)^2}{a^2} = 1$.

Example: $\dfrac{(x + 1)^2}{25} + \dfrac{(y - 2)^2}{9} = 1$

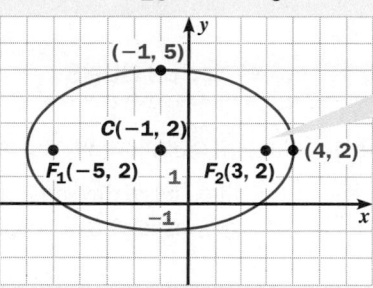

The foci lie on the major axis c units from the center. For an ellipse, $c = \sqrt{a^2 - b^2}$.

An equation of a **hyperbola** with major axis of length $2a$, minor axis of length $2b$, and center $C(h, k)$ is written in one of two ways:

- A *horizontal* hyperbola has equation $\dfrac{(x - h)^2}{a^2} - \dfrac{(y - k)^2}{b^2} = 1$ with asymptotes $y = \pm\dfrac{b}{a}(x - h) + k$.

- A *vertical* hyperbola has equation $\dfrac{(y - k)^2}{a^2} - \dfrac{(x - h)^2}{b^2} = 1$ with asymptotes $y = \pm\dfrac{a}{b}(x - h) + k$.

Example: $\dfrac{(y - 3)^2}{4} - \dfrac{(x - 5)^2}{9} = 1$

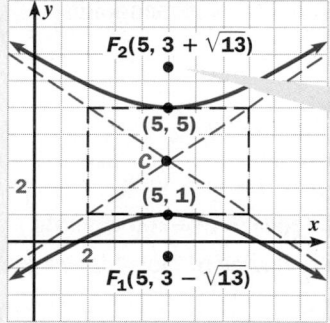

The foci lie on the major axis c units from the center. For a hyperbola, $c = \sqrt{a^2 + b^2}$.

SECTION 11.6

The general form of equations of all **conic sections**, including **degenerate conics**, is $Ax^2 + Bxy + Cy^2 + Dx + Ey + F = 0$. If $B = 0$, you can use the method of completing the square to write the equation in standard form. For example, you know the graph of $16x^2 - 25y^2 - 64x - 150y - 561 = 0$ is a hyperbola because the coefficients of x^2 and y^2 have opposite signs.

$$16x^2 - 25y^2 - 64x - 150y - 561 = 0$$

$$16(x^2 - 4x + \underline{?}) - 25(y^2 + 6y + \underline{?}) = 561 + \underline{?} + \underline{?}$$

Complete the square in x and y.

$$16(x^2 - 4x + 4) - 25(y^2 + 6y + 9) = 561 + 64 - 225$$

$$16(x - 2)^2 - 25(y + 3)^2 = 400$$

Add 64 and subtract 225 on both sides of the equation.

$$\dfrac{(x - 2)^2}{25} - \dfrac{(y + 3)^2}{16} = 1$$

This is an equation of the horizontal hyperbola with center $C(2, -3)$, vertices $V_1(-3, -3)$ and $V_2(7, -3)$, and asymptotes $y = \pm\dfrac{4}{5}(x - 2) - 3$.

Assessment

VOCABULARY QUESTIONS

For Questions 1–3, complete each paragraph.

1. When two lines have the same slope and different y-intercepts, the lines are ___?___ . Two lines are ___?___ to each other when their slopes are negative reciprocals.

2. A ___?___ consists of all points that are the same distance from a point F, called the ___?___ , and a line d, called the ___?___ .

3. The ___?___ of an ellipse is the line segment that contains the foci and has the two ___?___ as its endpoints.

SECTION 11.1

For each pair of points, find:

a. the distance between the points

b. the coordinates of the midpoint of the line segment connecting the points

4. $(1, 5), (2, 4)$ **5.** $(3, 2), (7, 5)$ **6.** $(-7, 0), (-3, 1)$

7. Given the point $P(-2, 6)$ and the line l with equation $y = \frac{3}{5}x - 4$, find an equation of the line through P and:

  a. parallel to l **b.** perpendicular to l

8. a. GEOMETRY If the endpoints of a diameter of circle P are $(-3, 4)$ and $(2, -2)$, find the coordinates of the center of P.

  b. What is the length of the diameter of circle P? What is the radius?

  c. Find an equation of the line that is tangent to the circle at $(-3, 4)$.

SECTIONS 11.2, 11.3, 11.4, and 11.5

9. Find an equation of the parabola with focus $F(4, -2)$ and directrix $d: x = -3$. Then graph the equation.

For each equation, identify the conic and state its important characteristics. Then graph the equation.

10. $(x + 2)^2 + (y - 5)^2 = 16$ **11.** $\dfrac{(x - 1)^2}{9} - \dfrac{(y - 6)^2}{16} = 1$

12. $\dfrac{(x + 2)^2}{36} + \dfrac{(y - 1)^2}{64} = 1$ **13.** $4(x + 1)^2 + 25(y + 7)^2 = 100$

14. **ASTRONOMY** Halley's comet travels around the sun in a path that is an ellipse. Assume that the center of the ellipse is placed at the origin of a coordinate plane and the foci are located on the x-axis. The distance between the vertices is 35.6 astronomical units (AU), and the distance between the foci is 34.4 AU.

    a. The sun is located at one of the foci of the ellipse. What are the possible coordinates of the sun on this graph?

    b. Write an equation of the ellipse that models the path of the comet. Then graph the equation.

15. **Writing** How is the equation of a horizontal hyperbola different from the equation of a vertical hyperbola? How are the equations alike?

Write an equation of each conic with the given characteristics.

16. a parabola with focus $(3, 2)$ and directrix $y = 6$

17. a circle with center $(0, 8)$ and radius 2

Write an equation of each graph.

18.

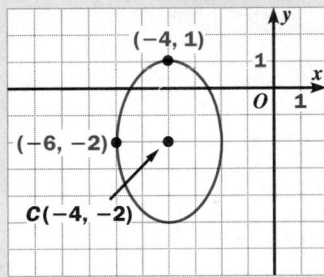

19.

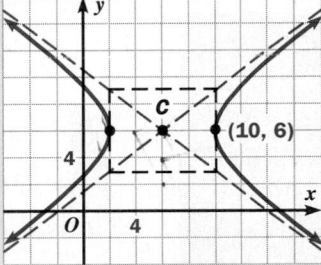

SECTION 11.6

For each equation of a conic, rewrite the equation in standard form and identify the conic.

20. $y^2 - 6y + 16x + 25 = 0$

21. $x^2 - 4y^2 - 2x + 16y - 19 = 0$

22. $x^2 + y^2 - 12y + 25 = 0$

23. $x^2 + 25y^2 - 6x - 100y + 84 = 0$

Identify and graph each equation of a degenerate conic.

24. $25x^2 - y^2 = 0$

25. $x^2 + y^2 + 2x + 2y + 2 = 0$

PERFORMANCE TASK

26. Make a poster to display the conic sections from this chapter. Highlight the important characteristics of each conic and give its equation. Describe the similarities and differences of the equations of the conics and of their graphs.

12 | Discrete Mathematics

Delivering POWER

Maria Rodriguez

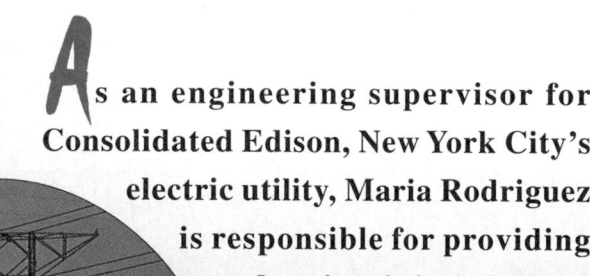

As an engineering supervisor for Consolidated Edison, New York City's electric utility, Maria Rodriguez is responsible for providing and maintaining all the power lines coming into the borough of Queens. The major facilities supplied include John F. Kennedy Airport, LaGuardia Airport, *The New York Times*, Riker's Island prison, hospitals, schools, factories, apartment buildings, and homes. "My main concern is that no one loses power," Rodriguez says. "If for some reason the lights go out, it's my job to get them back on as quickly as possible."

> **"My main concern is that no one loses power."**

"We study the entire system, looking for the weakest link in the chain."

Becoming an Engineer

Rodriguez has traveled far to reach a position of such great responsibility. Born in England and raised in Spain, she moved to the United States in 1980 and studied English for a year. Fluent in both Spanish and French, she had planned to become a translator until she entered college, when she discovered engineering.

After graduating from Columbia University in 1987 with a degree in electrical engineering, she participated in a management training program at Consolidated Edison. Five years and several promotions later, she was put in charge of delivering electric power to Queens.

Determining Peak Capacity

To ensure reliable electrical service, she has to anticipate peak power demands a year or more in advance. "We study the entire system, looking for the weakest link in the chain," she says. "Then we try to determine whether that link will be capable of withstanding the projected electrical load."

Too much electric current can overload power lines. "Pipes will burst if you try to make them carry too much water. Electrical equipment fails, too, when pushed beyond the limit," she explains.

Distributing Electricity

Electricity reaching the area stations in Queens is generated at a power plant and transmitted through high-voltage cables to switching stations, where the voltage is "stepped down," or reduced. The electricity then goes to area stations, where the voltage is reduced again. The area stations feed a network of transformers that distribute electricity to residential and commercial customers.

An Easy Way to Count

If Rodriguez needs to count the number of transformers branching out from the area stations in the diagram shown, she can use the *multiplication counting principle* to simplify her calculations. First, she counts the lines that leave the power plant. Then she counts the lines that leave each switching station. Next, she counts the lines that leave each area station. Finally, she multiplies these numbers together to determine the total number of transformers. This knowledge, in turn, helps her decide whether the system has the capacity to meet the growing need for electricity.

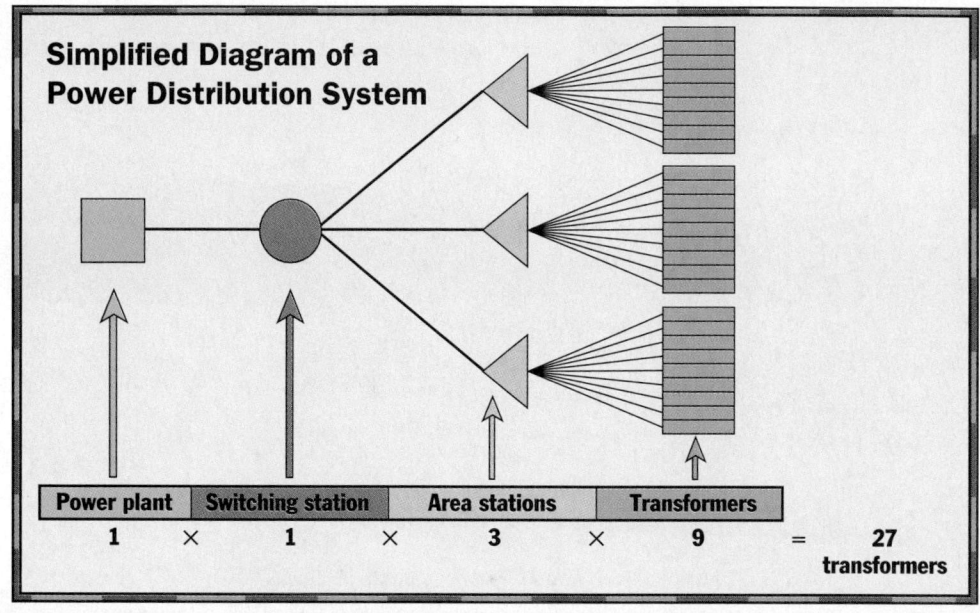

Simplified Diagram of a Power Distribution System

Power plant		Switching station		Area stations		Transformers		
1	×	1	×	3	×	9	=	27 transformers

Mathematics & Maria Rodriguez

In this chapter, you will learn more about how mathematics is used in electrical engineering.

Related Exercises

Section 12.2
• **Exercises 11–15**

Section 12.3
• **Exercises 12–18**

EXPLORE AND CONNECT

Maria Rodriguez stands at the Astoria Generating Station in Queens, New York.

1. Writing Suppose a second switching station is connected to the power plant in the diagram shown above. If the second switching station has the same subsequent connections as the first, how many transformers would there be? Explain.

2. Research Contact your local utility company to find out the cost of one kilowatt-hour of electricity for your school and for your home. Are these costs different? If so, what is the reason for this cost difference?

3. Project Determine the cost for using some household appliance during one year. Find out the wattage of the appliance and divide that figure by 1000 to convert it to kilowatts. Multiply your answer by the hours of usage for a year. Use the cost per kilowatt-hour from Question 2 to figure out the total cost for one year.

12.1

Coloring a Graph

Learn how to...

- **represent situations with graphs**
- **color graphs, trying to use as few colors as possible**

So you can...

- **solve problems that require sorting items into groups, such as coloring maps, scheduling exams, and planning aquariums**

Cartographers (mapmakers) are careful to use different colors to distinguish adjacent countries on a map. However, each additional color adds to the cost of printing the map. How can a cartographer color a map using as few colors as possible?

EXPLORATION
COOPERATIVE LEARNING

Investigating How to Color a Map

Work with a partner. You will need:
- colored pens or pencils

1 Trace or sketch the map of South America.

2 Using as few colors as possible, color the map so that no two countries sharing a border have the same color. How many colors do you need?

3 Compare your results with those of others in the class. Is there more than one way to group countries by color?

WATCH OUT!

In a graph, the position of each vertex does not matter. For example, these two graphs are equivalent:

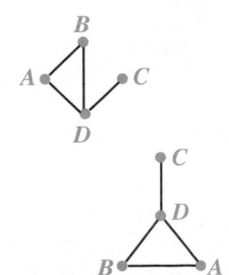

To simplify a map like the one of South America above, you can represent it with a **graph** consisting of **vertices** (plural of *vertex*) connected by **edges**.

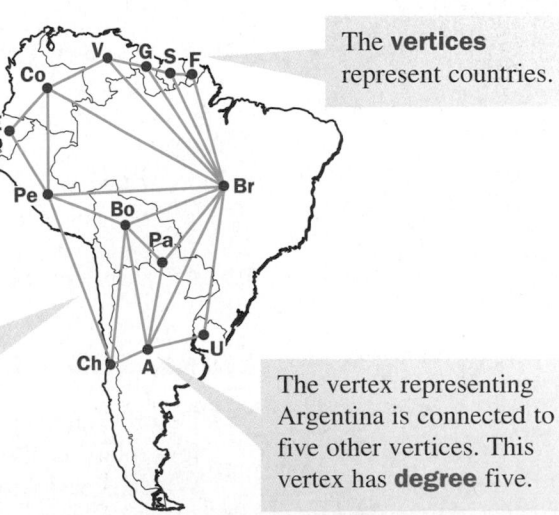

The **vertices** represent countries.

An **edge** connects two vertices if the corresponding countries share a border.

The vertex representing Argentina is connected to five other vertices. This vertex has **degree** five.

EXAMPLE 1   Application: Cartography

Color a map of South America. Try to use as few colors as possible.

SOLUTION

Use a graph to represent South America. Label each vertex with its degree.

Step 1 Choose a color, such as red, and apply it as shown.

First color **red** the vertex of highest degree (Brazil).

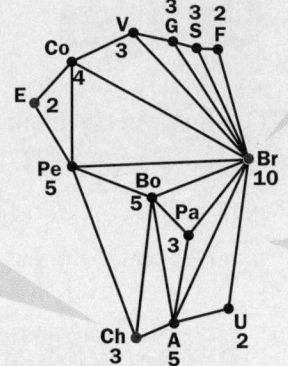

Then color **red** the vertex of next highest degree that does *not* share an edge with the first vertex.

Continue until all vertices are either red or adjacent to a red vertex.

Step 2 Choose another color, such as blue, and follow the procedure established in Step 1.

Color **blue** the uncolored vertex of highest degree. (Since there are three vertices of degree 5, choose one—say, Peru.)

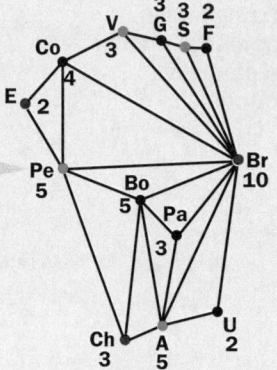

Step 3 Repeat using green and then yellow.

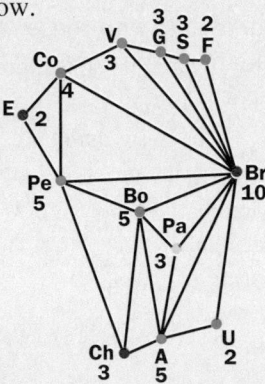

Step 4 Use the colored graph to color the map.

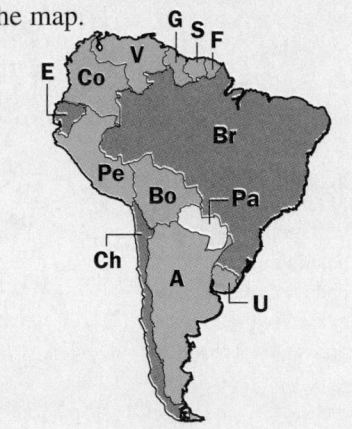

THINK AND COMMUNICATE

1. Describe another strategy for coloring a graph when you are trying to use as few colors as possible. Compare your strategy with the one in Example 1. Which is easier to use? Does one strategy use fewer colors than the other?

EXAMPLE 2   Application: Scheduling

Rosa Perez schedules final exams for Carver High School's summer school. The table lists the students in each class. Rosa Perez wants to ensure that no student is scheduled for more than one test at a time. How can she set up a testing schedule with the fewest time slots? Which exams can be given at the same time?

Algebra	Biology	Civics	Driver's Ed.	English	French
Marie-Ange	Jamarl	Ansu	Ansu	Eduardo	Eduardo
Marlon	James	Jamarl	Nancy	Marie-Ange	Marie-Ange
Mbuyi	Kiyana	James	Rachel	Parker	Melinda
Micah	Melinda	Marlon	Washington	Rachel	Micah
Tish	Wayne		Wayne	Tish	Washington

SOLUTION

Use a graph to represent the situation.

Represent each **class** by a **vertex**.

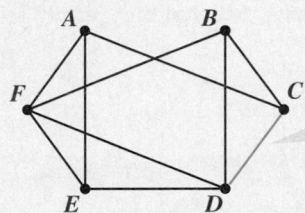

Draw **edges** between classes that **share one or more students**. The exams for these classes must *not* be given at the same time.

> **WATCH OUT!**
>
> Sometimes edges in a graph may cross each other. The intersections are not vertices.

Use a different color to represent each testing period. Use the procedure from Example 1 to color the graph.

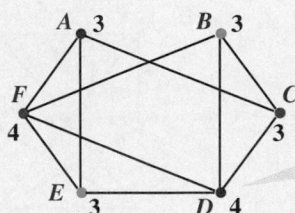

Two vertices have degree 4. Start by choosing one and coloring it **red**.

So, three testing times are needed: one for **Driver's Education** and **Algebra**, one for **French** and **Civics**, and one for **Biology** and **English**.

THINK AND COMMUNICATE

2. Explain why the two graphs at the right are equivalent.

3. Describe how to move the vertices of the graph in Example 2 so that no edges cross.

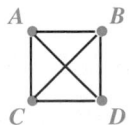
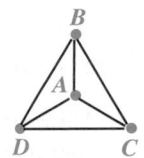

1. Which of the following graphs are equivalent? Explain your answer.

A. 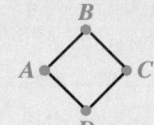 B. 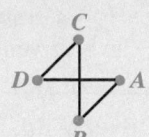 C. 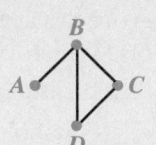 D.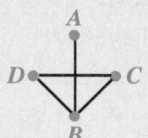

For Questions 2–4, look back at the graph on page 563. Give the degree of the vertex that represents each country.

2. Bolivia   3. Uruguay   4. Argentina

5. If you drew a graph like the one on page 563 for the United States, what would be the degree of your state?

6. A group of friends are making plans for the weekend. Tracy and Kim want to spend a day hiking; Tim wants to spend a day with his family; Tracy and Manuel want to spend a day working on a project for school; and Tim, Manuel, and Kim want to spend a day shopping.

   a. Explain how the graph at the left represents this situation.

   b. In two days, can everyone do what they want? How can you tell from the graph?

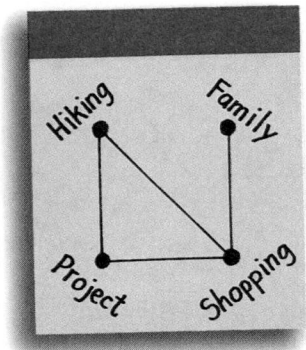

12.1 | **Exercises and Applications**

**Extra Practice
exercises on
page 767**

1. Which two graphs are equivalent? How do you know?

A. 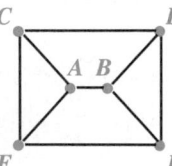 B. 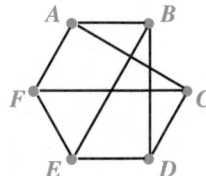 C.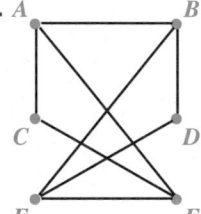

For each graph in Exercises 2–5:

a. **Find the degree of each vertex.**

b. **Draw an equivalent graph.**

c. **Color your graph so that vertices that share an edge are different colors.
   Try to use as few colors as possible.**

2. 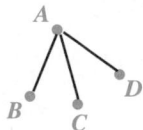   3. 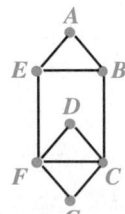   4. 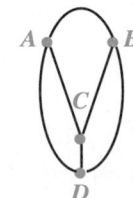   5.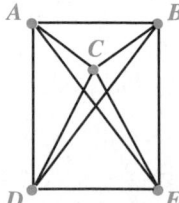

6. **a.** **CARTOGRAPHY** Use a graph to decide how to color the map at the right. Try to use as few colors as possible. Tasmania does not border any states or territories, so do not include it in your graph.

b. Trace or sketch the map of Australia and use your results from part (a) to color it. You can color Tasmania any color.

7. **Research** Suppose you represent a map of the United States with a graph as in Example 1 on page 564. Which vertex (or vertices) would have the highest degree?

8. **Writing** Look back at Example 2 on page 565.

a. If one student were enrolled in all six of the courses being offered, how many testing periods would be needed?

b. Explain why the graph at the right describes the situation in part (a). What happens if you try to use the method in Example 1 to color the graph?

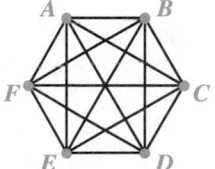

Connection ▶ **MARINE BIOLOGY**

Aquariums and pet stores must be careful when putting fish together In the same tank. For example, to minimize territorial disputes, a highly territorial species like the damselfish should be introduced to tanks after the other fish have had a chance to become established. Certain fish should never share a tank. The table below shows which fish are compatible and which are not.

9. **a.** Draw a graph in which each vertex represents a type of fish, and edges connect fish that are *incompatible*.

b. A graph is *planar* if it can be drawn without any edges crossing. Is your graph planar? Justify your answer.

10. Color your graph to decide how many tanks you need if you have all six types of fish named in the table. What grouping of fish does your graph represent?

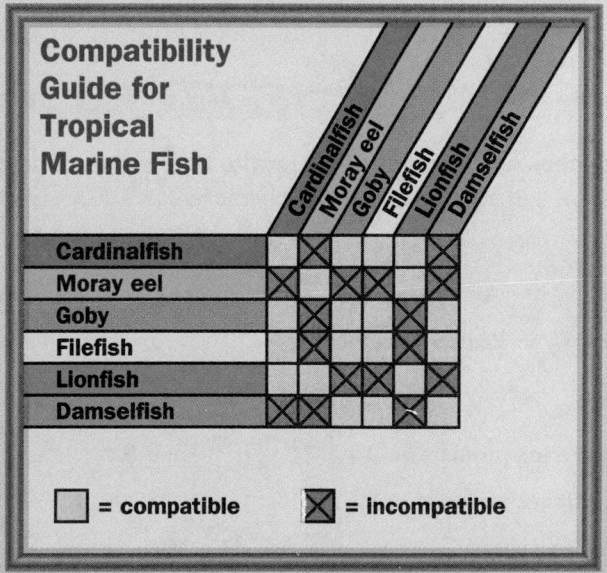

11. Your graph can be colored in several different ways. Give two more possible groupings of fish.

12. Find three types of fish that must be put into three different tanks. (*Hint:* Look for a triangle formed by three edges and three vertices in your graph.)

Compatibility Guide for Tropical Marine Fish

	Cardinalfish	Moray eel	Goby	Filefish	Lionfish	Damselfish
Cardinalfish		X	X	X		X
Moray eel	X		X		X	
Goby	X	X			X	
Filefish	X					X
Lionfish		X	X			X
Damselfish	X			X		

☐ = compatible   ☒ = incompatible

13. **Research** Look up the history of the *four-color problem*. Write a brief summary of your findings, including the origin of the problem and attempts to prove that all maps on a sphere can be colored with four or fewer colors.

14. **Open-ended Problem** Make up a map having six regions that can be colored using exactly:

   a. two colors   **b.** three colors   **c.** four colors

15. **Challenge** Make up a map that corresponds to the graph in Exercise 3. The vertices represent countries and the edges connect countries that share a border.

16. **CHEMISTRY** Many chemicals can be dangerous if allowed to come into contact with each other. For this reason, it is good practice to store incompatible chemicals in separate areas to avoid accidents. The table shows which chemicals in a high school lab are incompatible.

   a. Use a graph to decide how to store the chemicals. How many different storage areas are needed?

   b. If the lab did not have halogens and ethers, how many storage areas would be needed?

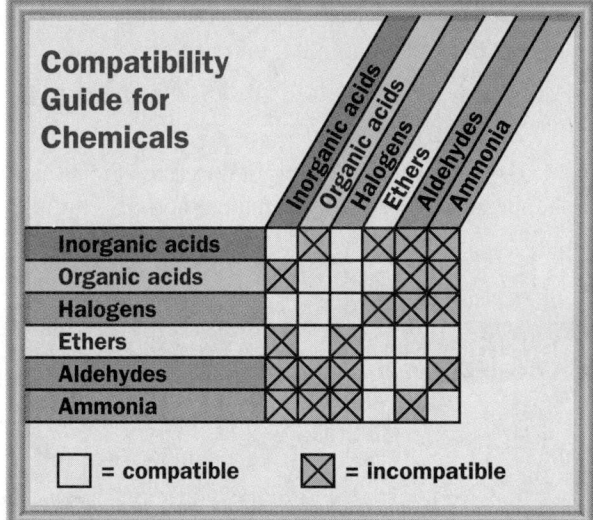

17. **Open-ended Problem** Choose a map showing several countries, states, or other subdivisions. Use a graph to color your map. Try to use as few colors as possible. If possible, show another way to color your map and explain how you found it.

SPIRAL REVIEW

For each equation of a conic, rewrite the equation in standard form, identify the conic, and state its important characteristics. *(Section 11.6)*

18. $y^2 + 2x^2 - 2y = 8$ **19.** $x^2 + y^2 + 3x = 4$

20. $2x^2 - 3y^2 + 6y = 9$ **21.** $y^2 + 10y + 2x = -24$

Solve. *(Section 8.3)*

22. $\sqrt{x} = 2$ **23.** $\sqrt{x - 3} = 6$ **24.** $\sqrt{4x - 1} = 5$

Find each product for $A = \begin{bmatrix} -1 & 0 & 2 \\ 3 & 5 & -4 \end{bmatrix}$ and $B = \begin{bmatrix} 2 & 1 \\ -1 & 0 \end{bmatrix}$. If the matrices cannot be multiplied, state that the product is *undefined*. *(Section 1.4)*

25. AB **26.** BA **27.** B^2

12.2 Directed Graphs and Matrices

Learn how to...

- **represent situations with directed graphs**
- **represent directed graphs with matrices**

So you can...

- **analyze situations involving direction, such as predator-prey relationships and electricity transfers**

A *food web* shows how energy, in the form of food, is transferred through an ecosystem. The illustration below shows a food web for polar seas.

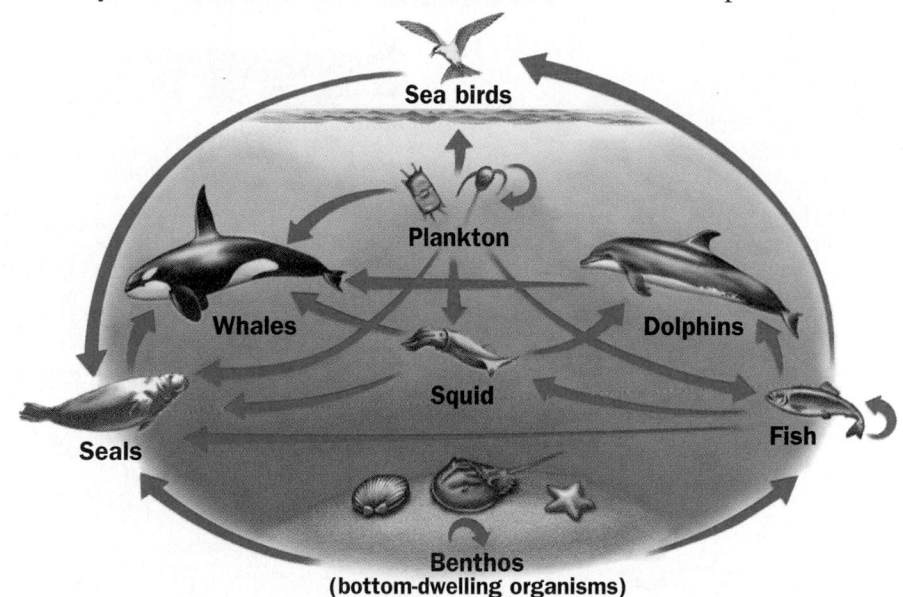

Sea birds

Plankton

Whales

Dolphins

Squid

Seals

Fish

Benthos
(bottom-dwelling organisms)

There are several kinds of whales in the polar seas. Bottlenose whales and sperm whales eat squid, killer whales eat dolphins and seals, and whalebone whales eat plankton.

THINK AND COMMUNICATE

1. In the food web above, what do the arrows represent?

2. Why is there an arrow from fish back to fish?

3. Which animals eat squid?

4. Seals eat benthos directly. They also eat benthos indirectly by eating fish that have eaten benthos. How else do seals eat benthos indirectly?

5. List the direct and indirect food sources for dolphins.

6. How would squid be affected if all of the plankton died?

7. Which animal is the most important food source for the rest of the animals? Explain.

You can represent the food web by a *directed graph* as shown at the right. A **directed graph** is a graph in which the **edges** are **arrows**.

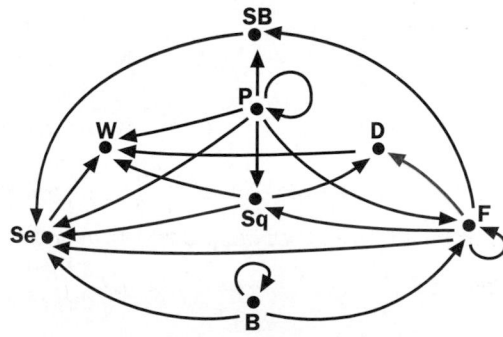

Earthworms eat decaying plant and animal matter in the soil. As they burrow through the ground, earthworms loosen and mix the soil, providing a better environment for plants.

EXAMPLE 1   Application: Ecology

Part of a forest ecosystem is described below. Draw a directed graph to represent the food web.

- Plants are eaten by insects, deer, earthworms, and songbirds.

- Insects are eaten by amphibians (frogs and toads), songbirds, raptors (hawks and owls), and each other.

- Amphibians are eaten by raptors.

- Earthworms are eaten by songbirds.

- Songbirds are eaten by raptors.

SOLUTION

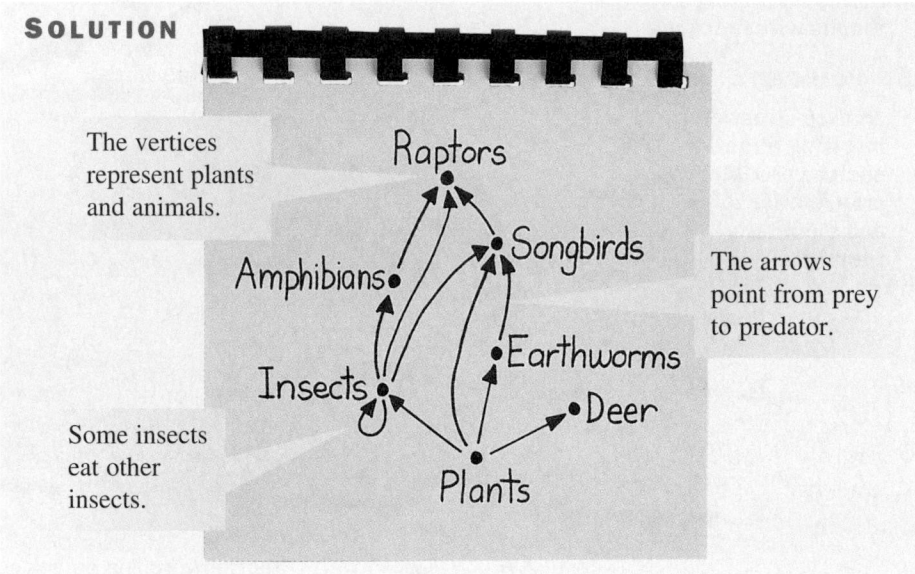

The vertices represent plants and animals.

The arrows point from prey to predator.

Some insects eat other insects.

THINK AND COMMUNICATE

8. In Example 1, are there any plants or animals that are not direct or indirect food sources for raptors? If so, what are they?

9. Plants (P) are an indirect food source for raptors (R) in several ways. Two are shown at the right.

   a. Describe at least two more ways that plants are a food source for raptors through two intermediate sources.

   b. Describe three ways that insects are a food source for raptors through one intermediate source.

   c. Are there any ways that plants are a food source for raptors through three intermediate sources?

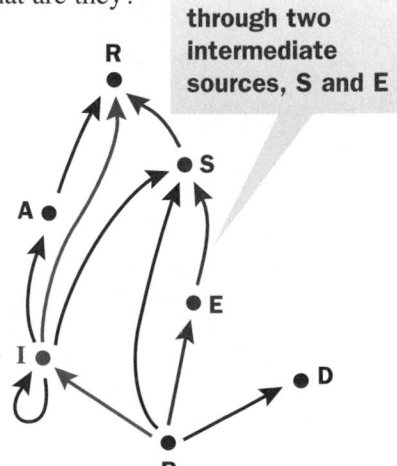

through two intermediate sources, S and E

through one intermediate source

If you represent a directed graph with a matrix, you can use a calculator or a computer to find direct and indirect relationships between vertices.

EXAMPLE 2

Represent the directed graph of the food web in Example 1 with a matrix.

SOLUTION

Let the rows and columns represent the vertices of the graph. Let an element of the matrix be 1 if the vertex represented by the row has an arrow going to the vertex represented by the column. Otherwise, let the element be 0.

Predator

There is an arrow from plants to deer.

There is no arrow from deer to plants.

Prey

$$
M = \begin{array}{c} \\ D \\ R \\ A \\ I \\ E \\ S \\ P \end{array}
\begin{array}{c} \begin{array}{ccccccc} D & R & A & I & E & S & P \end{array} \\
\left[\begin{array}{ccccccc}
0 & 0 & 0 & 0 & 0 & 0 & 0 \\
0 & 0 & 0 & 0 & 0 & 0 & 0 \\
0 & 1 & 0 & 0 & 0 & 0 & 0 \\
0 & 1 & 1 & 1 & 0 & 1 & 0 \\
0 & 0 & 0 & 0 & 0 & 1 & 0 \\
0 & 1 & 0 & 0 & 0 & 0 & 0 \\
1 & 0 & 0 & 1 & 1 & 1 & 0
\end{array} \right] \end{array} = M
$$

THINK AND COMMUNICATE

Questions 10 and 11 refer to both the matrix in Example 2 and the food web in Example 1.

10. In terms of the food web, what does it mean when an element of the matrix is 1? when an element is 0?

11. Find the sum of the elements in the last row of the matrix. What does the sum represent in terms of the food web? What does the sum of the elements in the second column represent?

The square of the matrix above tells you how many ways each plant or animal is a food source for the others through one intermediate source.

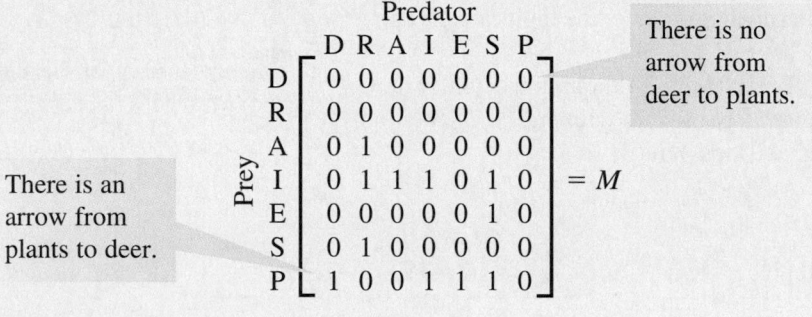

$M \times M = M^2$

Plants are eaten by **insects** and earthworms.

Insects and earthworms are eaten by **songbirds**.

$0 + 0 + 0 + 1 + 1 + 0 + 0 = 2$
Plants are eaten by songbirds indirectly in **two** ways, through **insects** and through earthworms.

EXAMPLE 3   Application: Ecology

Using the matrix in Example 2, find a matrix that gives the total number of ways each plant or animal is a food source for the others directly or through one intermediate source.

SOLUTION

Because M gives the number of ways that each plant or animal is a food source for the others directly, and M^2 gives the number of ways that each plant or animal is a food source for the others through one intermediate source, the totals are given by $M + M^2$.

Enter M into a graphing calculator or computer software with matrix calculation capabilities. Find $M + M^2$.

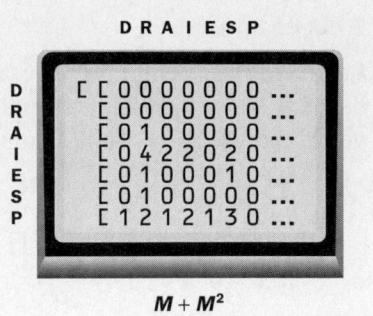

$M + M^2$

THINK AND COMMUNICATE

12. In how many ways are plants eaten by raptors directly or through one intermediate source? How can you tell from the matrix above? List the ways.

13. List the songbirds' food sources that are direct or through one intermediate source. How can you tell from the matrix above?

☑ CHECKING KEY CONCEPTS

Some of the mathematics courses offered at Fremont High School are listed below with their direct prerequisites.

Math C requires Math A.   **Math D requires Math A or B.**

Math E requires Math B, C, or D.   **Math F requires Math D or E.**

1. Which of the following directed graphs represents this situation?

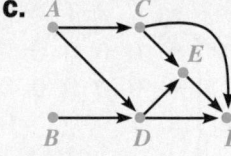

2. What do the arrows represent?

3. List the ways to take Math F with just two courses preceding it.

4. Write a matrix that represents the directed graph.

5. Square the matrix from Question 4. What do the elements represent?

12.2 Exercises and Applications

Extra Practice exercises on page 768

1. **MEDICINE** Six students at Ouray High School have come down with infectious mononucleosis in the past week. The school nurse, who suspects that the disease entered the school through a single student, is trying to trace the spread so that the disease can be contained. The nurse gathered the information at the right by interviewing each of the six students.

   a. Draw a directed graph representing possible transmissions from one student to another.

   b. List all possible pathways that the disease could have spread from Barbara to Katrina.

   c. Is it possible that the disease entered the school with only one of these six students and then spread to the others? Explain.

Student	Could have passed mononucleosis to:
Barbara	
Ezra	Ezra, Fatima, Ilana
Fatima	Fatima
Geoff	Katrina
Ilana	Ezra, Fatima
Katrina	Barbara, Fatima
	Ezra

Write a matrix representing each directed graph. Let an element be 1 if the row vertex has an edge directed to the column vertex. Otherwise, let the element be 0.

2.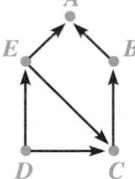

3.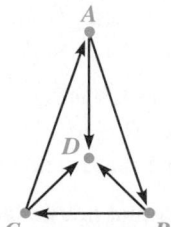

4.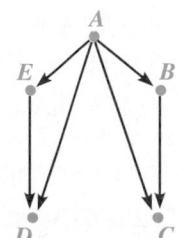

5.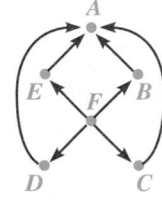

ECOLOGY The graph on page 570 represents part of a forest ecosystem. For Exercises 6–10, use the following additional information about the ecosystem.

- Plants are eaten by foxes, raccoons, and mice.
- Insects are eaten by snakes and mice.
- Snakes are eaten by raptors, foxes, and each other.
- Mice are eaten by raccoons, foxes, and snakes.
- Raccoons are eaten by foxes.

6. Using the information above, copy and extend the graph on page 570.

7. Represent the directed graph in Exercise 6 with a matrix.

8. **Technology** Find a matrix that gives the number of ways that each plant or animal is a food source for the others through one intermediate source.

9. **Technology** Find a matrix that gives the number of ways that each plant or animal is a food source for the others directly or through one intermediate source.

10. Which animal would be most affected if all of the mice died? Why? How can you tell from the matrix in Exercise 9?

Look back at the article on pages 560–562.

Electrical systems throughout the United States and Canada are connected together in large networks. During emergencies or times of high energy demand, one company may sell power to another company in the network. Rather than flowing directly from the source to the receiving station, the power divides and follows several paths of the network. Engineers such as Maria Rodriguez must monitor the transfer to make sure that no parts of the network are overloaded.

Suppose 100 megawatts (MW) of electricity must be sent from station A to station B within a network. Some constraints of the network are noted in the table.

Station	Can receive power from
B	A, C, D, E
C	E
D	A
E	D

11. Draw a directed graph that represents the possible paths of power transfer in the network.

12. If only 60 MW is transmitted directly from A to B, where does the additional 40 MW of electricity go initially?

13. If 10 MW flows from D to E and 5 MW flows from E to C, how much power flows from C to B, D to B, and E to B? Label your graph with all of the power transfers. Assume that no power is lost.

14. What is the most indirect path that power takes to get from A to B? How much power takes this path?

15. In how many ways can power flow from A to B through one intermediate station? through two? through three?

For Exercises 16 and 17, refer to the food web and graph on page 569.

16. **Technology** Find a matrix that gives the number of ways each plant or animal is a food source for the others directly or through one intermediate source. Use a graphing calculator or computer software with matrix calculation capabilities.

17. **Writing** Some environmentalists have been concerned about the impact of whaling on marine ecosystems. Explain how you can use the matrix from Exercise 16 to study how the polar marine ecosystem would be affected if the whale population were significantly reduced.

18. **Research** Draw a directed graph that summarizes the *hydrologic* (water) cycle. Explain what the vertices and arrows represent. Make a matrix to represent the information in your graph.

19. a. Challenge Represent the map of one-way streets with a 9×9 matrix M. The rows and columns should represent street intersections. What do the elements represent?

b. Use a graphing calculator or computer software with matrix calculation capabilities to find M^2, M^3, and M^4. What do they represent?

c. Can you drive from the intersection of First and C to the intersection of First and A without leaving the area shown in the map? How can you use matrices to find out?

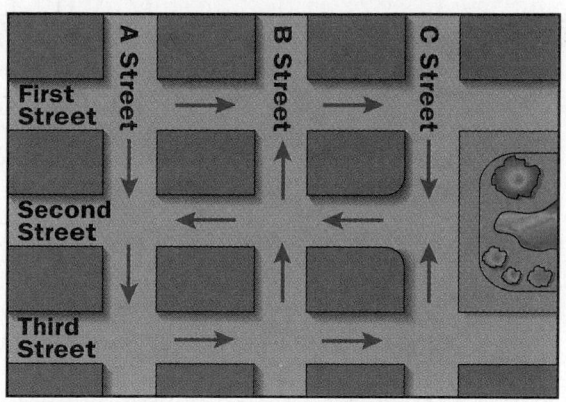

First Street / Second Street / Third Street / A Street / B Street / C Street

---

Connection ▶ SPORTS

To avoid ties when determining the regional champion for high school football, the Ohio High School Athletic Association rates Division I teams that play each other with a system similar to the following: Award the team one point for each win. Then add an additional point for each team that a defeated team has beaten.

BY THE WAY

Each school in the Ohio High School Athletic Association is assigned to a division based on the number of male students enrolled in grades 9–11.

Team	Won against
Angels	Diamonds, Cougars, Flames, Giants
Bison	Angels, Elks, Giants
Cougars	Diamonds, Bison, Elks
Diamonds	Bison
Elks	Angels, Diamonds, Flames, Giants
Flames	Diamonds, Cougars, Bison
Giants	Diamonds, Cougars, Flames

Suppose the results for a small region are shown in the table. The Diamonds receive 1 point for beating the Bison and 1 point for each of the Bison's wins, for a total of 4 points.

20. Draw a directed graph representing the information in the table. Can you tell who the champion would be? Explain.

21. a. Write a matrix M to represent the results for all 7 teams, using a 1 if the row team defeated the column team and a 0 otherwise.

b. What do the row sums of M represent?

22. a. Find M^2. What does it represent?

b. What do the row sums of M^2 represent?

23. a. Find $M + M^2$. What do the row sums represent?

b. Give the number of points awarded each team. Which team wins the championship?

24. Open-ended Problem Describe another method for choosing a regional champion. How can you use a graph or matrix with your method? Does your method give the same results as the one described above?

25. Open-ended Problem Draw a food web. Use a directed graph and matrices to analyze your food web. What would happen if one of the animals or plants were removed from the web?

26. Pete Morrill has several things he must do before leaving on a trip. Pete can't eat, pack, go to the store, or talk to his friend while he's showering. He can't pack or eat as he goes to the store. He also can't talk to his friend while he plans what to pack. Draw and color a graph to decide which of the six things Pete should do at the same time so that he can be ready to leave as soon as possible. *(Section 12.1)*

Find the sum of each series, if the sum exists. If a series does not have a sum, explain why not. *(Section 10.5)*

27. $3 + 9 + 27 + 81 + \cdots + 6561$

28. $\frac{1}{12} + \frac{1}{6} + \frac{1}{3} + \frac{2}{3} + \cdots + \frac{16}{3}$

29. $250 + (-50) + 10 + (-2) + \cdots$

30. $0.125 + 0.25 + 0.5 + 1 + \cdots$

ASSESS YOUR PROGRESS

VOCABULARY

graph (p. 563)
vertices (p. 563)
edges (p. 563)
degree of a vertex (p. 563)
directed graph (p. 569)

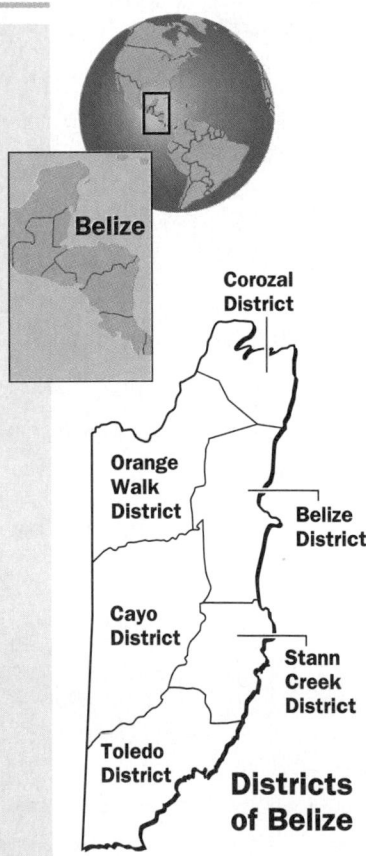

1. Draw a graph representing the map of Belize, a country in Central America. Label the degree of each vertex. Use your graph to color the map. Try to use as few colors as possible. *(Section 12.1)*

2. BUSINESS A company is planning meeting times for six committees. The following committees share members: 1 and 2, 1 and 3, 1 and 4, 2 and 5, 4 and 5, and 5 and 6. Use a graph to find the smallest number of meeting times that can be scheduled without conflict. *(Section 12.1)*

3. a. ECOLOGY Along the protected rocky shore of southern New England, plankton is eaten by mussels and barnacles; mussels are eaten by crabs, fish, and carnivorous snails; barnacles are eaten by crabs and fish; and crabs and carnivorous snails are eaten by fish. Draw a directed graph to represent this food web. *(Section 12.2)*

b. Write a matrix M to represent the graph, using a 1 if the row animal is a food source for the column animal, and a 0 otherwise.

c. Find M^2 and $M + M^2$. What does each matrix represent? Give specific examples.

4. Journal How are the graphs used in discrete mathematics different from the graphs you use in algebra? How are they alike?

Districts of Belize

12.3

Permutations

Will musicians ever run out of new tunes? In how many ways can the components of DNA be arranged? The answers to such questions involve counting large numbers of possibilities. As you will see in the Exploration, even designing the simplest of flags leads to numerous alternatives.

Learn how to...

- use the multiplication counting principle
- find the number of permutations of the elements of a set

So you can...

- count the possibilities in situations such as relay race strategy and protein synthesis

EXPLORATION
COOPERATIVE LEARNING

Investigating the Multiplication Counting Principle

Work with a partner.
You will need:
- paper of four different colors
- scissors

1 Cut a 1 in. by 3 in. strip from each of three colors of paper. Arrange the strips to form flags with two horizontal bars like those at the right.
List all of the different flags you can make. How many are there? This number is the product of what two consecutive integers?

Indonesia

2 Cut a 1 in. by 3 in. strip of the fourth color. List all of the two-bar flags you can make with four colors. How many are there? This number is the product of what two consecutive integers?

Burkina Faso

3 Look for a pattern in Steps 1 and 2. If you had one strip of each of ten different colors, how many different two-bar flags could you make?

Poland

Morocco

4 Cut another strip of each of the three original colors. How many different two-bar flags can you make from three colors if the bars may be the same color (as on the Moroccan flag) or different colors? Explain. This number is the square of what integer?

5 How many two-bar flags do you think can be made from four colors if repetition of colors is allowed? Explain.

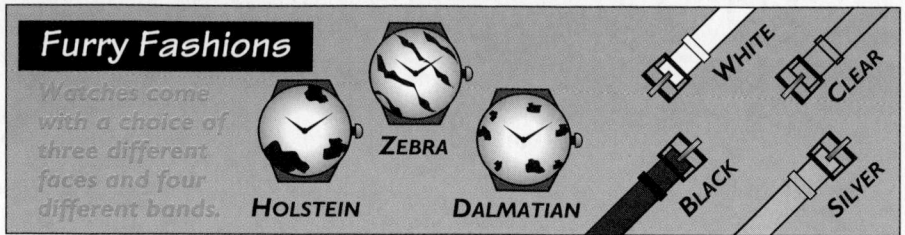

EXAMPLE 1   Application: Fashion

How many different *Furry Fashions* watches can you make?

Furry Fashions

Watches come with a choice of three different faces and four different bands.

HOLSTEIN   ZEBRA   DALMATIAN   WHITE   CLEAR   BLACK   SILVER

SOLUTION

Make a *tree diagram* to list the possibilities.

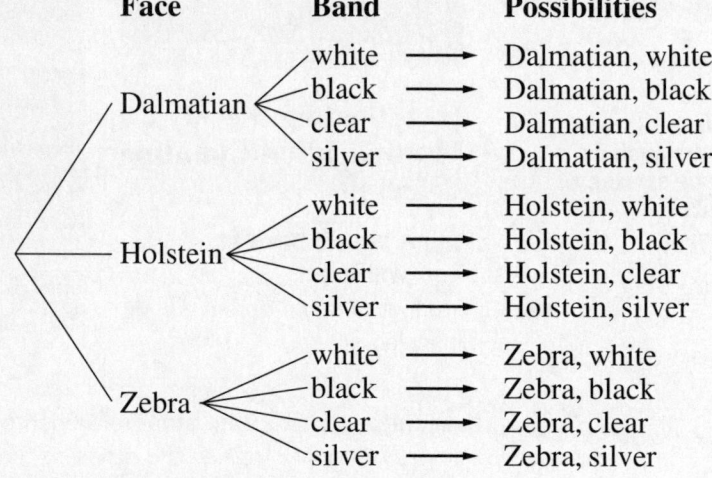

Face	Band	Possibilities
Dalmatian	white	Dalmatian, white
	black	Dalmatian, black
	clear	Dalmatian, clear
	silver	Dalmatian, silver
Holstein	white	Holstein, white
	black	Holstein, black
	clear	Holstein, clear
	silver	Holstein, silver
Zebra	white	Zebra, white
	black	Zebra, black
	clear	Zebra, clear
	silver	Zebra, silver

For each of the three choices of faces, there are four choices of bands. So there are $3 \times 4 = 12$ possible watches.

Example 1 suggests the following general principle.

> ### Multiplication Counting Principle
>
> If you must choose among m things, and for each of these you must choose among n more things, then you have a total of $m \times n$ possibilities.

THINK AND COMMUNICATE

1. How can you use the multiplication counting principle to tell how many two-bar flags you can make with 10 strips of different-colored paper?

2. How can you extend the multiplication counting principle to tell how many three-bar flags you can make with the 10 strips of paper?

EXAMPLE 2   Application: Sports

Short-track speed skaters occasionally compete in relay races. In planning the race strategy, a team's coach can arrange the four team members to skate in any order. How many arrangements are possible?

SOLUTION

Use the multiplication counting principle.

For the **first** skater, there are **4** choices.

For the **third** skater, **2** choices remain.

$$4 \quad \times \quad 3 \quad \times \quad 2 \quad \times \quad 1 \quad = \quad 24$$

For the **second** skater, **3** choices remain.

Only **1** choice remains for the **fourth** skater.

There are 24 possible orders in which the team can compete.

You can write $4 \times 3 \times 2 \times 1$ as 4! (read "four factorial"). In general, for any positive integer n, $n!$ is the product of all the integers from n down to 1.

$$n! = n(n - 1)(n - 2) \cdots (2)(1)$$

For consistency, 0! is defined to be equal to 1.

Permutations

An arrangement of the elements of a set is called a **permutation**. A set of four objects, such as a four-person relay race team, has 4! permutations. In general, the number of permutations of a set of n objects is given by $n!$.

EXAMPLE 3   Application: Sports

If the coach in Example 2 had filled the four slots from a pool of seven eligible skaters, how many four-member arrangements are possible?

SOLUTION

The **first** skater can be any of the **7**.

For the **third** skater, **5** choices remain.

$$7 \quad \times \quad 6 \quad \times \quad 5 \quad \times \quad 4 \quad = \quad 840$$

For the **second** skater, **6** choices remain.

For the **fourth** skater, **4** choices remain.

There are 840 permutations of four skaters chosen from seven.

In Example 3, the answer is the product of the first four factors of 7!. This is the same as $\dfrac{7!}{3!} = \dfrac{7 \cdot 6 \cdot 5 \cdot 4 \cdot \cancel{3} \cdot \cancel{2} \cdot \cancel{1}}{\cancel{3} \cdot \cancel{2} \cdot \cancel{1}}$ and suggests the following result.

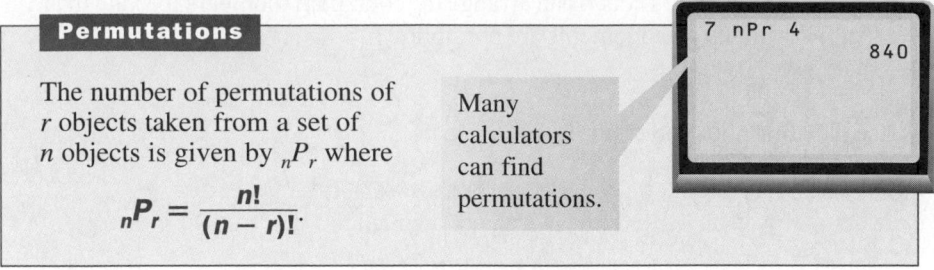

Permutations

The number of permutations of r objects taken from a set of n objects is given by $_nP_r$ where

$$_nP_r = \frac{n!}{(n-r)!}.$$

Many calculators can find permutations.

7 nPr 4
840

Distinguishable Permutations

When a set contains elements that are identical, some of the permutations look the same. For example, how many permutations are there of the letters in DAD? If you color the D's, you can arrange the 3 letters in **3!** different ways.

DAD   ADD   DDA

DAD   ADD   DDA

If you ignore the colors, you see **2!** copies of each "word" because there are **2!** ways to arrange the two D's.

So without color there are only $\dfrac{3!}{2!} = 3$ different permutations: DAD, ADD, and DDA.

Permutations with Repeated Elements

If a set of n elements has one element repeated q_1 times, another element repeated q_2 times, . . . , and the last element repeated q_k times, then the number of distinguishable permutations of the n elements is given by

$$\frac{n!}{q_1! q_2! \ldots q_k!}.$$

EXAMPLE 4

How many distinguishable permutations are there of the letters in MISSISSIPPI?

SOLUTION

Use the formula for permutations with repeated elements above. There are 11 letters, so $n = 11$. There are **one M ($q_1 = 1$)**, **four I's ($q_2 = 4$)**, **four S's ($q_3 = 4$)**, and **two P's ($q_4 = 2$)**.

$$\frac{11!}{1!4!4!2!} = \frac{11 \cdot 10 \cdot 9 \cdot \cancel{8} \cdot 7 \cdot \cancel{6} \cdot 5 \cdot \cancel{4} \cdot \cancel{3} \cdot \cancel{2} \cdot 1}{1 \cdot \cancel{4} \cdot \cancel{3} \cdot \cancel{2} \cdot 1 \cdot \cancel{4} \cdot \cancel{3} \cdot \cancel{2} \cdot 1 \cdot \cancel{2} \cdot 1}$$

$$= 11 \cdot 10 \cdot 9 \cdot 7 \cdot 5$$

$$= 34{,}650$$

There are 34,650 distinguishable permutations of the letters in MISSISSIPPI.

1. **a.** List the different meals you can choose from the menu at the right.

   b. Explain how you can use the multiplication counting principle to find the number of possible meals.

2. Write an expression for the number of ten-digit numerical codes that can be formed if:

   a. the digits must all be different   **b.** the digits may repeat

Simplify each expression.

3. $3!$ 4. $8!$ 5. $_8P_5$ 6. $_8P_3$

7. Which of the following represents the number of ways that eight runners can fill the top three slots in a race?

   A. $3!$ **B.** $8!$ **C.** $\dfrac{8!}{3!}$ **D.** $\dfrac{8!}{5!}$ **E.** $_8P_5$

Write an expression for the number of distinguishable permutations of the letters in each of the following six-letter words.

8. MATRIX   9. DIVIDE   10. BAOBAB

> *Dinner $6.99*
>
> *Entrees (choose one)*
> Spaghetti with Meatballs
> Chicken Enchiladas
>
> *Vegetables (choose one)*
> Corn on the Cob
> Green Beans
> Butternut Squash
>
> *Desserts (choose one)*
> Apple Pie
> Bread Pudding

12.3 | **Exercises and Applications**

Extra Practice exercises on page 768

Calculate each expression in Exercises 1–3.

1. **a.** $6!$ 2. **a.** $7!$ 3. **a.** $4!$

   b. $(2!)(3!)$ **b.** $2! + 5!$ **b.** $4(3!)$

Tell whether each statement is true for all positive integers m and n. Explain your answer.

4. $(m!)(n!) = (mn)!$ 5. $m! + n! = (m + n)!$ 6. $m! = m \cdot (m - 1)!$

Simplify.

7. $\dfrac{7!}{6!}$ 8. $\dfrac{8!}{4!4!}$ 9. $\dfrac{7!}{3!2!2!}$ 10. $\dfrac{(n + 2)!}{(n + 1)!}$

11. **FASHION** Janice has the clothes shown below, all of which coordinate. In how many ways can she create an outfit by choosing one of each item?

Look back at the article on pages 560–562.

Electrical engineers like Maria Rodriguez know that the components of an electrical circuit resist the flow of electricity. One such component, called a resistor, is shown below. The amount of resistance, measured in ohms, is indicated on the resistor using bands of color.

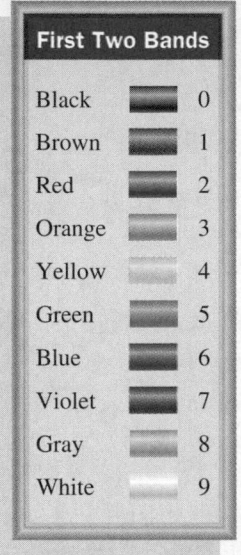

First Two Bands	
Black	0
Brown	1
Red	2
Orange	3
Yellow	4
Green	5
Blue	6
Violet	7
Gray	8
White	9

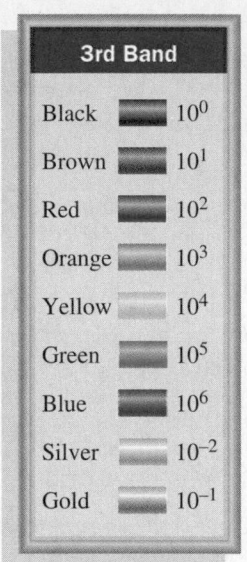

3rd Band	
Black	10^0
Brown	10^1
Red	10^2
Orange	10^3
Yellow	10^4
Green	10^5
Blue	10^6
Silver	10^{-2}
Gold	10^{-1}

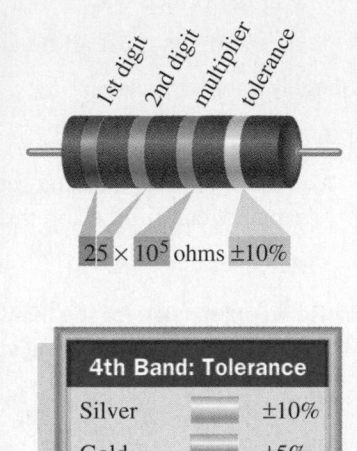

1st digit 2nd digit multiplier tolerance

25×10^5 ohms $\pm 10\%$

4th Band: Tolerance	
Silver	$\pm 10\%$
Gold	$\pm 5\%$
No Band	$\pm 20\%$

Find the resistance of each resistor.

12.  **13.** 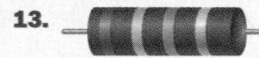 **14.**

15. How many different possible color sequences are there for the first three bands on a resistor?

16. How many different possible color sequences are there if tolerances are included? (Count "no band" as a color, since it still gives information.)

17. On some resistors a 5th color band indicates reliability. If the 5th band can be any of four colors, how many different color patterns can the resistor have?

18. **Challenge** Do all of the possible color patterns correspond to different resistance ratings? Give examples to explain your answer.

19. FORENSICS In the United Kingdom, a system called Photo-FIT (Facial Identification Technique) has been used by police to identify suspects. The basic five-section kit contains 195 hairlines, 99 eyes and eyebrows, 89 noses, 105 mouths, and 74 chins and cheeks. How many different faces can be constructed from the basic kit?

Simplify.

20. $_9P_5$ **21.** $_8P_1$ **22.** $_6P_2$ **23.** $_6P_4$

24. a. Open-ended Problem Describe a situation in which you would find $_8P_8$.

b. Describe a situation in which you would find $_8P_1$.

25. Adam saw a car speeding away from an accident scene. The car had a license plate from his state, which uses three letters followed by three digits. He remembers the first two letters and the first two digits. How many other cars could possibly share these same letters and digits?

26. You and five friends go to the movies. In how many ways can you line up to buy tickets?

27. SPORTS In baseball, the coach must plan the order in which the nine players will bat. How many possible batting orders are there?

28. Suppose 22 students are entering a classroom that has 22 desks.

a. How many possible seating arrangements are there? (*Note:* Assume that no one moves the desks.)

b. Some scientists estimate that the universe is about 10 billion years old. If the class had put itself into one of the seating arrangements each second since the beginning of the universe, when would the possibilities have been exhausted, or how much longer would it take?

29. SPORTS If 18 gymnasts are competing in an event, in how many ways can the gold, silver, and bronze medals be awarded?

30. SCHEDULING Jeanne Remsen, a health inspector, is making random checks of local restaurants. She has time today to visit four of nine restaurants on her list. In how many possible ways can her trips for the day be ordered?

31. SAT/ACT Preview In how many ways can a president, vice-president, treasurer, and recorder be chosen from a club of 11 members?

A. 14,641 **B.** 24 **C.** 7920 **D.** 1,663,200 **E.** 66

32. a. How many permutations of n objects are there?

b. What happens when you let $r = n$ in the formula for $_nP_r$?

c. What must be true for your answers to parts (a) and (b) to agree?

Give an expression for the number of distinguishable permutations of the letters of each word. Then evaluate the expression.

33. ROTOR **34.** HUMDRUM **35.** SYZYGY **36.** ONOMATOPOEIA

37. a. MUSIC In the key of G, the first 14 notes of "Twinkle, Twinkle, Little Star" include three G's, two A's, two B's, two C's, three D's, and two E's. Write and then evaluate an expression for the number of different melodies that can be formed from these 14 notes.

b. The first two measures of J. S. Bach's "Musette" from *English Suite No. 3* are shown at the right. How many different melodies can be formed from these 9 notes?

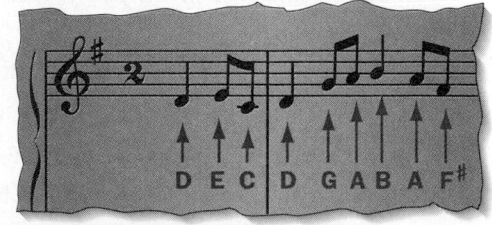

38. LITERATURE French poet Raymond Queneau's book *Cent mille milliards de poèmes* contains a sonnet on each of its 10 pages. The pages are cut so that any of the 14 lines of the sonnet can be turned independently, allowing any line of one sonnet to be combined with any line of another sonnet. How many different sonnets can result if the lines are structured so that all the possibilities make sense?

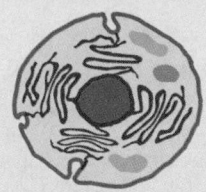

Special molecules called *RNA* tell your cells how to make proteins out of smaller molecules called *amino acids*. The RNA is made up of four kinds of *nucleotides*, which are often represented by the letters U, C, A, and G. Nucleotides are like letters in a sentence. Three nucleotides make a word, or *codon*. Each codon tells the cell to do something.

RNA

"Start making the protein. The first amino acid should be methionine."

"Add an alanine amino acid to the protein."

"Add a glycine amino acid."

"The protein is done. Stop."

AUG GCC UCU UGC AAA GGC UAU AGU AGU UAG

The cell always starts assembly with an AUG codon. This codon may also appear later in the sequence.

The codons UAA, UGA, and UAG tell the cell to stop assembly.

(Each codon except UAA, UGA, and UAG tells the cell to add an amino acid to the protein.)

39. How many possible codons are there? (*Note:* Nucleotides may repeat within a codon.)

40. Write an expression for the theoretical number of sequences of codons that can code for a protein 20 amino acids long. Remember that each sequence must begin with the "start" codon and must end with one of the "stop" codons, which do *not* add an amino acid.

ONGOING ASSESSMENT

41. Writing Explain why the number of ways to choose the top three winners from 10 contestants is $\frac{10!}{7!}$.

SPIRAL REVIEW

42. HOBBIES Five friends use ham radios to keep in touch. Api can receive broadcasts only from Destin. Ben can receive broadcasts from both Api and Destin. Charma can receive broadcasts from Api and Eddie. Destin can receive broadcasts from Eddie. And Eddie can receive broadcasts only from Ben. (*Section 12.2*)

a. Draw a directed graph representing this situation.

b. Can anyone send a broadcast that can be passed to all of the others with at most one intermediate stage? How you can use a matrix to find out?

Find the asymptotes of the graph of each function. Tell how the graph is related to a hyperbola with equation of the form $y = \frac{a}{x}$. (*Section 9.7*)

43. $f(x) = \dfrac{2}{x-1} + 3$ **44.** $g(x) = \dfrac{x}{x+2}$ **45.** $h(x) = \dfrac{2x+1}{x-4}$

12.4 Combinations

Learn how to...

- find the number of combinations of the elements of a set

So you can...

- count choices in which order is not important, such as the number of ways to plan a figure skating program

Lucky 3 Drawing
Three lucky winners will receive **$3000** each!

Grand Sweepstakes
1st prize: $30,000
2nd prize: $300
3rd prize: $30

In which contest would you rather be a winner? In the Lucky 3 Drawing, it matters only that you're in the top three, while in the Grand Sweepstakes, your exact position in the top three is important. There are more ways to award prizes for the Grand Sweepstakes than for the Lucky 3 Drawing. In the Exploration you'll see why.

EXPLORATION
COOPERATIVE LEARNING

Investigating Ways of Choosing Winners

Work in a group of 4 or 5 students.

Suppose only the members of your group enter the Grand Sweepstakes.

1 Each of you should list all of the ways that others in your group can win the second and third prizes if you are the first-prize winner.

2 As a group, count the number of ways to award the 3 prizes. Explain how to use permutations to check your answer. Write the answer in the form $_nP_r$.

Suppose only the members of your group enter the Lucky 3 Drawing.

3 Since each winner receives the same amount of money, the group of winners "John, Kia, and Ann," for example, is the same as "Ann, John, and Kia." In your lists from Step 1, how many times is each group of three winners listed? Explain why you can write your answer in the form $r!$.

4 Use the lists from Step 1 to count the number of ways to award the Lucky 3 prizes. How is your answer related to the answer in Step 2? Express the number of ways to choose the Lucky 3 winners in the form $\frac{_nP_r}{r!}$.

A selection made from a set when position is not important, as in the Lucky 3 Drawing, is called a **combination**. The symbol $_nC_r$ stands for the number of combinations of r items that can be chosen from a set of n items. The number of combinations of **three** Lucky 3 winners from a group of **five** entrants is $_5C_3$.

You can stack three of the four books in $_4P_3 = 24$ ways.

Suppose you are at a bookstore and have enough money to buy only three of the four books on your list. In how many ways can you choose three books? To find out, you could start by listing all the ways in which three of the four books can be stacked on the checkout counter.

T	T	M	M	L	L	M	M	L	L	W	W
M	L	L	T	T	M	L	W	W	M	M	L
L	M	T	L	M	T	W	L	M	W	L	M

L	L	W	W	T	T	W	W	T	T	M	M
W	T	T	L	L	W	T	M	M	W	W	T
T	W	L	T	W	L	M	T	W	M	T	W

Since it doesn't matter how you stack the books you buy, all of the permutations of **W**, **T**, and **M** are the same. There are $_3P_3 = 6$ permutations of **W**, **T**, and **M**.

So there are $\dfrac{_4P_3}{_3P_3} = \dfrac{24}{6} = 4$ combinations of three books that you can buy.

The example above suggests the following general formula.

Combinations

The number of combinations of r objects taken from a set of n objects is given by $_nC_r$ where

$$_nC_r = \frac{_nP_r}{_rP_r} = \frac{n!}{(n-r)!\,r!}$$

EXAMPLE 1 Application: Manufacturing

A quality control engineer in a manufacturing plant is inspecting five of each set of 100 resistors to make sure that they are within tolerance limits. How many possible sets of five resistors can be chosen from a set of 100?

SOLUTION

Note that the order in which the resistors are chosen is not important.

Method 1 Use the combinations formula with $n = 100$ and $r = 5$.

$$_{100}C_5 = \frac{100!}{(100-5)!\,5!}$$

$$= \frac{\overset{5}{\cancel{100}} \cdot \overset{33}{\cancel{99}} \cdot \overset{49}{\cancel{98}} \cdot 97 \cdot 96 \cdot \cancel{95!}}{\cancel{95!} \cdot \cancel{5} \cdot \cancel{4} \cdot \cancel{3} \cdot \cancel{2} \cdot 1}$$

$$= 75{,}287{,}520$$

Method 2 Use a calculator to find $_{100}C_5$.

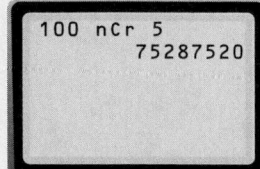

There are 75,287,520 sets of five resistors that can be chosen from a set of 100.

Suppose on a rainy afternoon you can choose to see one of 5 movies or read one of 3 books. You have a total of $5 + 3 = 8$ activities to choose from. But if you want to see one of 5 movies and then read one of 3 books, you have $5 \cdot 3 = 15$ possible ways to plan the afternoon.

When you have more than one choice to make, you must decide whether to use the multiplication counting principle or whether to add the possibilities.

EXAMPLE 2

Jasmine wants to buy at least 4 bagels of different flavors. How many different combinations of bagels can she buy?

SOLUTION

Jasmine can buy 4 *or* 5 bagels, so add the possibilities. You must consider purchases of 4 bagels and 5 bagels separately.

Step 1 Suppose Jasmine buys 4 different bagels.

$$_5C_4 = \frac{5!}{1!4!} = 5$$

Jasmine has **5** combinations of 4 flavors to choose from.

Step 2 Suppose Jasmine buys 5 different bagels.

$$_5C_5 = \frac{5!}{0!5!} = 1$$

Jasmine has **1** *additional* combination to choose from.

Step 3 Jasmine has $5 + 1 = 6$ different combinations of bagel flavors to choose from.

BAGEL Express

Today's Flavors:
plain
egg
raisin
blueberry
poppy seed

> **WATCH OUT!**
>
> Use the multiplication counting principle only when, for each choice from *m* things, you must choose from *n* more things. You shouldn't multiply in Step 3 because Jasmine will make one choice or the other, not both.

EXAMPLE 3

The school chorus has been invited to an international competition in Spain. How many different five-member fundraising committees can be formed from the 16 seniors, 14 juniors, and 10 sophomores in the chorus if the committee must have 2 seniors, 2 juniors, and 1 sophomore?

SOLUTION

The chorus must choose 2 seniors *and* 2 juniors *and* 1 sophomore for the committee, so use the multiplication counting principle.

| The two seniors can be chosen in $_{16}C_2$ ways. | The two juniors can be chosen in $_{14}C_2$ ways. | The sophomore can be chosen in $_{10}C_1$ ways. |

$$
\begin{aligned}
_{16}C_2 \cdot {}_{14}C_2 \cdot {}_{10}C_1 &= \frac{16!}{14!2!} \cdot \frac{14!}{12!2!} \cdot \frac{10!}{9!1!} \\
&= 120 \cdot 91 \cdot 10 \\
&= 109{,}200
\end{aligned}
$$

There are 109,200 possible committees.

For Questions 1–3, tell whether each situation describes *permutations* or *combinations*. Represent each number by an expression of the form $_nC_r$ or $_nP_r$.

1. the number of different three-filling burritos that you can choose if there are 9 fillings to choose from

2. the number of ways you can choose the four questions that you will answer on the test below

3. the number of orders in which you can answer four of the questions on the test below

> Name _____ Date _____
>
> **Answer four of the following six questions.**
>   **1.** Describa las comidas que le gustan.

4. Write an expression for the number of ways that you can choose one or two puppies from a litter of seven.

5. Write an expression for the number of committees of four girls and four boys that can be chosen from a class of 10 girls and 13 boys.

6. In the combinations formula on page 586, why does $\dfrac{_nP_r}{_rP_r} = \dfrac{n!}{(n-r)!r!}$?

7. Simplify each expression.

   a. $_7C_5$ **b.** $_7C_4$ **c.** $_7C_3$ **d.** $_{100}C_{98}$

12.4 | Exercises and Applications

Extra Practice exercises on page 768

Simplify each expression.

1. $_{12}C_5$ **2.** $_8C_4$ **3.** $_{16}C_4$ **4.** $_{16}C_{12}$

5. $_8C_0$ **6.** $_8C_8$ **7.** $_8C_1$ **8.** $_8C_7$

9. $_5C_3 + _5C_2$ **10.** $_6C_2 + _8C_2$ **11.** $_7C_5 \cdot _7C_2$ **12.** $_nC_1$

13. **MANUFACTURING** A quality control engineer is inspecting 10 of each set of 500 computers to make sure that they work properly. How many possible sets of 10 computers can be chosen from a set of 500?

14. Tyrone has four kinds of lettuce growing in his garden. How many different salads can he make using two of the types of lettuce?

Visual Thinking For Exercises 15 and 16, suppose eight pegs are arranged in a circle on a board. You can stretch a rubber band around the pegs as shown.

15. How many different triangles can be created? (*Note:* If two triangles have the same shape but are in different orientations, then consider them to be different triangles.)

16. How many quadrilaterals can be created?

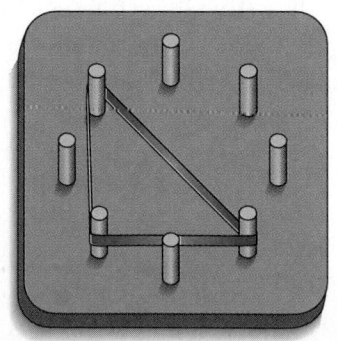

17. Karl wants either to go to the movies or to go bowling this evening. There are four movies and two kinds of bowling (tenpins and candlepins) to choose from. In how many different ways can Karl spend his evening?

18. Andrea wants to go bowling early in the evening and then see a movie later. She has two kinds of bowling and four movies to choose from. In how many ways can Andrea spend her evening?

19. Five bands that Sean likes have each released new music on both compact disc (CD) and cassette tape. Since tapes are less expensive than CDs, Sean plans to buy either three of the cassette tapes or two of the CDs. How many possible purchases does he have to choose from?

20. Trinh has roses, day lilies, clematis, lady's slippers, daisies, and poppies growing in her flower garden. She is making a bouquet to take to a neighbor. How many combinations of flowers can she have for the bouquet if she wants to include *at least* four different types of flowers?

21. **Open-ended Problem** For each expression, describe a situation in which the expression would be useful. Evaluate each expression and explain what the answer means in terms of the situation.

   a. $_6C_4 + {}_8C_4$ b. $_6C_4 \cdot {}_8C_4$ c. $_6P_4$

Connection SPORTS

Singles figure skating competitions are composed of two parts: the *short program*, which consists of required moves, and the *long program*, which skaters choreograph themselves. Each skater in a Ladies' Intermediate Freeskating competition must include all of the following moves in her short program:

Jumps:
   (1) an axel
   (2) a double loop or double salchow
   (3) a jump combination: a single jump and
       a double jump, or two double jumps

Spins:
   (4) a camel, sit, or upright spin
   (5) a spin combination

Step Sequences:
   (6) a straight line, circular, or serpentine step sequence

Double Salchow

Takeoff   Two revolutions   Landing

22. Skaters typically satisfy requirement (3) by choosing from six kinds of jumps: the salchow, axel, loop, toe loop, lutz, and flip. Each of these may be either a single or a double jump, but the double jump used to fulfill requirement (2) may not be used again in the jump combination. If the order of the jumps does not matter, and two double jumps of the same type may be used, how many different jump combinations may a skater use in her program to satisfy requirement (3)?

23. Suppose a skater has 36 choices for the spin combination to satisfy requirement (5). In how many ways can she choose all of the moves to include in her short program?

24. Explain how $_nC_r$ is related to $_nC_{n-r}$. Give an example.

25. SPORTS Suppose your physical education class has 20 students.

    a. In how many ways can a six-member volleyball team be chosen if players rotate so that all the positions are equivalent?

    b. How many baseball teams can be arranged given that each of the nine players plays a different position on the field?

    c. A basketball team consists of a center, two equivalent guards, and two equivalent forwards. In how many ways can a basketball team be chosen and positions assigned?

26. To open a combination lock like the one at the right, you dial a sequence of three numbers called the lock's *combination*. You turn the dial clockwise for the first number, counter-clockwise for the second, and clockwise again for the third.

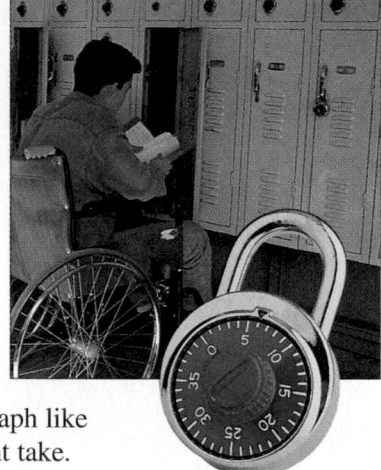

    a. How many *combinations* are possible for a lock like the one shown?

    b. Explain why a "combination" lock is misnamed.

27. COMPUTERS Before giving a computer a task, such as coloring a graph like those on page 566, computer scientists may predict how long it might take.

    a. In how many ways can you color a graph with *n* vertices using 4 colors? (For now, ignore the edges. Each vertex may be any color.)

    b. At most, how many edges must you check to be sure that no vertices that share an edge are the same color? (Consider the worst-case scenario in which each pair of the *n* vertices is connected by an edge. Assume that no two vertices are connected by more than one edge.)

    c. At most, how many edges must you check if you check every edge of every possible 4-color coloring of a graph with *n* vertices? What does it mean if each possible coloring of the graph has an edge that joins two vertices of the same color?

    d. Suppose a computer can check 1000 edges each second. At most, how long will the computer take to check every edge of every 4-color coloring of a graph with 10 vertices? 17 vertices?

28. ADVERTISING A national pizza chain had the following promotion: If you buy two pizzas, you may receive up to five toppings on each pizza free. In the television advertisement for this promotion, a child and an adult are waiting to buy pizza.

If the pizza chain offers 11 toppings, how did the child get the answer 1,048,576? Do you think the child is right? Explain your answer.

> **CHILD:** The possibilities are endless.
>
> **ADULT 1:** Five plus five is ten.
>   (The child does some quick calculations on a notepad.)
>
> **CHILD:** Actually, there are 1,048,576.
>
> **ADULT 1:** Well, ten was just a ballpark figure.
>
> **ADULT 2:** You got that right! Some ballpark!

A standard deck of playing cards contains 52 cards, with 13 cards in each of four suits: clubs, diamonds, hearts, and spades. Many games begin with one player dealing out a *hand* of cards to each player. The order of the cards within a hand is unimportant.

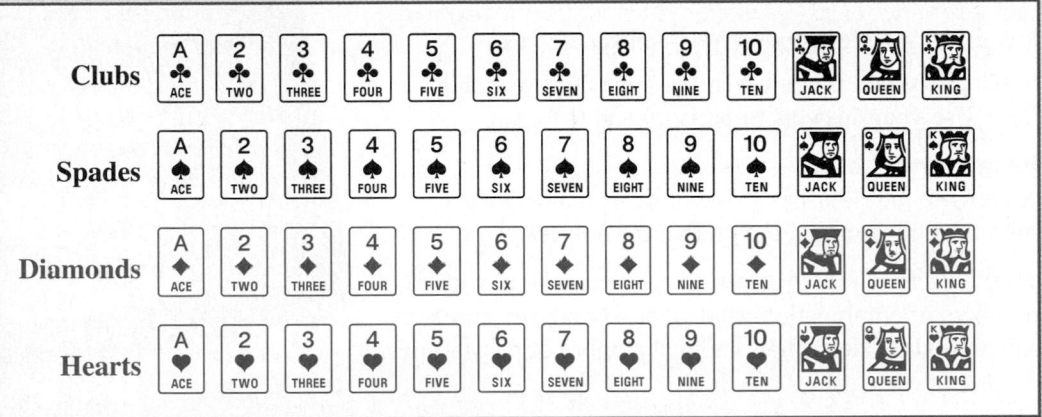

29. How many different 13-card hands can be dealt from a standard deck of cards?

30. In how many different orders can you play four cards from your 13-card hand?

31. How many 5-card hands contain a pair of aces? (*Hint:* The dealer must deal two of the four aces in the deck *and* three cards that are not aces.)

32. How many 5-card hands contain a pair of aces and three kings?

33. A hand with two cards of one value and three of another, such as in Exercise 32, is sometimes called a *full house*. A full house with 3 jacks and 2 fives is considered to be different from a full house with 3 fives and 2 jacks, but the suits of the cards are unimportant. How many different kinds of full houses are there?

34. How many 5-card hands contain a full house? Explain your answer. (*Note:* The two hands at the right are different.)

35. **Writing** How many different 5-card hands can be dealt from a standard deck of cards? Do you think it is likely that one of the 5-card hands in a 4-player game will be a full house? Explain.

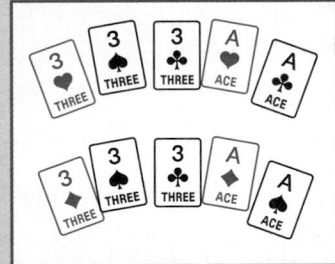

36. **FUNDRAISING** Many states sell lottery tickets to raise money for education and other state expenses. If the numbers on your ticket match the numbers the state draws at random, then you win. In some lottery games you get to choose six numbers from 1 to 40 with no numbers repeated. The order in which you choose the numbers does not matter. How many possible number combinations can you choose? Do you think it is likely that you will choose the same six numbers that the state will draw?

37. Challenge Some friends are at an Ethiopian restaurant that offers the specials shown. In how many ways can they choose four dishes to share if some of the dishes may be the same?

38. INDUSTRY Of 20 computer chips, a sample of two will be tested. Suppose that two of the 20 are defective and the other 18 are good.

   a. How many different samples can be selected?

   b. How many samples contain only good chips?

   c. How many samples contain only defective chips?

   d. How many samples contain at least one defective chip? (*Hint:* Use your answers to parts (a) and (b).)

39. Challenge Prove that $_{n+1}C_r = {}_nC_{r-1} + {}_nC_r$. (*Hint:* Rewrite the right-hand side of the equation with a common denominator and then factor the numerator.)

40. Writing Describe how to tell, for a given situation, when to add numbers of combinations and when to multiply numbers of combinations to determine the total number of possibilities.

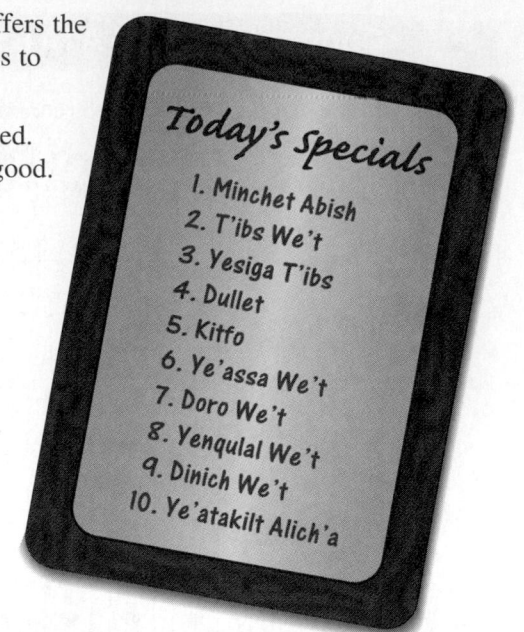

Today's Specials

1. Minchet Abish
2. T'ibs We't
3. Yesiga T'ibs
4. Dullet
5. Kitfo
6. Ye'assa We't
7. Doro We't
8. Yenqulal We't
9. Dinich We't
10. Ye'atakilt Alich'a

ONGOING ASSESSMENT

41. a. Writing How are combinations and permutations similar? How are they different? Use examples in your answer.

   b. Explain what $_nC_r$ and $_nP_r$ represent. How can you find $_nC_r$ if you know $_nP_r$? Use an example in your answer.

SPIRAL REVIEW

Simplify each expression. *(Section 12.3)*

42. $_5P_3$ **43.** $_6P_3$ **44.** $_{30}P_2$

45. $_8P_8$ **46.** $_8P_0$ **47.** $_{12}P_{10}$

Multiply. *(Section 9.2)*

48. $(5x + 2)(2x - 1)$ **49.** $(2t - 1)^2$ **50.** $(m - 3)(m^2 - 6m + 9)$

51. $(n - 2)^3$ **52.** $(n - 2)^4$ **53.** $(y + 1)^4$

Find an equation for each cubic function whose graph is shown. *(Section 9.4)*

54.

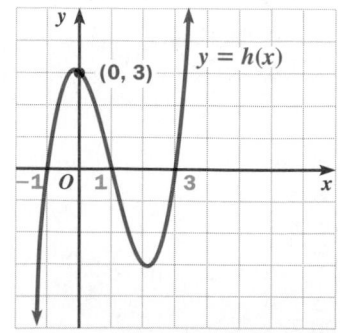

55.

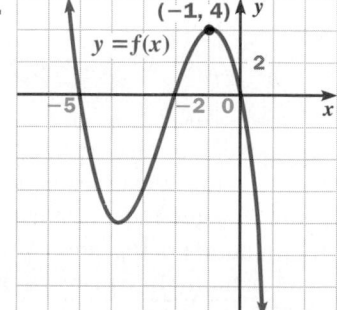

56.

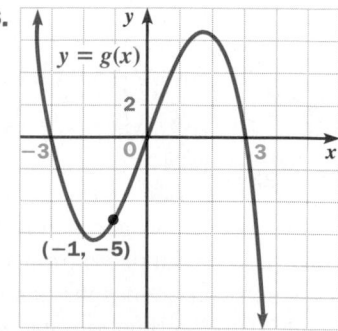

12.5 Pascal's Triangle

Learn how to...

- **apply the concept of combinations to Pascal's triangle**

So you can...

- **easily count the outcomes of sequences of events for which there are exactly two possibilities, such as left and right turns**

In the nineteenth century a device similar to the one shown was used to illustrate ideas about probability. A marble dropped into the device hits a sequence of pins and then falls into a slot at the bottom. The marble is equally likely to turn left or right when it hits each pin.

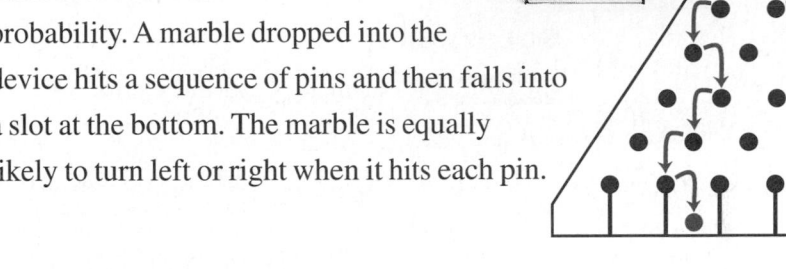

THINK AND COMMUNICATE

1. a. Express the paths of the red marble and the blue marble as sequences of left turns (L) and right turns (R).

b. Make a list of all possible paths that lead to the slot containing each of the two marbles shown.

c. For a falling marble to reach the slot that the blue marble is in, it must make exactly two right turns at any combination of two out of the six rows of pins that it falls through. What is $_6C_2$? How does this relate to your list from part (b)? What values of n and r in $_nC_r$ apply to the paths that lead to the slot where the red marble is?

d. Find the number of paths that lead to each of the other slots.

If you count the number of paths that a marble can follow to reach each pin, you get this result:

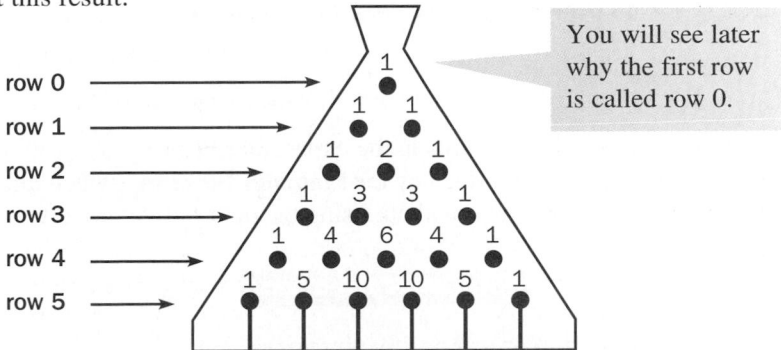

row 0 ⟶
row 1 ⟶
row 2 ⟶
row 3 ⟶
row 4 ⟶
row 5 ⟶

> You will see later why the first row is called row 0.

```
          1
        1   1
      1   2   1
    1   3   3   1
  1   4   6   4   1
1   5  10  10   5   1
```

This triangular display of numbers is called **Pascal's triangle**, named after French mathematician Blaise Pascal (1623–1662), who published it in 1653.

As you saw in *Think and Communicate* Question 1, term r in row n of Pascal's triangle, where r and n start at 0, is given by $_nC_r$.

THINK AND COMMUNICATE

2. Look closely at Pascal's triangle on the previous page to complete this statement: $_5C_4 = {}_4C_3 + {}_4C{\underline{\ ?\ }}$. Then verify that the statement is true.

3. Use the pattern in Question 2 to write any term $_nC_r$ ($0 < r < n$) of Pascal's triangle as a sum of terms from the preceding row. (*Hint:* The term $_nC_r$ is in row n of the triangle, so the preceding row is row $n - 1$.)

EXAMPLE 1

Generate row 7 of Pascal's triangle.

SOLUTION

Method 1 Calculate $_nC_r$ for each term.

Remember, r starts at 0.

$_7C_0$	$_7C_1$	$_7C_2$	$_7C_3$	$_7C_4$	$_7C_5$	$_7C_6$	$_7C_7$
1	7	21	35	35	21	7	1

Method 2 Add pairs of terms from the previous row.

These are the numbers you found in part (d) of *Think and Communicate* Question 1.

Each row begins and ends with 1.

Pascal's triangle is useful in *expanding*, or multiplying out, powers of binomials. Here are some expansions of the powers of $(a + b)$:

$$(a + b)^0 = 1$$
$$(a + b)^1 = a + b$$
$$(a + b)^2 = a^2 + 2ab + b^2$$
$$(a + b)^3 = a^3 + 3a^2b + 3ab^2 + b^3$$
$$(a + b)^4 = a^4 + 4a^3b + 6a^2b^2 + 4ab^3 + b^4$$
$$(a + b)^5 = a^5 + 5a^4b + 10a^3b^2 + 10a^2b^3 + 5ab^4 + b^5$$

THINK AND COMMUNICATE

4. What do you notice about the coefficients of the terms in the expansions? What do you notice about the exponents of a and b?

5. What do you think the expansion of $(a + b)^6$ is?

The patterns in the expansions of $(a + b)^0$, $(a + b)^1$, $(a + b)^2$, and so on are summarized by the **binomial theorem**, which allows you to expand any binomial without using the distributive property repeatedly.

Binomial Theorem

For any positive integer n,

$$(a + b)^n = {}_nC_0a^n + {}_nC_1a^{n-1}b + {}_nC_2a^{n-2}b^2 + \cdots + {}_nC_{n-1}ab^{n-1} + {}_nC_nb^n$$

EXAMPLE 2

Expand each binomial.

a. $(a + b)^7$

b. $(3p - 2q)^3$

SOLUTION

a. Use the binomial theorem. Let $n = 7$.

$$(a + b)^7 = {}_7C_0 a^7 + {}_7C_1 a^6 b + {}_7C_2 a^5 b^2 + {}_7C_3 a^4 b^3 + {}_7C_4 a^3 b^4 +$$
$$\qquad {}_7C_5 a^2 b^5 + {}_7C_6 ab^6 + {}_7C_7 b^7$$
$$= a^7 + 7a^6 b + 21a^5 b^2 + 35a^4 b^3 + 35a^3 b^4 + 21a^2 b^5 + 7ab^6 + b^7$$

b. Use the binomial theorem. Let $a = 3p$, $b = -2q$, and $n = 3$.

$$(3p - 2q)^3 = {}_3C_0 (3p)^3 + {}_3C_1 (3p)^2(-2q) + {}_3C_2 (3p)(-2q)^2 + {}_3C_3 (-2q)^3$$
$$= 3^3 p^3 + 3(3^2)(-2)p^2 q + 3(3)(-2)^2 pq^2 + (-2)^3 q^3$$
$$= 27p^3 - 54p^2 q + 36pq^2 - 8q^3$$

✓ CHECKING KEY CONCEPTS

1. Row 7 of Pascal's triangle is 1, 7, 21, 35, 35, 21, 7, 1. Find row 8.

2. Terms 7 and 8 of row 14 of Pascal's triangle are 3432 and 3003, respectively. Use these numbers to find ${}_{15}C_8$.

3. Suppose you toss a coin 10 times and get this sequence of heads (H) and tails (T): HTTHHHTTTH. How many such sequences of ten coin tosses show exactly five heads and five tails?

Give the values of *a*, *b*, and *n* that you would substitute in the binomial theorem to expand each power of a binomial.

4. $(x - y)^6$

5. $(4x + y)^{12}$

6. $(\pi p - 4q)^m$

12.5 Exercises and Applications

Extra Practice
exercises on
page 768

1. Find row 9 of Pascal's triangle.

2. Suppose you toss a coin 12 times and get this sequence of heads (H) and tails (T): HTTHHHTTTHHT. How many such sequences of twelve coin tosses show exactly six heads and six tails?

3. **Writing** Look back at the device shown at the top of page 593. What do you notice about the number of paths that involve one right turn out of six turns and the number of paths that involve five right turns out of six turns? Explain why Pascal's triangle has this symmetry. Compare paths with r right turns and $6 - r$ right turns.

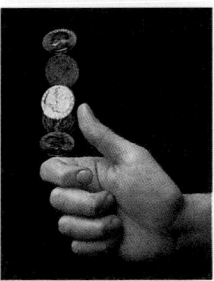

Use the binomial theorem to find the value of *k* for each of the following terms of a binomial expansion.

4. $_6C_4a^{6-k}b^k$ **5.** $_5C_ka^3b^2$ **6.** $_kC_4a^7b^4$ **7.** $_{14}C_6a^kb^{14-k}$

8. Find the coefficient of a^5b^4 in the expansion of $(a + b)^9$.

9. Find the coefficient of x^3y^8 in the expansion of $(x + y)^{11}$.

Use the binomial theorem to expand each power of a binomial.

10. $(x + y)^8$ **11.** $(x - y)^5$ **12.** $(3p + 4q)^4$ **13.** $(2y - 3z)^5$

14. $(a + 5b)^4$ **15.** $\left(m^2 - n\right)^7$ **16.** $\left(y - 2z^2\right)^5$ **17.** $\left(x^2 + y^3\right)^6$

18. a. Find the first three terms in the expansion of $(m + n)^{20}$.

    b. Find the last three terms in the expansion of $(m + n)^{20}$.

19. One term in the expansion of $(a + b)^n$ is $66a^{10}b^2$.

    a. What is the value of n?

    b. Express 66 in the form $_nC_r$.

    c. What is the next term of the expansion?

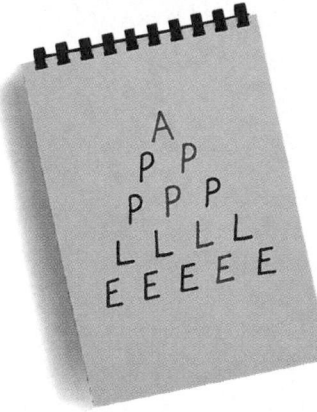

20. You can make a triangle out of a word by repeating the letters as shown in the arrangement of APPLE at the left. Then you can reconstruct the word by beginning at the top and choosing a letter in each row as you follow a path downward (such as the one shown in red) by turning right or left at each row.

    a. Explain why this situation is related to Pascal's triangle.

    b. How many ways are there to reconstruct APPLE?

    c. Express the number in part (b) as a sum of combinations.

    d. Make a word triangle from SUBTRACT. How many ways are there to reconstruct the word? To what sum of combinations does this correspond?

RECREATION Every day, Anchara jogs from her house at the intersection of 1st Street and Avenue A to the park entrance at the intersection of 6th Street and Avenue F. She runs so that she is always getting closer to the park, but she likes to take a different path as often as possible.

21. There are two paths from Anchara's house to 2nd and B: one passes through 1st and B, and one passes through 2nd and A. There are three paths from her house to 3rd and B: two paths pass through 2nd and B, and one passes through 3rd and A. Give a similar argument for the number of paths to 3rd and C.

22. Sketch a copy of the grid, and use the method in Exercise 21 to label each intersection with the number of different paths from Anchara's house to that intersection. What do you notice?

23. The distance from Anchara's house to the park is 10 blocks. To get to the park, how many times must she travel to a higher-numbered street, and how many times must she travel to a higher-lettered avenue? Use these facts to express the number of different paths to the park as a combination.

24. If Anchara also jogs back home from the park, how many different round-trip paths can she choose from? Explain.

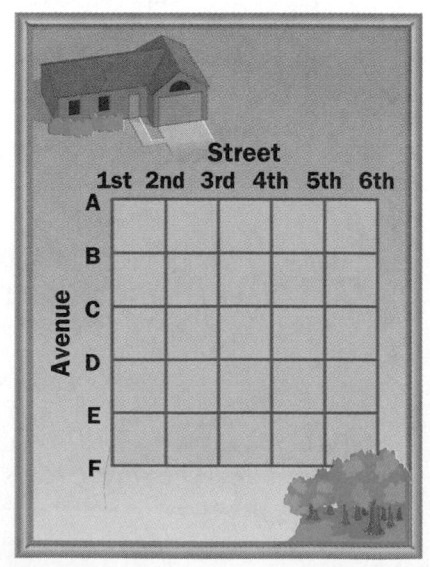

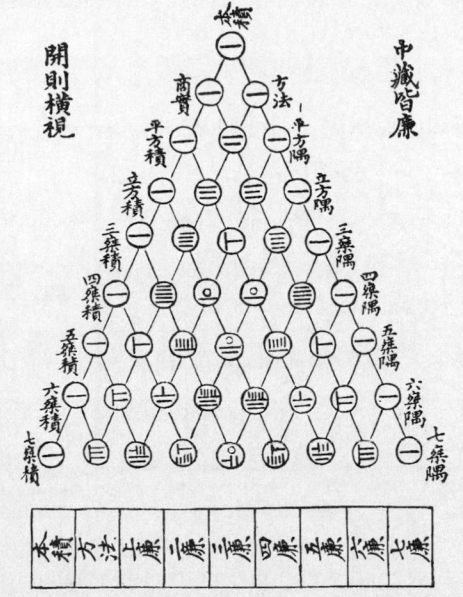

Connection ▶ **HISTORY**

Some 350 years before Pascal, Chinese mathematicians Yang Hui and Chu Shih Chieh gave a detailed treatment of what is now called Pascal's triangle. Their work was based on another Chinese work from about A.D. 1050.

25. Examine the symbols used in the Chinese triangle.

   a. What number does the symbol ⊦ represent?

   b. What number does the symbol ⚬ represent?

   c. How do you think the number 77 is written using the Chinese symbols? Explain your reasoning.

The Chinese discovered how to use the terms of the triangle to find roots of large numbers. Use the following version of the ancient Chinese method to complete Exercises 26 and 27.

To find an approximation for $\sqrt{465}$, first note that $21^2 < 465 < 22^2$, so $21 < \sqrt{465} < 22$. Then let $\sqrt{465} = 21 + d$, where d is the decimal part of the root $(0 < d < 1)$. So:

$$(21 + d)^2 = 465$$

$$21^2 + 2(21)d + d^2 = 465$$

    Use Pascal's triangle to expand $(21 + d)^2$.

$$441 + 42d + d^2 = 465$$

$$42d + d^2 = 24$$

$$d(d + 42) = 24$$

$$d = \frac{24}{d + 42}$$

    Solve for d in terms of itself.

If you let $d = 0$, you get $\sqrt{465} \approx 21 + \frac{24}{0 + 42} = 21\frac{4}{7}$, an overestimate of $\sqrt{465}$.

If you let $d = 1$, you get $\sqrt{465} \approx 21 + \frac{24}{1 + 42} = 21\frac{24}{43}$, an underestimate of $\sqrt{465}$.

26. Find an approximation for $\sqrt{301}$.

27. Challenge Find an approximation for $\sqrt[3]{623}$.

28. A diagonal of Pascal's triangle consists of the numbers $_nC_r$, where r is constant and n varies. You can generate one diagonal from another.

   a. Describe the numbers on the diagonal given by the terms $_nC_1$.

   b. The numbers on the diagonal given by the terms $_nC_2$ are called the *triangular numbers* because they give the total number of objects in a triangular arrangement of objects. Describe how to find the triangular numbers as sums of the terms in the diagonal given by the terms $_nC_1$.

   c. Find the sum of the first n terms along any diagonal. Where can that sum be found in Pascal's triangle? Do you think this always works? Why?

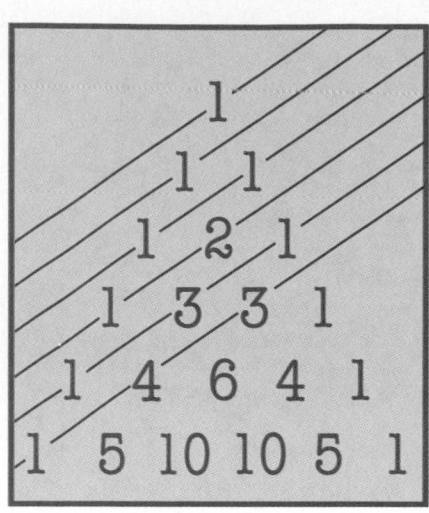

29. **Writing** Explain why the first and last numbers in a row of Pascal's triangle are always 1's.

30. The sums of the terms along the diagonals shown give a sequence called the *Fibonacci numbers*. Use Pascal's triangle to give the first 10 Fibonacci numbers.

31. **Research** Find out more about Pascal and Pascal's triangle. Describe some patterns that appear in the triangle that haven't been discussed in your class. Share your discoveries with your class.

32. a. **Technology** Use a graphing calculator or graphing software to graph $y = x^4$ and $y = x^4 + 4x^3 + 6x^2 + 4x + 1$. How is the second graph geometrically related to the first? Why?

    b. Repeat part (a) using $y = x^3$ and $y = x^3 - 6x^2 + 12x - 8$.

33. **Writing** Find the sum of the numbers in each of the first few rows of Pascal's triangle. Then give a formula for the sum of the numbers in row n. Explain why the sum of the numbers in a row doubles from row to row.

34. Substitute $a = 1$ and $b = 1$ into the binomial theorem. How do the results relate to the answers to Exercise 33?

35. a. **Challenge** Use the binomial theorem to find a general formula for term r (where $0 \le r \le n$) in the expansion of $(a + b)^n$.

    b. Use your formula to find term 11 of $(a - 3b)^{13}$.

    c. Use your formula to find term 6 of $(2m + n)^{12}$.

36. **Open-ended Problem** Think of a situation that you think can be modeled by Pascal's triangle. Explain your reasoning.

37. **SAT/ACT Preview** Which term is *not* part of the expansion of $(x + y)^7$?

    A. x^7 **B.** $21x^5y^2$ **C.** $7xy^6$ **D.** $35x^3y^3$ **E.** y^7

ONGOING ASSESSMENT

38. **Cooperative Learning** Work in groups of three or four. Each group should have several copies of the first 20 rows or so of Pascal's triangle. Begin by circling or coloring all of the odd numbers. The pattern represents a portion of the fractal called a *Sierpenski triangle*. (See page 464 for more information about a Sierpenski triangle.)

    a. On one copy of the triangle, circle or color all of the multiples of 3. What do you notice?

    b. On a different copy of the triangle, circle or color all of the multiples of 5. What do you notice?

    c. Repeat this process with multiples of other numbers. Does a pattern always seem to appear? If so, describe it.

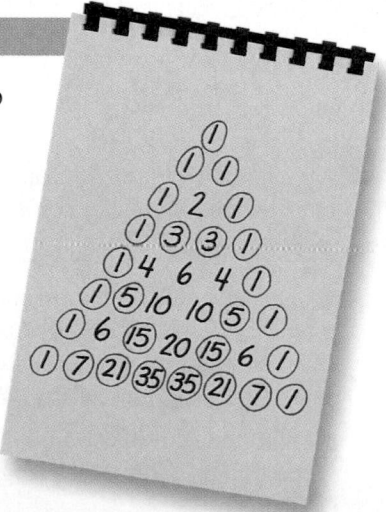

Simplify each expression. *(Section 12.4)*

39. $_1C_0$ **40.** $_7C_1$ **41.** $_7C_6$ **42.** $_{15}C_{12}$

43. The radius of a basketball is 12.0 cm. Find its surface area.
(Toolbox, page 802)

For Exercises 44–46, write a point-slope equation of the line passing through the given point and having the given slope. *(Section 2.3)*

44. point: $(1, 1)$
slope $= 2$

45. point: $(3, -2)$
slope $= 6$

46. point: $(-5, 5)$
slope $= -2$

ASSESS YOUR PROGRESS

VOCABULARY

permutation (p. 579) **Pascal's triangle** (p. 593)
combination (p. 585)

For Questions 1–6, simplify each expression. *(Sections 12.3 and 12.4)*

1. $_9C_1$ **2.** $_9P_1$ **3.** $_{10}C_4$

4. $_{10}P_4$ **5.** $_6C_3 + _6C_4$ **6.** $_5P_2 - _4P_2$

For Questions 7–9, give and evaluate an expression of the form $_nC_r$ or $_nP_r$ for each situation. *(Sections 12.3 and 12.4)*

7. the number of ways that five different jobs can be filled from among 40 employees

8. the number of different vanilla frozen-yogurt sundaes that can be created by choosing three of eight possible toppings

9. the number of ways that you can place 5 of your 20 compact discs into the slots of your stereo's CD changer

10. If you think of a rainbow as having seven different colors, how many possible rainbows can you paint if you painted the color bands in any order? *(Section 12.3)*

11. Niti plans to take two of four possible social science classes, one of three possible mathematics classes, one of four possible English classes, and two of six possible electives. How many possible combinations of classes can she take? *(Section 12.4)*

12. Find term 8 of row 11 of Pascal's triangle. *(Section 12.5)*

Use the binomial theorem to expand each power of a binomial. *(Section 12.5)*

13. $(j + k)^6$ **14.** $(2p - q)^5$ **15.** $(p + 0.5q)^3$

16. **Journal** Describe as many situations as you can in which you could use permutations to count possibilities during the course of a typical day. Then describe as many situations as you can in which you could use combinations.

Exploring the Possibilities

The average day presents many situations that call for choices. Various options and combinations present themselves. "Does the blue sweater go with this outfit?" "Would I rather have the entree with a baked potato or rice? And what type of dressing for the salad?" "Should I buy the two-door or the four-door model of this car?"

You can examine your options by writing them all down, but sometimes it helps to see the possibilities. By making a flip book, you can visualize all the combinations of choices.

PROJECT GOAL Make a flip book to show all the possibilities for a real-world situation involving choices.

Making Your Own Flip Book

Choose a real-world situation that has many options and where the options are divided into at least four categories. Make a flip book that displays every possible combination of the options. You will need a pad of paper, posterboard, and scissors.

The flip book below illustrates the choices for someone planning a dinner. Assemble your flip book in a similar manner.

1. USE a pad of paper or several sheets of paper stapled together.

2. DIVIDE the pad by cutting the paper so that there is a section for each category of options. Be sure to leave a border on the side where the sheets of paper are bound together.

3. MOUNT the pad on posterboard and put the name of each category on the side opposite the binding.

4. DRAW pictures (or clip and paste photos from magazines) to illustrate the options for each category. Remove any unused pieces of paper.

Writing a Report

Write a report about your flip book. Consider these questions:

• How many different combinations of options does your flip book display? How do you know?

• What are the advantages and drawbacks of using a flip book to display options?

• How might other people, such as a police sketch artist, a landscaper, or a fashion designer, use a flip book? Can you think of other areas in which a flip book might be useful?

| Cut vegetables | Salmon steak | Baked potato | Strawberries |
| Appetizer | Entree | Side dish | Dessert |

Extend your investigation by doing related projects that involve combinations. For example:

• Make drawings on strips of paper that you can put next to each other and slide back and forth to change, say, the expressions on a face or the clothes on a model.

• Make drawings on clear plastic overlays to explore options such as toppings on a pizza or disguises for a face.

• Create a colorful toy or game that a young child can play with to explore combinations.

Self-Assessment

Describe your understanding of combinations. In what ways do you feel that you are better able to recognize and examine situations involving choices?

12 | Review

Describe any exercises or ideas in this chapter that you found difficult but eventually understood. How did you resolve your difficulties? Now describe some things you still don't understand. Can you use some of the methods you used before to help you resolve these difficulties?

VOCABULARY

graph (p. 563)
vertices (p. 563)
edges (p. 563)
degree of a vertex (p. 563)

directed graph (p. 569)
permutation (p. 579)
combination (p. 585)
Pascal's triangle (p. 593)

SECTIONS | 12.1 *and* 12.2

You can use a **graph** to solve problems that require sorting items into groups, such as coloring a map of New England. The graph shown consists of **vertices**, representing states, connected by **edges**, representing a shared border. Each vertex is labeled with its **degree**.

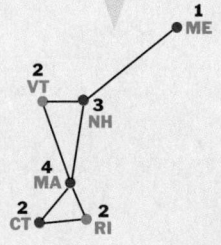

Using a procedure like that in Example 1 on page 564, you can assign colors.

You can use **directed graphs** and matrices to solve problems involving one-way relationships, such as predator-prey relationships.

Matrix *M* tells you how many ways each plant or animal is a direct food source for the others.

Matrix M^2 tells you how many ways each plant or animal is a food source for the others through one intermediate source.

The arrows point from prey to predator.

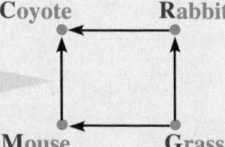

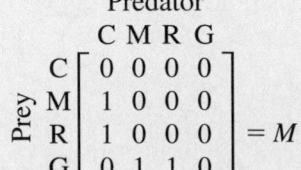

$$
\begin{array}{c}
\text{Predator} \\
\begin{array}{cc}
 & \begin{array}{cccc} C & M & R & G \end{array} \\
\begin{array}{c} \text{Prey} \end{array} \begin{array}{c} C \\ M \\ R \\ G \end{array} & \left[\begin{array}{cccc} 0 & 0 & 0 & 0 \\ 1 & 0 & 0 & 0 \\ 1 & 0 & 0 & 0 \\ 0 & 1 & 1 & 0 \end{array}\right] = M
\end{array}
\end{array}
$$

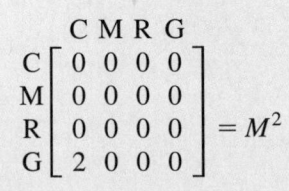

$$
\begin{array}{cc}
 & \begin{array}{cccc} C & M & R & G \end{array} \\
\begin{array}{c} C \\ M \\ R \\ G \end{array} & \left[\begin{array}{cccc} 0 & 0 & 0 & 0 \\ 0 & 0 & 0 & 0 \\ 0 & 0 & 0 & 0 \\ 2 & 0 & 0 & 0 \end{array}\right] = M^2
\end{array}
$$

SECTIONS 12.3 *and* 12.4

The multiplication counting principle states that if you must choose among *m* things, and for each of these you must choose among *n* more things, then you have a total of $m \times n$ possibilities. The multiplication counting principle leads to two important counting formulas.

	Permutation	Combination
Definition	an arrangement of items in which position is important	a selection of items in which position is *not* important
Formula	$_nP_r = \dfrac{n!}{(n-r)!}$	$_nC_r = \dfrac{_nP_r}{_rP_r} = \dfrac{n!}{(n-r)!\,r!}$
Example	In a club with 20 members, the number of ways to select 4 officers is: $_{20}P_4 = \dfrac{20!}{16!} = 116{,}280$	In a club with 20 members, the number of ways to select a committee of 4 is: $_{20}C_4 = \dfrac{20!}{16!\,4!} = 4845$

For any positive integer *n*, *n*! is the product of all the integers from *n* down to 1.

When a set contains elements that are identical, there are fewer distinguishable permutations. For example, the number of distinguishable arrangements of the letters in the word EEL is given by:

Divide by 2! because there are two E's.

$$\frac{3!}{2!} = 3$$

The three arrangements are: EEL, ELE, LEE.

SECTION 12.5

Pascal's triangle is formed by calculating $_nC_r$ $(0 \le r \le n)$ for term *r* in row *n*, or by adding pairs of terms from the previous row.

Pascal's triangle gives the coefficients when you expand powers of binomials.

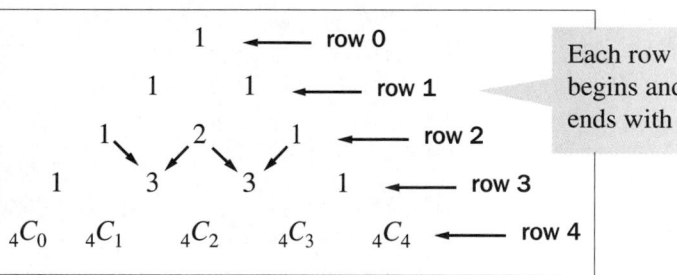

Each row begins and ends with a 1.

Binomial Theorem

For any positive integer *n*,

$$(a + b)^n = {_nC_0}a^n + {_nC_1}a^{n-1}b + {_nC_2}a^{n-2}b^2 + \cdots + {_nC_{n-1}}ab^{n-1} + {_nC_n}b^n$$

The binomial theorem gives the coefficients when you expand any power of any binomial, such as $(x + 3y)^4$.

$$
\begin{aligned}
(x + 3y)^4 &= {_4C_0}x^4 + {_4C_1}x^3(3y) + {_4C_2}x^2(3y)^2 + {_4C_3}x(3y)^3 + {_4C_4}(3y)^4 \\
&= x^4 + 4(3)x^3y + 6(3^2)x^2y^2 + 4(3^3)xy^3 + 3^4y^4 \\
&= x^4 + 12x^3y + 54x^2y^2 + 108xy^3 + 81y^4
\end{aligned}
$$

12 Assessment

VOCABULARY QUESTIONS

For Questions 1–3, complete each sentence.

1. On a graph, the _?_ of a vertex is five if the vertex is connected to five other vertices.

2. A(n) _?_ is a graph in which the edges are arrows.

3. A(n) _?_ is an arrangement of items in which position is important, and a(n) _?_ is a selection of items in which position is *not* important.

SECTIONS 12.1 *and* 12.2

For each graph in Questions 4–6:

a. Find the degree of each vertex.

b. Draw an equivalent graph.

c. Color your graph so that vertices that share an edge are different colors. Try to use as few colors as possible.

4. 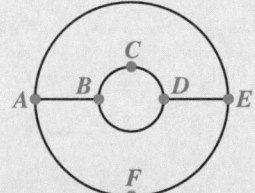   5. 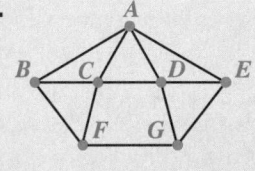   6.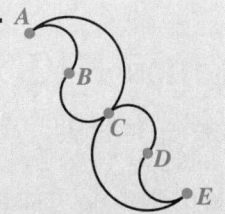

GAMES Some of the characters in a game are listed below along with their required companions.

- A healer requires an interpreter.
- A merchant requires a scribe.
- An advisor requires a merchant or an interpreter.
- A knight requires a healer.
- A prince requires a knight or an advisor.

7. Draw a directed graph to represent this situation.

8. What do the arrows in the graph represent?

9. Write a matrix that represents the directed graph.

10. Square the matrix from Question 9. What do the elements represent?

Simplify each expression.

11. $_7P_4$ **12.** $_7P_1$ **13.** $_6P_2$ **14.** $_6P_4$

15. $_7C_4$ **16.** $_7C_1$ **17.** $_5C_3 + _5C_4$ **18.** $_9C_3 + _9C_4$

19. GOVERNMENT Your Social Security number consists of nine digits from 0 to 9. The first three encode the location of your birth. For example, numbers beginning "001" originate in a certain area of New York. If "000" is not allowed for the first three digits, and "000000" is not allowed for the last six digits, how many possible nine-digit Social Security numbers are there? (*Hint:* Don't subtract the case of all 0 digits twice.)

20. Open-ended Problem Write two words with five letters each. One of your words should have a single letter repeated. Find the number of different five-letter permutations that can be made for each word. Are the numbers of permutations the same? Why or why not?

21. Find term 6 of row 12 of Pascal's triangle.

22. Find the coefficient of a^6b^2 in the expansion of $(a + b)^8$.

Use the binomial theorem to expand each power of a binomial.

23. $(x - y)^3$ **24.** $(2a + 2b)^4$ **25.** $(2y + z^2)^5$

26. One term in the expansion of $(a + b)^n$ is $78a^{11}b^2$. What is the next term of the expansion?

PERFORMANCE TASK

27. Draw a map of a fictitious continent with at least seven territories. Draw a graph that represents your map and use it to color your map so that territories that border each other are different colors. Write a story about the continent you created. Demonstrate what you have learned in this chapter by including mathematical explanations in your story, such as:

- a *directed graph* to show which territories export to other territories
- a *matrix* to represent your directed graph
- a *squared matrix* to show which territories export to the others through one intermediate territory
- the number of ways that all of the leaders of your territories can line up for a group photograph (a *permutation*)
- the number of handshakes if all the leaders shake hands after the group photograph (a *combination*)

Cumulative Assessment

CHAPTERS $10-12$

CHAPTER 10

Tell whether each sequence is *arithmetic*, *geometric*, or *neither*. Then write a formula for t_n and find the 10th term of the sequence.

1. $\dfrac{1}{2}, \dfrac{2}{3}, \dfrac{3}{4}, \dfrac{4}{5}, \ldots$ **2.** $2, 1, 0.5, 0.25, \ldots$ **3.** $5\sqrt{2}, 4\sqrt{2}, 3\sqrt{2}, 2\sqrt{2}, \ldots$

4. Find the arithmetic mean and the geometric mean of 0.9 and 1.6.

5. Write a recursive formula for each sequence in Questions 2 and 3.

Find the first 4 terms of each sequence.

6. $t_n = n(n + 1)$ **7.** $t_1 = 2; t_n = 2\left(t_{n-1}\right) + 1$

8. Open-ended Problem Make up an arithmetic sequence with a starting value of -8. Write a recursive formula and an explicit formula for your sequence.

Find the sum of each series, if the sum exists.

9. $2 + (-6) + 18 + (-54) + \cdots + 13{,}122$

10. $\displaystyle\sum_{n=1}^{\infty} 100(0.2)^{n-1}$ **11.** $\displaystyle\sum_{n=1}^{\infty} (2n - 3)$

12. a. Write the series of the first 100 positive even integers in sigma notation and find the sum.

     b. Find the value of n for which the sum of the first n positive even integers is 342.

13. Writing Explain how to write $0.727272\ldots$ as a fraction in simplest form.

CHAPTER 11

14. For the points $A(-4, -9)$ and $B(1, 3)$, find the distance between the points and an equation of the perpendicular bisector of $\overline{AB}$.

15. Show that the points $P(-5, -2)$, $Q(1, 2)$, $R(3, -1)$, and $S(-3, -5)$ are the vertices of a rectangle. Explain your reasoning.

For Questions 16–19, write an equation of each figure and then graph the equation.

16. a parabola with focus $(1, 4)$ and directrix $x = 3$

17. a circle with center $(-4, 0)$ and radius $2\sqrt{2}$

18. an ellipse with center $(1, -3)$, one vertex at $(1, 7)$, and one focus at $(1, 5)$

19. a hyperbola with center $(0, 0)$, one vertex at $(-3, 0)$, and one focus at $(-5, 0)$

20. Identify the vertex, the focus, and the directrix of the parabola with equation $y = \frac{1}{2}(x - 3)^2 + 5$.

21. Identify the center and the radius of the circle with equation $x^2 - 8x + 16 + y^2 = 4$.

22. Identify the center, the vertices, and the foci of the hyperbola with equation $\frac{(x - 2)^2}{64} - \frac{y^2}{225} = 1$. Find equations of the asymptotes.

23. Open-ended Problem Write an equation of a degenerate conic. Identify the conic and graph your equation.

24. Rewrite the equation $x^2 + 4y^2 + 6x + 5 = 0$ in standard form. Identify the conic, and state its important characteristics.

25. Writing Describe the placement and angle of a plane that intersects a double cone to form each of the conic sections.

CHAPTER 12

26. For the graph at the right, find the degree of each vertex, draw an equivalent graph, and color your graph so that vertices that share an edge must be different colors. Try to use as few colors as possible.

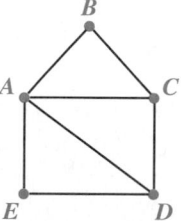

27. Writing Suppose a group of students have signed up to participate in volunteer activities, and some students want to participate in more than one activity. Describe how a graph can be used to find the least number of meeting times required to plan the volunteer activities.

28. COMPUTERS The table describes the connections in a computer network.

a. Draw a directed graph to represent the computer network.

b. Represent the directed graph in part (a) with a matrix.

c. **Technology** Use a graphing calculator or software with matrix calculation capabilities to determine which computers can send data to all the others either directly or through an intermediate computer.

Transmitting computer	Receiving computer(s)
A	B, C
B	D, E
C	A, D
D	E
E	B, C

Calculate each expression.

29. $\dfrac{9!}{5!}$

30. $_7P_4$

31. $_7C_4$

32. $\dfrac{n!}{(n - 2)!}$

33. In how many ways can five cards be randomly chosen and turned face up in a row from a standard deck of 52 cards?

34. A person who buys a punch card at an amusement center can choose 8 rides out of 12 and two activities from the following: miniature golf, bowling, and a water slide. How many different choices are possible?

35. Open-ended Problem Describe situations in which you would find $_4P_2$ and $_4C_2$. How are the two situations similar? How are they different?

36. Find the coefficients of x^7y^7 and x^9y^5 in the expansion of $(x + y)^{14}$.

13 | Probability

Probability in the courtroom

Robert Ward

*I*f you like drama, says attorney Robert Ward, there's nothing like a trial. "Courtrooms are a great place to study human behavior. You can see a full range of emotions in a single trial. For me, the attraction of trying cases is that it forces me to use the ideas and strategies I teach in the classroom. I get a firsthand chance to see how these principles actually play out in the courtroom."

"Someday, hopefully, DNA testing will make it easier for us to get criminals off the streets."

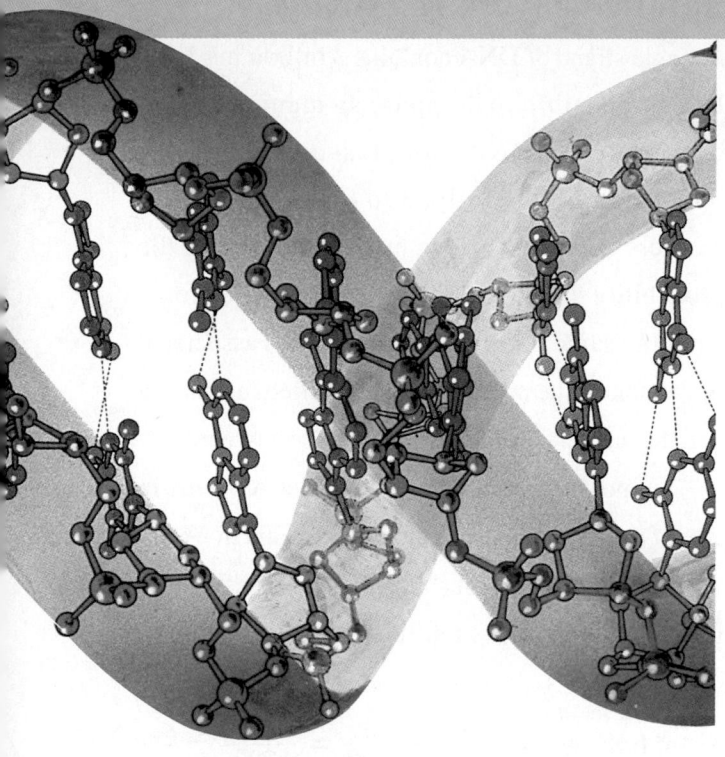

In Search of the Perfect Match

Scientists can analyze blood or hair samples by isolating the DNA contained within a cell nucleus. DNA is considered to be the chemical basis of heredity because it makes up the genes. Everyone in the world, except for identical twins, has a unique set of genes. The DNA found at a crime scene is compared with a DNA sample taken from the suspect to see if they match.

The DNA samples are compared at several well-known sites on the DNA chain. The genetic sequence found at each site is called a *marker*. If all the markers in the two samples "match up" (that is, look the same), then the two samples just may come from the same person.

Relevant Evidence

When he's not defending clients in court, Ward can be found teaching students the nuances of criminal law at New England School of Law. One of his classes is a course on evidence, which explores, among other things, the circumstances in which laboratory findings can be admitted in a trial.

In his class, Ward discusses the pros and cons of DNA testing, a relatively new technique hailed by some crime fighters as the greatest technological advance since fingerprinting. This method can be used when samples of, say, a perpetrator's blood or hair are found at the scene of a crime.

DNA Testing on Trial

DNA testing is based on the idea that the markers occur independently of one another in a strand of DNA. "Some critics have argued that DNA markers are not necessarily independent," Ward notes. "Questions have also been raised about the reliability of the tests themselves. In some cases, lab technicians may do the tests properly but still misread the results."

Given present uncertainties about the technique, Ward believes that DNA testing is most useful right now for upholding someone's innocence rather than proving his or her guilt. In time, he says, the technique should become more dependable as testing procedures improve and scientists reach a broader consensus on the methodology.

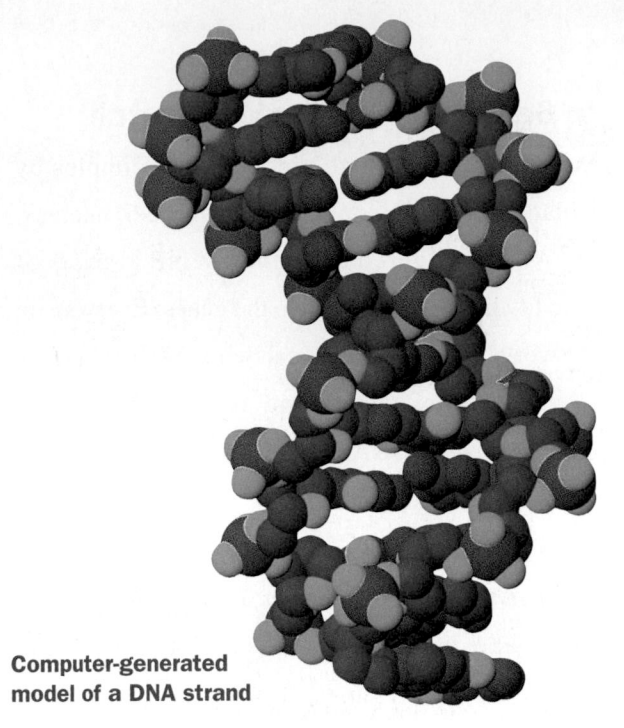

Computer-generated
model of a DNA strand

Determining the Probability

A single strand of DNA contains 3 billion markers. It is impossible to compare so many markers between two DNA samples using present technology. Typically, 3 to 6 distinguishing markers are compared. In order to determine the probability that two different DNA samples come from the same person, the probabilities associated with each of the markers are multiplied, assuming that the markers are *independent* of one another.

Suppose, for example, that marker A occurs in **2%** of the population and marker B occurs in **5%** of the population. The probability of two DNA samples having both markers is 1 in 1000:

$$\begin{array}{c}\textbf{probability}\\\textbf{of having}\\\textbf{both markers}\end{array} = \frac{2}{100} \times \frac{5}{100} = \frac{1}{1000}$$

marker **A**   marker **B**

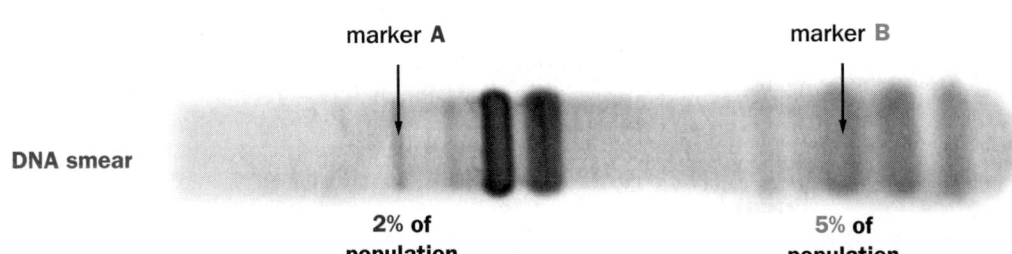

DNA smear

2% of
population

5% of
population

EXPLORE AND CONNECT

**Robert Ward reads
a legal brief in his
office.**

1. Writing Do you think the probability calculated above is low enough to convict a suspect if a match is made between the suspect's DNA and the DNA found at a crime scene? What do you think happens to the probability of a match if more markers are compared?

2. Research DNA is just one type of identifying evidence used in court trials. Find out what other kinds of identifying evidence are admitted in trials. Are there any problems associated with them?

3. Project Using information from your research, rank the different types of identifying evidence allowed in trials, based on your estimates of the probability of their accuracy. Explain your ranking.

**Mathematics
& Robert Ward**

In this chapter, you will learn more about how mathematics is related to legal issues.

**Related Examples
and Exercises**

Section 13.3
• **Example 3**
• **Exercises 10 and 11**

Section 13.4
• **Exercises 21–24**

13.1 Exploring Probability

During a Chicago Bulls basketball game on April 14, 1993, spectator Don Calhoun got a chance to win $1 million. All he had to do was make a 71 ft, three-quarter-court shot from the opposite free-throw line. Amazingly, Calhoun made the shot and became a rich man.

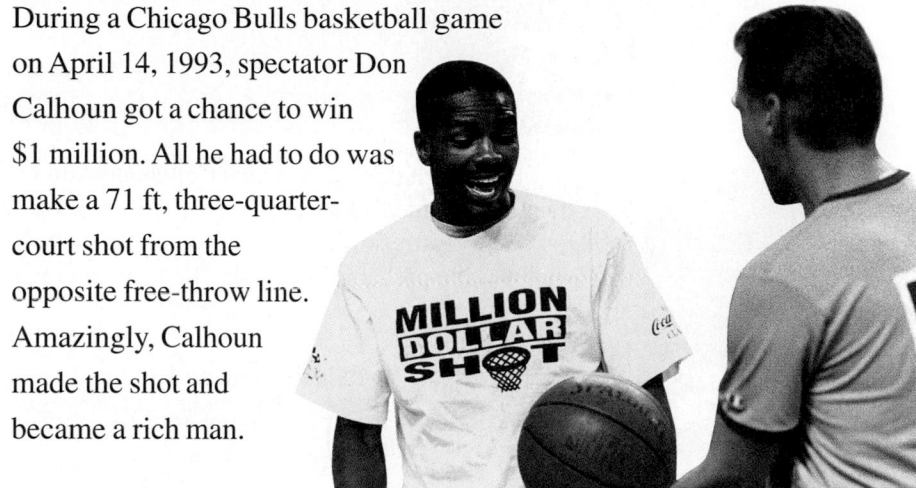

THINK AND COMMUNICATE

1. If Don Calhoun had taken 100 shots at a distance of 71 ft from the basket, how many do you think he would have made? How would you express Calhoun's chances of making a 71 ft shot?

2. A regulation free throw is shot 15 ft from the basket. What do you think your chances are of making a regulation free throw?

The **probability** of an event A, denoted $P(A)$, is the fraction of the time that A is expected to happen. For example, if you make **3** out of every **4** free throws you attempt, then the probability of your making a free throw is $\frac{3}{4}$, or 0.75.

For any event A, $0 \le P(A) \le 1$. The closer $P(A)$ is to 0, the *less* likely A is to happen. The closer $P(A)$ is to 1, the *more* likely A is to happen.

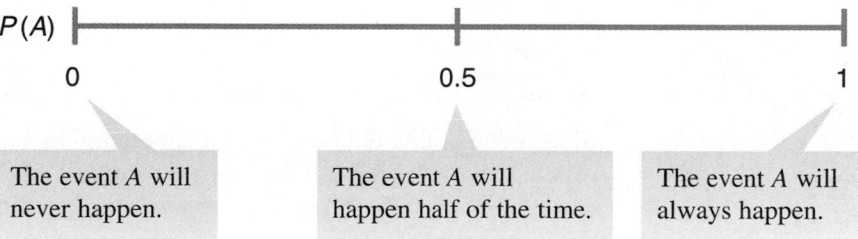

You can estimate your probability of making a free throw by performing an experiment in which you shoot, say, 50 free throws and record how many you make. The ratio of the number of free throws made to the number attempted is your *experimental probability* of making a free throw.

Suppose you perform an experiment consisting of a certain number of *trials* (such as free-throw attempts). The **experimental probability** of an event A (such as making a free throw) is given by:

$$P(A) = \frac{\text{number of trials where } A \text{ happens}}{\text{total number of trials}}$$

EXAMPLE 1   Application: Sports

Lorri Bauman holds the National Collegiate Athletic Association (NCAA) Division 1 women's basketball record for the most free throws made. While at Drake University during 1981–1984, she made **907** free throws in **1090** attempts. What was the probability of Lorri Bauman making a free throw?

SOLUTION

Use the formula for experimental probability.

$$P\left(\begin{array}{c}\text{Lorri Bauman}\\\text{makes a free throw}\end{array}\right) = \frac{\text{number of free throws made}}{\text{number of free throws attempted}}$$

$$= \frac{907}{1090}$$

$$\approx 0.832$$

The probability of Lorri Bauman making a free throw was about 0.832.

Lorri Bauman (55) prepares to take a shot.

Theoretical Probability

Suppose you want to find the probability of getting an odd number when you roll a six-sided die. There are **6** possible outcomes:

Of these outcomes, **3** correspond to getting an odd number:

Since the possible outcomes are equally likely to happen, it is reasonable to expect that the fraction of the time you will roll an odd number is the ratio of the number of odd outcomes to the number of possible outcomes:

$$P(\text{rolling an odd number}) = \frac{3}{6} = 0.5$$

This value, 0.5, is the *theoretical probability* of rolling an odd number.

Suppose a procedure (such as rolling a die) can result in n equally likely outcomes. If an event A (such as rolling an odd number) corresponds to m of these outcomes, then the **theoretical probability** of A is given by:

$$P(A) = \frac{m}{n}$$

EXAMPLE 2   Application: Sports

Every spring, the National Basketball Association (NBA) holds a lottery to determine the order in which its teams choose new members from among college players. In the 1993 lottery, each team that did not make the playoffs was assigned a certain number of lottery entries. Teams with poorer records received more entries.

Team (with 1992–93 win-loss record)	Number of entries
Dallas (11–71)	11
Minnesota (19–63)	10
Washington (22–60)	9
Sacramento (25–57)	8
Philadelphia (26–56)	7
Milwaukee (28–54)	6
Golden State (34–48)	5
Denver (36–46)	4
Miami (36–46)	3
Detroit (40–42)	2
Orlando (41–41)	1

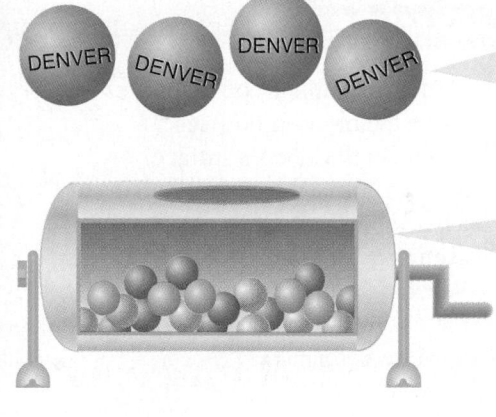

Each of a team's entries was represented by a ball with the team's name on it.

The balls were placed in a cylinder and were mixed well. The first ball drawn determined the team that chose first from the pool of college players.

What was the probability that Denver got to choose first in 1993?

SOLUTION

Each ball in the cylinder had an equally likely chance of being drawn first. Therefore:

$$P\left(\begin{array}{c}\text{Denver got}\\\text{to choose first}\end{array}\right) = \frac{\text{number of balls with Denver's name on them}}{\text{total number of balls}}$$

$$= \frac{4}{1 + 2 + 3 + \cdots + 11}$$

$$= \frac{4}{66}$$

$$= \frac{2}{33}, \text{ or about } 0.0606$$

The probability that Denver got to choose first was about 0.0606.

3. Orlando won the first pick in the NBA lottery in 1993 and chose Chris Webber of the University of Michigan. What was the probability of Orlando's winning?

4. a. Why is it possible to compute a team's probability of winning the first pick in the NBA lottery without first performing an experiment in which the lottery drawing is simulated many times?

b. Why *can't* you estimate a basketball player's probability of making a free throw without first having the player shoot many free throws?

Geometric Probability

Sometimes the probability of an event can be found by comparing the areas of geometric figures. The study of probability based on length, area, or volume is called **geometric probability**.

EXAMPLE 3   Application: Astronomy

Each year, millions of meteorites strike the moon, producing craters in the moon's surface. Many of these meteorites land in a *mare*—a large, flat plain that is seen as a dark patch from Earth. What is the probability that a meteorite striking the moon lands in Mare Serenitatis?

The area of Mare Serenitatis is approximately 125,000 mi².

The radius of the moon is about 1080 mi.

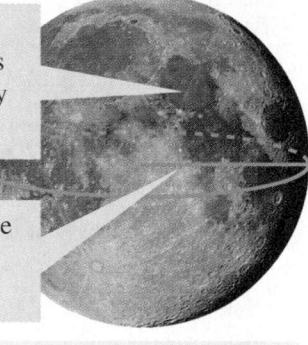

SOLUTION

It is reasonable to assume that each point on the moon has an equally likely chance of being hit by a meteorite. Therefore, the probability that a meteorite lands in Mare Serenitatis is just the fraction of the moon's surface area that the mare occupies:

$$P\left(\begin{array}{c}\text{a meteorite lands}\\\text{in Mare Serenitatis}\end{array}\right) = \frac{\text{area of Mare Serenitatis}}{\text{surface area of moon}}$$

$$\approx \frac{125{,}000}{4\pi(1080)^2}$$

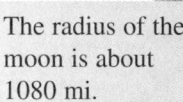 Use the formula **S.A. = $4\pi r^2$** for the surface area of a sphere.

$$\approx \frac{125{,}000}{14{,}657{,}415}$$

$$\approx 0.00853$$

The probability that a meteorite striking the moon lands in Mare Serenitatis is about 0.00853.

THINK AND COMMUNICATE

5. Out of every 1 million meteorites that strike the moon, how many would you expect to land in Mare Serenitatis? Explain.

☑ CHECKING KEY CONCEPTS

1. MANUFACTURING A quality control engineer randomly selects 250 hard disk drives from the assembly line of a computer manufacturing plant. The engineer finds that 8 of the drives are defective. Estimate the probability that a drive on the assembly line is defective.

2. Suppose you roll 2 six-sided dice, one white and one red. Find the probability that the number showing on the red die is greater than the number showing on the white die.

3. ASTRONOMY The area of the United States is about 3,618,770 mi². What is the probability that a meteorite striking Earth lands in the United States? (*Hint:* The radius of Earth is about 3963 mi.)

13.1 | **Exercises and Applications**

Extra Practice exercises on page 769

GAMES A standard deck of playing cards consists of 52 cards, with 13 cards in each of 4 *suits:* clubs, spades, diamonds, and hearts. *Face cards* are jacks, queens, and kings. Find the probability of choosing each type of card at random from a standard deck.

Clubs ♣ :	ace, 2, 3, 4, 5, 6, 7, 8, 9, 10, jack, queen, king
Spades ♠ :	ace, 2, 3, 4, 5, 6, 7, 8, 9, 10, jack, queen, king
Diamonds ♦ :	ace, 2, 3, 4, 5, 6, 7, 8, 9, 10, jack, queen, king
Hearts ♥ :	ace, 2, 3, 4, 5, 6, 7, 8, 9, 10, jack, queen, king

1. a 9 of diamonds

2. an ace

3. a heart

4. a face card

5. a red card

6. not a 3 of spades

7. not a club

8. a red or black card

9. an 11 of clubs

The targets shown are an equilateral triangle, a square, and a circle. Suppose a randomly thrown dart hits each target. Find the probability that the dart hits the target's shaded region. (The shaded regions are also formed from equilateral triangles, squares, and circles.)

10.
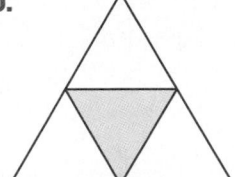

11.
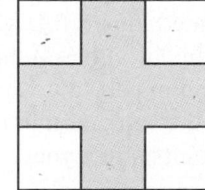

12.
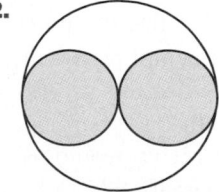

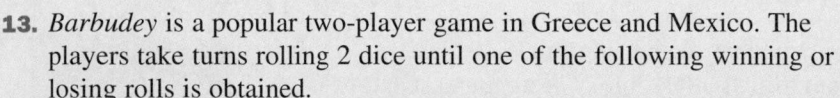

Connection ▶ GAMES

Did you know that the six-sided dice used in the United States today are almost identical to those used in China about 600 B.C. and in Egypt about 2000 B.C.? Games involving dice have been developed in virtually every part of the world.

13. *Barbudey* is a popular two-player game in Greece and Mexico. The players take turns rolling 2 dice until one of the following winning or losing rolls is obtained.

Ancient Egyptian dice

Winning rolls:

Losing rolls:

Find the probability of getting each type of roll.

a. a winning roll　　　　**b.** a losing roll　　　　**c.** neither a winning nor a losing roll

14. The Chinese game *kon mín yéung* is played with 6 six-sided dice. Players score points by rolling at least three of a kind.

　a. Cooperative Learning Work with a partner. You need 6 six-sided dice. Roll the dice 100 times. For each roll, record whether you get at least 3 of a kind. What is your experimental probability of getting at least 3 of a kind?

　b. Challenge Show that the theoretical probability of getting at least 3 of a kind when rolling 6 dice is $\dfrac{119}{324}$. How does this value compare with the experimental probability you obtained in part (a)?

15. Cooperative Learning You can use probability to approximate π. Work in a group of five students. You each need a graphing calculator.

　a. Suppose you randomly choose a point in the square region at the right. Show that:

$$P\left(\begin{array}{c}\text{the point lies in the} \\ \text{shaded quarter circle}\end{array}\right) = \frac{\pi}{4}$$

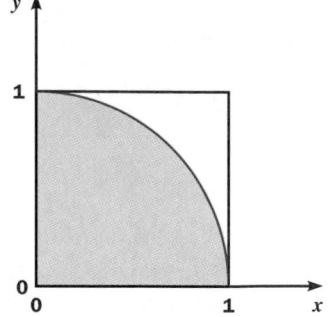

　b. 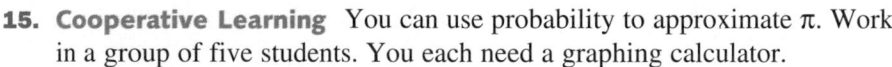 **Technology** You can generate a random point (x, y) in the square region by having your calculator produce two random numbers x and y between 0 and 1. Each of you should generate 20 such points and determine whether each point lies in the quarter circle. (*Hint:* Note that (x, y) lies in the quarter circle if and only if $x^2 + y^2 \le 1$.)

　c. Based on the 100 points your group generated in part (b), what is the experimental probability that a random point (x, y) lies in the quarter circle? Use this probability and the equation in part (a) to approximate π. How close is your approximation to 3.14, the value of π to two decimal places? How would using more random points affect your approximation?

16. **PHYSICS** A *neutron* is a subatomic particle that has no electric charge. When a beam of neutrons is directed at a piece of lead foil, almost all of the neutrons pass straight through the foil. This is because the lead atoms that make up the foil are mostly empty space. Sometimes, however, a neutron will strike a *nucleus*—the positively charged central region of an atom—and be deflected from its original path.

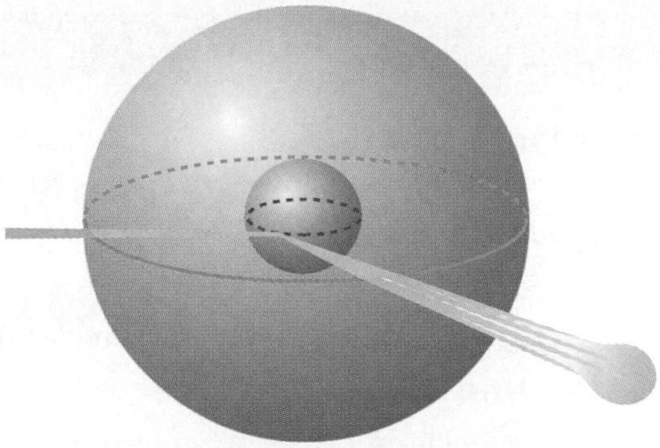

   a. The second diagram shows the cross sections of a lead atom and several neutrons. Complete this sentence: A neutron will strike the nucleus if the distance between the centers of the neutron and nucleus is less than   ? .

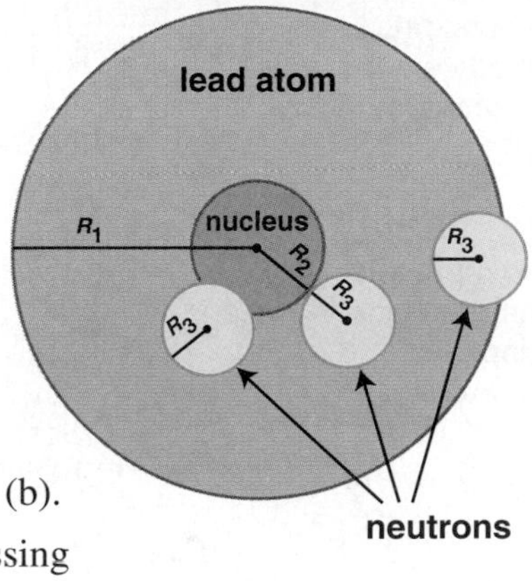

   b. In terms of R_1, R_2, and R_3, what is the probability that a neutron passing through a lead atom strikes the atom's nucleus? (*Hint:* Assume the path of the neutron's center intersects the atom's cross section at a random point.)

   c. The values (in meters) of R_1, R_2, and R_3 are as follows:
$$R_1 = 1.75 \times 10^{-10}$$
$$R_2 = 8.27 \times 10^{-15}$$
$$R_3 = 1.40 \times 10^{-15}$$
   Use these values to compute the probability expressed in part (b).

   d. Complete this sentence: About 1 out of every   ?   neutrons passing through a lead atom will strike the atom's nucleus.

17. **Cooperative Learning** Work with a partner. You will need a cup and a thumbtack.

   a. One of you should put the thumbtack into the cup, shake the cup, and "pour" the thumbtack onto a flat surface. The other should record whether the thumbtack lands "point up" or "point down." Repeat this process at least 20 times. What is the experimental probability that the thumbtack lands "point up"?

   b. How do you think the probability in part (a) would change if the thumbtack had a wider head? a longer point?

ONGOING ASSESSMENT

18. **Writing** In your own words, define experimental, theoretical, and geometric probability. Give examples of real-world situations where each type of probability is used.

SPIRAL REVIEW

Use the binomial theorem to expand each power of a binomial. (*Section 12.5*)

19. $(a + b)^3$ 20. $(x - y)^5$ 21. $(2u + 3v)^4$ 22. $\left(u^2 + v^3\right)^4$

Find all real and imaginary zeros of each function. Identify any double or triple zeros. (*Section 9.5*)

23. $f(x) = x^3 - 3x^2 - 7x + 12$ 24. $g(x) = 4x^4 - 4x^3 + 5x^2 - 4x + 1$

13.2 Working with Multiple Events

Learn how to...

- **find probabilities involving independent, mutually exclusive, and complementary events**

So you can...

- **solve problems about genetics, for example**

Many traits of plants and animals are passed from one generation to another through *genes*. Genes occur in pairs, and each pair controls a particular trait. One kind of pea plant has a pair of "height genes" that determines whether the plant is tall or short.

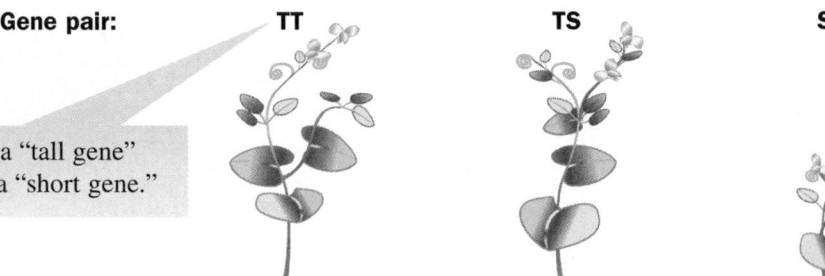

Gene pair: TT   TS   SS

Use T for a "tall gene" and S for a "short gene."

Plant height: **tall**   **tall**   **short**

Notice that a TS pair—a tall gene and a short gene—produces a tall plant. The tall gene is said to be *dominant* because it "overpowers" the short gene. The short gene is said to be *recessive*.

THINK AND COMMUNICATE

Suppose two pea plants, plant 1 and plant 2, both have a TS gene pair. The plants can be crossed to produce a new "child" plant. The height genes of the child consist of one height gene from each parent. A parent is equally likely to transmit its T or S gene.

	Plant 2	
	T	S
Plant 1 T	?	?
S	?	?

1. Copy and complete the table showing the possible gene pairs for the child plant.

2. Find each probability.

   a. $P\left(\dfrac{\text{plant 1}}{\text{transmits S gene}}\right)$ **b.** $P\left(\dfrac{\text{plant 2}}{\text{transmits S gene}}\right)$ **c.** $P\left(\dfrac{\text{child}}{\text{is short}}\right)$

   d. How are the probabilities in parts (a)–(c) related?

3. Find each probability.

   a. $P\left(\dfrac{\text{child has a}}{\text{TT gene pair}}\right)$ **b.** $P\left(\dfrac{\text{child has a}}{\text{TS gene pair}}\right)$ **c.** $P\left(\dfrac{\text{child}}{\text{is tall}}\right)$

   d. How are the probabilities in parts (a)–(c) related?

4. How can you use $P(\text{child is short})$ to find $P(\text{child is tall})$?

Think and Communicate Questions 2–4 illustrate the following types of events.

	•••• Multiple Events ••••	
Definition	**Rule**	**Example: Suppose two pea plants, each with TS height genes, are crossed to produce a third plant.**
Events A and B are **independent** if the occurrence of A does not affect whether B happens.	$P(A \text{ and } B) = P(A) \cdot P(B)$	A: One parent plant transmits an S gene. B: The other parent plant transmits an S gene. $P(A \text{ and } B) = \dfrac{1}{2} \cdot \dfrac{1}{2} = \dfrac{1}{4}$
Events A and B are **mutually exclusive** if they cannot both happen.	$P(A \text{ or } B) = P(A) + P(B)$	A: The child plant gets a TT gene pair. B: The child plant gets a TS gene pair. $P(A \text{ or } B) = \dfrac{1}{4} + \dfrac{1}{2} = \dfrac{3}{4}$
Events A and B are **complementary** if they are mutually exclusive and one of the events must happen.	$P(B) = 1 - P(A)$	A: The child plant is short. B: The child plant is tall. $P(B) = 1 - \dfrac{1}{4} = \dfrac{3}{4}$

EXAMPLE 1

Suppose a card is drawn at random from a standard deck of 52 playing cards. (See page 615 for a description of such a deck.) The card is put back in the deck, and a card is again drawn at random. Find the probability of each event.

a. The first card is a 7 of hearts, and the second card is a 2 of spades.

b. The first card is either a face card or an ace.

SOLUTION

a. Find $P(A \text{ and } B)$ where A is the event "the first card is a 7 of hearts" and B is the event "the second card is a 2 of spades."

$$P(A \text{ and } B) = P(A) \cdot P(B)$$

There is **one** 7 of hearts and **one** 2 of spades in a standard 52-card deck.

$$= \frac{1}{52} \cdot \frac{1}{52} \qquad \text{Events } A \text{ and } B \text{ are independent.}$$

$$= \frac{1}{2704}, \text{ or about } 0.000370$$

b. Find $P(A \text{ or } B)$ where A is the event "the first card is a face card" and B is the event "the first card is an ace."

$$P(A \text{ or } B) = P(A) + P(B) \qquad \text{Events } A \text{ and } B \text{ are mutually exclusive.}$$

There are **12** face cards and **4** aces in a standard 52-card deck.

$$= \frac{12}{52} + \frac{4}{52}$$

$$= \frac{16}{52}$$

$$= \frac{4}{13}, \text{ or about } 0.308$$

THINK AND COMMUNICATE

5. Look back at Example 1. Let A be the event "the first card is a 2," and let B be the event "the first card is a diamond." Find $P(A$ or $B)$ and $P(A) + P(B)$. Are these expressions equal? If not, why not?

6. Suppose Example 1 is changed so that the first card drawn is *not* put back in the deck before the second card is drawn. What effect, if any, does this have on the answer to part (a) of Example 1?

EXAMPLE 2 **Application: Games**

In the game of Monopoly®, a player sometimes has to "go to jail." The player can get out of jail by rolling doubles with a pair of six-sided dice. ("Doubles" is a roll such that both dice show the same number.) If the player doesn't roll doubles in three tries, the player must pay $50 to get out of jail. What is the probability of getting out of jail without paying $50?

SOLUTION

The desired probability is the same as P(rolling doubles in 3 tries). Let D_i be the event "doubles is rolled on the ith try," and let N_i be the event "doubles is not rolled on the ith try." Note that:

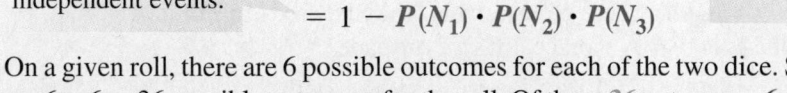

$$P\left(\begin{matrix}\text{rolling doubles}\\\text{in 3 tries}\end{matrix}\right) = 1 - P\left(\begin{matrix}\text{not rolling doubles}\\\text{in 3 tries}\end{matrix}\right)$$

> Rolling and not rolling doubles in 3 tries are complementary events.

N_1, N_2, and N_3 are independent events.

$$= 1 - P(N_1 \text{ and } N_2 \text{ and } N_3)$$

$$= 1 - P(N_1) \cdot P(N_2) \cdot P(N_3)$$

On a given roll, there are 6 possible outcomes for each of the two dice. So there are $6 \cdot 6 = 36$ possible outcomes for the roll. Of these **36** outcomes, **6** are doubles (two 1's, two 2's, . . . , two 6's). Therefore:

$$P(D_i) = \frac{6}{36} = \frac{1}{6} \text{ and } P(N_i) = 1 - P(D_i) = \frac{5}{6}$$

> D_i and N_i are complementary events.

It follows that:

$$P\left(\begin{matrix}\text{rolling doubles}\\\text{in 3 tries}\end{matrix}\right) = 1 - \left(\frac{5}{6}\right)\left(\frac{5}{6}\right)\left(\frac{5}{6}\right)$$

$$= 1 - \frac{125}{216}$$

$$= \frac{91}{216}, \text{ or about } 0.421$$

The probability of getting out of jail without paying $50 is about 0.421.

THINK AND COMMUNICATE

7. Look back at Example 2. Find each probability.

    a. $P(D_1)$ **b.** $P(N_1 \text{ and } D_2)$ **c.** $P(N_1 \text{ and } N_2 \text{ and } D_3)$

    d. Find the sum of the probabilities in parts (a)–(c). What do you notice? Explain why this result makes sense.

✓ CHECKING KEY CONCEPTS

Suppose a card is drawn at random from a standard deck of 52 playing cards. (See page 615 for a description of such a deck.) The card is put back in the deck, and a card is again drawn at random. Find the probability of each event.

1. The first card is a club, and the second card is a spade.

2. The first card is a heart or an 8 of spades.

3. The cards are not both jacks.

4. The first card is black, and the second card is a queen of hearts or a 2.

5. At least one of the cards is a diamond.

6. Exactly one of the cards is a diamond.

13.2 Exercises and Applications

Extra Practice exercises on page 769

1. **Writing** Explain the conditions under which each rule applies.

    a. $P(A \text{ and } B) = P(A) \cdot P(B)$ **b.** $P(A \text{ or } B) = P(A) + P(B)$ **c.** $P(B) = 1 - P(A)$

METEOROLOGY A meteorologist giving the weekend weather forecast says that there is a 30% chance of rain on Saturday and a 40% chance of rain on Sunday. Assuming these chances are correct and Sunday's weather is independent of Saturday's weather, find the probability of each event.

2. It doesn't rain on Saturday.

3. It rains on both Saturday and Sunday.

4. It rains on Saturday and doesn't rain on Sunday.

5. It doesn't rain at all over the weekend.

6. It rains on at least one day of the weekend.

7. It rains on exactly one day of the weekend.

8. **MOVIES** In the 1943 film *Casablanca*, casino owner Rick Blaine tells a roulette player to choose number 22. Incredibly, number 22 comes up twice in a row. What is the probability of this happening? (*Note:* In roulette, a ball spins around a wheel and lands in a slot on the wheel's edge. There are 36 slots numbered 1 through 36, plus two special slots labeled 0 and 00. The ball is equally likely to land in any slot.)

In Douglas Hofstadter's short story *The Tale of Happiton*, a mischievous demon writes a letter to the residents of a town called Happiton. The letter begins like this:

I've got some bad news and some good news for you. The bad first. You know your bell [in the courthouse clock] that rings every hour on the hour? Well, I've set it up so that each time it rings, there is exactly one chance in a hundred thousand—that is, $\frac{1}{100,000}$—that a Very Bad Thing will occur. The way I determine if that Bad Thing will occur is, I have this robot arm fling five dice and see if they all land with "7" on top.

9. The demon's dice are 20-sided. The sides of each die are numbered 0 through 9, with each number appearing on two opposite sides.

   a. What is the probability that a given die lands with "7" showing?

   b. Use your answer from part (a) to prove that the probability of all five dice landing with "7" showing is $\frac{1}{100,000}$, as stated in the passage.

10. In Hofstadter's story, Nellie Doobar, the mathematics teacher at Happiton High School, says that "the chances we'll make it through any eight-year period [without the Very Bad Thing happening] are almost exactly fifty-fifty."

   a. How many times will the dice be rolled in an eight-year period? (Assume 1 year = 365.25 days.)

   b. Find the probability that the Very Bad Thing happens during an eight-year period. Is your answer consistent with Nellie Doobar's remark?

   c. **Challenge** The "good news" referred to in the passage is that the demon will make the clock bell ring less often than once an hour provided the residents of Happiton write him postcards. If the demon receives p postcards on a given day, then the time t (in hours) between rings the next day will be:

   $$t = (1.00001)^p$$

   The population of Happiton is 20,000. How many postcards must each resident write per day for there to be only a 5% chance of the Very Bad Thing happening during an eight-year period?

11. **SPORTS** Elena, a member of her high school's basketball team, makes 80% of the free throws she attempts. Suppose Elena is fouled and gets to shoot a pair of free throws. Find the probability of each event.

   a. Elena makes both free throws.

   b. Elena makes at least 1 free throw.

   c. Elena makes exactly one free throw.

   d. Elena misses both free throws.

12. In some state lotteries, you buy a ticket and choose 6 numbers from among the integers 1, 2, 3, . . . , 49. (The order in which you choose the numbers doesn't matter.) You win all or a portion of the lottery if your numbers are the ones selected in that day's lottery drawing.

   a. If you buy a single lottery ticket, what is your probability of winning?

   b. Suppose you buy a lottery ticket every day for 60 years. What is your probability of winning at least once during that time? (Assume 1 year = 365.25 days.)

   c. **Writing** Based on your answer, do you think playing this type of lottery is a good idea? Explain.

13. **SAT/ACT Preview** What is the minimum number of times you must roll a six-sided die to have at least a 50% chance of getting a 1?

   A. 3   **B.** 4   **C.** 5   **D.** 6   **E.** 7

14. **ELECTRONICS** An automatic garage door opener includes a transmitter that you carry in your car and a receiver that you attach to your garage. On one brand of opener, both the transmitter and the receiver have 8 switches that you can set to either "on" or "off." The switches on the transmitter and the receiver must be set identically for the garage door to open.

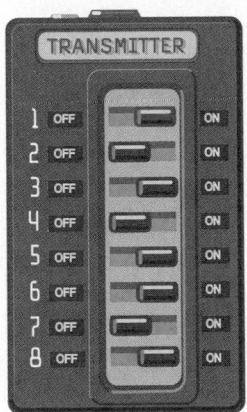

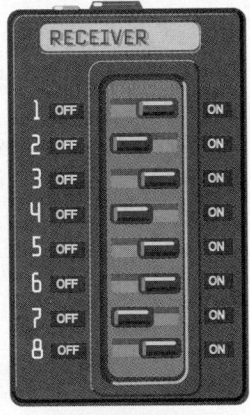

Setting 1

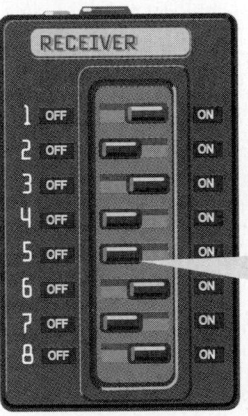

Setting 2

The position of switch 5 is different for the transmitter and the receiver, so the garage door *will not* open.

Setting 1 matches the transmitter's setting, so the garage door *will* open.

   a. Kaya bought the brand of garage door opener described above. In how many ways can she set the switches on the transmitter and the receiver so that the door opens?

   b. Suppose 10 houses on Kaya's street have her brand of garage door opener (including Kaya's house). What is the probability that at least one other opener has the same switch settings as Kaya's? Should Kaya feel confident that a neighbor won't accidentally open her garage door? Explain.

   c. Consider the probability that *any* 2 or more of the 10 houses in part (b) have the same switch settings for their garage door openers. Would you expect this probability to be higher or lower than the probability you found in part (b)? Check your answer by calculating this probability.

   d. **Challenge** Find the probability that at least 3 of the 10 houses in part (b) have the same switch settings for their garage door openers.

15. SPORTS In 1941, baseball player Joe DiMaggio of the New York Yankees got at least one hit in 56 consecutive games. What is the probability of this happening in any given sequence of 56 games? Assume DiMaggio batted an average of 4 times per game and had a 32.5% chance of getting a hit during each at-bat.

16. Visual Thinking The rectangular target shown contains two overlapping colored regions—the red square *PQSU* and the blue triangle *QRT*. Suppose a randomly thrown dart hits the target. Let *A* be the event "the dart hits square *PQSU*," and let *B* be the event "the dart hits triangle *QRT*."

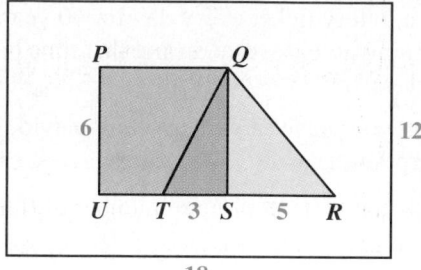

a. Writing Explain why the rule

$$P(A \text{ or } B) = P(A) + P(B)$$

on page 619 does not apply in this situation.

b. Find $P(A \text{ or } B)$ by calculating the ratio of the area of quadrilateral *PQRU* to the area of the entire rectangular target. Show that:

$$P(A \text{ or } B) = P(A) + P(B) - P(A \text{ and } B)$$

c. Suppose a six-sided die is rolled. Let *A* be the event "the number showing on the die is even," and let *B* be the event "the number showing on the die is greater than 4." Find $P(A \text{ or } B)$, $P(A)$, $P(B)$, and $P(A \text{ and } B)$. Show that these probabilities satisfy the equation in part (b), but not the equation in part (a).

d. Writing Explain why the equation in part (b) is a generalization of the equation in part (a).

ONGOING ASSESSMENT

17. Open-ended Problem Give examples of independent, mutually exclusive, and complementary events. Write and solve at least two probability problems involving the events you chose.

SPIRAL REVIEW

18. The sequence of numbers and letters on a California license plate has the form

d–LLL–DDD

where *d* is a digit from 1 to 9, each *L* is any letter except "O," and each *D* is any digit from 0 to 9. Find the probability that the three letters on a California license plate spell "CAT." *(Section 13.1)*

Find the sum of each series. *(Sections 10.4 and 10.5)*

19. $2 + 7 + 12 + 17 + \cdots + 87$

20. $\displaystyle\sum_{n=1}^{\infty} 17(-0.25)^{n-1}$

13.3

Using Conditional Probability

Learn how to...

- **find conditional probabilities**

So you can...

- **make judgments about evidence presented in court cases, for example**

Most movies are screened by test audiences before they open at your local theater. This helps movie studios predict whether their movies will be successful. It also allows studios to target advertising toward groups (such as men or women) that seem most receptive to a particular movie.

EXPLORATION
COOPERATIVE LEARNING

Conducting a Movie Survey

Work as an entire class.
You will need:
- small slips of paper

1 Your class should choose a movie that most of you have seen.

2 Each of you should have a slip of paper. On your slip, write whether you are male or female and whether you liked or disliked the movie. If you haven't seen the movie, write "no opinion."

3 Collect all the slips of paper. Tally the results of the survey in a table like the one shown in the photo.

Questions

1. Suppose a student is randomly selected from your class. Let *A* be the event "the student is female," and let *B* be the event "the student liked the movie."

   a. Find *P*(*A* and *B*) and *P*(*A*) • *P*(*B*).

   b. Are the events *A* and *B* independent? Explain.

2. Suppose a student is randomly selected from the *males* in your class. What is the probability that the student liked the movie? disliked the movie? had no opinion about the movie?

It is often important to know the probability of an event under restricted conditions. For example, a physician may need to know the probability that a patient has the flu even though the patient has no fever. You can write this probability as:

$$P\left(\begin{array}{c|c} \text{the patient} & \text{the patient} \\ \text{has the flu} & \text{has no fever} \end{array}\right)$$

You read this as "the probability that the patient has the flu, **given that** the patient has no fever."

In general, the **conditional probability** $P(B|A)$ is the probability of event B, given that event A has occurred.

EXAMPLE 1

Susan's class performed the Exploration on the previous page. Each student was asked whether he or she liked the movie *Jurassic Park*. The results of the survey are shown in the table.

	Liked	**Disliked**	**No opinion**	**Total**
Males	11	3	2	16
Females	9	4	6	19
Total	20	7	8	35

Find each probability for a randomly selected student in Susan's class.

a. $P\left(\begin{array}{c|c} \text{student liked} & \text{student} \\ \text{Jurassic Park} & \text{is male} \end{array}\right)$ **b.** $P\left(\begin{array}{c|c} \text{student} & \text{student disliked} \\ \text{is female} & \text{Jurassic Park} \end{array}\right)$

SOLUTION

a. There are **16** male students. Of these students, **11** liked *Jurassic Park*. Therefore:

$$P\left(\begin{array}{c|c} \text{student liked} & \text{student} \\ \text{Jurassic Park} & \text{is male} \end{array}\right) = \frac{11}{16}, \text{ or about } 0.688$$

b. There are **7** students who disliked *Jurassic Park*. Of these students, **4** are female. Therefore:

$$P\left(\begin{array}{c|c} \text{student} & \text{student disliked} \\ \text{is female} & \text{Jurassic Park} \end{array}\right) = \frac{4}{7}, \text{ or about } 0.571$$

THINK AND COMMUNICATE

1. Suppose a student is randomly selected from Susan's class. Let A be the event "the student liked *Jurassic Park*," and let B be the event "the student is male."

  a. Find $P(A \text{ and } B)$ and $P(A) \cdot P(B)$. Are the events A and B independent? Explain.

  b. Find $P(A) \cdot P(B|A)$. What do you notice?

In Section 13.2, you saw that $P(A \text{ and } B) = P(A) \cdot P(B)$ provided the events A and B are independent. Using conditional probability, you can generalize this equation so that it applies to *any* events A and B.

Conditional Probability

For any events A and B:

$$P(A \text{ and } B) = P(A) \cdot P(B|A) \quad \text{and} \quad P(B|A) = \frac{P(A \text{ and } B)}{P(A)}$$

EXAMPLE 2

Royce randomly chooses a marble from the jar shown. He places the marble in his pocket, then randomly chooses a second marble from the jar. Find the probability that the first marble is red and the second marble is green.

SOLUTION

Let R be the event "a red marble is chosen" and G be the event "a green marble is chosen."

Step 1 Make a *probability tree diagram* showing the possible outcomes of each stage of the experiment and the probabilities of these outcomes.

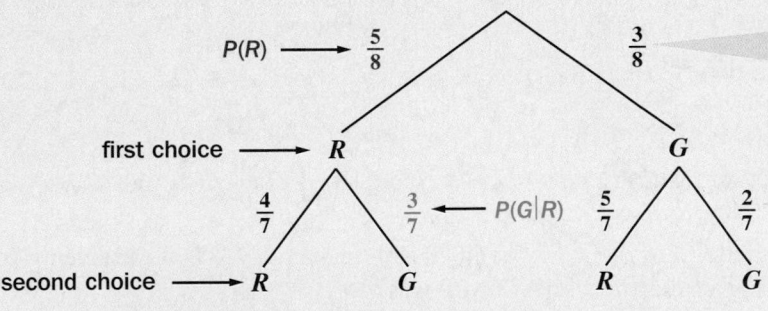

3 of the 8 marbles are green.

If the first marble chosen is green, then 2 of the remaining 7 marbles are green.

Step 2 Find $P(R \text{ and } G)$ using a formula for conditional probability.

$$P(R \text{ and } G) = P(R) \cdot P(G|R)$$

$$= \frac{5}{8} \cdot \frac{3}{7}$$

Read these probabilities from the tree diagram.

$$= \frac{15}{56}, \text{ or about } 0.268$$

The probability of choosing red first and green second is about 0.268.

THINK AND COMMUNICATE

2. Suppose an outcome of an experiment, such as choosing two marbles from a jar, corresponds to a path through a probability tree diagram. How is the probability of the outcome related to the probabilities assigned to the branches of the path?

EXAMPLE 3   Interview: Robert Ward

Attorneys like Robert Ward have to make sure that jurors understand how to interpret evidence based on probability. Consider the following problem:

You are a juror in a case involving a nighttime hit-and-run accident by a taxicab. Two cab companies, one with green cabs and one with blue cabs, operate in your city. You are told that:

- Of the cabs in your city, 85% are green and 15% are blue.

- A witness identified the cab as blue.

- In reenactments of the accident, the witness correctly identified the color of the cab 80% of the time.

What is the probability that the cab involved in the accident was blue, given the witness's statement?

SOLUTION

Step 1 Use symbols to represent the different events.

G: The cab was green. G_w: The witness says the cab was green.

B: The cab was blue. B_w: The witness says the cab was blue.

Step 2 Make a probability tree diagram representing the situation.

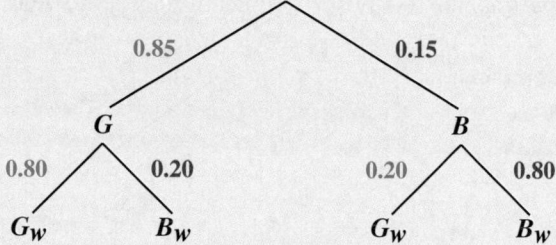

Step 3 Find $P(B|B_w)$, the probability that the cab involved in the accident was blue, given that the witness says it was blue.

$$P(B|B_w) = \frac{P(B \text{ and } B_w)}{P(B_w)}$$

Use a formula for conditional probability.

There are two paths through the tree diagram that lead to B_w. **Add** the probabilities associated with each path.

$$= \frac{P(B \text{ and } B_w)}{P(G \text{ and } B_w) + P(B \text{ and } B_w)}$$

$$= \frac{(0.15)(0.80)}{(0.85)(0.20) + (0.15)(0.80)}$$

$$\approx 0.414$$

The probability that the cab was blue, given the witness's statement, is about 0.414.

THINK AND COMMUNICATE

3. The problem in Example 3 is sometimes called the *juror's fallacy*. Why do you think this name is used?

Suppose two marbles are taken at random from the jar shown. The first marble is *not* put back in the jar before the second marble is taken. Find the probability of each event.

1. The second marble is green, given that the first marble is red.

2. The first marble is red, and the second marble is green.

3. The first marble is green, and the second marble is red.

4. Both marbles are red.

5. Both marbles are green.

6. The second marble is red.

13.3 | **Exercises and Applications**

Extra Practice exercises on page 769

The table gives the majors of students at a small technical college.

	Freshmen	Sophomores	Juniors	Seniors	Total
Architecture	50	30	40	25	145
Business	60	55	45	30	190
Engineering	40	35	50	55	180
Total	150	120	135	110	515

Suppose a student from the college is selected at random. Find each probability.

1. $P(\text{sophomore})$

2. $P(\text{architecture major})$

3. $P(\text{engineering major} \mid \text{freshman})$

4. $P(\text{architecture major} \mid \text{senior})$

5. $P(\text{freshman} \mid \text{engineering major})$

6. $P(\text{junior} \mid \text{business major})$

7. $P(\text{business major} \mid \text{junior or senior})$

8. $P(\text{business or engineering major} \mid \text{sophomore})$

9. LITERATURE In Michael Crichton's novel *Congo*, a team of scientists searches for the legendary Lost City of Zinj in Africa. At one point, the scientists consider parachuting into the jungles of Zaire.

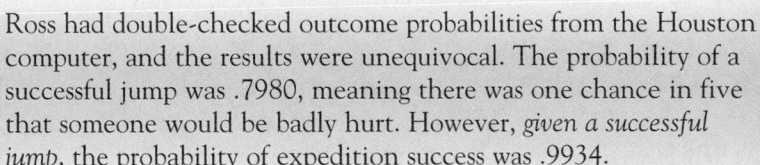

> Ross had double-checked outcome probabilities from the Houston computer, and the results were unequivocal. The probability of a successful jump was .7980, meaning there was one chance in five that someone would be badly hurt. However, *given a successful jump*, the probability of expedition success was .9934.

What is the probability that the scientists make a successful jump and then complete their expedition successfully?

Robert Ward

Look back at the article on pages 608–610.

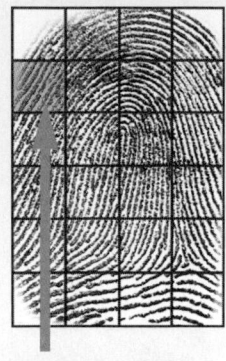

corresponding squares

Attorneys like Robert Ward often rely on fingerprints to place a suspect at the scene of a crime. Fingerprint evidence is considered very strong because each person's fingerprints are assumed to be unique. This was established mathematically by the British scientist Sir Francis Galton in 1892.

Galton wanted to estimate the probability that two fingerprints from two randomly chosen people match. He divided each of two hypothetical fingerprints into 24 square sections. He then defined these three events:

   A: The same number of ridges pass through two corresponding squares.

   B: The ridges adjacent to two corresponding squares have the same "general course."

   C: Two corresponding squares match.

10. Galton claimed that:

$$P(C) = P(C \mid (A \text{ and } B)) \cdot P(A \text{ and } B)$$

He assumed *A* and *B* are independent events, and estimated that

$P(A) = \dfrac{1}{2^{1/3}}$, $P(B) = \dfrac{1}{2^{1/6}}$, and $P(C \mid (A \text{ and } B)) = \dfrac{1}{2}$. Find the probability that two corresponding squares match.

11. Galton also assumed that matches between corresponding squares are independent events. Based on this assumption and your answer to Exercise 10, what is the probability of a complete match between two fingerprints from two randomly chosen people?

Suppose two cards are drawn at random from a standard deck of 52 playing cards. (See page 615 for a description of such a deck.) The first card is *not* put back in the deck before the second card is drawn. Find the probability of each event.

12. The second card is a 4, given that the first card is an ace.

13. The second card is a king, given that the first card is a king.

14. The first card is a club, and the second card is a diamond.

15. Both cards are black.

16. The first card is a queen, and the second card is a face card.

17. The second card is a 9.

18. **HEALTH** The ELISA test is used for diagnosing HIV (the human immunodeficiency virus, which causes AIDS). The accuracy of this test is such that 99.3% of people who are HIV-positive will test positive, and 99.99% of people who are HIV-negative will test negative. It has been estimated that about 0.7365% of the population in North America is HIV-positive. What is the probability that a randomly selected person in North America is HIV-negative, given that the person tests positive?

19. MARKET RESEARCH In a focus group designed to test the appeal of a new TV sitcom, 80% of the women and 60% of the men said they plan to watch the sitcom. Women made up 55% of the focus group. If you assume the composition and preferences of the focus group are representative of the TV audience as a whole, what is the probability that a person watching the sitcom is female?

ONGOING ASSESSMENT

20. Open-ended Problem Write a conditional probability exercise based on the table in Example 1. Then solve your exercise.

SPIRAL REVIEW

21. Suppose a six-sided die is rolled twice. Find the probability of each event. *(Section 13.2)*

  a. The first roll is a 1 or a 4.

  b. The first roll is a 5, and the second roll is an even number.

  c. The sum of the two rolls is less than 11.

Use the binomial theorem to expand each power of a binomial. *(Section 12.5)*

22. $(x + y)^4$ **23.** $(x - y)^3$ **24.** $(a + 2b)^5$ **25.** $\left(s^2 + t^4\right)^3$

ASSESS YOUR PROGRESS

VOCABULARY

probability (p. 611)   **independent events** (p. 619)
experimental probability (p. 612)   **mutually exclusive events** (p. 619)
theoretical probability (p. 613)   **complementary events** (p. 619)
geometric probability (p. 614)   **conditional probability** (p. 626)

1. Suppose a randomly thrown dart hits the circular target shown. Find the probability that the dart hits the target's square shaded region. *(Section 13.1)*

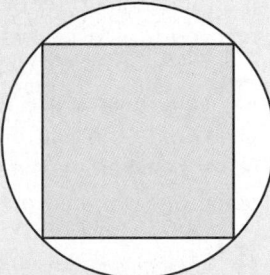

2. HISTORY In a game popular in France during the seventeenth century, a player tried to roll a six-sided die four times without getting a 6. What is the probability that a player wins this game? *(Section 13.2)*

3. MANUFACTURING Seven percent of the welds made on an automobile assembly line are defective. An X-ray machine correctly rejects 90% of the defective welds and correctly accepts 95% of the good welds. Find the probability that an accepted weld is defective. *(Section 13.3)*

4. Journal Explain why the equation $P(A \text{ and } B) = P(A) \cdot P(B|A)$ is a generalization of the equation $P(A \text{ and } B) = P(A) \cdot P(B)$. Give an example of two events A and B that satisfy the first equation but not the second.

13.4 Binomial Distributions

Do you think you could pass a true/false test in a language you can't read? Suppose you are taking a six-question true/false test and decide to guess each answer by flipping a coin. How likely are you to get all of the questions right? How likely are you to get more than half of the questions right? You will examine this situation experimentally in the Exploration.

Learn how to...
- **find the probability distribution for a binomial experiment**

So you can...
- **make predictions in cases where there are many trials, each with two possible outcomes, such as guessing answers on a true/false test**

EXPLORATION
COOPERATIVE LEARNING

Guessing on a True/False Test

Work with your class.
You will need:
- a coin

1 Flip a coin to guess the answer to each question on the test shown. If the coin comes up heads, answer that question *True*. If it comes up tails, answer *False*.

Esperanto Geography Test
Tell whether each statement is *True* or *False*.

1. Pli da homoj loĝas en Kalifornio ol en la tuta Aŭstralio.
2. La ĉefa rikoltaĵo en Ukrajno estas tritiko.
3. Kenjo iĝis sendependa en 1963.
4. Kalkuto estas la ĉefurbo de Bharato.
5. Ŝanhajo estas lando en Azio.
6. Malavio troviĝas sur la okcidenta bordo de granda laĝo.

2 What is the theoretical probability of answering the first question correctly? of answering all six correctly?

3 Your teacher will tell you the answers to the test. Find your score by counting the number of questions you answered correctly.

4 As a class, make a relative frequency histogram of everyone's scores. What is the general shape of the histogram?

5 Based on the histogram, what is the probability of answering all of the questions correctly? Does this agree with the probability you calculated in Step 2? Why or why not?

6 Based on the histogram, what is the probability of answering four or more questions correctly? of answering three or fewer questions correctly? Is one situation theoretically more likely than the other? Why?

BY THE WAY

Esperanto is an international language invented by Ludwik Lazar Zamenhof, a Polish opthamologist, in 1887. It is the most widely used of the more than 200 universal languages that have been proposed since the 1600s.

In Steps 5 and 6 of the Exploration, you used your class results to calculate probabilities experimentally. You can also calculate the probabilities theoretically, as Example 1 illustrates.

EXAMPLE 1

Find the probability of getting exactly 4 out of 6 right on the following two types of six-question tests if you guess the answer to each question randomly.

a. a true/false test

b. a multiple-choice test, with 5 possible responses to each question

SOLUTION

a. The probability of getting any particular question right (or wrong) is $\frac{1}{2}$, so the probability of guessing a particular sequence of answers—such as right, right, wrong, right, wrong, right—is $\left(\frac{1}{2}\right)^6$. There are $_6C_4$ different sequences that contain exactly 4 right answers.

$$P(4 \text{ out of 6 right}) = {}_6C_4 \cdot \left(\frac{1}{2}\right)^6$$

$$\approx 0.234$$

```
(6nCr4)*(1/2)^6
            .234375
```

b. The probability of getting a particular question **right** is $\frac{1}{5}$. The probability of getting a particular question **wrong** is $1 - \frac{1}{5} = \frac{4}{5}$.

$$P(4 \text{ out of 6 right}) = {}_6C_4 \cdot \left(\frac{1}{5}\right)^4 \cdot \left(\frac{4}{5}\right)^2$$

There are $_6C_4$ different sequences that contain exactly 4 right answers.

$$= 15 \cdot \frac{1}{625} \cdot \frac{16}{25}$$

$$\approx 0.0154$$

The probability of guessing a particular sequence of **4 right** and **2 wrong** answers is $\left(\frac{1}{5}\right)^4 \cdot \left(\frac{4}{5}\right)^2$.

THINK AND COMMUNICATE

1. For each test in Example 1, what is $P(2 \text{ out of 6 } wrong)$? Explain.

2. Based on part (b) of Example 1, explain how to find $P(5 \text{ out of 6 right})$.

3. Based on parts (a) and (b) of Example 1, what can you say about $P(4 \text{ out of 6 right})$ on a six-question test as the number of answer choices increases?

Flipping a coin is an example of a *binomial experiment* because there are two possible results, heads or tails. In a binomial experiment, the two mutually exclusive outcomes are often called *success* and *failure*. For each trial, $P(\text{success}) = p$ and $P(\text{failure}) = 1 - p$. The probability of k successes in n independent trials is given by the following formula.

$$P(k \text{ successes in } n \text{ trials}) = {}_nC_k \cdot p^k \cdot (1 - p)^{n - k}$$

There are ${}_nC_k$ different sequences of trials that contain exactly k successes.

The probability of a particular sequence of k successes and $n - k$ failures is $p^k \cdot (1 - p)^{n - k}$.

This formula determines a theoretical **binomial distribution**. The distribution of your class's scores in the Exploration is an experimental binomial distribution.

EXAMPLE 2 | Application: Architecture

An architect is designing a new lecture hall for a university. Each seat will have a writing desk attached to one of the arms. Left-handed students usually prefer to have the desk on their left. The architect wants to know the number of left-handed desks needed so that each left-handed student in a class of 180 students can have a left-handed desk. About 10% of the general population is left-handed.

a. Write a formula and make a histogram giving the binomial distribution for this situation.

b. Find the probability that more than 25 students in the class will be left-handed.

SOLUTION

a. Substitute $n = 180$ and $p = 0.1$ into the formula for a binomial distribution.

$$P(k \text{ left-handed students}) = {}_{180}C_k \cdot (0.1)^k \cdot (0.9)^{180 - k}$$

Find $P(k$ left-handed students) for each possible value of k and make a histogram.

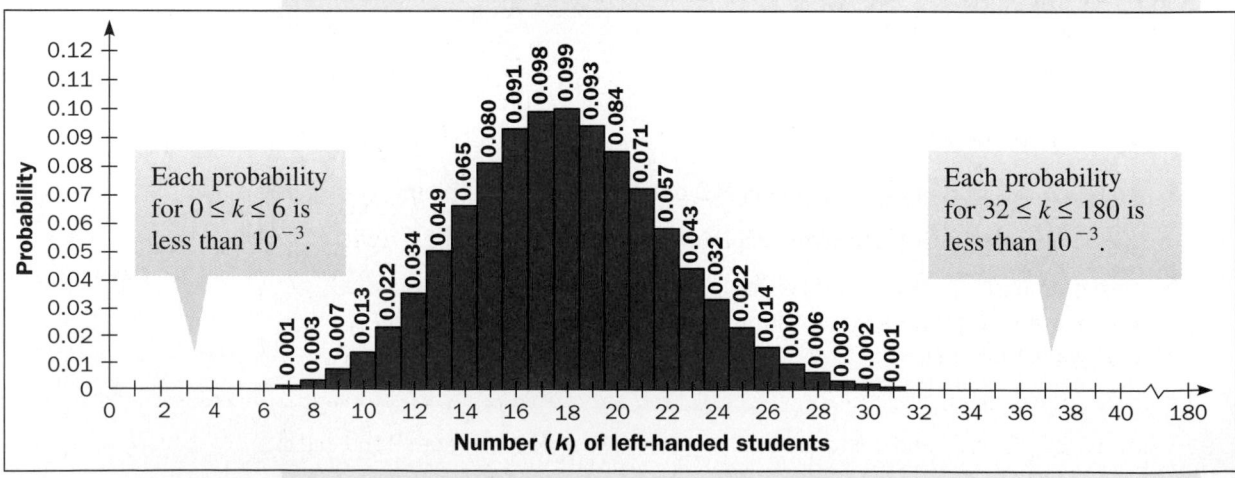

Each probability for $0 \le k \le 6$ is less than 10^{-3}.

Each probability for $32 \le k \le 180$ is less than 10^{-3}.

b. The probability that more than 25 students in the class will be left-handed is the sum of the probabilities that 26, 27, 28, 29, . . . , or 180 students will be left-handed.

$P(\text{more than } 25) \approx 0.014 + 0.009 + 0.006 + 0.003 + 0.002 + 0.001$
$= 0.035$

The probability that more than 25 students will be left-handed is about 0.035.

☑ CHECKING KEY CONCEPTS

1. Are you likely to guess more questions correctly on a 20-question true/false test or on a 20-question multiple-choice test? Why?

2. What is the probability of answering exactly 3 questions correctly on a 5-question true/false test if you guess each answer randomly?

3. Each of the 5 questions on a multiple-choice test has 4 possible answers. Find $P(\text{exactly } 4 \text{ correct})$ if you guess each answer randomly.

4. A binomial experiment has 10 trials where each trial has a 40% chance of success. Explain how to find the probability of at least 8 successes.

5. Why is the distribution of your class's scores in the Exploration on page 632 an example of a binomial distribution? Write a formula to describe the theoretical distribution.

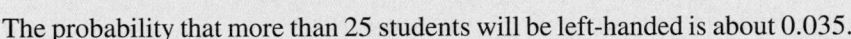

13.4 **Exercises and Applications**

Extra Practice exercises on page 770

1. Writing Write the binomial probability formula, and explain how each part of the formula relates to a binomial experiment.

Write a formula to describe the probability distribution for a binomial experiment with *n* trials, each with probability *p* of success. Then find the probability that the experiment will have exactly *k* successful trials.

2. $n = 5, p = 0.4; k = 2$

3. $n = 10, p = \frac{1}{2}; k = 7$

4. $n = 9, p = 0.02; k = 1$

5. $n = 7, p = 10\%; k = 6$

6. $n = 12, p = \frac{1}{4}; k = 0$

7. $n = 1, p = 70\%; k = 1$

8. Suppose you flip a coin to answer each of the ten questions on a true/false quiz. What is the probability that you will get:

a. all the answers right?

b. half the answers right?

c. 9 or 10 answers wrong?

9. About 10% of the population is left-handed. What is the probability that a class of 30 students will have:

a. no left-handed students?

b. exactly 1 left-handed student?

c. exactly 2 left-handed students?

d. more than 2 left-handed students?

10. Look back at the description of a standard deck of playing cards on page 615.

  a. If you draw a card at random from a standard deck of cards, what is the probability of getting a heart? a non-heart?

  b. Suppose you draw a card at random from each of two standard decks of cards. Copy and complete the table.

Number of hearts	0	1	2	3 or more
Probability	?	?	?	?

11. **Writing** Suppose you roll a die 3 times and count how many fives occur. Is this a binomial experiment? Explain.

12. a. If you toss a coin twice, what is the probability of getting 2 heads? 1 head? 0 heads?

  b. If the coin from part (a) is slightly bent so that it comes up heads 60% of the time, what is the probability of getting 2 heads? 1 head? 0 heads?

13. **MARKETING** A telemarketer has found that 20% of the calls made result in a sale.

  a. If the telemarketer makes 10 calls, what is the probability that half of them result in a sale?

  b. What minimum number of calls is needed for the probability of at least one sale to be greater than 99%?

14. **DRIVER'S EDUCATION** The written test for a regular Massachusetts driver's license has 20 multiple-choice questions. You must answer at least 14 questions correctly to pass the test. If you can always narrow the choices down so that you have a 50% chance of guessing the answer to each question correctly, what is the probability that you will pass? Do you think the minimum score for passing the test should be changed? Explain.

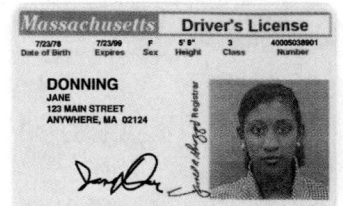

Describe how *P*(*k* successes in *n* trials, each with probability *p* of success) is related to *P*(*j* successes in *m* trials, each with probability *q* of success) for each set of values of *n*, *k*, *p*, *m*, *j*, and *q*. Explain.

15. $n = 7, k = 4, p = 0; m = 7, j = 2, q = 0$

16. $n = 9, k = 3, p = \frac{1}{2}; m = 9, j = 6, q = \frac{1}{2}$

17. $n = 5, k = 1, p = \frac{1}{5}; m = 5, j = 4, q = \frac{4}{5}$

18. $n = 8, k \geq 4, p = 0.1; m = 8, j \leq 4, q = 0.9$

19. **Cooperative Learning** Work with a partner. One of you should do part (a) and the other should do part (b). Work together on part (c).

  a. Shake 4 coins in a plastic cup or other container and "pour" them out. Write down how many of the 4 coins come up heads. Repeat the experiment 20 times and make a relative frequency histogram.

  b. Make a histogram of a theoretical binomial distribution with $n = 4$ and $p = 0.5$.

  c. **Writing** Compare your results from parts (a) and (b). Discuss the reasons behind any similarities and differences.

20. **Challenge** In the 1991 National Hockey League (NHL) playoffs, Minnesota beat Chicago in a best-of-seven series. Suppose Minnesota had a 45% chance of beating Chicago in any particular game of the series. The winner of the series is the first team to win four games. What was the probability that Minnesota would win the series?

In criminal court cases, which require a unanimous decision, it is often difficult for a jury to reach a decision. Suppose the probability that any given jury member thinks the defendant is guilty is a constant, p. This constant could be called the "appearance of guilt" based on the evidence and arguments given by lawyers like Robert Ward. You will use two models to explore how this probability and the jury size affect the chances of a guilty verdict.

Look back at the article on pages 608–610.

21. **Technology** The *Friedman* model of jury decision-making assumes that each jury member votes independently and that the defendant is declared guilty if all members vote for conviction or not guilty if all members vote for acquittal. Otherwise, there is a "hung jury."

a. Suppose the vote of each jury member is a binomial trial with probability p of a vote for conviction. Express the probabilities of conviction by a 6-member jury and by a 12-member jury as functions of p.

b. Graph both functions from part (a) using a graphing calculator or graphing software. Compare the probability of a conviction for a 12-person jury with that for a 6-person jury.

22. **Spreadsheets** The *Walbert* model assumes that the majority will convince the minority. If the jury is initially split evenly, the *Walbert* model states that there is a 50% chance of eventual conviction.

a. According to the *Walbert* model, a 6-member jury will always vote for conviction if 4, 5, or 6 members initially think the defendant is guilty, and half the time if just 3 members believe the defendant is guilty. If $p = 0.9$, what is the probability of a conviction?

b. Use the *Walbert* model to express the probability of a conviction as a function of p if a jury has 6 members. Use a spreadsheet to find this probability for $p = 0, 0.1, 0.2, \ldots, 1$.

c. If a jury has 12 members, use the *Walbert* model to express the probability of a conviction as a function of p. Use a spreadsheet to find this probability for the same values of p as in part (b).

d. Compare the results of parts (b) and (c). Which size jury is more likely to convict the defendant for each value of p?

23. **Open-ended Problem** The *Walbert* model is more realistic than the *Friedman* model, but it still differs from an actual jury. How is the *Walbert* model unrealistic? Can you suggest any improvements? How do you think your changes would affect the probability of a conviction?

24. **Writing** Do you think juries in criminal cases should have 6 members or 12 members? Why?

Before launching a new product, a manufacturing company will conduct various marketing surveys to see how well the product will compete in the existing market. For example, a new food product will be taste-tested, consumer reactions to different marketing approaches will be charted, and so on.

25. Suppose people at a shopping mall are asked to look at an advertisement and then answer some questions about it. Out of the first ten shoppers surveyed independently, 7 said they thought the claims seemed believable. What is the probability of getting this result if the proportion of the general population believing the ad is 50%?

26. In a blind taste test, 11 out of 15 people said they preferred a new formulation for a company's diet cola. What is the probability of getting at least this many positive responses if the average consumer really has no preference between the two?

27. a. Write an expression for the probability of getting k out of 15 positive responses to a new diet cola if you assume that people in general have no preference.

b. Of the 15 people polled, let $x =$ the number who indicated a preference for the new cola. Use the expression from part (a) to find $P(x \geq k)$ for each value of k from 0 to 15. Make a histogram of these probabilities as a function of k.

c. In part (b), for what value of k is $P(x \geq k)$ first less than 10%? first less than 5%? At what point do you think it is safe to conclude that people really prefer the new formulation to the old?

ONGOING ASSESSMENT

28. Writing Explain what a binomial experiment is and how to find binomial distributions. Include some examples.

SPIRAL REVIEW

29. Suppose $P(A) = 0.8$ and $P(A \text{ and } B) = 0.32$. Find $P(B|A)$. *(Section 13.3)*

Find the mean and the standard deviation of each data set. *(Section 6.5)*

30. 125, 170, 250, 270, 144, 185, 176, 100, 220, 160

31. 0.18, 0.25, 0.63, 0.40, 0.26, 0.25, 0.20

Graph each inequality. *(Section 7.4)*

32. $y < 3x + 6$ **33.** $5x - 4y \geq 10$ **34.** $y > 2.7x - 1.3$

13.5 Normal Distributions

Learn how to...

- recognize a normal distribution
- find probabilities involving normally distributed data

So you can...

- find probabilities involving the heights of women, for example

Toolbox p. 790
Mean, Median, and Mode

Chances are that most of the women you see during a typical week are all close to the same height, and that very few are either much shorter or much taller. This is due to the fact that women's heights are *normally* distributed. The histogram below shows the distribution of the heights of 1000 women.

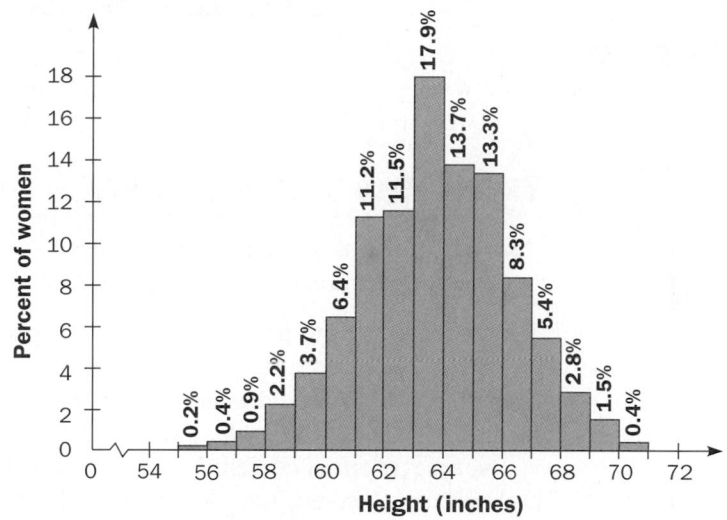

THINK AND COMMUNICATE

1. Does the histogram show a *symmetric distribution* or a *skewed distribution*?

2. Estimate the mean, median, and mode of the data in the histogram.

3. The height data have a mean of 64 in. and a standard deviation of 2.5 in. Estimate the percent of the data that lies within:

   a. one standard deviation of the mean (that is, between $64 - 2.5 = 61.5$ in. and $64 + 2.5 = 66.5$ in.)

   b. two standard deviations of the mean

   c. three standard deviations of the mean

4. Imagine a smooth curve that comes reasonably close to passing through the midpoints of the tops of the bars in the histogram. What shape does the curve have?

Rather than working with a large set of data, mathematicians sometimes use a curve that approximates the shape of a distribution. In particular, when a histogram has a **normal distribution** like the one above, it is modeled with an equation whose graph is a bell-shaped curve.

The *area under the curve* (that is, the area between the curve and the horizontal axis) for some interval on the horizontal axis gives an approximation of the percent of data that lies within the interval.

The graph shown below is a normal distribution with mean $\bar{x}$ and standard deviation σ.

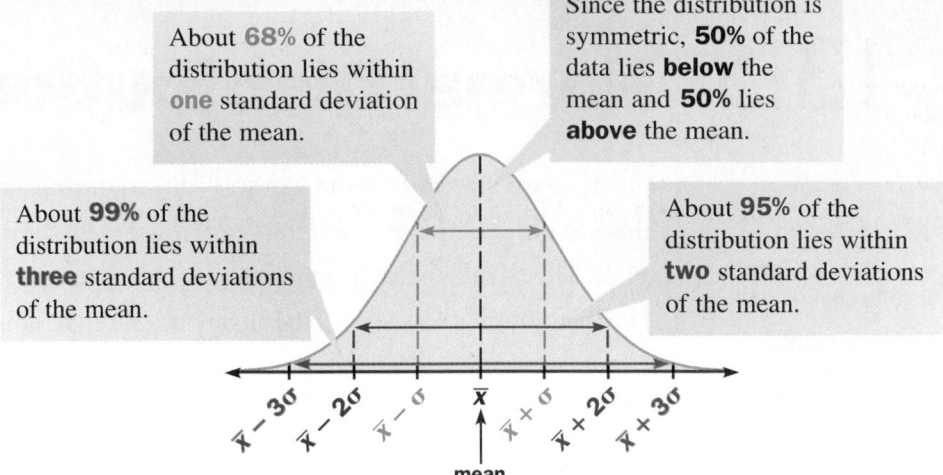

About **68%** of the distribution lies within **one** standard deviation of the mean.

Since the distribution is symmetric, **50%** of the data lies **below** the mean and **50%** lies **above** the mean.

About **99%** of the distribution lies within **three** standard deviations of the mean.

About **95%** of the distribution lies within **two** standard deviations of the mean.

$\bar{x} - 3\sigma$ $\bar{x} - 2\sigma$ $\bar{x} - \sigma$ $\bar{x}$ $\bar{x} + \sigma$ $\bar{x} + 2\sigma$ $\bar{x} + 3\sigma$

mean

THINK AND COMMUNICATE

5. What percent of the data in a normal distribution lies between the mean and one standard deviation above the mean? Explain.

You can interpret the percent of data that lies within a given interval as a probability. For example, your chance of selecting any one data value from a set of normally distributed data and having it lie within one standard deviation of the mean is about 68%, or 0.68.

Because a normal curve is a *model* for a histogram of actual data, probabilities obtained from the curve are theoretical, and probabilities obtained from the data are experimental.

EXAMPLE 1

Find the theoretical probability that the height of a randomly selected woman is between 61.5 in. and 69 in. tall. Use the fact that the mean of the data is 64 in. and the standard deviation is 2.5 in.

SOLUTION

Step 1 Find how far above or below the mean the endpoints of the given interval are.

61.5 in. − 64 in. = −2.5 in. ⟵ **1** standard deviation **below** the mean

69 in. − 64 in. = **5 in.** ⟵ **2** standard deviations **above** the mean

Step 2 Use the mean to rewrite the interval as two separate intervals. Let x be the height in inches of a randomly selected woman.

$$P(61.5 \text{ in.} < x < 69 \text{ in.}) = P(61.5 \text{ in.} < x < 64 \text{ in.}) + P(64 \text{ in.} < x < 69 \text{ in.})$$

$$= 0.34 + 0.475$$

$$= 0.815$$

The probability that a woman's height is between 61.5 in. and 69 in. is 0.815.

The Standard Normal Distribution

The **standard normal distribution** is the normal distribution with mean 0 and standard deviation 1. You can use the *standard normal table* shown below to find the probability that a randomly selected data value is less than a given number of standard deviations from the mean. The table is a refinement of the percents given on the previous page.

To use the table, you need to convert a given x-value from a normal distribution with mean $\bar{x}$ and standard deviation σ to a **z-score** given by:

Subtract the mean from the given value.

$$z = \frac{x - \bar{x}}{\sigma}$$

Divide by the standard deviation.

A z-score gives the number of standard deviations that x lies from the mean. Obtaining a z-score for a data value is called *standardizing* the value.

z	.0	.1	.2	.3	.4	.5	.6	.7	.8	.9
−3	.0013	.0010	.0007	.0005	.0003	.0002	.0002	.0001	.0001	.0000+
−2	.0228	.0179	.0139	.0107	.0082	.0062	.0047	.0035	.0026	.0019
−1	.1587	.1357	.1151	.0968	.0808	.0668	.0548	.0446	.0359	.0287
−0	.5000	.4602	.4207	.3821	.3446	.3085	.2743	.2420	.2119	.1841
0	.5000	.5398	.5793	.6179	.6554	.6915	.7257	.7580	.7881	.8159
1	.8413	.8643	.8849	.9032	.9192	.9332	.9452	.9554	.9641	.9713
2	.9772	.9821	.9861	.9893	.9918	.9938	.9953	.9965	.9974	.9981
3	.9987	.9990	.9993	.9995	.9997	.9998	.9998	.9999	.9999	1.0000−

This means "slightly more than 0."

The probability that a data value is less than 1.9 standard deviations above the mean is **0.9713**.

This means "slightly less than 1."

EXAMPLE 2   Application: Biology

Baboon skulls can be classified by their dental structure. A fossilized baboon skull with a third premolar of length 9.0 mm was discovered in Angola. In genus *Papio*, the third premolar has a mean length of 8.18 mm with a standard deviation of 0.47 mm. If you assume molar lengths are normally distributed, what is the probability of one having a length of at least 9.0 mm? What does this probability suggest about the baboon?

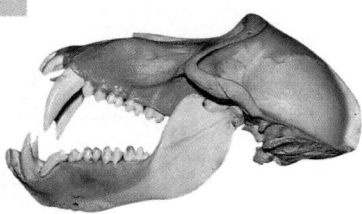

SOLUTION

Standardize the observed value so you can use the standard normal table.

$$P(\text{length} \geq 9.0) = 1 - P(\text{length} < 9.0)$$

Use the complementary probability.

$$= 1 - P\left(z < \frac{9.0 - 8.18}{0.47}\right)$$

Standardize the value 9.0.

$$\approx 1 - P(z < 1.7)$$

$$\approx 1 - 0.9554$$

Use the standard normal table.

$$= 0.0446$$

The probability that an animal of this genus has a third premolar at least 9.0 mm long is about 0.045. This small probability suggests that the premolar may not be from genus *Papio*.

6. Look at the standard normal table on the previous page. How is $P(z < k)$ related to $P(z < -k)$ for some positive constant k? Explain this relationship in terms of the area under the standard normal curve.

7. Describe how you can use the standard normal table to solve Example 1.

8. Use the standard normal table to show that about 68% of the data in a normal distribution lies within one standard deviation of the mean, about 95% lies within two standard deviations, and about 99% lies within three standard deviations.

☑ **CHECKING KEY CONCEPTS**

1. At a large university, the heights of male students are normally distributed with mean 175 cm and standard deviation 10 cm.

   a. Sketch a normal curve to show the distribution of male student heights. Label the mean. Also label the heights that are one, two, and three standard deviations from the mean.

   b. About what percent of the male students have heights between 155 cm and 195 cm? between 165 cm and 205 cm?

2. The following values are from a normal distribution with mean 25 and standard deviation 15. Standardize each value.

   a. 25   **b.** 10   **c.** 55   **d.** 30   **e.** 50

3. Suppose the scores on a national mathematics test are normally distributed with mean 500 and standard deviation 100. What percent of students who took the test scored below 420? above 690?

13.5 Exercises and Applications

Extra Practice exercises on page 770

1. Find the probability that a randomly chosen x-value lies within each interval if x comes from a normal distribution with mean $\bar{x}$ and standard deviation σ. Explain how you found each answer.

   a. $\bar{x} - 2\sigma < x < \bar{x}$ **b.** $\bar{x} < x < \bar{x} + 3\sigma$ **c.** $\bar{x} - 2\sigma < x < \bar{x} + 3\sigma$

2. **BOTANY** The *guayule* plant, which grows in the southwestern United States and in Mexico, is one of several plants that can be used as a source of rubber. Suppose that in a large group of *guayule* plants, the heights of the plants are normally distributed with mean 12 in. and standard deviation 2 in.

   a. What percent of the *guayule* plants are taller than 16 in.?

   b. What percent of the *guayule* plants are between 10 in. and 14 in. tall?

   c. What percent of the *guayule* plants are between 8 in. and 12 in. tall?

   d. What percent of the *guayule* plants are no taller than 6 in.?

Connection · HISTORY

Adolphe Quetelet was a Belgian scholar who demonstrated that normal distributions can be used to model many different kinds of data. In one of his studies he measured the chests of 5738 Scottish soldiers. To check whether the data were normally distributed, Quetelet compared theoretical and experimental probabilities.

Chest (in.)	33	34	35	36	37	38	39	40	41	42	43	44	45	46	47	48
Frequency	3	18	81	185	420	749	1073	1079	934	658	370	92	50	21	4	1

3. The mean of the data is $\bar{x} = 39.8$ in. and the standard deviation is $\sigma = 2.05$ in. If you assume that the data are normally distributed, what is the theoretical probability for each interval?

   a. $\bar{x} - \sigma < x < \bar{x}$
   b. $\bar{x} < x < \bar{x} + 2\sigma$
   c. $\bar{x} - 3\sigma < x < \bar{x} - \sigma$

4. Use the table to find the experimental probability for each interval in Exercise 3.

5. **Writing** Do you think the data are normally distributed? Why or why not? You may want to test other intervals before deciding.

6. **MANUFACTURING** Boxes of cereal are filled by a machine. Tests of the machine's accuracy show that the amount of cereal in each box varies. The weights are normally distributed with mean 20 oz and standard deviation 0.25 oz.

   a. Sketch a normal curve to show the distribution of weights. Label the mean. Also label the weights that are one, two, and three standard deviations from the mean.

   b. What percent of the boxes weigh more than 20 oz?

   c. What percent of the boxes weigh less than 19.5 oz?

7. **Cooperative Learning** Work with a partner.

   a. Ask at least 30 people to each draw a line segment 5 in. long without using a ruler. Measure each line segment to the nearest $\frac{1}{16}$ in. and display the data in a histogram.

   b. Find the mean and the standard deviation of the data. What percent of your data values are within one standard deviation of the mean? within two standard deviations of the mean? within three standard deviations of the mean? Do you think the data are normally distributed? Explain.

8. The following values are from a normal distribution with mean 17 and standard deviation 4. Standardize each value.

   a. 17   **b.** 21   **c.** 15   **d.** 26   **e.** 30

9. The scores on a statewide mathematics test were normally distributed with mean 400 and standard deviation 40. Each score is reported as a *percentile*, the percent of people who scored less than or equal to that score. José's score was 550. What was his percentile score?

10. a. **Technology** Use a graphing calculator or graphing software to

graph $f(x) = \frac{1}{\sqrt{2\pi}} e^{-x^2/2}$. Use a viewing window with $-4 \le x \le 4$

and $0 \le y \le 1$.

b. The graph in part (a) is the standard normal curve. What should the total area under this curve be? Why? Some calculators and graphing software can *integrate* to find the area under the curve, denoted by $\int f(x)\, dx$. Check your answer by finding $\int f(x)\, dx$ for $-4 \le x \le 4$.

c. What value would you expect for $\int f(x)\, dx$ when x is between -1 and 1? between -2 and 2? Explain. Check your answers.

11. Open-ended Problem Think of a real-world variable that might have a normal distribution. Gather at least 30 data values. Are the data normally distributed? Write a brief report about your project. Be sure to justify your conclusion.

Connection ▶ MEDICINE

Pediatricians check a child's height and weight, against standard tables to track his or her rate of growth. Measurements of weight for each age group tend to be normally distributed, and each child tends to remain at the same place within the distribution as she or he grows. If a child's location within the distribution changes significantly, the pediatrician may want to find out why.

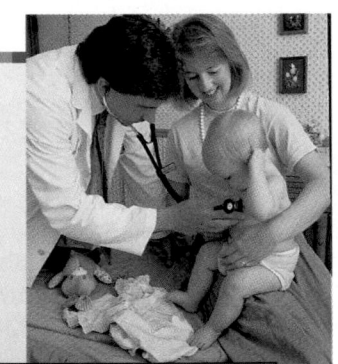

Distributions of Weights for Girls Aged 2 Years to 13 Years												
Age (years)	2	3	4	5	6	7	8	9	10	11	12	13
Weight (kg)	11.9	14.1	15.9	17.7	19.5	21.8	24.8	28.3	32.5	36.9	41.4	45.9
Standard deviation	1.34	1.88	2.37	2.98	3.77	4.68	5.90	7.36	8.81	10.33	11.79	12.83

12. Make a scatter plot of mean weight versus age. Over which year does the average girl gain the most weight?

13. About what percent of each group weighs more than 19 kg?

    a. 5-year-old girls   **b.** 6-year-old girls   **c.** 7-year-old girls

14. a. The table at the right gives Cathy's weights at different ages. At each age, what percent of girls in Cathy's age group weigh less than she does?

    b. At 10 years old Cathy weighed 31.9 kg, and at 13 years old she weighed 47.8 kg. Compare her growth through age 13 with the typical growth pattern of girls.

Age (years)	Weight (kg)
2	11.55
3	13.43
4	14.35
5	15.87

15. Writing How is the standard normal distribution like other normal distributions? What is special about the standard normal distribution?

16. What is the probability that you will answer exactly 3 questions correctly on a 4-question true/false test if you guess each answer randomly? *(Section 13.4)*

17. A company sells *FunBags* that each contain a random assortment of 10 colored erasers. If 20% of the erasers manufactured for the *FunBags* are green, what is the probability that a given *FunBag* will contain at least 3 green erasers? *(Section 13.4)*

Simplify each expression. *(Section 12.4)*

18. $_5C_1$ **19.** $_7C_4$ **20.** $_{22}C_0$ **21.** $_{19}C_{17}$

22. Are $\triangle ABC$ and $\triangle ADE$ similar? Explain. *(Toolbox, page 801)*

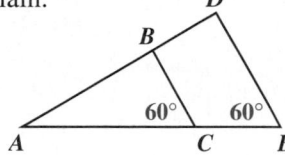

ASSESS YOUR PROGRESS

VOCABULARY

binomial distribution (p. 634)   **standard normal distribution** (p. 641)
normal distribution (p. 639)   **z-score** (p. 641)

1. Suppose you roll a die 6 times and count how many times the die shows a one. Find $P(\text{exactly 3 ones})$ and $P(\text{at least 2 ones})$. *(Section 13.4)*

2. MANUFACTURING There is a 7% chance that any given capacitor is defective. If an inspector examines 20 capacitors, what is the probability that at least one will be defective? *(Section 13.4)*

3. AGRICULTURE The pumpkins that a farmer harvested varied in size, having diameters that were normally distributed with mean diameter 12 in. and standard deviation 3 in. *(Section 13.5)*

     a. About what percent of the pumpkins were larger than 15 in. in diameter?

     b. About what percent were smaller than 6 in. in diameter?

4. MANUFACTURING A manufacturer produces bags of pretzels marked "Net weight 16 oz." The machine is set to fill each bag with 16.3 oz, but the actual weights are normally distributed with mean 16.3 oz and standard deviation 0.18 oz. What is the probability that a given bag will contain less than 16 oz of pretzels? *(Section 13.5)*

5. Journal Compare the binomial and normal distributions. How are they alike? How are they different? Give examples of each.

A Matter of Taste

If you've ever gone grocery shopping, you know that generic foods are generally less expensive than their brand-name counterparts. However, many shoppers believe that "buying generic" means sacrificing quality. Is there really a difference? You can find out by doing some market research.

PROJECT GOAL Determine whether people can distinguish a generic beverage from one or more brand-name beverages.

Performing an Experiment

1. CHOOSE a beverage, such as cola or orange juice, that you want to test. Buy a generic version and one or more brand-name versions of the beverage.

2. FIND a person to survey. Have the person taste each beverage *without* letting him or her see the beverage containers. You may want to use small paper cups to hold beverage samples.

3. ASK the person to identify the generic beverage. Record whether the person is correct.

4. REPEAT Steps 2 and 3 with other people. You should survey a total of 20 people.

Evaluating Your Results

If the responses of the people you surveyed were totally random (that is, if the people were only guessing when they tried to identify the generic beverage), then the probability that r of the 20 people responded correctly is

$$P(r) = {}_{20}C_r \cdot p^r \cdot (1 - p)^{20 - r}$$

where p is the probability of a correct guess.

1. What kind of probability distribution does the formula for $P(r)$ determine?

2. What is the value of p for your experiment? How is this value related to the number of beverages you tested? Explain.

In a binomial experiment like this one (with more than just a few trials), the value of $P(r)$ is small for any value of r. So $P(r)$ alone is not a good indicator of the likelihood that the experiment's outcome is due to chance. A better indicator is the probability of getting *at least r* correct responses:

$$P(\text{at least } r) = P(r) + P(r + 1) + \cdots + P(20)$$

3. Calculate the value of $P(\text{at least } r)$ for the value of r from your experiment. Does this probability suggest that chance determined the outcome of the experiment? Explain your reasoning.

Being very cautious, scientists and mathematicians *presume* that chance determines the outcome of an experiment unless there is strong evidence to the contrary. In this case, "strong evidence" might mean that the value of $P(\text{at least } r)$ is less than 0.1 or even 0.05 (for extra-cautious types).

4. Given the requirement for strong evidence stated above, can you conclude that people really can distinguish the generic beverage from the brand-name beverage(s)?

Writing a Report

Write a report summarizing your results. Include the data from your experiment and your answers to the questions above. If you wish, you can extend your project by conducting a survey that compares generic and brand-name versions of a different product, such as peanut butter or paper towels.

Self-Assessment

After completing this project, how comfortable are you with the ideas of probability as they relate to a scientific experiment? What ideas do you understand well? What ideas are still unclear?

13 Review

STUDY TECHNIQUE

Read the title of each section in the chapter. Without looking at the section's contents, write a summary of the section and at least three questions that test the section's main objectives.

VOCABULARY

probability (p. 611)
experimental probability (p. 612)
theoretical probability (p. 613)
geometric probability (p. 614)
independent events (p. 619)
mutually exclusive events (p. 619)

complementary events (p. 619)
conditional probability (p. 626)
binomial distribution (p. 634)
normal distribution (p. 639)
standard normal distribution (p. 641)
z-score (p. 641)

SECTION | 13.1

Number on die	Number of times rolled
1	8
2	11
3	9
4	5
5	7
6	10

The **probability** of an event A, denoted $P(A)$, is the fraction of the time that A is expected to happen. There are several types of probability.

For example, the table shows the results of an experiment in which a six-sided die is rolled **50** times. You can use these results to find the **experimental probability** of rolling a number less than 3:

$$P\left(\begin{array}{c}\text{rolling a number}\\ \text{less than 3}\end{array}\right) = \frac{8 + 11}{50} = \frac{19}{50}, \text{ or about } 0.380$$

To find the corresponding **theoretical probability**, compare the number of possible rolls with the number of rolls less than 3:

6 possible rolls ⟶

2 rolls less than 3 ⟶

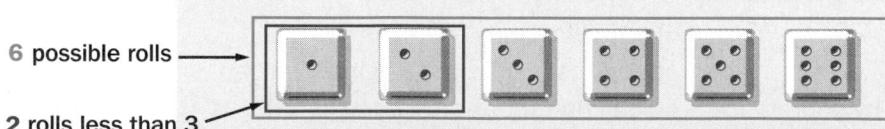

$$P\left(\begin{array}{c}\text{rolling a number}\\ \text{less than 3}\end{array}\right) = \frac{2}{6} = \frac{1}{3}, \text{ or about } 0.333$$

A **geometric probability** is a probability based on length, area, or volume. For example, suppose a randomly thrown dart hits the target shown. Since the area of each wedge is one sixth the area of the circle,

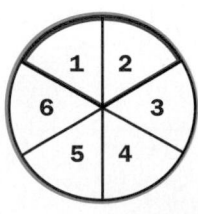

$$P\left(\begin{array}{c}\text{dart hitting}\\ \text{wedge 1 or wedge 2}\end{array}\right) = \frac{2}{6} = \frac{1}{3}, \text{ or about } 0.333$$

Two events, A and B, are:

- **independent** if the occurrence of A does not affect whether B happens.
- **mutually exclusive** if A and B cannot both happen.
- **complementary** if A and B are mutually exclusive and either A or B *must* happen.

To illustrate these definitions, suppose that two marbles are chosen at random—one from jar 1 and one from jar 2—and consider the following pairs of events and corresponding rules.

A: The marble chosen from jar 1 is **red**.
B: The marble chosen from jar 2 is **red**.

These events are independent.

$$P(A \text{ and } B) = P(A) \cdot P(B) = \frac{4}{9} \cdot \frac{2}{7} = \frac{8}{63}, \text{ or about } 0.127$$

C: The marble chosen from jar 1 is **green**.
D: The marble chosen from jar 1 is **blue**.

These events are mutually exclusive.

$$P(C \text{ or } D) = P(C) + P(D) = \frac{3}{9} + \frac{2}{9} = \frac{5}{9}, \text{ or about } 0.556$$

E: The marble chosen from jar 2 is **green**.
F: The marble chosen from jar 2 is *not* **green**.

These events are complementary.

$$P(F) = 1 - P(E) = 1 - \frac{4}{7} = \frac{3}{7}, \text{ or about } 0.429$$

The **conditional probability** $P(B|A)$ is the probability of event B, given that event A has occurred. For example, suppose two marbles are removed one at a time from jar 1. Let A be the event "the first marble is **blue**," and let B be the event "the second marble is **blue**." Then $P(B|A) = \frac{1}{8}$, or 0.125.

Jar 1

Jar 2

A *binomial experiment* consists of n independent trials, each of which has two possible outcomes, *success* and *failure*. If $P(\text{success}) = p$ and $P(\text{failure}) = 1 - p$, then the probability that k of the n trials are successes is given by the formula:

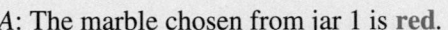

$$P(k \text{ successes in } n \text{ trials}) = {}_nC_k \cdot p^k \cdot (1 - p)^{n - k}$$

This formula defines a **binomial distribution**.

For example, the probability of getting 4 threes in 20 rolls of a six-sided die is $P(4) = {}_{20}C_4 \cdot \left(\frac{1}{6}\right)^4 \cdot \left(\frac{5}{6}\right)^{16}$, or about 0.202.

A **normal distribution** is characterized by a bell-shaped curve. In a normal distribution, about **68%** of the data lies within **1** standard deviation of the mean, about **95%** of the data lies within **2** standard deviations of the mean, and about **99%** of the data lies within **3** standard deviations of the mean.

13 Assessment

VOCABULARY QUESTIONS

For Questions 1–3, complete each sentence.

1. If events A and B are ___?___, then $P(A \text{ and } B) = P(A) \cdot P(B)$.

2. The ___?___ distribution has mean 0 and standard deviation 1.

3. If a procedure can result in n equally likely outcomes, and an event A corresponds to m of these outcomes, then $\frac{m}{n}$ is called the ___?___ of A.

SECTION 13.1

Find the probability of choosing each type of card at random from a standard deck of 52 playing cards. (See page 615 for a description of such a deck.)

4. a 6 of hearts   5. a 6   6. a heart

7. a black card   8. a red face card   9. a 15 of clubs

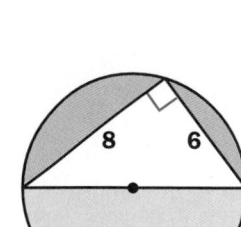

10. **GAMES** In a certain game that uses 5 six-sided dice, you get points for rolling a "full house." A full house consists of three dice showing one number and two dice showing a different number (for example, three 1's and two 4's as shown).

    a. Jorge rolls a set of 5 six-sided dice 80 times and gets 4 full houses. What is his experimental probability of rolling a full house?

    b. What is the theoretical probability of rolling a full house?

Suppose a randomly thrown dart hits the circular target shown. Find the probability that the dart lands in each region.

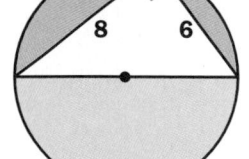

11. the blue region   12. the white region   13. the red region

SECTIONS 13.2 *and* 13.3

14. A soda manufacturer puts a message on the underside of 10% of its bottle caps indicating that the purchaser has won a free bottle of soda. Yolanda buys two bottles of the manufacturer's soda. Find the probability that Yolanda wins the following:

    a. two bottles of soda   **b.** no bottles of soda

    c. at least one bottle of soda   **d.** exactly one bottle of soda

15. SPORTS On July 28, 1991, Dennis Martinez of the Montreal Expos pitched a "perfect game" against the Los Angeles Dodgers. This means that he prevented all 27 batters he faced from getting on base. If you assume each batter had a 30% chance of getting on base, what was the probability of Martinez pitching a perfect game that day?

16. POLITICS Two candidates, one Democratic and one Republican, are running for a seat in the United States Senate. Recent polls predict that the Republican will receive 53% of the vote. Of voters who say they will vote Republican, 82% support a balanced-budget amendment to the Constitution. Of voters who say they will vote Democratic, only 59% support such an amendment. What is the probability that a voter who supports a balanced-budget amendment votes Republican?

SECTIONS 13.4 *and* 13.5

17. Matt is taking a history quiz that consists of 6 multiple-choice questions. Each question has 4 choices. Since Matt didn't study for the quiz, he is forced to guess each answer randomly. What is the probability that he gets:

a. all the answers right?   **b.** none of the answers right?

c. exactly half the answers right?   **d.** at least half the answers right?

18. Open-ended Problem A company hires 10 engineers—8 men and 2 women—from a large pool of equally qualified applicants. Of the applicants, 65% are men and 35% are women. Do you think the company's hiring decisions indicate a bias against women? Give a mathematical justification for your answer.

19. MANUFACTURING A manufacturer makes ball bearings that must have diameters between 20.0 mm and 21.0 mm. Quality control tests indicate that the diameters of the bearings produced are normally distributed with mean 20.4 mm and standard deviation 0.4 mm.

a. What percent of the ball bearings have diameters less than 20.0 mm?

b. What percent of the ball bearings have diameters greater than 21.0 mm?

c. What percent of the ball bearings have diameters that are *not* between 20.0 mm and 21.0 mm?

PERFORMANCE TASK

20. In Section 13.1, you found probabilities associated with the dice games *barbudey* and *kon mín yéung*. Choose another game and analyze some probabilities associated with playing this game. Your analysis should incorporate ideas from at least three of the five sections in this chapter.

14 | Triangle Trigonometry

It's a JUNGLE *in there*

Johnpaul Jones

Deep in the still greenness of *Tiger River*, a three-acre replica of a Southeast Asian jungle at the San Diego Zoo, live tapirs, pythons, crocodiles, purple herons, gold-crested mynas, and rare Sumatran tigers. The exhibit was designed to feel like a patch of tropical rain forest in Indonesia. Johnpaul Jones, a senior partner in the architectural and landscape design firm of Jones & Jones, and his partners pioneered the design of exhibits like *Tiger River*.

> "The idea we started, putting wild animals and wild nature together, has traveled to zoos around the world."

Knowing the Animals

In addition to examining the site, the staff studied each of the animal species destined for *Tiger River*. "We figured out the height, length, and weight of every animal and came up with diagrams showing what each wild animal was capable of doing in terms of jumping, running, and leaping," Jones says.

Creating Safe Barriers

With information about the site and the animals, Jones could design safe, natural barriers to keep the tigers and other animals within their new habitats. "We could put a pool in the foreground of the tigers' space and make it deep enough so that a running tiger couldn't put its feet on the bottom and leap out. Then we could make the barrier above the water line lower so people could get closer. At all of the various barriers around the tiger exhibit, trigonometry played a major role in creating a better place for the wild animal and for the visitor," explains Jones.

Developing Natural Environments

Designing each animal exhibit for *Tiger River* meant solving two basic problems: creating an environment that looks and feels like the animals' natural wild habitat, and providing a safe, close viewing space for visitors. In the case of *Tiger River*, Jones and his colleagues first studied the zoo site, which had been a canyon occupied by African birds and antelope. Because the existing canyon would eventually house rain forest creatures, he decided to install pools of water that would seem to flow continuously throughout the canyon, representing the Southeast Asian rain forest floor.

Analyzing a Tiger's Leap

Jones analyzed the leaping capabilities of the tiger to design the barriers for its habitat. A tiger's initial leaping velocity v_0 can be broken down into vertical and horizontal components. The tiger's vertical velocity, $v_0 \sin \theta$ (θ is pronounced THAY *tuh*), and its horizontal velocity, $v_0 \cos \theta$, correspond to the tiger's upward and forward motion in the diagram.

These velocity components are used in parametric equations that define the path of the tiger as it leaps through the air. The factors $\sin \theta$ and $\cos \theta$ are examples of *trigonometric ratios* that you will study in this chapter.

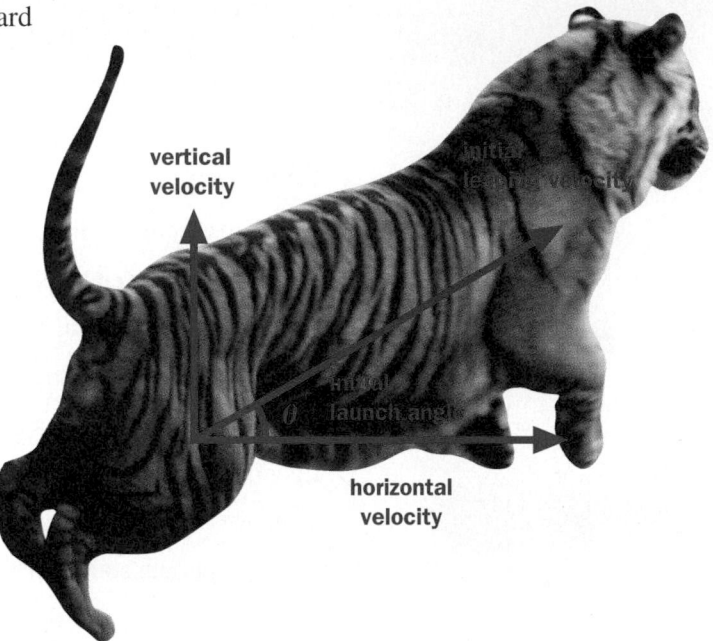

vertical velocity

initial leaping velocity

θ initial launch angle

horizontal velocity

1. Writing Suppose a tiger's initial leaping velocity is 1.5 ft/s. Use a calculator to evaluate $1.5 \sin \theta$ and $1.5 \cos \theta$ for a launch angle θ of 25°, 35°, and 45°. What happens to the vertical velocity, $1.5 \sin \theta$, and the horizontal velocity, $1.5 \cos \theta$, as the launch angle increases?

2. Research Choose an animal from the wild. Find out about the animal's natural habitat and its physical capabilities. Write a paragraph describing the habitat's climate, landscape, and vegetation. Write another paragraph about the animal's height, weight, and body structure, along with any leaping, flying, or running abilities.

3. Project Describe and sketch a zoo exhibit for the wild animal you chose to research in Question 2. Share your exhibit with your classmates and find ways to combine exhibits to house various wild animals.

Mathematics
& Johnpaul Jones

In this chapter, you will learn more about how mathematics is related to designing animal parks.

Related Exercises

Section 14.1
• Exercises 24 and 25

Section 14.2
• Exercises 16 and 17

14.1 Using Sine and Cosine

Trigonometry is a branch of mathematics used in fields such as astronomy, architecture, surveying, aviation, and navigation. The term "trigonometry" comes from two Greek words, *trigonos* and *metron,* meaning "triangle measurement." Trigonometry can be used to find an unknown angle measure or an unknown length of a side of a triangle.

Learn how to...
- **find the sine and cosine of an acute angle**

So you can...
- **find the length of a ladder needed to reach a given height, for example**

Toolbox p. 801
Triangle Relationships

EXPLORATION
COOPERATIVE LEARNING

Finding Ratios of Side Lengths in Similar Right Triangles

Work with a partner.
You will need:
- graph paper
- a protractor

1 Draw a segment along a horizontal grid line on your graph paper. Label one endpoint *A.*

2 Use a protractor to draw a 60° angle at *A.*

3 Draw segments along vertical grid lines to form right triangles, such as △*ABC,* △*ADE,* and △*AFG* shown.

Questions

1. Why are △*ABC,* △*ADE,* and △*AFG* similar triangles?

2. Cut a strip of graph paper to use as a ruler. Use your ruler to find the value of the following ratio for △*ABC* and for △*ADE:*

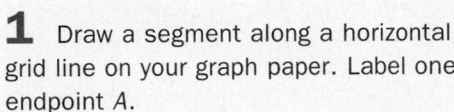
$$\frac{\text{length of leg opposite } \angle A}{\text{length of hypotenuse}}$$
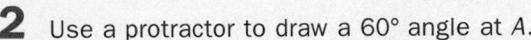

   What do you notice?

3. Make a conjecture about the value of the ratio in Question 2 for any triangle similar to △*ABC.* To test your conjecture, find the value of this ratio for △*AFG.*

In the Exploration you found that corresponding ratios of side lengths in similar right triangles do *not* depend on the lengths of the sides. These ratios depend *only* on the shape of the triangles as determined by the measures of the acute angles. These constant ratios are so important they are given names.

WATCH OUT!

Sometimes the lengths of sides of triangles are named with the lowercase form of the capital letter at the vertex of the opposite angle.

Sine and Cosine of an Angle

In right $\triangle ABC$, the **sine** of $\angle A$, which is written "sin A," is given by

$$\sin A = \frac{\text{length of leg opposite } \angle A}{\text{length of hypotenuse}} = \frac{a}{c}$$

and the **cosine** of $\angle A$, which is written "cos A," is given by

$$\cos A = \frac{\text{length of leg adjacent to } \angle A}{\text{length of hypotenuse}} = \frac{b}{c}$$

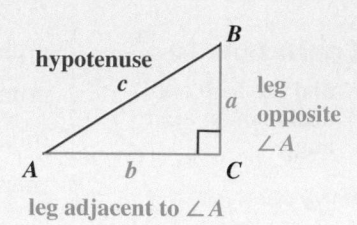

EXAMPLE 1

For $\triangle ABC$ below, find sin A and cos A.

SOLUTION

$$\sin A = \frac{\text{opposite}}{\text{hypotenuse}} = \frac{36}{39} \approx 0.9231$$

$$\cos A = \frac{\text{adjacent}}{\text{hypotenuse}} = \frac{15}{39} \approx 0.3846$$

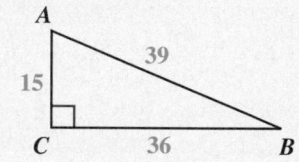

EXAMPLE 2　Application: Archaeology

Between A.D. 1000 and 1300, the Anasazi people lived in cliff dwellings in the southwestern part of the United States. The doors to the cliff dwellings opened onto balconies that were reached by climbing ladders.

Suppose the ladder shown rests on the ground and extends 3 ft above the balcony. How long is the ladder?

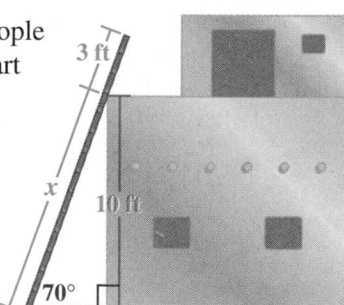

BY THE WAY

The Anasazi built and inhabited a cliff dwelling with almost 200 rooms in what is now Colorado. Known as "Cliff Palace," it is located in Mesa Verde National Park, a United Nations world heritage site.

SOLUTION

Use the sine ratio to find the unknown side length.

$$\sin 70° = \frac{\text{opposite}}{\text{hypotenuse}} = \frac{10}{x}$$

$$x = \frac{10}{\sin 70°} \approx \frac{10}{0.9397} \approx 11$$

The ladder is about $11 + 3 = 14$ ft long.

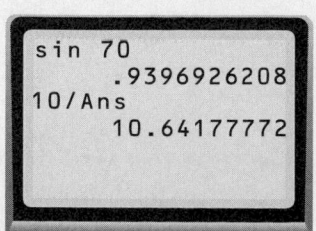

All angles do not have measures that are an integral number of degrees, such as the 70° angle in Example 2. The measures of some angles involve fractions of a degree, such as 74.3125°. Angle measures written in this form are said to be in *decimal degrees*.

Angle measures can also be written using *degrees* (°), *minutes* ('), *and seconds* ("), such as 74°18'45". Just as a circle is divided into 360 degrees, a degree is divided into 60 minutes, and a minute is divided into 60 seconds.

$$1° = 60' \qquad 1' = 60''$$

You can use the equations above to convert an angle measure from degrees, minutes, seconds to decimal degrees. For example, consider 74°18'45":

$$74°18'45'' = 74° + 18\left(\frac{1}{60}°\right) + 45\left(\frac{1}{3600}°\right) = 74.3125°$$

Each minute is $\frac{1}{60}$ of a degree.

Each second is $\frac{1}{60}\left(\frac{1}{60}\right) = \frac{1}{3600}$ of a degree.

EXAMPLE 3 Application: Astronomy

Thousands of years before the invention of telescopes, satellites, computers, and other tools of modern astronomy, people used trigonometry to calculate the sizes and distances of the moon and the sun.

Use the diagram below to estimate the distance between the surface of Earth and the moon using trigonometry. In the diagram, the moon is directly above point P and is just visible from point Q. Points P and Q are on the equator. Use 3963 mi for the radius of Earth at the equator.

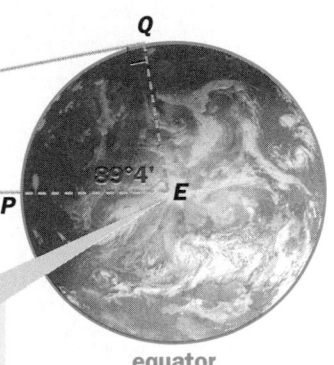

Earth

moon

M

P ---- $89°4'$ ---- E

equator

The measure of $\angle E$ is the difference in longitude between points P and Q on the surface of Earth.

SOLUTION

Find the distance between E, the center of Earth, and M, the moon. Then subtract the radius of Earth.

$$\cos 89°4' = \frac{\text{adjacent}}{\text{hypotenuse}}$$

Substitute **3963** for EQ, the radius of Earth at the equator.

$$= \frac{3963}{EM}$$

$$EM = \frac{3963}{\cos 89°4'}$$

$$\approx \frac{3963}{0.01629}$$

$$\approx 243{,}000$$

```
cos 89°4'
          .0162890193
3963/Ans
          243292.7321
```

The distance between the surface of Earth and the moon is about $243{,}000 - 3{,}963 \approx 239{,}000$ mi.

Inverse Sine and Cosine Functions

For a right triangle, sine and cosine are functions whose domain consists of all angle measures between 0° and 90°. What do you think the range of each function is?

THINK AND COMMUNICATE

1. a. What is the longest side of a right triangle?

b. What does your answer to part (a) tell you about the ratios $\dfrac{\text{opposite}}{\text{hypotenuse}}$ and $\dfrac{\text{adjacent}}{\text{hypotenuse}}$?

c. What does your answer to part (b) tell you about the range of sine and cosine?

The inverses of the sine and cosine functions are written "$\sin^{-1}$" and "$\cos^{-1}$." The inverse sine function is formed by reversing the domain and range of the sine function. Similarly, the inverse cosine function is formed by reversing the domain and range of the cosine function.

For a right triangle, the **domain** of sine and cosine consists of **angle measures between 0° and 90°**, and the **range** consists of **ratios between 0 and 1**.

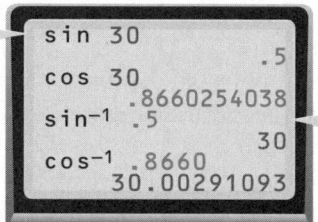

```
sin 30
                  .5
cos 30
         .8660254038
sin-1 .5
                  30
cos-1 .8660
         30.00291093
```

For a right triangle, the **domain** of inverse sine and inverse cosine consists of **ratios between 0 and 1**, and the **range** consists of **angle measures between 0° and 90°**.

You can use the inverse sine or cosine to find the measure of an angle whose sine or cosine is known. In this and certain other situations, it is convenient to refer to the sine or cosine of an angle without reference to its vertex label. In such cases, the Greek letter θ (read "theta") is often used to name the angle.

EXAMPLE 4

Find θ when $\sin \theta = 0.4731$. Express the answer in two ways:

a. in decimal degrees, to the nearest tenth of a degree

b. in degrees, minutes, and seconds, to the nearest second

SOLUTION

To find θ, use inverse sine.

$$\theta = \sin^{-1} 0.4731$$

a. A calculator gives θ as 28.2° to the nearest tenth of a degree.

b. Converting θ to degrees, minutes, and seconds, you get 28°14′9″ to the nearest second.

```
sin-1 0.4731
         28.23571314
Ans▸ DMS
         28°14'8.567"
```

Many calculators will convert decimal degrees to degrees, minutes, and seconds (DMS).

☑ CHECKING KEY CONCEPTS

1. For △ABC with right ∠C, write sin B and cos B in terms of the side lengths a, b, and c.

2. Find θ to the nearest tenth of a degree when cos θ = 0.7859.

3. Given θ = 32°19′, find sin θ and cos θ to four decimal places.

4. A kite is on a 50 yard string that makes a 50° angle with the ground.

   a. Make a sketch to represent the situation described. Label the sides of the triangle formed by the kite, the string, and the ground.

   b. Find the height of the kite above the ground.

14.1 Exercises and Applications

Extra Practice exercises on page 770

Find sin A and cos A for each right triangle. Express answers in decimal form.

1.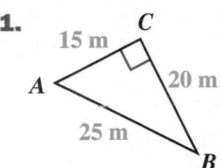

2.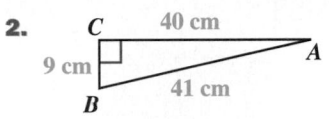

3.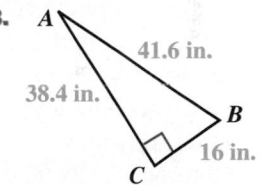

4–6. Find sin B and cos B for each of the triangles in Exercises 1–3. Compare these ratios with the ones you found in Exercises 1–3. What do you notice?

7. **EXERCISING** When Carol read these directions for an exercise, she thought, "How can I know when I have lifted my shoulders 30°? Wouldn't it be easier if I knew how *high* to lift my shoulders instead?"

   Suppose Carol's shoulder-to-hip length is 28 in. How high should she lift her shoulders to do this exercise correctly?

> CRUNCHES: Legs bent; arms back, supporting head and neck; chin up. Lift shoulders and back about 30°. (Lifting up more than 30° activates back muscles, thus negating abdominal work.)

For each trigonometric ratio, find θ. Express the answer in two ways:

a. in decimal degrees, to the nearest tenth of a degree

b. in degrees, minutes, and seconds, to the nearest second

8. sin θ = 0.9532

9. cos θ = 0.7248

10. sin θ = 0.2361

11. cos θ = 0.2854

12. sin θ = 0.5377

13. sin θ = 0.4444

14. cos θ = 0.7800

15. sin θ = 0.8580

Complete each statement using <, >, =, ≤, or ≥. Assume that ∠A and ∠B are both acute angles.

16. If ∠A > ∠B, then sin A ⎯?⎯ sin B.

17. If ∠A > ∠B, then cos A ⎯?⎯ cos B.

Recall from Chapter 5 that a *projectile* is an object that is thrown or launched. The horizontal position *x*, in feet, and the vertical position *y*, in feet, of a projectile *t* seconds after it is launched at an angle θ can be found using the parametric equations

$$x = (v_0 \cos \theta)t + x_0$$
$$y = -16t^2 + (v_0 \sin \theta)t + y_0$$

where v_0 is the launch velocity in feet per second, and x_0 and y_0 are the initial horizontal and vertical positions, respectively, in feet. This model has its limitations, since it does not take into account factors such as air resistance, which affect a projectile's movement.

In Exercises 18 and 19, assume the origin of the coordinate system is at (x_0, y_0).

18. Suppose a golfer hits a golf ball 167 yd using a 7-iron. If the launch angle is 45° and the launch velocity is 127 ft/s, for how many seconds is the golf ball in the air?

BY THE WAY

Golfers use two types of clubs, woods and irons. Both woods and irons are numbered 1 through 9 depending on the angle of the surface that contacts the ball. This angle affects how far away the ball lands.

19. In a court case, a track coach was called as an expert witness. The coach testified that it was impossible for the defendant to have jumped the 13 ft gap between the roof of one building and the roof of another building that was one story (about 10 ft) lower. The reason given was that the world record for the standing broad jump is 12 ft.

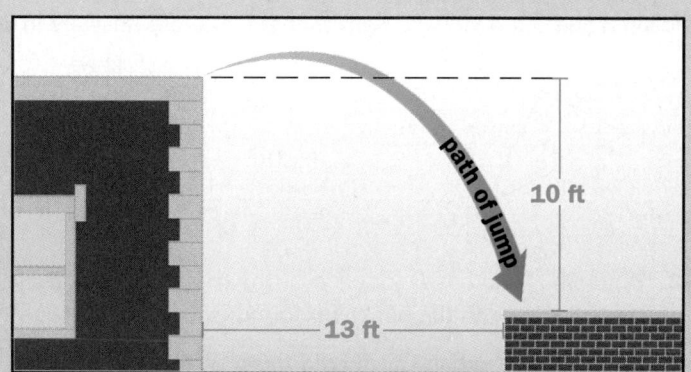

path of jump

10 ft

13 ft

a. **Technology** Graph the parametric equations $x = (v_0 \cos \theta)t$ and $y = -16t^2 + (v_0 \sin \theta)t$ on a graphing calculator. Experiment to find the value of θ that maximizes the horizontal distance that a projectile travels for any given value of v_0. (*Hint:* Choose any value for v_0 and then vary the value of θ.)

b. Estimate the launch velocity for the world record standing broad jump by completing Steps 1–3.

   Step 1 Express the time in the air for a standing broad jump in terms of v_0 and θ. (*Hint:* Solve $y = -16t^2 + (v_0 \sin \theta)t$ for *t* when $y = 0$. Why?)

   Step 2 Express the length of a broad jump in terms of v_0 and θ. (*Hint:* Substitute the nonzero value of *t* from Step 1 into $x = (v_0 \cos \theta)t$.)

   Step 3 Express the length of the longest possible standing broad jump as a function of the launch velocity. (*Hint:* Use the value of θ you found in part (a).)

c. **Technology** On a graphing calculator, graph the path of the jumper in the situation described. Determine whether the conclusion drawn by the expert witness is correct. (*Hint:* Is $x > 13$ when $y = -10$?)

Find the measure of ∠ A in decimal degrees, to the nearest tenth of a degree.

20.
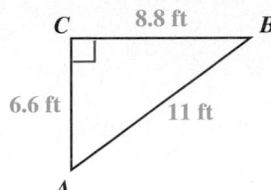
C 8.8 ft B
6.6 ft 11 ft
A

21.
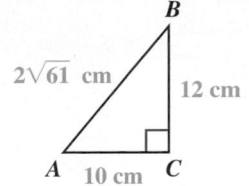
B
$2\sqrt{61}$ cm 12 cm
A 10 cm C

22.
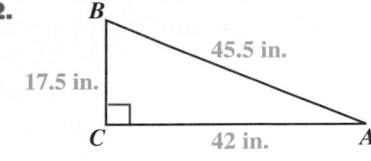
B
45.5 in.
17.5 in.
C 42 in. A

23. ASTRONOMY For the triangle formed by Earth E, Venus V, and the sun S, $\angle E$ reaches its maximum value, $47°$, when $\overline{EV} \perp \overline{VS}$. Use the diagram to find each distance. (The distance between Earth and the sun is about 93,000,000 mi.)

a. distance between Venus and the sun

b. distance between Venus and Earth

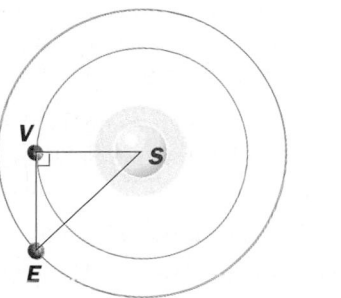

INTERVIEW Johnpaul Jones

Look back at the article on pages 652–654.

The moat that surrounds the Sumatran tiger habitat in Johnpaul Jones's Tiger River exhibit benefits both the human visitors to the zoo and the tigers. In warm climates like that of San Diego, tigers may spend hours lying in the water to escape the heat and the flies. Combined with a wall that slants inward toward the base, the moat helps to keep the tigers inside the simulated jungle.

Use the parametric equations for projectile motion given on page 660.

24. Suppose that the observation deck for the Sumatran tiger habitat were on the same level as the water in the moat. If the launch angle and velocity for the *longest* possible leap by a Sumatran tiger are $20°$ and 19 mi/h, respectively, what should be the minimum width of the moat?

25. Suppose that the tiger habitat were not surrounded by a moat, and that the wall below the observation deck were straight, not slanted. If the launch angle and velocity for the *highest* possible leap by a Sumatran tiger are $30°$ and 24 mi/h, respectively, what should be the minimum height of the wall?

26. PARASAILING This sport is a cross between hang-gliding and water-skiing. The rider wears a harness fitted with a special parachute that acts like an airplane wing. As the tow line connecting the rider to the boat is let out, the parachute lifts the rider up into the air.

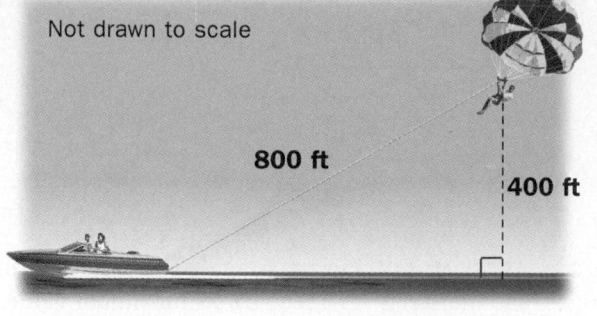

Not drawn to scale

800 ft

400 ft

Suppose the tow line is let out to its maximum length of 800 ft. What is the measure of the angle between the tow line and the surface of the water when the rider is at a height of 400 ft?

27. Challenge Without using a calculator, find $\sin \theta$ and $\cos \theta$ for each value of θ. Explain how you found your answers.

a. $\theta = 45°$ (*Hint:* Sketch a right triangle with a 45° angle. How are the lengths of the two shorter sides related?)

b. $\theta = 30°$ (*Hint:* Sketch an equilateral triangle and draw an altitude.)

c. $\theta = 60°$

28. SAT/ACT Preview For right $\triangle ABC$, which equation can you use to solve for c?

A. $\cos 72.8° = \dfrac{34.2}{c}$ **B.** $\sin 72.8° = \dfrac{34.2}{c}$

C. $\sin 72.8° = \dfrac{c}{34.2}$ **D.** $\cos 72.8° = \dfrac{c}{34.2}$

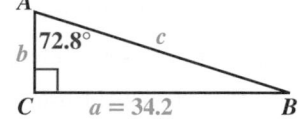

E. none of these

29. Writing Use the diagram and a calculator to describe how the value of $\sin A$ changes with the measure of $\angle A$ in a right triangle when the length of the hypotenuse remains constant.

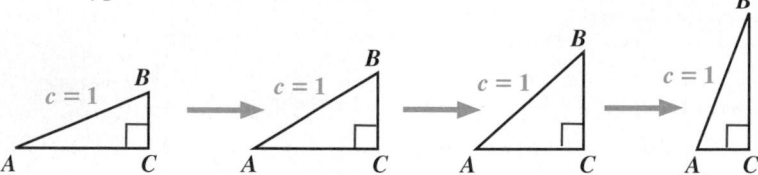

30. Suppose the donations received at a charity event form a normal distribution. If the mean donation was \$55 and the standard deviation was \$10, what percent of donations were below \$50? above \$75? (*Section 13.5*)

Find the geometric mean for each pair of numbers. (*Section 10.2*)

31. 17, 6

32. 1.8, 0.5

33. $\dfrac{1}{2}, \dfrac{3}{8}$

Solve each proportion. (*Toolbox, p. 785*)

34. $\dfrac{x}{16} = \dfrac{3}{4}$

35. $\dfrac{18}{x} = \dfrac{x}{4}$

36. $\dfrac{2}{3} = \dfrac{10}{x+1}$

Section

14.2 Using Tangent

In addition to the sine and cosine ratios, you can use the *tangent* ratio to find the measures of the sides and angles of a right triangle.

Learn how to...

- find the tangent of an acute angle

So you can...

- find the length of a runway when you know the height of a plane and the angle of descent, for example

---

Tangent of an Angle

In right $\triangle ABC$, the **tangent** of $\angle A$, which is written "tan A," is given by

$$\tan A = \frac{\text{length of leg opposite } \angle A}{\text{length of leg adjacent to } \angle A} = \frac{a}{b}$$

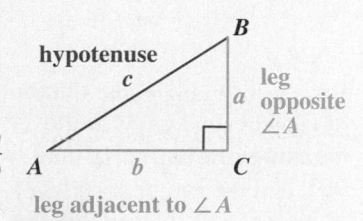

EXPLORATION

COOPERATIVE LEARNING

Estimating the Distance Across Your Classroom

Work with a partner.
You will need:

- a ruler, yardstick, or meterstick

1 Stand at one end of the classroom holding the ruler. Your partner should stand at the other end of the room.

2 Hold the ruler about 12 in. in front of your eyes and line up the "0" end of the ruler with the top of your partner's head.

3 Spot the floor at your partner's feet along the ruler. What are your partner's "ruler height" and actual height?

4 Copy and complete the diagram below. Include the measurements from Step 3.

Questions

1. What is the value of tan θ using the small triangle?

2. Let d be the unknown distance across the room. Using the large triangle, write an expression involving d for the value of tan θ.

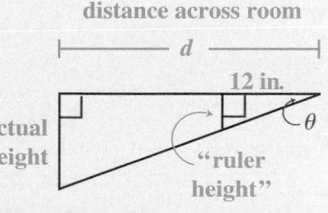

3. Use your answers from Questions 1 and 2 to write an equation. Solve for d. (You may want to use the ruler to find the actual distance across the room and compare it with your calculated result.)

14.2 Using Tangent **663**

THINK AND COMMUNICATE

Use the definition of tangent given at the top of the previous page.

1. What is the value of tan *A* for a right △*ABC* with the given leg lengths?

    a. *a* = 1, *b* = 10   **b.** *a* = 10, *b* = 10   **c.** *a* = 100, *b* = 10

2. For a right triangle, tangent is a function whose domain consists of all angle measures between 0° and 90°. What do you think is the range of the tangent function for a right triangle?

The inverse of the tangent function is written "tan⁻¹." The inverse tangent function is formed by reversing the domain and range of the tangent function.

For a right triangle, the **domain** of tangent consists of **angle measures between 0° and 90°**, and the **range** consists of **all nonnegative numbers**.

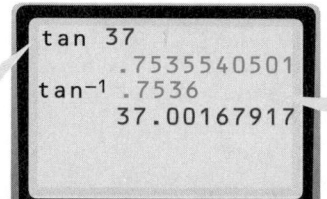

```
tan 37
        .7535540501
tan⁻¹ .7536
        37.00167917
```

For a right triangle, the **domain** of inverse tangent consists of **all nonnegative numbers**, and the **range** consists of **angle measures between 0° and 90°**.

Surveyors sometimes use a professional instrument called a *transit* to measure angles and then use trigonometry to calculate distances.

EXAMPLE 1   Application: Surveying

Surveyors use trigonometry to calculate distances that would otherwise be difficult to find, such as the distance between two houses located across a lake from each other, as shown in the diagram. What is this distance?

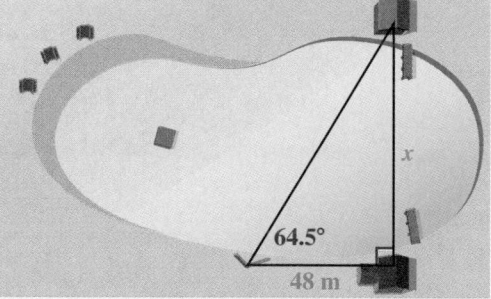

SOLUTION

Let *x* = the distance in meters across the lake. Write an equation involving *x* and a trigonometric ratio.

```
tan 64.5
        2.096543599
Ans*48
        100.6340928
```

$$\frac{x}{48} = \tan 64.5°$$

$$x = 48 \tan 64.5°$$

$$x \approx 48(2.0965) \approx 100.6$$

The distance between the two houses is about 101 m.

THINK AND COMMUNICATE

3. In Example 1, what would you do to find the other acute angle in the diagram? What would you do to find the hypotenuse?

4. Use a calculator to find $\tan^{-1}\left(\frac{101}{48}\right)$. Does this agree with the angle given in the diagram for Example 1?

Finding the lengths of *all* sides and the measures of *all* angles of a triangle is called **solving a triangle**.

EXAMPLE 2

Solve right △ABC.

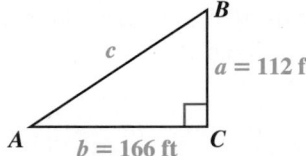

SOLUTION

Step 1 The lengths of the two legs of the triangle are known. To find ∠A, use inverse tangent.

$$\tan A = \frac{\text{opposite}}{\text{adjacent}} = \frac{112}{166}$$

$$\angle A = \tan^{-1}\left(\frac{112}{166}\right) \approx 34°$$

```
tan⁻¹ (112/166)
          34.00749242
```

Step 2 To find ∠B, use the fact that the sum of angle measures in a triangle is 180°.

$$\angle A + \angle B + \angle C = 180°$$
$$34° + \angle B + 90° \approx 180°$$
$$\angle B \approx 56°$$

Step 3 Now use the Pythagorean theorem (or a trigonometric ratio) to find *c*, the length of the hypotenuse.

$$c^2 = 166^2 + 112^2$$
$$c = \sqrt{40,100}$$
$$c \approx 200.2$$

> **Toolbox p. 801**
> *Triangle Relationships*

Therefore, $\angle A \approx 34°$, $c \approx 200$ ft, and $\angle B \approx 56°$.

Angles of Elevation and Depression

In the Exploration on page 663, you measured an **angle of depression** from your eye level to your partner's feet. There is an angle of equal measure, called the **angle of elevation**, from your partner's feet up to your eyes.

Angles of depression and elevation are often used with trigonometry to calculate distances in navigation, aviation, and surveying.

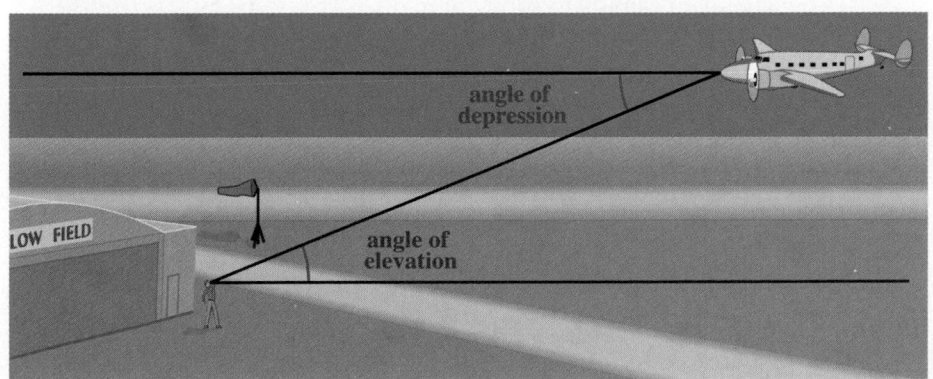

EXAMPLE 3   Application: Aviation

Suppose an airplane begins its descent to a runway from an altitude of 10,000 ft.

a. If the airplane is still 75,000 ft (about 14 mi) from the runway as measured along the ground, what is its angle of descent (the angle of depression)?

b. What is the length of the runway if the angle of elevation from the far end of the runway to the airplane is 6.6°?

SOLUTION

Not drawn to scale

a.

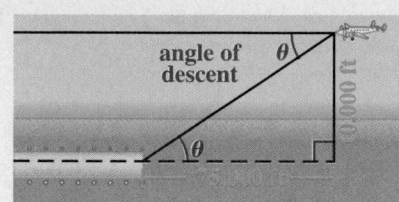

angle of descent θ

b.

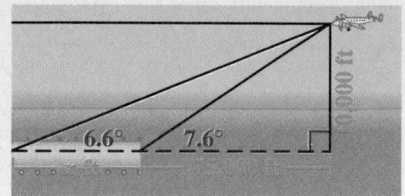

6.6°　7.6°

Use the fact that the measures of the angle of elevation and the angle of depression are the same to find the airplane's angle of descent using the tangent ratio.

$$\tan \theta = \frac{10,000}{75,000}$$

$$\theta = \tan^{-1}\left(\frac{10,000}{75,000}\right)$$

$$\approx 7.59°$$

The angle of descent is about 7.6°.

> Multiply both sides by $75,000 + x$ and divide both sides by 0.1157.

The distance from the airplane to the far end of the runway as measured along the ground is $75,000 + x$, where x is the length in feet of the runway.

$$\tan 6.6° = \frac{10,000}{75,000 + x}$$

$$0.1157 \approx \frac{10,000}{75,000 + x}$$

$$75,000 + x \approx \frac{10,000}{0.1157}$$

$$x \approx 86,430 - 75,000$$

$$= 11,430$$

The runway is about 11,000 ft long.

☑ CHECKING KEY CONCEPTS

1. a. For right $\triangle ABC$ shown, state an equation you can use to solve for a.

  b. Solve $\triangle ABC$.

2. In right $\triangle RST$, $\angle R$ is a right angle, $t = 3$ ft, and $s = 4$ ft. Find $\angle S$.

3. Use your knowledge of geometry to explain why the angle of elevation ($\angle 1$) and the angle of depression ($\angle 2$) between two objects at different heights are equal in measure.

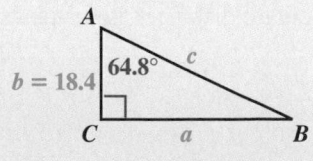

$b = 18.4$　64.8°　c

C　a　B

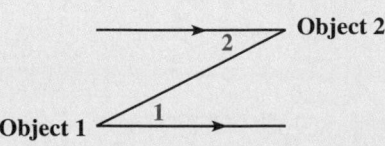

Object 2

Object 1

14.2 **Exercises and Applications**

Extra Practice
exercises on
page 770

1. Refer to △*XYZ*.

   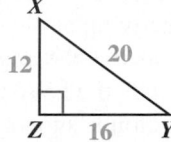

   a. Find tan *Y*.   **b.** Find tan *X*.

   c. How are your answers to parts (a) and (b) related?

2. Find each value.

   a. tan 32.8°   **b.** tan 28°41′   **c.** tan 89.1°   **d.** tan 45°

3. **SAT/ACT Preview** For right △*ABC*, which equation
   can you use to solve for *a*?

   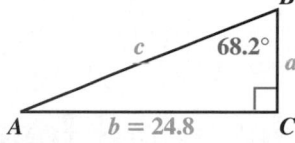

   A. $\tan 68.2° = \dfrac{24.8}{a}$ **B.** $\tan 68.2° = \dfrac{a}{24.8}$

   C. $\tan 21.8° = \dfrac{24.8}{a}$ **D.** $\sin 21.8° = \dfrac{a}{24.8}$

   E. none of these

4. **AVIATION** The pilot of an airplane finds that the angle of depression to the
   start of a runway is 52°34′.

   a. If the altitude of the airplane is 1250 m, how far is the airplane from the
   runway as measured along the ground?

   b. If the angle of depression to the other end of the runway is 36°, how
   long is the runway to the nearest meter?

Solve △*ABC* using the given measures.

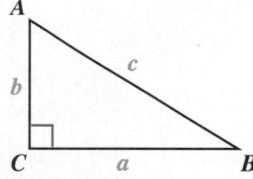

5. $\angle A = 19°, b = \sqrt{12}$ 6. $c = 32, \angle B = 72°$

7. $b = 12, c = 19.5$ 8. $a = 114, \angle B = 39.3°$

9. $\tan A = \dfrac{1}{3}, a = 7$ 10. $\angle B = 62.8°, c = 18.4$

11. **ASTRONOMY** On the moon, when the sun is at an angle of 30° to the
    horizon, the rim of a crater casts a 120 ft long shadow at the bottom of
    the crater. Assuming the sides of the crater are vertical, find the depth of
    the crater.

12. **GEOMETRY** For any line that has a positive slope and intersects the *x*-axis,
    let θ be the angle that the line makes with the *x*-axis. (Measure θ counter-
    clockwise from the *x*-axis to the line, so that 0° < θ < 90°.) Discuss how θ
    is related to the slope of the line.

13. **Cooperative Learning** Work with a partner. You may want to use
    graph paper.

    a. Find the perimeter of rectangle *ABCD* with vertices *A*(3, 7), *B*(5, 5),
    C(1, 1), and *D*(−1, 3).

    b. Find the measures of ∠ *DAC* and ∠ *CAB* formed by drawing diagonals
    $\overline{AC}$ and $\overline{BD}$. Describe your method.

    c. Find the measure of the acute angle formed by the intersection of the
    diagonals. Describe your method.

14. METEOROLOGY Pilots flying at night need to know the *cloud ceiling*, the lowest height of a mass of clouds. Pilots cannot take off or land if this ceiling is too low. To determine the cloud ceiling at night, a meteorologist shines a searchlight, located a known distance from a weather station, directly up at the clouds. The meteorologist then measures the angle of elevation to the spot of light and finds the height of the clouds. Find the height of the clouds shown in the diagram.

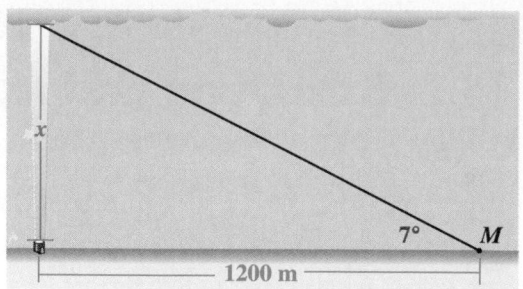

Not drawn to scale

15. Refer to △*LMN*. Name an angle whose tangent is $\sqrt{3}$. Name an angle whose tangent is $\frac{1}{\sqrt{3}}$. Is there an angle of △*LMN* whose tangent is $\frac{1}{2}$? Explain.

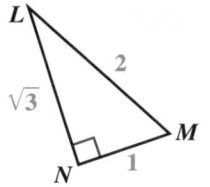

INTERVIEW **Johnpaul Jones**

Look back at the article on pages 652–654.

Johnpaul Jones and his staff designed a marsh **aviary,** *or bird enclosure, for the San Diego Zoo. The drawing below is similar to an architectural drawing of the aviary proposed by Jones and his staff. Zoo visitors would observe the birds from behind a railing and a harpwire barrier.*

The roof is designed to be high enough to give birds flying room, but not so high that birds can fly out of sight from viewers.

Harpwire is strung between roof supports.

Tubular steel posts hold up the nylon netting roof.

Not drawn to scale

16. Suppose a person whose eye level is at 6.5 ft stands at the railing. Some of the person's view of the sky would be blocked by the overhang as shown.

  a. What would be the vertical distance between the person's eyes and the overhang?

  b. What should be the measure of ∠1?

17. Suppose the person in Exercise 16 is 60 ft from a point where a tubular steel post is to be placed, and suppose that the bottom of the post is at the person's eye level.

  a. What should be the measure of ∠2 (the angle of elevation determined by the overhang)?

  b. How tall should the post be for the person to just see the top of it?

  c. Would someone shorter still find this an acceptable height for the pole? Explain.

Throughout history, people have constructed sundials (or "shadow clocks") to tell time. In the 13th century, a Moroccan scientist named Abu'l Hasan first introduced the idea of pointing the *gnomon*, or shadow pole, toward the North Star. In this way, the hour lines on a sundial could be positioned so that consecutive lines measure equal intervals of elapsed time.

A sundial can be constructed using the directions shown.

The angle of the gnomon should be the same as the latitude where the sundial is located.

Place the gnomon on the north-south line.

The noon shadow points directly north. Place the mark for 12:00 noon on the north-south line. The hour angle for noon is 0°.

The positions of the hour marks around the sundial depend on the latitude where the sundial is located.

18. To position the hour marks around the sundial, you need to determine the shadow angles that correspond to each hour. These angles depend on the sundial's latitude and can be found by this formula:

tan (shadow angle) = sin (latitude) × tan (hour angle)

a. Each hour is $\frac{1}{24}$ of a day. Each day, Earth makes a revolution of 360°, so *each hour* Earth turns through an angle of $\frac{1}{24} \cdot 360°$, or 15°. At 2:00 P.M., the hour angle is 30°. What is the hour angle for 5:00 P.M.?

b. What shadow angle corresponds to 5:00 P.M. in New York City, which has a latitude of 41°?

19. Research What is the latitude where you live? If you build a sundial like the one shown, what is the shadow angle on your sundial at 5:00 P.M.?

20. Challenge From point *A*, at an elevation of about 11,750 ft above sea level, suppose the angle of elevation to the top of Mt. Everest is about 8.7°. From point *B*, at the same elevation as *A* but 18,325 ft farther away from the mountain than *A*, suppose the angle of elevation is about 7.5°. How high is Mt. Everest?

Not drawn to scale

B 7.5° *A* 8.7°

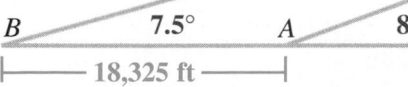

├── 18,325 ft ──┤

21. Writing Use ratios to help you explain this alternative definition of tan A:

$$\tan A = \frac{\sin A}{\cos A}$$

SPIRAL REVIEW

22. SPORTS It takes only about 5 min to travel up the *Extreme Access* chair lift at the Copper Mountain Resort ski area in Colorado. *(Section 14.1)*

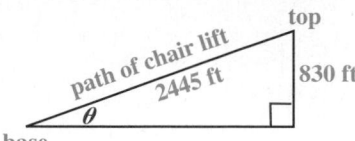

path of chair lift
2445 ft
top
830 ft
θ
base

a. How fast does the chair lift travel in feet per minute? in miles per hour?

b. Find θ, the measure of the angle of elevation from the base of the chair lift to the top of the lift.

23. Tell whether each sequence is *geometric* or *arithmetic*. State the common difference or common ratio for each sequence. *(Section 10.2)*

a. 35, 125, 215, 305, . . .   **b.** 0.5234, −0.5234, 0.5234, −0.5234, . . .

24. Name the vertex and the line of symmetry for the graph of the equation $y = 3x^2 + 6x + 2$. *(Section 5.2)*

ASSESS YOUR PROGRESS

VOCABULARY

sine (p. 656)
cosine (p. 656)
tangent (p. 663)

solving a triangle (p. 665)
angle of depression (p. 665)
angle of elevation (p. 665)

1. For $\triangle ABC$ with right $\angle C$, find $\cos A$ if $a = 5$ m and $b = 12$ m. *(Section 14.1)*

2. NAVIGATION An observer at the top of a lighthouse 100 m above sea level spots a boat at sea. *(Section 14.2)*

a. The angle of depression of the boat is 28°. How far is the boat from the shore?

b. Suppose a second boat, in line with the observer and the first boat, is sighted far away at an angle of depression of 12°. How far apart are the two boats?

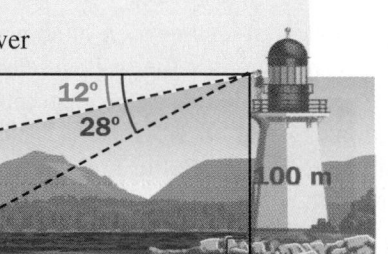

12°
28°
100 m
x
Not drawn to scale

3. Journal Give specific examples of situations where you can use trigonometry to find unknown side lengths or angle measures of right triangles.

14.3 Angles of Rotation

Learn how to...

- **find the trigonometric functions of an angle of rotation**

So you can...

- **convert courses to angles of rotation and find the coordinates of a plane's position, for example**

The direction that an airplane or ship travels relative to north is called its *course*. Courses can be any angle between 0° and 360°. The navigator of a plane or ship measures angles in a *clockwise* direction from north.

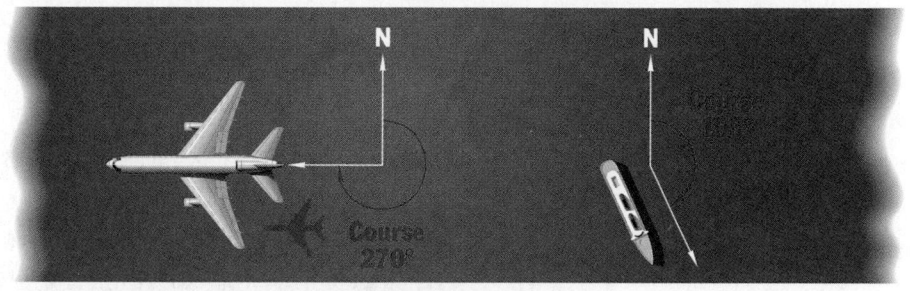

THINK AND COMMUNICATE

1. a. What course corresponds to a ship traveling northwest? southeast?

b. In general, what mathematical relationship holds for the courses of two ships traveling in opposite directions?

2. If a plane flies from Chicago to St. Louis on a course of 205° as shown, what is its return course? Explain.

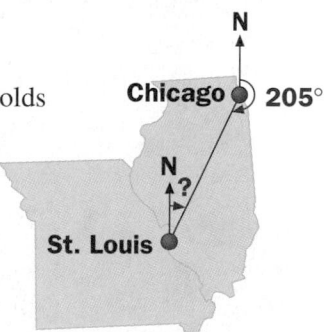

Mathematicians create angles between 0° and 360° in the coordinate plane by rotating a ray in a *counterclockwise* direction from the positive *x*-axis, which is called the **initial side** of an angle. The other ray that forms the angle is its **terminal side**. This is known as the **standard position** of an angle.

This is a **Quadrant III angle** because the terminal side lies in Quadrant III.

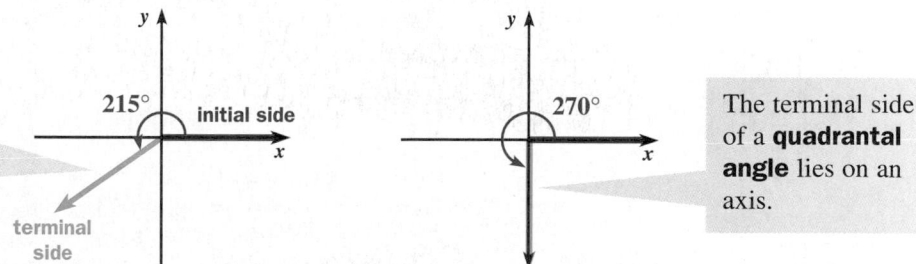

The terminal side of a **quadrantal angle** lies on an axis.

THINK AND COMMUNICATE

3. Name a Quadrant II angle and a Quadrant IV angle.

4. Name a quadrantal angle other than the one shown above.

Trigonometric Functions

The *trigonometric functions* of an angle θ between $0°$ and $360°$ are defined by placing θ in standard position in a coordinate plane and choosing a point P on the terminal side of θ. P must be a point other than the origin.

Dropping a perpendicular from P to the x-axis forms a right triangle.

The hypotenuse has length $r = \sqrt{x^2 + y^2}$, which is always positive.

The legs of the triangle have "lengths" that can be positive, negative, or 0.

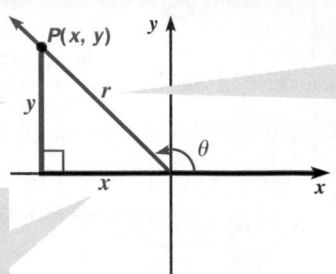

The trigonometric functions of θ are defined in terms of the x- and y-coordinates of P together with r, the distance between P and the origin.

WATCH OUT!

The ratio $\frac{y}{x}$ is undefined when $x = 0$, so $\tan \theta$ is undefined when $\theta = 90°$ or $\theta = 270°$.

Trigonometric Functions

If θ is an angle in standard position with the point $P(x, y)$ on its terminal side and if r is the distance between P and the origin, then:

$$\sin \theta = \frac{y}{r} \qquad \cos \theta = \frac{x}{r} \qquad \tan \theta = \frac{y}{x}$$

EXAMPLE 1

Find the sine, cosine, and tangent of each angle.

a. The terminal side of θ_1 passes through $(-2, 2)$.

b. The measure of θ_2 is $295°$.

SOLUTION

a. Find the value of r and then use the definitions given above.

$$r = \sqrt{(-2)^2 + 2^2} = \sqrt{8} = 2\sqrt{2}$$

$$\sin \theta_1 = \frac{y}{r} = \frac{2}{2\sqrt{2}} = \frac{1}{\sqrt{2}} \approx 0.7071$$

$$\cos \theta_1 = \frac{x}{r} = \frac{-2}{2\sqrt{2}} = \frac{-1}{\sqrt{2}} \approx -0.7071$$

$$\tan \theta_1 = \frac{y}{x} = \frac{2}{-2} = -1$$

b. Use a calculator set in degree mode.

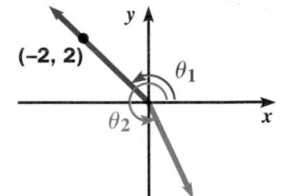

```
sin 295
        -.906307787
cos 295
        .4226182617
tan 295
        -2.144506921
```

Therefore:

$$\sin 295° \approx -0.9063$$

$$\cos 295° \approx 0.4226$$

$$\tan 295° \approx -2.145$$

Notice that the signs of the *x*- and *y*-coordinates of the point in part (a) of Example 1 determined whether the sine, cosine, and tangent values of the angle were positive or negative. The trigonometric functions can be positive or negative depending on the quadrant in which the terminal side lies.

The table at the right and the coordinate plane below show which trigonometric functions are positive in each quadrant.

I	II	III	IV
all	sine	tangent	cosine

Quadrant II:

$x < 0, y > 0$
$90° < \theta < 180°$
$\sin \theta > 0$
$\cos \theta < 0$
$\tan \theta < 0$

Quadrant I:

$x > 0, y > 0$
$0° < \theta < 90°$
$\sin \theta > 0$
$\cos \theta > 0$
$\tan \theta > 0$

Quadrant III:

$x < 0, y < 0$
$180° < \theta < 270°$
$\sin \theta < 0$
$\cos \theta < 0$
$\tan \theta > 0$

Quadrant IV:

$x > 0, y < 0$
$270° < \theta < 360°$
$\sin \theta < 0$
$\cos \theta > 0$
$\tan \theta < 0$

You may find it helpful to remember this sentence:
All scholars take calculus.

The first letter of each word tells you which trigonometric functions are positive in each quadrant:
A for all (Q. I)
s for sine (Q. II)
t for tangent (Q. III)
c for cosine (Q. IV)

EXAMPLE 2   **Application: Aviation**

After leaving an airport, a plane travels on a course of 160° for 300 mi.

a. What angle of rotation is equivalent to the plane's course?

b. Write and solve two equations relating the plane's *x*- and *y*-coordinates to the angle of rotation found in part (a).

c. How far has the plane traveled in an east-west direction? in a north-south direction?

SOLUTION

a. Imagine the airport at the origin of a coordinate plane. The positive *y*-axis represents north, and the positive *x*-axis represents east. Make a sketch of the course and the corresponding angle of rotation as shown. The terminal side of a **160° course** is in Quadrant IV. The **angle of rotation** is 270° + 20° = **290°**.

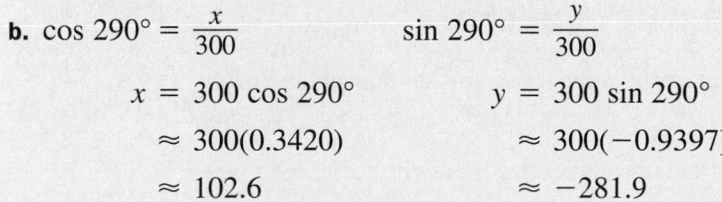

b. $\cos 290° = \dfrac{x}{300}$ $\sin 290° = \dfrac{y}{300}$

$x = 300 \cos 290°$ $y = 300 \sin 290°$

$\approx 300(0.3420)$ $\approx 300(-0.9397)$

≈ 102.6 ≈ -281.9

The coordinates of the plane are approximately $(103, -282)$.

c. The plane has traveled about 103 mi east and about 282 mi south.

THINK AND COMMUNICATE

5. In part (a) of Example 2, why was 20° added to 270° to get the angle of rotation?

Consider an angle θ with its terminal side in Quadrant I. If you reflect that terminal side over the axes, you will get other angles whose trigonometric values are related to those for θ, as shown below.

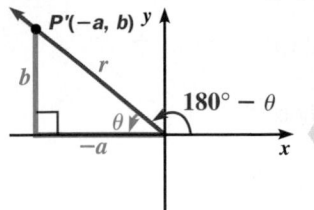

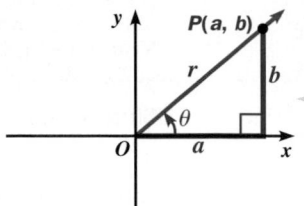

The x- and y-coordinates of each reflected point P' are the **same as or the opposite of** the coordinates of P.

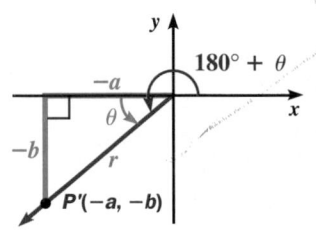

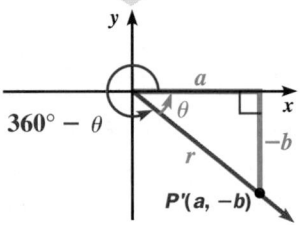

EXAMPLE 3

Find all angles θ between 0° and 360° such that $\sin \theta = 0.5878$.

SOLUTION

Find a Quadrant I angle using inverse sine on a calculator.

$\theta = \sin^{-1} 0.5878 \approx 36°$

Another angle with the same sine value is in Quadrant II.

$180° - \theta \approx 180° - 36° = 144°$

So, $\theta \approx 36°$ or $\theta \approx 144°$.

```
sin⁻¹ 0.5878
       36.00104446
sin (180-36)
       .5877852523
```

☑ CHECKING KEY CONCEPTS

1. In what quadrant is the terminal side of each angle?

   a. 150°   **b.** 252°   **c.** 76°

2. An angle θ is in standard position with the given point P on its terminal side. Find $\sin \theta$, $\cos \theta$, and $\tan \theta$.

   a. $P(-12, -5)$ **b.** $P\left(1, \sqrt{2}\right)$ **c.** $P(24, -7)$

3. Find all angles θ between 0° and 360° such that $\tan \theta = 2.246$.

14.3 Exercises and Applications

Extra Practice
exercises on page 770

Sketch each angle of rotation in a coordinate plane.

1. 30°   **2.** 150°   **3.** 210°   **4.** 330°

5. A snorkeler swam away from an anchored boat on a course of 25°. What course should the snorkeler use to swim back to the boat?

6. a. Name all the quadrantal angles from 0° to 360°.

b. What quadrantal angles correspond to courses that are exactly north, south, east, and west?

c. Find the sine and cosine of each quadrantal angle you listed in part (a).

7. DANCING In ballroom dancing, some of the moves are described as "quarter turns."

a. What is the measure of a quarter turn?

b. If a pair of dancers has completed three quarter turns, through what angle have they rotated?

8. ARCHITECTURE The Space Needle is a 605 ft tall tower built for the 1962 World's Fair in Seattle, Washington. The dining section of a circular restaurant near the top of the Space Needle rotates, completing one revolution each hour.

a. Suppose your seat in the restaurant is near a window. Through what angle have you rotated in 40 min when you move from a view of Mt. Rainier to a view of Puget Sound?

b. Suppose your seat at the restaurant has rotated through an angle of 84°. How long have you been sitting there?

An angle θ is in standard position with the given point P on its terminal side. Find sin θ, cos θ, and tan θ.

9. $P(8, 15)$ **10.** $P\left(3, -3\sqrt{3}\right)$ **11.** $P(-4, -3)$ **12.** $P(-100, 100)$

For each angle θ, use a calculator to find sin θ, cos θ, and tan θ.

13. $\theta = 192°$ **14.** $\theta = 302°$ **15.** $\theta = 111.6°$ **16.** $\theta = 24°16'45''$

17. Writing Suppose an angle θ has its terminal side in Quadrant IV. Explain in your own words how you know that tan θ is negative.

18. What is the relationship between sin θ and sin $(180° + \theta)$? Explain.

19. Find two values of θ for each trigonometric value below. Round your answers to the nearest degree.

a. sin $\theta = 0.8660$ **b.** cos $\theta = 0.7660$ **c.** tan $\theta = 2.475$

20. Use the results of Exercise 19 to predict two values of θ for each trigonometric value given. Check your answers with a calculator.

a. sin $\theta = -0.8660$ **b.** cos $\theta = -0.7660$ **c.** tan $\theta = -2.475$

21. NAVIGATION After leaving a harbor, a ship travels on a course of 190° for 98 mi.

   a. What angle of rotation is equivalent to the ship's course?

   b. Write and solve two equations relating the ship's x- and y-coordinates to the angle of rotation found in part (a).

   c. How far has the boat traveled in an east-west direction? in a north-south direction?

22. Challenge A ship travels 38 mi on a course of θ degrees for 1 h. Find formulas for the distance d traveled north or south as a function of θ. Consider different quadrants.

Connection ▶ ROBOTICS

Robots are useful in such diverse fields as manufacturing, office automation, and space and underwater exploration. A robot is sometimes controlled by a human operator, who gives the robot instructions to follow. These instructions can take the form of turn angles and distances.

Suppose a robot can move only forward and must first turn its body to face the direction it needs to travel. The diagram shows the obstacles that the robot must avoid while moving around a room.

23. Suppose the robot travels from its current location A to another location B by following the given set of instructions. Each of the turns described is *counterclockwise* from the direction that the robot faces at any given time.

   a. Copy the diagram. Sketch the robot and obstacles. Determine the robot's location after carrying out the instructions.

   b. If the robot can make only *clockwise* turns, how would the instructions to get from A to B change?

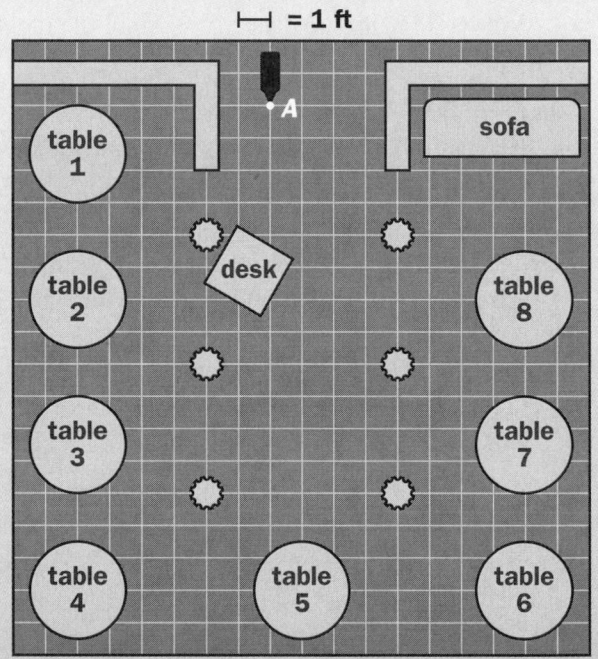

```
From location A
turn 27°, move 4.5 ft;
turn 333°, move 4.0 ft;
turn 288°, move 6.5 ft;
   to reach location B.
```

24. Sometimes it is important for a robot to complete instructions in the least amount of time or to travel the shortest distance possible.

   a. Open-ended Problem Choose what you think is a shorter path from location A to location B. Write a set of instructions using counterclockwise turns to get the robot from A to B.

   b. Find the total distance traveled for your path in part (a) and the path described in Exercise 23. Is your new path shorter than the one in Exercise 23?

25. **Cooperative Learning** Work in a group of three students. You will design a treasure map using a protractor and a compass.

a. Find or draw a map of some open area such as the schoolyard, the town green, a golf course, or a local park.

b. Put a starting point on your map and secretly choose a location for the treasure.

c. Write directions to the treasure in terms of courses and distances with no fewer than four moves and no more than eight moves.

d. Exchange your map and directions with another group. Use the process shown in Example 2 on page 673 to separate each course and distance you receive into north-south and east-west components. Then combine all the north-south components and all the east-west components to find out the location of the treasure on the map you receive.

e. Compare your group's answer with the treasure's location chosen by the group that created the map. Were you successful in finding the treasure?

26. **Research** Find out about a famous journey in history or a treasure map in literature. Discuss how to get to the goal using courses and distances. Then separate these courses and distances into north-south and east-west components.

ONGOING ASSESSMENT

27. **Writing** Describe how you can find all angles θ between $0°$ and $360°$ whose tangent is $\frac{3}{4}$. Include sketches of the angles in a coordinate plane.

SPIRAL REVIEW

28. Solve each triangle. *(Section 14.2)*

a.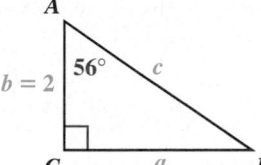

b.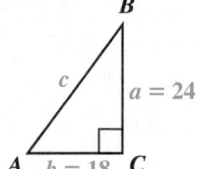

29. Find the area of each triangle described. *(Toolbox, page 802)*

a. base length = 17 ft; height = 24 ft

b. base length = $3\sqrt{2}$ cm; height = $\sqrt{6}$ cm

30. Suppose you roll a six-sided die three times. What is the probability of getting at least one 6? *(Section 13.2)*

14.4 Finding the Area of a Triangle

Learn how to...

- **find the area of a triangle given two side lengths and the measure of the included angle**

- **find the area of a sector**

So you can...

- **find the area of the Bermuda Triangle, for example**

The map below shows the Bermuda Triangle, a large region of the Atlantic Ocean where many ships and planes have mysteriously disappeared. A trigonometric ratio can help you to find the area of this region.

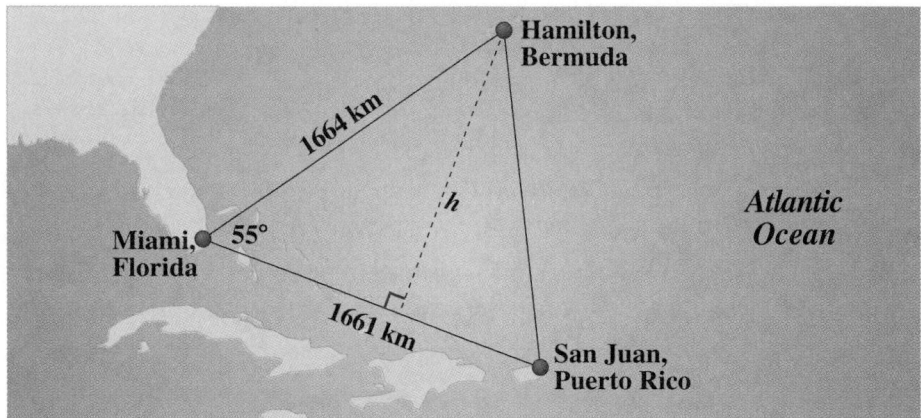

Hamilton, Bermuda

1664 km

h

Miami, Florida 55°

1661 km

San Juan, Puerto Rico

Atlantic Ocean

THINK AND COMMUNICATE

1. a. Write and solve an equation involving a trigonometric ratio to find the height *h* of the Bermuda Triangle.

  b. Use the value of *h* you found in part (a) to find the area of the Bermuda Triangle.

2. Write an expression for the area of $\triangle XYZ$ in terms of *x*, *y*, and the measure of $\angle Z$. Use a diagram to explain your reasoning.

You can find the area of any triangle if you know the lengths of two sides and the measure of the included angle.

Area of a Triangle

The area *K* of $\triangle ABC$ is given by any of the following formulas:

$$K = \frac{1}{2}bc \sin A$$

$$K = \frac{1}{2}ac \sin B$$

$$K = \frac{1}{2}ab \sin C$$

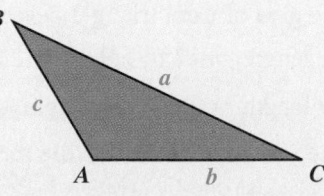

EXAMPLE 1

Find the area of △*XYZ*.

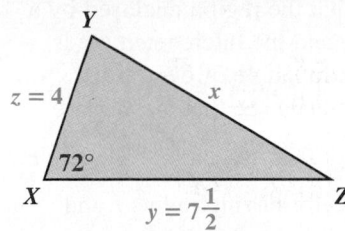

SOLUTION

Since you know the lengths of two sides and the measure of the included angle, you can find the area using $K = \frac{1}{2}yz \sin X$.

$$K = \frac{1}{2}\left(7\frac{1}{2}\right)(4) \sin 72°$$

> Substitute $7\frac{1}{2}$ for y, **4** for z, and **72°** for X.

$$\approx 14.3$$

The area of △*XYZ* is about 14.3 square units.

THINK AND COMMUNICATE

3. Complete parts (a)–(c) to show that the area formula given on the previous page applies to an obtuse triangle as well as an acute triangle.

a. Express the height *h* of △*ABC* in terms of one or more of the labeled parts of △*ABD*.

b. Use your answer to part (a) to express the area *K* of △*ABC* in terms of the labeled parts of the diagram.

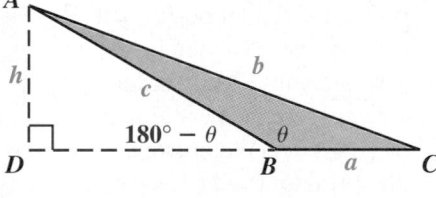

c. What is the relationship between sin (180° − θ) and sin θ? Use your answer to simplify the area formula you wrote for part (b).

EXAMPLE 2

Find the area of △*RST* where *t* = 11.4, *r* = 8.6, and ∠*S* = 125°.

SOLUTION

Step 1 Make a sketch. You can see that the given angle is between the two given sides.

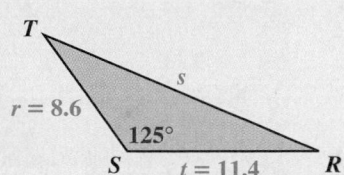

Step 2 Use an area formula.

$$K = \frac{1}{2}tr \sin S$$

$$= \frac{1}{2}(11.4)(8.6) \sin 125°$$

$$\approx 40.2$$

> The area of △*RST* is **about 40.2 square units**.

Area of a Sector

Recall from geometry that the region enclosed by a central angle of a circle and the intercepted arc is called a *sector*. The central angle of the shaded sector shown is 90°. You can see that the area of the sector is $\frac{90}{360}$, or $\frac{1}{4}$, of the total area of the circle.

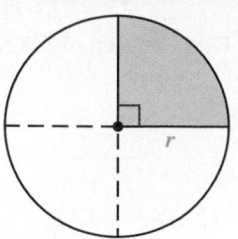

In general, for any sector having radius r and central angle θ:

$$\frac{\text{area of the sector}}{\text{area of the circle}} = \frac{\theta}{360}$$

$$\text{area of the sector} = \frac{\theta}{360} \cdot (\text{area of the circle})$$

$$\mathbf{\text{area of the sector} = \frac{\theta}{360} \cdot \pi r^2}$$

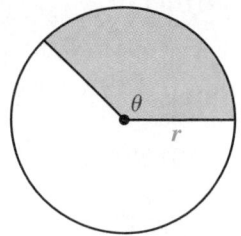

THINK AND COMMUNICATE

4. Explain the significance of the "360" in the formula above.

5. What values of θ can be substituted into the formula?

EXAMPLE 3 | **Application: Landscaping**

A water sprinkler positioned in the corner of a yard is set to rotate only 120° in order to keep the pavement around the grass dry. Water from the sprinkler reaches as far as 15 ft away. What is the area of the yard that is watered by the sprinkler?

sprinkler

SOLUTION

Substitute **120°** for θ and **15** for r into the formula for the area of a sector.

$$\text{area of sector} = \frac{\theta}{360}\,\pi r^2$$

$$\text{area of yard watered} = \frac{120}{360}\,\pi(15)^2$$

$$= 75\pi$$

$$\approx 240$$

About 240 ft² of the yard is watered by the sprinkler.

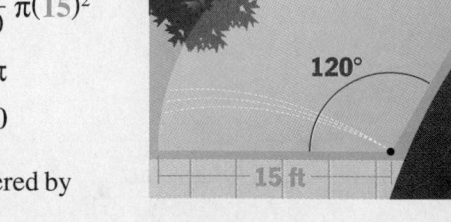

120°

15 ft

☑ CHECKING KEY CONCEPTS

Find the area of each figure.

1.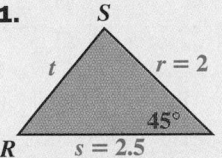

2.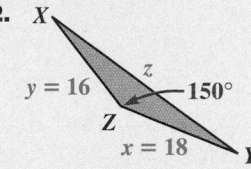

3.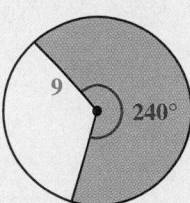

Find the area of △ABC.

4. $a = \frac{1}{2}$, $b = 12$, $\angle C = 40°$ **5.** $b = 10$, $c = 3.9$, $\angle A = 145°$

Find the area of each sector having central angle θ and radius r.

6. $\theta = 60°$, $r = \frac{3}{5}$ **7.** $\theta = 144°$, $r = \sqrt{6}$ **8.** $\theta = 110°$, $r = 1.5$

9. Writing In your own words, state the formula for the area of a triangle given on page 678.

14.4 **Exercises and Applications**

Extra Practice exercises on page 771

For Exercises 1–9, find the area of △DEF. If there is not enough information, explain.

1.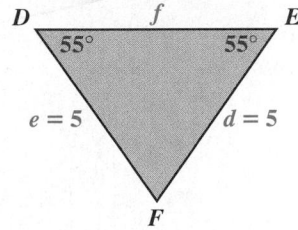

2.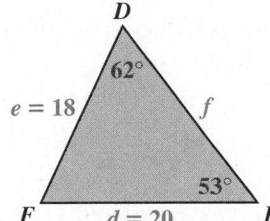

3.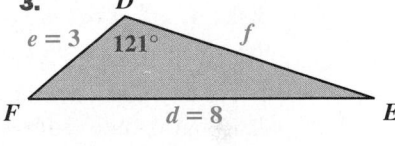

4. $e = 18.2$, $d = 9.4$, $\angle F = 91°$ **5.** $f = 32$, $e = 40$, $\angle D = 115°$ **6.** $d = 6$, $e = 11$, $\angle E = 30°$

7. $d = 9$, $f = 8\frac{1}{2}$, $\angle E = 57°$ **8.** $f = 13$, $e = 4$, $\angle F = 135°40'$ **9.** $f = 10$, $d = 9.7$, $\angle D = 85°$

COOKING One kind of pastry is shaped by rolling a triangular piece of dough from the wide end, as shown. The long sides of the triangle measure about $4\frac{1}{2}$ in. each, and the angle between them measures $22\frac{1}{2}°$.

10. Let $t =$ the thickness of a triangular piece of dough before it is rolled. Express the volume of the dough in terms of t.

11. Challenge How many pieces of pastry can be made from $1\frac{1}{2}$ cups of dough when the dough is $\frac{1}{8}$ in. thick? (*Hint:* 1 cup ≈ 14.4 in.³)

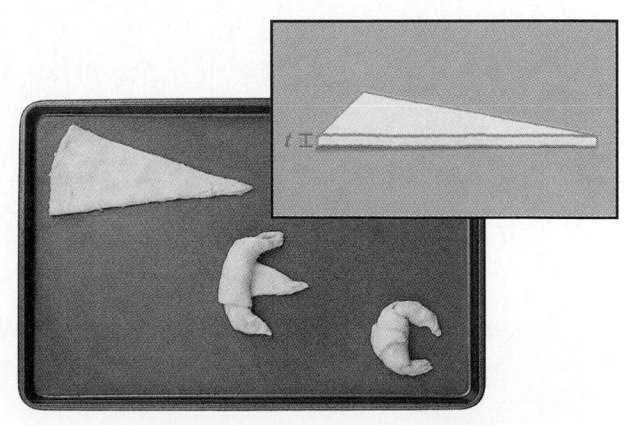

Find the area of each sector having central angle θ and radius r.

12. $\theta = 180°$, $r = \dfrac{5}{8}$

13. $\theta = 280°$, $r = 3\dfrac{1}{4}$

14. $\theta = 72°$, $r = 2.5$

15. $\theta = 22°30'$, $r = 12$

16. $\theta = 3°36'$, $r = 5$

17. $\theta = 225°$, $r = 1.6$

Find the area of each shaded sector.

18.

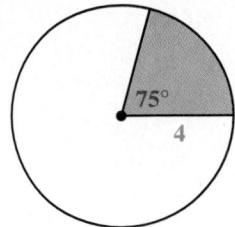

19.

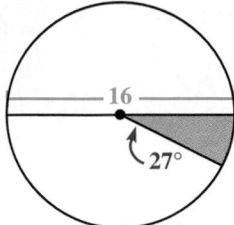

20.

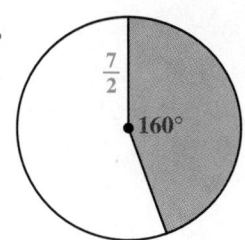

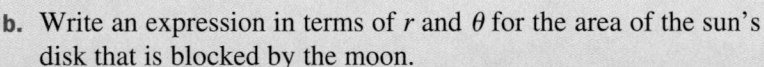

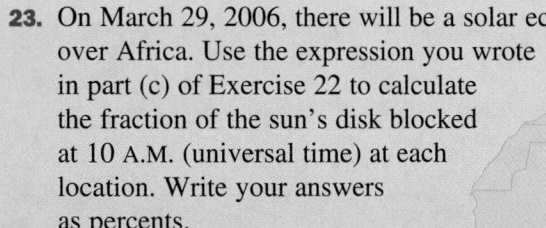

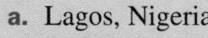

Connection ▶ ASTRONOMY

During a solar eclipse, the moon blocks some or all of your view of the sun. You can use the areas of sectors and triangles to find the fraction of the sun that is blocked by the moon at any given moment during an eclipse.

21. a. Write an expression for the area of sector $XABC$.

    b. Write an expression for the area of $\triangle XAC$.

    c. The region bounded by $\overline{AC}$ and arc ABC is called a *segment* of the circle with center X. Write an expression for the area of segment ABC. Explain your reasoning.

22. a. From Earth, the sun and the moon appear to be circular disks of about the same size. What does this tell you about the areas of segments ADC and ABC?

    b. Write an expression in terms of r and θ for the area of the sun's disk that is blocked by the moon.

    c. Write an expression in terms of θ for the *fraction* of the sun's disk blocked by the moon. Explain how you found your answer.

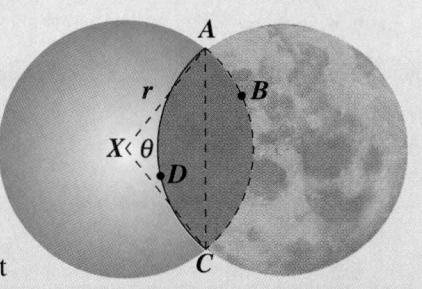

sun   moon

23. On March 29, 2006, there will be a solar eclipse over Africa. Use the expression you wrote in part (c) of Exercise 22 to calculate the fraction of the sun's disk blocked at 10 A.M. (universal time) at each location. Write your answers as percents.

    a. Lagos, Nigeria

    b. Lake Chad in Chad

    c. Cairo, Egypt

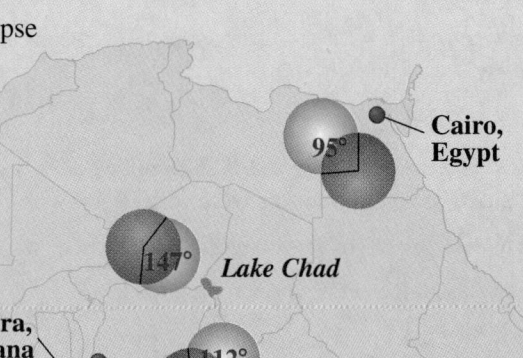

24. Writing On March 29, 2006, at about 9:15 A.M. (universal time), the moon will completely block the sun in Accra, Ghana. What will the value of θ be in this case? Explain your answer.

25. HISTORY All the Native American people of the Great Plains used *tipis* for shelter. Tipis were made of wooden poles slanted toward a center point and covered by buffalo hides sewn together in a shape similar to a sector.

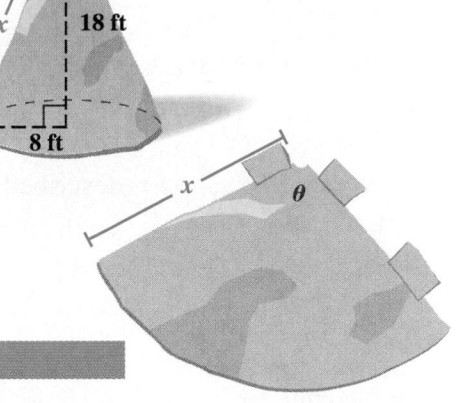

    a. What is the length *x* from the ground to where the poles meet? This is the radius of the buffalo hide.

    b. Find the length of the circular edge of the buffalo hide.

    c. Challenge Find the central angle θ of the buffalo hide.

$$\left(\textit{Hint: } \frac{\text{length of circular edge of the sector}}{\text{circumference of the circle}} = \frac{\theta}{360}\right)$$

    d. Find the surface area of the tipi.

ONGOING ASSESSMENT

26. Open-ended Problem Draw a triangle and a sector where each has an area of 40 square units.

SPIRAL REVIEW

Find two measures of ∠A for each trigonometric value. *(Section 14.3)*

27. $\sin A = 0.8192$ **28.** $\tan A = 3.271$ **29.** $\cos A = 0.8988$

Tell whether *y* varies inversely with *x*. If so, state the constant of variation. *(Section 9.6)*

30. $y = -\dfrac{5}{x}$ **31.** $xy = 6$ **32.** $y = \dfrac{8}{x-4}$ **33.** $x = \dfrac{12}{y}$

ASSESS YOUR PROGRESS

VOCABULARY

initial side of an angle (p. 671)   **standard position of an angle** (p. 671)
terminal side of an angle (p. 671)

Sketch each angle of rotation in a coordinate plane. Then use a calculator to find sin θ, cos θ, and tan θ. *(Section 14.3)*

1. $\theta = 40°$ **2.** $\theta = 225°$ **3.** $\theta = 308°$ **4.** $\theta = 95°$

Each point *P* lies on the terminal side of an angle θ in standard position. Find sin θ, cos θ, and tan θ. *(Section 14.3)*

5. $P(6, 2)$ **6.** $P(4, -1)$ **7.** $P(-7, 9)$ **8.** $P(-5, -8)$

Find the area of △*JKL*. *(Section 14.4)*

9. $j = 22, l = 17, \angle K = 38°$ **10.** $k = 0.75, l = 1.2, \angle J = 165°$

Find the area of each sector having central angle θ and radius *r*. *(Section 14.4)*

11. $\theta = 330°, r = 60$ **12.** $\theta = 200°, r = \sqrt{3}$ **13.** $\theta = 54°, r = 15$

14. Journal List the important ideas you learned in Sections 14.3 and 14.4. Tell whether each was easy or difficult for you to understand.

The Law of Sines

Learn how to...

- use the law of sines to solve a triangle

So you can...

- find the distance of a ship from the Rock of Gibraltar, for example

In his book *Shakl al-qita*, the Persian mathematician al-Tusi (1201–1274) described a relationship that is true for any triangle. To derive this law, first set the three expressions for area shown below equal to each other.

$$K = \frac{1}{2}bc \sin A \qquad K = \frac{1}{2}ac \sin B \qquad K = \frac{1}{2}ab \sin C$$

$$\frac{1}{2}bc \sin A = \frac{1}{2}ac \sin B = \frac{1}{2}ab \sin C$$

Multiply by $\frac{2}{abc}$.

$$\frac{2}{abc}\left(\frac{1}{2}bc \sin A\right) = \frac{2}{abc}\left(\frac{1}{2}ac \sin B\right) = \frac{2}{abc}\left(\frac{1}{2}ab \sin C\right)$$

$$\frac{\sin A}{a} = \frac{\sin B}{b} = \frac{\sin C}{c}$$

Notice that each ratio involves only one angle and its opposite side.

Law of Sines

For any $\triangle ABC$:

$$\frac{\sin A}{a} = \frac{\sin B}{b} = \frac{\sin C}{c}$$

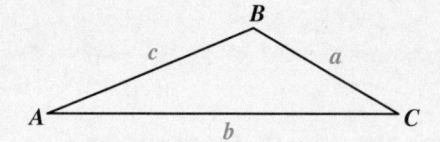

EXAMPLE 1 Application: Navigation

A ship at S is sighted simultaneously from Point Almina at A and from the Rock of Gibraltar at G, as shown. Find the ship's distance from the Rock of Gibraltar.

SOLUTION

You need to find $\angle S$ before you can use the law of sines.

$\angle S = 180° - \angle A - \angle G$

$\quad = 180° - 32° - 37° = 111°$

Use the fact that $\frac{\sin S}{s} = \frac{\sin A}{a}$.

$$\frac{\sin 111°}{15} = \frac{\sin 32°}{a}$$

$$a = \frac{15 \sin 32°}{\sin 111°}$$

$$a \approx 8.5$$

The ship is about 8.5 mi from the Rock of Gibraltar.

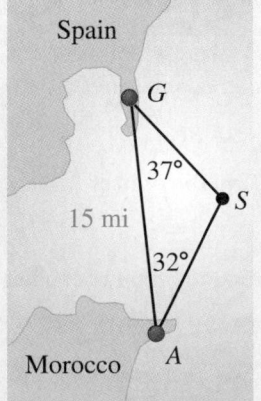

You can use the law of sines when you know two side lengths of a triangle and the measure of a non-included angle. In the Exploration below, you will investigate whether this information determines a unique triangle.

EXPLORATION
COOPERATIVE LEARNING

Constructing a Triangle

Work with a partner.
You will need:
- a ruler
- a protractor
- a compass

1 Draw $\overrightarrow{AX}$ at least 15 cm long. Then draw $\overline{AC}$ so that $\angle CAX = 30°$ and $AC = b = 8$ cm.

2 Place the compass point at C. Choose a radius from the table below and draw an arc as shown.

3 Each point (if any) where the arc meets $\overrightarrow{AX}$ should be labeled B to complete a $\triangle ABC$. Determine the number of such triangles that can be drawn with the given radius.

4 Copy and complete the table by repeating Steps 1–3 for each radius r.

Radius r (cm)	Number of triangles
2	?
4	?
6	?
8	?
10	?

Questions

1. Find the height h of $\triangle ABC$ when $\angle A = 30°$ and $b = 8$ cm. Do all the triangles you constructed have the same height? Explain.

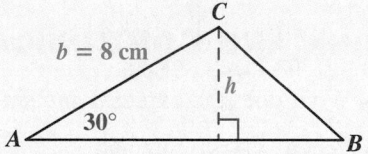

2. How many triangles are possible when $\overrightarrow{AX}$ is tangent to the arc drawn with the compass? For which value(s) of r does this happen?

3. a. For which value(s) of r does the arc intersect $\overrightarrow{AX}$ twice?

  b. Express your answer to part (a) as an inequality in terms of b, h, and r.

4. How many triangles are possible when the arc intersects $\overrightarrow{AX}$ exactly once? For which value(s) of r does this happen?

5. Repeat Steps 1–4 above using $\angle A = 120°$. Under what circumstances do you get a triangle? don't you get a triangle?

The Exploration showed that two sides and a non-included angle do not always determine a unique triangle. This is called the *ambiguous case*. All the possible outcomes of the ambiguous case are shown below.

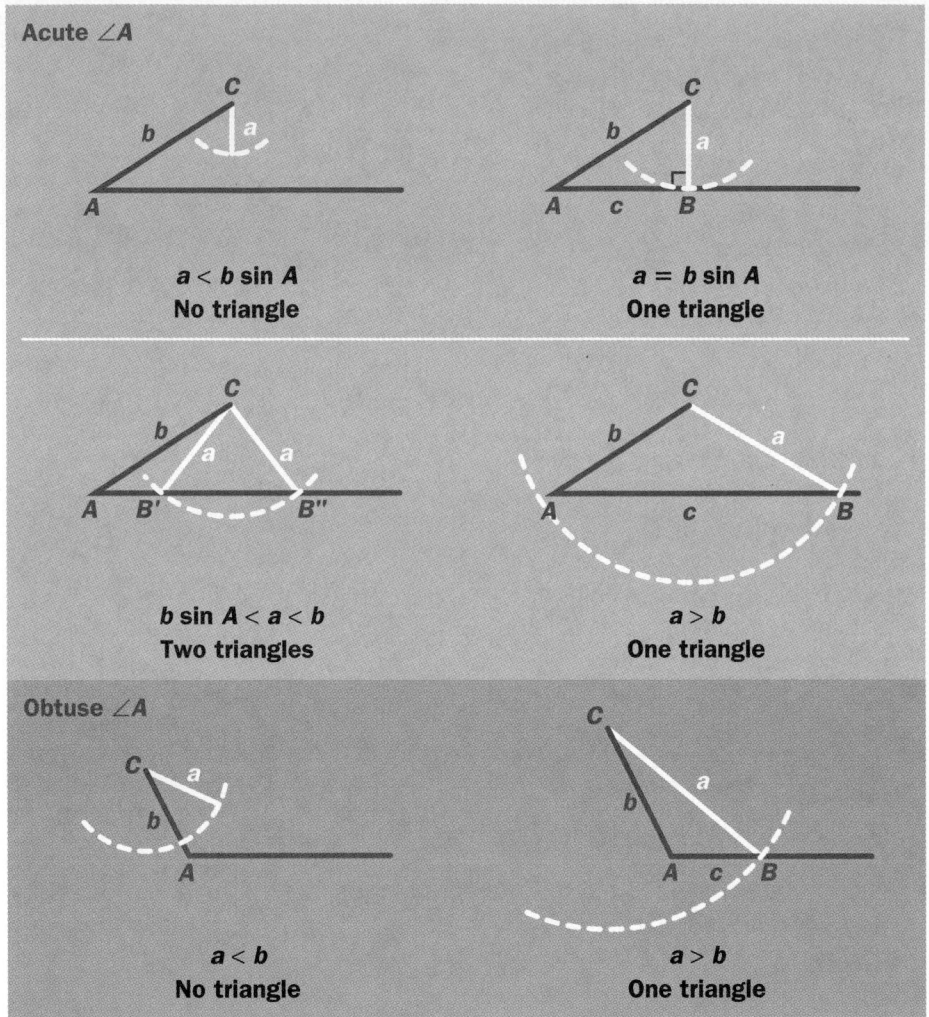

Acute ∠A

$a < b \sin A$
No triangle

$a = b \sin A$
One triangle

$b \sin A < a < b$
Two triangles

$a > b$
One triangle

Obtuse ∠A

$a < b$
No triangle

$a > b$
One triangle

THINK AND COMMUNICATE

1. What does the expression $b \sin A$ represent for the triangles above?

2. When ∠A is acute and $a < b \sin A$, why can't any triangles be formed?

3. What kind of triangle is formed when ∠A is acute and $a = b \sin A$?

4. How are ∠AB′C and ∠AB″C related in the diagram for the situation where ∠A is acute and $b \sin A < a < b$?

5. **a.** How many obtuse angles can a triangle have?

   b. What must be true about the length of the side opposite an obtuse angle of a triangle in comparison with the lengths of the other two sides? How does this explain the conditions and diagrams for the situation where ∠A is obtuse?

6. Based on the diagrams above, explain why the case where the measures of two sides and a non-included angle are known is called the *ambiguous* case.

EXAMPLE 2

For $\triangle ABC$, $\angle A = 38°$, $a = 14$, and $b = 21$. Find the measure of $\angle B$.

SOLUTION

Step 1 Check to see how many triangles are possible.

$b \sin A = 21 \sin 38°$
≈ 12.9

$12.9 < 14 < 21$

Compare a with $b \sin A$ and with b.

Since $b \sin A < a < b$, two triangles are possible: $\triangle ACB'$ and $\triangle ACB''$.

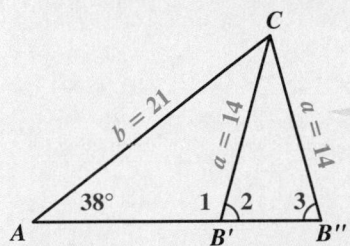

Note that $\triangle B'CB''$ is isosceles, so $\angle 2 = \angle 3$. Therefore, since $\angle 1$ and $\angle 2$ are supplementary, $\angle 1$ and $\angle 3$ are also supplementary.

Step 2 Use the fact that $\dfrac{\sin A}{a} = \dfrac{\sin B}{b}$.

$$\frac{\sin 38°}{14} = \frac{\sin B}{21}$$

$$\frac{21 \sin 38°}{14} = \sin B$$

$$0.9235 \approx \sin B$$

$\angle B \approx 67.4°$ or $\angle B \approx 180° - 67.4° = 112.6°$

```
sin 38
        .6156614753
21*Ans/14
        .923492213
sin⁻¹ Ans
        67.44208077
```

EXAMPLE 3

For $\triangle JKL$, $\angle J = 128°$, $j = 32$, and $k = 20$. Find the measure of $\angle K$.

SOLUTION

Step 1 Check to see how many triangles are possible. Since $\angle J$ is obtuse and $j > k$, one triangle is possible.

Note that $\angle K$ must be acute since $\angle J$ is obtuse and a triangle can have only one obtuse angle.

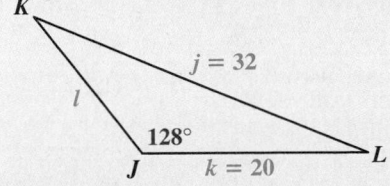

Step 2 Use the fact that $\dfrac{\sin K}{k} = \dfrac{\sin J}{j}$.

$$\frac{\sin K}{20} = \frac{\sin 128°}{32}$$

$$\sin K = \frac{20 \sin 128°}{32}$$

$$\sin K \approx 0.4925$$

$$\angle K \approx 29.5°$$

Solve each △*ABC*. If there are two solutions, give both. If there is no solution, explain why.

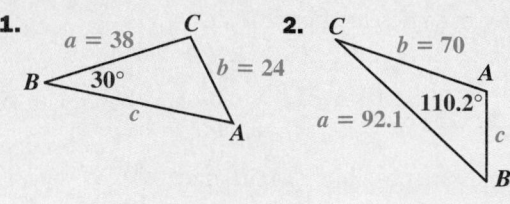 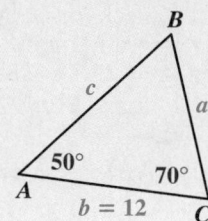

1.   **2.**   **3.**

4. $a = 16$, $b = 19.2$, $\angle A = 81°$ **5.** $b = 14$, $\angle A = 100°$, $\angle C = 32°$

6. $a = 7.9$, $b = 9.9$, $\angle A = 53°$ **7.** $a = 21\frac{2}{5}$, $b = 24\frac{1}{3}$, $\angle A = 59°$

14.5 | **Exercises and Applications**

Extra Practice exercises on page 771

Solve each triangle.

1. 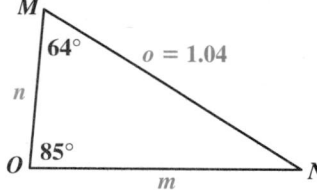   **2.** 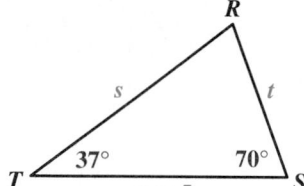   **3.**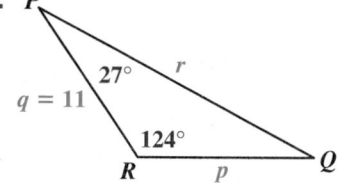

4. △*DEF* with $f = 30$, $\angle D = 50°$, $\angle E = 25°$ **5.** △*XYZ* with $y = 18$, $\angle X = 84°$, $\angle Z = 39°$

6. △*JKL* with $j = 7.25$, $\angle K = 116°$, $\angle L = 44°$ **7.** △*ABC* with $a = 4\frac{3}{5}$, $\angle B = 28°$, $\angle C = 76°$

8. Cooperative Learning Play three rounds of the following game with two other students. The object is to make the greatest number of triangles.

 • Have the person to your left choose a measure for $\angle A$.

 • Have the person to your right choose a length for b.

 • You should choose a length for a. You earn one point for each △*ABC* that can be made with the given side lengths and angle measure.

 a. What are the possible scores you can get in one turn?

 b. What kind of angle should the person to your left choose to make sure that you will not earn more than one point during your turn?

 c. How should you choose a when $\angle A$ is obtuse? when $\angle A$ is acute?

 d. Writing Describe a strategy for the game that maximizes the points you earn. Try out your strategy by playing two or three more rounds with your partners. Did your partners use similar strategies? What happens when you all use the same strategy?

Solve each △ABC. If there are two solutions, give both. If there is no solution, explain why.

9.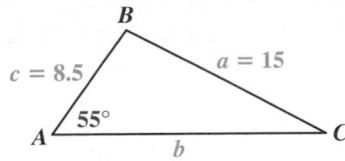

10.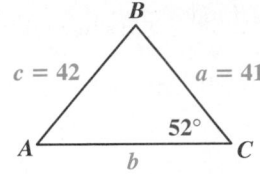

11.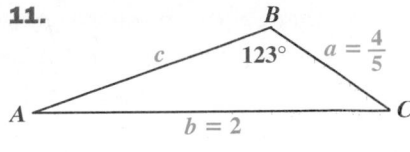

12. $a = 34$, $b = 42$, $\angle A = 47°$

13. $\angle A = 56°$, $a = 20$, $b = 12$

14. $\angle A = 102°$, $b = 27$, $a = 41$

15. $\angle C = 137.1°$, $b = 97.2$, $c = 72$

16. $a = 12$, $b = 8$, $\angle B = 44°$

17. $a = 17.4$, $c = 21.3$, $\angle C = 92°$

Connection   LITERATURE

In the C.S. Forester novel *The African Queen*, the mechanic Mr. Allnut and his employer Miss Rose travel along a river on a dilapidated boat during World War I. In this passage from the novel, the enemy captain tries to capture the boat as it travels past him.

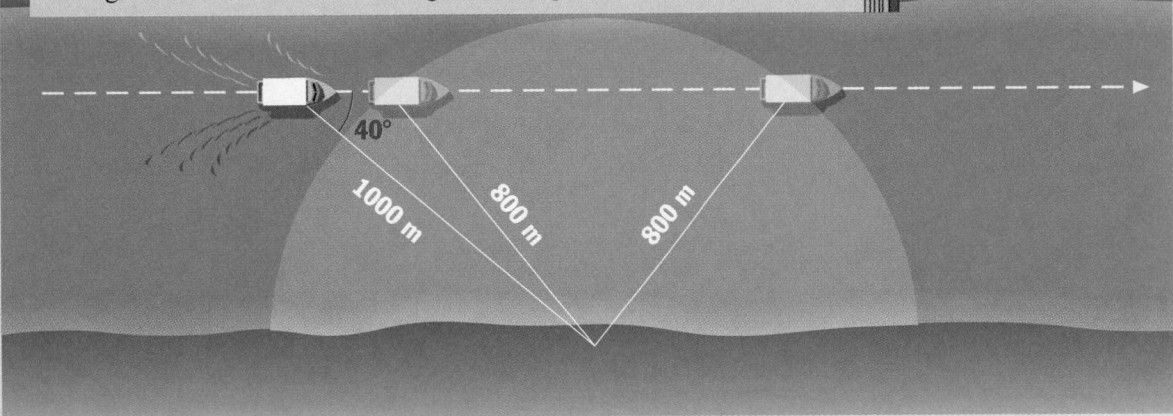

> They were right in the eye of the sun now, and the glare off the water made the foresight indistinct. It was very easy to lose sight of the white awning of the boat as he aimed.
>   A thousand metres was a long range for a Martini rifle with worn rifling. He fired, reloaded, fired again, and again, and again.

Suppose that when the boat was 1000 m from the captain, the angle formed by the direction of the boat and the line of sight from the boat to the enemy captain was 40°. Assume the range of the Martini rifle was 800 m.

18. How far did the boat travel before the captain's fire could reach it?

19. What distance did the boat have to travel through the captain's fire?

20. **Visual Thinking** Where should the boat travel to minimize its exposure to the gunfire? Explain.

21. **Challenge** Suppose the boat traveled a path 100 m closer to the bank of the river where the captain stood. How far would the boat have to travel through the captain's fire?

The map below represents part of the Great Trigonometrical Survey of India in the nineteenth century. The surveyors began with one known distance and used triangle trigonometry to find other distances. Complete Exercises 22–25 to see how measurements were found from these surveys.

22. Find the distances between Ghirya and Manoli and between Ghirya and Valvan.

23. Find the distance between Adhúr and Kumbhárli and between Manoli and Kumbhárli.

24. Use two different triangles to find the distance between Manoli and Mirya. Compare your answers.

25. a. Find the distance between Ghirya and Adhúr going through Mirya.

b. **Writing** Do you think the distance you found in part (a) is a good approximation of the straight-line distance between Ghirya and Adhúr? Explain.

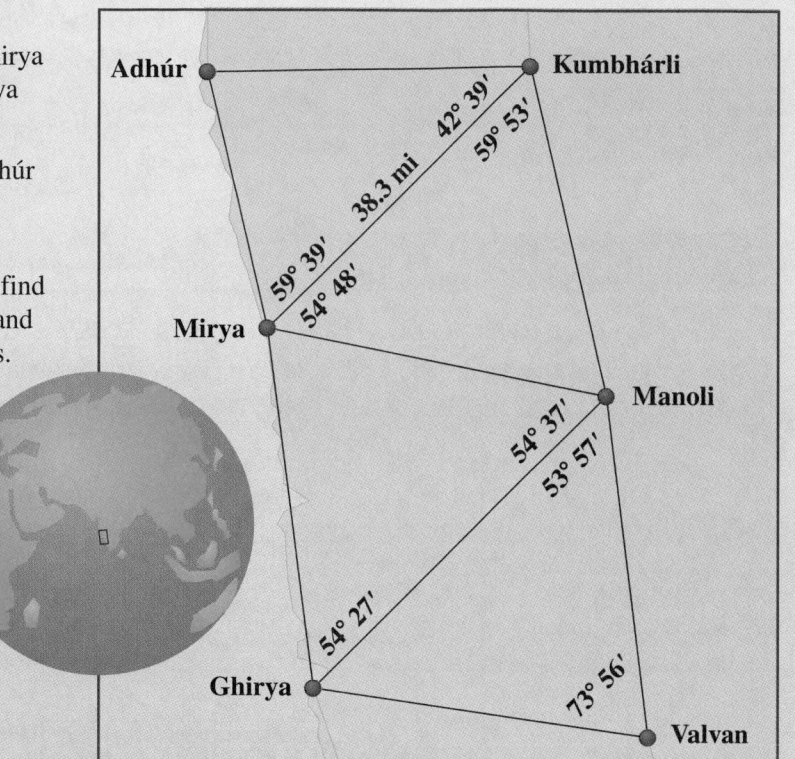

ONGOING ASSESSMENT

26. Open-ended Problem Suppose $a = 8$ and $\angle A = 62°$ in $\triangle ABC$. Find a value for b so that $\angle B$ has:

    a. two possible values   **b.** one possible value   **c.** no possible values

---

SPIRAL REVIEW

Find the area of $\triangle XYZ$. *(Section 14.4)*

27. $x = 4$, $y = 6$, $\angle Z = 50°$

28. $y = 22$, $z = 38$, $\angle X = 160°$

29. $x = 5.6$, $z = 9.2$, $\angle Y = 97°$

30. $x = \dfrac{8}{9}$, $y = \dfrac{3}{4}$, $\angle Z = 34°$

Write as a logarithm of a single number or expression. *(Section 4.3)*

31. $3 \log_a 4$

32. $5 \log_b 6 - \log_b 12$

33. $-2 \log_b 5$

34. $\log_a 8 + 2 \log_a 4$

35. $2 \ln m^2 - \ln m^4$

36. $\log_c 3 + \log_c 9$

14.6 | The Law of Cosines

Learn how to...

- use the law of cosines to solve a triangle

So you can...

- find the step angle of dinosaur tracks, for example

You can think of the hands of a clock as two sides of a triangle whose third side connects the ends of the hands. The shape of the triangle changes as the hands move and the angle they form changes.

Los Angeles

Chicago

New York

THINK AND COMMUNICATE

On each clock shown above, the minute hand is 4 in. long and the hour hand is 2.5 in. long. Use this information for Questions 1–4.

1. What is the measure of the angle formed by the hands of the Chicago clock? How far apart are the ends of the hands?

2. Compare the Los Angeles clock with the Chicago clock.

   a. Is the measure of the angle formed by the hands *greater than* or *less than* the measure of the angle for the Chicago clock?

   b. Is the distance between the ends of the hands *greater than* or *less than* the distance for the Chicago clock?

3. Compare the New York clock with the Chicago clock by repeating parts (a) and (b) of Question 2.

4. Can you find the distance between the ends of the hands of the Los Angeles clock or the New York clock? Why or why not?

You already know how to use the Pythagorean theorem to find the third side of a right triangle. The *law of cosines* is a general rule you can apply to any triangle.

Law of Cosines

For any $\triangle ABC$:
$$a^2 = b^2 + c^2 - 2bc \cos A$$
$$b^2 = a^2 + c^2 - 2ac \cos B$$
$$c^2 = a^2 + b^2 - 2ab \cos C$$

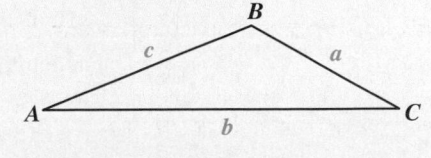

To derive the law of cosines, draw the altitude $\overline{CD}$ of any $\triangle ABC$ and let $CD = h$. Then let $AD = x$. It follows that $DB = c - x$.

Use the Pythagorean theorem to find two different expressions for h^2.

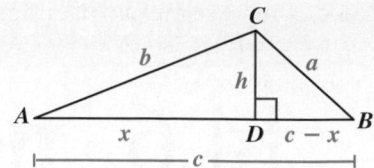

Right $\triangle ACD$

$$x^2 + h^2 = b^2$$
$$h^2 = b^2 - x^2$$

Right $\triangle BCD$

$$h^2 + (c - x)^2 = a^2$$
$$h^2 = a^2 - (c - x)^2$$

Equate the two expressions for h^2.

$$b^2 - x^2 = a^2 - (c - x)^2$$
$$b^2 - x^2 = a^2 - c^2 + 2cx - x^2$$
$$b^2 = a^2 - c^2 + 2cx$$
$$a^2 = b^2 + c^2 - 2cx$$

Since $\cos A = \dfrac{x}{b}$, $x = b \cos A$. Substitute **$b \cos A$** for **x**.

$$a^2 = b^2 + c^2 - 2c(\boldsymbol{b \cos A})$$
$$a^2 = b^2 + c^2 - 2bc \cos A$$

THINK AND COMMUNICATE

5. What happens to the law of cosines when the included angle is 90°?

6. Look back at the clocks shown on page 691. Use the law of cosines to find the distance between the ends of the hands of the Los Angeles clock and between the ends of the hands of the New York clock.

EXAMPLE 1

Find the missing side length e for $\triangle DEF$ below.

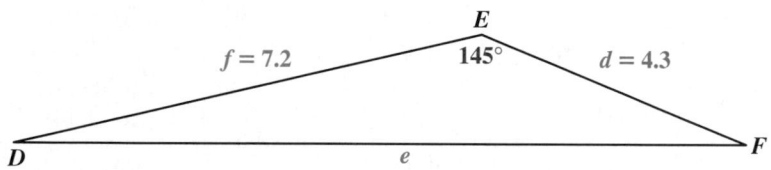

SOLUTION

You know the lengths of two sides and the measure of the included angle, so you can use the law of cosines.

Substitute **4.3** for d, **7.2** for f, and **145°** for **E**.

$$e^2 = d^2 + f^2 - 2df \cos E$$
$$e^2 = (4.3)^2 + (7.2)^2 - 2(4.3)(7.2) \cos \mathbf{145°}$$
$$e = \sqrt{(4.3)^2 + (7.2)^2 - 2(4.3)(7.2) \cos 145°}$$
$$\approx 11.0$$

```
cos 145
        -.8191520443
√(4.3²+7.2²-
2(4.3*7.2*Ans))
       11.00235859
```

The length e is about 11 units.

You can also use the law of cosines to find angle measures when only the side lengths of a triangle are known. Start with the formula $a^2 = b^2 + c^2 - 2bc \cos A$ and solve for cos A:

$$a^2 = b^2 + c^2 - 2bc \cos A$$

$$a^2 - b^2 - c^2 = -2bc \cos A$$

$$\frac{b^2 + c^2 - a^2}{2bc} = \cos A$$

You can derive the following formulas for cos B and cos C in the same way.

$$\frac{a^2 + c^2 - b^2}{2ac} = \cos B \qquad \frac{a^2 + b^2 - c^2}{2ab} = \cos C$$

EXAMPLE 2   Application: Paleontology

The tracks shown at the left were made by a carnivorous dinosaur during the Cretaceous period in what is now Texas. The tracks are on display at the American Museum of Natural History in New York City.

The *pace lengths*, a and c, and the *stride length*, b, were measured from the tracks themselves. Use the measurements shown in the diagram to find $\angle B$, the *step angle*.

$c = 158.7$ cm B $a = 162.6$ cm

A $b = 315$ cm C

SOLUTION

You know the three side lengths, so you can use the law of cosines to find the angle.

$$\cos B = \frac{a^2 + c^2 - b^2}{2ac}$$

$$\cos B = \frac{(162.6)^2 + (158.7)^2 - 315^2}{2(162.6)(158.7)}$$

Substitute **162.6** for **a**, **158.7** for **c**, and **315** for **b**.

$$\cos B \approx -0.9223$$

$$\angle B \approx \cos^{-1}(-0.9223)$$

$$\angle B \approx 157°$$

The step angle is about 157°.

You can use the table below to help you remember when to use the law of sines and when to use the law of cosines.

Information given	Law to use	Information to find
angle-angle-side	law of sines	remaining sides*
side-side-angle	law of sines	remaining side and angles
angle-side-angle	law of sines	remaining sides*
side-angle-side	law of cosines	remaining side and one angle*
side-side-side	law of cosines	all three angles

Remember: This is the ambiguous case, where the given information may lead to 0, 1, or 2 triangles.

*You can find the remaining angle by using the fact that the sum of the measures of the angles of a triangle is 180°.

☑ CHECKING KEY CONCEPTS

Find the missing side length of each triangle.

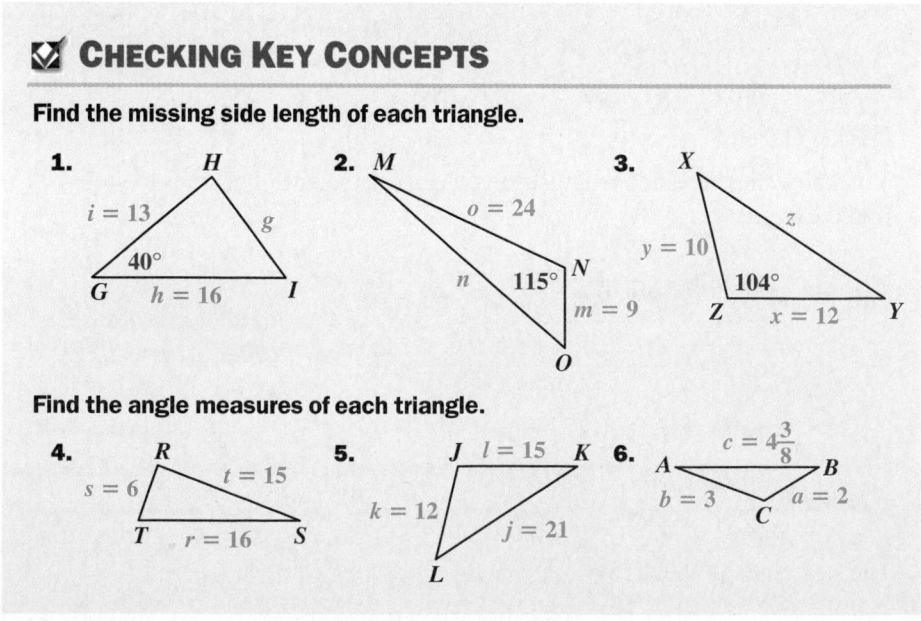

1. $i = 13$, $40°$, G, $h = 16$, I, g, H

2. M, $o = 24$, n, $115°$, N, $m = 9$, O

3. X, z, $y = 10$, $104°$, Z, $x = 12$, Y

Find the angle measures of each triangle.

4. R, $s = 6$, $t = 15$, T, $r = 16$, S

5. J, $l = 15$, K, $k = 12$, $j = 21$, L

6. A, $c = 4\frac{3}{8}$, B, $b = 3$, C, $a = 2$

14.6 Exercises and Applications

Extra Practice exercises on page 771

For Exercises 1–9, find the missing side length of each triangle.

1.
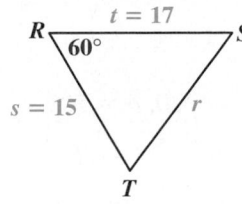
$t = 17$, R $60°$ S, $s = 15$, r, T

2.
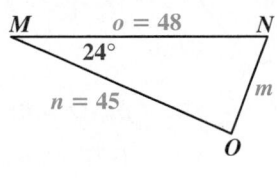
M $o = 48$ N, $24°$, $n = 45$, m, O

3.
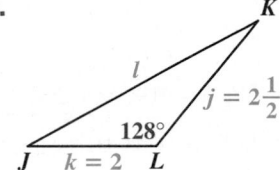
K, l, $j = 2\frac{1}{2}$, $128°$, J $k = 2$ L

4. $\triangle MNO$ with $m = 1$, $o = 7$, $\angle N = 75°$

5. $\triangle XYZ$ with $x = 10$, $y = 11$, $\angle Z = 83°$

6. $\triangle ABC$ with $a = 0.47$, $b = 0.47$, $\angle C = 96°$

7. $\triangle LKJ$ with $k = 20.5$, $l = 29.2$, $\angle J = 15°$

8. $\triangle RST$ with $s = \frac{8}{3}$, $t = \frac{3}{4}$, $\angle R = 172°$

9. $\triangle DEF$ with $d = 6\frac{1}{2}$, $f = 9\frac{1}{4}$, $\angle E = 154°$

10. Writing Chandra used the law of cosines to solve $\triangle ABC$ as shown at the right. Describe Chandra's reasoning. Then apply the same reasoning to side-side-angle cases that lead to two triangles and no triangle.

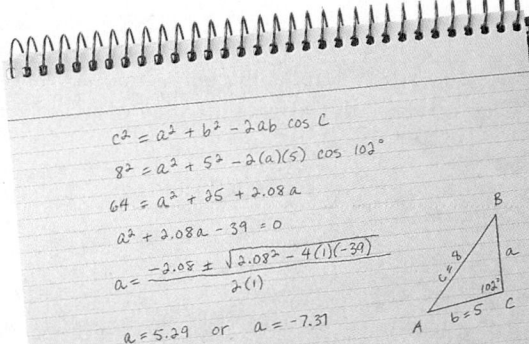

$$c^2 = a^2 + b^2 - 2ab \cos C$$
$$8^2 = a^2 + 5^2 - 2(a)(5) \cos 102°$$
$$64 = a^2 + 25 + 2.08a$$
$$a^2 + 2.08a - 39 = 0$$
$$a = \frac{-2.08 \pm \sqrt{2.08^2 - 4(1)(-39)}}{2(1)}$$
$$a = 5.29 \quad \text{or} \quad a = -7.37$$
$$\boxed{a = 5.29}$$

$c = 8$, $b = 5$, $102°$, a

Connection ▶ CHEMISTRY

A water molecule is made up of one atom of oxygen and two atoms of hydrogen. The bonds between the oxygen atom and each hydrogen atom form an angle that makes the molecule triangular. As water freezes, the shape of the triangle changes. (*Note:* All measurements are given in picometers (pm), with 1 pm = 10^{-12} m.)

liquid water
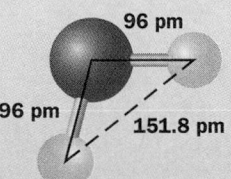
96 pm, 96 pm, 151.8 pm

ice
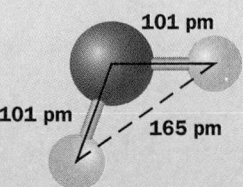
101 pm, 101 pm, 165 pm

11. Find the angle formed by the two bonds in a molecule of liquid water.

12. Find the angle formed by the two bonds in a molecule of ice.

13. Research Find out which is less dense, *liquid water* or *ice*. How do your answers to Exercises 11 and 12 support this fact?

For Exercises 14–22, find the angle measures of each triangle.

14.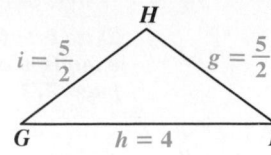

15.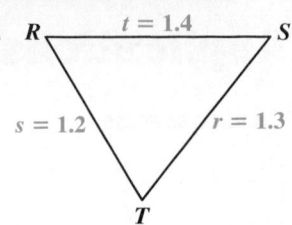

16.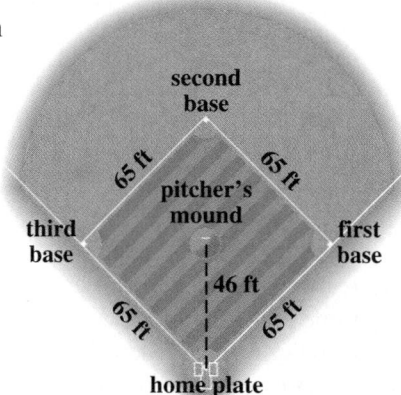

17. $\triangle XYZ$ with $x = 20$, $y = 19$, $z = 2$

18. $\triangle MNO$ with $m = 245$, $n = 720$, $o = 595$

19. $\triangle PQR$ with $p = 8.0$, $q = 14.6$, $r = 6.7$

20. $\triangle DEF$ with $d = 7.5$, $e = 6.3$, $f = 11.9$

21. $\triangle JKL$ with $j = 4\frac{2}{9}$, $k = 5\frac{1}{3}$, $l = 2\frac{7}{8}$

22. $\triangle GHI$ with $g = \frac{17}{2}$, $h = \frac{13}{3}$, $i = \frac{34}{5}$

23. SPORTS The three bases and home plate of a slow-pitch softball field form a 65 ft by 65 ft square. The pitcher's mound lies between home plate and second base, 46 ft from home plate.

a. Find the distance between the pitcher's mound and second base.

b. Find the distance between the pitcher's mound and first base.

c. **Writing** How do the distances between the pitcher's mound and each base differ? Do you think this affects the play of the game? Explain.

d. **Research** Find out the dimensions of a baseball field. Repeat parts (a)–(c) using these measurements.

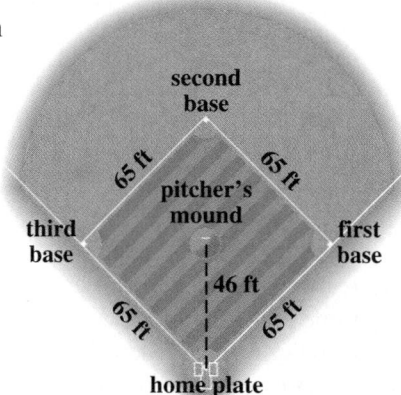

Connection ▶ ART

In 1974, the town of Vegreville in Alberta, Canada, honored the Ukrainian settlers of the region by building a giant *pysanki*, or Ukrainian decorated egg. The egg is made of 2208 equilateral triangles and 524 three-pointed stars. Each star consists of three isosceles triangles with 12 in. legs and bases that form equilateral triangles of different sizes.

24. Suppose the point of a three-pointed star is 20°. How long is each side of the equilateral triangle in the center of the star?

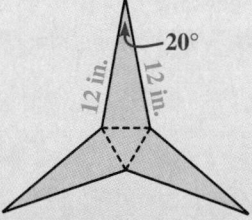

25. To make a three-pointed star so that the equilateral triangle in the center is 8 in. on each side, what must be the angle measure of each point of the star?

26. Open-ended Problem Draw a star like the ones shown. Give lengths and angle measures for all four triangles that make up the star.

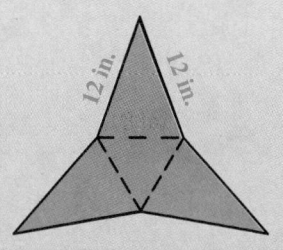

27. Challenge A pilot flies a plane 3580 km on a 146° course to get from Caracas to Brasília. On a second flight, the pilot flies 3780 km on a 298° course to get from Brasília to Quito. What course and distance must the pilot fly to return directly to Caracas from Quito?

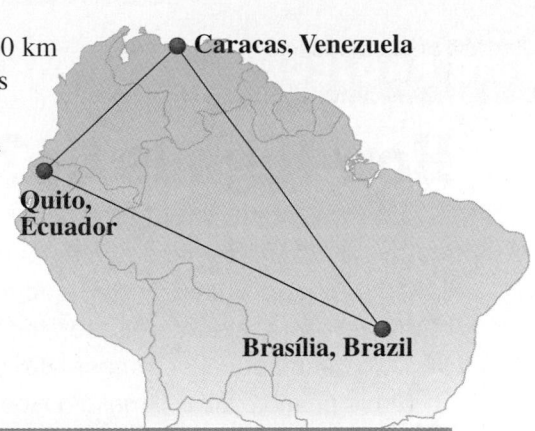

28. Writing Explain how you can use the law of cosines to draw accurately a triangle with side lengths 5, 6, and 7.

Find each missing side length and angle measure of △ABC. *(Section 14.5)*

29. $b = 7.4$, $\angle A = 36°$, $\angle B = 60°$ **30.** $c = 2$, $\angle A = 104°$, $\angle C = 28°$

31. $a = 4.25$, $b = 8.5$, $\angle A = 30°$ **32.** $b = 4$, $c = 11$, $\angle C = 149°$

Find sin θ and cos θ for each value of θ. *(Section 14.3)*

33. $\theta = 135°$ **34.** $\theta = 120°$ **35.** $\theta = 90°$ **36.** $\theta = 45°$

37. $\theta = 330°$ **38.** $\theta = 270°$ **39.** $\theta = 225°$ **40.** $\theta = 210°$

ASSESS YOUR PROGRESS

Solve △XYZ. If there are two solutions, give both. If there is no solution, explain why. *(Sections 14.5 and 14.6)*

1. 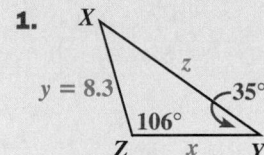   **2.** 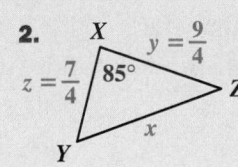   **3.**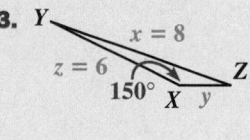

4. $y = 9.75$, $z = 11$, $\angle Y = 57°$ **5.** $x = 40$, $y = 29$, $z = 58$

6. $x = 9$, $z = 7$, $\angle Z = 165°$ **7.** $y = 13.7$, $\angle Y = 43°$, $\angle Z = 28°$

8. $x = 8.3$, $y = 6.05$, $z = 2.6$ **9.** $x = 14.5$, $y = 8$, $\angle Y = 35°$

10. $x = 13$, $y = 19$, $\angle Y = 25°$ **11.** $y = 12.5$, $z = 5$, $\angle X = 124°$

12. $x = \frac{3}{8}$, $z = \frac{3}{4}$, $\angle X = 30°$ **13.** $x = 6\frac{2}{5}$, $z = 6\frac{3}{5}$, $\angle X = 72°$

14. Journal Is the *law of sines* or the *law of cosines* easier for you to remember and use? Explain.

How High Is Up?

A surveyor can determine positions and elevations with the help of a tool called a *transit*, which can be used to measure angles. After taking some measurements, a surveyor uses trigonometry to calculate distances that would be impractical or impossible to measure directly.

In this project, you will make a type of transit. Then you will explore and evaluate two trigonometric methods of calculating the height of a tall object.

PROJECT GOAL Determine the height of a tall object using two trigonometric methods.

Making and Using a Transit

Work with a partner. You will need a protractor, a drinking straw, a piece of string, a small weight (such as a key or a washer), and masking tape. Assemble your transit as shown.

To measure the angle of elevation to some high point, view the point through the straw. Have your partner record the angle of elevation while you view the point.

For each of the following methods, you will need a tape measure.

Tape a straw to the base of a protractor.

Tie a string to a weight. Attach the string to the center mark (or hole) on the base of the protractor.

The angle of elevation is equal to the measure of the angle formed by the string and the 90° mark.

Method 1: The Direct Approach

Choose a tall object whose height you wish to measure. Stand some distance away from the object, and look through your transit at the top of the object. Draw a diagram of the situation, and record your data on the diagram.

1. MEASURE and record the angle of elevation of the top of the object.

2. MEASURE and record your distance from the bottom of the object.

3. USE a trigonometric ratio to calculate the object's height above eye level.

4. ADD the height at eye level to the height in Step 3 to find the object's total height.

Method 2: The Indirect Approach

Even when it's not possible to measure the distance between observer and object, it's still possible to measure the height of the object. To do this, you must measure the angle of elevation from two points, *A* and *B*, that lie on a line running directly to the object. Draw a diagram like the one shown, and record your data on the diagram.

1. **MEASURE** and record the angle of elevation of the top of the object from each of points *A* and *B*.

2. **MEASURE** and record the distance between *A* and *B*.

3. **CALCULATE** the measure of ∠*ABC*.

4. **CALCULATE** the measure of ∠*ACB*.

5. **USE** the law of sines to find *BC*.

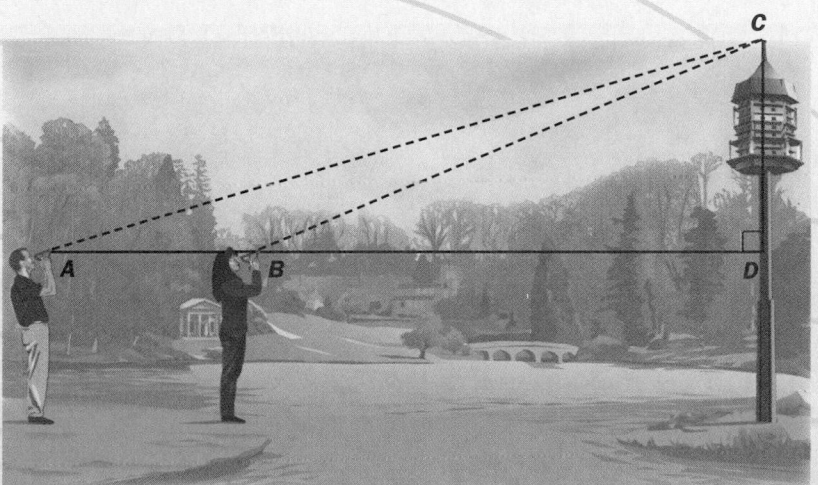

6. **USE** a trigonometric ratio to calculate *CD*, and add the height at eye level to find the total height of the object.

Writing a Report

Write a report summarizing your results. Include your data, calculations, and a completed diagram for each method used. Compare your results from the first and second methods, and try to account for any differences. Why might someone prefer to use one method rather than the other?

Here are some ideas for extending your project:

- Give a geometric argument to explain why the angle between the 90° mark on the protractor and the position of the string on your transit is the same as the angle of elevation.

- Apply the methods of this project to calculate the height of a tall object whose height is already known from some reliable source, and compare your results with the known height.

Self-Assessment

Describe any problems that you had completing the project. How well were you able to apply the trigonometric formulas needed to calculate some of the distances? What formulas, if any, still give you trouble?

14 | Review

Explain a section of the chapter to someone who is not in the class: a parent, a brother or sister, or a friend. Trying to teach someone else what you know often helps to clarify your own understanding.

VOCABULARY

sine (p. 656)

cosine (p. 656)

tangent (p. 663)

solving a triangle (p. 665)

angle of depression (p. 665)

angle of elevation (p. 665)

initial side of an angle (p. 671)

terminal side of an angle (p. 671)

standard position of an angle (p. 671)

SECTIONS | 14.1 *and* 14.2

You can find the unknown lengths of sides or measures of angles in a right triangle by using the **sine**, **cosine**, and **tangent** ratios.

Sine, Cosine, and Tangent Ratios

In right $\triangle ABC$, the sine, cosine, and tangent of $\angle A$ are given by:

$$\sin A = \frac{\text{length of leg opposite } \angle A}{\text{length of hypotenuse}}$$

$$\cos A = \frac{\text{length of leg adjacent to } \angle A}{\text{length of hypotenuse}}$$

$$\tan A = \frac{\text{length of leg opposite } \angle A}{\text{length of leg adjacent to } \angle A}$$

Example:

hypotenuse $c = 17$

leg opposite $\angle A$ $a = 8$

$b = 15$

leg adjacent to $\angle A$

$$\sin A = \frac{8}{17}$$

$$\cos A = \frac{15}{17}$$

$$\tan A = \frac{8}{15}$$

You can measure angles in decimal degrees or in degrees, minutes, and seconds. For example, the angle measure $54.68°$ is the same as the measure $54°40'48''$.

You can find the measure of an angle θ in a right triangle if you know its sine, cosine, or tangent value by using the inverse trigonometric functions. For example, if $\tan \theta = 1.256$, then $\theta = \tan^{-1} 1.256 \approx 51.5°$.

SECTION 14.3

If θ is an angle in **standard position** with the point $P(x, y)$ on its **terminal side** and if r is the distance between P and the origin, then the **trigonometric functions** of θ are defined by:

$$\sin\theta = \frac{y}{r} \quad \cos\theta = \frac{x}{r} \quad \tan\theta = \frac{y}{x}$$

For any Quadrant I angle θ, there are three other angles between 90° and 360° having the **same** trigonometric values as θ **or opposite** trigonometric values.

Examples:

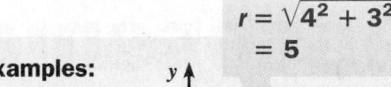

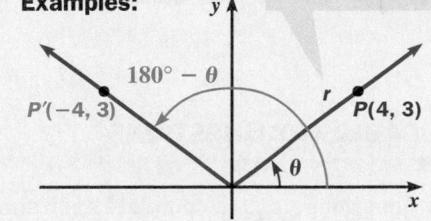

$$\sin\theta = \frac{3}{5} \qquad \sin(180° - \theta) = \frac{3}{5}$$

$$\cos\theta = \frac{4}{5} \qquad \cos(180° - \theta) = -\frac{4}{5}$$

$$\tan\theta = \frac{3}{4} \qquad \tan(180° - \theta) = -\frac{3}{4}$$

SECTIONS 14.4, 14.5, and 14.6

You can find the area K of $\triangle ABC$ by using any of these formulas:

$$K = \frac{1}{2}bc\sin A \quad K = \frac{1}{2}ac\sin B \quad K = \frac{1}{2}ab\sin C$$

Example: The area of $\triangle ABC$ where $b = 5$, $c = 6$, and $\angle A = 120°$ is:

$$K = \frac{1}{2}(5)(6)\sin 120° \approx 15(0.8660) \approx 13$$

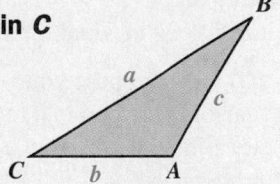

When you know the measures of some sides or angles in $\triangle ABC$, you can use the *law of sines* or the *law of cosines* to **solve the triangle** to find any unknown measures.

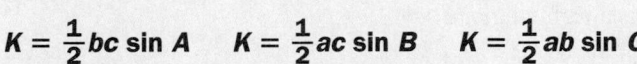

Law of Sines

$$\frac{\sin A}{a} = \frac{\sin B}{b} = \frac{\sin C}{c}$$

Example: If $a = 10$, $b = 8$, and $\angle A = 42°$ for $\triangle ABC$, use the law of sines to find the measure of $\angle B$.

$$\frac{\sin 42°}{10} = \frac{\sin B}{8}$$

$$\sin B = \frac{8\sin 42°}{10} \approx 0.5353$$

$$\angle B = \sin^{-1} 0.5353 \approx 32.4°$$

Law of Cosines

$$a^2 = b^2 + c^2 - 2bc\cos A$$

$$b^2 = a^2 + c^2 - 2ac\cos B$$

$$c^2 = a^2 + b^2 - 2ab\cos C$$

Example: If $a = 5$, $b = 12$, and $\angle C = 125°$ for $\triangle ABC$, use the law of cosines to find c.

$$c^2 = 5^2 + 12^2 - 2(5)(12)\cos 125°$$

$$\approx 237.8$$

$$c \approx \sqrt{237.8} \approx 15.4$$

By solving these formulas for cos A, cos B, and cos C, you can find $\angle A$, $\angle B$, and $\angle C$ when a, b, and c are known.

Assessment

VOCABULARY QUESTIONS

For Questions 1 and 2, complete each sentence.

1. In a right $\triangle ABC$, the ratio $\dfrac{\text{length of leg opposite } \angle A}{\text{length of leg adjacent to } \angle A}$ is the value of the __?__ of $\angle A$.

2. When you __?__, you find the measures of all the sides and all the angles of the triangle.

SECTIONS 14.1 *and* 14.2

For △*ABC* shown at the right, find each value.

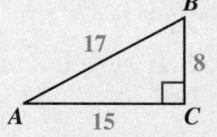

3. $\sin A$ **4.** $\cos A$ **5.** $\tan B$

6. In right $\triangle DEF$, $\angle F$ is a right angle, $d = 12$ in., and $f = 24$ in. Find the measure of $\angle D$.

7. BOTANY Suppose you want to estimate the height of a several-hundred-year-old redwood in a redwood forest in northern California. The angle of elevation is 82.4° from a point 35 ft from the base of the tree. How tall is the redwood?

SECTION 14.3

8. NAVIGATION After leaving an airport, a plane travels on a course of 250° for 100 mi.

  a. What angle of rotation is equivalent to the plane's course?

  b. Write and solve equations relating the plane's *x*- and *y*-coordinates to the angle of rotation found in part (a).

  c. How far has the plane traveled in an east-west direction? in a north-south direction?

9. An angle θ is in standard position with point $P(-4, -5)$ on its terminal side. Find $\sin \theta$, $\cos \theta$, and $\tan \theta$.

10. Find all the angles θ between 0° and 360° such that $\cos \theta = 0.2345$.

Find the area of each triangle.

11.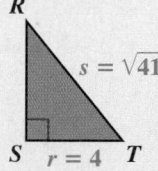
$s = \sqrt{41}$, $r = 4$

12.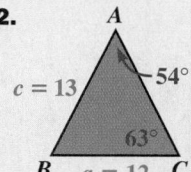
$c = 13$, $54°$, $63°$, $a = 12$

13.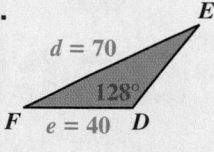
$d = 70$, $128°$, $e = 40$

14. GEOMETRY Find the area of the stop sign, which is a regular octagon. (*Hint:* Divide the sign into eight triangles. What is the area of each triangle?)

16.2 in.

Find the area of each sector.

15.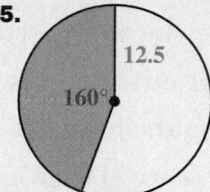
12.5, 160°

16.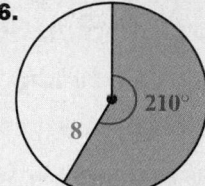
210°, 8

17.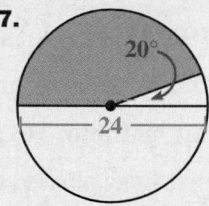
20°, 24

Solve each △ABC. If there are two solutions, give both. If there is no solution, explain why.

18. $a = 40$, $b = 32$, $\angle A = 115°20'$

19. $a = 71$, $b = 84$, $\angle A = 160°$

20. $a = 6$, $b = 6\sqrt{2}$, $\angle C = 45°$

21. $b = 25$, $c = 50$, $\angle B = 15°$

22. $a = 54$, $c = 112$, $\angle C = 120°$

23. $a = 60$, $b = 50$, $c = 66$

24. Writing Describe the situations in which you can use the law of sines to solve a triangle. Also, describe the situations in which you can use the law of cosines to solve a triangle.

PERFORMANCE TASK

25. Choose two angle measures and a side length, or choose two side lengths and an angle measure. Indicate whether the single side (or angle) chosen is included between the two angles (or sides) chosen.

a. How many triangles can you form with your chosen values?

b. If your answer to part (a) was "none," choose new values until you can form at least one triangle.

c. Solve your triangle(s).

d. Find the area(s) of your triangle(s).

15 | Trigonometric Functions

Making beautiful Music

Joe Lopez

In 1962 Joe Lopez started Zaz Recording Studio when he purchased a one-track tape recorder and a small, rundown building on the outskirts of San Antonio, Texas. He'd already spent countless hours in studios as a guest bass player, and he was ready to use what he'd learned to go into business for himself. Today, Lopez heads a family-run operation that is the largest Hispanic-owned recording complex in the Southwest.

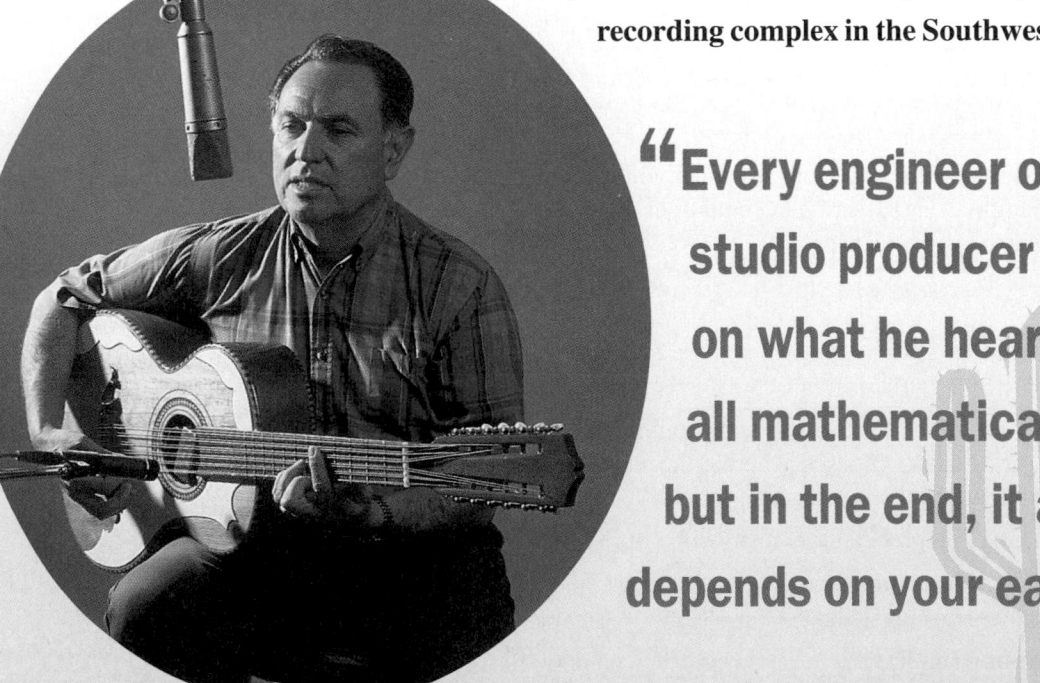

"Every engineer or studio producer relies on what he hears. It's all mathematical, yes, but in the end, it all depends on your ears."

A Blend of Musical Worlds

Lopez, a San Antonio native, is no stranger to success. In the mid-1950s his band, "Los Guadalupanos," was one of the most popular *conjunto* groups in south Texas. *Conjunto* music features the accordion and *bajo sexto,* a Mexican twelve-string guitar.

This musical form developed in the late 19th century along the Texas-Mexico border when German, Czech, and Polish immigrants brought their accordions, along with polkas, waltzes, and mazurkas, to local dances. *Conjunto* is now regarded as one of this country's purest forms of folk music, and in San Antonio, it's as popular as chili.

Changing Tracks

Lopez gradually moved from creating music to recording it. He now owns Joey Records International, which occupies six acres of land and includes two 24-track recording studios, cassette and vinyl production facilities, a talent agency, and several record labels featuring more than 300 artists.

But Lopez still spends most of his time in the studio. "It takes a lot of coordination to create a beautiful sound," he says. "My heart's always in what I'm doing, and I never have a dull day in the studio. You get in there with the musicians and you create something. That's why music is so beautiful. You try to do the best you can."

The Science behind the Sound

When Lopez started in the recording business, he hung heavy curtains and tacked thousands of empty egg cartons to the walls of his makeshift studio for soundproofing. "What we used for acoustics then is nothing like what we use today. But when I hear those tapes, they still sound great," he says.

The old building was eventually torn down, and Lopez consulted architectural acoustics experts from around the country to design the studios with higher ceilings, concrete walls and floors, and foam insulation that is shaped like egg cartons. Thick glass divides the musicians' booths and allows them to see and keep time with each other. The result, Lopez says, is state-of-the-art.

Nick Villarreal (accordion) and Rogelio Valdéz (guitar) prepare to record with Joe Lopez.

Music to the Ears

Sound waves are generated by any object that is vibrating. Sound waves move in all directions, bouncing off the floor, ceiling, and walls. The ear receives a mixture of direct sound waves and their reflections. Acoustics is used in the music studio to minimize unwanted reflections and other noise in recordings.

The image of a sound wave can be displayed by an electronic instrument called an *oscilloscope*. Pictured below are oscilloscope traces of a C-note played at three different volumes on a guitar. Each sound wave establishes a repetitive pattern, which can be modeled by *sine functions*. You will learn more about the graphs of sine, cosine, and tangent functions in this chapter.

Eva Ybarra, known as "the queen of the accordion," plays *conjunto* music in San Antonio, Texas.

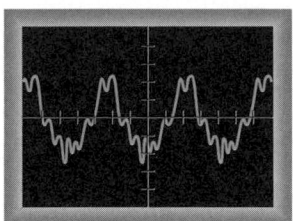

Guitar C-note at low volume

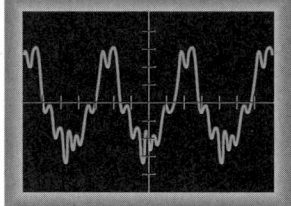

Guitar C-note at normal volume

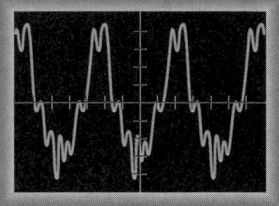

Guitar C-note at high volume

EXPLORE AND CONNECT

Joe Lopez plays his *bajo sexto* at his recording studio.

1. Writing Describe the similarities and differences you see between the three oscilloscope traces of the guitar's C-note. What happens to the wave as the note is played louder? softer? Do you think a C-note played by another instrument would look the same on the oscilloscope? Why or why not?

2. Project Find a recording of one song, by the same artist, on a vinyl record, cassette tape, and compact disc (CD). Listen to the three recordings, and rank them according to their sound quality. Explain your ranking. Describe the similarities and differences you notice between the three different recordings.

3. Research There are two methods of recording sound—analog and digital. Find out how each of these methods works. For each method, what is a sound wave converted to and how is it stored? Which recording formats are typically used for analog recordings? for digital recordings?

Mathematics & Joe Lopez

In this chapter, you will learn more about how mathematics is related to music and recording.

Related Examples and Exercises

Section 15.2
• Example 3
• Exercises 43–45

Section 15.3
• Exercises 24–27

15.1

Sine and Cosine on the Unit Circle

Learn how to...

- **work with sine and cosine as functions**

- **extend the domain of sine and cosine**

So you can...

- **solve problems, such as finding the height of a seat on a Ferris wheel as a function of time**

In Chapter 14 you used sine and cosine to find missing parts of triangles. The sine and cosine functions are not restricted to triangles, however. They are also used to describe many types of *periodic,* or repeating, phenomena, such as the motion of a Ferris wheel.

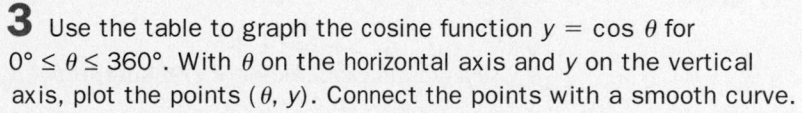

EXPLORATION

COOPERATIVE LEARNING

Graphing Sine and Cosine as Functions

Work with a partner.
You will need:
- a scientific calculator
- graph paper

1 Copy the table. Use a calculator to complete the table.

2 Use the table to graph the sine function $y = \sin \theta$ for $0° \le \theta \le 360°$. With θ on the horizontal axis and y on the vertical axis, plot the points (θ, y). Then connect the points with a smooth curve.

Angle θ	sin θ	cos θ
0°	?	?
30°	?	?
60°	?	?
90°	?	?
120°	?	?
150°	?	?
180°	?	?
210°	?	?
240°	?	?
270°	?	?
300°	?	?
330°	?	?
360°	?	?

3 Use the table to graph the cosine function $y = \cos \theta$ for $0° \le \theta \le 360°$. With θ on the horizontal axis and y on the vertical axis, plot the points (θ, y). Connect the points with a smooth curve.

Questions

1. How are the two graphs alike? How are they different?

2. How many values of θ for $0° \le \theta \le 360°$ satisfy each of the following?

    a. $\sin \theta = 0.5$ **b.** $\cos \theta = 0.5$

    c. $\cos \theta = -1$ **d.** $\sin \theta = -0.8$

When an object moves on a circular path, you can use **sine** and **cosine** to determine the object's position. For example, consider an object moving counterclockwise around a *unit circle* (a circle of radius 1, centered at the origin).

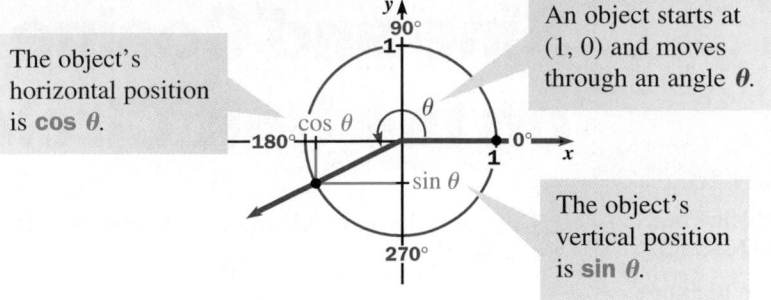

The object's horizontal position is **cos θ**.

An object starts at (1, 0) and moves through an angle **θ**.

The object's vertical position is **sin θ**.

THINK AND COMMUNICATE

1. Use the definitions of sine and cosine on page 672 to explain why the coordinates of the object in the diagram above are (cos θ, sin θ).

2. How does the unit circle tell you the ranges of the sine and cosine functions?

Graphs of Sine and Cosine for 0° ≤ θ ≤ 360°

The graphs of $y = \sin \theta$ and $y = \cos \theta$ for $0° \le \theta \le 360°$ are shown below.

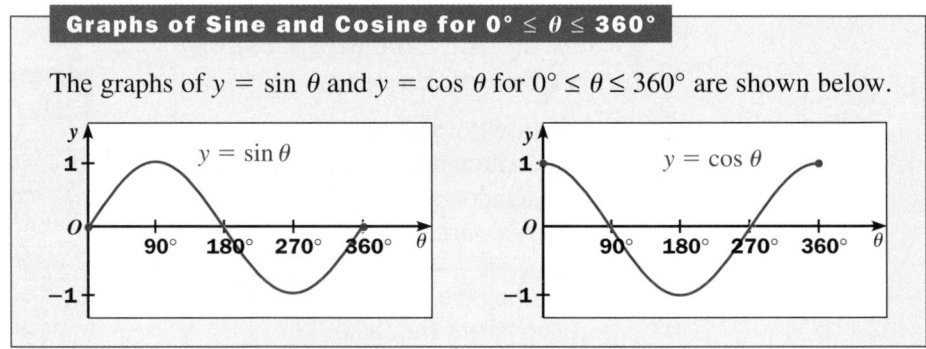

EXAMPLE 1

Find all angles **θ** for 0° ≤ **θ** ≤ 360° that satisfy each equation. Give answers to the nearest degree.

a. $\sin \theta = 0.7000$　　　　　　**b.** $\cos \theta = -0.4518$

SOLUTION

Use a graphing calculator or graphing software set in degree mode.

a. Graph $y = \sin x$ and $y = 0.7000$, and find the intersections of the two graphs.

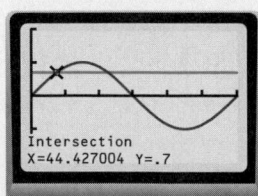

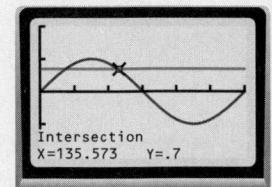

$\sin \theta = 0.7000$ for
θ ≈ 44° and **θ ≈ 136°**.

b. Graph $y = \cos x$ and $y = -0.4518$, and find the intersections of the two graphs.

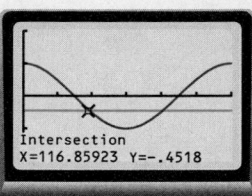

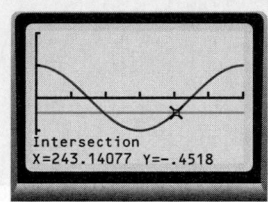

$\cos \theta = -0.4518$ for $\theta \approx \mathbf{117°}$ and $\theta \approx \mathbf{243°}$.

THINK AND COMMUNICATE

3. Use a calculator to find $\sin^{-1} 0.7000$ and $\cos^{-1} (-0.4518)$. How do these values relate to the solutions found in Example 1?

4. Why does the equation $\sin \theta = 2$ have no solution?

Extending Sine and Cosine's Domain

An object moving on the unit circle might keep moving beyond one revolution counterclockwise ($\theta > 360°$). It might also move clockwise ($\theta < 0°$).

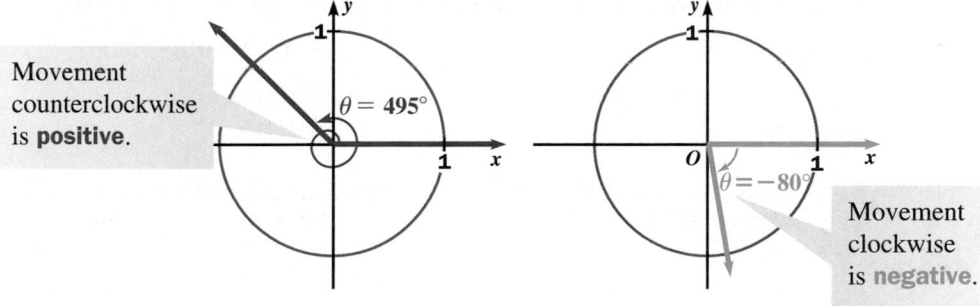

Movement counterclockwise is **positive**.

$\theta = \mathbf{495°}$

$\theta = -80°$

Movement clockwise is **negative**.

You can extend the domain of the sine and cosine functions to include any angle, positive or negative.

Graphs of Sine and Cosine for All Angles θ

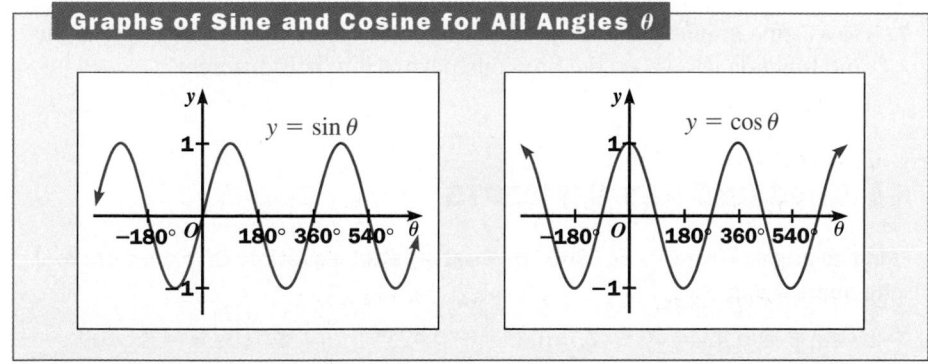

THINK AND COMMUNICATE

5. Compare the complete graphs of sine and cosine shown above to the graphs for $0° \le \theta \le 360°$ shown on the previous page. What do you notice?

EXAMPLE 2 Application: Amusement Park Rides

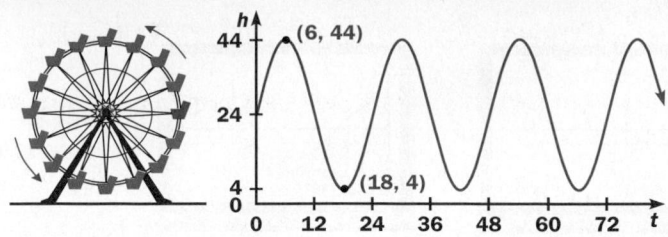

(6, 44)

(18, 4)

The modern Ferris wheel developed from the pleasure wheel, which probably originated in Eastern Europe or the Near East. The first pleasure wheel may have been a water wheel children clung to for rides.

The Ferris wheel shown has a radius of 20 ft. The center of the wheel is 24 ft above the ground, and the wheel rotates at a constant speed of 15°/s. The height h, in feet, of the red seat above the ground is given as a function of time t, in seconds, by:

$$h = 24 + 20 \sin (15° \cdot t)$$

Find h when:

a. $t = 1$ **b.** $t = 30$

> When $t = 0$, the red seat is in the position shown in the diagram above.

SOLUTION

Evaluate the function for each value of t.

> Use a calculator to find a sine or cosine value you don't know.

a. $h = 24 + 20 \sin (15° \cdot \mathbf{1})$
     $= 24 + 20 \sin (15°)$
     $\approx 24 + 20(0.2588)$
     ≈ 29

After 1 s, the red seat is at a height of about 29 ft.

b. $h = 24 + 20 \sin (15° \cdot \mathbf{30})$
     $= 24 + 20 \sin (450°)$
     $= 24 + 20(1)$
     $= 44$

After 30 s, the red seat is at a height of 44 ft.

THINK AND COMMUNICATE

6. In Example 2, can you get the same h-value for two different t-values? Explain.

7. How is the graph of the height function in Example 2 like the graph of the sine function shown on the previous page? How is it different?

☑ CHECKING KEY CONCEPTS

Find all angles θ for $0° \le \theta \le 360°$ that satisfy each equation. Give answers to the nearest degree.

1. $\sin \theta = 0.8244$ **2.** $\sin \theta = -0.4257$ **3.** $\cos \theta = 0.2209$

4. $\cos \theta = -0.3000$ **5.** $\sin \theta = \sin (-135°)$ **6.** $\cos \theta = \sin 395°$

Use the height function in Example 2 to find the height of the red seat for each value of t.

7. $t = 5$ **8.** $t = 45$ **9.** $t = 60$

15.1 Exercises and Applications

Extra Practice exercises on page 771

Find all angles θ for 0° ≤ θ ≤ 360° that satisfy each equation. Give answers to the nearest degree.

1. $\sin \theta = -1$
2. $\cos \theta = -1$
3. $\sin \theta = 0$
4. $\cos \theta = 0$

5. $\sin \theta = 0.5000$
6. $\cos \theta = -0.5000$
7. $\sin \theta = 0.4580$
8. $\cos \theta = 0.3223$

9. $\sin \theta = -0.5000$
10. $\sin \theta = 1$
11. $\cos \theta = -0.3250$
12. $\cos \theta = 0.9140$

Evaluate each expression.

13. $\sin 390°$
14. $\cos 450°$
15. $\cos (-75°)$
16. $\sin (-69°)$

17. $\sin 1350°$
18. $\cos (-480°)$
19. $\cos 1000°$
20. $\sin (-10°)$

Connection ▸ CAROUSELS

For a typical carousel, or merry-go-round, a horse makes a complete up-and-down movement 4.8 times for each revolution of the carousel, and the horse is raised 6 in. above and 6 in. below its center position. The up-and-down motion of the horse can be modeled by the function

$$h = 6 \cos (4.8\theta)$$

where h is the horse's displacement from its center position and θ is the angle turned by the carousel.

The brown horse started at its highest position at $\theta = 0°$.

Find the value of h for each value of θ.

21. $\theta = 90°$
22. $\theta = 360°$
23. $\theta = 400°$

24. **Open-ended Problem** Do you think that the function $h = 6 \cos (4.8\theta)$ holds for $\theta < 0°$? What would a negative value of θ indicate about the motion of the carousel?

The height of the horse can also be expressed as a function of time, t, in minutes, assuming the speed of the carousel is constant. If the carousel makes 4 revolutions per minute, then:

Toolbox p. 793
Dimensional Analysis

$$\theta = \frac{360°}{1 \text{ revolution}} \cdot \frac{4 \text{ revolutions}}{1 \text{ min}} \cdot t$$

25. Write an equation for h as a function of t.

26. Use the height function you found in Exercise 25 to find the value of h for each value of t.

   a. $t = 1$ b. $t = 5$

   c. $t = 0.5$ d. $t = 0.25$

   e. $t = 2$ f. $t = -1$

If $y = 4 - \cos \theta$, find the value of y for each value of θ.

27. $60°$ **28.** $120°$ **29.** $420°$ **30.** $-180°$

If $y = 3 + \cos 3\theta$, find the value of y for each value of θ.

31. $60°$ **32.** $120°$ **33.** $420°$ **34.** $-180°$

35. **Technology** Graph $y = 3 + 2 \cos \theta$ on a graphing calculator or graphing software.

    a. What is the maximum value of the function?

    b. What is the minimum value of the function?

    c. What values of θ for $0 \le \theta \le 360°$ give the maximum and minimum values of the function?

36. SAT/ACT Preview If $A = \sin(-40°)$ and $B = -\sin 40°$, then:

    A. $A > B$ **B.** $A = B$ **C.** $A < B$

    D. relationship cannot be determined

ONGOING ASSESSMENT

37. a. Open-ended Problem Name a positive angle and a negative angle that have the same sine value and the same cosine value. How are the two angles related to each other?

    b. Are there other positive or negative angles that have the same sine value and cosine value as the two angles you named? Explain.

SPIRAL REVIEW

Solve each triangle. *(Section 14.6)*

38.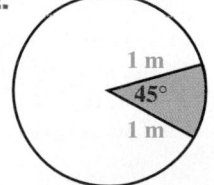

39.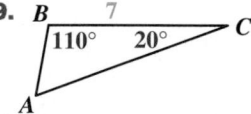

40.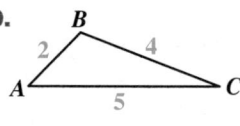

41. A circle has a radius of 1 in. *(Toolbox, page 802)*

    a. What is the circumference of the circle?

    b. An arc of the circle is intercepted by a 270° central angle. What is the length of the arc?

Find the area of each shaded region. *(Section 14.4)*

42. **43.** 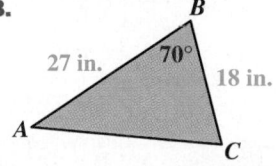 **44.**

15.2 Measuring in Radians

So far you have measured angles in degrees. People who use the sine and cosine functions for modeling often use *radians,* rather than degrees, to measure angles.

Learn how to...

- **convert between degree measure and radian measure**
- **find the sine and cosine of angles given in radians**

So you can...

- **explore models of periodic phenomena, such as music**

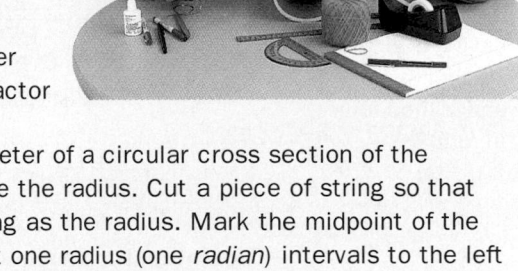

EXPLORATION

COOPERATIVE LEARNING

Making a Radian String

Work with a partner.
You will need:

- a cylinder
- a string
- tape
- a ruler
- a marker
- a protractor

SET UP Measure the diameter of a circular cross section of the cylinder and then calculate the radius. Cut a piece of string so that it is about 14 times as long as the radius. Mark the midpoint of the string, and make marks at one radius (one *radian*) intervals to the left and right of the midpoint. You may want to use a different color in each direction.

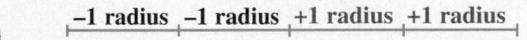

| –1 radius | –1 radius | +1 radius | +1 radius |

1 Attach the midpoint of your "radian string" to your cylinder. This is the origin.

2 Wrap the string around the cylinder. The counterclockwise direction is positive, and the clockwise direction is negative.

Questions

Use your string and cylinder to answer the following questions.

1. About how many radians are in a full circle?

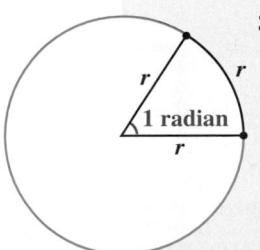

2. Compare your answer to Question 1 with the answer found by a group that used a different sized cylinder. Does the size of the cylinder affect the answer? Explain.

3. How are the points at 2 radians and −2 radians geometrically related to each other? Does this relationship hold for any two points that are at ±*r* radians from the origin?

The **radian** measure of an angle is the length of the unit circle's arc intercepted by the angle. Since the circumference of the unit circle is 2π, you have the following relationship between radians and degrees:

In this diagram,

$\theta = 135° = \dfrac{3\pi}{4}$ radians.

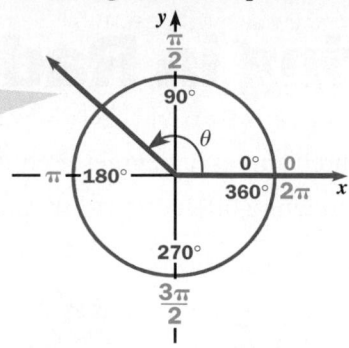

2π **radians = 360°**

1 radian $= \dfrac{180°}{\pi}$ 1° $= \dfrac{\pi}{180}$ **radians**

EXAMPLE 1

a. Convert 236° to radians.

b. Convert $\dfrac{\pi}{4}$ radians to degrees.

If an angle measure does not have a degree symbol, it is understood to be in radians.

SOLUTION

a. $236° = 236 \cdot \dfrac{\pi}{180}$ radians

$\approx 1.31\pi$ radians

≈ 4.12

b. $\dfrac{\pi}{4} = \dfrac{\pi}{4} \cdot \dfrac{180°}{\pi}$

$= \dfrac{180°}{4}$

$= 45°$

The sine and cosine functions can be defined for radians as well as degrees.

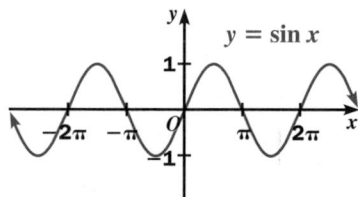

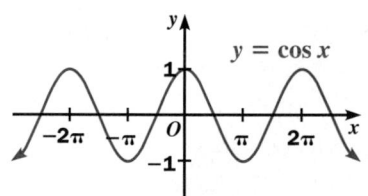

In general, this book will use x to represent angles measured in radians and θ to represent angles measured in degrees.

EXAMPLE 2

a. Find $\cos\dfrac{\pi}{2}$.

b. Find $\sin 3$.

SOLUTION

WATCH OUT!

In part (b), if your calculator gives 0.0523, it is set in degree mode. Remember: $\sin 3 \neq \sin 3°$

a. Since $\dfrac{\pi}{2}$ represents a quarter of the way around the unit circle, $\cos\dfrac{\pi}{2}$ is the x-coordinate of the point $(0, 1)$.

$$\cos\dfrac{\pi}{2} = 0$$

b.

sin 3
      .1411200081

Set a calculator to radian mode and find sin 3.
sin 3 ≈ 0.1411

Periodic Phenomena

The tides rising and falling, the moon waxing and waning, and a pendulum swinging back and forth are all examples of *periodic* phenomena, because the events repeat themselves in a regular cycle.

Many periodic phenomena can be modeled with sine or cosine functions. Radians are usually used when describing or discussing periodic phenomena that are not obviously related to angles, such as musical tones.

EXAMPLE 3 Interview: Joe Lopez

When Joe Lopez records music, he is recording small pressure changes in the air. If the atmospheric pressure is 15.000 pounds per square inch (psi), a loud noise might cause the pressure to vary between 14.999 psi and 15.001 psi.

A pure musical tone will cause the air pressure to vary in a way that can be modeled with a sine function. The variation in pressure for a soft Concert A is given by

$$P = \left(5 \cdot 10^{-4}\right) \sin \left(2\pi \cdot 440t\right)$$

where P is the deviation from the average atmospheric pressure in psi, and t is the time in seconds.

a. Graph the function. Find the maximum deviation in pressure.

b. Find the time between consecutive maximum P-values.

SOLUTION

a. Graph $y = \left(5 \cdot 10^{-4}\right) \sin \left(2\pi \cdot 440x\right)$ using a graphing calculator or graphing software, and use trace to find the maximum value of y.

b. Use trace to find the values of x at the first and second maximum values of y, and then subtract the two x-values.

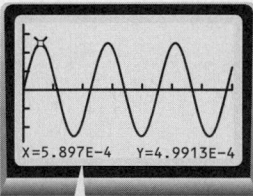

X=5.897E-4   Y=4.9913E-4

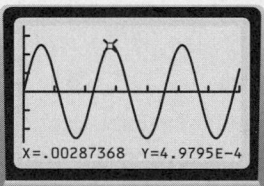

X=.00287368   Y=4.9795E-4

Find the difference between the x-values shown here and on the screen in part (a).

The maximum deviation from the average pressure is about **$5 \cdot 10^{-4}$ psi**.

$$x_2 - x_1 \approx 0.0029 - 0.0006$$
$$\approx 0.0023$$

The time between consecutive maximum P-values is about 0.0023 s.

THINK AND COMMUNICATE

1. Maximum P-values occur at about 0.0006 s and 0.0029 s. When does the next maximum P-value occur?

☑ CHECKING KEY CONCEPTS

Convert each angle measure from radians to degrees (to the nearest degree).

1. $\dfrac{5\pi}{2}$ **2.** $-\dfrac{3\pi}{4}$ **3.** 3.86 **4.** 8.95

Convert each angle measure from degrees to radians (to the nearest hundredth of a radian).

5. $-200°$ **6.** $540°$ **7.** $30°$ **8.** $311°$

Evaluate each expression.

9. $\sin \dfrac{3\pi}{2}$ **10.** $\cos(-3\pi)$ **11.** $\cos \dfrac{4\pi}{3}$ **12.** $\sin \dfrac{\pi}{2}$

15.2 Exercises and Applications

Extra Practice exercises on page 772

Convert each angle measure from radians to degrees (to the nearest degree).

1. 1 **2.** $\dfrac{5\pi}{8}$ **3.** $\dfrac{5\pi}{4}$ **4.** -2.5

5. 3π **6.** $-\pi$ **7.** 7π **8.** -6.5

9. 5.89 **10.** 7.25π **11.** -0.5π **12.** -13.27

13.  **Technology** Use a graphing calculator set in parametric, simultaneous, and radian mode. Use $0 \le t \le 2\pi$, $-2 \le x \le 7$, and $-3 \le y \le 3$. Have the calculator increment t by 0.05. Graph each pair of parametric equations below. Describe what happens on your calculator screen. Explain the two graphs in terms of what you learned in this section.

a. $x_1 = \cos t$ $x_2 = t$ **b.** $x_1 = \cos t$ $x_2 = t$

     $y_1 = \sin t$ $y_2 = \sin t$ $y_1 = \sin t$ $y_2 = \cos t$

Convert each angle measure from degrees to radians. Express each answer as a decimal to the nearest hundredth of a radian, and in terms of π.

14. $45°$ **15.** $-180°$ **16.** $210°$ **17.** $720°$

18. $-90°$ **19.** $-65°$ **20.** $20°$ **21.** $390°$

22. $-700°$ **23.** $87°$ **24.** $-35°$ **25.** $405°$

26. GEOMETRY Steve's racing bicycle has tires with a diameter of 27 in. Joan's mountain bike has tires with a diameter of 24 in. If both bikes are wheeled forward 4π radians, how far (in inches) will each bike travel? Explain why the two bikes travel different distances, even though the tires move through the same number of radians.

racing bicycle   **mountain bicycle**

Evaluate each expression.

27. $\sin\left(-\dfrac{\pi}{2}\right)$ **28.** $\cos\dfrac{7\pi}{4}$ **29.** $\sin 8\pi$ **30.** $\cos\left(-\dfrac{2\pi}{3}\right)$

31. $\sin \pi$ **32.** $\cos 3\pi$ **33.** $\cos 0.82$ **34.** $\sin(-2.18)$

35. $\cos(-5.34)$ **36.** $\cos 2.8\pi$ **37.** $\sin 2$ **38.** $\sin 6.32$

Connection ▸ SOLAR POWER

The amount of sunlight striking Earth at a particular latitude changes through the year because the orientation of Earth's axis relative to the sun changes. The table shows the variation in the average solar radiation falling on a solar panel lying flat on the ground at latitude 40° north.

The heliostat field and solar tower at Sandia National Laboratories in NM (above) and the Solar One Power Plant in Dagget, CA (below) collect solar energy.

Month	Average daily solar radiation (BTU/ft²)
March	1182
April	1584
May	1862
June	1920
July	1862
August	1584
September	1182
October	838
November	524
December	463
January	524
February	838

39. Make a scatter plot of the data. Put time t, in months since March, on the horizontal axis, and the average daily solar radiation s, in BTU per square foot, on the vertical axis. Connect the points with a smooth curve.

   a. What function does your graph resemble? What are the maximum and minimum values?

   b. Visual Thinking Extend your graph one year backward (letting t be negative) and one year forward. Explain what you did.

40. Open-ended Problem How do you think your graph in Exercise 39 would change for a location closer to the equator? Explain your reasoning.

41. **Technology** The data in the table can be modeled by this equation:

$$s = 1190 + 730 \sin\left(\frac{\pi}{6}t\right)$$

Use a graphing calculator or graphing software to graph the function and compare it with the graph you made in Exercise 39. How well does the model fit the data?

42. Open-ended Problem Given what you learned above, when are solar collectors most useful at latitude 40° north? What might someone do to gather more energy from sunlight?

Look back at the
article on pages
704–706.

*When Joe Lopez records music, the varying pressure of the sound wave
against the microphone is transformed into varying electrical voltage.
He then records this electrical signal.*

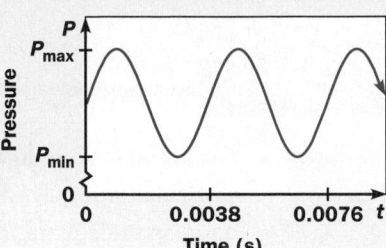

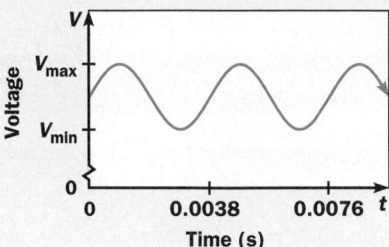

43. Visual Thinking The graphs show the air pressure over time of
a pure musical tone, Concert C, and the voltage over time as the
tone is transformed into an electrical signal. Describe the
similarities and differences between the two graphs.

44. Look at the graph of pressure over time. How much time passes
before the pattern repeats itself?

45. Open-ended Problem What do you think happens to the
graph of pressure over time when the music gets louder? softer?
What do you think happens to the graph of voltage over time?

ONGOING ASSESSMENT

46. Open-ended Problem Choose four angles, one in each of the four
quadrants. Give the radian measure, the sine value, and the cosine value
of each angle.

SPIRAL REVIEW

**Find all angles θ for $0° \leq \theta \leq 360°$ that satisfy each equation. Give answers to the
nearest degree.** *(Section 15.1)*

47. $\sin \theta = 0.5218$ **48.** $\cos \theta = -0.9900$ **49.** $\sin \theta = 0.4050$

50. **Technology** Use a graphing calculator or graphing software to
graph the function $y = 5 + 2 \sin \theta$. *(Section 15.1)*

   a. What is the maximum value of the function?

   b. What is the minimum value of the function?

51. Suppose you roll a 20-sided die, marked with the numbers 1 through 20.
What is the probability of getting a 1 or a 2? *(Section 13.1)*

Exploring Amplitude and Period

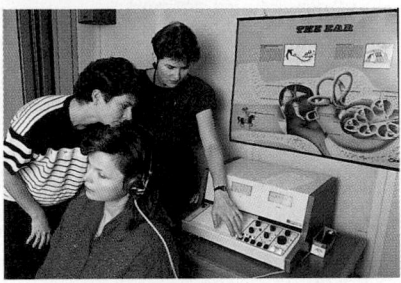

Learn how to...

- **graph equations of these forms:**
  $y = a \sin bx + k$
  $y = a \cos bx + k$

So you can...

- **understand periodic phenomena, such as the pure-tone sounds used in hearing tests**

An audiologist tests your hearing by changing the pressure of pure-tone sounds to find the lowest pressure you can hear at various frequencies.

An audiologist uses a frequency of 1000 cycles per second (cps) first, increasing the sound pressure level until you can hear the tone. The graphs below show the changes in air pressure associated with tones at sound pressure levels of 20, 25, and 30 decibels (db).

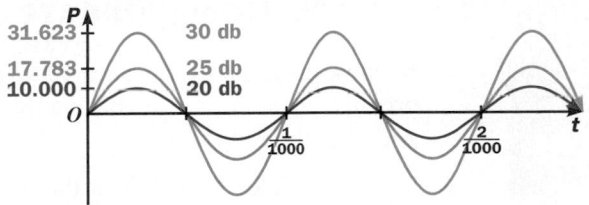

After using the frequency of 1000 cps, the audiologist may use other frequencies, such as 500 cps (a lower pitch) and 2000 cps (a higher pitch).

The unit of sound pressure on the vertical axis is the smallest sound pressure audible to most humans, approximately 3.53×10^{-9} atmosphere.

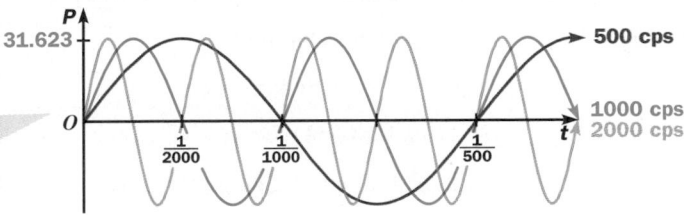

THINK AND COMMUNICATE

1. In the first set of graphs, the frequency of the tone is constant but the pressure level varies. How are the graphs alike? How are they different?

2. The equations $P = 10.000 \sin 2000\pi t$, $P = 17.783 \sin 2000\pi t$, and $P = 31.623 \sin 2000\pi t$ describe the first set of graphs. How does changing the value of a affect the graph of the equation $y = a \sin bx$?

3. In the second set of graphs, the pressure level is constant but the frequency varies. How are the graphs alike? How are they different?

4. The equations $P = 31.623 \sin 1000\pi t$, $P = 31.623 \sin 2000\pi t$, and $P = 31.623 \sin 4000\pi t$ describe the second set of graphs. How does changing the value of b affect the graph of the equation $y = a \sin bx$?

The values of a and b affect the *amplitude* and *period* of the graphs of $y = a \sin bx$ and $y = a \cos bx$ as shown below.

The **amplitude** is

$$\frac{M - m}{2}$$

where M and m are the maximum and minimum values of the function.

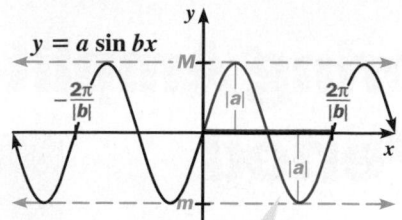

$y = a \sin bx$

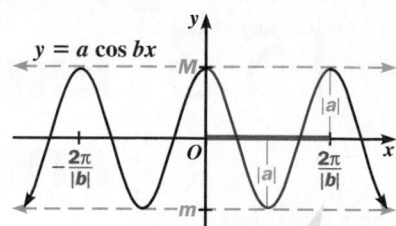

$y = a \cos bx$

A **cycle** is the smallest portion of the graph that repeats.

The **period** is the length of one cycle.

In general, the graphs of $y = a \sin bx$ and $y = a \cos bx$ have an amplitude of $|a|$ and a period of $\dfrac{2\pi}{|b|}$. Throughout this chapter, only positive values of b will be used.

THINK AND COMMUNICATE

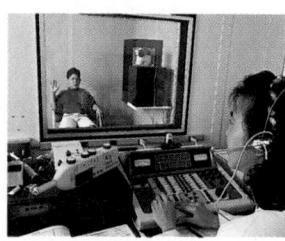

5. Suppose an audiologist uses a 10 db pure-tone sound with the sound pressure equation $P = 3.162 \sin 16{,}000\pi t$. What are the amplitude and period of the sound?

6. If the test subject cannot hear the sound in Question 5, how should the audiologist change it? What effect would this have on the equation?

EXAMPLE 1

Graph each equation.

a. $y = 2 \sin 4\pi x$

b. $y = 3.5 \cos 0.5x$

SOLUTION

a. The amplitude is 2.

The period is $\dfrac{2\pi}{4\pi} = \dfrac{1}{2}$.

The graph rises 2 units above and falls 2 units below the x-axis.

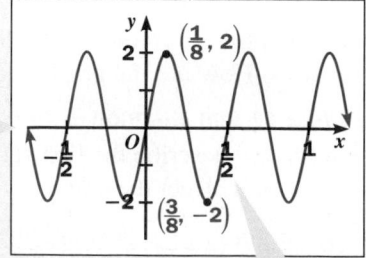

b. The amplitude is 3.5.

The period is $\dfrac{2\pi}{0.5} = 2\pi \cdot 2 = 4\pi$.

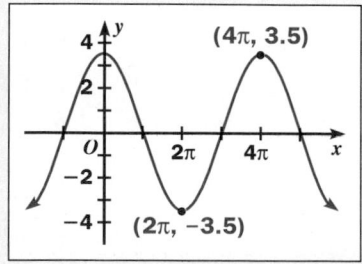

The graph completes one cycle in $\dfrac{1}{2}$ unit.

7. In the same coordinate plane, graph each pair of functions. How are the graphs geometrically related?

a. $y = \sin x$

$y = -\sin x$

b. $y = \cos x$

$y = -\cos x$

8. Suppose the value of b in the equation $y = a \sin bx$ is between 0 and 1. Is the period of the graph *greater than* or *less than* 2π?

EXAMPLE 2

Find an equation for each graph.

a.
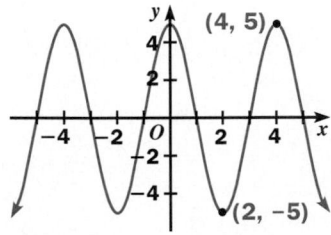

b.
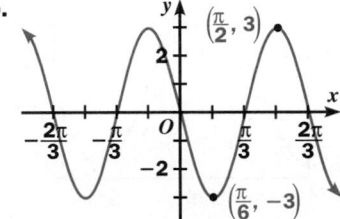

SOLUTION

a. The curve resembles the graph of a cosine function.

The amplitude is 5, so $a = 5$.

The period is 4, so $\dfrac{2\pi}{b} = 4$ and $b = \dfrac{2\pi}{4} = \dfrac{\pi}{2}$.

An equation for the graph is $y = 5 \cos \dfrac{\pi}{2}x$.

b. The curve resembles the graph of a sine function reflected over the x-axis.

The amplitude is 3, so $a = -3$.

The period is $\dfrac{2\pi}{3}$, so $\dfrac{2\pi}{b} = \dfrac{2\pi}{3}$ and $b = 3$.

An equation for the graph is $y = -3 \sin 3x$.

> Use a negative value for a because of the reflection.

Vertical Translation

If you translate the graph of a sine or cosine function vertically by k units, its equation will have the form $y = a \sin bx + k$ or $y = a \cos bx + k$.

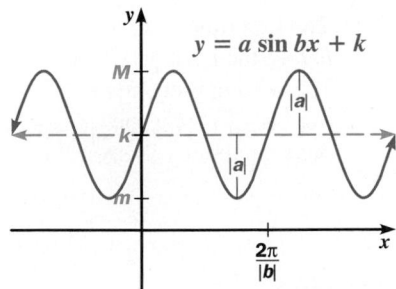

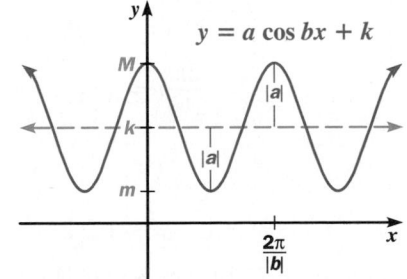

> The line $y = k$ is called the *axis*. The value of k is the average of M and m.

EXAMPLE 3 Application: Amusement Park Rides

The Navy Pier Ferris wheel in Chicago, Illinois, has an axle that is 80 ft above the ground. The wheel has a radius of 70 ft. After passengers are seated, the wheel makes one complete turn in 210 s.

a. Suppose you are in the seat at the bottom when the wheel begins to turn. Draw a graph of your height above the ground as a function of time.

b. Write an equation for your height above the ground as a function of time.

c. Find the first two times you are 100 ft above the ground.

SOLUTION

a.

This graph looks like the graph of a cosine function reflected over its axis.

b. $h = -70 \cos \dfrac{2\pi}{210}t + 80$

The radius is 70 ft.

The period is 210 s.

The axle is 80 ft above the ground.

c. **Method 1** Use a graphing calculator or graphing software.

Graph $y = -70 \cos \dfrac{2\pi}{210}x + 80$
and $y = 100$ and find the points of intersection of the graphs.

You are at 100 ft about 62 s and about 148 s after the wheel begins to turn.

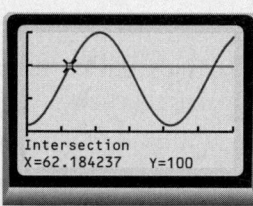

Intersection
X=62.184237   Y=100

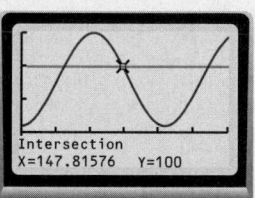

Intersection
X=147.81576   Y=100

Method 2 Solve the equation using algebra.

$$-70 \cos \frac{2\pi}{210}t + 80 = 100$$

$$-70 \cos \frac{2\pi}{210}t = 20$$

$$\cos \frac{2\pi}{210}t = -\frac{20}{70}$$

$$\frac{2\pi}{210}t = \cos^{-1}\left(-\frac{20}{70}\right)$$

$$t = \frac{210}{2\pi} \cos^{-1}\left(-\frac{20}{70}\right)$$

$$t \approx 62.2$$

The first time is 43 s *before* the peak at 105 s, so the second time is 43 s *after* the peak, at about 148 s.

You reach 100 ft when $t \approx 62$ s and when $t \approx 148$ s.

1. **a.** In the same coordinate plane, graph $y = 2 \sin 4\pi x$ and $y = 4 \sin 2\pi x$.

   b. Compare the amplitudes and periods of the two graphs.

2. **a.** Draw the graph of a cosine function that has period $\frac{\pi}{2}$, amplitude 3, and a vertical translation of 5 units.

   b. What is an equation for the graph in part (a)?

15.3 **Exercises and Applications**

Extra Practice exercises on page 772

Graph each function. State the amplitude and period of each graph.

1. $y = 2 \cos 2\pi x$

2. $y = \frac{1}{2} \sin \frac{1}{2}x$

3. $y = -\frac{1}{2} \sin 3x$

4. $y = 2 \cos \frac{3\pi}{2}x$

5. $y = -\cos \frac{\pi}{2}x$

6. $y = 4 \sin 4\pi x$

7. $y = 4 \sin 4x$

8. $y = -\frac{1}{3} \cos 2x$

9. $y = -\frac{1}{3} \sin \frac{1}{4}x$

10. $y = 3 \cos \pi x$

11. $y = 2 \sin 2.5x$

12. $y = 2 \cos 4x$

Write an equation for each graph.

13.

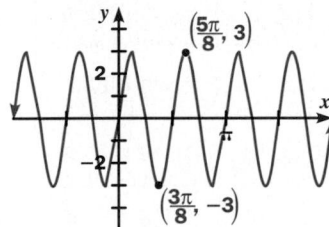

14.

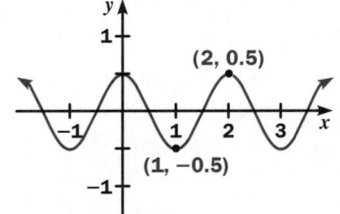

15.

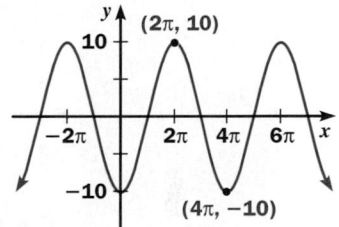

Write a sine function with the given maximum M, minimum m, and period.

16. $M = 5, m = -5,$ period $= \frac{1}{4}$

17. $M = 7, m = 3,$ period $= 3\pi$

18. $M = 6, m = -9,$ period $= 1$

19. $M = -4, m = -10,$ period $= \frac{\pi}{2}$

Match each equation with its graph.

20. $y = 2 + \frac{1}{2} \cos 2x$

21. $y = -1 - 3 \sin 2x$

22. $y = 2 - \cos \frac{\pi}{2}x$

A.

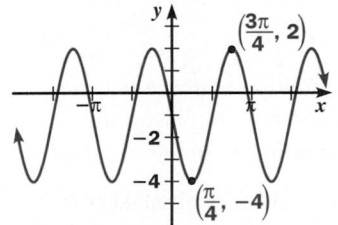

B.

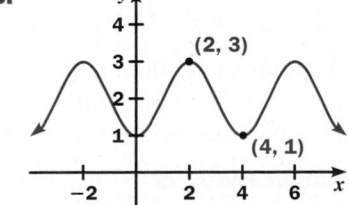

C.

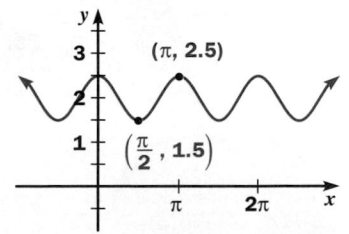

23.  **Technology** Use a graphing calculator or graphing software. Set it to radian mode and use a viewing window that shows $0 \le x \le 2\pi$ and $-1.5 \le y \le 1.5$.

   a. Graph $y = \sin x$, $y = \sin 1.5x$, $y = \sin 2x$, and $y = \sin 2.5x$. How many cycles appear between 0 and 2π along the x-axis for each graph?

   b. How many cycles do you expect will appear between 0 and 2π in the graphs of $y = \sin 1.25x$, $y = \sin 1.75x$, and $y = \sin 2.25x$? Check your predictions by graphing.

   c. How many cycles will appear between 0 and 2π in a graph with equation of the form $y = \sin bx$? in a graph with equation of the form $y = \cos bx$?

Joe Lopez tunes his *bajo sexto* using harmonics.

Look back at the article on pages 704–706.

INTERVIEW Joe Lopez

One way to tune a guitar is to use **harmonics,** *the ringing tones you hear if you pluck a string while lightly touching the string at points one half, one third, or one fourth of the way along its length. To find the exact points to touch, you can use the 12th, 7th, and 5th frets as guides.*

The plucked string vibrates this way.

String (frequency)	Fundamental (open string)	1st harmonic (12th fret)	2nd harmonic (7th fret)	3rd harmonic (5th fret)
E $\left(82\frac{1}{2}\text{ cps}\right)$	$P = \sin 165\pi t$	$P = \sin 330\pi t$	$P = \sin 495\pi t$	$P = \sin 660\pi t$
A (110 cps)	$P = \sin 220\pi t$	$P = \sin 440\pi t$	$P = \sin 660\pi t$	$P = \sin 880\pi t$
D $\left(146\frac{2}{3}\text{ cps}\right)$	$P = \sin 293\frac{1}{3}\pi t$	$P = \sin 586\frac{2}{3}\pi t$	$P = \sin 880\pi t$	?
G $\left(195\frac{5}{9}\text{ cps}\right)$	$P = \sin 391\frac{1}{9}\pi t$	$P = \sin 782\frac{2}{9}\pi t$	?	?
B $\left(247\frac{1}{2}\text{ cps}\right)$	$P = \sin 495\pi t$	?	?	?
E (330 cps)	?	?	?	?

5th fret ——
7th fret ——
12th fret ——

The table shows sound pressure equations for the harmonic tones of a guitar tuned in the key of A.

24. Look at the first two rows of the table. How do the equations of the 1st, 2nd, and 3rd harmonics compare with the fundamental equations?

25. Copy and complete the table.

26. Strings that vibrate according to the same equation sound the same. Explain how the first four strings can be tuned by comparing the 3rd harmonic of one string with the 2nd harmonic of the next string.

27. Open-ended Problem Explain how you might tune the last two strings based on the harmonics of one of the first four strings.

Electronic Distance Measurement (EDM) is a surveying technique. Because of its accuracy, it is used to monitor dams so that small but potentially disastrous changes can be recognized early and repaired. EDM is also used for tunneling through mountains or below water. To measure the distance between two points *A* and *B* by EDM, an electromagnetic wave of known period is transmitted from *A* and reflected back from a prism at *B*. A receiver at *A* analyzes the amplitude of the wave when it returns.

28. Suppose previous measurements show that the distance from *A* to *B* is between 430 m and 440 m. If you transmit a wave from *A* to *B* and back to *A*, how far might it travel?

29. Writing Explain how using a wave with a period of 20 m can help you find the exact distance between *A* and *B*.

30. Write an equation for a sine wave with period 20 m and amplitude *a*.

31. Suppose that when a sine wave with period 20 m returns to *A*, the receiver detects that the wave height is **0.6*a***, as shown in the diagram at the right. What are the two possible distances the wave has traveled when going from *A* to *B* and back to *A*?

32. Suppose the receiver also detects that the wave height is *decreasing* when the wave reaches the receiver. Which of your two answers from Exercise 31 is the actual distance the wave has traveled? What is the actual distance between *A* and *B*?

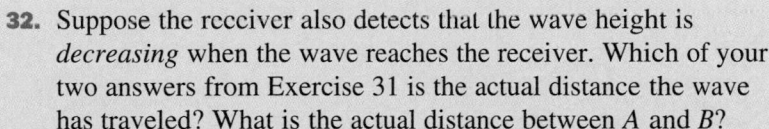

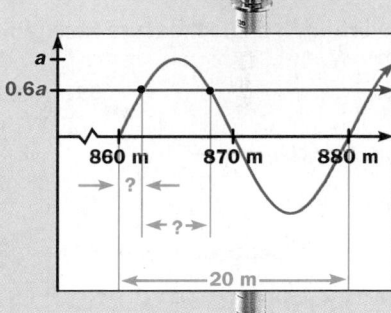

33. SAT/ACT Preview A sine function has a maximum value of 20, a minimum value of -10, and period 4π. Which equation describes this function?

A. $y = 20 - 10 \sin 4x$ **B.** $y = 5 - 15 \sin \frac{1}{2}x$ **C.** $y = 15 \sin \frac{1}{2}x - 5$

D. $y = 20 - 10 \sin \frac{1}{2}x$ **E.** none of these

34. Consider the function $y = a \sin bx + k$.

   a. Give expressions in terms of *a*, *b*, and *k* for *M* and *m*, the maximum and minimum values of the function.

   b. Use your results from part (a) to show that the definition of amplitude applies to vertically translated graphs of sine functions. That is,

$$\frac{M - m}{2} = |a|$$

35. Open-ended Problem In the same coordinate plane, sketch the graphs of two sine functions, one with twice the amplitude and half the period of the other. Write equations for the two graphs.

36. Challenge Find three different cosine equations whose graphs include the points $(0, 2)$ and $(3, 0)$.

37. Cooperative Learning Work in a group of four students. Prepare your results on a poster or overhead transparency.

a. In the same coordinate plane, graph $y = a \cos x$ for three different positive values of a, showing at least one period. In a different color, graph three functions with negative values of a. Label the intersections of the graphs with the y-axis.

b. In the same coordinate plane, graph $y = \cos bx$ for three different positive values of b, showing at least one period. Label the intersections of each graph with the x-axis.

c. Write a summary statement that describes the effects of varying a and b in the equations $y = a \cos x$ and $y = \cos bx$.

SPIRAL REVIEW

Convert each angle measure from radians to degrees. *(Section 15.2)*

38. $\dfrac{\pi}{6}$

39. -3π

40. $\dfrac{9}{4}\pi$

Multiply. *(Section 8.4)*

41. $(4 + i)(2 + 3i)$

42. $(6 - 3i)^2$

43. $(1 + i)(1 - i)$

Graph each function. *(Section 9.7)*

44. $y = \dfrac{1}{x - 1}$

45. $y = \dfrac{1}{x + 2} + 4$

46. $y = \dfrac{1}{x + 7} - 3$

ASSESS YOUR PROGRESS

VOCABULARY

sine (p. 708)   **cycle** (p. 720)
cosine (p. 708)   **period** (p. 720)
radian (p. 714)   **amplitude** (p. 720)

Find all angles θ for $0° \leq \theta \leq 360°$ that satisfy each equation. Give answers to the nearest degree. *(Section 15.1)*

1. $\sin \theta = 0.7589$

2. $\cos \theta = 0.8234$

3. $\sin \theta = -0.5239$

Evaluate each expression. *(Section 15.2)*

4. $\cos 4.1$

5. $\cos (-1.5)$

6. $\sin (-315°)$

Graph each function. State the amplitude and period of each graph. *(Section 15.3)*

7. $y = 2 \sin \dfrac{\pi}{2}x + 4$

8. $y = -\dfrac{1}{2} \cos \dfrac{1}{2}x$

9. Journal Explain how to write an equation of the form $y = a \sin bx + k$ if you know the maximum and minimum values, the amplitude, and the period of a sine function.

15.4 Exploring Phase Shifts

Learn how to...

- **graph equations of these forms:**
  $y = a \sin b(x - h) + k$
  $y = a \cos b(x - h) + k$

- **write equations for sine and cosine graphs that have a phase shift**

So you can...

- **describe periodic phenomena that have a phase shift, such as the motions of pistons in an engine**

The combustion of fuel causes the pistons of a car engine to move up and down. Connecting rods link the pistons to the throws of a crankshaft, transforming the vertical motion of the pistons into a circular motion that powers the wheels.

piston

connecting rod

throw

crankshaft

Because the throws are placed at 120° angles around the crankshaft, the pistons must be fired "out of phase" (that is, at different times) for the engine to run smoothly.

The graphs below show the heights of the pistons above the crankshaft in a 6-cylinder engine as a function of time. There are only three graphs, because the pistons are paired. Pistons 1 and 6 move together, as do pistons 2 and 5, and pistons 3 and 4.

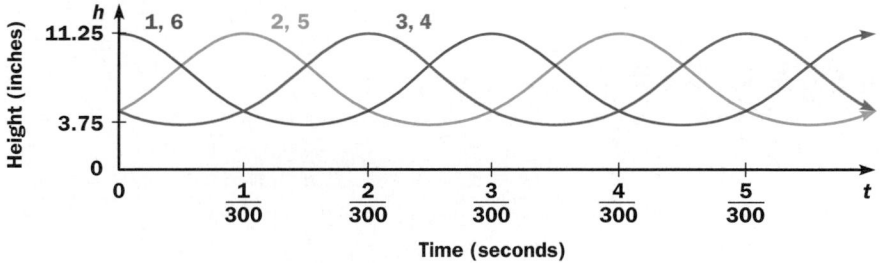

BY THE WAY

To help the crankshaft rotate smoothly, a film of oil about 0.002 inch thick is placed between the crankshaft and the surrounding bearings.

THINK AND COMMUNICATE

1. Based on the graphs, how long does it take the crankshaft to rotate once?

2. At what times are pistons 1 and 6 at their maximum height? pistons 2 and 5? pistons 3 and 4?

3. Describe how the graphs of the piston heights are geometrically related to each other.

4. Suppose the equation $h = f(t)$ describes the height of piston 1 as a function of time t. Complete each sentence.

   a. An equation for piston 2 is $h = f(t - \underline{\ ?\ })$.

   b. An equation for piston 3 is $h = f(t - \underline{\ ?\ })$.

A horizontal translation of a periodic function is called a **phase shift**.

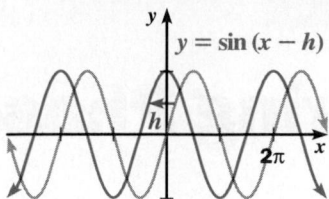

If *h* < 0, the graph is
shifted to the **left**.

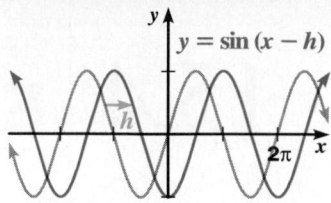

If *h* > 0, the graph is
shifted to the **right**.

THINK AND COMMUNICATE

5. Refer to the piston height graphs on the previous page. Compared with piston 1, what is the phase shift for piston 2? for piston 3?

6. a. What is the smallest phase shift that will move a sine graph onto a cosine graph? Express your answer in radians.

   b. Complete this equation: $\cos x = \sin (x - \underline{\ ?\ })$.

EXAMPLE 1

Graph each equation.

a. $y = 3 \cos (x - \pi)$

b. $y = \cos \left(3x - \dfrac{\pi}{2}\right)$

SOLUTION

Rewrite each equation in the form $y = a \cos b(x - h)$.

WATCH OUT!

In part (b), the phase shift is *not* $\dfrac{\pi}{2}$.

a. $y = 3 \cos (x - \pi)$

$\quad = 3 \cos 1(x - \pi)$

$a = 3 \quad b = 1 \quad h = \pi$

Think of shifting the graph of $y = 3 \cos x$ (shown in gray) to the right π units.

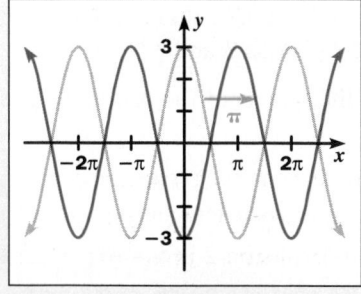

b. $y = \cos \left(3x - \dfrac{\pi}{2}\right)$

$\quad = 1 \cos 3\left(x - \dfrac{\pi}{6}\right)$

$a = 1 \quad b = 3 \quad h = \dfrac{\pi}{6}$

Think of shifting the graph of $y = \cos 3x$ (shown in gray) to the right $\dfrac{\pi}{6}$ units.

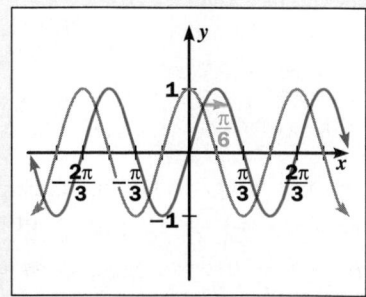

EXAMPLE 2

Write a sine function for the blue graph shown below.

SOLUTION

Step 1 Find the amplitude and period of the graph.

The amplitude is 2, so $a = 2$.

The period is 6, so $\frac{2\pi}{b} = 6$, and $b = \frac{\pi}{3}$.

There is no vertical shift for this graph.

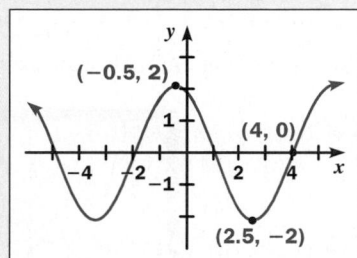

Step 2 Draw a helping graph using a sine function with the values of a and b from Step 1.

The graph of $y = 2 \sin \frac{\pi}{3}x$ is shown in gray.

The blue graph is shifted 2 units to the left of the gray graph.

A sine function that describes the blue graph is $y = 2 \sin \frac{\pi}{3}(x + 2)$.

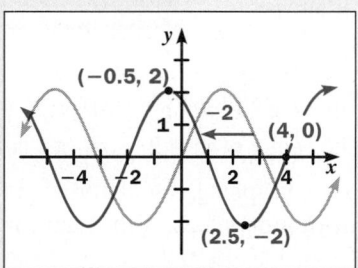

Combining Sine and Cosine Functions

When you add sine or cosine functions, the resulting functions can be periodic, but their graphs may be more complicated than the graphs of simple sine and cosine functions.

In general, the **sum** of the functions $y = f(x)$ and $y = g(x)$ is the function $y = f(x) + g(x)$.

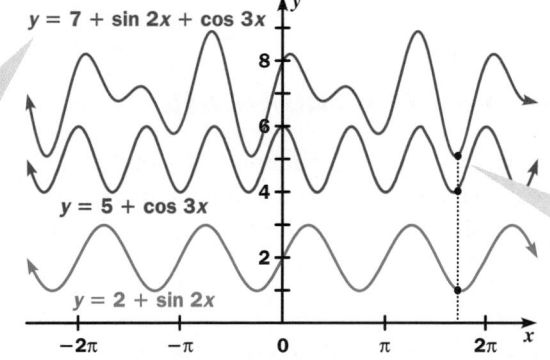

The y-coordinate of this point is the **sum of the y-coordinates** of the points with the same x-coordinate on the green and blue graphs.

THINK AND COMMUNICATE

7. What is the period of each of the three functions shown above?

8. Is the *period of the sum* function (shown in red) equal to the *sum of the periods* of the two original functions (shown in green and blue)?

EXAMPLE 3

In the same coordinate plane, graph the functions $y = 4 + \cos \frac{2\pi}{3}x$, $y = 2 + \sin \pi x$, and their sum. Find the period and range of each function.

SOLUTION

Use a graphing calculator or graphing software.

The period of
$$y = 4 + \cos \frac{2\pi}{3}x$$
is **3**.

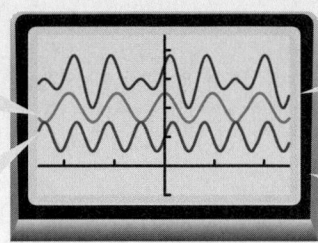

The period of
$$y = 2 + \sin \pi x$$
is **2**.

The period of the **sum** of the functions is **6**.

On some calculators, you can enter
$$y_3 = y_1 + y_2$$
to graph the sum of two functions.

The range of $y = 4 + \cos \frac{2\pi}{3}x$ is $3 \le y \le 5$.

The range of $y = 2 + \sin \pi x$ is $1 \le y \le 3$.

Use a graphing calculator or graphing software to find the minimum and maximum values of the sum function.

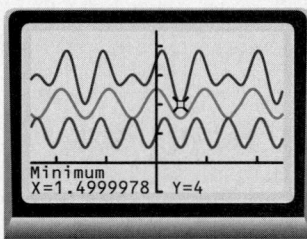

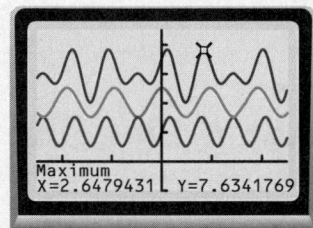

The minimum is exactly 4.   The maximum is about 7.63.

The range of $y = 6 + \sin \pi x + \cos \frac{2\pi}{3}x$ is about $4 \le y \le 7.63$.

THINK AND COMMUNICATE

9. In Example 3, the maximum values of the given functions (graphs shown in green and blue) are 5 and 3, but the maximum value of the sum function (graph shown in red) is *not* $5 + 3 = 8$. Explain why not.

☑ CHECKING KEY CONCEPTS

1. Graph the functions $y = \sin\left(x - \frac{\pi}{2}\right)$ and $y = \sin\left(x + \frac{\pi}{2}\right)$. How do they compare with the graph of $y = \sin x$?

2. Write a cosine function that is equivalent to $y = \sin \pi(x - 3)$.

3. Graph the functions $y = 2 \sin 2x$, $y = \sin(2x - \pi)$, and their sum. Find the period and range of each function.

Graph each function. Show at least one period.

1. $y = 3 \sin\left(x - \dfrac{\pi}{2}\right)$ 2. $y = 2 \cos\left(x + \dfrac{\pi}{2}\right)$ 3. $y = 4 \sin \dfrac{\pi}{4}(x + 5)$ 4. $y = 10 \cos \dfrac{\pi}{6}(x - 7)$

5. $y = \sin 3\left(x + \dfrac{\pi}{9}\right)$ 6. $y = 1.5 \cos 2\left(x - \dfrac{3\pi}{4}\right)$ 7. $y = -2 \sin \pi\left(x - \dfrac{3}{4}\right)$ 8. $y = -5 \cos \dfrac{\pi}{2}(x + 2.5)$

Write a sine function for each graph.

9.
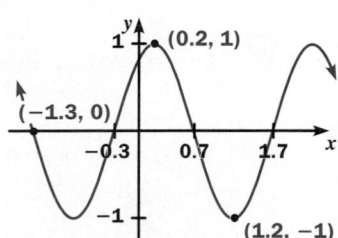

10.
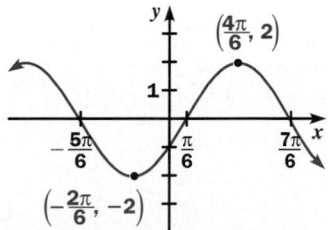

11.
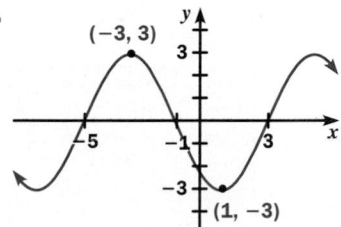

12. **Open-ended Problem** Write a cosine function for one of the graphs in
Exercises 9–11.

Write a cosine function for each graph.

13.
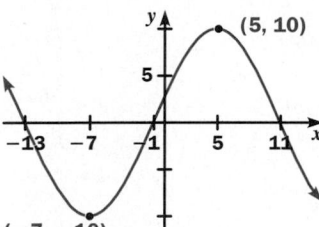

14.
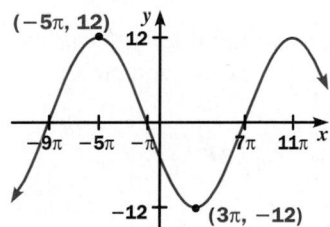

15.
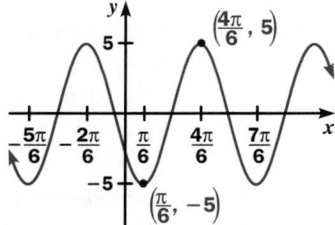

16. **Open-ended Problem** Write a sine function for one of the graphs in
Exercises 13–15.

17. **AUTOMOTIVE MECHANICS** The pistons in an engine force the crank pins
to rotate in a circle around the center of the crankshaft. The graphs show the
heights of crankpins 1, 2, and 3 relative to the axle as a function of time.

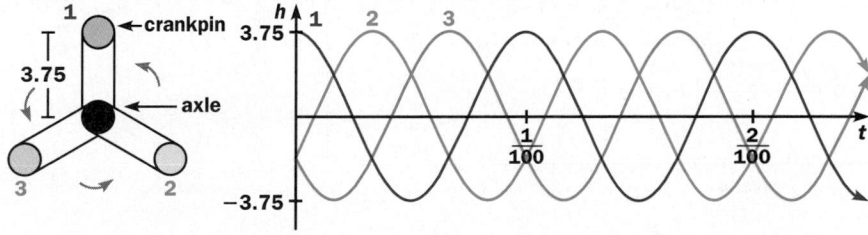

a. Explain why the periods of the graphs are $\dfrac{1}{100}$ s if the engine speed is
6000 rev/min.

b. Write a cosine function for the height of crank pin 1.

c. Use phase shifts to write equations for crank pins 2 and 3.

Architecture protects people from environments in which they would not survive unsheltered. An igloo, for instance, maintains inside air temperatures near or above freezing while outside temperatures are well below freezing. An adobe house protects its inhabitants from extreme heat in the daytime (by absorbing the sun's energy) as well as from uncomfortably low temperatures at night (by radiating the absorbed energy).

The diagrams below show the temperature patterns over a 24-hour period inside and outside an igloo and an adobe house.

Igloo Temperatures

18. Suppose t represents hours after midnight. Explain why the temperature outside the igloo, T_o, can be modeled by this equation:

$$T_o = -20 + 10 \sin \frac{\pi}{12}(t - 8)$$

19. Write sine functions like the one in Exercise 18 for T_c, T_s, and T_f, the temperatures at the ceiling, sleeping platform, and floor of the igloo. How are the equations alike? How are they different?

Adobe House Temperatures

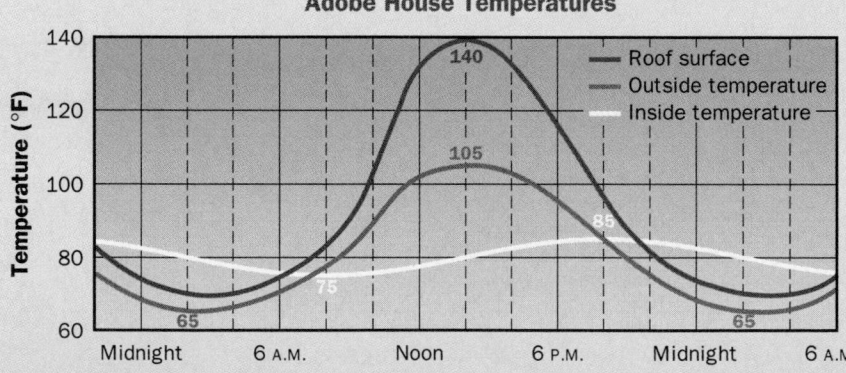

20. Write sine functions to model the temperatures inside and outside the adobe house.

21. Explain why the functions in Exercise 20 have different phase shifts.

22. Writing Explain why a sine function is not as good a model for the adobe house roof temperature as it is for the inside temperature.

23. Investigation Draw a graph of $y = \sin x$ using the domain $-4\pi \le x \le 4\pi$. On a piece of tracing paper, draw a graph of $y = \cos x$ using the same domain. Place the cosine graph over the sine graph and shift it until the two graphs coincide. Using different phase shifts, write three cosine functions that are equivalent to $y = \sin x$.

 Technology In the same coordinate plane, graph each pair of functions and their sum. Find the period and range of each function.

24. $y = \sin x$
$y = 2 \sin x$

25. $y = 3 + \cos x$
$y = -3 + \cos x$

26. $y = 5 \sin \frac{1}{2}x$

$y = -2 \sin \frac{1}{2}x$

27. $y = 2 + \sin \pi x$
$y = 4 + \sin \frac{\pi}{2}x$

28. $y = 1 + \sin \pi x$
$y = 3 + \sin \frac{\pi}{2}\left(x - \frac{1}{2}\right)$

29. $y = 5 + 2 \cos \frac{2\pi}{3}x$
$y = -5 + 2 \cos \frac{\pi}{2}x$

30. a. **Technology** Use a graphing calculator or graphing software to graph $y = (\cos x)^2$, $y = (\sin x)^2$, and $y = (\cos x)^2 + (\sin x)^2$ in the same coordinate plane.

b. Explain why what you saw in part (a) is reasonable, using what you know about the coordinates of points on the unit circle.

31. MUSIC If you simultaneously play two musical notes that are very close in frequency (measured in cycles per second), you may hear "beats," which are loud-soft pulsations in the sound. The diagram shows how adding two sounds with frequencies of 15 cps and 10 cps creates a wave form that has 5 beats per second.

a. **Technology** Copy and complete the table. Use a graphing calculator or graphing software with this viewing window: $-0.5 \le x \le 0.5$ and $-7 \le y \le 7$.

15 cps
Loud
5 beats per second
soft
10 cps

Notes	Equations	Sum	Beats/sec
15 cps 10 cps	$y = \sin 30\pi x + 5$ $y = \sin 20\pi x - 5$	$y = \sin 30\pi x + \sin 20\pi x$	5
14 cps 10 cps	$y = \sin 28\pi x + 5$ $y = \sin 20\pi x - 5$	$y = \sin 28\pi x + \sin 20\pi x$	?
13 cps 10 cps	$y = \sin 26\pi x + 5$ $y = \sin 20\pi x - 5$	?	?
12 cps 10 cps	$y = ?$ $y = \sin 20\pi x - 5$	?	?
11 cps 10 cps	$y = ?$ $y = \sin 20\pi x - 5$	?	?

The constants $+5$ and -5 are in the equations to help separate the graphs so that you can see the sum better.

b. What happens to the number of beats per second as two musical notes get closer in frequency? How is the number of beats related to the frequencies of the two notes?

Technology Use a graphing calculator or graphing software to graph each function. Find its maximum value by using trace.

32. $y = \sin x + \cos x$

33. $y = 3 \sin x + 4 \cos x$

34. $y = 4 \sin x + 3 \cos x$

35. $y = 5 \sin x + 12 \cos x$

36. $y = 8 \sin x + 6 \cos x$

37. $y = 2 \sin x + 5 \cos x$

38. Based on your results in Exercises 32–37, write a conjecture about the maximum value of the function $y = a \sin x + b \cos x$. Check your conjecture by testing other values of a and b.

39. Writing Describe how changing the values of a, b, h, and k affects the graph of $y = a \cos b(x - h) + k$. Use drawings in your explanation.

ONGOING ASSESSMENT

40. Cooperative Learning Work in a group of four students.

a. Suppose each of you is in a different seat on the Ferris wheel shown when it begins to turn counterclockwise. It takes 160 s to make one complete turn. Draw a graph of your height above the ground as a function of time.

b. Write a sine function for your graph in part (a). What is the phase shift for your graph compared with an unshifted sine graph?

c. Combine your results with those of the others in your group on a single graph, labeling each graph with an appropriate equation.

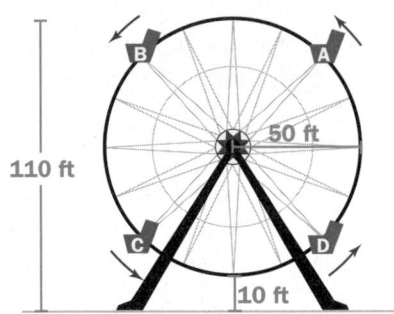

SPIRAL REVIEW

Write a cosine function with the given maximum *M*, minimum *m*, and period. *(Section 15.3)*

41. $M = 8$, $m = 2$, period $= 6$

42. $M = -1$, $m = -5$, period $= \pi$

Find the vertical asymptotes of the graph of each function. *(Section 9.8)*

43. $y = \dfrac{1}{(x - 2)(x - 5)}$

44. $y = \dfrac{5}{(x + 3)(x - 4)}$

45. $y = \dfrac{2}{x^2 - 36}$

Find tan *A* for each right triangle. *(Section 14.2)*

46.

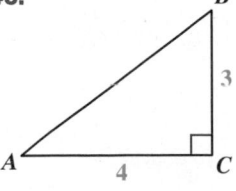

47.

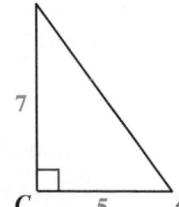

48.

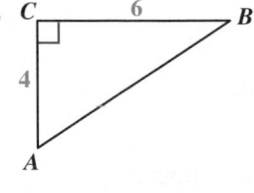

15.5 The Tangent Function

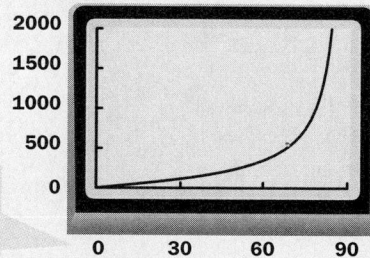

Learn how to...

• graph the tangent function

So you can...

• solve problems, such as finding the height of a balloon

More than two thousand years ago, Chinese children made the first hot air balloons from empty eggshells. They heated the air inside the eggshells by burning small pieces of dried plants placed inside the eggshells. The air inside a modern hot air balloon is heated with a propane burner.

EXAMPLE 1 **Application: Hot Air Ballooning**

A hot air balloon is rising as shown, where h is the height of the balloon (in feet), and θ is the angle between the ground and a line drawn from an observer to the balloon. The balloon is rising above a point 200 ft from the observer.

a. Write an equation for h as a function of θ.

b. Find the value of h when $\theta = 25°$.

c. Graph h as a function of θ.

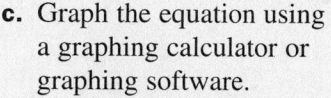

h

θ

200 ft

SOLUTION

a. From Section 14.2 you know that $\tan \theta = \dfrac{h}{200}$. So $h = 200 \tan \theta$.

b. $h = 200 \tan \theta$

$\quad = 200 \tan 25°$

$\quad = 200(0.4663)$

$\quad \approx 93$

> Use a calculator to find a tangent value you don't know.

The balloon is at a height of about 93 ft when $\theta = 25°$.

c. Graph the equation using a graphing calculator or graphing software.

> The domain of the height function is $0° \le \theta < 90°$.

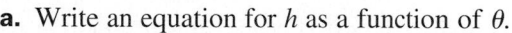

THINK AND COMMUNICATE

1. What is the balloon's height when $\theta = 0°$? Why does this make sense?

2. What happens to the height of the balloon as θ approaches 90°? How is this shown in the graph?

As with sine and cosine, you can define **tangent** on the unit circle. The domain of the tangent function can be extended to all angles except those for which tangent is undefined, and the angles can be expressed in degrees or radians.

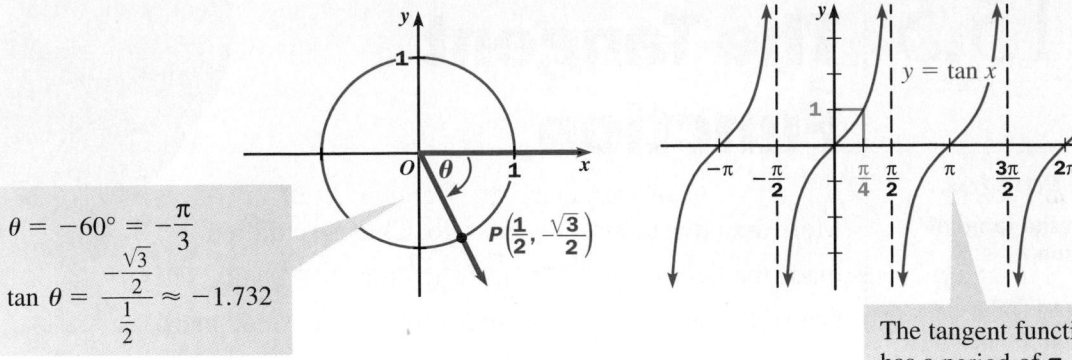

$\theta = -60° = -\dfrac{\pi}{3}$

$\tan \theta = \dfrac{-\dfrac{\sqrt{3}}{2}}{\dfrac{1}{2}} \approx -1.732$

The tangent function has a period of π.

The tangent function is undefined at odd multiples of $\dfrac{\pi}{2}$ (that is, at $x = \dfrac{(2n + 1)\pi}{2}$, where n is an integer). The graph of $y = \tan x$ has vertical asymptotes at these x-values.

THINK AND COMMUNICATE

3. Explain why $y = \tan x = \dfrac{\sin x}{\cos x}$ is undefined at odd multiples of $\dfrac{\pi}{2}$.

4. Compare the period of the tangent function with the periods of the sine and cosine functions. Why is there a difference?

The graph of the tangent function $y = \tan x$ can be shifted and stretched vertically and horizontally. Examples are shown below.

Vertical stretch: $y = 2 \tan x$

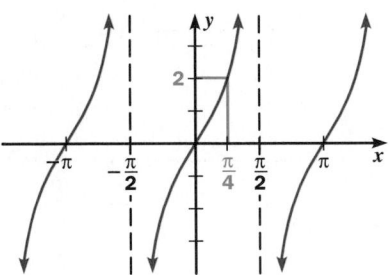

Horizontal stretch: $y = \tan 2x$

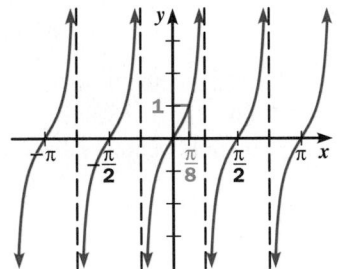

Horizontal shift: $y = \tan \left(x - \dfrac{\pi}{4} \right)$

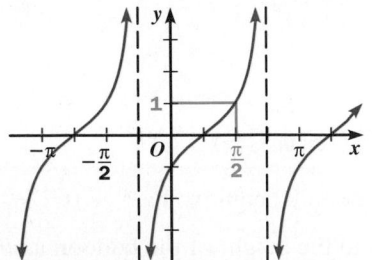

Vertical shift: $y = \tan x + 1$

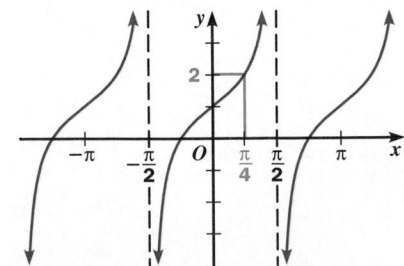

The general form of the equation for a tangent function is $y = a \tan b(x - h) + k$, where a, b, h, and k determine vertical and horizontal stretches and shifts.

EXAMPLE 2

Graph the function $y = 3 \tan \dfrac{x}{2}$.

SOLUTION

The graph will involve a vertical stretch and a horizontal stretch of the basic tangent graph.

Step 1 Find the asymptotes of the graph. You want to know when $\dfrac{x}{2}$ equals odd multiples of $\dfrac{\pi}{2}$.

$$\frac{x}{2} = \frac{(2n + 1)\pi}{2}$$
$$x = (2n + 1)\pi$$

So asymptotes occur when x is an odd multiple of π.

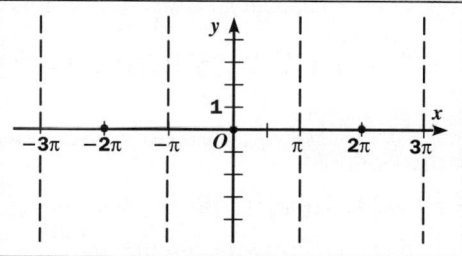

Step 2 The graph is not shifted vertically ($k = 0$), so the x-intercepts of the basic tangent graph are not shifted vertically. The x-intercepts remain halfway between the asymptotes, at even multiples of π.

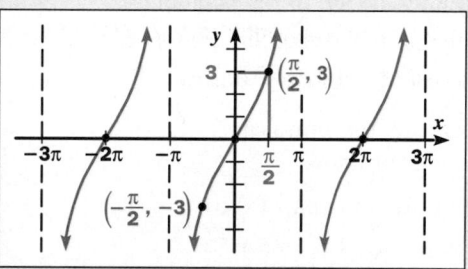

Step 3 The factor 3 stretches the basic tangent graph vertically.

For $y = \tan x$, $y = 1$ when $x = \dfrac{\pi}{4}$, midway between the x-intercept at 0 and the asymptote at $x = \dfrac{\pi}{2}$.

For $y = 3 \tan \dfrac{x}{2}$, $y = 3$ when $x = \dfrac{\pi}{2}$, midway between the

x-intercept at 0 and the asymptote at $x = \pi$.

☑ CHECKING KEY CONCEPTS

Evaluate each expression.

  1. $\tan 38°$ **2.** $\tan (-5°)$ **3.** $\tan 0.75\pi$ **4.** $\tan \left(-\dfrac{2\pi}{3}\right)$

Graph each function.

  5. $y = -\tan x$ **6.** $y = \tan x - 2$ **7.** $y = \tan \left(x + \dfrac{\pi}{3}\right) + \dfrac{1}{2}$

15.5 Exercises and Applications

Extra Practice
exercises on
page 772

Evaluate each expression.

1. $\tan \pi$

2. $\tan \dfrac{\pi}{6}$

3. $\tan \dfrac{11\pi}{6}$

4. $\tan (-30°)$

5. $\tan \dfrac{3\pi}{8}$

6. $\tan \left(-\dfrac{7\pi}{4}\right)$

7. $\tan 520°$

8. $\tan \dfrac{13\pi}{16}$

9. $\tan 0.18$

10. $\tan 95°$

11. $\tan 0$

12. $\tan (-3.10)$

13. **Writing** Explain why the graph of the tangent function has asymptotes, while the graphs of the sine and cosine functions do not.

14. **GEOMETRY** In the diagram, $\overleftrightarrow{AB}$ is tangent to the unit circle at $A(1, 0)$ and $\angle AOB = \theta$.

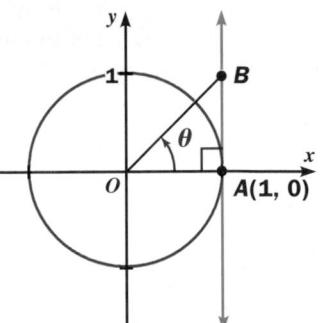

   a. Use trigonometry to express the length of $\overline{AB}$ in terms of θ.

   b. **Writing** Use your answer to part (a) to explain the relationship between the tangent line and the tangent function.

   c. If θ increases steadily, does $\tan \theta$ increase steadily? (*Hint:* Find the tangent of 0°, 10°, 20°, and so on.)

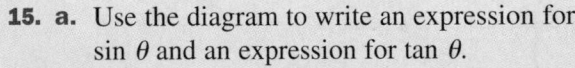

Connection ▶ SPORTS

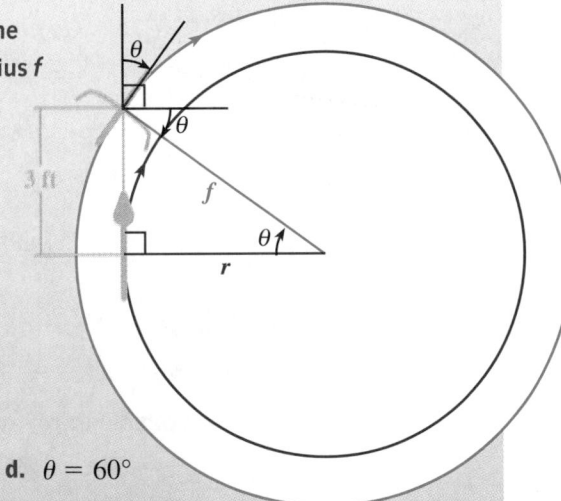

The more you turn the handlebars of your bicycle, the smaller the circle followed by each tire. The diagram shows the turning radius f of the front wheel and the turning radius r of the rear wheel. Suppose the length of the bike frame between the centers of the wheels is 3 ft and the handlebars are turned an angle θ.

15. a. Use the diagram to write an expression for $\sin \theta$ and an expression for $\tan \theta$.

   b. Use the equations you found in part (a) to express f and r as functions of θ. Use these equations for Exercises 16–18.

16. Find the values of f and of r for each turning angle.

   a. $\theta = 0°$ b. $\theta = 30°$ c. $\theta = 45°$ d. $\theta = 60°$

17. **Challenge** For what approximate values of θ will f and r have close to the same value? $\left(\textit{Hint: } \text{Remember that } \tan \theta = \dfrac{\sin \theta}{\cos \theta}.\right)$

18. **Writing** Some trick riders spin their bikes around quickly, with the handlebars at $\theta = 90°$. What are the front and rear turning radii for this value of θ? Why does this make sense? Describe what the bike does.

Graph each function. Show at least one period.

19. $y = 3 \tan \dfrac{x}{4}$ **20.** $y = \tan \dfrac{\pi}{2}(x - 1)$ **21.** $y = \tan 4x - 2$

ONGOING ASSESSMENT

22. Writing Write an equation of a tangent function of the form $y = a \tan bx$. Then graph your equation. Write an explanation of each step you take.

SPIRAL REVIEW

23. Write a cosine function that is equivalent to $y = 2 \sin x$. *(Section 15.4)*

Evaluate each expression. *(Section 15.1)*

24. $\sin 45°$ **25.** $\cos 450°$ **26.** $\cos (-45°)$ **27.** $\sin (-450°)$

ASSESS YOUR PROGRESS

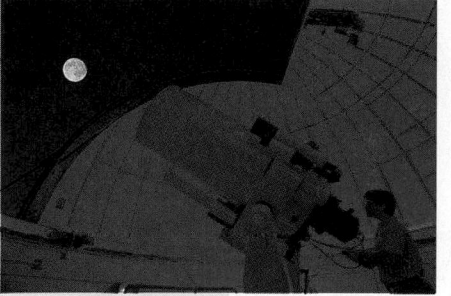

VOCABULARY

phase shift (p. 728) **tangent** (p. 736)

Graph each function. Show at least one period. *(Section 15.4)*

1. $y = -3 \sin \left(x + \dfrac{\pi}{6}\right)$ **2.** $y = \cos \left(x + \dfrac{\pi}{2}\right) - 1$

3. ASTRONOMY In 1725 James Bradley discovered that there is a small shift in the observed position of a star due to the movement of Earth. This *aberration angle* is greatest when the true direction of the star is perpendicular to the direction of Earth's motion around the sun.

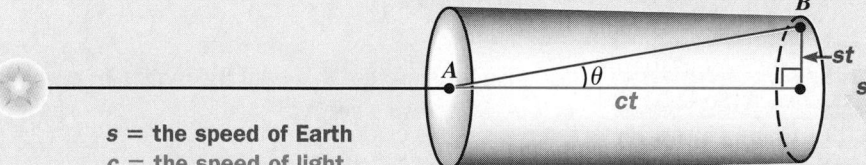

s = the speed of Earth
c = the speed of light

The diagram shows that light enters the telescope at *A*, but while it travels the distance *ct*, the telescope has moved a distance *st*, so the light hits the focal plane at *B*. The telescope must be tilted along the red line to view the star, and the star appears to be shifted by θ from its true position.

a. Use the diagram to write an expression for *s* in terms of *c* and θ.

b. Find *s*, given that $c = 3.00 \times 10^8$ m/s and $\theta = 0.00569°$. *(Section 15.5)*

Graph each function over two periods. *(Section 15.5)*

4. $y = \tan x - 2$ **5.** $y = 2 \tan 2x$ **6.** $y = \tan (2x - \pi)$

7. Journal Consider the three general equations below:

$y = a \sin b(x - h) + k$ $y = a \cos b(x - h) + k$ $y = a \tan b(x - h) + k$

Explain what effect the values of *a*, *b*, *h*, and *k* have on the graph of each function.

Modeling Ups and Downs

Suppose a weight is attached to a spring, pulled down, and released. As you can see from the diagram below, the position of the weight as it moves up and down is a periodic function of time.

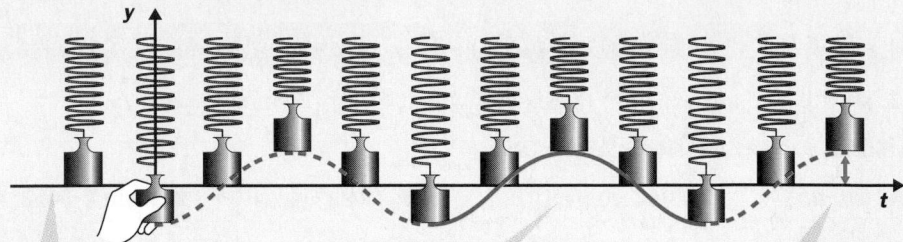

The line $y = 0$ represents the weight's **equilibrium position**, that is, its position before being pulled down and released.

One **oscillation** of the weight is one complete motion up and down.

The **amplitude** of the weight's motion is the maximum displacement of the weight from its equilibrium position.

The motion of the weight is an example of *simple harmonic motion*. An object undergoes simple harmonic motion if its displacement from its equilibrium position can be modeled by a function of the form

$$y = a \cos bt$$

where a and b are constants.

PROJECT GOAL Use a cosine function to model the motion of a weight as it oscillates on a spring.

Experimenting with a Spring

Suspend a weight from a spring, and find an equation modeling the motion of the weight when it is pulled down and released. Work in a group of four students. Your group will need a spring, a paper clip, some masking tape, a 1 lb weight, some string, a meterstick, and a stopwatch.

1. **BEND** a paper clip and tape it to the edge of a desk. Hook one end of the spring to the paper clip and use string to tie the weight to the other end.

2. **TAPE** the meterstick next to the spring and weight as shown.

3. **STRETCH** the spring some distance d, using the meterstick to measure the distance stretched.

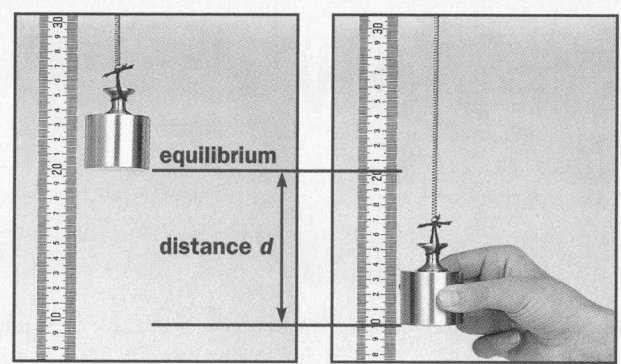

equilibrium

distance d

4. **RELEASE** the weight and use the stopwatch to measure the time it takes the weight to complete five oscillations. Divide this time by 5 to calculate the time of one oscillation.

5. **REPEAT** Step 4 two more times using the same stretch distance d. Find T, the average of your three oscillation times.

Modeling Oscillation

1. What is the relationship between your values of d and T and the values of a and b in the equation $y = a \cos bt$?

2. Write an equation of the form $y = a \cos bt$ for the oscillating weight.

3. Graph your equation from Question 2, and use your graph to predict the time required for the weight to complete four oscillations. Test your prediction.

4. What does your equation assume about the amplitude of the oscillating weight as time passes? Do you think this assumption is accurate?

Writing a Report

Write a report about your experiment. Describe your procedures, and present your data and the equation of your model. Also include these items:

• a graph of your model

• a comparison of your model with your experimental results

• an evaluation of your model's strengths and weaknesses

An equation of the form $y = ae^{-kt} \cos bt$ (where $k > 0$) more accurately models the weight's displacement as a function of time. This is because the factor e^{-kt} *damps*, or decreases, the amplitude of the oscillations as time passes. You can extend your report by finding an equation of the form $y = ae^{-kt} \cos bt$ for your oscillating weight.

Self-Assessment

Describe your understanding of the usefulness of trigonometric functions in modeling. What is it about the graphs of these functions that makes them useful? What are some other types of situations that might be modeled by trigonometric functions?

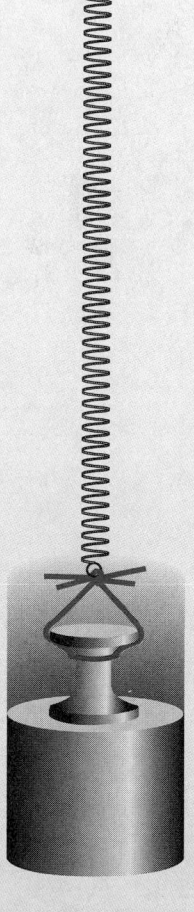

15 Review

Make a test study sheet. Write down definitions, equations, diagrams, or hints that will help you prepare for a test. Even if your teacher does not allow you to use your study sheet during the test, you will benefit from having organized and summarized your notes.

VOCABULARY

sine (p. 708)
cosine (p. 708)
radian (p. 714)
amplitude (p. 720)

cycle (p. 720)
period (p. 720)
phase shift (p. 728)
tangent (p. 736)

SECTIONS 15.1 *and* 15.2

You can measure angles in degrees or **radians**.

$$360 \text{ degrees } = 2\pi \text{ radians}$$

$$1 \text{ degree } = \frac{\pi}{180} \text{ radians}$$

$$\frac{180}{\pi} \text{ degrees } = 1 \text{ radian}$$

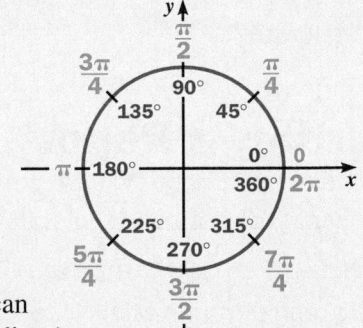

The domain of the **sine** and **cosine** functions can include angles with measures less than 0° (0 radians) or greater than 360° (2π radians).

Graph of $y = \sin \theta$
(θ measured in degrees)

Graph of $y = \cos x$
(x measured in radians)

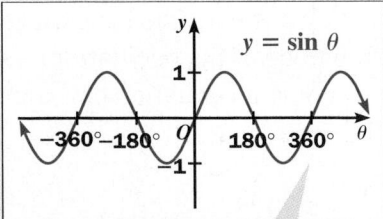

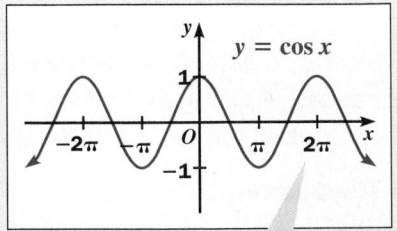

The period is 360°.

The period is 2π.

Functions of the form $y = a \sin b(x - h) + k$ and $y = a \cos b(x - h) + k$ have graphs with **amplitude** $|a|$, **period** $\dfrac{2\pi}{|b|}$, **phase shift** h, and vertical shift k.

Example:

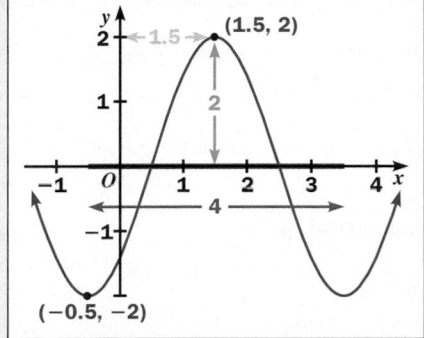

The amplitude is **2**.

$$y = 2 \cos \frac{2\pi}{4}(x - 1.5)$$

The period is **4**.

The phase shift is 1.5.

You can use trigonometric functions to model periodic phenomena, such as the mean monthly temperature in Columbia, South Carolina.

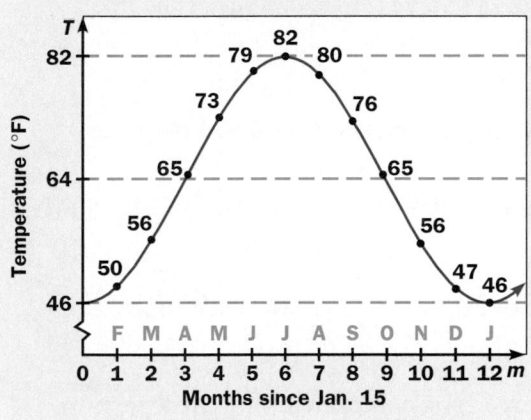

The amplitude is 18.

$$T = -18 \cos \frac{2\pi}{12}m + 64$$

The graph has a vertical shift of 64, because the temperature varies around a mean of 64°F.

The period is 12.

As with the sine and cosine functions, the domain of the **tangent** function can be extended.

The function $y = \tan x$ is undefined when x is an odd multiple of $\dfrac{\pi}{2}$.

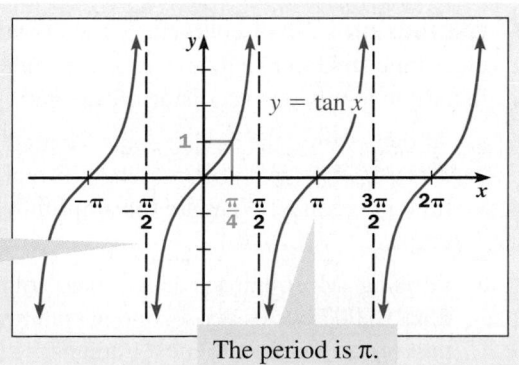

The period is π.

15 Assessment

VOCABULARY QUESTIONS

For Questions 1–3, complete each sentence.

1. There are $360°$ or 2π __?__ in a full circle.

2. The graph of $y = 3 \sin 4(x - 2)$ has a(n) __?__ of 3, a(n) __?__ of $\frac{\pi}{2}$, and a(n) __?__ of 2.

3. The __?__ and __?__ functions have a period of 2π.

SECTIONS 15.1 *and* 15.2

4. The equation $h = 30 + 25 \sin (2° \cdot t)$ gives the height h (in meters) of a Ferris wheel seat as a function of the time t (in seconds) since the ride began. Find h when:

   a. $t = 10$ **b.** $t = 45$ **c.** $t = 90$

Find all angles θ for $0° \le \theta \le 360°$ that satisfy each equation. Give answers to the nearest degree.

   5. $\sin \theta = 0.7315$ **6.** $\cos \theta = -0.7315$ **7.** $\cos \theta = 0.8250$

Evaluate each expression.

   8. $\sin 720°$ **9.** $\cos (-270°)$ **10.** $\sin (-450°)$

   11. $\cos (-\pi)$ **12.** $\sin \pi$ **13.** $\cos 16\pi$

14. **Writing** Explain how to convert an angle measurement from degrees to radians, and from radians to degrees. Show some examples.

SECTIONS 15.3 *and* 15.4

15. **ENGINEERING** To discourage pigeons from nesting in subway stations, electronic soundmaking devices are installed that emit a pitch that is inaudible to humans yet disturbs pigeons.

   a. At one setting, the device emits a sound with the equation $P = 316,200 \sin 44,000\pi t$, where P represents pressure and t represents time in seconds. What are the amplitude and period of the equation's graph?

   b. **Writing** Most humans cannot hear sounds with frequencies greater than 16,000 cycles per second. Explain why the sound described in part (a) is inaudible to most humans.

Write a cosine function with the given maximum *M*, minimum *m*, and period.

16. $M = 7$, $m = -7$, period $= \pi$ **17.** $M = 4$, $m = -6$, period $= 2$

Open-ended Problem Write a sine or cosine function for each graph.

18.

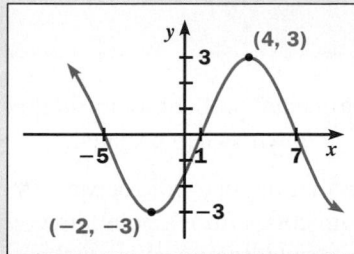

19.

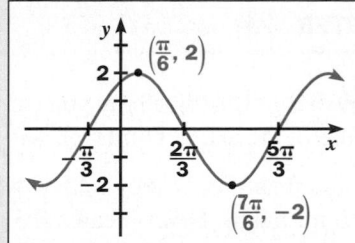

  Technology In the same coordinate plane, graph each pair of functions and their sum. Find the period and range of each function.

20. $y = \sin x$
    $y = 2 \cos x$

21. $y = \cos \frac{1}{2}x$
    $y = 2 \sin x$

22. $y = 3 \sin \pi x$
    $y = \cos 2\pi x$

SECTION 15.5

Graph each function. Show at least two periods.

23. $y = 4 \tan x$ **24.** $y = \tan \pi x$ **25.** $y = 2 \tan \left(x - \frac{\pi}{2}\right)$

26. Open-ended Problem Write a tangent function with period 3π and a phase shift of your choice. Graph your function and show its horizontal intercepts and vertical asymptotes.

27. ASTRONOMY If you push a stick into the ground so it points directly at the sun at noon, then wait a while, the shadow it casts will point east.

   a. Use the diagram to explain why the length l of the shadow cast by a stick of height h is $l = h \tan \theta$.

   b. What is the length of the shadow cast by an 18 in. stick when $\theta = 0°$? $\theta = 30°$? $\theta = 60°$? $\theta = 90°$?

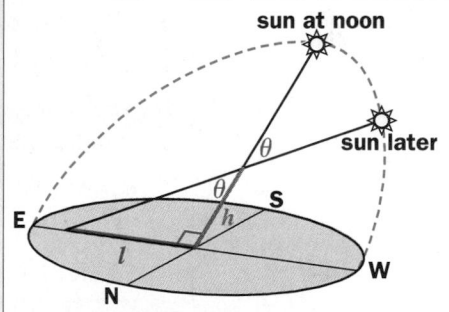

A stick pointed directly at the sun later casts a shadow pointing east.

PERFORMANCE TASK

28. a. Research a phenomenon that can be modeled using a sine function.

   b. Find the maximum value, minimum value, and period of the function. Write an equation for the function. Graph your equation.

   c. Prepare a poster or overhead transparency that explains the phenomenon and shows how your function models it.

Cumulative Assessment
CHAPTERS $13-15$

CHAPTER 13

1. **Writing** Describe how you can use experimental probability to find the likelihood that a thumbtack will land "point down" when dropped.

2. A sunken ship is known to be within a circular region of the ocean with a 10 mi radius. Divers search 0.5 mi in all directions from a point directly below a salvage ship anchored within the circular region. What is the probability that the divers find the sunken ship on the first try?

Suppose there is a 2% chance that an accident will occur in a certain factory during any month. Find the probability of each event.

3. at least one accident occurs during three consecutive months

4. accidents occur in exactly two months during one year

For Questions 5 and 6, suppose that a new machine produces 75% of the rivets made in a factory. Of these rivets, 2% are defective. An older machine produces the rest of the rivets. Of these, 5% are defective. Find the probability of each event.

5. a randomly chosen rivet is from the new machine and is not defective

6. a randomly chosen rivet is from the older machine and is defective

7. There are 8 *bit*s in a computer *byte*. If each bit in a particular byte is equally likely to be on or off, find the probability that 6 or more bits are on.

8. **Open-ended Problem** Sketch two normal distributions that have the same standard deviation but different means.

9. The scores on a quiz are normally distributed with mean 72.5 and standard deviation 7.5. About what percent of the scores are:

   a. over 80?   b. under 92?   c. between 65 and 75?

CHAPTER 14

10. In right $\triangle ABC$, $a = 12$, $b = 35$, and $c = 37$.

    a. Find the sine, cosine, and tangent of $\angle A$ and of $\angle B$. Express your answers in decimal form.

    b. Find the measure of $\angle A$ and of $\angle B$. Express your answers in decimal degrees, to the nearest tenth of a degree, and in degrees, minutes, and seconds, to the nearest second.

11. The wires that support an 800 ft tower extend from the top of the tower to the ground and form 70° angles with the ground. Find the distance from the base of the tower to the point at which a wire is anchored to the ground.

12. **Writing** Explain why the sine and the cosine of an acute angle must be less than 1, but the tangent of an acute angle may be greater than 1. (*Hint:* Sketch a right triangle.)

13. An angle θ is in standard position with the point $P(-\sqrt{2}, -\sqrt{2})$ on its terminal side. Find $\sin \theta$, $\cos \theta$, and $\tan \theta$.

14. Find all angles θ between $0°$ and $360°$ such that $\cos \theta = \dfrac{3}{5}$.

15. **NAVIGATION** A ship traveled on a course of $280°$ for 50 km.

   a. What angle of rotation is equivalent to the ship's course?

   b. How far did the ship travel in an east-west direction? in a north-south direction?

16. **Open-ended Problem** Give an example of the ambiguous case for which no triangle can be formed. Explain why no triangle can be formed.

Solve △*ABC*. If there are two solutions, give both. If there is no solution, explain why.

17. $\angle A = 52°$, $\angle B = 45°$, $a = 8$ 18. $\angle B = 50°$, $b = 16$, $c = 20$

CHAPTER 15

Evaluate each expression.

19. $\tan(-135°)$ 20. $\cos 1.47$ 21. $\sin\left(-\dfrac{4\pi}{3}\right)$ 22. $\sin 150°$

23. Find the maximum and minimum values of the function $y = 2 - \sin 2\theta$. What values of θ for $0° \le \theta \le 360°$ give the maximum and minimum values?

24. Convert $\dfrac{10\pi}{9}$ from radians to degrees (to the nearest degree).

25. Convert $-84°$ from degrees to radians. Express your answer as a decimal to the nearest hundredth of a radian, and as a multiple of π.

Graph each function. Show at least one period. State the amplitude and period of each graph.

26. $y = 3 \sin(-2x)$ 27. $y = -\dfrac{3}{2} \cos \dfrac{\pi}{2}x$ 28. $y = -\cos(x - \pi)$

29. a. Write an equation for a cosine function with period $\dfrac{\pi}{3}$, maximum value 1, and minimum value -5.

   b. **Open-ended Problem** Choose a phase shift to apply to the graph of the equation you wrote in part (a). Write a function that describes the graph after the phase shift.

   c. How is the graph of the function from part (b) different from the graph of the function from part (a)?

30. **Writing** Explain how to use technology to find the period and the maximum value of a function of the form $y = a \sin x + b \cos x$.

31. Graph at least two periods of the function $y = \tan \dfrac{1}{2}\left(x - \dfrac{\pi}{2}\right)$.

Contents of Student Resources

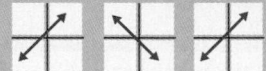

Extra Practice

CHAPTER 1

For Exercises 1–5, use the histogram below, which gives the circulation of Sunday newspapers in the United States in selected years. *Section 1.1*

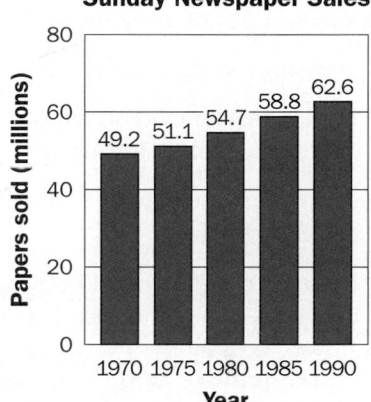

Sunday Newspaper Sales

1. Find the average 5-year increase in Sunday newspaper sales between 1970 and 1990.

2. Find the average 5-year percent increase in Sunday newspaper sales between 1970 and 1990.

3. Using a constant growth model, estimate the Sunday newspaper sales in the years 1995 and 2000.

4. Repeat Exercise 3 using a proportional growth model.

5. Which model do you think provides a more accurate estimate? Give a reason for your answer.

For Exercises 6–8, use the table below, which lists the expenditures for public elementary and secondary schools in the United States from 1985 to 1993.
Section 1.2

Year	1985	1986	1987	1988	1989	1990	1991	1992	1993
Expenditure (billions)	137	149	161	173	193	212	229	243	258

6. **a.** Write a linear function to model the annual expenditures for public elementary and secondary schools from 1985 to 1993. Let $x =$ the number of years since 1985.

   b. Repeat part (a) with an exponential function.

7. **Technology** Use a graphing calculator or graphing software to graph your functions from Exercise 6 along with the data from the table. Does one model fit the data better? If so, which one?

8. Predict the expenditures for public elementary and secondary schools in 2000 and in 2010 using:

   a. the linear function from part (a) of Exercise 6

   b. the exponential function from part (b) of Exercise 6

Use matrices _A_ and _B_ to evaluate each matrix expression. *Section 1.3*

$$A = \begin{bmatrix} 4 & 0 & -1 \\ 2 & -3 & 1 \end{bmatrix} \qquad B = \begin{bmatrix} 2 & -6 & 10 \\ 0 & 8 & -4 \end{bmatrix}$$

9. $A + B$

10. $B - A$

11. $3A$

12. $2A - B$

13. $A - 2B$

14. $A + \frac{1}{2}B$

15. $-4A + 3B$

16. $A - \frac{3}{2}B$

Use matrices *P*, *Q*, *R*, and *S* to find each product. If the matrices cannot be multiplied, state that the product is *undefined*. *Section 1.4*

$$P = \begin{bmatrix} 2 & 3 \\ 0 & -1 \\ 4 & 1 \end{bmatrix} \qquad Q = \begin{bmatrix} 4 & 3 & -2 \\ 1 & 12 & 0 \end{bmatrix} \qquad R = \begin{bmatrix} 3 & 1 & -2 \\ -5 & 0 & 4 \\ -1 & 2 & 6 \end{bmatrix} \qquad S = \begin{bmatrix} 3 & -1 \\ 1 & 4 \end{bmatrix}$$

17. *PQ*

18. *QP*

19. *QR*

20. *RQ*

21. *PS*

22. *SP*

23. *SQ*

24. *RP*

For Exercises 25–28, describe how you would simulate each situation. *Section 1.5*

25. guessing the right answer to a multiple-choice question with 4 choices

26. finding that a randomly chosen student in a school is left-handed, if 1 out of every 10 people is left-handed

27. getting on base in a given at-bat in baseball, if the player gets on base 2 out of every 5 at-bats

28. having a rainstorm start between 12:00 noon and 3:00 P.M., if the storm arrives at some time during a 24-hour day

 Technology **Write a calculator formula that will generate a random number satisfying the given condition.** *Section 1.5*

29. a positive integer less than 7

30. one of the integers 5–12

31. one of the integers 2, 3, 4, 5, and 6

32. one of the integers 0–7

CHAPTER 2

For each equation, tell whether *y* varies directly with *x*. If so, graph the equation. *Section 2.1*

1. $y = \dfrac{x}{2}$

2. $y = 4x - 1$

3. $y = \dfrac{4}{3x}$

4. $y = -\dfrac{4}{5}x$

5. $y = 5x$

6. $y = 3x - 2$

7. $y = 2$

8. $y = 4x - 9$

For each table, write a direct variation equation relating the two variables. Use the equation to find the missing value. *Section 2.1*

9.

x	y
3	4.5
4	6
9	?
14	21

10.

x	y
6	2.7
?	3.6
12	5.4
22	9.9

11.

x	y
6	61.8
15	154.5
24	247.2
32	?

Find the slope-intercept equation of the line passing through each pair of points. *Section 2.2*

12. (1, 4) and (3, 8)

13. (−1, 5) and (1, 11)

14. (6, −1) and (7, −4)

15. (−3, 2) and (5, −10)

16. (−2, 1) and (−4, −9)

17. (−13, −9) and (5, 3)

Graph each equation. *Section 2.2*

18. $y = 4 - 3x$

19. $y = -2 + \frac{3}{2}x$

20. $y = -4$

21. $y = 3 + \frac{1}{3}x$

22. $y = 5 - \frac{3}{4}x$

23. $y = -1 - \frac{2}{3}x$

For Exercises 24–29:

a. Write a point-slope equation of the line passing through the given point and having the given slope.

b. Graph the equation.

c. Write the equation in slope-intercept form. *Section 2.3*

24. point: $(4, 1)$
slope $= -2$

25. point: $(-3, 2)$
slope $= \frac{1}{3}$

26. point: $(5, -6)$
slope $= \frac{3}{4}$

27. point: $(6, -2)$
slope $= 2$

28. point: $(-1, 5)$
slope $= -3$

29. point: $(9, 2)$
slope $= \frac{2}{3}$

Find a point-slope equation of the line passing through each pair of points. *Section 2.3*

30. $(2, 4)$ and $(3, 7)$

31. $(-1, 5)$ and $(4, 3)$

32. $(7, -2)$ and $(5, 4)$

33. $(1, 6)$ and $(-3, -2)$

34. $(9, -5)$ and $(-1, 3)$

35. $(-1, -4)$ and $(-6, 5)$

For Exercises 36–39, use the table below, which lists the number of cable television subscribers in the United States in selected years. *Section 2.4*

Year	1982	1983	1985	1986	1988	1990	1991	1993
Subscribers (millions)	29.3	34.1	39.8	42.2	48.6	54.9	55.8	58.8

36. Make a scatter plot of the data and draw a line of fit.

37. Find an equation of the line. Let $x =$ the number of years since 1982.

38. What does the slope of the line from Exercise 37 represent?

39. Use the equation from Exercise 37 to predict the number of cable television subscribers in the year 2000.

For Exercise 40, use the table below, which lists the 1993 assets and incomes of 8 major life insurance companies in the United States. *Section 2.5*

Assets (billions)	67.5	53.6	51.5	48.4	47.3	44.1	43.7	40.0
Income (billions)	3.2	8.9	6.5	5.3	3.7	5.3	8.4	9.9

40. **Technology** Use a graphing calculator or statistical software to make a scatter plot of the data.

a. **Writing** Is the correlation between a life insurance company's assets and its income strong or weak? Explain.

b. Find the correlation coefficient for the data.

Tell whether you would expect the correlation between the two quantities to be positive, negative, or about zero. *Section 2.5*

41. a student's travel time to and from school and his or her average grade in Spanish class

42. number of hours of daylight in a certain city in a given month and the average high temperature in that city for that month

Graph each pair of parametric equations for the given restriction on *t*. *Section 2.6*

43. $x = 2t$
$y = -t$
$t \le 0$

44. $x = -3t$
$y = t + 1$
$t \ge -1$

45. $x = -2 + t$
$y = -4 + 4t$
$0 \le t \le 3$

46. $x = 1 + \frac{1}{2}t$
$y = 2t$
no restriction on *t*

CHAPTER 3

Write each expression as a power of 2. *Section 3.1*

1. $2 \cdot 2 \cdot 2 \cdot 2 \cdot 2$

2. $8 \cdot 16$

3. $4 \cdot 2^6$

4. $2^3 \cdot 2^7$

Evaluate each expression when *x* = 6. *Section 3.1*

5. $10(2^x)$

6. $320\left(\frac{1}{2}\right)^x$

7. $0.125(2^x)$

8. $1600\left(\frac{1}{2}\right)^x$

Tell whether each equation represents growth that is *linear*, *exponential*, or *neither*. *Section 3.1*

9. $y = 5(2^x)$

10. $y = 5x^2$

11. $y = (5^2)x$

12. $y = 2(5^x)$

Simplify. *Section 3.2*

13. 4^{-3}

14. 5^0

15. 3^{-5}

16. 6^{-4}

Simplify using the properties of exponents. *Section 3.2*

17. $5^{1/3} \cdot 5^{2/3}$

18. $3^{5/2} \cdot 3^{1/2}$

19. $4^{7/2} \cdot 4^{-1/2}$

20. $6^{13/4} \cdot 6^{-5/4}$

21. $\dfrac{25^{1/4}}{25^{3/4}}$

22. $\left(16^{5/2}\right)^{1/5}$

23. $\dfrac{49^{5/2}}{49^2}$

24. $\left(27^{4/3}\right)^{-1/4}$

For each function:

a. Find the *y*-intercept of the graph.

b. Tell whether the graph represents *exponential growth* or *exponential decay*.

c. Sketch the graph. *Section 3.3*

25. $y = 7(1.5)^x$

26. $y = (3.5)^x$

27. $y = 6(0.9)^x$

28. $y = 4^x$

29. $y = 2(0.75)^x$

30. $y = 0.5(1.8)^x$

Sketch the graph of $y = 5 \cdot 2^x$ and its reflection in each axis, and give an equation for each reflection. *Section 3.3*

31. the *y*-axis

32. the *x*-axis

Suppose a bank offers interest compounded continuously. Use the formula $A = Pe^{rt}$ to find the value of \$1000 after 10 years at each interest rate. *Section 3.4*

33. 5%   **34.** 7.5%   **35.** 4.25%   **36.** 9.75%

37. The growth of railways in the United States from 1830 to 1920 can be modeled by the logistic function

$$M(t) = \frac{303.8}{1 + e^{-0.0687(t - 1891.85)}}$$

where $M(t)$ = railway mileage in thousands and t = year. *Section 3.4*

  a. **Technology** Use a graphing calculator or graphing software to graph the function.

  b. Evaluate the function when $t = 1900$.

  c. Find the year when railway mileage reached 162 thousand miles.

Write an exponential function that passes through each pair of points. *Section 3.5*

38. $(0, 6), (1, 8)$ **39.** $(0, 4), (2, 9)$ **40.** $(0, 50), (2, 8)$ **41.** $(2, 10), (3, 25)$

For Exercises 42–45, use the table, which shows the population growth for the city of Austin, Texas, from 1960 to 1990. *Section 3.5*

Year	1960	1970	1980	1990
Population	186,545	253,539	345,496	465,648

42. Use the population values for 1960 and 1970 to write an exponential function to represent the growth. Let x = number of years since 1960.

43. **Technology** Use a graphing calculator or statistical software to perform an exponential regression. What function do you get?

44. Compare your exponential functions from Exercises 42 and 43. Which function gives the closest values to the data for 1980 and 1990?

45. Use the model you chose in Exercise 44 to predict the population of Austin in the year 2000.

CHAPTER 4

For each function:

a. **Graph the function and its inverse in the same coordinate plane.**

b. **Find an equation for the inverse.** *Section 4.1*

1. $f(x) = -4x$ **2.** $g(x) = -2x + 3$ **3.** $y = -5x + 7$ **4.** $h(x) = 3x - 1$

5. $y = -\frac{5}{4}x$ **6.** $y = \frac{2}{5}x$ **7.** $f(x) = -\frac{3}{2}x - 6$ **8.** $y = \frac{1}{4}x + 7$

Write each equation in logarithmic form. *Section 4.2*

9. $5^3 = 125$ **10.** $49^{1/2} = 7$ **11.** $\left(\frac{2}{3}\right)^2 = \frac{4}{9}$ **12.** $\left(\frac{1}{3}\right)^{-1} = 3$

Write each equation in exponential form. *Section 4.2*

13. $\log_{1/2} 8 = -3$ **14.** $\log_{64} 2 = \dfrac{1}{6}$ **15.** $\log_{0.36} 0.6 = \dfrac{1}{2}$ **16.** $\log_{3/2} \dfrac{27}{8} = 3$

Evaluate each logarithm. *Section 4.2*

17. $\log_2 128$ **18.** $\log_{1/3} 81$ **19.** $\ln 1$ **20.** $\log \sqrt{10}$

Write each expression in terms of $\log_5 x$ and $\log_5 y$. *Section 4.3*

21. $\log_5 \dfrac{x^3}{y^2}$ **22.** $\log_5 (xy)^7$ **23.** $\log_5 y\sqrt{x}$ **24.** $\log_5 \dfrac{1}{x^4 y^7}$

Write as a logarithm of a single number or expression. *Section 4.3*

25. $\log_2 x - \log_2 y$ **26.** $3 \log_a 10$ **27.** $2 \log_b 3 + \dfrac{1}{2} \log_b 25$

28. $4 \log_8 p + 3 \log_8 q$ **29.** $\dfrac{2}{3} \ln 8 - 3 \ln 2$ **30.** $6 \log x^2 - 2 \log x^5 - \log x^3$

Solve each equation. Round each answer to the nearest hundredth. *Section 4.4*

31. $2e^{5x} = 12$ **32.** $e^{0.7x} = 15$ **33.** $10^{x+1} = 73$ **34.** $10^{4x} = 88$

35. $6(2.5)^x = 27$ **36.** $\left(\dfrac{3}{2}\right)^x = 20$ **37.** $5e^{2x} - 7 = e^{2x}$ **38.** $12 - 3(10^x) = 1$

Evaluate each logarithm. Round each answer to the nearest hundredth. *Section 4.4*

39. $\log_3 8$ **40.** $\log_7 15$ **41.** $\log_5 28$ **42.** $\log_{12} 0.52$

43. $\log_6 \dfrac{3}{4}$ **44.** $\log_{0.8} 47$ **45.** $\log_{1/4} 3$ **46.** $\log_{5/6} \dfrac{2}{3}$

Write y as an exponential function of x. *Section 4.5*

47. $\log y = 1.8x + 2.4$ **48.** $\log y = 0.3x - 1.5$ **49.** $\ln y = -0.6x + 3.7$ **50.** $\ln y = 4.2x + 0.98$

Write y as a power function of x. *Section 4.5*

51. $\log y = 0.76 \log x - 2.4$ **52.** $\log y = 1.7 - 0.2 \log x$ **53.** $\log y = 5.8 \log x + 0.4$

54. $\ln y = 6.2 \ln x + 3.6$ **55.** $\ln y = -1.09 \ln x + 5.82$ **56.** $\ln y = -4.6 + 2 \ln x$

CHAPTER 5

Match each graph with its equation. *Section 5.1*

1. 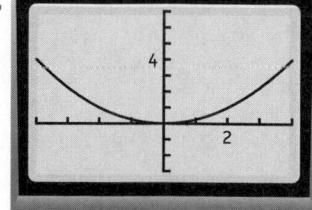   **2.** 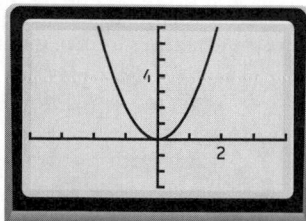   **3.**

A. $y = 2x^2$ **B.** $y = \dfrac{1}{4}x^2$ **C.** $y = 8x^2$

Solve each equation. Give solutions to the nearest tenth when necessary. *Section 5.1*

4. $x^2 = 28$

5. $5x^2 = 45$

6. $-\frac{1}{3}x^2 = -75$

7. $\frac{2}{5}x^2 = 20$

8. $-27x^2 = -3$

9. $3x^2 - 45 = 0$

10. $-6x^2 = -72$

11. $x^2 - 10 = 20$

Each graph has an equation of the form $y = ax^2$. Find a. *Section 5.1*

12.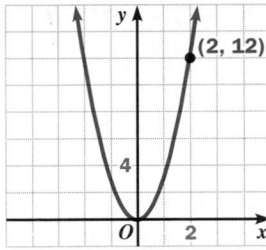

13.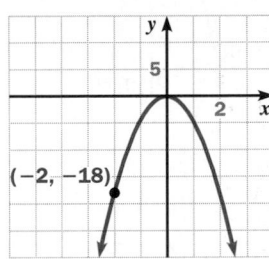

14.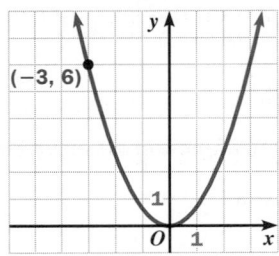

State whether each function has a *maximum value* or a *minimum value*. Then find that value. *Section 5.2*

15. $f(x) = 3(x - 5)^2 - 2$

16. $f(x) = -1.7(x + 11)^2 + 4$

17. $f(x) = 9(x - 3)^2 - 5$

18. $f(x) = -0.6(x - 8)^2 + 13$

19. $f(x) = 5(x + 9)^2 - 10$

20. $f(x) = -0.8(x - 4)^2 + 7$

For each function:

a. Find an equation of the line of symmetry for the graph of the function.

b. Find the coordinates of the vertex.

c. Tell whether the graph *opens up* or *opens down*.

d. Sketch the graph. *Section 5.2*

21. $y = -2(x + 3)^2$

22. $y = \frac{1}{3}(x - 5)^2 + 1$

23. $y = -4(x + 1)^2 + 3$

24. $y = \frac{3}{2}(x - 6)^2 - 5$

25. $y = -\frac{2}{3}(x + 4)^2 - 2$

26. $y = 3(x + 5)^2 - \frac{5}{2}$

Write an equation in the form $y = a(x - h)^2 + k$ for each parabola shown. *Section 5.2*

27.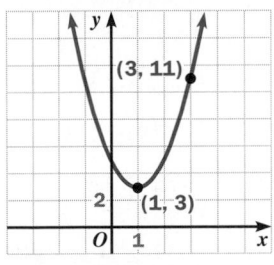

28.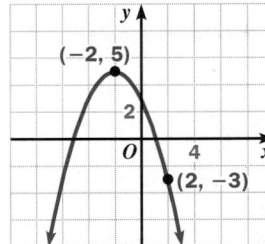

29.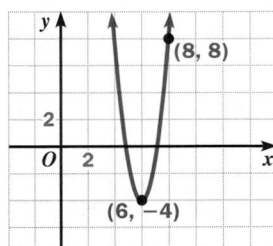

For each equation:

a. Find the *x*-intercept(s).

b. Find the vertex.

c. Sketch the graph. *Section 5.3*

30. $y = (x - 1)(x + 3)$

31. $y = -2(x - 2)(x - 5)$

32. $y = \frac{1}{3}(x + 8)(x - 4)$

33. $y = -1.5(x + 3)(x + 7)$

34. $y = (2x - 6)(x - 2)$

35. $y = (12 - 3x)(x - 2)$

Write an equation for each graph. *Section 5.3*

36.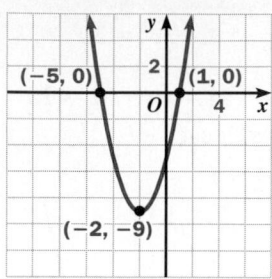

37.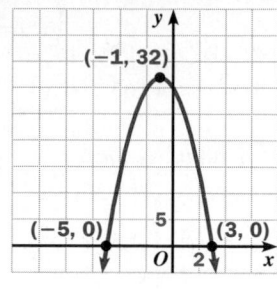

38.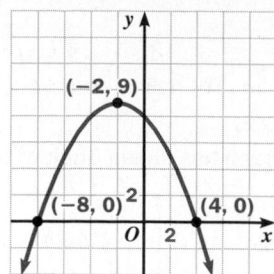

Write each function in vertex form. *Section 5.4*

39. $y = x^2 - 6x + 8$

40. $y = x^2 + 10x + 31$

41. $y = x^2 - 8x$

42. $y = -x^2 + 3x - 1$

43. $y = 2x^2 - 8x + 5$

44. $y = 4x^2 + 6x - 7$

State whether each function has a *maximum value* or a *minimum value*. Then find that value. *Section 5.4*

45. $y = 8x^2 - 4x + 3$

46. $y = -5x^2 - 10x + 2$

47. $y = -x^2 + 6x - 10$

48. $y = 2x^2 + 14x + 5$

49. $y = \dfrac{3}{4}x^2 - 6x$

50. $y = -2x^2 - 8x - 1$

Find the solution(s) of each quadratic equation. Use the quadratic formula. *Section 5.5*

51. $x^2 - 4x - 12 = 0$

52. $2x^2 + 3x - 2 = 0$

53. $5x^2 - 9x + 2 = 0$

54. $3x^2 - 7x + 2 = 0$

55. $2x^2 + x - 5 = 0$

56. $4x^2 + 12x + 9 = 0$

57. $-x^2 + 4x - 1 = 0$

58. $6x^2 - x - 5 = 0$

59. $x^2 + 12x + 8 = 0$

Tell whether each equation has *two solutions*, *one solution*, or *no solution*. *Section 5.5*

60. $x^2 - 3x + 5 = 0$

61. $2x^2 - 7x - 4 = 0$

62. $x^2 + 6x + 3 = 0$

63. $4x^2 - 20x + 25 = 0$

64. $3x^2 - 2x + 5 = 0$

65. $\dfrac{1}{4}x^2 - 3x + 9 = 0$

Write each quadratic function as a product of factors. *Section 5.6*

66. $y = x^2 - 7x - 18$

67. $y = 2x^2 - 9x + 10$

68. $y = 3x^2 - 2x - 5$

69. $y = 25x^2 - 49$

70. $y = 4x^2 - 20x + 9$

71. $y = 5x^2 + 17x - 12$

72. $y = 2x^2 + 19x + 35$

73. $y = 3x^2 + 16x - 12$

74. $y = x^2 - 8x + 12$

Find the *x*-intercept(s) of the graph of each function. *Section 5.6*

75. $y = x^2 - 12x + 35$

76. $y = x^2 + 6x - 40$

77. $y = 2x^2 - 5x + 3$

78. $y = 0.3x^2 + 1.1x - 0.4$

79. $y = 8x - 10x^2$

80. $y = 5x^2 - 17x + 6$

Use factoring and the properties of logarithms to solve for *x*. Be sure to check your answers. *Section 5.6*

81. $\log x + \log (x - 3) = 1$

82. $\log_3 (x - 6) + \log_3 (x + 2) = 2$

83. $\log_4 (3x + 2) + \log_4 x = 2$

84. $\log_2 x + \log_2 (x - 4) = 5$

CHAPTER 6

Tell whether the data that can be gathered about each variable are *categorical* or *numerical*. Then describe the categories or numbers. *Section 6.1*

1. the batting average of a baseball player

2. the country of origin of a car

3. a commuter's mode of transportation

4. a person's annual salary

Each member of the junior class was asked to name his or her favorite subject. The graphs show the results of the survey in two different ways. *Section 6.1*

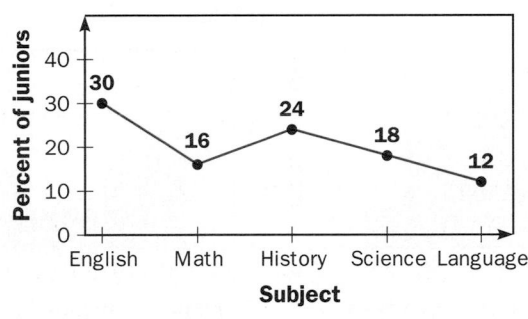
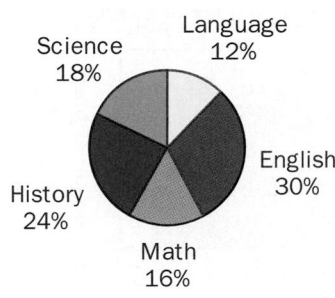

5. Tell which of the two graphs is a better way to display the data. Explain your reasoning.

6. **Open-ended** Display the data using another method. Explain why the graph you created is an appropriate way to organize the data.

For Exercises 7–10, tell why each question may be biased. Then describe what changes you would make to improve the question. *Section 6.2*

7. "Do you think this city should re-elect Mayor Fong, who is in favor of lower taxes and more public recreation facilities?"

8. "Are you in favor of repealing Proposition 17?"

9. "How many times a year do you visit the public library?"

10. "Do you favor building a dam that will bring low-cost electricity to the region?"

For Exercises 11–14, suppose a survey is to be conducted to find out the average number of hours students spend doing homework each night. Identify the type of sample and tell if the sample is biased. Explain your reasoning. *Section 6.3*

11. The survey will be given out to the three homerooms nearest the principal's office.

12. The names of 20 students will be chosen by a computer, which will select students by randomly generating 20 identification numbers.

13. Students will be asked to volunteer to participate in the survey. The first 30 students to volunteer will be selected.

14. Every fourth student on the school computer's list of all students will be selected to participate in the survey.

A magazine surveyed its subscribers to find out their annual incomes and collected the data listed below (in thousands of dollars). Use the data to answer Exercises 15–17. *Section 6.4*

   25, 36, 18, 23.5, 19, 44, 34, 32, 28, 29.5, 24, 26, 29.5, 25, 33, 52, 56, 33, 30, 27

15. Find the median of the data and make a box plot.

16. Find the mean of the data and make a histogram.

17. Is the median or the mean a better measure of the center of the distribution? Explain.

Use the graphs below to answer Exercises 18–20. *Section 6.5*

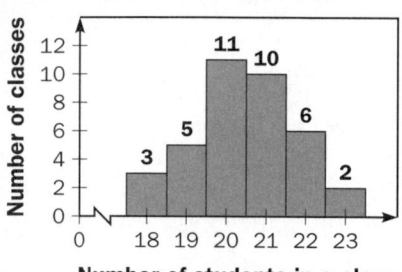
Number of students in a class

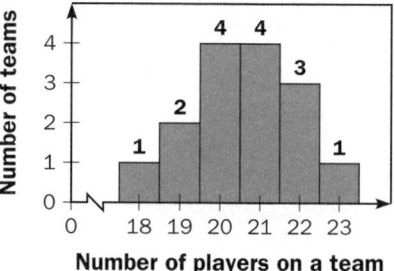
Number of players on a team

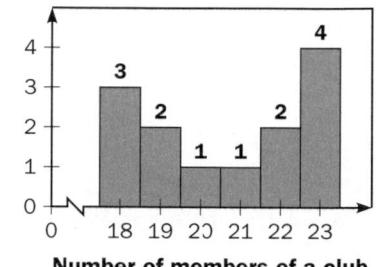
Number of members of a club

18. Find the mean of each data set. Compare the three means.

19. Find the standard deviation of each data set. Compare the standard deviations.

20. For each data set, what percent of the values fall within one standard deviation of the mean? within two standard deviations of the mean?

In a random telephone survey, only 18% of those questioned could name the current United States Secretary of State. *Section 6.6*

21. a. If 2540 people responded to the survey, what is the margin of error?

   b. Find an interval that is likely to contain the proportion of the entire United States population that knows the name of the Secretary of State.

22. Suppose the margin of error of a similar survey is 1.5%, based on the sample size. What is the sample size in this case?

In a poll conducted by a potential advertiser, 324 out of 700 people questioned said they planned to watch the Super Bowl game. *Section 6.6*

23. What percent of the entire population of the United States would be likely to watch the Super Bowl game, based on the sample data?

24. What is the margin of error for this poll?

CHAPTER 7

Solve each system of equations. *Section 7.1*

1. $4x - y = 5$
$2x + y = 13$

2. $-2x + 5y = 9$
$x = y - 3$

3. $6x + 1.4y = 5$
$x + y = -3$

4. $x + 3y = 23$
$5x + 2y = -2$

Technology Use a graphing calculator or graphing software to solve each system of equations. *Section 7.1*

5. $y = x^2 - 5$
$y = 4 - x$

6. $y = 3x - x^2$
$y = 4x - 7$

7. $y = x^2 + 3x - 10$
$y = 1 - 2x$

8. $y = 2(0.6)^x$
$y = -3x + 5$

9. $y = 5(2)^x$
$y = 4x + 8$

10. $y = 4 - x^2$
$y = x^2 - 9$

Solve each system of equations. State whether the system has *one solution*, *infinitely many solutions*, or *no solution*. *Section 7.2*

11. $2x + 3y = 10$
$3x + 2y = 5$

12. $5x - 25y = 8$
$4x - 20y = 8$

13. $4x - 2y = 8$
$6x - 3y = 12$

14. $3x - 11y = 36$
$11x - 3y = 20$

15. $2.5x + 3.2y = 31$
$1.5x - 2.6y = -4$

16. $6x + 15y = 28$
$7x + 17.5y = 24$

Rewrite each system in matrix form, and solve for *x* and *y*. *Section 7.3*

17. $11x - 8y = 48$
$15x - 14y = 33$

18. $1.3x + 2.7y = 18$
$2.5x - 3.5y = 17$

19. $5x + 13y = 40.8$
$9x - 4y = 2.2$

20. $7x + 15y = 103$
$21x - 9y = 65$

21. $9x + y = -13.5$
$1.6x + y = 0$

22. $-3x + y = 7.4$
$-6.2x + y = -19$

Use matrices to solve each system for *x*, *y*, and *z*. *Section 7.3*

23. $3x - 2y + 5z = 11$
$7x + 8y - 4z = 10$
$-5x + y + 2z = 7$

24. $y - 6z = 5$
$8x + 3y = 21$
$5x - y + 2z = 18$

25. $2.4x + 1.8y - 3.5z = 6.8$
$0.6x - 2.7y - 4.8z = 3.1$
$7.3x + 5.6y + 3.4z = 9.2$

Graph each inequality. *Section 7.4*

26. $y < -2x + 7$

27. $y \geq x - 4$

28. $y \leq 5$

29. $y < \frac{1}{2}x - 3$

30. $y \geq 1.4x - 2.6$

31. $y > -\frac{4}{3}x + 3$

32. $3x + 2y \leq 6$

33. $2x - 5y < 8$

Find an inequality that defines each shaded region. *Section 7.4*

34.
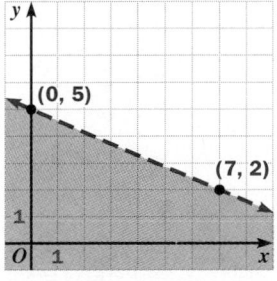

35.
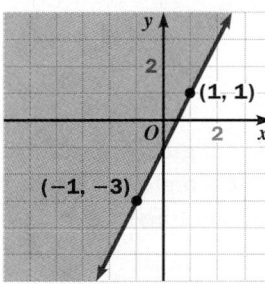

36.
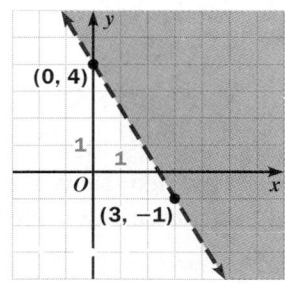

Graph each system of inequalities. *Section 7.5*

37. $y < 3x - 8$
$x < 2$

38. $y \geq -\frac{5}{3}x + 4$
$y < 2x - 7$

39. $y > \frac{1}{2}x + 1$
$y < 5$

40. $y \leq -2x + 9$
$y > \frac{3}{4}x - 2$

Write a system of inequalities defining each shaded region. *Section 7.5*

41.

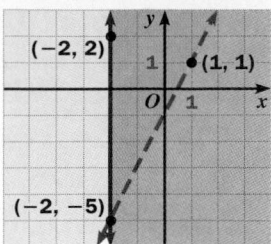

42.

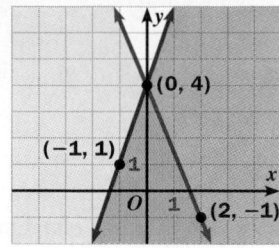

43.

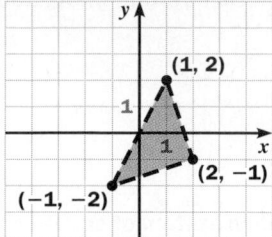

Graph each system of inequalities. *Section 7.5*

44. $y > -3$
$y \le -\frac{1}{2}x + \frac{3}{2}$
$y > x$

45. $y \ge 2x - 9$
$x + y < 3$
$y < 5 - \frac{1}{3}x$

46. $y \le \frac{3}{4}x + 3$
$y > \frac{3}{4}x - 2$
$x < 4$

47. $x + 2y \le 8$
$x - y \ge 5$
$2x + y \ge -5$
$2y - 5x < 8$

48. Ana Lopez wants to mix seaweed meal and mushroom compost to make a fertilizer. These ingredients contain nitrogen (N) and phosphorus (P) as shown in the table. Ana needs a mixture that will contain at least 0.63 kg of N and at least 0.21 kg of P. *Section 7.6*

	Percent N	Percent P
Seaweed meal	2.8	0.2
Mushroom compost	0.7	0.3

a. Let x = number of kilograms of seaweed meal, and let y = number of kilograms of mushroom compost in the mixture. Write a system of inequalities to represent the constraints.

b. Graph the feasible region. Label the corner points.

c. Suppose seaweed meal costs $.40 per kilogram and mushroom compost costs $.15 per kilogram. How much of each ingredient should Ana use in order to minimize the cost of the mixture?

CHAPTER 8

Graph each function. State the domain and range. *Section 8.1*

1. $y = \sqrt{x + 2}$

2. $y = \sqrt{2 - x}$

3. $y = \sqrt{x} - 2$

4. $y = \sqrt{x + 2} - 2$

5. $y = 3\sqrt{x} - 3$

6. $y = \sqrt{4 - x} + 2$

Technology Use a graphing calculator or graphing software to graph the left and right sides of each equation as separate functions. Decide whether each statement is *always true*, *sometimes true*, or *never true*. *Section 8.1*

7. $\sqrt{x^2 - 4} = x - 2$

8. $\sqrt{9x^2} = |3x|$

9. $2\sqrt{x} + 3 = \sqrt{2x + 3}$

10. $\sqrt{25x^6} = 5x^3$

11. $\sqrt{x - 3} = \sqrt{3 - x}$

12. $\sqrt{x^3} = \sqrt{x} \cdot \sqrt{x^2}$

Evaluate each radical expression, or state that it is *undefined*. *Section 8.2*

13. $\sqrt[3]{-1}$

14. $\sqrt{0.0009}$

15. $\sqrt[4]{-16}$

16. $\sqrt[3]{1,000,000}$

17. $\sqrt[5]{-32}$

18. $\sqrt{-\frac{1}{49}}$

19. $\sqrt[4]{\frac{81}{16}}$

20. $\sqrt[3]{\frac{1}{125}}$

State the domain and range of each function. *Section 8.2*

21. $\sqrt{x-5}$

22. $\sqrt[3]{x+4}$

23. $\sqrt[4]{x}-3$

24. $\sqrt[4]{2x+8}$

In Exercises 25–32, assume all variables are restricted to nonnegative values. Simplify each expression. *Section 8.2*

25. $\sqrt{18}$

26. $\sqrt[3]{-250}$

27. $\sqrt[4]{48}$

28. $\sqrt[3]{135}$

29. $\sqrt{5x^8}$

30. $\sqrt[5]{96x^{15}}$

31. $\sqrt{9x^{16}}$

32. $\sqrt[3]{8x^{27}}$

Solve. *Section 8.3*

33. $\sqrt{x}=9$

34. $\sqrt{2x-1}=5$

35. $\sqrt{15-x}=6$

36. $\sqrt{x+11}=8$

37. $\sqrt{\dfrac{2x}{5}}=8$

38. $3+\sqrt{x-2}=6$

39. $\sqrt{\dfrac{x-5}{2}}=7$

40. $\sqrt{2x-3}-1=4$

Solve by raising both sides of each equation to the same power. *Section 8.3*

41. $\sqrt[4]{3x+5}=1$

42. $\sqrt[3]{5-2x}=5$

43. $\sqrt[5]{\dfrac{1}{3}x+8}=2$

44. $\sqrt[4]{4x-7}=3$

Solve using factoring. Check to eliminate extraneous solutions. *Section 8.3*

45. $\sqrt{x+2}=x-4$

46. $\sqrt{3x-8}=x-6$

47. $\sqrt{5x+9}=x-1$

48. $\sqrt{89-2x}=x+5$

49. $\sqrt{7x+15}=x+3$

50. $\sqrt{34-3x}=2x-1$

Perform the indicated operation. *Section 8.4*

51. $(3-7i)+(2+5i)$

52. $(2+3i)-(-5+6i)$

53. $(-11+2i)+(-3-9i)$

54. $(2-5i)(3+8i)$

55. $(7-2i)^2$

56. $(-6+5i)(-6-5i)$

Use complex conjugates to write each quotient in $a+bi$ form. *Section 8.4*

57. $\dfrac{5-2i}{3+i}$

58. $\dfrac{3+8i}{4-i}$

59. $\dfrac{-6+7i}{2+3i}$

60. $\dfrac{9-5i}{3i}$

Solve using the quadratic formula. Check your solutions. *Section 8.4*

61. $x^2-4x+5=0$

62. $x^2+8x+20=0$

63. $-3x^2+6x-30=0$

Plot each pair of complex numbers and their sum in the complex plane. Find their magnitudes. *Section 8.5*

64. $2,-5i$

65. $3+2i,3-2i$

66. $-1+5i,1-3i$

67. $-3-6i,2-i$

68. $5,-2+7i$

69. $4+2i,-3+3i$

Plot each pair of complex numbers and their product in the complex plane. Find their magnitudes. *Section 8.5*

70. $3+i,3-i$

71. $-2+3i,1+i$

72. $-3,4-i$

73. $5-3i,2-i$

A complex number and its *opposite* have a sum of 0. Find the opposite of each complex number. *Section 8.5*

74. $4i$

75. $3+2i$

76. $5-7i$

77. $-3-2i$

For each table, tell whether the given group property holds. If a group property does not hold, give a counterexample. *Section 8.6*

78. commutative

+	0	1	2	3
0	0	1	2	3
1	1	2	3	0
2	2	3	0	1
3	3	0	1	2

79. closure

×	1	2	3
1	1	3	3
2	2	4	6
3	3	6	9

80. identity

+	α	β	γ	δ
α	α	β	γ	δ
β	β	α	δ	γ
γ	γ	δ	α	β
δ	δ	β	γ	α

If possible, give an example of a number that satisfies each description. *Section 8.6*

81. integral and complex

82. whole and imaginary

83. real and rational

CHAPTER 9

Use synthetic substitution to evaluate each polynomial for the given value of the variable. *Section 9.1*

1. $x^3 - 3x^2 + 2x - 5; x = 5$

2. $2x^3 + 4x^2 - 5x + 7; x = -4$

3. $u^4 + 2u^3 - u + 10; u = -2$

4. $-3r^4 + 5r^3 + 12r^2 - 11; r = 3$

Add or subtract, as indicated. *Section 9.1*

5. $(5x^3 - 3x^2 + 4x - 8) + (-2x^3 + 3x^2 - 7x + 15)$

6. $(-4x^5 + x - 10) + (7x^5 + 5x^2 + 4)$

7. $(3x^4 - 5x^3 + 4) - (-4x^4 - 2x^2 + 7)$

8. $(-2x^5 + x^3 - 7x + 8) - (5x^5 - 3x^4 - 5)$

Multiply or divide, as indicated. *Section 9.2*

9. $(5x + 3)(2x - 7)$

10. $(v - 3)(v^2 + 3v + 9)$

11. $(y - 2)(5y^2 - y + 4)$

12. $(n^2 + n - 2)(n^2 - n + 2)$

13. $(a + 2)(a - 1)(a + 1)$

14. $(x - 4)^3$

15. $\dfrac{2x^2 - 5x - 3}{x - 2}$

16. $\dfrac{3k^2 - 5k}{k + 3}$

17. $\dfrac{4c^3 - 3c^2 + c - 1}{c - 3}$

18. $\dfrac{y^3 - 40}{y - 4}$

19. $\dfrac{2b^3 - 9b^2 + 4}{2b - 5}$

20. $\dfrac{4x^3 + 7x^2 - 2x - 8}{x^2 + 3x - 1}$

 Technology Use a graphing calculator or graphing software to graph each polynomial function. For each function:

a. Describe the end behavior using infinity notation.

b. Find all local maximums and local minimums. *Section 9.3*

21. $f(x) = -2x^2 + 8x - 5$

22. $f(x) = -x^4 + 4x^3 - 5$

Find the zeros of each function. *Section 9.4*

23. $f(x) = 3(x + 2)(x - 1)(x - 5)$

24. $f(x) = (2x - 5)(x + 2)(x + 7)$

25. $g(x) = x^3 - 4x^2 + x + 6$

26. $g(x) = x^3 + 5x^2 - 4x - 20$

Find an equation for each cubic function whose graph is shown. *Section 9.4*

27.

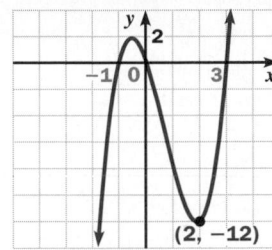

28.

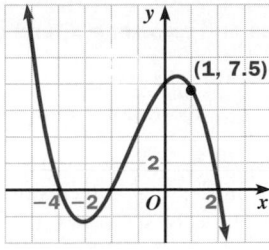

29.

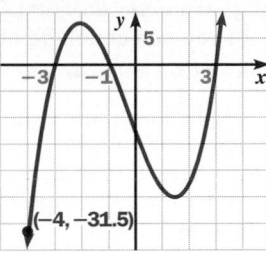

 Technology Use a graphing calculator or graphing software to approximate all real zeros of each function to the nearest hundredth. *Section 9.5*

30. $f(x) = x^3 - 4x^2 - 5x + 3$

31. $f(x) = x^3 - 6x^2 + 5x + 7$

For each function:

a. List the possible rational zeros of the function.

b. Find all real and imaginary zeros of the function. Identify any double or triple zeros. *Section 9.5*

32. $f(x) = 2x^3 - 11x^2 + 17x - 6$

33. $f(x) = 10x^4 - 33x^3 + 33x^2 - 7x - 3$

 Technology Use a graphing calculator or graphing software to graph each equation. For each equation, tell whether *y* varies inversely with *x*. If so, state the constant of variation. *Section 9.6*

34. $y = -\dfrac{6}{x}$

35. $xy = 8$

36. $x = \dfrac{4}{y}$

37. $y = -9x$

Tell whether you think the two quantities show inverse variation. Explain your reasoning. *Section 9.6*

38. the time it takes to drive a fixed distance and the average speed for the trip

39. the number of pounds of pears in a grocery bag and the price of the bag

Find an equation of each hyperbola. *Section 9.7*

40.

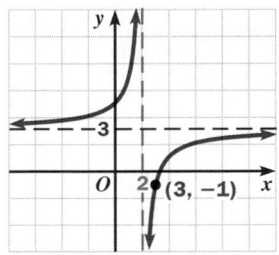

41.

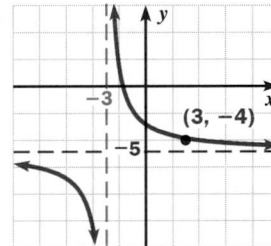

42.

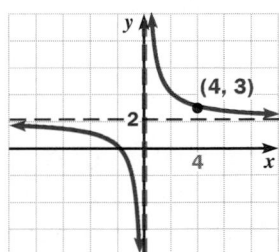

For each function:

a. Find the asymptotes of the function's graph.

b. Tell how the function's graph is related to a hyperbola with equation of the form $y = \dfrac{a}{x}$. *Section 9.7*

43. $y = \dfrac{2x + 4}{x - 3}$

44. $y = \dfrac{-5x - 2}{x + 2}$

45. $y = \dfrac{3x - 6}{x - 5}$

For Exercises 46–48, suppose $f(x) = \dfrac{p(x)}{q(x)}$, where $p(x)$ and $q(x)$ are polynomials.

State what you can conclude about the degrees and zeros of $p(x)$ and $q(x)$ from the given information about the graph. *Section 9.8*

46. has vertical asymptotes at $x = 3$ and $x = -5$ and a horizontal asymptote at $y = 0$

47. has vertical asymptotes at $x = -2$, $x = 2$, and $x = 4$ and has no horizontal asymptote

48. has a vertical asymptote at the y-axis and the line $y = \dfrac{2}{3}$ as a horizontal asymptote

For each function:

a. Find the vertical asymptotes of the function's graph.

b. Describe the function's end behavior using infinity notation. *Section 9.8*

49. $f(x) = \dfrac{-x}{x^2 - 3x - 4}$

50. $f(x) = \dfrac{x^3 + 5}{15x^2 + 2x - 1}$

51. $g(x) = \dfrac{-3x + 8}{x^3 - 2x^2 + x}$

52. $g(x) = \dfrac{6x^3}{x^3 + 4x^2 - x - 4}$

53. $h(x) = \dfrac{x^2 - 3x - 10}{2x^3 - 3x^2 + 6x - 9}$

54. $h(x) = \dfrac{5x^4 + 7}{x^2 - 16}$

Solve each equation. *Section 9.9*

55. $\dfrac{2}{x} + \dfrac{5}{2x} = 3$

56. $\dfrac{4}{x} - \dfrac{3}{x + 1} = 1$

57. $\dfrac{6}{x} - \dfrac{2}{x(x - 3)} = 1$

58. $\dfrac{y + 2}{y - 1} + \dfrac{2}{y + 1} = 3$

59. $\dfrac{4}{n - 2} = \dfrac{4}{(n + 3)(n - 2)}$

60. $\dfrac{2r}{r - 1} + \dfrac{5}{r + 5} = 4$

CHAPTER 10

Find the next 4 terms of each sequence. *Section 10.1*

1. $2, -4, 8, -16, \ldots$

2. $5, 8, 11, 14, \ldots$

3. $11, 101, 1001, 10{,}001, \ldots$

4. $1, 3, 6, 10, \ldots$

5. $50, 5, 0.5, 0.05, \ldots$

6. $-15, -8, -1, 6, \ldots$

7. $1, \dfrac{1}{4}, \dfrac{1}{9}, \dfrac{1}{16}, \ldots$

8. $\dfrac{3}{4}, \dfrac{4}{5}, \dfrac{5}{6}, \dfrac{6}{7}, \ldots$

9. $\dfrac{1}{9}, 9\dfrac{1}{9}, 99\dfrac{1}{9}, 999\dfrac{1}{9}, \ldots$

Write a formula for t_n. *Section 10.1*

10. $\dfrac{1}{5}, \dfrac{3}{10}, \dfrac{9}{15}, \dfrac{27}{20}, \ldots$

11. $17, 13, 9, 5, \ldots$

12. $3, 3\sqrt{2}, 6, 6\sqrt{2}, \ldots$

13. $\dfrac{1}{2}, \dfrac{1}{4}, \dfrac{1}{6}, \dfrac{1}{8}, \ldots$

14. $\sqrt{6}, 3\sqrt{6}, 5\sqrt{6}, 7\sqrt{6}, \ldots$

15. $\dfrac{4}{5}, -\dfrac{8}{5}, \dfrac{16}{5}, -\dfrac{32}{5}, \ldots$

Tell whether each sequence is *arithmetic*, *geometric*, or *neither*. *Section 10.2*

16. $1, 1.2, 1.04, 1.008, \ldots$

17. $-3, 6, -12, 24, \ldots$

18. $\dfrac{1}{3}, 1, \dfrac{5}{3}, \dfrac{7}{3}, \ldots$

19. $5, 43, 81, 119, \ldots$

20. $-\dfrac{4}{5}, -\dfrac{7}{5}, -2, -\dfrac{13}{5}, \ldots$

21. $0.4, 0.44, 0.444, \ldots$

22. $3, 3i, -3, -3i, \ldots$

23. $6, 1.2, 0.24, 0.048, \ldots$

24. $-2, -3.5, -5, -6.5, \ldots$

For each pair of numbers:

a. Find the arithmetic mean.

b. Find the geometric mean. *Section 10.2*

25. $4, 121$

26. $2, 18$

27. $5, 80$

28. $\dfrac{1}{3}, 75$

29. $0.45, 0.05$

30. $7, 29$

31. $2, \dfrac{81}{2}$

32. $6i, 54i$

Write a recursive formula for each sequence. *Section 10.3*

33. $3, 15, 75, 375, \ldots$

34. $16, 9, 2, -5, \ldots$

35. $\dfrac{5}{2}, \dfrac{5}{4}, \dfrac{5}{8}, \dfrac{5}{16}, \ldots$

36. $t_n = -0.5n + 3$

37. $t_n = \dfrac{8}{3} \cdot \dfrac{1}{2^n}$

38. $t_n = 4n - 11$

For each sequence:

a. Find the first 4 terms.

b. Write an explicit formula. *Section 10.3*

39. $t_1 = 0.2$
$t_n = -5t_{n-1}$

40. $t_1 = -10$
$t_n = t_{n-1} + 7$

41. $t_1 = 32$
$t_n = -\dfrac{1}{2}t_{n-1}$

42. $t_1 = 4$
$t_n = i(t_{n-1})$

Find the sum of each series. *Section 10.4*

43. $-1 + 5 + 11 + 17 + \cdots + 89$

44. $38 + 35 + 32 + 29 + \cdots + (-13)$

45. $\dfrac{3}{2} + 5 + \dfrac{17}{2} + 12 + \cdots + 33$

46. $-5.6 + (-4.8) + (-4) + \cdots + 2.4$

Expand each series. *Section 10.4*

47. $\displaystyle\sum_{n=1}^{7} (n + 3)$

48. $\displaystyle\sum_{n=1}^{5} (3n^2 - 6)$

49. $\displaystyle\sum_{n=1}^{6} (5 - 10n^3)$

Write each series in sigma notation and find the sum. *Section 10.4*

50. $9 + 13 + 17 + 21 + \cdots + 149$

51. $-87 + (-62) + (-37) + \cdots + 113$

52. $28 + 25 + 22 + 19 + \cdots + (-41)$

53. $\dfrac{1}{3} + \dfrac{7}{3} + \dfrac{13}{3} + \dfrac{19}{3} + \cdots + \dfrac{91}{3}$

Find the sum of each series. *Section 10.5*

54. $162 + 108 + 72 + 48 + \cdots + \dfrac{128}{9}$

55. $\dfrac{1}{4} + \dfrac{1}{2} + 1 + 2 + \cdots + 64$

56. $\dfrac{5}{4}\sqrt{3} + 5\sqrt{3} + 20\sqrt{3} + \cdots + 1280\sqrt{3}$

57. $250 + (-50) + 10 + (-2) + \cdots + 0.016$

Find the sum of each series, if the sum exists. If a series does not have a sum, explain why not. *Section 10.5*

58. $54 + 36 + 24 + 16 + \cdots$

59. $-6 + 2 + \left(-\dfrac{2}{3}\right) + \dfrac{2}{9} + \cdots$

60. $\dfrac{100}{7} + \dfrac{10}{7} + \dfrac{1}{7} + \dfrac{1}{70} + \cdots$

61. $\displaystyle\sum_{n=1}^{\infty} 4\left(\dfrac{3}{4}\right)^n$

62. $\displaystyle\sum_{n=1}^{\infty} 3(1.1)^n$

63. $\displaystyle\sum_{n=1}^{\infty} 8\left(-\dfrac{3}{5}\right)^n$

CHAPTER 11

For each pair of points, find:

a. the distance between the points

b. an equation of the perpendicular bisector of the line segment connecting the points *Section 11.1*

1. $(2, 3), (8, 9)$
2. $(-2, 5), (6, 1)$
3. $(1, -7), (-5, 11)$
4. $(-1, 9), (-5, 14)$

Find the value(s) of *k* so that the given points are *n* units apart. *Section 11.1*

5. $(7, -3), (-5, k); n = 13$
6. $(4, k), (-8, 3); n = 15$
7. $(k, 1.9), (0.8, 4.3); n = 2.5$

For Exercises 8–10:

a. Find an equation of the parabola with the given characteristics.

b. Graph your equation from part (a). *Section 11.2*

8. vertex $(3, 0)$; directrix $y = -1$
9. vertex $(2, -7)$; focus $\left(\frac{7}{4}, -7\right)$
10. directrix $y = 2$; focus $(5, 4)$

Name the vertex, focus, and directrix of a parabola with the given equation. Sketch the parabola with its focus and directrix. You may want to use a graphing calculator or graphing software to check your work. *Section 11.2*

11. $y = -2x^2$
12. $y = \frac{1}{8}x^2 - 1$
13. $y = -\frac{1}{12}(x + 4)^2 + 3$
14. $x = \frac{1}{2}(y + 3)^2 - 2$

For each equation of a circle, identify the center and the radius. Then graph the equation. *Section 11.3*

15. $(x + 3)^2 + (y - 5)^2 = 16$
16. $x^2 + (y - 4)^2 = 49$
17. $x^2 - 6x + 9 + y^2 = 10$

Write an equation of the circle with the given center *C* and radius *r*. *Section 11.3*

18. $C(2, 0); r = 3$
19. $C(-1, 8); r = \sqrt{7}$
20. $C(3, -4); r = 2\sqrt{3}$

21. $C\left(-\sqrt{5}, -1\right); r = 6$
22. $C(0, k); r = 4$
23. $C(p, q); r = p$

Identify the center, the vertices, and the foci of each ellipse with the given equation. Tell whether the ellipse is *horizontal* or *vertical*. Then graph the equation. *Section 11.4*

24. $\dfrac{x^2}{36} + \dfrac{y^2}{81} = 1$
25. $\dfrac{x^2}{49} + \dfrac{y^2}{25} = 1$
26. $\dfrac{(x + 2)^2}{36} + \dfrac{(y - 5)^2}{100} = 1$

Write an equation of each ellipse. *Section 11.4*

27.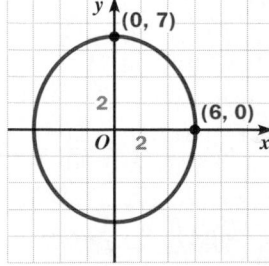

28.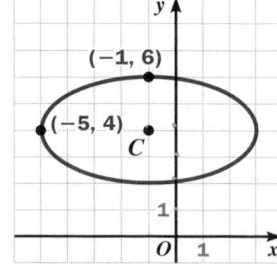

29.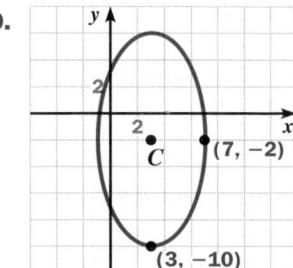

Identify the center, the vertices, and the foci of each hyperbola with the given equation. Tell whether the hyperbola is *horizontal* or *vertical*. Find equations of the asymptotes. Then graph the equation. *Section 11.5*

30. $\dfrac{y^2}{4} - \dfrac{x^2}{25} = 1$

31. $\dfrac{x^2}{49} - \dfrac{y^2}{36} = 1$

32. $\dfrac{(x-4)^2}{36} - \dfrac{(y-1)^2}{81} = 1$

33. $\dfrac{(y-2)^2}{9} - \dfrac{(x+3)^2}{49} = 1$

34. $16(y+4)^2 - (x+1)^2 = 64$

35. $25(x+2)^2 - 9(y-5)^2 = 225$

Write an equation for each hyperbola with the given characteristics. Then sketch its graph. *Section 11.5*

36. asymptotes: $y = \pm\dfrac{3}{4}x$; foci $(0, 10)$ and $(0, -10)$

37. center $(1, 3)$; vertex $(7, 3)$; focus $(10, 3)$

38. center $(2, -4)$; vertex $(2, 1)$; focus $(2, 9)$

For each equation of a conic, rewrite the equation in standard form, identify the conic, and state its important characteristics. *Section 11.6*

39. $4x^2 - 8x - y^2 - 6y - 9 = 0$

40. $x^2 - 8x + 7 + 9y^2 = 0$

41. $x^2 + 4y - 6x + 17 = 0$

42. $x^2 + 4y^2 - 24y - 72 = 0$

Identify and graph each equation of a degenerate conic. *Section 11.6*

43. $x^2 + 9y^2 = 0$

44. $x^2 - 4x + 4 - y^2 = 0$

45. $x^2 - 25y^2 = 0$

46. $y^2 + 2y + 1 + x^2 = 0$

CHAPTER 12

For each graph:

a. **Find the degree of each vertex.**

b. **Draw an equivalent graph.**

c. **Color your graph so that vertices that share an edge are different colors. Try to use as few colors as possible.** *Section 12.1*

1.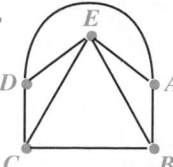

2.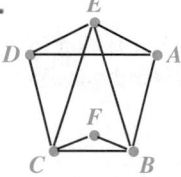

3.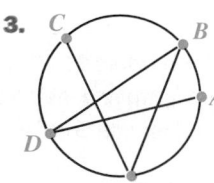

A map of the Rocky Mountain states is shown at the right. Use this map in Exercises 4 and 5. *Section 12.1*

4. Draw a graph with each vertex corresponding to one of the 6 states in the map.

5. Label the vertices of the graph to show how you could color the map with as few colors as possible.

Write a matrix representing each directed graph. Let an element be **1** if the row
vertex has an edge directed to the column vertex. Otherwise let the element be **0**.
Section 12.2

6.

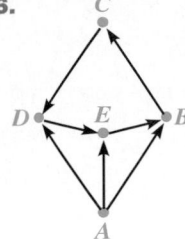

7.

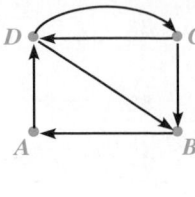

8.

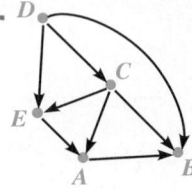

For each matrix, draw a directed graph. Assume that a **1** means that the row vertex
has an edge directed to the column vertex, and a **0** means no such edge exists.
Section 12.2

9.
$$\begin{array}{c} \\ A \\ B \\ C \\ D \end{array} \begin{array}{cccc} A & B & C & D \\ \begin{bmatrix} 0 & 0 & 1 & 0 \\ 1 & 0 & 0 & 1 \\ 1 & 1 & 0 & 0 \\ 1 & 0 & 1 & 0 \end{bmatrix} \end{array}$$

10.
$$\begin{array}{c} \\ A \\ B \\ C \\ D \\ E \end{array} \begin{array}{ccccc} A & B & C & D & E \\ \begin{bmatrix} 0 & 0 & 0 & 0 & 1 \\ 1 & 0 & 0 & 1 & 0 \\ 1 & 0 & 0 & 0 & 1 \\ 0 & 0 & 1 & 0 & 1 \\ 0 & 1 & 0 & 0 & 0 \end{bmatrix} \end{array}$$

11.
$$\begin{array}{c} \\ A \\ B \\ C \\ D \end{array} \begin{array}{cccc} A & B & C & D \\ \begin{bmatrix} 0 & 1 & 0 & 1 \\ 0 & 0 & 1 & 0 \\ 1 & 0 & 0 & 0 \\ 0 & 1 & 1 & 0 \end{bmatrix} \end{array}$$

Calculate each expression. *Section 12.3*

12. $\dfrac{7!}{5!}$

13. $(3!)(6!)$

14. $\dfrac{8!}{2!\,2!\,4!}$

15. $\dfrac{n!}{(n-2)!}$

Simplify each expression. *Section 12.3*

16. $_5P_2$

17. $_7P_3$

18. $_6P_6$

19. $_8P_2$

Give an expression for the number of distinguishable permutations of the letters of
each word. Then evaluate the expression. *Section 12.3*

20. ALMANAC

21. MINIMUM

22. NONSENSE

Simplify each expression. *Section 12.4*

23. $_8C_5$

24. $_9C_3$

25. $_{15}C_2$

26. $_{12}C_8$

A frozen yogurt shop offers **11** different toppings that can be added to a frozen
yogurt sundae. How many possibilities are there if you order the given number of
toppings combined on one sundae? *Section 12.4*

27. one topping

28. two toppings

29. three toppings

For each expression:

a. Find *n* such that $(a + b)^n$ has a term containing the given powers of *a* and *b*.

b. Find the coefficient of the term in the binomial expansion $(a + b)^n$. *Section 12.5*

30. a^5b^3

31. ab^5

32. a^3b^6

33. a^4b^4

Use the binomial theorem to expand each power of a binomial. *Section 12.5*

34. $(x + y)^7$

35. $(x - y)^8$

36. $(x + 3y)^5$

37. $(2x - 3y)^6$

38. $(x^2 + y)^4$

39. $\left(x^3 - y^2\right)^7$

CHAPTER 13

The targets shown consist of a large square divided into smaller, congruent squares. Suppose a randomly thrown dart hits each target. Find the probability that the dart hits the target's shaded region. *Section 13.1*

1.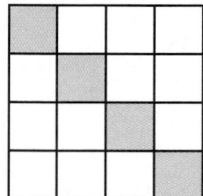

2.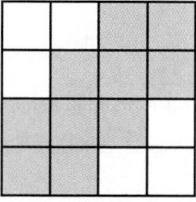

3.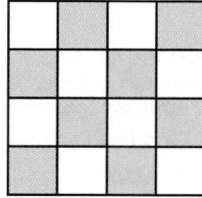

Find the probability of choosing each type of card at random from a standard deck of 52 playing cards. (See page 615 for a description of such a deck.) *Section 13.1*

4. a diamond

5. an ace

6. a black card

7. not a heart

8. not the 5 of clubs

9. a red six

10. not a face card

11. a jack or queen

12. a card neither black nor red

Suppose a six-sided die is rolled and a fair coin is flipped. Find the probability of each event. *Section 13.2*

13. the die shows 6, and the coin comes up heads

14. the die shows 2, 3, or 5, and the coin comes up tails

15. the die does not show 4, and the coin comes up heads or tails

A baseball player's batting average is the probability that the player will get a hit in his or her next official at-bat. Suppose a player batting .250 is followed by a player batting .320, and both have official at-bats. Find the probability of each event. *Section 13.2*

16. both players get hits

17. exactly one of the two players gets a hit

18. at least one player gets a hit

19. neither player gets a hit

Suppose two cards are drawn at random from a standard deck of 52 playing cards. (See page 615 for a description of such a deck.) The first card is *not* put back in the deck before the second card is drawn. Find the probability of each event. *Section 13.3*

20. the first card is a face card, and the second card is not a face card

21. the second card is a heart, given that the first card is a spade

22. the second card is a 10, given that the first card is a queen

23. at least one of the cards is a diamond

24. neither card is a face card

Write a formula to describe the probability distribution for a binomial experiment with *n* trials, each with probability *p* of success. Then find the probability that the experiment will have exactly *k* successful trials. *Section 13.4*

25. $n = 6, p = 0.25; k = 2$
26. $n = 4, p = 40\%; k = 3$
27. $n = 10, p = \frac{4}{5}; k = 5$

Suppose 10% of the population are left-footed kickers. What is the probability that, on a soccer team of 11 players, exactly *k* players are left-footed kickers? *Section 13.4*

28. $k = 2$
29. $k = 5$
30. $k = 0$
31. $k = 11$

The following values are from a normal distribution with mean 24 and standard deviation 5. Standardize each value. *Section 13.5*

32. 29
33. 20
34. 16
35. 35

A group of test scores is normally distributed with mean 500 and standard deviation 100. Find the percent of scores in the group that satisfy each of the given conditions. *Section 13.5*

36. below 650
37. above 700
38. between 390 and 540

CHAPTER 14

For each trigonometric ratio, find θ. Express the answer in two ways:

a. in decimal degrees, to the nearest tenth of a degree

b. in degrees, minutes, and seconds, to the nearest second *Section 14.1*

1. $\sin \theta = 0.5623$
2. $\cos \theta = 0.8920$
3. $\sin \theta = 0.4337$

4. $\cos \theta = 0.0682$
5. $\sin \theta = 0.1706$
6. $\cos \theta = 0.3625$

Find the measure of $\angle A$ in decimal degrees, to the nearest tenth of a degree.
Section 14.1

7.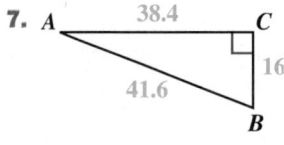
8.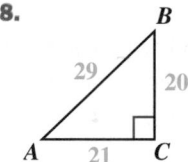
9.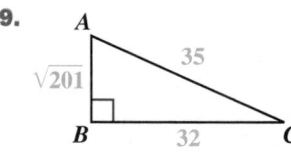

Find each value. *Section 14.2*

10. $\tan 62°$
11. $\tan 47° \, 33'$
12. $\tan 25.6°$
13. $\tan 60°$

Solve $\triangle ABC$ using the given measures. *Section 14.2*

14. $a = 14; b = 23$
15. $c = 17; \angle A = 38°$

16. $\tan B = \frac{4}{3}; c = 42$
17. $\sin A = \frac{7}{25}; b = 18$

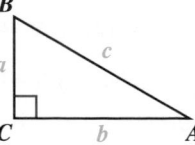

An angle θ is in standard position with the given point *P* on its terminal side. Find $\sin \theta$, $\cos \theta$, and $\tan \theta$. *Section 14.3*

18. $P(6, 2.5)$
19. $P(7, -24)$
20. $P(-4.5, 6)$
21. $P(-6\sqrt{2}, -7)$

For each angle θ, use a calculator to find sin θ, cos θ, and tan θ. *Section 14.3*

22. $\theta = 220°$

23. $\theta = 114°$

24. $\theta = 310°$

25. $\theta = 67°$

26. $\theta = 248°$

27. $\theta = 153°$

For Exercises 28–33, find the area of $\triangle ABC$. If there is not enough information, explain why. *Section 14.4*

28. $a = 14.7, b = 8.5, \angle C = 42°$

29. $b = 24.6, c = 17.3, \angle A = 37°$

30. $b = 13, c = 45.6, \angle B = 17°$

31. $a = 9, c = 11, \angle B = 56°$

32. $b = 25, c = 18, \angle A = 22°$

33. $a = 9.5, b = 16, \angle A = 29°$

Find the area of each sector having central angle θ and radius r. *Section 14.4*

34. $\theta = 90°; r = \dfrac{7}{4}$

35. $\theta = 240°; r = \dfrac{15}{8}$

36. $\theta = 108°; r = 4.5$

37. $\theta = 265°; r = 14$

38. $\theta = 7.5°; r = 8$

39. $\theta = 72°; r = 3.6$

Solve each $\triangle DEF$. *Section 14.5*

40. $d = 15, \angle D = 33°, \angle E = 65°$

41. $e = 26, \angle D = 21°, \angle F = 49°$

42. $f = 27, \angle D = 18°, \angle E = 115°$

43. $d = 8.5, \angle E = 29°, \angle F = 73°$

44. $e = 4\dfrac{2}{5}, \angle E = 127°, \angle F = 15°$

45. $f = 11, \angle D = 34°, \angle F = 67°$

Solve each $\triangle ABC$. If there are two solutions, give both. If there is no solution, explain why. *Section 14.5*

46. $b = 26, c = 17, \angle C = 33°$

47. $a = 19, b = 28, \angle A = 105°$

48. $a = 8, b = 34, \angle A = 40°$

49. $b = 12, c = 9.5, \angle C = 26°$

50. $a = 29, b = 14, \angle B = 30°$

51. $a = 23, c = 12, \angle A = 53°$

Find the missing side length of each triangle. *Section 14.6*

52.

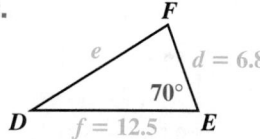

53.

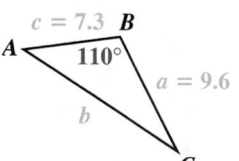

54.

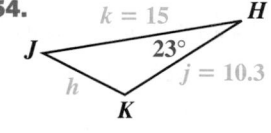

Find the angle measures of each $\triangle XYZ$. *Section 14.6*

55. $x = 17, y = 10, z = 14$

56. $x = 38, y = 24, z = 31$

57. $x = 4.6, y = 5.2, z = 8.3$

58. $x = \dfrac{13}{5}, y = \dfrac{7}{2}, z = \dfrac{25}{8}$

CHAPTER 15

Find all angles θ for $0° \le \theta \le 360°$ that satisfy each equation. Give answers to the nearest degree. *Section 15.1*

1. $\cos \theta = 1$

2. $\sin \theta = 0.4$

3. $\sin \theta = -0.8660$

4. $\cos \theta = 0.5$

5. $\sin \theta = 0.8660$

6. $\cos \theta = 0.7675$

7. $\cos \theta = -0.6250$

8. $\sin \theta = -0.1128$

9. $\cos \theta = 0.2145$

10. $\sin \theta = 0.8164$

11. $\sin \theta = -0.5722$

12. $\cos \theta = -0.1700$

Evaluate each expression. *Section 15.2*

13. $\cos 170°$ **14.** $\sin 380°$ **15.** $\sin(-110°)$ **16.** $\cos(-55°)$

17. $\sin 156°$ **18.** $\cos(-450°)$ **19.** $\cos 230°$ **20.** $\sin(-25°)$

Convert each angle measure from radians to degrees (to the nearest degree). *Section 15.2*

21. $\dfrac{3\pi}{2}$ **22.** $\dfrac{5\pi}{6}$ **23.** 4.2 **24.** $-\dfrac{\pi}{3}$

25. -3.2 **26.** 4π **27.** $\dfrac{3\pi}{8}$ **28.** -5.6

Convert each angle measure from degrees to radians. Express each answer as a decimal to the nearest hundredth of a radian, and in terms of π. *Section 15.2*

29. $60°$ **30.** $-45°$ **31.** $150°$ **32.** $450°$

33. $195°$ **34.** $226°$ **35.** $-130°$ **36.** $144°$

Graph each function. State the amplitude and the period of each graph. *Section 15.3*

37. $y = 4\cos 2x$ **38.** $y = -2\sin \dfrac{2}{3}x$ **39.** $y = \dfrac{3}{4}\cos 3x$ **40.** $y = -\dfrac{4}{3}\sin\left(-\dfrac{2\pi}{3}x\right)$

Write a sine function with the given maximum M, minimum m, and period. *Section 15.3*

41. $M = 4$, $m = -4$, period $= 2$ **42.** $M = 5$, $m = -1$, period $= \dfrac{2\pi}{3}$

43. $M = 3$, $m = -7$, period $= 5\pi$ **44.** $M = -2$, $m = -5$, period $= \dfrac{1}{4}$

Graph each function. Show at least one period. *Section 15.4*

45. $y = 2\sin\left(x - \dfrac{\pi}{3}\right)$ **46.** $y = -3\cos\dfrac{\pi}{2}(x - 5)$ **47.** $y = \dfrac{5}{4}\sin 2\left(x + \dfrac{2\pi}{3}\right)$ **48.** $y = 1.5\cos\dfrac{\pi}{4}(x + 1)$

Write a sine function for each graph. *Section 15.4*

49.
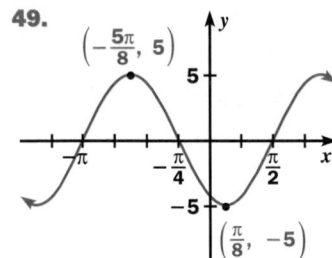

50.
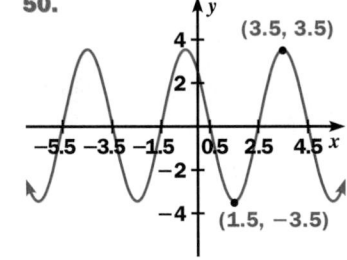

51.
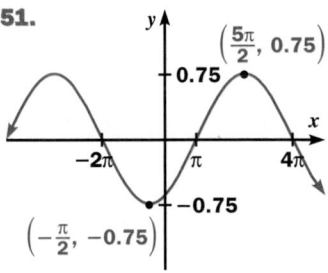

Evaluate each expression. *Section 15.5*

52. $\tan\dfrac{\pi}{4}$ **53.** $\tan 360°$ **54.** $\tan\dfrac{\pi}{3}$ **55.** $\tan 135°$

56. $\tan 38°$ **57.** $\tan\left(-\dfrac{7\pi}{6}\right)$ **58.** $\tan 0.62$ **59.** $\tan\pi$

Graph each function over two periods. *Section 15.5*

60. $y = \dfrac{1}{2}\tan\dfrac{x}{3}$ **61.** $y = 2\tan\dfrac{x}{2} + 3$ **62.** $y = -\dfrac{1}{3}\tan\dfrac{\pi}{4}(x - 2)$

Toolbox

NUMBER OPERATIONS AND PROPERTIES

Numbers and Number Lines

The *real numbers* consist of positive numbers, negative numbers, and zero. You can graph real numbers on a *number line*.

The *coordinate* of point *A* is 3.

Arrows show that the line and the numbers continue without end in both directions.

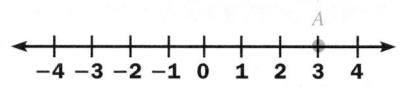

The *graph* of $x = 3$ is point *A*.

The set of real numbers includes each of the following sets of numbers:

whole numbers: 0, 1, 2, 3, . . .

integers: . . . , -3, -2, -1, 0, 1, 2, 3, . . .

positive integers: 1, 2, 3, . . .

negative integers: -1, -2, -3, . . .

The three dots mean that the list continues.

EXAMPLE

Graph the numbers $-2\frac{1}{4}$, -2.5, and -2 on a number line. Then list the numbers in order from least to greatest.

SOLUTION

Numbers on a number line increase from left to right.

From least to greatest, the numbers are -2.5, $-2\frac{1}{4}$, and -2.

PRACTICE

Refer to the number line below. Name the point with each coordinate.

1. -0.5

2. 1

3. $-\dfrac{4}{3}$

4. $\dfrac{9}{4}$

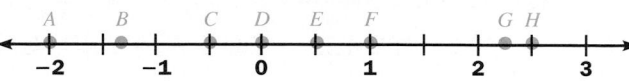

Refer to the number line above. Name the coordinate of each point.

5. *D*

6. *A*

7. *H*

8. *E*

9. Refer to the number line for Exercises 1–8. Name each point whose coordinate is:

    a. positive   **b.** negative   **c.** a whole number   **d.** a real number

Graph each set of numbers on a separate number line. Then list the numbers in order from least to greatest.

10. $\frac{3}{2}, -2, -0.25$ **11.** $2\frac{1}{3}, 3\frac{1}{2}, 1\frac{2}{3}$ **12.** $-\frac{3}{4}, -1.45, -1\frac{1}{4}$

Properties of Real Numbers

The following properties are useful when you are calculating with numbers or simplifying algebraic expressions.

Property	Example	Summary
Commutative $a + b = b + a$ $ab = ba$	$-2 + 3 = 3 + (-2)$ $-2(3) = 3(-2)$	You can add or multiply numbers in any order without changing the result.
Associative $(a + b) + c = a + (b + c)$ $(ab)c = a(bc)$	$(-8 + 6) + 5 = -8 + (6 + 5)$ $(-8 \cdot 6) \cdot 5 = -8 \cdot (6 \cdot 5)$	When you add or multiply three or more numbers, you can regroup the numbers without changing the result.
Distributive $a(b + c) = ab + ac$	$7(60 + 2) = 7(60) + 7(2)$	When a sum is multiplied by a number, you can distribute the multiplication to each of the numbers being added.
Absolute Value $\lvert x \rvert = x$ if $x \geq 0$ $\lvert x \rvert = -x$ (the opposite of x) if $x < 0$	$\lvert 8 \rvert = 8$ $\lvert -8 \rvert = -(-8) = 8$	The absolute value of a number is its distance from 0 on a number line. This distance is always a nonnegative number.

EXAMPLE 1

Tell what property is used in the following: $2(4a) = (2 \cdot 4)a = 8a$.

SOLUTION

The factors have been regrouped. The associative property is used.

PRACTICE

Tell what property is used in each lettered step.

1. a. $7\left(\lvert -3 \rvert + \lvert 5 \rvert\right) = 7(3 + 5)$
   b. $\qquad\qquad = 7(3) + 7(5)$
         $\qquad\qquad = 21 + 35$

2. a. $(4a)(7b) = (4a \cdot 7)b$
   b. $\qquad\quad = (7 \cdot 4a)b$
   c. $\qquad\quad = (7 \cdot 4)ab$
        $\qquad\quad = 28ab$

3. a. $7x + (8 + 3x) = 7x + (3x + 8)$
   b. $\qquad\qquad\quad = (7x + 3x) + 8$
   c. $\qquad\qquad\quad = (7 + 3)x + 8$
          $\qquad\qquad\quad = 10x + 8$

You can show that a statement is false in general if you can find a specific example, called a *counterexample*, for which the statement is false.

EXAMPLE 2

Tell whether each property is true for all real numbers *a* and *b*. Give an example to support your answer.

a. $0 \cdot a = 0$

b. $a \div b = b \div a$

SOLUTION

a. True; the product of any number and 0 is 0; for example, $0 \cdot 7 = 0$.

b. False; for example, $2 \div 10 = 0.2$ and $10 \div 2 = 5$.

PRACTICE

Tell whether each property is true for all nonzero real numbers *a, b, c,* and *d*. Give an example to support your answer.

4. $(a + b)(a - b) = a^2 - b^2$

5. $a^2 + b^2 = (a + b)^2$

6. $a - b = a + (-b)$

7. $a(b - c) = ab - ac$

8. $1 \cdot a = 1$

9. $a \div b = a \cdot \dfrac{1}{b}$

10. $\dfrac{a}{b} + \dfrac{c}{d} = \dfrac{a + c}{b + d}$

11. $\dfrac{a}{b} \cdot \dfrac{c}{d} = \dfrac{ac}{bd}$

12. $|a| - |b| = |a - b|$

13. $0 + a = 0$

Percent Change

Recall that $n\%$ means $\dfrac{n}{100}$. For example, $54\% = \dfrac{54}{100} = 0.54$. To calculate a percent change, use the following formula:

$$\text{percent change} = \frac{\text{amount of change}}{\text{original amount}} \cdot 100$$

EXAMPLE

a. A stock that once sold for $10 per share now sells for $12 per share. Find the percent change in the price.

b. A bracelet, originally priced at $40, is on sale for $30. Find the percent change in the price.

SOLUTION

a. $\text{percent change} = \dfrac{\text{new price} - \text{old price}}{\text{old price}} \cdot 100$

$$= \frac{12 - 10}{10} \cdot 100$$

$$= 20$$

There is a 20% increase in the price.

b. $\text{percent change} = \dfrac{\text{new price} - \text{old price}}{\text{old price}} \cdot 100$

$$= \frac{30 - 40}{40} \cdot 100$$

$$= -25$$

There is a 25% decrease in the price.

Find the percent change. Round answers to the nearest tenth.

1. The Goldsteins bought a house in 1990 for $124,000. They sold the house in 1995 for $128,500.

2. A car purchased for $12,400 is worth only $9810 one year later.

3. The price of lettuce was $.69/lb during the summer and $1.39/lb during the following winter.

Exponents and Powers

The expression b^n is called a *power*. The number b is called the *base*, and the number n is called the *exponent*. The expression means you use b as a factor n times.

This is read "3 to the 4th power."

exponent

base

$$3^4 = 3 \cdot 3 \cdot 3 \cdot 3 = 81$$

4 *factors* of 3

Here are some properties of exponents for $b > 0$:

Property	Summary
Product property: $b^m \cdot b^n = b^{m+n}$	When you multiply powers with the same base, add the exponents.
Quotient property: $\dfrac{b^m}{b^n} = b^{m-n}$	When you divide powers with the same base, subtract the exponents.
Power property: $(b^m)^n = b^{mn}$	When you raise a power to a power, multiply the exponents.
Zero exponent: $b^0 = 1$	When you raise any nonzero base to the 0 power, you get 1.
Negative exponent: $b^{-n} = \dfrac{1}{b^n}$	A nonzero base raised to a negative power is the same as the reciprocal of the base raised to the opposite power.

EXAMPLE

Simplify using properties of exponents.

a. $(-5)^3$

b. $\dfrac{2^8}{2^5}$

c. $\left(x^4\right)^2$

SOLUTION

a. $(-5)^3 = (-5)(-5)(-5)$

$\qquad = -125$

b. $\dfrac{2^8}{2^5} = 2^{8-5}$

$\qquad = 2^3$

$\qquad = 8$

c. $\left(x^4\right)^2 = x^{4 \cdot 2}$

$\qquad = x^8$

Simplify using properties of exponents.

1. 4^3 **2.** $(-2)^5$ **3.** 12^0 **4.** $(-3)^{-4}$

5. $3^2 \cdot 3^3$ **6.** $5^0 \cdot 5^4$ **7.** $2^7 \cdot 2^{-7}$ **8.** $5^{-3} \cdot 5^2$

9. $a^7 \cdot a^3$ **10.** $\dfrac{b^8}{b^3}$ **11.** $c^{-3} \cdot c^{-2}$ **12.** $\dfrac{d^2}{d^5}$

13. $\left(a^2\right)^3$ **14.** $\left(b^{-3}\right)^2$ **15.** $\left(c^5\right)^0$ **16.** $\left(d^{-1}\right)^{-1}$

Square Roots

A number x is a *square root* of a number y if $x^2 = y$. For example, 5 and -5 are the square roots of 25 because $5^2 = 25$ and $(-5)^2 = 25$. Every positive number has two square roots.

25 has two square roots. ⟶ $\sqrt{25} = 5$ and $-\sqrt{25} = -5$ $\sqrt{0} = -\sqrt{0} = 0$ ⟵ 0 has only one square root.

EXAMPLE 1

Simplify.

a. $-\sqrt{81}$

b. $\sqrt{\dfrac{4}{9}}$

SOLUTION

a. $81 = 9^2$, so $-\sqrt{81} = -9$

b. $\dfrac{4}{9} = \left(\dfrac{2}{3}\right)^2$, so $\sqrt{\dfrac{4}{9}} = \dfrac{2}{3}$

Here are two properties of square roots for a and b not both negative and $b \neq 0$:

Property	Summary
Product property: $\sqrt{ab} = \sqrt{a} \cdot \sqrt{b}$	The square root of the product of a and b equals the product of the square roots of a and b.
Quotient property: $\sqrt{\dfrac{a}{b}} = \dfrac{\sqrt{a}}{\sqrt{b}}$	The square root of the quotient of a and b equals the quotient of the square roots of a and b.

EXAMPLE 2

Simplify.

a. $\sqrt{48}$

b. $\sqrt{3} \cdot \sqrt{6}$

c. $\dfrac{\sqrt{3}}{\sqrt{27}}$

SOLUTION

a. $\sqrt{48} = \sqrt{16 \cdot 3}$

$\qquad = \sqrt{16} \cdot \sqrt{3}$

$\qquad = 4\sqrt{3}$

b. $\sqrt{3} \cdot \sqrt{6} = \sqrt{3 \cdot 6}$

$\qquad = \sqrt{3 \cdot 3 \cdot 2}$

$\qquad = 3\sqrt{2}$

c. $\dfrac{\sqrt{3}}{\sqrt{27}} = \sqrt{\dfrac{3}{27}}$

$\qquad = \sqrt{\dfrac{1}{9}}$

$\qquad = \dfrac{1}{3}$

Find the square roots of each number. If the number has no real square roots, write *none*.

1. 49   **2.** -100 **3.** $\dfrac{1}{64}$ **4.** 0.16

Simplify, if possible.

5. $\sqrt{400}$ **6.** $-\sqrt{1.69}$ **7.** $\sqrt{-1}$ **8.** $\sqrt{\dfrac{100}{121}}$

9. $\sqrt{125}$ **10.** $-\dfrac{1}{2}\sqrt{72}$ **11.** $0.7\sqrt{300}$ **12.** $\sqrt{10}\cdot\sqrt{6}$

13. $3\sqrt{2}\cdot4\sqrt{3}$ **14.** $\dfrac{\sqrt{12}}{\sqrt{3}}$ **15.** $\dfrac{\sqrt{6}}{\sqrt{8}}$ **16.** $\dfrac{\sqrt{8}}{\sqrt{18}}$

17. Is the statement $\sqrt{a^2} = a$ true for all values of a? Explain your answer.

Order of Operations

When you simplify a numerical expression, you must use the *order of operations*, a set of rules that guarantees that an expression has just one value.

> **Order of Operations**
>
> **1.** First simplify expressions inside parentheses or other grouping symbols.
> **2.** Then evaluate powers.
> **3.** Next, do multiplications and divisions in order from left to right.
> **4.** Last, do additions and subtractions in order from left to right.

EXAMPLE

Simplify each expression.

a. $3(4 - 6) + 2$ **b.** $\dfrac{1 - 7^2}{2(5 - 1)}$ **c.** $6^3 \div 4 \cdot 5 + 3 - 4 \div 2$ **d.** $\left|-2 \div 5^2\right|$

SOLUTION

a.
$$3(4 - 6) + 2 = 3(-2) + 2$$
$$= -6 + 2$$
$$= -4$$

b. Simplify the numerator and the denominator before dividing.

$$\dfrac{1 - 7^2}{2(5 - 1)} = \dfrac{1 - 49}{2(4)}$$

        The fraction bar acts like parentheses:
$$\left(1 - 7^2\right) \div \left(2(5 - 1)\right)$$

$$= \dfrac{-48}{8}$$

$$= -6$$

c. $6^3 \div 4 \cdot 5 + 3 - 4 \div 2 = 216 \div 4 \cdot 5 + 3 - 4 \div 2$ Evaluate powers.

$$= 54 \cdot 5 + 3 - 2$$

Multiply and divide in order from left to right.

$$= 270 + 3 - 2$$

$$= 273 - 2$$

$$= 271$$

Add and subtract in order from left to right.

d. Simplify the expression inside the absolute value bars first.

$$\left| -2 \div 5^2 \right| = \left| -2 \div 25 \right|$$

$$= \left| -0.08 \right|$$

The absolute value of a negative number is positive.

$$= 0.08$$

Simplify each expression.

1. $9 - 5 - 6 + 1$ **2.** $9 - (5 - 6) + 1$ **3.** $2 \cdot 5 + 7 \cdot 6$ **4.** $2(5 + 7 \cdot 6)$

5. $2(5 + 7) \cdot 6$ **6.** $\left| 2 - 5 \right| + \left| 5 - 2 \right|$ **7.** $9 - 6^2 \div 2 + 4$ **8.** $9 - (6^2 \div 2 + 4)$

9. $(9 - 6^2) \div (2 + 4)$ **10.** $(9 - 6^2) \div 2 + 4$ **11.** $\dfrac{1 - 5^3}{(1 + 3)^2}$ **12.** $\dfrac{7 + 3 \cdot 11}{1 + 8 \div 2}$

Rational Numbers and Irrational Numbers

The numbers that you graph on a number line are *real numbers*. Every real number is either *rational* or *irrational*.

A *rational number* is a number that can be written as a ratio of two integers or as a decimal that repeats or ends.

An *irrational number* cannot be written as a ratio of two integers. When written as a decimal, an irrational number neither repeats nor ends.

Tell whether each number belongs to the *whole numbers*, the *integers*, the *rational numbers*, the *irrational numbers*, and the *real numbers*.

a. $-\sqrt{4}$ **b.** $2.777\ldots$ **c.** $\sqrt{5}$

SOLUTION

a. $-\sqrt{4} = -2$, so $-\sqrt{4}$ is an integer, a rational number $\left(-2 = -\dfrac{2}{1} \right)$, and a real number.

b. $2.777\ldots$ is a decimal that repeats, so it is a rational number and a real number.

c. $\sqrt{5} = 2.236067977\ldots$, which is a decimal that neither repeats nor ends, so it is an irrational number and a real number.

Tell whether each number belongs to the *whole numbers*, the *integers*, the *rational numbers*, the *irrational numbers*, and the *real numbers*. (See page 773.)

1. 3.2

2. $\sqrt{9}$

3. $1.234567891011\ldots$

4. 0

5. π

6. $-\dfrac{1}{4}$

7. $8.535353\ldots$

8. $\sqrt{2}$

Name a real number that fits each description. If there is no such number, write *not possible*.

9. an integer that is not a whole number

10. a real number that is not rational

11. a real number that is not positive or negative

12. a number that is rational and irrational

OPERATIONS WITH VARIABLE EXPRESSIONS

Evaluating Variable Expressions

A *variable expression* is an expression formed using variables, numbers, and operation symbols. To *evaluate* a variable expression, substitute a value for each variable and simplify the resulting expression using the *order of operations*.

EXAMPLE

Evaluate each expression when $a = -5$ and $b = 2$.

a. $\dfrac{a+b}{2}$

b. $a^3 - 2a + 4$

c. $7(a+b)^2 - a$

SOLUTION

a. $\dfrac{a+b}{2} = \dfrac{-5+2}{2}$

$= \dfrac{-3}{2}$ ◁ Simplify the numerator first.

$= -1.5$

b. $a^3 - 2a + 4 = (-5)^3 - 2(-5) + 4$ ◁ Evaluate powers.

$= -125 - 2(-5) + 4$

◁ Multiply and divide.

$= -125 + 10 + 4$

$= -111$ ◁ Add and subtract.

c. $7(a+b)^2 - a = 7(-5+2)^2 - (-5)$

$= 7(-3)^2 - (-5)$ ◁ Simplify inside parentheses first.

$= 7(9) - (-5)$

$= 63 - (-5)$

$= 63 + 5$ ◁ Remember: Subtracting is the same as adding the opposite.

$= 68$

Evaluate each expression when $x = 4$.

1. $\frac{1}{2}x - 4$ **2.** $x^2 - 5x - 2$ **3.** $3(7 - x) + 5$ **4.** $x^3 - 5x^2$

5. $3(x - 6)^2 + 3x$ **6.** $47 + 7.25x$ **7.** $\frac{6}{x - 1}$ **8.** $x + \frac{2x + 7}{3}$

Evaluate each expression when $r = -7$ and $s = 5$.

9. $-4(r + s) - 5$ **10.** $9 - 3r + s^2$ **11.** $9 - 3(r + s^2)$ **12.** $9 - (3r + s^2)$

13. $(r + 2s)^3$ **14.** $-6rs + rs^2$ **15.** $\frac{3r - 4}{2s}$ **16.** $\frac{r^2 + rs}{r + s}$

Simplifying Variable Expressions

For a variable expression like $3x - 2 + x$, the parts that are joined by plus signs or minus signs are called *terms*. Terms with the same variable parts, such as $3x$ and x, are called *like terms*.

To write an expression in *simplest form*, find an *equivalent expression* that has no parentheses and has all like terms combined. In simplest form, the expression $3x - 2 + x$ becomes $4x - 2$.

Simplify.

a. $-(1 - 5x + x^2)$ **b.** $(t^2 - 5t) + (-t^2 + 6t + 3)$

SOLUTION

a. $\begin{aligned} -(1 - 5x + x^2) &= -1(1 - 5x + x^2) \\ &= -1(1) - (-1)(5x) + (-1)(x^2) \\ &= -1 - (-5x) + (-x^2) \\ &= -1 + 5x - x^2 \end{aligned}$ Use the distributive property.

b. $\begin{aligned} (t^2 - 5t) + (-t^2 + 6t + 3) &= (t^2 - t^2) + (-5t + 6t) + 3 \\ &= 0t^2 + (-5 + 6)t + 3 \\ &= t + 3 \end{aligned}$

    Group like terms.
    Use the distributive property.
    Remember: $1 \cdot t = t$.

Simplify.

1. $2(7t - 4) - 8t$ **2.** $-5(k + 1) + 4$ **3.** $-(a - 3b + 2)$ **4.** $-(x^2 + 7) + 7$

5. $3(-4 - 2j) + 3j$ **6.** $12 - 4(-a + 4)$ **7.** $5x + 3y - 4x + y$ **8.** $-a + 2ab - 3b + 7a$

9. $m^2 + 2m + 5 - 3m$ **10.** $(a - 5b) + (4b + 7)$ **11.** $(a - 5b) - (4b + 7)$ **12.** $a - (5b - 4b + 7)$

13. $2(3 - 5x) - (x - 1)$ **14.** $-1.5(4 + 2k) + (3k - 10)$ **15.** $(3y^2 - 4y + 1) + (y^2 + y - 5)$

16. $-(3d^2 - 7d) + 5(1 + d)$ **17.** $(6 - n + n^2) - (n^3 - 7n - 8)$ **18.** $3(c^2 + 5cd + 4d^2) - (c^2 - d^2)$

Multiplying Variable Expressions

To find the product of two variable expressions, use the *distributive property*. (See page 774.)

EXAMPLE

Multiply.

a. $-t(2t - 3)$

b. $(x - 3)(4x + 1)$

SOLUTION

a. $-t(2t - 3) = -t(2t) - (-t)(3)$

$\qquad\qquad\quad = (-1 \cdot 2)(t \cdot t) - (-3t)$

$\qquad\qquad\quad = -2t^2 + 3t$

b. $(x - 3)(4x + 1) = x(4x + 1) - 3(4x + 1)$ Use the distributive property.

$\qquad\qquad\qquad\quad = [x(4x) + x(1)] - [3(4x) + 3(1)]$ Use it again.

$\qquad\qquad\qquad\quad = (4x^2 + x) - (12x + 3)$

$\qquad\qquad\qquad\quad = 4x^2 + x - 12x - 3$

$\qquad\qquad\qquad\quad = 4x^2 - 11x - 3$

PRACTICE

Multiply.

1. $b(2b - 7)$ **2.** $2m(m + 4)$ **3.** $-h(5 - h)$ **4.** $(r + 2)(r - 1)$

5. $(p + 4)(-p + 7)$ **6.** $(s - 4)(s + 4)$ **7.** $(2c + 3)(c + 3)$ **8.** $(5z - 1)(z - 2)$

9. $(8 - 3w)(4 + w)$ **10.** $(3j - 1)(2j - 3)$ **11.** $(9 - 4y)(1 + 3y)$ **12.** $(2k + 5)(5k + 2)$

13. $(a - 5)^2$ **14.** $(m + 3)^2$ **15.** $(2s + 7)^2$ **16.** $(5j - 1)^2$

Factoring Variable Expressions

The distributive property can be read in two ways:

$$a(b + c) = ab + ac \qquad\qquad ab + ac = a(b + c)$$

Distribute a to b and c.   Factor a from ab and ac.

Distributing changes a product into a sum. Factoring changes a sum into a product by finding common factors. Each process is the reverse of the other.

To factor a sum, first find the factors of each addend. Then look for the *greatest common factor* (GCF).

Rewrite each expression in factored form.

a. $12a + 16b$

b. $3z^2 - 5z$

SOLUTION

a. $12a + 16b = (2 \cdot 2 \cdot 3 \cdot a) + (2 \cdot 2 \cdot 2 \cdot 2 \cdot b)$ Factor each term.

$\qquad\qquad = (4 \cdot 3a) + (4 \cdot 4b)$

$\qquad\qquad\qquad\qquad\qquad$ The GCF of the terms is $2 \cdot 2$, or 4.

$\qquad\qquad = 4(3a + 4b)$

b. $3z^2 - 5z = (3 \cdot z \cdot z) - (5 \cdot z)$

$\qquad\qquad = z(3z - 5)$ The GCF is z.

PRACTICE

Rewrite each expression in factored form.

1. $2t + 8$ **2.** $-6m + 9$ **3.** $r^2 - 5r$ **4.** $2v^2 + 7v$

5. $18j + 6k$ **6.** $-9d - 15f$ **7.** $-4g + 15g^2$ **8.** $bc + bd$

9. $21a^2 + 49$ **10.** $5c + 25c^2$ **11.** $-2pt + 3st$ **12.** $20x^2 - 50y^2$

Adding and Subtracting Rational Expressions

A *rational expression* is an expression that can be written as a fraction. If two rational expressions have the same denominator, you can add or subtract them by adding or subtracting the numerators. If the expressions have different denominators, rewrite them using the *least common denominator* (LCD).

EXAMPLE

Simplify.

a. $\dfrac{4}{x - 2} - \dfrac{2x}{x - 2}$

b. $\dfrac{4}{x - 2} + 2$

SOLUTION

a. $\dfrac{4}{x - 2} - \dfrac{2x}{x - 2} = \dfrac{4 - 2x}{x - 2}$ Subtract the numerators.

$\qquad\qquad\qquad = \dfrac{-2(x - 2)}{x - 2}$ Factor the numerator.

$\qquad\qquad\qquad = -2$

b. $\dfrac{4}{x - 2} + 2 = \dfrac{4}{x - 2} + \dfrac{2(x - 2)}{x - 2}$ Rewrite using the LCD $x - 2$.

$\qquad\qquad\qquad = \dfrac{4}{x - 2} + \dfrac{2x - 4}{x - 2}$ Use the distributive property.

$\qquad\qquad\qquad = \dfrac{4 + 2x - 4}{x - 2}$ Add the numerators.

$\qquad\qquad\qquad = \dfrac{2x}{x - 2}$

Simplify.

1. $\dfrac{1}{t} + \dfrac{5}{t}$ 2. $\dfrac{1}{t} - \dfrac{5}{t}$ 3. $\dfrac{2n}{n+1} + \dfrac{3}{n+1}$ 4. $\dfrac{2n}{n+1} - \dfrac{3}{n+1}$

5. $\dfrac{3y}{y-1} - \dfrac{3}{y-1}$ 6. $\dfrac{1}{4k} + \dfrac{3}{4k}$ 7. $\dfrac{5}{x} + \dfrac{3}{2x}$ 8. $\dfrac{5}{x} - \dfrac{3}{2x}$

9. $\dfrac{1}{3} + \dfrac{1}{2x-1}$ 10. $\dfrac{4}{5z} + 6$ 11. $1 - \dfrac{1}{r+2}$ 12. $2 + \dfrac{1}{2x-1}$

13. $\dfrac{c}{c+4} + 1$ 14. $\dfrac{c}{c+4} - 1$ 15. $\dfrac{1}{y} + \dfrac{1}{y+1}$ 16. $\dfrac{2}{m+2} - \dfrac{1}{m-2}$

LINEAR EQUATIONS AND INEQUALITIES

Solving Linear Equations

An *equation* is a mathematical sentence that says that two expressions are equal. A linear equation with one variable can be written in the form $ax + b = c$, where a, b, and c are real numbers and $a \neq 0$.

You *solve* an equation by finding each value of the variable that makes the equation true. To do this, undo operations to get the variable alone on one side. It's always a good idea to check your solutions in the original equation.

EXAMPLE

Solve each linear equation.

a. $\dfrac{x-4}{3} = -2$

b. $7(x+2) = 4x + 20$

SOLUTION

a. $\dfrac{x-4}{3} = -2$

$x - 4 = -6$ ⟵ Multiply both sides by 3.

$x = -6 + 4$ ⟵ Add 4 to both sides.

$x = -2$

Check

$\dfrac{-2-4}{3} \stackrel{?}{=} -2$

$\dfrac{-6}{3} \stackrel{?}{=} -2$

$-2 = -2$ ✔

b. $7(x+2) = 4x + 20$

$7x + 14 = 4x + 20$ ⟵ Use the distributive property.

$3x + 14 = 20$ ⟵ Subtract 4x from both sides.

$3x = 6$ ⟵ Subtract 14 from both sides.

$x = 2$ ⟵ Divide both sides by 3.

Check

$7(2+2) \stackrel{?}{=} 4(2) + 20$

$7(4) \stackrel{?}{=} 8 + 20$

$28 = 28$ ✔

Solve each linear equation.

1. $\frac{1}{8} + k = \frac{7}{8}$

2. $4z = -8$

3. $-12 = -16r$

4. $n - 5 = 2.4$

5. $7a + 3 = -18$

6. $9 - z = z + 5$

7. $\frac{x}{4} - 3 = 1$

8. $2 - \frac{y}{8} = 0$

9. $\frac{v - 2}{9} = 2$

10. $\frac{8 - s}{3} = s$

11. $4(h + 3) = h$

12. $-2(3x + 1) = 4x$

13. $\frac{r}{5} + 7 = 3r$

14. $\frac{3b + 5}{4} = b$

15. $2(y + 5) = 6 - 3y$

16. $\frac{4j}{5} + 7.2 = 5.6$

Ratios and Proportions

When one number is divided by another, the quotient is called a *ratio*. Ratios can be expressed in several equivalent ways:

$$a \div b \qquad \frac{a}{b} \qquad a \text{ to } b \qquad a : b$$

When two ratios are set equal, you get an equation that is called a *proportion*. To solve a proportion, use the *means-extremes property*:

$$\text{If } \frac{a}{b} = \frac{c}{d}, \text{ then } ad = bc.$$

Solve each proportion.

a. $\frac{3}{2x} = \frac{15}{8}$

b. $\frac{x - 1}{x + 4} = 6$

SOLUTION

a. $\frac{3}{2x} = \frac{15}{8}$

$3 \cdot 8 = 2x \cdot 15$

$24 = 30x$

$\frac{24}{30} = x$

$0.8 = x$

b. $\frac{x - 1}{x + 4} = \frac{6}{1}$

$1(x - 1) = 6(x + 4)$

$x - 1 = 6x + 24$

$-1 = 5x + 24$

$-25 = 5x$

$-5 = x$

Solve each proportion.

1. $\frac{x}{4} = \frac{5}{12}$

2. $\frac{9}{x} = \frac{-2}{7}$

3. $\frac{10}{9} = \frac{3x}{5}$

4. $\frac{7}{15} = \frac{21}{5x}$

5. $\frac{4t - 1}{5} = \frac{3}{4}$

6. $\frac{5}{6} = \frac{3k + 7}{8}$

7. $\frac{14}{3} = \frac{x - 6}{2x + 5}$

8. $\frac{12}{3 - r} = \frac{16}{2r + 1}$

Use a proportion to solve each problem.

9. In 1980, Janice Brown flew the first solar-powered aircraft 6 mi in 22 min near Marana, Arizona. During that flight, about how far did she travel in the first 2 min?

10. Carlos Rodriguez bought a 3 lb roast beef on sale for $5.67. If he had bought a $3\frac{1}{2}$ lb roast instead, how much would the roast have cost?

11. On a map of Pennsylvania, 1 in. represents 10 mi. The distance between Philadelphia, Pennsylvania, and Allentown, Pennsylvania, is 49 mi. How far apart are these cities on the map?

Identities and False Statements

Sometimes when you solve an equation you may get a puzzling result: a statement that is always true or a statement that is never true.

EXAMPLE

Solve each linear equation.

a. $7x - 9 = -\frac{1}{2}(18 - 14x)$

b. $7x - 9 = x + 6x$

SOLUTION

a. $7x - 9 = -\frac{1}{2}(18 - 14x)$

$7x - 9 = -9 + 7x$

$-9 = -9$

The equation $-9 = -9$ is an *identity*, because it is always true. When solving an equation leads to an identity, the equation has all real numbers as solutions.

b. $7x - 9 = x + 6x$

$7x - 9 = 7x$

$-9 = 0$

The equation $-9 = 0$ is a *false statement* since $-9 \neq 0$. When solving an equation leads to a false statement, the equation has no solution.

PRACTICE

Solve each equation. If all real numbers are solutions, write *all real numbers*. If there is no solution, write *no solution*.

1. $x - 8 = -(8 - x)$

2. $x - 8 = x + 3$

3. $x - 8 = 4 - x$

4. $2t - 5 = 5 - 2t$

5. $2t - 5 = 5 + 2t$

6. $2t - 5 = 2(t - 2.5)$

7. $3(2a + 3) = 9 + 6a$

8. $3(2a + 3) = 6(a + 1)$

9. $3(2a + 3) = 2(a - 3)$

10. $20 - 8b = 4(2b - 1)$

11. $\frac{3}{4}h + 8 = 2$

12. $\frac{5 + 2r}{2} = r + 6$

13. $23z - 46 = -46$

14. $-18 + 9x = -9(2 - x)$

15. $0.4k + 1 = 0.02(50 + 20k)$

Solving Linear Inequalities

You solve a linear inequality such as $\frac{2}{3}n - 5 > 1$ in much the same way that you solve a linear equation. One important difference is this: When you multiply or divide both sides of an inequality by a negative number, you must reverse the direction of the inequality sign.

EXAMPLE

Solve each inequality. Graph the solution on a number line.

a. $\frac{2}{3}n - 5 > 1$

b. $-3(z + 2) \leq 7 + z$

SOLUTION

a. $\qquad \frac{2}{3}n - 5 > 1$

$\frac{2}{3}n - 5 + 5 > 1 + 5 \qquad$ Add 5 to both sides.

$\qquad \frac{2}{3}n > 6$

$\frac{3}{2} \cdot \frac{2}{3}n > \frac{3}{2} \cdot 6 \qquad$ Multiply both sides by the reciprocal of $\frac{2}{3}$.

$\qquad n > 9$

The solution is all real numbers greater than 9.

The *open* circle shows that 9 is not a solution.

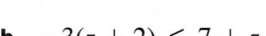

b. $-3(z + 2) \leq 7 + z \qquad$ Use the distributive property.

$\quad -3z - 6 \leq 7 + z \qquad$ Add 6 to both sides.

$\qquad -3z \leq 13 + z$

$\qquad \qquad \qquad$ Subtract z from both sides.

$\qquad -4z \leq 13$

$\quad \frac{-4z}{-4} \geq \frac{13}{-4} \qquad$ Divide both sides by -4. *Reverse* the direction of the inequality sign.

$\qquad z \geq -3.25$

The *solid* circle shows that -3.25 is a solution.

The solution is all real numbers greater than or equal to -3.25.

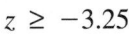

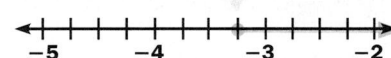

PRACTICE

Solve each inequality. Graph the solution on a number line.

1. $2x < 18$

2. $y - 4 \geq -7$

3. $-4k < -8$

4. $\frac{1}{3}z \leq -1$

5. $2 - p > -3$

6. $r + 7 \leq 5$

7. $11 + 2c > 35$

8. $3 - 7x \leq 17$

9. $1.2d + 1 < 4$

10. $5h - 6 \geq 2h$

11. $-\frac{1}{3}t + 2 < t$

12. $-(s - 3) \leq 3$

13. $4 - 2g > 3g - 6$

14. $4(x + 3) \geq -12$

15. $-6(w - 4) < 2w$

16. $5(1 - m) \leq -10m$

Rewriting Equations and Formulas

Sometimes you need to solve an equation or a formula for a particular variable. For example, if you want to use a graphing calculator to graph the equation $2x + 3y = -6$, you need to solve the equation for y first.

EXAMPLE

Solve $2x + 3y = -6$ for y.

SOLUTION

$$2x + 3y = -6$$

$$3y = -6 - 2x$$ Subtract $2x$ from both sides.

$$\frac{1}{3}(3y) = \frac{1}{3}(-6 - 2x)$$ Multiply both sides by the reciprocal of 3.

$$y = -2 - \frac{2}{3}x$$

PRACTICE

Solve each equation or formula for the indicated variable.

1. $5x - y = 15$ for y **2.** $A = \frac{1}{2}bh$ for b **3.** $A = \pi r^2$ for r **4.** $A = \frac{1}{2}h(b_1 + b_2)$ for b_1

5. $x = \frac{2}{3}y + 8$ for y **6.** $P = 2l + 2w$ for l **7.** $ax + by = c$ for y **8.** $x = \frac{180(n - 2)}{n}$ for n

PROBABILITY AND STATISTICS

Experimental Probability

Probability is a ratio between 0 and 1 that measures how likely it is that an *event* will occur. If an event has a probability close to 0, it is unlikely to occur. If an event has a probability close to 1, it is almost certain to occur.

EXAMPLE 1

Express each probability as a decimal between 0 and 1, inclusive.

a. a 30% chance   **b.** a 1 in 5 chance

SOLUTION

a. Write the percent as a decimal.

$$30\% = 0.3$$

b. Write the probability as a fraction and convert it to a decimal.

$$\frac{1}{5} = 1 \div 5 = 0.2$$

You can calculate the *experimental probability* of an event by finding the ratio of the number of times an event occurs to the number of times an experiment is performed.

EXAMPLE 2

In the 1994 National Football League season, quarterback Troy Aikman of the Dallas Cowboys completed 233 of 361 passes. Based on this performance, find the probability that Aikman completed his first pass in the 1995 season.

SOLUTION

$$\text{probability} = \frac{\text{number of completed passes}}{\text{number of attempted passes}} = \frac{233}{361} \approx 0.65$$

PRACTICE

Express each probability as a decimal between 0 and 1, inclusive.

1. a 75% chance   **2.** a 1 in 3 chance   **3.** a 3 in 25 chance

4. a 53% chance   **5.** a 4 out of 4 chance   **6.** a 7 in 8 chance

7. It rains in San Diego, California, an average of about 42 days a year. Find the probability that it will rain in San Diego on any given day.

8. A quality control engineer at the Best Manufacturing Company selected and tested a sample of 500 batteries and found that 6 were defective.

    a. Find the probability that a randomly chosen battery is defective.

    b. Find the probability that a randomly chosen battery is not defective.

9. Registered voters were asked in a survey about the proposed construction of a new highway. The survey found that 132 voters are in favor of the construction, 102 are opposed, and 16 have no opinion.

    a. Find the probability that a voter is in favor of the construction.

    b. Suppose that 60% of voters must vote in favor of the construction in order for it to be approved. Is it likely that the construction will be approved? Explain.

Significant Digits

Because measuring tools, such as rulers and thermometers, are limited in their ability to measure precisely, measurements must be recorded with a certain number of *significant digits*. You can determine the number of significant digits using the following guidelines:

- All nonzero digits are significant. For example, 467, 46.7, and 4.67 all have three significant digits.

- For a decimal, any zeros that appear after the last nonzero digit or between two nonzero digits are significant, but any zeros that appear before the first nonzero digit are not significant. For example, 0.04067, 4067, and 46.70 all have four significant digits.

- For a whole number, any zeros that appear between two nonzero digits are significant. Unfortunately, you cannot tell whether any zeros after the last nonzero digit are significant, so you should assume that they are not significant (unless you know otherwise). For example, 47, 470, and 4700 all have two significant digits.

When you perform calculations using measurements having various numbers of significant digits, a general rule to follow is to give the result with no more significant digits than the measurement with the *fewest* number of significant digits. (*Note:* Many science books require that you follow more specific rules than the one given here.)

Also, when a calculation involves multiple operations, carry *all* digits through the calculation and then round the result to the appropriate number of significant digits.

EXAMPLE

Simplify. Write your answers with the appropriate number of significant digits.

a. $4.37 - 2.6$

b. 2.05×30

SOLUTION

a. $4.37 - 2.6 = 1.77$

≈ 1.8

Write the answer with two significant digits because 2.6 has two significant digits.

b. $2.05 \times 30 = 61.5$

≈ 60

Write the answer with one significant digit because 30 has one significant digit.

PRACTICE

Simplify. Write your answers with the appropriate number of significant digits.

1. $3.7 + 1.6$

2. $3.75 - 1.6$

3. 3.75×1.125

4. $3.7 \div 1.125$

5. $30.78 + 1.5$

6. $30.0 - 1.5$

7. 370×1.875

8. $3.78 \div 2.0$

9. $2.057 + 0.38$

10. $2.05 - 0.0375$

11. 0.057×0.03

12. $20.057 \div 4.03$

Mean, Median, and Mode

The mean, the median, and the mode are types of averages for a set of data. To find the *mean*, add the data values and divide by the number of data values. To find the *median*, write the data values in order from least to greatest and find the middle number. If there is an even number of data values, then the median is the mean of the two middle values. The *mode* is the value that appears most often. There may be no mode or more than one mode.

EXAMPLE

Find the mean, the median, and the mode of Brandon's test scores.

Brandon's Test Scores
78, 83, 88, 90, 90, 93

SOLUTION

$$\text{mean} = \frac{78 + 83 + 88 + 90 + 90 + 93}{6} = \frac{522}{6} = 87$$

$$\text{median} = \frac{88 + 90}{2} = 89$$

Since there is an even number of scores, the median is the mean of the middle two scores.

$$\text{mode} = 90$$

The score of 90 appears most often in the data set.

The tables below show the final standings for the teams in each division of the National Football Conference for the 1995 season. Use the tables for Exercises 1 and 2.

Eastern Division		
Team	W	L
Dallas	12	4
Philadelphia	10	6
Washington	6	10
New York Giants	5	11
Arizona	4	12

Central Division		
Team	W	L
Green Bay	11	5
Detroit	10	6
Chicago	9	7
Minnesota	8	8
Tampa Bay	7	9

Western Division		
Team	W	L
San Francisco	11	5
Atlanta	9	7
New Orleans	7	9
St. Louis	7	9
Carolina	7	9

1. Find the mean, the median, and the mode for the number of wins (W) for all the teams in each division.

   a. Eastern Division   b. Central Division   c. Western Division

2. Find the mean, the median, and the mode for the number of wins for all the teams in the National Football Conference. How do these values compare with the values you found for the divisions?

3. The television ratings for the first 20 football Super Bowl games are listed below. Find the mean, the median, and the mode for the data.

   41.1, 36.8, 36.0, 39.4, 39.9, 44.2, 42.7, 41.6, 42.4, 42.3,
   44.4, 47.2, 47.1, 46.3, 44.4, 49.1, 48.6, 46.4, 46.4, 48.3

Data Displays

A *bar graph* displays data that fall into distinct categories.

EXAMPLE 1

The bar graph shows the total number of people who voted in the 1994 national election in the United States.

a. About how many people voted in the 1994 national election?

b. About 20% of the United States population live in the Northeast. Does the number of votes cast in the Northeast accurately represent the population?

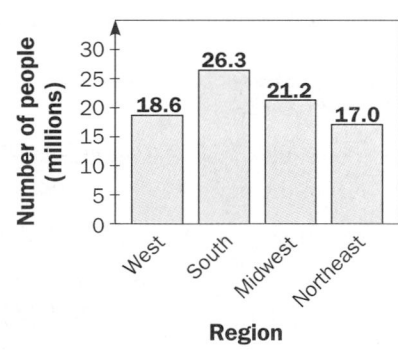

SOLUTION

a. $18.6 + 26.3 + 21.2 + 17.0 = 83.1$

   About 83.1 million people voted in the 1994 national election.

b. Find the percent of all votes that came from the Northeast: $\frac{17.0}{83.1} \approx 20.5\%$.

   About 20% of the people who voted lived in the Northeast. The region was accurately represented.

A *circle graph* shows how data relate to a whole and to each other.

EXAMPLE 2

The manager of a car dealership decides to record the colors of the cars that customers buy to see which colors are most popular. The data collected are shown in the table. Draw a circle graph to display the data.

Most Popular Car Colors	
Color	**Number of cars**
Red	11
Blue	9
White	14
Silver/Gray	5
Black	2
Other	9
Total	50

SOLUTION

Step 1 Write the result for each category as a fraction of all the data collected. Then write each fraction as a decimal.

Red: $\dfrac{11}{50} = 0.22$

Blue: $\dfrac{9}{50} = 0.18$

White: $\dfrac{14}{50} = 0.28$

Silver/gray: $\dfrac{5}{50} = 0.1$

Black: $\dfrac{2}{50} = 0.04$

Other: $\dfrac{9}{50} = 0.18$

Step 2 Multiply each decimal from Step 1 by 360° to find the degree measure for each *sector*, or wedge, of the circle graph.

$0.22 \times 360° \approx 79°$

$0.18 \times 360° \approx 65°$

$0.28 \times 360° \approx 101°$

$0.1 \times 360° = 36°$

$0.04 \times 360° \approx 14°$

$0.18 \times 360° \approx 65°$

Step 3 Draw a circle. Use a protractor to divide the circle into sectors of the appropriate size. Convert each decimal from Step 1 to a percent and label each sector appropriately.

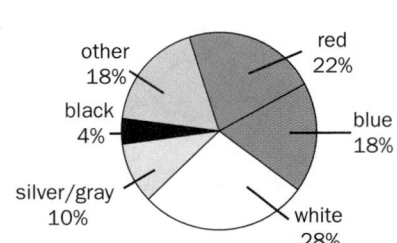

A *line graph* shows an amount and a direction of change in data over a period of time.

EXAMPLE 3

The line graph shows the number of teens, aged 16−19, who had jobs during the period 1990−1994.

a. About how many teens had jobs in 1992?

b. How does the graph show that the number of teens with jobs increased more from 1993 to 1994 than from 1992 to 1993?

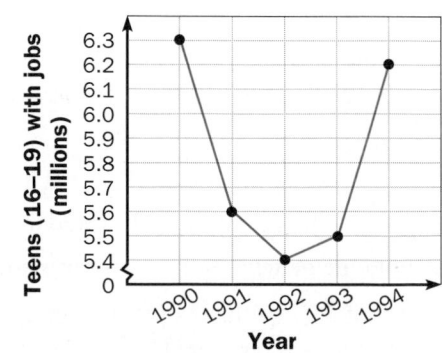

SOLUTION

a. About 5.4 million teens had jobs in 1992.

b. The line segment between 1993 and 1994 has a greater slope than the line segment between 1992 and 1993.

PRACTICE

Use the bar graph shown to answer Exercises 1–3.

1. Which denomination is in circulation an average of 2 years? 4 years?

2. What is the difference in the average circulation time of a $10 bill and a $50 bill?

3. Which denominations are in circulation the longest? Why do you think this is so?

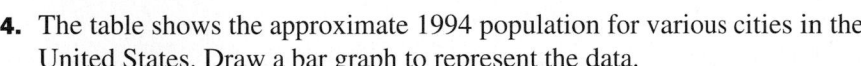

4. The table shows the approximate 1994 population for various cities in the United States. Draw a bar graph to represent the data.

City	New York	Los Angeles	Chicago	Houston	Philadelphia
Population	7,300,000	3,500,000	2,800,000	1,700,000	1,600,000

5. In 1993, 32.8% of the cars sold in the United States were compact or subcompact cars, 43.3% were midsize, 11.1% were large, and the rest were luxury cars. Draw a circle graph to represent the data.

6. A survey conducted in Chicago, Illinois, in 1990 showed that 1,700,857 workers commuted by car, 826,767 used public transportation, and 256,103 used another method. Draw a circle graph to represent the data.

7. a. The table at the right shows the number of students in high schools in the United States from 1960 to 1990. Draw a line graph to represent the data.

   b. State a conclusion you can make from the graph.

Year	Enrollment (millions)
1960	13.0
1970	19.7
1980	18.0
1990	15.3

Dimensional Analysis

When you want to convert from one unit of measure to another, you multiply by an appropriate *conversion factor* and cancel the units of measurement as if they were numbers. This method is called *dimensional analysis*.

EXAMPLE

Convert 55 mi/h to feet per minute.

SOLUTION

Set up the conversion factors so that miles and hours will cancel, leaving feet in the numerator and minutes in the denominator.

$$\frac{55 \text{ mi}}{1 \text{ h}} \times \frac{? \text{ ft}}{1 \text{ mi}} \times \frac{1 \text{ h}}{? \text{ min}} = \frac{? \text{ ft}}{1 \text{ min}}$$

$$\frac{55 \text{ mi}}{1 \text{ h}} \times \frac{5280 \text{ ft}}{1 \text{ mi}} \times \frac{1 \text{ h}}{60 \text{ min}} = \frac{? \text{ ft}}{1 \text{ min}}$$

> **1 h = 60 min**
> **1 mi = 5280 ft**

$$\frac{55 \text{ mi}}{1 \text{ h}} \times \frac{5280 \text{ ft}}{1 \text{ mi}} \times \frac{1 \text{ h}}{60 \text{ min}} = 4840 \text{ ft/min}$$

Use dimensional analysis for each conversion.

1. A car is traveling 80 km/h. Find this speed in meters per second.

2. Adam is taking a vacation in Mexico and wants to exchange $1250 for Mexican pesos. On the day he goes to the bank, one United States dollar is worth 3.10 pesos. How many pesos did Adam get?

3. Sound travels at a speed of about 3650 m/s through brick. Find this speed in feet per second. (*Hint:* 1 ft ≈ 0.3048 m)

4. The average fuel efficiency rate of Mona's car is about 24 mi/gal. The car's gas tank holds 15 gal of gas. If Mona pays $1.30 per gallon for gas, find the approximate cost of the gas in dollars per mile.

GRAPHS OF POINTS AND EQUATIONS

Graphing Points in a Coordinate Plane

A *coordinate plane* consists of a horizontal *x-axis* and a vertical *y-axis* that intersect at a point called the *origin*, labeled O. The axes divide the coordinate plane into four *quadrants* as shown.

Each point in a coordinate plane is associated with an *ordered pair* (a, b) of real numbers. The first number, a, is the *x-coordinate*. The second number, b, is the *y-coordinate*.

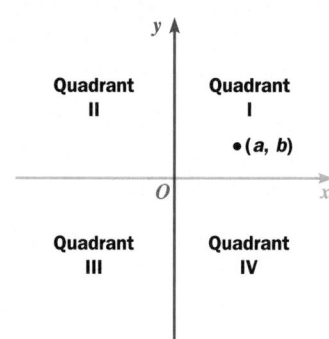

EXAMPLE

Graph the points $A(-3, 2)$ and $B(2, 0)$ in a coordinate plane. Name the quadrant, if any, in which each point lies.

SOLUTION

To graph $A(-3, 2)$, start at the origin and move left 3 units and **up 2 units**. The point is in Quadrant II.

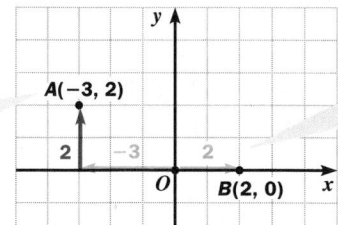

To graph $B(2, 0)$, start at the origin and move right 2 units and **vertically 0 units**. The point is not in any quadrant.

PRACTICE

Graph the following points in the same coordinate plane. Name the quadrant, if any, in which each point lies.

1. $C(1, 6)$ 2. $D(-4, -5)$ 3. $E(0, 3)$ 4. $F(2, -4)$

5. $G(-5, 0)$ 6. $H(-3, 2.5)$ 7. $I\left(\dfrac{14}{3}, \dfrac{14}{3}\right)$ 8. $J(0, 0)$

Graphing an Equation

You can use a coordinate plane to graph an equation.

EXAMPLE

Graph $y = |x|$.

SOLUTION

Step 1 Make a table of values. Be sure to include both positive and negative values of x.

$y = |x|$	
x	y
-3	$|-3| = 3$
-2	$|-2| = 2$
-1	$|-1| = 1$
0	$|0| = 0$
1	$|1| = 1$
2	$|2| = 2$
3	$|3| = 3$

Step 2 Draw a coordinate grid. Plot the ordered pairs (x, y) from the table. Connect the plotted points.

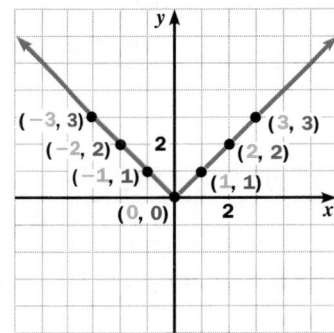

PRACTICE

Graph each equation.

1. $y = x$

2. $y = -4x - 5$

3. $y = |2x|$

4. $y = 3|x|$

5. $y = x^2$

6. $y = -\frac{1}{2}x^2$

7. $y = 2^x$

8. $y = \left(\frac{1}{2}\right)^x$

Making a Scatter Plot

A *scatter plot* is a graph that shows the relationship between two sets of data.

EXAMPLE

The table gives the number of VCRs owned by people in the United States for each year from 1985 to 1990. Make a scatter plot of the data.

SOLUTION

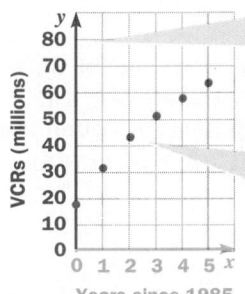

Years since 1985

Step 1 Draw a first-quadrant coordinate grid. Use the horizontal axis for years since 1985 and the vertical axis for number of VCRs.

Step 2 Plot each data pair from the table.

Years since 1985	Number of VCRs (in millions)
0	18
1	31
2	43
3	51
4	58
5	63

Make a scatter plot of the data in each table.

1.

Energy Used While In-Line Skating					
Time skating (min)	3	7	12	18	25
Energy used (Cal)	29	67	115	172	238

2.

Plain Cheese Pizzas at Doug's Pizza Palace					
Diameter (in.)	7	10	12	14	16
Price ($)	2.49	6.49	8.49	11.49	14.49

TRANSFORMATIONS IN THE COORDINATE PLANE

A *transformation* of a geometric figure is a change in the position or size of the figure. The result of a transformation is called an *image*. Some common transformations are *translations*, *reflections*, and *dilations*.

Translating a Figure

A *translation* is a transformation that moves each point of a figure the same distance in the same direction. A translation neither flips nor turns a figure, and it does not change the figure's size or shape.

EXAMPLE

a. Translate △*ABC* (shown below) right 9 units and down 5 units. Give the coordinates of the image's vertices.

b. Write a formula representing the translation in part (a).

SOLUTION

a.

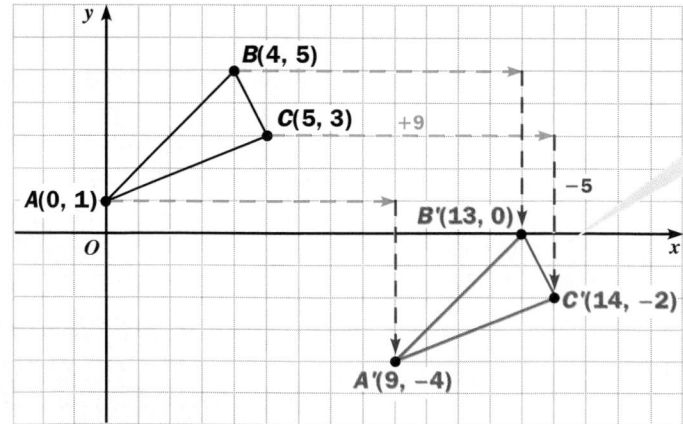

Move each vertex of △*ABC*
right 9 units and **down 5 units**
to get new vertices *A′*, *B′*, and *C′*.
Connect these vertices to form
△*A′B′C′*, the image of △*ABC*.

To find the coordinates of *C′*, add 9
to the *x*-coordinate of *C* and
subtract 5 from the *y*-coordinate
of *C*: **(5 + 9, 3 − 5)**, or **(14, −2)**

b. Under the translation "right 9 units and down 5 units," each point *P*(*x*, *y*) is moved to the point *P′*(*x* + 9, *y* − 5). You can write *P*(*x*, *y*) → *P′*(*x* + 9, *y* − 5).

Let △*DEF* be the triangle with vertices *D*(−3, 2), *E*(4, 4), and *F*(2, 0). For each of Exercises 1–6, draw △*DEF* and its image under the given translation. Write a formula representing the translation.

1. right 4 units

2. down 5 units

3. left 5 units and up 2 units

4. right 1 unit and up 1 unit

5. right 7.5 units and down 6 units

6. left 4 units and down 9 units

Reflecting a Figure

A *reflection* is a transformation that flips a figure over a line, called the *line of reflection*. A reflection does not change the size or shape of a figure.

EXAMPLE 1

Reflect △*ABC* shown on the previous page over each line.

a. the *y*-axis

b. the *x*-axis

SOLUTION

First reflect each vertex of △*ABC* over the given line to get new vertices *A*′, *B*′, and *C*′. Then connect these vertices to form △*A*′*B*′*C*′, the image of △*ABC* under the given reflection.

a.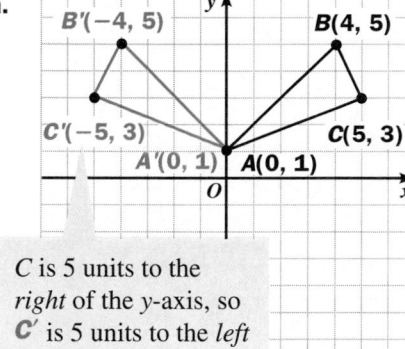

C is 5 units to the *right* of the *y*-axis, so *C*′ is 5 units to the *left* of the *y*-axis.

b.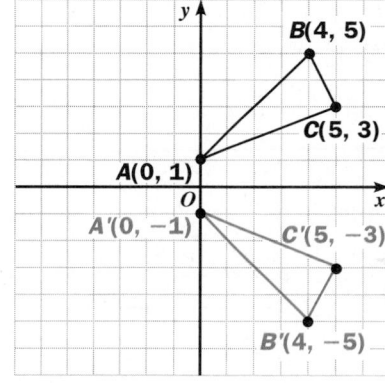

C is 3 units *above* the *x*-axis, so *C*′ is 3 units *below* the *x*-axis.

If a figure is reflected over a line and the image coincides with the original figure, then the line is called a *line of symmetry* for the figure.

EXAMPLE 2

Find the number of lines of symmetry of an equilateral triangle.

SOLUTION

The diagram shows that an equilateral triangle has 3 lines of symmetry. Each line of symmetry passes through a vertex and is perpendicular to the side opposite the vertex.

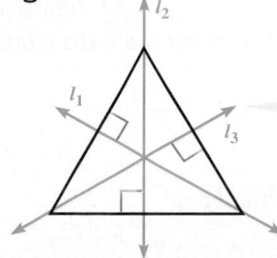

Note that if you fold a figure along a line of symmetry, the figure's edges coincide.

Let △*GHI* be the triangle with vertices *G*(0, 1), *H*(3, 3), and *I*(5, 0). For each of Exercises 1–4, draw △*GHI* and its image when △*GHI* is reflected over the given line.

1. the *x*-axis   **2.** the *y*-axis   **3.** $x = 6$ **4.** $y = -2$

Give the minimum number of lines of symmetry for each type of figure.

5. a rectangle   **6.** a square   **7.** a parallelogram

8. an isosceles triangle   **9.** a parabola   **10.** a circle

Dilating a Figure

A *dilation* of a figure is a transformation whose image is similar to the original figure. The ratio of any length in the image to the corresponding length in the original figure is called the *scale factor* of the dilation.

EXAMPLE

Dilate △*ABC* shown on page 796. Use *P*(8, 1) as the *center* of the dilation and a scale factor of 2.

SOLUTION

Step 1 Draw **rays** from *P* through the vertices of △*ABC*.

Step 2 Locate point *A'* on $\overrightarrow{PA}$ so that *PA'* = 2(*PA*), since the scale factor is 2. Locate points *B'* and *C'* similarly.

Step 3 Draw △*A'B'C'*, the image of △*ABC*.

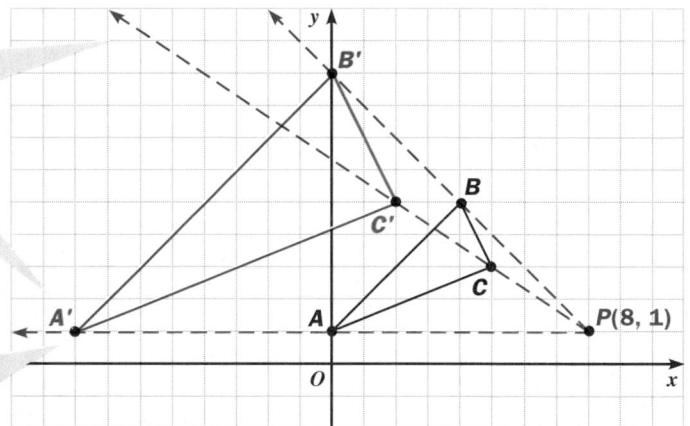

In general, when the scale factor of a dilation is greater than 1 (as in the example above), the image is larger than the original figure. When the scale factor is between 0 and 1, the image is smaller than the original figure.

PRACTICE

Let △*JKL* be the triangle with vertices *J*(−6, 0), *K*(−6, 4), and *L*(0, 2). For each of Exercises 1–4, draw △*JKL* and its image under the dilation with the given center and scale factor.

1. center: *P*(0, 0); scale factor: 2   **2.** center: *P*(0, 0); scale factor: $\frac{1}{2}$

3. center: *P*(−2, −4); scale factor: $\frac{3}{4}$ **4.** center: *P*(−8, 6); scale factor: 3

LOGIC AND GEOMETRIC RELATIONSHIPS

Logical Statements

A *conditional statement* can be written in the form "if p, then q." The phrase represented by p is the *hypothesis* and the phrase represented by q is the *conclusion*.

You write the *converse* of a conditional statement by interchanging the hypothesis and the conclusion. The converse of "if p, then q" is "if q, then p."

If-then statements are either true or false. If you find an instance where the hypothesis is true and the conclusion is false, the statement in general is false.

EXAMPLE 1

For parts (a) and (b), use the statement "If $b = \sqrt{a}$, then $b^2 = a$."

a. Identify the hypothesis and the conclusion. Then tell whether the statement is *true* or *false*.

b. Write the converse. Then tell whether the converse is *true* or *false*.

SOLUTION

a. The hypothesis is "$b = \sqrt{a}$." The conclusion is "$b^2 = a$." Since $(\sqrt{a})^2 = a$, the statement is true.

b. The converse is "If $b^2 = a$, then $b = \sqrt{a}$." If $b^2 = a$, then b can equal either $\sqrt{a}$ or $-\sqrt{a}$, so the statement is not necessarily true. The converse is false.

A statement and its converse can be combined into one statement, called a *biconditional*, using the phrase "*if and only if*." The truth of one statement in a biconditional implies the truth of the other statement.

For example, the biconditional "$b^2 = a$ if and only if $b = \pm\sqrt{a}$" is a true statement, but the biconditional "$b^2 = a$ if and only if $b = \sqrt{a}$" is false.

EXAMPLE 2

Tell whether the statement is *true* or *false*: "A number is divisible by 4 if and only if the number is divisible by 2."

SOLUTION

One conditional statement is "If a number is divisible by 2, then the number is divisible by 4." Since 10 is divisible by 2 but not divisible by 4, this conditional statement is false.

The other conditional statement, which is the converse of the first, is "If a number is divisible by 4, then the number is divisible by 2." Since every number that is divisible by 4 is an even number, the converse is true.

Since divisibility by 2 does not imply divisibility by 4, the biconditional is false.

For Exercises 1–6:

a. Identify the hypothesis and the conclusion of each conditional statement. Then tell whether the statement is *true* or *false*.

b. Write the converse of each statement. Then tell whether the converse is *true* or *false*. Explain.

c. Write a biconditional using each statement and its converse. Then tell whether the biconditional is *true* or *false*. Explain.

1. If a figure is a rectangle, then it is a quadrilateral.

2. If $a = 0$ or $b = 0$, then $ab = 0$.

3. If a triangle is equilateral, then it is isosceles.

4. If $a > b$, then $a + c > b + c$.

5. If two lines intersect, then they are not parallel.

6. If $x^2 = 9$, then $x = 3$.

Angle Relationships

A *right angle* has a measure of 90°, and a *straight angle* has a measure of 180°. An angle with a measure between 0° and 90° is an *acute angle*. An angle with a measure between 90° and 180° is an *obtuse angle*.

Two angles whose measures have a sum of 90° are *complementary angles*. Two angles whose measures have a sum of 180° are *supplementary angles*. Two angles whose sides form two pairs of opposite rays are *vertical angles*. Vertical angles have the same measure.

EXAMPLE

For the diagram shown, find the values of x and y.

SOLUTION

Since $\angle POS$ and $\angle QOR$ are vertical angles, their measures are equal. Therefore, $x = 75$.

Since $\angle POS$ and $\angle SOR$ are supplementary angles, $75 + y = 180$. Therefore, $y = 105$.

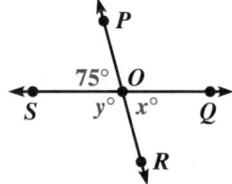

PRACTICE

Find the values of x and y in each diagram.

1.
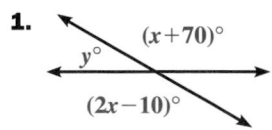

2.
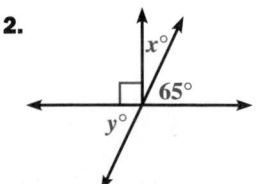

3.
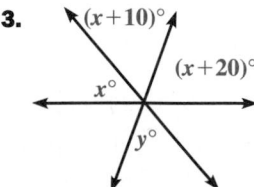

Tell if each statement is *always true, sometimes true,* or *never true*.

4. Two angles that are supplementary are acute angles.

5. Two angles that are right angles are supplementary.

6. Two angles that are vertical angles are complementary.

Triangle Relationships

The sum of the measures of the angles of a triangle is 180°. Two triangles are *congruent* if they have the same size and shape. If two triangles have the same shape but are not necessarily the same size, the two triangles are *similar*. If two pairs of corresponding angles are congruent, the triangles are similar.

The Pythagorean theorem and its converse tell you that $\triangle ABC$ is a right triangle if and only if $a^2 + b^2 = c^2$.

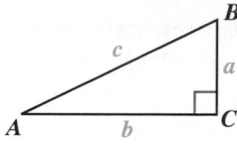

EXAMPLE

Use the diagram at the right.

a. Find the value of x.

b. Find the value of y.

SOLUTION

a. The sum of the measures of the angles of a triangle is 180°.

$$90 + 60 + x = 180$$
$$x = 30$$

b. Use the Pythagorean theorem.

$$y^2 + (2\sqrt{3})^2 = 4^2$$
$$y^2 + 12 = 16$$
$$y^2 = 4$$
$$y = 2$$

Use the positive square root because y is a length.

PRACTICE

Find the value of x in each diagram.

1.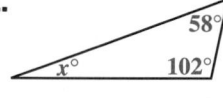

2.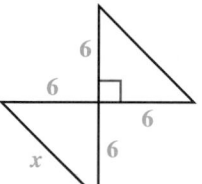

3.

4.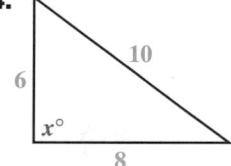

Polygons

A *polygon* is the union of several line segments that are joined end to end so as to completely enclose an area. A polygon with all angles of equal measure is *equiangular*. A polygon with all sides of equal length is *equilateral*. A polygon that is both equiangular and equilateral is a *regular polygon*.

EXAMPLE

Complete each statement with one of the words *always*, *sometimes*, or *never*. Explain your reasoning.

a. An isosceles triangle is _?_ equilateral.

b. A rectangle is _?_ a parallelogram.

SOLUTION

a. All isosceles triangles have two congruent sides, but only some have three congruent sides. Therefore, an isosceles triangle is *sometimes* equilateral.

b. All rectangles are quadrilaterals that have parallel opposite sides, so a rectangle is *always* a parallelogram.

**Complete each statement with one of the words *always*, *sometimes*, or *never*.
Explain your reasoning.**

1. A hexagon is <u>?</u> a regular polygon.

2. A quadrilateral <u>?</u> has five sides.

3. An isosceles triangle is <u>?</u> a right triangle.

4. An obtuse triangle is <u>?</u> equiangular.

5. A square is <u>?</u> a regular polygon.

6. A rectangle is <u>?</u> an equiangular quadrilateral.

Perimeter, Circumference, Area, and Volume

Refer to the *Tables, Properties, and Formulas* section on pages 817–825 for a list
of specific formulas for perimeter, circumference, area, surface area, and volume.

EXAMPLE

Use the rectangular prism shown at the right.

a. Find the surface area.

b. Find the volume.

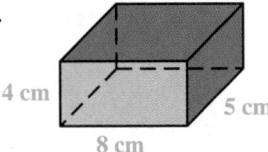

SOLUTION

a. surface area of the prism = twice the areas of the 3 faces shown

$$= 2(8 \cdot 5 + 4 \cdot 8 + 4 \cdot 5)$$

$$= 184$$

The surface area of the prism is 184 cm².

b. volume of the prism = **area of a base × height of the prism**

$$= \mathbf{8 \cdot 5 \cdot 4}$$

$$= 160$$

The volume of the prism is 160 cm³.

PRACTICE

Find the perimeter and area of each figure.

1. a trapezoid with height 4 cm, bases with lengths 3 cm and 9 cm, and both
   legs with length 5 cm

2. a right triangle with legs of lengths 18 ft and 15 ft

3. a rectangle with length 9 m and width 3.5 m

Find the circumference and area of each circle. Round to the nearest hundredth.

4. a circle with radius 3

5. a circle with diameter 12

6. a circle with radius $\sqrt{5}$

Find the surface area and volume of each figure. Round to the nearest hundredth.

7.
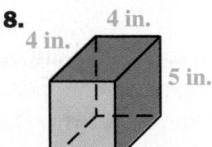

8.
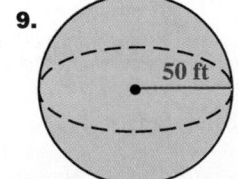

9.

10.
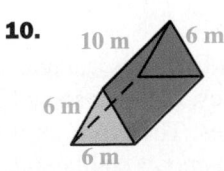

Technology Handbook

This handbook introduces you to the basic features of most graphing calculators. Check your calculator's instruction manual for specific keystrokes and any details not provided here.

PERFORMING CALCULATIONS

The Keyboard

Look closely at your calculator's keyboard. Notice that most keys serve more than one purpose. Each key is labeled with its primary purpose, and labels for any secondary purposes appear somewhere near the key. You may need to press `2nd`, `SHIFT`, or `ALPHA` to use a key for a secondary purpose.

On the TI-82, for example, the `x²` key can be used as follows:

- Press `x²` to square a number.
- Press `2nd` and then `x²` to take a square root.
- Press `ALPHA` and then `x²` to type the letter I.

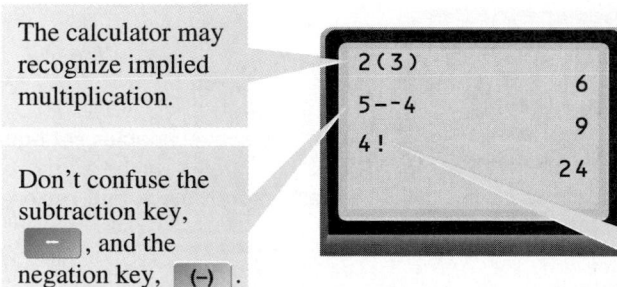

The Home Screen

Your calculator has a *home screen* where you can do calculations. You can usually enter a calculation on a graphing calculator just as you would write it on a piece of paper.

Below are some things to remember as you enter calculations on your graphing calculator.

The calculator may recognize implied multiplication.

Don't confuse the subtraction key, `–`, and the negation key, `(–)`.

```
2(3)
              6
5 – – 4
              9
4 !
             24
```

You may need to press `MATH` to access some operations, such as evaluating a factorial.

Use your calculator to find the value of each expression.

1. $-3 - 9$
2. $\sqrt{12.25}$
3. $\sqrt[3]{2.744}$
4. $6!$
5. 8^5
6. $_5C_2$
7. $\tan 75°$
8. $\left| e^{-1} - 2\pi \right|$

DISPLAYING GRAPHS

Entering and Graphing a Function

To graph a function, enter its equation in the form $y = f(x)$. If variables other than x and y are used in the equation, replace the independent variable with x and the dependent variable with y.

For example, to enter the equation $-3r + 5s = 10$, first rewrite it as $-3x + 5y = 10$ (assuming r is the independent variable and s is the dependent variable). Then solve for y to get $y = \frac{3}{5}x + 2$. The graph of this equation is shown.

Use parentheses. If you enter $y = 3/5x + 2$, the calculator may interpret the equation as $y = \frac{3}{5x} + 2$.

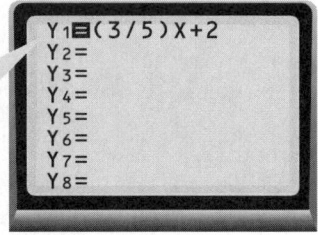

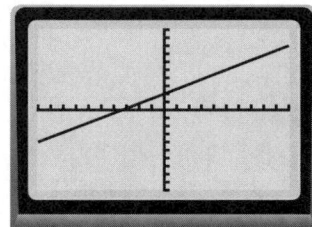

Be sure to use parentheses when groupings are implied by radical signs or fraction bars. For example, enter $y = \sqrt{x + 1} - 2$ as $y = \sqrt{\,}(x + 1) - 2$, and enter $y = \frac{1}{x + 1}$ as $y = 1/(x + 1)$.

The Viewing Window

When graphing, think of the calculator screen as a *viewing window* that lets you look at a portion of the coordinate plane.

On many calculators, the *standard window* shows values from -10 to 10 on both the x- and y-axes. You can adjust the viewing window by pressing WINDOW or RANGE and entering new values for the window variables.

The interval $-10 \leq x \leq 10$ will be shown on the x-axis.

The interval $-10 \leq y \leq 10$ will be shown on the y-axis.

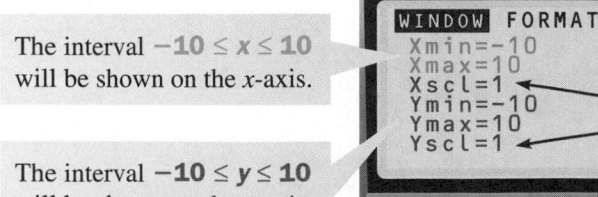

With the scale variables set equal to 1, tick marks will be 1 unit apart on both the x- and y-axes.

Squaring the Viewing Window

In a *square* viewing window, the distance between consecutive integers is the same on both the *x*- and *y*-axes. This causes graphs to appear undistorted. For example, the circle with equation $x^2 + y^2 = 49$ is shown below in both the standard window and a square window.

Standard Viewing Window

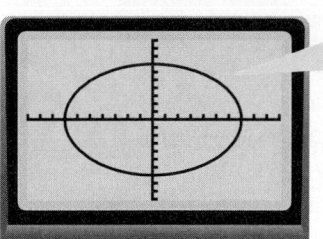

A circle looks like an ellipse in the standard window.

Square Viewing Window

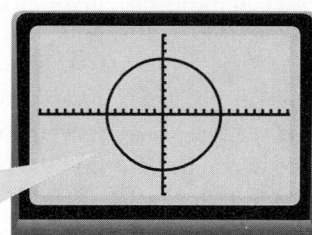

A circle appears undistorted in a square window.

Your calculator may have a feature that automatically produces a square viewing window. If not, you may be able to create one using the fact that the ratio of the screen's height to its width is often about 2 to 3. Just choose values for the window variables that make the "length" of the *y*-axis about two thirds the "length" of the *x*-axis:

$$\text{Ymax} - \text{Ymin} \approx \frac{2}{3}(\text{Xmax} - \text{Xmin})$$

The *Zoom* Feature

To magnify part of a graph, you can use your calculator's *zoom* feature. One common way of zooming involves putting a *zoom box* around the portion of the graph you want to magnify.

Fix one corner of the zoom box. Then fix the opposite corner.

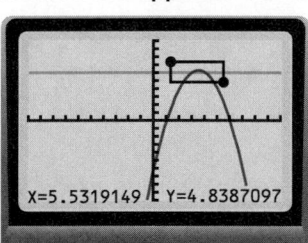

The calculator draws what is inside the zoom box at full-screen size.

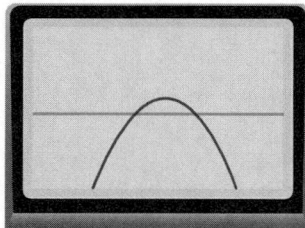

CALCULATOR PRACTICE

Enter and graph each function. Use the standard viewing window.

9. $-6x + 2y = 2$ **10.** $y = -x^2 - x + 6$ **11.** $y = e^x$ **12.** $y = |x - 2|$

13. Graph the perpendicular lines $y = x - 1$ and $y = -x + 7$ in the same coordinate plane. Adjust the viewing window so that the lines look perpendicular. Then zoom in on the point where the lines intersect.

EXAMINING GRAPHS

The *Trace* Feature

After a graph is displayed, you can use your calculator's *trace* feature. When you press TRACE, a flashing cursor appears on the graph. The x- and y-coordinates of the cursor's location are shown at the bottom of the screen. You can move the cursor along the graph by pressing ◄ and ► .

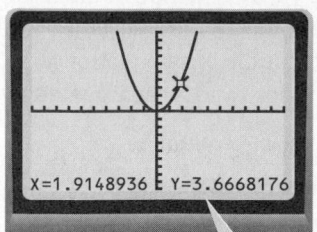

X=1.9148936 Y=3.6668176

The trace cursor is at the point (1.9148936, 3.6668176) on the graph of $y = x^2$.

For example, consider the equation $y = 64x + 2100$, which gives the total pressure y (in pounds per square foot) on a diver working at a depth of x ft. You can use trace to find the depth at which the pressure reaches 4000 lb/ft^2.

Graph $y = 64x + 2100$. Press TRACE, and move the cursor along the graph until $y \approx 4000$.

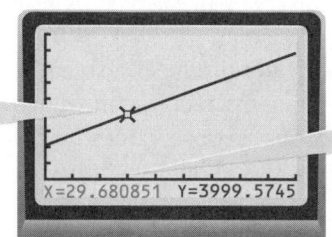

X=29.680851 Y=3999.5745

The corresponding water depth is about **30 ft**.

Friendly Windows

As you press ► while tracing along a graph, the x-coordinate may increase in "unfriendly" increments. For example, if you use the standard window on the TI-82, the x-increment, $\triangle$X, is 0.21276596.

Your calculator may allow you to control $\triangle$X directly. If not, you can control $\triangle$X indirectly by choosing an appropriate Xmax for a given Xmin. On the TI-82, for example, choose Xmax so that:

$$\text{Xmax} = \text{Xmin} + 94(\triangle X)$$

This number depends on the calculator you are using.

Suppose you want Xmin to equal **−5** and $\triangle$X to equal **0.1**. Then:

$$\text{Xmax} = -5 + 94(0.1) = 4.4$$

Setting Xmax equal to **4.4** gives a "friendly window" in which the trace cursor's x-coordinate increases by **0.1** each time you press ► .

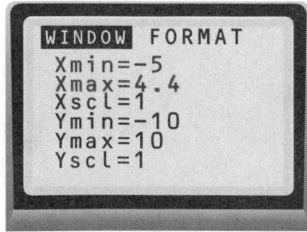

WINDOW FORMAT
Xmin=-5
Xmax=4.4
Xscl=1
Ymin=-10
Ymax=10
Yscl=1

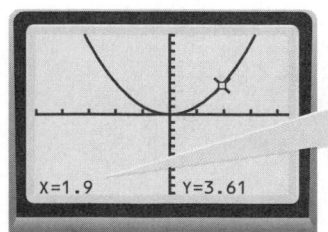

X=1.9   Y=3.61

Using trace with the window settings shown gives "friendly" x-coordinates like 1.9.

Calculating Coordinates of Interest

Some graphing calculators have a *calculate* feature that you can use to find coordinates of interest on a graph. On the TI-82, you access the calculate menu, shown below, by pressing [2nd] and then [TRACE] after you have graphed a function.

To evaluate the function for a given value of x, select *value* from the calculate menu.

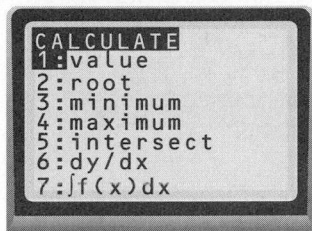

Enter the *x*-value.

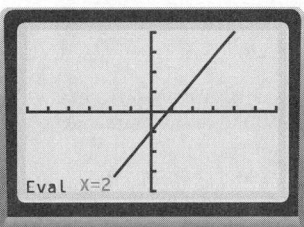

The calculator displays the corresponding *y*-value.

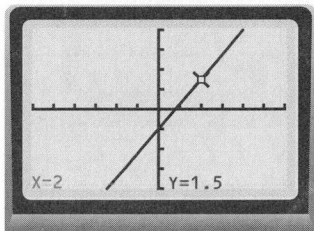

To find a zero of a function, select *root* from the calculate menu.

Use the cursor to specify lower and upper bounds for the zero.

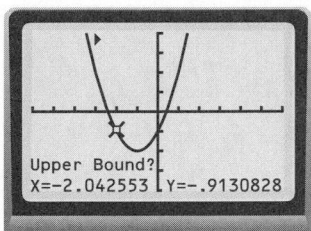

Then guess the location of the zero to help the calculator find it more quickly.

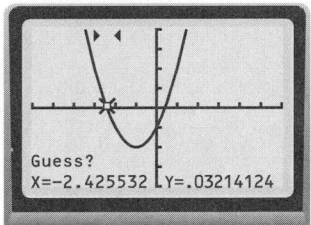

The zero is the *x*-coordinate displayed by the calculator.

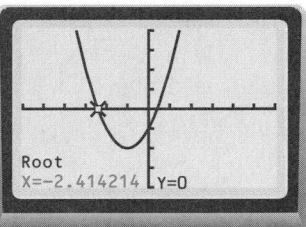

To find a local minimum or local maximum of a function, select *minimum* or *maximum* from the calculate menu.

Use the cursor to specify lower and upper bounds for the minimum or maximum.

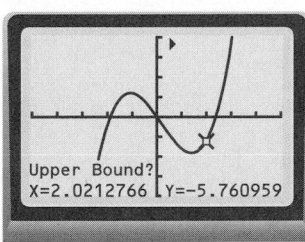

Then guess the location of the minimum or maximum.

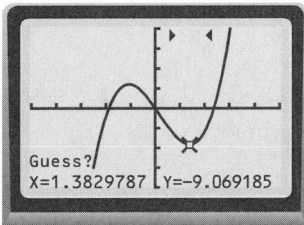

The minimum or maximum is the *y*-coordinate displayed by the calculator.

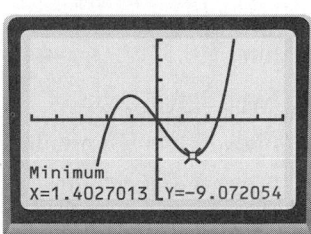

To find a point where two graphs intersect, select *intersect* from the calculate menu.

Use 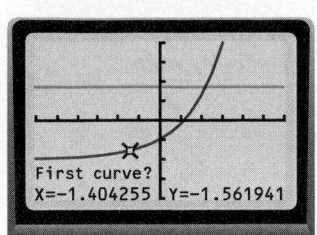 and ▼ to choose the two graphs.

Use ◄ and ► to move the cursor near a point of intersection.

The calculator displays the coordinates of the point of intersection.

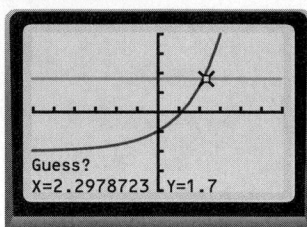

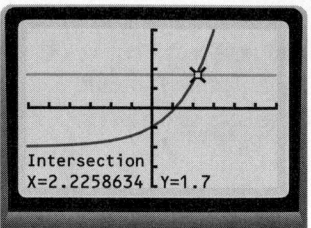

The *Table* Feature

Instead of using the trace and calculate features to find the coordinates of points on a function's graph, you may want to create a table of function values. Some calculators have a *table* feature that allows you to do this. For example, the table below shows values of the function $f(x) = x^3 + 2x^2 - 5x - 10$.

The minimum x-value is 0, and the x-increment is 1. Many calculators let you change these settings.

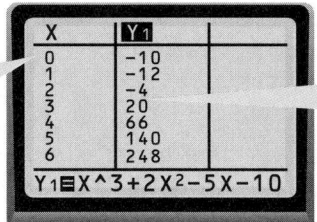

Since the y-values change sign between $x = 2$ and $x = 3$, a zero of f lies between these x-values.

CALCULATOR PRACTICE

14. Graph $y = \dfrac{6}{x^2 + 1}$. Create a friendly window such that Xmin $= -4$ and $\triangle X = 0.1$. Use trace to find the value of y when $x = -0.8$.

15. Let $f(x) = x^4 - 7x^2 + x + 5$. Use the calculate feature to complete parts (a)–(d).

   a. Find $f(0.5)$.

   b. Find all real zeros of f.

   c. Find all local minimums and local maximums of f.

   d. Find all real solutions of the equation $f(x) = 9$.

16. Let $g(x) = x - \sqrt{x + 8}$. Make a table of values for g. The minimum x-value in the table should be 0, and the x-increment should be 1.

   a. Use the table to find $g(6)$.

   b. By repeatedly adjusting the table's minimum x-value and x-increment, find a zero of g to the nearest hundredth.

COMPARING GRAPHS

Using a List to Graph a Family of Functions

Your graphing calculator may allow you to enter a *list* as part of an equation. The graph of such an equation is a family of curves, rather than a single curve. The calculator draws one curve for each number in the list.

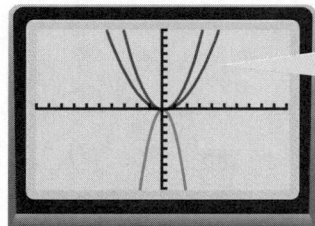

$Y_1 = \{1, -3, 0.5\}X^2$ gives the family of functions $y = 1x^2$, $y = -3x^2$, and $y = 0.5x^2$.

$Y_2 = X^2 + \{5, 1, -6\}$ gives the family of functions $y = x^2 + 5$, $y = x^2 + 1$, and $y = x^2 - 6$.

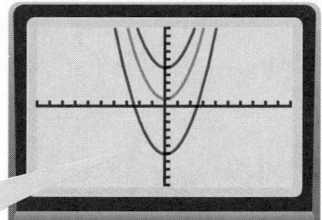

Graphing a Function and Its Inverse

Some graphing calculators have a *draw inverse* feature that lets you graph the inverse of a function without entering the inverse's equation. For example, suppose you want to graph $y = 2x + 4$ and its inverse.

Enter the equation $y = 2x + 4$.

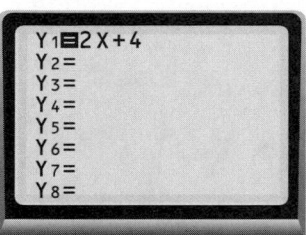

Using draw inverse, tell the calculator to graph the inverse of $y = 2x + 4$.

The calculator graphs $y = 2x + 4$ and its inverse in the same coordinate plane.

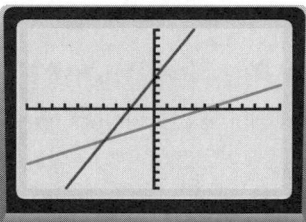

CALCULATOR PRACTICE

17. Describe how to use a list to graph the family of functions $y = x$, $y = -x$, $y = \frac{1}{2}x$, and $y = 2x$.

18. Graph each family of functions. (Set your calculator to radian mode. Adjust the viewing window so that the intervals $-2\pi \le x \le 2\pi$ and $-4 \le y \le 4$ are shown on the axes. Use $\text{Xscl} = \frac{\pi}{2}$ and $\text{Yscl} = 1$.)

    a. $y = \{1, -3, 0.5\}\sin x$ **b.** $y = \cos x + \{0, 2, -2\}$

19. Use the draw inverse feature to graph each function and its inverse.

    a. $y = -3x + 2$ **b.** $y = 2^x$ **c.** $y = -\sqrt{x}$

GRAPHING IN DIFFERENT MODES

Graphing Parametric Equations

Suppose an airplane flying at an altitude of **15,000 ft** is preparing to land. Once the plane begins its descent to the runway, its altitude decreases at a rate of **20 ft/s**. The plane's horizontal speed is **250 ft/s**.

You can use parametric equations to describe the plane's horizontal position x and vertical position y after t seconds:

$$x = 250t$$

$$y = 15{,}000 - 20t$$

To graph these equations on your graphing calculator, switch to *parametric mode* (denoted by "par" on some calculators). Then complete these steps.

Set appropriate intervals for t, x, and y.	Enter the parametric equations.	Graph the equations. You can use trace to find the plane's position as a function of time.

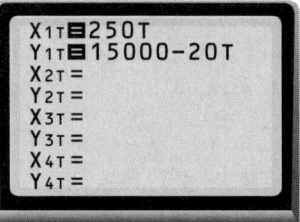
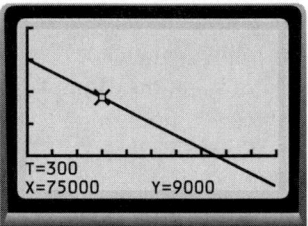

Graphing Sequences

Many graphing calculators have a *sequence mode* (often denoted by "seq") that lets you graph sequences defined either explicitly or recursively. For example, suppose you want to graph the first 10 terms of this sequence:

$$t_1 = 2$$

$$t_n = 0.6(t_{n-1}) + 3$$

On the TI-82, use the following steps. Note that you must replace t_n with u_n or v_n when entering the sequence.

Enter the starting value t_1 and the minimum and maximum values of n.	Enter the recursion equation.	Graph the sequence in dot mode. Use trace to find values of t_n for different values of n.

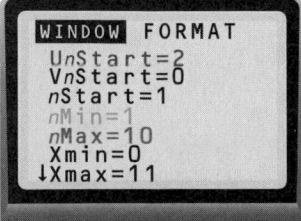
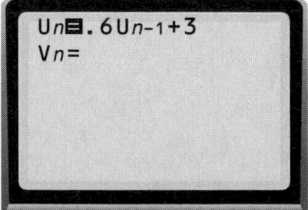
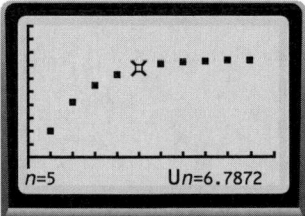

20. Graph the parametric equations that describe the position of the airplane as it descends for landing. (See the top of the previous page.) How long does it take for the plane to reach the ground?

21. Graph the first 10 terms of each sequence.

    a. $t_1 = 10$
       $t_n = t_{n-1} - 1$

    b. $t_1 = 1$
       $t_n = 1.25(t_{n-1})$

    c. $t_1 = 6$
       $t_n = 0.7(t_{n-1}) + 0.2$

USING MATRICES

Adding and Subtracting Matrices

You can use a graphing calculator to add or subtract matrices that have the same dimensions. Consider the tables below, which show the numbers of left- and right-handed males and females in two algebra classes.

Class 1	Left	Right	Both
Males	2	8	1
Females	1	13	0

Class 2	Left	Right	Both
Males	3	16	0
Females	2	9	2

To get a single table containing data for *both* classes, enter each table as a matrix and find the sum of the two matrices.

Use matrix [A] for class 1. The dimensions of [A] are 2 × 3.

```
MATRIX[A]  2 x 3
[2    8    1    ]
[1    13   0    ]



2,3=0
```

Use matrix [B] for class 2. This matrix is also 2 × 3.

```
MATRIX[B]  2 x 3
[3    16   0    ]
[2    9    2    ]



2,3=2
```

From the home screen, add the two matrices.

```
[A]+[B]
        [[5 24 1]
         [3 22 2]]
```

Multiplying a Matrix by a Scalar

Look at matrix *A* above showing the numbers of males and females in class 1 who are left- and right-handed. To find the *percent* of students who fall into each category, first divide each matrix entry by the total number of students in the class (25). Then multiply the resulting entries by 100. You can perform these operations quickly using scalar multiplication on a graphing calculator, as shown at the right.

```
(1/25)[A]
   [[.08 .32 .04]
    [.04 .52 0  ]]
Ans*100
        [[8 32 4]
         [4 52 0]]
```

Finding the Product of Two Matrices

You can multiply two matrices when the number of columns in the first matrix is the same as the number of rows in the second matrix.

For example, suppose you want to find this product: $\begin{bmatrix} 9 & 4 \\ 3 & 1 \\ 2 & 8 \\ 1 & 5 \end{bmatrix} \begin{bmatrix} 4 & 2 & 0 \\ 3 & 0 & 2 \end{bmatrix}$

A has **2** columns and *B* has **2** rows, so the product *AB* is defined.

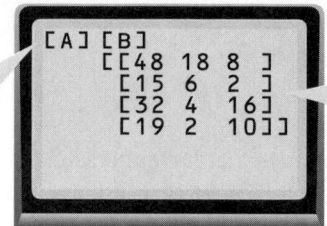

```
[A] [B]
   [[48  18   8 ]
    [15   6   2 ]
    [32   4  16]
    [19   2  10]]
```

Since *A* is a **4 × 2** matrix and *B* is a **2 × 3** matrix, *AB* is a **4 × 3** matrix.

Finding the Inverse of a Matrix

If a square matrix *A* has an inverse, you can use a graphing calculator to find it. This feature is useful when you want to solve a system of linear equations. For example, suppose you want to solve this system:

$$\begin{array}{rcl} 2x - 5y + 3z &=& 35 \\ 7x - 2y - 4z &=& -11 \\ -x \quad\quad + z &=& 5 \end{array}$$

$$\begin{bmatrix} 2 & -5 & 3 \\ 7 & -2 & -4 \\ -1 & 0 & 1 \end{bmatrix} \begin{bmatrix} x \\ y \\ z \end{bmatrix} = \begin{bmatrix} 35 \\ -11 \\ 5 \end{bmatrix}$$

$\uparrow$ $\uparrow$ $\uparrow$
A   *X*   *B*

This equation, $AX = B$, has solution $X = A^{-1}B$.

Enter the matrices *A* and *B*. Then calculate $A^{-1}B$.

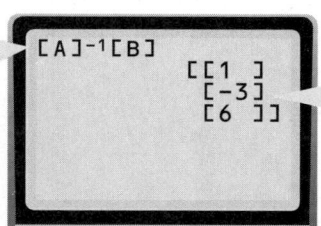

```
[A]-1[B]
       [[1 ]
        [-3]
        [6 ]]
```

The solution of the system of equations is $x = 1$, $y = -3$, and $z = 6$.

CALCULATOR PRACTICE

Use matrices A, B, C, and D to evaluate each matrix expression. If an operation cannot be performed, write *undefined*.

$$A = \begin{bmatrix} 3 & 4 \\ 6 & 7 \end{bmatrix} \quad B = \begin{bmatrix} 5 & 11 \\ -3 & -7 \end{bmatrix} \quad C = \begin{bmatrix} 9 & 3 \\ 12 & 4 \end{bmatrix} \quad D = \begin{bmatrix} 2 & 0 & -8 & 1 \\ 5 & 7 & 3 & 9 \end{bmatrix}$$

22. $A + B$ **23.** $B - C$ **24.** $2A$ **25.** AB

26. $C + D$ **27.** CD **28.** A^{-1} **29.** D^{-1}

STATISTICS AND PROBABILITY

Histograms, Box Plots, and Line Graphs

Many graphing calculators can display histograms, box plots, and line graphs of data that you enter. For example, suppose you want to make a histogram of the following employee commuting times from page 260:

15, 25, 30, 40, 60, 40, 25, 15, 15, 20, 30, 75, 20, 15, 65, 40,
30, 20, 15, 10, 20, 35, 45, 35, 30, 20, 10, 55, 25, 15, 5, 45, 20,
10, 5, 15, 30, 30, 50, 15, 5, 10, 45, 30, 20, 20, 10, 30, 20, 30

The steps used to enter data and perform statistical functions vary greatly from calculator to calculator. On the TI-82, follow these steps:

Press **STAT**, and select the "Edit..." option. Enter the data values into a list, such as L1.

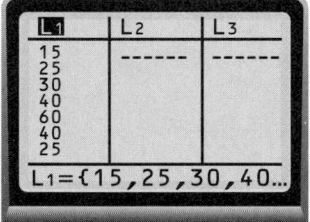

Tell the calculator to display a **histogram** using list **L1**. Set the frequency option to **1**.

Set up a good viewing window. Then press **GRAPH** to display the histogram.

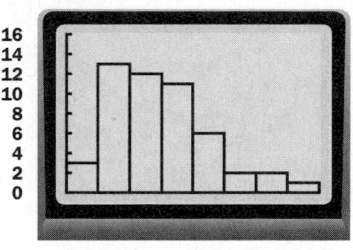

Mean and Standard Deviation

You can use a scientific or graphing calculator to find the mean and standard deviation of a set of data values. Some calculators give two types of standard deviation—the *sample standard deviation S* and the *population standard deviation* σ. In this book, the population standard deviation is always used.

For example, you can calculate the mean and standard deviation of the employee commuting times shown above.

Select the 1-variable statistics option from the menu of statistics functions.

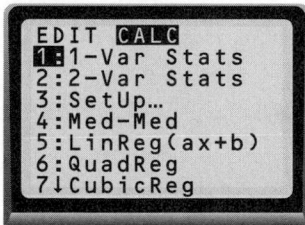

Specify the list containing the data (in this case, **L1**).

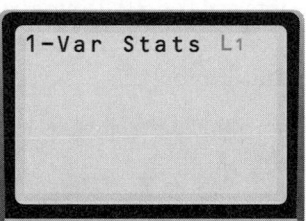

The calculator displays the **mean** and **standard deviation** of the data.

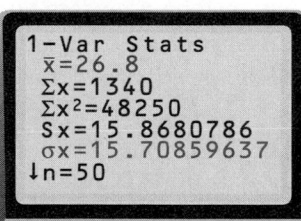

Scatter Plots and Curve Fitting

Your graphing calculator may have a curve-fitting feature that lets you fit a line, parabola, or other type of curve to a scatter plot. For example, suppose you want to make a scatter plot and find a line of fit for the following bicycle data from page 66:

Years since 1965	0	5	10	15	20	25
Number of bicycles (in millions)	21	36	43	62	79	95

On the TI-82, follow these steps:

Enter the years since 1965 in list L1. Enter the numbers of bicycles in list L2.

Use the linear regression feature to find an equation of the least-squares line.

Graph the **data pairs** and the **regression equation** in the same coordinate plane.

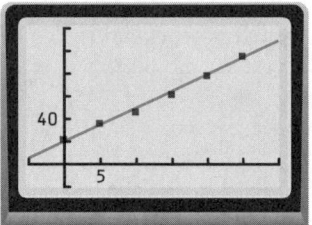

Generating Random Numbers

Most scientific and graphing calculators will generate random numbers between 0 and 1. On the TI-82, this is accomplished using the *rand* feature, as shown below.

On the TI-82, you can find the rand feature by pressing **MATH**.

You get a new random number each time you press **ENTER**.

You can transform a random number *x* in the interval $0 < x < 1$ into a random number in any interval. For example, suppose you want to simulate rolling a six-sided die by generating random integers from 1 to 6.

Use this formula. In general, the expression **int(n * rand) + m** generates random integers from m to $m + n - 1$.

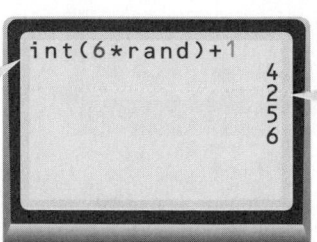

Press **ENTER** to simulate each roll of the die.

30. Students in a history class obtained these scores on an exam:

75, 88, 63, 79, 90, 52, 81, 68, 84, 77,
92, 75, 64, 70, 57, 89, 66, 96, 75, 83

   a. Make a histogram of the scores. Adjust the viewing window so that the intervals $50 \le x \le 100$ and $0 \le y \le 8$ are shown. Use Xscl = 10 and Yscl = 1.

   b. Find the mean and standard deviation of the scores.

31. Use the data on swimming given in the table.

   a. Find an equation of the least-squares line for the data.

   b. Graph the data pairs (x, y) and the equation of the least-squares line in the same coordinate plane.

32. Explain how you can use a calculator to generate random integers from 10 to 25.

Men's Winning Times in Olympic 400 m Freestyle Swimming	
x = years since 1960	y = time (seconds)
0	258.3
4	252.2
8	249.0
12	240.27
16	231.93
20	231.31
24	231.23
28	226.95
32	225.00

PROGRAMMING YOUR CALCULATOR

You may be able to program your graphing calculator to carry out mathematical procedures that are not built in. The following program for the TI-82 determines the number of real or imaginary solutions of a quadratic equation $ax^2 + bx + c = 0$. It also finds these solutions.

```
:Input "ENTER A:",A:Input "ENTER B:",B:
 Input "ENTER C:",C
:B²−4AC→D:−B/(2A)→R:(√ abs D)/(2A)→S
:ClrHome
:If D>0:Then:Disp "TWO REAL:",R+S,R−S
:Else
:If D=0:Then:Disp "ONE REAL:",R
:Else
:Disp "TWO IMAGINARY:","R+SI AND R−SI","WHERE"
:Output(4,1,"R="):Output(4,3,R):Output(5,1,"S="):
 Output(5,3,S)
:End:End
```

You can use this program to solve $x^2 + 4x + 13 = 0$, for example.

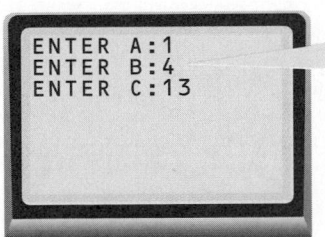

ENTER A:1
ENTER B:4
ENTER C:13

Enter the equation's coefficients.

The solutions of $x^2 + 4x + 13 = 0$ are $-2 \pm 3i$.

TWO IMAGINARY:
R+SI AND R−SI
WHERE
R=−2
S=3

33. Enter the program on the previous page into your calculator. (If you don't have a TI-82, you will have to modify some of the commands.) Use the program to solve each quadratic equation.

    a. $x^2 + 3x - 18 = 0$ **b.** $25x^2 - 20x + 4 = 0$ **c.** $16x^2 = 8x - 33$

34. Write a program for your calculator that solves linear equations of the form $ax + b = 0$.

USING A SPREADSHEET

Suppose a high school volunteer club sells T-shirts to raise money. The club raises $3000 in 1995 and projects that this amount will increase by 25% per year. In what year will the amount raised exceed $7000?

You can solve this problem using a computer spreadsheet. A spreadsheet is made up of cells named by a column letter and a row number, such as A3 or B4. You can enter a label, a number, or a formula into a cell.

row numbers   column letters

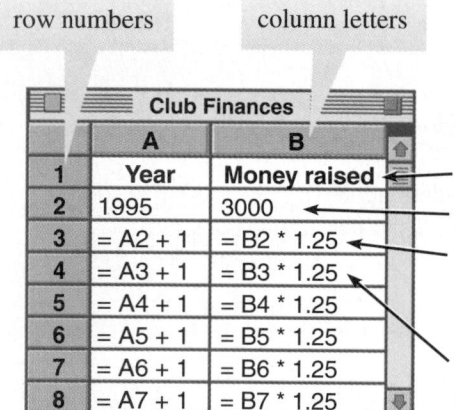

Club Finances

	A	B
1	Year	Money raised
2	1995	3000
3	= A2 + 1	= B2 * 1.25
4	= A3 + 1	= B3 * 1.25
5	= A4 + 1	= B4 * 1.25
6	= A5 + 1	= B5 * 1.25
7	= A6 + 1	= B6 * 1.25
8	= A7 + 1	= B7 * 1.25

Cell B1 contains the label "Money raised."

Cell B2 contains the number 3000.

Cell B3 contains the formula "= B2 * 1.25." This formula tells the computer to take the number in cell B2, multiply it by 1.25, and store the result in cell B3.

Instead of typing a formula into each cell individually, you can use the spreadsheet's *copy* or *fill down* command.

The computer replaces all formulas in the spreadsheet with calculated values, as shown at the right. You can see that the amount of money raised by the volunteer club will **exceed $7000** in 1999.

Club Finances

	A	B
1	Year	Money raised
2	1995	3000.00
3	1996	3750.00
4	1997	4687.50
5	1998	5859.38
6	1999	7324.22
7	2000	9155.27
8	2001	11444.09

35. Suppose you have $600 in a savings account that pays 5% annual interest. Use a spreadsheet to answer parts (a) and (b).

    a. How much money will you have in your account after 10 years if you make no deposits or withdrawals?

    b. How many full years must you wait before the value of your account exceeds $800?

Table of Measures

Time

60 seconds (s) = 1 minute (min)
60 minutes = 1 hour (h)
24 hours = 1 day
7 days = 1 week
4 weeks (approx.) = 1 month

$\left.\begin{array}{l}\text{365 days} \\ \text{52 weeks (approx.)} \\ \text{12 months}\end{array}\right\}$ = 1 year

10 years = 1 decade
100 years = 1 century

Metric

Length

10 millimeters (mm) = 1 centimeter (cm)

$\left.\begin{array}{l}\text{100 cm} \\ \text{1000 mm}\end{array}\right\}$ = 1 meter (m)

1000 m = 1 kilometer (km)

Area

100 square millimeters = 1 square centimeter
(mm^2) (cm^2)
$10,000\ cm^2$ = 1 square meter (m^2)
$10,000\ m^2$ = 1 hectare (ha)

Volume

1000 cubic millimeters = 1 cubic centimeter
(mm^3) (cm^3)
$1,000,000\ cm^3$ = 1 cubic meter (m^3)

Liquid Capacity

1000 milliliters (mL) = 1 liter (L)
1000 L = 1 kiloliter (kL)

Mass

1000 milligrams (mg) = 1 gram (g)
1000 g = 1 kilogram (kg)
1000 kg = 1 metric ton (t)

Temperature — Degrees Celsius (°C)

0°C = freezing point of water
37°C = normal body temperature
100°C = boiling point of water

United States Customary

Length

12 inches (in.) = 1 foot (ft)

$\left.\begin{array}{l}\text{36 in.} \\ \text{3 ft}\end{array}\right\}$ = 1 yard (yd)

$\left.\begin{array}{l}\text{5280 ft} \\ \text{1760 yd}\end{array}\right\}$ = 1 mile (mi)

Area

144 square inches $(in.^2)$ = 1 square foot (ft^2)
$9\ ft^2$ = 1 square yard (yd^2)

$\left.\begin{array}{l}43,560\ ft^2 \\ 4840\ yd^2\end{array}\right\}$ = 1 acre (A)

Volume

1728 cubic inches $(in.^3)$ = 1 cubic foot (ft^3)
$27\ ft^3$ = 1 cubic yard (yd^3)

Liquid Capacity

8 fluid ounces (fl oz) = 1 cup (c)
2 c = 1 pint (pt)
2 pt = 1 quart (qt)
4 qt = 1 gallon (gal)

Weight

16 ounces (oz) = 1 pound (lb)
2000 lb = 1 ton (t)

Temperature — Degrees Fahrenheit (°F)

32°F = freezing point of water
98.6°F = normal body temperature
212°F = boiling point of water

Table of Symbols

Symbol		Page	Symbol		Page
$\begin{bmatrix} 1 & 0 \\ 0 & 1 \end{bmatrix}$	matrix	16	$_nP_r$	number of permutations of r objects taken from n objects	580
$a_{2,1}$	element in first column and second row of a matrix A	16	$_nC_r$	number of combinations of r objects taken from n objects	585
x_1	x sub 1	46	$P(A)$	probability of event A	611
$f(x)$	f of x, or the value of f at x	60	$P(B \mid A)$	probability of event B given that event A has occurred	626
b^{-n}	$\frac{1}{b^n}, b \neq 0$	101			
e	irrational number, about 2.718	116	sin	sine	656
f^{-1}	inverse of function f	142	cos	cosine	656
$\log_b x$	base-b logarithm of x	149	$\circ$ $'$ $''$	degrees, minutes, and seconds	657
$\log x$	base-10 logarithm of x	150	$\sin^{-1}$	inverse sine	658
$\ln x$	base-e logarithm of x	150	$\cos^{-1}$	inverse cosine	658
$\pm$	plus-or-minus	187	θ	theta; name of an angle, or measure of an angle	658
$\sqrt{a}$	the nonnegative square root of a	187			
$\bar{x}$	x-bar; the mean of $x_1, x_2, \ldots, x_n$	268	tan	tangent	663
σ	sigma; the standard deviation of a data set	268	$\tan^{-1}$	inverse tangent	664
			$\cdot$	multiplication, times ($\times$)	774
A^{-1}	inverse of matrix A	304	$\lvert a \rvert$	absolute value of a	774
$\sqrt[n]{a}$	nth root of a	345	a^n	nth power of a	776
π	pi; irrational number, about 3.14	358	$\neq$	is not equal to	786
i	$\sqrt{-1}$	359	$>$	is greater than	787
			$<$	is less than	787
$\lvert a + bi \rvert$	the magnitude of a complex number $a + bi$	366	$\geq$	is greater than or equal to	787
$-a$	opposite of a	373	$\leq$	is less than or equal to	787
$\frac{1}{a}$	reciprocal of a, $a \neq 0$	374	$\approx$	is approximately equal to	790
			(x, y)	ordered pair	794
∞	infinity	405	$\angle A$	angle A, or measure of $\angle A$	800
Σ	summation	490	$\triangle ABC$	triangle ABC	801
$!$	factorial	579			

Properties of Algebra

Properties of Real Numbers

Let a, b, and c be real numbers.

Group Properties of Real Numbers under Addition, p. 373

- **Closure:** $a + b$ is a real number.
- **Identity:** $a + 0 = 0 + a = a$. 0 is the identity for addition.
- **Inverse:** There is a real number $-a$, called the *opposite*, or *additive inverse*, of a, such that $a + (-a) = 0$.
- **Associative:** $(a + b) + c = a + (b + c)$
- **Commutative:** $a + b = b + a$

Group Properties of Real Numbers under Multiplication, p. 374

- **Closure:** ab is a real number.
- **Identity:** $a \cdot 1 = 1 \cdot a = a$. 1 is the identity for multiplication.
- **Inverse:** If a is a nonzero real number, then there is a real number $\frac{1}{a}$, called the *reciprocal*, or *multiplicative inverse*, of a, such that $a \cdot \frac{1}{a} = 1$.
- **Associative:** $(ab)c = a(bc)$
- **Commutative:** $ab = ba$

Field Properties of Real Numbers under Addition and Multiplication, p. 377:
The field properties include all of the properties listed above as well as the *distributive property*:

- **Distributive:** $a(b + c) = ab + ac$

Properties of Exponents

Let b, m, and n be positive real numbers.

Zero Exponent, p. 776: $b^0 = 1$

Negative Exponent, p. 101: $b^{-n} = \frac{1}{b^n}$

Product of Powers, p. 94: $b^m \cdot b^n = b^{m+n}$

Quotient of Powers, p. 94: $\frac{b^m}{b^n} = b^{m-n}$

Power of a Power, p. 94: $(b^m)^n = b^{mn}$

Properties of Exponents and Radicals

Let a, b, and c be positive real numbers, and let m and n be positive integers.

Relationship between Exponents and Radicals, pp. 102, 346: $b^{1/n} = \sqrt[n]{b} = c$ if and only if $b = c^n$

Power or Root of a Product, p. 346

$(ab)^m = a^m \cdot b^m$ and $\sqrt[m]{ab} = \sqrt[m]{a} \cdot \sqrt[m]{b}$

Power or Root of a Quotient, p. 346

$\left(\frac{a}{b}\right)^m = \frac{a^m}{b^m}$ and $\sqrt[m]{\frac{a}{b}} = \frac{\sqrt[m]{a}}{\sqrt[m]{b}}$

Power or Root of a Power, p. 346

$\left(b^m\right)^n = b^{mn} = b^{nm} = \left(b^n\right)^m$ and

$b^{n/m} = \sqrt[m]{b^n} = \left(\sqrt[m]{b}\right)^n$

Properties of Logarithms

Let M, N, b, and c be positive real numbers with $b \neq 1$ and $c \neq 1$.

Relationship between Exponents and Logarithms, p. 149: $\log_b M = N$ if and only if $b^N = M$

Logarithm of a Product, p. 156

$\log_b MN = \log_b M + \log_b N$

Logarithm of a Quotient, p. 156

$\log_b \frac{M}{N} = \log_b M - \log_b N$

Logarithm of a Power, p. 156

$\log_b M^k = k \log_b M$ for any real k

Change of Base, p. 163: $\log_b M = \frac{\log_c M}{\log_c b}$

Visual Glossary of Graphs

Constant

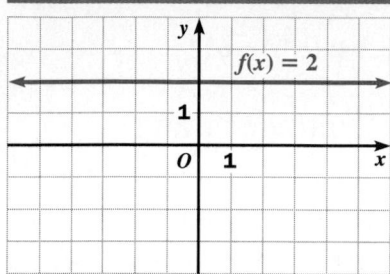

$f(x) = 2$

Direct Variation

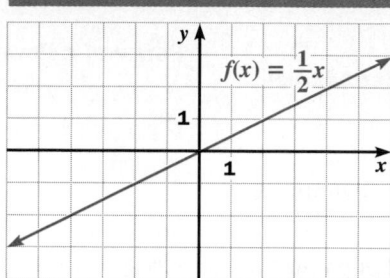

$f(x) = \frac{1}{2}x$

Linear

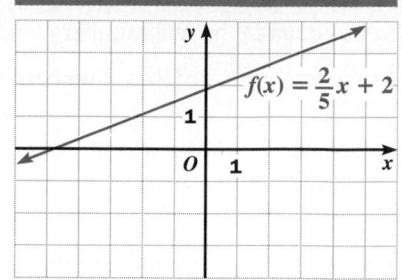

$f(x) = \frac{2}{5}x + 2$

Exponential Growth

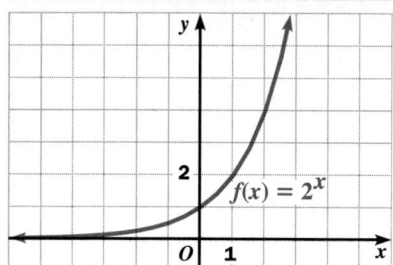

$f(x) = 2^x$

Exponential Decay

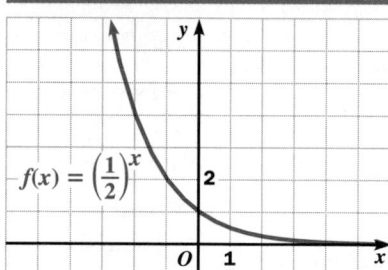

$f(x) = \left(\frac{1}{2}\right)^x$

Exponential/Logarithmic (inverse functions)

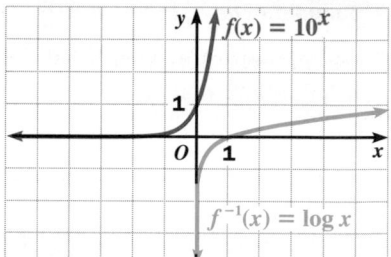

$f(x) = 10^x$

$f^{-1}(x) = \log x$

Quadratic

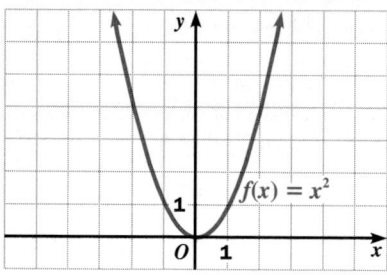

$f(x) = x^2$

Quadratic/Square Root (inverse functions)

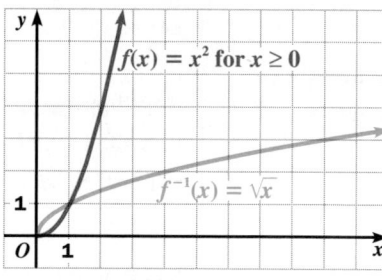

$f(x) = x^2$ for $x \geq 0$

$f^{-1}(x) = \sqrt{x}$

Cubic/Cube Root (inverse functions)

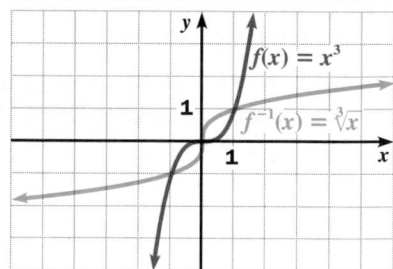

$f(x) = x^3$

$f^{-1}(x) = \sqrt[3]{x}$

Polynomial (even degree)

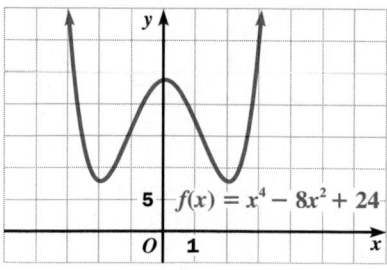

$f(x) = x^4 - 8x^2 + 24$

Polynomial (odd degree)

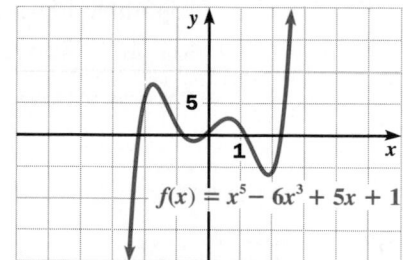

$f(x) = x^5 - 6x^3 + 5x + 1$

Rational

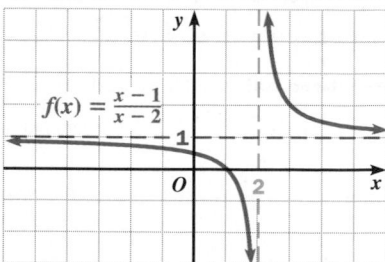

$f(x) = \frac{x-1}{x-2}$

Sine	Cosine	Tangent

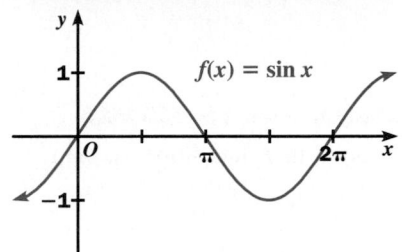
$f(x) = \sin x$

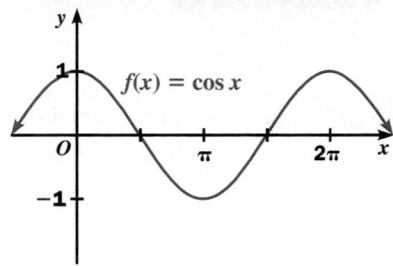
$f(x) = \cos x$

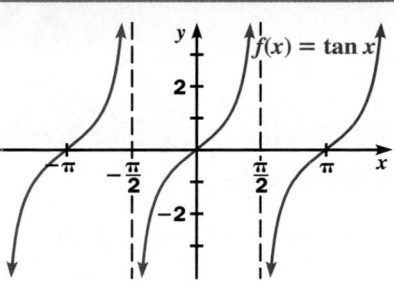
$f(x) = \tan x$

Transformations

Translation	Vertical Stretch	Reflection over x-axis

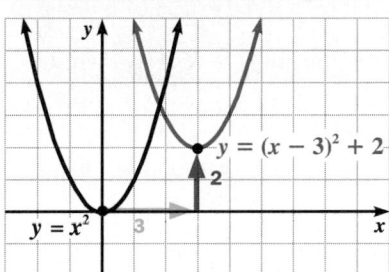
$y = (x - 3)^2 + 2$
$y = x^2$

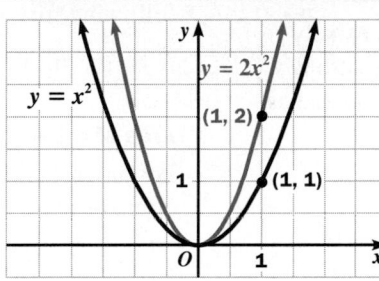
$y = 2x^2$
$y = x^2$
$(1, 2)$
$(1, 1)$

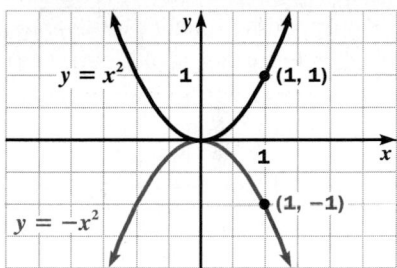
$y = x^2$
$(1, 1)$
$y = -x^2$
$(1, -1)$

Conic Sections

Circle	Horizontal Ellipse	Vertical Ellipse

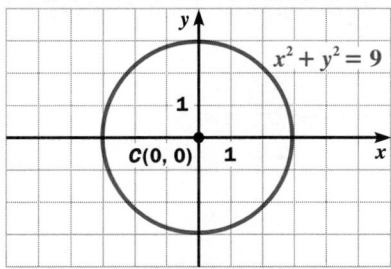
$x^2 + y^2 = 9$
$C(0, 0)$

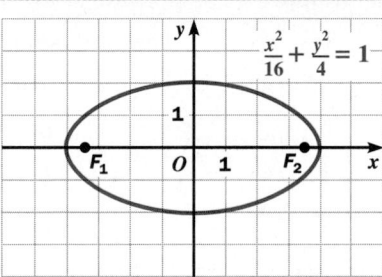
$\dfrac{x^2}{16} + \dfrac{y^2}{4} = 1$
F_1 F_2

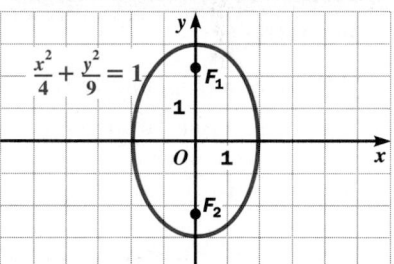
$\dfrac{x^2}{4} + \dfrac{y^2}{9} = 1$
F_1
F_2

Horizontal Parabola	Horizontal Hyperbola	Vertical Hyperbola

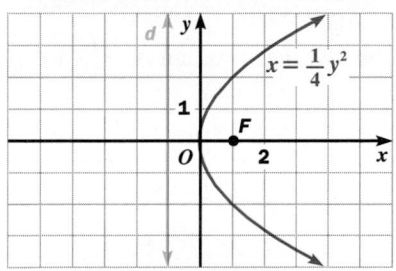
$x = \dfrac{1}{4}y^2$
d
F

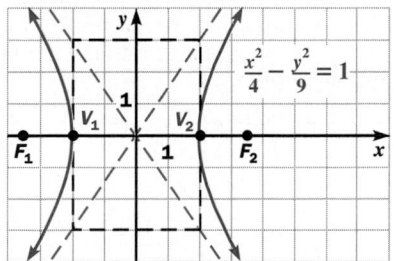
$\dfrac{x^2}{4} - \dfrac{y^2}{9} = 1$
F_1 V_1 V_2 F_2

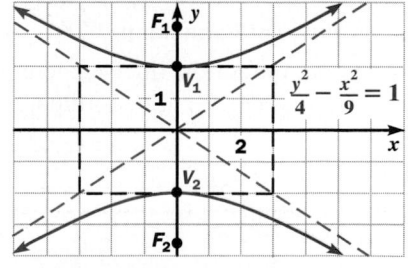
$\dfrac{y^2}{4} - \dfrac{x^2}{9} = 1$
F_1 V_1 V_2 F_2

Formulas

Formulas and Theorems about Equations

The Quadratic Formula, p. 216

The solutions of any quadratic equation in the form $ax^2 + bx + c = 0$ are:

$$x = \frac{-b + \sqrt{b^2 - 4ac}}{2a} \text{ and } x = \frac{-b - \sqrt{b^2 - 4ac}}{2a}$$

Fundamental Theorem of Algebra, p. 421

A polynomial function of degree n has exactly n complex zeros, provided each double zero is counted as 2 zeros, each triple zero is counted as 3 zeros, and so on.

Rational Zeros Theorem, p. 421

Let f be a polynomial function with integral coefficients. Then the only possible rational zeros of f are $\frac{p}{q}$ where p is a divisor of the constant term of $f(x)$ and q is a divisor of the leading coefficient.

Formulas from Coordinate Geometry

The Slope of a Line, pp. 47, 53

The slope of a line through the points (x_1, y_1) and (x_2, y_2) is $\frac{\text{vertical change}}{\text{horizontal change}} = \frac{y_2 - y_1}{x_2 - x_1}$.

The Distance Formula, p. 514

The distance d between two points (x_1, y_1) and (x_2, y_2) is:

$$d = \sqrt{(x_2 - x_1)^2 + (y_2 - y_1)^2}$$

The Midpoint Formula, p. 514

The midpoint of the line segment joining the points (x_1, y_1) and (x_2, y_2) has coordinates:

$$\left(\frac{x_1 + x_2}{2}, \frac{y_1 + y_2}{2} \right)$$

Parallel and Perpendicular Lines, pp. 515, 516

If line l_1 has slope a_1 and line l_2 has slope a_2, then:

$$l_1 \parallel l_2 \text{ if and only if } a_1 = a_2$$
$$l_1 \perp l_2 \text{ if and only if } (a_1)(a_2) = -1$$

Formulas from Trigonometry

Area of a Triangle, p. 678

The area K of a $\triangle ABC$ is given by any of the following formulas:

$$K = \frac{1}{2}bc \sin A \quad K = \frac{1}{2}ac \sin B \quad K = \frac{1}{2}ac \sin C$$

Area of a Sector, p. 680

For any sector determined by a central angle θ in a circle of radius r:

$$\text{area of the sector} = \frac{\theta}{360} \cdot \pi r^2$$

Law of Sines, p. 684

For any $\triangle ABC$:

$$\frac{\sin A}{a} = \frac{\sin B}{b} = \frac{\sin C}{c}$$

Law of Cosines, p. 691

For any $\triangle ABC$:

$$a^2 = b^2 + c^2 - 2bc \cos A$$
$$b^2 = a^2 + c^2 - 2ac \cos B$$
$$c^2 = a^2 + b^2 - 2ab \cos C$$

Converting Degree Measure to Radian Measure, p. 714

Since there are $360°$ and 2π radians in a full circle, use the fact that $1° = \frac{\pi}{180}$ radians to convert degrees to radians, and use the fact that $1 \text{ radian} = \frac{180°}{\pi}$ to convert radians to degrees.

Formulas from Statistics

The Mean of a Data Set, p. 790

For the data $x_1, x_2, x_3, \ldots, x_n$, the mean of the n data values, $\bar{x}$, is given by this formula:

$$\bar{x} = \frac{x_1 + x_2 + \cdots + x_n}{n}$$

The Standard Deviation of a Data Set, p. 268

Standard deviation measures the variability of data. For the data $x_1, x_2, x_3, \ldots, x_n$, where $\bar{x}$ is the mean of the data values and n is the number of data values, the standard deviation σ is given by this formula:

$$\sigma = \sqrt{\frac{(x_1 - \bar{x})^2 + (x_2 - \bar{x})^2 + \cdots + (x_n - \bar{x})^2}{n}}$$

The Margin of Error for a Sample Proportion, p. 274

When a random sample of size n is taken from a large population, the sample proportion has a margin of error approximated by this formula:

$$\text{margin of error} \approx \pm \frac{1}{\sqrt{n}}$$

Formulas from Probability

Experimental Probability, p. 612

Suppose an experiment is performed that consists of a certain number of trials. The experimental probability of an event A is:

$$P(A) = \frac{\text{number of trials where } A \text{ happens}}{\text{total number of trials}}$$

Theoretical Probability, p. 613

Suppose a procedure can result in n equally likely outcomes. If an event A corresponds to m of these outcomes, then the theoretical probability of A is:

$$P(A) = \frac{m}{n}$$

Multiple Events, p. 619

- **Independent Events:** The occurrence of event A does not affect whether event B happens.

$$P(A \text{ and } B) = P(A) \cdot P(B)$$

- **Mutually Exclusive Events:** Events A and B cannot both happen.

$$P(A \text{ or } B) = P(A) + P(B)$$

- **Complementary Events:** Events A and B are mutually exclusive and one of the events must happen.

$$P(B) = 1 - P(A)$$

Conditional Probability, p. 627

For any events A and B:

$$P(A \text{ and } B) = P(A) \cdot P(B \mid A)$$

$$P(B \mid A) = \frac{P(A \text{ and } B)}{P(A)}$$

Binomial Distribution, p. 634

In a binomial experiment, there are two mutually exclusive outcomes, often called *success* and *failure*. For each trial, $P(\text{success}) = p$ and $P(\text{failure}) = 1 - p$. The probability of k successes in n independent trials is given by this formula:

$$P\left(\begin{matrix} k \text{ successes} \\ \text{in } n \text{ trials} \end{matrix}\right) = {}_nC_k \cdot p^k \cdot (1 - p)^{n-k}$$

The probability values for $k = 0, 1, 2, \ldots, n$ form a binomial distribution.

Normal Distribution, p. 640

In a normal distribution:

- 50% of the data lies below the mean and 50% of the data lies above the mean.

- about 68% of the distribution lies within one standard deviation of the mean.

- about 95% of the distribution lies within two standard deviations of the mean.

- about 99% of the distribution lies within three standard deviations of the mean.

z-score, p. 641

The x-values in a normal distribution with mean $\bar{x}$ and standard deviation σ can be converted to a z-score by this formula:

$$z = \frac{x - \bar{x}}{\sigma}$$

A z-score gives the number of standard deviations that x lies from the mean. Obtaining a z-score for a data value is called *standardizing* the value.

Formulas from Discrete Mathematics

Explicit Formulas for Arithmetic and Geometric Sequences, p. 473

- The nth term of an arithmetic sequence with common difference d is given by this explicit formula:
$$t_n = t_1 + d(n - 1)$$

- The nth term of a geometric sequence with common ratio r is given by this explicit formula:
$$t_n = t_1 \cdot r^{n-1}$$

Arithmetic and Geometric Means, p. 475

- The term x between any two terms a and b of an arithmetic sequence is given by:
$$x = \frac{a+b}{2}$$

- The term x between any two terms a and b of a geometric sequence is given by:
$$x = \sqrt{ab}$$

Recursive Formulas for Arithmetic and Geometric Sequences, p. 481

- The nth term of an arithmetic sequence with common difference d is given by this recursive formula:
$$t_1 = \text{starting value}$$
$$t_n = t_{n-1} + d$$

- The nth term of a geometric sequence with common ratio r is given by this recursive formula:
$$t_1 = \text{starting value}$$
$$t_n = r(t_{n-1})$$

Sum of a Finite Arithmetic Series, p. 489

The sum S_n of the n terms of a finite arithmetic series $t_1 + t_2 + t_3 + t_4 + \cdots + t_n$ is:
$$S_n = \frac{n}{2}(t_1 + t_n)$$

Sum of a Finite Geometric Series, p. 496

The sum S_n of the finite geometric series $t_1 + t_1 \cdot r + t_1 \cdot r^2 + \cdots + t_1 \cdot r^{n-1}$ where $r \neq 1$ is:
$$S_n = \frac{t_1(1 - r^n)}{1 - r}$$

Sum of an Infinite Geometric Series, p. 498

The sum S of the infinite geometric series $t_1 + t_1 \cdot r + t_1 \cdot r^2 + t_1 \cdot r^3 + t_1 \cdot r^4 + \cdots$ where $|r| < 1$ is:
$$S = \frac{t_1}{1 - r}$$

The Multiplication Counting Principle, p. 578

If a choice is to be made among m things, and for each of those choices another choice is to be made among n things, then there are a total of $m \times n$ possibilities to choose from.

Permutations, p. 580

The number of permutations of r objects taken from a set of n objects is given by ${}_nP_r$ where
$${}_nP_r = \frac{n!}{(n - r)!}.$$

Permutations with Repeated Elements, p. 580

If a set of n elements has one element repeated q_1 times, another element repeated q_2 times, . . . , and the last element repeated q_k times, then the number of distinguishable permutations of the n elements is given by $\dfrac{n!}{q_1! q_2! \cdots q_k!}.$

Combinations, p. 586

The number of combinations of r objects taken from a set of n objects is given by ${}_nC_r$ where
$${}_nC_r = \frac{{}_nP_r}{{}_rP_r} = \frac{n!}{(n - r)! r!}.$$

The Binomial Theorem, p. 594

For any positive integer n,
$$(a + b)^n = {}_nC_0 a^n + {}_nC_1 a^{n-1}b +$$
$${}_nC_2 a^{n-2}b^2 + \cdots + {}_nC_{n-1}ab^{n-1} + {}_nC_n b^n$$

Formulas from Geometry

To find the perimeter P of any plane figure made up of line segments, find the sum of the lengths of each side of the figure.

Formulas for the areas of several plane figures as well as the volumes and surface areas of several space figures are given on the next page.

For the formulas below, let A = area, C = circumference, V = volume, B = the area of the base of a space figure, S.A. = the (total) surface area of a space figure, and L.S.A. = the lateral surface area of a space figure. Note that $\pi \approx 3.14$.

Parallelogram

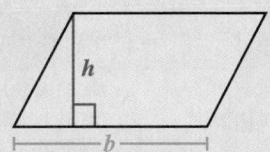

A = base $\times$ height

$A = bh$

Triangle

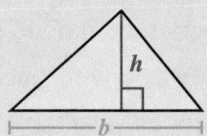

$A = \dfrac{1}{2} \times$ base $\times$ height

$A = \dfrac{1}{2}bh$

Trapezoid

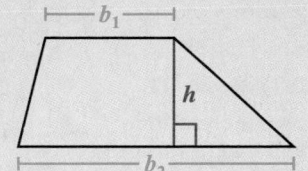

$A = \dfrac{1}{2} \times$ sum of bases $\times$ height

$A = \dfrac{1}{2}(b_1 + b_2)h$

Circle

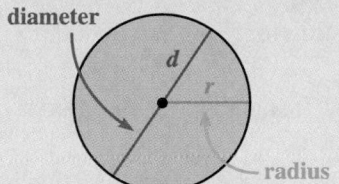

$C = 2\pi r$ or $C = \pi d$

$A = \pi r^2$

Right Rectangular Prism

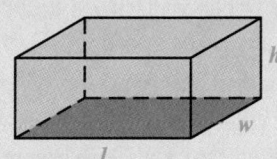

V = area of base $\times$ height

$V = Bh = lwh$

S.A. $= 2(lw + wh + lh)$

Right Cylinder

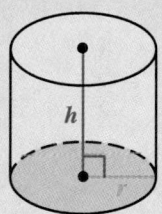

V = area of base $\times$ height

$V = Bh = \pi r^2 h$

L.S.A. $= 2\pi rh$

S.A. $= 2\pi r^2 + 2\pi rh$

Right Regular Pyramid

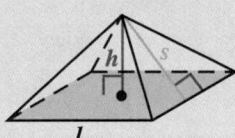

$V = \dfrac{1}{3} \times$ area of base $\times$ height

$V = \dfrac{1}{3}Bh$

L.S.A. $= \dfrac{1}{2} \times \begin{matrix} \text{perimeter} \\ \text{of base} \end{matrix} \times \begin{matrix} \text{slant} \\ \text{height} \end{matrix}$

For a base with n sides:

L.S.A. $= \dfrac{1}{2}nls$

Right Circular Cone

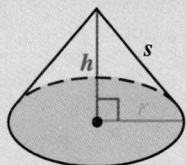

$V = \dfrac{1}{3} \times$ area of base $\times$ height

$V = \dfrac{1}{3}Bh = \dfrac{1}{3}\pi r^2 h$

L.S.A. $=$
$\dfrac{1}{2} \times \begin{matrix} \text{circumference} \\ \text{of base} \end{matrix} \times \begin{matrix} \text{slant} \\ \text{height} \end{matrix}$

L.S.A. $= \dfrac{1}{2}(2\pi r)s = \pi rs$

S.A. $= \pi r^2 + \pi rs$

Sphere

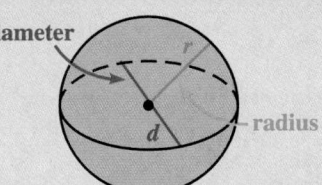

$V = \dfrac{4}{3}\pi r^3$ or $V = \dfrac{1}{6}\pi d^3$

S.A. $= 4\pi r^2$ or   S.A. $= \pi d^2$

1

Solving Inequalities

Learn how to...

• **solve inequalities involving different types of functions**

So you can...

• **solve problems about investing, for example**

Financial planners recommend that you invest money you won't need for many years in stocks. The table at the right shows why. Over the long run, stocks have increased in value more rapidly than other types of investments.

Investment Data (1926–1995)	
Type of investment	Average annual return
Stocks	10.5%
Bonds	5.2%
Cash reserves	3.7%

EXAMPLE 1 Application: Investing

Alma Ramos puts $5000 in a mutual fund investing in stocks and $6000 in a mutual fund investing in bonds.

a. Write equations giving the predicted values $s(t)$ and $b(t)$ of Alma's stock and bond investments, respectively, after t years.

b. Predict when Alma's stock investment will be worth more than her bond investment.

SOLUTION

The growth factor for each function is 1 plus the average annual return expressed as a decimal.

a. Use the **initial amounts invested** and the average annual returns in the table to write $s(t)$ and $b(t)$ as exponential functions of t.

$$s(t) = 5000(1 + 0.105)^t \qquad s(t) = 5000(1.105)^t$$
$$b(t) = 6000(1 + 0.052)^t \qquad b(t) = 6000(1.052)^t$$

b. Use a spreadsheet and the equations from part (a) to calculate $s(t)$ and $b(t)$ for different values of t. Find the t-values for which $s(t) > b(t)$.

The inequality $s(t) > b(t)$ is true when $t \geq 4$.

Alma's Investments			
	A	**B**	**C**
1	t	$s(t)$	$b(t)$
2	0	5000.00	6000.00
3	1	5525.00	6312.00
4	2	6105.13	6640.22
5	3	6746.16	6985.52
	4	7454.51	7348.76
7	5	8237.23	7730.90
8	6	9102.14	8132.90

Alma's stock investment will be worth more than her bond investment after 4 years.

BY THE WAY

Although stocks make good long-term investments, they can be risky in the short run. For example, in October of 1987, stocks lost more than 20% of their value in a single day.

THINK AND COMMUNICATE

1. a. **Technology** Use a spreadsheet to find the values of $s(t)$ and $b(t)$ in Example 1 for $t = 3.0, 3.1, 3.2, \ldots, 4.0$. Write a solution of $s(t) > b(t)$ that is more precise than the solution given in Example 1.

   b. Describe how you can use a spreadsheet to make the solution of $s(t) > b(t)$ as precise as desired.

In Example 1, you used a spreadsheet to solve an inequality. You can also solve inequalities using graphs.

EXAMPLE 2

Solve $x^2 - 6x + 11 < 5$.

SOLUTION

The solution is the set of x-values for which the graph of $y = x^2 - 6x + 11$ lies below the graph of $y = 5$. First find where these graphs intersect, as shown.

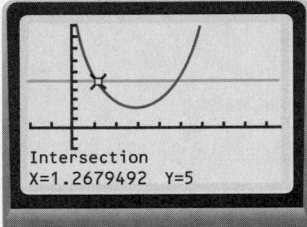

Intersection
X=1.2679492   Y=5

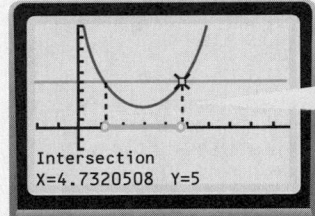

Intersection
X=4.7320508   Y=5

The graph of $y = x^2 - 6x + 11$ lies below the graph of $y = 5$ when $1.27 < x < 4.73$.

The solution is $1.27 < x < 4.73$.

EXAMPLE 3

Solve $\sqrt{2x - 5} \geq \sqrt{x + 3} - 1$.

SOLUTION

The inequality has the form $y_1 \geq y_2$ where $y_1 = \sqrt{2x - 5}$ and $y_2 = \sqrt{x + 3} - 1$. This inequality is equivalent to $y_1 - y_2 \geq 0$, which you can solve by graphing the function $y_3 = y_1 - y_2$ and finding the x-values for which the graph lies on or above the x-axis.

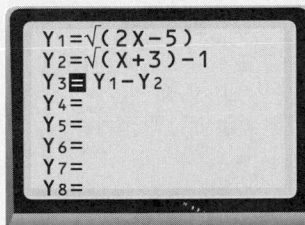

Y1=√(2X−5)
Y2=√(X+3)−1
Y3◼Y1−Y2
Y4=
Y5=
Y6=
Y7=
Y8=

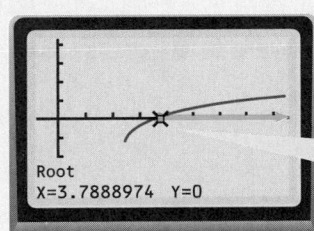

Root
X=3.7888974   Y=0

The x-intercept of the graph of y_3 is about 3.79. The graph lies on or above the x-axis when $x \geq 3.79$.

The solution is $x \geq 3.79$.

THINK AND COMMUNICATE

2. Use the graphs in Example 2 to solve $x^2 - 6x + 11 \geq 5$.

3. Consider the inequality $y_1 < y_2$ where y_1 and y_2 are as defined in Example 3. Explain why the solution of this inequality is *not* $x < 3.79$. Then solve the inequality.

Another way to solve an inequality is to graph the associated *truth function* using a graphing calculator or graphing software. The value of an inequality's truth function is 1 for x-values that satisfy the inequality and 0 for x-values that do not satisfy the inequality.

EXAMPLE 4

Solve $\dfrac{3}{x + 2} \leq \dfrac{1}{x - 2}$.

SOLUTION

Graph the inequality's truth function. On most graphing calculators and graphing software, you can define the functions $y_1 = \dfrac{3}{x + 2}$ and $y_2 = \dfrac{1}{x - 2}$ and then enter the truth function as $y_3 = (y_1 \leq y_2)$.

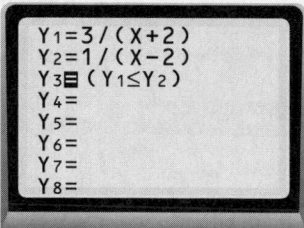

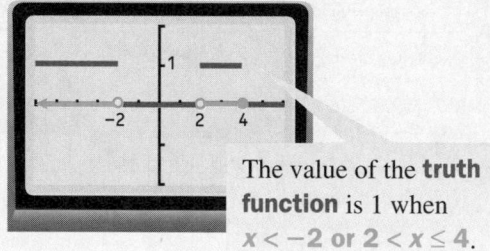

The value of the **truth function** is 1 when $x < -2$ or $2 < x \leq 4$.

The solution is $x < -2$ or $2 < x \leq 4$.

THINK AND COMMUNICATE

4. In Example 4, how do you know that 4 is a solution of the given inequality, but -2 and 2 are not solutions?

Exercises and Applications

Technology **Solve each inequality. Round all numbers in your answers to the nearest hundredth.**

1. $2^x \geq 8$

2. $(0.5)^{x + 1} \leq 16$

3. $6e^{0.17x} < 34$

4. $100(1.09)^x > 250(1.03)^x$

5. $(0.77)^x > 0$

6. $\ln x \geq 2$

7. $3 - \ln x \geq \sqrt{2}$

8. $\log (|x| + 1) < 0$

9. $\log_3 x \leq \log_2 x$

10. **INVESTING** Look back at Example 1 and the table of investment data on page 826. Suppose Alma Ramos has $3000 in cash reserves in addition to her stock and bond investments. Predict when Alma's stock investment will be worth more than her bond investment and cash reserves combined.

11. **a.** Sketch the graph of $y = \ln x$. Use the graph to explain why $\ln b > \ln a$ if and only if $b > a$, where a and b are any positive numbers.

   b. Use the result from part (a) to solve the inequality $e^{2x} > 8$ algebraically. (*Hint:* Take the natural logarithm of both sides.)

   c. Challenge Solve the inequality $s(t) > b(t)$ in Example 1 algebraically.

 Technology **Solve each inequality. Round all numbers in your answers to the nearest hundredth.**

12. $x^2 - 7x + 10 \leq 0$ 13. $2x^2 - 3x - 9 < -4$ 14. $3x^2 - 20x + 35 > 5$

15. $-0.7x^2 + 2.3x + 3.6 \leq 1.8$ 16. $-x^2 + 2x + 4 \geq 3x + 8$ 17. $-2x^2 - 7x > x^2 + 8x + 6$

18. **ENGINEERING** The fuel efficiency E of an average car, measured in miles per gallon, can be modeled by the equation

$$E = -0.0177s^2 + 1.48s + 3.39$$

where s is the speed of the car in miles per hour. For what speeds is an average car's fuel efficiency at least 25 mi/gal?

 Technology **Solve each inequality. Round all numbers in your answers to the nearest hundredth.**

19. $\sqrt{x - 1} \geq 2$ 20. $\sqrt{3x - 10} \geq \sqrt{x + 4}$ 21. $6.5 - \sqrt{x} < x$

22. $\sqrt{x + 1} \leq \sqrt{x} + 1$ 23. $\sqrt{2x} > \sqrt{5x - 4} - 2$ 24. $\sqrt[3]{x} + 3 < 2\sqrt{x}$

25. **OCEANOGRAPHY** The speed s of a wave depends on the depth d of the water through which it travels according to the equation

$$s = \sqrt{35.28d}$$

where d is in kilometers and s is in kilometers per minute. For what depths is wave speed between 7 km/min and 10 km/min?

 Technology **Solve each inequality. Round all numbers in your answers to the nearest hundredth.**

26. $\dfrac{1}{x} \leq 3$ 27. $\dfrac{5}{x + 2} > 1$ 28. $\dfrac{x + 4}{x - 2} \geq 0$

29. $\dfrac{x^2 + 1}{x - 1} < 0$ 30. $\dfrac{2}{x} < x + 1$ 31. $\dfrac{1}{x - 4} \geq \dfrac{x}{x - 6}$

32. **DENTISTRY** The pH level in your mouth t minutes after eating sugary foods can be modeled by this equation:

$$\text{pH} = \frac{65t^2 - 204t + 2340}{10t^2 + 360}$$

During what time interval after eating is the pH level in your mouth below 5?

33. Open-ended Problem Write an inequality. Solve the inequality using at least two of the methods presented in this appendix.

2

Translating Graphs of Exponential and Logarithmic Functions

Learn how to...

- **recognize how the parameters *h* and *k* affect the domains, ranges, and graphs of $y = b^{x-h} + k$ and $y = \log_b (x - h) + k$**

So you can...

- **translate graphs of exponential and logarithmic functions**

Look at the graph of the exponential function $y = 2^x$.

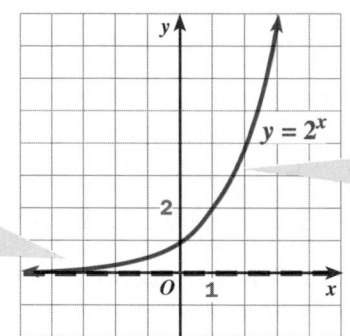

The **x-axis** is a horizontal *asymptote* of the graph. An asymptote is a line that a graph approaches more and more closely.

$y = 2^x$

The domain of $y = 2^x$ is all real numbers. The range is $y > 0$.

In each of the coordinate planes below, the graph of $y = 2^x$ is shown along with graphs of some related functions.

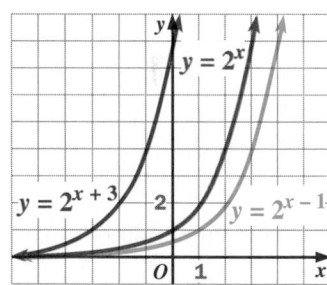

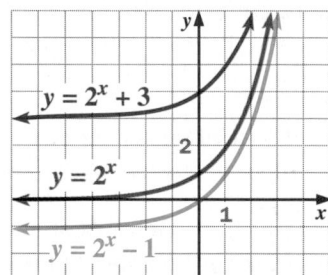

THINK AND COMMUNICATE

Use the graphs shown above to help you answer Questions 1 and 2.

1. a. Give the domains, ranges, and asymptotes for $y = 2^{x-1}$ and $y = 2^{x+3}$.

    b. How are the graphs of $y = 2^{x-1}$ and $y = 2^{x+3}$ geometrically related to the graph of $y = 2^x$?

2. a. Give the domains, ranges, and asymptotes for $y = 2^x - 1$ and $y = 2^x + 3$.

    b. How are the graphs of $y = 2^x - 1$ and $y = 2^x + 3$ geometrically related to the graph of $y = 2^x$?

3. What do you think are the domain, range, and asymptote(s) for $y = 2^{x-1} + 3$? How do you think the graph of $y = 2^{x-1} + 3$ is geometrically related to the graph of $y = 2^x$? Explain your reasoning.

You can generalize the results you obtained in the *Think and Communicate* questions as follows.

Exponential Functions of the Form $f(x) = b^{x-h} + k$

An exponential function $f(x) = b^{x-h} + k$ has these properties:

- The domain of f is all real numbers, and the range is $y > k$.
- The graph of f is the graph of $y = b^x$ translated horizontally h units and vertically k units. The function $y = b^x$ is called the *parent function* of f.
- The line $y = k$ is a horizontal asymptote of the graph of f.

EXAMPLE 1

a. Give the domain, range, and parent function of $f(x) = \left(\frac{1}{2}\right)^{x+1} - 3$.

b. Sketch the graphs of f and its parent function in the same coordinate plane. Identify any asymptotes of the graph of f.

SOLUTION

a. First write the equation for f in the form $f(x) = b^{x-h} + k$:

$$f(x) = \left(\frac{1}{2}\right)^{x-(-1)} + (-3)$$

For this function, $h = -1$ and $k = -3$.

The domain of f is all real numbers, and the range is $y > -3$.

The parent function of f is $y = \left(\frac{1}{2}\right)^x$.

b.

The graph of f is the graph of $y = \left(\frac{1}{2}\right)^x$ translated **left 1 unit** and **down 3 units**.

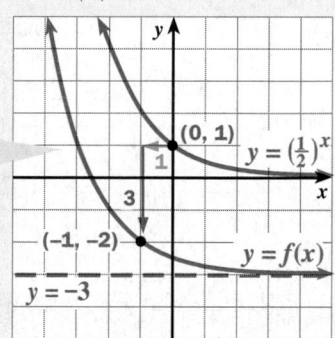

The line $y = -3$ is a horizontal asymptote of the graph of f.

THINK AND COMMUNICATE

4. In Example 1, explain how you know that the direction of the translation is left and down (rather than right and up, for example).

5. Tell how the graph of $y = b^{x-h} + k$ is geometrically related to the graph of $y = b^x$ if:

   a. $h > 0$ and $k > 0$ **b.** $h > 0$ and $k < 0$

   c. $h < 0$ and $k > 0$ **d.** $h < 0$ and $k < 0$

You can also translate graphs of logarithmic functions. The equations of the translated graphs have the form given below.

Logarithmic Functions of the Form $f(x) = \log_b (x - h) + k$

A logarithmic function $f(x) = \log_b (x - h) + k$ has these properties:

- The domain of f is $x > h$, and the range is all real numbers.
- The graph of f is the graph of $y = \log_b x$ translated horizontally h units and vertically k units. The function $y = \log_b x$ is the parent function of f.
- The line $x = h$ is a vertical asymptote of the graph of f.

EXAMPLE 2

a. Give the domain, range, and parent function of $f(x) = \log_2 (x - 3) + 4$.

b. Sketch the graphs of f and its parent function in the same coordinate plane. Identify any asymptotes of the graph of f.

SOLUTION

a. Note that the function $f(x) = \log_2 (x - 3) + 4$ is already in the form $f(x) = \log_b (x - h) + k$ with $h = 3$ and $k = 4$. The domain of f is $x > 3$, and the range is all real numbers. The parent function of f is $y = \log_2 x$.

b.

The graph of f is the graph of $y = \log_2 x$ translated **up 4 units** and **right 3 units**.

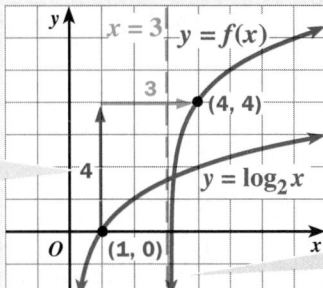

The line $x = 3$ is a vertical asymptote of the graph of f.

Exercises and Applications

For each function f:

a. Give the domain, range, and parent function of f.

b. Sketch the graphs of f and its parent function in the same coordinate plane. Identify any asymptotes of the graph of f.

1. $f(x) = 2^{x-3}$

2. $f(x) = 3^x + 1$

3. $f(x) = 4^{x-2} + 3$

4. $f(x) = e^{x+3} - 5$

5. $f(x) = \left(\dfrac{1}{2}\right)^{x+4} + 2$

6. $f(x) = \left(\dfrac{1}{3}\right)^{x-1} - 4$

7. $f(x) = \log_2 (x + 4)$

8. $f(x) = \log_2 x - 3$

9. $f(x) = \log_3 (x - 1) + 6$

10. $f(x) = \log_5 (x + 2) + 1$

11. $f(x) = \log (x - 3) - 1$

12. $f(x) = \ln (x + 5) - 2$

Selected Answers for Appendices

APPENDIX 1

Pages 828–829 Exercises and Applications

1. $x \geq 3$ **3.** $x < 10.20$ **5.** all real numbers **7.** $0 < x \leq 4.88$
9. $x \geq 1$ **13.** $-1 < x < 2.5$ **15.** $x \leq -0.65$ or $x \geq 3.94$
17. $-4.56 < x < -0.44$ **19.** $x \geq 5$ **21.** $x > 4.40$
23. $0.8 \leq x < 8$ **25.** $1.39 < d < 2.83$ **27.** $-2 < x < 3$
29. $x < 1$ **31.** $2 \leq x \leq 3$ or $4 < x < 6$ **33.** Answers may
vary. An example is given. Consider the inequality
$\sqrt{4x + 1} > 3$. One way to solve this inequality is to graph
$y = \sqrt{4x + 1}$ and $y = 3$ in the same coordinate plane and find
the x-values for which the graph of $y = \sqrt{4x + 1}$ lies above
the graph of $y = 3$. A second way to solve the inequality is to
graph its truth function, $y = (\sqrt{4x + 1} > 3)$, and find the
x-values for which the truth function equals 1. The solution
of the inequality is $x > 2$.

APPENDIX 2

Page 832 Exercises and Applications

1. a. domain: all real numbers; range: $y > 0$; parent
function: $y = 2^x$

b.
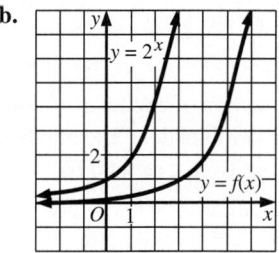

A horizontal asymptote of the graph of f is the x-axis.

3. a. domain: all real numbers; range: $y > 3$; parent
function: $y = 4^x$

b.
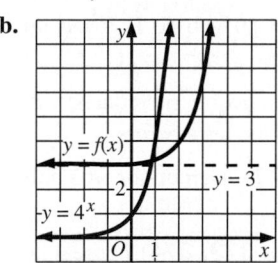

A horizontal asymptote of the graph of f is $y = 3$.

5. a. domain: all real numbers; range: $y > 2$; parent
function: $y = \left(\dfrac{1}{2}\right)^x$

b.
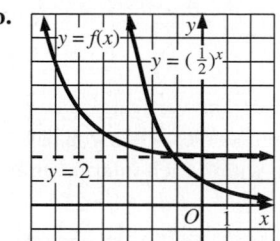

A horizontal asymptote of the graph of f is $y = 2$.

7. a. domain: $x > -4$; range: all real numbers; parent
function: $y = \log_2 x$

b.
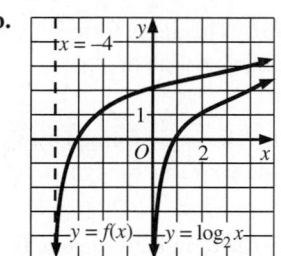

A vertical asymptote of the graph of f is $x = -4$.

9. a. domain: $x > 1$; range: all real numbers; parent
function: $y = \log_3 x$

b.
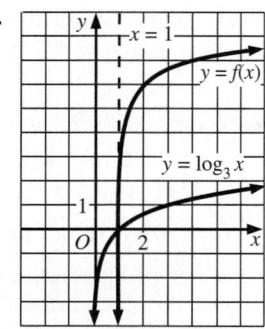

A vertical asymptote of the graph of f is $x = 1$.

11. a. domain: $x > 3$; range: all real numbers; parent
function: $y = \log x$

b.
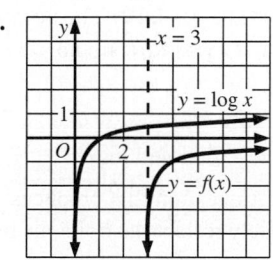

A vertical asymptote of the graph of f is $x = 3$.

Glossary

absolute frequency (p. 260) The vertical axis of a histogram shows absolute frequencies when you can read the actual count of data within each interval on the horizontal axis.

amplitude (p. 720) One half the difference between the maximum and minimum values of a sine or cosine function. The functions $y = a \sin bx$ and $y = a \cos bx$ have an amplitude of $|a|$.

angle of depression (p. 665) The angle formed by a horizontal line at eye level and your line of sight to an object lower than you are.

angle of elevation (p. 665) The angle formed by a horizontal line at eye level and your line of sight to an object higher than you are.

arithmetic mean (p. 475) The term between any two terms of an arithmetic sequence. If a, x, b form an arithmetic sequence, then $x = \dfrac{a + b}{2}$.

arithmetic sequence (p. 473) A sequence where the difference between any term and the term before it is constant.

arithmetic series (p. 488) The indicated sum of an arithmetic sequence.

associative property (pp. 373, 374) For real numbers a, b, and c, $(a + b) + c = a + (b + c)$ and $(ab)c = a(bc)$.

asymptote (p. 427) A line that a graph approaches more and more closely.

base of a logarithmic function (p. 149) The number b, which must be positive and not equal to 1, in the logarithmic function $y = \log_b x$.

base-b logarithm of x (p. 149) The power to which the base b must be raised to equal x, denoted $\log_b x$.

biased question (p. 245) A question that produces responses that do not accurately reflect the opinions or actions of the respondents.

biased sample (p. 251) A sample that overrepresents or underrepesents part of the population.

binomial distribution (p. 634) In a binomial experiment with n independent trials, each having probability p of success, the values of $_nC_k \cdot p^k \cdot (1 - p)^{n-k}$ for $k = 0, 1, 2, \ldots, n$ form a binomial distribution of probabilities for the possible outcomes of the experiment.

box plot (p. 259) A display that shows the median, the quartiles, and the extremes of a numerical data set.

categorical data (p. 238) Data that are names or labels.

center (pp. 528, 536, 543) A point used to define certain conic sections. See also *circle*, *ellipse*, and *hyperbola*.

circle (p. 528) The set of all points in a plane that are the same distance from a point called the *center* of the circle.

closure property (pp. 373, 374) For real numbers a and b, the numbers $a + b$ and ab are also real.

cluster sample (p. 250) A sample chosen as a group.

combination (p. 585) A selection of elements of a set. Position is not important in a combination. The number of combinations of r objects taken from a set of n objects is given by $_nC_r$ where $_nC_r = \dfrac{n!}{(n-r)!r!}$.

common difference (p. 473) In an arithmetic sequence, the constant difference between any term and the term before it.

common logarithm (p. 150) The base-10 logarithm of N ($N > 0$), denoted $\log N$.

common ratio (p. 473) In a geometric sequence, the constant ratio of any term to the term before it.

commutative group (p. 373) A group for which the commutative property holds.

commutative property (pp. 373, 374) For real numbers a and b, $a + b = b + a$ and $ab = ba$.

complementary events (p. 619) Two mutually exclusive events, one of which must happen. For complementary events A and B, $P(B) = 1 - P(A)$.

complex conjugates (p. 361) Two complex numbers of the form $a + bi$ and $a - bi$.

complex numbers (p. 359) The set of numbers that consists of all real and all imaginary numbers.

complex plane (p. 365) A coordinate plane where each point (a, b) represents the complex number $a + bi$. The complex plane has a horizontal real axis and a vertical imaginary axis.

conditional probability (p. 626) The probability of event B, given that event A has happened, denoted $P(B|A)$. For any events A and B, $P(B|A) = \dfrac{P(A \text{ and } B)}{P(A)}$.

conic section (p. 550) A parabola, circle, ellipse, or hyperbola produced by slicing a double cone with a plane.

constant of variation (pp. 46, 427) The nonzero constant a in the direct variation function $y = ax$ or in the inverse variation function $y = \dfrac{a}{x}$.

constant term (p. 390) The term in a polynomial that does not have a variable or that has a variable with 0 as the exponent.

convenience sample (p. 250) A sample chosen because it is easy to obtain.

corner-point principle (p. 323) For a polygonal feasible region, the solution of a linear programming problem will be at a corner point of the region.

correlation coefficient (p. 72) A number, denoted r, between -1 and $+1$ that measures how well data points line up.

cosine (pp. 656, 708) In a right $\triangle ABC$, the cosine of $\angle A$, which is written "cos A," is given by the equation $\cos A = \dfrac{\text{length of leg adjacent to } \angle A}{\text{length of hypotenuse}}$. For a unit circle, $\cos \theta$ is the x-coordinate of the point where the terminal side of an angle θ in standard position intersects the circle.

cubic function (p. 413) A polynomial function of degree 3.

cycle (p. 720) For a periodic graph, the smallest portion that repeats.

degenerate conic (p. 551) A point, a line, or a pair of intersecting lines produced when a plane intersects a double cone at its vertex.

degree of a polynomial (p. 390) The greatest exponent of the polynomial.

degree of a vertex (p. 563) In discrete mathematics, the number of other vertices connected to that vertex by edges.

dependent system (p. 299) A system of equations that has infinitely many solutions.

dependent variable (p. 10) A variable whose value is determined by the value of some other variable.

dimensions of a matrix (p. 16) The number of rows r and the number of columns c in a matrix, written $r \times c$.

direct variation (p. 46) The relationship between two variables in which the ratio of the variables is constant. If $\dfrac{y}{x} = a$, or $y = ax$, where a is a nonzero constant, then y varies directly with x.

directed graph (p. 569) In discrete mathematics, a graph in which the edges are arrows.

directrix (p. 521) See *parabola*.

discriminant (p. 217) The value of $b^2 - 4ac$ in the quadratic formula. It tells you how many solutions the quadratic equation $ax^2 + bx + c = 0$ has.

domain of a function (p. 60) The set of values of the independent variable for which a function is defined.

double zero of a polynomial function (p. 420) A number k is a double zero of a polynomial function f if $(x - k)^2$ is a factor of $f(x)$.

doubling time (p. 108) The time it takes for the initial value of an exponential growth function to double.

e (p. 116) The irrational number e, approximately equal to 2.718, is often used as the base of exponential functions. As the value of n increases, the value of $\left(1 + \dfrac{1}{n}\right)^n$ approaches e.

edge of a graph (p. 563) In discrete mathematics, a connecting link between two vertices of a graph. Edges may have arrows.

element of a matrix (p. 16) A number in a matrix.

ellipse (p. 536) The set of all points in a plane such that the sum of the distances between each point and two fixed points (called the *foci*) is constant. The midpoint of the segment whose endpoints are the foci is called the *center* of the ellipse.

end behavior (p. 405) The behavior of a function $f(x)$ as x takes on large positive and large negative values.

experimental probability (p. 612) For an event A, the ratio of the number of trials where A happens to the total number of trials performed in an experiment.

explicit formula (p. 480) A formula that expresses the nth term, t_n, of a sequence as a function of its position number n.

exponential decay function (p. 96) A function of the form $y = ab^x$ where $a > 0$ and $0 < b < 1$. Its graph *falls* from left to right.

exponential growth function (p. 96) A function of the form $y = ab^x$ where $a > 0$ and $b > 1$. Its graph *rises* from left to right.

extraneous solution (p. 165) A solution that does not satisfy the original equation.

factoring (p. 222) For a quadratic expression of the form $ax^2 + bx + c$ where a, b, and c are integers, factoring means finding (if possible) integers k, l, m, and n such that $ax^2 + bx + c = (kx + m)(lx + n)$.

feasible region (p. 323) A graph of the solution of the system of inequalities representing all the given constraints in a linear programming problem.

finite sequence (p. 467) A sequence that has a last term.

focus, foci (pp. 521, 536, 543) A fixed point (focus) or two fixed points (foci) used to define certain conic sections. See also *parabola*, *ellipse*, and *hyperbola*.

function (p. 10) A pairing in which there is exactly one output value for each input value.

function notation (p. 60) The notation $y = f(x)$, which is read "y equals f of x" and which means that y is a function of x.

geometric mean (p. 475) The term between any two terms of a geometric sequence. If a, x, b form a geometric sequence and a and b are both positive, then $x = \sqrt{ab}$.

geometric probability (p. 614) A probability based on length, area, or volume.

geometric sequence (p. 473) A sequence where the ratio of any term to the term before it is constant.

geometric series (p. 495) The indicated sum of a geometric sequence.

graph (p. 563) In discrete mathematics, a collection of points, called *vertices,* connected by links, called *edges.*

group (p. 373) A number set and an operation form a group if the set is closed under the operation and the identity, inverse, and associative properties hold.

half-life (p. 109) The time it takes for the initial value of an exponential decay function to be halved.

half-planes (p. 310) A line divides the coordinate plane into two regions called half-planes.

histogram (p. 260) A graph that displays the overall shape of the distribution of numerical data. You create a histogram by dividing the data into intervals of equal width and drawing a bar over each interval so that the height indicates the frequency of the data within that interval.

hyperbola (pp. 427, 543) The set of all points in a plane such that the difference of the distances between each point and two fixed points (called *foci*) is constant. The midpoint of the segment whose endpoints are the foci is called the *center* of the hyperbola. The graph of an inverse variation equation is a hyperbola.

identity matrix (p. 304) A square matrix that has 1's on the diagonal running from the upper left to the lower right and 0's everywhere else. For any square matrix A and identity matrix I with the same dimensions, $AI = IA = A$.

identity property (pp. 373, 374) For a real number a, $a + 0 = 0 + a = a$ and $a \cdot 1 = 1 \cdot a = a$. The identity for addition is 0, and the identity for multiplication is 1.

imaginary axis (p. 365) The vertical axis of the complex plane.

imaginary numbers (p. 359) All numbers of the form $a + bi$, where a and b are real numbers and $b \neq 0$.

inconsistent system (p. 299) A system of equations that has no solution.

independent events (p. 619) Two events such that the occurrence of one event does not affect whether the other event happens. For independent events A and B, $P(A \text{ and } B) = P(A) \cdot P(B)$.

independent variable (p. 10) A variable whose value determines the value of some other variable.

infinite sequence (p. 467) A sequence that continues without end.

initial side of an angle (p. 671) See *standard position of an angle*.

intercept form (p. 200) The form $y = a(x - p)(x - q)$ for a quadratic function, where p and q are the x-intercepts of the function's graph.

interquartile range (IQR) (p. 266) The difference between the upper quartile and the lower quartile for a set of numerical data.

inverse matrix (p. 304) For a square matrix A, the inverse of A, denoted A^{-1}, is the matrix that, when multiplied by A, gives the identity matrix I: $A^{-1}A = A(A^{-1}) = I$. Not all square matrices have inverses.

inverse of a function f (p. 142) A function, denoted f^{-1}, that satisfies this property: $f^{-1}(b) = a$ if and only if $f(a) = b$. Sometimes the domain of f must be restricted in order for f^{-1} to exist.

inverse property (pp. 373, 374) For every real number a, there is a real number $-a$ such that $a + (-a) = 0$. For every nonzero real number a, there is a real number $\frac{1}{a}$ such that $a\left(\frac{1}{a}\right) = 1$.

inverse variation (p. 427) The relationship between two variables in which the product of the variables is constant. If $xy = a$, or $y = \frac{a}{x}$, where a is a nonzero constant, then y varies inversely with x.

leading coefficient (p. 406) The coefficient of the highest power of the variable in a polynomial.

least-squares line (p. 67) A line of fit that minimizes the sum of the squares of the vertical distances of a given set of data points from the line.

like terms (p. 392) Terms of polynomials that have the same power of the variable.

line of fit (p. 65) A line that lies as close as possible to all the points in a scatter plot. It does not necessarily have to pass through any of the points.

line of symmetry (p. 186) For a graph, a line that divides the graph so that the part of the graph on one side of the line is a reflection of the part on the other side.

linear programming (p. 323) A method for finding the best solution to a problem that involves maximizing or minimizing the value of some linear function of two variables when the values of the variables are constrained to some feasible region.

local maximum (p. 407) The y-coordinate of a turning point that is higher than all nearby points.

local minimum (p. 407) The y-coordinate of a turning point that is lower than all nearby points.

logarithmic function (p. 149) A function that is the inverse of the exponential function $f(x) = b^x$, denoted $f^{-1}(x) = \log_b x$.

logistic growth (p. 118) A type of limited growth that is basically exponential at first and then levels off.

lower extreme (p. 259) In a set of numerical data, the least value.

lower quartile (p. 259) In a set of numerical data, the median of the lower half of the data.

magnitude of a complex number (p. 366) In a complex plane, the distance between $(0, 0)$ and (a, b), denoted $|a + bi|$.

major axis (pp. 536, 543) For an ellipse or a hyperbola, the line segment whose endpoints are the vertices.

margin of error (p. 274) An indication of the expected variability in a sample proportion when a random sample is taken from a large population. For a sample of size n, it is approximated by this formula:

$$\text{margin of error} \approx \pm \frac{1}{\sqrt{n}}$$

mathematical model (p. 4) A table, a graph, an equation, a function, or an inequality that describes a real-world situation.

matrix (p. 16) A group of numbers arranged in rows and columns.

matrix multiplication (p. 23) The product of an $m \times n$ matrix A and an $n \times p$ matrix B is the $m \times p$ matrix C such that each element $c_{i,j}$ of C is obtained by adding the products of corresponding elements as you go across the ith row of A and down the jth column of B.

maximum value of a quadratic function (p. 195) The value k of the function $y = a(x - h)^2 + k$ when $a < 0$.

median (p. 259) The middle number when you put data in order from least to greatest. When the number of data items is even, the median is the mean of the two middle numbers.

minimum value of a quadratic function (p. 195) The value k of the function $y = a(x - h)^2 + k$ when $a > 0$.

minor axis (pp. 536, 543) For an ellipse or a hyperbola, a line segment that has its midpoint at the center of the conic and is perpendicular to the major axis.

mutually exclusive events (p. 619) Two events that cannot both happen. For mutually exclusive events A and B, $P(A \text{ or } B) = P(A) + P(B)$.

natural logarithm (p. 150) The base-e logarithm of N ($N > 0$), denoted $\ln N$.

negative correlation (p. 72) For two variables x and y, the correlation coefficient r is negative ($-1 \le r < 0$) when y tends to decrease as x increases.

normal distribution (p. 639) A distribution of data whose graph is a bell-shaped curve that reaches its maximum height at the mean.

numerical data (p. 238) Data that are counts or measurements.

outlier (p. 266) A data value whose distance from the nearer quartile is more than 1.5 times the interquartile range.

parabola (pp. 186, 521) The set of all points in the plane such that each point is the same distance from a point F, called the *focus*, and a line d, called the *directrix*. The graph of a quadratic function is a parabola.

parameter (p. 79) A variable, usually denoted t, upon which two other variables depend.

parametric equations (p. 79) Equations that express two variables in terms of a third variable, called the *parameter*.

Pascal's triangle (p. 593) A triangular display of numbers that gives the coefficients when you expand $(a + b)^n$ for $n = 0, 1, 2, \ldots$.

period (p. 720) For a function $y = f(x)$ whose graph repeats, the length of the interval on the x-axis covered by the smallest part of the graph that repeats. The functions $y = a \sin bx$ and $y = a \cos bx$ have a period of $\frac{2\pi}{|b|}$. The function $y = a \tan bx$ has a period of $\frac{\pi}{|b|}$.

permutation (p. 579) An arrangement of the elements of a set. Position is important in a permutation. The number of permutations of r objects taken from a set of n objects is given by ${}_nP_r$ where ${}_nP_r = \dfrac{n!}{(n - r)!}$.

phase shift (p. 728) A horizontal translation of a periodic function.

point-slope equation (p. 59) An equation of the form $y = y_1 + a(x - x_1)$ where (x_1, y_1) is a point on a line and a is the slope of the line.

polynomial (p. 390) A term or a sum of terms, where each term is the product of a real-number coefficient and a variable with a whole-number exponent.

population (p. 239) A complete group from which a sample is taken.

positive correlation (p. 72) For two variables x and y, the correlation coefficient r is positive ($0 < r \le 1$) when y tends to increase as x increases.

power function (p. 170) A function of the form $y = ax^b$ where a and b are constants.

probability (p. 611) The fraction of the time that an event A is expected to happen, denoted $P(A)$.

pure imaginary number (p. 358) The square root of any negative real number. A pure imaginary number can be written in the form bi, where $b \ne 0$.

quadratic formula (p. 216) The formula $x = \dfrac{-b \pm \sqrt{b^2 - 4ac}}{2a}$, which gives the solutions of any quadratic equation in the form $ax^2 + bx + c = 0$.

quadratic function (p. 186) A function that involves squaring the independent variable. Quadratic functions can be written in standard form, $y = ax^2 + bx + c$; intercept form, $y = a(x - p)(x - q)$; or vertex form, $y = a(x - h)^2 + k$.

radian (p. 714) On a unit circle, the radian measure of an angle in standard position is the length of the arc that is intercepted by the angle. Radians are related to degrees by the fact that there are 2π radians or $360°$ in a full circle, so 1 radian $= \dfrac{180°}{\pi}$.

radical function (p. 344) A function that involves taking the nth root of the independent variable.

radius (p. 528) For a circle, the distance between the center of the circle and any point on the circle.

random sample (p. 250) A sample chosen so that each person or object has an equally likely chance of being selected.

range of a data set (p. 266) The difference between the maximum and minimum values in a set of numerical data.

range of a function f (p. 60) The set of all values of $f(x)$, where x is in the domain of f.

rational equation (p. 450) An equation that contains only polynomials or quotients of polynomials.

rational function (p. 435) A function of the form $f(x) = \dfrac{p(x)}{q(x)}$ where $p(x)$ and $q(x)$ are polynomials.

real axis (p. 365) The horizontal axis of the complex plane.

real numbers (p. 358) The set of numbers that includes all rational and all irrational numbers.

recursion (p. 481) The process of repeatedly applying the same operation(s) to a starting value and to each successive answer.

recursive formula (p. 481) A formula for a sequence that includes two parts: a starting value and a recursion equation for the nth term, t_n, as a function of t_{n-1}.

relative frequency (p. 260) The vertical axis of a histogram shows relative frequencies when the count of the data within each interval on the horizontal axis is expressed as a percent of all the data.

sample (p. 239) A subset of a population.

sample proportion (p. 273) The fraction of categorical data that fall into a particular category.

sample size (p. 274) The number of items in a sample.

sampling distribution (p. 274) A histogram that shows the distribution of sample proportions when many samples of the same size are taken from the same population of categorical data.

scalar (p. 18) A number by which a matrix is multiplied.

scalar multiplication (p. 18) The process of multiplying each element of a matrix by a scalar.

self-selected sample (p. 250) A sample of people who have volunteered to be surveyed.

sequence (p. 465) An ordered list of numbers or figures.

series (p. 488) The indicated sum when the terms of a sequence are added.

simulation (p. 29) An experiment that is used to model a situation and make predictions.

sine (pp. 656, 708) In a right $\triangle ABC$, the sine of $\angle A$, which is written "sin A," is given by the equation $\sin A = \dfrac{\text{length of leg opposite } \angle A}{\text{length of hypotenuse}}$. For a unit circle, $\sin \theta$ is the y coordinate of the point where the terminal side of an angle θ in standard position intersects the circle.

skewed distribution (p. 260) A distribution that is not symmetric.

slope (p. 47) The ratio $\dfrac{\text{vertical change}}{\text{horizontal change}}$ for two points on a line. If (x_1, y_1) and (x_2, y_2) are two points on a line, then the slope a of the line is given by:
$$a = \frac{y_2 - y_1}{x_2 - x_1}$$

slope-intercept form (p. 53) The form $y = ax + b$ for a linear function, where a is the slope and b is the vertical intercept of the function's graph.

solving a triangle (p. 665) Finding the lengths of all sides and the measures of all angles of a triangle.

square root (p. 187) A number x is a square root of a number y if it satisfies the equation $x^2 = y$. A positive number y has two square roots, written $\sqrt{y}$ and $-\sqrt{y}$.

standard deviation (p. 268) A measure of the variability in a set of numerical data, denoted σ. For the data $x_1, x_2, x_3, \ldots, x_n$ with mean $\bar{x}$, the standard deviation is given by:
$$\sigma = \sqrt{\frac{(x_1 - \bar{x})^2 + (x_2 - \bar{x})^2 + \cdots + (x_n - \bar{x})^2}{n}}$$

standard form of a quadratic function (p. 206) The form $y = ax^2 + bx + c$.

standard form of a polynomial (p. 390) The form of a polynomial where exponents decrease from left to right.

standard normal distribution (p. 641) The normal distribution with mean 0 and standard deviation 1.

standard position of an angle (p. 671) An angle in standard position is formed by rotating a ray, called the *terminal side*, in a counterclockwise direction from the positive x-axis, called the *initial side*.

stratified random sample (p. 250) A sample chosen by dividing the population into groups and randomly selecting people or objects from each group.

symmetric distribution (p. 260) A distribution whose histogram is approximately symmetrical about a vertical line passing through the interval with the greatest frequency.

synthetic substitution (p. 391) A method of evaluating a polynomial for any value of the variable.

system of equations (p. 292) Two or more equations involving the same variables.

system of inequalities (p. 315) Two or more inequalities involving the same variables.

systematic sample (p. 250) A sample chosen using a pattern, such as every fourth person in line.

---

tangent (pp. 663, 736) In a right $\triangle ABC$, the tangent of $\angle A$, which is written "tan A," is given by the equation $\tan A = \dfrac{\text{length of leg opposite } \angle A}{\text{length of leg adjacent to } \angle A}$. For a unit circle, $\tan \theta$ is the ratio of the y-coordinate to the x-coordinate of the point where the terminal side of an angle θ in standard position intersects the circle.

term of a sequence (p. 465) An element of a sequence, denoted t_n where n is the position number.

terminal side of an angle (p. 671) See *standard position of an angle.*

theoretical probability (p. 613) For a procedure that can result in n equally likely outcomes and for an event A that corresponds to m of these outcomes, the theoretical probability of A is given by $P(A) = \dfrac{m}{n}$.

triple zero of a polynomial function (p. 420) A number k is a triple zero of a polynomial function f if $(x - k)^3$ is a factor of $f(x)$.

turning point (p. 407) A point that is higher or lower than all nearby points on the graph of a function.

---

upper extreme (p. 259) In a set of numerical data, the greatest value.

upper quartile (p. 259) In a set of numerical data, the median of the upper half of the data.

---

vertex form (p. 193) The form $y = a(x - h)^2 + k$ for a quadratic function, where (h, k) is the vertex of the function's graph.

vertex of a graph (p. 563) In discrete mathematics, a point used to represent some object, such as a country on a map or an animal in a food web.

vertex of a parabola (pp. 186, 521) The point where the line of symmetry crosses the parabola.

vertices of a hyperbola (p. 543) The points of intersection of the hyperbola and the line containing the foci.

vertices of an ellipse (p. 536) The points of intersection of the ellipse and the line containing the foci.

vertical intercept (p. 53) The value of the dependent variable when the value of the independent variable is 0. On a graph of a function, the vertical intercept indicates where the graph crosses the vertical axis.

---

x-intercept (p. 199) The x-coordinate of any point where a graph crosses the x-axis.

---

z-score (p. 641) The number of standard deviations that a given data value lies from the mean. The z-score for a data value x from a normal distribution with mean $\bar{x}$ and standard deviation σ is given by $z = \dfrac{x - \bar{x}}{\sigma}$.

zero of a polynomial function (p. 413) For a polynomial function f, each solution of the equation $f(x) = 0$.

Index

Absolute frequency, 260
Absolute value, 774
ACT/SAT Preview *See* SAT/ACT Preview.
Acute angle, 800
Addition
  clock, 376
  of complex numbers, 360, 366
  of first *n* terms
    arithmetic series, 489
    geometric series, 496
  group properties of real numbers under, 373
  of infinite geometric series, 498
  of linear equations to solve a system, 297
  of matrices, 17
  of polynomials, 392
  repeated, 480
  sequence of partial sums, 493
  sigma notation, 490
  of sine or cosine function, 729
Algebra, fundamental theorem of, 421
Algebra tiles, 206–207, 221–222
Algorithms *See* Discrete mathematics.
Amplitude, 720, 740
Angle
  acute, 800
  cosine of, 656
  degree measures of, 657
  of depression, 665
  of elevation, 665
  initial side of, 671
  measuring in the coordinate plane, 671
  negative, 709
  obtuse, 800
  positive, 709
  quadrantal, 671
  radian measures of, 713–714
  right, 800
  of rotation, 673
  sine of, 656
  standard position of, 671
  straight, 800
  tangent of, 663

  terminal side of, 671
Angles
  complementary, 800
  congruent, 801
  supplementary, 800
  vertical, 800
Applications
  advertising, 292, 590
  agriculture, 134, 313, 645
  amusement park rides, 710, 722
  archaeology, 656
  architecture, 478, 489, 509, 514, 539, 634, 675
  art, 307, 325
  astronomy, 126, 190, 350, 523, 525, 559, 614, 615, 657, 661, 667, 739, 745
  automotive mechanics, 731
  aviation, 666, 667, 673
  banking, 108, 112, 117, 124, 137
  bicycling, 185, 398
  biochemistry, 476
  biology, 63, 71, 118, 122, 125, 129, 170, 175, 198, 254, 345, 349, 394, 428, 468, 477, 641
  bookmaking, 93–95
  botany, 642, 702
  business, 3, 4, 75, 144, 201–202, 203, 265, 322, 323, 325, 576
  camping, 24
  canoeing, 22
  carpentry, 541
  cartography, 313, 564, 567
  census, 100
  chemistry, 117, 121, 137, 181, 448, 467, 568
  computers, 590, 607
  consumer economics, 275, 276, 302, 310, 321, 324, 332, 439
  cooking, 162, 303, 306, 681
  dancing, 675
  dentistry, 444
  driver's education, 636
  driving, 224, 225
  earth science, 157, 159, 267, 268
  ecology, 169, 570, 572, 573, 576
  economics, 68, 135, 259, 501
  education, 276, 417

  electricity, 233
  electronics, 623
  employment, 276
  energy, 425
  engineering, 744
  entertainment, 15, 21
  exercising, 659
  fashion, 578, 581
  finance, 302
  fitness, 333
  flag-making, 518
  forensics, 582
  forestry, 145
  fundraising, 591
  games, 604, 615, 620, 650
  gardening, 448
  genealogy, 508
  geology, 318
  geometry, 48, 57, 99, 128, 189, 225, 347, 461, 471, 485, 517, 518, 519, 558, 667, 703, 716, 738
  government, 77, 219, 256, 605
  health, 71, 125, 630
  heredity, 618
  history, 418, 631, 683
  hobbies, 584
  home repair, 429, 459
  hot air ballooning, 735
  hydraulics, 432
  industry, 12, 592
  landscaping, 680
  language arts, 484, 552
  literature, 583, 629
  manufacturing, 10, 11, 48, 66, 67, 137, 314, 586, 588, 615, 631, 643, 645, 651
  marketing, 636
  market research, 246, 631
  medicine, 239, 314, 482–483, 573
  metallurgy, 449, 451
  meteorology, 73, 621, 668
  mountain climbing, 151
  movies, 621
  music, 475, 583, 733
  mythology, 98
  navigation, 670, 676, 684, 702, 747

oceanography, 61, 460

packaging, 411

parasailing, 662

paleontology, 693

personal finance, 20, 51, 166, 291, 327

physics, 79, 194, 208–209, 210, 233, 286, 332, 384, 426, 494, 617

physiology, 109, 129, 407

politics, 651

polling, 275

population, 154, 161, 436, 458

precision skating, 45

psychology, 123, 396

recreation, 146, 298, 596

scheduling, 565, 583

snowmaking, 105

social science, 294

space travel, 412

sports, 12, 48, 63, 89, 114, 127, 136, 218, 285, 287, 294, 339, 357, 453, 459, 496, 579, 583, 590, 612, 613, 622, 624, 651, 670, 696

surveying, 664

temperature scales, 55

transportation, 260

wind chill, 355

See also Connections.

Area

as function of folds of paper, 95

of a circle, 802, 825

of quadrilaterals, 825

of a sector, 680

of squares in a Koch construction, 502

of a triangle, 678, 825

under a standard normal curve, 639

Arithmetic mean, 475

Arithmetic sequence, 473, 481

Arithmetic series

sum of finite, 489

Assessment

Assess Your Progress, 15, 35, 64, 83, 114, 129, 161, 175, 205, 227, 257, 279, 308, 327, 357, 379, 411, 425, 440, 453, 487, 503, 527, 542, 553, 576, 599, 631, 645, 670, 683, 697, 726, 739

Chapter Assessment, 40–41, 88–89, 134–135, 180–181, 232–233, 284–285, 332–333, 384–385, 458–459, 508–509, 558–559, 604–605, 650–651, 702–703, 744–745

Cumulative Assessment, 136–137, 286–287, 460–461, 606–607, 746–747

Journal, 15, 35, 64, 83, 114, 129, 161, 175, 205, 227, 257, 279, 308, 327, 357, 379, 411, 425, 440, 453, 487, 503, 527, 542, 553, 576, 599, 631, 645, 670, 683, 697, 726, 739

Ongoing, 8, 15, 21, 28, 35, 51, 57, 64, 71, 77, 83, 99, 106, 113, 122, 129, 147, 154, 161, 167, 175, 191, 198, 205, 213, 220, 227, 243, 249, 256, 265, 272, 278, 296, 302, 308, 314, 321, 327, 343, 350, 356, 364, 370, 379, 396, 403, 410, 418, 425, 432, 440, 448, 453, 471, 479, 487, 494, 503, 519, 527, 534, 542, 549, 553, 568, 576, 584, 592, 598, 617, 624, 631, 638, 645, 662, 670, 677, 683, 690, 697, 712, 718, 726, 734, 739

Performance Task, 41, 89, 135, 181, 233, 285, 333, 385, 459, 509, 559, 605, 651, 703, 745

Self-Assessment, 37, 85, 131, 177, 229, 281, 329, 381, 455, 505, 555, 601, 647, 699, 741

See also Projects, portfolio.

Associative properties, 373–374

Asymptote, 427, 441–442, 830–832

Average *See* Mean.

Axis

imaginary, 365

real, 365

Bar graph, 2, 3–4, 238, 791

Base of an exponent, 776

Biconditional, 799

Binomial distribution, 634

Binomial expansion, 594

Binomial experiment, 634

Binomial theorem, 594

Biographical note *See* History of mathematics *and* Interview.

Box plot

extremes, 259

median, 259

quartiles, 259

using manipulatives, 258

By the Way, 4, 6, 7, 10, 12, 18, 33, 48, 56, 61, 66, 75, 93, 95, 103, 116, 117, 128, 141, 151, 153, 157, 167, 169, 175, 187, 194, 211, 218, 219, 239, 242, 245, 251, 253, 263, 264, 265, 266, 270, 314, 318, 320, 323, 339, 346, 355, 357, 391, 396, 401, 402, 409, 410, 417, 418, 421, 425, 427, 431, 438, 444, 453, 469, 470, 476, 485, 493, 496, 501, 514, 523, 569, 570, 575, 579, 582, 593, 598, 614, 618, 628, 630, 632, 637, 656, 657, 660, 664, 666, 669, 684, 693, 710, 715, 727, 733

Calculator, graphing,

box plot, 261, 813

combination formula, 586

degree mode, 672, 714

evaluating logarithms, 164

exponential regression, 125, 256, 329

finding correlation coefficient for linear model, 72–73, 172

graphing, 46, 97, 107, 117, 118, 163, 164, 169, 189, 192, 199, 207, 222, 229, 277, 278, 292, 297, 328–329, 338, 344, 352, 358, 404, 405, 406, 407, 408, 412, 419, 426, 433, 436, 443, 444, 483, 529–530, 531, 538, 545, 633, 644, 660, 708, 709, 712, 715, 716, 717, 718, 722, 724, 730, 733, 734, 735, 745

histogram, 261, 813

linear regression, 67, 73, 122, 147, 169, 328

matrices, 25, 28, 304, 305, 306, 308, 395, 403, 453, 572

Data analysis

analyzing a survey, 280–281

categorical data, 238

collecting data, 8, 21, 36, 50, 62, 71, 77, 84, 130, 167, 176, 228–229, 250, 328, 380, 454, 504, 646, 698, 741

comparing models and data, 4, 5, 9, 37, 45–46, 94–96, 102, 118, 123–125, 141, 168–171, 328–329, 381, 741

distribution of data set, 260–261, 266

fitting models to data, 3–5, 9–11, 16–18, 22–25, 29–31, 37, 46, 65–68, 72–74, 85, 96, 100, 101, 109, 117, 118, 123–125, 131, 135, 169, 171, 176–177, 211, 229, 256, 307, 328–329, 740–741

frequency table, 261

numerical data, 238

organizing data, 16, 37, 85, 131, 177, 229, 258–261, 281, 329, 381, 455, 505, 647, 699, 741

sample, 239, 250–252, 273–274

using box plots, 258–259, 261

using histograms, 238, 260–261, 632, 639

using matrices, 16–18, 22–25

using spreadsheets, 4, 45, 51, 97, 99, 100, 131, 173, 191, 197, 381, 413, 428, 505, 637

using a table to find a pattern, 4, 9, 37, 52, 84, 115, 130, 170, 229, 381, 515

See also Graphing, Projects, portfolio, *and* Technology Handbook.

Decay factor, 96, 109

Decimal

repeating, 499

See also Toolbox.

Degenerate conics, 551

Degree of a polynomial, 390, 406

Depression, angle of, 665

Deviation, standard, 268–269

Difference *See* Subtraction.

Digits, significant, 789–790

Dilation, 187, 798

Dimensional analysis, 793

Direct variation, 45–46

Directed graph

edges of, 569

representing with a matrix, 571

vertices of, 570

Discrete mathematics

coloring a graph, 563–565

complementary events, 619

conditional probability, 626–627

constraints in linear programming, 322–324

corner-point principle, 323–324

degree of a vertex, 563

directed graph, 569

edges of a graph, 563

edges of a directed graph, 569

equivalent graphs, 563

experimental probability, 611–612

fractals, 365, 369, 462–464, 479, 502, 598

geometric probability, 614

independent events, 619

linear programming, 322–324

multiple events, 619

mutually exclusive events, 619

Pascal's triangle, 593–594

planar graphs, 567

probability tree diagram, 496, 627–628

series, 488–490, 495–499

sequences, 465–467, 472–475, 480–483, 493, 500

theoretical probability, 612–613

vertices of a graph, 563

vertices of a directed graph, 570

Discriminant, 217

Distance formula, 512, 514

Distribution

normal, 639–641

skewed, 260

standard normal, 641

symmetric, 260

Distributive property, 377, 398, 774, 782

Division

of complex numbers, 361

of polynomials, 399–400

by proportional growth factor, 101

Domain, 60

Doubling time, 108

e, 115–116

Elevation, angle of, 665

Ellipse

drawing, 535

foci of, 536

horizontal, 537

major axis, 536

minor axis, 536

standard form of equation for, 537

vertical, 537

vertices of, 536

End behavior, 405–406, 443–444

Enrichment *See* Challenge exercises.

Equation

direct variation, 45–46

exponential, 10–11, 92, 93–96, 107–110, 116–117, 123–125, 131, 148–149, 162–163, 171

exponential form of logarithmic, 149

factoring polynomial, 206–207, 221–224

finding integral zeros of polynomial, 415

intercept form of quadratic, 199–202

inverse variation, 427

linear, 10–11, 46–47, 52–54, 58–59, 78–80, 141–142, 297, 784

logarithmic, 149, 164–165

parametric, 78–80

point-slope form of linear, 58–59

polynomial, 186, 206–209, 211, 212, 412–415, 419–422

quadratic, 193, 199–202, 206–209

rational, 449–451

recursion, 481

rewriting, 788

slope-intercept form of linear, 52, 53

solving, 784

solving with algebra tiles, 206–207, 221–222

standard form of circle, 529

standard form of ellipse, 537

standard form of hyperbola, 544

standard form of linear, 297

standard form of polynomial, 390

Scalar multiplication of matrices, 18

Scale factor of a dilation, 798

Scatter plot, 9, 52, 65–67, 72–73, 85, 169, 170, 177, 229, 309, 328, 381, 795

Scientific calculator *See* Calculator, scientific.

Sector
area of, 680
of a circle graph, 792

Self-Assessment, 37, 85, 131, 177, 229, 281, 329, 381, 455, 505, 555, 601, 647, 699, 741

Sequence
arithmetic, 473, 481
domain of, 466
Fibonacci, 485, 493
finite, 467
geometric, 473, 481
infinite, 467
of partial sums, 493, 500
range of, 466
recursion, 481
terms of, 465–466

Series
arithmetic, 488–490
compact form, 490
expanded form, 490
geometric, 495–499
infinite geometric, 497–499
sum of finite arithmetic, 489
sum of finite geometric, 496
sum of infinite geometric, 498

Sets, 371–375

Sierpinski pattern, 464, 479, 598

Sigma notation, 490

Significant digits, 789–790

Similar triangles, 655, 801

Simplest form of a variable expression, 781

Simulation, 29–31

Sine of an angle, 656

Sine function
amplitude of, 720, 740
combining with cosine functions, 729, 730
graph of, 707–709, 720
inverse of, 658
on the unit circle, 708

Sines, Law of, 684, 694

Skewed distribution, 260, 639

Slope of a line
definition of, 47
of a perpendicular line, 515–516
positive and negative, 73

Slope-intercept form of a linear equation, 53–54, 65

Solving equations *See* Equation.

Spatial reasoning *See* Manipulatives *and* Visual thinking.

Spiral review *See* Review.

Spreadsheet, 4, 45, 51, 97, 99, 100, 131, 169, 170, 173, 191, 197, 369, 381, 413, 428, 493, 500, 505, 637
See also Technology Handbook.

Square roots
graph of function, 338
inequalities involving, 827
properties of, 187, 777

Standard deviation, 268

Standard normal distribution, 641

Standard form
of cubic equation, 413
for equation of a circle, 529
for equation of an ellipse, 537
for equation of a hyperbola, 544
for equation of a parabola, 521
of linear equation, 297
of polynomial equation, 390
of quadratic equation, 206

Statistics
binomial distribution, 634
binomial experiment, 634
Chebyshev's rule, 270
interquartile range (IQR), 266
margin of error, 274
mean, 3, 475, 513
median, 259
normal distribution, 639–641
outlier, 266
population, 239
range, 60, 266
sample, 239, 250–252, 273–274
sampling distribution, 273–274
simulation, 29–31
standard deviation, 268
standard normal distribution, 641
writing survey questions, 244–246
z-score, 641
See also Data analysis *and* Sample.

Straight angle, 800

Study techniques, 38, 86, 132, 178, 230, 282, 330, 382, 456, 506, 556, 602, 648, 700, 742

Subscript notation, 466

Subset, 376

Subtraction
of complex numbers, 360
of matrices, 17
of polynomials, 392
of system of linear equations, 297, 299–300

Sum *See* Addition.

Supplementary angles, 800

Surface area
of a cylinder, 410, 454–455, 825
of a sphere, 190, 825
of space figures, 825

Survey
analyzing, 280–281
biased questions, 245
conducting, 646–647
writing questions for, 244–245
See also Sample.

Symmetric distribution, 260, 639

Symmetry, line of, 186, 528, 536, 543, 797

Synthetic substitution, 391

Systems of inequalities, 290, 315–317

Systems of linear equations
dependent, 299–300
inconsistent, 299–300
infinitely many solutions, 299–300
no solution, 299–300
solving by adding or subtracting, 297, 299, 300
solving by graphing, 294, 297
solving with matrices, 303–305
solving by substitution, 292–293

Table
frequency, 261
as mathematical model, 4
recording data in, 4, 9, 29, 30, 35, 37, 45, 48, 52, 66, 84, 130, 229, 515
See also Spreadsheet.

Tangent of an angle, 663

Tangent function
  graph of, 736
  inverse of, 664
Technology Handbook, 803–816
Technology exercises, 13, 14, 15,
      28, 34, 51, 57, 68, 70, 75, 76,
      77, 82, 83, 89, 97, 111, 120,
      121, 122, 126, 127, 129, 135,
      147, 160, 161, 163, 165, 169,
      172, 173, 181, 189, 191, 197,
      207, 211, 222, 256, 277, 278,
      293, 294, 302, 305, 306, 307,
      308, 332, 343, 355, 369, 395,
      403, 408, 409, 410, 411, 416,
      417, 423, 424, 425, 430, 438,
      439, 446, 447, 448, 453, 461,
      493, 500, 525, 534, 548, 573,
      574, 575, 598, 607, 616, 637,
      644, 660, 712, 716, 717, 718,
      724, 733, 734, 745
Term of a variable expression, 781
Testing *See* Assessment.
Theoretical probability, 612–613
Toolbox, 773–802
Transformation
  dilation, 798
  reflection, 797
  translation, 796
Translation
  for exponential functions, 830–831
  for logarithmic functions, 832
  of hyperbolas, 433
  of parabolas, 192–193
  phase shift, 728, 736
Tree diagram, 496, 627, 628
Triangles
  area of, 678
  congruent, 801
  constructing, 685
  similar right, 655
  solving, 665
Trigonometric ratios, 654, 655, 663
Trigonometric functions, 672,
      707–709, 729, 735–737,
      740–741
Truth function, 828
Turning point, 186, 404–407

Unit circle, 708, 736

Variables, 10
Variable expression
  definition of, 780
  factoring, 782
  term of, 781
Variation
  constant of, 46, 427
  direct, 45–46
  inverse, 427
Vertex
  of a directed graph, 570
  form of quadratic equation, 193
  of a graph, 563
  of a hyperbola, 543
  of a parabola, 186
Vertical angles, 800
Vertical axis, 238, 260
Vertical intercept, 53
Visual thinking, 50, 69, 70, 97, 167,
      219, 255, 296, 354, 368, 369,
      371–372, 395, 409, 430, 477,
      491, 588, 624, 689, 717, 718
  See also Manipulatives.
Vocabulary, 15, 35, 38, 64, 83, 86,
      114, 129, 132, 161, 175, 178,
      205, 227, 230, 257, 279, 282,
      308, 327, 330, 357, 379, 382,
      411, 425, 440, 453, 456, 487,
      503, 506, 527, 542, 553, 556,
      576, 599, 602, 631, 645, 648,
      670, 683, 700, 726, 739, 742
  See also Glossary.
Volume
  of a cylinder, 410, 454–455, 802,
    825
  of a rectangular prism, 397, 802,
    825
  of a sphere, 402, 802, 825
  of space figures, 825

Whole number, 773
Writing
  Exercises, 6, 8, 13, 14, 15, 19, 27,
      28, 31, 33, 40, 51, 55, 57, 64,
      68, 70, 71, 75, 82, 83, 98,
      104, 111, 120, 121, 129, 135,
      146, 147, 152, 153, 154, 158,

159, 160, 161, 166, 171, 173,
175, 180, 189, 196, 197, 198,
203, 205, 211, 219, 227, 232,
241, 247, 248, 254, 255, 256,
257, 262, 264, 265, 270, 277,
279, 285, 286, 287, 295, 302,
314, 317, 318, 319, 327, 332,
343, 354, 356, 364, 370, 376,
378, 384, 395, 396, 403, 408,
410, 418, 425, 432, 436, 439,
448, 452, 460, 461, 468, 477,
478, 479, 485, 501, 502, 509,
518, 525, 526, 533, 541, 548,
549, 559, 567, 574, 584, 591,
592, 595, 598, 606, 607, 617,
621, 624, 635, 636, 638, 643,
645, 654, 662, 670, 675, 677,
681, 682, 688, 690, 695, 696,
697, 699, 703, 725, 732, 734,
738, 739, 744, 746, 747
  See also Journal, Projects,
    introductory, and Projects,
    portfolio.

x-axis, 427
x-coordinate, 794
x-intercept, 199–202, 413

y-axis, 427
y-coordinate, 794
y-intercept, 53

Zero
  of cubic function, 413
  double, 420
  finding integral zeros of
    polynomial functions, 415
  finding real zeros of polynomial
    functions, 419–420
  rational zeros theorem, 421
  triple, 420
Zero exponent property, 776
z-score, 641

Credits

Cover Images

Front cover Earth Imaging/Tony Stone Images/Chicago, Inc.(tl); Dana Berry, Wright Center Tufts University(m); Seth Shostak, Science Photo Library/Photo Researchers, Inc. (m background); Roger Ressmeyer—©1989 CORBIS(r background); NASA (l background) **Back cover** Roger Ressmeyer—©1989 CORBIS; NASA (background)

Chapter Opener Writers

Linda Borcover interviewed Johnpaul Jones (pp. 652–654); Melissa Burdick Harmon interviewed Vera Rubin (pp. 386–388); Yleanna Martinez interviewed Norbert Wu (pp. 42–44), Ednaly Ortiz (pp. 138–140), Jhane Barnes (pp. 462–464); Steve Nadis interviewed Twyla Lang (pp. xxxiv–2), Finn Strong (pp. 90–92), Mark Thomas (pp. 182–184), Donna Cox (pp. 234–236), Gina Oliva (pp. 288–290), Ronald Toomer (pp. 334–336), Kija Kim (pp. 510–512), Maria Rodriguez (pp. 560–562), Robert Ward (pp. 608–610).

Acknowledgements

371 Excerpt from *Smilla's Sense of Snow* by Peter Høeg, translation by Tina Nunnally. Copyright © 1993 by Farrar, Straus & Giroux, Inc. Reprinted by permission of Farrar, Straus & Giroux, Inc. **484** Excerpts and illustrations from *Bringing the Rain to Kapiti Plain*, retold by Verna Aardema. Illustrations by Beatriz Vidal. Copyright © 1981 by Verna Aardema, text, copyright © 1981 by Beatriz Vidal, illustrations. Used by permission of Dial Books, a division of Penguin Books USA.

Stock Photography

i EarthImaging/Tony Stone Images/Chicago, Inc.(tl); Dana Berry, Wright Center Tufts University(m); Seth Shostak, Science Photo Library/Photo Researchers, Inc.(m background); Roger Ressmeyer—©1989 CORBIS(r background); **iv** Ted Thai/Time Inc.(t, m); **v** ©1993 Charles Lindsay/Mo Yung Productions(b); **vi** James King-Holmes, Science Photo Library/Photo Researchers, Inc.(t); photo by Alessandro Sanvito. Artifacts from the collection of Museo Archeologico di Napoli, Italy.(m); **viii** ©The Stock Market(t); courtesy Mark Thomas/General Motors(b); **ix** ©John Eastcott/Yvamomativk/Photo Researchers, Inc.(t); **x** courtesy Gina Oliva; **xi** Daniel Forster Photography(tl, tr); ©1995 Chad Slattery(b); **xii** Katherine Lambert; **xiii** based on an illustration from *Navajo Architecture: Forms, History, Distributions*, by Jett and Spencer ©1981 University of Arizona Press(t); Jerry Jacka Photography(m); courtesy Jhane Barnes(b); David Austen/Stock Boston(t); Tom Herde/Boston Globe(b); **xv** courtesy Consolidated Edison Company, New York(b); **xvi** ©Ken Eward, Science Source/Photo Researchers, Inc.(t); **xvii** Comstock Photofile Limited, Toronto(t); **xviii** ©Dan McCoy/Rainbow(t); Robert W. Parvin(b); **xx** courtesy Jhane Barnes(bl); Craig Fujii/Detroit Free Press, Inc.(br); **xxi** ©Norbert Wu(tl, m); ©1993 Bob Cranston/Mo Yung Productions(tr, bl); ©1993 Charles Lindsay/Mo Yung Productions(br); **xxii** Michael Rosenfeld/ Tony Stone Images/Chicago, Inc.; **xxiv** Benn Mitchell/The Image Bank(t); **xxv** Royal Tyrell Museum/Alberta Community Development(tl);

Julie Sheffield/Hannibal Courier Post(bl); **xxvii** ©Allsport/Chris Cole; **xxviii** Rhoda Sidney/Leo De Wys(b); **xxix** ©Michael Marten, Science Photo Library/Photo Researchers, Inc.(tl); Thomas Braise/Tony Stone Images/Chicago, Inc.(tr); © 1995 Chris Arend/Alaska Stock Images(bl); **xxxi** (m)Bill Gallery/Stock Boston(tl, bl); ©Mark Marten, Science Source, Photo Researchers, Inc.(tr); Galen Rowell/Mountain Light(br); **xxxii** Bruce Kliewe/Picture Cube; **xxxiii** ©P. Gontier, Photo Researchers, Inc.(t); courtesy Jhane Barnes(br); **1** Bonwire/Owen Franken/Stock Boston(tr, far r); **9** Superstock(t); ©The Stock Market/Chris Sorensen(l); ©Leonard Lessin/Peter Arnold, Inc.(b); **12** David Young Wolff/Tony Stone Images/Chicago; **13** Reuters/The Bettmann Archive(t); Ted Thai/Time Inc.(m, b); **16** Philip & Karen Smith/Tony Stone Images/Chicago, Inc.; **19** Jeff Hetler/Stock Boston(far l); Kevin R. Morris/Tony Stone Images/Chicago, Inc.(ml); ©Norm Thomas, Photo Researchers, Inc.(mr); Tony Stone Images/Chicago, Inc.(far r); **21** Michael Melford/The Image Bank(t); CMCD, Inc.(b); **22** Lambert/Archive Photos **24** Rhoda Sidney/Leo de Wys(tl); **27** Hall Puckett/Houston Astros Baseball Club; **32** The Topps Company, Inc.; **33** Trans. No. 4971 (4) (Photo by Don Eiler) Courtesy Department of Library Services, American Museum of Natural History(l); Nevada Historical Society(r); **42** ©Norbert Wu; **43** ©1993 Bob Cranston/Mo Yung Productions(tr); ©Norbert Wu(bl); **44** ©1993 Charles Lindsay/Mo Yung Productions(tr); **44** ©1993 Bob Cranston/Mo Yung Productions(bl); **48** ©Allsport/Shaun Botterill; **49** ©1993 Charles Lindsay/Mo Yung Productions(tr); **50** Tony Freeman/PhotoEdit; **56** Renee Lynn/Tony Stone Images/Chicago, Inc.(l); Daryl Balfour/Tony Stone Images/Chicago, Inc. (tr); ©Animals Animals/Richard Packwood(br); **58** Superstock; **62** Greg Vaugh/Tony Stone Images/Chicago, Inc.(br); ©NYNEX Information Resources Company. Printed by permission of NYNEX Information Resources Company.(overlay); **66** Ed Pritchard/Tony Stone Images/Chicago, Inc.; **69** Julie Sheffield/Hannibal Courier Post;—courtesy Hannibal Courier Post(b) **70** David Burnett/Contact Press Images(l); Tony Freeman/PhotoEdit(r); **72** Bachmann/Uniphoto(tl); Fotoworld/The Image Bank(tm); Harald Sund/The Image Bank(tr); David Ball/Picture Cube(bl); Colin Molyneux/The Image Bank(bm); ©Arvind Garg, Photo Researchers, Inc.(br); **75** courtesy World Data Center A for Glaciology, National Snow and Ice Data Center(br); **76** Darrell Gulin/Tony Stone Images/Chicago, Inc.(tl); Paul H. Humann(tr); ©Animals Animals/Raymond A. Mendez(bl); ©Animals Animals/Zig Leszczynski(br); **81** Superstock; **82** ©The Stock Market/Frank Rossotto; **90–91** courtesy Finn Strong, River Tank Systems(m); **90** ©E.R. Degginger, The National Audubon Society Collection/Photo Researchers, Inc.(bl); courtesy Finn Strong, River Tank Systems(bm); **91** ©Jeff Lepore, The National Audubon Society Collection/Photo Researchers, Inc.(t); ©Rod Planck, The National Audubon Society Collection/Photo Researchers, Inc.(m); **98** K. Tarusov/ITAR-TASS/Sovfoto/Eastfoto(br); **100** ©Bob Daemmrich; **103** Archive Photos; **105** John Coletti/Picture Cube(t); Michael Dwyer/Stock Boston(m); Superstock(b); **106** Richard Pasley/Stock Boston(t); UPI/Bettmann Archive(b); **113** Roger Ressmeyer—©1995 CORBIS(l); ©James King-Holmes, Science Photo Library/Photo Researchers, Inc.(tr); photo by Alessandro Sanvito. Artifacts from the collection of Museo Archeologico di Napoli, Italy.(br); **114** Rob Crandall;

Images/Chicago, Inc.(t); Superstock(b); **733** courtesy of Emil Fries Piano Hospital and Training Center; **735** Superstock **738** John Kelly/The Image Bank; **739** ©The Stock Market/Bob Krist.

592(t), **593**, **710**(tl), **711**, **740**, **741**
Patrice Rossi **159**
David Shepherd Design **48**, **55**, **69**, **74**, **82**, **119**, **146**, **166**, **168**(l)
Doug Stevens **92**, **224**, **292**, **301**(b), **302**, **669**, **671**, **680**, **689**

Assignment Photography

Kindra Clineff **553**
Jeffrey Dunn **vi**(b), **90**(br), **91**(b), **92**(t, b), **112**(l, r)
Charlotte Fiorito **iv**(b), **xxiv** (bl, br), **xxxiv**, **1**(tl, bl), **2**(t, b), **6**, **26**
Ken Karp **495**(r)
John Marshall **vii**, **xx**(tl), **x**, **138–139**, **140**, **146**(l), **159**
RMIP/Richard Haynes **v**(t), **xv**(t), **xvi**(b), **xxiii**, **xxvi**, **xxvii**(br, bl), **xxviii**(tr), **xxix**(r), **3**, **12**(tl, tr), **20**, **29**, **31**, **36**, **37**, **46**, **51**, **52**, **53**, **65**(tr), **69**(b), **84**, **85**, **93**(bl, br), **94**, **95**, **98**(tl, tr, m), **99**, **107**, **111**, **115**(t, b), **120**(t), **123**(t), **127**(t), **128**(t, ml, m, mr), **130**(l, r), **131**(r), **148**, **155**, **158**, **160**(b), **162**, **165**, **167**, **176**(b), **177**, **183**(tr), **191**(r), **192**, **195**, **199**, **203**, **206**, **209**, **214**, **216**, **219**, **220**(tr), **221**, **225**, **228**, **229**, **237**, **244**, **249**, **250**, **258**, **273**, **278**, **280**, **297**, **299**(b), **324**(tl), **337**, **343**, **344**, **350**(m), **355**(b), **377**, **378**(br), **380-381**, **404**, **419**, **433**(l, m), **436**, **441**, **454**(bl), **455**(bl), **462-464**(background), **472**, **480**, **486**, **497**, **501**(l, br), **504**, **505**(t), **513**, **515**, **520**, **526**(tl, m, tr), **535**(t, bl, br), **541**(b), **547**, **555**, **560**(t), **563**, **566**, **570**(r), **573**, **577**, **585**, **586**, **590**(tr), **593**(t), **596**, **598**(b), **600**, **601**, **608**(b), **610**(b), **625**, **632**, **637**(t), **646**, **647**, **655**, **663**, **673**, **677**, **681**, **685**, **691**, **695**(t), **698**, **707**(b), **713**, **726**(tl), **740**, **741**
Jeff Reinking **xvii**(b), **653**(t), **654**(b), **661**, **668**
Tony Scarpetta **455**(tr)
Robert Schoen **xxxiii**(l), **510**(b), **511**(t), **512**
Bill Wiegand **ix**(b), **234**(b), **236**(b), **255**(tr), **271**(t)

Illustrations

Arnold Bombay **341**, **563**, **564**, **567**(t), **576**, **584**
Hannah Bonner **198**, **204**(b), **296**(t), **303**, **307**, **468**
Steve Cowden **xxvii**, **xxx**, **45**, **54**, **78**, **79**, **81**, **208**, **212**, **298**, **301**(t), **315**(t), **326**, **328**, **336**, **409**, **424**, **432**, **569**, **589**, **659**, **661**(b), **662**, **668**(b), **670**, **699**
Dartmouth Publishing **vii**, **2**, **4**, **6**(t), **12**, **24**, **25**, **45**, **62**, **84**, **93**, **97**, **99**(tl, mr), **100**, **131**, **169**, **170**, **173**, **191**, **197**, **204**(t, b), **239**, **242**, **247**, **248**, **253**, **256**(b), **264**(t), **271**, **275**, **276**(t), **351**(background), **369**(t), **371**, **377**, **381**, **413**, **428**, **454**(b), **490**, **493**, **500**, **505**, **518**, **590**, **598**
DLF Group **vii**, **xxxii**, **1**, **8**, **22**, **56**, **69**, **114**, **118**(tl), **154**, **168**(l), **172**, **534**
Eureka Cartography **176**(tm), **313**, **315**(b), **320**, **501**, **602**
Harvard Design & Mapping **512**, **517**, **532**
Phillip Jones **727**
Piotr Kaczmarek **xxv**, **14**, **49**, **296**(b), **313**, **354**, **431**, **446**(b), **657**, **661**(t)
Joe Klim **426**
Andrea Maginnis **351**
Precision Graphics **xii**, **xxi**, **xxxii**, **5**, **6**(b), **7**, **8**(b), **9**, **13**, **15**, **29**, **30**, **33**(b), **35**, **44**, **61**, **77**, **101**, **125**(r), **147**, **226**, **227**, **237**, **239**, **243**, **246**, **252**, **254**, **256**(t), **263**, **264**(t), **270**, **273**, **276**(b), **277**, **280**, **281**, **290**, **291**, **310**, **349**, **370**, **388**, **417**, **430**, **439**(m), **514**, **528**, **533**, **539**, **541**, **548**, **550**, **552**, **567**(b), **568**, **582**, **596**, **626**, **656**, **664**, **665**, **666**, **668**(t), **683**, **696**, **717**, **732**, **738**(b)
Ann Raymond **347**, **359**, **364**, **372**, **376**(t), **402**, **410**, **447**, **448**, **562**, **565**, **570**, **575**, **577**, **578**, **581**, **583**, **586**, **587**, **588**,

Selected Answers

CHAPTER 1

Page 5 Checking Key Concepts

1. about 1,178,000 people **3.** Yes; the number of degrees (in thousands) predicted by the constant growth model for the second and third years are 921 and 1049 and by the proportional growth model are about 912 and 1049. Both sets of estimates are reasonable, since they are close to the actual values for those years.

Pages 5–8 Exercises and Applications

1. Prediction estimates may vary due to rounding. Increases are rounded to the nearest dollar or the nearest percent.

a.

Average Daily Hospital Cost			
Year	Daily charge ($)	$ Increase	% Increase
1975	134	—	—
1980	245	111	83%
1985	460	215	88%
1990	687	227	49%

b. about $870 **c.** about $1200; It is much higher. **3.** Prediction estimates may vary due to rounding. Answers may vary. An example is given. I think a reasonable estimate is about 2300 scarves. This is based on the constant growth model, which I think is a reasonable model. **5.** Prediction estimates may vary due to rounding. **a.** constant growth model: about $6.10; proportional growth model: about $8.86 **b.** the estimate obtained from the constant growth model; Both estimates will be off because of the fact that the minimum wage did not increase at all for 9 years. **c.** 1979–1981: The estimate with the constant growth model would be slightly higher; the estimate with the proportional growth model would be slightly lower. 1974–1976: The estimate with the constant growth model would be significantly lower; the estimate with the proportional growth model would be about the same. **7, 9.** Estimates may vary due to rounding. Answers may vary. Two examples are given for each item. The first example is based on a constant growth model, the second on a proportional growth model. **7.** about $13,126; about $14,100 **9.** about $3028; about $3250 **11.** total expense; tuition **13. a.** about 43% **b.** about $42,954; about $61,424 **c.** No; the actual statistics are significantly lower than the estimates. Answers may vary. An example is given. The difference between a master's degree and a doctorate may not have as significant an effect on income as the difference between lower education levels.

21.

x	y
0	6
1	3
2	0
3	−3

23. 3.4

Page 12 Checking Key Concepts

1. number of workers; years since 1987 **3.** $y = 7147(1.374)^x$

Pages 12–15 Exercises and Applications

1. Marya; Jon-Paul **3.** 1985; 1980; Jon-Paul's data begins in 1985, while Marya's begins in 1980. **5.** 3000; Jon-Paul's (the linear model) **7.** about 39,000 teams **11. a.** Answers may vary slightly due to rounding. Let x = the number of four-year periods since 1960 and y = time in seconds.
(a) $y = 258.3 - 4.16x$ (b) $y = 258.3(0.983)^x$
(c) linear exponential

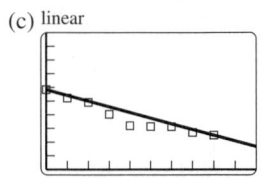 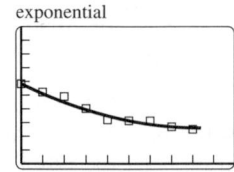

Answers may vary. An example is given. Yes; the data points are reasonably near both graphs. (d) linear: about 220.86 s, about 216.70 s; exponential: about 221.36 s, about 217.60 s **17.** Answers may vary due to rounding. Examples are given. **a.** about $22.2 billion **b.** about $22.4 billion **19.** −1 **21.** 13.5; 13; 13

Page 15 Assess Your Progress

2. Answers may vary due to rounding. Examples are given. In parts (a) and (b), let x = the number of decades since 1951 and y = the cost in dollars of a movie ticket. **a.** $y = 0.50 + 1.1x$ **b.** $y = 0.50(1.78)^x$ **c.**

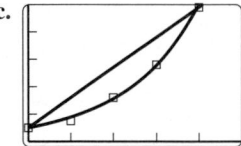

The exponential model fits the data almost exactly.

d. about $6 (linear model); about $8.93 (exponential model), which would probably be rounded to $9.00

Page 18 Checking Key Concepts

1. $\begin{bmatrix} 5 & 4 \\ 6 & 5 \end{bmatrix}$ **3.** $\begin{bmatrix} -3 & 2 \\ 6 & -1 \end{bmatrix}$ **5.** $\begin{bmatrix} 2 & 4 & 4 \\ 18 & -26 & 18 \\ 3 & -20 & 2 \end{bmatrix}$

Pages 19–21 Exercises and Applications

1. 2×4 **3.** No; the matrices do not have the same dimensions, so they do not have corresponding elements.

5. $\begin{bmatrix} 14 & 6 \\ 19 & 18 \\ 11 & -17 \end{bmatrix}$ **7.** $\begin{bmatrix} -2 & 10 \\ -19 & 6 \\ 9 & 9 \end{bmatrix}$ **9.** $\begin{bmatrix} -26 & -44 \\ 68 & -23 \end{bmatrix}$ **11.** undefined

13. $A + B$; $\begin{bmatrix} 2,497,800 & 136,000 & 696,150 & 18,160 \\ 2,989,800 & 5,075 & 674,225 & 62,050 \\ 394,250 & 835,000 & 90,500 & 12,430 \end{bmatrix}$

15. The dimensions would be 5 × 4; the dimensions would be 3 × 5 (or 5 × 5 if Nebraska and Missouri were added as well).

17. a. $\begin{bmatrix} -1 & 1 & 5 & 3 \\ -3 & 2 & 2 & -3 \end{bmatrix}$

b.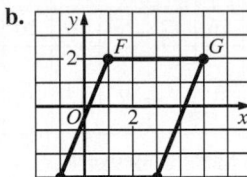

EFGH is *ABCD* moved 1 unit in the negative *x*-direction and 3 units in the negative *y*-direction.

19. enlarged by a factor of 3

21. position moved 1 unit in the negative *x*-direction and 3 units in the positive *y*-direction

23. The element in row *m*, column *n* of *A* + *B* is $a_{m,n} + b_{m,n}$. The element in row *m*, column *n* of *B* + *A* is $b_{m,n} + a_{m,n}$, which is equal to $a_{m,n} + b_{m,n}$.

$$A + B = \begin{bmatrix} 3+(-9) & 1+10 & -2+(-3) \\ -1+0 & 5+6 & 6+1 \\ 4+14 & 13+7 & 0+(-8) \end{bmatrix} = \begin{bmatrix} -6 & 11 & -5 \\ -1 & 11 & 7 \\ 18 & 20 & -8 \end{bmatrix};$$

$$B + A = \begin{bmatrix} -9+3 & 10+1 & -3+(-2) \\ 0+(-1) & 6+5 & 1+6 \\ 14+4 & 7+13 & -8+0 \end{bmatrix} = \begin{bmatrix} -6 & 11 & -5 \\ -1 & 11 & 7 \\ 18 & 20 & -8 \end{bmatrix}$$

25. The element in row *m*, column *n* of *r*(*A* + *B*) is $r(a_{m,n} + b_{m,n})$. The element in row *m*, column *n* of *rA* + *rB* is $ra_{m,n} + rb_{m,n}$, which is equal to $r(a_{m,n} + b_{m,n})$.

$$r(A + B) = r\begin{bmatrix} -6 & 11 & -5 \\ -1 & 11 & 7 \\ 18 & 20 & -8 \end{bmatrix} = \begin{bmatrix} -6r & 11r & -5r \\ -r & 11r & 7r \\ 18r & 20r & -8r \end{bmatrix};$$

$$rA + rB = \begin{bmatrix} 3r & r & -2r \\ -r & 5r & 6r \\ 4r & 13r & 0 \end{bmatrix} + \begin{bmatrix} -9r & 10r & -3r \\ 0 & 6r & r \\ 14r & 7r & -8r \end{bmatrix} = \begin{bmatrix} -6r & 11r & -5r \\ -r & 11r & 7r \\ 18r & 20r & -8r \end{bmatrix}$$

31. Answers may vary due to rounding. Let *x* = the number of decades since 1960 and *y* = the number (in millions) of cable users. $y = 0.65(4.53)^x$ **33.** 1 **35.** 48 **37.** False; for example, 5 − 1 = 4, but 1 − 5 = −4. **39.** True for every real number *a* except 0; for example, $7 \cdot \frac{1}{7} = 1$. ($\frac{1}{0}$ is not defined.)

41. False; for example, $|-5| + |5| = 5 + 5 = 10$, while $|-5 + 5| = |0| = 0$.

Page 25 **Checking Key Concepts**

1. a. $\begin{bmatrix} 8 & 6 & -2 \\ 18 & 20 & -4 \end{bmatrix}$ **b.** undefined

3. a. $\begin{bmatrix} 16 & -23 & 3 \\ 47 & -61 & -3 \\ 45 & -57 & -1 \end{bmatrix}$ **b.** $\begin{bmatrix} -32 & -14 & 22 \\ -1 & -8 & 11 \\ 27 & 4 & -6 \end{bmatrix}$ **5.** *RS* and *SR*; No.

Pages 25–28 **Exercises and Applications**

1. *A*: 3 × 5; *B*: 5 × 2; *AB*: 3 × 2 **3.** *LM*: Yes; 8 × 6. *ML*: No, the number of columns in *M* does not equal the number of rows in *L*. **5.** *LM*: No; the number of columns in *L* does not equal the number of rows in *M*. *ML*: Yes; 3 × 5.

7. $\begin{bmatrix} 8 & -3 \\ 5 & -10 \end{bmatrix}$ **9.** $\begin{bmatrix} -24 \\ -10 \end{bmatrix}$ **11.** $\begin{bmatrix} -4 & -7 & -4 \\ 6 & 30 & -7 \end{bmatrix}$ **13.** undefined

15.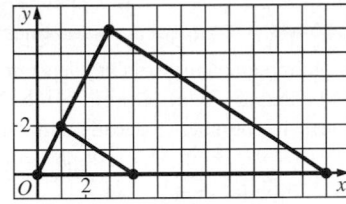

$S = \begin{array}{c} \text{Day 1} \\ \text{Day 2} \end{array}\begin{bmatrix} 40 & 45 & 12 & 28 & 25 & 65 \\ 32 & 52 & 24 & 16 & 18 & 45 \end{bmatrix};$

$$T = \begin{array}{r} \\ \text{Class} \\ \text{Plain} \\ \text{2-color sorority} \\ \text{Sorority} \\ \text{Earrings} \\ \text{Shirts} \end{array}\begin{array}{c} \$ \\ \begin{bmatrix} 25 \\ 20 \\ 25 \\ 30 \\ 15 \\ 12 \end{bmatrix} \end{array}; \quad \begin{array}{l} S \text{ has dimensions } 2 \times 6 \text{ and } T \text{ has} \\ \text{dimensions } 6 \times 1. \end{array}$$

b. $\begin{array}{c} \\ \text{Day 1} \\ \text{Day 2} \end{array}\begin{array}{c} \$ \\ \begin{bmatrix} 4195 \\ 3730 \end{bmatrix} \end{array};$ total sales for the two days

23. a. $\begin{bmatrix} -510 & 90 & 60 & 696 \\ -1071 & 653 & -18 & 444 \\ -810 & 450 & 0 & 432 \end{bmatrix}$

b. An error statement results. *TS* cannot be computed because the number of columns of *T* and the number of rows of *S* are not equal. **25.** sometimes true **27.** always true

31. $\begin{bmatrix} -\frac{3}{2} & 0 \\ -3 & -6 \end{bmatrix}$ **33.** $\begin{bmatrix} 1 & 2 \\ -1 & -1 \end{bmatrix}$ **35.** 0.25 **37.** 0.01

39. $x \le -9$ **41.** $x \ge -\frac{3}{5}$

Page 32 **Checking Key Concepts**

Answers may vary. Examples are given. In each case, repeat the simulation a number of times, say 10, and compute the average. **1.** Roll a die and let rolling a 6 represent getting the particular sticker. Count the number of trials needed to roll a 6. **3.** Use the random number feature on a calculator. Let random numbers greater than or equal to 0.2 represent individual stickers and numbers less than 0.2 represent team stickers. Count the number of trials needed to generate a number less than 0.2.

Pages 32–35 **Exercises and Applications**

1, 3. Answers may vary. Examples are given. In each case, repeat the simulation a number of times, say 10, and compute the average.

1. Use the random number feature on a calculator. Let the first digit of each random number represent the number of correct answers guessed. Count the number of trials needed to generate a number greater than or equal to 7.   **3.** Roll a die. Let rolling a 1 or a 2 represent a hit. Count the number of trials needed to roll a 1 or a 2.   **5.** B   **7.** Answers may vary. Sample trial results are given.   **a.** 7, 15, 18, 10, 9, 9, 13, 13, 16, 15
b. for the preceding sample, 18 packs; 7 packs; about 13 packs
9. a. Simulation method may vary. For example, toss a coin. Let heads represent painted side up and tails represent painted side down. To simulate each player's turn, toss the coin 12 times. The player gets 1 point if heads come up exactly five times.   **b.** Answers may vary.   **c.** Answers may vary. An example is given. On each turn, each player has an equal chance of getting 1 point. However, I think the player who throws first has a slight advantage. Consider, for example, if both players score 1 point on each turn, Player 1 will win.

17. $\begin{bmatrix} 14 \\ 3 \end{bmatrix}$ **19.** $\begin{bmatrix} 7 & -5 & 10 \\ -3 & 4 & -7 \\ -4 & -12 & 16 \end{bmatrix}$

21.

x	y
-3	3
-2	2
-1	1
0	0
1	-1
2	-2
3	-3

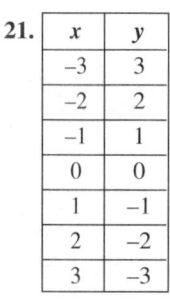 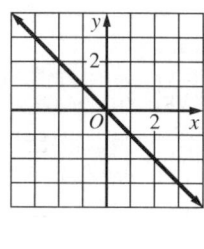

23.

x	y
-3	11
-2	9
-1	7
0	5
1	3
2	1
3	-1

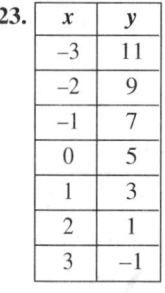 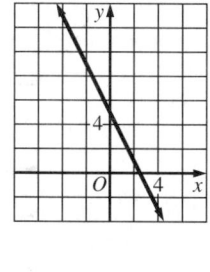

Page 35   Assess Your Progress

1. $\begin{bmatrix} 1 & 6 & 1 \\ -3 & 4 & 1 \end{bmatrix}$ **2.** $\begin{bmatrix} 9 & -6 & 6 \\ 15 & -3 & -9 \\ 0 & 12 & 3 \end{bmatrix}$ **3.** undefined   **4.** undefined

5. $\begin{bmatrix} -3 & -6 & 0 \\ 6 & -5 & -5 \end{bmatrix}$ **6.** $\begin{bmatrix} 6 & -8 & 3 \\ -4 & 21 & -5 \end{bmatrix}$ **7.** $\begin{bmatrix} 29 & -14 \\ 3 & -3 \end{bmatrix}$

8. undefined   **9.** $\begin{bmatrix} 1 & 2 \\ -8 & 5 \\ 19 & -9 \end{bmatrix}$ **10.** undefined

11. a. Simulation methods may vary. An example is given. For each employee, use int(6*rand)+1 to generate random numbers. Let 1 represent the employee's leaving the company. Generate random numbers until you get a 1. The number of tries is the number of years the employee stays at the company. Sample trial results are given.

Employee	1	2	3	4	5	6	7	8	9	10
No. of years employee stays	6	12	4	2	11	5	3	5	12	5

Employee	11	12	13	14	15	16	17	18	19	20
No. of years employee stays	7	6	14	5	2	1	3	6	11	5

b. for the preceding sample, about 6 years; 9 employees; 5 employees

Pages 40–41   Chapter 1 Assessment

1. function; independent variable; dependent variable
2. matrix; element   **3. a.** 4.9 million   **b.** 22%
c, d. Let x = the number of decades since 1970.
c. $y = 20.0 + 4.9x$; about 34.7 million   **d.** $y = 20.0(1.22)^x$; about 36.3 million   **4. a.** $y = 24.7 + 2.3x$ **b.** $y = 24.7(1.09)^x$
c. Amounts are in millions of dollars.

Year	Actual Amount Spent	Amount predicted by linear function	Amount predicted by exponential function
1986	24.7	24.7	24.7
1987	27.1	27.0	26.9
1988	29.4	29.3	29.3
1989	31.6	31.6	32.0

d. The values predicted by the linear model are slightly closer to the actual values than are the values predicted by the exponential model. However, both models are reasonable.

5. $\begin{bmatrix} 8 & -2 \\ -2 & 0 \\ 7 & 31 \end{bmatrix}$ **6.** undefined   **7.** $\begin{bmatrix} 7 & -18 \\ -8 & 5 \\ -7 & 69 \end{bmatrix}$ **8.** $\begin{bmatrix} 53 \\ 6 \\ 25 \end{bmatrix}$

9. $\begin{bmatrix} 3 & 0 \\ -6 & 3 \end{bmatrix}$ **10.** undefined   **11.** 2×4 **12.** -2 **13.** 3×4

14. a. Let three numbers, say 1, 2, and 3, represent the letter A. Let two numbers, say 4 and 5, represent the letter B, and let the remaining number represent the letter C. Roll the die until a number representing each of the three letters comes up. Record the number of rolls it took. Repeat a number of times, say 20, and calculate the average number of rolls.
b. 7 or 8 bottle caps

CHAPTER 2

Page 47   Checking Key Concepts

1. a. Yes; the ratio of y and x is constant; $\frac{y}{x} = 3.2$.   **b.** Yes; the ratio of y and x is constant; $\frac{y}{x} = -3.2$.   **c.** No; the ratio of y and x is not constant. The graph of $y = 3.2x + 1$ does not pass through the origin.   **3. a.** Yes; $y = \frac{1}{2}x$.   **b.** No.

Pages 48–51   Exercises and Applications

1. Yes; the common ratio is about 62.5; $y = 62.5x$.   **3.** No.
5. No.   **7.** No.   **9.** No.   **11.** No.

13. Let l = the number of laps and d = the distance of the race in meters; $l = 0.0025d$; 7.5 laps. **19.** Yes; $y = \frac{2}{3}x$. **21.** Yes; $y = -3x$. **23.** Answers may vary. about 50 Cal **25. a.** Let x = the number of hours spent exercising and y = the number of calories used; $y = 900x$; 900 Cal/h. **b.** Yes; the constant would be less for a person who weighs less than 150 lb, because it requires less energy to move a smaller mass. **31.** the basic monthly cost of the account; the cost per check

Page 54 Checking Key Concepts

1. 2   **3.** 0   **5.** $-\frac{3}{7}; \frac{1}{4}$ **7.** $y = -\frac{4}{3}x + 4$

Pages 55–57 Exercises and Applications

1. 3; 7   **3.** 1; 0   **5.** $y = -2$ **7.** $y = \frac{1}{2}x - 1.5$

9. Let x = the number of gal of water in the pitcher and y = the total weight in pounds of the pitcher and water; $y = 8.3x + 1.2$.

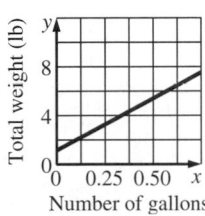

11. Let x = the number of wrong answers and y = the test score; $y = -5x + 100$.

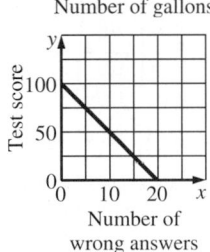

13. Let x = the number of miles driven and y = the number of gallons of gas in the tank; $y = -\frac{1}{15}x + 16$.

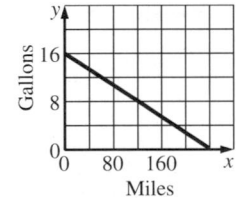

15.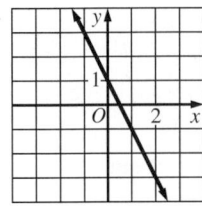

17.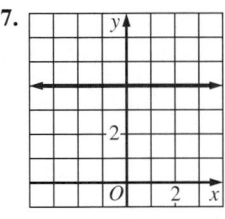

19.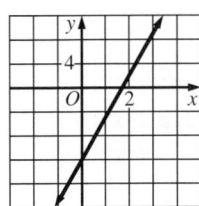

27. B   **31.** A   **33.** $-\frac{3}{2}$

Page 61 Checking Key Concepts

1. $y = 3 + (x - 2)$ **3.** $y = 3 - \frac{3}{2}x$ **5.** $y = 3 + (x + 2)$ or $y = 2 + (x + 3)$ **7. a.** -1 **b.** $x \geq -2; f(x) \geq -5$

Pages 61–64 Exercises and Applications

1. a. $y = 2 + (x - 3)$

b.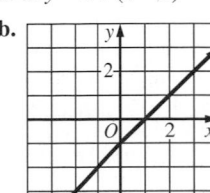

c. $y = x - 1$

3. a. $y = 1 - \frac{2}{3}(x - 3)$

b.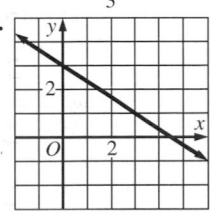

c. $y = -\frac{2}{3}x + 3$

5. a. $s = 1484 + 0.017(d - 1000)$ **b.** about 1501 m/s; about 4803 m   **c.** The domain is $1000 \leq d \leq 10{,}924$ since the function describes the relation at depths below approximately 1000 m and the deepest point is 10,924 m. Using the domain, you can find that the range is $1484 \leq s \leq 1565$. (The speed at 10,924 m is $1652.708 \approx 1653$ m/s.)
11. all real numbers; all real numbers   **13.** all real numbers; 2
15. a. $t = 390 - 0.5(w - 8)$ **b.** about 394 s or 6 min 34 s
17. $y = 4 + 2(x - 3)$ or $y = -4 + 2(x + 1)$
19. $y = 1 + \frac{1}{2}(x - 1)$ or $y = \frac{1}{2}(x + 1)$ **21.** $x > 0; f(x) > -2$
23. $x \geq 14; f(x) \geq 420.8$ **27.** $y = -\frac{2}{3}x - 3$ **29.** Let x = the number of years since 1993 and y = the number of families with preschoolers; $y = 11 - x$.   **31.** 4   **33.** $-\frac{8}{5}$

Page 64 Assess Your Progress

1. a. For every data pair, the ratio $\frac{\text{distance}}{\text{time}} \approx 55$. Since the ratio is constant, distance varies directly with time.
b. Estimates may vary; about 9 h.

2.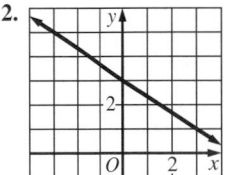

3. $f(x) \geq -1$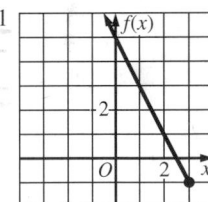

Page 68 Checking Key Concepts

1. Answers may vary. Examples are given.

a.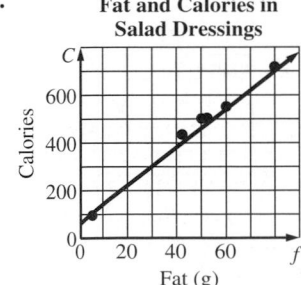

Fat and Calories in Salad Dressings

b. $C = 8f + 62$
c. about 302 g

1. Answers may vary. Examples are given.

b. Let x = the number of years since 1958 and y = the number of hours worked per day to pay taxes; $y = 0.02x + 2.2$; the increase per year.

c. about $3\frac{1}{4}$ h

a.

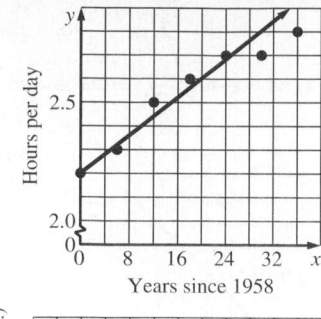

Working to Pay Taxes

Hours per day / Years since 1958

3. Answers may vary. **a.** Examples are given.

b. Let x = the date in April, 1993 and y = the height of the river above its banks; $y = 0.45x + 0.2$.

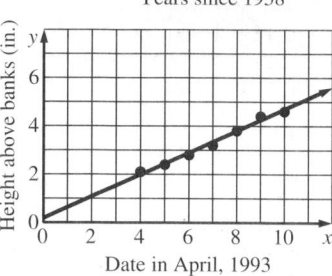

Height above banks (in.) / Date in April, 1993

5. a.

Year	District	m	a
1982	2	26,427	346
1982	4	15,904	282
1982	8	42,448	223
1986	2	15,671	293
1986	4	36,276	360
1986	8	36,710	306
1990	2	700	151
1990	4	11,529	–349
1990	8	26,047	160

b. Line of fit may vary.

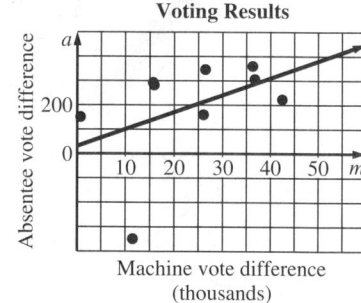

Voting Results

Absentee vote difference / Machine vote difference (thousands)

9. Answers may vary. Examples are given.

b. Let w = weight in kg and c = oxygen consumption in mL/min; $c = 7w + 43$.

c. about 218 mL/min

a.

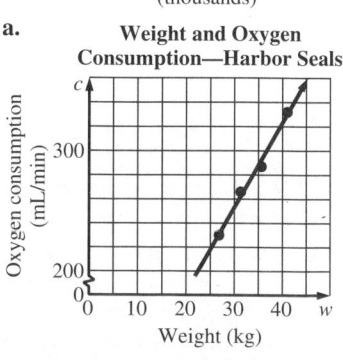

Weight and Oxygen Consumption—Harbor Seals

Oxygen consumption (mL/min) / Weight (kg)

13. $y = 3 - 4.2(x - 7)$ **15.** $y = -3 + 0.5(x + 2)$ **17.** Yes.
19. Yes.

1. -0.9 **3.** -0.3

1, 3. Estimates may vary.   **1.** about -0.9 **3.** about 1
5. positive   **7.** about zero

13. a.

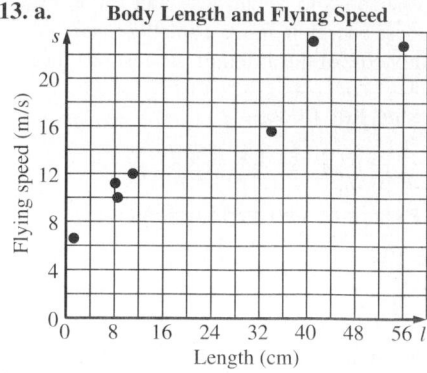

Body Length and Flying Speed

Flying speed (m/s) / Length (cm)

Estimates may vary; about 0.9.
b. 0.95
c. 0.95; They are the same. No.

17. a.

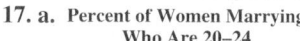

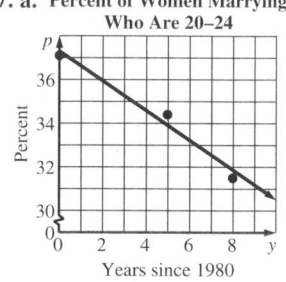

Percent of Women Marrying Who Are 20–24

Percent / Years since 1980

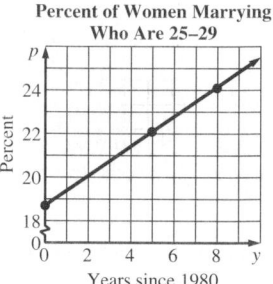

Percent of Women Marrying Who Are 25–29

Percent / Years since 1980

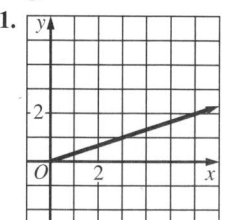

Percent of Women Marrying Who Are 65 and Older

Percent / Years since 1980

b. Answers may vary. Examples are given. Let y = the number of years since 1980 and p = the percent of women marrying who are the given age. 20–24: $p = -0.68y + 37.30$; 25–29: $p = 0.68y + 18.71$; 65 and older: $p = 1$ **c.** 20–24: about 30.5% in 1990, about 23.7% in 2000; 25–29: about 25.5% in 1990, about 32.3% in 2000; 65 and older: 1% in 1990, 1% in 2000   **19.** $y = -2$ **21.** 1   **23.** 4

1. 3; -12 **3.** The equations are not defined for $t < 5$.
5. $y = 2x + 1$; $0 \le x \le 6$

1.

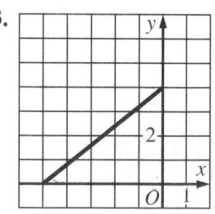

3.

5. $y = \frac{1}{3}x$; $x \ge 0$ **7.** $y = \frac{4}{5}x + 4$; $-5 \le x \le 0$ **11. a.** 280 s after deploying the parachute   **b.** $x = 12t$, $y = 3000 - 10t$, $20 \le t \le 300$ **17. a.** Let x = the number of years since 1990 and y = the number of computer books sold; $y = 560(1.15)^x$.

b. about 979 books   **19.** $y = 7 - \frac{12}{5}(x + 3)$ or
$y = -5 - \frac{12}{5}(x - 2)$

Page 83   Assess Your Progress

1. Lines of fit may vary.

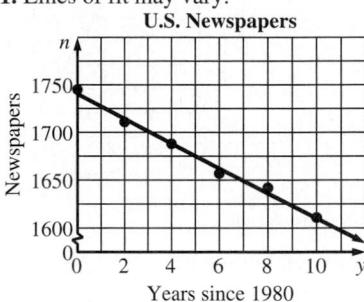

U.S. Newspapers

2. –0.996; strong; negative
3. a. 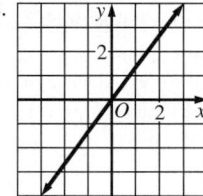   **b.** $y = -x + 3$; $x \geq 2$

Pages 88–89   Chapter 2 Assessment

1. direct variation   **2.** slope-intercept; slope; y-intercept
3. least-squares line; correlation coefficient   **4.** No.
5. Yes. 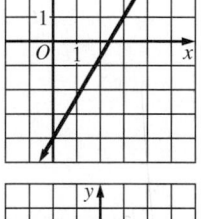   **6.** No.   **7. a.** $p = \frac{5}{12}s$
b. $11\frac{1}{4}$ **8.** 6; 0   **9.** –5; 2
10. 0; –3

11. 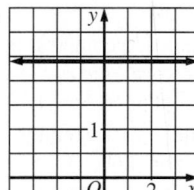   **12.**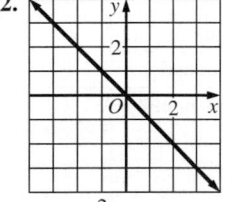

13. 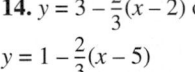   **14.** $y = 3 - \frac{2}{3}(x - 2)$ or
$y = 1 - \frac{2}{3}(x - 5)$
15. $y = -3 + 2(x - 1)$ or
$y = 1 + 2(x - 3)$
16. $y = 1 - \frac{1}{4}(x - 1)$ or
$y = 2 - \frac{1}{4}(x + 3)$

17. a. $x \leq 3$; $y \geq -1$ **b.**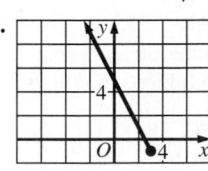

18. a.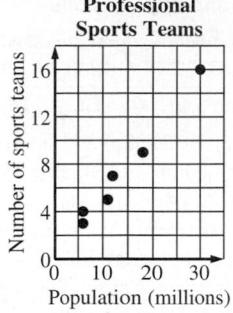

Professional Sports Teams

b. positive; strong
c. 0.99
d. Answers may vary. The least-squares line is $y = 0.519x + 0.151$.
e. Answers may vary. The estimate based on the least-squares line is about 4 teams; that is higher than the actual figure.

20. a. 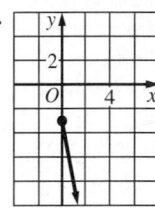   **21. a.**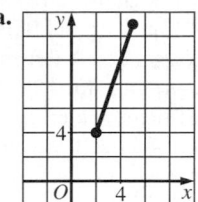

b. $y = -5x - 3$; $x \geq 0$ **b.** $y = 3x - 2$; $2 \leq x \leq 5$
22. a.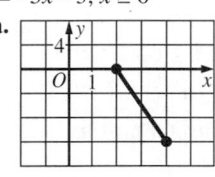

b. $y = -6x + 12$; $2 \leq x \leq 4$

CHAPTER 3

Page 96   Checking Key Concepts

1. 128 leaves   **3.** 512 leaves

5. a.

Number of folds	Page width (cm)	Page height (cm)	Page area (cm²)
0	38	51	1938
1	25.5	38	969
2	19	25.5	484.5
3	12.75	19	242.25
4	9.5	12.75	121.125
5	6.375	9.5	60.5625
6	4.75	6.375	30.28125

b. Let n = the number of folds and A = the page area in square centimeters; $A = 1938\left(\frac{1}{2}\right)^n$.

Pages 97–99   Exercises and Applications

1. 4 folds   **3.** 3 folds   **5.** 2^4 **7.** 2^6 **9.** 160   **11.** 200
13. linear   **15.** neither   **29.** $y = x + 1$ **31.** Yes.   **33.** No.

Page 104   Checking Key Concepts

1, 3. Answers are rounded to two decimal places.   **1.** 5.31; the predicted population in millions in 1800   **3.** about 9.53; the predicted population in millions in 1820   **5.** The growth rate is an estimate. Also, the proportional growth rates in Column D of the spreadsheet are not all equal, only approximately equal. This indicates that the exponential equation models the data only approximately.   **7.** 4   **9.** 16   **11.** about 4.64

Pages 104–106   Exercises and Applications

1. The growth is exponential, not linear.   **3, 5, 7.** Answers are rounded to two decimal places.   **3.** 22.94 million

5. 7.57 million   **7.** 447.38 million   **11.** $\frac{1}{16}$ **13.** $\frac{1}{7}$ **15.** 36

17. 1   **19.** 16   **21.** 2   **23.** 5   **25.** 1   **27.** Let d = the number of decades after 1959 and v = the number of vehicles per day in thousands; $v = 75(1.36)^d$.   **29.** Answer may vary due to rounding. about 119,000 vehicles per day   **35. a.** Answers may vary depending on rounding and how the independent variable is defined. An example is given. Let d = the number of degrees above zero and a = the air flow per gallon of water in ft³/min; $a = 3.0(1.06)^d$.   **b.** Answers may vary due to rounding; about 4.0 ft³/min; about 7.2 ft³/min; about 12.9 ft³/min   **43.** 0.2   **45.** The graphs have the same slope but different y-intercepts. The graph of $y = 6 + 2x$ is the graph of $y = 3 + 2x$ shifted 3 units in the positive y-direction.
47. The equation is simply a point-slope equation of the same line. The graphs are identical.

Page 110   Checking Key Concepts

1. exponential growth   **3.** exponential growth   **5.** 3   **7.** –1
9. –2.5   **11.** 4 (half-life)

Pages 110–114   Exercises and Applications

1. a. 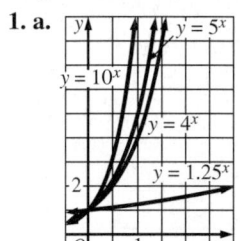   **3. a.**

b. $y = 10^x$; $y = (1.25)^x$
c. The y-intercept of each of the graphs is 1.

b. $y = -10^x$; $y = -5^x$;
$y = -4^x$; $y = -(1.25)^x$

5. a. 1
b. exponential growth
c.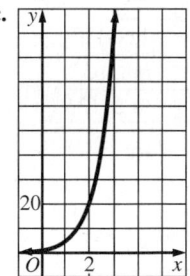

7. a. 2.5
b. exponential decay
c.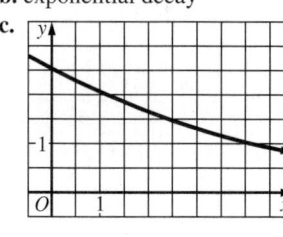

9. a. For $b = 1$, the graph of $y = ab$ is a horizontal line through $(0, a)$.   **b.** For $a = 0$, the graph of $y = ab^x$ is the x-axis.
11. C   **13.** A   **17.** 12 years   **19.** 8 years   **21.** 4 years
25, 27. Answers are given to two decimal places.   **25.** 0.74
27. 0.30   **29, 31.** Estimates may vary.   **29.** about 3700 years ago   **31.** about 10.4%   **35.** about 10.6 million   **37.** Estimates may vary. An example is given. about 35 years   **39.** $\frac{1}{9}$

41. a. Lines of fit may vary.
b. Answers may vary. An example is given. Let P = percent of water and C = the number of calories; $C = -4.4P + 435$.
c. Estimates may vary; about 31 calories.

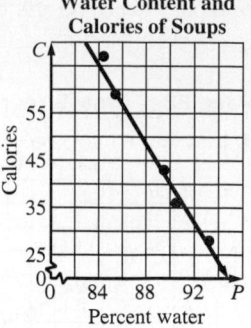

Water Content and Calories of Soups

Page 114   Assess Your Progress

1. 16   **2.** $\frac{1}{16} = 0.0625$ **3.** 24   **4.** Let T = the number of teams; $T = 64\left(\frac{1}{2}\right)^n$.   **5.** $2^{3/4} \approx 1.68$ **6.** 7   **7.** 2   **8.** about 267 accidents   **9.** about 241 accidents   **10.** about 327 accidents
11. Estimates may vary. An example is given. in about $13\frac{1}{2}$ years

Page 119   Checking Key Concepts

1. $1032.52; the annual interest rate that would yield the same amount after one year   **3. a.** about 0.2501 g
b. about 0.1251 g   **c.** about 0.0625 g   **5. a.** 100   **b.** When t is very large, $e^{-0.1t} \approx 0$, so $L(t) \approx 2000$.   **c.** Estimates may vary. An example is given. about 29.4   **d.** none

Pages 119–122   Exercises and Applications

1, 3. Answers are given to two decimal places.   **1.** 2.53%
3. 7.79%   **5.** $1454.99   **7.** $2585.71   **9.** 2   **11.** 2.704814
13. 2.718146   **15.** 2.718280   **17. a.** Graphs may vary.
b. Answers may vary. See table in part (c).
c. Estimates may vary.

Doubling Your Money in a Year		
Compounding	n	Rate
annually	1	100%
semiannually	2	82.8%
quarterly	4	75.7%
monthly	12	71.4%
daily	365	69.4%

As the frequency of compounding increases, the interest rate needed to double your money drops. However, the rate is very high even for daily compounding. It would be extremely unusual to find an investment that offered an annual interest rate of 69.4%.   **e.** Estimates may vary; about 69.3%.
19. Estimates and equations may vary depending on rounding. An example is given. about 5.8 years; $E(t) = 179.2\left(\frac{1}{2}\right)^{t/5.8}$
27. $A(t) = 100(0.995)^t$ **33.** $a > 0$ and $b > 1$
35. Let y = the number of years since 1985 and P = public funds in billions of dollars; $P = 0.12y + 1.32$
37. $\begin{bmatrix} 39 & -8 & -10 \\ 96 & 48 & -40 \end{bmatrix}$

Page 125 Checking Key Concepts

1. $y = 5\left(\dfrac{3}{5}\right)^x$ **3.** $y = 3 \cdot 4^x$

Pages 126–129 Exercises and Applications

1. $y = 5\left(\dfrac{8}{5}\right)^x$ **3.** $y = 10\left(\dfrac{4}{5}\right)^x$ **5.** $y = 1.804(1.185)^x$

7. $y = 12.642(0.944)^x$

17. Let x = the number of degrees Celsius above zero and y = the oxygen absorbed in mL/L; $y = 10.015(0.98)^x$.
21. a. Let t be the number of hours after 1:00 P.M. and $P(t)$ = the number of bacteria; $P(t) = 30e^{0.116t}$. **b.** about 48 **c.** about 27 **d.** Substitute 2.75 for t in the equation.

23. a.

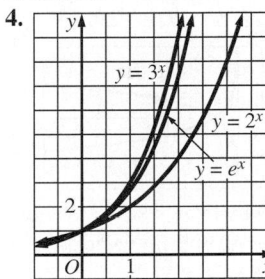

b. $y = -3x + 7;\ x \le 2$

Page 129 Assess Your Progress

1. $2385.04 **2.** $2394.25 **3.** $2394.43

4.

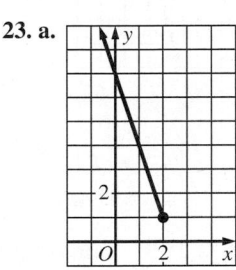

Summaries may vary. All three graphs have the same y-intercept, 1. All three are exponential, with equations of the form $y = b^x$. The greater the value of b, the steeper the graph.
5. a. Answers may vary.
b. $y = 4.975(1.046)^x$
c. about 181.7 cm

Pages 134–135 Chapter 3 Assessment

1. exponential growth; exponential decay **2.** e **3.** 2
4. $\dfrac{1}{6\sqrt{6}}$ or $\dfrac{\sqrt{6}}{36}$ **5.** $\dfrac{1}{49}$ **6.** 1 **7. a.** $E = 179.2(0.89)^{w/52}$
b. about 142 eggs per year **8. a.** $s = 30,000(1.05)^n$
b. 14.2 years **9. a.** $y = 10\left(\dfrac{1}{2}\right)^{x/11.2}$ **b.** about 8.84 g
10. a. $3136.62 **b.** $4738.15 **c.** $11,225.04 **11. a.** $E(t) = 50(0.970)^t$ **b.** about 27 errors **13.** $y = 4.43(1.17)^x$ **14.** $y = 7.40(1.04)^x$ **15.** $y = 0.943(2.12)^x$ **16. a.** $y = 0.766(2.17)^x$
b. $36.9 trillion **17. a.** Let x = the number of years since 1990 and y = the CPI; $y = 130.7(1.04)^x$. **b.** about 120.8; about 141.4

CHAPTERS 1–3

Pages 136–137 Cumulative Assessment

1. Let y = the winning time in seconds; $y = 44.6 - 0.367x$;
$y = 44.6(0.992)^x$. **5.** undefined **7.** $\begin{bmatrix} 11 & -7 & 0 \\ 0 & -6 & -2 \end{bmatrix}$ **9.** No.

11. Yes.

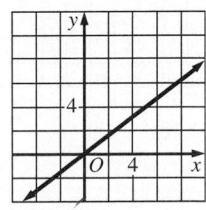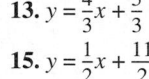

13. $y = \dfrac{4}{3}x + \dfrac{5}{3}$

15. $y = \dfrac{1}{2}x + \dfrac{11}{2}$

17.

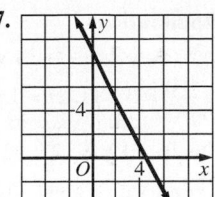

19. a. $E = 38 + 1(A - 7)$ or $E = 40 + 1(A - 9)$ **b.** 41
23. 3 **25.** 1

27. a.

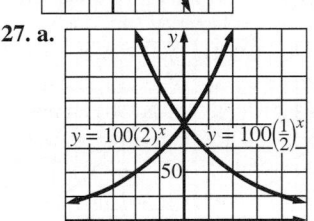

29. a. about 17.7 g
b. about 20.0 s; $A(t) = 50\left(\dfrac{1}{2}\right)^{t/20.0}$

CHAPTER 4

Page 145 Checking Key Concepts

1. 2 **3.** No.

Pages 145–147 Exercises and Applications

1. a.

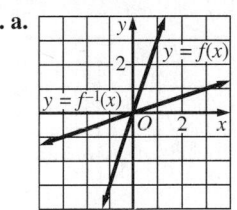

b. $f^{-1}(x) = \dfrac{1}{3}x$

3. a.

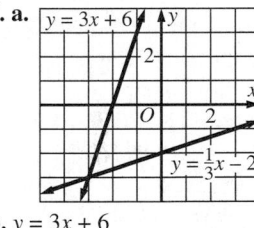

b. $y = -4x$

5. a.

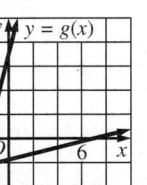

7. a.

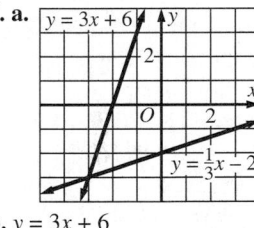

b. $y = 3x + 6$

b. $g^{-1}(x) = \dfrac{1}{4}x - \dfrac{7}{4}$

9. E **11.** For C, the domain is the nonnegative real numbers and the range is the real numbers greater than or equal to 84,000. For S, the domain is the real numbers greater than or equal to 84,000 and the range is the nonnegative real numbers. The domain of C is the range of S and the range of C is the domain of S. **17. a.** $y = 1.849x - 28.60$; average fuel economy; number of years since 1980 **b.** 2007 **19. a.** $y = 50,100(1.26)^x$ **b.** about 318,000 m³ **c.** about 25,000 m³
21. $y = 4 - 2(x + 3)$ or $y = -2 - 2x$ **23.** $563.75

Page 152 Checking Key Concepts

1. $f^{-1}(x) = \log_2 x$ **3.** $y = \ln x$ **5.** $\log_8 1 = 0$ **7.** $6^3 = 216$
9. $27^{1/3} = 3$ **11.** $\dfrac{2}{3}$ **13.** 2.16 **15.** -0.49

Pages 152–154   Exercises and Applications

1. a. $x > 1$ **b.** $0 < x < 1$ **c.** 1 **d.** $x \le 0$ **3.** $\log_2 8 = 3$

5. $\ln 1 = 0$ **7.** $\log_{64} \frac{1}{4} = -\frac{1}{3}$ **9.** $\log_{4/5} \frac{64}{125} = 3$ **11.** $2^4 = 16$

13. $2^{-5} = \frac{1}{32}$ **15.** $\left(\frac{1}{6}\right)^2 = \frac{1}{36}$ **17.** $64^{2/3} = 16$ **19.** 2 **21.** -3

23. 0 **25.** $\frac{3}{4}$ **31.** 1.08 **33.** 1.87

35. a.

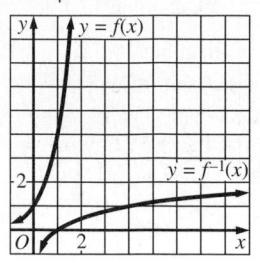

37. a.

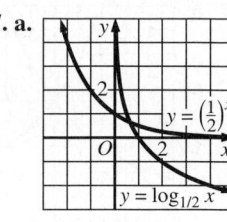

b. $y = \log_{1/2} x$

b. $f^{-1}(x) = \log_4 x$

39. a. $t = 40 \ln \dfrac{P}{19.1}$; the number of years after 1990 that the population reaches a given population P **b.** 2008   **c.** 1980

45. a.

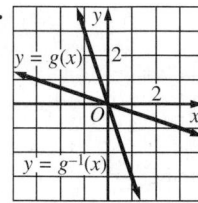

47. a.

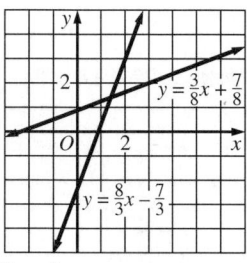

b. $g^{-1}(x) = -3x$

b. $y = \frac{8}{3}x - \frac{7}{3}$

49. neither   **51.** exponential

Page 158   Checking Key Concepts

1. $3 \log_2 p$ **3.** $2 \log_2 p - 4 \log_2 r$ **5.** $\log_a 32$ **7.** $\log_a \dfrac{x^9}{y^2}$

Pages 158–161   Exercises and Applications

1. $4 \log_7 p$ **3.** $2 \log_7 q + 3 \log_7 r$ **5.** $-\log_7 q$ **7.** $9 \log_7 p + 11 \log_7 q - 7 \log_7 r$ **11.** $\log_a \frac{1}{7}$ **13.** $\log_5 \dfrac{u^6}{v^{10}}$ **15.** $\log x^{32}$

17. $\log_b 9$ **23.** $y - x$ **25.** $2x + y$ **27. a.** about 26%   **29.** A

31. substitution; the product property of exponents; definition of logarithm; substitution   **35.** $\log_4 16 = 2$ **37.** 9.16

Page 161   Assess Your Progress

1. a.

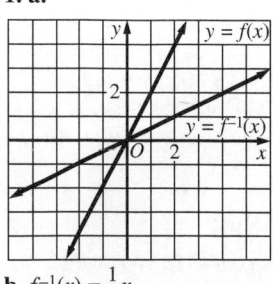

2. a.

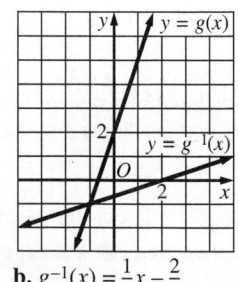

b. $f^{-1}(x) = \frac{1}{2}x$

b. $g^{-1}(x) = \frac{1}{3}x - \frac{2}{3}$

3. a.

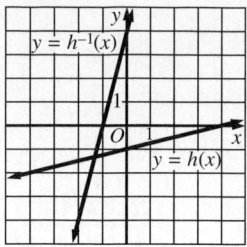

4. a.

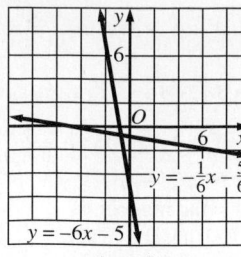

b. $h^{-1}(x) = 4x + 4$

b. $y = -\frac{1}{6}x - \frac{5}{6}$

5. 2 **6.** $\frac{1}{2}$ **7.** -4 **8.** $\frac{5}{11}$ **9. a.** $t = 91.7 \ln \dfrac{P}{33.9}$ **b.** 1997

10. $3 \log p + 4 \log q - 5 \log r$ **11.** $\log_b 8$

Page 165   Checking Key Concepts

1. 1.10 **3.** 0.86 **5.** 2.32 **7.** 1.28 **9.** 32 **11.** 8

Pages 165–167   Exercises and Applications

1. 1.95 **3.** 0.17 **5.** 0.42 **7.** 0.82 **13.** 0.32 **15.** 1.56

17. -1.63 **19.** 0.40 **23.** 272.99 **25.** 5 **27.** 2 **29.** 10

31. a.

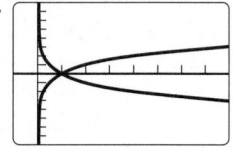

b.

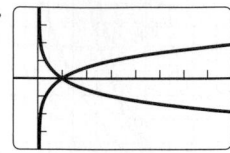

c.

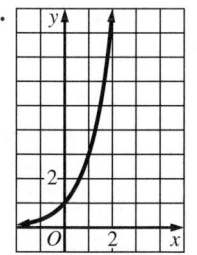

d. Each graph is the reflection of the other over the x-axis.

e. $\log_{1/b} x = -\log_b x$

33. $\log_a 63$

35.

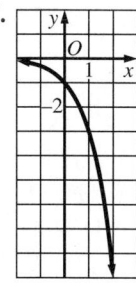

37.

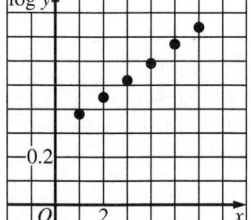

Page 171   Checking Key Concepts

1. a.

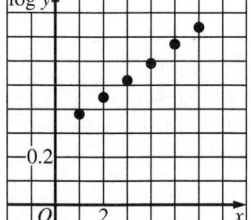

b.

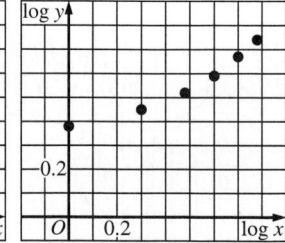

c. exponential function; $y = 2.01(1.18)^x$

3. a. 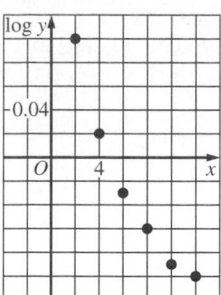 **b.**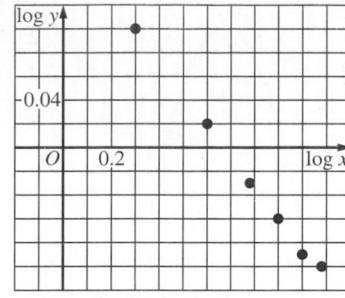

c. power function; $y = 1.50x^{-0.260}$

Pages 171–175 Exercises and Applications

3. $y = 6310(1.58)^x$ **5.** $y = 0.811(0.912)^x$
7. $y = 2.29(0.0380)^x$ **11.** $y = 219x^{0.720}$ **13.** $y = 0.914x^{-0.61}$
15. $y = 4.39x^{-1}$

17. a. 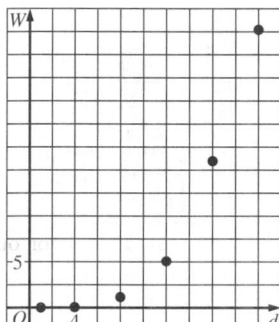 **b.**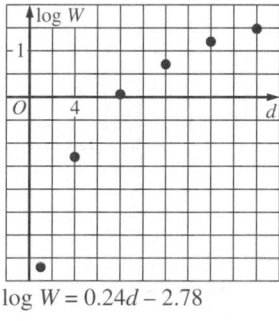

$\log W = 0.24d - 2.78$

No; power functions and exponential functions have similarly shaped graphs.
21. D **23.** Parksosaurus: about 176 kg; Struthiomimus: about 391 kg; Allosaurus: about 5740 kg; Tyrannosaurus: about 19,500 kg **27.** 0.94 **29.** –0.85 **31.** 9; 4
33. –36; –16

Page 175 Assess Your Progress

1. 1.79 **2.** 2.26 **3.** 8.60 **4.** 64 **5.** no solution **6. a.** $W = 4.83(1.14)^t$ **b.** $W = 5.19t^{0.365}$ **c.** The exponential function is the better model since the correlation coefficient for the exponential function, 0.9963581796, is closer to 1 than the correlation coefficient for the power function, 0.9550944564.

Pages 180–181 Chapter 4 Assessment

1. inverse; reflection **2.** common; natural; base-5; change-of-base
3. a. 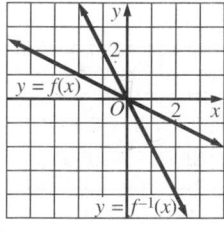 **4. a.**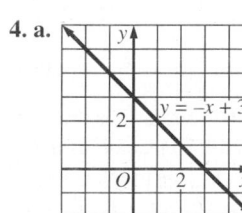

b. $f^{-1}(x) = -2x$ **b.** $y = -x + 3$

5. a.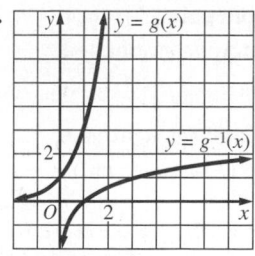

b. $g^{-1}(x) = 2 \log x$
7. a. The equation for the function and the equation for its inverse are the same. **8. a.** Let t = time in hours after the candle is lit and h = the height of the candle in inches; $h = 10 - 0.25t$ **b.** $t = 40 - 4h$ **c.** 16 h after the candle is lit
9. $\log_2 64 = 6$ **10.** $\ln \frac{1}{e} = -1$ **11.** $\log_{0.5} 8 = -3$ **12.** $4^{0.5} = 2$

13. $e^0 = 1$ **14.** $\left(\frac{1}{9}\right)^{-3/2} = 27$ **15.** $\frac{1}{3}$ **16.** –2 **17.** 4 **18.** –2
19. If $0 < x < 1$, then $\log x < 0$. Raising 10 to any positive power produces a number greater than 1. **20.** $3^2 = 9$ and $3^3 = 27$; Since $9 < 15 < 27$, $2 < \log_3 15 < 3$. **21.** $\log_5 p + 2 \log_5 q$ **22.** $1 - \log_5 r$ **23.** $\frac{1}{4} \log_5 r - 3 \log_5 p - \log_5 q$

24. $\ln 2$ **25.** $\log_a u^3 v^2$ **26.** $\log_b 25$ **28.** 4.52 **29.** 0.59
30. –0.22 **31.** 61 **32.** 1.70 **33.** –0.61 **34.** –0.32
35. Answers may vary. Examples are given. In the examples, x = years since 1980 and y = price in thousands.

a.

Type of function	Equation	Correlation coefficient
exponential	$y = 59.8(1.05)^x$	0.9965490038
power	$y = 42.8x^{0.353}$	0.9986221616

b. The power function is the better model since the correlation coefficient for the power function, 0.9986221616, is closer to 1 than the correlation coefficient for the exponential function, 0.9965490038. **c.** about $103,000; The prediction is very close to the actual median price.

CHAPTER 5

Page 188 Checking Key Concepts

1. 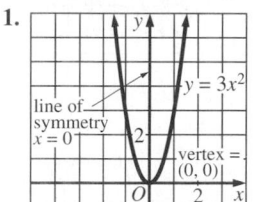 **3.** ±5 **5.** ±3.5 **7.** ±15.8

Pages 188–191 Exercises and Applications

1. B **3.** C **7.** ±3.5 **9.** ±0.5 **11.** ±9 **13.** ±2.6 **15.** 5
17. a. $r^2 = \frac{V}{\pi h} = \frac{75}{\pi}$ **b.** about 4.9 in. **25. a.** 9.25×10^{-10}
b. about 13,000,000 mi **29.** $y = 20.0x^2$ **31.** They all have the same slope, 3. Each graph has a different vertical intercept.

6. a.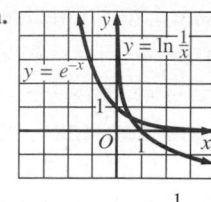

b. $y = -\ln x$ or $y = \ln \frac{1}{x}$

Page 196   Checking Key Concepts

1. $y = -2(x - 1)^2 - 2$ **3.** down; maximum value: -2
5. $-3.7, -0.3$

Pages 196–198   Exercises and Applications

1. B   **3.** A   **9.** 1 (maximum)   **11.** -4 (minimum)
13. -2 (minimum)
15. The vertex is $(-3, -3)$.
The line of symmetry
is $x = -3$.
The parabola opens down.

17. The vertex is $(-2, 0)$.
The line of symmetry is
$x = -2$.
The parabola opens down.

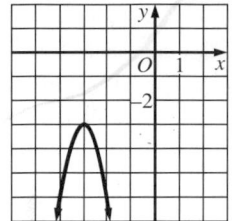 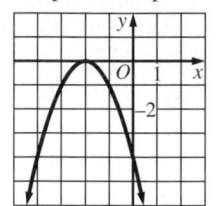

19. The vertex is $\left(-4, \frac{1}{2}\right)$.
The line of symmetry is $x = -4$.
The parabola opens down.

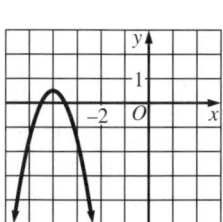

21. $y = 2(x + 2)^2 + 4$ **23.** $y = -3(x - 5)^2 + 2$ **29.** ± 3.2
31. $3(x + 2)$ **33.** $x(5x + 2)$ **35.** $g^{-1}(x) = \frac{1}{3}x - 5$

Page 202   Checking Key Concepts

1. a. $-7; 6$ **b.** $\left(-\frac{1}{2}, -253.5\right)$ **c.**
3. 72

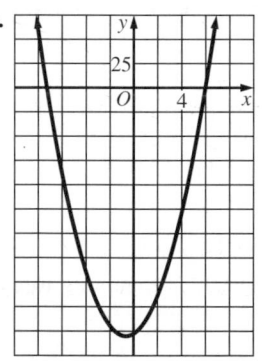

Pages 203–205   Exercises and Applications

1. a. $1; -3$
b. $(-1, -16)$
c.

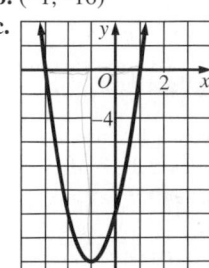

3. a. $4; -4$
b. $(0, 16)$
c.

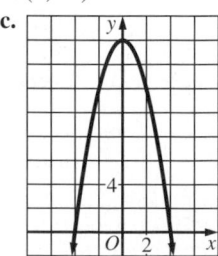

5. a. $6; -4$ **b.** $(1, 125)$
c.

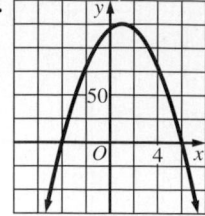

9. $y = -6(x - 4)(x - 6)$
17. D   **23.** ± 5
25. $-1.9; 0.9$ **27.** $\log_2 x^6$

Page 205   Assess Your Progress

1–3. a.

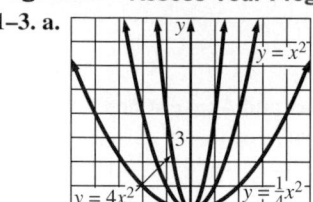

1. b. ± 3.5
2. b. ± 1.7
3. b. ± 6.9
4. $y = -3(x - 3)^2 + 7$
5. $y = \frac{1}{2}(x + 1)^2 + 4$
6. $y = 2(x - 5)^2 - 2$
7. $y = \frac{1}{8}(x + 6)^2 - 1$

8. $y = 3(x + 9)(x + 4)$

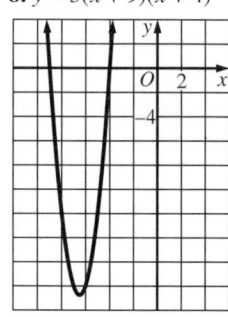

9. $y = -6(x - 3)(x - 4)$

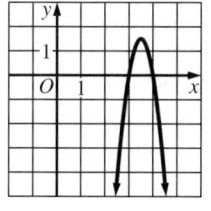

10. $y = 4(x - 3)(x + 3)$

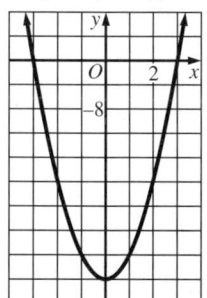

11. $y = -10(x - 2)(x + 2)$

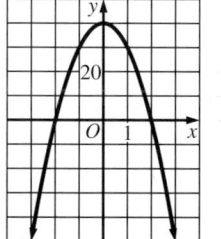

Page 209   Checking Key Concepts

1. a. 49   **b.** $\frac{25}{4}$ **c.** 0.01   **3.** 11

Pages 210–213   Exercises and Applications

1. $y = (x + 2)^2 - 1$ **3.** $y = \left(x - \frac{7}{2}\right)^2 - \frac{97}{4}$ **5.** $y = (x + 3)^2 - 9$

7. $y = 2\left(x - \frac{5}{4}\right)^2 - \frac{49}{8}$ **9. a.** $h = -16t^2 + 24t + 3$ **b.** 12 ft

c. 0.75 s after it is thrown   **d.** 1.6 s after it is thrown
11, 13. Explanations may vary. Examples are given.
11. A; This graph is the only one in which the parabola
opens down.   **13.** B; The vertex form of this equation is
$y = 2(x + 1)^2$, so the vertex of the graph is $(-1, 0)$.
17. a. $y = 0.06(x - 33.33)^2 + 20.33$

b. 20.33; The minimum oxygen consumption of which the parrot is capable is about 20.33 mL/g · h. This is achieved by flying at a speed of about 33.33 km/h.   **25.** maximum; 20

27. maximum; 12   **29.** maximum; 11   **31.** minimum; $\frac{5}{24}$

33. maximum; 21   **37.** $y = -\frac{3}{8}x(x - 8)$ **39.** $y = \frac{3}{4}(x - 3)(x - 7)$

41. ± 2 **43.** ± 1.5 **45.** $f^{-1}(x) = 0.4x + 3.12$

47. $h^{-1}(x) = -\frac{1}{2}x + 2$

Page 218   Checking Key Concepts
1. a. 1.25 s   **b.** 0.5 s   **c.** 9 ft

Pages 218–220   Exercises and Applications
Answers may vary slightly due to rounding.   **1.** −5.2; 1.2
3. −0.6; 1   **5.** −5; 3   **7.** −3.4; 0.9   **9.** −1.1; 2.4   **11. a.** 32.9°
b. 324.6 ft   **13.** one solution   **15.** no solution
17. two solutions   **19.** no x-intercepts   **21.** 1   **23.** −0.4; 2.4
31. 2001   **37.** (−1, −3)   **39.** ± 5 **41.** ± 12 **43.** 2.31

Page 224   Checking Key Concepts
1. $y = (x - 9)(x - 9)$ **3.** $y = 2(3x - 1)(x + 2)$ **5.** 18 mi/h

Pages 225–227   Exercises and Applications
1. $y = (x - 4)(x - 1)$ **3.** $y = (x + 1)(x - 4)$
5. $y = (3x + 2)(3x + 2)$ **7.** $y = (4x - 5)(x + 2)$
9. $y = 4(3x - 5)(3x + 5)$ **13, 15, 17, 19.** Answers are rounded to the nearest tenth when necessary.   **13.** −6   **15.** 1; 3
17. $-4\frac{1}{2}$; $1\frac{2}{3}$ **19.** −5; 0.6

21. a. graph:
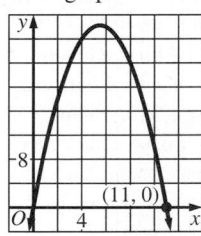
algebra: $0 = 11x - x^2$; $0 = -x(x - 11)$; $x = 0$ or $x = 11$
arithmetic: Add terms until the sum is zero: $10 + 8 + 6 + 4 + 2 + 0 + (-2) + (-4) + (-6) + (-8) + (-10) = 0$; there are 11 terms.   **b.** The answers are the same; preferences may vary. An example is given. I think the algebraic method was the easiest since the function was so easy to write as a product of factors.
c. Conjecture: The sum of the first x terms of the numbers $12 + 10 + 8 + 6 + \cdots$ is given by the function $S(x) = 13x - x^2$. It takes 13 terms for the function to get the sum of zero.
27. 0; 6   **29.** $\frac{7}{6}$ **31.** 0.4; 7.6   **33.** −0.7; 1.1   **35.** about \$780

37. 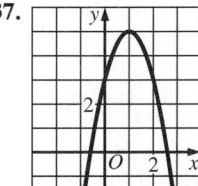   **39.**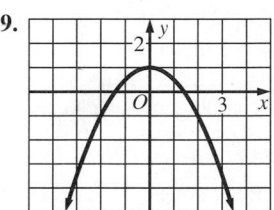

Page 227   Assess Your Progress
1. $y = 15(x + 0.3)^2 - 1.35$ **2.** $y = (x + 6)^2 - 29$
3. $y = 4(x - 2.5)^2 - 28$ **4.** −1.5; −0.5   **5.** no x-intercepts
6. ± 3.2 **7.** $y = (x + 4)(x + 3)$ **8.** $y = (3x + 1)(x - 2)$
9. $y = (2x - 1)(x + 7)$

Pages 232–233   Chapter 5 Assessment
1. quadratic; vertex; minimum; parabola; vertex; line of symmetry   **2.** standard; discriminant; x-intercepts
3.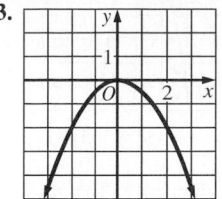
4. 3
5. 19.4 ft
8. a. $y = 2(x + 1)^2 - 7$
b. −2.87; 0.87

9. $y = -2(x - 2)(x - (-1))$
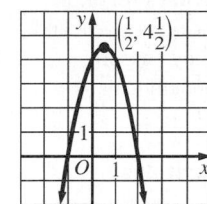
10. a. $y = \frac{1}{3}(x + 1)(x - 5)$
b. $y = -x(x + 4)$
11. a. 12.25 ft   **b.** 1.5 s; Answers may vary. An example is given. I used the quadratic formula to solve the equation $-16t^2 + 20t + 6 = 0$.

12. $y = 2\left(x + \frac{9}{4}\right)^2 - \frac{57}{8}$ **13.** $y = -(x - 3)^2 + 12$ **14.** $y = \frac{1}{4}(x - 4)^2 + 7$ **15. a.** $P = -16(I - 3.75)^2 + 225$ **b.** (3.75, 225); The maximum power, 225 watts corresponds to a current of 3.75 amperes.   **16.** 2.5   **17.** no x-intercepts   **18.** $-\frac{2}{3}$; 6
19. 35 mi/h   **20.** $y = (x - 4)(x - 2)$ **21.** $y = (3x - 5)(2x + 1)$
22. $y = (4x + 3)(x + 6)$

CHAPTER 6

Page 240   Checking Key Concepts
1. The graph on the left is a bar graph displaying the percents of class members who own a dog, a cat, a bird, another pet, or no pet. The data are categorical. The graph on the right is a histogram displaying the percentages of those students who are cat owners and have 1, 2, or 3 cats. The data are numerical.   **3.** Estimates may vary; about 14,000 cat owners.
5. Estimates may vary; about 22,000 cats.

Pages 240–243   Exercises and Applications
Opinions may vary as to whether some data are numerical or categorical. Classifications other than those given in the following answers should be accepted if they can be reasonably justified.   **1.** numerical; whole numbers between, say, 100 and 300   **3.** categorical; names of colors   **5.** No; the data are categorical, so a bar graph would be more appropriate.
7. Yes; the data are categorical.   **9. a.** histogram
b. Estimates may vary. about 72 in. (6 ft)   **c.** heights in inches: 72, 72, 65, 72, 74, 72, 74, 72, 74, 73, 68, 72, 70, 74, and 71; $71\frac{2}{3}$ in. (5 ft $11\frac{2}{3}$ in.); The estimate given in part (b) is fairly close.   **13. a.** all cars of a given model; 100 of the cars
b. numerical   **c.** histogram   **15. a.** all radio listeners in a city between the ages of 12 and 25; those selected for the survey
b. categorical   **c.** bar graph or circle graph   **17. a.** all eligible voters; the voters who are surveyed   **b.** categorical
c. bar graph or circle graph   **25.** −3; 1   **27.** −9; −4

29. 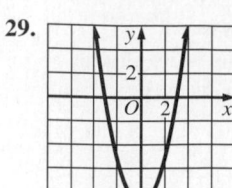 **31.**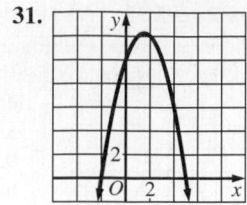

Page 247 Checking Key Concepts

1, 3. Answers may vary. Examples are given. **1.** The question assumes that the respondent knows all the candidates' platforms and supports one. You could provide a list of the key issues, each with all the candidates' positions given, and ask respondents to check which positions they support. **3.** The first statement encourages the respondent to reply positively. Omit the first sentence.

Pages 247–249 Exercises and Applications

1, 3, 5, 7. Answers may vary. Examples are given. **1.** Weight is a sensitive issue and people may not wish to answer or may not answer truthfully. The information should be collected anonymously. **3.** Unless all the people surveyed are familiar with Japanese food, they may have no information about the dishes. A reasonable question might describe each of the dishes and ask respondents to rank them in order of preference. **5.** The question assumes that the respondent is familiar with the facts of the case. Any presentation of the facts (as interpreted by the pollster) may be biased as well. It might be best then to ask the question as given only to those who reply affirmatively to the question, "Are you familiar with the facts of the Carter case?" **7.** I think questions A, C, and D do not need to be rewritten if question B is rephrased, "Do you think a student can maintain good grades while holding down an after-school job?" **11.** Answers may vary. Examples are given. (1) categorical; It will help determine whether there is an equal amount of interest among males and females. (2) categorical; It will help determine whether a banquet should be held at all. (3) categorical; It will help determine what type of banquet should be held. (4) numerical; It will help determine what type of banquet students can afford. **17, 19.** Survey questions may vary. Examples are given. **17.** Have you ever used the train service? If so, how often? How often do you anticipate using the service in the future? Which hours do you usually use the service? If service were expanded to include more early morning and late evening trains, do you think you would use the service during the new hours? If so, at what times and what routes? How often do you think you would use the new service? People who live or work in the area served by the trains should be surveyed. **19.** Have you watched any shows on this station on Thursday night during the past month? Give the frequency with which you have watched each of the Thursday shows listed below during the past month. Give the frequency with which you anticipate watching each of the Thursday shows listed below during the next month. People within the service area of the station should be surveyed. **21.** numerical; The data might be expressed in miles per gallon, which can vary widely.

23. either categorical (data that classify members as family members, individual members, patrons, students, and so on) or numerical (data that indicate membership totals for given museums) **25.** 0.25

Page 253 Checking Key Concepts

1. systematic sample; all customers of the store; The sample is biased unless it is done over a time period covering all the store's hours. Doing the survey at a particular time of day or on a particular day of the week will underrepresent customers who shop at other times. **3.** stratified random sample; all students of the school; The sample may be somewhat biased if the number of students in each grade varies too widely. It would be better to find the percent of students in each grade and randomly choose students from the grade to match this percentage.

Pages 253–256 Exercises and Applications

1. a. Ms. Rose: stratified random sample; Mr. Champine: cluster sample; Mr. Santanella: random sample; Mrs. Kim: convenience sample **b.** Answers may vary. An example is given. I would use Mr. Santanella's method. It is random and gives all students an equal chance of participating. **3.** convenience or cluster sample; people in the viewing area; Yes; sports fans are overrepresented. **5.** systematic sample; all customers of the taxicab company; probably not; Choosing every tenth rider in each taxicab should produce a fairly representative sample. **7.** systematic sampling **9.** cluster sampling **11. a.** cluster sampling **b.** Answers may vary. An example is given. I think no one sampling method will always produce the best results; under different circumstances, different methods will produce the most representative samples. **13.** Estimates may vary; field on the left: about 70%; field on the right: about 50%. **15.** Answers may vary. An example is given. Plants are numbered from 1 to 800, starting at the top of the first column and continuing from top to bottom in each column. every plant whose number is a multiple of 80 **21, 23.** Answers may vary. Examples are given. **21. a.** The first sentence encourages an exaggerated response because the respondents may want to look good to the pollster. **b.** numerical; histogram **23. a.** The wording of the question implies the respondent eats ice cream and has a favorite flavor. However, the question is not biased if possible responses include "none" and "I don't eat ice cream." **b.** categorical; bar graph **25. a.** Answers may vary. An example is given. Let x = the number of 30-year periods since period 1839–1868 and y = the number of new stamps issued. If the first and last data points are used, then the function is $y = 88(1.95)^x$; the equation fits the data reasonably well. **b.** $y = 95.5(1.86)^x$; Answers may vary. The fit is similar.

Page 257 Assess Your Progress

1. numerical; nonnegative decimal numbers **2.** categorical; the names of the month **3.** categorical; examples: small, medium, and large **4.** numerical; usually numbers between 0.0 and 6.0 **5.** Yes. **6.** No; the data are categorical and should be displayed in a bar graph or a circle graph. **7, 8.** Answers may vary. Examples are given. **7.** The first sentence encourages a positive response. The first sentence should be omitted.

8. The question is too vague. A better question would be, "How many times a week do you floss between your teeth?" **9.** stratified random sample; residents of the county; Yes; people without phones or with unlisted phone numbers are underrepresented. **10.** self-selected sample; customers of the store; Yes; only customers (and perhaps people who are not even regular customers) who want to win a gift certificate are represented.

Page 262 Checking Key Concepts

1. Width of intervals may vary. The graph on the left is an absolute frequency histogram. The graph on the right is a relative frequency histogram. The distribution is skewed.

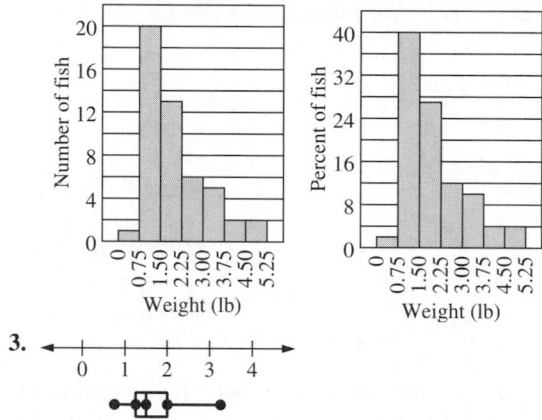

3.
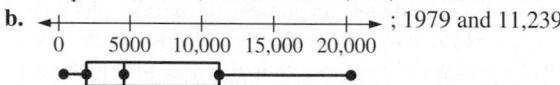

Both box plots have the same lower extreme and lower quartile. The median, upper quartile, and upper extreme of the first data set are greater than those for the second. The first harvest was better. In that harvest, half of the fish were over 1.75 lb, with 25% over 2.75 lb, and the largest fish weighed 5 lb. In the second harvest, 75% of the fish were under 2 lb and the largest fish weighed only 3.25 lb.

Pages 262–265 Exercises and Applications

3. skewed **7. a.** median: 4588.5; upper quartile: 11,239; lower quartile: 1979; maximum: 20,320; minimum: 345
b. ; 1979 and 11,239
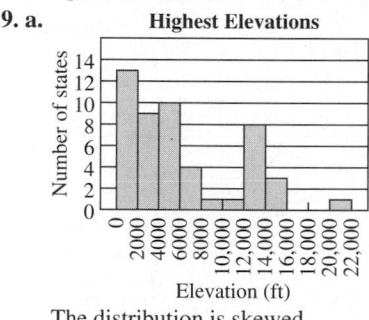

c. The median changes to 4393, the upper quartile is 9994, the lower quartile is 1965, and the maximum is 14,494.
9. a.

Highest Elevations

b. Estimates will vary; about 5000 ft.
c. mean: 6162.0; median: 4588.5
d. The median is a better measure of the "middle." The data value 20,320 raises the mean and makes the mean less descriptive of the "middle."

The distribution is skewed.

11. The intervals on the horizontal axis are not of equal width.

The data should be divided into intervals of equal width.
15. Since at least 11 territories had become states by 1788, it indicates that the lower quartile is about 1788. Since the lower extreme is 1787, the lower half of the box plot will be very condensed.
17.

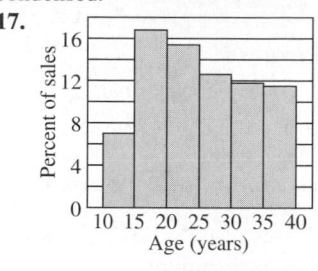

skewed distribution; The shape of the histogram is not symmetric.
21. systematic; people at the spring play; Yes; people who are not at the spring play are not represented.
23. $3\sqrt{6}$ **25.** 10

Page 269 Checking Key Concepts

1.
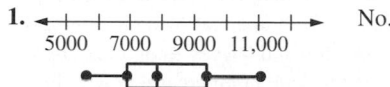
No.

3. values between 6570.4 and 9640.8; 63.0%

Pages 270–272 Exercises and Applications

1. 12 **3.** There are none. **5.** 15; 20.4; The estimate is about 0.7σ. **7.** 6; 9.3; The estimate is about 0.6σ.
9.
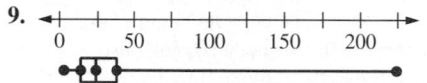

range: 217; IQR: 24; outliers: 79, 89, 97, 161, 224
11. 90%, 96% **13.** about 3.3 standard deviations **17.** The mean for each data set is 16. **19.** 0.7; 1.2; 2.5; The standard deviations vary greatly, with that of the third data set much higher than the others. **25.** $x \le 8$ **27.** $x < 2$

Page 276 Checking Key Concepts

1. Mirdik: 56%; Wong: 44% **3.** between 52% and 60%
5. It would be ±3%.

Pages 276–279 Exercises and Applications

1. between 51.3% and 55.3% **3.** No; you need to know the sample size. **7.** The sample is self-selected and, therefore, biased. **9.** No. Answers may vary. Examples are given. Even if the intervals did not overlap (and they do), they refer to percents of two different populations. You would have to calculate actual numbers based on population figures. Also, the survey was not scientific.
11. a.

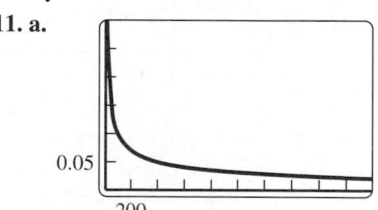

Sample size	Margin of error
100	0.1
200	0.0707
400	0.05
800	0.0354
1600	0.025
2400	0.0204

b. The larger the sample, the smaller the margin of error.
c. Using the margin of error formula, you could not take a large enough sample to have zero error, since there is no

number n for which $\frac{1}{\sqrt{n}} = 0$. In actuality, however, the margin of error would be zero if the sample consisted of the entire population.   **13.** ± 1.1 (to two significant digits)

17. $8.46; \approx 2.90$ **19.** $y = 4 + \frac{3}{5}(x - 1)$ or $y = 1 + \frac{3}{5}(x + 4)$

Page 279   Assess Your Progress

1. a. 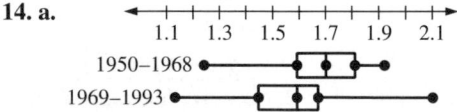   **b.** 6 years old; 8 years old

3. a. 7; 2; There are no outliers.   **b.** 1.8   **4. a.** between 48% and 56%   **b.** No.

Pages 284–285   Chapter 6 Assessment

1. histogram; box plot; categorical   **2.** systematic; random   **3.** absolute frequencies; relative frequencies   **4.** categorical; names of the species   **5.** may be numerical (usually numbers between 0 and 100 or between 0 and 4.0) or categorical (letter grades A through F)   **6.** numerical; numbers of kilowatt-hours   **7.** Yes; the data are categorical.   **8.** No; the data are categorical. The data should be displayed in a bar graph or circle graph.   **9–10.** Answers may vary. Examples are given.   **9.** Unless the question is asked anonymously, it may not elicit an honest response since it involves sensitive personal information. The information should be collected anonymously.   **10.** The first sentence encourages a negative response, since it appeals to a person's respect for individual freedoms. The first sentence should be eliminated.   **11.** random; all those in the area covered by the types of numbers generated (for example, with the appropriate area codes); Yes; those without phones are underrepresented   **12.** systematic; all students in the class; No (unless seating was not assigned randomly).   **13.** cluster; all students in the school; Yes; students who are not members of the student council are underrepresented (while students who are members are overrepresented).

14. a.

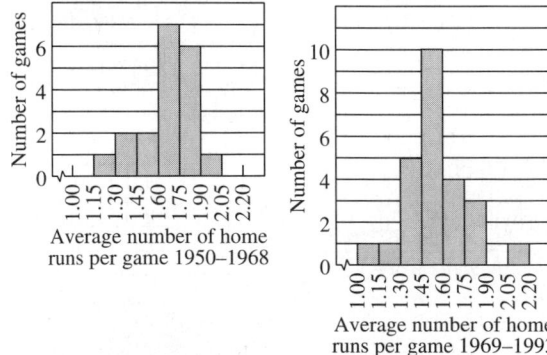

b. 1.23 and 1.7 or 1.57 and 1.81 or 1.7 and 1.91
c. 1.15 and 1.56 or 1.425 and 1.67 or 1.56 and 2.12
15. Intervals may vary. Examples are given.

16 a. 1950-1968: 0.68; 0.24; no outliers; 1969-1993: 0.97; 0.25; 2.12   **b.** 1950-1968: 0.18; 1969-1993: 0.19

The distribution for the 1950–1968 data is skewed; the distribution for the 1969–1993 data is roughly symmetric.
17. Answers may vary. An example is given. Since the upper quartile for the later data is about the same as the median for the earlier data, it appears the average number of home runs per game decreased after 1969. I think it is reasonable to assume that the changes described were at least partially responsible.

18 a. 75%   **b.** about $\pm 11\%$; between 64% and 86%

CHAPTERS 4–6

Pages 286–287   Cumulative Assessment

1. a.

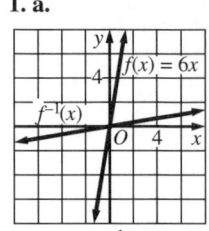

3. a.

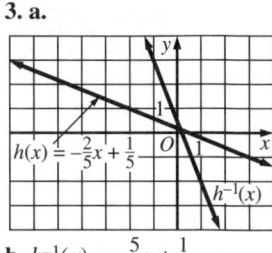

b. $f^{-1}(x) = \frac{1}{6}x$ **b.** $h^{-1}(x) = -\frac{5}{2}x + \frac{1}{2}$

5. $\log_{81} \frac{1}{9} = -\frac{1}{2}$ **7.** -2 **9.** -2.59 **11.** $\log_a \frac{1}{9}$ **13.** 0.02

15. $y = 10(6.31)^x$ **17.** Answers may vary slightly due to differences in technology or to rounding.   **a.** $y = 31.5(1.03)^x$
b. $y = 14.5x^{0.489}$

19. Summaries may vary. The graph is a parabola that opens down. Its vertex is at $(1, 0)$ and it is wider than the graph of $y = x^2$.   **21.** ± 2.2
23. 1.9; 5.6   **25.** $y = 2(x + 4)^2 + 1$
27. $y = -\frac{1}{8}(x + 5)(x - 3)$ **29.** -2

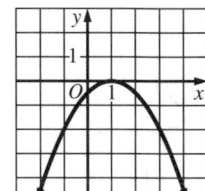

31. $y = (x + 8)(x - 8)$ **33.** $y = (3x - 4)^2$
35. Answers may vary. Examples are given.   **a.** contributors to the station   **b.** self-selected; Yes; it underrepresents listeners who do not contribute.   **c.** numerical (for example, how many listeners listen to a particular type of show); categorical (types of programs that are most popular)   **d.** histogram; bar graph; circle graph

39. a.

Conference	Minimum	Lower quartile	Median	Upper quartile	Maximum
Eastern	9	18	22	27	30
Western	16	16.5	18.5	24	33

b.

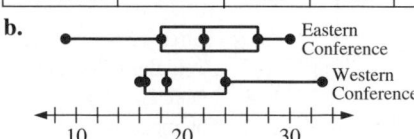

c. Eastern Conference: 21; 9; no outliers; Western Conference: 17; 7.5; no outliers.   **d.** Eastern Conference: 5.62; Western Conference: 5.25; The standard deviation is greater for the Eastern Conference data.

CHAPTER 7

Page 293   Checking Key Concepts

1. To solve a system of equations means to find the coordinates of the point(s) where the graphs of the equations intersect (that is, the values of the variables that make all the equations of the system true).   **3.** $(1.5, 14.5)$ **5.** $(3, 39)$ **7.** $(2, 4.5)$

Pages 293–296   Exercises and Applications

1. $(2, 4)$ **3.** $(3, 4)$ **5.** $(2, 11)$ **7.** $\left(\dfrac{60}{41}, \dfrac{156}{41}\right)$ **9. a.** $(5, 19)$; The y_1-value and the y_2-value that correspond to the x-value 5 are the same.   **11.** 75 times   **13.** $(2.02, 120), (13.7, 295)$ **15.** $(-0.303, -0.908), (3.303, 9.908)$ **17.** $(-1, -3), (9, 97)$ **19.** $(-3.32, -5), (3.32, -5)$ **21.** $(2.98, -53)$ **23.** $(-1.79, 0.417),$ $(2.79, 9.58)$ **35.** no solution   **37.** no solution

Page 300   Checking Key Concepts

1, 3, 5. Answers may vary. All the systems may be solved by graphing. Examples of algebraic solution methods are given. **1.** Add.   **3.** Use substitution.   **5.** Multiply both sides of the first equation by 3 and both sides of the second equation by -2, then add.   **7, 9.** Answers may vary. Examples are given. **7.** $x - y = 7, x + y = 9$ **9.** $2x - 3y = -7, 3x + 2y = -4$

Pages 300–302   Exercises and Applications

1. $(2, -1)$ **3.** $\left(-1, 5\frac{1}{2}\right)$ **5.** $(-40, -27)$ **7.** $(2, 0)$ **9.** Values that satisfy one equation always satisfy the other; infinitely many.   **11.** no solution   **13.** $\left(-40\frac{5}{6}, 26\frac{52}{63}\right)$; one solution   **15.** no solution   **17. a.** $10d_1 = 2d_2$ **b.** $d_1 + d_2 = 14$ **c.** $2\frac{1}{3}$ in.; $11\frac{2}{3}$ in.   **23.** $(3, 21)$ **25.** $(3, 4)$

27. $\begin{bmatrix} 8 & 11 \\ -11 & -2\frac{1}{2} \end{bmatrix}$ **29.** $\begin{bmatrix} -2 & 10 \\ -19 & 6 \\ 9 & 9 \end{bmatrix}$

Page 305   Checking Key Concepts

1. $\begin{bmatrix} 14 & -13 \\ -5 & -1 \end{bmatrix} \begin{bmatrix} x \\ y \end{bmatrix} = \begin{bmatrix} -6 \\ 2 \end{bmatrix}$; $(0.253, 0.734)$

3. $\begin{bmatrix} 81.2 & 17.4 \\ 4.6 & -12.8 \end{bmatrix} \begin{bmatrix} x \\ y \end{bmatrix} = \begin{bmatrix} 0.8 \\ 12.8 \end{bmatrix}$; $(0.208, -0.925)$

5. $(4.63, -1.03, -0.567)$

Pages 306–308   Exercises and Applications

3. $\begin{bmatrix} 5 & -12 \\ 14 & -7 \end{bmatrix} \begin{bmatrix} x \\ y \end{bmatrix} = \begin{bmatrix} 13 \\ 34 \end{bmatrix}$; $(2.38, -0.90)$ **5. a.** 3 batches of rolls, 2 batches of muffins   **b.** There is no way Charlie can use up exactly 8 cups of buttermilk and 6 eggs using the given recipes.   **7.** $(0.29, 6.12)$ **9.** $(-0.70, 8.35)$ **11.** $(3, -1)$ **13.** $(-0.26, 0.12)$ **17.** $(1.94, -6.43, 1.62)$ **19, 21.** Coefficients may vary due to rounding. Examples are given. **19.** $y = 1.42x^2 + 2.37x - 1.2$ **21.** $y = 0.19x^2 - 1.5x + 3.31$ **29.** $\left(\dfrac{15}{4}, -\dfrac{5}{4}\right)$; one solution   **31.** $x \geq -\dfrac{3}{2}$ **33.** $x > \dfrac{1}{2}$ **35.** $y = (2x + 1)(x - 5)$

Page 308   Assess Your Progress

1. $(-0.35, 0.25), (2.85, 16.3)$ **2.** $\left(-\dfrac{9}{5}, 3\right)$ **3.** $\left(0, \dfrac{3}{5}\right)$ **4.** $(0.96, 1.27)$; one solution   **5.** no solution   **6.** $(3, -4)$; one solution   **7.** $y = -0.35x^2 - 3.65x - 4.8$

Page 312   Checking Key Concepts

1. 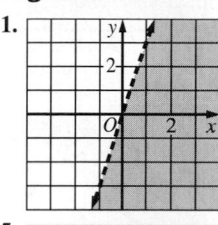   **3.**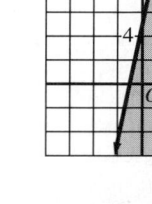

5.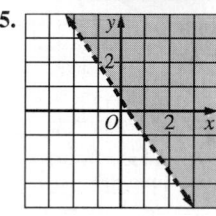

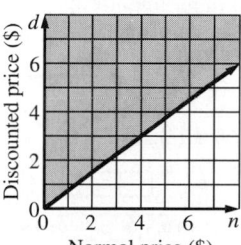

7. Let n = the normal price and d = the discounted price; $d \geq 0.75n$.

Pages 312–314   Exercises and Applications

1. Let i = the number of infants and s = the number of supervisors; $s \geq \dfrac{1}{3}i$.

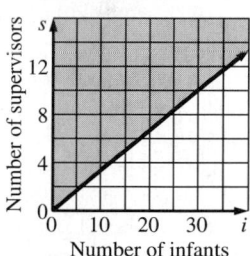

3. 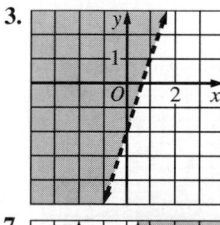   **5.**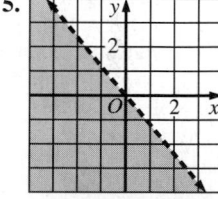

7. 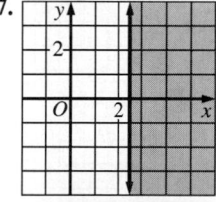   **9.**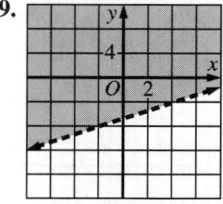

11. 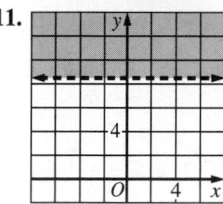   **13.**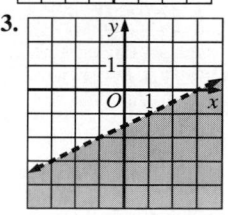

15. $x \geq -4$ **17.** $y \leq -\frac{1}{3}x + \frac{11}{3}$

23. Let f = the number of female ostriches and e = the number of eggs; $e \leq 70f$.

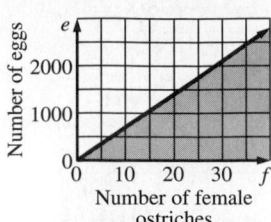

25. Let x = the blood pressure in the arm and y = the blood pressure in the ankle; $y \geq 0.9x$. (Both x and y are measured in milligrams of mercury.)

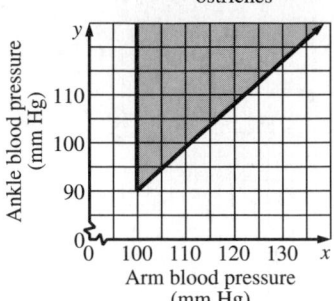

29. (15.97, 5.26)

31. a.

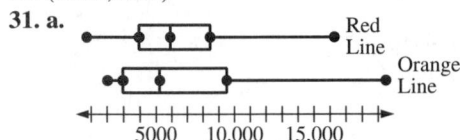

Page 317   Checking Key Concepts

1. 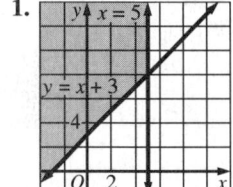   **3.**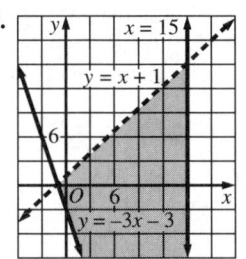

Pages 317–321   Exercises and Applications

1. 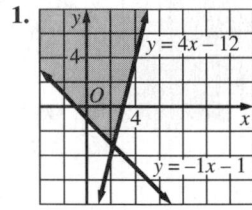   **3.**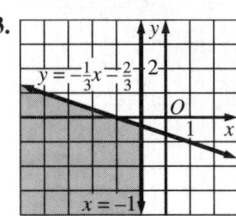

5. 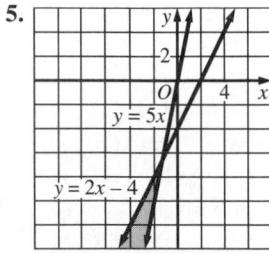   **7.**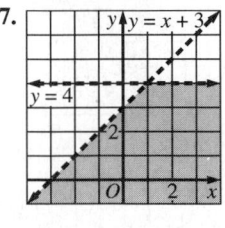

11. $x \leq 0, y \geq 3, y \leq \frac{1}{2}x + 4$ **13.** $x \geq 0, y \geq 0, y \leq -x + 4$

15. Lines of fit may vary. Least-squares lines are shown.

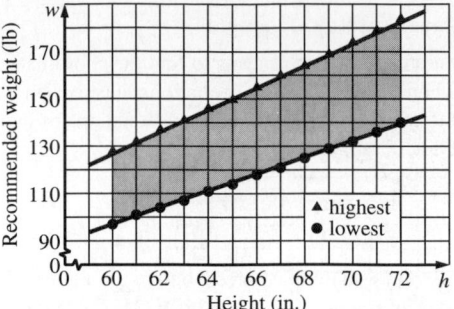

Height (in.)

23. 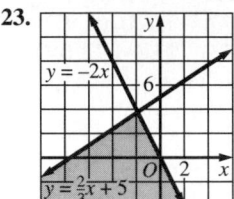   **25.**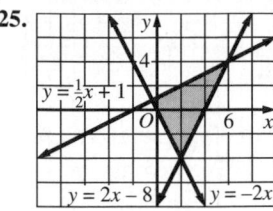

27. 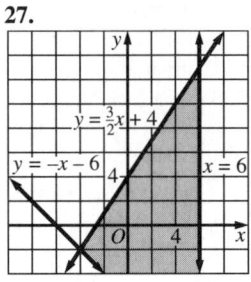   **29.**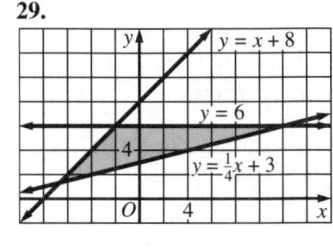

31. $x \leq 5, y \leq 3, y \geq -\frac{1}{3}x - \frac{4}{3}, y \leq 2x + 1$

33. $x \geq 0, y \geq 0, y \leq 20, y \leq x + 8, y \leq -\frac{5}{2}x + 80$

37. 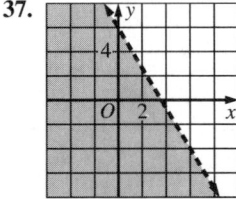   **39.**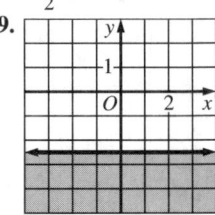

41. (5, 5)   **43.** (−15, 4)   **45.** 1, −4   **47.** ±1

Page 325   Checking Key Concepts

1. Let x = the number of oatmeal cookies and y = the number of chocolate chip cookies; $x \geq 24, y \geq 24, y \geq -x + 120,$ $y \leq -x + 144.$

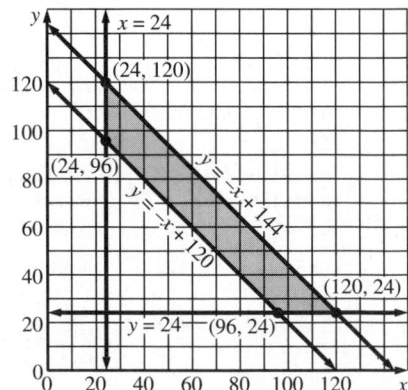

3. 96 oatmeal cookies, 24 chocolate chip cookies

Pages 325–327 Exercises and Applications

1.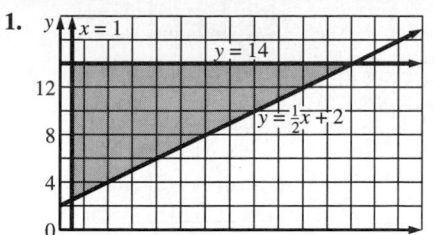

3.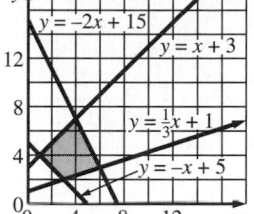

5. $176 **7.** $48 **9.** $10.50
11. $15

13. a. $x \geq 4$, $y \geq 0$, $y \leq -\dfrac{5}{4}x + 10$

b. 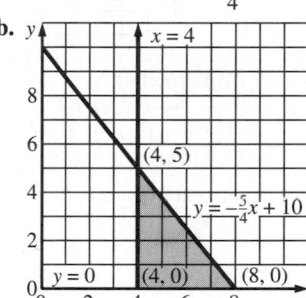 **c.** 4 bowls, 5 plates

19. $102 **21.** $32 **23.** $f^{-1}(x) = \dfrac{1}{2}x - \dfrac{7}{2}$
25. $h^{-1}(x) = \dfrac{1}{15}x + \dfrac{1}{5}$

27. Answers may vary. An example is given. The first part of the question might encourage a negative answer. I would omit that part of the question as well as the word "another."

Page 327 Assess Your Progress

1.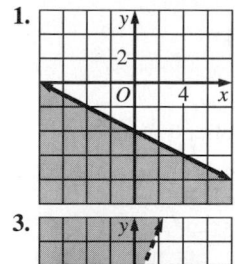

2.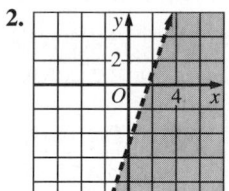

3.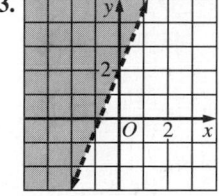

4. Let g = gross income in dollars and m = mortgage amount in dollars; $m \leq 0.28g$.

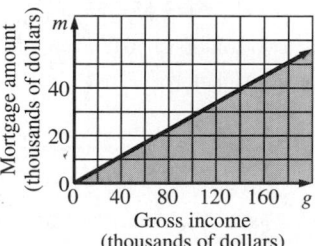

5.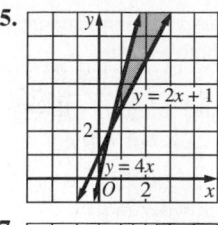

6.

7.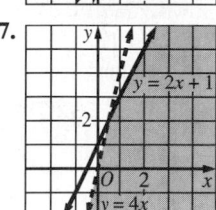

8.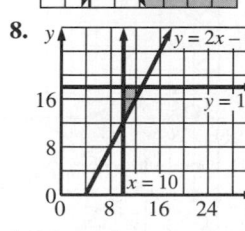

$426

9.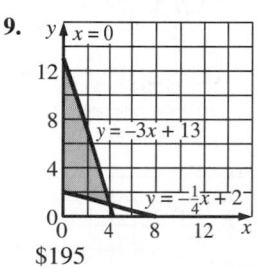

$195

10.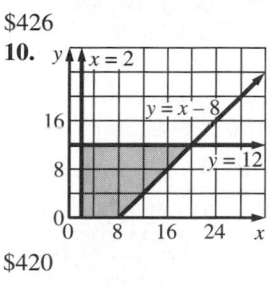

$420

Pages 332–333 Chapter 7 Assessment

1. inconsistent; dependent **2.** inverse; identity matrix
3. (8.07, 44.9) **4.** (0.118, 1.94); one solution **5.** no solution
6. (1.02, 0.408); one solution **7.** no solution
8. $\left(\dfrac{3}{2}, -\dfrac{1}{2}\right)$; one solution **9.** $\left(-2, -\dfrac{2}{3}\right)$; one solution
11. a. Let h = the number of hours of use and C = the cost in dollars; $C = 0.0018h + 12$ (fluorescent) and $C = 0.0072h + 0.50$ (incandescent). **b.** (2130, 15.83); If bulbs are used for 2130 h, the cost is the same for fluorescent and incandescent bulbs. **12. a.** $h = -15t^2 + 52.5t + 212.5$ **b.** about 258 ft

13.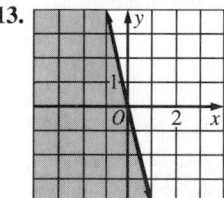

14.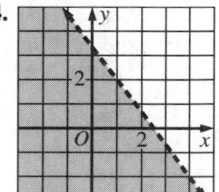

15.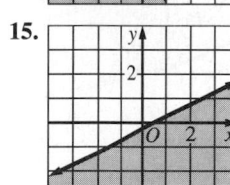

16. $x \geq 0$; $y \geq 0$; $y \leq \dfrac{1}{2}x + 2$;
$y \leq -2x + 12$
17. $y > -3$; $y < 2$; $y > -\dfrac{5}{2}x - 8$;
$y > \dfrac{5}{3}x - \dfrac{14}{3}$

18. a.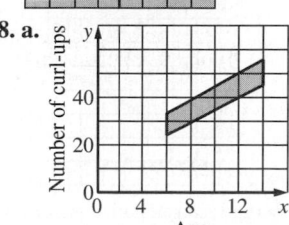

b. With the exception of the upper boundary, the region represents all scores for boys between 6 and 14 that achieve or exceed the limits for the National Award and fall short of the limits for the Presidential

Award; every point on the upper boundary represents an age and number of curl-ups that qualify for both awards.

19. a. Let b = the amount in dollars invested in the balanced fund and g = the amount in dollars invested in the growth fund; $b + g \geq 12{,}000$; $b + g \leq 20{,}000$; $b \geq \frac{2}{3}g$; $b \leq 10{,}000$.

b.

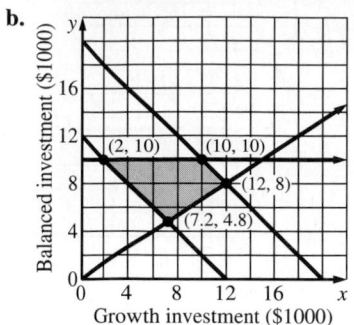

c. $12,000 in the growth fund and $8,000 in the balanced fund

CHAPTER 8

Page 340 Checking Key Concepts

1. a. $v = \sqrt{19.6h}$

b.

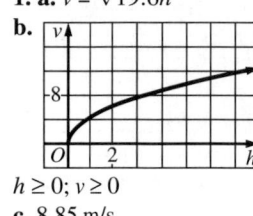

$h \geq 0$; $v \geq 0$

c. 8.85 m/s

3.

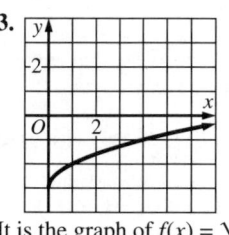

It is the graph of $f(x) = \sqrt{x}$ translated down 3 units.

Pages 340–343 Exercises and Applications

1. a, b.

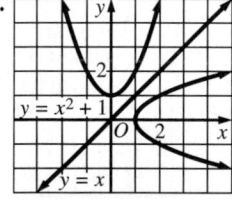

c. No; there are two outputs for each input except 1.
d. $y = \sqrt{x-1}$ and $y = -\sqrt{x-1}$; $x \geq 1$

3. $2\sqrt{2x}$ **5.** $5\sqrt{x}$ **7.** $10\sqrt{x}$ **11.** C **13.** F **15.** A

21.

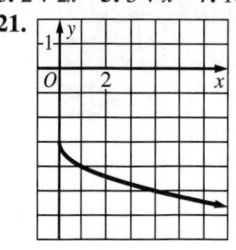

nonnegative numbers; $y \leq -3$

23.

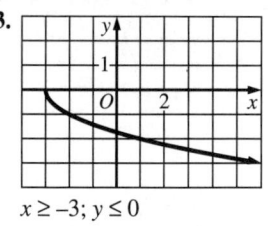

$x \geq -3$; $y \leq 0$

25.

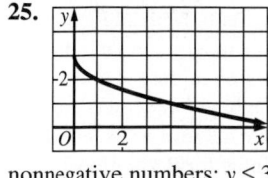

nonnegative numbers; $y \leq 3$

27.

$x \geq 1$; nonnegative numbers

29.

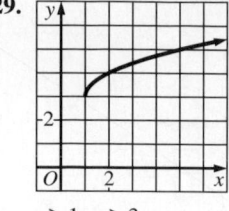

$x \geq 1$; $y \geq 3$

31.

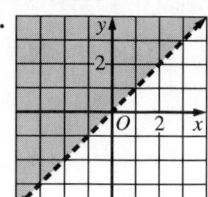

$x \geq -3$; $y \geq -1$

33. never **35.** sometimes **37.** sometimes **39.** always
43. 10 **45.** 5

47.

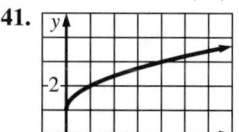

49.

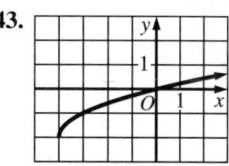

Page 347 Checking Key Concepts

1. all numbers; all numbers **3.** all numbers; all numbers
5. undefined **7.** –3 **9.** –4 **11.** $3\sqrt[3]{10}$
13. a. $r = \sqrt[3]{\frac{3V}{\pi}}$ **b.** 9.85 cm

Pages 347–350 Exercises and Applications

1. 15 in. **3.** 24 in. **5.** undefined **7.** 10 **9.** $\frac{1}{2}$ **11.** 5
13. 10 **15.** 2 **17. a.** $d = \sqrt[3]{\frac{6V}{\pi}}$ **b.** 12.4 cm
19. all numbers; all numbers **21.** nonnegative numbers;
nonnegative numbers **23. a.** $\sqrt{z}$ **b.** $\sqrt[3]{p^2}$ **c.** $\sqrt{q^3}$
d. $\sqrt[10]{y^7}$ **25.** $\sqrt[6]{x} \cdot \sqrt[6]{x} = x^{1/6} \cdot x^{1/6} = x^{1/6 + 1/6} = x^{1/3} = \sqrt[3]{x}$
27. $x^{m/n} = x^{m(1/n)} = (x^m)^{1/n} = \sqrt[n]{x^m}$

41.

43.

nonnegative numbers; $y \geq 1$ $x \geq -4$; $y \geq -2$
45. 10

Page 353 Checking Key Concepts

1. 12 **3.** –3 **5.** 20 **7.** –4 **9.** 3

Pages 353–357 Exercises and Applications

1. 49 **3.** 61 **5.** 18 **7.** 15 **9.** 1 **11.** no solution **13.** 173
15. –35 **19.** 27.5 m/s; 33.9 m/s; 45.0 m/s; 59.8 m/s
21. Wind velocity increases. **23.** 7 **25.** –2; –5 **27.** 3
29. 9 **31.** –2 **33.** –5 **37.** 4 **39.** 2.22 **45.** 0; 2 **47.** 0; 1
49. 0.791 **51.** 1.55 **55.** $5\sqrt{2}$ **57.** $2\sqrt[4]{25} = 2\sqrt{5}$
59. two solutions **61.** no solution **63.** $\left(-\frac{4}{7}, -\frac{2}{7}\right)$
65. $(1, 3)$; $(-1, 5)$

Page 357   Assess Your Progress

1. a.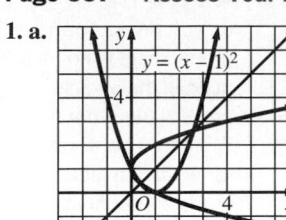

b. $y = 1 + \sqrt{x}$; $y = 1 - \sqrt{x}$, $x \geq 0$

c. The reflection of the graph is not the graph of a function since each input value has two output values. The reflection is the combined graph of two functions.

2. $T = 47.3w^{2/3}$ **3.** 659; 692.5; 725; 784; 841; 896.5; 950; 1019; 1086; 1182.5   **4.** no solution   **5.** $\dfrac{3 + \sqrt{13}}{2} \approx 3.30$ **6.** $\pm 2\sqrt{6} \approx \pm 4.90$

Page 362   Checking Key Concepts

1. $i\sqrt{7}$ **3.** $10i\sqrt{3}$ **5.** $14 + 5i$ **7.** $5 + 6i$ **9.** $12 + 8i$ **11.** $11 + 23i$ **13.** $\dfrac{11}{25} + \dfrac{2}{25}i$ **15.** $7 \pm 2i$

Pages 362–364   Exercises and Applications

1. $6i$ **3.** $-2i\sqrt{7}$ **5, 7.** Answers may vary. Examples are given.   **5.** $x^2 + 16 = 0$ **7.** $x^2 + 2 = 0$ **9.** $12 + 5i$ **11.** $18 + 3i$ **13.** $-2 + 4i$ **15.** $10 + 4i$ **17.** $-11 + 4i$ **19.** 0 **21.** $20i$ **23.** $10 + 15i$ **25.** $-45 + 18i$ **27.** $19 + 13i$ **29.** $10 + 5i$ **31.** 25 **33.** $\dfrac{21}{65} + \dfrac{38}{65}i$ **35.** $-\dfrac{9}{25} + \dfrac{13}{25}i$ **43.** $7 \pm i$ **45.** $-3 \pm i\sqrt{3}$ **47.** $-\dfrac{1}{2} \pm \dfrac{\sqrt{3}}{2}i$ **51.** 66 **53.** $\dfrac{21}{2}$ **55.** $y = 3.28(1.25)^x$ **57.** $\sqrt{74} \approx 8.60$ **59.** $4\sqrt{2} \approx 5.66$

Page 367   Checking Key Concepts

1.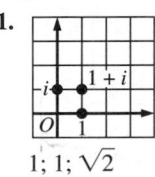
$1; 1; \sqrt{2}$

3.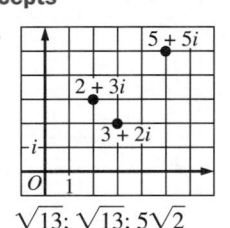
$\sqrt{13}; \sqrt{13}; 5\sqrt{2}$

5.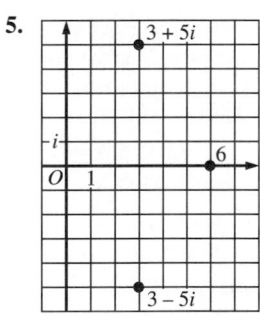
$\sqrt{34}; \sqrt{34}; 6$

7.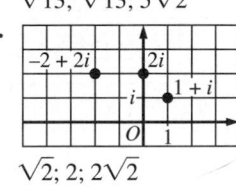
$\sqrt{2}; 2; 2\sqrt{2}$

9.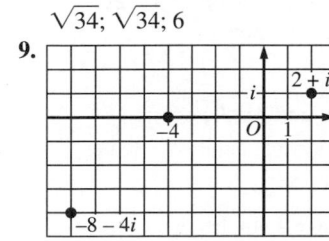
$4; \sqrt{5}; 4\sqrt{5}$

11.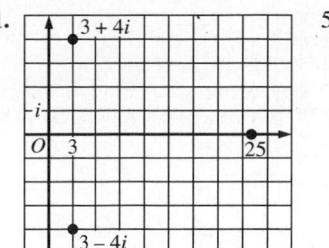
$5; 5; 25$

13.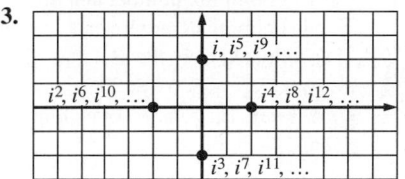

Answers may vary. The sequence with $t_n = i^n$ is the repeating sequence $i, -1, -i, 1, i, -1, -i, 1, \ldots$.

Pages 368–370   Exercises and Applications

1, 3. Choices of numbers with the same magnitude may vary. Examples are given.   **1.** $1 + 2i$ and $1 - 2i$; $\sqrt{5}$; $-1 + 2i$ and $-1 - 2i$ **3.** $-2 - 3i$ and $2 - 3i$; $\sqrt{13}$; $2 + 3i$ and $-2 + 3i$

5.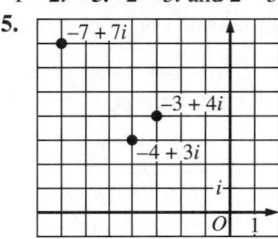
$5; 5; 7\sqrt{2}$

7.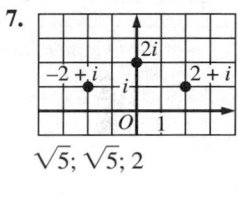
$\sqrt{5}; \sqrt{5}; 2$

9.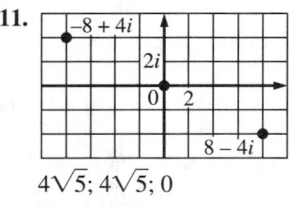
$\sqrt{5}; \sqrt{34}; 5$

11.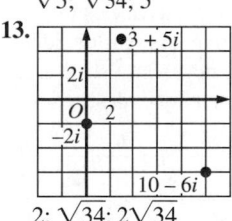
$4\sqrt{5}; 4\sqrt{5}; 0$

13.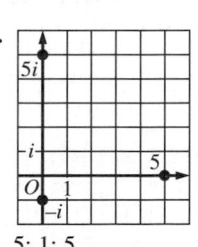
$2; \sqrt{34}; 2\sqrt{34}$

15.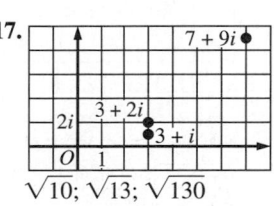
$5; 1; 5$

17.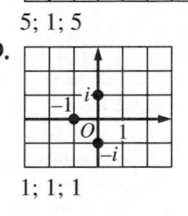
$\sqrt{10}; \sqrt{13}; \sqrt{130}$

19.
$1; 1; 1$

21. $5 - 2i$ **23.** $-5 - 6i$ **25.** $4 - 7i$ **27.** $-a - bi$

29.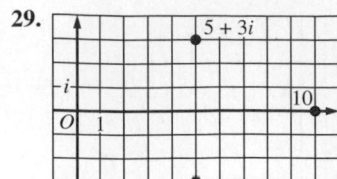

31.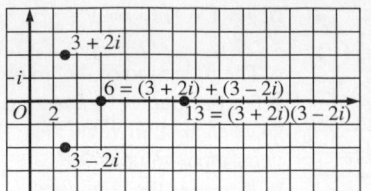

33.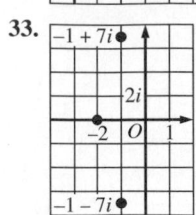

35. Answers may vary. An example is given. A complex number and its complex conjugate are images of each other reflected over the real axis. **53.** 16.8; 7.1 **55.** 51.8; 25.5 **57.** Answers may vary. An example is given. $x^2 + 9 = 0$

59. $\begin{bmatrix} 7 & 3 \\ 13 & 5 \end{bmatrix}$ **61.** $\begin{bmatrix} 11 & 3 \\ 56 & 27 \end{bmatrix}$

Page 375 Checking Key Concepts

1. whole numbers, integers, rational numbers, real numbers, complex numbers **3.** imaginary numbers, complex numbers **5.** not possible **7.** Answers may vary. An example is given. π **9.** Yes; do not form a group; there is no identity and the inverse property does not hold. **11.** No; do not form a group; the set is not closed under addition.

Pages 375–379 Exercises and Applications

1. imaginary numbers, complex numbers **3.** irrational numbers, real numbers, complex numbers **5.** not possible **7.** Answers may vary. An example is given. 7 **9.** No; do not form a group; the set is not closed under multiplication. **11.** Yes; form a commutative group; the set is closed under multiplication and all the group properties as well as the commutative property hold. **13.** No; do not form a group; the set is not closed under multiplication. **15.** Yes; form a commutative group; the set is closed under multiplication and all the group properties as well as the commutative property hold. **17.** Yes; do not form a group; there is no identity and the inverse property does not hold. **19.** No; do not form a group; the set is not closed under multiplication. **21. a.** closure, identity, inverse, associative, commutative **b.** closure, identity, associative **25.** Counterexamples may vary. No; $\alpha \dagger \delta = \alpha$ but $\gamma \dagger \alpha = \gamma$, so there is not a unique identity.

27.

×	A	B	C	D
A	D	C	B	A
B	C	D	A	B
C	B	A	D	C
D	A	B	C	D

29.

@	□	●	▽	▲
□	▲	□	●	▽
●	□	●	▽	▲
▽	●	▽	▲	□
▲	▽	▲	□	●

37. a. No. **b.** Yes. **c.** Yes. **39.** RF **41.** TRF **43.** I **45.** RF

57. 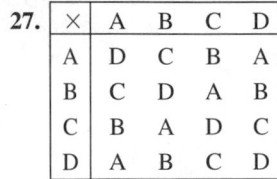 **59.** $y = (2x - 5)(x + 3)$ **61.** $y = (2x - 7)(2x + 7)$ **63.** $5x - 15$

Page 379 Assess Your Progress

1. $-1 + 5i$ **2.** 58 **3.** $39 - 80i$

4. Answers may vary. An example is given. **5.** $\sqrt{153}$

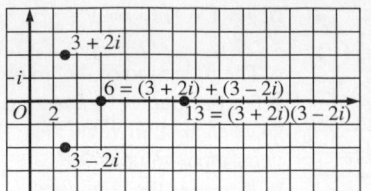

6. Yes; do not form a group; there is no identity and the inverse property does not hold. **7.** Yes; do not form a group; the inverse property does not hold.

Pages 384–385 Chapter 8 Assessment

1. radical function **2.** complex plane; magnitude **3.** closure **4.** nonnegative numbers; $y \geq -4$ **5.** nonnegative numbers; $y \geq 1$ **6.** $x \geq 2$; $y \geq 3$ **7.** all real numbers; all real numbers **8.** $x \geq -5$; $y \geq 1$ **9.** nonnegative numbers; $y \geq 2$ **10.** $z^{1/9}$ **11.** $y^{3/5}$ **12.** $y^{7/4}$ **13.** $\sqrt[3]{t^4}$ **14.** $\sqrt[8]{b^3}$ **15.** $\sqrt[7]{p^5}$ **16.** 4 **17.** 13 **18.** 2 **20. a.** 33.6 m/s **b.** 32.8 m/s **c.** 14.1 m/s **21.** $3 - 2i$ **22.** $9 - 3i$ **23.** 15 **24.** $12 - 9i$ **25.** $-2 + 14i$ **26.** $\frac{38}{29} - \frac{21}{29}i$ **27–32.** Examples of numbers with the same magnitude may vary. **27.** $\sqrt{17}$; $1 - 4i$ and $-4 + i$ **28.** $\sqrt{58}$; $7 - 3i$ and $-7 + 3i$ **29.** $4\sqrt{2}$; $4 - 4i$ and $-4 - 4i$ **30.** $\sqrt{5}$; $2 + i$ and $-2 + i$ **31.** $\sqrt{29}$; $5 - 2i$ and $-5 - 2i$ **32.** $\sqrt{10}$; $3 - i$ and $-3 - i$

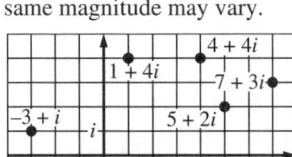

33. 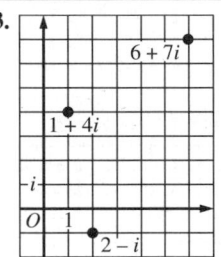 **34.**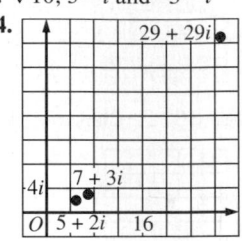

35. 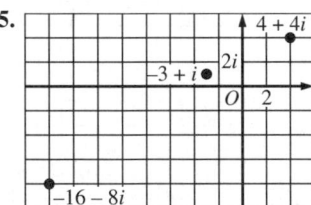 **36.** Yes; form a commutative group; the set is closed under addition and all the group properties as well as the commutative property hold. **37.** No. **38.** No.

39. Yes. **40.** No; for example A Δ C = C, but C Δ A = D. **41.** No; for example, $3 \times 7 = 1$, which is not in the set.

CHAPTER 9

Page 393 Checking Key Concepts

1. No; the exponents are not whole numbers. **3.** Yes; $-0.94y^5 - 3.8y^4 - 1.6y^2 + 0.22y$; 5 **5.** 7 **7.** $8x^3 + 8x^2 + 4x + 9$

Pages 393–396 Exercises and Applications

1. No; the exponent 2.9 is not a whole number. **3.** Yes; $t^2 + 6t - 1$; 2 **5.** Yes; $-0.6u^{10} - \sqrt{2}u^8 + 5u^7 - \frac{3}{2}u^4 + \pi$; 10 **7.** No; 2^r is not the product of a real number and a whole-number power of r. **9.** 143 **11.** -2 **13.** 221 **15.** -1 **21.** $16x^3 + 9x^2 + 9x + 13$

23. $9.1y^5 - 2.2y^4 - 4.8y^3 - 10.8y^2 + 8.5y + 4$
25. $-4\sqrt{2}m^3 - 7\sqrt{5}m^2 + 23m$ **27.** $-x^3 - 10x^2 + 4$
29. $-v^3 + \frac{1}{14}v^2 + \frac{27}{20}v - \frac{59}{36}$ **31.** $0.29t^5 - 0.78t^4 + 2.59t^2 -$
$2.1t + 3.52$ **35.** A **41.** Yes; do not form a group; the
inverse property does not hold. **43.** Yes; do not form a
group; the inverse property does not hold. **45.** 4 **47.** $\frac{3}{2}$
49. $x^2 + x - 12$ **51.** $36n^2 + 12n + 1$

Page 400 Checking Key Concepts
1. $3x^2 + x - 10$ **3.** $u^3 - 4u^2 - 6u - 1$ **5.** $x + 4 + \frac{2}{x+2}$
7. $y^2 + 7y + 1 + \frac{4}{y-3}$

Pages 401–403 Exercises and Applications
1. $8x^2 + 26x - 7$ **3.** $-6t^5 + 12t^4 + 2t^3$ **5.** $u^3 + 2u^2 + 2u + 1$
7. $6m^3 - 31m^2 - 40m + 16$ **9.** $40n^3 + 66n^2 - 31n - 42$
11. $w^4 - 12w^3 + 54w^2 - 108w + 81$ **19.** $x + 2 + \frac{4}{x+1}$
21. $3t + 13 + \frac{26}{t-2}$ **23.** $y^2 + 4y - 2 + \frac{5}{y+4}$

41.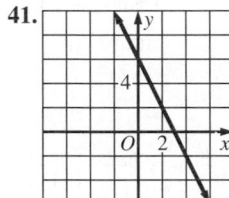
25. $6x^2 + 9x + 4$
27. $2x^3 - 5x^2 + 11x - 12 + \frac{18}{x+2}$
29. $4n - 1 + \frac{12n+4}{2n^2+7n-5}$
33. 47 **35.** 47 **37.** (3, 1)
39. (−2, 1, −3)

43. 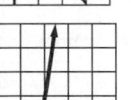 **45.**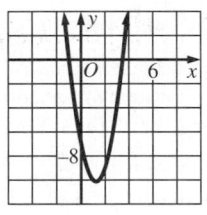

Page 408 Checking Key Concepts
1. A **3.** B
5. 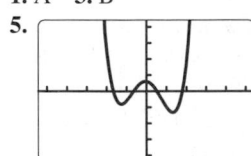 **7.**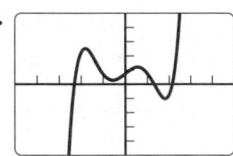

local minimums: -4.06 and -6.64   local minimums: 1.20 and −4.10;
local maximum: 3.03   local maximums: 10.1 and 4.80

Pages 408–411 Exercises and Applications
3.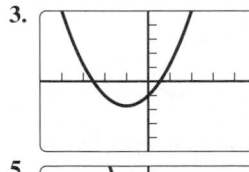
a. As $x \to +\infty$, $f(x) \to +\infty$ and
as $x \to -\infty$, $f(x) \to +\infty$.
b. local minimum: −7

5.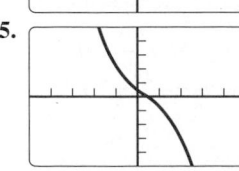
a. As $x \to +\infty$, $f(x) \to -\infty$ and
as $x \to -\infty$, $f(x) \to +\infty$.
b. no local maximum,
no local minimum

7.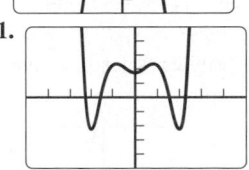
a. As $x \to +\infty$, $g(x) \to +\infty$ and
as $x \to -\infty$, $g(x) \to -\infty$.
b. local maximums: 2.57 and 0;
local minimums: −3.44 and
−13.44

9.
a. As $x \to +\infty$, $h(x) \to -\infty$ and
as $x \to -\infty$, $h(x) \to -\infty$.
b. local maximum: 5

11.
a. As $x \to +\infty$, $f(x) \to +\infty$ and
as $x \to -\infty$, $f(x) \to +\infty$.
b. local minimums: −9 and 7;
local maximum: 9.52

21. $y = \frac{1}{3}(x+5) - 3$ or $y = \frac{1}{3}(x-4)$ **23.** 3; 4 **25.** $-\frac{3}{4} \pm \frac{\sqrt{17}}{4}$

Page 411 Assess Your Progress
1. $-2x^5 + 8x^4 + x^3 - 13x + 9$; 5 **8. a.** $V = x(24 - 2x)(18 - 2x)$
2. 41 **3.** $3x^3 - 3x^2 - 7$ **b.**
4. $4y^4 + 9y^3 - 8y^2 - y - 9$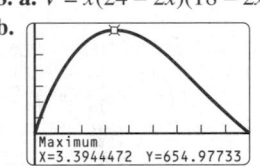
5. $8u^3 + 10u^2 + 13u + 35$
6. $4x^2 - 9x + 18 - \frac{28}{x+2}$
7. As $x \to +\infty$, $f(x) \to -\infty$ about 3.39 in.; about 655 in.^{3}
and as $x \to -\infty$, $f(x) \to -\infty$.

Page 416 Checking Key Concepts
1. 0.66 **3.** $f(x) = 2(x-1)(x-2)(x-3)$ **5.** 2; 4; 5 **7.** −1; 2; 6

Pages 416–418 Exercises and Applications
1. 3 **3.** −2.62; −0.38; 1.50 **5.** −1.44; 1.27; 4.38 **7.** $f(x) =$
$(x-1)(x-3)(x-6)$ **9.** $f(x) = 5x(x+2)(x-2)$ **13.** −10; 3; 7
15. −1; 1; 4 **17.** −9; $\frac{1-\sqrt{5}}{2}$; $\frac{1+\sqrt{5}}{2}$ **19.** 6; $\frac{2-\sqrt{14}}{2}$;
$\frac{2+\sqrt{14}}{2}$ **21.** 1989 **23.** $g(x) = 24\left(x + \frac{3}{2}\right)\left(x + \frac{4}{3}\right)\left(x + \frac{5}{4}\right)$
25. $f(x) = (x+7)(x-1-\sqrt{5})(x-1+\sqrt{5})$ **31.** As $x \to +\infty$,
$g(x) \to -\infty$ and as $x \to -\infty$, $g(x) \to -\infty$. **33.** $\frac{1}{25}$ **35.** 3
37. $\pm 4i$ **39.** $2 \pm 3i$

Page 423 Checking Key Concepts
1. 2.00 **3.** −2.98; −1.87; −0.70; 0.63; 1.42 **5. a.** ± 1; ± 3;
$\pm\frac{1}{2}$; $\pm\frac{3}{2}$ **b.** −1; $\frac{1}{2}$; 3 **7. a.** ± 1; ± 2; ± 4; ± 8
b. 2 (double zero); $1 \pm i$

Pages 423–425 Exercises and Applications
1. −3.67; 0.35; 2.32 **3.** −2.24; 2.76 **5.** −2 (double zero);
−0.41; 0.41; 1 **7.** −2.21; −1.67; −0.59; 0.64; 1.57; 2.26
9. a. ± 1; ± 13 **b.** 1, $2 \pm 3i$ **11. a.** ± 1; ± 2; ± 4; ± 8;
$\pm\frac{1}{3}$; $\pm\frac{2}{3}$; $\pm\frac{4}{3}$; $\pm\frac{8}{3}$ **b.** $\frac{2}{3}$; −2 (double zero) **13. a.** ± 1; ± 2;
± 4; ± 5; ± 8; ± 10; ± 20; ± 40 **b.** −4; −1; 2; 5 **15. a.** ± 1;
$\pm\frac{1}{2}$; $\pm\frac{1}{4}$; $\pm\frac{1}{8}$; $\pm\frac{1}{16}$ **b.** $-\frac{1}{4}$ (double zero); $\pm i$ **17. a.** ± 1; ± 2

b. -1 (triple zero); $\dfrac{5 \pm \sqrt{17}}{2}$ **21.** D
25. $f(x) = -2x(x + 7)(x + 5)$ **27.** No.   **29.** No.

Page 425   Assess Your Progress

1. about 9.98 m/s   **2.** 1; -2; 3   **3.** -4; $2 \pm \sqrt{2}$
4. $\dfrac{5}{3}$; $\dfrac{-3 \pm i\sqrt{7}}{4}$ **5.** $-\dfrac{1}{2}$; 3; $\pm 2i$

Page 429   Checking Key Concepts

1. Yes; 5.   **3.** No.   **5.** No.   **7.** No.

Pages 430–432   Exercises and Applications

1. Yes; 2.   **3.** No.   **7.** Yes; 5.
5. No.

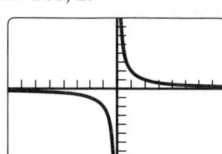   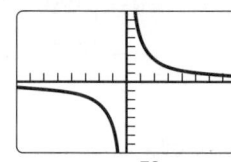

9. Yes; let l = length and w = width; $lw = 72$, so $l = \dfrac{72}{w}$ or
$w = \dfrac{72}{l}$.   **11.** No; let n = the number of gallons of gasoline
used, d = the distance driven in miles, and let M represent the
car's gasoline usage in mi/gal (a constant); $n = \dfrac{d}{M}$; this is a
direct variation.   **23.** $\dfrac{1}{3}$; $-\dfrac{1}{2}$; 4   **25.** $\dfrac{1}{2}$ (double zero); $-2 \pm i$
27. translated 1 unit right   **29.** translated 1 unit right and
3 units up

Page 437   Checking Key Concepts

1. D   **3.** B

Pages 437–440   Exercises and Applications

1. Yes; $2x + 3$ and $5x - 1$ are both polynomials.   **3.** No; $2^t + 1$
is not a polynomial.   **7.** $f(x) = \dfrac{4}{x} + 2$ **9.** $f(x) = \dfrac{6}{x + 2} + 4$
11. $f(x) = \dfrac{-12}{x + 8} - 10$ **17. a.** $x = \dfrac{3}{2}$; $y = 4$ **b.** It is the graph of
$y = \dfrac{13}{x}$ translated $\dfrac{3}{2}$ units right and 4 units up.   **19. a.** $x = -2$;
$y = 3$ **b.** It is the graph of $y = \dfrac{5}{x}$ translated 2 units left and
3 units up.   **21. a.** $x = -\dfrac{1}{4}$; $y = -5$ **b.** It is the graph of $y = \dfrac{2}{x}$
translated $\dfrac{1}{4}$ unit left and 5 units down.   **23. a.** $x = \dfrac{4}{5}$; $y = -6$
b. It is the graph of $y = -\dfrac{5}{x}$ translated $\dfrac{4}{5}$ unit right and 6 units
down.   **27.** Yes; 1.5.   **29.** No.   **31.** $x + 7 + \dfrac{3}{x + 1}$
33. $4x^2 - x - 1 - \dfrac{5}{2x - 1}$

Page 440   Assess Your Progress

1. Yes; $y = \dfrac{-12}{x}$.   **2.** 20 min   **3.** $f(x) = \dfrac{2}{x - 1} - 3$ **4.** $x = -\dfrac{2}{3}$; $y =$
2; It is the graph of $y = \dfrac{3}{x}$ translated $\dfrac{2}{3}$ unit left and 2 units up.

Page 445   Checking Key Concepts

1. a. $x = -7$; $x = 4$ **b.** $f(x) \to 0$ as $x \to \pm\infty$.   **3. a.** $x = \dfrac{17}{4}$
b. $g(x) \to -\infty$ as $x \to -\infty$ and $g(x) \to +\infty$ as $x \to +\infty$.

Pages 445–448   Exercises and Applications

1. C   **3.** B   **7. a.** $x = -5$; $x = 2$ **b.** $f(x) \to 0$ as $x \to \pm\infty$.
9. a. $x = \dfrac{9}{5}$ **b.** $h(x) \to +\infty$ as $x \to -\infty$ and $h(x) \to -\infty$ as
$x \to +\infty$.   **11. a.** $x = -4$; $x = 2$; $x = 3$ **b.** $g(x) \to -1$ as
$x \to \pm\infty$.   **19.** It is the graph of $y = \dfrac{5}{x}$ translated 6 units left
and 3 units down.   **21.** It is the graph of $y = \dfrac{-2}{x}$ translated $\dfrac{1}{2}$
unit right and 7 units down.   **23.** 6   **25.** 4

Page 451   Checking Key Concepts

1. 4   **3.** 1; 15   **5.** $16\dfrac{2}{3}$ oz

Pages 451–453   Exercises and Applications

1. 2   **3.** no solution   **5.** $\dfrac{7}{3}$; 3   **7.** 6   **9.** no solution   **11.** $-\dfrac{1}{4}$
17. $13 - 4i$ **19.** $8 + 6i$ **21.** $y = x^2 - x$

Page 453   Assess Your Progress

1. a. $x = 1$; $x = 5$ **b.** $f(x) \to 0$ as $x \to \pm\infty$.   **2. a.** $x = -3$
b. $g(x) \to +\infty$ as $x \to -\infty$ and $g(x) \to -\infty$ as $x \to +\infty$.
3. a. $x = -5$; $x = 2$ **b.** $h(x) \to +\infty$ as $x \to \pm\infty$.   **4. a.** $x = 0$;
$x = -3$; $x = 3$ **b.** $f(x) \to 3$ as $x \to \pm\infty$.   **5.** -1 **6.** 2
7. 5 consecutive hits

Pages 458–459   Chapter 9 Assessment

1. degree; leading coefficient; constant term   **2.** inverse variation; constant of variation   **3.** -32 **4.** $-2x^3 + 13x^2 - 11x + 1$
5. $4u^4 + 3u^3 - 7u^2 + 11u - 21$ **6.** $12y^3 - 10y^2 + 23y - 7$
7. $5x^2 - 6x + 9 - \dfrac{32}{2x + 3}$ **9.** local minimums: 3.92 and -11;
local maximum: 9.58   **11.** 1956   **12.** $f(x) = 2x(x + 2)(x - 3)$
13. 4; $3 + 2\sqrt{2} \approx 5.83$; $3 - 2\sqrt{2} \approx 0.17$
14. $-\dfrac{3}{2}$ (double zero); $1 \pm i\sqrt{3}$
15. a. No; the products dl are not constant.

b.

d (in.)	l (in.)	A (in.²)
$\dfrac{1}{8}$	1440	$\dfrac{\pi}{256}$
$\dfrac{1}{4}$	360	$\dfrac{\pi}{64}$
$\dfrac{3}{8}$	160	$\dfrac{9\pi}{256}$
$\dfrac{1}{2}$	90	$\dfrac{\pi}{16}$

Yes; for all corresponding values of l and A, $lA = \dfrac{45\pi}{8}$.

c. $l = \dfrac{45\pi}{8A}$ **16.** horizontal asymptote: $y = 5$; vertical asymptote: $x = -\dfrac{1}{6}$; The graph is the graph of $y = \dfrac{2}{x}$ translated $\dfrac{1}{6}$ unit
left and 5 units up.   **17.** $x = -3$ and $x = \dfrac{1}{4}$; As $x \to \pm\infty$,
$h(x) \to +\infty$.   **18.** at least 10 three-point shots   **19.** -1

CHAPTERS 7–9

Pages 460–461   Cumulative Assessment

1. $(2, -3)$; one solution   **3.** $(10, 4)$; one solution

5. $\begin{bmatrix} 0.3 & -2.1 & 4.5 \\ 5.2 & 0.2 & -3.7 \\ -1.8 & -3.0 & -1.6 \end{bmatrix} \begin{bmatrix} x \\ y \\ z \end{bmatrix} = \begin{bmatrix} 12.4 \\ 4.0 \\ 9.2 \end{bmatrix}$; $(1.41, -4.27. 0.669)$

7. 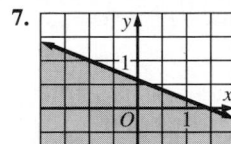   **9.**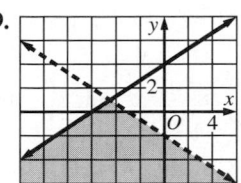

11. 4 tapes; 4 CDs

13. a.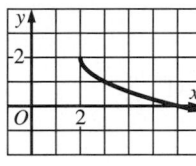

$x \geq 2$; $y \leq 2$

15. 5

17. $\pm 4i\sqrt{2}$

19.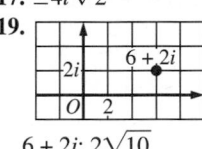

$6 + 2i$; $2\sqrt{10}$

21. $\dfrac{24}{25} + \dfrac{7}{25}i$ **25.** $1.2x^3 - x^2 + 10x - 13$

29. $y = -\dfrac{1}{2}(x + 2)(x - 1)(x - 4)$ **31.** possible rational zeros:

$\pm1; \pm2; \pm5; \pm10; \pm25; \pm50$; zeros: $-1; 5; -1 \pm 3i$ **33.** Yes; 40.   **35.** $x = -3$; $x = 2$; As $x \to \pm\infty$, $g(x) \to 0$.   **37.** 0; 3

CHAPTER 10

Page 468   Checking Key Concepts

1. a. $\dfrac{1}{25}, \dfrac{1}{36}, \dfrac{1}{49}, \dfrac{1}{64}$ **b.** $t_n = \dfrac{1}{n^2}$ **3. a.** $\sqrt{5}, \sqrt{6}, \sqrt{7}, 2\sqrt{2}$

b. $t_n = \sqrt{n}$ **5.** 65,536

Pages 468–471   Exercises and Applications

1. 1024, 4096, 16,384, 65,536   **3.** 7, 72, 9, 144   **5.** 55,555, 555,555, 5,555,555, 55,555,555   **7.** 183, 243, 313, 393
9. $\sqrt{243}, 27, \sqrt{2187}, 81$ **11.** Yes; U.S. presidential elections are held every four years.   **13.** Yes; on most streets, house lots are numbered by consecutive even or consecutive odd integers. (There are exceptions to this rule.)   **17.** 8; 125; 1728   **19.** $\dfrac{5}{2}; \dfrac{50}{17}; \dfrac{60}{19}$ **21.** 16; 169; 1156   **23.** 16, 100, 576
25. $-1; 1; -1$ **27.** Logs are rounded to two decimal places. $\log 16 \approx 1.20$; $\log 40 \approx 1.60$; $\log 96 \approx 1.98$ **29.** $t_n = 6n + 37$
31. $t_n = \dfrac{3}{5^n}$ **33.** $t_n = 109 - 6n$ **35.** $t_n = \dfrac{1}{2n-1}(-1)^{n+1}$
43. finite; The circumference of the vase is finite, so the pattern can only contain a finite number of terms.   **45.** infinite; The sequence 1, 3, 5, 7, 9, … continues without end.   **47.** finite; $\dfrac{1}{128} = 0.0078125$ **57.** $-4; 7$ **59.** -2 **61.** 729   **63.** $2\sqrt[5]{3}$

65. 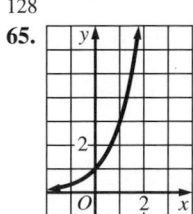   **67.**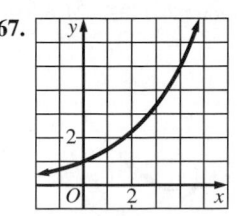

Page 476   Checking Key Concepts

1. a. arithmetic   **b.** 51   **3. a.** arithmetic   **b.** -7 **5. a.** neither
b. 121   **7. a.** 14.5   **b.** 10   **9. a.** $2\sqrt{2}$ **b.** $\sqrt{6}$

Pages 476–479   Exercises and Applications

1. arithmetic   **3.** neither   **5.** geometric   **7.** neither
9. geometric   **11.** arithmetic   **13. a.** (1) 8; (4) 1; (8) -41;
(11) -0.03 **b.** (1) 112; (4) 45; (8) -404; (11) -1 **17. a.** 36.5
b. 24   **19. a.** 10.5   **b.** $4\sqrt{5}$ **21. a.** 3   **b.** $\sqrt{8.51}$
23. a. 2.02   **b.** 0.4   **25. a.** neither; The sequence is 1, 6, 12, 18, 24, … .   **b.** $t_1 = 1$ and for $n \geq 2$, $t_n = 6(n - 1)$.
35. a. 1, 3, 9, 27   **b.** geometric; Each term is 3 times the preceding term.   **c.** $t_n = 3^{n-1}$ **37. a.** $\dfrac{\sqrt{3}}{4}, \dfrac{3\sqrt{3}}{16}, \dfrac{9\sqrt{3}}{64}, \dfrac{27\sqrt{3}}{256}$
b. geometric; Each term is $\dfrac{3}{4}$ times the precedingterm.
39. $15\sqrt{3}$ **41.** 435   **43.** $\pm\sqrt{2}$; 5

Page 483   Checking Key Concepts

1. 0, 6, 12, 18   **3.** $-2, -7, -12, -17$ **5.** 6.2, 3.1, 1.55, 0.775
7. $t_1 = 4; t_n = 7(t_{n-1})$ **9.** $t_1 = 19; t_n = t_{n-1} - 6$ **11.** $t_1 = \dfrac{1}{3}$;
$t_n = (t_{n-1})^2$

Pages 483–487   Exercises and Applications

1. a. $4, -2, 1, -\dfrac{1}{2}$ **b.** $t_n = 4\left(-\dfrac{1}{2}\right)^{n-1}$ **3. a.** 4.5, 27, 162, 972
b. $t_n = 4.5 \cdot 6^{n-1}$ **5. a.** 9, 9, 9, 9   **b.** $t_n = 9$ **7. a.** 0, 6, 12, 18   **b.** $t_n = 6(n - 1)$ or $t_n = 6n - 6$ **11.** $t_1 = 1.1; t_n = t_{n-1} + 1.1$
13. $t_1 = 50; t_n = \dfrac{1}{2}(t_{n-1})$ **15.** $t_1 = 3; t_n = (t_{n-1})^2$ **17.** $t_1 = 3$;
$t_n = \sqrt{5}(t_{n-1})$ **19.** $t_1 = 2; t_n = t_{n-1} + 3$ **21.** $t_1 = 1$;
$t_n = t_{n-1} + 1$ **27.** $t_1 = 6; t_2 = 7; t_n = 2(t_{n-1} + t_{n-2})$
33. $12\sqrt{3}$ **35.** $2\sqrt{6}\sqrt{10}$ **37.** Answers may vary.
Justifications are given for both. Numerical data may involve measurement such as with shirts sized by neck measurements. Categorical data may involve clothing such as sweaters grouped into sizes such as small, medium, and large.   **39.** 75
41. $-\dfrac{55}{14} \approx -3.93$

Page 487   Assess Your Progress

1. $-19, -25, -32$; neither   **2.** 128, -256, 512; geometric
3. 40, 49, 58; arithmetic   **4.** 15; $4\sqrt{11}$ **5.** 31.7; 7.5   **6.** $3\sqrt{3}$;
$\sqrt{15}$ **7.** explicit: $t_n = -7 + (-6)(n - 1)$ or $t_n = -6n - 1$;
recursive: $t_1 = -7; t_n = t_{n-1} - 6$ **8.** explicit: $t_n = 3 \cdot \left(\dfrac{2}{3}\right)^{n-1}$;
recursive: $t_1 = 3; t_n = \dfrac{2}{3}(t_{n-1})$ **9.** explicit: $t_n = $
$19 + 3(n - 1)$ or $t_n = 3n + 16$; recursive: $t_1 = 19; t_n = t_{n-1} + 3$

Page 491   Checking Key Concepts

1. a. 21 terms   **b.** 1260   **3. a.** 13 terms   **b.** -39
5. $4 + 32 + 108 + 256 + 500 + 864$ **7.** $1 + 4 + 7 + 10 + 13$
9. $\displaystyle\sum_{n=1}^{4} 3 \cdot 4^{n-1}$ **11.** $\displaystyle\sum_{n=1}^{9} (28 - 2n)$

Pages 491–494   Exercises and Applications

1. 3240   **3.** 15,150   **5.** 19,943   **9.** 50   **11.** 68

13. $-11 + (-10) + (-9) + (-8) + (-7) + (-6)$

15. $3 + (-11) + (-49) + (-123) + (-245) + (-427) + (-681)$

17. $\sum_{n=1}^{41} (2.3n - 2.3)$; 1886 **19.** $\sum_{n=1}^{18} \left(\frac{7}{2}n + \frac{9}{2}\right)$; $679\frac{1}{2}$

21. $\sum_{n=1}^{33} (7.3 - 0.2n)$; 128.7 **25. a.** 91 oranges **b.** $\sum_{n=1}^{12} n^2$

27. B **31.** $t_1 = 14$; $t_n = t_{n-1} - 8$ **33.** $\begin{bmatrix} -42 & -27 & 19 \\ -33 & -18 & 11 \\ -21 & -27 & 26 \end{bmatrix}$

Page 499   Checking Key Concepts

1. 728 **3.** -255 **5.** The sum does not exist because the series is infinite with $r = 4$ and $|4|$ is not less than 1. **7.** 1.5

9. $\sum_{n=1}^{6} 2 \cdot 3^{n-1}$ **11.** $\sum_{n=1}^{8} 3(-2)^{n-1}$

Pages 500–503   Exercises and Applications

1. 508 **3.** 2,222,222,222,222 **5.** 7.9921875 **7.** 61.4

9. 2.33×10^{20} **11.** 265,720 employees **13.** 18 **15.** $117\frac{1}{3}$

17. The sum does not exist because the series is infinite with $r = 1\frac{1}{3}$ and $\left|1\frac{1}{3}\right|$ is not less than 1. **19.** 6.25 **21.** The sum does not exist because the series is infinite with $r = -1$ and $|-1|$ is not less than 1. **25.** $\frac{8}{9}$ **27.** $\frac{4}{99}$ **29.** $\frac{368}{99}$ **31.** $\frac{1058}{999}$

33. $-\frac{151}{3330}$ **35.** $-\frac{359}{33}$ **45.** $\sum_{n=1}^{42} (-14n + 197) = -4368$

47. $\left(-14, 6\frac{3}{4}\right)$ **49.** $\left(\frac{26}{31}, \frac{94}{31}\right)$ **51.** $-\frac{1}{5}$ **53.** 5

Page 503   Assess Your Progress

1. a. geometric **b.** 8 terms **c.** 1275 **2. a.** arithmetic **b.** 8 terms **c.** $\frac{1}{2}$ **3. a.** geometric **b.** 7 terms **c.** 117,186 **4.** does not exist **5.** 125

6. $\sum_{n=1}^{8} (-17n + 131)$; 436 **7.** $\sum_{n=1}^{\infty} 3\left(\frac{2}{3}\right)^{n-1}$; 9

8. $6 + 11 + 16 + 21 + 26 + 31 + 36 + 41 + 46$; 234
9. $0.3 + 2.1 + 14.7 + 102.9 + 720.3 + 5042.1 + 35,294.7$; 41,177.1 **10.** $0.8 + 0.64 + 0.512 + 0.4096 + \ldots$; 4

Pages 508–509   Chapter 10 Assessment

1. geometric; common ratio; arithmetic; arithmetic mean
2. finite; series **3.** 125, 216, 343, 512; neither **4.** 1, -1, 1, -1; geometric **5.** $-13, -18, -23, -28$; arithmetic **6.** $t_n = n^3$
7. $t_n = (-1)^{n-1}$ **8.** $t_n = 7 - 5(n - 1)$ or $t_n = -5n + 12$ **9.** $\frac{12}{13}$
10. $\frac{1}{531,441}$ **11.** 58 **12. a.** 20.5 **b.** 20 **13. a.** 9 **b.** $\sqrt{77}$
14. a. 0.75 **b.** 0.6 **15. a.** 100, -10, 1, -0.1
b. $t_n = 100(-0.1)^{n-1}$ **16. a.** 100, 99.9, 99.8, 99.7
b. $t_n = 100 - 0.1(n - 1)$ or $t_n = -0.1n + 100.1$ **17. a.** 3, $3\sqrt{2}$, 6, $6\sqrt{2}$ **b.** $t_n = 3\left(\sqrt{2}\right)^{n-1}$ **18.** $t_1 = 7$; $t_n = t_{n-1} - 5$

19. $t_1 = \frac{1}{3}$; $t_n = \frac{1}{3}(t_{n-1})$ **20.** $t_1 = 3$; $t_n = t_{n-1} + 5$ **21.** $t_n = 2^n$; $t_1 = 2$, $t_n = 2(t_{n-1})$; geometric; Each term is twice the preceding term. **23. a.** arithmetic **b.** 16 terms

c. $\sum_{n=1}^{16} (-8n + 28)$ **d.** -640 **24. a.** geometric **b.** 8 terms

c. $\sum_{n=1}^{8} 2 \cdot 5^{n-1}$ **d.** 195,312 **25. a.** geometric **b.** 6 terms

c. $\sum_{n=1}^{6} 8\left(-\frac{1}{2}\right)^{n-1}$ **d.** $5\frac{1}{4}$ **26. a.** arithmetic **b.** 17 terms

c. $\sum_{n=1}^{17} (0.5n + 6.5)$ **d.** 187 **27.** 20 rows; 59 seats

28. a. $\frac{1}{2} + 4 + \frac{27}{2} + 32 + \frac{125}{2}$; $112\frac{1}{2}$

b. $1000 - 250 + 62.5 - 15.625 + \ldots$; 800 **30.** $\frac{35}{1998}$

CHAPTER 11

Page 516   Checking Key Concepts

1. a. $3\sqrt{2} \approx 4.24$ **b.** (2.5, 6.5) **3. a.** $\sqrt{13} \approx 3.61$
b. $(-4.5, -1)$

Pages 516–519   Exercises and Applications

1. a. $(1.5, -4)$ **b.** $\frac{4}{5}$ **c.** $-\frac{5}{4}$ **d.** $y + 4 = -\frac{5}{4}(x - 1.5)$ or $y = -\frac{5}{4}x - \frac{17}{8}$ **3. a.** 2.5 **b.** $y = \frac{3}{4}x - \frac{21}{16}$ **5. a.** $\sqrt{197} \approx 14.04$
b. $y = -14x - 60.5$ **7. a.** $\sqrt{106} \approx 10.30$ **b.** $y = -\frac{5}{9}x + \frac{2}{9}$
9. a. $\sqrt{183.38} \approx 13.54$ **b.** $y = -\frac{107}{83}x - \frac{2118}{415}$ **11. a.** $\frac{\sqrt{1445}}{8} \approx$
4.75 **b.** $y = -38x - \frac{979}{16}$ **13. a.** $\sqrt{460.85} \approx 21.47$ **b.** $y =$
$\frac{17}{214}x + \frac{8177}{4280}$ **15.** 7 or 15 **17.** 14 or -34 **19.** $-3 \pm 4\sqrt{2}$;
about 2.66 or -8.66 **21.** Answers may vary. An example is given. $QR = RS = ST = TQ = 5\sqrt{2}$. The slope of $\overline{QR}$ and $\overline{ST}$ is 1, and the slope of $\overline{RS}$ and $\overline{TQ}$ is -1. The adjacent sides of $QRST$ are perpendicular. Therefore, $QRST$ is a rhombus with four right angles, or a square. **25. a.** $10 + 2\sqrt{34} \approx 21.66$
b. Proofs may vary. An example is given. $AB = CD = \sqrt{34}$, and $BC = AD = 5$. Since both pairs of opposite sides are congruent, $ABCD$ is a parallelogram. **29. a.** (1.5, 2.5), (2.5, 6.5),
(6.5, 5.5), (5.5, 1.5) **b.** (2, 4.5), (4.5, 6), (6, 3.5), (3.5, 2)
c. $4\sqrt{34}$; $4\sqrt{17}$; $4\sqrt{8.5}$; The perimeter of the next square should be $4\sqrt{4.25}$. To check the prediction, find two vertices of the next square, for example, (3.25, 5.25) and (5.25, 4.75). These give a side length of $\sqrt{4.25}$, and a perimeter of $4\sqrt{4.25}$,
as predicted. **35.** $\sum_{n=1}^{7} 6(3)^{n-1} = 6558$

37. a.

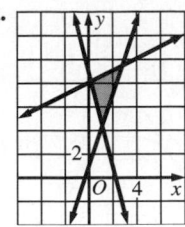

b. 55.40 **c.** $\frac{97}{7} \approx 13.86$

39. parabola that opens up and has vertex $(3, -4)$

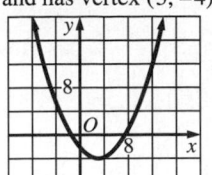

Page 524   Checking Key Concepts

1. The farther apart the directrix and focus are, the wider the parabola becomes. **3.** $y - \frac{3}{2} = \frac{1}{6}(x - 1)^2$

5. $x - 1 = -\frac{1}{12}(y + 1)^2$

Pages 525–527   Exercises and Applications

3. A

5. a. $y = -\frac{1}{12}x^2$

b.

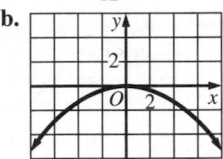

7. a. $y = \frac{1}{4}x^2$

b.

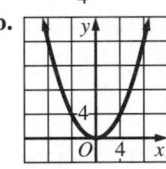

9. a. $x - 2 = -\frac{1}{24}(y - 6)^2$

b.

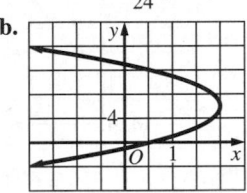

11. about 1.8 m

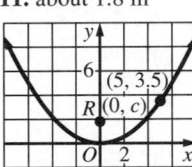

13. about 27.8 m

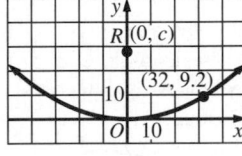

15. vertex: $(0, 0)$; focus: $\left(0, \frac{1}{12}\right)$;

directrix: $y = -\frac{1}{12}$

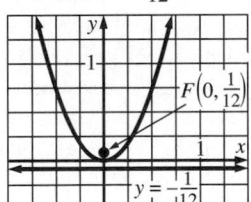

17. vertex: $(-1, 4)$; focus: $(-1, 6)$; directrix: $y = 2$

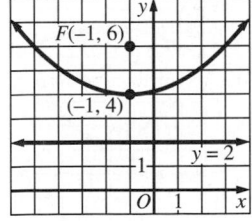

19. vertex: $(2, 1)$; focus: $\left(2\frac{1}{16}, 1\right)$;

directrix: $x = 1\frac{15}{16}$

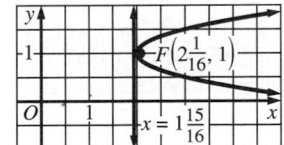

21. vertex: $(0, 2)$;

focus: $\left(0, 2\frac{1}{4}\right)$;

directrix: $y = 1\frac{3}{4}$

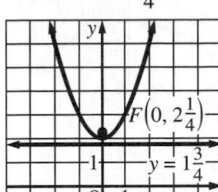

23. vertex: $\left(-\frac{3}{4}, 0\right)$;

focus: $\left(\frac{1}{4}, 0\right)$;

directrix: $x = -1\frac{3}{4}$

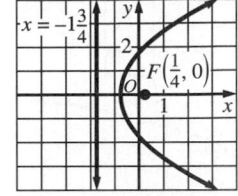

31. a. $8\sqrt{13} \approx 28.84$ **b.** $(4, -3)$ **33.** $\left(\frac{15}{4}, \frac{5}{4}\right)$

Page 527   Assess Your Progress

1. a. $\sqrt{145} \approx 12.04$ **b.** $(1, 3.5)$ **2. a.** $\sqrt{185} \approx 13.60$

b. $(0.5, -1)$ **3. a.** $y = -\frac{4}{5}x + 1$ **b.** $y = \frac{5}{4}x - \frac{37}{4}$

4. vertex: $(-3, 1)$; focus: $\left(-2\frac{3}{4}, 1\right)$; directrix: $x = -3\frac{1}{4}$

5. vertex: $(-5, -6)$; focus: $(-5, -1)$; directrix: $y = -11$

6. $y - 1 = \frac{1}{12}(x - 3)^2$ **7.** $x + 2\frac{1}{2} = -\frac{1}{14}y^2$

Page 531   Checking Key Concepts

1. $C(6, 4)$; $r = 12$

3. $C(-5, 0)$; $r = 6$

5. $(x + 2)^2 + (y - 3)^2 = 16$

9. $(0, -2)$

7.

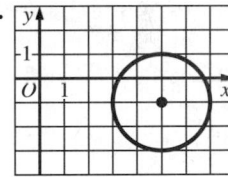

Pages 532–534   Exercises and Applications

1. $C(4, 2)$; $r = 5$

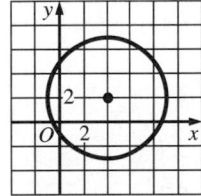

3. $C(0, 1)$; $r = 2$

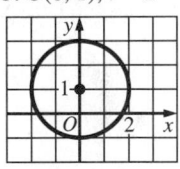

5. $C(-1, -1)$; $r = \sqrt{10} \approx 3.16$

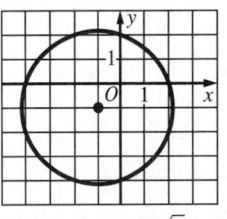

7. $C(0, -3)$; $r = \sqrt{2} \approx 1.41$

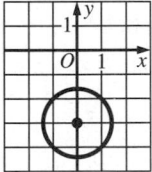

9. $C(0, 1)$; $r = 2\sqrt{2} \approx 2.83$

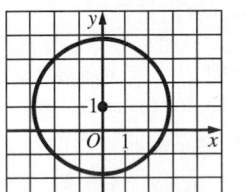

11. $x^2 + y^2 = 1$; 1 mi

13. Yes; $(0.9)^2 + (1.2)^2 < (1.6)^2$

15. $x^2 + y^2 = 49$

17. $(x + 1)^2 + (y - 6)^2 = 9$

19. $x^2 + (y - \sqrt{3})^2 = 121$

21. $(x - a)^2 + (y - b)^2 = 121$

27. Yes. **29.** Yes.

33. $(x - 3)^2 + (y - 1)^2 = 25$

35. about (3.46, 3) and (−3.46, 3)

41. $y = \frac{1}{4}(x - 3)^2$

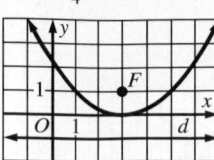

43. $y + 1 = -\frac{1}{8}(x - 4)^2$

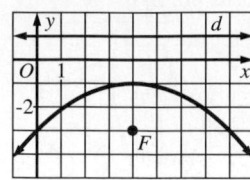

45.

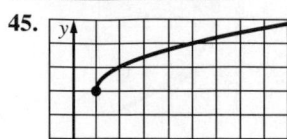

$x \geq 1; y \geq 4$

Page 539 Checking Key Concepts

1. (6, −4); (0, −4) and (12, −4)   **3.** $\frac{(x-6)^2}{100} + \frac{(y+4)^2}{64} = 1$

5. center: (−1, 3); vertices: (−9, 3)
and (7, 3); foci: (−1 − 2$\sqrt{15}$, 3) and
(−1 + 2$\sqrt{15}$, 3); horizontal

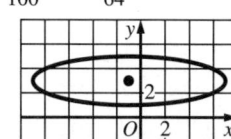

Pages 540–542 Exercises and Applications

1. center: (0, 0); vertices: (8, 0)
and (−8, 0); foci: ($\sqrt{39}$, 0) and
(−$\sqrt{39}$, 0) ≈ (±6.24, 0);
horizontal

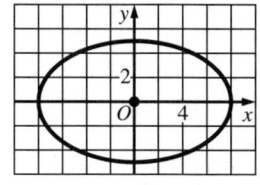

3. center: (0, 1); vertices: (2$\sqrt{2}$, 1)
and (−2$\sqrt{2}$, 1) ≈ (±2.83, 1); foci:
($\sqrt{6}$, 1) and (−$\sqrt{6}$, 1) ≈ (±2.45, 1);
horizontal

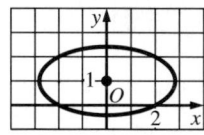

5. center: (−8, −1); vertices: (−8, 9) and
(−8, −11); foci: (−8, −1 ± 5$\sqrt{3}$) ≈
(−8, 7.66) and (−8, −9.66); vertical

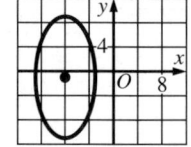

7. $\frac{x^2}{4} + \frac{y^2}{16} = 1$ **9.** $\frac{(x+3)^2}{16} + \frac{(y+1)^2}{9} = 1$

21. $\frac{(x-5)^2}{4} + \frac{(y+3)^2}{9} = 1$ **23.** $\frac{(x+2)^2}{49} + \frac{(y-4)^2}{9} = 1$

27. $(x + 2)^2 + (y + 1)^2 = 36$

29. $y = 2x$

31. $y = -\frac{b}{a}x$

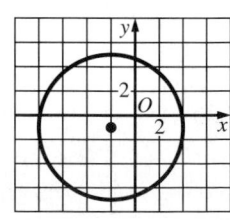

Page 542 Assess Your Progress

1. center: (−1, 5); radius: 4   **2.** center: (3, 0); radius: 5

3. (3, 5) and $\left(-\frac{77}{41}, \frac{45}{41}\right)$ ≈ (−1.88, 1.10)   **4.** center: (0, 5);

vertices: (−8, 5) and (8, 5); foci: (−2$\sqrt{7}$, 5) and (2$\sqrt{7}$, 5);
horizontal

SA27 Selected Answers

5. center: (−3, 5); vertices: (−3, 19) and (−3, −9);
foci: (−3, 5 + 4$\sqrt{6}$) and (−3, 5 − 4$\sqrt{6}$); vertical

6. $x^2 + (y + 5)^2 = 16$

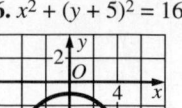

7. $\frac{(x+1)^2}{9} + \frac{(y-2)^2}{25} = 1$

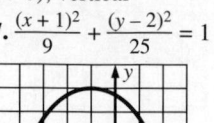
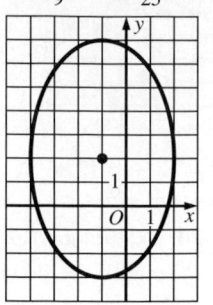

Page 546 Checking Key Concepts

1. (−2, 5); (−2, 10) and (−2, 0)   **3.** $\frac{(y-5)^2}{16} - \frac{(x+2)^2}{9} = 1$

5. center: (6, 4); vertices: (2, 4) and
(10, 4); foci: (6 ± $\sqrt{65}$, 4) ≈ (14.06, 4)
and (−2.06, 4); horizontal;
asymptotes: $y = \pm\frac{7}{4}(x - 6) + 4$

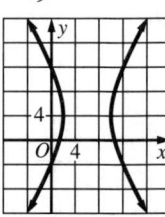

Pages 547–549 Exercises and Applications

1. center: (0, 0); vertices: (±6, 0);
foci: (±$\sqrt{61}$, 0) ≈ (±7.81, 0);
horizontal; asymptotes: $y = \pm\frac{5}{6}x$

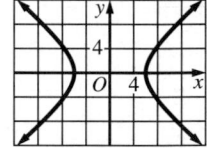

3. center: (0, 3); vertices: (±9, 3);
foci: (±3$\sqrt{10}$, 3) ≈ (±9.49, 0);
horizontal; asymptotes:
$y = \pm\frac{1}{3}x + 3$

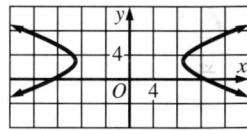

5. center: (5, −7); vertices: (5, −2) and
(5, −12); foci: (5, −7 ± 5$\sqrt{5}$) ≈
(5, 4.18) and (5, −18.18); vertical;
asymptotes: $y = \pm\frac{1}{2}(x - 5) - 7$

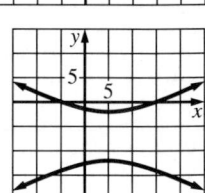

7. center: (−4, 0); vertices: (−4, ±3);
foci: (−4, ±$\sqrt{58}$) ≈ (−4, ±7.62);
vertical; asymptotes: $y = \pm\frac{3}{7}(x + 4)$

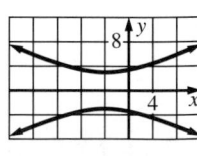

9. center: (0, 0); vertices:
(0, ±6$\sqrt{3}$) ≈ (0, 10.39);
foci: (0, ±$\sqrt{183}$) ≈ (0, ±13.53);
vertical; asymptotes: $y = \pm1.2x$

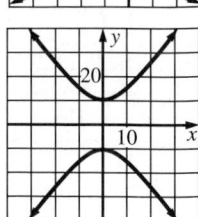

11. C   **13.** A

19. a.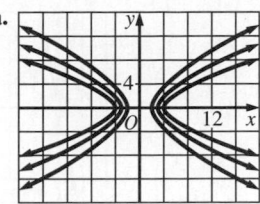

b. The vertices move out farther from the center (the origin), and the hyperbola becomes narrower.

c. Answers may vary. An example is given, using the equations $\frac{x^2}{4} - \frac{y^2}{4} = 1$, $\frac{x^2}{4} - \frac{y^2}{9} = 1$, and $\frac{x^2}{4} - \frac{y^2}{16} = 1$.

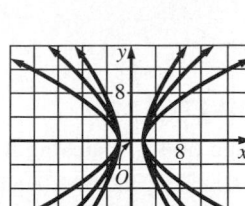

As the value of b increases, the hyperbola widens, while the vertices remain fixed.

21. $\frac{x^2}{9} - \frac{y^2}{16} = 1$

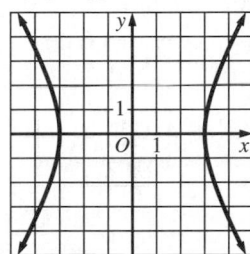

23. $\frac{x^2}{36} - \frac{y^2}{64} = 1$

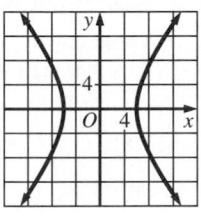

25. $\frac{(x-1)^2}{36} - \frac{(y+4)^2}{64} = 1$

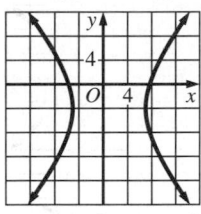

29. $\frac{(x+6)^2}{16} + \frac{(y+1)^2}{1} = 1$

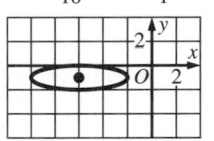

31. $\frac{(x-2)^2}{169} + \frac{(y+3)^2}{25} = 1$

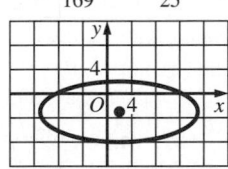

33. $y = (x-6)^2 - 4$; $y = (x-8)(x-4)$

Page 552 Checking Key Concepts

1. $(x-3)^2 + \left(y + \frac{1}{2}\right)^2 = \frac{25}{4}$; a circle with center $\left(3, -\frac{1}{2}\right)$ and radius $\frac{5}{2}$ **3.** a point, (0, 0), representing a degenerate ellipse; The graph is the point (0, 0).

Pages 552–553 Exercises and Applications

1. $\frac{(x+5)^2}{4} - \frac{(y+2)^2}{9} = 1$; a horizontal hyperbola with center (−5, −2), vertices (−3, −2) and (−7, −2), and asymptotes $y = \pm\frac{3}{2}(x+5) - 2$ **3.** $y = 6(x+1)^2$; a parabola, opening upward, with vertex (−1, 0) and axis $x = -1$

5. $x^2 + \left(y + \frac{3}{2}\right)^2 = \frac{25}{4}$; a circle with center $\left(0, -\frac{3}{2}\right)$ and radius $\frac{5}{2}$

13. a degenerate parabola; the line $y = -x$

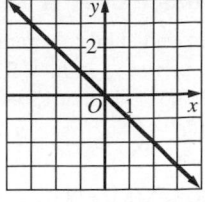

15. a degenerate ellipse; the point (2, 0)

17. a degenerate hyperbola; the intersecting lines $y = \frac{1}{2}x + \frac{3}{2}$ and $y = -\frac{1}{2}x + \frac{1}{2}$

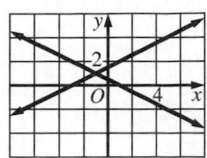

19. center: (−1, 3); vertices: (6, 3) and (−8, 3); foci: $(-1 \pm \sqrt{74}, 3) \approx$ (7.60, 3) and (−9.60, 3); horizontal; asymptotes: $y = \pm\frac{5}{7}(x+1) + 3$

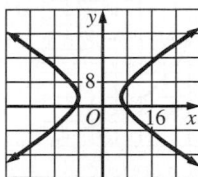

21. 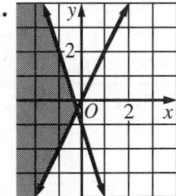 **23.**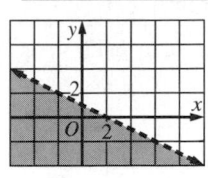

Page 553 Assess Your Progress

1. center: (0, 5); vertices: (±9, 5); foci: $(\pm 3\sqrt{13}, 5) \approx (10.82, 5)$; horizontal; asymptotes: $y = \pm\frac{2}{3}x + 5$

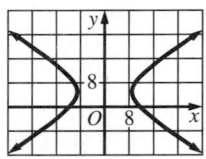

2. center: (−2, 2), vertices: (−2, 7) and (−2, −3); foci: $(-2, 2 \pm \sqrt{29}) \approx$ (−2, 7.39) and (−2, −3.39); vertical; asymptotes: $y = \pm\frac{5}{2}(x+2) + 2$

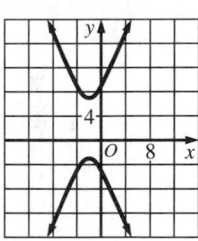

3. $(x+7)^2 + (y-4)^2 = 64$; a circle with center (−7, 4) and radius 8 **4.** $\frac{(x+2)^2}{9} - \frac{(y-5)^2}{9} = 1$; a horizontal hyperbola with center (−2, 5), vertices (1, 5) and (−5, 5), and asymptotes $y = \pm(x+2) + 5$

5. a degenerate circle; the point (0, 0)

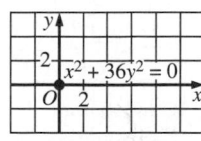

6. a degenerate hyperbola; the intersecting lines $y = \frac{1}{6}x$ and $y = -\frac{1}{6}x$

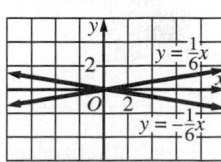

1. parallel; perpendicular   2. parabola; focus; directrix
3. major axis; vertices   **4. a.** $\sqrt{2} \approx 1.41$ **b.** (1.5, 4.5)
5. a. 5   **b.** (5, 3.5)   **6. a.** $\sqrt{17} \approx 4.12$ **b.** (–5, 0.5)

7. a. $y = \frac{3}{5}x + \frac{36}{5}$ **b.** $y = -\frac{5}{3}x + \frac{8}{3}$ **8. a.** $\left(-\frac{1}{2}, 1\right)$

b. $\sqrt{61} \approx 7.81$; $\frac{1}{2}\sqrt{61} \approx 3.91$ **c.** $y = \frac{5}{6}x + \frac{13}{2}$

9. $x - \frac{1}{2} = \frac{1}{14}(y + 2)^2$

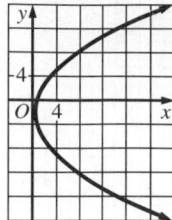

10. a circle with center (–2, 5) and radius 4

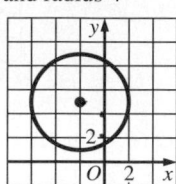

11. a horizontal hyperbola with center (1, 6), vertices (4, 6) and (–2, 6), foci (6, 6) and (–4, 6), and asymptotes $y = \pm\frac{4}{3}(x - 1) + 6$

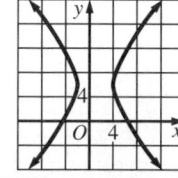

12. a vertical ellipse with center (–2, 1), vertices (–2, –7) and (–2, 9), and minor axis with endpoints (–8, 1) and (4, 1)

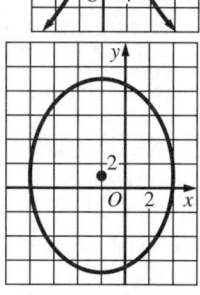

13. a horizontal ellipse with center (–1, –7), vertices (–6, –7) and (4, –7), and minor axis with endpoints (–1, –9) and (–1, –5)

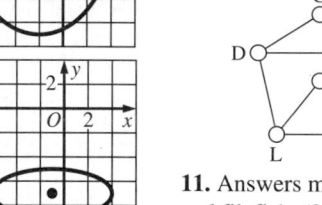

14. a. (17.2, 0) or (–17.2, 0)

b. $\frac{x^2}{316.84} + \frac{y^2}{21} = 1$

16. $y - 4 = -\frac{1}{8}(x - 3)^2$

17. $x^2 + (y - 8)^2 = 4$

18. $\frac{(x + 4)^2}{4} + \frac{(y + 2)^2}{9} = 1$

19. $\frac{(x - 6)^2}{16} - \frac{(y - 6)^2}{9} = 1$

20. $x + 1 = -\frac{1}{16}(y - 3)^2$; a parabola that opens left, has vertex (–1, 3), and axis $y = 3$ **21.** $\frac{(x - 1)^2}{4} - \frac{(y - 2)^2}{1} = 1$; a horizontal hyperbola with center (1, 2), vertices (3, 2) and (–1, 2), and asymptotes $y = \pm\frac{1}{2}(x - 1) + 2$ **22.** $x^2 + (y - 6)^2 = 11$; a circle with center (0, 6) and radius $\sqrt{11} \approx 3.32$ **23.** $\frac{(x - 3)^2}{25} + \frac{(y - 2)^2}{1} = 1$; a horizontal ellipse with center (3, 2), vertices (8, 2) and (–2, 2), and minor axis with endpoints (3, 3) and (3, 1)

24. a degenerate hyperbola; intersecting lines: $y = 5x$ and $y = -5x$

25. a degenerate circle; the point (–1, –1)

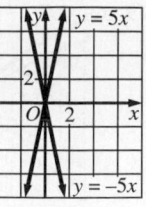

CHAPTER 12

Page 566   Checking Key Concepts

1. Graphs A and B are equivalent. Answers may vary. An example is given. Each graph consists of edges connecting A and B, B and C, C and D, and D and A. Graphs C and D are equivalent.   **3.** 2   **5.** Answers may vary. An example is given. Vermont would have degree 3.

Pages 566–568   Exercises and Applications

1. A and B; In both graphs, each vertex is connected in the same way to the other vertices.   **3, 5.** Answers to parts (b) and (c) may vary. Examples are given.
3. a. A, 2; B, 3; C, 4; D, 2; E, 3; F, 4; G, 2
b, c. Three colors are needed.

5. a. A, 4; B, 4; C, 4; D, 4; E, 4
b, c. Five colors are needed.

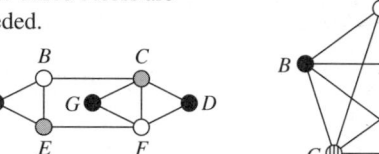

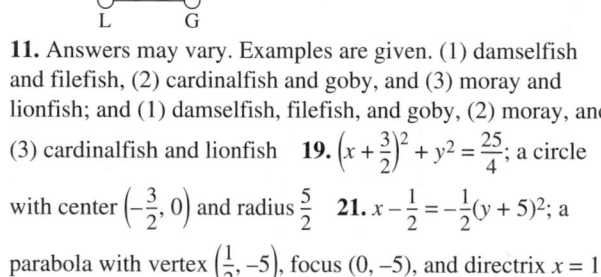

9, 11. Answers may vary. Examples are given.
9. a.

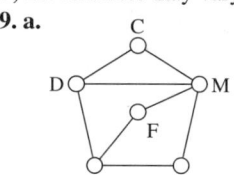

b. Yes, as shown in the graph in part (a).

11. Answers may vary. Examples are given. (1) damselfish and filefish, (2) cardinalfish and goby, and (3) moray and lionfish; and (1) damselfish, filefish, and goby, (2) moray, and (3) cardinalfish and lionfish   **19.** $\left(x + \frac{3}{2}\right)^2 + y^2 = \frac{25}{4}$; a circle with center $\left(-\frac{3}{2}, 0\right)$ and radius $\frac{5}{2}$ **21.** $x - \frac{1}{2} = -\frac{1}{2}(y + 5)^2$; a parabola with vertex $\left(\frac{1}{2}, -5\right)$, focus (0, –5), and directrix $x = 1$

23. 39   **25.** undefined   **27.** $\begin{bmatrix} 3 & 2 \\ -2 & -1 \end{bmatrix}$

Page 572   Checking Key Concepts

1. B   **3.** A $\rightarrow$ D $\rightarrow$ F, B $\rightarrow$ D $\rightarrow$ F, and B $\rightarrow$ E $\rightarrow$ F

5.

$$\begin{array}{c} \\ A \\ B \\ C \\ D \\ E \\ F \end{array} \begin{array}{c} \quad A\ B\ C\ D\ E\ F \\ \begin{bmatrix} 0 & 0 & 0 & 0 & 2 & 1 \\ 0 & 0 & 0 & 0 & 1 & 2 \\ 0 & 0 & 0 & 0 & 0 & 1 \\ 0 & 0 & 0 & 0 & 0 & 1 \\ 0 & 0 & 0 & 0 & 0 & 0 \\ 0 & 0 & 0 & 0 & 0 & 0 \end{bmatrix} \end{array}$$; the number of ways a course can be taken with two prerequisites

1. a.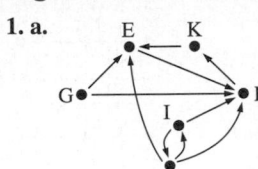

b. Barbara → Fatima → Katrina, Barbara → Ezra → Fatima → Katrina, Barbara → Ilana → Fatima → Katrina

c. No. Explanations may vary. An example is given. Ilana could have gotten the disease only from Barbara, and Barbara could have gotten it only from Ilana. Since Geoff could not have gotten the disease from either Barbara or Ilana, it is not possible that the disease began with one of these six students.

3.
$$\begin{array}{c c}& \begin{array}{cccc} A & B & C & D \end{array}\\ \begin{array}{c} A\\ B\\ C\\ D \end{array} & \left[\begin{array}{cccc} 0 & 1 & 0 & 1\\ 0 & 0 & 1 & 1\\ 1 & 0 & 0 & 1\\ 0 & 0 & 0 & 0 \end{array}\right]\end{array}$$

5.
$$\begin{array}{c c}& \begin{array}{cccccc} A & B & C & D & E & F \end{array}\\ \begin{array}{c} A\\ B\\ C\\ D\\ E\\ F \end{array} & \left[\begin{array}{cccccc} 0 & 0 & 0 & 0 & 0 & 0\\ 1 & 0 & 0 & 0 & 0 & 0\\ 1 & 0 & 0 & 0 & 0 & 0\\ 1 & 0 & 0 & 0 & 0 & 0\\ 1 & 0 & 0 & 0 & 0 & 0\\ 0 & 1 & 1 & 1 & 1 & 0 \end{array}\right]\end{array}$$

7.
$$\begin{array}{c c}& \begin{array}{cccccccccccc} D & R & A & I & E & S & P & Sn & M & Rc & F \end{array}\\ \begin{array}{c} D\\ R\\ A\\ I\\ E\\ S\\ P\\ Sn\\ M\\ Rc\\ F \end{array} & \left[\begin{array}{cccccccccccc} 0 & 0 & 0 & 0 & 0 & 0 & 0 & 0 & 0 & 0 & 0 & 0\\ 0 & 0 & 0 & 0 & 0 & 0 & 0 & 0 & 0 & 0 & 0 & 0\\ 0 & 1 & 0 & 0 & 0 & 0 & 0 & 0 & 0 & 0 & 0 & 0\\ 0 & 1 & 1 & 1 & 1 & 0 & 1 & 0 & 1 & 1 & 0 & 0\\ 0 & 0 & 0 & 0 & 0 & 1 & 0 & 0 & 0 & 0 & 0 & 0\\ 0 & 1 & 0 & 0 & 0 & 0 & 0 & 0 & 0 & 0 & 0 & 0\\ 1 & 0 & 0 & 1 & 1 & 1 & 0 & 0 & 1 & 1 & 1 & 1\\ 0 & 1 & 0 & 0 & 0 & 0 & 0 & 1 & 0 & 0 & 1 & 1\\ 0 & 0 & 0 & 0 & 0 & 0 & 0 & 1 & 0 & 1 & 1 & 1\\ 0 & 0 & 0 & 0 & 0 & 0 & 0 & 0 & 0 & 0 & 0 & 1\\ 0 & 0 & 0 & 0 & 0 & 0 & 0 & 0 & 0 & 0 & 0 & 0 \end{array}\right]\end{array}$$

9.
$$\begin{array}{c c}& \begin{array}{cccccccccccc} D & R & A & I & E & S & P & Sn & M & Rc & F \end{array}\\ \begin{array}{c} D\\ R\\ A\\ I\\ E\\ S\\ P\\ Sn\\ M\\ Rc\\ F \end{array} & \left[\begin{array}{cccccccccccc} 0 & 0 & 0 & 0 & 0 & 0 & 0 & 0 & 0 & 0 & 0 & 0\\ 0 & 0 & 0 & 0 & 0 & 0 & 0 & 0 & 0 & 0 & 0 & 0\\ 0 & 1 & 0 & 0 & 0 & 0 & 0 & 0 & 0 & 0 & 0 & 0\\ 0 & 5 & 2 & 2 & 0 & 2 & 0 & 4 & 2 & 1 & 2\\ 0 & 1 & 0 & 0 & 0 & 1 & 0 & 0 & 0 & 0 & 0\\ 0 & 1 & 0 & 0 & 0 & 0 & 0 & 0 & 0 & 0 & 0\\ 1 & 2 & 1 & 2 & 1 & 3 & 0 & 2 & 2 & 2 & 3\\ 0 & 2 & 0 & 0 & 0 & 0 & 0 & 2 & 0 & 0 & 2\\ 0 & 1 & 0 & 0 & 0 & 0 & 0 & 2 & 0 & 1 & 3\\ 0 & 0 & 0 & 0 & 0 & 0 & 0 & 0 & 0 & 0 & 1\\ 0 & 0 & 0 & 0 & 0 & 0 & 0 & 0 & 0 & 0 & 0 \end{array}\right]\end{array}$$

21. a.
$$M = \begin{array}{c c}& \begin{array}{ccccccc} A & B & C & D & E & F & G \end{array}\\ \begin{array}{c} A\\ B\\ C\\ D\\ E\\ F\\ G \end{array} & \left[\begin{array}{ccccccc} 0 & 0 & 1 & 1 & 0 & 1 & 1\\ 1 & 0 & 0 & 0 & 1 & 0 & 1\\ 0 & 1 & 0 & 1 & 1 & 0 & 0\\ 0 & 1 & 0 & 0 & 0 & 0 & 0\\ 1 & 0 & 0 & 1 & 0 & 1 & 1\\ 0 & 1 & 1 & 1 & 0 & 0 & 0\\ 0 & 0 & 1 & 1 & 0 & 1 & 0 \end{array}\right]\end{array}$$

b. The row sums represent the total number of games won by the row team.

23. a.
$$M + M^2 = \begin{array}{c c}& \begin{array}{ccccccc} A & B & C & D & E & F & G & Total \end{array}\\ \begin{array}{c} A\\ B\\ C\\ D\\ E\\ F\\ G \end{array} & \left[\begin{array}{ccccccc|c} 0 & 3 & 3 & 4 & 1 & 2 & 1 & 14\\ 2 & 0 & 2 & 3 & 1 & 3 & 3 & 14\\ 2 & 2 & 0 & 2 & 2 & 1 & 2 & 11\\ 1 & 1 & 0 & 0 & 1 & 0 & 1 & 4\\ 1 & 2 & 3 & 4 & 0 & 3 & 2 & 15\\ 1 & 3 & 1 & 2 & 2 & 0 & 1 & 10\\ 0 & 3 & 2 & 3 & 1 & 1 & 0 & 10 \end{array}\right]\end{array}$$

The row sums represent the total number of points awarded to the row team. **b.** See the last column in the matrix in part (a). The Elks win the championship. **27.** 9840

29. $\dfrac{625}{3} \approx 208.33$

Page 576   Assess Your Progress

1. Answers may vary. An example is given.

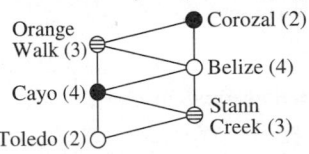

Three colors are needed. Color groupings: Corozal and Cayo, Belize and Toledo, Orange Walk and Stann Creek.

2. Two meeting times are needed.

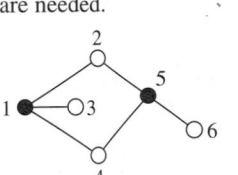

3. a.

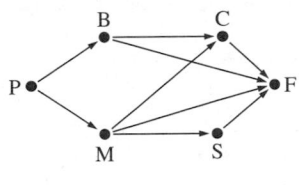

b. $M =$
$$\begin{array}{c c}& \begin{array}{cccccc} F & C & S & B & M & P \end{array}\\ \begin{array}{c} F\\ C\\ S\\ B\\ M\\ P \end{array} & \left[\begin{array}{cccccc} 0 & 0 & 0 & 0 & 0 & 0\\ 1 & 0 & 0 & 0 & 0 & 0\\ 1 & 0 & 0 & 0 & 0 & 0\\ 1 & 1 & 0 & 0 & 0 & 0\\ 1 & 1 & 1 & 0 & 0 & 0\\ 0 & 0 & 0 & 1 & 1 & 0 \end{array}\right]\end{array}$$

c. $M^2 =$
$$\begin{array}{c c}& \begin{array}{cccccc} F & C & S & B & M & P \end{array}\\ \begin{array}{c} F\\ C\\ S\\ B\\ M\\ P \end{array} & \left[\begin{array}{cccccc} 0 & 0 & 0 & 0 & 0 & 0\\ 0 & 0 & 0 & 0 & 0 & 0\\ 0 & 0 & 0 & 0 & 0 & 0\\ 1 & 0 & 0 & 0 & 0 & 0\\ 2 & 0 & 0 & 0 & 0 & 0\\ 2 & 2 & 1 & 0 & 0 & 0 \end{array}\right]\end{array}$$

M^2 gives the number of ways that each prey is a food source for the column predator through one intermediate source. Answers may vary. An example is given. Barnacles are an intermediate food source for fish through crabs.

$$M + M^2 = \begin{array}{c c}& \begin{array}{cccccc} F & C & S & B & M & P \end{array}\\ \begin{array}{c} F\\ C\\ S\\ B\\ M\\ P \end{array} & \left[\begin{array}{cccccc} 0 & 0 & 0 & 0 & 0 & 0\\ 1 & 0 & 0 & 0 & 0 & 0\\ 1 & 0 & 0 & 0 & 0 & 0\\ 2 & 1 & 0 & 0 & 0 & 0\\ 3 & 1 & 1 & 0 & 0 & 0\\ 2 & 2 & 1 & 1 & 1 & 0 \end{array}\right]\end{array}$$

$M + M^2$ gives the total number of ways that each prey is a food source for the column predator either directly or through one intermediate source. Answers may vary. An example is given. Barnacles are a food source for fish in two ways, both directly and indirectly through crabs.

Page 581   Checking Key Concepts

1. a. spaghetti/corn/pie, spaghetti/corn/pudding, spaghetti/beans/pie, spaghetti/beans/pudding, spaghetti/squash/pie, spaghetti/squash/pudding, chicken/corn/pie, chicken/corn/pudding, chicken/beans/pie, chicken/beans/pudding, chicken/squash/pie, chicken/squash/pudding **b.** There are 2 choices of entrees, 3 choices of vegetables, and 2 choices of desserts, so there are $2 \times 3 \times 2 = 12$ choices of meals. **3.** 6 **5.** 6720

7. D **9.** $\dfrac{6!}{2!\ 2!\ 1!\ 1!}$

Pages 581–584   Exercises and Applications

1. a. 720 **b.** 12 **3. a.** 24 **b.** 24 **7.** 7 **9.** 210 **11.** 96 ways **21.** 8 **23.** 360 **25.** 259 other cars **27.** 362,880 batting orders **31.** C

33. $\dfrac{5!}{2!\,2!\,1!} = 30$ **35.** $\dfrac{6!}{3!\,1!\,1!\,1!} = 120$ **43.** $x = 1$ and $y = 3$; $y = \dfrac{2}{x}$ is translated 1 unit right and 3 units up. **45.** $x = 4$ and $y = 2$; $y = \dfrac{9}{x}$ is translated 4 units right and 2 units up.

Page 588   Checking Key Concepts

1. combinations; $_9C_3$ **3.** permutations; $_6P_4$ **5.** $_{10}C_4 \cdot _{13}C_4$ **7. a.** 21 **b.** 35 **c.** 35 **d.** 4950

Pages 588–592   Exercises and Applications

1. 792 **3.** 1820 **5.** 1 **7.** 8 **9.** 20 **11.** 441 **13.** $_{500}C_{10} \approx$ 2.458×10^{20} sets **15.** 56 triangles **17.** 6 ways **19.** 20 possible purchases **23.** 35,640 ways **25. a.** 38,760 ways **b.** $_{20}P_9 \approx 6.095 \times 10^{10}$ ways **c.** 465,120 ways **29.** 635,013,559,600 different hands **31.** 103,776 hands **33.** 156 different kinds of full houses **43.** 120 **45.** 40,320 **47.** 239,500,800 **49.** $4t^2 - 4t + 1$ **51.** $n^3 - 6n^2 + 12n - 8$ **53.** $y^4 + 4y^3 + 6y^2 + 4y + 1$ **55.** $f(x) = -x(x + 5)(x + 2)$

Page 595   Checking Key Concepts

1. 1, 8, 28, 56, 70, 56, 28, 8, 1 **3.** 252 sequences **5.** $a = 4x$; $b = y$; $n = 12$

Pages 595–599   Exercises and Applications

1. 1, 9, 36, 84, 126, 126, 84, 36, 9, 1 **5.** 2 **7.** 8 **9.** 165 **11.** $x^5 - 5x^4y + 10x^3y^2 - 10x^2y^3 + 5xy^4 - y^5$ **13.** $32y^5 - 240y^4z + 720y^3z^2 - 1080y^2z^3 + 810yz^4 - 243z^5$ **15.** $m^{14} - 7m^{12}n + 21m^{10}n^2 - 35m^8n^3 + 35m^6n^4 - 21m^4n^5 + 7m^2n^6 - n^7$ **17.** $x^{12} + 6x^{10}y^3 + 15x^8y^6 + 20x^6y^9 + 15x^4y^{12} + 6x^2y^{15} + y^{18}$ **19. a.** 12 **b.** $_{12}C_2$ **c.** $220a^9b^3$ **37.** D **39.** 1 **41.** 7 **43.** $576\pi \approx 1810$ cm^2 **45.** $y + 2 = 6(x - 3)$

Page 599   Assess Your Progress

1. 9 **2.** 9 **3.** 210 **4.** 5040 **5.** 35 **6.** 8 **7.** $_{40}P_5 = 78,960,960$ **8.** $_8C_3 = 56$ **9.** $_{20}P_5 = 1,860,480$ **10.** 5040 possible rainbows **11.** 1080 possible combinations of classes **12.** 165 **13.** $j^6 + 6j^5k + 15j^4k^2 + 20j^3k^3 + 15j^2k^4 + 6jk^5 + k^6$ **14.** $32p^5 - 80p^4q + 80p^3q^2 - 40p^2q^3 + 10pq^4 - q^5$ **15.** $p^3 + 1.5p^2q + 0.75pq^2 + 0.125q^3$

Pages 604–605   Chapter 12 Assessment

1. degree **2.** directed graph **3.** permutation; combination **4–6. b, c.** Answers may vary. Examples are given.
4. a. A, 3; B, 3; C, 2; D, 3; E, 3; F, 2 **5. a.** A, 4; B, 3; C, 4; D, 4; E, 3; F, 3; G, 3
b, c.

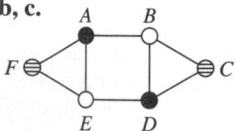

b, c.

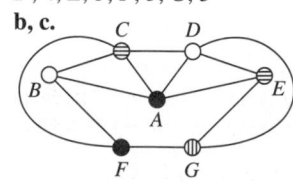

6. a. A, 2; B, 2; C, 4; D, 2; E, 2 **b, c.**

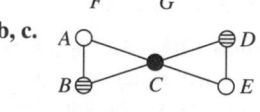

7–10. Answers may vary. Examples are given.
7. In the diagram, H = healer, I = interpreter, M = merchant, S = scribe, A = advisor, K = knight, and P = prince. The arrows are drawn from a person who requires another to a person who is required, although the arrows can be drawn in the opposite direction.

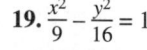

8. The arrows indicate which characters are required by others.
9.

$$
\begin{array}{c}
\text{Character required} \\
\begin{array}{l}
\text{Character} \\
\text{requiring another}
\end{array}
\begin{array}{c}
\text{P A M S K H I} \\
\begin{bmatrix}
0 & 1 & 0 & 0 & 1 & 0 & 0 \\
0 & 0 & 1 & 0 & 0 & 0 & 1 \\
0 & 0 & 0 & 1 & 0 & 0 & 0 \\
0 & 0 & 0 & 0 & 0 & 0 & 0 \\
0 & 0 & 0 & 0 & 0 & 1 & 0 \\
0 & 0 & 0 & 0 & 0 & 0 & 1 \\
0 & 0 & 0 & 0 & 0 & 0 & 0
\end{bmatrix}
\begin{array}{l}
\text{P} \\ \text{A} \\ \text{M} \\ \text{S} \\ \text{K} \\ \text{H} \\ \text{I}
\end{array}
\end{array}
\end{array}
$$

or

$$
\begin{array}{c}
\text{Character} \\
\text{requiring another}
\end{array}
\begin{array}{c}
\text{Character required} \\
\begin{array}{c}
\text{P A M S K H I} \\
\begin{bmatrix}
0 & 0 & 0 & 0 & 0 & 0 & 0 \\
1 & 0 & 0 & 0 & 0 & 0 & 0 \\
0 & 1 & 0 & 0 & 0 & 0 & 0 \\
0 & 0 & 1 & 0 & 0 & 0 & 0 \\
1 & 0 & 0 & 0 & 0 & 0 & 0 \\
0 & 0 & 0 & 0 & 1 & 0 & 0 \\
0 & 1 & 0 & 0 & 0 & 1 & 0
\end{bmatrix}
\begin{array}{l}
\text{P} \\ \text{A} \\ \text{M} \\ \text{S} \\ \text{K} \\ \text{H} \\ \text{I}
\end{array}
\end{array}
\end{array}
$$

10.

$$
\begin{array}{c}
\text{Character required} \\
\begin{array}{l}
\text{Character} \\
\text{requiring another}
\end{array}
\begin{array}{c}
\text{P A M S K H I} \\
\begin{bmatrix}
0 & 0 & 1 & 0 & 0 & 1 & 1 \\
0 & 0 & 0 & 1 & 0 & 0 & 0 \\
0 & 0 & 0 & 0 & 0 & 0 & 0 \\
0 & 0 & 0 & 0 & 0 & 0 & 0 \\
0 & 0 & 0 & 0 & 0 & 0 & 1 \\
0 & 0 & 0 & 0 & 0 & 0 & 0 \\
0 & 0 & 0 & 0 & 0 & 0 & 0
\end{bmatrix}
\begin{array}{l}
\text{P} \\ \text{A} \\ \text{M} \\ \text{S} \\ \text{K} \\ \text{H} \\ \text{I}
\end{array}
\end{array}
\end{array}
$$

or

$$
\begin{array}{c}
\text{Character} \\
\text{requiring another}
\end{array}
\begin{array}{c}
\text{Character required} \\
\begin{array}{c}
\text{P A M S K H I} \\
\begin{bmatrix}
0 & 0 & 0 & 0 & 0 & 0 & 0 \\
0 & 0 & 0 & 0 & 0 & 0 & 0 \\
1 & 0 & 0 & 0 & 0 & 0 & 0 \\
0 & 1 & 0 & 0 & 0 & 0 & 0 \\
0 & 0 & 0 & 0 & 0 & 0 & 0 \\
1 & 0 & 0 & 0 & 0 & 0 & 0 \\
1 & 0 & 0 & 0 & 1 & 0 & 0
\end{bmatrix}
\begin{array}{l}
\text{P} \\ \text{A} \\ \text{M} \\ \text{S} \\ \text{K} \\ \text{H} \\ \text{I}
\end{array}
\end{array}
\end{array}
$$

The elements of the squared matrix tell you which characters are required indirectly through one intermediate character.
11. 840 **12.** 7 **13.** 30 **14.** 360 **15.** 35 **16.** 7 **17.** 15 **18.** 210 **19.** $10^9 - 10^6 - 10^3 + 1 = 998,999,001$ **21.** $_{12}C_6 = 924$ **22.** 28 **23.** $x^3 - 3x^2y + 3xy^2 - y^3$ **24.** $16a^4 + 64a^3b + 96a^2b^2 + 64ab^3 + 16b^4$ **25.** $32y^5 + 80y^4z^2 + 80y^3z^4 + 40y^2z^6 + 10yz^8 + z^{10}$ **26.** $_{13}C_3a^{10}b^3 = 286a^{10}b^3$

CHAPTERS 10–12

Pages 606–607   Cumulative Assessment

1. neither; $t_n = \dfrac{n}{n+1}$; $\dfrac{10}{11}$ **3.** arithmetic; $t_n = (6 - n)\sqrt{2}$; $-4\sqrt{2}$
5. (2) $t_1 = 2$; $t_n = \dfrac{1}{2}t_{n-1}$; (3) $t_1 = 5\sqrt{2}$; $t_n = t_{n-1} - \sqrt{2}$
7. 2, 5, 11, 23 **9.** 9842 **11.** no sum **15.** $\overleftrightarrow{PQ}$ and $\overleftrightarrow{SR}$ both have slope $\dfrac{2}{3}$; $\overleftrightarrow{PS}$ and $\overleftrightarrow{QR}$ both have slope $-\dfrac{3}{2}$. Then $\overleftrightarrow{PQ} \perp \overleftrightarrow{QR}$, $\overleftrightarrow{PQ} \perp \overleftrightarrow{PS}$, $\overleftrightarrow{SR} \perp \overleftrightarrow{PS}$, and $\overleftrightarrow{SR} \perp \overleftrightarrow{QR}$, so $\angle P$, $\angle Q$, $\angle R$, and $\angle S$ are right angles and $PQRS$ is a rectangle.
17. $(x + 4)^2 + y^2 = 8$ **19.** $\dfrac{x^2}{9} - \dfrac{y^2}{16} = 1$

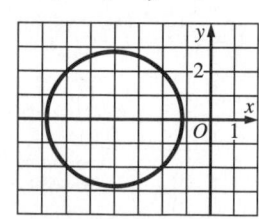

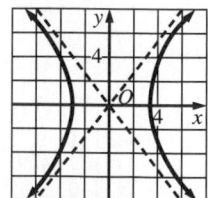

21. $(4, 0)$; 2 **29.** 3024 **31.** 35 **33.** 311,875,200 ways

CHAPTER 13

Page 615 Checking Key Concepts

1. $\frac{8}{250}$, or 0.032 **3.** $\frac{3,618,770}{4\pi(3963)^2}$, or about 0.0183

Pages 615–617 Exercises and Applications

1. $\frac{1}{52}$, or about 0.0192 **3.** $\frac{1}{4}$, or 0.25 **5.** $\frac{1}{2}$, or 0.5

7. $\frac{3}{4}$, or 0.75 **9.** 0 **11.** $\frac{5}{9}$, or about 0.556

19. $a^3 + 3a^2b + 3ab^2 + b^3$ **21.** $16u^4 + 96u^3v + 216u^2v^2 + 216uv^3 + 81v^4$ **23.** $4, \frac{-1 - \sqrt{13}}{2}, \frac{-1 + \sqrt{13}}{2}$

Page 621 Checking Key Concepts

1. $\frac{1}{16}$, or 0.0625 **3.** $\frac{168}{169}$, or about 0.994 **5.** $\frac{7}{16}$, or 0.4375

Pages 621–624 Exercises and Applications

3. 0.12 **5.** 0.42 **7.** 0.46 **11. a.** 0.64 **b.** 0.96 **c.** 0.32
d. 0.04 **13.** B **15.** about 0.000002192 **19.** 801

Page 629 Checking Key Concepts

1. $\frac{4}{9}$, or about 0.444 **3.** $\frac{4}{15}$, or about 0.267

5. $\frac{2}{15}$, or about 0.133

Pages 629–631 Exercises and Applications

1. $\frac{24}{103}$, or about 0.233 **3.** $\frac{4}{15}$, or about 0.267

5. $\frac{2}{9}$, or about 0.222 **7.** $\frac{15}{49}$, or about 0.306

13. $\frac{1}{17}$, or about 0.0588 **15.** $\frac{25}{102}$, or about 0.245

17. $\frac{1}{13}$, or about 0.0769 **19.** about 0.620

21. a. $\frac{1}{3}$, or about 0.333 **b.** $\frac{1}{12}$, or about 0.0833

c. $\frac{11}{12}$, or about 0.917 **23.** $x^3 - 3x^2y + 3xy^2 - y^3$
25. $s^6 + 3s^4t^4 + 3s^2t^8 + t^{12}$

Page 631 Assess Your Progress

1. $\frac{2}{\pi}$, or about 0.637 **2.** $\frac{625}{1296}$, or about 0.482
3. about 0.00786

Page 635 Checking Key Concepts

1. true/false test; On a true/false test, the probability of getting any question right is $\frac{1}{2}$, while on a multiple-choice test, the probability of guessing any question right from among 3 answers is $\frac{1}{3}$, from among 4 answers is $\frac{1}{4}$, and so on.

3. $\frac{15}{1024}$, or about 0.0146 **5.** For each trial, there are two possible mutually exclusive results: correct or incorrect. Each trial is also independent, so the distribution is binomial; $_6C_k\left(\frac{1}{2}\right)^6$, where k is the number correct.

Pages 635–638 Exercises and Applications

3. $_{10}C_k(0.5)^{10}$; about 0.117 **5.** $_7C_k(0.1)^k(0.9)^{7-k}$; 6.3×10^{-6}
7. $_1C_k(0.7)^k(0.3)^{1-k}$; 0.7 **9. a.** about 0.0424 **b.** about 0.141
c. about 0.228 **d.** about 0.589 **15.** $P(k$ successes) and $P(j$ successes) both equal 0, since k and j are both positive, and there is a 0% chance of success (certain failure) on every trial of either experiment. **17.** $P(k$ successes) and $P(j$ successes) are the same (about 0.410). This is because the first experiment asks for the probability of 1 success in 5 trials with a 20% chance of success. You can restate this as the probability of 4 failures in 5 trials with an 80% chance of failure. This is what the second experiment describes, redefining "failure" as "success." **29.** 0.4 **31.** mean: 0.31; standard deviation: about 0.146 **33.**

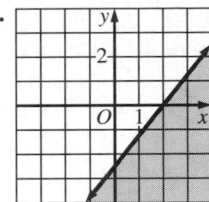

Page 642 Checking Key Concepts

1. a.

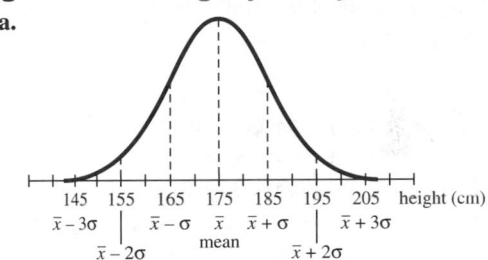

b. 95%; 84% **3.** about 21.2%; about 2.87%

Pages 642–645 Exercises and Applications

1. Answers may vary. Examples are given. **a.** about 0.475; Because about 95% of the distribution lies within two standard deviations of the mean, half of that lies between the mean and two standard deviations below the mean. **b.** about 0.495; Because about 99% of the distribution lies within three standard deviations of the mean, half of that lies between the mean and three standard deviations above the mean. **c.** about 0.976; The probability can be found by subtracting the probability for a z-score of up to -2 from the probability for a z-score of up to 3: $0.9987 - 0.0228 = 0.9759$ **9.** 99.99% **13.** Answers may vary depending upon when rounding is done, and if interpolation is used. Examples are given. **a.** 34% **b.** 54% **c.** 73%
17. about 0.322 **19.** 35 **21.** 171

Page 645 Assess Your Progress

1. about 0.0536; about 0.263 **2.** about 0.766 **3. a.** 15.9%
b. 2.28% **4.** about 0.0446

Pages 650–651 Chapter 13 Assessment

1. independent **2.** standard normal distribution **3.** probability
4. $\frac{1}{52}$, or about 0.0192 **5.** $\frac{1}{13}$, or about 0.0769 **6.** $\frac{1}{4}$, or 0.25
7. $\frac{1}{2}$, or 0.5 **8.** $\frac{3}{26}$, or about 0.115 **9.** 0 **10. a.** $\frac{1}{20}$, or 0.05

b. $\frac{25}{648}$, or about 0.0386 **11.** $\frac{1}{2}$, or 0.5 **12.** $\frac{24}{25\pi}$, or

about 0.306 **13.** $0.5 - \frac{24}{25\pi}$, or about 0.194 **14. a.** 0.01

b. 0.81 **c.** 0.19 **d.** 0.18 **15.** about 6.57×10^{-5}

16. about 0.610 **17. a.** about 0.000244 **b.** about 0.178
c. about 0.132 **d.** about 0.169 **19. a.** about 15.9%
b. about 6.68% **c.** about 22.6%

CHAPTER 14
Page 659   Checking Key Concepts

1. $\sin B = \frac{b}{c}$; $\cos B = \frac{a}{c}$ **3.** $\sin \theta = 0.5346$; $\cos \theta = 0.8451$

Pages 659–662   Exercises and Applications

1. $\sin A = 0.8000$; $\cos A = 0.6000$ **3.** $\sin A = 0.3846$;
$\cos A = 0.9231$ **5.** $\sin B = 0.9756$; $\cos B = 0.2195$ **7.** 14 in.
9. a. 43.5° **b.** 43°32′52″ **11. a.** 73.4° **b.** 73°25′2″
13. a. 26.4° **b.** 26°23′6″ **15. a.** 59.1° **b.** 59°5′34″
17. < **21.** 50.2° **31.** 10.1 **33.** 0.43 **35.** $x = \pm 6\sqrt{2}$

Page 666   Checking Key Concepts

1. a. $\tan 64.8° = \frac{a}{18.4}$ **b.** $\angle B = 25.2°$; $a \approx 39.1$; $c \approx 43.2$

3. They are alternate interior angles and are therefore
congruent.

Pages 667–670   Exercises and Applications

1. a. $\frac{3}{4}$ **b.** $\frac{4}{3}$ **c.** They are reciprocals. **3.** A

5, 7, 9. Side lengths are rounded to the nearest tenth.
5. $\angle B = 71°$; $a \approx 1.2$; $c \approx 3.7$ **7.** $\angle A \approx 52.0°$; $\angle B \approx 38.0°$;
$a \approx 15.4$ **9.** $\angle A \approx 18.4°$; $\angle B \approx 71.6°$; $b = 21$; $c \approx 22.1$
11. about 69 ft **13. a.** $12\sqrt{2} \approx 17$ **b, c.** Methods may vary.
Examples are given. **b.** $\angle DAC \approx 26.6°$; $\angle CAB \approx 63.4°$;
$\triangle DAC$ and $\triangle CAB$ are right triangles with right angles at

D and B. So, $\angle DAC = \tan^{-1}\left(\frac{\sqrt{8}}{\sqrt{32}}\right) \approx 26.6°$ and $\angle CAB =$

$\tan^{-1}\left(\frac{\sqrt{32}}{\sqrt{8}}\right) \approx 63.4°$. **c.** 53.2°; By continuing the process in

part (b), you can find that two of the angles of each acute triangle have measures of 63.4°. The sum of the angles of a triangle is 180°, so the measure of the third angle is about 53.2°.
15. $\angle M$; $\angle L$; No; the tangent is the ratio of the legs of a right
triangle, but $\overline{LM}$ is the hypotenuse. **23. a.** arithmetic; common difference = 90 **b.** geometric; common ratio = –1

Page 670   Assess Your Progress

1. $\frac{12}{13}$ **2. a.** about 188 m **b.** about 282 m

Page 674   Checking Key Concepts

1. a. II **b.** III **c.** I **3.** about 66°, about 246°

Pages 675–677   Exercises and Applications

1. 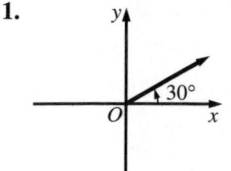 **3.**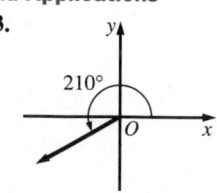

5. 205° **7. a.** 90° **b.** 270° **9.** 0.8824; 0.4706; 1.875
11. –0.6; –0.8; 0.75 **13.** –0.2079; –0.9781; 0.2126
15. 0.9298; –0.3681; –2.526 **19. a.** 60°; 120° **b.** 40°; 320°

c. 68°; 248° **21. a.** 260° **b.** $\cos 260° = \frac{x}{98}$; about –17;

$\sin 260° = \frac{y}{98}$; about –97 **c.** about 17 mi; about 97 mi

29. a. 204 ft² **b.** $3\sqrt{3}$ cm², or about 5.2 cm²

Page 681   Checking Key Concepts

1. about 1.8 square units **3.** about 170 square units
5. about 11.2 square units **7.** about 7.5 square units

Pages 681–683   Exercises and Applications

1. about 11.7 square units **3.** not enough information; need
the measure of included angle **5.** about 580 square units
7. about 32.1 square units **9.** not enough information; need
the measure of included angle **13.** about 25.8 square units
15. about 28.3 square units **17.** about 5 square units
19. about 15.1 square units **27.** 55°, 125° **29.** 26°, 334°
31. Yes; 6. **33.** Yes; 12.

Page 683   Assess Your Progress

1. 0.6428; 0.7660; 0.8391 **2.** –0.7071; –0.7071; 1

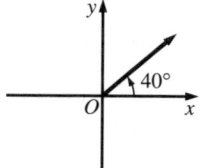 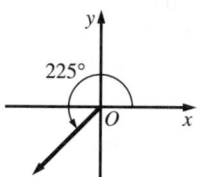

3. –0.7880; 0.6157; –1.280 **4.** 0.9962; –0.0872; –11.43

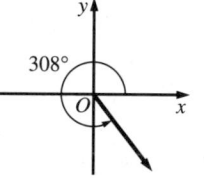 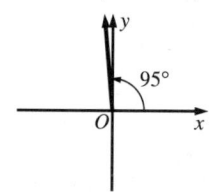

5. 0.3162; 0.9487; 0.3333 **6.** –0.2425; 0.9701; –0.25
7. 0.7894; –0.6139; –1.286 **8.** –0.8480; –0.5300; 1.6
9. about 115 square units **10.** about 0.12 square units
11. about 10,367 square units **12.** about 5.2 square units
13. about 106 square units

Page 688   Checking Key Concepts

1. $\angle A \approx 52.3°$; $\angle C \approx 97.7°$; $c \approx 47.6$ **3.** $\angle B = 60°$;
$a \approx 10.6$; $c \approx 13.0$ **5.** $\angle B = 48°$; $a \approx 18.6$; $c \approx 10.0$
7. $\angle B \approx 77.1°$, $\angle C \approx 43.9°$; $c \approx 17.3$; $\angle B \approx 102.9°$;
$\angle C \approx 18.1°$; $c \approx 7.8$

Pages 688–690   Exercises and Applications

1. $\angle N = 31°$; $m \approx 0.94$; $n \approx 0.54$ **3.** $\angle Q = 29°$; $r \approx 19$;
$p \approx 10$ **5.** $\angle Y = 57°$; $x \approx 21$; $z \approx 14$ **7.** $\angle A = 76°$;
$b \approx 2.2$; $c = 4.6$ **9.** $\angle B \approx 97.3°$; $\angle C \approx 27.7°$; $b \approx 18.2$
11. $\angle A \approx 19.6°$; $\angle C \approx 37.4°$; $c \approx 1.4$ **13.** $\angle B \approx 29.8°$;
$\angle C \approx 94.2°$; $c \approx 24$ **15.** no solution; The side opposite the
obtuse angle is not the longest. **17.** $\angle A \approx 54.7°$; $\angle B \approx 33.3°$;
$b \approx 11.7$ **27.** about 9.2 square units **29.** about 25.6 square
units **31.** $\log_a 64$ **33.** $\log_b \frac{1}{25}$ **35.** $\ln 1 = 0$

Page 694   Checking Key Concepts
1. about 10   3. about 17   5. $\angle J \approx 101.5°$; $\angle K \approx 34.0°$; $\angle L \approx 44.4°$

Pages 695–697   Exercises and Applications
1. about 16   3. about 4.05   5. about 14   7. about 10.8
9. about 15.4   15. $\angle R \approx 59.4°$; $\angle S \approx 52.6°$; $\angle T \approx 68.0°$
17. $\angle X \approx 117.4°$; $\angle Y \approx 57.5°$; $\angle Z \approx 5.1°$ 19. $\angle P \approx 7.3°$; $\angle Q \approx 166.6°$; $\angle R \approx 6.1°$ 21. $\angle J \approx 52.0°$; $\angle K \approx 95.6°$; $\angle L \approx 32.4°$ 25. 38.9°   29. $\angle C = 84°$; $a \approx 5.0$; $c \approx 8.5$
31. $\angle B = 90°$; $\angle C = 60°$; $c \approx 7.36$ 33. $\frac{\sqrt{2}}{2} \approx 0.7071$; $-\frac{\sqrt{2}}{2} \approx -0.7071$ 35. 1; 0   37. -0.5; $\frac{\sqrt{3}}{2} \approx 0.8660$
39. $-\frac{\sqrt{2}}{2} \approx -0.7071$; $-\frac{\sqrt{2}}{2} \approx -0.7071$

Page 697   Assess Your Progress
1. $\angle X = 39°$; $x \approx 9.1$; $z \approx 13.9$ 2. $\angle Y \approx 55.3°$; $\angle Z \approx 39.7°$; $x \approx 2.73$ 3. $\angle Y \approx 8.0°$; $\angle Z \approx 22.0°$; $y \approx 2$ 4. $\angle X \approx 51.9°$, $\angle Z \approx 71.1°$, $x \approx 9.15$; $\angle X \approx 14.1°$, $\angle Z \approx 108.9°$, $x \approx 2.84$
5. $\angle X \approx 39.3°$, $\angle Y \approx 27.3°$; $\angle Z \approx 113.4°$ 6. no solution; The side opposite the obtuse angle is not the longest side.
7. $\angle X \approx 109°$; $x \approx 19.0$; $z \approx 9.4$ 8. $\angle X = 144.2°$, $\angle Y = 25.2°$; $\angle Z = 10.5°$ 9. no solution; $y < x \sin y$, so no triangle can be formed.   10. $\angle X \approx 16.8°$, $\angle Z \approx 138.2°$, $z \approx 30$
11. $\angle Y \approx 40.8°$; $\angle Z \approx 15.2°$; $x \approx 15.8$ 12. $\angle Y = 60°$; $\angle Z = 90°$; $y = \frac{3\sqrt{3}}{8} \approx 0.65$ 13. $\angle Y \approx 29.3°$, $\angle Z \approx 78.7°$, $y \approx 3.3$; $\angle Y \approx 6.7°$, $\angle Z \approx 101.3°$, $y \approx 0.79$

Pages 702–703   Chapter 14 Assessment
1. tangent   2. solve a triangle   3. $\frac{8}{17}$ 4. $\frac{15}{17}$ 5. $\frac{15}{8}$ 6. 30°
7. about 262 ft   8. a. 200°   b. $\cos 200° = \frac{x}{100}$; $x \approx -94$; $\sin 200° = \frac{y}{100}$; $y \approx -34$ c. about 94 mi west; about 34 mi south   9. -0.7809; -0.6247; 1.25   10. 76.4°, 283.6°
11. 10 square units   12. about 69 square units
13. about 597 square units   14. about 742 in.²
15. about 218 square units   16. about 117 square units
17. about 201 square units   18. $\angle B \approx 46°18'$; $\angle C \approx 18°22'$; $c \approx 14$ 19. no solution; The side opposite the obtuse angle is not the longest side.   20. $\angle A = 45°$; $\angle B = 90°$; $c = 6$
21. $\angle A \approx 133.8°$, $\angle C \approx 31.2°$, $a \approx 70$; $\angle A \approx 16.2°$, $\angle C \approx 148.8°$, $a \approx 27$ 22. $\angle A \approx 24.7°$; $\angle B \approx 35.3°$; $b \approx 75$ 23. $\angle A \approx 60.4°$; $\angle B \approx 46.5°$; $\angle C \approx 73.1°$

CHAPTER 15
Page 710   Checking Key Concepts
1. 56°, 124°   3. 77°, 283°   5. 225°, 315°   7. about 43 ft
9. 24 ft

Pages 711–712   Exercises and Applications
1. 270°   3. 0°, 180°, 360°   5. 30°, 150°   7. 27°, 153°
9. 210°, 330°   11. 109°, 251°   13. 0.5   15. 0.2588   17. -1
19. 0.1736   27. 3.5   29. 3.5   31. 2   33. 2

35.
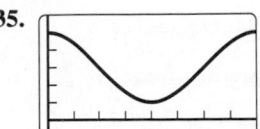
a. 5   b. 1   c. maximum: 0°, 360°; minimum: 180°
39. m $\angle A = 50°$; $AB \approx 3$; $AC \approx 9$ 41. a. 2π in.
b. $\frac{3\pi}{2}$ in.   43. 228 in.²

Page 716   Checking Key Concepts
1. 450°   3. 221°   5. -3.49 7. 0.52   9. -1 11. -0.5

Pages 716–718   Exercises and Applications
1. 57°   3. 225°   5. 540°   7. 1260°   9. 337°   11. $-90°$
13. a. The first pair of equations graphs the unit circle counterclockwise from (1, 0). As each point is plotted, the second set of equations plots the sine value of the corresponding angle.   b. The first pair of equations again plots the unit circle counterclockwise from (1, 0). As each point is plotted on the unit circle, the second pair of equations plots the cosine function of the corresponding angle.   15. -3.14; $-\pi$
17. 12.57; 4π 19. -1.13; $-\frac{13\pi}{36}$ 21. 6.81; $\frac{13\pi}{6}$ 23. 1.52; $\frac{29\pi}{60}$
25. 7.07; $\frac{9\pi}{4}$ 27. -1 29. 0   31. 0   33. 0.6822   35. 0.5872
37. 0.9093   47. 31°; 149°   49. 24°; 156°   51. $\frac{1}{10}$

Page 723   Checking Key Concepts
1. a.
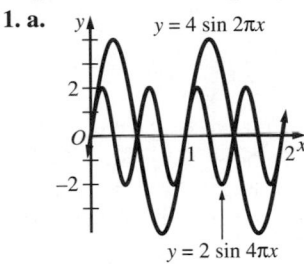
b. $y = 2 \sin 4\pi x$ has amplitude 2 and period 0.5; $y = 4 \sin 2\pi x$ has amplitude 4 and period 1.

Pages 723–726   Exercises and Applications
1. amplitude: 2; period: 1   3. amplitude: $\frac{1}{2}$; period: $\frac{2\pi}{3}$
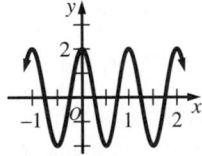
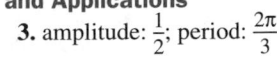
5. amplitude: 1; period: 4   7. amplitude: 4; period: $\frac{\pi}{2}$
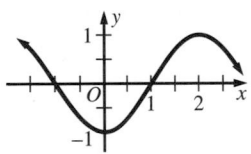
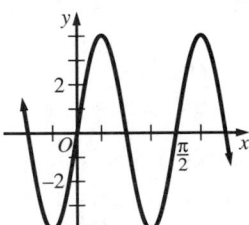
9. amplitude: $\frac{1}{3}$; period: 8π
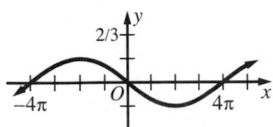

11. amplitude: 2; period: $\frac{4\pi}{5}$

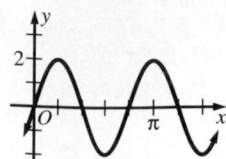

13. $y = 3 \sin 4x$

15. $y = -10 \cos \frac{1}{2}x$

17. $y = 2 \sin \frac{2}{3}x + 5$

19. $y = 3 \sin 4x - 7$ **21.** A

23. a. 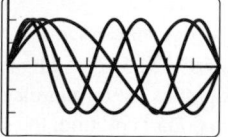 1 cycle; 1.5 cycles; 2 cycles; 2.5 cycles   **b.** 1.25 cycles; 1.75 cycles; 2.25 cycles
c. $y = \sin bx$: $|b|$ cycles; $y = \cos bx$: $|b|$ cycles

33. B   **45.**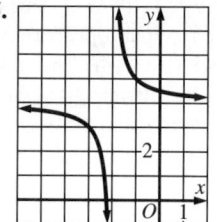
39. $-540°$
41. $5 + 14i$
43. 2

Page 726   Assess Your Progress

1. $49°, 131°$ **2.** $35°, 325°$ **3.** $328°, 212°$ **4.** -0.5748
5. 0.0707 **6.** 0.7071

7.   **8.**

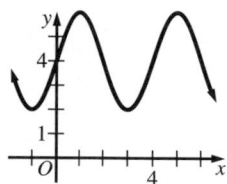   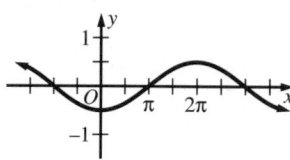

amplitude: 2; period: 4   amplitude: $\frac{1}{2}$; period: 4π

Page 730   Checking Key Concepts

1.

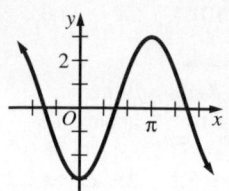

The graph of $y = \sin\left(x - \frac{\pi}{2}\right)$ is the graph of $y = \sin x$ shifted $\frac{\pi}{2}$ to the right, while the graph of $y = \sin\left(x + \frac{\pi}{2}\right)$ is the graph of $y = \sin x$ shifted $\frac{\pi}{2}$ to the left.

3.

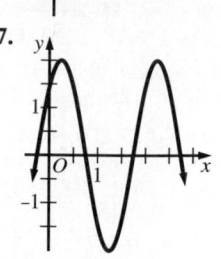

Function	Period	Range
$y = 2 \sin 2x$	π	$-2 \le y \le 2$
$y = \sin(2x - \pi)$	π	$-1 \le y \le 1$
sum	π	$-1 \le y \le 1$

Pages 731–734   Exercises and Applications

1.

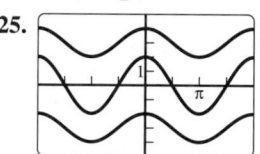

3.

7.

5.

9, 11, 13, 15. Answers may vary. Examples are given.

9. $y = \sin \pi(x + 0.3)$ **11.** $y = 3 \sin \frac{\pi}{4}(x - 3)$

13. $y = 10 \cos \frac{\pi}{12}(x - 5)$ **15.** $y = 5 \cos 2\left(x + \frac{\pi}{3}\right)$

17. a. 6000 rev/min = 100 rev/s, and since the period is the reciprocal of the frequency, the period is $\frac{1}{100}$ s.

b. $y = 3.75 \cos 200\pi x$ **c.** 2: $y = 3.75 \cos 200\pi\left(x - \frac{1}{300}\right)$;

3: $y = 3.75 \cos 200\pi\left(x - \frac{1}{150}\right)$ **23.** Answers may vary. An

example is given. $y = \cos\left(x - \frac{\pi}{2}\right)$; $y = \cos\left(x + \frac{3\pi}{2}\right)$;

$y = \cos\left(x - \frac{5\pi}{2}\right)$

25.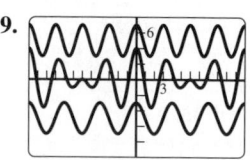

Function	Period	Range
$y = 3 + \cos x$	2π	$2 \le y \le 4$
$y = -3 + \cos x$	2π	$-4 \le y \le -2$
sum	2π	$-2 \le y \le 2$

27.

Function	Period	Range
$y = 2 + \sin \pi x$	2	$1 \le y \le 3$
$y = 4 + \sin \frac{\pi}{2}x$	4	$3 \le y \le 5$
sum	4	$4.24 \le y \le 7.76$

29.

Function	Period	Range
$y = 5 + 2 \cos \frac{2\pi}{3}x$	3	$3 \le y \le 7$
$y = -5 + 2 \cos \frac{\pi}{2}x$	4	$-7 \le y \le -3$
sum	12	$-3.61 \le y \le 4$

33. 5 **35.** 13 **37.** 5.39 **41.** Answers may vary. An example is given. $y = 3 \cos \frac{\pi}{3}x + 5$ **43.** $x = 2$; $x = 5$
45. $x = -6$; $x = 6$ **47.** 1.4

Page 737 Checking Key Concepts

1. 0.7813 **5.** 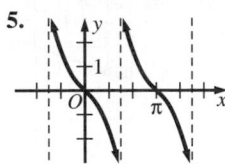 **7.**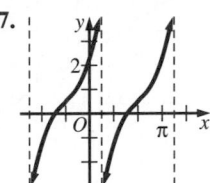
3. -1

Pages 738–739 Exercises and Applications

1. 0
3. -0.5774
5. 2.4142
7. -0.3640
9. 0.1820
11. 0

19. 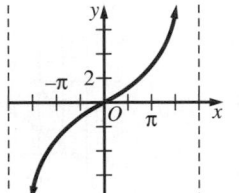 **21.**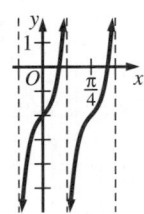

23. Answers may vary. An example is given.
$y = 2 \cos \left(x - \frac{\pi}{2}\right)$ **25.** 0 **27.** -1

Page 739 Assess Your Progress

1. 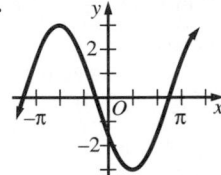 **2.**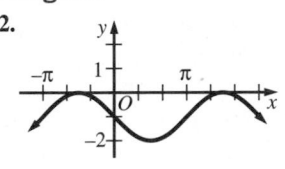

3. a. $s = c \tan \theta$ **b.** $s = 29{,}800 \frac{\text{m}}{\text{s}}$

4. 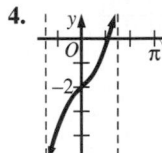 **5.** 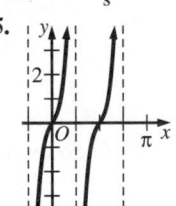 **6.**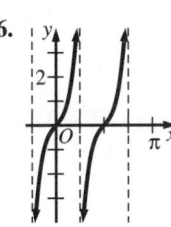

Pages 744–745 Chapter 15 Assessment

1. radians **2.** amplitude, period, phase shift **3.** sine, cosine
4. a. 39 m **b.** 55 m **c.** 30 m **5.** 47°, 133° **6.** 137°, 223°
7. 34°, 326° **8.** 0 **9.** 0 **10.** -1 **11.** -1 **12.** 0 **13.** 1
15. a. amplitude: 316,200; period: $\frac{1}{22{,}000}$ **16, 17.** Answers
may vary. Examples are given. **16.** $y = 7 \cos 2x$
17. $y = 5 \cos \pi x - 1$

20.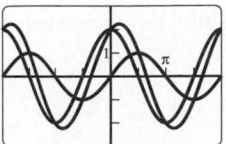

Function	Period	Range
$y = \sin x$	2π	$-1 \le y \le 1$
$y = 2 \cos x$	2π	$-2 \le y \le 2$
sum	2π	$-2.24 \le y \le 2.24$

21.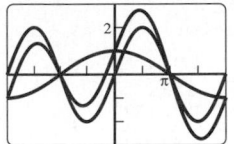

Function	Period	Range
$y = \cos \frac{1}{2}x$	4π	$-1 \le y \le 1$
$y = 2 \sin x$	2π	$-2 \le y \le 2$
sum	4π	$-2.74 \le y \le 2.74$

22.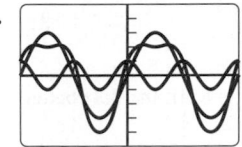

Function	Period	Range
$y = 3 \sin \pi x$	2	$-3 \le y \le 3$
$y = \cos 2\pi x$	1	$-1 \le y \le 1$
sum	2	$-4 \le y \le 2.13$

23. 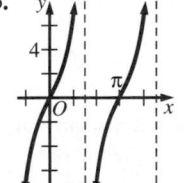 **24.** **25.**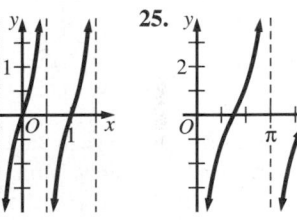

27. a. By the diagram, $\tan \theta = \frac{l}{h}$, and thus $l = h \tan \theta$.
b. 0, 10.4 in., 31.2 in., and (when $\theta = 90°$) undefined

CHAPTERS 13–15

Pages 746–747 Cumulative Assessment

3. about 0.0588 **5.** 0.735 **7.** about 0.145
9. a. about 0.1587 **b.** about 0.995 **c.** about 0.4592
11. about 291 ft **13.** -0.7071; -0.7071; 1 **15. a.** 170°
b. 49.2 km; 8.68 km **17.** $\angle C = 83°$; $b \approx 7.18$; $c \approx 10.1$
19. 1 **21.** about 0.8660 **23.** 3; 1; maximum: $\theta = 135°$ and
$\theta = 315°$; minimum: $\theta = 45°$ and $\theta = 225°$ **25.** about -1.47
or $-\frac{7\pi}{15}$

27.

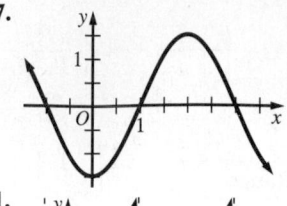

amplitude: $\frac{3}{2}$; period: 4

31.

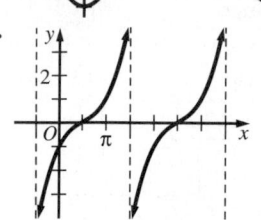

EXTRA PRACTICE

Pages 749–750   Chapter 1

1, 3, 5. Answers may vary. Examples are given.
1. 3.35 million   **3.** 66.0 million; 69.3 million
5. The constant growth model provides a more accurate estimate, because the lower estimates provided by this model seem to reflect the fact that sales did not increase as much from 1985–1990 as from 1980–1985.
7.

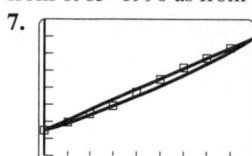

Answers may vary. An example is given. The linear model appears to fit the data better.

9. $\begin{bmatrix} 6 & -6 & 9 \\ 2 & 5 & -3 \end{bmatrix}$ **11.** $\begin{bmatrix} 12 & 0 & -3 \\ 6 & -9 & 3 \end{bmatrix}$ **13.** $\begin{bmatrix} 0 & 12 & -21 \\ 2 & -19 & 9 \end{bmatrix}$

15. $\begin{bmatrix} -10 & -18 & 34 \\ -8 & 36 & -16 \end{bmatrix}$ **17.** $\begin{bmatrix} 11 & 42 & -4 \\ -1 & -12 & 0 \\ 17 & 24 & -8 \end{bmatrix}$ **19.** $\begin{bmatrix} -1 & 0 & -8 \\ -57 & 1 & 46 \end{bmatrix}$

21. $\begin{bmatrix} 9 & 10 \\ -1 & -4 \\ 13 & 0 \end{bmatrix}$ **23.** $\begin{bmatrix} 11 & -3 & -6 \\ 8 & 51 & -2 \end{bmatrix}$ **25, 27.** Answers may

vary. Examples are given.   **25.** Draw cards at random from a standard deck of 52 cards, letting each suit represent one of the four answer choices, and choosing a particular suit to represent a right answer.   **27.** Generate a random number from 1–5 on a calculator by using the formula "int(5*rand)+1." Let "1" and "2" represent getting on base.   **29.** int(6*rand)+1
31. int(5*rand)+2

Pages 750–752   Chapter 2

1. Yes.   **3.** No.   **5.** Yes.   **7.** No.

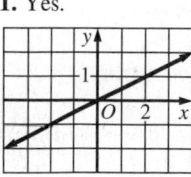   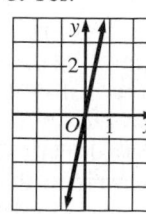

9. $y = 1.5x$; 13.5   **11.** $y = 10.3x$; 329.6

13. $y = 3x + 8$ **15.** $y = -\frac{3}{2}x - \frac{5}{2}$ **17.** $y = \frac{2}{3}x - \frac{1}{3}$
19.

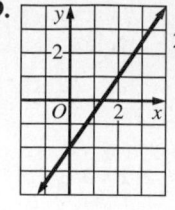   **21.**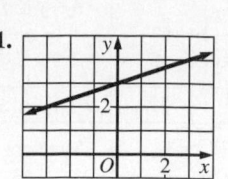

23.

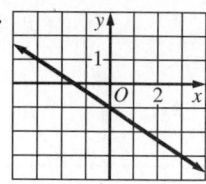

25. a. $y = 2 + \frac{1}{3}(x + 3)$
b.

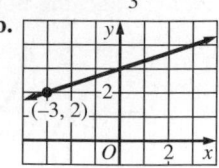

c. $y = \frac{1}{3}x + 3$

27. a. $y = -2 + 2(x - 6)$ **29. a.** $y = 2 + \frac{2}{3}(x - 9)$
b.

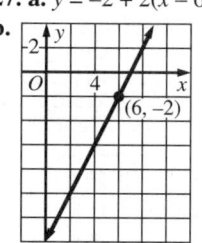   **b.**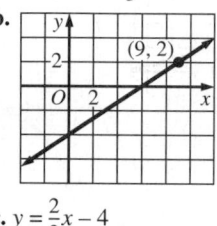

c. $y = 2x - 14$ **c.** $y = \frac{2}{3}x - 4$
31. $y = 5 - \frac{2}{5}(x + 1)$ or $y = 3 - \frac{2}{5}(x - 4)$ **33.** $y = 6 + 2(x - 1)$ or
$y = -2 + 2(x + 3)$ **35.** $y = -4 - \frac{9}{5}(x + 1)$ or $y = 5 - \frac{9}{5}(x + 6)$
37, 39. Answers may vary. Examples are given.
37. $y = 2.7x + 31$ **39.** about 80 million   **41.** about zero
43.   **45.**

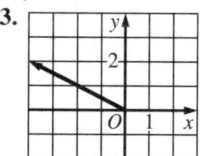   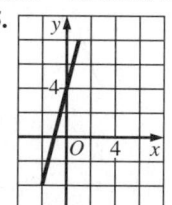

Pages 752–753   Chapter 3

1. 2^5 **3.** 2^8 **5.** 640   **7.** 8   **9.** exponential   **11.** linear
13. $\frac{1}{64}$ **15.** $\frac{1}{243}$ **17.** 5   **19.** 64   **21.** $\frac{1}{5}$ **23.** 7
25. a. 7   **27. a.** 6
b. exponential growth   **b.** exponential decay
c.   **c.**

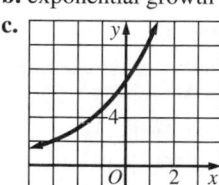   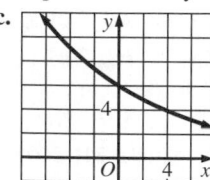

29. a. 2
b. exponential decay
c.

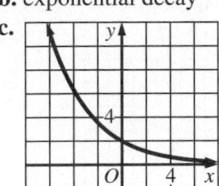

33. $1648.72   **35.** $1529.59

37. a.

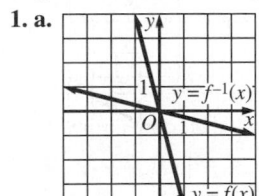

b. about 193,000 mi
c. in 1893, shortly before the turn of 1894

39. $y = 4\left(\frac{3}{2}\right)^x$

41. $y = 1.6\left(\frac{5}{2}\right)^x$ **43.** $y = 186{,}810(1.03101)^x$

45. Predictions should range from 633,746 to 636,540.

Pages 753–754   Chapter 4

1. a.

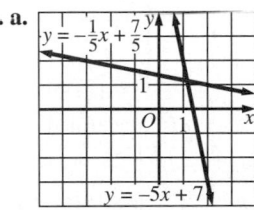

b. $f^{-1}(x) = -\frac{1}{4}x$

3. a.

b. $y = -\frac{1}{5}x + \frac{7}{5}$

5. a.

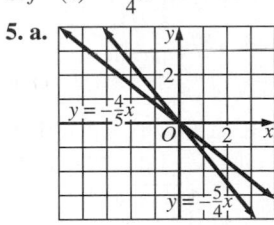

b. $y = -\frac{4}{5}x$

7. a.

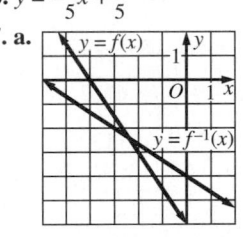

b. $f^{-1}(x) = -\frac{2}{3}x - 4$

9. $\log_5 125 = 3$ **11.** $\log_{2/3} \frac{4}{9} = 2$ **13.** $\left(\frac{1}{2}\right)^{-3} = 8$

15. $(0.36)^{1/2} = 0.6$ **17.** 7   **19.** 0   **21.** $3\log_5 x - 2\log_5 y$

23. $\frac{1}{2}\log_5 x + \log_5 y$ **25.** $\log_2 \frac{x}{y}$ **27.** $\log_b 45$ **29.** $\ln \frac{1}{2}$

31. 0.36   **33.** 0.86   **35.** 1.64   **37.** 0.28   **39.** 1.89   **41.** 2.07

43. −0.16   **45.** −0.79   **47.** $y = 251(63.1)^x$

49. $y = 40.4(0.549)^x$ **51.** $y = 0.00398x^{0.76}$ **53.** $y = 2.5x^{5.8}$

Pages 754–756   Chapter 5

1. C   **3.** A   **5.** ±3   **7.** $\pm 5\sqrt{2} \approx \pm 7.1$ **9.** $\pm\sqrt{15} \approx \pm 3.9$
11. $\pm\sqrt{30} \approx \pm 5.5$ **13.** −4.5   **15.** minimum value; −2
17. minimum value; −5   **19.** minimum value; −10

21. a. $x = -3$ **b.** $(-3, 0)$
c. opens down
d.

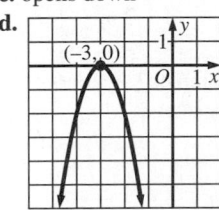

23. a. $x = -1$ **b.** $(-1, 3)$
c. opens down
d.

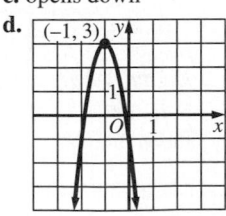

25. a. $x = -4$ **b.** $(-4, -2)$
c. opens down
d.

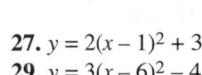

27. $y = 2(x - 1)^2 + 3$
29. $y = 3(x - 6)^2 - 4$

31. a. 2, 5   **b.** (3.5, 4.5)
c.

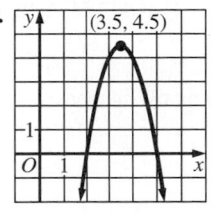

33. a. −3, −7   **b.** (−5, 6)
c.

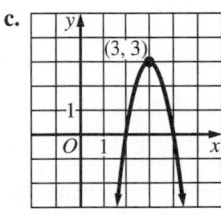

35. a. 2, 4   **b.** (3, 3)
c.

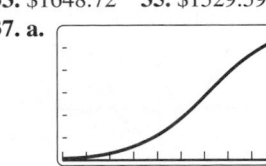

37. $y = -2(x + 5)(x - 3)$
39. $y = (x - 3)^2 - 1$
41. $y = (x - 4)^2 - 16$
43. $y = 2(x - 2)^2 - 3$
45. minimum value; 2.5
47. maximum value; −1
49. minimum value; −12

51. −2, 6   **53.** $\frac{9 \pm \sqrt{41}}{10} \approx 0.26,\ 1.54$

55. $\frac{-1 \pm \sqrt{41}}{4} \approx -1.85,\ 1.35$ **57.** $2 \pm \sqrt{3} \approx 0.27,\ 3.73$

59. $-6 \pm 2\sqrt{7} \approx -11.29,\ -0.71$ **61.** two solutions   **63.** one
solution   **65.** one solution   **67.** $y = (2x - 5)(x - 2)$
69. $y = (5x - 7)(5x + 7)$ **71.** $y = (5x - 3)(x + 4)$ **73.** $y =$
$(3x - 2)(x + 6)$ **75.** 5, 7   **77.** $1, \frac{3}{2}$ **79.** $0, \frac{4}{5}$ **81.** 5   **83.** 2

Pages 757–758   Chapter 6

1, 3. Descriptions may vary. Examples are given.   **1.** numerical; A batting average can vary from 0.0 to 1.0, with values around 0.2 to 0.25 being common.   **3.** categorical; Some of the modes are train, car, bus, bicycle, and taxi.   **5.** Answers may vary. An example is given. the circle graph; The circle graph allows you to see more easily how each choice is related to the results as a whole.   **7, 9.** Answers may vary. Examples are given.   **7.** This is a leading question, since most people would like lower taxes and public recreation facilities. The question could be ended after "Do you think this city should re-elect Mayor Fong?"   **9.** Respondents might inflate their answers to look good in the eyes of the questioner. The question would be best if it could be answered anonymously.
11, 13. Answers may vary. Examples are given.
11. convenience sample; biased; The students in a few specific homerooms might not represent students as a whole.
13. self-selected sample; biased; Students who volunteer might spend more time doing homework than others.
15. median: $29,500

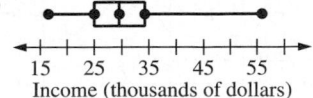

Income (thousands of dollars)

17. Answers may vary. An example is given. The median is a better measure of the center because the two highest incomes skew the mean toward the right, which makes it less representative of most of the incomes. **19.** 1st set: about 1.3; 2nd set: about 1.3; 3rd set: about 2.0; The standard deviation of the 3rd set is much greater. **21. a.** $\pm 2\%$ **b.** 16–20%
23. about 46.3%

Pages 758–760   Chapter 7

1. (3, 7)   **3.** (2, –5)   **5.** about (–3.54, 7.54) and (2.54, 1.46)
7. about (1.65, –2.31) and (–6.65, 14.31)
9. about (–1.58, 1.67) and (1.48, 13.90)
11. (–1, 4); one solution   **13.** infinitely many solutions
15. (6, 5); one solution

17. $\begin{bmatrix} 11 & -8 \\ 15 & -14 \end{bmatrix}\begin{bmatrix} x \\ y \end{bmatrix} = \begin{bmatrix} 48 \\ 33 \end{bmatrix}$; (12, 10.5)

19. $\begin{bmatrix} 5 & 13 \\ 9 & -4 \end{bmatrix}\begin{bmatrix} x \\ y \end{bmatrix} = \begin{bmatrix} 40.8 \\ 2.2 \end{bmatrix}$; (1.4, 2.6)

21. $\begin{bmatrix} 9 & 1 \\ 1.6 & 1 \end{bmatrix}\begin{bmatrix} x \\ y \end{bmatrix} = \begin{bmatrix} -13.5 \\ 0 \end{bmatrix}$; about (–1.82, 2.92)

23. about (0.30, 2.49, 3.01)   **25.** about (1.20, 0.58, –0.82)

27. 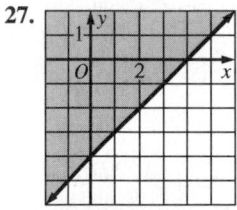   **29.**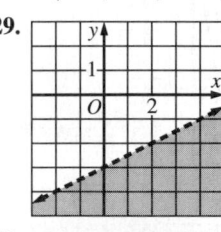

31. 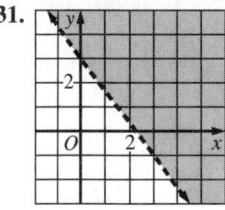   **33.**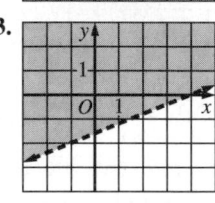

35. $y \geq 2x - 1$
37. 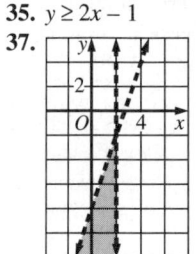   **39.**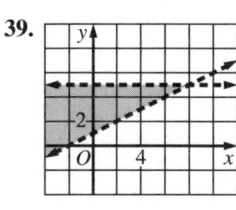

41. $x \geq -2$ **43.** $y < 2x$
$\quad\ y > 2x - 1$ $\quad\ y < -3x + 5$
$\qquad\qquad\qquad\qquad\ y > \frac{1}{3}x - \frac{5}{3}$

45. 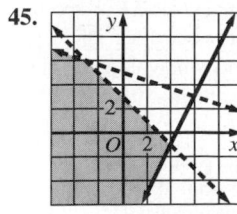   **47.**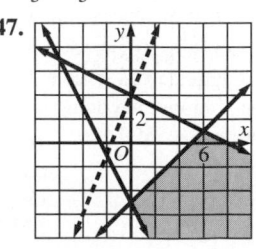

Pages 760–762   Chapter 8

1. domain: $x \geq -2$;   **3.** domain: $x \geq 0$;
range: $y \geq 0$ range: $y \geq -2$

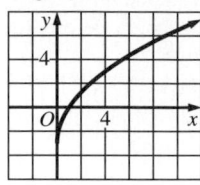   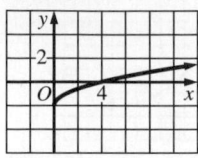

5. domain: $x \geq 0$;
range: $y \geq -3$ **7.** sometimes true

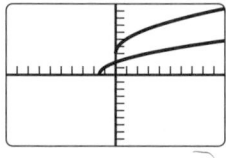   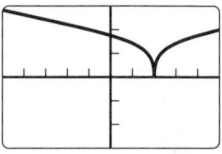

9. never true   **11.** sometimes true

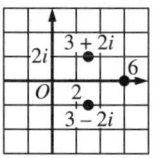   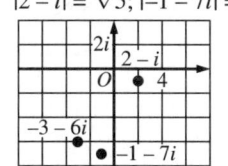

13. –1   **15.** undefined   **17.** –2   **19.** $\frac{3}{2}$ **21.** domain: $x \geq 5$;
range: $y \geq 0$ **23.** domain: $x \geq 0$; range: $y \geq -3$ **25.** $3\sqrt{2}$

27. $2\sqrt[4]{3}$ **29.** $x^4\sqrt{5}$ **31.** $3x^8$ **33.** 81   **35.** –21   **37.** 160
39. 103   **41.** $-\frac{4}{3}$ **43.** 72   **45.** 7   **47.** 8   **49.** –2, 3
51. $5 - 2i$ **53.** $-14 - 7i$ **55.** $45 - 28i$ **57.** $\frac{13}{10} - \frac{11}{10}i$
59. $\frac{9}{13} + \frac{32}{13}i$ **61.** $2 \pm i$ **63.** $1 \pm 3i$

65. $|3 + 2i| = \sqrt{13}$;   **67.** $|-3 - 6i| = 3\sqrt{5}$;
$|3 - 2i| = \sqrt{13}$; $|6| = 6$ $|2 - i| = \sqrt{5}$; $|-1 - 7i| = 5\sqrt{2}$

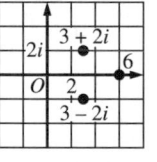   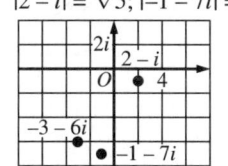

69. $|4 + 2i| = 2\sqrt{5}$;   **71.** $|-2 + 3i| = \sqrt{13}$;
$|-3 + 3i| = 3\sqrt{2}$; $|1 + i| = \sqrt{2}$;
$|1 + 5i| = \sqrt{26}$ $|-5 + i| = \sqrt{26}$

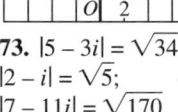   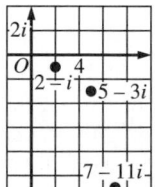

73. $|5 - 3i| = \sqrt{34}$;
$|2 - i| = \sqrt{5}$;
$|7 - 11i| = \sqrt{170}$

75. $-3 - 2i$ **77.** $3 + 2i$ **79.** No; for example, $3 \times 2 = 6$, but 6 is not in the set $\{1, 2, 3\}$.   **81.** any integer
83. any rational number

Pages 762–764 Chapter 9

1. 55 **3.** 12 **5.** $3x^3 - 3x + 7$ **7.** $7x^4 - 5x^3 + 2x^2 - 3$
9. $10x^2 - 29x - 21$ **11.** $5y^3 - 11y^2 + 6y - 8$
13. $a^3 + 2a^2 - a - 2$ **15.** $2x - 1 - \dfrac{5}{x-2}$

17. $4c^2 + 9c + 28 + \dfrac{83}{c-3}$ **19.** $b^2 - 2b - 5 - \dfrac{21}{2b-5}$

21.

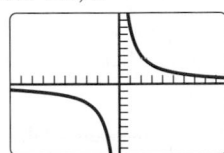

a. As $x \to -\infty, f(x) \to -\infty$.
As $x \to +\infty, f(x) \to -\infty$.
b. local (and global)
maximum: (2, 3)

23. –2, 1, 5 **25.** –1, 2, 3 **27.** $y = 2x(x+1)(x-3)$ **29.** $y = \dfrac{3}{2}(x+3)(x+1)(x-3)$ **31.** –0.71, 2.14, 4.57 **33. a.** $\pm1, \pm3,$

$\pm\dfrac{1}{2}, \pm\dfrac{3}{2}, \pm\dfrac{1}{5}, \pm\dfrac{3}{5}, \pm\dfrac{1}{10}, \pm\dfrac{3}{10}$ **b.** $-\dfrac{1}{5}, \dfrac{3}{2}, 1$ (double root)

35. Yes; 8.

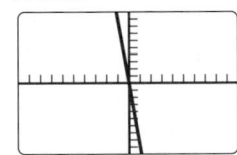

37. No.

39. No; let n = the number of pounds in the bag, c = the total cost of the bag, and k = the fixed price per pound. Then $nk = c$, or $\dfrac{c}{n} = k$, which is a direct variation equation.

41. $y = \dfrac{6}{x+3} - 5$ **43. a.** $x = 3$ and $y = 2$ **b.** The graph is the graph of $y = \dfrac{10}{x}$ translated 3 units to the right and 2 units up.

45. a. $x = 5$ and $y = 3$ **b.** The graph is the graph of $y = \dfrac{9}{x}$ translated 5 units to the right and 3 units up. **47.** degree $p(x) >$ degree $q(x)$; degree $q(x) \geq 3$; roots of $q(x)$: –2, 2, 4
49. a. $x = 4$ and $x = -1$ **b.** As $x \to \pm\infty, f(x) \to 0$.
51. a. $x = 0$ and $x = 1$ **b.** As $x \to \pm\infty, g(x) \to 0$. **53. a.** $x = \dfrac{3}{2}$
b. As $x \to \pm\infty, h(x) \to 0$. **55.** $\dfrac{3}{2}$ **57.** 4, 5 **59.** –2

Pages 764–765 Chapter 10

1. 32, –64, 128, –256 **3.** 100,001, 1,000,001, 10,000,001, 100,000,001 **5.** 0.005, 0.0005, 0.00005, 0.000005
7. $\dfrac{1}{25}, \dfrac{1}{36}, \dfrac{1}{49}, \dfrac{1}{64}$ **9.** $9999\dfrac{1}{9}, 99,999\dfrac{1}{9}, 999,999\dfrac{1}{9}, 9,999,999\dfrac{1}{9}$
11, 13, 15. Answers may vary. Examples are given. **11.** $t_n = -4n + 21$ **13.** $t_n = \dfrac{1}{2n}$ **15.** $t_n = \dfrac{1}{5}(-2)^{n+1}$ **17.** geometric
19. arithmetic **21.** neither **23.** geometric **25. a.** 62.5
b. 22 **27. a.** 42.5 **b.** 20 **29. a.** 0.25 **b.** 0.15 **31. a.** $21\dfrac{1}{4}$
b. 9 **33.** $t_1 = 3; t_n = 5(t_{n-1})$ **35.** $t_1 = \dfrac{5}{2}; t_n = \dfrac{1}{2}(t_{n-1})$
37. $t_1 = \dfrac{4}{3}; t_n = \dfrac{1}{2}(t_{n-1})$ **39. a.** 0.2, –1, 5, –25

b. $t_n = 0.2(-5)^{n-1}$ **41. a.** 32, –16, 8, –4 **b.** $t_n = 32\left(-\dfrac{1}{2}\right)^{n-1}$
43. 704 **45.** 172.5 **47.** $4 + 5 + 6 + 7 + 8 + 9 + 10$
49. $-5 - 75 - 265 - 635 - 1245 - 2155$

51, 53. Answers may vary. Examples are given.
51. $\displaystyle\sum_{n=1}^{9} (25n - 112) = 117$ **53.** $\displaystyle\sum_{n=1}^{16}\left(2n - \dfrac{5}{3}\right) = 245\dfrac{1}{3}$

55. $127\dfrac{3}{4}$ **57.** 208.336 **59.** –4.5 **61.** 12 **63.** –3

Pages 766–767 Chapter 11

1. a. $6\sqrt{2}$ **b.** $y = -x + 11$
3. a. $6\sqrt{10}$ **b.** $y = \dfrac{1}{3}x + \dfrac{8}{3}$

5. 2 or –8
7. 1.5 or 0.1

9. a. $x - 2 = -(y + 7)^2$
b.

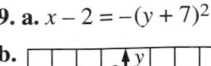
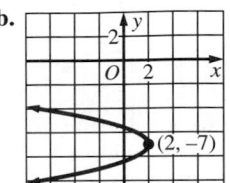

11. vertex: (0, 0); focus: $\left(0, -\dfrac{1}{8}\right)$;
directrix: $y = \dfrac{1}{8}$

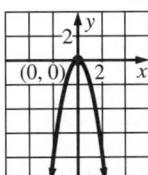

13. vertex: (–4, 3);
focus: (–4, 0);
directrix: $y = 6$

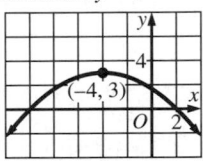

15. center: (–3, 5); radius: 4

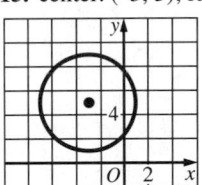

17. center: (3, 0);
radius: $\sqrt{10}$

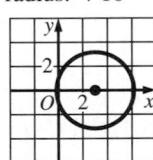

19. $(x + 1)^2 + (y - 8)^2 = 7$ **21.** $(x + \sqrt{5})^2 + (y + 1)^2 = 36$
23. $(x - p)^2 + (y - q)^2 = p^2$
25. center: (0, 0); vertices: (±7, 0);
foci: (±2√6, 0); horizontal

27. $\dfrac{x^2}{36} + \dfrac{y^2}{49} = 1$

29. $\dfrac{(x-3)^2}{16} + \dfrac{(y+2)^2}{64} = 1$

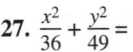
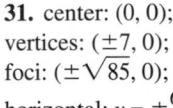

31. center: (0, 0);
vertices: (±7, 0);
foci: (±√85, 0);
horizontal; $y = \pm\dfrac{6}{7}x$

33. center: (–3, 2);
vertices: (–3, 5), (–3, –1);
foci: (–3, 2 ± √58); vertical;
$y = \pm\dfrac{3}{7}(x + 3) + 2$

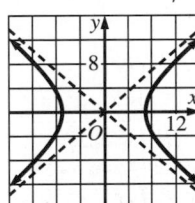
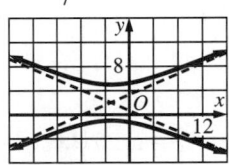

35. center: $(-2, 5)$; vertices: $(1, 5)$, $(-5, 5)$; foci: $(-2 \pm \sqrt{34}, 5)$; horizontal; $y = \pm\frac{5}{3}(x + 2) + 5$

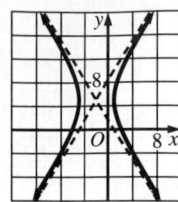

37. $\frac{(x-1)^2}{36} + \frac{(y-3)^2}{45} = 1$

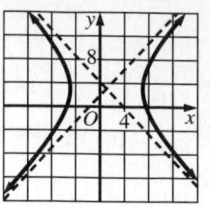

39. $\frac{(x-1)^2}{1} - \frac{(y+3)^2}{4} = 1$; horizontal hyperbola with center $(1, -3)$, vertices $(2, -3)$ and $(0, -3)$, foci $(1 \pm \sqrt{17}, -3)$, and asymptotes $y = \pm2(x - 1) - 3$ **41.** $y + 2 = -\frac{1}{4}(x - 3)^2$; parabola that opens down with vertex $(3, -2)$, focus $(3, -3)$, and directrix $y = -1$ **43.** degenerate ellipse; The graph is the point $(0, 0)$. **45.** degenerate hyperbola; The graph is the pair of lines $y = \frac{1}{5}x$ and $y = -\frac{1}{5}x$.

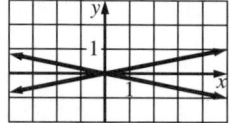

Pages 767–769   Chapter 12

1–3. b, c. Answers may vary. Examples are given.
1. a. A: 3; B: 3; C: 3; D: 3; E: 4   **b.**
c. 3 colors: E red, A and C green, B and D blue

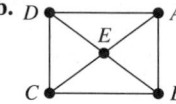

3. a. A: 3; B: 4; C: 3; D: 4; E: 4   **b.**
c. 4 colors: E red, D blue, B green, A and C yellow

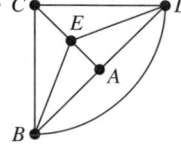

5. Answers may vary. An example is given. 3 colors: UT and MT red, ID and CO green, NV and WY blue

7.
$$\begin{array}{c} \\ A \\ B \\ C \\ D \end{array} \begin{array}{cccc} A & B & C & D \\ \left[\begin{array}{cccc} 0 & 0 & 0 & 1 \\ 1 & 0 & 0 & 0 \\ 0 & 1 & 0 & 1 \\ 0 & 1 & 1 & 0 \end{array}\right] \end{array}$$

9, 11. Answers may vary. Examples are given.
9. 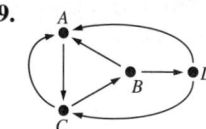   **11.**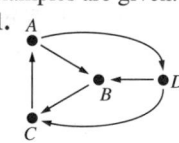

13. 4320   **15.** $n(n - 1)$, or $n^2 - n$ **17.** 210   **19.** 56
21. $\frac{7!}{3!2!} = 420$ **23.** 56   **25.** 105   **27.** 11   **29.** 165   **31. a.** 6
b. 6   **33. a.** 8   **b.** 70   **35.** $x^8 - 8x^7y + 28x^6y^2 - 56x^5y^3 + 70x^4y^4 - 56x^3y^5 + 28x^2y^6 - 8xy^7 + y^8$ **37.** $64x^6 - 576x^5y + 2160x^4y^2 - 4320x^3y^3 + 4860x^2y^4 - 2916xy^5 + 729y^6$
39. $x^{21} - 7x^{18}y^2 + 21x^{15}y^4 - 35x^{12}y^6 + 35x^9y^8 - 21x^6y^{10} + 7x^3y^{12} - y^{14}$

Pages 769–770   Chapter 13

1. $\frac{1}{4}$, or 0.25   **3.** $\frac{1}{2}$, or 0.5   **5.** $\frac{1}{13}$, or about 0.0769

7. $\frac{3}{4}$, or 0.75   **9.** $\frac{1}{26}$, or about 0.0385   **11.** $\frac{2}{13}$, or about 0.154
13. $\frac{1}{12}$, or about 0.0833   **15.** $\frac{5}{6}$, or about 0.833   **17.** 0.41
19. 0.51   **21.** $\frac{13}{51}$, or about 0.255   **23.** $\frac{15}{34}$, or about 0.441
25. $_6C_k(0.25)^k(0.75)^{6-k}$; 0.297   **27.** $_{10}C_k\left(\frac{4}{5}\right)^k\left(\frac{1}{5}\right)^{10-k}$; 0.0264
29. 0.00246   **31.** 1×10^{-11} **33.** -0.8 **35.** 2.2   **37.** 2.28%

Pages 770–771   Chapter 14
1. a. $34°$ **b.** $34°12'54''$ **3. a.** $26°$ **b.** $25°42'9''$ **5. a.** $10°$
b. $9°49'22''$ **7.** $22.6°$ **9.** $66.1°$ **11.** 1.0932   **13.** 1.7321
15. $\angle B = 52°$; $a \approx 10.5$; $b \approx 13.4$ **17.** $\angle A \approx 16.3°$; $\angle B \approx 73.7°$; $a = 5.25$; $c = 18.75$ **19–27.** Answers are given in the order $\sin \theta$; $\cos \theta$; $\tan \theta$.   **19.** $-\frac{24}{25} = -0.96$; $\frac{7}{25} = 0.28$; $-\frac{24}{7} \approx -3.4286$ **21.** $-\frac{7}{11} \approx -0.6364$; $-\frac{6\sqrt{2}}{11} \approx -0.7714$; $\frac{7}{6\sqrt{2}} \approx 0.8250$ **23.** 0.9135; -0.4067; -2.2460 **25.** 0.9205; 0.3907; 2.3559   **27.** 0.4540; -0.8910; -0.5095 **29.** about 128.1 square units   **31.** about 41.0 square units   **33.** not enough information; Given a and b, you need to know the measure of the included angle, $\angle C$.   **35.** about 7.36 square units   **37.** about 453 square units   **39.** about 8.14 square units   **41.** $\angle E = 110°$; $d \approx 9.92$; $f \approx 20.9$ **43.** $\angle D = 78°$; $e \approx 4.21$; $f \approx 8.31$ **45.** $\angle E = 79°$; $d \approx 6.68$; $e \approx 11.7$ **47.** no solution; $\angle A$ is obtuse and $a < b$.   **49.** $\angle A \approx 120.4°$; $\angle B \approx 33.6°$; $a \approx 18.7$; or $\angle A \approx 7.6°$; $\angle B \approx 146.4°$; $a \approx 2.9$ **51.** $\angle B \approx 102.4°$; $\angle C \approx 24.6°$; $b \approx 15.9$ **53.** about 28.1   **55.** $\angle X \approx 88.6°$; $\angle Y \approx 36.0°$; $\angle Z \approx 55.4°$ **57.** $\angle X \approx 30.0°$; $\angle Y \approx 34.4°$; $\angle Z \approx 115.6°$

Pages 771–772   Chapter 15
1. $0°, 360°$ **3.** $240°, 300°$ **5.** $60°, 120°$ **7.** $129°, 231°$
9. $78°, 282°$ **11.** $215°, 325°$ **13.** -0.9848
15. -0.9397 **17.** 0.4067   **19.** -0.6428 **21.** $270°$ **23.** $241°$
25. $-183°$ **27.** $68°$ **29.** 1.05; $\frac{\pi}{3}$ **31.** 2.62; $\frac{5\pi}{6}$ **33.** 3.40; $\frac{13\pi}{12}$
35. -2.27; $-\frac{13\pi}{18}$

37. amplitude: 4; period: π

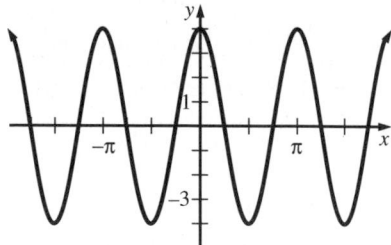

39. amplitude: $\frac{3}{4}$; period: $\frac{2\pi}{3}$

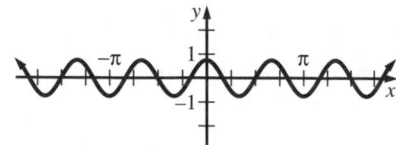

41. $y = 4 \sin \pi x$ **43.** $y = 5 \sin \frac{2}{5}x - 2$

45.

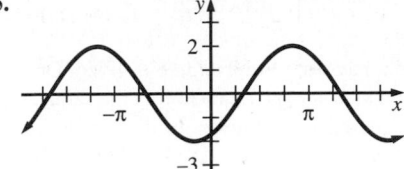

47.

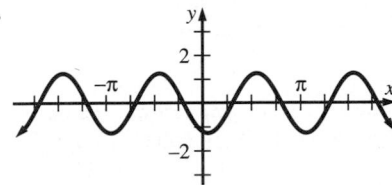

49, 51. Answers may vary. Examples are given.

49. $y = 5 \sin \frac{4}{3}\left(x - \frac{\pi}{2}\right)$

51. $y = 0.75 \sin \frac{1}{3}(x - \pi)$ **53.** 0

55. -1 **57.** $-\frac{\sqrt{3}}{3} \approx -0.5774$ **59.** 0

61.

TOOLBOX

Pages 773–774   Numbers and Number Lines
1. C **2.** F **3.** B **4.** G **5.** 0   **6.** -2 **7.** $2\frac{1}{2}$ **8.** $\frac{1}{2}$

9. a. E, F, G, H **b.** A, B, C **c.** D, F **d.** A, B, C, D, E, F, G, H

10.

$-2, -0.25, \frac{3}{2}$

11.

$1\frac{2}{3}, 2\frac{1}{3}, 3\frac{1}{2}$

12.

$-1.45, -1\frac{1}{4}, -\frac{3}{4}$

Pages 774–775   Properties of Real Numbers
1. a. absolute value property   **b.** distributive property
2. a. associative property   **b.** commutative property
c. associative property   **3. a.** commutative property
b. associative property   **c.** distributive property
4–13. Examples given may vary.   **4.** True; $(2 + 3)(2 - 3) = 5(-1) = -5$; $2^2 - 3^2 = 4 - 9 = -5$.   **5.** False; $1^2 + 2^2 = 5$, but $(1 + 2)^2 = 9$.   **6.** True; $3 - 5 = -2$; $3 + (-5) = -2$.   **7.** True; $4(1 - 3) = 4(-2) = -8$ and $4(1) - 4(3) = 4 - 12 = -8$.   **8.** False; $1 \cdot 13 = 13$.   **9.** True; $12 \div 3 = 4$ and $12 \cdot \frac{1}{3} = 4$.   **10.** False; $\frac{1}{2} + \frac{3}{4} = \frac{5}{4}$ and $\frac{1 + 3}{2 + 4} = \frac{2}{3}$.   **11.** True; $\frac{1}{2} \cdot \frac{3}{4} = \frac{3}{8}$ and $\frac{1 \cdot 3}{2 \cdot 4} = \frac{3}{8}$.
12. False; $|3| - |7| = 3 - 7 = -4$ and $|3 - 7| = |-4| = 4$.
13. False; $0 + (-27) = -27$.

Page 776   Percent Change
1. 3.6% increase   **2.** 20.9% decrease   **3.** 101.4% increase

Page 777   Exponents and Powers
1. 64   **2.** -32 **3.** 1   **4.** $\frac{1}{81}$ **5.** 243   **6.** 625   **7.** 1   **8.** $\frac{1}{5}$
9. a^{10} **10.** b^5 **11.** $\frac{1}{c^5}$ **12.** $\frac{1}{d^3}$ **13.** a^6 **14.** $\frac{1}{b^6}$ **15.** 1   **16.** d

Page 778   Square Roots
1. ± 7 **2.** none   **3.** $\pm \frac{1}{8}$ **4.** ± 0.4 **5.** 20   **6.** -1.3 **7.** not possible   **8.** $\frac{10}{11}$ **9.** $5\sqrt{5}$ **10.** $-3\sqrt{2}$ **11.** $7\sqrt{3}$ **12.** $2\sqrt{15}$
13. $12\sqrt{6}$ **14.** 2   **15.** $\frac{\sqrt{3}}{2}$ **16.** $\frac{2}{3}$ **17.** No; the statement is true only if $a \geq 0$. If $a < 0$, $\sqrt{a^2} = -a$.

Page 779   Order of Operations
1. -1 **2.** 11   **3.** 52   **4.** 94   **5.** 144   **6.** 6   **7.** -5 **8.** -13
9. -4.5 **10.** -9.5 **11.** -7.75 **12.** 8

Page 780   Rational Numbers and Irrational Numbers
1. the rational numbers, the real numbers   **2.** the whole numbers, the integers, the rational numbers, the real numbers
3. the irrational numbers, the real numbers   **4.** the whole numbers, the integers, the rational numbers, the real numbers
5. the irrational numbers, the real numbers   **6.** the rational numbers, the real numbers   **7.** the rational numbers, the real numbers   **8.** the irrational numbers, the real numbers
9, 10. Examples may vary.   **9.** any negative integer   **10.** π
11. 0   **12.** not possible

Page 781   Evaluating Variable Expressions
1. -2 **2.** -6 **3.** 14   **4.** -16 **5.** 24   **6.** 76   **7.** 2   **8.** 9   **9.** 3
10. 55   **11.** -45 **12.** 5   **13.** 27   **14.** 35   **15.** -2.5 **16.** -7

Page 781   Simplifying Variable Expressions
1. $6t - 8$ **2.** $-5k - 1$ **3.** $-a + 3b - 2$ **4.** $-x^2$ **5.** $-12 - 3j$
6. $-4 + 4a$ **7.** $x + 4y$ **8.** $6a + 2ab - 3b$ **9.** $m^2 - m + 5$
10. $a - b + 7$ **11.** $a - 9b - 7$ **12.** $a - b - 7$ **13.** $7 - 11x$
14. -16 **15.** $4y^2 - 3y - 4$ **16.** $-3d^2 + 12d + 5$
17. $-n^3 + n^2 + 6n + 14$ **18.** $2c^2 + 15cd + 13d^2$

Page 782   Multiplying Variable Expressions
1. $2b^2 - 7b$ **2.** $2m^2 + 8m$ **3.** $-5h + h^2$ **4.** $r^2 + r - 2$
5. $-p^2 + 3p + 28$ **6.** $s^2 - 16$ **7.** $2c^2 + 9c + 9$ **8.** $5z^2 - 11z + 2$
9. $32 - 4w - 3w^2$ **10.** $6j^2 - 11j + 3$ **11.** $9 + 23y - 12y^2$
12. $10k^2 + 29k + 10$ **13.** $a^2 - 10a + 25$ **14.** $m^2 + 6m + 9$
15. $4s^2 + 28s + 49$ **16.** $25j^2 - 10j + 1$

Page 783   Factoring Variable Expressions
1. $2(t + 4)$ **2.** $3(-2m + 3)$ or $-3(2m - 3)$ **3.** $r(r - 5)$
4. $v(2v + 7)$ **5.** $6(3j + k)$ **6.** $-3(3d + 5f)$ **7.** $g(-4 + 15g)$ or $-g(4 - 15g)$ **8.** $b(c + d)$ **9.** $7(3a^2 + 7)$ **10.** $5c(1 + 5c)$
11. $t(-2p + 3s)$ or $-t(2p - 3s)$ **12.** $10(2x^2 - 5y^2)$

Page 784   Adding and Subtracting Rational Expressions
1. $\frac{6}{t}$ **2.** $-\frac{4}{t}$ **3.** $\frac{2n + 3}{n + 1}$ **4.** $\frac{2n - 3}{n + 1}$ **5.** 3   **6.** $\frac{1}{k}$ **7.** $\frac{13}{2x}$ **8.** $\frac{7}{2x}$
9. $\frac{2x + 2}{3(2x - 1)}$ **10.** $\frac{4 + 30z}{5z}$ **11.** $\frac{r + 1}{r + 2}$ **12.** $\frac{4x - 1}{2x - 1}$ **13.** $\frac{2c + 4}{c + 4}$
14. $-\frac{4}{c + 4}$ **15.** $\frac{2y + 1}{y(y + 1)}$ **16.** $\frac{m - 6}{(m + 2)(m - 2)}$

Pages 785   Solving Linear Equations
1. $\frac{3}{4}$ **2.** -2 **3.** $\frac{3}{4}$ **4.** 7.4   **5.** -3 **6.** 2   **7.** 16   **8.** 16   **9.** 20
10. 2   **11.** -4 **12.** $-\frac{1}{5}$ **13.** $2\frac{1}{2}$ **14.** 5   **15.** $-\frac{4}{5}$ **16.** -2

Pages 785–786　Ratios and Proportions

1. $1\frac{2}{3}$　**2.** $-31\frac{1}{2}$　**3.** $1\frac{23}{27}$　**4.** 9　**5.** $1\frac{3}{16}$　**6.** $-\frac{1}{9}$　**7.** $-3\frac{13}{25}$

8. $\frac{9}{10}$　**9.** $\frac{6}{11}$ mi ≈ 0.55 mi　**10.** $6.62　**11.** 4.9 in.

Page 786　Identities and False Statements

1. all real numbers　**2.** no solution　**3.** 6　**4.** $2\frac{1}{2}$

5. no solution　**6.** all real numbers　**7.** all real numbers

8. no solution　**9.** $-3\frac{3}{4}$　**10.** $1\frac{1}{2}$　**11.** -8　**12.** no solution

13. 0　**14.** all real numbers　**15.** all real numbers

Page 787　Solving Linear Inequalities

1. $x < 9$

2. $y \geq -3$

3. $k > 2$

4. $z \leq -3$

5. $p < 5$

6. $r \leq -2$

7. $c > 12$

8. $x \geq -2$

9. $d < 2.5$

10. $h \geq 2$

11. $t > 1\frac{1}{2}$

12. $s \geq 0$

13. $g < 2$

14. $x \geq -6$

15. $w > 3$

16. $m \leq -1$

Page 788　Rewriting Equations and Formulas

1. $y = 5x - 15$　**2.** $b = \frac{2A}{h}$　**3.** $r = \sqrt{\frac{A}{\pi}}$　**4.** $b_1 = \frac{2A}{h} - b_2$ or

$b_1 = \frac{2A - b_2 h}{h}$　**5.** $y = \frac{3}{2}x - 12$　**6.** $l = \frac{P - 2w}{2}$ or $l = \frac{P}{2} - w$

7. $y = \frac{c - ax}{b}$　**8.** $n = \frac{360}{180 - x}$

Page 789　Experimental Probability

1. 0.75　**2.** 0.333 (to three decimal places)　**3.** 0.12　**4.** 0.53
5. 1　**6.** 0.875　**7.** 0.12　**8. a.** 0.012　**b.** 0.988　**9. a.** 0.528
b. No; even if those who now have no opinion decide to support the construction, only 59.2% of the voters would vote in favor of construction.

Page 790　Significant Digits

1. 5.3　**2.** 2.2　**3.** 4.22　**4.** 3.3　**5.** 32　**6.** 29　**7.** 690
8. 1.9　**9.** 2.4　**10.** 2.01　**11.** 0.002　**12.** 4.98

Page 791　Mean, Median, and Mode

1. a. 7.4; 6; no mode　**b.** 9; 9; no mode　**c.** 8.2; 7; 7
2. 8.2; 8; 7; Answers may vary. An example is given. Only the Western Division values have a mode; it is the same as that for the entire conference. The mean for the conference data is higher than that for the Eastern Division, lower than that for the Central Division, and the same as that for the Western Division. The median for the conference data is higher than those for the Eastern and Western Divisions, and lower than that for the Central Division. Overall, the values for the conference data are most similar to those for the Western Division.　**3.** 43.73; 44.3; 44.4, 46.4

Page 793　Data Displays

1. $5 bill; $20 bill　**2.** 6 years　**3.** $50 bills and $100 bills; Answers may vary. An example is given. I think the bills are probably handled much less and so wear out less quickly than more commonly used bills.

4.

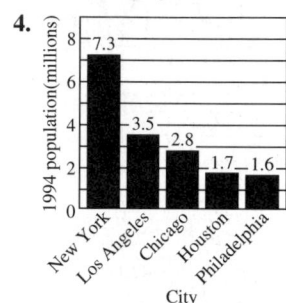

5.

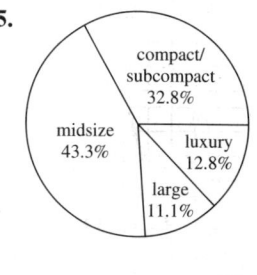

6.

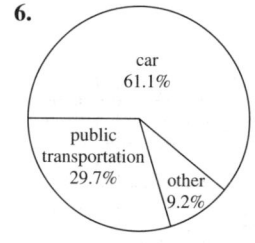

7. a.

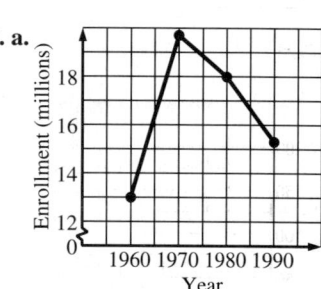

b. Answers may vary. An example is given. It appears that enrollment peaked in 1970 and has been declining since.

Page 794　Dimensional Analysis

1. $22\frac{2}{9}$ m/s　**2.** 3875 pesos　**3.** about 11,975 ft/s
4. about $0.05 per mile

Page 794　Graphing Points on a Coordinate Plane

1–8.

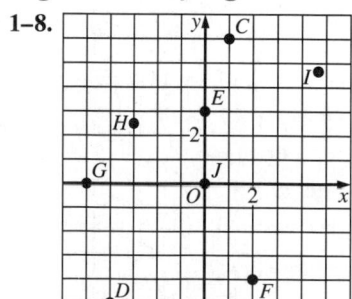

1. I　**2.** III　**3.** none
4. IV　**5.** none　**6.** II
7. I　**8.** none

Page 795   Graphing an Equation

1.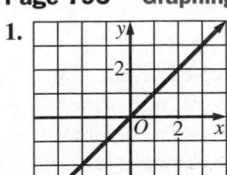

2.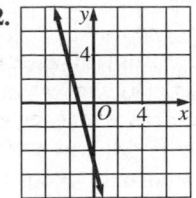

3.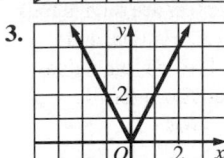

4.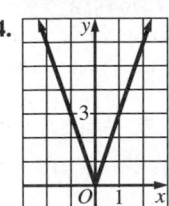

5.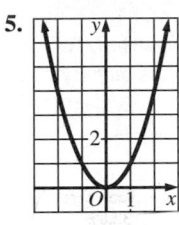

6.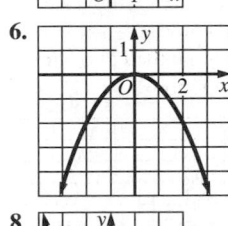

7.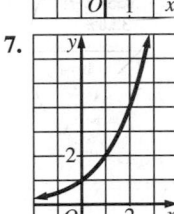

8.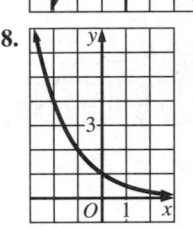

Page 796   Making a Scatter Plot

1.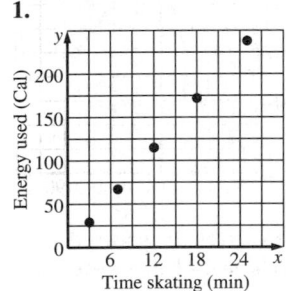

2.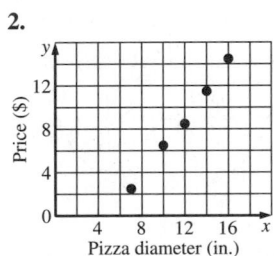

Page 797   Translating a Figure

1.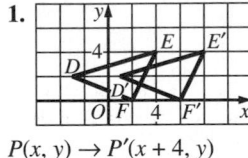

$P(x, y) \rightarrow P'(x + 4, y)$

2.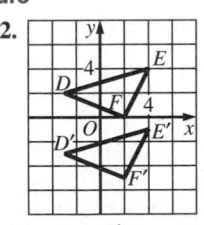

$P(x, y) \rightarrow P'(x, y - 5)$

3.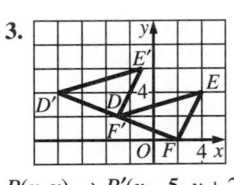

$P(x, y) \rightarrow P'(x - 5, y + 2)$

4.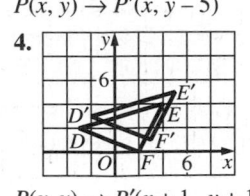

$P(x, y) \rightarrow P'(x + 1, y + 1)$

5.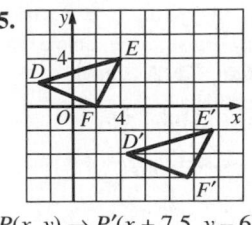

$P(x, y) \rightarrow P'(x + 7.5, y - 6)$

6.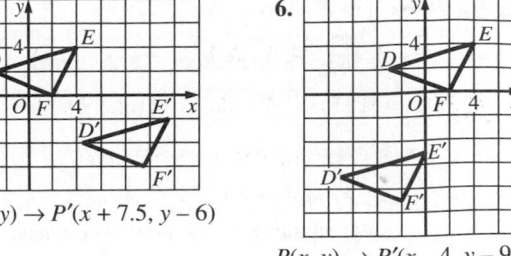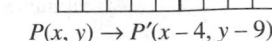

$P(x, y) \rightarrow P'(x - 4, y - 9)$

Page 798   Reflecting a Figure

1.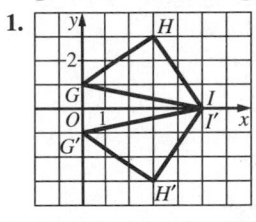

2.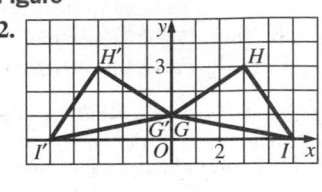

3.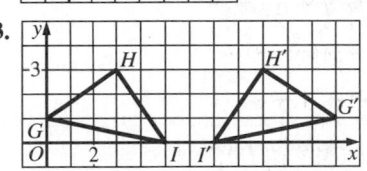

4.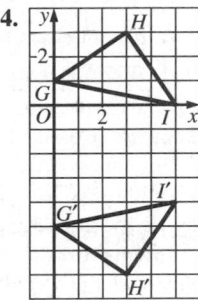

5. two lines of symmetry   **6.** four lines of symmetry   **7.** no lines of symmetry **8.** one line of symmetry   **9.** one line of symmetry   **10.** infinitely many lines of symmetry

Page 798   Dilating a Figure

1.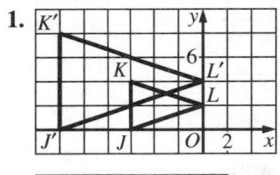

2.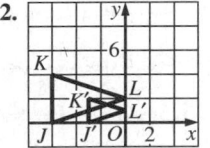

3.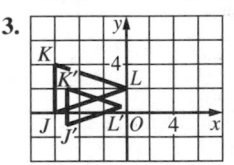

4.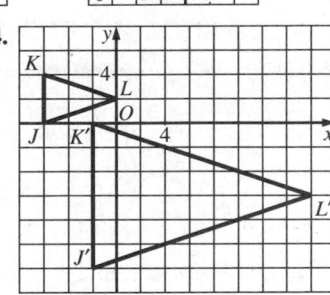

Page 800   Logical Statements

1. a. a figure is a rectangle; it is a quadrilateral; True.
b. If a figure is a quadrilateral, then it is a rectangle. False; a non-square rhombus is a quadrilateral that is not a rectangle.
c. A figure is a rectangle if and only if it is a quadrilateral. False; one of the conditionals is false.   **2. a.** $a = 0$ or $b = 0$; $ab = 0$; True.   **b.** If $ab = 0$, then $a = 0$ or $b = 0$. True; if the product of two numbers is zero, at least one of the numbers is 0.   **c.** $ab = 0$ if and only if $a = 0$ or $b = 0$. True; both of the conditionals are true.

3. a. a triangle is equilateral; it is isosceles; True. **b.** If a triangle is isosceles, then it is equilateral. False; for example, an isosceles right triangle is not equilateral. **c.** A triangle is equilateral if and only if it is isosceles. False; one of the conditionals is false. **4. a.** $a > b$; $a + c > b + c$; True. **b.** If $a + c < b + c$, then $a < b$; True. **c.** $a + c < b + c$ if and only if $a < b$. True; both of the conditionals are true. **5. a.** Two lines intersect; they are not parallel. True. **b.** If two lines are not parallel, then they intersect. False; Parallel lines are lines that lie in the same plane and do not intersect. **c.** Two lines intersect if and only if they are not parallel. False; One of the conditionals is false. **6. a.** $x^2 = 9$; $x = 3$; False. **b.** If $x = 3$, then $x^2 = 9$. True; $3^2 = 9$. **c.** $x^2 = 9$ if and only if $x = 3$. False; one of the conditionals is false.

Page 800 Angle Relationships

1. 80; 30 **2.** 25; 65 **3.** 50; 60 **4.** never true **5.** always true **6.** sometimes true

Page 801 Triangle Relationships

1. 20 **2.** $6\sqrt{2}$ **3.** 12 **4.** 90

Page 802 Polygons

1. sometimes; A hexagon may or may not be equiangular and equilateral. **2.** never; A quadrilateral has four sides. **3.** sometimes; A right triangle with two 45° angles is an isosceles triangle. **4.** never; If a triangle had three obtuse angles, the sum of the measures of the angles would be greater than 180°. **5.** always; Every square is both equiangular and equilateral. **6.** always; Every rectangle is a quadrilateral with four 90° angles.

Page 802 Perimeter, Circumference, Area, and Volume

1. 22 cm; 24 cm^2 **2.** $33 + 3\sqrt{61}$ ft $\approx$ 56.43 ft; 135 ft^2 **3.** 25 m; 31.5 m^2 **4.** 18.85; 28.27 **5.** 37.70; 113.10 **6.** 14.05; 15.71 **7.** 3518.58 cm^2; 10,053.10 cm^3 **8.** 112 in.^{2}; 80 in.^{3} **9.** 31,415.93 ft^2; 523,598.78 ft^3 **10.** 211.18 m^2; 155.88 m^3

TECHNOLOGY HANDBOOK

Pages 804–815 Calculator Practice

1. –12 **2.** 3.5 **3.** 1.4 **4.** 720 **5.** 32,768 **6.** 10 **7.** 3.732 **8.** 5.92

9.
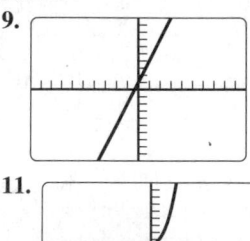

10.
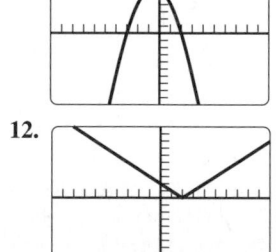

11.

12.

13. To make the lines $y = x - 1$ and $y = -x + 7$ look perpendicular, use a square viewing window like the one shown below. Zoom windows may vary.
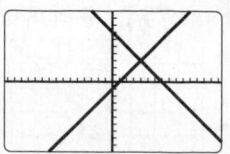

14. 3.66 **15. a.** 3.81 **b.** 2.39, 1, –0.813, –2.58 **c.** local minimums: –5.40, –9.14; local maximum: 5.04 **d.** 2.68, –2.80 **16. a.** 2.26 **b.** 3.37 **17.** Enter the equation $Y_1 = \{1,-1,0.5,2\}X$.

18. a.
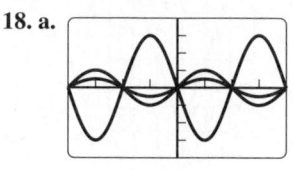
b.
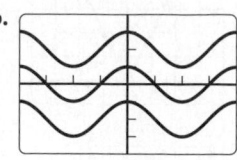

19. a.
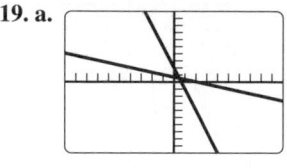
b.
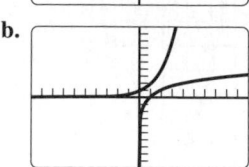

c.
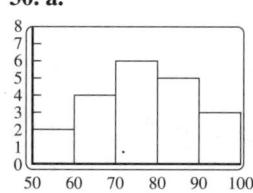

20. 750 s, or 12.5 min

21. a.
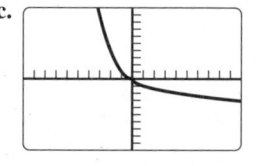

b.
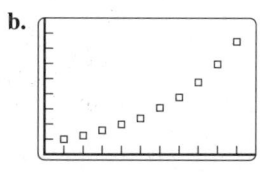
c.
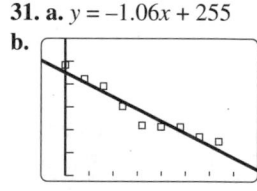

22. $\begin{bmatrix} 8 & 15 \\ 3 & 0 \end{bmatrix}$ **23.** $\begin{bmatrix} -4 & 8 \\ -15 & -11 \end{bmatrix}$ **24.** $\begin{bmatrix} 6 & 8 \\ 12 & 14 \end{bmatrix}$ **25.** $\begin{bmatrix} 3 & 5 \\ 9 & 17 \end{bmatrix}$

26. undefined **27.** $\begin{bmatrix} 33 & 21 & -63 & 36 \\ 44 & 28 & -84 & 48 \end{bmatrix}$ **28.** $\begin{bmatrix} -2.33 & 1.33 \\ 2 & -1 \end{bmatrix}$

29. undefined

30. a.
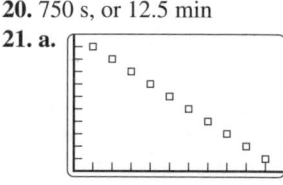

31. a. $y = -1.06x + 255$
b.
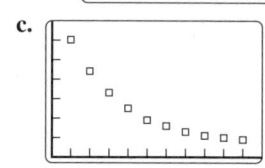

b. mean: 76.2; standard deviation: 11.8

32. Answers will depend on the type of calculator used. On the TI-82, enter the expression "int(16*rand)+10."

33. a. 3, –6 **b.** 0.4 **c.** $0.25 \pm 1.41i$ **34.** The specific commands in the program will depend on the calculator used. The following program is for the TI-82:
:Input "ENTER A:",A:Input "ENTER B:",B
:Disp "SOLUTION:",–B/A

Page 816 Spreadsheet Practice

35. a. $977.34 **b.** 6 years